Ecology and Conservation of North American Sea Ducks

STUDIES IN AVIAN BIOLOGY

A Publication of The Cooper Ornithological Society

www.crcpress.com/browse/series/crcstdavibio

Studies in Avian Biology is a series of works published by The Cooper Ornithological Society since 1978. Volumes in the series address current topics in ornithology and can be organized as monographs or multi-authored collections of chapters. Authors are invited to contact the series editor to discuss project proposals and guidelines for preparation of manuscripts.

See complete series list on page [583].

Volume 46
Studies in Avian Biology
Cooper Ornithological Society

Ecology and Conservation of North American Sea Ducks

EDITED BY

Jean-Pierre L. Savard
Environment Canada, Québec City, QC, Canada

Dirk V. Derksen
U.S. Geological Survey, Anchorage, AK, USA

Dan Esler
U.S. Geological Survey, Anchorage, AK, USA

John M. Eadie
University of California, Davis, CA, USA

ASSISTED BY

Vanessa W. Skean
U.S. Geological Survey, Anchorage, AK, USA

CRC Press
Taylor & Francis Group
Boca Raton London New York

CRC Press is an imprint of the
Taylor & Francis Group, an **informa** business

Cover photo by Jeff Coats: Long-tailed Duck on Barnegat Inlet, New Jersey.

PERMISSION TO COPY

CRC Press
Taylor & Francis Group
6000 Broken Sound Parkway NW, Suite 300
Boca Raton, FL 33487-2742

First issued in paperback 2017

© 2015 by Cooper Ornithological Society
CRC Press is an imprint of Taylor & Francis Group, an Informa business

No claim to original U.S. Government works

ISBN-13: 978-1-4822-4897-5 (hbk)
ISBN-13: 978-1-138-57579-0 (pbk)

Visit the Taylor & Francis Web site at
http://www.taylorandfrancis.com

and the CRC Press Web site at
http://www.crcpress.com

CONTENTS

Editors / vii

Contributors / ix

Foreword / xiii

Preface / xv

Introduction / xvii

1 • STATUS AND TRENDS OF NORTH
AMERICAN SEA DUCKS: REINFORCING
THE NEED FOR BETTER MONITORING / 1
Timothy D. Bowman,
Emily D. Silverman,
Scott G. Gilliland, and Jeffery B. Leirness

2 • PHYLOGENETICS, PHYLOGEOGRAPHY,
AND POPULATION GENETICS OF NORTH
AMERICAN SEA DUCKS (TRIBE: MERGINI) / 29
Sandra L. Talbot, Sarah A. Sonsthagen,
John M. Pearce, and Kim T. Scribner

3 • POPULATION DYNAMICS OF SEA DUCKS:
USING MODELS TO UNDERSTAND THE
CAUSES, CONSEQUENCES, EVOLUTION,
AND MANAGEMENT OF VARIATION IN
LIFE HISTORY CHARACTERISTICS / 63
Paul L. Flint

4 • INFECTIOUS DISEASES, PARASITES, AND
BIOLOGICAL TOXINS IN SEA DUCKS / 97
Tuula E. Hollmén and J. Christian Franson

5 • BREEDING COSTS, NUTRIENT
RESERVES, AND CROSS-SEASONAL
EFFECTS: DEALING WITH DEFICITS IN
SEA DUCKS / 125
Ray T. Alisauskas and Jean-Michel Devink

6 • CONTAMINANTS IN SEA DUCKS: METALS,
TRACE ELEMENTS, PETROLEUM,
ORGANIC POLLUTANTS, AND RADIATION / 169
J. Christian Franson

7 • FORAGING BEHAVIOR, ECOLOGY,
AND ENERGETICS OF SEA DUCKS / 241
Ramūnas Žydelis and Samantha E. Richman

8 • VARIATION IN MIGRATION STRATEGIES
OF NORTH AMERICAN SEA DUCKS / 267
Margaret R. Petersen and
Jean-Pierre L. Savard

9 • REMIGIAL MOLT OF SEA DUCKS / 305
Jean-Pierre L. Savard and
Margaret R. Petersen

10 • SITE FIDELITY, BREEDING HABITATS,
AND THE REPRODUCTIVE STRATEGIES
OF SEA DUCKS / 337
Mark L. Mallory

11 • BREEDING SYSTEMS, SPACING
BEHAVIOR, AND REPRODUCTIVE
BEHAVIOR OF SEA DUCKS / 365
John M. Eadie and Jean-Pierre L. Savard

12 • HARVEST OF SEA DUCKS IN NORTH
AMERICA: A CONTEMPORARY
SUMMARY / 417
Thomas C. Rothe, Paul I. Padding,
Liliana C. Naves, and Gregory J. Robertson

13 • HABITATS OF NORTH AMERICAN SEA
DUCKS / 469
    Dirk V. Derksen, Margaret R. Petersen,
    and Jean-Pierre L. Savard

14 • CONSERVATION OF NORTH AMERICAN
SEA DUCKS / 529
    W. Sean Boyd, Timothy D. Bowman,
    Jean-Pierre L. Savard, and Rian D. Dickson

15 • CONCLUSIONS, SYNTHESIS,
AND FUTURE DIRECTIONS / 561
    Daniel Esler, Paul L. Flint,
    Dirk V. Derksen, Jean-Pierre L. Savard,
    and John M. Eadie

    Appendix A: North American Sea
    Ducks / 569

    Index / 573

    Studies in Avian Biology / 583

EDITORS

Jean-Pierre L. Savard is a Scientist Emeritus with Environment Canada, Quebec, Canada. Dr. Savard earned his PhD from the University of British Columbia, where he conducted research on the territorial behavior, nesting success, and brood survival of Barrow's Goldeneye, Common Goldeneye, and Bufflehead ducks. He has published numerous papers focused on the biology and population dynamics of sea ducks, including Common Eider, Harlequin Duck, Surf Scoter, Black Scoter, Long-tailed Duck, Bufflehead, Barrow's Goldeneye, Common Goldeneye, Hooded Merganser, and Red-breasted Merganser.

Dirk V. Derksen is retired from the U.S. Geological Survey, Alaska Science Center, Anchorage, Alaska, where he served as Chief of migratory bird, terrestrial mammal, and genetics research over a 26-year period. Dr. Derksen earned his PhD from Iowa State University and conducted his dissertation research on Adelie penguins in Antarctica. He studied the habitat ecology of waterbirds on the North Slope of Alaska and published a suite of papers detailing the freshwater and marine wetland and terrestrial areas important for Spectacled Eiders, King Eiders, and Long-tailed Ducks.

Daniel Esler is a Research Wildlife Biologist with the U.S. Geological Survey, Alaska Science Center, Anchorage, Alaska, where he leads ecological studies of nearshore marine systems along the Pacific Coast of North America. Dr. Esler examined the effects of the 1989 *Exxon Valdez* oil spill on the demography of Harlequin Ducks during winter in Prince William Sound, Alaska, for his PhD at Oregon State University. He has published extensively on the biology of sea ducks, including Steller's Eider, Spectacled Eider, Harlequin Duck, Surf Scoter, White-winged Scoter, Black Scoter, and Barrow's Goldeneye.

John M. Eadie is the Dennis G. Raveling Professor in Waterfowl Biology and Chair of the Department of Wildlife, Fish & Conservation Biology, University of California, Davis, California. Dr. Eadie earned his PhD in zoology at the University of British Columbia, where he researched the reproductive ecology and behavior of Barrow's and Common Goldeneyes. His numerous publications cover a diversity of behavioral and ecological aspects of sea duck species, including Steller's Eider, Harlequin Duck, Surf Scoter, Bufflehead, Barrow's Goldeneye, and Common Goldeneye.

CONTRIBUTORS

RAY T. ALISAUSKAS
Wildlife Research Division
Science and Technology Branch
Environment Canada
Prairie and Northern Wildlife Research
  Centre
115 Perimeter Road
Saskatoon, SK S7N 0X4, Canada
ray.alisauskas@ec.gc.ca

TIMOTHY D. BOWMAN
U.S. Fish and Wildlife Service
Division of Migratory Bird Management
1011 East Tudor Road
Anchorage, AK 99503, USA
tim_bowman@fws.gov

W. SEAN BOYD
Wildlife Research Division
Science and Technology Branch
Environment Canada
5421 Robertson Road
Delta, BC V4K 3N2, Canada
sean.boyd@ec.gc.ca

DIRK V. DERKSEN
U.S. Geological Survey
Alaska Science Center
4210 University Drive
Anchorage, AK 99508, USA
dderksen@usgs.gov

JEAN-MICHEL DEVINK
Stantec Consulting Inc.
100-75 24th Street East
Saskatoon, SK S7K 0K3, Canada
jeanmichel.devink@stantec.com

RIAN D. DICKSON
Black Duck Biological
1103 Hadfield Avenue
Victoria, BC V9A 5N6, Canada
riandel@gmail.com

JOHN M. EADIE
University of California at Davis
One Shields Avenue
1053 Academic Surge Building
Davis, CA 95616, USA
jmeadie@ucdavis.edu

DANIEL ESLER
U.S. Geological Survey
Alaska Science Center
4210 University Drive
Anchorage, AK 99508, USA
desler@usgs.gov

PAUL L. FLINT
U.S. Geological Survey
Alaska Science Center
4210 University Drive
Anchorage, AK 99508, USA
pflint@usgs.gov

J. CHRISTIAN FRANSON
U.S. Geological Survey
National Wildlife Health Center
6006 Schroeder Road
Madison, WI 53711, USA
jfranson@usgs.gov

SCOTT G. GILLILAND
Environment Canada
Canadian Wildlife Service
17 Waterfowl Lane
Sackville, NB A1N 4T3, Canada
scott.gilland@ec.gc.ca

TUULA E. HOLLMÉN
Alaska SeaLife Center
and
University of Alaska Fairbanks
School of Fisheries and Ocean Sciences
PO Box 1329
Seward, AK 99664, USA
tuulah@alaskasealife.org

JEFFERY B. LEIRNESS
University of Delaware
and
U.S. Fish and Wildlife Service
11510 American Holly Drive
Laurel, MD 20708, USA
jeffery_leirness@fws.gov

MARK L. MALLORY
Acadia University
33 Westwood Drive
Wolfville, NS B4P 2R6, Canada
mark.mallory@acadiau.ca

LILIANA C. NAVES
Alaska Department of Fish and Game
Division of Subsistence
333 Raspberry Road
Anchorage, AK, 99518, USA
liliana.naves@alaska.gov

PAUL I. PADDING
U.S. Fish and Wildlife Service
Division of Migratory Bird Management
11510 American Holly Drive
Laurel, MD 20708, USA
paul_padding@fws.gov

JOHN M. PEARCE
U.S. Geological Survey
Alaska Science Center
4210 University Drive
Anchorage, AK 99508, USA
jpearce@usgs.gov

MARGARET R. PETERSEN
U.S. Geological Survey
Alaska Science Center
4210 University Drive
Anchorage, AK 99508, USA
mrpetersen@usgs.gov

SAMANTHA E. RICHMAN
Long Point Waterfowl
P.O. Box 160
Port Rowan, ON N0E 1M0, Canada
cruciger7@gmail.com

GREGORY J. ROBERTSON
Wildlife Research Division
Science and Technology Branch
Environment Canada
6 Bruce Street
Mount Pearl, NL A1N 4T3, Canada
greg.robertson@ec.gc.ca

THOMAS C. ROTHE
Alaska Department of Fish and Game
Division of Wildlife Conservation (retired)
11828 Broadwater Drive
Eagle River, AK 99577, USA
tom.halcyon@gmail.com

JEAN-PIERRE L. SAVARD
Environment Canada
Wildlife Research Division
801-1550 Av d'Estimauville
Québec, QC G1J 0C3, Canada
jean-pierre.savard@ec.gc.ca

KIM T. SCRIBNER
Michigan State University
480 Wilson Road
East Lansing, MI 48824, USA
scribne3@msu.edu

EMILY D. SILVERMAN
U.S. Fish and Wildlife Service
Division of Migratory Bird Management
11510 American Holly Drive
Laurel, MD 20708, USA
emily_silverman@fws.gov

SARAH A. SONSTHAGEN
U.S. Geological Survey
Alaska Science Center
4210 University Drive
Anchorage, AK 99508, USA
ssonsthagen@usgs.gov

SANDRA L. TALBOT
U.S. Geological Survey
Alaska Science Center
4210 University Drive
Anchorage, AK 99508, USA
stalbot@usgs.gov

RAMŪNAS ŽYDELIS
DHI-Denmark
Agern Allé 5
DK-2970 Hørsholm, Denmark
rzy@dhigroup.com

FOREWORD

Regardless of one's experience with sea ducks, the name seems to bring an immediate vision of fast-flying, goal-oriented ducks in tight formation flying low over cold waters: eiders, scoters, Long-tailed Ducks, goldeneyes, or mergansers. There is a feeling of excitement behind the binoculars, camera, bola, arrow, or gun. And it is true: most sea ducks are found in the northern hemisphere, high latitude, cold water, near-shore marine, bay, or large lakes and rivers. The richness and abundance of animal resources in cold waters is well known, and the diving waterfowl can access them not only in freezing temperatures but also at great depths. During the breeding season, in which the air may be sharp, the newly hatched young have dense down suitable for their tundra or rugged shoreline nest site. Islands are favored nesting areas, and some species or populations are colonial, some clustered, and others solitary—dictated in part by habitat. But that distribution pattern is not without variation for some species like goldeneyes, Red-breasted Mergansers, and Harlequin Ducks, which nest in cool forested areas along inland lakes and streams. Tree-hole nest sites are common among Buffleheads, Common Mergansers, and especially Hooded Mergansers, which may nest in southern wooded swamps and even use nest sites serially shared with distant relatives, such as Wood Ducks. To gain perspective, the tribe Mergini contains 21 recent species worldwide, of which 19 are still living. Remarkably, these birds are separated with little controversy into 10 genera, reflecting dramatic morphological and ecological differentiation. Of the 21 species, 19 (18 living) are from the northern hemisphere and only 2 (1 living) have distributions in the southern hemisphere. The last known specimen of the northeastern Labrador Duck was taken in 1878, before we learned much about their breeding range and habits. Although they were abundant in coastal Labrador at some seasons and shot by east coast duck hunters as far south as the Long Island, early observers suggested that they may have nested in the St. Lawrence Estuary. Recent DNA studies suggest that their tribal relationship is still uncertain. The Chinese or Scaly-sided Merganser survives in small numbers (perhaps 2,500 total); its distribution in several countries complicates surveys but enhances chances for survival. At least three North American species or distinct populations are also threatened, as will become clear in several chapters. There were two species of sea ducks in the southern hemisphere, where I have searched for them in vain. The Auckland Islands Merganser—seemingly now extinct—was known only from a few specimens taken at the mouths of small estuaries and along small streams on Auckland Island and also from skeletal remains elsewhere that indicate a once-larger range. The Brazilian Merganser, a rare and isolated resident of subtropical forested rivers of Brazil, Argentina, and perhaps Paraguay, nests in tree holes. The discovery of new breeding populations in Brazil is encouraging, where estimates are now in the range of 200–250 individuals. However, the status of former populations in Argentina is uncertain.

One ponders the *why* of this distribution pattern. Obviously, a few pioneers made it south, seemingly derived from the Common Merganser stock, which are forest lake and stream birds. But North American species that choose seashores may winter well south along both coasts, and a few make it to the Gulf Coast. Remarkably, as I was writing this, an e-mail arrived stating that the Common Eider has been accepted as the 631st species recorded in Texas. But these coastal species have not pioneered into or beyond the tropics. There is, however, another tribe of cold-water ducks, unrelated to Mergini, that form an ecological counterpart of northern eiders: the steamer ducks of Patagonia and the Falkland Islands. Climatically less pressured to move to warmth in winter, three of the four species are flightless. Most of the species and genera of Mergini considered in this volume occur in both North America and Eurasia, but the emphasis in this treatise is on North America with the direct involvement of this unique set of authors intent on studying a fascinating group of ducks. While they produced original and important results from personal field and laboratory studies, their work, like that of all scientists, is based on the work of many early and current European scientists who have lived closely with sea ducks for many years. Collectively, all strive for the same goal: understanding the group, their adaptations, their limitations in facing societal stresses, and ultimately their conservation. Arctic tundra– and northern island–based societies have utilized and managed products from these birds—down, eggs, and meat—for centuries via strategic harvesting of down to minimize the impact on egg success, creation of shelters (ranging from canopies to man-made holes) to reduce the high rate of egg loss to aerial predators, and focus on postbreeding season harvest such as the renowned eider pass near Barrow, Alaska.

But times have changed dramatically, and seas, rivers, lakes, and associated nesting habitats are not what they once were because of climactic events and impacts of human activities. Moreover, the problems that impact sea ducks and other seabirds impact all of the society as well, demanding that we address such integrative ecological, political, and conservation problems collectively using approaches and analyses that transcend species or even generic features. The results reflect the multispecies thinking that stimulated the cooperative Sea Duck Joint Venture (SDJV). These summaries are truly current and will demand rewriting of some current textbooks, old or traditional laws, and government policies at all levels. The chapters will also present goals and approaches to guide future research. By necessity then, this work will provide a must-read foundation for resource ecologists working in associated disciplines: survival of native customs and societal needs; hunting and related harvest policies; soil erosion and riverine water quality; streamside forest removal; oil and chemical pollution at sea and in-stream; coal, tar sand, and oil extraction; marine and inland riverine transport problems; wind and geothermal energy development; and management of both inland and coastal fisheries as well as marine mammal resources. Although global warming has impacted sea duck habitat many times in geologic history, coastlines have never been under great pressure from human developments—with no reductions in sight. How to identify and understand such current problems, impending threats, possible solutions, and future research directions is discussed in this book. After reading this book, I think you will agree that our knowledge of sea duck ecology has significantly grown.

MILTON W. WELLER
Kerrville, Texas

PREFACE

People have long been fascinated by sea ducks because of their spectacular courtship displays, brightly colored and patterned plumage, remarkable variation in life history traits, remote northern breeding locations, unique migrations, and the many mysteries that remain about their ecology and behavior. However, the past decade has witnessed a dramatic increase in the interest and attention directed toward the 15 species of North American sea ducks, which include four species of eiders, the Harlequin Duck, three species of scoters, Long-tailed Duck, Bufflehead, two species of goldeneyes, and three species of mergansers that breed largely in arctic and subarctic habitats across the continent. We were motivated to produce this book, in large part, by the conservation concerns associated with simultaneous declines in several sea duck species and populations. Additionally, we wanted to synthesize the considerable research that has recently been conducted on this small tribe of ducks. These studies form the basis for identifying factors that may be limiting population recovery and will help to develop a conservation strategy for sea ducks in North America. We thus felt it was imperative to capture the current state of knowledge and to offer future research directions for researchers, managers, students, conservationists, and avian enthusiasts. *Ecology and Conservation of North American Sea Ducks* is a collection of 15 linked chapters focused on population dynamics, ecology, and behavior of sea ducks in North America. When we outlined this volume, we brought together researchers and managers who had field and laboratory experiences with all members of the sea duck tribe Mergini in North America. This in turn led to collaborations among specialists from different geographic areas of the continent to develop comprehensive treatises on key life history and management topics important for the conservation of sea duck species and their diverse marine and freshwater habitats. The chapter authors have combined the summaries of published accounts with new data, as well as presented additional analyses, models, and interpretations to advance our understanding of sea duck population dynamics and genetics, infectious diseases and parasites, breeding costs and cross-seasonal effects, contaminant burdens, foraging behavior and energetics, migration strategies, molt ecology, breeding systems and reproductive behavior, harvest history and contemporary trends, and habitat affinities and dynamics, and to document conservation concerns. We hope this volume of the *Studies in Avian Biology* series will stimulate further interest and research on the sea ducks of North America and worldwide.

JEAN-PIERRE L. SAVARD
Québec City, Québec

DIRK V. DERKSEN
Anchorage, Alaska

DANIEL ESLER
Anchorage, Alaska

JOHN M. EADIE
Davis, California

Introduction

Dirk V. Derksen, Jean-Pierre L. Savard, Daniel Esler, and John M. Eadie

Sea ducks breed across the entire continent of North America. Different species nest as far north as eastern Greenland, westward through the Aleutian Islands of Alaska, and south into the Canadian Prairie Provinces, Rocky Mountains, Mississippi River Valley, and Atlantic Coastal States. Sea ducks have diverse migration strategies, ranging from species that travel thousands of kilometers to the near-sedentary populations of Common Eider (*Somateria mollissima sedentaria*) residing in Hudson Bay. As a result, wintering locations vary greatly within the tribe from arctic marine polynyas to the Great Lakes of the midcontinent and south to temperate ocean waters of the Gulf of Mexico and along the east and west coasts of the Baja Peninsula.

The waterfowl tribe Mergini includes 21 species (Table I.1). Our focus in this volume of *Studies in Avian Biology* is on the 15 extant species of sea ducks indigenous to North America, not including the Labrador Duck (*Camptorhynchus labradorius*), which became extinct by 1875 (Livezey 1995, Chilton 1997, Kear 2005).

The tribe Mergini is a fascinating group of waterfowl because of the unique and diverse ecology, behavior, and life histories of these species. Sea ducks reach sexual maturity between two and three years of age and have long life spans and low annual recruitment. All sea ducks dive to forage, and their diet is largely animal matter, including mollusks, crustaceans, echinoderms, polychaete worms, insects, fishes, and fish eggs, although some species also consume smaller quantities of aquatic plants and algae. Sea ducks exhibit dimorphic plumage and spectacular courtship displays (Myres 1959, Johnsgard 1965). In most species, males are the larger sex and more numerous in the population than females.

Nearly all sea ducks rely on marine habitats for most of the annual cycle. Nine North American species are found exclusively in salt water habitats during winter. Four species—White-winged Scoter (*Melanitta fusca*), Long-tailed Duck (*Clangula hyemalis*), Bufflehead (*Bucephala albeola*), and Common Goldeneye (*B. clangula*)—occur mostly in marine waters during winter but are known to exploit freshwater systems during winter. Only two species—Hooded Merganser (*Lophodytes cucullatus*) and Common Merganser (*Mergus merganser*)—occur primarily in freshwater habitats during winter (Table I.1). Some sea duck species nest exclusively on freshwater wetlands and streams at a considerable distance from the sea, while other members of the tribe breed on freshwater sites, barrier islands, and other habitats that are in close proximity to oceans and estuaries.

Within the complex of eider species, Steller's Eiders (*Polysticta stelleri*) are of special interest because the North American breeding population is small and currently restricted to two discrete areas in western and northern Alaska (Flint and Herzog 1999, Petersen et al. 2000, Fredrickson 2001). Spectacled Eiders (*Somateria fischeri*) are remarkable birds because their pelagic wintering locations are far offshore in the pack ice of the mid-Bering Sea (Petersen et al. 1999). King Eiders (*Somateria spectabilis*) breed at especially high northern latitudes, and their coastal

TABLE I.1

Common name	Scientific name[a]	Distribution	Focal species	Breed	Winter	Molt
Steller's Eider	*Polysticta stelleri*	NA, EU	x	FW, SW	SW	SW
Spectacled Eider	*Somateria fischeri*	NA, RU	x	FW, SW	SW	SW
King Eider	*Somateria spectabilis*	NA, EU, RU	x	FW	SW	SW
Common Eider	*Somateria mollissima*	NA, IC, EU, RU	x	FW, SW	SW	SW
Harlequin Duck	*Histrionicus histrionicus*	NA, IC	x	FW	SW	SW
Labrador Duck	*Camptorhynchus labradorius*	NA	Extinct	?	SW	?
Surf Scoter	*Melanitta perspicillata*	NA	x	FW	SW	SW
White-winged Scoter	*Melanitta fusca*	NA, EU, RU	x	FW	FW, SW	SW
Common Scoter	*Melanitta nigra*	EU, RU, IC		FW	SW	SW
Black Scoter	*Melanitta americana*	NA	x	FW	SW	SW
Long-tailed Duck	*Clangula hyemalis*	NA, EU, RU	x	FW, SW	FW, SW	FW, SW
Bufflehead	*Bucephala albeola*	NA	x	FW	FW, SW	FW, SW
Common Goldeneye	*Bucephala clangula*	NA, EU, RU	x	FW	FW, SW	FW, SW
Barrow's Goldeneye	*Bucephala islandica*	NA, IC	x	FW	SW	FW, SW
Smew	*Mergellus albellus*	EU, RU		FW	FW	FW, SW
Hooded Merganser	*Lophodytes cucullatus*	NA	x	FW	FW	FW
Auckland Islands Merganser	*Mergus australis*	AI	Extinct	?	?	?
Brazilian Merganser	*Mergus octosetaceus*	BR, AR		FW	FW	FW
Common Merganser	*Mergus merganser*	NA, EU, RU	x	FW	FW	FW
Red-breasted Merganser	*Mergus serrator*	NA, EU, RU	x	FW, SW	SW	SW
Scaly-sided Merganser	*Mergus squamatus*	RU, KO, CH		FW	SW	FW

NOTES: NA, North America; IC, Iceland; EU, Europe; RU, Russia; BR, Brazil; AR, Argentina; AI, Auckland Islands; KO, Korea; CH, China; FW, freshwaters; SW, salt waters; Sea ducks addressed in this volume are identified as focal species (x).

[a] Taxonomy follows American Ornithologists' Union (1998), Kear (2005), Chesser et al. (2010), and Harrop et al. (2013).

migrations are considered spectacular among sea ducks because of the protracted passage of large flocks along ice-free ocean leads (Suydam 2000). Common Eiders nest in colonies and are perhaps more closely tied to marine habitats than any other sea duck (Goudie et al. 2000, Bédard et al. 2008). Harlequin Ducks (*Histrionicus histrionicus*) are river specialists that exploit fast-flowing, fluvial systems for breeding in two widely separated areas of North America (Robertson and Goudie 1999), where Atlantic and Pacific populations are thought to be genetically isolated (Scribner et al., Michigan State University, unpubl. report). All three of the North American scoter species inhabit Nearctic waters and exhibit long-distance migrations from coastal winter habitats to breeding areas (Bordage and Savard 1995, Brown and Fredrickson 1997, Savard et al. 1998). Surf Scoters (*Melanitta perspicillata*) and White-winged Scoters are largely dependent on the continental boreal forest biome during the breeding period, whereas Black Scoters (*M. americana*) breed more commonly in tundra and taiga habitats in Alaska and Canada. Long-tailed Ducks breed at high latitudes up to 80°N and have unique and complex molt patterns with three distinct plumages, in contrast to other sea ducks that exhibit only two plumages (Robertson and Savard 2002). The genus *Bucephala*, which includes the Bufflehead and two species of goldeneyes, is well known for cavity nesting, nest parasitism, and strong intra- and interspecific territorial behavior (Gauthier 1993; Eadie et al. 1995, 2000). The three North American species of mergansers use a great diversity of marine and freshwater habitats across the continent. Common Mergansers breed on rivers and large lakes and like their congener, the Red-breasted Merganser (*Mergus serrator*), are reviled by some fishermen because their diet is largely piscivorous (Mallory and Metz 1999, Titman 1999). The Hooded Merganser is restricted in distribution to North America, where the species nests in natural cavities as well as man-made boxes and commonly lays its eggs in the nests of conspecifics and other cavity-nesting ducks (Dugger et al. 1994).

Why are sea ducks of interest in North America? Sea ducks have been important as sources of food and material to northern communities for centuries and remain so to this day. North American Inuit continue to use the down and eggs of Common Eiders, and in northern Québec and in the St. Lawrence River Estuary, eiderdown is harvested commercially (Bédard et al. 2008). Arctic and subarctic villages at northern latitudes across the continent have a long history of sea duck harvest throughout the year (Phillips 1925, 1926; Bellrose 1976; Gilliland et al. 2009; Naves 2012). Similarly, sea ducks have historically been harvested in large numbers by commercial interests. The practice has been discontinued along the northeastern coast of the United States, but the commercial take of Common Eiders still occurs in west Greenland (Merkel 2004, 2010; Gilliland et al. 2009). Sport hunters also have a long tradition of taking sea ducks as part of the regulated harvest in Canada and the United States (Caithamer et al. 2000). Now, more than ever, sea ducks are enjoyed by increasing numbers of birding enthusiasts and conservationists who contribute to our knowledge of their status, trends, and distribution through citizen-based observation networks and surveys (Bond et al. 2007, Greenwood 2007, Sullivan et al. 2009).

Early in the twentieth century, little was known about most sea ducks in North America, with the exception of the Common Eider (Phillips 1925, 1926). Not much had changed by 1951 (Bent 1951) or even 1976 when Bellrose (1976) wrote, "Of all the ducks in North America, the Surf Scoter has the dubious distinction of being the least studied." Of course, the primary reasons for a lack of information on the status and biology of sea ducks were that most species occurred in remote breeding and wintering areas, and waterfowl management placed greater emphasis on the biology of geese and dabbling ducks (Phillips and Lincoln 1930).

Importantly, with limited availability of population data, uncertainty existed concerning the status of most sea duck species in North America. Nevertheless, as early as the 1920s, John C. Phillips, who wrote A Natural History of the Ducks,

identified declines of some species and populations of sea ducks, especially Common Eiders of the Atlantic region during the era of market hunting and feather trade (Phillips 1925, 1926; Austin 1932). It is now clear that subsistence hunting and harvest of eggs and down from the Common Eider were the primary factors limiting the Greenland breeding population (Hansen 2002, Merkel 2004) and that recent protection measures have resulted in population recovery (Merkel 2010). Similarly, colonies of Common Eiders along the New England coast, Canadian Maritime Provinces, and the Gulf of St. Lawrence have rebounded with increased protection (Todd 1963, Bellrose 1976).

In 1986, the Committee on the Status of Endangered Wildlife in Canada listed the eastern population of Harlequin Duck as endangered. Subsequently, the eastern populations of Harlequin Duck and Barrow's Goldeneye were determined to be species of concern (Robert et al. 2000, Thomas and Robert 2001). In the United States, sea ducks have been listed under the authority of the U.S. Endangered Species Act: the global population of Spectacled Eider was listed as threatened in 1993 (U.S. Fish and Wildlife Service 1996), and in 1994, the North American breeding population of Steller's Eider was determined to be threatened (U.S. Fish and Wildlife Service 2002). Collectively, these listings stimulated research interest in these four sea duck species as well as other members of the tribe Mergini, which for decades had largely been overlooked with regard to collecting fundamental population data through surveys because of the emphasis placed on hunted species of waterfowl that were more popular with the public. During the same period, Goudie et al. (1994) reviewed the status of sea ducks in the North Pacific and recommended a suite of management changes to reduce mortality. The authors also stressed the need for long-term ecological studies and basic surveys of sea ducks as essential for making informed management decisions. Importantly, the North American Waterfowl Management Plan (NAWMP) was updated in 1994 with immediate objectives to improve the understanding of population status, production, harvest, and factors affecting mortality and survival of sea ducks (U.S. Fish and Wildlife Service and the Canadian Wildlife Service 1994). Thereafter, Petersen and Hogan (1996) presented a paper at the 7th International Waterfowl Symposium that summarized the status of sea ducks in North

America, identified factors that may have been limiting sea ducks, and proposed studies to collect information critical to develop species-specific management plans. In 1997, a coalition of the US and Canadian federal, state, provincial, and nongovernment conservation organizations proposed the establishment of the Sea Duck Joint Venture (SDJV). In 1998, the NAWMP Committee approved the creation of the SDJV to "promote development of short- and long-term information gathering programs to determine basic parameters of sea duck populations, such as delineation of ranges and subunits, abundance and trends, production, harvest, and survival rates" (Sea Duck Joint Venture Management Board 2008). To achieve these goals, the SDJV relies on international partnerships and is guided by detailed strategic and implementation plans (Sea Duck Joint Venture Management Board 2008, Sea Duck Joint Venture 2013).

Much has been accomplished to fill the enormous gaps in our knowledge of North American sea ducks since Canada and the United States took actions to list and fund studies of the eastern populations of Harlequin Duck and Barrow's Goldeneye, the world population of Spectacled Eider, and the North American breeding population of Steller's Eider. Likewise, funding provided by the SDJV for priority monitoring and research has resulted in a much clearer picture of the status and biology of North American sea ducks. Inspired by the recent rapid increase in our knowledge of sea duck ecology, we conceived this volume of Studies in Avian Biology to summarize and synthesize what we now know about the tribe Mergini in North America and highlight the remaining data gaps that must be addressed to facilitate continued efforts to conserve this fascinating group of waterfowl.

This book is composed of 15 interrelated chapters that focus on many aspects of the ecology of members of the tribe Mergini that occur in North America (Table I.1). The chapters are organized conceptually so as to not duplicate other efforts that have been prepared on a species-by-species basis such as the species accounts in the Birds of North America series, and to compare and contrast the ecological attributes of birds in the tribe. To enhance the interpretation of our results, conclusions, and recommendations, we have drawn on research conducted on the same species as well as other sea duck populations outside of North America. To create

this volume, 27 subject experts developed in-depth treatments of sea duck population status and trends, phylogeography and phylogenetics, population dynamics and demography, diseases, reproductive energetics and cross-seasonal effects, contaminant burdens, foraging behavior and energetics, migration strategies, molt ecology, reproductive strategies and behavior, harvest, habitat affinities, and conservation issues. We hope that the collection of integrated chapters will serve as a primary reference for researchers, managers, teachers, and students interested in detailed treatments of the biology, ecology, and conservation of this unique and widespread tribe of marine-dependent waterfowl.

We, the editors of Ecology and Conservation of North American Sea Ducks, would like to thank the following reviewers who contributed significantly to the quality of the chapters in this book: Courtney Amundson, Rebecca Bentzen, Tim Bowman, Jan Bustnes, Susan De La Cruz, Kathy Dickson, Lynne Dickson, John Elliott, George Finney, Anthony Fox, Chris Franson, Scott Gilliland, Matthieu Guillemain, Jerry Hupp, Sam Iverson, Kim Jaatinen, Don Kraege, Phil Lavretsky, Tyler Lewis, Scott McWilliams, Steffen Oppel, John Pearce, Jeffrey Peters, Abby Powell, Hannu Pöysä, Andy Ramey, Austin Reed, Joel Schmutz, Caz Taylor, Jonathan Thompson, Caroline Van Hemert, Mark Wayland, Caitlin Wells, Heather Wilson, and Elise Zipkin. Leslie Holland-Bartels and Carl Markon, U.S. Geological Survey, Alaska Science Center, provided administrative and financial support for this project. Vanessa Skean was invaluable as editorial assistant on this project. We also express our gratitude to Brett Sandercock, Series Editor for Studies in Avian Biology, for the expert guidance and oversight in editing the chapters that comprise this volume. Chuck Crumly and the staff at CRC Press, Taylor & Francis Group, provided professional and timely support in bringing this volume to publication. Funding support for the publication of this book was provided by the Sea Duck Joint Venture of the North American Waterfowl Management Plan, the Alaska Science Center of the U.S. Geological Survey, Environment Canada, and the University of California at Davis.

LITERATURE CITED

American Ornithologists' Union. 1998. Check-list of North American birds, 7th edn. American Ornithologists' Union, Washington, DC.

Austin, O. L. Jr. 1932. The birds of Newfoundland Labrador. No. VII. Nuttall Ornithological Club, Cambridge, MA.

Bédard, J., A. Nadeau, J.-F. Giroux, and J.-P. L. Savard. 2008. Eiderdown: characteristics and harvesting procedures. Société Duvetnor Ltée and Canadian Wildlife Service, Environment Canada, Québec Region, Québec, Canada.

Bellrose, F. C. 1976. Ducks, geese and swans of North America. The Wildlife Management Institute, Stackpole Books, Harrisburg, PA.

Bent, A. C. 1951. Life histories of North American wild fowl (Vol. 2). Dover Publications, New York, NY.

Bond, A. L., P. W. Hicklin, and M. Evans. 2007. Daytime spring migrations of scoters (Melanitta spp.) in the Bay of Fundy. Waterbirds 30:566–572.

Bordage, D., and J.-P. L. Savard. 1995. Black Scoter (Melanitta nigra). in A. Poole and F. Gill (editors), The birds of North America (No. 177). The Academy of Natural Sciences, Philadelphia, PA and the American Ornithologists' Union, Washington, DC.

Brown, P. W., and L. H. Fredrickson. 1997. White-winged Scoter (Melanitta fusca). in A. Poole and F. Gill (editors), The birds of North America (No. 274). The Academy of Natural Sciences, Philadelphia, PA and the American Ornithologists' Union, Washington, DC.

Caithamer, D. F., M. Otto, P. I. Padding, J. R. Sauer, and G. H. Haas. 2000. Sea ducks in the Atlantic flyway: population status and a review of special hunting seasons. Office of Migratory Bird Management, United States Fish and Wildlife Service, Laurel, MD.

Chesser, R. T., R. C. Banks, F. K. Barker, C. Cicero, J. L. Dunn, A. W. Kratter, I. J. Lovette et al. 2010. Fifty-first supplement to the American Ornithologists' Union check-list of North American birds. Auk 127:726–744.

Chilton, G. 1997. Labrador Duck (Camptorhynchus labradorius). in A. Poole and F. Gill (editors), The birds of North America (No. 307). The Academy of Natural Sciences, Philadelphia, PA and the American Ornithologists' Union, Washington, DC.

Dugger, B. D., K. M. Dugger, and L. H. Fredrickson. 1994. Hooded Merganser (Lophodytes cucullatus). in A. Poole and F. Gill (editors), The birds of North America (No. 98). The Academy of Natural Sciences, Philadelphia, PA and the American Ornithologists' Union, Washington, DC.

Eadie, J. M., M. L. Mallory, and H. G. Lumsden. 1995. Common Goldeneye (Bucephala clangula). in A. Poole and F. Gill (editors), The birds of North America (No. 170). The Academy of Natural Sciences, Philadelphia, PA and the American Ornithologists' Union, Washington, DC.

Eadie, J. M., J.-P. L. Savard, and M. L. Mallory. 2000. Barrow's Goldeneye (Bucephala islandica). in A. Poole and F. Gill (editors), The birds of North America (No. 548). The Academy of Natural Sciences, Philadelphia, PA and the American Ornithologists' Union, Washington, DC.

Flint, P. L., and M. P. Herzog. 1999. Breeding of Steller's Eiders, Polysticta stelleri, on the Yukon-Kuskokwim Delta, Alaska. Canadian Field-Naturalist 113:306–308.

Fredrickson, L. H. 2001. Steller's Eider (Polysticta stelleri). in A. Poole and F. Gill (editors), The birds of North America (No. 571). The Academy of Natural Sciences, Philadelphia, PA and the American Ornithologists' Union, Washington, DC.

Gauthier, G. 1993. Bufflehead (Bucephala albeola). in A. Poole and F. Gill (editors), The birds of North America (No. 67). The Academy of Natural Sciences, Philadelphia, PA and the American Ornithologists' Union, Washington, DC.

Gilliland, S. G., H. G. Gilchrist, R. F. Rockwell, G. J. Robertson, J.-P. L. Savard, F. Merkel, and A. Mosbech. 2009. Evaluating the sustainability of harvest among northern Common Eiders Somateria mollissima borealis in Greenland and Canada. Wildlife Biology 15:24–36.

Goudie, R. I., S. Brault, B. Conant, A. V. Kondratyev, M. R. Petersen, and K. Vermeer. 1994. The status of sea ducks in the north Pacific Rim: toward their conservation and management. Transactions of the North American Wildlife and Natural Resources Conference 59:27–49.

Goudie, R. I., G. J. Robertson, and A. Reed. 2000. Common Eider (Somateria mollissima). in A. Poole and F. Gill (editors), The birds of North America (No. 546). The Academy of Natural Sciences, Philadelphia, PA and the American Ornithologists' Union, Washington, DC.

Greenwood, J. J. D. 2007. Citizens, science and bird conservation. Journal of Ornithology 148:S77–S124.

Hansen, K. 2002. A farewell to Greenland's wildlife. Gads Forlag et Narayana Press, Gylling, Copenhagen, Denmark.

Harrop, A. H. J., J. M. Collinson, S. P. Dudley, C. Kehoe, and the British Ornithologists' Union Records Committee. 2013. The British list: a check-list of birds of Britain (8th edn). Ibis 155:635–676.

Johnsgard, P. 1965. Handbook of waterfowl behavior: tribe mergini (Sea Ducks). University of Nebraska, Lincoln, NE.

Kear, J. 2005. Ducks, geese and swans (Vol. 2). Oxford University Press, Oxford, UK.

Livezey, B. C. 1995. Phylogeny and evolutionary ecology of modern seaducks (Anatidae: Mergini). Condor 97: 233–255.

Mallory, M., and K. Metz. 1999. Common Merganser (Mergus merganser). in A. Poole and F. Gill (editors), The birds of North America (No. 442).

The Academy of Natural Sciences, Philadelphia, PA and the American Ornithologists' Union, Washington, DC.

Merkel, F. R. 2004. Impact of hunting and gillnet fishery on wintering eiders in Nuuk, southwest Greenland. Waterbirds 27:469–479.

Merkel, F. R. 2010. Evidence of recent population recovery in Common Eiders breeding in western Greenland. Journal of Wildlife Management 74:1869–1874.

Myres, M. T. 1959. Display behavior of Bufflehead, scoters and goldeneyes at copulation. Wilson Bulletin 71:159–168.

Naves, L. C. 2012. Alaska migratory bird subsistence harvest estimates, 2010. Alaska Migratory Bird Co-Management Council. Technical Paper No. 376, Alaska Department of Fish and Game, Anchorage, AK.

Petersen, M. R., J. B. Grand, and C. P. Dau. 2000. Spectacled Eider (Somateria fischeri). in A. Poole and F. Gill (editors), The birds of North America (No. 547). The Academy of Natural Sciences, Philadelphia, PA and the American Ornithologists' Union, Washington, DC.

Petersen, M. R., and M. E. Hogan. 1996. Seaducks: a time for action. Seventh International Waterfowl Symposium, Memphis, TN.

Petersen, M. R., W. W. Larned, and D. C. Douglas. 1999. At-sea distributions of Spectacled Eiders (Somateria fischeri): a 120 year-old mystery resolved. Auk 116:1009–1020.

Phillips, J. C. 1925. A natural history of the ducks (Vol. 3). Houghton-Mifflin Company, Boston, Massachusetts, USA.

Phillips, J. C. 1926. A natural history of the ducks (Vol. 4). Houghton-Mifflin Company, Boston, MA.

Phillips, J. C., and F. C. Lincoln. 1930. American waterfowl: their present situation and the outlook for their future. Houghton-Mifflin Company, Boston, MA.

Robert, M., R. Benoit, and J.-P. L. Savard. 2000. COSEWIC status report on the Barrow's Goldeneye (Bucephala islandica) eastern population in Canada. Committee on the Status of Endangered Wildlife in Canada, Canadian Wildlife Service, Ottawa, Ontario, Canada.

Robertson, G. J., and R. I. Goudie. 1999. Harlequin Duck (Histrionicus histrionicus). in A. Poole and F. Gill (editors), The birds of North America (No. 547). The Academy of Natural Sciences, Philadelphia, PA and the American Ornithologists' Union, Washington, DC.

Robertson, G. J., and J.-P. L. Savard. 2002. Long-tailed Duck (Clangula hyemalis). in A. Poole and F. Gill (editors), The birds of North America (No. 651).

The Academy of Natural Sciences, Philadelphia, PA and the American Ornithologists' Union, Washington, DC.

Savard, J.-P. L., D. Bordage, and A. Reed. 1998. Surf Scoter (Melanitta perspicillata). in A. Poole and F. Gill (editors), The birds of North America (No. 363). The Academy of Natural Sciences, Philadelphia, PA and the American Ornithologists' Union, Washington, DC.

Sea Duck Joint Venture. 2013. Sea duck joint venture implementation plan for April 2013 through March 2016. Report of the Sea Duck Joint Venture, U.S. Fish and Wildlife Service, Anchorage, Alaska and Environment Canada, Sackville, New Brunswick, Canada.

Sea Duck Joint Venture Management Board. 2008. Sea duck joint venture strategic plan 2008–2012. U.S. Fish and Wildlife Service, Anchorage, Alaska, USA and Canadian Wildlife Service, Sackville, New Brunswick, Canada.

Sullivan, B. L., C. L. Wood, M. J. Iliff, R. E. Bonney, D. Fink, and S. Kelling. 2009. eBird: a citizen-based bird observation network in the biological sciences. Biological Conservation 142:2282–2292.

Suydam, R. S. 2000. King Eider (Somateria spectabilis). in A. Poole and F. Gill (editors), The birds of North America (No. 491). The Academy of Natural Sciences, Philadelphia, PA and the American Ornithologists' Union, Washington, DC.

Thomas, P. W., and M. Robert. 2001. Updated COSEWIC status report of the eastern Canada Harlequin Duck (Histrionicus histrionicus). Committee on the Status of Endangered Wildlife in Canada, Canadian Wildlife Service, Ottawa, Ontario, Canada.

Titman, R. D. 1999. Red-breasted Merganser (Mergus serrator). in A. Poole and F. Gill (editors), The birds of North America (No. 443). The Academy of Natural Sciences, Philadelphia, PA and the American Ornithologists' Union, Washington, DC.

Todd, W. E. C. 1963. Birds of the Labrador Peninsula and adjacent areas. University of Toronto Press, Toronto, Ontario, Canada.

U.S. Fish and Wildlife Service. 1996. Spectacled Eider recovery plan. Anchorage, AK.

U.S. Fish and Wildlife Service. 2002. Steller's Eider recovery plan. Fairbanks, AK.

U.S Fish and Wildlife Service and Canadian Wildlife Service. 1994. 1994 update to the North American waterfowl management plan. Environment Canada, U.S. Department of Interior, and de Desarrollo Social Mexico, Washington, DC.

CHAPTER ONE

Status and Trends of North American Sea Ducks*

REINFORCING THE NEED
FOR BETTER MONITORING

*Timothy D. Bowman, Emily D. Silverman,
Scott G. Gilliland, and Jeffery B. Leirness*

Abstract. The value of existing waterfowl survey data for assessing the status and trends of North American sea duck populations is limited due to short time series, insufficient geographic coverage, improper timing, and species identification problems. Despite these shortcomings, contemporary data provide insights into the status of several sea duck populations. In this chapter, we synthesize available information on population status and trends in abundance for sea ducks and recommend efforts that could improve our ability to monitor sea duck populations. The Alaska breeding population of Spectacled Eiders is currently stable (Arctic Coastal Plain) or increasing (Yukon-Kuskokwim Delta) in numbers. Steller's Eiders (*Polysticta stelleri*) wintering in Alaska have declined since the early 1990s. Spectacled and Steller's Eiders remain below historic levels and are listed as threatened in the United States. In western North America, King Eiders (*Somateria spectabilis*) declined substantially between the mid-1970s and mid-1990s; recent data suggest regional differences, but a stable population overall. There is insufficient information on trend for King Eiders in eastern North America. An assessment of trends for Pacific Common Eiders (*S. mollissima v-nigra*) is based on limited information, but data suggest that this subspecies declined substantially in northern parts of its range in the 1980s to the early 2000s. Recent regional trend estimates note declines in central arctic Canada and northwestern Alaska, and stable to increasing numbers in other parts of Alaska. Population trends for American Common Eiders (*S. dresseri*) are variable range-wide, with apparent increases in northern parts of their range and decreases in southern parts. Trends for Hudson Bay (*S. m. sedentaria*) and Northern subspecies (*S. m. borealis*) of Common Eiders are uncertain. The population trajectories of the three scoter species (*Melanitta* spp.) are also not well understood, but available data suggest that, as a group, scoters decreased from the 1980s to the early 2000s, with greater declines noted in the northern boreal forest and northern prairies than in Alaska, and with overall increases since about 2004. Black Scoters on Pacific breeding areas have declined significantly since the mid-1970s, but have increased in number in the last decade. There is no measurable trend for eastern Black Scoters. The limited data for Long-tailed Ducks (*Clangula hyemalis*) suggest long-term declines in the parts of their range that are surveyed, with

* Bowman, T. D., E. D. Silverman, S. G. Gilliland, and J. B. Leirness. 2015. Status and trends of North American sea ducks: Reinforcing the need for better monitoring. Pp. 1–28 in J.-P. L. Savard, D. V. Derksen, D. Esler, and J. M. Eadie (editors). Ecology and conservation of North American sea ducks. Studies in Avian Biology (no. 46), CRC Press, Boca Raton, FL.

more stable numbers in recent years. Buffleheads (*Bucephala albeola*), goldeneyes (*B. clangula* and *B. islandica*) and mergansers (*Mergus* spp. and *Lophodytes cucullatus*) have increased, although lack of differentiation among species of both goldeneyes and mergansers prohibits reliable species-specific evaluations. Numbers of Harlequin Ducks (*Histrionicus histrionicus*) along the Atlantic coast are increasing, while the Pacific trend is unknown. Data suggest that, for the 22 populations of North America sea ducks currently recognized as distinct or allopatric, 11 populations appear to be stable or have increased in abundance over the last 10–20 years, and two populations are apparently declining. Data are insufficient to determine status for the remaining nine populations. Reliable information about population status and trends requires surveys designed with specific consideration of sea duck distribution and phenology. We recommend increasing observer training, incorporating detection adjustments, and using aerial photography to improve the accuracy of species identification and abundance estimation. Management agencies in the United States and Canada must devote greater resources to monitoring sea ducks if they wish to better inform harvest management, focus habitat conservation efforts on areas of greatest importance to sea ducks, and effectively evaluate their management actions.

Key Words: aerial and nest surveys, Christmas Bird Count, population trends, species identification.

Estimates of trends in waterfowl abundance are critical to understand population status, to determine the causes of changes in population size, and to assess the need for, and response to, management actions. In North America, measures of status are typically derived from breeding ground surveys or, for a few species, winter inventories. However, estimates of abundance and population trends for sea ducks are generally poor or lacking altogether. These species inhabit vast, remote breeding areas, and molting and wintering birds often gather on large lakes, coastal waters, and far at sea—habitats that are difficult, costly, and risky to survey. Even the distributions of sea duck species are poorly understood. For some species, population assessment is further complicated by the fact that their ranges cross international boundaries into countries that do not adequately monitor sea duck populations including Russia, Japan, Mexico, and Greenland.

Many previous and ongoing waterfowl surveys are of limited value for monitoring sea duck populations because they do not cover large portions of breeding ranges or offshore wintering areas, are not optimally timed to capture peak counts of breeding birds, or only identify sea ducks to genus or species groups such as scoters, goldeneyes, or mergansers. Indices of abundance from such surveys are often highly variable and of limited utility in evaluating trends (Stott and Olson 1972, Heusmann 1999).

Currently, only two sea duck populations are adequately monitored on a routine basis in North America: Spectacled Eiders (*Somateria fischeri*) and the Pacific population of Black Scoters (*Melanitta americana*). Within North America, both species breed only in Alaska. In 1993, Spectacled Eiders were listed as threatened under the U.S. Endangered Species Act, prompting development of surveys of this species on its breeding areas. A Pacific Black Scoter survey encompassing most of its breeding area was initiated in 2004, evolving from several years of design and operational adjustments informed by satellite telemetry studies that identified proper timing and appropriate geographic coverage.

Despite the limitations of most waterfowl surveys for sea duck assessment, several large-scale, long-term monitoring programs provide information on sea duck populations. Most notable is the Waterfowl Breeding Population and Habitat Survey (WBPHS), which is conducted annually by the U.S. Fish and Wildlife Service (USFWS) and the Canadian Wildlife Service. There are also a number of rigorous smaller-scale, shorter-duration surveys that contribute to our understanding of sea duck status.

In this chapter, we drew upon a variety of data sources to examine population status and trends for sea ducks. The timing and methods of these surveys can differ substantially, so it is not possible to directly compare the abundance estimates they produce nor resolve contradictory information about population trends. For example, spring breeding season surveys enumerate adult breeders, while late summer molting surveys count primarily adult males, and winter surveys enumerate both adults and young following the fall migration. Because they are measuring different demographic components of the population, population estimates and trends among these various surveys may differ.

In addition, most surveys do not incorporate an adjustment for incomplete detection of birds or nests. A limited number of surveys have employed double-observer techniques to estimate detection (Raven and Dickson 2006; Stehn and Platte 2012; J. B. Fischer and R. A. Stehn, U.S. Fish and Wildlife Service, unpubl. report). Double-observer methods account for perception bias of observers missing birds that are within view, but not availability bias if birds are present but out of view under water or in vegetation (Williams et al. 2002). Surveys from helicopters have low perception bias for sea ducks (estimates from Eastern Canada are 0.97, N. Plante and D. Bordage, Canadian Wildlife Service, unpubl. data, and 0.98, Gilliland et al. 2009b) and surveys in the WBPHS protocols correct counts from fixed-wing platforms using paired helicopter data, but these corrections are subject to additional availability bias, because the fixed-wing airplanes and helicopters do not fly at the same time, nor do they cover exactly the same survey area, and sample sizes are often small (Smith 1995; E. D. Silverman, pers. obs.). Given the infrequent use of detection correction in sea duck surveys, and the limits of the existing methods, most estimates of abundance should be considered indices and not necessarily comparable across surveys. Thus, aside from the annotated summary in the electronic Appendix (www.crcpress.com/product/isbn/9781482248975), we have generally refrained from reporting numerical estimates of abundance for most sea duck populations and include indices only if they provide perspective on the relative size or order of magnitude of surveyed populations or subpopulations.

In general and unlike most goose and some duck populations, sea ducks have not been classified into distinct management units based on survey or banding data. Nonetheless, Black Scoters, Harlequin Ducks (Histrionicus histrionicus), and Barrow's Goldeneyes (Bucephala islandica) are well documented as having distinct populations in eastern and western North America, which should be considered separately for purposes of conservation and management. Currently, the Sea Duck Joint Venture recognizes 22 distinct populations for the 15 species of sea ducks in North America (Sea Duck Joint Venture Management Board 2008).

We use data from surveys that cover a large geographic range such as multiple provinces and states or entire coastlines, a significant portion of a species range, or a long time series (>10 years). When such survey data were lacking, we have also presented less detailed information and analyses from local or periodic surveys. We chose not to consider data from surveys that, for various reasons, do a relatively poor job of monitoring sea ducks. For example, several surveys do not adequately cover offshore habitats used by some sea ducks; these include the Mid-Winter Waterfowl Survey (Heusmann 1999), the Breeding Bird Survey (Sauer et al. 2013), and the Atlantic Sea Duck Survey conducted from 1990 to 2002 (D. F. Caithamer et al., U.S. Fish and Wildlife Service, unpubl. report). Because we have access to the WBPHS data, we included some original analyses for these data; all other estimates and summaries were obtained from existing reports, publications, or pers. comms. with researchers.

DATA SOURCES

Waterfowl Breeding Population and Habitat Survey

A large-scale study of North American waterfowl breeding habitats was initiated by the USFWS and its partners in 1947 (Crissey 1949). These fixed-wing aerial surveys evolved into the WBPHS with the current design established in Alaska in 1966 and in Canada and the US prairies in 1974 and with the current survey protocol mostly unchanged since 1974. The survey is divided into two broad areas: (1) the midcontinent area, which covers waterfowl habitats in the northern Yukon, the Northwest Territories, Alberta, Saskatchewan, Manitoba, and western Ontario in Canada, and in Alaska, Montana, and the Dakotas in the United States (hereafter, midcontinent and Alaska); and (2) the eastern survey area, which includes parts of Ontario, Quebec, Newfoundland and Labrador, and the Maritime provinces in Canada, and New York and Maine in the United States (hereafter, eastern; Figure 1.1). Historically, the survey was focused on estimating population sizes for prairie dabbling ducks. As a result, the survey is optimally timed for these species and is too early for sea ducks. Also, effort is concentrated in the prairie-pothole region, which constitutes only

Figure 1.1. Transects surveyed by the WBPHS: the midcontinent and Alaska transects are thick solid lines; the eastern transects are dashed lines. Stratum 50, which is considered part of the midcontinent survey, but analyzed here with the eastern survey data, is indicated by thin solid lines.

a fraction of the habitat types used by sea ducks during the breeding season. Smith (1995) provides a description of the survey protocol. Survey observers record the abundance of Buffleheads (Bucephala albeola), Long-tailed Ducks (Clangula hyemalis), and species groups of goldeneyes, mergansers, eiders, and scoters on 200 m wide strip transects.

Eastern Waterfowl Survey

The Eastern Waterfowl Survey, conducted by the Canadian Wildlife Service, was developed primarily to survey American Black Ducks (Anas rubripes), but it has also monitored several species of inland breeding sea ducks since 1990 (M. C. Bateman et al., Canadian Wildlife Service, unpubl. data). All waterfowl are recorded within square plots located in parts of Quebec, Ontario, New Brunswick, Nova Scotia, Newfoundland, and Labrador (Figure 1.2). The current protocol involves surveying 304, 5 km × 5 km, plots by helicopter on a rotational basis with about half of all plots surveyed in a given year. The survey covers only a small portion of breeding ranges for several eastern sea duck populations but provides species-specific estimates of trend and population

size for the three merganser and three scoter species that occur there. The survey is well timed for mergansers, Buffleheads, and Common Goldeneyes (Bucephala clangula) but overlaps only with the early part of the breeding period for scoters.

Yukon–Kuskokwim Delta Aerial Goose–Duck–Waterbird Survey

Although this aerial transect survey was designed primarily to estimate breeding populations of geese on the Yukon–Kuskokwim Delta, Alaska, ducks are also recorded. The survey is flown by the USFWS in early June using fixed-wing aircraft and covers most of the high-density waterfowl habitats in the coastal zone of the Yukon–Kuskokwim Delta, Alaska (Figure 1.3). A backseat observer records only ducks and other waterbirds. Since 1988, survey techniques and observers have been relatively consistent and indices of breeding duck populations are available for every year except 2011 (R. M. Platte and R. A. Stehn, U.S. Fish and Wildlife Service, unpubl. report). This survey covers the primary breeding areas and provides reliable estimates of Common Eiders (Somateria

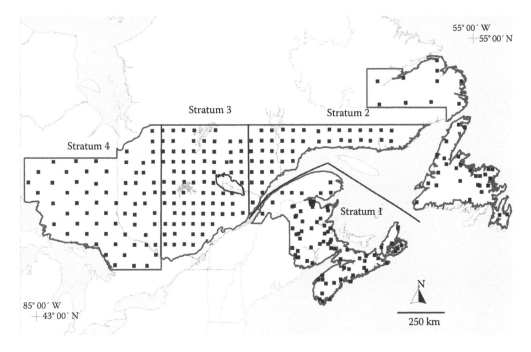

Figure 1.2. Plot locations for the Eastern Waterfowl Breeding Survey.

Figure 1.3. Alaskan survey areas for the Arctic Coastal Plain Breeding Waterfowl Survey (gray) and Pacific Black Scoter Breeding Survey (black) and location of Yukon–Kuskokwim Delta.

mollissima) and Spectacled Eiders but is of limited use for Long-tailed Ducks and Black Scoters, which are more abundant in areas further inland.

Yukon–Kuskokwim Delta Nest Survey

Since 1985, the USFWS has conducted ground-based sampling to monitor nest populations of waterfowl and waterbirds on the coastal zone of the Yukon–Kuskokwim Delta, Alaska (Figure 1.3). Two

to four biologists search all nesting habitat within randomly located plots and record data for all nests found. This survey is currently designed to optimize estimates of nesting Spectacled Eider but also provides indices of local nesting populations of geese and other duck and waterbird species (J. B. Fischer and R. A. Stehn, unpubl. report). Counts of nests are adjusted for incomplete detection, and species-specific estimates from the nest plot survey are expanded to the entire Yukon–Kuskokwim

Delta using data from the concurrent aerial survey (R. M. Platte and R. A. Stehn, unpubl. report).

Alaska Arctic Coastal Plain Breeding Waterfowl Survey

This breeding population survey, initiated in 1992 and restricted to northern Alaska, is important because it is one of the few long-running surveys conducted on arctic breeding grounds (Figure 1.3). It is a fixed-wing stratified transect survey conducted annually by USFWS. The survey was redesigned in 2007 to combine two previously independent surveys: an eider-specific survey (W. W. Larned et al., U.S. Fish and Wildlife Service, unpubl. report) and a more general waterfowl survey (E. J. Mallek et al., U.S. Fish and Wildlife Service, unpubl. report). Data were reanalyzed to provide indices of breeding population size (R. A. Stehn et al., unpubl. report). The survey is appropriately timed (mid-June) and covers extensive breeding habitat for several species of sea ducks including Long-tailed Ducks, King Eiders (*Somateria spectabilis*), and Spectacled Eiders.

Pacific Black Scoter Breeding Survey

From 2004 to 2012, an aerial transect survey was flown annually (except 2011) in western Alaska to estimate breeding population size for Pacific Black Scoters (R. A. Stehn and R. M. Platte, U.S. Fish and Wildlife Service, unpubl. report). This survey was designed based on data from previous reconnaissance surveys and satellite telemetry studies to cover most (>80%) of the breeding range of Pacific Black Scoters (Figure 1.3) and is timed appropriately for breeding scoters. The survey is intended to provide precise estimates of breeding population size for Pacific Black Scoters. Using fixed-wing aircraft, two observers record all scoters, scaup (*Aythya* spp.), and Long-tailed Ducks. Most scoters are identified to species, and >95% are identified as Black Scoters. On a subset of transects, a double-observer technique is employed in most years to estimate observer-specific detection rates, which are used to adjust estimates (Magnussen et al. 1978, Pollock and Kendall 1987, Graham and Bell 1989).

British Columbia Coastal Waterbird Survey

Bird Studies Canada implemented a winter, shore-based survey in the Strait of Georgia in 1999 (Crewe et al. 2012). Observers count all species of waterbirds

at more than 200 predefined sites between December and February each year. Although the survey covers a relatively small geographic area, it is the only survey that provides trend information for several species of sea ducks not covered by other surveys in that region. Annual mean counts per site are used as indices of abundance and to estimate variance and trends in abundance.

Audubon Christmas Bird Counts

The National Audubon Society has conducted Christmas Bird Counts (CBC) annually since 1900. Between December 14 and January 5, volunteer teams count all birds in a count circle with a diameter of 24 km along assigned routes. The benefits of CBC data are its long time series, continental scope, and ability to examine species-specific trends, which can complement aerial survey data for species that are recorded only to species groups such as mergansers or goldeneyes and for species that are difficult to count during aerial surveys such as Hooded Mergansers (*Lophodytes cucullatus*) and Harlequin Ducks. The drawbacks to CBC data are that they are derived from nonrandom samples with variable effort, based on volunteer data, and do not cover offshore habitats and areas in the far north. We restricted summaries of CBC data to 1974–2011 to be consistent with analyses of WBPHS and only to sea duck species with more freshwater and near-coastal winter distributions. Data derived from CBC online (National Audubon Society 2011) were reported as birds observed per party hour; an effort to standardize for variable effort among years.

METHODS

WBPHS Trend Analysis

We used count data from the WBPHS to estimate trends in sea duck abundance. We fit trend models independently to the midcontinent and Alaska data from 1974 to 2012 and to the eastern data from 1996 to 2012. Although midcontinent survey work was initiated in 1947 and the eastern survey in 1990, we limited the trend analysis for the midcontinent and Alaska survey areas to post-1973 data because the survey protocols, transect locations, and effort allocation were variable prior to 1974. Likewise, the complete eastern survey area has been consistently surveyed only since 1996.

We did not combine observations from the two areas for the trend estimation because of the large discrepancy in the time span of the available data and because some species have distinct eastern and western breeding populations. We included the data from western Ontario, usually considered part of the midcontinent, in the eastern analysis because the area was not flown between 1974 and 1984 and is contiguous with the eastern area. Although WBPHS analyses typically employ visibility corrections based on helicopter surveys, we did not adjust the WBPHS counts by visibility because the corrections used for sea ducks are based on small data sets collected before GPS technology allowed the helicopter crews to track the actual airplane path and the corrections vary little over the regions where sea ducks are observed and are not annually adjusted.

Trend Estimates

Trends were estimated by fitting cubic polynomial regression models to the total indicated birds (TIBs, Smith 1995) counted on all survey transects by year. Models were fit assuming normal (Gaussian) errors and included an offset for total number of miles flown per year. If the P-value for the coefficient of the highest-order polynomial was greater than 0.05, this term was dropped and the model was updated. The process was repeated as necessary to find the simplest, adequate model for TIBs per year (cubic, polynomial, or linear). We examined the residuals from each resulting model using autocorrelation and partial autocorrelation plots and found the assumption of independent observations to be reasonable; probability plots indicated that the normality assumption was also reasonable.

For cases where the quadratic or cubic model was selected, we estimated the annual trend for each survey year as the value of the derivative of the fitted curve in that year and determined the associated 95% confidence interval for each annual trend estimate. All analyses were completed using R version 3.0.1 (R Development Core Team 2013) and ArcGIS version 10.1 (Environmental Systems Research Institute 2012).

Scoter Trends by Species

In 1993, the protocol in the Alaska area was modified so that all scoters within the innermost 100 m of each transect were identified to species. These data provided a consistent index of species-specific abundance and species composition across the Alaskan survey area between 1993 and 2012. We used these records to estimate species-specific scoter trends for the Alaska survey area (including also Old Crow Flats in the Northwest Territories) from 1993 to 2012. We also divided the Alaska survey area into two regions based on the scoter species composition: the Bering and Chukchi Sea coastal survey areas, where Black Scoters are more common, and the interior and Gulf of Alaska survey areas, where most White-winged Scoters (*Melanitta fusca*) and Surf Scoters (*M. perspicillata*) occur. We then separately fit trend models to the scoter TIBs from 1974 to 2012 for these regions to differentiate Black Scoter population trends from those of the other two species.

RESULTS

Spectacled Eider

The majority of the worldwide population of Spectacled Eiders breed in arctic Russia. A single aerial survey, conducted over 3 years, 1993–1995, provided a minimum population index of 146,000 birds for the arctic Russia breeding population (Hodges and Eldridge 2001; R. A. Stehn et al., unpubl. report). In North America, there are two smaller breeding populations of Spectacled Eiders, both in Alaska, on the Yukon–Kuskokwim Delta and Arctic Coastal Plain. Both the Alaskan and Russian populations winter together among the pack ice of the northern Bering Sea near St. Lawrence Island (Petersen et al. 1999).

Almost all eiders recorded during the WBPHS are observed on the Yukon–Kuskokwim Delta, Alaska (Figure 1.4). Eiders are not differentiated to species during this survey, and indices represent a mix of Spectacled, Common, and Steller's Eiders (*Polysticta stelleri*). However, Spectacled Eiders were formerly the more common species on the Delta; thus, the >90% decline documented between the 1970s and early 1990s is thought to have been driven largely by declines in Spectacled Eiders (Stehn et al. 1993). Anecdotal information indicated that populations in the other two primary breeding areas, Russia and Alaska Arctic Coastal Plain, also declined, along with the much smaller breeding population on St. Lawrence Island in the Bering Sea (U.S. Fish and Wildlife Service 1996).

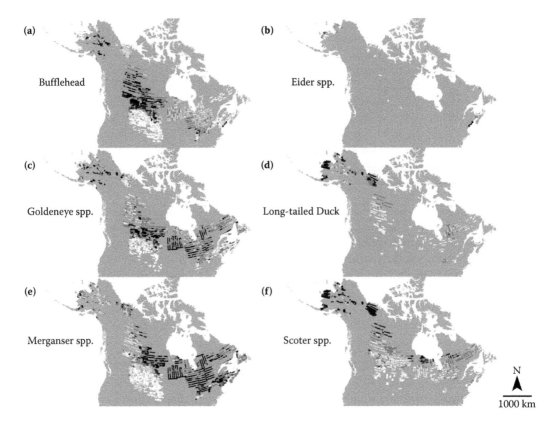

Figure 1.4. Average segment density (TIBs per km) for (a) Bufflehead (*B. albeola*), (b) eiders (*Somateria* spp. and *P. stelleri*), (c) goldeneyes (*B. clangula* and *B. islandica*), (d) Long-tailed Duck (*C. hyemalis*), (e) mergansers (*Mergus* spp. and *L. cucullatus*), and (f) scoters (*Melanitta* spp.) from the WBPHS for 1974–2012 in the midcontinent and Alaska and for 1996–2012 in the east. Segments are colored according to log density, low to high: white–light gray–dark gray–black. Segments where no birds were present (zero density) are not shown.

All Spectacled Eider breeding populations were listed as threatened in 1993 because of documented population declines.

Surveys of the only known wintering area of this species, presumed to represent the world population, were conducted in 1997, 1998, 2009, and 2010. Photographic surveys estimated a total population ranging from about 305,000 to 375,000 birds (Petersen et al. 1999; W. W. Larned et al., unpubl. report).

Surveys for breeding population trend were developed for the two Alaska breeding populations. The Yukon–Kuskokwim Delta Nest Survey counts are used in conjunction with the Yukon–Kuskokwim Delta Aerial Goose–Duck–Waterbird Survey to provide an annual estimate of the Yukon–Kuskokwim Delta breeding population. Recent estimates are about 6,000 nests with an increasing population trend since the early 1990s (J. B. Fischer and R. A. Stehn, unpubl. report). The Alaska Arctic

Coastal Plain Breeding Waterfowl Survey estimates approximately 6,400 total birds (not adjusted for incomplete detection) with a stable population since surveys began in 1992 (R. A. Stehn et al., unpubl. report).

Steller's Eider

The vast majority of Steller's Eiders in the Pacific originate from breeding areas in eastern Siberia, with only remnant or small breeding populations in western and northern Alaska. In 1997, the Alaska breeding population was listed as a threatened species under the U.S. Endangered Species Act based on a substantial decrease in abundance, reduction in breeding range, and vulnerability of the remaining Alaska breeding population to extirpation. Worldwide, Steller's Eiders are listed as vulnerable in the IUCN Red List of Threatened Species (BirdLife International 2012). An extensive survey of the Russian Far East in 1993–1995 reported more than 129,000 birds in the Pacific population (Hodges and Eldridge 2001).

Steller's Eiders have nearly disappeared as a breeding species from the Yukon–Kuskokwim Delta where there were perhaps several thousand breeding prior to the 1960s; only a few nests have been found there in recent years and the current population probably numbers fewer than a dozen (Kertell 1991, Flint and Herzog 1999). Numbers of breeding Steller's Eiders on the Arctic Coastal Plain are highly variable, with highest densities around the Barrow area (Obritschkewitsch and Ritchie 2011; Safine 2012; U.S. Fish and Wildlife Service, unpubl. report). Although ground and aerial surveys estimate several hundred birds occur in mid-June on the Arctic Coastal Plain, Alaska, in most years, the number of observations are few and highly variable (Obritschkewitsch and Ritchie 2011, Safine 2012). For example, the aerial survey index for indicated breeding birds resulted in an uninformative estimate (−1.5% per year; 90% CI: −7.3 to 5.4) for 1989–2013 growth rate (R. A. Stehn, U.S. Fish and Wildlife Service, unpubl. data).

Most Steller's Eiders breeding in eastern Russia and Alaska molt and winter along the Alaska Peninsula, the southwest coast of Alaska, and Aleutian Islands and migrate north to breeding areas during spring along the coast of western Alaska. A spring aerial survey for Steller's Eiders begun in 1992 provided an index to population size of birds migrating northward in coastal habitats in southwest Alaska (W. W. Larned, unpubl. report). This survey has yielded counts ranging from 55,000 to 138,000 birds, with an average of about 82,000 birds. The survey indicated a long-term average annual decline of 2.3% but a stable estimate from 2003 to 2011. Although the survey was subject to several potential biases and the estimates imprecise, it represented the only long-term data set specifically targeting population trend for Steller's Eiders in the Pacific. That said, Steller's Eiders are opportunistically counted during two other aerial surveys targeting Emperor Geese (*Chen canagica*), one in fall (E. J. Mallek and C. P. Dau, unpubl. report) and one in spring (C. P. Dau and E. J. Mallek, U.S. Fish and Wildlife Service, unpubl. report). Those two surveys are conducted in similar areas of Alaska, but at different times, and both surveys indicate similar rates of decline over the same period beginning in 1992 (spring survey: −3.5% per year; fall survey: −4.2% per

year). The spring aerial survey was replaced in 2012 by a photographic aerial survey of molting Steller's Eiders during fall on the primary molting areas in southwest Alaska, which yielded an index of 50,400 molting Steller's Eiders, primarily males (H. M. Wilson et al., U.S. Fish and Wildlife Service, unpubl. report).

Common Eider

Common Eiders breed extensively throughout Canada and Alaska, inhabiting arctic and subarctic marine environments most of the year and breeding in near-coastal wetlands or on coastal islands. There are four subspecies of Common Eiders in North America: Pacific (*S. mollissima v-nigra*), American (*S. m. dresseri*), Hudson Bay (*S. m. sedentaria*), and Northern (*S. m. borealis*). Survey methods for Common Eiders vary throughout their range, and most surveys of Common Eiders are specific to a particular subspecies based on geography. Status and trends for each of the four subspecies are discussed separately below.

Pacific Common Eider

Pacific Common Eiders nest on islands and a few mainland areas throughout the western and central Canadian Arctic from the Yukon coast to Queen Maud Gulf and north to include Victoria and Banks islands (Barry 1986, Cornish and Dickson 1997). Within Alaska, the largest breeding aggregations of Pacific Common Eiders have been found along the coastlines of the Aleutian Islands, Yukon–Kuskokwim Delta, Northwest Alaska, in the vicinity of the Seward Peninsula, and barrier islands of the Chukchi and Beaufort seas. Birds from breeding areas in northern Alaska and Canada winter primarily along the Chukotka Peninsula and near St. Lawrence Island in the northern Bering Sea (Petersen and Flint 2002, Dickson 2012a), and birds from the Yukon–Kuskokwim Delta winter along the north coast of Bristol Bay (Petersen and Flint 2002). Common Eiders breeding in the Aleutians Islands, Alaska, remain there during winter (M. R. Petersen, U.S. Geological Survey, pers. comm.). There has been no systematic effort to census the entire population (U.S. Fish and Wildlife Service, unpubl. report).

In the western Canadian Arctic, Barry (1986) estimated a total of 81,500 breeders, while spring migration data from 1993 suggested a total Canadian population of 90,500 birds (Alexander et al. 1997). This number is consistent with estimates of spring migrants passing Point Barrow (Suydam et al. 2000) from roughly the same time period. Numbers of Common Eiders breeding in the Bathurst Inlet area of Nunavut, Canada, declined by 43%–50% from nearly 17,000 to <10,000 over a 13-year period between 1995 and 2007–2008 (Raven and Dickson 2009).

Spring migration counts at Point Barrow, Alaska, which sample both Alaska Arctic Coastal Plain and western Canadian Arctic Common Eiders, suggested a decline of more than 50% between 1976 and 1996 (Woodby and Divoky 1982, Suydam et al. 2000), but the spring migration count in 2003 (about 120,000 birds; Suydam et al. 2004) increased 70% above the 1996 count, suggesting that the Canadian Arctic and Alaska Arctic Coastal Plain populations may have partially rebounded since the mid-1990s. The trends obtained from these migration counts should be viewed with caution, as considerable variation may be attributed to sampling methods and environmental variables (Day et al. 2004, Quakenbush et al. 2009). Aerial surveys specifically targeting Common Eiders in near-shore waters and along barrier islands of Alaska's Arctic Coastal Plain indicated a stable population of about 2400 birds from 1999 to 2009 (C. P. Dau and K. S. Bollinger, U.S. Fish and Wildlife Service, unpubl. report). Similar surveys of coastal areas between the Yukon Delta and the Arctic Coastal Plain in 2008 and 2009 estimated 4,000–5,000 Common Eiders (K. S. Bollinger and R. M. Platte, U.S. Fish and Wildlife Service, unpubl. report), which was 37% fewer than estimated in 1992 using similar survey methods (W. W. Larned et al., U.S. Fish and Wildlife Service, unpubl. report).

Nest surveys have been conducted by ground crews on the Yukon–Kuskokwim Delta since 1985 (J. B. Fischer and R. A. Stehn, unpubl. report). An average of approximately 3,500 Common Eider nests were estimated (after correcting for undetected nests; J. B. Fischer and R. A. Stehn, unpubl. report) on the Yukon–Kuskokwim Delta during the most recent 5-year period (2008–2012). The 25-year trend is increasing (7.5% per year; 90% CI: 5.0–10.1). An increasing trend (3.5% per year; 90% CI: 1.6–5.5) was also estimated from an aerial survey of the coastal zone of the Yukon–Kuskokwim Delta from 1988 to 2012 (R. M. Platte and R. A. Stehn, unpubl. report).

In the Aleutian Islands, various surveys conducted periodically since the 1970s provide minimal estimates for the region. Combining the most recent counts for all islands, about 25,000–30,000 Common Eiders reside in the Aleutians (U.S. Fish and Wildlife Service, unpubl. report). Reliable long-term trend data are nearly nonexistent for most of the Aleutians, as is the case for other large sections of the breeding range for the Pacific Common Eider.

Based on all available data throughout their range, it appears that overall Pacific Common Eiders are below historic levels, but with considerable uncertainty in the estimates and regional differences in trends.

American Common Eider

The American Common Eider subspecies breeds on islands in coastal waters from Labrador to Massachusetts and winters from Newfoundland to Rhode Island, with greatest numbers wintering in Maine and Massachusetts. While abundance for some segments of the population has been monitored regularly such as the St. Lawrence Estuary, there has been no comprehensive, range-wide survey of American Common Eider abundance and no estimates of overall trend in abundance exist.

Along the Lower North Shore of the Gulf of St. Lawrence, Quebec, ground-based colony counts have been made almost every 5 years since 1925 (Rail and Chapdelaine 2002, Rail and Cotter 2007) and indicate strong positive growth over the last 30 years (Figure 1.5), with a current estimate of about 17,000–20,000 breeding pairs for this region. Little information on numbers or trend exists for Newfoundland and Labrador. The number of male eider ducks counted on the Labrador coast during the breeding season increased from 8,800 in 1980 (Lock 1986) to 18,000 males in 1994—an increase of about 5% per year (S. G. Gilliland, Canadian Wildlife Service, unpubl. data). Similar growth has been observed across several archipelagos in northern Newfoundland with an average increase of 5% per year between 1988 and 2006 and a current index of about 6,000 males (S. G. Gilliland, unpubl. data).

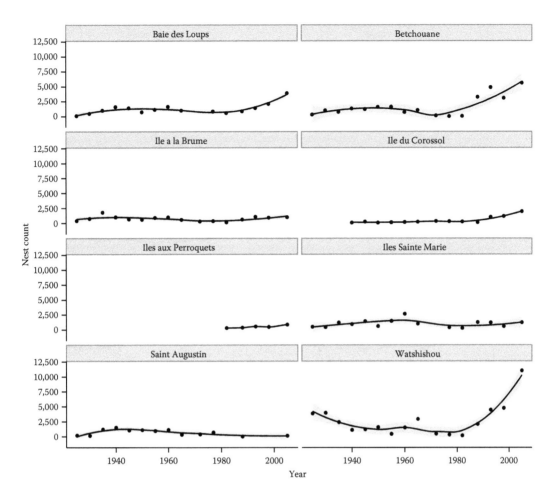

Figure 1.5. Number of American Common Eider (*S. m. dresseri*) nests detected in seabird colonies on Migratory Bird Sanctuaries along the Lower North Shore of the Gulf of St. Lawrence, Quebec, 1925–2005. Fitted local polynomial regressions (solid curve) with 95% confidence intervals (gray shading) for fitted values are also plotted. (From Rail and Chapdelaine 2002, Rail and Cotter 2007.)

In the St. Lawrence Estuary, Duvetnor and the Société Protectrice des Eiders de l'Estuaire have collected eiderdown from various colonies, which has provided valuable information on colony size for this segment of the population. The number of eiders breeding in the estuary was stable from the mid-1960s to the late 1990s (Figure 1.6). Epidemics of avian cholera occurred in eider colonies in the estuary in 1976, 1985, and, most recently, in 2002, when an estimated 6,000 breeding females died (Joint Working Group on the Management of the Common Eider 2004). The number of nests in the estuary has not increased since the 2002 epidemic, and current estimates are between 20,000 and 30,000 nests.

In southwestern New Brunswick, a comprehensive aerial survey of breeding areas in the Bay of Fundy has been conducted biannually since 1991 and indicates an overall decline of 3.1% per year between 1991 and 2012 (P = 0.003; K. Conner, New Brunswick Department of Natural Resources, unpubl. data; Figure 1.7). The number of males detected on the survey was stable until the late 1990s then declined at a rate of about 5% per year since 2000 (P = 0.027, Figure 1.7).

In coastal Maine, Common Eiders were nearly extirpated in the early 1900s but rebounded in response to protective measures (Krohn et al. 1992, Allen 2000). By 2000, an estimated 29,000 pairs of eiders nested in 320 colonies. In 2012, the population estimate was revised to 22,740 pairs nesting on 312 islands, based on updated data for several colonies,

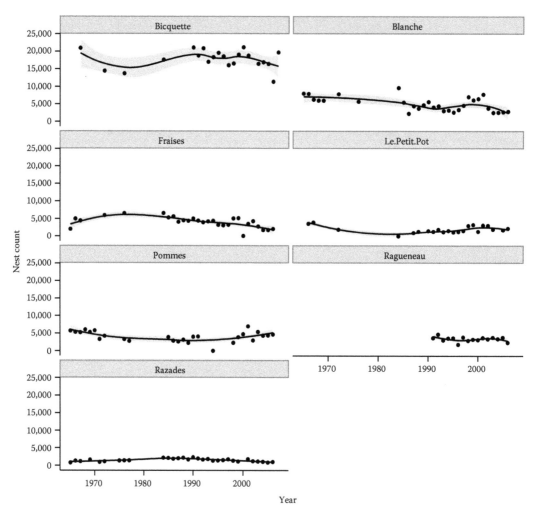

Figure 1.6. Number of American Common Eider (*S. m. dresseri*) nests detected in the largest colonies of the St. Lawrence Estuary, Quebec, during eiderdown collections, 1965–2007. Fitted local polynomial regressions (solid curve) with 95% confidence intervals (gray shading) for fitted values are also plotted. (From Duvetnor and Société Protectrice des Eiders de l'Estuaire, unpubl. data.)

although no comprehensive coast-wide survey has been conducted (R. B. Allen, Maine Department of Inland Fisheries and Wildlife, unpubl. data).

Numbers of American Common Eiders in the southern part of their breeding range appeared to peak in the late 1990s; however, data from colonies in the St. Lawrence Estuary, Quebec, and Bay of Fundy, New Brunswick, suggest this segment of the population has declined since then. Although there are no data available to evaluate trends, long-term banding programs in Nova Scotia and Maine have experienced increasing difficulty capturing breeding females at several major colonies

suggesting their numbers have been declining in these regions during the same period (R. B. Allen and R. Milton, Nova Scotia Department of Natural Resources, pers. comm.). In contrast, numbers of American Common Eiders in the northern part of their breeding range appear to have increased throughout the latter part of the twentieth and early twenty-first centuries.

Hudson Bay Common Eider

Hudson Bay Common Eiders are resident to Hudson Bay, where it is logistically difficult and expensive to conduct surveys for this subspecies. The breeding population was estimated to

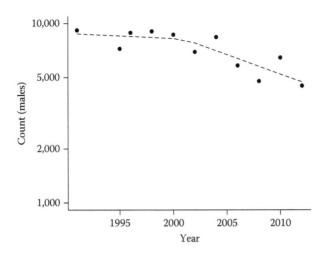

Figure 1.7. Number of male American Common Eiders (*S. m. dresseri*) counted during aerial breeding season surveys of southwestern Bay of Fundy, New Brunswick, between 1991 and 2012. Broken line represents fitted linear models broken at 2000. (From K. Connor, New Brunswick Department of Natural Resources, Fredericton, New Brunswick, Canada, unpubl. data.)

be 83,000 pairs in the 1980s (Nakashima and Murray 1988). The wintering population was estimated at about 255,000 birds in 2006 (Gilliland et al. 2008a). Ground-based colony surveys in the Belcher Islands were conducted twice, once between 1985 and 1989 as part of Nakashima and Murray's (1988) survey and in 1997 following a large winter kill in 1991–1992 (Robertson and Gilchrist 1998); a 75% decline in numbers of nesting eiders was documented. There is no recent estimate of trend for Hudson Bay Common Eiders.

Northern Common Eider

Northern Common Eiders breed in the Eastern Canadian Arctic, as well as in west Greenland. About 75% of the Canadian breeding population overwinter primarily in Greenland, with smaller numbers wintering in coastal Newfoundland and along the North Shore of the Gulf of St. Lawrence (Gilliland et al. 2009a). Obtaining reliable estimates of abundance during winter is complicated because they occur in three countries, including Canada, Greenland, and the French islands of St. Pierre and Miquelon, and mix to an unknown degree with other subspecies of Common Eiders (*S. m. dresseri*, Gilliland and Robertson 2009, Gilliland et al. 2009a).

An island colony at East Bay, Southampton Island, Nunavut, has been monitored since 1996. The colony was stable from 1996 to 2004, then increased to a maximum of 9,400

nests (G. Gilchrist, Environment Canada, pers. comm.) apparently in response to harvest reductions in Greenland that were implemented in 2002–2004 (Gilliland et al. 2008b, Merkel et al. 2008). Avian cholera was detected in 2005 resulting in a loss of a large portion of the colony in 2006 (Descamps et al. 2009). Cholera was still present on the colony in 2013 when the colony was about 2,300 nests, and recovery now appears to be restricted by frequent depredation by polar bears (*Ursus maritimus*, Iverson et al. 2014).

In northern Labrador, Northern Common Eiders appear to be doing well. In 1980, Lock (1986) estimated about 6,700 bred along the north coast of Labrador, and their numbers have been increasing at about 5% per year from 1980 to 2006 (Chaulk 2009). Numbers of breeding eiders along the central and southern coast of Ungava Bay have been stable to increasing between 1980 and 2000 (Chapdelaine et al. 1986, Falardeau et al. 2003). However, fewer eiders were detected breeding across northwestern Ungava Bay during the same time period, suggesting a significant decline of breeding eiders in that region (Falardeau et al. 2003).

Using photographic counts and ratio and regression estimators, Bordage et al. (1998) surveyed wintering areas in the Gulf of St. Lawrence, off Newfoundland, and off the French Islands of St. Pierre and Miquelon in 2003 and have since repeated the survey on a 3-year cycle. Estimates were 204,000 (SE = 23,000), 176,000 (SE = 8,000), and 204,800 (SE = 22,500) eiders in 2003, 2006, and 2009, respectively, suggesting that the segment of the population wintering in Canada has been relatively stable over the last 10 years (C. Lepage and S. G. Gilliland, Canadian Wildlife Service, unpubl. data).

King Eider

No comprehensive range-wide survey for King Eiders has been undertaken; only intermittent or regional surveys provide insights into abundance and population trends. Monitoring is complicated by the fact that many King Eiders that breed in western Canada and Alaska winter off the east coast of Russia (Phillips et al. 2006, Dickson 2012b), while many King Eiders breeding in Eastern Canada winter along the west coast of Greenland (Mosbech et al. 2006). The Sea Duck Joint Venture recognizes two populations of King Eiders in North America: a western arctic population and an eastern arctic population, although the dividing line between the two is not well documented and there likely is an area of overlap (Mehl et al. 2004, Dickson 2012b, Sea Duck Joint Venture 2013).

Western North America

King Eiders migrating past Point Barrow, Alaska, in spring represent the northern Alaska and western Canadian Arctic populations and have been counted there at periodic intervals between 1976 and 2003. From 1976 to 1996, counts suggested a decline of more than 50%, from 800,000 to 350,000 (Suydam et al. 2000), although a spring 2003 count estimated about 362,000 King Eiders (Suydam et al. 2004), suggesting the population may have stabilized since 1996. As noted for Pacific Common Eiders, these migration counts and apparent trends may be biased due to variation in sampling methods and environmental variables (Day et al. 2004, Quakenbush et al. 2009).

Systematic aerial surveys of breeding areas on western Victoria Island, Northwest Territories, documented a 50% decline in King Eider abundance from over 70,000 in 1992–1994 to 33,000 in 2004–2005 (Raven and Dickson 2006). On Alaska's Arctic Coastal Plain, an average total bird index of 21,000 birds was estimated for the most

recent 5-year period, 2008–2012. This survey indicates an increasing population with a long-term (1986–2012) average annual growth rate of 3.1% (90% CI: 2.1–4.1) and a growth rate of 2.4% (90% CI: 1.1–3.7) for the most recent 10-year period (2003–2012; R. A. Stehn et al., unpubl. report).

Eastern North America

No range-wide or contemporary trend information exists for King Eiders in eastern North America. A substantial decrease in numbers of wintering and molting King Eiders in Greenland, most of which breed in Canada, suggests that the eastern arctic population is declining (Environment Canada 2013). A significant decrease in numbers of King Eiders was observed in the Rasmussen Lowlands of Nunavut, a relatively small breeding area, between 1975–1976 and 1994–1995 (Gratto-Trevor et al. 1998).

Scoters

The WBPHS is the only long-term, broad-scale survey in which scoters have been consistently counted, although the survey is not optimally timed for scoters, which arrive relatively late on their breeding grounds. The annual counts are variable and may not accurately represent the breeding population in a given area. Based on satellite telemetry studies (De La Cruz et al. 2009, Sea Duck Joint Venture 2012), the midcontinent and Alaska areas include a substantial portion of the continental breeding range of White-winged and Surf Scoters. Highest densities of scoters are observed in the northern boreal and Alaska areas (Figure 1.4). During this survey, scoter species (Black, Surf, and White-winged) are typically recorded only to genera, so species-specific trends are difficult to determine. Nevertheless, the long time series is valuable for understanding overall trends of scoters, and some species-specific conclusions can be drawn based on known differences in breeding habitats and from observations recorded to species on Alaska areas since 1993.

WBPHS data for the midcontinent and Alaska region suggest a decreasing trend from the early 1980s to early 2000s, and increasing trend since about 2004 (Figure 1.8) with somewhat smaller declines in interior Alaska (~2% average annual decline in counted birds between the early 1980s

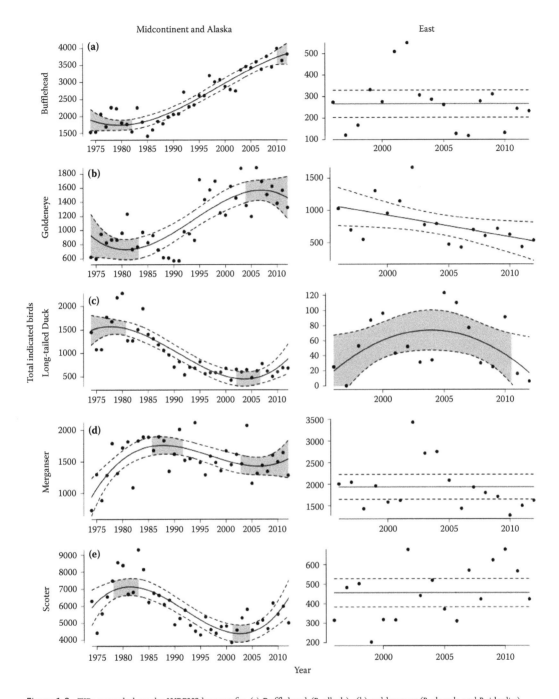

Figure 1.8. TIBs recorded on the WBPHS by year for (a) Bufflehead (*B. albeola*), (b) goldeneyes (*B. clangula* and *B. islandica*), (c) Long-tailed Duck (*C. hyemalis*), (d) mergansers (*Mergus* spp. and *L. cucullatus*), and (e) scoters (*Melanitta* spp.). Separate plots are shown for the midcontinent and Alaska (first column) and eastern (second column) survey areas. Fitted polynomial models (solid curve) with 95% confidence intervals for the fitted values (dashed curves) are also plotted. Gray shading indicates years when the estimated annual trend (change in TIBs per year) was not significantly different from zero (P-value > 0.05).

and early 2000s) than in northern boreal and prairie regions (~5% average annual decline). In the eastern strata of the WBPHS, counts are highly variable, with no discernible trend since 1996 (Figure 1.8). In what follows, we present species-specific summaries where data permit.

Black Scoter

Based on satellite telemetry studies of Black Scoter in both western (Bowman et al. 2008) and eastern North America (S. G. Gilliland, pers. comm.), it is clear that there are two distinct populations of Black Scoters: one that breeds in Alaska and one that breeds in Eastern Canada. We refer to these populations as Pacific and Atlantic Black Scoters, respectively.

Pacific Black Scoter

The WBPHS covers part of the breeding range of Pacific Black Scoters, mainly in western Alaska (Figure 1.4). Aerial survey and ground observations have indicated that nearly all scoters in the Alaska tundra strata are Black Scoters. Data from WBPHS for the Alaska tundra suggest a long-term (1974–2012) downward trend for Black Scoters, although no observable trend over all of the Alaska survey area in the last 20 years (Figure 1.9).

The Pacific Black Scoter survey, initiated in 2004 specifically to monitor this breeding population, indicates an increasing trend from 2004 to 2012. This survey should provide a more reliable indicator of trend than the WBPHS, although a cautious interpretation is required because the time series for this survey is relatively short. This survey employs an adjustment for incomplete detection, so that the index could be interpreted as an estimate of the breeding population size in the survey area. The average population size from 2004 to 2012 was about 140,000 indicated total birds and 133,000 indicated breeding birds (Stehn and Platte 2012). These estimates do not include breeding or nonbreeding birds that occur outside the survey area, so the numbers should be considered a minimum estimate of range-wide population size.

Atlantic Black Scoter

Estimates of abundance do not exist for the entire population and there are no reliable data to evaluate long-term trends for Atlantic Black Scoters.

The WBPHS covers little of the breeding range of Atlantic Black Scoters and the lack of observations recorded to species precludes any species-specific assessment in eastern North America.

Surf Scoter

Satellite telemetry studies (De La Cruz et al. 2009, Sea Duck Joint Venture 2012) suggest that there may be two separate populations of Surf Scoters in North America—one in eastern North America, and one in western North America, although further study is needed to confirm this (Sea Duck Joint Venture 2012). In western North America, the scoter species-specific data (i.e., positive identifications) from the WBPHS indicate a decreasing trend for Surf Scoters in Alaska from 1993 to 2012 (Figure 1.9). In the Strait of Georgia, British Columbia, shore-based winter waterbird surveys indicated no trend in Surf Scoters over the 1999–2011 period (Crewe et al. 2012). Based on aerial surveys during winter in Puget Sound, Washington, Surf Scoters have declined by 37% from 1999 to 2013 with recent indices of about 40,000 birds (J. R. Evenson, Washington Department of Fish and Wildlife, pers. comm.). In eastern Canada, the Eastern Waterfowl Survey indicates a stable population of Surf Scoters from 1990 to 2012 (Figure 1.10).

White-winged Scoter

Using only the scoter species-specific data with positive identifications from the WBPHS, White-winged Scoters appear to have an increasing trend in Alaska from 1993 to 2012 (Figure 1.9). In 2001, an aerial survey specifically targeting scoters was initiated at Yukon Flats National Wildlife Refuge, an area that supports the highest densities of scoters in interior Alaska. About 99% of the scoters are positively identified as White-winged Scoters. The population has been stable for the 10-year period beginning 2001 (N. Guldager et al., U.S. Fish and Wildlife Service, unpubl. report). Winter aerial surveys in Puget Sound, Washington, suggest that White-winged Scoters have declined by 35% from 1999 to 2013 (~12,000 birds; J. R. Evenson, pers. comm.).

Long-tailed Duck

Estimates of abundance for Long-tailed Ducks in North America are difficult to ascertain due to

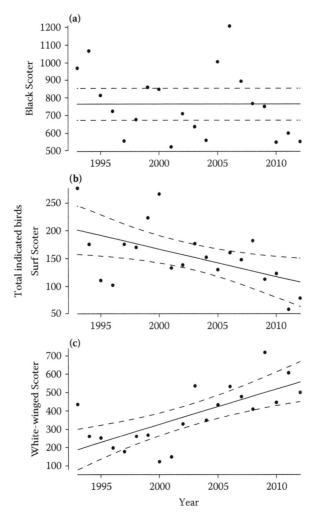

Figure 1.9. TIBs recorded for (a) Black Scoter (*M. americana*), (b) Surf Scoter (*M. fusca*), and (c) White-winged Scoter (*M. perspicillata*) in the innermost 100 m of the survey transects for the Alaska survey area of the WBPHS, 1993–2012. In those years, all scoters within the innermost 100 m of each transect were identified to species. Fitted polynomial models (solid lines) with 95% confidence intervals for the fitted values (dashed curves) are shown.

their vast breeding range including areas that are largely outside survey boundaries. Further, satellite telemetry has revealed that many Long-tailed Ducks that breed in western North America winter along the coast of Asia (Petersen et al. 2003; B. Bartzen et al., Canadian Wildlife Service, unpubl. report). A rough estimate of one million birds in North America has been used previously, but without a strong basis (Robertson and Savard 2002). Long-tailed Ducks remain one of the most poorly surveyed species of sea ducks in North America.

The WBPHS covers only a small part of the North American breeding range for Long-tailed Ducks, primarily on the Yukon–Kuskokwim Delta, Alaska, and northern Northwest Territories (Figure 1.4).

Data indicate substantial declines from about 1980 to 2002, and an increasing trend from the most recent 6 years (2007–2012; Figure 1.8). Few Long-tailed Ducks are encountered in the eastern survey area of the WBPHS, and no discernible trend is evident since 1996 (Figure 1.8).

Aerial transect surveys of breeding birds on western Victoria Island in central arctic Canada indicate a decline from a mean population estimate of 21,100 for 1992–1994 to 14,900 for 2004–2005 (Raven and Dickson 2006). The Yukon–Kuskokwim Delta Aerial Goose–Duck–Waterbird Survey, which also covers a fairly small proportion of the continental population of Long-tailed Ducks, suggests a stable population from

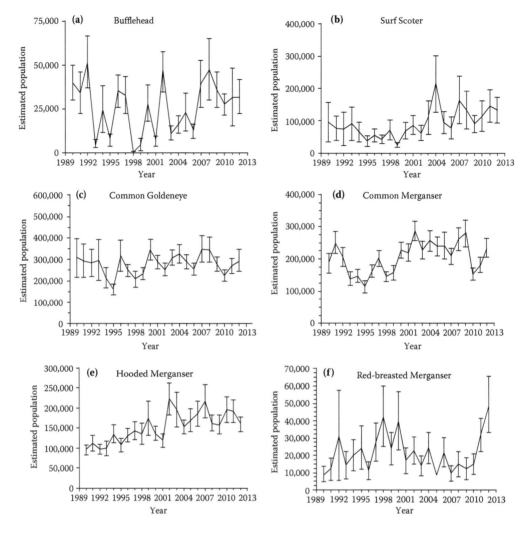

Figure 1.10. Population estimates (mean and SE) for (a) Bufflehead (*B. albeola*), (b) Surf Scoter (*M. fusca*), (c) Common Goldeneye (*B. clangula*), (d) Common Merganser (*M. merganser*), (e) Hooded Merganser (*L. cucullatus*), and (f) Red-breasted Merganser (*M. serrator*) in the Eastern Waterfowl Survey area, 1990–2012. Figures represent results from the helicopter surveys only. (From Environment Canada, Population status of migratory game birds in Canada, November 2012, Canadian Wildlife Service Migratory Birds Regulatory Report, No. 37, Canadian Wildlife Service, Ottawa, Ontario, Canada, 2013.)

1988 to 2012 (−0.5% per year; 90% CI: −1.8 to 0.9) but a declining trend for the most recent 10-year period 2003–2012 (−7.2% per year; 90% CI: −10.9 to −3.3) and a recent index of about 4,200 TIBs (R. M. Platte and R. A. Stehn, unpubl. report).

An extensive survey of the Alaska Arctic Coastal Plain, begun in 1986, estimated an average population index (uncorrected for incomplete detection) of 47,000 birds from 2008 to 2012, long-term (1986–2012) stability (−0.1% per year; 90% CI: −1.6 to 0), and an increasing trend (2.5% per year; 90% CI: 0.1–5.0) in recent years (2003–2012;

R. A. Stehn et al., unpubl. report), consistent with the positive trajectory observed in the WBPHS from 2007 to 2012. Overall, while several surveys suggest long-term declines in populations of Long-tailed Ducks, more recent data (mid-2000s to 2012) suggest stabilization or slight increases, albeit at levels below historical estimates.

Harlequin Duck

There are two distinct populations of Harlequin Ducks in North America—a relatively small population that resides primarily in Eastern Canada and a much larger population that breeds from Alaska south through the far western provinces and northwest states (Robertson and Goudie 1999). The western population winters in coastal areas from Alaska to California. Harlequin Ducks are a challenge to survey from aircraft during the breeding season because they are difficult to see in riverine breeding habitats; thus, most population estimates are based on winter counts.

Eastern North America

In Canada, the eastern population is designated as a species of Special Concern under the federal Species at Risk Act, and under provincial legislation, it is listed as endangered in New Brunswick and Nova Scotia and vulnerable in Newfoundland, Labrador, and Quebec. In the United States, Harlequin Ducks are listed as a threatened species in Maine.

In eastern North America, the breeding range of Harlequin Ducks covers northern Quebec and Labrador east to Newfoundland and south into northern New Brunswick. This population was formerly thought to contain two distinct segments based on where birds winter, although recent satellite telemetry studies suggest there may be some intermixing between the two population segments (Chubbs et al. 2008, Robert et al. 2008). Birds breeding in northern Quebec and Labrador molt and winter primarily along the southwest coast of Greenland, whereas birds in southern parts of the breeding range typically molt along the southwest Labrador coast and other areas to the south. The southern segment of the breeding population winters in southern Newfoundland, the Maritime Provinces, and New England states, mostly in Maine.

There are no range-wide abundance estimates for the breeding population. Abundance estimates for the eastern population are primarily based on counts of males at molting areas and counts of birds at wintering sites. Recent surveys suggest there are between 3,000 and 3,500 birds wintering in eastern North America (Boyne 2008, Mittelhauser 2008) and 5,000–10,000 birds wintering in Greenland, many of which originated from breeding areas in Eastern Canada (Boertmann 2008).

Harlequin Ducks have been counted at key wintering sites in Atlantic Canada at various intervals from 2001 to 2013. These counts suggest the number of Harlequin Ducks has been rapidly increasing in Atlantic Canada during the last 12 years (8.6% per year; Figure 1.11). However, these surveys have similar drawbacks to CBC data because they rely on nonrandom samples and counts were not standardized within or among sites. Also, refinement of survey effort and increasing observer experience may be partially responsible for some of the observed increases.

Western North America

There are no range-wide surveys that provide robust estimates of population size or long-term trend data for Harlequin Ducks in western North America. CBC data show a steadily increasing trend along the Pacific coast since 1974 (Niven et al. 2004, National Audubon Society 2011). Most other data on trend are from relatively small geographic areas.

In Prince William Sound, Alaska, winter surveys conducted in most years between 1997 and 2009 suggest an increasing trend for Harlequin Duck densities (Rosenberg et al. 2013). Densities also appeared slightly higher in 2007–2009 compared with 1972–1973. In 1989, the *Exxon Valdez* oil spill had a significant impact on the Harlequin Duck population in Prince William Sound, and recovery has been slow (Iverson and Esler 2010). The most recent survey in 2009 estimated 12,500 Harlequin Ducks in surveyed areas of Prince William Sound—a minimal estimate of total population size because not all areas were surveyed.

On Kodiak Island, Alaska, several bays with a combined population of at least 3,000 Harlequin Ducks were surveyed from 1994–1997 and 2004–2007. The mean population remained stable between these time periods, with the exception of one bay that showed decreased numbers suspected to be the result of high hunting pressure (D. Zwiefelhofer, U.S. Fish and Wildlife Service, unpubl. report).

The majority of the Pacific population of Harlequin Ducks winter in the Aleutian Islands in Alaska. Byrd et al. (1992) estimated about 147,000 harlequins in the Aleutian Islands in the early 1990s. A subjective estimate of 600,000 to 1 million Harlequin Ducks in the Aleutians for the period 1967–1969 was put forth by Bellrose (1976). There are no reliable data to evaluate trend

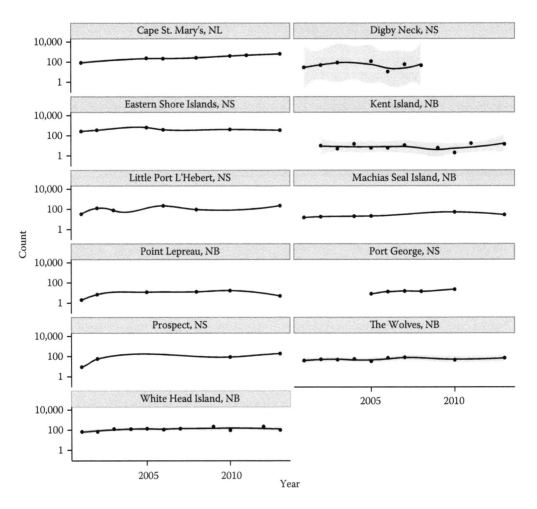

Figure 1.11. Annual trends in number of Harlequin Ducks (H. histrionicus) detected at key wintering locations in Atlantic Canada from 2001 to 2013. Fitted local polynomial regressions (solid curve) with 95% confidence intervals (gray shading) for fitted values are also plotted. (From Canadian Wildlife Service, Ottawa, Ontario, Canada, unpubl. data.)

in the Aleutian Islands. In the Strait of Georgia, British Columbia, shore-based winter waterbird surveys indicated a significant declining trend (−2.6% per year) for Harlequin Ducks for the period 1999–2011 (Crewe et al. 2012).

Goldeneyes

Common Goldeneyes and Barrow's Goldeneyes are difficult to identify to species during aerial surveys and are often lumped into a generic goldeneye category, including during the WBPHS. Based on the location of survey transects and the respective ranges of Common and Barrow's Goldeneyes, data from the WBPHS are likely more indicative of trends for Common Goldeneyes, which breed

widely across the continent (Figure 1.4). Few of the midcontinent and Alaska WBPHS transects are located in the range of Barrow's Goldeneyes. In surveyed areas, the trend from the early 1980s to early 2000s is clearly positive with no significant trend in the 5–10 years prior to, and after, that period (Figure 1.8). In the eastern portion of the WBPHS, goldeneyes showed a downward trend from 1996 to 2012 (Figure 1.8).

In Puget Sound, Washington, goldeneyes have declined slightly from 1996 to 2013 (J. R. Evenson, Washington Department of Fish and Wildlife, unpubl. data). While both species of goldeneyes occur there, species-specific trends cannot be ascertained because species composition has been estimated only in recent years (2008–2013).

Common Goldeneye

CBC data show a steadily increasing trend for Common Goldeneyes since 1974 (Niven et al. 2004, National Audubon Society 2011). In Eastern Canada, the Eastern Waterfowl Survey indicates no trend for Common Goldeneyes from 1990 to 2012 (Figure 1.10). In coastal British Columbia, shore-based winter waterbird surveys in the Strait of Georgia indicated no trend in Common Goldeneyes over the 1999–2011 period (Crewe et al. 2012).

Barrow's Goldeneye

Two distinct populations of Barrow's Goldeneyes occur in North America. The eastern population is relatively small and breeding is restricted largely to Quebec (Robert et al. 2000, Savard and Robert 2013), although there have been incidental observations of Barrow's Goldeneyes along the Quebec–Labrador border in northeast Quebec and in insular Newfoundland suggesting their breeding range may be somewhat larger (S. G. Gilliland, pers. comm.). In Canada, the eastern population of Barrow's Goldeneyes is designated as a species of Special Concern under the federal Species at Risk Act, and under provincial legislation, it is listed as endangered in New Brunswick and Nova Scotia and vulnerable in Newfoundland and Labrador and Quebec. In the eastern United States, Barrow's Goldeneyes are listed as a threatened species in Maine due to low numbers and vulnerability to extirpation. The Pacific population is more numerous and breeds over a much larger area from British Columbia into western Alberta and as far north as central Alaska.

Eastern North America

Barrow's Goldeneyes primarily winter along the St. Lawrence Estuary and Gulf coasts, with small concentrations of birds occurring in New Brunswick, Prince Edward Island, Nova Scotia, and Maine. Robert and Savard (2006) estimated the wintering population to be about 5200 birds. A helicopter survey of the major overwintering sites in Quebec has been conducted at regular 3–5 year intervals since 1999. The survey suggests the population is stable, but the results are imprecise and are considered inadequate for monitoring trends (M. Robert, Canadian Wildlife Service, pers. comm.). Thus, there are few data on trends for eastern Barrow's

Goldeneyes during winter, and no data on trends for breeding populations.

Western North America

No range-wide surveys provide estimates of population size or long-term trends. Most available data are from relatively small geographic areas. In Prince William Sound, Alaska, winter surveys conducted in most years between 1997 and 2009 suggested a stable population for Barrow's Goldeneyes (Rosenberg et al. 2013). In coastal British Columbia, shore-based winter waterbird surveys in the Strait of Georgia indicated a significant declining trend (−4.3% per year) for Barrow's Goldeneyes for the period 1999–2011 (Crewe et al. 2012).

Bufflehead

Buffleheads are widely distributed across most of the northern United States including Alaska, as well as in Canada from western Quebec to British Columbia, with higher densities in western North America. They winter along the entire ice-free portions of the Atlantic and Pacific coasts as well as inland waterways of the southern United States and Mexico. The best large-scale, consistent survey for Buffleheads in North America is the WBPHS. The timing of the WBPHS for Buffleheads is excellent (Smith 1995), and surveyed areas coincide well with their breeding range (Figure 1.4). In the midcontinent and Alaska survey areas, the trend since the early 1980s has been decidedly upward (Figure 1.8). CBC data show a steadily increasing trend continentally for Buffleheads since 1974 (Niven et al. 2004, National Audubon Society 2011). Data from the eastern area suggest a stable population since 1996 (Figure 1.8). The Eastern Waterfowl Survey shows highly variable numbers of Buffleheads with no obvious trend from 1990 to 2012 (Figure 1.10). Buffleheads have been stable in Puget Sound, Washington, from 1996 to 2013 (index = 50,000 birds, uncorrected for incomplete detection; J. R. Evenson, unpubl. data).

Mergansers

Mergansers are not identified to species during the WBPHS, but data for the generic merganser category represent some unknown proportion of Common Mergansers (*Mergus merganser*) and Red-breasted Mergansers (*M. serrator*); both species occur in surveyed areas and their ranges overlap (Figure 1.4). Hooded Mergansers are included in the merganser category, but few are counted because they are difficult to detect in the wooded habitats where they occur during the breeding season. The midcontinent and Alaska region of the WBPHS shows an increasing trend from the early 1970s to mid-1980s, with relatively stable populations since then (Figure 1.8). In the east, the WBPHS shows no trend since 1996 (Figure 1.8).

Common Merganser

Common Mergansers breed in forested areas across North America from interior Alaska to Newfoundland, south to the New England and upper Great Lake states, and most of the northwestern states. The species winters widely across the western, central, and northeast United States, mostly in freshwater environments. CBC data show a steadily increasing trend for Common Mergansers continentally since 1974 (Niven et al. 2004, National Audubon Society 2011). The Eastern Waterfowl Survey shows highly variable numbers with no obvious trend from 1990 to 2012 (Figure 1.10). In the Strait of Georgia, British Columbia, shore-based winter waterbird surveys indicated no trend in Common Mergansers over the 1999–2011 period (Crewe et al. 2012).

Red-breasted Merganser

Red-breasted Mergansers (M. serrator) breed across the subarctic and arctic regions of Alaska and Canada, and as far south as the Great Lakes in eastern North America, and winter primarily along the Atlantic and Pacific coasts and Great Lakes. Based on CBC data, Red-breasted Mergansers have been stable since 1974 on a continental scale (Niven et al. 2004, National Audubon Society 2011). Numbers of Red-breasted Mergansers in the Eastern Waterfowl Survey are highly variable with no obvious trend from 1990 to 2012 (Figure 1.10). Shore-based winter waterbird surveys in the Strait of Georgia, British Columbia, indicated no trend in Red-breasted Mergansers over the 1999–2011 period (Crewe et al. 2012).

Hooded Merganser

Hooded Mergansers have two largely disjunct breeding regions in North America; one includes nearly all of the eastern US and southeastern Canadian provinces, with another segment centered in the Pacific northwestern states and British Columbia. Primary wintering areas include the southeastern United States and coastal areas from Texas to the New England states, as well as coastal British Columbia and Pacific northwestern states. CBC data suggest that Hooded Mergansers have steadily increased in both eastern and western North America since 1974 (National Audubon Society 2011). In Eastern Canada, the Eastern Waterfowl Survey shows increasing numbers from 1990 to 2012 (Figure 1.10).

DISCUSSION

Existing waterfowl surveys provide limited information to estimate abundance, relative densities, and population trends for many species of sea ducks. There is an urgent need for surveys that will provide accurate indices of population size for long-term monitoring and robust detection of trends for all sea ducks. To be most useful, new surveys must cover appropriate geographical areas, be properly timed, and address the particular methodological challenges of enumerating sea ducks.

The full extent of breeding, wintering, molting, and staging areas remains to be described for many species, and current monitoring efforts focus on reconnaissance surveys to fill geographic gaps in our knowledge of seasonal distribution (Sea Duck Joint Venture 2013). Information from these surveys can be complemented by data from birds outfitted with satellite transmitters, which provide valuable insights on areas missed by reconnaissance work, the timing of breeding and migration, annual site fidelity, and the identity of regionally distinct subpopulations (Sea Duck Joint Venture 2012). Such information will allow waterfowl managers to design appropriately scaled and timed surveys for sea ducks and possibly identify areas that could provide indices representative of larger continental populations.

Effective sea duck surveys must address the particular challenges that sea ducks pose during enumeration. Three groups of sea ducks are especially prone to misidentification: scoters (Black, Surf, and White-winged), goldeneyes (Common and Barrow's), and the two large merganser species (Common and Red-breasted). These species are often identified only to genus, and even species-specific counts may not reflect the true composition of the community if the probability of correct identification varies among species. There are also identification problems among the black and white species that include goldeneyes, mergansers, Buffleheads, and sometimes also scaup and Ring-necked Ducks (*Aythya collaris*). Species misidentification and identification to genera complicate detection estimation and can lead to biased population estimates.

Sea ducks are also regularly found in large mixed species flocks during the nonbreeding season (Silverman et al. 2013) and are observed in multispecies groups more commonly than other waterfowl during breeding surveys (S. G. Gilliland, pers. comm.). Large aggregations introduce counting errors and the potential for bias (Bordage et al. 1998), which may be compounded when similar species co-occur.

One approach to addressing these challenges is to ensure observers are well trained in sea duck identification and practiced in counting large groups. Rigorous presurvey training and the dissemination of an appropriate aerial field guide and other training materials would reduce errors and help to standardize survey counts. When possible, presurvey training should include simultaneous aerial photography and visual observations to provide feedback to observers and help correct misidentification problems and adjust count biases. High-definition photography could also be employed on surveys as a means of sampling for species composition and for enumerating birds in large flocks. Aerial photography can also aid in the development of detection corrections, possibly including availability bias. More work using double-observer and distance sampling methods will also improve our ability to estimate detection probabilities and understand the factors that affect detection.

Despite the limitations of many surveys for monitoring sea ducks, some general trends across species are notable. In particular, some species that exhibited declines throughout the 1980s and 1990s (scoters, Long-tailed Ducks, some eider populations) appear to have stabilized or increased in recent years, albeit below historic levels. While this is encouraging, the factors driving population dynamics are still unknown for most species. Recent research suggests that sea ducks may be responding to changes in the ocean environment (Zipkin et al. 2010, Flint 2013), with survey counts correlating with major decadal shifts in the Pacific marine environment (Flint 2013). Notably, many of the species we classify as stable to increasing in recent years, including Spectacled Eiders, western King Eiders, Pacific Common Eiders, Pacific Black Scoters, and Long-tailed Ducks, are the more marine-dependent species of Mergini and inhabit the North Pacific and Bering Sea (Table 1.1).

A decade ago, available evidence suggested that 10 of 15 species of sea ducks in North America were declining (Sea Duck Joint Venture Management Board 2008). The situation appears more optimistic today. Population trajectories appear to be stable or increasing for 11 of the 22 sea duck populations recognized by the Sea Duck Joint Venture, and only 2 populations appear to be declining (Table 1.1). Data are insufficient or lacking for the other nine populations. That said, characterizations of trends for several populations are based on data that are less than robust, and some species are currently at levels below numbers observed in the most recent two to four decades.

There is growing interest in using population indices to better assess and inform harvest management and to focus habitat conservation efforts on areas of greatest importance to sea ducks (see Chapter 14, this volume). Accurate estimates of sea duck population trends would improve managers' ability to make decisions about these and other conservation priorities. Until management agencies in the United States and Canada devote greater resources to monitoring sea ducks, conservation efforts will be hampered by the lack of reliable data. Monitoring surveys also need to be complimented by other data sources. Additional sea duck banding, where practical, along with continued studies involving satellite telemetry, genetics (Sonsthagen et al. 2011; Chapter 2, this volume), and stable isotope techniques (Mehl et al. 2004, 2005), should be encouraged to better delineate populations and support the development of robust monitoring programs. Considering the increasing impact of humans on northern sea duck habitats, including expansion of resource development in boreal forest and arctic marine areas, increased shipping of materials including oil, and large-scale oceanic changes due to global warming (Caldeira and Wickett 2003; Descamps et al. 2011, 2012; Benoit 2012; Chapter 14, this

TABLE 1.1
Summary of contemporary (10–20 Years) population status and trends for North American sea ducks.

Species	Recent trend	Status relative to historical levels	Confidence in trend
Bufflehead	Stable to increasing	Above	High
Hooded Merganser	Stable to increasing	Above	High
Common Goldeneye	Stable to increasing	Above	Medium
Common Merganser	Stable to increasing	Above	Medium
Red-breasted Merganser	Stable to increasing	Above	Medium
Spectacled Eider	Stable to increasing	Below	High
Pacific Black Scoter	Stable to increasing	Below	Medium
Western King Eider	Stable to increasing	Below	Low
Long-tailed Duck	Stable to increasing	Below	Low
Eastern Harlequin Duck	Stable to increasing	Unknown	High
Eastern Barrow's Goldeneye	Stable	Unknown	Low
Steller's Eider	Decline	Below	Medium
American Common Eider	Decline	Below	Low
Surf Scoter	Unknown	Below	
White-winged Scoter	Unknown	Below	
Eastern King Eider	Unknown	Unknown	
Hudson Bay Common Eider	Unknown	Unknown	
Northern Common Eider	Unknown	Unknown	
Pacific Common Eider	Unknown	Unknown	
Atlantic Black Scoter	Unknown	Unknown	
Western Harlequin Duck	Unknown	Unknown	
Pacific Barrow's Goldeneye	Unknown	Unknown	

volume), it is becoming ever more important to accurately determine sea duck abundance and distribution and establish long-term monitoring programs, particularly for boreal and arctic species including eiders, scoters, and Long-tailed Ducks.

ACKNOWLEDGMENTS

Such an extensive compilation would not have been possible without the dedicated work of pilots and observers too numerous to name, who have contributed over the years to the WBPHS and other surveys. We thank the following people for sharing information and providing insight into surveys mentioned in this chapter: R. Stehn, R. Platte, and W. Larned. Thanks to R. Platte for assistance in creating figures illustrating survey areas and D. Groves for providing speciated scoter counts for the Alaskan portion of the WBPHS.

LITERATURE CITED

Alexander, S. A., D. L. Dickson, and S. W. Westover. 1997. Spring migration of eiders and other waterbirds in offshore areas of the western Arctic. in D. L. Dickson (editor), King and Common Eiders of the Western Canadian Arctic. Canadian Wildlife Service Occasional Paper, No. 94. Canadian Wildlife Service, Ottawa, ON.

Allen, R. B. 2000. Common Eider assessment. Maine Department of Inland Fisheries and Wildlife, Bangor, ME.

Barry, T. W. 1986. Eiders of the western Canadian Arctic. Pp. 74–80 in A. Reed (editor), Eider ducks in Canada. Canadian Wildlife Service Report Series, No. 47. Canadian Wildlife Service, Ottawa, ON.

Bellrose, F. C. 1976. Ducks, geese, and swans of North America. Stackpole Books, Harrisburg, PA.

Benoit, L. 2012. Resource development in northern Canada. Report of the Standing Committee on Natural Resources. 41st Parliament, 1st Session. Public Works and Government Services, Ottawa, ON.

BirdLife International. [online]. 2013. *Polysticta stelleri*. in IUCN 2014. IUCN Red List of Threatened Species. Version 2014.2. <www.iucnredlist.org> (1 September 2014).

Boertmann, D. 2008. Harlequin Ducks in Greenland. Waterbirds 31:4–7.

Bordage, D., N. Plante, A. Bourget, and S. Paradis. 1998. Use of ratio estimators to estimate the size of Common Eider populations in winter. Journal of Wildlife Management 62:185–192.

Bowman, T. D., J. L. Schamber, W. S. Boyd, D. H. Rosenberg, D. Esler, M. J. Petrula, and P. L. Flint. 2008. Characterization of annual movements, distribution and habitat use of Pacific Black Scoters. P. 24 in Third North American Sea Duck Conference, 10–14 November 2008, Québec, QC.

Boyne, A. W. 2008. Harlequin Ducks in the Canadian Maritime provinces. Waterbirds 31:50–57.

Byrd, G. V., J. C. Williams, and A. Durand. 1992. The status of Harlequin Ducks in the Aleutian Islands, Alaska. U.S. Fish and Wildlife Service Report. Prepared for the Harlequin Duck Symposium, Moscow, Idaho, April 1992.

Caldeira, K., and M. E. Wickett. 2003. Oceanography: anthropogenic carbon and ocean pH. Nature 425:365.

Chapdelaine, G., A. Bourget, W. B. Kemp, D. J. Nakashima, and D. J. Murray. 1986. Population d'Eider à duvet près des côtes du Québec septentrional. Pp. 39–50 in A. Reed (editor), Eider Ducks in Canada. Canadian Wildlife Service Report Series, No. 47. Canadian Wildlife Service, Ottawa, ON.

Chaulk, K. G. 2009. Suspected long-term population increases in Common Eiders, *Somateria mollissima*, on the mid-Labrador coast, 1980, 1994, and 2006. Canadian Field-Naturalist 123:304–308.

Chubbs, T. E., P. G. Trimper, G. W. Humphries, P. W. Thomas, L. T. Elson, and D. K. Laing. 2008. Tracking seasonal movements of adult male Harlequin Ducks from central Labrador using satellite telemetry. Waterbirds 31:173–182.

Cornish, B. J., and D. L. Dickson. 1997. Common Eiders nesting in the western Canadian Arctic. in D. L. Dickson (editor), King and Common Eiders of the Western Canadian Arctic. Canadian Wildlife Service Occasional Paper, No. 94. Canadian Wildlife Service, Ottawa, ON.

Crewe, T., K. Barry, P. Davidson, and D. Lepage. 2012. Coastal waterbird population trends in the Strait of Georgia 1999–2011: results from the first 12 years of the British Columbia Coastal Waterbird Survey. British Columbia Birds 22:8–35.

Crissey, W. F. 1949. Waterfowl populations and breeding conditions—Summer 1949. Special Scientific Report—Wildlife, No. 2. U.S. Fish and Wildlife Service, Washington, DC.

Day, R. H., J. R. Rose, A. K. Prichard, R. J. Blaha, and B. A. Cooper. 2004. Environmental effects on the fall migration of eiders at Barrow, Alaska. Marine Ornithology 32:13–24.

De La Cruz, S. E. W., J. Y. Takekawa, M. T. Wilson, D. R. Nysewander, J. R. Evenson, D. Esler, W. S. Boyd, and D. H. Ward. 2009. Surf Scoter spring migration routes and chronology: a synthesis of Pacific coast studies. Canadian Journal of Zoology 87:1069–1086.

Descamps, S., M. R. Forbes, H. G. Gilchrist, O. P. Love, and J. Bety. 2011. Avian cholera, post-hatching survival and selection on hatch characteristics in a long-lived bird, the Common Eider *Somateria mollissima*. Journal of Avian Biology 42:39–48.

Descamps, S., H. G. Gilchrist, J. Bety, E. I. Buttler, and M. R. Forbes. 2009. Costs of reproduction in a long-lived bird: large clutch size is associated with low survival in the presence of a highly virulent disease. Biology Letters 5:278–281.

Descamps, S., S. Jenouvrier, H. G. Gilchrist, and M. R. Forbes. 2012. Avian cholera, a threat to the viability of an Arctic seabird colony? PLoS One 7:e29659.

Dickson, D. L. 2012a. Seasonal movement of Pacific Common Eiders breeding in arctic Canada. Technical Report Series, No. 521, Canadian Wildlife Service, Edmonton, AB.

Dickson, D. L. 2012b. Seasonal movement of King Eiders breeding in western arctic Canada and northern Alaska. Technical Report Series, No. 520, Canadian Wildlife Service, Edmonton, AB.

Environment Canada. 2013. Population status of migratory game birds in Canada, November 2012. Canadian Wildlife Service Migratory Birds Regulatory Report, No. 37. Canadian Wildlife Service, Ottawa, ON.

Falardeau, G., J. F. Rail, S. Gilliland, and J-P. L. Savard. 2003. Breeding Survey of Common Eiders along the West Coast of Ungava Bay, in summer 2000, and a supplement on other nesting aquatic birds. Technical Report Series, No. 405, Canadian Wildlife Service, Quebec Region, Sainte-Foy, QC.

Flint, P. L. 2013. Changes in size and trends of North American sea duck populations associated with North Pacific oceanic regime shifts. Marine Biology 160:59–65.

Flint, P. L., and M. P. Herzog. 1999. Breeding of Steller's Eiders, *Polysticta stelleri*, on the Yukon–Kuskokwim Delta, Alaska. Canadian Field-Naturalist 113:306–308.

Gilliland, S. G., H. G. Gilchrist, D. Bordage, C. Lepage, F. R. Merkel, A. Mosbech, B. Letournel, and J-P. L. Savard. 2008a. Winter distribution and abundance of Common Eiders in the Northwest

Atlantic and Hudson Bay. P. 95 in Third North American Sea Duck Conference, 10–14 November 2008, Quebec, QC.

Gilliland S. G., H. G. Gilchrist, R. F. Rockwell, G. J. Robertson, J-P. L. Savard, F. Merkel, and A. Mosbech. 2009a. Evaluating the sustainability of harvest among northern Common Eiders *Somateria mollissima borealis* in Greenland and Canada. Wildlife Biology 15:24–36.

Gilliland, S. G., C. Lepage, J.-P. Savard, D. Bordage, G. J. Robertson, and E. Reed. [online]. 2009b. Sea Duck Joint Venture Project number 115: developmental surveys for breeding scoters in eastern North America. Progress report, November 2012 Sea Duck Joint Venture. <http://seaduckjv.org/studies/pro3/pr115.pdf> (15 November 2013).

Gilliland, S. G., and G. J. Robertson. 2009. Composition of eiders harvested in Newfoundland. Northeastern Naturalist 16:501–518.

Gilliland, S. G., G. J. Robertson, H. G. Gilchrist, S. Descamps, R. F. Rockwell, J-P.L. Savard, A. Mosbech, and F. R. Merkel. 2008b. Applying demographic modelling techniques to support sea duck conservation: the continuing case of the northern Common Eider. P. 45 in Third North American Sea Duck Conference, 10–14 November 2008, Quebec, QC.

Graham, A., and R. Bell. 1989. Investigating observer bias in aerial survey by simultaneous double-counts. Journal of Wildlife Management 53:1009–1016.

Gratto-Trevor, C. L., V. H. Johnston, and S. T. Pepper. 1998. Changes in shorebird and eider abundance in the Rasmussen Lowlands, NWT. Wilson Bulletin 110:316–325.

Heusmann, H. W. 1999. Let's get rid of the midwinter waterfowl inventory in the Atlantic Flyway. Wildlife Society Bulletin 27:559–565.

Hodges, J. I., and W. D. Eldridge. 2001. Aerial surveys of eiders and other waterbirds on the eastern Arctic coast of Russia. Wildfowl 52:127–142.

Iverson, S. A., and D. Esler 2010. Harlequin Duck population injury and recovery dynamics following the 1989 Exxon Valdez oil spill. Ecological Applications 7:1993–2006.

Iverson, S. A., H. G. Gilchrist, P. A. Smith, A. J. Gaston, and M. R. Forbes. 2014. Longer ice-free seasons increase the risk of nest depredation by polar bears for colonial breeding birds in the Canadian Arctic. Proceedings of the Royal Society B 281:art20133128.

Joint Working Group on the Management of the Common Eider. 2004. Québec management plan for the Common Eider *Somateria mollissima dresseri*. A special publication of the Joint Working Group on the Management of the Common Eider, Québec, QC.

Kertell, K. 1991. Disappearance of the Steller's Eider from the Yukon–Kuskokwim Delta, Alaska. Arctic 44:177–187.

Krohn, W. B., P. O. Corr, and A. E. Hutchinson. 1992. Status of American Eider Ducks with special reference to northern New England. Fish and Wildlife Research 12. U.S. Fish and Wildlife Service, Orono, ME.

Lock, A. R. 1986. A census of Common Eiders breeding in Labrador and the Maritime provinces. Pp. 30–38 in A. Reed (editor), Eider Ducks in Canada. Canadian Wildlife Service Report Series, No. 47. Canadian Wildlife Service, Ottawa, ON.

Magnussen, W. E., G. J. Caughley, and G. C. Grigg. 1978. A double-survey estimate of population size from incomplete counts. Journal of Wildlife Management 43:174–176.

Mehl, K. R., R. T. Alisauskas, K. A. Hobson, and D. K. Kellett. 2004. To winter east or west? Heterogeneity in winter philopatry in a central-arctic population of King Eiders. Condor 106:241–251.

Mehl, K. R., R. T. Alisauskas, K. A. Hobson, and F. R. Merkel. 2005. Linking breeding and wintering areas of King Eiders: making use of polar isotopic gradients. Journal of Wildlife Management 69:1297–1304.

Merkel, F. R., A. Mosbech, H. G. Gilchrist, and S. Descamps. 2008. Recent population trends of Common Eiders breeding in northwest Greenland as derived from a community-based monitoring program. P. 72 in Third North American Sea Duck Conference, 10–14 November 2008, Quebec, QC.

Mittelhauser, G. H. 2008. Harlequin Ducks in the eastern United States. Waterbirds 31:58–66.

Mosbech, A., R. S. Danø, F. Merkel, C. Sonne, G. Gilchrist, and A. Flagstad. 2006. Use of satellite telemetry to locate key habitats for King Eiders *Somateria spectabilis* in West Greenland. Pp. 769–776 in G. C. Boere, C. A. Galbraith, and D. A. Stroud (editors), Waterbirds around the world. The Stationery Office, Edinburgh, U.K.

Nakashima, D. J., and D. J. Murray. 1988. The Common Eider (*Somateria mollissima sedentaria*) of eastern Hudson Bay: a survey of nest colonies and Inuit ecological knowledge. Environmental Studies Revolving Funds Report, No. 201. Ottawa, ON.

National Audubon Society. [online]. 2011. The Christmas bird count historical results. <http://www.christmasbirdcount.org> (27 September 2013).

Niven, D. K., J. R. Sauer, and G. S. Butcher. 2004. Population trends of North American sea ducks based on Christmas bird count and breeding bird survey data. P. 101 in Second North American Sea Duck Conference, 7–11 November 2005, Annapolis, MD.

Obritschkewitsch, T., and R. J. Ritchie. 2011. Steller's Eider Surveys Near Barrow, Alaska, 2010. ABR, Inc., Fairbanks, AK.

Petersen, M. R., and P. L. Flint. 2002. Population structure of Pacific Common Eiders breeding in Alaska. Condor 104:780–787.

Petersen, M. R., W. W. Larned, and D. C. Douglas. 1999. At-sea distribution of Spectacled Eiders: a 120-year-old mystery solved. Auk 116:1009–1020.

Petersen, M. R., B. J. McCaffery, and P. L. Flint. 2003. Post-breeding distribution of Long-tailed Ducks Clangula hyemalis from the Yukon–Kuskokwim Delta, Alaska. Wildfowl 54:103–113.

Phillips, L. M., A. N. Powell, and E. A. Rexstad. 2006. Large-scale movements and habitat characteristics of King Eiders throughout the nonbreeding period. Condor 108:887–900.

Pollock, K. H., and W. L. Kendall. 1987. Visibility bias in aerial surveys: a review of procedures. Journal of Wildlife Management 51:502–510.

Quakenbush, L. T., R. S. Suydam, R. Acker, M. Knoche, and J. Citta. 2009. Migration of king and Common Eiders past Point Barrow, Alaska, during summer/fall 2002 through spring 2004: population trends and effects of wind. Final Report Outer Continental Shelf Study Minerals Management Service 2009–036. University of Alaska, Fairbanks, AK.

Rail, J.-F., and G. Chapdelaine. 2002. Quinzième inventaire des oiseaux marins dans les refuges de la Côte-Nord: techniques et résultats détaillés. Technical report Series, No. 392, Canadian Wildlife Service, Québec Region, Sainte-Foy, Québec, QC (in French).

Rail, J.-F., and R. C. Cotter. 2007. Sixteenth census of seabird populations in the sanctuaries of the North Shore of the Gulf of St. Lawrence, 2005. Canadian Field-Naturalist 121:287–294.

Raven, G. H., and D. L. Dickson. 2006. Changes in distribution and abundance of birds on western Victoria Island from 1992–1994 to 2004–2005. Canadian Wildlife Service Technical Report Series, No. 456. Edmonton, AB.

Raven, G. H., and D. L. Dickson. 2009. Surveys of Pacific Common Eiders (Somateria mollissima v-nigra) in the Bathurst Inlet area of Nunavut, 2006–2008. Canadian Wildlife Service Technical Report Series, No. 503. Canadian Wildlife Service, Edmonton, AB.

Robert, M., D. Bordage, J-P. L. Savard, G. Fitzgerald, and F. Morneau. 2000. The breeding range of the Barrow's Goldeneye in eastern North America. Wilson Bulletin 112:1–7.

Robert, M., G. H. Mittelhauser, B. Jobin, G. Fitzgerald, and P. Lamothe. 2008. New insights on Harlequin Duck population structure in eastern North America as revealed by satellite telemetry. Waterbirds 31:159–172.

Robert, M., and J-P. L. Savard. 2006. The St. Lawrence River Estuary and Gulf: a stronghold for Barrow's Goldeneyes wintering in eastern North America. Waterbirds 29:437–450.

Robertson, G. J., and H. G. Gilchrist. 1998. Evidence for population declines among Common Eiders breeding in the Belcher Islands, Northwest Territories. Arctic 51:378–385.

Robertson, G. J., and R. I. Goudie. 1999. Harlequin Duck (Histrionicus histrionicus). in A. Poole and F. Gill (editors), The Birds of North America, No. 651. The Academy of Natural Sciences, Philadelphia, PA and the American Ornithologists' Union, Washington, DC.

Robertson, G. J., and J.-P. L. Savard. 2002. Long-tailed Duck (Clangula hyemalis). in A. Poole and F. Gill (editors), The Birds of North America, No. 651. The Academy of Natural Sciences, Philadelphia, PA and the American Ornithologists' Union, Washington, DC.

Rosenberg, D. H., M. J. Petrula, D. D. Hill, and A. M. Christ. 2013. Harlequin Duck population dynamics: measuring recovery from the Exxon Valdez oil spill. Exxon Valdez Oil Spill Restoration Project Final Report (Restoration Project 040407). Alaska Department of Fish and Game, Division of Wildlife Conservation, Anchorage, AK.

Safine, D. E. 2012. Breeding ecology of Steller's and Spectacled Eiders nesting near Barrow, Alaska, 2012. U.S. Fish and Wildlife Service Technical Report, Fairbanks, AK.

Sauer, J. R., W. A. Link, J. E. Fallon, K. L. Pardieck, and D. J. Ziollowski Jr. 2013. The North American Breeding Bird Survey 1966–2011: summary analysis and species accounts. North American Fauna 79:2–32.

Savard, J.-P. L., and M. Robert. 2013. Relationships among breeding, molting and wintering areas of adult female Barrow's Goldeneyes (Bucephala islandica) in eastern North America. Waterbirds 36:34–42.

Sea Duck Joint Venture. [online]. 2012. Atlantic and Great Lakes Sea Duck migration study. Progress report November 2012. <http://seaduckjv.org/atlantic_migration_study> (accessed February 12, 2013).

Sea Duck Joint Venture. 2013. Sea Duck Joint Venture implementation plan for April 2013 through March 2016. Report of the Sea Duck Joint Venture. U.S. Fish and Wildlife Service, Anchorage, AK.

Sea Duck Joint Venture Management Board. 2008. Sea Duck Joint Venture Strategic Plan 2008–2012. U.S. Fish and Wildlife Service, Anchorage, AK; Canadian Wildlife Service, Sackville, NB.

Silverman, E. D., D. T. Saalfeld, J. B. Leirness, and M. D. Koneff. 2013. Wintering sea duck distribution along the Atlantic coast of the United States. Journal of Fish and Wildlife Management 4:178–198.

Smith, G. W. 1995. A critical review of the aerial and ground surveys of breeding waterfowl in North America. National Biological Service, Biological Science Report, No. 5. National Technical Information Service, Springfield, MA.

Sonsthagen, S. A., S. L. Talbot, K. T. Scribner, and K. G. McCracken. 2011. Multilocus phylogeography of Common Eiders (*Somateria mollissima*) breeding in North America and Scandinavia. Journal of Biogeography 38:1368–1380.

Stehn, R. A., C. P. Dau, B. Conant, and W. I. Butler, Jr. 1993. Decline of Spectacled Eiders nesting in western Alaska. Arctic 46:264–277.

Stehn, R. A., and R. M. Platte. 2012. Pacific Black Scoter breeding survey. Progress report to Sea Duck Joint Venture, September 2012. U.S. Fish and Wildlife Service, Anchorage, AK.

Stott, R. S., and D. P. Olson. 1972. An evaluation of waterfowl surveys on the New Hampshire coastline. Journal of Wildlife Management 36:468–477.

Suydam, R., D. L. Dickson, J. B. Fadely, and L. T. Quakenbush. 2000. Population declines of King and Common Eiders of the Beaufort Sea. Condor 102:219–222.

Suydam, R., L. Quakenbush, and M. Knoche. 2004. Status of King and Common Eiders migrating past Point Barrow, Alaska. P. 28 in 10th Alaska Bird Conference and Workshops, 15–19 March 2004, Anchorage, AK.

U.S. Fish and Wildlife Service. 1996. Spectacled Eider recovery plan. Anchorage, AK.

Williams, B. K., J. D. Nichols, and M. J. Conroy. 2002. Analysis and management of animal populations. Academic Press, New York, NY.

Woodby, D. A., and G. J. Divoky. 1982. Spring migration of eiders and other waterbirds at Point Barrow, Alaska. Arctic 35:403–410.

Zipkin, E. F., B. Gardner, A. T. Gilbert, A. F. O'Connell, J. A. Royle, and E. D. Silverman. 2010. Distribution patterns of wintering sea ducks in relation to the North Atlantic Oscillation and local environmental characteristics. Oecologia 163:893–902.

CHAPTER TWO

Phylogenetics, Phylogeography, and Population Genetics of North American Sea Ducks (Tribe: Mergini)*

Sandra L. Talbot, Sarah A. Sonsthagen, John M. Pearce, and Kim T. Scribner

Abstract. Many environments occupied by North American sea ducks are remote and difficult to access, and as a result, detailed information about life history characteristics that drive population dynamics within and across species is limited. Nevertheless, progress on this front during the past several decades has benefited by the application of genetic technologies, and for several species, these technologies have allowed for concomitant tracking of population trends and genetic diversity, delineation of populations, assessment of gene flow among metapopulations, and understanding of migratory connectivity between breeding and wintering grounds.

This chapter provides an overview of phylogenetic, phylogeographic, and population genetics studies of North American sea duck species, many of which have sought to understand the major and minor genetic divisions within and among sea duck species, and most of which have been conducted with the understanding that the maintenance of genetic variation in wild sea duck populations is fundamental to the group's long-term persistence.

Key Words: fossil, genetic, glacial refugia, molecular ecology, phylogenetic, phylogeography, taxonomy, systematics.

The maintenance of genetic variation in wild populations of Arctic, subarctic, and temperate species, including sea ducks, is fundamental to the long-term persistence of regional biodiversity. Variability provides opportunities for species to respond to novel challenges, such as changing environmental conditions and emerging pathogens. When individual species decline in abundance and their geographic distributions contract, genetic variability can also erode. However, we have not yet developed a clear understanding of how genetic variability is spatially and temporally distributed, particularly in Arctic and subarctic regions of North America, which provides breeding and in some cases wintering habitat for 15 extant sea duck species. Further, we lack understanding about how genetic variation, and therefore evolutionary potential, is generated and maintained for most high-latitude species. The advent of new technologies and analytical approaches now provides opportunities to generate more comprehensive views of genetic variation and can provide the necessary foundation for diverse theoretical and applied endeavors,

* Talbot, S. L., S. A. Sonsthagen, J. M. Pearce, and K. T. Scribner. 2015. Phylogenetics, phylogeography, and population genetics of North American sea ducks (Tribe: Mergini). Pp. 29–61 in J.-P. L. Savard, D. V. Derksen, D. Esler, and J. M. Eadie (editors). Ecology and conservation of North American sea ducks. Studies in Avian Biology (no. 46), CRC Press, Boca Raton, FL.

Systematics integrates across all other sub-disciplines within the biological sciences by charting the evolutionary record necessary to discover, categorize, and interpret biodiversity. Without this foundation, we cannot ask pertinent questions related to species and their evolutionary responses. Excluding the more charismatic species, our knowledge of the systematics of the vast majority of Arctic and subarctic organisms, including sea ducks, remains meager. Within systematics, two particular disciplines are closely interlinked: *phylogenetics* (the study of evolutionary relationships) and *taxonomy* (the generation of a standard nomenclature). Concurrently, knowledge emerging from these two disciplines contributes to the broader issues of viability, sustainability, and availability of critical biotic diversity. Therefore, systematics is an integral feature informing the management and conservation of Arctic species (Kutz et al. 2009), including sea ducks.

Phylogenetics is the study of the evolutionary relationships among various groups of organisms, based on evolutionary similarities and differences in existing characteristics. Above the level of populations, phylogenetics becomes a history of speciation, with species regarded as independent lineages. Phylogenetic trees reconstruct history and constrain explanations about the emergence and distribution of biodiversity. *Taxonomy* is the theory and practice of classifying organisms (Mayr 1969). While early taxonomic studies relied almost exclusively on morphological variation, in recent decades, diversity has been queried using genetics techniques, which have provided increasingly refined views of geographic variation. Classification systems provided by taxonomy are foundational for efficient funding prioritization as well as rigorous project design. A poorly developed taxonomic framework can have serious negative conservation consequences (Mace 2004). When a taxonomic framework accurately reflects evolutionary relationships among organisms, the

taxonomy provides more than a universally accepted name or common language for scientists and managers. By presenting the species' history, systematic and taxonomic research generates direct connections linking ecology, evolution, and biogeography, affording a predictive framework within which to investigate and identify emerging risks such as hybridization and disease (Schrag and Wiener 1995, Brooks and Hoberg 2006, Seehausen et al. 2007).

ranging in this case from recovering the history of diversification and extinction of sea ducks to developing robust projections for the long-term persistence of sea ducks in North America.

This chapter provides an overview of the systematic studies of North American sea ducks, including phylogenetic, phylogeographic, and population genetics relationships. We emphasize that an understanding of how genetic diversity is shaped in deeper evolutionary time is essential to predicting future responses and persistence of sea ducks in high-latitude habitats (Box 2.1).

SYSTEMATICS, TAXONOMY, AND PHYLOGENETIC RELATIONSHIPS IN MERGINI

The fossil record of waterfowl is more complete than many other avian families (Olson and Feduccia 1980, Box 2.2), leading to paleontological reconstructions that placed a major radiation of the waterfowl within the last 60 million years (Delacour and Mayr 1945, Olson and Feduccia 1980). Delacour and Mayr (1945) considered the sea ducks, tribe Mergini, to have emerged after tribe Oxyurini, the Stiff-tailed Ducks, sharing most common ancestry with members of the tribe Cairinini (or Cairininae), the perching ducks (Figure 2.1), although taxonomists disagree over the classification within Cairinini (Johnsgard 2010). Given the largely northern hemisphere origin of Mergini, Howard (1964) and Weller (1964) proposed a northern hemisphere origin for the tribe.

Sea ducks are generally considered to comprise up to 20 extant species worldwide (Delacour and Mayr 1945, Johnsgard 1975, Livezey 1995), largely distributed across northern hemisphere habitats, including six mergansers (*Lophodytes cucullatus, Mergellus albellus, Mergus merganser, M. serrator, M. squamatus,* and *M. octosetaceus,* the sole species

Paleontology. New paleontological findings continue to provide insight into the evolution of sea ducks and factors contributing to their population demography and, sometimes, extinction. For example, the goose-sized, flightless sea duck, *Chendytes lawi,* thought to be most closely related to the eiders (*Somateria;* Mosimann and Martin 1975) and once common off the California coast, was once thought to have gone extinct at the Pleistocene/Holocene transition. However, its survival well into the Holocene was verified in 1976 when bones attributed to *C. lawi,* Carbon dated (^{14}C) to between 5,400 and 3,800 ybp, were found in archaeological sites north of Santa Cruz (Morejohn 1976). Morejohn (1976) speculated the species bred on offshore islands, providing immunity to predation by humans, and this was confirmed by Guthrie (1992), who reported immature individuals and egg shells found in late Pleistocene sites on San Miguel Island, one of the Channel Islands. Morejohn (1976) and others (Jones et al. 2008) suggest that *C. lawi* became extinct between 3,000 and 2,200 ybp. Extinction corresponded to the development of watercraft by native Californians that increased efficiency in accessing remote islands, islets, and offshore rocks used as breeding colonies by the birds (Steadman and Martin 1984). Jones et al. (2008) proposed that *C. lawi* was hunted by humans for at least 8,000 years before it was driven to extinction.

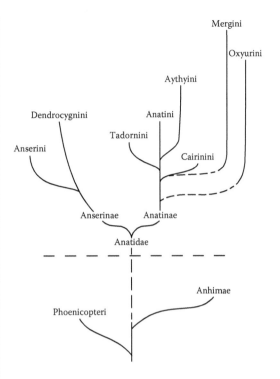

Figure 2.1. Hypothesized relationships among subfamilies and tribes of the Anatidae. (After Delacour and Mayr 1945.)

a species complex, *S. mollissima*-gp. Likewise, the Palearctic *M. nigra* and the Nearctic *M. americana* are sometimes grouped into a single Black Scoter complex (*Melanitta nigra*-gp.) and *M. fusca* and *M. deglandi* into a White-winged Scoter complex (*M. fusca*-gp.). The recently extinct Mergini include the enigmatic Labrador Duck (*Camptorhynchus labradorius*) and the Auckland Islands Merganser (*Mergus australis*), both of which went extinct sometime during the nineteenth- and twentieth-century transition (Hahn 1963). Of the extant species, 15 occupy habitats in North America.

distributed only in the southern hemisphere), five or six scoters (*Melanitta nigra, M. americana, M. perspicillata, M. fusca, M. deglandi,* and *M. stejnegeri*), four eiders (*Somateria mollissima, S. spectabilis, S. fischeri,* and *Polysticta stelleri*), two goldeneyes and Bufflehead (*Bucephala clangula, B. islandica,* and *B. albeola*), the Harlequin Duck (*Histrionicus histrionicus*), and the Long-tailed Duck (*Clangula hyemalis*). Some researchers (Livezey 1995) elevate some or all of the four to seven subspecies of Common Eider (*S. mollissima mollissima, S. m. faeroeensis, S. m. islandica, S. m. borealis, S. m. sedentaria, S. m. dresseri,* and *S. m. v-nigrum*) into

Morphology and Behavior

In the 10th edition of *Systema Naturae,* Linnaeus (1758) divided the class Aves into six orders—Accipitres, Picae, Anseres, Grallae (Scolopaces in previous editions), Gallinae, and Passeres, based largely on morphological characteristics, heavily emphasizing bill and foot characteristics. His third order, Anseres, comprised 11 genera largely made up of web-footed, short-legged waterbirds: *Anas, Mergus, Procellaria, Diomedea, Pelecanus, Phaeton, Alca, Columbus, Larus, Sterna,* and *Rynchops.* In the Linnaean arrangement, waterfowl comprised *Anas* (within which 39 species were recognized) and *Mergus* (within which 5 species were recognized). Waterfowl (*Anas* and *Mergus*) were not segregated into distinct subcategories. Within *Anas*, which included 13 sea ducks, four groups were identified according to certain morphological characters: (1) bill gibbous (with bulbous protuberances at the base), (2) bill equal at base, (3) certain feathers recurved, or (4) crested. Under Linnaeus' schema, White-winged Scoter (*M. fusca*), Black Scoter (*M. nigra*), and King Eider (*S. spectabilis*) are placed in the first group within *Anas* and Surf Scoter (*M. perspicillata*) and Common Eider in the second group. *Mergus* included only mergansers. Contemporaneous eighteenth-century and subsequent nineteenth-century taxonomic treatments of waterfowl provided diverse groupings of sea ducks (Illiger 1811, Cuvier 1817, Vigors 1825) often retaining the mergansers within a group separated from the remainder of the sea duck species. Cuvier (1817) grouped the Macreuses (scoters), Garrots (goldeneyes, Harlequin and Long-tailed Ducks), eiders, and Milloiuns (pochards) into a duck subgenus. Recognizing that variations in the trachea and syrinx of avian species, which are greatly simplified in waterfowl (Johnsgard 1961), can provide a means to judge taxonomic relationships, Yarrell (1827) examined tracheal characteristics and linked the true ducks (including species within the genera *Tadorna* [shelducks] and *Anas* [gadwalls, shovelers, Garganey, teal]) to the Harlequin Duck, Long-tailed Duck, and goldeneyes via eiders and scoters. Yarrell (1827) also noted similarities between the mergansers and goldeneyes based on sternum and stomach as well as trachea.

Twentieth-century taxonomists continued to revise earlier taxonomies. In their revision of the classification of Anatidae, Delacour and Mayr (1945) erected the tribe Mergini, separating the pochards (*Netta* and *Aythya*) from the remaining diving ducks and merging the mergansers (*Mergus*, *Mergellus*, and *Lophodytes*) and the goldeneyes (*Bucephala*) with the other sea ducks into a monophyletic group (Figure 2.1). Mergini was characterized by the following life history and ecological characteristics: (1) a sea-dwelling habit and feeding primarily on animal food obtained during diving; (2) strong, hooked bills; (3) moderately heavy wing loading and rapid flight; (4) brightly colored eclipse plumage in males, with metallic coloration generally restricted to the head

region; (5) sexual maturity at 2–3 years of age; (6) elaborate displays in males; (7) largely northern distribution; and (8) bold gray and white patterning with distinctive capped head appearance in downy young. Delacour and Mayr (1945) considered perching ducks (Cairinini) to be Mergini's closest relatives, based on similarities in nesting habits and tail morphology, although Johnsgard (1960a) noted differences between the Cairinini and Mergini with respect to behavior, characteristics of downy young, and ability to hybridize.

Although previous to their treatment, eiders were placed with scoters, goldeneyes, and other diving ducks, Delacour and Mayr (1945) gave only provisional assignment of eiders to Mergini, citing differences in eider tracheal anatomy and appearance of downy young. In-depth studies of tracheal anatomy, however, led Humphrey (1955, 1958) to conclude that Mergini comprises two unrelated groups of birds, of which some members independently evolved similar diving adaptations. His research suggested that the eiders and Harlequin Duck shared common, uniform features in tracheal anatomy, whereas the remainder of the tribe possessed variable tracheas, ranging from rudimentary (scoters) to more complex shapes with membranaceous fenestrae (the remainder of the tribe). Humphrey (1955, 1958) supported the placement of eiders into a separate tribe (Somateriini) more closely allied to the Anatini than the remainder of Mergini, largely based on tracheal structure of male eiders, plumage patterns of females, coloration of downy young, food habits, and foraging behavior of eider ducks. Subsequent studies of female courtship and copulatory behavior in 11 of the 18 existing species of Mergini led Myres (1959) to reject Humphrey's hypothesis of a close relationship between eiders and the Anatini. To date, however, investigators have largely agreed with the *Bucephala–Mergus* assemblage and the composition of the tribe Mergini (Livezey 1995), but within-tribe relationships continue to be debated. For example, relationships between eiders and the rest of Mergini (Humphrey 1955, 1958; Delacour 1969; Cramp and Simmons 1977), and the placement of Harlequin Duck and Smew (*M. albellus*) within the tribe, remained unresolved. There is also disagreement regarding the separation of Mergini from other diving ducks (Phillips 1925, 1926; Peters 1931; Livezey 1995) and the perching ducks (Cairinini, Delacour and Mayr 1945, Johnsgard 1960b).

Livezey (1995) performed cladistics analysis of 137 morphological characters (skeletal, tracheal, and natal and definitive plumage) across 25 sea duck lineages to assess phylogenetic relationships within Mergini. To polarize characters and assess intertribal relationships, he also included data from several species of Anatini, Oxyurini, and Aythyini. His analyses are as follows: (1) Eiders (*Polysticta* and *Somateria*) are placed into a monophyletic group that is sister to all other Mergini and (2) *Somateria* is placed as monophyletic, with Spectacled Eider, *S. fischeri*, as the basal taxon of the congeners and Steller's Eider, *P. stelleri*, as basal in the eider clade. Among the non-eider clades, (3) Harlequin Ducks are placed closest to the eider group, followed (in order of descending relationship from eiders) by *Melanitta* and the extinct *Camptorhynchus*, *Clangula*, *Bucephala* and *Mergellus*, *Lophodytes*, and *Mergus*; (4) Black Scoters (here, *M. nigra* and *M. americana*) comprise the sister group of the other scoters; (5) Buffleheads (*B. albeola*) are the sister group to the goldeneyes, and (6) *Lophodytes* (Hooded Merganser, *L. cucullata*) is the sister group to the *Mergus*; and (7) Smew are placed with the goldeneyes. Livezey (1995) suggested that among attributes characterizing sea ducks, the clearest patterns include primary preference for nest site (selection for terrestrial nest sites considered a primitive character, with preference for nest cavities in goldeneyes and mergansers considered a derived character), frequency of semicolonial nesting, and method of diving, followed by variation in clutch size and relative clutch mass, sexual size dimorphism, and frequency of interspecific brood parasitism. Moderate phylogenetic constraint was observed in the attributes of sexual dichromatism and migratory habit.

Molecular

Phylogenetic studies of sea ducks based on molecular data are still few and typically do not include all species comprising Mergini. The composition of integumental lipids suggested moderate distance between Common Eider and other genera comprising Mergini (Jacob and Glaser 1975, Jacob 1982), and feather proteins differentiated eiders from the rest of the Mergini (Brush 1976). Analyses of 13 proteins suggested a separation of *Bucephala* from *Melanitta* and *Clangula* (Patton and Avise 1986), but analyses of 25 presumptive protein loci of 40 waterfowl species, including 12

species of sea duck, separated Black Scoters from the other *Melanitta* (White-winged and Surf Scoter) and grouped them with the Long-tailed Duck in an ancestral position relative to the remainder of Mergini (Oates and Principato 1994). The Wagner distance phenogram also placed *Histrionicus* with *Lophodytes* (Oates and Principato 1994). DNA hybridization studies conducted in the late 1980s (Madsen et al. 1988, Sibley and Ahlquist 1990, Sibley and Monroe 1990) precluded insights into intratribe relationships, as only a single member of Mergini (*Melanitta*) was included. Donne-Goussé et al. (2002) analyzed sequence data from the mitochondrial genome in 45 waterfowl species representing 24 waterfowl genera to infer phylogenetic relationships among Anseriformes and included six sea duck species (Black Scoter, Smew, Hooded Merganser, Red-breasted Merganser [*M. serrator*], Long-tailed Duck, and Common Eider). Their analyses placed these representatives of Mergini, as well as, surprisingly, *Callonetta leucophrys* (Ringed Teal; sometimes placed with the dabbling ducks but of unresolved phylogenetic relationship; Bulgarella et al. 2010), within a single clade that was sister to a clade comprising Cairinini, Anatini, and Aythyini. The mergansers, including *Mergellus*, were placed in a monophyletic clade, and *Somateria* was placed close to *Clangula* within the Mergini clade.

Gonzalez et al. (2009) included 12 species of sea duck in a phylogenetic analysis of Anatidae, based on mitochondrial DNA (mtDNA) cytochrome *b* and ND2 genes; their analysis also placed Mergini in a single clade, sister to the Aythyini and Anatini, with Cairinini (represented by *Aix sponsa*, *A. galericulata*, and *Cairina moschata*) sister to Tadornini. *Clangula* is placed as the basal species in the monophyletic clade Mergini. The two eider species analyzed (Common and King Eiders) were placed within a single clade sister to a clade comprising scoters, goldeneyes, and mergansers. *Callonetta* did not group with Mergini in this analysis. Solovyeva and Pearce (2011) and Liu et al. (2012) also used mtDNA control region and cytochrome *b* sequences to examine phylogenetic relationships among several mergansers. Solovyeva and Pearce (2011) analyzed data from Common Mergansers (*M. merganser*) from North America and Europe/Russia and Hooded, Red-breasted, and Scaly-sided Mergansers; their analyses placed Red-breasted and Hooded Mergansers in a single clade. This classification differed from Liu et al. (2012), who placed Common

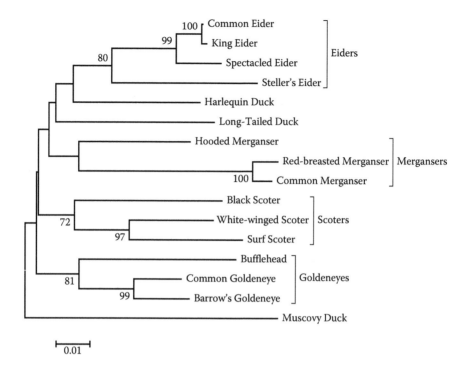

Figure 2.2. Phylogenetic reconstruction of all 15 North American sea duck species assayed from 549 bp of mtDNA cytochrome *b* sequence, using a minimum evolution distance approach. Bootstrap support values are provided at relevant nodes (>70% support).

and Red-breasted Mergansers in a single clade, sister to a clade comprising Smew and Hooded Mergansers, concluding that both *Lophodytes* and *Mergellus* belong within *Mergus*.

In an effort to provide a more comprehensive evaluation of the evolutionary relationships among the 15 extant North American sea duck species, we obtained sequence data from 549 base pairs of the mtDNA cytochrome *b* gene and used the distance approach to generate a phylogenetic tree, following methods similar to those outlined in Pearce et al. (2004). Unfortunately, the phylogenetic reconstructions failed to resolve intragenera relationships, with most nodes receiving <70% bootstrap support (Figure 2.2). *Somateria* eiders comprised a single clade that was sister to Steller's Eider. Buffleheads were basal to Common (*B. clangula*) and Barrow's Goldeneyes (*B. islandica*), and Black Scoter was basal to White-winged and Surf Scoters. The scoters and mergansers both formed monophyletic clades. However, the relationships of the Harlequin Duck and Long-tailed Duck, relative to the other sea duck species, remain unresolved. Additional data are needed to provide sufficient resolution to infer higher-level phylogenetic relationships within Mergini. We suggest that future assessments that query the mitochondrial genome apply whole mitogenomic data to the question of Mergini phylogeny to provide greater resolution. Further, whole genomic (or reduced representational genomic) next-generation sequencing technologies, which simultaneously generate whole mitogenomes, are becoming increasingly affordable for nonmodel organisms (Miller et al. 2012), and application of these technologies to generate a wide range of markers (mtDNA, autosomal, and sex linked) to phylogenetic investigations of sea ducks will likely clarify a number of these unresolved relationships.

PHYLOGEOGRAPHY OF NORTH AMERICAN SEA DUCKS

General Genetic Patterns

The application of comparative population genetic approaches to the examination of multiple and widespread species can be used to determine whether each taxon has been similarly influenced by historical isolating mechanisms

Phylogeography. Phylogeography, the assessment of the spatial distribution of genealogical lineages within species (Avise 2000, Knowles 2009, Hickerson et al. 2010), bridges phylogenetic (macroevolution) and population genetic (microevolution) analyses (see Box 2.1). Phylogeography employs our ability to generate DNA data from individuals across multiple species and ranges, allowing us to acquire fundamental insights about evolutionary origins, concordant genetic patterns of different species within a single ecosystem, location of refugia, historical demography, biogeographic barriers, temporal niche conservatism, and evolutionarily significant units (ESUs). Deep knowledge of such biotic and abiotic attributes will ultimately facilitate better predictions about the range of future responses of species, such as sea ducks.

(Avise 2000). Recent comparative population genetic studies of avian species have led to several general conclusions: (1) sympatric taxa distributed across similar regions exhibit a range of phylogeographic patterns (Zink et al. 2001, Qu et al. 2010, Humphries and Winker 2011) and (2) levels of population genetic structure are not necessarily correlated with taxonomic similarity; that is, closely related taxa do not always share similar phylogeographic patterns or levels of genetic diversity and gene flow among populations (Gómez-Diaz et al. 2006, Friesen et al. 2007). Thus, comparisons among species can aid in the identification of isolating barriers (Klicka et al. 2011) and their demographic impacts on populations (Hewitt 2000, Hansson et al. 2008). Genetic information can also be used to infer aspects of species biology, such as flexibility of life history traits, and thus the response to past and future changes in climate (Qu et al. 2010). Studies that investigate phylogeographic relationships using both mitochondrial and nuclear markers, which differ in mode and tempo of inheritance, can also uncover sex-biased life history characteristics, such as dispersal and philopatry (Scribner et al. 2001, Sonsthagen et al. 2011, Peters et al. 2012).

Assessments of neutral genetic variation within North American sea duck species, based on data from both the mtDNA and neutral nuclear markers, have been reported for all the eiders (Scribner et al. 2001; Pearce et al. 2004, 2005b; McKinnon et al. 2006; Sonsthagen et al. 2007, 2009, 2010, 2011, 2013), Harlequin Duck (Lanctot et al. 1999), Long-tailed Duck (Humphries and Winker 2011), and mergansers (Pearce 2008; Pearce et al. 2008, 2009b; Fishman 2010; Solovyeva and Pearce 2011; Peters et al. 2012). Additionally, genetic assessments have been made based on a single mtDNA locus for Red-breasted Mergansers (Pearce et al. 2009a), Buffleheads and goldeneyes (Pearce et al. 2014). No comprehensive molecular-based comparative phylogeographic studies including all North American species within Mergini have been published. Nevertheless, three general patterns have emerged from these single-species or group studies.

Genetic analyses of Steller's, Spectacled, and King Eiders (Scribner et al. 2001; Pearce et al. 2004, 2005b) occupying ice-dominated western and eastern Beringian habitats and in some cases high-latitude habitats in Canada reflect a signal of high variability but low to absent population structuring, particularly at the nuclear genome (Figure 2.2). This pattern is observed also in other highly migratory high-latitude vertebrate species (Dalén et al. 2005; Sonsthagen et al. 2012a; E. Peacock et al., U.S. Geological Survey, unpubl. data), in Beringian scoters (S. L. Talbot and J. M. Pearce, U.S. Geological Survey, unpubl. data) and Long-tailed Ducks (Humphries and Winker 2011; R. E. Wilson et al., U.S. Geological Survey, unpubl. data). Exceptions to this general observation include Common Eiders, which show high levels of population structuring relative to the other eiders (Tiedemann et al. 2004, Sonsthagen et al. 2011).

A second pattern, largely associated with high-latitude boreal nesting sea ducks, is a disjunct North American distribution observed for lineages within some species, including Harlequin Ducks (Scribner et al. 1998) and possibly White-winged Scoters (S. L. Talbot, unpubl. data). This east–west disjunct has been observed in other avian species, including forest obligate accipiters (Hull and Girman 2005, Sonsthagen et al. 2012b, Bayard de Volo et al. 2013). In the Sharp-shinned Hawk (*Accipiter striatus*), this pattern has been attributed to westward expansion following the expansion

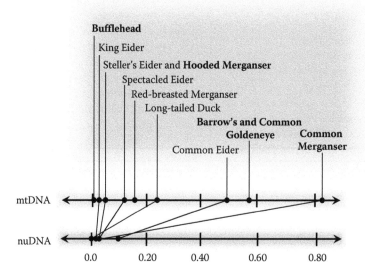

Figure 2.3. Estimates of genetic differentiation for 11 sea duck species, as measured by F_{ST} from large geographic scale studies in North America. Cavity-nesting species are shown in bold. For six species, both mitochondrial (mt) and nuclear (nu) DNA results are shown to illustrate the lower levels of population differentiation observed with nuDNA markers in sea duck species. From left to right, species names represent results from Bufflehead (Pearce et al. 2014), King Eider (Pearce et al. 2004), Steller's Eider (Pearce et al. 2005b), Hooded Merganser (Pearce et al. 2008), Spectacled Eider (Scribner et al. 2001), Red-breasted Merganser (Pearce et al. 2009b), Long-tailed Duck (R. E. Wilson et al., U.S. Geological Survey, unpubl. data), Common Eider (Sonsthagen et al. 2011), Barrow's Goldeneye (Pearce et al. 2014), Common Goldeneye (Pearce et al. 2014), and Common Merganser (Pearce et al. 2009a, b). Studies with $F_{ST} > 0.2$ represent cases of significant population structuring (Category II from Avise 2000). For mtDNA, values are based on control region sequences except for King Eider and Steller's Eider, which are from cytochrome b. (Adapted from Pearce et al. 2014.)

of forest during the Holocene Thermal Optimum (or Hypsithermal; Hull and Girman 2005), although this explanation does not explain the pattern in at least one sea duck species, the Harlequin Duck (S. L. Talbot et al., unpubl. data).

A third pattern is associated with cavity nesters. Larger-bodied, cavity-nesting waterfowl species apparently exhibit greater levels of population differentiation than smaller-bodied congeners (Pearce et al. 2014; Figure 2.3). Seven sea ducks are either obligate or semi-obligate cavity-nesting species and all are secondary cavity nesters, relying upon naturally occurring cavities from tree decay and breakage or on excavator species that create holes into trees. Nest-box studies of cavity-nesting sea ducks (Gauthier 1990, 1993; Eadie et al. 1995, 2000) have documented high levels of nest (breeding) site fidelity, a possible indicator of population structure if accompanied by philopatric behavior. However, patterns of fidelity may be driven by variables other than cavity availability, such as competition, food requirements, brood habitat,

and body size (Boyd et al. 2009). For example, despite high levels of breeding site fidelity, Pearce et al. (2008) found little evidence for population genetic structure in Hooded Mergansers, which likely are not limited by nest cavity availability (Denton et al. 2012). In contrast, another cavity-nesting sea duck, the Common Merganser, exhibited a high degree of population genetic structure across North America (Figure 2.3). These findings led to the hypothesis that population structure among cavity-nesting ducks could be influenced by body size and cavity competition (Pearce et al. 2009b). Among cavity-nesting species of waterfowl, Common Mergansers have the largest body size, requiring larger cavities that may be rare in some forested landscapes (Vaillancourt et al. 2009). As a result, Common Mergansers may exhibit greater fidelity and population structure than smaller-bodied congeners such as the Hooded Merganser. Thus, there may be a positive relationship between body size and level of population structure among cavity-nesting waterfowl due to the greater abundance of smaller cavities and rarity of large cavities. Consistent with this prediction was the finding that Buffleheads, the smallest cavity-nesting species, exhibited the lowest level of population differentiation relative to intermediate-sized Barrow's and Common Goldeneyes (Pearce et al. 2014).

Impact of Historical Climate Change on Genetic Structure

It is increasingly evident that environmental processes associated with Quaternary cycling have resulted in predictable sequences of diversification relative to geography (Avise 2000, Hewitt 2004). While species at lower latitudes experienced fragmentation and local population contraction/expansion through climate cycles, they did not drastically shift ranges. On the other hand, temperate and high-latitude species, including sea ducks, likely experienced range shifts that resulted in more shallow coalescent times due to recent speciation or a loss of historical diversity as glacial cycles wiped the genetic slate clean (Hewitt 1996). In general, vagile species may exhibit broad regional differentiation (Sonsthagen et al. 2012a), although some widespread vagile species, such as Common Eiders, still exhibit relatively fine-scaled phylogeographic structure (Sonsthagen et al. 2007, 2009, 2011, 2013). It is now well established that fluctuating climatic conditions associated with these Pleistocene glacial cycles (Hewitt 2004) greatly influenced the distribution of species in northern latitudes and their genetic diversity. Over 20 Pleistocene glacial cycles have been recorded (Williams et al. 1998), each resulting in major range shifts and fluctuations in population demography of northern latitude species. During glacial stadials, many high-latitude species apparently retracted to southern refugia, recolonizing northward following the retreat of the Laurentide and Cordilleran ice sheets in North America (Avise 2000), and the Fennoscandian ice sheet following climate amelioration in Europe (Hewitt 2001, Schmitt 2007). Nevertheless, much of Eastern Europe, Siberia, and North America remained virtually ice-free throughout these cycles, potentially providing refugia for tundra-obligate species (Hope et al. 2011), or ice-obligate species that occupy adjacent terrestrial habitats for a portion of their annual cycle. Beringia was a large refugium during glacial advances, and multiple species apparently recolonized the Arctic from Beringia after the ice sheets receded (Hultén 1937). However, Beringia was itself recolonized by species expanding out of glacial refugia, mostly from the south (Stewart et al. 2010), and thus this region, which spans eastern Russia, Alaska, and western Canada to the MacKenzie Highlands, may have played a central role in the diversification of populations of Arctic biota (Rand 1954, MacPherson 1965). The Bering Isthmus that connected Eurasia and North America during glacial periods facilitated transcontinental migration of plants and animals, homogenizing lineages there. Separation of Eurasia and North America by the Bering Strait is not generally reflected in genetic analyses, suggesting that this recurring barrier to terrestrial species dispersal (most recently formed 11,000 years ago) has imposed only minor influence on genetic structure or divergence within many free-living terrestrial organisms (Galbreath and Cook 2004).

There is also considerable evidence for additional high Arctic glacial refugia in North America, with divergent lineages and high genetic diversity observed in populations of a number of taxa occupying the high Canadian Arctic and Greenland (Tremblay and Schoen 1999, Abbott et al. 2000, Fedorov and Stenseth 2002, Waltari and Cook 2005). Refugia typically harbor divergent DNA lineages within species that reflect their long-term isolation (Fedorov et al. 2003), and expansion of populations from glacial refugia has left predictable genetic patterns in recently colonized regions (Lessa et al. 2003). Genetic data analyzed under the principles of the coalescent theory have enabled researchers to investigate historical species distribution and demography, aiding in the identification and location of glacial refugia (Lessa et al. 2003, Waltari and Cook 2005, Sonsthagen et al. 2011).

Based on current distribution of waterfowl species, Ploeger (1968) postulated the relative importance of these ice-free areas as potential refugia for Arctic waterfowl (Anatidae) during the last Pleistocene glacial period. Proposed high Arctic refugia that harbored migratory North American species include Beringia, the Canadian Arctic Archipelago, northern Greenland, Spitsbergen Bank near Svalbard, and northwest Norway. Proposed temperate refugia included Newfoundland, western Greenland, Iceland, and Western Europe. Convergence in genetic signatures of population expansion across multiple Arctic species has provided insights into the locations of proposed refugia and their relative importance as historical reservoirs of species genetic diversity. To date, molecular data have inferred and, in some cases, substantiated Beringia, Canadian Arctic Archipelago, and western Greenland as ice-free refugia for many Arctic vertebrates (Holder et al. 1999, 2000; Fedorov and Stenseth 2002; Fedorov et al. 2003; Waltari and Cook 2005) including sea ducks (Spectacled Eider, Scribner et al. 2001; Common Eider, Sonsthagen et al. 2011; Common Merganser, Pearce et al. 2009a; and goldeneyes, Pearce et al. 2014). Additionally, the mtDNA haplotype network for Barrow's Goldeneye shows a central placement of Alaska haplotypes as a possible origin of subsequent genetic diversity across the current range of the species (Figure 2.6a).

Impact of Contemporary Factors on Population Structure

It is axiomatic that the ability of a population to respond to current ecological and evolutionary forces is partially dependent upon the maintenance of genetic diversity. Populations that undergo large reductions in the number of effective breeders are expected to exhibit reduced levels of genetic variation (Jackson et al. 2013). Extinction risk increases in small, isolated populations due to negative effects associated with genetic drift, such as the erosion of quantitative genetic variation required for adaptive evolution (Westemeier et al. 1998, England et al. 2003) and inbreeding depression (Mills and Smouse 1994). These expectations have been corroborated empirically (Spielman et al. 2004, see Frankham et al. 2002 for review). Prior knowledge that a population has endured a recent, severe population decline can help managers anticipate problems, such as decreased reproductive fitness, reduced survival, and increased susceptibility to disease, even though current population sizes may suggest no problems. Population history is often unknown, but past fluctuations in population numbers can be detected using genetic data that can help to infer how both recent and historical demographic, ecological, and genetic histories of species interact to affect persistence (Cornuet and Luikart 1996). Such assessments are crucial to management prescriptions applied to recovering populations of threatened and endangered species (Brown Gladden et al. 1999), such as the Steller's Eider (Pearce et al. 2005b).

Effective Population Size, Homozygosity, and Heterozygosity

Populations of sea ducks, especially those at the southern periphery of distributions and those in the high Arctic, such as the Steller's Eider, could become reduced in size or further isolated as a result of climate warming. Smaller populations may experience higher inbreeding, increasing the possibility that highly deleterious recessive alleles are expressed (Hedrick and Kalinowski 2000). Isolation has profound effects on genetic variability and ultimately the ability of a species to withstand environmental or biotic challenges. Wright's (1931, 1938) concept of effective population size (N_e) is a fundamental parameter in many population models that can be used to monitor populations that have experienced decreases (Johnson et al. 2010). A recent review of indices of genetic diversity in bird populations found significant relationships between measures of mtDNA control region variation and populations size (Jackson et al. 2013). Coalescent analyses can also offer inferences into how past populations have responded to environmental change and thus offer insights into the possible magnitude of future population changes (Hope et al. 2011).

Peripheral Populations

Although peripheral populations may have reduced levels of variability, these populations often harbor unique alleles that can comprise a significant portion of the genetic variability maintained by individual species. Hence, isolated or peripheral populations become key players in the long-term persistence of species. Novel genetic variability in peripheral populations may increase the adaptive potential necessary for species to respond to novel challenges. At northern distributional margins, conserving evolutionary processes in peripheral populations on the edge of species distribution may be important as population shifts in response to climate change (Lessica and Allendorf 1995, Hampe and Petit 2005, Gibson et al. 2009). For example, the Alaskan population of the Steller's Eider represents a peripheral population relative to the core population in Siberia (Pearce et al. 2005b).

Behavior

As a group, waterfowl show strong female-biased natal and breeding site fidelity (Greenwood 1980,

Anderson et al. 1992), and this behavior is thought to characterize sea ducks. Fidelity to sites or to specific habitat types (e.g., cavities for cavity nesters) can lead to particular patterns of population structure, and in particular, female bias in philopatry (natal site fidelity) coupled with male-biased dispersal can lead to discrepancies in patterns of population structure assayed using different classes of molecular markers. In species with high levels of female philopatry and male-biased gene flow, population structure assayed using maternally inherited loci, such as mitochondrial genes, should exceed levels assayed using biparentally inherited nuclear markers, such as microsatellite loci (Zink and Barrowclough 2008). As a result, assessments of the magnitude of spatial variation in gene frequencies in sea duck populations have largely taken advantage of markers that differ in mode of inheritance, since this approach can be used to assess levels of sex-biased dispersal and the geographic scale over which such dispersal occurs (Scribner et al. 2001). Nevertheless, care must be taken to determine whether the observed pattern is due to male bias in dispersal or the smaller effective population size and faster sorting of the mtDNA relative to nuclear genes (Peters et al. 2012). Clarification of behavioral and molecular sources of discordance between molecular markers is especially valuable if rates of survival, or other demographic parameters, vary spatially (Peters et al. 2012). This issue is particularly true for sea ducks since demographic and movement data are sparse for this group relative to other waterfowl.

POPULATION RELATIONSHIPS ACROSS NORTH AMERICA

Phylogenetic studies investigate macroevolutionary processes that lead to differentiation and speciation, whereas population genetic studies focus on understanding microevolutionary processes. An understanding of microevolutionary processes, such as gene flow and genetic drift, is critical when monitoring distribution and abundance of sea duck species across North America because effective management of populations requires accurate identification and delineation of populations. Additionally, information is needed on population connectivity between breeding and wintering areas. Genetic and band-recovery information have both been useful for investigating

populations and migratory patterns of sea ducks and for understanding the relationship among populations via gene flow. For example, Pearce et al. (2014) found high concordance for genetic and band-recovery information in terms of identifying population segments, but the combination also allowed an understanding of connectivity between breeding and wintering areas that would not have been possible with genetic data alone. Similarly, Pearce et al. (2009b) established through the use of both banding and genetic data that genetic signatures of admixed groups of molting male mergansers on Kodiak Island, Alaska, suggest a pattern where some males undertake long-distance migrations from breeding areas in the contiguous United States to Alaska.

Uncovering the origin of taxa and population trajectories can benefit from the use of fossils. Fossil evidence can be used to investigate both macro- and microevolutionary processes by providing information about the origins of species and verifying times of occupations for study taxa relative to changing environments (Jones et al. 2008). Incorporation of DNA extracted from fossils (ancient DNA, or aDNA) can be a powerful method for investigating past population fluctuations, and recent advances in genomic technologies is facilitating the expanded use of aDNA to study changes in populations over even greater temporal scales (Lindqvist et al. 2010, Miller et al. 2012). Relatively few studies have investigated microevolutionary processes in extant populations within sea duck species in North America, and to our knowledge, none have incorporated aDNA into population-level investigations, despite the relative abundance of sea duck fossils. Here, we provide an overview of individual species' origin and microevolutionary characteristics, as suggested by fossil evidence and recent population genetic studies. Species nomenclature follows Chesser et al. (2013).

Eiders

Origins

A *Somateria* specimen was found in middle-Oligocene deposits in central Kazakhstan and dated at ca. 28 million years before present (ybp, Kurochkin 1976). A fossil ascribed to Common Eider (S. cf. *mollissima*) was described from northern Norway dating to ca. 115,000 ybp (Lauritzen et al. 1996). A pelvic bone ascribed to Common Eider was found at the Champlain Sea site in Québec, Canada, dated at 10,300 ± 100 ybp (Harington and Occhietti 1980). Late Pleistocene fossils of Common Eider have been reported from Alaska, Ireland, Norway, and Denmark (Brodkorb 1964), and early Pleistocene to late Holocene fossils were found in the Old Crow Basin Yukon Territory, Canada (Fitzgerald 1991). There are limited records of the other eider species. Fossils ascribed to King Eider were found in prehistoric deposits along the coast of Labrador (Fitzhugh 1978), with Pleistocene deposits in the Ural Mountains (Potapova 1990). Fossil evidence of Spectacled Eiders is limited to Alaska (Friedmann 1934, 1935). The first Palearctic fossil record of a Steller's Eider, dating to between 9,100 and 12,400 ybp, was reported from a late glacial settlement on the Bay of Biscay, Spain (Elorza 2005). Subfossil material from Steller's Eiders has been reported from Sweden (Liljegren 1977) and Alaska (Brodkorb 1964).

Phylogenetic analysis based on the mtDNA cytochrome b gene (Figure 2.2) suggests Common and King Eiders are sister species relative to Spectacled Eiders, with Steller's Eiders basal to the other eider species; this is concordant with morphological analyses (Livezey 1995). Subspecific designations have only been described for Common Eiders. Presumed hybridization has been reported between King and Common Eiders (Pettingill 1959) and between Steller's and Common Eiders (Forsman 1995).

Spectacled Eider

Spectacled Eiders breed in large but patchily distributed areas along coastal western and northern Alaska and Arctic Russia. Population sizes in Alaska declined precipitously during the last decade of the twentieth century (Stehn et al. 1993). Because Spectacled Eiders from different portions of the species' breeding range aggregate in sea ice polynyas during the nonbreeding season, it was hypothesized that Spectacled Eiders were panmictic, that is, belonged to a single, randomly breeding population. Scribner et al. (2001) used molecular markers with contrasting modes of inheritance to evaluate the degree of spatial genetic structuring among breeding populations of Spectacled Eiders. Specifically, population-level structuring was observed between breeding populations, based on mtDNA control region sequences, but levels of structuring were lower

based on analyses of biparentally inherited and sex-linked (Z-specific) nuclear microsatellite loci. Estimates of effective population size were used to assess levels of gene flow among breeding populations and suggested that gene flow between breeding regions differed substantially between males and females (i.e., male-biased gene flow), a pattern seen in other sea ducks (Pearce et al. 2005a, b; Sonsthagen et al. 2011; Peters et al. 2012).

King Eider

The King Eider has a Holarctic distribution, nesting in coastal Arctic tundra in northern Alaska, Canada, Greenland, Svalbard, and Russia and wintering throughout the Bering Sea and Pacific Ocean near Russia and Alaska, the Atlantic coast of Canada and the United States, in southern Greenland, and along the coast of Norway. Ploeger (1968) concluded that the lack of morphological distinction among King Eiders of North America was due to the species' nomadic nature and sharing of common wintering areas that eventually led to dispersal among populations restricted to different Pleistocene refugia. Despite the lack of morphological differentiation, however, two breeding populations—western and eastern Arctic—are generally recognized in North America (Suydam et al. 2000). These two regions are delineated based on geographic isolation of wintering areas in the Pacific and Atlantic oceans, corroborated to some extent by banding (Salomonsen 1968, 1979; Lyngs 2003) and satellite telemetry data (Dickson 2012). Some level of nest site fidelity has been observed within King Eiders in the Central Canadian Arctic (Kellett 1999). This result, coupled with geographic isolation of wintering areas, suggests that gene flow mediated by females might be reduced, facilitating population- or regional-level structure. However, population genetics research of King Eiders in North America that analyzed fragment data from nuclear biparentally inherited microsatellite loci and sequence data from the cytochrome b gene of the maternally inherited mtDNA genome failed to uncover a signal of spatial genetic structure (Pearce et al. 2004). Results from nested clade analyses, mismatch distributions, and coalescence-based analyses suggested that historical population growth and high levels of gene flow may have homogenized gene frequencies among both nesting and wintering areas over time. Nevertheless, the presence of unique haplotypes among birds wintering in Greenland may reflect some level of reduction of gene flow between the western and eastern Arctic, consistent with the banding and stable isotope data.

Common Eider

Common Eiders have a circumpolar distribution and inhabit coastal marine systems (Goudie et al. 2000) throughout the Holarctic, breeding from northern Alaska, northeast United States, Canada, Greenland, Iceland, Scandinavia, northern Europe, and Russia (with the exception of north central Russia). Common Eiders show a propensity to nest colonially and form large aggregations in nonbreeding months. Populations vary in their degree of migratory behavior, with some populations remaining resident (Belcher and Aleutian islands), while others are long-distant migrants (>2,300 km; Goudie et al. 2000). Typically, populations overwinter as far north as open water persists, and facultative movements induced by advancing pack ice or freeze-up have been observed (Goudie et al. 2000). Similar to other sea ducks, numbers of Common Eiders declined >50% in the mid-1970s–1990s (Cooch 1965, Robertson and Gilchrist 1998, Suydam et al. 2000, Merkel 2004) and in some areas >90% (Yukon–Kuskokwim Delta (YK Delta), Alaska; Stehn et al. 1993). Recent estimates within North America and Greenland suggest that populations have stabilized and some areas may be increasing in number (Hipfner et al. 2002, Chaulk et al. 2005, Quakenbush et al. 2009, Merkel 2010, Chapter 1, this volume).

Female Common Eiders are highly philopatric, whereas males disperse among populations that share wintering grounds. Common Eiders are one of the few Holarctic waterfowl species in which morphology is sufficiently variable that six to seven subspecies have been described (Goudie et al. 2000). Subspecific designations are based on morphology (Palmer 1976) and appear to correspond to overwintering areas (Ploeger 1968, Goudie et al. 2000). Recent molecular work uncovered partitions in the nuclear genetic markers that are concordant with overwintering areas and therefore subspecific designations (Sonsthagen et al. 2011), supporting the inferences that both sexes display some degree of winter site fidelity (Spurr and Milne 1976). Further, phylogeographic analyses indicate that Common Eiders were restricted to four areas during the Last Glacial Maximum (LGM): Belcher Islands, Newfoundland Bank, northern Alaska, and Svalbard (Sonsthagen et al. 2011). Three of these areas coincide with previously identified glacial refugia: Newfoundland Bank, Beringia, and Spitsbergen Bank. Gene flow and clustering analyses indicated that the Beringian refugium contributed little to the Common Eider postglacial colonization of North America, whereas the Canadian, Scandinavian, and southern Alaskan postglacial colonization is likely to have occurred in a stepwise fashion from the same glacial refugium (Figure 2.4; Sonsthagen et al. 2011). Furthermore, suture zones identified at MacKenzie River, western Alaska/Aleutians, and Scandinavia coincide with those identified for other Arctic vertebrates, suggesting that these regions were strong geographic barriers limiting dispersal from Pleistocene refugia (Sonsthagen et al. 2011).

Common Eiders are unusual among sea ducks and other high-latitude avian species as they exhibit fine-scale spatial genetic structuring for both mtDNA and nuclear markers (Tiedemann et al. 1999, 2004; Sonsthagen et al. 2007, 2009, 2013). High levels of structure in mtDNA were observed among colonies in the Aleutian Islands, Baltic Sea, Beaufort Sea, and YK Delta ($\Phi_{ST} = 0.135$–0.866, $F_{ST} = 0.074$–0.709). Significant, but lower, levels of structure were detected among Baltic Sea and Beaufort Sea colonies at microsatellite loci ($F_{ST} = 0.009$–0.029). Furthermore, molecular data indicate that Common Eider females nest in kin groups, which increases microgeographic genetic structure within populations via the formation of spatially close associations among individuals that share matrilineages (McKinnon et al. 2006, Sonsthagen et al. 2010).

Steller's Eider

The Steller's Eider, the smallest of the four eider species, breeds in coastal tundra and Arctic Russia and Alaska (Fredrickson 2001). A majority of the estimated 200,000 Steller's Eiders nest in Russia, with a small number nesting in two disjunct regions in North America: the YK Delta of western Alaska and the western North Slope of Alaska. Following nesting, Steller's Eiders undergo a flightless molt along the

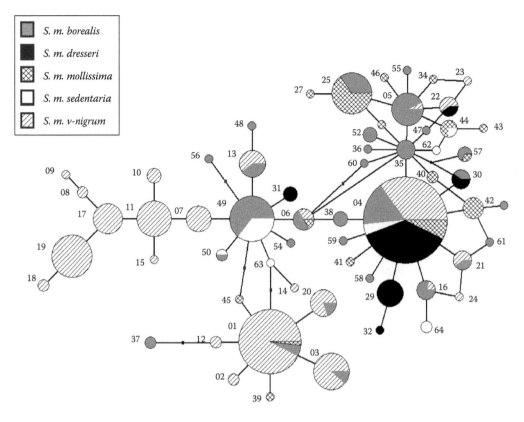

Figure 2.4. Unrooted 95% parsimony network illustrating relationships of 64 mtDNA control region haplotypes from Common Eiders (*S. mollissima*), with the size of the circle node corresponding to the frequency of each haplotype (numbered). Small black squares indicate intermediate ancestral haplotypes that were not sampled (Sonsthagen et al. 2011).

Alaska Peninsula, then winter throughout the Aleutian Islands eastward to Cook Inlet, Alaska (Fredrickson 2001). Banding data suggest that Steller's Eiders from multiple wintering groups aggregate along the Alaska Peninsula during molt (Dau et al. 2000). Pearce et al. (2005b) employed biparentally inherited nuclear microsatellite loci and mtDNA cytochrome *b* sequence data to infer levels of population differentiation and population exchange across the range of Steller's Eider. They found that levels of population differentiation were low within, but high between, Pacific and Atlantic nesting areas, but only for mtDNA. Summary statistics for mtDNA indicated that Steller's Eider went through a population expansion; however, they remained spatially unstructured, possibly because an insufficient amount of time has passed since expansion to allow for behavioral influences, such as philopatry, to arise between populations that now winter in different ocean basins (Pearce et al. 2005b).

Scoters

Origins

The earliest fossil assigned to the genus *Melanitta* (*M. ceruttii*; Chandler 1990, an extinct species) is reported from the San Diego Formation near San Diego, California, and dating to the late Pliocene (~2–2.5 million ybp; in North America, the Late Blancan). The earliest records for the three extant North American species date to the Pleistocene (Brodkorb 1964, Howard 1964). Fitzgerald (1991) reports fossils from White-winged Scoters from late Illinoian sediments (>140,000 ybp) and upper interlake beds (dated to >14,000–25,000 ybp), as well as Surf Scoter fossils from the Sangamon (dated to >54,000 ybp) and river terrace (dated to between ca. 1,700 and 2,400 ybp) sediments from the Old Crow River, Yukon. Fossils from Black Scoters, which do not currently occupy the region, were also recovered from river terraces and sandbars from the Old

Crow River Basin and dated to the early to late Holocene (Fitzgerald 1991). Black Scoter fossils dating from the late Pleistocene to late Holocene have also been recovered from the Pribilof Islands, Aleutian Islands, Kodiak Island, Cape Denbigh, and Cape Prince of Wales (Brodkorb 1964). White-winged Scoters are reported from Nearctic Pleistocene deposits on the west (southern California: Howard 1933, Brodkorb 1964, Oregon: Jehl 1967) and east coasts (Maryland: Wetmore 1973).

White-winged and Surf Scoters

Morphological (Livezey 1995) and molecular data from the mtDNA cytochrome *b* gene (Figure 2.2) consistently place Surf and White-winged Scoters as sister taxa and support current nomenclature that place these with Velvet Scoters within the subgenus *Melanitta*, which is sister to the subgenus *Oidemia*, which contain Black Scoters (see the following). Some researchers consider White-winged Scoters to be conspecific with the Velvet Scoter of Europe (Brown and Fredrickson 1997), with White-winged Scoters placed in a North American/eastern Asian subspecies group (*M. fusca deglandi*) and Velvet Scoters within a European subspecies group (*M. fusca fusca*). Others including Collinson et al. (2006) consider morphological distinctiveness sufficient to consider the Velvet and White-winged Scoters to be distinct species. No geographic variation and no subspecies are described for Surf Scoters (Savard et al. 1998). Baird et al. (1884) and Ball (1934) report hybridization between White-winged and Surf Scoters, Gray (1958) and Palmer (1976) found possible hybridization between White-winged Scoters and Common Goldeneyes, and Gundmundsson (1932) noted a possible hybridization between a male White-winged Scoter and a female Common Eider. No genetic evidence of hybridization has been reported, however.

White-winged Scoters breed across the high-latitudes of North America, from western Ontario northwest across Canada into northern Alaska (Palmer 1976), and across northern Asia, east of the Yenisey Basin, wintering in temperate regions along the Great Lakes, northern coasts of the northern United States and southern coasts of Canada, and south to China in Asia. Surf Scoters breed in Alaska and Canada and, like White-winged Scoters, winter in more temperate zones.

Black Scoter

Black Scoters are considered a subarctic species (Bordage and Savard 2011), and occur in two apparently disjunct North American populations. The western (Bering/Pacific) population occurs mostly in Alaska, where the species nests on coastal alpine habitats and alpine lakes from southwestern and northwestern Alaska, the Alaska Peninsula, and Kodiak Island eastward to the central Alaskan Range (Gabrielson and Lincoln 1959; Hodges 1991; B. Conant and D. J. Groves, U.S. Fish and Wildlife Service, unpubl. report). The eastern (Atlantic) population nests largely in northern Québec (Bordage and Savard 2011), although breeding records or pairs have been reported from Newfoundland, Labrador, and the Northwest Territories. Black Scoters nesting in the Pacific region winter along the Aleutian and Pribilof islands and extend southeast, although with decreasing abundance (Root 1988) to Baja California. In the Atlantic region, Black Scoters winter from Newfoundland south to Florida and in interior lakes such as the Great Lakes and the St. Lawrence River (Johnsgard 1975, Godfrey 1986, Root 1988). There are no morphological differences evident between the Pacific and Atlantic groups, and no subspecies have been described. Black and Common Scoters (*M. nigra*-gp.), previously considered conspecific or allospecific (Collinson et al. 2006, Sangster 2009, Chesser et al. 2010), are considered sister species relative to Surf and White-winged Scoters (Livezey 1995). Breeding distribution of Black and Common Scoters is largely allopatric, but the two species come into contact at the lower Lena River valley of Russia (Cramp and Simmons 1977)—a well-known contact zone (Abbott and Brochmann 2003)—but no hybrids have been reported.

Population and phylogeographic studies employing mtDNA cytochrome *b* sequences and microsatellite analyses are ongoing (S. L. Talbot, unpubl. data). Preliminary analyses of the same suite of 11 biparentally inherited microsatellite data from the western portion of the range of all three scoter species suggest Black Scoters have higher levels of diversity (observed heterozygosity = 0.59) relative to White-winged and Surf Scoters (observed heterozygosity of 0.50 and 0.49, respectively) and that each species has low levels of population differentiation within the region (S. L. Talbot, unpubl. data). Microsatellite loci clearly differentiate the three

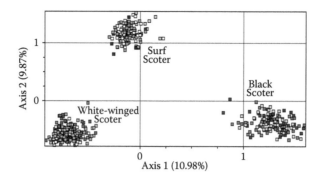

Figure 2.5. Factorial correspondence analysis performed on allele frequencies at 11 biparentally inherited microsatellite loci in three scoter species (Black Scoter, n = 130; White-winged Scoter, n = 209; Surf Scoter, n = 143) samples from across each species' western North American range. Within each species, different gray-scale values represent different sampling locales. Analysis includes (1) Black Scoters sampled from Alaska, the Pacific Northwest, and the east coast of North America; (2) White-winged Scoters sampled from Alaska and the Pacific Northwest, the Northwest Territory of Canada, and the east coast of North America; and (3) Surf Scoters sampled from Alaska and the Pacific Northwest and the east coast of North America.

species, and there is little suggestion of hybridization (Figure 2.5), although additional sampling may be needed to uncover rare hybridization events. Analyses of a nesting population of Black Scoters in Aropuk Lake, Alaska (S. L. Talbot, unpubl. data), failed to uncover any signature of preference for nesting near kin, as observed for Common Eiders (McKinnon et al. 2006, Sonsthagen et al. 2010). Preliminary analyses of White-winged Scoters from the western portion of their range relative to samples from Europe suggest this species might show an east–west disjunct in the distribution of alleles, although this requires further analyses (S. L. Talbot, unpubl. data).

Goldeneyes and Buffleheads

Origins

Fossils ascribed to *Bucephala* date to the Middle Miocene of Mátraszölos, Hungary, and the Late Pliocene of Chilhac, France (*B. cereti*); the Late Miocene/Early Pliocene of Bone Valley, USA (*B. ossivalis*); the Late Pliocene of California (*B. fossilis*); and the Early Pleistocene of Central Europe (*B. angustipes*) and Turkey (*Bucephala* spp.; Louchart et al. 1998).

Goldeneyes and Buffleheads

No subspecies are described within the three species of *Bucephala* across their North American range, although Bufflehead shows some geographic variation in juvenile plumage (Erskine 1992). Barrow's Goldeneye is monotypic (Eadie et al. 2000); the Common Goldeneye in North America is larger and has a thicker bill than the species in Eurasia (Eadie et al. 1995) but shows little morphological variation across North America. Common Goldeneyes reportedly hybridize readily with other species; hybrids have been reported between Common Goldeneye and Barrow's Goldeneye (Martin and DiLabio 1994), Hooded Merganser, Common Merganser, Smew, Pochard (*Aythya ferina*), Greater Scaup (*A. marila*) and, in captivity, other species (Ball 1934, Gray 1958, Palmer 1976, Panov 1989). Hybridization between wild Barrow's and Common Goldeneyes has been supported by DNA evidence (Eadie et al. 2000). Hybridization is reported between Buffleheads and Hooded Mergansers (Marcisz 1981). Gauthier (1993) hypothesizes that the small size and distinctive behavior of the Buffleheads prevent hybridization between Buffleheads and goldeneyes.

Pearce et al. (2014) examined the population structure of three cavity-nesting waterfowl species distributed across much of North America: Barrow's Goldeneye, Common Goldeneye, and Bufflehead. Patterns of population structure were assayed, using both variation in mtDNA control region sequences and band-recovery data for the same species and geographic regions. Results between data types were highly congruent with the hypothesis derived from the larger waterfowl literature, that larger-bodied, cavity-nesting species exhibit greater levels of population

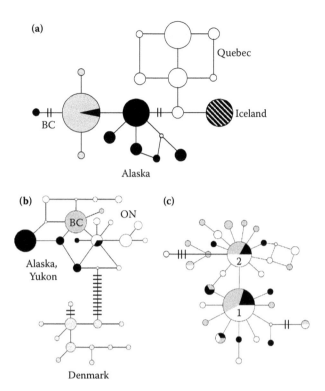

Figure 2.6. MtDNA haplotype networks for (a) Barrow's Goldeneye (black for Alaska, gray for British Columbia, white for Quebec, and diagonal for Iceland), (b) Common Goldeneye (black for Alaska/Yukon, gray for British Columbia, white for Ontario, light gray for Denmark), and (c) Bufflehead (black for Alaska/Yukon/Northwest Territories, gray for western North America, and white for eastern North America). A single site substitution links each circle except where bars are present, which denote multiple substitutions between haplotypes. Circles are drawn proportionally to observed number of each haplotype. The smallest open circles in each network represent inferred haplotypes that were not sampled. Numbers within larger circles correspond to the haplotype number. (Adapted from Pearce et al. 2014.)

differentiation than smaller-bodied congeners. Populations were structured for Barrow's and Common Goldeneyes, which are intermediate sized, but not for Buffleheads, the smaller-sized cavity nester. Further, Pearce et al. (2014) found evidence for discrete Old and New World populations of Common Goldeneyes and for differentiation of regional groups of both goldeneye species in Alaska, the Pacific Northwest, and the eastern coast of North America (Figure 2.6).

Mergansers

Origins

Fossils of the genus *Mergus* are described from the Middle Miocene Calvert Formation (*M. miscellus*, dated to approximately 14 million years ago) of Virginia (Alvarez and Olson 1978) and from the Early Pleistocene (*Mergus connectens*, 1–2 million years ago) of central and Eastern Europe (Mlíkovský 2002). Fossils from the Late Pliocene of Texas have been attributed to *Lophodytes* (Miller and Bowman 1956), and fossils from the Late Pleistocene from various sites, such as Georgia (Steadman 2005) and Kansas (Stettenheim 1958), have been ascribed to Hooded Mergansers.

Phylogenetic analysis of morphological and plumage characters among the sea ducks placed the mergansers (both *Mergus* and *Lophodytes*) in a monophyletic group, sister to the goldeneyes (Livezey 1995) and the Smew of Eurasia. The Hooded Merganser is the basal merganser species and the Red-breasted Merganser is in a sister species relationship with the Scaly-sided Merganser (*M. squamatus*) of Eurasia, relative to the Common Merganser. Phylogenetic analyses employing data from diverse mtDNA genes have been equivocal. Analyses based on 2103 base pairs (bp) of the mitochondrial genome from 45 waterfowl species placed North American mergansers (Red-breasted and Hooded) and the Smew of Eurasia in a monophyletic group within the tribe Mergini and sister to the goldeneyes and Bufflehead (Donne-Goussé et al. 2002). Liu et al. (2012) analyzed the entire mitogenome of the Scaly-sided Merganser and compared homologous sequences from concatenated control region and cytochrome *b* sequence data with those of three *Mergus* species (Common, Red-breasted, and Scaly-sided Mergansers), the Hooded Merganser, and the Smew. Their analyses placed the *Mergus*, *Mergellus*, and *Lophodytes* within a single monophyletic group relative to the Muscovy Duck (*Cairina moschata*), Redhead (*Aythya americana*), and Mallard (*Anas platyrhynchos*); they did not include the closely allied *Bucephala* in their analyses. All phylogenetic analyses were well supported and within the mergansers; the Scaly-sided Merganser was basal to the sister relationship of the Red-breasted and Common Mergansers, whereas the Smew and Hooded Merganser comprise a monophyletic clade. Other analyses of two mtDNA genes (1045 bp of the cytochrome *b* gene and 1041 bp of the NADH dehydrogenase subunit 2 gene) from among 121 species of Anseriformes placed the three North American mergansers in a monophyletic group relative to the Smew, with Hooded Merganser basal to the Common and Red-breasted Merganser (Gonzalez et al. 2009). Solovyeva and Pearce (2011) present comparative analyses of North American and Eurasian mergansers, which included phylogenetic analyses based on 405 bp of a portion of the mtDNA control region (domain 1) from a greater number of individuals representing the three North American mergansers and the Scaly-sided Merganser than previously analyzed. Their analyses placed the Scaly-sided Merganser (n = 38) basal to the different forms of Common Mergansers in North America (n = 92) and Europe (Goosander, n = 27), relative to sequences from three *Bucephala* outgroup species. The Red-breasted and Hooded Mergansers occurred as well-defined sister species. Levels of genetic diversity within North American populations of *Mergus* were found to be higher than within the Eurasian Scaly-sided Merganser, a species considered to be one of the most threatened sea duck species in the Palearctic region.

Like most waterfowl species, the Common Merganser appears to exhibit a female bias in philopatry, a characteristic that leads to population-level structure (Avise 2000, Peters et al. 2012). However, a number of waterfowl species, including the Hooded Merganser (Pearce et al. 2008), demonstrate a wide range of mtDNA differentiation that are at times inconsistent with predictions

based solely on female philopatry (Cronin et al. 1996, Scribner et al. 2003, Peters and Omland 2007). Thus, the assumption of female philopatry, an important life history trait that can drive population differentiation over time and be an indication of limited species dispersal and thus vulnerability, should be investigated alongside other supporting evidence (Pearce and Talbot 2006). Avian species with a high population structure based on genetic and morphological measures are often differentiated into subspecies (Wilson et al. 2013); however, among the North American mergansers, subspecific designations have only been described for the Common Merganser (Mallory and Metz 1999).

Common Merganser

The Common Merganser is a Holarctic, cavity-nesting species that occupies boreal forests (Mallory and Metz 1999). Mark-recovery analyses based on banding data and satellite telemetry studies suggest geographic and sex-specific variation in migratory tendency in North America (Pearce et al. 2005a, Pearce and Petersen 2009) and Europe (Little and Furness 1985, Hatton and Marquiss 2004). Hefti-Gautschi et al. (2009) and Pearce et al. (2009b) showed that breeding groups within both North America and Europe were structured for mtDNA, but not nuclear DNA, suggesting males disperse among genetically differentiated breeding areas (Peters et al. 2012). Pearce et al. (2009a) examined mtDNA data from both males and females sampled from multiple wintering grounds to analyze migratory patterns and population structure among breeding and wintering North American Common and Red-breasted Mergansers. Common Mergansers breeding in western and eastern North America were strongly differentiated at both the regional level and at finer scales, suggestive of limited female-mediated gene flow. No evidence of secondary contact among historically isolated groups of Common Mergansers was observed. North American samples clustered into three major mtDNA haplotype groups: (1) Beringia, including samples from interior and western Alaska, the Aleutian Islands, and Magadan, Russia; (2) south central and southeastern Alaska and British Columbia; and (3) Washington, western Ontario, and eastern North America. Samples from Scotland and Denmark formed a fourth group. The south central Alaska/British Columbia phylogroup appears to have derived from the Beringian group, which is more closely allied to North America than Asia. This phylogenetic pattern differs from a number of other Beringian avian taxa that show closer affinities with Asian lineages (Zink et al. 1995, 2006; Drovetski et al. 2004), although admixture of Palearctic and Nearctic lineages near Vladivostok may signify the presence of a suture zone in the Russian Far East. Microsatellite and nuclear intron sequence data did not reflect strong differentiation observed in mtDNA among regions within continents, but showed strong differences between Eurasia and North America (Hefti-Gautschi et al. 2009, Pearce et al. 2009a, Peters et al. 2012).

The pattern of population-level structure in Common Mergansers contrasted with patterns observed in Red-breasted Mergansers, a ground-nesting species, which exhibited lower levels of population structure across their North American range. Because population structure could not be attributed to limited migration of breeding female Common Mergansers during nonbreeding seasons, Pearce et al. (2009a) hypothesized that interspecific differences in mtDNA patterns may stem from factors relating to nesting ecology, body size, and responses to historical climate change (see Pearce et al. 2014).

Red-breasted Merganser

Like the Common Merganser, the Red-breasted Merganser has a Holarctic distribution but, unlike the Common Merganser, is a ground-nesting species occupying both tundra and boreal forest areas (Titman 1999). The Red-breasted Merganser is the only species within the cavity-nesting clade that nests on the ground, apparently arising from an ancestral group of the cavity-nesting Mergus species (Livezey 1995). Pearce et al. (2009a) hypothesized that ground-nesting behavior arose from competitive avoidance among closely related cavity-nesting sea duck species, such as with the Common Merganser and goldeneyes. Additionally, ground nesting may have facilitated a northward expansion into interior and coastal tundra areas where opportunities for cavity nesting are few. mtDNA analyses in the Red-breasted Merganser revealed decreased levels of population structuring compared to the Common Merganser,

high haplotype diversity and low nucleotide diversity, and a starlike haplotype network, all of which are expected with population expansions (Kvist et al. 1999). Consequently, these results support the hypothesis that Red-breasted Mergansers colonized northward along both the eastern and western coasts of North America, following amelioration of climate and expanding boreal forests during the Holocene.

Hooded Merganser

The Hooded Merganser is restricted to North America and resident throughout the year in broad areas of the Pacific Northwest of British Columbia, Canada, the northwestern United States, and much of the eastern half of the continental United States. Additionally, the species is known to breed across the southern portion of the eastern Canadian Provinces (Dugger et al. 2009). In addition to areas where it is resident, the species winters in freshwater habitats of the western United States and along the Gulf Coast.

As with all merganser species, the population size of the Hooded Merganser is thought to be stable or even increasing (Dugger et al. 2009, Flint 2013, Chapter 1, this volume). Population stability is supported by genetic analyses conducted by Pearce et al. (2008), who estimated that the North American population of Hooded Mergansers began expanding approximately 60,000 years ago (range is 25,000–157,000) based on control region mtDNA. Haplotype networks resemble an expanding population with little evidence for population subdivision, likely due to high dispersal tendency (see Pearce et al. 2014). Despite a lack of broad-scaled subdivision, Pearce et al. (2008) found some evidence for fine-scale structure, suggesting a role of female philopatry.

MONOTYPIC SEA DUCKS

Harlequin Duck

Origins

The genus *Histrionicus* is considered monotypic and is known from fossils recovered from Middle to Late Miocene deposits in Oregon (*H. shotwelli*; Brodkorb 1961). The fossil history of the species is scant; until recently, it was known only from the Pleistocene of North America, although a closely related taxon is reported from the Pliocene of Lee Creek Mine, North Carolina (Olson and Rasmussen 2001). In 2012, a nearly complete coracoid of a Harlequin Duck was described from Middle Pleistocene fluvial deposits in Casal Selce, near Rome, Italy (Pavia and Bedetti 2013:877), although the authors suggest that, based on the species' current "sedentary habit … in the Western Palearctic," the fossil was likely to represent a vagrant. Researchers assume that the Harlequin Duck had a continuous range throughout North America prior to the last glacial period (Brodkorb 1964, Ploeger 1968, Hobson and Driver 1989, Olson and Rasmussen 2001). The extinction of the central portion of the species' range may have occurred when the Laurentide ice sheet separated eastern from western Nearctic populations—west Greenland, Iceland, and Newfoundland providing a refugia for eastern populations (Ploeger 1968). Nevertheless, Bennike (1997) cited the harsh conditions present during the LGM to refute the hypothesis that Greenland represented a glacial refugium; however, the concept is still debated (Bennike et al. 1999). Brodeur et al. (2002) consider the migration route of Harlequin Ducks from breeding grounds at southern latitudes in Labrador and Quebec to wintering areas in eastern portions of the United States to possibly reflect a southeasterly origin from an eastern North American refuge.

Livezey (1995) placed Harlequin Ducks as basal in the Mergini phylogeny, based on morphological, plumage, and behavioral characteristics, and considered Harlequin Ducks to be more closely allied with other sea ducks than with eiders. Hybridization with other waterfowl species has not been reported. Harlequin Ducks occupy an ecological niche unique among sea ducks: the species uses fast-flowing rivers and streams for breeding. The species is characterized by a disjunct Holarctic breeding distribution, comprising two widely separated breeding populations, the western (Pacific) Arctic and subarctic and eastern (Atlantic) Arctic and subarctic (Robertson and Goudie 1999). This broad-scale population delineation is based on the geographic isolation of both breeding and wintering areas associated with watersheds draining into the Pacific and Atlantic oceans; there are no known breeding populations between the eastern seaboard of North America and northwestern Wyoming and

western Montana. Nevertheless, although Brooks (1915) described two subspecies corresponding to Pacific and Atlantic birds (H. h. pacificus and H. h. histrionicus, respectively), there is no evidence of morphological differentiation between the two (Palmer 1976), and no subspecies are currently recognized. Preliminary genetic analyses suggest low levels of population differentiation but support the regional distinction of eastern and western Harlequin Duck populations (Scribner et al. 1998; K. T. Scribner, University of Michigan, unpubl. report).

Few genetic data are available regarding the extent and level of broad-scale movement among breeding locales or migratory routes or the degree of fidelity to breeding, molting, and wintering areas by Harlequin Ducks. However, an increasing number of banding and satellite telemetry studies are leading to a greater understanding of the species' seasonal movement patterns and distribution. For instance, short-term field studies suggest that individual Harlequin Ducks demonstrate a high degree of fidelity to certain breeding, molting, and wintering areas (Anderson et al. 1992, Robertson and Cooke 1999, Robertson and Goudie 1999, Robertson et al. 2000, Iverson et al. 2004), suggesting demographic independence of population segments. Juvenile Harlequin Ducks migrate with adult females to wintering areas (Regehr et al. 2001), but male dispersal and evidence for local-scale gene flow has been observed among wintering Harlequin Ducks (Lanctot et al. 1999, Cooke et al. 2000). Studies assessing microgeographic movement of breeding birds in the Pacific populations (Cassirer and Groves 1994, Reichel and Genter 1996, Bruner 1997) suggest breeding females and subadults move among watersheds within the breeding range. Nevertheless, fidelity by philopatric females (and associated young) to specific wintering areas, followed by pair formation, may lead to genetic differentiation of wintering populations, as seen in Common Eiders (Sonsthagen et al. 2011).

In western North America, wintering populations are found along exposed coastlines of Alaska, British Columbia, Washington and Oregon, and south to California and Baja California (Howell and Webb 1995). In Alaska, migratory distances can be short (Dzinbal 1982, Crowley 1994), whereas satellite telemetry studies of eastern North American birds suggest wider-ranging migratory movements (Brodeur et al. 2002). Genetic studies of breeding and wintering populations in western North America failed to reject the null hypothesis of panmixia among populations comprising the western portion of the species' range (exclusive of Russia, Brown 1998, Lanctot et al. 1999). However, mark-recapture data within this region suggested wintering aggregations may be demographically independent at a much finer scale than has been observed genetically (Iverson et al. 2004, but see Pearce and Talbot 2006).

Eastern birds may comprise at least two distinct wintering populations that may have originated from a Pleistocene glacial refuge in western Greenland and one south of the Laurentide ice sheet in eastern Canada or United States (Brodeur et al. 2002). Banding and satellite telemetry studies indicate that Harlequin Ducks in the wintering population of eastern North America nest in northern New Brunswick, the Gaspé Peninsula in Québec, southern Labrador, and Newfoundland (Brodeur et al. 2002). The wintering population of Greenland nests in Greenland, northern Québec (Nunavik), northern Labrador, and Nunavut. A large portion of the eastern North American population winters in coastal Maine (Vickery 1988; Mittelhauser 1991, 1993). The historical population size of Harlequin Ducks on the eastern coast of North America is considered to have been smaller than western populations, and increased hunting, disturbance, and habitat loss are the primary factors determining number of breeding Harlequin Ducks on the east coast (MacCallum 2001).

Although genetic studies reported by Brown (1998) and Lanctot et al. (1999) suggest the western population is characterized by little microgeographic structuring at different temporal scales, little is known about the population genetic structure and gene flow among the two broadly separated regions containing populations of Harlequin Ducks, either within or between the wintering populations of the western and eastern Arctic and subarctic. Female nest site fidelity, which is thought to characterize some sea duck species in the Arctic (Kellett 1999), and the geographic isolation of Harlequin Duck wintering areas suggest that limited female gene flow may occur between the western and eastern Arctic, which could lead to genetic differentiation of these populations. However, some data also suggest that a fairly broad-scale interchange can potentially link the two populations. A female banded as a duckling

in Wyoming in August of 1994 was harvested in Alaska in January of 1998 (U.S. Geological Survey Bird Banding Laboratory). Nevertheless, preliminary analyses (Scribner et al. 1998; K. T. Scribner, University of Michigan, unpubl. report) suggest that, like White-winged Scoters as well as Common Eiders, Harlequin Ducks show an east–west disjunct genetic population structure.

Long-tailed Duck

Origins

A congener of the Long-tailed Duck has been described from the Middle Miocene of Hungary (Gál et al. 1999), and in North America, middle Pleistocene fossils (between 54,000 and >140,000 ybp) were recovered from the Old Crow River basin in the Yukon Territory, Canada (Fitzgerald 1991). Late Pleistocene fossils of Long-tailed Ducks have been reported from Oregon (Howard 1964), Florida (Ligon 1965), Utah (Emslie and Heaton 1987), Alaska (Brodkorb 1964), and Arctic Canada (McGhee 1979). Ploeger (1968) postulated that the Long-tailed Duck survived the last glaciation in a high Arctic refugium, but the earliest Holocene record of the extant species in the Arctic dates to around 6500 BP from the Canadian High Arctic (Stewart and Hourston-Wright 1990). T. H. Heaton (University of South Dakota, pers. comm.) reports a Long-tailed Duck fossil humerus, recovered from On Your Knees Cave on Prince of Wales Island in southeastern Alaska's Alexander Archipelago and dated at 39,500 ybp, prior to the LGM. Prince of Wales Island and surrounding outer islands of the Alexander Archipelago and the Haida Gwaii Archipelago are thought to have comprised a Pleistocene refugium for certain vertebrate species (Heaton et al. 1996), including birds (Barry and Tallmon 2010).

Livezey (1995) placed Long-tailed Ducks close to the goldeneye and Bufflehead group (*Bucephala* spp. and *Mergellus*) and merganser group (*Mergus* spp. and *Lophodytes*), based on 137 morphological and plumage characters. No geographic variation has been reported, and no subspecies are recognized (Ploeger 1968, Palmer 1976). No wild hybrids have been reported.

Long-tailed Ducks have a circumpolar distribution, breeding throughout northern Canada, Alaska, Siberia, the Kola Peninsula, Svalbard, Greenland, Iceland, and Fennoscandia and wintering primarily along the subarctic and temperate coastlines, although in North America they also winter in the Great Lakes region (Robertson and Savard 2002). The world population size has been estimated at ~8 million birds (Rose and Scott 1997), and thus Long-tailed Ducks are among the most abundant of sea ducks. However, Long-tailed Ducks, like other sea ducks, have experienced population declines due to unknown factors, subsequent to the 1970s, and in some areas up to 50% (Hodges et al. 1996, Dickson and Gilchrist 2002, Bower 2009). In North America, the population has apparently remained stable during the past decade (Platte and Stehn 2011, Larned et al. 2012). Moreover, based on 12 microsatellite loci, R. E. Wilson et al., U.S. Geological Survey (unpubl. data), were unable to detect a genetic signature of a demographic bottleneck in most western North American Arctic populations, although they did detect a signature of recent (within the past hundred generations) population growth among Long-tailed Ducks occupying the YK Delta of Alaska.

Satellite telemetry data and isotope analyses of North American Arctic populations demonstrate that female Long-tailed Ducks from the same breeding population winter in multiple areas (Petersen et al. 2003, Lawson 2006), where they may intermix and form pair bonds with males from other breeding populations. As a result, although females show some degree of philopatry (Schamber et al. 2009), as well as breeding and wintering site fidelity (Alison 1975, Sea Duck Joint Venture 2012), substantial gene flow is likely to be mediated by males. This conjecture is supported by the absence of population structure observed across nesting populations along the northern Arctic coast of Alaska and central Canada, based on data from the nuclear genome (12 microsatellite loci), coupled with significant population structuring across the same region, and based on the maternally inherited mitochondrial genome (R. E. Wilson et al., U.S. Geological Survey, unpubl. data). Humphries and Winker (2011) also failed to uncover significant differentiation between Alaskan and Russian Long-tailed Ducks, based on nuclear markers (amplified fragment length polymorphisms or AFLPs), although greater structuring was observed at mtDNA. Lower or no population differentiation in nuclear markers coupled with significant structure in mtDNA is consistent with the pattern found in other sea ducks (Figure 2.3), such as Spectacled and Common Eiders (Scribner et al. 2001, Sonsthagen et al. 2011), as well as some other waterfowl species (Lavretsky et al. 2014). However, the significant differentiation at mtDNA was also based, at least in part, on the presence of two divergent clades, one showing signs of admixture among all breeding locales and one primarily composed of Long-tailed Ducks sampled from the YK Delta. This finding may reflect the isolation of two refugial populations, with subsequent expansion and admixture in the widespread clade. Data from eastern populations of Long-tailed Ducks are needed to determine whether coastal Arctic breeding populations form a panmictic population, as exhibited in the King Eider (Pearce et al. 2004), or whether there is a central Arctic contact zone between western and eastern populations (Common Eider; Sonsthagen et al. 2011; Harlequin Duck, Scribner and Talbot, unpubl. data). As well, data are lacking from European populations, precluding a broad-scale phylogeographic assessment of the species.

CONCLUSIONS AND PERSPECTIVES

Maintenance of genetic variation in wild sea duck populations is fundamental to the group's long-term persistence, and a first step in maintaining genetic variation is to understand the levels of genetic variability as well as how that variation is partitioned among species and populations (Figure 2.6). Genetic approaches will allow for concomitant tracking of population trends and genetic diversity, population delineation for effective monitoring, determination of gene flow to define metapopulations, and migratory connectivity between breeding and wintering grounds. Pearce et al. (2014) found that genetic and band-recovery information yielded remarkably similar results in regard to population delineation and distribution during winter. A combined approach may be useful for other less characterized sea duck populations, such as those on the eastern coast of North America, Greenland, Iceland, and throughout Eurasia.

Clearly, there is still much work to do in regard to an overall sea duck phylogeny as evolutionary relationships among species remain problematic (Box 2.4), or unresolved. Effective conservation and management of contemporary population characteristics and future potential for biodiversity requires this understanding of the major and minor divisions within and among species.

BOX 2.4

Labrador Duck. Advances in molecular methods that allow researchers to obtain DNA from small quantities of biological materials facilitate the assay of museum-preserved eggs and other specimens. Thus, Chilton and Sorenson (2007) amplified mtDNA (12S, ND2, and control region genes) from nine putative Labrador Duck eggs in an attempt to clarify the breeding distribution of the species. Sequences from six of the eggs were consistent matches with sequences of the Red-breasted Merganser, one with a Common Eider, and two with Mallard.

Climate warming is substantially changing the distribution and population dynamics of terrestrial organisms, particularly at high latitudes. Population responses include adapting to new conditions, tracking climate shifts into new range that may lead to new zones of contact between species, or even the possibility of local extirpation or extinction. To forecast the impact of climate-induced perturbations, an essential first step is to develop an understanding of how high-latitude species such as sea ducks have been structured by past episodes of dynamic environmental change. Because high-latitude environments used by sea ducks are remote and difficult to access, there is limited information available about many of these essential factors for the majority of species, but progress has been made.

The recognition of biodiversity has typically been in the form of subspecies, a taxonomic designation not typically ascribed to sea ducks, but since the 1980s, other descriptors such as *evolutionary significant units* (ESUs) (Ryder 1986) or *designatable units* (DUs; Green 2005, Committee on the Status of Endangered Wildlife in Canada 2010) have been emphasized. In some cases, such units have been uncovered or corroborated as part of phylogeographic or population genetics analyses (Sonsthagen et al. 2011). However, for high-latitude sea ducks, population-level data have contributed more commonly to the resolution of shallower population structure and *management units* (MUs; Moritz 1994). Landscape genetic techniques, which can resolve population structuring across different geographic scales at finer scales than phylogenetics (Manel et al. 2003), may provide valuable approaches for researchers seeking to better understand how microevolutionary processes generate structure within sea ducks.

Another role of conservation genetics, and one that has not yet been applied to sea ducks, is to understand the adaptive potential of species exposed to environmental challenges such as climate change. New genomics approaches are now allowing researchers to map associations between adaptive genome regions and environmental gradients in space and time (Hemmer-Hansen et al. 2014). Recent advances in functional genomics are revolutionizing genetic analyses of natural populations but have yet to be applied to sea ducks.

ACKNOWLEDGMENTS

M. Fowler and J. Fristensky assisted with graphics, and V. Skean assisted with editing. Three anonymous reviewers provided valuable comments to an earlier draft of this chapter. Genbank Accession numbers associated with sequences listed in Figure 2.2 include: Common Eider (AF515264), King Eider (AY124077), Spectacled Eider (AY244332), Steller's Eider (AY244333), Harlequin Duck (KM501447), Long-tailed Duck (KM501453), Hooded Merganser (EU585650), Red-breasted Merganser (KM501448), Common Merganser (KM501449), Black Scoter (KM501452), White-winged Scoter (KM501450), Surf Scoter (KM501451), Bufflehead (EU585633), Common Goldeneye (EU585634), Barrow's Goldeneye (pending), and Muskovy Duck (L08385).

LITERATURE CITED

Abbott, R. J., and C. Brochmann. 2003. History and evolution of the arctic flora: in the footsteps of Eric Hultén. Molecular Ecology 12:299–313.

Abbott, R. J., L. C. Smith, R. I. Milne, R. M. M. Crawford, K. Wolff, and J. Balfour. 2000. Molecular analysis of plant migration and refugia in the Arctic. Science 289:1343–1346.

Alison, R. M. 1975. Breeding biology and behavior of the Oldsquaw (*Clangula hyemalis* L.). Ornithological Monographs 18:1–52.

Alvarez, R., and S. L. Olson. 1978. A new merganser from the Miocene of Virginia (Aves: Anatidae). Proceedings of the Biological Society of Washington 91:522–532.

Anderson, M. G., J. M. Rhymer, and F. C. Rohwer. 1992. Philopatry, dispersal, and the genetic structure of waterfowl populations. Pp. 365–395 in B. D. J. Batt, A. D. Afton, M. G. Anderson, C. D. Ankney, D. H. Johnson, J. A. Kadlec, and G. L. Krapu (editors), Ecology and management of breeding waterfowl. University of Minnesota Press, Minneapolis, MN.

Avise, J. C. 2000. Phylogeography: the history and formation of species. Harvard University Press, Cambridge, MA.

Baird, A. F., T. M. Brewer, and R. Ridgway. 1884. The water birds of North America. Memoirs of the Museum of Comparative Zoology at Harvard College (vols. 12 and 13). Little, Brown, and Company, Boston, MA.

Ball, S. C. 1934. Hybrid ducks, including descriptions of two crosses of Bucephala and Lophodytes. Peabody Museum of Natural History Bulletin 3:1–26.

Barry, P., and D. A. Tallmon. 2010. Genetic differentiation of a subspecies of Spruce Grouse (Falcipennis canadensis isleibi) in an endemism hotspot. Auk 127:617–625.

Bayard de Volo, S., R. T. Reynolds, S. A. Sonsthagen, S. L. Talbot, and M. F. Antolin. 2013. Phylogeography, gene flow, and population history of Northern Goshawks in North America. Auk 130:342–354.

Bennike, O. 1997. Quaternary vertebrates from Greenland: a review. Quaternary Science Reviews 16:899–909.

Bennike, O., S. Björck, J. Böcker, L. Hansen, J. Heinemeier, and B. Wohifarth. 1999. Early Holocene plant and animal remains from north-east Greenland. Journal of Biogeography 26:667–677.

Bordage, D., and J.-P. L. Savard. 2011. Black Scoter (Melanitta americana). in A. Poole and F. Gill (editors), The Birds of North America, No. 177. The Academy of Natural Sciences, Philadelphia, PA.

Bower, J. L. 2009. Changes in marine bird abundance in the Salish Sea: 1975 to 2007. Marine Ornithology 37:9–17.

Boyd, W. S., B. D. Smith, S. A. Iverson, M. R. Evans, J. E. Thompson, and S. Schneider. 2009. Apparent survival, natal philopatry, and recruitment of Barrow's Goldeneyes (Bucephala islandica) in the Cariboo-Chilcotin region of British Columbia, Canada. Canadian Journal of Zoology 87:337–345.

Brodeur, S., J.-P. L. Savard, M. Robert, P. Laporte, P. Lamothe, R. D. Titman, S. Marchand, S. Gilliland, and G. Fitzgerald. 2002. Harlequin Duck Histrionicus histrionicus population structure in eastern Nearctic. Journal of Avian Biology 33:127–137.

Brodkorb, P. 1961. Birds from the Pliocene of Juntura, Oregon. Journal of the Florida Academy of Sciences 24:169–184.

Brodkorb, P. 1964. Catalogue of fossil birds: part 2 (Anseriformes through Galliformes). Bulletin of the Florida State Museum, Biological Sciences Series 8:195–335.

Brooks, D. R., and E. P. Hoberg. 2006. Systematics and emerging infectious diseases: from management to solution. Journal of Parasitology 92:426–429.

Brooks, W. S. 1915. Notes on birds from east Siberia and arctic Alaska. Bulletin of the Museum of Comparative Zoology 59:361–413.

Brown, M. E. 1998. Population genetic structure, philopatry and social structure in three Harlequin Duck (Histrionicus histrionicus) breeding subpopulations. Thesis, University of California, Davis, Davis, CA.

Brown, P. W., and L. H. Fredrickson. 1997. White-winged Scoter (Melanitta fusca). in A. Poole and F. Gill (editors), The Birds of North America, No. 274. The Academy of Natural Sciences, Philadelphia, PA.

Brown Gladden, J. G., M. M. Ferguson, M. K. Friesen, and J. W. Clayton. 1999. Population structure of North American beluga whales (Delphinapterus leucas) based on nuclear DNA microsatellite variation and contrasted with the population structure revealed by mitochondrial DNA variation. Molecular Ecology 8:347–363.

Bruner, H. J. 1997. Habitat use and productivity in the central Cascade Range of Oregon. Thesis, Oregon State University, Corvallis, OR.

Brush, A. H. 1976. Waterfowl feather proteins: analysis of use in taxonomic studies. Journal of Zoology 179:467–498.

Bulgarella, M., M. D. Sorenson, J. L. Peters, R. E. Wilson, and K. G. McCracken. 2010. Phylogenetic relationships of Amazonetta, Speculanas, Lophonetta, and Tachyeres: four morphologically divergent duck genera endemic to South America. Journal of Avian Biology 41:186–199.

Cassirer, E. F., and C. R. Groves. 1994. Ecology of Harlequin Ducks in northern Idaho. Idaho Department of Fish and Game, Boise, ID.

Chandler, R. M. 1990. Fossil birds of the San Diego formation, late Pliocene, Blancan, San Diego County, CA. Ornithological Monographs 44:73–161.

Chaulk, K. G., G. J. Robertson, B. T. Collins, W. A. Montevecchi, and B. C. Turner. 2005. Evidence of recent population increases in Common Eiders breeding in Labrador. Journal of Wildlife Management 69:805–809.

Chesser, R. T., R. C. Banks, F. K. Barker, C. Cicero, J. L. Dunn, A. W. Kratter, I. J. Lovette et al. 2010. Fifty-first supplement to the American Ornithologists' Union check-list of North American birds. Auk 127:726–744.

Chesser, R. T., R. C. Banks, F. K. Barker, C. Cicero, J. L. Dunn, A. W. Kratter, I. J. Lovette et al. 2013. Fifty-fourth supplement to the American Ornithologists' Union check-list of North American birds. Auk 130:558–571.

Chilton, G., and M. D. Sorenson. 2007. Genetic identification of eggs purportedly from the extinct Labrador Duck. Auk 124:262–268.

Collinson, M., D. T. Parkin, A. G. Knox, G. Sangster, and A. J. Helbig. 2006. Species limits within the genus *Melanitta*, the scoters. British Birds 99:183–201.

Committee on the Status of Endangered Wildlife in Canada. [online]. 2010. Committee on the Status of Endangered Wildlife in Canada. <www.cosewic.gc.ca> (27 March 2014).

Cooch, F. G. 1965. Breeding biology and management of the Northern Eider (*Somateria mollissima*) in the Cape Dorset area, Northwest Territories. Wildlife Management Bulletin Series 2. Canadian Wildlife Service, Edmonton, AB.

Cooke, F., G. J. Robertson, C. M. Smith, R. I. Goudie, and W. S. Boyd. 2000. Survival, emigration, and winter population structure of Harlequin Ducks. Condor 102:137–144.

Cornuet, J. M., and G. Luikart. 1996. Description and power analysis of two tests for detecting recent population bottlenecks from allele frequency data. Genetics 144:2001–2014.

Cramp, S., and K. L. Simmons. 1977. Handbook of the birds of Europe, the Middle East and North Africa: the birds of the Western Palearctic (vol. 1). Oxford University Press, Oxford, U.K.

Cronin, M. A., J. B. Grand, D. Esler, D. V. Derksen, and K. T. Scribner. 1996. Breeding populations of Northern Pintails have similar mitochondrial DNA. Canadian Journal of Zoology 74:992–999.

Crowley, D. W. 1994. Breeding habitat of Harlequin Ducks in Prince William Sound, Alaska. Thesis, Oregon State University, Corvallis, OR.

Cuvier, G. 1817. Le règne animal distribué d'après son organisation, pour servir de base à l'histoire naturelle des animaux et d'introduction à l'anatomie comparée. Tome I. Les mammifères et les oiseaux. [The animal kingdom arranged according to its organization as a basis in natural history of animals and introduction to comparative anatomy. Volume 1. Mammals and birds]. Deterville: Paris (in French).

Dalén, L., E. Fuglei, P. Hersteinsson, C. M. O. Kapel, J. D. Roth, G. Samelius, M. Tannerfeldt, and A. Angerbjörn. 2005. Population history and genetic structure of a circumpolar species: the arctic fox. Biological Journal of the Linnean Society 84:79–89.

Dau, C. P., P. L. Flint, and M. R. Petersen. 2000. Distribution of recoveries of Steller's Eiders banded on the lower Alaska Peninsula, Alaska. Journal of Field Ornithology 71:541–548.

Delacour, J. 1969. The waterfowl of the world (vol. 3). Country Life, London, U.K.

Delacour, J., and E. Mayr. 1945. The family Anatidae. Wilson Bulletin 57:3–55.

Denton, J. C., C. L. Roy, G. J. Soulliere, and B. A. Potter. 2012. Current and projected abundance of potential nest sites for cavity-nesting ducks in the hardwoods of the North Central United States. Journal of Wildlife Management 76:422–432.

Dickson, D. L. 2012. Seasonal movement of King Eiders breeding in western Arctic Canada and northern Alaska. Canadian Wildlife Service Technical Report Series, No. 520. Canadian Wildlife Service, Edmonton, AB.

Dickson, D. L., and H. G. Gilchrist. 2002. Status of marine birds on the Southern Beaufort Sea. Arctic 55:46–58.

Donne-Goussé, C., V. Laudet, and C. Hännia. 2002. A molecular phylogeny of Anseriformes based on mitochondrial DNA analysis. Molecular Phylogenetics and Evolution 23:339–356.

Drovetski, S. V., R. M. Zink, S. Rohwer, I. V. Fadeev, E. V. Nesterov, I. Karogodin, E. A. Koblik, and Y. A. Red'kin. 2004. Complex biogeographic history of a Holarctic passerine. Proceedings of the Royal Society B 271:545–551.

Dugger, B. D., K. M. Dugger, and L. H. Fredrickson. 2009. Hooded Merganser (*Lophodytes cucullatus*). in A. Poole (editor), The Birds of North America, No. 98. The Academy of Natural Sciences, Philadelphia, PA.

Dzinbal, K. A. 1982. Ecology of Harlequin Ducks in Prince William Sound, Alaska, during summer. Thesis, Oregon State University, Corvallis, OR.

Eadie, J. M., M. L. Mallory, and H. G. Lumsden. 1995. Common Goldeneye (*Bucephala clangula*). in A. Poole and F. Gill (editors), The Birds of North America, No. 170. The Academy of Natural Sciences, Philadelphia, PA.

Eadie, J. M., J.-P. L. Savard, and M. L. Mallory. 2000. Barrow's Goldeneye (*Bucephala islandica*). in A. Poole and F. Gill (editors), The Birds of North America, No. 548. The Academy of Natural Sciences, Philadelphia, PA.

Elorza, M. 2005. First palearctic fossil record of *Polysticta stelleri* (Pallas) 1769. Munibe Antropologia-Arkeologia 57:297–301.

Emslie, S. D., and T. H. Heaton. 1987. The late Pleistocene avifauna of Crystal Ball Cave, Utah. Journal of the Arizona-Nevada Academy of Science 21:53–60.

England, P. R., G. H. R. Osler, L. M. Woodworth, M. E. Montgomery, D. A. Briscoe, and R. Frankham. 2003. Effects of intense versus diffuse population bottlenecks on microsatellite genetic diversity and evolutionary potential. Conservation Genetics 4:595–604.

Erskine, A. J. 1992. Atlas of breeding birds of the Maritime Provinces. Nimbus Publications and the Nova Scotia Museum, Halifax, NS.

Fedorov, V. B., M. Jaarola, A. V. Goropashnaya, and J. A. Cook. 2003. Phylogeography of lemmings (*Lemmus*): no evidence for postglacial colonization of Arctic from the Beringian refugium. Molecular Ecology 12:725–731.

Fedorov, V. B., and N. C. Stenseth. 2002. Glacial survival of the Norwegian lemming (*Lemmus lemmus*) in Scandinavia: inference from mitochondrial DNA variation. Proceedings of the Royal Society B 268:809–814.

Fishman, D. J. 2010. Philopatry and the spatial structuring of kin in the Red-breasted Merganser (*Mergus serrator*). Thesis, McGill University, Montreal, QC.

Fitzgerald, G. R. 1991. Pleistocene ducks of the Old Crow basin, Yukon Territory, Canada. Canadian Journal of Earth Science 28:1561–1571.

Fitzhugh, W. W. 1978. Maritime archaic cultures of the central and northern Labrador Coast. Arctic Anthropology 15:61–95.

Flint, P. L. 2013. Changes in the size and trends of North American sea duck populations associated with North Pacific oceanic regime shifts. Marine Biology 160:59–65.

Forsman, D. 1995. A presumed hybrid of Steller's Eider × Common Eider in Norway. Birding World 8:138.

Frankham, R., J. D. Ballou, and D. A. Briscoe. 2002. Introduction to conservation genetics. Cambridge University Press, Cambridge, U.K.

Fredrickson, L. H. 2001. Steller's Eider (*Polystica stelleri*). Pp. 1–24 in A. Poole and F. Gill (editors), The Birds of North America, No. 571. The Academy of Natural Sciences, Philadelphia, PA.

Friedmann, H. 1934. Bird bones from old Eskimo ruins in Alaska. Journal of the Washington Academy of Sciences 24:230–237.

Friedmann, H. 1935. Avian bones from prehistoric ruins on Kodiak Island, Alaska. Journal of the Washington Academy of Sciences 25:44–51.

Friesen, V. L., T. M. Burg, and K. D. McCoy. 2007. Mechanisms of population differentiation in seabirds. Molecular Ecology 16:1765–1785.

Gabrielson, I. N. and F. C. Lincoln. 1959. Birds of Alaska. Stackpole Books, Harrisburg, PA.

Gál, E., J. Hír, E. Kessler, and J. Kókay. 1999. Középsõmiocén õsmaradványok, a Mátraszõlõs, Rákóczikápolna alatti útbevágásból. I. A Mátraszõlõs 1. Lelõhely. [Middle Miocene fossils from the sections at the Rákóczi chapel at Mátraszõlõs. Locality Mátraszõlõs I.] Folia Historico Naturalia Musei Matraensis 23:33–78 (in Hungarian).

Galbreath, K. E., and J. A. Cook. 2004. Genetic consequences of Pleistocene glaciations for the tundra vole (*Microtus oeconomus*) in Beringia. Molecular Ecology 13:135–148.

Gauthier, G. 1990. Philopatry, nest-site fidelity, and reproductive performance in Buffleheads. Auk 107:126–132.

Gauthier, G. 1993. Bufflehead (*Bucephala albeola*). in A. Poole and F. Gill (editors), The Birds of North America, No. 67. The Academy of Natural Sciences, Philadelphia, PA.

Gibson, S. Y., R. C. Van der Marel, and B. M. Starzomski. 2009. Climate change and conservation of leading-edge peripheral populations. Conservation Biology 23:1369–1373.

Godfrey, W. E. 1986. Birds of Canada, revised ed. Natural Sciences Museum of Canada, Ottawa, ON.

Gómez-Diaz, E., J. González-Solis, M. A. Peinado, and R. D. M. Page. 2006. Phylogeography of the *Calonectris* shearwaters using molecular and morphometric data. Molecular Phylogenetics and Evolution 41:322–332.

Gonzalez, J., H. Düttmann, and M. Wink. 2009. Phylogenetic relationships based on two mitochondrial genes and hybridization patterns in Anatidae. Journal of Zoology 279:310–318.

Goudie, R. I., G. J. Robertson, and A. Reed. 2000. Common Eider (*Somateria mollissima*). in A. Poole and F. Gill (editors), The Birds of North America, No. 546. The Academy of Natural Sciences, Philadelphia, PA.

Gray, A. P. 1958. Bird hybrids: a checklist with bibliography. Commonwealth Agricultural Bureaux, Farnham Royal, U.K.

Green, D. M. 2005. Designatable units for status assessment of endangered species. Conservation Biology 19:1813–1820.

Greenwood, P. J. 1980. Mating systems, philopatry and dispersal in birds and mammals. Animal Behaviour 28:1140–1162.

Gundmundsson, F. 1932. Beobachtungen an isländichen Eiderenten (*Somateria m. mollissima*). Sonderdruck aus Beiträge zur Fortpflanzungsbiologie der Vögel 8:85–97 (in German).

Guthrie, R. D. 1992. A late Pleistocene avifauna from San Miguel Island, California. Natural History Museum of Los Angeles County Science Series 36:319–327.

Hahn, P. 1963. Where is that vanished bird? An index to the known specimens of the extinct and near extinct North American species. Royal Ontario Museum and University of Toronto Press, Toronto, ON.

Hampe, A., and R. J. Petit. 2005. Conserving biodiversity under climate change: the rear edge matters. Ecology Letters 8:461–467.

Hansson, B., D. Hasselquist, M. Tarka, P. Zehtindjiev, and S. Bensch. 2008. Postglacial colonization patterns and the role of isolation and expansion in driving diversification in a passerine bird. PLoS One 3:e2794.

Harington, C. R., and S. Occhietti. 1980. Pleistocene eider duck (*Somateria* cf. *mollissima*) from Champlain Sea deposits near Shawinigan, Québec. Géographie Physique et Quaternaire 34:239–245.

Hatton, P. L., and M. Marquiss. 2004. The origins of molting Goosanders on the Eden Estuary. Ringing and Migration 22:70–74.

Heaton, T. H., S. L. Talbot, and G. F. Shields. 1996. An ice age refugium for large mammals in the Alexander Archipelago, southeastern Alaska. Quaternary Research 46:186–192.

Hedrick, P. W., and S. T. Kalinowski. 2000. Inbreeding depression in conservation biology. Annual Review of Ecology and Systematics 31:139–162.

Hefti-Gautschi, Z. B., M. Pfunder, L. Jenni, V. Keller, and H. Ellegren. 2009. Identification of conservation units in the European *Mergus merganser* based on nuclear and mitochondrial DNA markers. Conservation Genetics 10:87–99.

Hemmer-Hansen, J., N. O. Therkildsen, D. Meldrup, and E. E. Nielsen. 2014. Conserving marine biodiversity: insights from life-history trait genes in Atlantic cod (*Gadus morhua*). Conservation Genetics 15:213–228.

Hewitt, G. M. 1996. Some genetic consequences of ice ages, and their role in divergence and speciation. Biological Journal of the Linnean Society 58:247–276.

Hewitt, G. M. 2000. The genetic legacy of the Quaternary ice ages. Nature 405:907–913.

Hewitt, G. M. 2001. Speciation, hybrid zones and phylogeography—or seeing genes in space and time. Molecular Ecology 10:537–549.

Hewitt, G. M. 2004. Genetic consequences of climatic oscillations in the Quaternary. Philosophical Transactions of the Royal Society of London B 359:183–195.

Hickerson, M. J., B. C. Carstens, J. Cavender-Bares, K. A. Crandall, C. H. Graham, J. B. Johnson, L. Rissler, P. F. Victoriano, and A. D. Yoder. 2010. Phylogeography's past, present, and future: 10 years after Avise 2000. Molecular Phylogenetics and Evolution 54:291–301.

Hipfner, J. M., H. G. Gilchrist, A. J. Gaston, and D. K. Cairns. 2002. Status of Common Eiders, *Somateria mollissima*, nesting in the Digges Sound region, Nunavut. Canadian Field-Naturalist 116:22–25.

Hobson, K. A., and J. C. Driver. 1989. Archaeological evidence for the use of the Strait of Georgia by marine birds. Pp. 168–173 in K. Vermeer and R. W. Butler (editors), The ecology and status of marine and shoreline birds in the Strait of Georgia, British Columbia. Canadian Wildlife Service Special Publication, Ottawa, ON.

Hodges, J. I. 1991. Alaska water fowl production surveys. U.S. Fish Wildlife Service, Juneau, AK.

Hodges, J. I., J. G. King, B. Conant, and H. A. Hanson. 1996. Aerial surveys of waterbirds in Alaska 1957–94: population trends and observer variability. Informal Technical Report 4. National Biological Service Juneau, AK.

Holder, K., R. Montgomerie, and V. L. Friesen. 1999. A test of the glacial refugium hypothesis using patterns of mitochondrial and nuclear DNA sequence variation in Rock Ptarmigan (*Lagopus mutus*). Evolution 53:1936–1950.

Holder, K., R. Montgomerie, and V. L. Friesen. 2000. Glacial vicariance and historical biogeography of Rock Ptarmigan (*Lagopus mutus*) in the Bering Region. Molecular Ecology 9:1265–1278.

Hope, A. G., E. Waltari, V. B. Fedorov, A. V. Goropashnaya, S. L. Talbot, and J. A. Cook. 2011. Persistence and diversification of the Holarctic shrew, *Sorex tundrensis* (Family Soricidae), in response to climate change. Molecular Ecology 20:4346–4370.

Howard, H. 1933. Bird remains from an Indian shellmound near Point Mugu, California. Condor 35:235.

Howard, H. 1964. Fossil Anseriformes. Pp. 233–326 in J. Delacour (editor), Waterfowl of the world (vol. 4). County Life, London, U.K.

Howell, S. N. G., and S. Webb. 1995. A guide to the birds of Mexico and Northern Central America. Oxford University Press, Oxford, U.K.

Hull, J. M., and D. J. Girman. 2005. Effects of Holocene climate change on the historical demography of migrating Sharp-shinned Hawks (*Accipiter striatus velox*) in North America. Molecular Ecology 14:159–170.

Hultén, E. 1937. Outline of the history of Arctic and boreal biota during the Quaternary period. Aktiebolaget Thule, Stockholm, Sweden.

Humphrey, P. S. 1955. The relationships of the seaducks (Tribe Mergini). Dissertation, University of Michigan, Ann Arbor, MI.

Humphrey, P. S. 1958. Classification and systematic position of the eiders. Condor 60:129–135.

Humphries, E. M., and K. Winker. 2011. Discord reigns among nuclear, mitochondrial, and phenotypic estimates of divergence in nine lineages of trans-Beringian birds. Molecular Ecology 20:573–583.

Illiger, J. 1811. Prodromus systematis mammalium et avium. Sumptibus C. Salfield, Berlin, Germany (in German).

Iverson, S. A., D. Esler, and D. J. Rizzolo. 2004. Winter philopatry of Harlequin Ducks in Prince William Sound, Alaska. Condor 106:711–715.

Jackson, H., B. J. T. Morgan, and J. J. Groombridge. 2013. How closely do measures of mitochondrial DNA control region diversity reflect recent trajectories of population decline in birds? Conservation Genetics 14:1291–1296.

Jacob, J. 1982. Integument lipide-ihre cheische Struktur und ihre Bedeutung als systematisches Merkmal in der Zoologie. Funckt. Biol. Med. 1:83–90 (in German).

Jacob, J., and A. Glaser. 1975. Chemotaxonomy of Anseriformes. Biochemical Systematics and Ecology 2:215–220.

Jehl Jr., J. R. 1967. Pleistocene birds from Fossil Lake, Oregon. Condor 69:24–27.

Johnsgard, P. A. 1960a. Classification and evolutionary relationships of the sea ducks. Condor 62:426–433.

Johnsgard, P. A. 1960b. Comparative behavior of the Anatidae and its evolutionary implications. Wildfowl 11:31–45.

Johnsgard, P. A. 1961. Tracheal anatomy of the Anatidae and its taxonomic significance. Wildfowl Trust Annual Report 12:58–69.

Johnsgard, P. A. 1975. Waterfowl of North America. Indiana University Press, Bloomington, IN.

Johnsgard, P. A. [online]. 2010. Waterfowl of North America: perching ducks Tribe Cairinini. Pp. 160–180 in Waterfowl of North America, revised ed. <http://digitalcommons.unl.edu/biosciwaterfowlna/10> (27 March 2014).

Johnson, J. A., S. L. Talbot, G. K. Sage, K. K. Burnham, J. W. Brown, T. L. Maechtle, W. S. Seegar, M. A. Yates, B. Anderson, and D. P. Mindell. 2010. The use of genetics for the management of a recovering population: temporal assessment of migratory Peregrine Falcons in North America. PLoS One 5:e14042.

Jones, T. L., J. F. Porcasi, J. M. Erlandson, H. Dallas Jr., T. A. Wake, and R. Schwarderer. 2008. The protracted Holocene extinction of California's flightless sea duck (Chendytes lawi) and its implications for the Pleistocene overkill hypothesis. Proceedings of the National Academy of Sciences USA 105:4105–4108.

Kellett, D. 1999. Causes and consequences of variation in nest success of King Eiders (Somateria spectabilis) at Karrak Lake, Northwest Territories. Thesis, University of Saskatchewan, Saskatoon, SK.

Klicka, J., G. M. Spellman, K. Winker, V. Chua, and B. T. Smith. 2011. A phylogeographic and population genetic analysis of a widespread, sedentary North American bird: the Hairy Woodpecker (Picoides villosus). Auk 128:346–362.

Knowles, L. L. 2009. Statistical phylogeography. Annual Review of Ecology and Systematics 40:593–612.

Kurochkin, E. N. 1976. A survey of the paleogene birds of Asia. Smithsonian Contributions to Paleobiology 27:75–86.

Kutz, S. J., E. J. Jenkins, A. M. Veitch, J. Ducrocq, L. Polley, B. Elkin, and S. Lair. 2009. The arctic as a model for anticipating, preventing, and mitigating climate change impacts on host-parasite interactions. Veterinarian Parasitology 163:217–228.

Kvist, L., M. Ruokonen, J. Lumme, and M. Orell. 1999. The colonization history and present-day population structure of the European Great Tit (Parus major major). Heredity 82:495–502.

Lanctot, R., B. Goatcher, K. Scribner, S. Talbot, B. Pierson, D. Esler, and D. Zwiefelhofer. 1999. Harlequin Duck recovery from the Exxon Valdez Oil Spill: a population genetics perspective. Auk 116:781–791.

Larned, W., P. Stehn, and R. Platte. 2012. Waterfowl breeding population survey Arctic coastal plain, Alaska 2011. U.S. Fish and Wildlife Service, Anchorage, AK.

Lauritzen, S. E., H. Nese, R. W. Lie, A. Lauritsen, and R. Løvlie. 1996. Interstadial/interglacial fauna from Norcemgrotta, Kjøepsvik, northern Norway. Pp. 89–92 in S. E. Lauritzen (editor), Climate change: the karst record. Karst Waters Institute Special Publication 2. Karst Waters Institute, Charles Town West, VA.

Lavretsky, P., K. G. McCracken, and J. L. Peters. 2014. Phylogenetics of a recent radiation in the Mallards and allies (Aves: Anas): inferences from a genomic transect and multispecies coalescent. Molecular Phylogenetics and Evolution 70:402–411.

Lawson, S. L. 2006. Comparative reproductive strategies between Long-tailed Ducks and King Eiders at Karrak Lake, Nunavut: use of energy resources during the nesting season. Thesis, University of Saskatchewan, Saskatoon, SK.

Lessa, E. P., J. A. Cook, and J. L. Patton. 2003. Genetic footprints of demographic expansion in North America, but not Amazonia, following the Late Pleistocene. Proceedings of the National Academy of Sciences 100:10331–10334.

Lessica, P., and F. W. Allendorf. 1995. When are peripheral populations valuable for conservation? Conservation Biology 9:753–760.

Ligon, J. D. 1965. A Pleistocene avifauna from Haile, Florida. Bulletin of the Florida State Museum, Biological Sciences Series 10:127–158.

Liljegren, R. 1977. Subfossila fågelfynd. Anser 16:17–25 (in Swedish).

Lindqvist, C., S. C. Schuster, Y. Sun, S. L. Talbot, J. Qi, A. Ratan, L. P. Tomsho et al. 2010. Complete mitochondrial genome of a Pleistocene jawbone unveils the origin of polar bear. Proceedings of the National Academy of Sciences USA 107:5053–5057.

Linnaeus, C. 1758. Systema naturae per regna tria naturae, 10th ed. reformata (vol. 1). Stocholm, Sweden. [Systema naturae. A photographic facsimile of the first volume of the tenth edition (1758)] British Museum of Natural History, London, U.K., 1939 (in Swedish).

Little, B., and R. W. Furness. 1985. Long distance molt migration by British Goosanders, *Mergus merganser*. Ringing and Migration 6:77–82.

Liu, G., L. Zhou, and C. Gu. 2012. Complete sequence and gene organization of the mitochondrial genome of the Scaly-sided Merganser (*Mergus squamatus*) and phylogeny of some Anatidae species. Molecular Biology Reports 39:2139–2145.

Livezey, B. C. 1995. Phylogeny and evolutionary ecology of modern seaducks (Anatidae: Mergini). Condor 97:233–255.

Louchart, A., C. Mourer-Chauviré, E. Guleç, F. C. Howell, and T. D. White. 1998. L'avifaune de Dursulu, Turquie, Pléistocène inférieur: climat, environement et biogéographie. [The bird life of Kursunlu, Turkey, Pleistocene climate, environment and biography] Comptes Rendus de l'Académie des Sciences Series IIA 327:341–346 (in French).

Lyngs, P. 2003. Migration and winter ranges of birds in Greenland: an analysis of ringing recoveries. Dansk Ornitologisk Forenings Tidsskrift 97:1–168.

MacCallum, B. 2001. Status of the Harlequin Duck (*Histrionicus histrionicus*) in Alberta. Wildlife Status Report, No. 36. Alberta Sustainable Resource Development, Fisheries and Wildlife Management Division, and Alberta Conservation Association, Edmonton, AB.

Mace, G. 2004. The role of taxonomy in species conservation. Philosophical Transactions of the Royal Society of London B 359:711–719.

MacPherson, A. H. 1965. The origin of diversity in mammals of the Canadian arctic tundra. Systematic Zoology 14:153–173.

Madsen, C. S., K. P. McHugh, and S. R. De Kloet. 1988. A partial classification of waterfowl (Anatidae) based on single-copy DNA. Auk 105:452–459.

Mallory, M., and K. Metz. 1999. Common Merganser (*Mergus merganser*). in A. Poole and F. Gill (editors), The Birds of North America, No. 442. The Academy of Natural Sciences, Philadelphia, PA.

Manel, S., M. K. Schwartz, G. Luikart, and P. Taberlet. 2003. Landscape genetics: combining landscape ecology and population genetics. Trends in Ecology and Evolution 18:189–197.

Marcisz, W. J. 1981. A presumed Bufflehead × Hooded Merganser hybrid in Illinois. American Birds 35:340–341.

Martin, P. R., and B. M. DiLabio. 1994. Natural hybrids between the Common Goldeneye, *Bucephala clangula*, and Barrow's Goldeneye, *B. islandica*. Canadian Field-Naturalist 108:195–198.

Mayr, E. 1969. Principles of systematic zoology. McGraw-Hill, New York.

McGhee, R. 1979. The palaeoeskimo occupations of Port Refuge, high arctic Canada. Archaeological Survey of Canada, National Museums of Canada, Ottawa, ON.

McKinnon, L., H. G. Gilchrist, and K. T. Scribner. 2006. Genetic evidence for kin-based female social structure in Common Eiders. Behavioral Ecology 17:614–621.

Merkel, F. R. 2004. Evidence of population decline in Common Eiders breeding in western Greenland. Arctic 57:27–36.

Merkel, F. R. 2010. Evidence of recent population recovery in Common Eiders breeding in western Greenland. Journal of Wildlife Management 78:1869–1874.

Miller, A. H., and R. I. Bowman. 1956. Fossil birds of the late Pliocene of Cita Canyon, Texas. Wilson Bulletin 68:38–46.

Miller, W., S. C. Schuster, A. J. Welch, A. Ratan, O. C. Bedoya-Reina, F. Zhao, H. L. Kim et al. 2012. Polar and brown bear genomes reveal ancient admixture and demographic footprints of past climate change. Proceedings of the National Academy of Sciences USA 109:E2382–E2390.

Mills, L. S., and P. E. Smouse. 1994. Demographic consequences of inbreeding in remnant populations. American Naturalist 144:412–431.

Mittelhauser, G. 1991. Harlequin Ducks at Acadia National Park and coastal Maine, 1988–91. Island Research Center, College of the Atlantic, Bar Harbor, ME.

Mittelhauser, G. 1993. Status of Harlequin Ducks, 1993. Report to the endangered and nongame wildlife grants program. Maine Department of Inland Fisheries and Wildlife, Bangor, ME.

Mlíkovský, J. 2002. Early Pleistocene birds of Stránská skála, Czech Republic: 2. Absolon's cave. Sylvia 38:19–28.

Morejohn, G. V. 1976. Evidence of the survival to recent times of the extinct flightless duck *Chendytes lawi* Miller. Smithsonian Contributions to Paleobiology 27:207–211.

Moritz, C. 1994. Defining 'evolutionarily significant units' for conservation. Trends in Ecology and Evolution 9:373–375.

Mosimann, J. E., and P. S. Martin. 1975. Simulating overkill by Paleoindians. American Scientist 63:303–313.

Myres, M. T. 1959. The behavior of the sea-ducks and its value in the systematics of the tribes Mergini and Somateriini, of the family Anatidae. Dissertation, University of British Columbia, Vancouver, BC.

Oates, D. W., and J. Principato. [online] 1994. Genetic variation and differentiation of North American waterfowl (Anatidae). Nebraska Game and Parks Commission Staff Research Publications 15:1–166. <http://digitalcommons.unl.edu/nebgamestaff/15> (27 March 2014).

Olson, S. L., and A. Feduccia. 1980. *Presbyornis* and the origin of the Anseriformes (Aves: Charadriomorphae). Smithsonian Contributions to Zoology 323:1–24.

Olson, S. L., and P. C. Rasmussen. 2001. Miocene and Pliocene birds from the Lee Creek Mine, North Carolina. in C. E. Ray and D. J. Bohaska (editors), Geology and paleontology of the Lee Creek Mine, North Carolina, IL. Smithsonian Contributions to Paleobiology 90, Smithsonian Institute, Washington, DC.

Palmer, R. S. 1976. Handbook of North American birds (vol. 3). Waterfowl. Yale University Press, New Haven, CT.

Panov, E. N. 1989. Natural hybridization and ethological isolation in birds. Nauka Publishing House, Moscow, Russia.

Patton, J. C., and J. C. Avise. 1986. Evolutionary genetics of birds IV: rates of protein divergence in waterfowl (Anatidae). Genetica 68:129–143.

Pavia, M., and C. Bedetti. 2013. The presence of Harlequin Duck Histrionicus histrionicus (Linnaeus 1758) in the Middle Pleistocene of Italy. Journal of Ornithology 154:875–878.

Pearce, J. M. 2008. Site fidelity: definition, measurement and implications for population structure using mark-recapture, genetic, and comparative data in Hooded, Red-breasted, and Common Mergansers. Dissertation, University of Alaska, Fairbanks, AK.

Pearce, J. M., P. Blums, and M. Lindberg. 2008. Site fidelity is an inconsistent determinant in population structure in the Hooded Merganser (Lophodytes cucullatus): evidence from mark-recapture, genetic and comparative data. Auk 125:711–722.

Pearce, J. M., J. M. Eadie, J.-P. L. Savard, T. K. Christensen, J. Berdeen, E. Taylor, S. Boyd, and S. L. Talbot. 2014. Comparative population structure of cavity-nesting sea ducks. Auk 131:195–207.

Pearce, J. M., K. G. McCracken, T. K. Christensen, and Y. N. Zhuralev. 2009a. Migratory patterns and population structure among breeding and wintering Red-breasted Mergansers (Mergus serrator) and Common Mergansers (M. merganser). Auk 126:784–798.

Pearce, J. M., and M. R. Petersen. 2009. Post-fledging movements of juvenile Common Mergansers (Mergus merganser) in Alaska as inferred by satellite telemetry. Waterbirds 32:133–137.

Pearce, J. M., J. A. Reed, and P. L. Flint. 2005a. Geographic variation in survival and migratory tendency among North American Common Mergansers. Journal of Field Ornithology 76:109–216.

Pearce, J. M., and S. L. Talbot. 2006. Demography, genetics and the value of mixed messages. Condor 108:474–479.

Pearce, J. M., S. L. Talbot, M. R. Petersen, and J. R. Rearick. 2005b. Limited genetic differentiation among breeding, molting, and wintering groups of the threatened Steller's Eider: the role of historic and contemporary factors. Conservation Genetics 6:743–757.

Pearce, J. M., S. L. Talbot, B. J. Pierson, M. R. Petersen, K. T. Scribner, D. L. Dickson, and A. Mosbech. 2004. Lack of spatial genetic structure among nesting and wintering King Eiders. Condor 106:229–240.

Pearce, J. M., D. Zwiefelhofer, and N. Maryanski. 2009b. Mechanisms of population heterogeneity among molting Common Mergansers on Kodiak Island, Alaska: implications for genetic assessments of migratory connectivity. Condor 111:283–293.

Peters, J. L. 1931. Check-list of the Birds of the World (vol. 1). Harvard University Press, Cambridge, MA.

Peters, J. L., K. A. Bolender, and J. M. Pearce. 2012. Behavioral vs. molecular sources of conflict between nuclear and mitochondrial DNA: the role of male-biased dispersal in a Holarctic sea duck. Molecular Ecology 21:3562–3575.

Peters, J. L., and K. E. Omland. 2007. Population structure and mitochondrial polyphyly in North American Gadwalls (Anas strepera). Auk 124:444–462.

Petersen, M. R., B. J. McCaffery, and P. L. Flint. 2003. Moult and winter distribution of Long-tailed Ducks breeding on the Yukon-Kuskokwim Delta. Wildfowl 54:129–139.

Pettingill Jr., O. S. 1959. King Eiders mated with Common Eiders in Iceland. Wilson Bulletin 71:205–207.

Phillips, J. C. 1925. A natural history of the ducks (vol. 3). Houghton Mifflin, Boston, MA.

Phillips, J. C. 1926. A natural history of the ducks (vol. 4). Houghton Mifflin, Boston, MA.

Platte, P. M., and R. A. Stehn. 2011. Abundance and trend of waterbirds on Alaska's Yukon-Kuskokwim Delta coast based on 1988–2010 aerial surveys. U.S. Fish and Wildlife Service, Anchorage, AK.

Ploeger, P. L. 1968. Geographical differentiation in arctic Anatidae as a result of isolation during the last glacial period. Ardea 56:1–59.

Potapova, O. R. 1990. Ostatki ptits iz pleystotsenovykh otlozheniy Medvezh'yey peshchery na severnom urale. [Remains of birds from Pleistocene deposits in the Medvezh'ey cave in the northern Urals]. Trudy Zoologicheskogo Instituta Akademii Nauk SSSR 212:135–153 (in Russian).

Qu, Y., F. Lei, R. Zhang, and X. Lu. 2010. Comparative phylogeography of five avian species: implications for Pleistocene evolutionary history in the Qinghai-Tibetan plateau. Molecular Ecology 19:338–351.

Quakenbush, L. T., R. S. Suydam, R. Acker, M. Knoche, and J. Citta. 2009. Migration of King and Common Eiders past Point Barrow, Alaska, during summer/fall 2002 through spring 2004: population trends and effects of wind. Final Report. OCS Study MMS 2009–036. Coastal Marine Institute, University of Alaska, Fairbanks, AK.

Rand, A. L. 1954. The Ice Age and mammal speciation in North America. Arctic 7:31–35.

Regehr, H. M., C. M. Smith, B. Arquilla, and F. Cooke. 2001. Post-fledging broods of migratory Harlequin Ducks accompany females to wintering areas. Condor 103:408–412.

Reichel, J. D., and D. L. Genter. 1996. Harlequin Duck surveys in western Montana: 1995. Montana Natural Heritage Program, Helena, MT.

Robertson, G. J., and F. Cooke. 1999. Winter philopatry in migratory waterfowl. Auk 116:20–34.

Robertson, G. J., F. Cooke, R. I. Goudie, and W. S. Boyd. 2000. Spacing patterns, mating systems, and winter philopatry in Harlequin Ducks. Auk 117:299–307.

Robertson, G. J., and H. G. Gilchrist. 1998. Evidence of population declines among Common Eiders breeding in the Belcher Islands, Northwest Territories. Arctic 51:378–385.

Robertson, G. J., and R. I. Goudie. 1999. Harlequin Duck (Histrionicus histrionicus). in A. Poole and F. Gill (editors), The Birds of North America, No. 466. The Academy of Natural Sciences, Philadelphia, PA.

Robertson, G. J., and J.-P. L. Savard. 2002. Long-tailed Duck (Clangula hyemalis). in A. Poole and F. Gill (editors), The Birds of North America, No. 651. The Academy of Natural Sciences, Philadelphia, PA.

Root, T. 1988. Atlas of wintering North American birds: an analysis of Christmas Bird Count data. University of Chicago Press, Chicago, IL.

Rose, P. M., and D. A. Scott. 1997. Waterfowl population estimates, 2nd ed. Wetlands International Publication 44, Wageningen, Netherlands.

Ryder, O. A. 1986. Species conservation and systematic: the dilemma of subspecies. Trends in Ecology and Evolution 1:9–10.

Salomonsen, F. 1968. The moult migration. Wildfowl 19:5–24.

Salomonsen, F. 1979. Thirteenth preliminary list of recoveries abroad of birds ringed in Greenland. Danks Ornitologisk Forenings Tidsskrift 73:191–206.

Sangster, G. 2009. Acoustic differences between the scoters Melanitta nigra nigra and M. n. americana. Wilson Journal of Ornithology 121:696–702.

Savard, J.-P. L., D. Bordage, and A. Reed. 1998. Surf Scoter (Melanitta perspicillata). in A. Poole and F. Gill (editors), The Birds of North America, No. 363. The Academy of Natural Sciences, Philadelphia, PA.

Schamber, J. L., P. L. Flint, J. B. Grand, H. M. Wilson, and J. A. Morse. 2009. Population dynamics of Long-tailed Ducks breeding on the Yukon-Kuskokwim Delta, Alaska. Arctic 62:190–200.

Schmitt, T. 2007. Molecular biogeography of Europe: Pleistocene cycles and postglacial trends. Frontiers in Zoology 4:11.

Schrag, S. J., and P. Wiener. 1995. Emerging infectious disease: what are the relative roles of ecology and evolution? Trends in Ecology and Evolution 10:319–324.

Scribner, K. T., R. Lanctot, S. Talbot, B. Pierson, and K. M. Dickson. 1998. Genetic variation among populations of Harlequin Ducks (Histrionicus histrionicus): are eastern populations distinct? A Final Report to Region 5, U.S. Fish and Wildlife Service, Anchorage, AK.

Scribner, K. T., M. R. Petersen, R. L. Fields, S. L. Talbot, J. M. Pearce, and R. K. Chesser. 2001. Sex-biased gene flow in Spectacled Eiders (Anatidae): inferences from molecular markers with contrasting modes of inheritance. Evolution 55:2105–2115.

Scribner, K. T., S. L. Talbot, J. M. Pearce, B. J. Pierson, K. S. Bollinger, and D. V. Derksen. 2003. Phylogeography of Canada Geese (Branta canadensis) in western North America. Auk 120:889–907.

Sea Duck Joint Venture. [online] 2012. Atlantic and Great Lakes sea duck migration study. Progress Report. Washington, DC. <http://seaduckjv.org/atlantic_migration_study.html> (27 March 2014).

Seehausen, O., G. Takimoto, D. Roy, and J. Jokela. 2007. Speciation reversal and biodiversity dynamics with hybridization in changing environments. Molecular Ecology 17:30–44.

Sibley, C. G., and J. E. Ahlquist. 1990. Phylogeny and classification of birds: a study in molecular evolution. Yale University Press, New Haven, CT.

Sibley, C. G., and B. L. Monroe Jr. 1990. Distribution and taxonomy of birds of the world. Yale University Press, New Haven, CT.

Solovyeva, D. V., and J. M. Pearce. 2011. Comparative mitochondrial genetics of North American and Eurasian mergansers with an emphasis on the endangered Scaly-sided Merganser (Mergus squamatus). Conservation Genetics 12:839–844.

Sonsthagen, S. A., C. V. Jay, A. S. Fischbach, G. K. Sage, and S. L. Talbot. 2012a. Spatial genetic structure and asymmetrical gene flow within the Pacific walrus. Journal of Mammalogy 93:1512–1524.

Sonsthagen, S. A., R. N. Rosenfield, J. Bielefeldt, R. K. Murphy, A. C. Stewart, W. E. Stout, T. G. Driscoll, M. A. Bozek, B. L. Sloss, and S. L. Talbot. 2012b. Genetic and morphological divergence among Cooper's Hawk populations breeding in north-central and western North America. Auk 129:427–437.

Sonsthagen, S. A., S. L. Talbot, R. B. Lanctot, and K. G. McCracken. 2010. Do Common Eiders nest in kin groups? Microgeographic genetic structure in a philopatric sea duck. Molecular Ecology 19:647–657.

Sonsthagen, S. A., S. L. Talbot, R. A. Lanctot, K. Scribner, and K. McCracken. 2009. Hierarchical spatial genetic structure of Common Eiders (Somateria mollissima) breeding along a migratory corridor. Auk 126:744–754.

Sonsthagen, S. A., S. L. Talbot, and K. G. McCracken. 2007. Genetic characterization of Common Eiders (Somateria mollissima) breeding on the Yukon–Kuskokwim Delta, Alaska. Condor 109:879–894.

Sonsthagen, S. A., S. L. Talbot, K. Scribner, and K. McCracken. 2011. Multilocus phylogeography and population structure of Common Eiders breeding in North America and Scandinavia. Journal of Biogeography 38:1368–1380.

Sonsthagen, S. A., S. L. Talbot, R. E. Wilson, M. R. Petersen, J. C. Williams, G. V. Byrd, and K. G. McCracken. 2013. Genetic structure of the Common Eider in the Western Aleutian Islands prior to fox eradication. Condor 115:28–39.

Spielman, D., B. W. Brook, and R. Frankham. 2004. Most species are not driven to extinction before genetic factors impact them. Proceedings of the National Academy of Sciences USA 101:15261–15264.

Spurr, E., and H. Milne. 1976. Adaptive significance of autumn pair formation in Common Eider Somateria mollissima (L.). Ornis Scandinavica 7:85–89.

Steadman, D. W. 2005. Late Pleistocene birds from Kingston Saltpeter Cave, southern Appalachian Mountains, Georgia. Bulletin of the Florida Museum of Natural History 45:231–248.

Steadman, D. W., and P. S. Martin. 1984. Quaternary extinctions: a prehistoric revolution. Pp. 466–467 in P. S. Martin and R. G. Kelin (editors), University of Arizona Press, Tucson, AZ.

Stehn, R. A., C. P. Dau, B. Conant, and W. I. Butler Jr. 1993. Decline of Spectacled Eiders nesting in western Alaska. Arctic 46:264–277.

Stettenheim, P. 1958. Bird fossils from the Late Pleistocene of Kansas. Wilson Bulletin 70:197–199.

Stewart, J. R., A. M. Lister, I. Barnes, and L. Dalén. 2010. Refugia revisited: individualistic responses of species in space and time. Proceedings of the Royal Society B 277:661–671.

Stewart, T. G., and J. Hourston-Wright. 1990. 6500 bp Oldsquaw duck (Clangula hyemalis) from Northern Ellesmere Island, Arctic Archipelago, Canada. Arctic 43:239–243.

Suydam, R. S., D. L. Dickson, J. B. Fadely, and L. T. Quakenbush. 2000. Population declines of King and Common Eiders of the Beaufort Sea. Condor 102:219–222.

Tiedemann, R., K. B. Paulus, M. Scheer, K. G. von Kistowski, K. Skirnisson, D. Bloch, and M. Dam. 2004. Mitochondrial DNA and microsatellite variation in the Eider Duck (Somateria mollissima) indicate stepwise postglacial colonization of Europe and limited current long-distance dispersal. Molecular Ecology 13:1481–1494.

Tiedemann, R., K. G. von Kistowski, and H. Noer. 1999. On sex-specific dispersal and mating tactics in the Common Eider Somateria mollissima as inferred from the genetic structure of breeding colonies. Behaviour 136:1145–1155.

Titman, R. D. 1999. Red-breasted Merganser (Mergus serrator). in A. Poole and F. Gill (editors). The Birds of North America, No. 443. The Academy of Natural Sciences, Philadelphia, PA.

Tremblay, N. O., and D. J. Schoen. 1999. Molecular phylogeography of Dryas integrifolia: glacial refugia and postglacial recolonization. Molecular Ecology 8:1187–1198.

Vaillancourt, M.-A., P. Drapeau, M. Robert, and S. Gauthier. 2009. Origin and availability of large cavities for Barrow's Goldeneye (Bucephala islandica), a species at risk inhabiting the Eastern Canadian boreal forest. Avian Conservation and Ecology 4:art6.

Vickery, P. D. 1988. Distribution and population status of Harlequin Ducks (Histrionicus histrionicus) wintering in eastern North America. Wilson Bulletin 100:119–126.

Vigors, N. A. 1825. Observations on the natural affinities that connect the orders and families of the birds. Transactions of the Linnean Society of London 14:395–517.

Waltari, E., and J. A. Cook. 2005. Historical demographics and phylogeography of arctic hares (Lepus): genetic signatures test glacial refugial hypotheses. Molecular Ecology 14:3005–3016.

Weller, M. W. 1964. Distribution and species relationships. Pp. 108–120 in J. Delacour (editor), The waterfowl of the world (vol. 4). Country Life, London, U.K.

Westemeier, R. L., J. D. Brawn, S. A. Simpson, T. L. Esker, R. W. Jansen, J. W. Walk, E. L. Kershner, J. L. Bouzat, and K. N. Paige. 1998. Tracking the long-term decline and recovery of an isolated population. Science 282:1695–1698.

Wetmore, A. 1973. A Pleistocene record for the White-winged Scoter. Auk 90:910–911.

Williams, M., D. Dunkerley, P. De Deckker, P. Kershaw, and J. Chappel. 1998. Quaternary environments, 2nd ed. Oxford University Press, New York, NY.

Wilson, R. E., J. L. Peters, and K. G. McCracken. 2013. Genetic and phenotypic divergence between low- and high-altitude populations of two recently diverged Cinnamon Teal subspecies. Evolution 67:170–184.

Wright, S. 1931. Evolution in Mendelian populations. Genetics 16:97–159.

Wright, S. 1938. Size of population and breeding structure in relation to evolution. Science 87:430–431.

Yarrell, W. 1827. Observations on the trachea of birds; with descriptions and representations of several not hitherto figured. Transactions of the Linnean Society London 15:378–391.

Zink, R. M., and G. F. Barrowclough. 2008. Mitochondrial DNA under siege in avian phylogeography. Molecular Ecology 17:2107–2121.

Zink, R. M., A. E. Kessen, T. V. Line, and R. C. Blackwell-Rago. 2001. Comparative biogeography of some aridland bird species. Condor 103:1–10.

Zink, R. M., A. Pavlova, S. Rohwer, and S. V. Drovetski. 2006. Barn Swallows before barns: population histories and intercontinental colonization. Proceedings of the Royal Society B 273:1245–1251.

Zink, R. M., S. Rohwer, A. V. Andreev, and D. L. Dittmann. 1995. Trans-Beringian comparisons of mitochondrial DNA differentiation in birds. Condor 97:639–649.

CHAPTER THREE

Population Dynamics of Sea Ducks*

USING MODELS TO UNDERSTAND THE CAUSES, CONSEQUENCES, EVOLUTION, AND MANAGEMENT OF VARIATION IN LIFE HISTORY CHARACTERISTICS

Paul L. Flint

Abstract. In this chapter, I explore population dynamics of sea ducks by developing population models. In determining which life history characteristics had the greatest influence on future population dynamics, adult female survival consistently had the highest sensitivity and elasticity and this result was robust across a wide range of life history parameter values. Conversely, retrospective models consistently found that the majority of annual variation in λ was associated with variation in productivity. Stochastic models that are based on process variation and incorporate correlations among life history parameters are the most useful for visualizing the probability of achieving a desired management outcome. Effective management targets both the mean and the variance of parameters and takes advantage of correlations among life history parameters. Example models demonstrate that sea duck species can achieve equal fitness using a variety of survival and productivity combinations. Sea duck populations will tend to have long time lags in terms of responding to management actions. Understanding the role of density-dependent population regulation is critical for effective sea duck management and conservation.

Key Words: correlated vital rates, density-dependence, demography, deterministic, elasticity, matrix model, prospective, retrospective, sensitivity, stochastic.

Population dynamics are defined by changes in the number of individuals and age composition of populations, as well as the biological and environmental processes influencing those changes. In this chapter, I focus on ways in which populations are affected by variation in birth and death rates, as opposed to immigration and emigration. Population dynamics provides the means to interpret life history characteristics relative to one another. That is, the interpretation of any one

---

* Flint, P. L. 2015. Population dynamics of sea ducks: Using models to understand the causes, consequences, evolution, and management of variation in life history characteristics. Pp. 63–96 in J.-P. L. Savard, D. V. Derksen, D. Esler, and J. M. Eadie (editors). Ecology and conservation of North American sea ducks. Studies in Avian Biology (no. 46), CRC Press, Boca Raton, FL.

life history characteristic value must be made in the context of all the other life history characteristics. For example, if a given population has nesting success of 40%, is that good or bad? The answer depends on all of the other demographic characteristics of the population. In some cases, 40% nesting success would lead to population growth; in others, the population would decline.

Population models provide a quantitative structure for evaluating and interpreting values of various life history parameters relative to one another. By understanding the interactions among life history parameters, it may be possible to identify and assess the effects of potential management actions, environmental perturbations, or other constraints on vital rates and subsequent changes in population size. Given that multiple species of sea ducks have experienced long-term fluctuations in population size, understanding underlying demographic processes is important for defining management priorities. While factors leading to variation in population dynamics of many sea ducks remain poorly known, development of population models has proven useful for evaluating and predicting population change in several cases (Schamber et al. 2009, Coulson 2010, Iverson and Esler 2010, Wilson et al. 2012). Reliable determination of factors affecting population change in other sea ducks will almost certainly require application of some form of population modeling, and this chapter is intended to provide background for understanding and accomplishing this approach.

WHAT IS A POPULATION?

A population is a group of animals of any given species that share a common suite of demographic parameters. That is, individuals of a population must share resources in terms of time and space such that they would have the same expectation of survival and fecundity. For highly migratory waterfowl species like sea ducks, population definition can be somewhat problematic (Robertson and Cooke 1999, Cooke et al. 2000, Esler 2000). For the purposes of examining population dynamics or developing a population model, should sea ducks be defined in terms of breeding populations or wintering populations? For some species, these two definitions would be equivalent; for others, they would be starkly different. On the Yukon–Kuskokwim Delta, breeding Long-tailed Ducks

showed fidelity to specific study areas, implying a distinct breeding population (Schamber et al. 2009). However, birds from one breeding area wintered over a wide range from British Columbia, Canada, to the Kamchatka Peninsula in Russia (Petersen et al. 2003). Thus, for a large portion of the year, this breeding population is not sympatric and likely experiences substantial variation in life history parameters such as adult survival. Some species may winter in consistent areas and show wintering site fidelity but may breed over a large range. For example, the entire global population of Spectacled Eiders winter in one specific area in the Bering Sea but breed in three separate groups in Yukon–Kuskokwim Delta, Arctic Coastal Plain of Alaska, and Northern Siberia (Scribner et al. 2001). Thus, the wintering population would be expected to have considerable heterogeneity in productivity. Last, within a single species, there may be clearly definable population structure. Petersen and Flint (2002) found that Common Eiders breeding in different parts of Alaska remained essentially segregated throughout the annual cycle. Accordingly, it is clear that the definition of a population may vary depending upon the specific question at hand. For many of the concepts of population dynamics and population models to apply, the primary determinants of population change have to be births and deaths, as opposed to immigration and emigration. Thus, if the dynamics of a potential population are primarily driven by immigration and emigration, the geographic frame of reference may need to shift to a larger scale, within which the collective rates of births and deaths drive overall dynamics.

The issue of population definition relative to modeling of population dynamics represents a trade-off in terms of balancing the potential effects of immigration and emigration against heterogeneity in life history parameters. For example, if a population is declared to be a small number of birds breeding in a specific area, there will likely be little heterogeneity in life history parameters, but there may be substantial effects of immigration and emigration. Conversely, if a population is declared to be the entire global distribution of a species, then obviously there can be no immigration or emigration, but there is almost certainly substantial variation in life history parameters within this population. To successfully model population dynamics, careful consideration of the definition of the modeled population is essential.

HOW DO WE DESCRIBE POPULATION DYNAMICS?

Lambda (abbreviated as λ in equations) is one of the primary terms used in discussion of population dynamics. Lambda is a demographic statistic that summarizes the overall trajectory of a population. Lambda can be used to predict future population size using simple multiplication. That is, population size (N) at time $t + n$ (i.e., N_{t+n}) is simply population size at time (t) × lambda^n (i.e., $N_t \times \lambda^n$). Lambda is scaled in relative terms and is independent of population size (e.g., a population with a lambda of 1.10 is increasing at a rate of 10% per year). As such, a population of 100 individuals and a population of 10,000 individuals can both have a lambda of 1.10. Obviously, the larger population is growing in greater absolute numbers each year, but the two populations are comparable on a relative scale. As such, lambda represents the balance between mortality and recruitment.

Understanding and describing population dynamics require synthesis of productivity and survival data for a given population. This synthesis is accomplished through development of a population model. Building such a model requires detailed data on all relevant life history parameters for a population. Unfortunately, complete data sets of all life history parameters do not exist for most avian species, including sea ducks (Heppell et al. 2000, Sæther and Bakke 2000). Thus, it is impossible to generate and compare independent population models for multiple species. In lieu of comprehensive data for a given population, I developed population models for a generic sea duck for the purposes of this chapter. I utilized model structures that captured general patterns of life history characteristics for most sea duck species (Chapter 10, this volume). In general, sea ducks have relatively high survival with low, variable productivity, and most new recruits do not breed until several years of age.

One of the primary uses of a population model is to predict an overall numerical trend. Thus, demographic data are used to estimate the population trend. However, in the absence of complete demographic data, this process can be reversed. That is, for the models presented in this chapter, I frequently assumed that populations were numerically stable and, based on that assumption, estimated the combination of life history parameter values that yielded this result. As such, I was primarily interested in the inherent patterns of population dynamics as opposed to trend predictions specific to any given population. I demonstrate the degree of variability in results across ranges of parameter estimates applicable to a variety of sea duck species. Although these models are not specific to any population, results indicate patterns that are generalizable to most sea duck species. As confirmation that results are generally applicable to most sea duck species, I demonstrate that these results are robust to variation in specific parameter estimates.

MODELING FEMALES ONLY OR BOTH SEXES

As noted previously, a key component of modeling is to define a population that is primarily influenced by births and deaths as opposed to immigration and emigration. However, dispersal rates in most sea ducks vary by sex (Swennen 1990, Scribner et al. 2001). Adult females tend to have high site fidelity, and juvenile females tend to show high natal philopatry, whereas males have a much higher likelihood of dispersal (Swennen 1990). Accordingly, a female-only model is far less likely to violate assumptions regarding immigration and emigration and therefore more accurately reflect true population dynamics.

In waterfowl, adult female survival is usually lower than that of adult males, due to differences in reproductive investment (Johnson et al. 1992). However, survival of male sea ducks has rarely been estimated. Flint et al. (2000b) reported that male Steller's Eiders had lower apparent annual survival than females, but more recent analyses suggest that these estimates of male survival were biased low by emigration. The general pattern of lower survival for females is supported for sea ducks by sex ratio data based on adult plumage characteristics, which indicate either equal or male-biased sex ratios (Anderson and Timken 1972; Swennen et al. 1979; Bolen and Chapman 1981; Iverson et al. 2004, 2006; Caron and Paton 2007; Lehikoinen et al. 2008; Robertson 2008). Given assumed or observed male-biased sex ratios in waterfowl populations, models typically only consider females as their vital rates will determine population dynamics. If males were to become the limiting sex, not all females who would otherwise breed would be successful in acquiring a mate. A two-sex model could be developed and the proportion of adult males could be used as a factor affecting breeding propensity of females. That is, one could assume that all adult males breed, and the total number of available males then determines how many females attempt to reproduce in a given year. However, situations in which numbers of males actually limit breeding propensity are sufficiently rare that I consider female-only models for the remainder of this chapter.

POPULATION MODELING BASICS

The simplest form of a closed population model is the balance equation:

$1 -$ annual mortality rate $+$ annual recruitment rate $=$ annual rate of population change (λ).

Accordingly, when populations are stable, annual mortality must equal annual recruitment. If recruitment exceeds mortality, populations will increase and vice versa. North American numbers of many species of sea ducks have declined (Flint 2013); thus, for these populations, during the periods of decline, annual mortality must have exceeded recruitment. With data on population trends and estimates of annual survival of breeding aged birds, overall recruitment can be estimated. However, it can be difficult to make predictions with this type of model as all productivity parameters are grouped together until age of first breeding or recruitment, precluding assessment of potential effects of individual life history parameters on lambda. As such, this type of model may not be an efficient use of all available data such as when one or more estimates of productivity components exist. Nevertheless, Heppell et al. (2000) demonstrated that even these simple models do a good job of generally describing behavior of populations, unless there is significant individual variation in life history values.

PROSPECTIVE POPULATION PROJECTION

Prospective population projection requires the development of a stage- or age-based population model from which potential future changes in population trajectory can be estimated from changes in the life history input parameters. These models serve as a useful indicator of overall population dynamics. Such models are probably the most common

population models used in wildlife biology. These models utilize life history parameters typically collected in field situations and synthesize them into a numerical summary of a population's dynamics. The key is that these models allow interpretation of individual life history parameters relative to all other parameters. The simplest of these models do not allow input parameters to vary and thus are termed deterministic models. I will start with simple deterministic models of population dynamics and generate increasingly complex models, describing benefits and drawbacks of each permutation.

DETERMINISTIC SEA DUCK MATRIX POPULATION MODEL

I developed a general, age-structured, deterministic population model that fits the life history of most sea duck species. While the specific model structure for age of first reproduction and other parameters will differ slightly from this basic model formulation for some species, I assess the robustness of conclusions across a range of biologically plausible model structures and parameter inputs later in the chapter. I developed a female-only-based model, assuming that all females that would otherwise attempt to breed would find a mate. As such, I assumed that females were controlling population dynamics and that there were surplus males in the population.

An age-structured population can be represented using a diagram that depicts life cycle stages and transitions among stages (Figure 3.1). In this generalized sea duck model, there are four distinct stages. Three of these stages are based on distinct age classes (i.e., 1, 2, 3+) and one is a progeny stage (i.e., 0). The solid-lined arrows in Figure 3.1 indicate transition probabilities among age classes or the age-specific survival rates. The dashed lines indicate age-specific fertilities. Fertilities represent total production of female ducklings as the product of breeding propensity, clutch size (adjusted for a 1:1 sex ratio), nesting success, and duckling survival. In the diagram presented here, it was assumed that 1-year-old females do not breed, and thus, there is no fertility line for the 1-year-old age class. Given this definition of fertilities, first-year survival (i.e., S0) represents survival from fledging until 1 year old. Conversely, the other survival rates (i.e., S1 and S2) represent survival from the time an individual joins one age class until it graduates to the next age class.

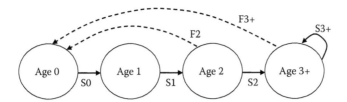

Figure 3.1. Diagrammatic representation of a typical sea duck population age-structured model with a postbreeding census. Circles represent the age classes. Age 0 represents progeny and age 3+ represents all individuals 3 years old or older. Solid-lined arrows represent age-specific survival probabilities (i.e., S0, S1, S2, and S3+) and dashed lines represent age-specific fertilities (i.e., F2 and F3+). Age 3+ has two survival transitions representing individuals graduating from age 2 into age 3 and the proportion of individuals already in age class 3+ that remain in the class (i.e., the proportion that does not die).

The duration of S0 and the other survival rates differs, as S0 does not include the period of egg laying, incubation, and brood rearing. As such, S0 does not represent a true annual (12 month) survival rate, but instead covers the ~10-month period from fledging until the start of breeding the following year. The age class 3+ includes all individuals that were 3 years old or older. This age class has two relevant survival transitions: (1) S2 that represents individuals graduating from age class 2 into age class 3 and (2) S3+ that represents the proportion of individuals already in age class 3+ that remain in age class 3+ or the proportion of individuals already in age class 3+ that do not die. This model structure assumes that there is no age-specific variation in survival and productivity within the 3+ age class.

Reproduction in sea ducks occurs once annually, during a relatively short window in summer. Seasonality simplifies models of overall population dynamics by allowing a birth-pulse structure. When constructing population models, the modeler must decide when the model will be paused for the number of individuals present in the population to be counted. The counts will serve as the basis for determining annual rates of population change and potential responses to management scenarios. I developed a prebreeding census model, which counts the number of individuals by age class just prior to onset of reproduction in a given year. This model formulation closely matches available survey data, which are commonly counted early in the nesting period (Wilson et al. 2012). Using a prebreeding census, the onset of reproduction also defines the time at which individuals graduate from one age class into the next. Under a prebreeding census model structure, there is no opportunity to count the number of individuals in the age class 0 because the young have not yet hatched, and accordingly, fecundity is defined as the number of young hatched in year i that survive until the onset of breeding in year i + 1 (Morris and Doak 2002). Thus, individuals are recruited into age class 1 when they are just under 1 year old. Given that the age class 0 is never counted in a prebreeding census, the overall model is functionally simplified to three age classes (Figure 3.2). The only thing different about this model formulation compared to the previous model is that juvenile survival or the probability that a fledged duckling survives to the prebreeding census in the following year (denoted as S0) is modeled as a component of fecundity. In the previous model formulation (Figure 3.1), fertility was the number of ducklings fledged. In this prebreeding census format, fecundity

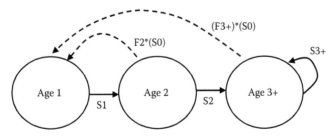

Figure 3.2. Reduced version of the age-structured population represented in Figure 3.1 as modeled with a prebreeding census. Age class 0 is dropped from consideration as the size of the 0 age class is never determined. Instead, the juvenile survival rate (i.e., S0) is included as a fecundity parameter.

refs to the total number of female recruits to the 1-year-old stage of the model on a per female basis. Thus, a fecundity of 0.5 implies that, on average, each female in that stage class produced 0.5 female recruits that ultimately survive to join the 1-year-old age class. I followed definitions of Morris and Doak (2002) where fecundity refers to fledged ducklings and fertility refers to 1-year-old recruits.

INITIAL MODEL PARAMETERS

With the possible exception of some populations of Common Eiders, there are no sea ducks for which the entire suite of life history parameters required to build a comprehensive population model has been estimated. As such, it is impossible to build a single-population model from existing data for most species. The basic population model I have developed applies to what I have termed a *generic* sea duck. The basic model structure was designed to fit the known age of first reproduction for eiders and scoters. Parameter estimates for fertility and survival are generally based on published estimates in combination with some unpubl. data. This basic model would likely fit most sea duck populations in the broad sense, but not fitting any population specifically. I use this general model to demonstrate the basic characteristics of sea duck population dynamics and then demonstrate that results generated from this model are robust over a wide range of parameter estimates and model formulations.

In developing this model, I assumed that female sea ducks start breeding at 2–3 years of age. That is, no birds attempted to nest at 1 year of age. I assumed that 40% of the 2-year-old females attempted to breed and that almost all birds age 3 or older attempted to breed (97%). I allowed clutch size to vary with age and assigned clutch sizes of 3.5 eggs for 2-year-old females and 4.5 eggs for all older birds. I assumed a sex ratio at hatch of 1:1. Following the logic that young birds have lower productivity, I structured model inputs such that young, inexperienced birds breeding for the first time at age 2 would tend to have lower nesting success (28.5%) than older birds (51.3%; Ost and Steele 2010). I also assumed that a small proportion of older 3+-year-old females may renest or lay continuation clutches following loss of their first clutch and set this renesting probability

at 0 for 2-year-old birds and 11.4% for older females. Renests were assumed to have the same probability of nesting success as all other nests. Duckling survival was defined as the probability of a hatched duckling surviving to fledging and was set at 42.8% for all females. Juvenile survival (S0) was defined as the probability of a fledged duckling surviving until 1 year of age and was set at 40%.

Survival was defined as the proportion of females surviving from the census in year i to the census in year i + 1. I set this value at 85%. I applied this annual survival rate to all birds ≥ age 1. I also included a survival component in the fecundity estimation that is defined as breeding season survival. The relevance is that females that die during the breeding season were assigned a fertility of 0. Thus, not only does the female die, but all her dependent offspring (eggs or ducklings) are also assumed to perish. I set breeding season survival at 0.99.

Using the data and assumptions listed earlier, I developed a Leslie style matrix model (Caswell 2001). The advantage of a Leslie matrix model is matrix algebra is used to track the number of individuals of each age class in the population. Further, the known properties of these matrices in terms of eigenvalues and eigenvectors can be interpreted in a biological fashion and provide convenient analytical solutions for several aspects of population dynamics, including lambda, stable age distributions, and reproductive values. The Leslie matrix is presented as a square matrix of survival and fertility values with cells defined as in Figure 3.3. The zero cells in this matrix represent biologically impossible transitions. For example, a female in the 2-year-old age class cannot survive for a year and remain in the 2-year-old age class because, by definition, this bird would now be 3 years old. Therefore, zeros in the second row force all birds to graduate to the next highest age class. The zero in the first column

$$\begin{bmatrix} \text{Fertility of} & \text{Fertility of} & \text{Fertility of} \\ \text{1–yr old females} & \text{2–yr old females} & \text{>3–yr old females} \\ \text{Survival from 1 to} & & \\ \text{2–yr old females} & 0 & 0 \\ & \text{Survival from 2 to} & \text{Survival of} \\ 0 & \text{3–yr old females} & \text{>3–yr old females} \end{bmatrix}$$

Figure 3.3. Definitions of cell values for a Leslie matrix.

indicates that a 1-year-old female cannot graduate directly to the 3-year-old age class; that is, she has to pass through the 2-year-old age class first. As such, these zeros in the matrix are required in age-structured models.

In this model, fecundity of 1-year-old females is defined as the product of breeding propensity × breeding season survival × clutch size × sex ratio × nesting success × duckling survival × juvenile survival. However, because breeding propensity of 1-year-old females was fixed at 0, the overall fertility is 0. Fecundity of 2-year-old females was calculated as the product of breeding propensity × breeding season survival × clutch size × sex ratio × nesting success × duckling survival × juvenile survival. Fecundity of >3-year-old females starts with the same equation as for 2-year-old females and adds the following term to account for productivity from renests: (1 nesting success) × renesting probability × clutch size × sex ratio × nesting success × duckling survival × juvenile survival. Accordingly, the actual Leslie matrix for the model is given in Figure 3.4. The three columns represent the three age classes. The top row indicates the age-specific fertilities, and the next two rows indicate the age-specific survival probabilities.

This set of conditions was selected as biologically representative for sea ducks; the parameter estimates fall within the observed range of values and they yield a stable population (i.e., $\lambda = 1.0$). For any given set of life history parameters, the proportion of the population that occurs within each stage can be estimated. Further, under any deterministic set of time invariant conditions, the stable age distribution is fixed and does not vary through time. In this example, the stable age distribution is 15% 1-year-olds, 12.8% 2-year-olds, and 72.2% birds ≥3 years of age. The distribution corresponds with patterns observed for many populations of sea ducks, which are primarily composed of adult birds (Bolen and Chapman 1981; Smith et al. 2001; Iverson et al. 2004, 2006; Robertson 2008; Wilson et al. 2012).

$$\begin{bmatrix} 0 & 0.034 & 0.203 \\ 0.85 & 0 & 0 \\ 0 & 0.85 & 0.85 \end{bmatrix}$$

Figure 3.4. Input Leslie matrix for a hypothetical sea duck population.

It is also typical of K-selected species that are long-lived with relatively low reproductive rates. A similar pattern is expected for males, but with the same level of productivity under the assumption of a 1:1 sex ratio (Swennen et al. 1979) and with somewhat higher levels of annual survival that might produce a surplus of males in the population, the expected stable age distribution of males would show an even higher proportion of older birds.

Under the scenario used to define this population, the stable age distribution also can be used to estimate the proportion of the population that does not attempt to breed in a given year. Recall that this model uses a prebreeding census where the accounting occurs just prior to the breeding season. It is assumed that 1-year-old birds never breed, and only 40% of the 2-year-old females breed. In this example, 22.7% of the prebreeding population of females would be considered nonbreeders: 15% + (12.8% × 0.6%). Limited information from a handful of long-term sea duck data sets indicates this pattern likely holds true in several, if not most, sea duck species (Boyd et al. 2009). For example, in >12 years of marking Spectacled Eiders on the Yukon–Kuskokwim Delta, 1-year-old birds were never observed on the nesting grounds, and only a fraction of the 2-year-old birds were observed (P. Flint, U.S. Geological Survey, unpubl. data), leading to the conclusion that nonbreeding birds likely remained at sea. If this is the case, almost a quarter of the population is not present on the breeding grounds in a given year (Coulson 1984).

Given this basic deterministic model, it is possible to predict population responses to perturbation of life history parameters. This approach can be thought of as the predicted response of the population to management actions or environmental changes. Perturbations can be considered in two ways: (1) absolute and (2) relative changes in life history parameters.

ELASTICITY: PROPORTIONAL CHANGES IN LIFE HISTORY PARAMETERS

The term elasticity is used to describe the relative change in lambda resulting from a relative change in life history parameters. Elasticity analyses have several unique attributes. First, the matrix cell elasticities always sum to 1.0.

$$\begin{bmatrix} 0 & 0.003 & 0.115 \\ 0.118 & 0 & 0 \\ 0 & 0.115 & 0.650 \end{bmatrix}$$

Figure 3.5. Elasticity estimates for model components of a hypothetical sea duck population.

Thus, elasticities can be directly compared to one another. Elasticities are estimated by direct manipulation of the Leslie matrix parameters and measurement of the effect of the manipulation on lambda. Given a matrix model, we can estimate the elasticity for each nonzero cell of the matrix (Figure 3.5). Consider the cell in the lower right corner that represents elasticity of survival for females ≥3 years of age. The interpretation of this value is that a 1% increase in adult female survival of birds ≥3 years of age would yield a 0.64% increase in lambda. In this example, the survival rate was 0.85 and lambda was 1.0. Thus, the expected effect of increasing adult female survival by 1% from 0.85 to 0.859 (i.e., 0.85 × 1.01) would result in a lambda of 1.0064. However, as mentioned previously, elasticities are additive. Therefore, overall elasticity of survival versus productivity can be assessed by summing elasticities for all cells of the matrix that contain survival or productivity estimates. Accordingly, the overall elasticity of survival for females ≥1 year old is 0.882 (i.e., 0.1179 + 0.1146 + 0.6495), whereas the overall elasticity of productivity is 0.118 (i.e., 0.0033 + 0.1146). Accordingly, relative changes in survival will have far greater effects on sea duck populations compared to relative changes in productivity. Alternatively, to achieve the same result in terms of the change in lambda from potential management actions, a relative change in productivity would have to be 7.5 times greater than a relative change in survival.

To assess the robustness of these conclusions across a range of survival and productivity values, I kept this same age-specific fecundity structure and constrained lambda to equal 1.0 and then varied survival between 0.5 and 1.0. For each value of survival, I iteratively solved for the overall level of productivity that yielded a lambda of 1.0. I varied productivity by adjusting all life history parameters by an equal proportion. At relatively low levels of survival (0.5) and associated high levels of productivity, elasticity for

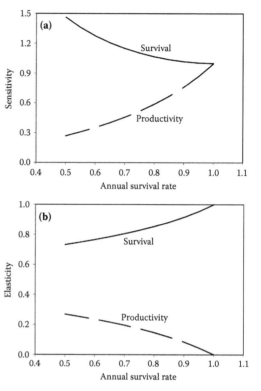

Figure 3.6. (a) Sensitivity and (b) elasticity values for a range of survival and productivity values. All pairs of survival and productivity yield a lambda of 1.0. Dashed lines represent productivity; solid lines represent survival. The overall pattern of sensitivities and elasticities remained constant across the biologically plausible range of values.

survival was three times higher than elasticity of productivity (Figure 3.6). The ratio of elasticities becomes more divergent as survival increases. Ultimately, this process ends in the biologically unrealistic scenario where survival equals 1.0 and productivity is zero—adults live forever but never reproduce. Thus, when lambda is held constant, the conclusion that survival has the greatest relative effect on population dynamics of sea ducks is robust across a wide range of potential life history parameter values.

Next, I considered the robustness of the conclusion that adult survival had the greatest relative effect when lambda differed from 1.0. Starting with this same matrix model, I first varied relative levels of productivity while holding survival constant. Thus, as productivity varied, lambda also varied. I considered a range of lambda values from 0.9 to 1.1 (populations increasing or decreasing by 10% per year). Under all scenarios, elasticity of survival remained dominant (Figure 3.7). Next, I repeated this process holding productivity constant and varying survival rate. Under this regime, there was little variation in elasticity values. However, because of the difference in elasticity values, obtaining the same range of variation in lambda required larger proportional changes in productivity compared to survival. Because of this difference in ranges, nonlinear trends in elasticities resulting from changes in productivity were apparent (Figure 3.7).

Overall, the conclusion that elasticity of survival is dominant was robust across a broad range of sea duck life history parameter values and overall population trajectories. This conclusion appeared to be primarily driven by the specifics of the structure of the matrix models. Sæther and Bakke (2000) concluded that the positive relationship between adult survival and elasticity was a function of the model structure as mediated by age of first reproduction. Species with delayed age of first reproduction and increasing age-specific fecundities will be more K-selected by definition and hence will always show this pattern of overall elasticities. Indeed, this conclusion has been found consistently in the few sea duck population models that have been published (Schamber et al. 2009, Iverson and Esler 2010, Wilson et al. 2012).

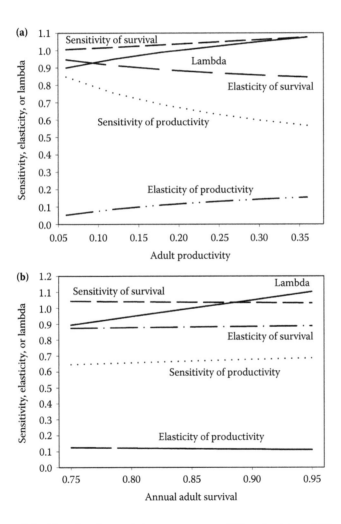

Figure 3.7. Sensitivity and elasticity values for populations across a range of (a) productivity and (b) survival values. Lambda in these simulations varies from 0.9 to 1.1. Overall, the relative pattern of sensitivities and elasticities did not vary across this range of parameter inputs.

LOWER-LEVEL ELASTICITIES

Individual elements in a matrix can be decomposed into more specific components of a species life history. For instance, as mentioned earlier, fecundity of a given age class is a product of a suite of parameters including breeding propensity, clutch size, nesting success, and duckling survival. Elasticities of each of these parameters that are a part of a matrix element are referred to as lower-level elasticities. They represent the relative effect of a proportional change in any one life history parameter (e.g., clutch size, nest success, rather than the composite parameter fertility, of which these lower-level parameters are components) on lambda. Lower-level elasticities have the same interpretation as matrix element level elasticities; however, lower-level elasticities are not constrained to sum to 1.0. The pattern of lower-level elasticities for the generic model shows the overall dominance of the effect of survival on lambda and lower and relatively uniform effects of individual productivity parameters (Figure 3.8). The equality of lower-level elasticities for productivity parameters is primarily a function of the structure of the fertility equation. For example, in estimating recruitment, the probability of a hatched duckling being recruited is estimated as duckling survival × juvenile survival. Because the matrix element is estimated as the product of two probabilities, a proportional change in one parameter is equivalent to a proportional change in the other. Thus, the multiplicative theorem of mathematics dictates that purely multiplicative components of the fertility equations will have equal lower-level elasticities. The observed slight variation in lower-level elasticities for fertility components is due to differences in age-specific fertilities that are not simply multiplicative. Accordingly, a similar pattern in lower-level elasticities would be expected for all sea duck species.

SENSITIVITY: ABSOLUTE CHANGES IN LIFE HISTORY PARAMETERS

While elasticities are used to contrast proportional changes in life history parameters, sensitivities measure the effect of absolute changes in parameter values. Sensitivities are not constrained to sum to 1, and therefore, interpretation of any given sensitivity can only be made relative to other sensitivities estimated from the same model. Like elasticities, sensitivities are estimated by direct manipulation of the model estimates and calculation of the effect on lambda (Figure 3.9). Consider the cell in the lower right corner, which represents the sensitivity of ≥3-year-old female survival. The interpretation of this value is that a 1% point increase in adult female survival (i.e., birds ≥3 years of age) would yield a 0.76% point increase in lambda. In this example, the survival rate was 0.85 and lambda was 1.0. Thus, the expected effect

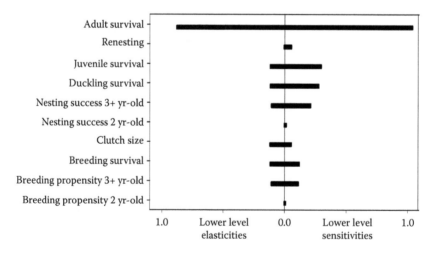

Figure 3.8. Lower-level sensitivities and elasticities for a hypothetical sea duck species. Lower-level responses predict the effect of a relative (elasticity) or absolute (sensitivity) change in one of the lower-level parameters on the population growth rate (lambda). For any species with a life history similar to a sea duck, changes in adult female survival will always yield the largest effect on lambda.

$$\begin{bmatrix} 0 & 0.100 & 0.568 \\ 0.139 & 0 & 0 \\ 0 & 0.135 & 0.764 \end{bmatrix}$$

Figure 3.9. Matrix parameter sensitivities for a hypothetical sea duck population.

of increasing adult female survival by 1% point from 0.85 to 0.86 (i.e., 0.85 + 0.01) would result in a lambda of 1.0076. While sensitivities do not sum to 1.0, they are still additive. Therefore, the overall sensitivity of survival versus productivity can be assessed by summing the sensitivity for any cell of the matrix that contains survival or productivity estimates. Accordingly, the overall sensitivity of survival is 1.038 (i.e., 0.139 + 0.139 + 0.764), whereas the overall sensitivity of productivity is 0.668 (i.e., 0.100 + 0.568). Absolute changes in survival will have greater effects on sea duck populations compared to changes of a similar magnitude in productivity. To think of this another way, in order to achieve the same result from potential management actions, an absolute change in productivity would have to be 1.6 times greater (i.e., 1.038/0.668) than a change in survival, to produce the same change in lambda.

To assess how robust this result is across a range of survival and productivity values, I kept the same age-specific fecundity structure and constrained lambda to equal 1.0 and then varied survival between 0.5 and 1.0. For each value of survival, I iteratively solved for the overall level of productivity that yielded a lambda of 1.0. I varied productivity by adjusting all life history parameters by an equal proportion. At relatively low levels of survival (i.e., 0.5) and associated high levels of productivity, sensitivity for survival was 5.4 times higher than the sensitivity of productivity (Figure 3.6). The ratio of sensitivities converged toward 1.0 as survival increased. Ultimately, this example ends in the biologically unrealistic scenario where survival equals 1.0 and productivity is zero. In this case, sensitivities of survival and productivity are equal at 1.0. Thus, the conclusion that survival has the greatest absolute effect on population dynamics of sea ducks is robust across a wide range of life history parameter values, although the dominance (relative influence) of survival with regard to lambda declines as survival rate increases.

Next, I considered how robust this conclusion is when lambda differs from 1.0. Starting with the same matrix model, I first varied relative levels of productivity holding survival constant. Thus, as productivity varied, lambda also varied. I considered a range of lambda values from 0.9 to 1.1 (populations increasing or decreasing by 10% per year). Under all scenarios, sensitivity of survival remained dominant (Figure 3.7). Next, I repeated this process holding productivity constant and varying survival rate. Under this regime, there was little variation in sensitivity values (Figure 3.7).

Overall, the conclusion that sensitivity of survival is higher than sensitivity of fertility appears to be robust to realistic variation in input parameters and lambda values. In support of this, Sæther and Bakke (2000) also found that the sensitivity of survival tended to increase as survival rate increased across a range of species. Thus, species like sea ducks, with relatively high survival rates, will always have a high sensitivity of survival. Like elasticities, this result is related to the matrix model structure and, as such, is generally applicable to all sea duck species.

The concept of matrix sensitivities also can be extended to individual demographic parameters or *lower-level* parameters used to parameterize the matrix model. The pattern of lower-level sensitivities for our generic model also shows the overall dominance of the effect of survival on lambda (Figure 3.8). Lower-level sensitivities of individual productivity parameters were more variable with the greatest effects for nesting success, duckling survival, and juvenile survival; intermediate effects for clutch size, breeding survival, and breeding propensity of adults; and low effects for breeding propensity of second-year birds and renesting probability.

COMPARISON OF ELASTICITY AND SENSITIVITY VALUES

Elasticity and sensitivity values, while obviously related, provide somewhat different perspectives on various components of population dynamics. Both analyses indicate that changes (either relative or absolute) in survival would have a larger effect than changes in productivity on future rates of population change of sea ducks. The main difference between elasticities and sensitivities is related to issues of scale. For example, nesting success is

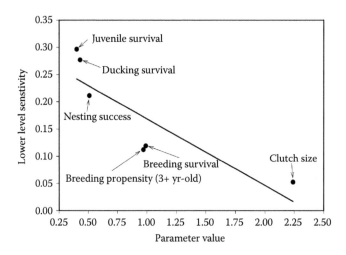

Figure 3.10. Negative relationship between lower-level sensitivity values and parameter estimates for a suite of parameters with approximately equivalent lower-level elasticity values.

constrained by definition to be between 0 and 1. Conversely, clutch size for sea ducks might vary from 4 to 12 eggs (e.g., Common Eiders to mergansers). Changing these two parameters by the same absolute amount (e.g., 0.5) represents a large proportional change in nesting success but a relatively minor change in clutch size. Thus, when comparing parameters measured on different scales (e.g., nesting success and clutch size), elasticity values may be more interpretable. However, as noted in the lower-level elasticity analyses, when parameters are multiplicative in the development of the fertility estimate, then they all have the same elasticity and, as such, elasticities are uninformative. In this case, parameters such as nesting success, duckling survival, and juvenile survival are all measured on the same scale and sensitivities can be more useful in a management context.

There is a general pattern within lower-level sensitivities that can be useful in a management context. That is, for groups of parameters with similar elasticity values, sensitivity values are negatively related to parameter values. That is, the parameters with the lowest estimated values will have the highest sensitivities (Figure 3.10). For example, all fertility parameters have essentially the same lower-level elasticity estimates. Therefore, if duckling survival is 10% and nesting success is 30% for a given population, then duckling survival will have a higher sensitivity than nesting success. Thus, it has been suggested that lower-level sensitivity analyses can be used to detect potential bottleneck parameters that may

be have a large effect on population dynamics simply because the estimates are low. An example was provided in recent models of Long-tailed Ducks by Schamber et al. (2009), King Eiders by Bentzen and Powell (2012), and Common Eiders by Wilson et al. (2012). In these cases, elasticities for most reproductive parameters were substantially lower than for survival; however, sensitivity of duckling survival approached the sensitivity of survival. This result occurred because estimates of duckling survival were low (i.e., <10%). When estimates like this are low, even small absolute changes in parameters are proportionally large (i.e., a 1% point absolute change in duckling survival would be 10% relative change). Accordingly, lower-level sensitivity values can be useful in discriminating among parameters with equivalent elasticity values and identifying aspects of species life history that may be potential bottlenecks. Such parameters with high sensitivities might be logical targets for management actions.

STOCHASTIC MATRIX POPULATION MODELS

The models presented so far have demonstrated dynamics of sea duck populations under deterministic conditions. While these models allowed for age-specific variation in survival and reproductive parameters, they considered population dynamics in the absence of environmental variation or stochasticity. Obviously, this is not realistic as all populations are subject to variation in life history parameters through both natural and anthropogenic influences. Importantly, Benton and Grant (1996) demonstrated that optimal life history strategies can vary between stochastic and deterministic projections. Thus, failure to consider stochasticity may lead to unrealistic models and hence incorrect conclusions. Further, the goal of most management actions is to have an effect on life history parameters, and variation in parameters may be indicative of management potential. Stochastic models can be used to demonstrate the dynamics of sea duck populations under realistic levels of variation in life history parameters.

In the context of population dynamics, it is important to consider the type or source of variance included in a stochastic model. For example, Schamber et al. (2009) presented nesting success estimates for Long-tailed Ducks that varied from <1% to 79% across sites and years. However, each of these individual point estimates of nesting success has some degree of uncertainty as indicated by their respective standard errors. As such, the variation among these individual points represents the total variation where total variation equals process (or true environmental variation) plus sampling variation (Gould and Nichols 1998). Assuming no sampling error, then process variation would equal total variation. Given an overall total variance from a series of estimates in combination with some measure of error for each individual point estimate, it is possible to decompose the total variance into process and sampling. The simple method of estimating the process variance is to average the sampling variance of each of the individual estimates, then subtract this average from the total variance yielding a naïve estimate of the process variance ignoring potential covariance (Gould and Nichols 1998, Morris and Doak 2002). There are other methods of estimating the process variance including Bayesian and random effects models, which may be more appropriate for specific applications (Burnham and White 2002, Morris and Doak 2002, White et al. 2008), but the key point is that the source of error must be considered when using stochastic models (Gould and Nichols 1998).

In the context of stochastic models, there are three potential types of variance that could be considered: total, sampling, and process. Which of these sources of variance should be used in a stochastic model depends on the goal of the modeling process. If the goal is to assess the true biological range of variation in population trajectories, then stochastic models should only include process variation. Thus, a stochastic lambda and its associated confidence interval based on process variance of input parameters would represent the expected trajectories of the actual population under realized environmental variation. In many ways, a model based solely on process variance represents the best description of a population under real conditions and as such probably is the most common model formulation. However, if the goal of the stochastic model involves statistical comparison of lambda to a fixed value or other estimates, then the model must include sampling variation and should utilize estimates of total variation. For example, if I wanted to assess if my modeled population was significantly different from a stable population where lambda is 1, I would need to model total variance to accurately represent the degree of statistical uncertainty in my parameter estimates. As such, models based on process variance represent the best estimate of what a population is really doing, whereas models based on total variance represent the realized ability to measure and predict population processes. Models based entirely on process variance while ignoring sampling variance should not attempt to infer statistical significance of the estimates. For the remainder of this chapter, I am interested in the true biological variation in sea duck populations and therefore restrict all stochastic models to inclusion of only process variance.

In applying estimates of process variation for my theoretical model, I used the relative pattern of variation found in field studies of Spectacled Eiders (B. Grand, Auburn University, unpubl. data). These overall patterns of variation are generally supported in the literature. Benton and Grant (1996) report that the coefficients of variation (CV) for fecundity rates were approximately twice as high as the CV for survival across a range of species. For parameters that were logically constrained between 0 and 1.0 (i.e., all rates), I used a beta distribution. For clutch size, I used a normal distribution. I selected values of the standard deviation that yielded logical ranges of parameter estimates given published literature for sea ducks (Table 3.1). Given these estimates of process variation for each of the life history parameters, I reran the model under stochastic conditions using Monte Carlo resampling methods (Caswell 2001). During each iteration of the model, I randomly selected from a distribution for each

TABLE 3.1

Input values used for stochastic trials of a generalized sea duck population model.

Input variable	Mean	Standard deviation	Lower CI	Upper CI
Breeding propensity 2 yr-old	0.40	0.05	0.30	0.50
Breeding propensity 3+ yr-old	0.97	0.05	0.83	1.00
Breeding season survival	0.99	0.01	0.96	0.99
Clutch size/2	2.24	0.01	0.91	3.59
Nesting success 2 yr-old	0.28	0.20	0.02	0.73
Nesting success 3+ yr-old	0.51	0.20	0.14	0.89
Duckling survival	0.42	0.20	0.09	0.81
Juvenile survival	0.40	0.10	0.21	0.61
Renesting probability 3+ yr-old	0.12	0.05	0.03	0.23
Adult female survival	0.85	0.05	0.74	0.93

parameter, using the same mean value as I used in the deterministic trial and various levels of process variation. Thus, each year of the simulation has a different lambda, but the overall average lambda is still 1.0. Under this scenario, it is obvious that there is a wide range of potential population trajectories. Some populations increased substantially, while others decreased. In looking at Figure 3.11, the average lambda is 1.0 with a normal error distribution, yet over 50-year projections, it is clear that there is an overall average negative trend. In these projections, the initial population size is multiplied by each year's simulated lambda in succession. Imagine a scenario where a population starts with 1,000 individuals. In year 1, lambda is 0.9 (i.e., a 10% decline in population size), yielding a year-1

population estimate of 900 individuals. Then in year 2, lambda is 1.10 (i.e., a 10% increase in population size) yielding a year 2 estimate of 990 individuals. In this simple 2-year example, the average lambda is 1.0 (i.e., [0.9 + 1.1]/2) yet the population declined slightly (i.e., 1%). This demonstrates one of the important properties of population growth in a stochastic environment; that is, realized population growth in a stochastic environment is less than the average lambda (Caswell 2001). The degree of this bias is related to the magnitude of the variance. That is, the greater the stochastic variation, the greater the difference between deterministic growth potential and realized stochastic growth. Thus, all else being equal, more consistent environmental conditions that lead to less process variation in life

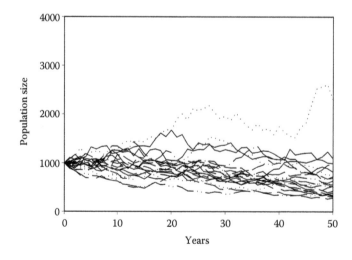

Figure 3.11. Population projections under stochastic conditions for a population with a mean matrix projection of stability ($\lambda = 1.0$). The average stochastic projection is always less than the mean lambda.

history parameters will result in less total variation in population trajectories and somewhat higher average rates of population growth.

Even though average lambda was 1.0, under stochastic conditions, more than half of modeled populations declined. Further, even for simulated populations that ultimately grew and were at higher levels at the end than at the start of the simulation, most showed periods when they declined for variable amounts of time. This result brings up an important consideration regarding measurement of life history parameters for use in population projection. When life history parameters are measured at only a single (or a few) location(s), over relatively short periods of time, there is a reasonable chance that the population was sampled during a nonrepresentative period of time. For example, Chaulk et al. (2004) found that clutch size of Common Eiders varied among years and across archipelagos, and Coulson (1999) showed that clutch size declined through time at a single site. Similarly, Pearce et al. (2005) reported that annual survival rate of Common Mergansers varied among banding locations. Hence, extrapolations of population dynamics based on parameter estimates from one or a few years or sites will not be representative of future overall population performance (Doak et al. 2005). Thus, long-term, multilocation studies are needed, ideally replicated across a population's range, before confidence can be placed in projected population trajectories.

The key difference in interpreting deterministic and stochastic population modeling results is functionally the same as considering the mean of a parameter and its variance. Deterministic population models do a good job of describing basic population dynamics in terms of lambda, matrix level estimates of sensitivities and elasticities, and lower-level sensitivities and elasticities. Inclusion of stochastic variation has little effect on estimates of sensitivity and elasticity (Caswell 2001). Further, deterministic models allow a clear understanding of the potential response of populations to management actions. For example, using a deterministic model, it is easy to see the potential effect of altering life history parameters through management actions. However, stochastic models provide a much more realistic view of the probability of actually realizing a predicted effect, given natural variation that exists. That is, even when a deterministic model predicts population stability, many of the predicted trajectories from a stochastic model show declines. Similarly, when deterministic models show positive effects of potential management actions, stochastic models can help to understand the probability of actually observing the desired effect. For example, managers may increase nesting success, but there is still some chance that a population may decline at a higher rate. Thus, the deterministic model can demonstrate the potential effect of management actions, but the stochastic model projections demonstrate the probability of realizing that effect.

CORRELATED VARIATION IN LIFE HISTORY PARAMETERS

In the previous examples of stochastic models, I included random variation in life history parameters. In these examples, all life history parameters were assumed to be independent of one another. However, in reality, there are likely various levels of correlation among different subsets of life history parameters (Benton and Grant 1996). In this section, I consider positive and negative correlations in life history parameter values both within and among years and the influences these may have on sea duck population dynamics. The key point is that failure to consider potential correlations among life history parameters can misrepresent conclusions from elasticity and sensitivity analyses (van Tienderen 2000, Iles 2012). For example, in the case of positive correlations, the response of the population will be underestimated by modeling projections. Conversely, if negative correlations exist among life history characteristics, population response may be overestimated.

Positive Correlations

Positive correlations in life history parameters are likely fairly common for sea ducks. They can result from annual variation in habitat conditions or annual changes in potential predator populations. Flint et al. (2006b) demonstrated a positive correlation between duckling survival and duckling growth rate of Spectacled Eiders and hypothesized that this correlation was caused by annual variation in habitat conditions. Thus, when conditions were favorable, ducklings both survived at a higher rate and grew more quickly. If high growth rates of ducklings lead to higher first-year survival, as has been found in Common Eiders (Christensen 1999), then habitat conditions would result in a positive correlation between duckling survival and first-year survival. Erikstad and Tveraa (1995) found a weak positive relationship between clutch size and nesting success of Common Eiders, likely resulting from positive relationships between body condition and both parameters. Of course, the converse also appears to be true; an apparent overwinter food shortage resulted in reductions in breeding propensity and clutch size of Common Eiders in the following year (Oosterhuis and Van Dijk 2002). Yoccoz et al. (2002) reported a positive correlation between clutch size and annual survival for Common Eiders. Positive correlations among life history parameters have two primary effects on population dynamics. First, positive correlations will tend to increase the range of stochastic variation among population projections. Second, positive correlations will result in underestimation of the realized effects of perturbations of individual parameters (Iles 2012). For example, I previously discussed lower-level sensitivities and elasticities. However, these values are calculated by examining change in lambda based on a change in a given parameter while holding all other parameters constant (van Tienderen 2000). If correlated parameters exist, then simple sensitivities and elasticities will underestimate the effect of given changes in a single life history component.

Positive Correlations in Parameters within Years

Within-year positive correlations among life history parameters tend to magnify the level of variation in lambda. This pattern creates the boom–bust scenario of productivity (Hario and Rintala 2006). Good years result in high levels of productivity and recruitment, whereas poor years result in little or no productivity and recruitment. I simulated stochastic population growth with a mean lambda of 1.0, but correlated the variation in life history parameters. I simulated estimates of nesting success of 2-year-old and 3+-year-old females that covaried positively and estimates of duckling survival that were positively correlated with nesting success. Thus, I simulated a situation where potential predators of both eggs and ducklings varied through time. Under this scenario, variation in population trajectories increased dramatically (Figure 3.12). The degree to which positive correlations influence variation in lambda depends on the degree of variation in life history parameters, the correlation in life history parameters, and the elasticities of the correlated parameters. As noted previously, larger fluctuations in life history parameters will have greater effects on lambda. Stronger correlations among parameters will tend to increase this effect (Iles 2012). Last, strong correlations among parameters with high elasticities will have a far greater effect compared to correlations among parameters with low elasticities. Accordingly, positive correlations

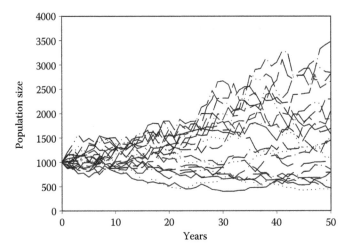

Figure 3.12. Population projections under stochastic conditions for a population with a mean matrix projection of stability ($\lambda = 1.0$) and positively correlated fertility parameters. The positive correlation increases the mean projection as well as the variation among trajectories (compare to Figure 3.11).

between annual survival and reproductive parameters will have the greatest influence on variation in sea duck population variation.

Positive Correlations in Parameters among Years

Among-year positive correlations in life history parameters can occur for a variety of reasons. For sea ducks, a common cause is related to population dynamics of potential predator species. In many areas, arctic or red foxes are the dominant predators of sea duck nests. However, fox populations at high latitudes tend to fluctuate with microtine abundance (Roth 2003, Samelius 2006). Similarly, Iles (2012) found a negative trend in Common Eider nesting success as availability of snow geese as alternative prey declined over time. Thus, when fox populations are high and alternative prey are low, sea ducks may suffer a series of years with low productivity. Conversely, after fox populations decline due to either food limitation or epizootics, such as rabies, sea ducks may experience a series of years with relatively high productivity.

Further, broad-scale habitat conditions may vary in somewhat predictable patterns. Examples include the Pacific Decadal Oscillation, the North Atlantic Oscillation, and the Arctic Oscillation, which tend to alter broad scale weather patterns over periods of years to decades. Such oscillations may have consistent effects on survival and recruitment of sea duck populations through influences on both winter and breeding habitat conditions (Lehikoinen et al. 2006, D'Alba et al. 2010, Descamps et al. 2010, Mehlum 2012). Such broad-scale environmental perturbations would tend to create inherent correlations among life history parameters over long time scales. Positive among-year correlations in life history parameters have the effect of creating short-term trends that deviate from long-term projections. Thus, a truly stable population will show multiyear periods of consistent increase and decrease.

Negative Correlations

Negative parameter correlations occur when increases in one or more parameters are correlated with decreases in others. For example, Pehrsson and Nyström (1988) found a negative correlation between nesting success and duckling survival for Long-tailed Ducks. They concluded that increasing nest success did not result in increased recruitment because duckling abundance was limited by prey availability in brood-rearing habitat. Similarly, Hario and Rintala (2006) found a negative relationship between population size and duckling survival in Common Eiders in Finland. In other cases, negative correlations are associated with life history trade-offs. A classic example is the clutch size–egg size trade-off. When resources available for egg production are limiting, females laying small eggs would tend to lay a larger clutch than females laying large eggs (Rohwer 1988). If egg size confers a survival advantage to ducklings, this pattern represents a negative correlation in life history parameters.

Negative Correlations in Parameters within Years

These examples of density dependence and life history trade-offs represent negative correlations of parameters within years. The general effect of within-year negative correlations of parameters is a reduction in annual variation in population trends, in other words, a reduction in stochastic variation in lambda. Of course, realized effects of potential negative correlations depend on the degree of correlation and the sensitivities of the parameters. Negative correlations in parameters also will result in over-estimation of sensitivities and elasticities. Thus, realized effects of a single parameter manipulation will be less than anticipated because they will be somewhat canceled by negative response of other parameters.

Negative Correlations in Parameters among Years

Alternatively, there can be negative correlations in life history parameter values among years. One example of this would be the trade-off between current reproductive effort and future reproductive performance. Yoccoz et al. (2002) found that breeding propensity was lower in future years for female Common Eiders laying large clutches. Coulson (1984) suggested that Common Eiders were likely to skip a year of breeding following their first reproductive attempt. Descamps et al. (2009) found lower survival for females laying large clutches when avian cholera was present. Similarly, Common Goldeneyes that successfully fledged young tended to have a slightly smaller clutch size and delayed date of nest initiation in the subsequent year (Milonoff et al. 2004). Cooke et al. (1995) found that, like sea ducks, some snow geese first breed at 2 years of age, while others delay to >3 years of age. Further, the birds that attempted to breed, as 2 year olds, had a lower likelihood of breeding at 3 years of age (Viallefont et al. 1995) and a similar pattern has been shown in other birds (Weimerskirch 1990, Wooler et al. 1990, Barbraud and Weimerskirch 2005). Thus, attempted breeding in year i resulted in reduced breeding propensity in year i + 1. This pattern would create an overall negative relationship among a single productivity parameter among years (breeding propensity). A negative correlation can even occur in time lags >1 year. In developing this model, I assumed that young females breeding for the first time at age 2 would tend to have lower nesting success than adult birds. Under a stochastic situation of high annual variation in productivity, we would expect high annual variation in population age structure (Flint et al. 2006b). Specifically, two years following a season of high productivity, we would expect a high proportion of 2-year-old females to attempt to breed (Flint et al. 2006a). If these birds tend to have lower overall success, then we would expect a drop in nesting success two years after a pulse of high productivity. Time lags would tend to create cycles in the overall population trends where increases in population size lead to decreases in the average fertility rate.

Negative correlations among parameters reduce the degree of variation in lambda and associated variation in population projections (Wisdom et al. 2000). Reduced variation is the opposite effect of positive correlations among parameters. Obviously, the overall effect of negative correlations among parameters depends on both the strength of the correlations and the elasticities of correlated parameters.

IMPORTANT ISSUES RELATIVE TO CORRELATED PARAMETERS

The combination of stochastic population models and the potential for correlated parameters emphasizes the importance of input parameters being used in modeling efforts. Studies must be of sufficient duration to measure temporal variation and ideally replicated across space such that true process variation in parameters can be estimated (Hario and Rintala 2006, Coulson 2010). Rarely do individual studies record or study all relevant life history parameters simultaneously. In many cases, some parameters are assumed to have certain values, estimates are obtained from other species, or estimates are obtained from other populations or perhaps different periods of study. In these cases, there is no opportunity to estimate levels of correlation among parameters. Limited available data from the few studies that have examined correlated parameters suggest both positive and negative correlations are likely for sea duck populations. Thus, even when stochastic models are used, failure to consider correlated parameters may yield less than ideal results regarding variation in rates of population growth. Generally, positive correlations will increase stochastic variation in lambda and create multiyear patterns or cycles in population trajectories. Conversely, negative correlations will decrease stochastic variation in lambda and create short-term or annual oscillations in sea duck population trajectories.

RETROSPECTIVE POPULATION MODELING

Thus far, all of the modeling results presented would be classified as prospective projections. That is, I considered the potential dynamics of populations in the future. However, it can also be informative to consider the history of population dynamics using a retrospective approach. Retrospective analyses consider how populations have varied in the past and how these fluctuations in population size varied in relation to life history parameters. The advantage of retrospective analyses is that given the variation has already occurred, we can be more definitive in assigning sources of variation. Direct retrospective analyses of population dynamics are possible when detailed data on populations are recorded for a series of years (Jónsson et al. 2013). For example, if population size is estimated on an annual basis concurrent with estimates of life history parameters, then a correlation analysis can determine relationships between annual variation in life history parameters and subsequent variation in population size. Coulson (2010) examined a 48-year data set that included estimates of population size and life history parameters of Common Eiders and concluded that population size was highly responsive to small changes in adult survival and that long-term trends in abundance were related to variation in adult survival. Coulson (2010) also concluded that substantial variation in recruitment was responsible for short-term fluctuations in numbers, but had no appreciable effect on long-term trends in abundance. Conversely, Hario et al. (2009) found no relationship between survival and long-term population trends and concluded that reductions in recruitment best explained population declines. Unfortunately, such long-term data sets are rare for sea ducks, limiting the use of retrospective analyses of population dynamics.

Retrospective variation in population dynamics was modeled using the same parameter data used in the prospective stochastic population models of a generic sea duck. I used the same estimates of life history parameters and their associated process variance and matrix model estimates of sensitivities and elasticities. Given these data, historic variation in lambda resulting from any given parameter can be estimated as either the (elasticity^2) × (coefficient of process variance^2) where the coefficient was calculated as the process variance divided by the (parameter mean) or the (sensitivity^2) × (process variance, Coulson et al. 2005). Both metrics yield equivalent estimates of the relative contribution of each parameter to historic variation in lambda. The logic is that for a parameter to have had a major influence on historic variation in lambda, that parameter had to both vary and have a relatively high elasticity/sensitivity. Thus, parameters with high elasticities but little process variation will not have much effect on variation in lambda (e.g., adult survival in K-selected species).

Much like sensitivity and elasticity analyses, the concepts of variance decomposition can be applied at different levels. When we apply this process to the Leslie matrix cells, it becomes clear that fertility of adult females is the dominant

factor resulting in variation in lambda (Table 3.2). Summing fertility and survival values makes it clear that annual variation in productivity has a far greater effect on annual variation of lambda compared to survival. Proportionately, variation in fertility accounts for 86% of the annual variation in lambda. Thus, in retrospective analyses, variation in fertility has had a far greater effect on variation in lambda than variation in survival.

Variance decomposition also can be considered with regard to lower-level parameters of fertility. Because lower-level estimates of sensitivity and elasticity do not sum to a fixed value and only have relevance to one another in a relative sense, the variance decomposition approach is only useful to examine relative contributions of the fertility parameters. Using lower-level sensitivity values and individual parameter variances, I estimated the proportional variance decomposition of the fertility parameters (Table 3.3).

The interpretation of these results is functionally nested. That is, the matrix level analyses suggest that most of the annual fluctuation in lambda results from variation in fertility. The lower-level analysis suggests that most of the variation in fertility results from variation in just a few parameters. Duckling survival accounted for the highest proportion of this variation, followed by adult female nesting success, clutch size, and juvenile survival. If there are inherent positive correlations among some of these fertility parameters, then the lower-level sensitivities are underestimated and the associated contribution to the variance decomposition is also underestimated.

The interesting pattern here is that prospective population projection identified survival as having the greatest effect on future population dynamics. Conversely, the retrospective variance decomposition analyses suggest that historic variation in productivity determines most of the annual variation in lambda. Thus, it

TABLE 3.2

Leslie matrix variance decomposition values.

			Totals	%
0	0.000399	0.007253	0.007652	0.859502
0.0000422	0	0		
0	0.000042	0.001167	0.001251	0.140498
			0.008903	

TABLE 3.3
Variance decomposition of lower-level fertility parameters.

Parameter	Lower-level sensitivity	Process variance	Variance decomposition	Proportion of total variance (%)
Breeding propensity 2 yr-old	0.008	0.003	1.82×10^{-7}	<1
Breeding propensity 3+ yr-old	0.111	0.002	2.71×10^{-5}	<1
Breeding season survival	0.118	<0.001	1.27×10^{-6}	<1
Clutch size/2	0.052	0.470	0.001	19
Nesting success 2 yr-old	0.012	0.039	5.5×10^{-6}	<1
Nesting success 3+ yr-old	0.208	0.039	0.002	25
Duckling survival	0.276	0.039	0.003	43
Juvenile survival	0.292	0.010	0.001	12
Renesting probability 3+ yr-old	0.054	0.003	7.53×10^{-6}	<1

would seem that these two analyses are reaching opposite conclusions. However, these two analyses are designed to measure different effects (Caswell 2000). The fact that they identify different life history components as being the most important is not unusual (Ehrlén and van Groenendael 1998, Coulson et al. 2005, Iles 2012). Pfister (1998) examined results from population models across a range of species and found a negative relationship between parameter process variance and elasticity values. Thus, individual parameters with high elasticity values, such as adult female survival in our sea duck model, tend to have low levels of process variation. The converse is also true of parameters with low elasticity values, which tend to have higher levels of process variation. Pfister (1998) argued that natural selection would tend to remove variation in traits with high elasticity values. That is, at the individual parameter level, elasticity is the functional equivalent of fitness (Grant and Benton 2000). Thus, selection will act to remove heritable variation in fitness parameters. Survival is obviously related to fitness as a bird must be alive to reproduce and likely has some level of heritable variation. Conversely, productivity is the combination of numerous factors, many of which are largely influenced by extrinsic, environmental factors (de Kroon et al. 2000). Thus, most of the process variation in productivity is likely environmental variation and as such is not as responsive to selection pressure. For K-selected sea ducks, prospective analyses will likely always identify adult female survival as having the highest

sensitivity/elasticity, with most of the historic variation in population trajectories occurring as a result of variation in productivity.

USING MODELS TO EXAMINE POTENTIAL SEA DUCK MANAGEMENT SCENARIOS

There are several examples where models have been developed to examine specific management issues for sea duck populations. Gilliland et al. (2009) developed a model to examine sustainability of commercial harvest for Common Eiders. Bentzen and Powell (2012) modeled King Eider population dynamics in relation to a hypothetical oil spill. Iverson and Esler (2010) modeled Harlequin Duck population recovery following the Exxon Valdez oil spill. Krementz et al. (1997) used a simple model to consider harvest of White-winged Scoters in relation to fall age ratios. Robertson (2008) adapted basic modeling approaches to utilize winter age ratio data as input values for fecundity. Last, for breeding sea ducks, Wilson et al. (2012) and Iles (2012) modeled Common Eiders, and Schamber et al. (2009) modeled dynamics of Long-tailed Ducks to identify potential population limiting parameters and explain long-term population trends. Thus, to a limited extent, population models, with varying levels of assumptions and caveats, have been developed to aid in the management of sea duck species. In general, the results from these specific models fit with my results for a generic sea duck.

Thus far, I have demonstrated the basic pattern of population dynamics of sea ducks. Now, I explore potential management scenarios and demonstrate potential effects on sea duck population dynamics.

Sensitivities and elasticities are built upon average parameters, and thus they clearly describe the effects of altering mean life history parameters. For example, Stien et al. (2010) removed predators and raised nest success of Common Eiders by 30% at one site; given our elasticities, this might equate to a 3.4% increase in lambda (0.30 × 0.11). In our deterministic model, elasticity of survival was 0.88. Thus, increasing mean survival by 1% (i.e., mean survival × 1.01) would yield a 0.88% change in lambda. It is important to realize that the viability of management actions to change survival rates is likely asymmetrical (de Kroon et al. 2000). That is, there may be few or no options for increasing survival, but there are almost certainly multiple options for decreasing survival such as by increasing harvest levels. Accordingly, interpretation of elasticities for management is somewhat dependent upon the overall management goal.

I also demonstrated that variation in life history parameters leads to variation in lambda. All else being equal, populations with higher variation in lambda have lower levels of population growth or higher levels of population decline. Thus, it follows that ideal management actions would alter both the mean and the variance in life history parameters simultaneously. If the goal is to increase lambda, then actions that simultaneously reduce parameter variation and increase mean parameter values will have a double benefit on the effect. In fact, management actions targeted solely on reducing parameter variation could have exactly this effect. Imagine a scenario where nesting success varies over a wide range of values, centered on some mean value (Figure 3.13). A management action designed to reduce the frequency of years with low nesting success would have the simultaneous effect of reducing the variance and increasing the mean. This management action might be habitat manipulation to avoid extreme conditions or a directed predator removal program that only removes predators when numbers are high—assuming high predator density leads to low nesting success. The goal of such a program would be to reduce the frequency of years with low productivity as observed by Reed et al. (2007) for Common Eiders. Here, I demonstrate the effect of a management action that eliminates the occurrence of low values of nesting success. I start with a random normal distribution of logit transformed estimates. I then progressively eliminate low estimates of nesting success from this distribution. The simultaneous effect of such a hypothetical management

action is an increase in the mean parameter estimate and a decrease in the variance (Figure 3.13). This simulation demonstrates the potential utility of managing variation in a directional fashion. The potential increase in mean parameter value and the reduction in process variation would both have positive effects on lambda.

NONLINEAR RESPONSES OF POPULATIONS TO MANAGEMENT ACTIONS

Sensitivities and elasticities are calculated for a certain set of conditions, a given balance between survival and productivity. Sensitivities and elasticities are estimated by making small perturbations to the matrix and measuring the response. Small perturbations are used because the larger the perturbation, the greater the change in lambda. The response of the population is frequently not linear across different values of lambda (de Kroon et al. 2000). Thus, the greater the potential change in lambda initiated by a management action, the less accurate the prediction of the response (Mills et al. 1999). Given basic sea duck population dynamics, substantial changes in lambda can occur by changes in survival with little effect on matrix level sensitivities or elasticities. However, comparable changes in lambda resulting from changes in fertility may alter values of sensitivities and elasticities, while the overall patterns or ranking remain consistent (Ehrlén et al. 2001; Figure 3.7). Accordingly, effectiveness of a given management action may increase or decrease as the population begins to respond to that management action. Last, the actual values for sensitivities and elasticities are dependent upon both input parameter values and estimated lambda. As such, sensitivities and elasticities should not be thought of as values appropriate for a given species or population, but as a set of values appropriate for a given population at a particular point in time. While the overall ranking of parameter elasticities will remain consistent, the actual elasticity values can vary substantially for a population that is increasing compared to a population that is decreasing (de Kroon et al. 2000).

MODELING THE EXTREMES OF LIFE HISTORY OF SEA DUCKS

So far, I have based my discussion on the model developed for a hypothetical sea duck life history with average demographic rates. Now, I develop two models

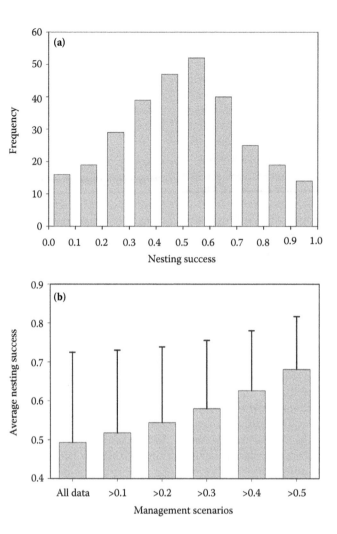

Figure 3.13. (a) A hypothetical distribution of nest success values. (b) The mean estimate of nesting success and the associated variance under a range of management scenarios. The six scenarios include no management action (noted as *All data*) and a range of cases where the frequency of years with low levels of nesting success is eliminated. That is, the distribution in (a) is truncated by eliminating progressive proportions of the lower tail. An example of this type of action would be removal of nest predators in years when predator populations are particularly abundant, thereby reducing the occurrence of years of low nesting success. Such management actions have the effect of both increasing the mean parameter estimate and reducing process variation.

to demonstrate the extremes of the high survival–low productivity and low survival–high productivity continuum for sea ducks, using Common Eiders and Common Mergansers as example species.

Common Eider Model

The population model for Common Eiders demonstrates the dynamics of a species that is long lived with low reproductive output. I assumed that only a small proportion of females attempted to breed at age 2 (i.e., 20%), with 42% of females attempting to breed at age 3 and 97% of females >3 attempt

to breed each year. I used a clutch size of four eggs and adult female survival of 90% (Coulson 1984, Wilson et al. 2007). Nesting success, duckling survival, and juvenile survival were the same as used in our general model. These parameters result in a population with a lambda of 1.0 with elasticities of survival at 0.918 and fecundity at 0.082.

Common Merganser Model

The population model for Common Mergansers demonstrates the dynamics of a sea duck with relatively low survival and high productivity.

I assumed higher rates of breeding propensity for young birds with 40% of females attempting to breed at age 2 and 98% of females 3+ years old attempt to breed each year. I set clutch size at 11 eggs and assumed that cavity nests would have much higher levels of nesting success of 60% for 2-year-old and 80% for 3+-year-old females (Evans et al. 2002). Adult female survival was set at 60% (Pearce et al. 2005). These parameters result in a population with a lambda of 1.0 with elasticities of survival at 0.755 and fecundity at 0.245.

COMPARISON AMONG THESE MODELS

Common Eiders and Common Mergansers likely represent the two extremes of life history strategy within sea ducks, mergansers being a high-reproductive output species and eiders being a high-survival species. In spite of these large differences in life history inputs, the overall pattern of lower-level sensitivities and elasticities is similar (Figure 3.14).

In comparing models for the two species, it appears that there is an inherent trade-off between survival and productivity. Thus, I explore potential changes in productivity and ask what comparable change in survival would have to occur to offset these changes. Starting with Common Eiders, I can ask the question "Why don't eiders lay larger clutches?" To answer this question using the models, I compare the relative effect of a one-egg change in clutch size to the corresponding change in survival. For eiders, a one-egg increase in clutch size is a 25% increase given an average four-egg clutch. The elasticity of clutch size is 0.082 making the relative effect of a one-egg increase in clutch size a 2.1% change in lambda. The trade-off with survival then asks the question "How much of a reduction in survival would offset a 2.1% increase in lambda resulting from the change in clutch size?" The elasticity of survival was 0.918, thus $0.021/0.918 = 2.3\%$ reduction in survival. Thus, survival becomes 0.88, calculated as $0.90 - (0.9 \times 0.023)$. Given that Common Eiders rely on stored reserves for egg production, attempting to lay additional eggs may place them below a nutritional threshold such that they suffer lower survival than predicted in this hypothetical situation. Accordingly, eiders laying an additional egg (all else being equal) may have annual survival rates <0.88 and therefore would be less fit.

Alternatively, I can ask: "Why don't Common Mergansers lay fewer eggs?" A reduction in clutch size for Common Mergansers of one egg is a 9.1% reduction in clutch size. Elasticity of clutch size is 0.249 making the relative effect of a one-egg reduction in clutch size equate to a 2.3% reduction in lambda. The elasticity of survival was 0.755, thus $0.023/0.755 = 3.1\%$ increase in survival. In the case of mergansers, survival would have to increase to 0.62 to offset the reduction in clutch size. Given that mergansers do not rely on stored reserves to produce eggs, it is unlikely that a one-egg reduction in clutch size would result in a 2%

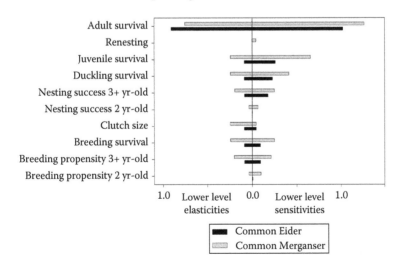

Figure 3.14. Lower-level sensitivities and elasticities from Common Eider and Common Merganser models. Both models were constrained to $\lambda = 1.0$. In spite of differences in model structure (i.e., age at first breeding) and substantial differences in life history parameters, the overall patterns of sensitivities and elasticities are similar.

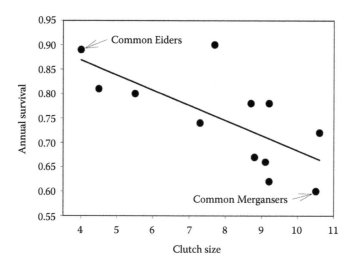

Figure 3.15. Negative relationship between adult female survival and productivity (as indexed by clutch size) for 12 of the 15 sea duck species. Three species were not plotted as estimates of survival were not available. $R^2 = 0.49$, $P = 0.01$.

point increase in survival. Thus, reducing clutch size for mergansers would likely decrease fitness.

The apparent trade-off between survival and productivity among species, as demonstrated by this example with two species, may exist within sea ducks in general. In fact, looking across the sea duck species, there is an overall negative relationship between rates of annual survival and clutch size as an index of productivity (Figure 3.15). Considering just survival and productivity, I can hypothesize the shape of the overall fitness gradient that must exist. A species having high survival and high productivity would clearly have the highest fitness. At lower levels of survival and productivity, there is a wide range of survival and productivity pairs that would yield similar levels of fitness (Figure 3.16). Of course, species with both low survival and low productivity would likely go extinct quickly. Interestingly, the patterns of variation observed among species in survival and productivity, as predicted by the regression equation, are similar to the patterns of elasticity predicted by the generic model. The relative range of variation in survival between the eider and merganser models is 30%, and the range of variation in clutch size is 262%. The ratio of these ranges (i.e., 8.7) is similar to the ratio of the elasticities of survival and productivity (~7.5). Accordingly, the trade-off between survival and productivity observed among species is close to what would be predicted based on relative effects, and as such, all populations would be predicted to

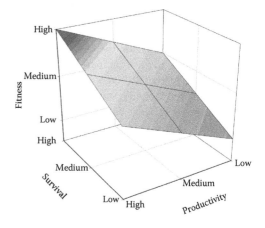

Figure 3.16. Hypothetical fitness gradient for sea ducks. The highest fitness could be achieved with both high survival and high productivity. A broad array of combinations of survival and productivity could be used to achieve medium levels of fitness.

have a similar lambda, all else being equal. Thus, the hypothesized fitness plane has all species with a similar overall fitness that is realized by different combinations of survival and productivity (Figure 3.17).

For any given species, it is clear that an increase in fitness could be obtained with an increase in either survival or productivity. In other words, the best possible scenario for a sea duck would be to have the survival of a Common Eider and the productivity of a merganser. Why doesn't such a sea duck species exist? Likely because some environmental factor is

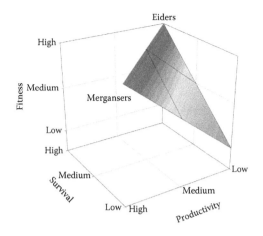

Figure 3.17. Apparent realized fitness gradient for sea ducks. There are no sea duck species that show both high survival and high productivity implying that such a combination is functionally unachievable. Sea duck species are distributed along a gradient of survival and productivity values (as indexed by clutch size) with eiders and mergansers at the extremes.

always limiting and the limiting factor likely varies among species. For eiders, productivity appears to be limiting. They simply cannot carry sufficient reserves to produce more offspring. Because forage is not readily available on breeding grounds during the period of nest initiation, they cannot increase their investment in the clutch. Given this constraint, selection pressure would appear to have maximized survival with more lifetime chances to reproduce, rather than high investment in any given year. For mergansers, the opposite appears to be true—survival may be limiting. Given that survival is controlled by extrinsic factors, selection appears to have modified the only parameters with remaining heritable variation, those related to productivity. The effect of selection is to remove heritable genetic variation from a population. Thus, one can assess the degree of variation in a given parameter, or suite of parameters, in order to draw inference relative to previous selection pressures. Total among female process variance in a given parameter is the sum of genetic and environmental variation. Separating genetic from environmental variation requires controlled studies examining mother–daughter correlations. However, total phenotypic variation sets the upper limit to potential heritable variation, such that when environmental variation is zero, all phenotypic variation is genetic. Thus, it is reasonable to assume that parameters with relatively low phenotypic variation likely have relatively low genetic

variation as well. Certainly, this would have to be true when variance differs by orders of magnitude.

Given the variance components described earlier, I predict that there is a positive relationship between process variance for a given parameter and the degree of heritable genetic variation in that parameter. Pfister (1998) and Sæther and Bakke (2000) found negative relationships between parameter variance and elasticity values. Morris and Doak (2004) demonstrated that the techniques used by Pfister (1998) to estimate the relationship between variance and elasticity were biased, but that correcting the bias only makes the negative relationships stronger. Combining my prediction regarding process variation and heritable genetic variation with these results would result in a negative relationship between the amount of heritable genetic variation and the elasticity of parameters. Thus, parameters with high elasticity values would tend to have low levels of heritable variation. If this hypothesis were true, then elasticity values would be positively correlated with the intensity of historic selection pressure on life history components (Benton and Grant 1996, Pfister 1998, Grant and Benton 2000). In other words, parameters with high elasticity values would tend to have little heritable variation because high selection pressure would be expected to remove any variation that may have existed. Accordingly, I would expect more genetic variation in survival of mergansers than Common Eiders. Conversely, I would expect more genetic variation in productivity of eiders than in mergansers. As such, elasticity values may provide some clues regarding relative selection pressures on life history traits and their subsequent evolution.

DENSITY-DEPENDENCE

All of the models considered thus far have been density independent. In these density-independent models, the life history parameter values are independent of population size. The degree to which sea duck populations behave in a density-dependent fashion is unknown. Dabbling and diving ducks nesting in the prairie pothole region of North America have been shown to behave in a density-dependent fashion at a broad scale (Murray et al. 2010). That is, rates of population growth decline as population size increases. However, it is unclear when or how this density-dependent mechanism might work (Murray et al.

2010). The true causes of density dependence could occur during breeding, molting, staging, or winter. For example, cavity nesting species have frequently been thought to be limited by breeding sites (Pöysä and Pöysä 2002). Coulson (2010) showed a weak negative relationship between survival and population size that may be indicative of density-dependence in local breeding populations of Common Eiders. Further, Coulson (1984) showed that both clutch size and breeding propensity decreased during a period when the breeding population was increasing. Hario and Rintala (2006) demonstrated that ducking survival was negatively correlated with population size and concluded that this was evidence of density dependence operating during the brood-rearing period. Rönkä et al. (2011) showed positive relationships between productivity and subsequent population size for Common Eiders and Common Mergansers, but no such relationship for Common Goldeneyes, implying that of these three species, only goldeneyes may have been constrained by density dependence. Hartman et al. (2013) showed evidence for time-lagged density dependence in Velvet Scoters. Thus, there is evidence for density dependence in sea ducks, but the strength of the effect varies as well as the affected life history parameter.

Given that most sea duck populations have declined (Flint 2013; Chapter 1, this volume), it might be assumed that populations are below carrying capacity and thus any effects of density dependence should be reduced through time. If so, the density-independent models developed in this chapter would reasonably apply. In this scenario, population declines may be caused by a density-independent source such as predation effects on survival or productivity. Conversely, sea duck populations may have declined because carrying capacity has declined. Flint (2013) found that continental scale changes in population size and trend of sea ducks in North America were correlated with timing of oceanic regime shifts in the North Pacific. Flint (2013) hypothesized that regime shifts represent rapid changes in functional carrying capacities of marine habitats that ultimately influence both survival and productivity of sea duck species. As such, survival and/or productivity are reduced as a direct result of population density, and this process will continue until a new carrying capacity threshold is established. For example, Coulson (2013) suggested that a

long-term decline in survival of Common Eiders resulted from a deterioration in the local environment. Therefore, sea duck populations may have declined because carrying capacity declined, and, as such, even declining populations may still be experiencing density-dependent effects.

Density-dependent models are relatively simple adaptations of density-independent models where survival and/or productivity are scaled relative to population size. The problem is such scaling requires some estimate of carrying capacity that is difficult if not impossible to estimate for sea ducks, given our current lack of knowledge of factors that set carrying capacity. Further, scaling becomes even more problematic if carrying capacity is changing through time. In density-dependent models, sensitivities and elasticities are only relevant for a given population size. As such, the relative effect of a given perturbation is not a single number, but a relationship with population size. Density dependence tends to reduce the variation among population projections resulting from stochastic models (Hartman et al. 2013). This result occurs because large populations have a low likelihood of further large increases, whereas small populations tend to increase quickly. Thus, density dependence reduces variation relative to a density-independent population and forces populations toward an equilibrium size. Grant and Benton (2000) demonstrated that sensitivities and elasticities estimated from density-independent models are reasonably valid for density-dependent populations as long as these populations are not at carrying capacity. Once a population reaches its carrying capacity equilibrium size, further increases in population size cannot be maintained by perturbation of life history parameters and can only occur by increasing carrying capacity. For populations at equilibrium size, increases in life history parameters are either very short term or immediately counteracted by negative correlations with other parameters (Grant and Benton 2000). The potential effects of negative changes in life history parameters for populations at carrying capacity may be approximated by sensitivities and elasticities, but such changes are frequently associated with correlated responses in other parameters that minimize the rate of population decline.

Last, density dependence in sea duck populations may be partially confounded with stochastic environmental effects on life history parameters (Rönkä et al. 2005, Hartman et al. 2013). In other words, sea ducks may experience periodic density dependence as a function of stochastic variation in carrying capacity. For example, sea ducks commonly winter in areas with seasonal ice cover, and the period and extent of ice coverage vary among years (Robertson and Gilchrist 1998, Petersen and Douglas 2004). In these wintering areas, sea ducks may consume a considerable proportion of benthic prey (Lewis et al. 2007, Bilcher et al. 2011). Under the assumption that in most years ice cover is not limiting access to forage, then in good years, there is no density dependence. However, in years of severe ice conditions, food may be limiting and survival may decline (Robertson and Gilchrist 1998, Camphuysen et al. 2002) or productivity may be reduced (Mehlum 1991, 2012; Oosterhuis and Van Dijk 2002; Lehikoinen et al. 2006; D'Alba et al. 2010; Chaulk and Mahoney 2012). Under this scenario, the magnitude of the negative effect that occurs in these severe conditions, such as mortality or reduction in breeding effort, would likely be correlated with population size. Positive correlations between mortality and population size, or negative correlations between productivity and population size, are the definitions of density dependence. Thus, sea duck populations may periodically experience density dependence because they live in such highly dynamic environments. Unfortunately, detection of these periodic density-dependent events is difficult, and these effects are likely labeled as stochastic process variation in life history parameters. In fact, the high interannual variation in survival and nonbreeding observed by Coulson (1984, 2013) may be a periodic density-dependent effect resulting from stochastic variation in conditions. However, Hartman et al. (2013) were able to demonstrate separate effects of density dependence and environmental variation for Velvet Scoters demonstrating that these two factors are not always linked.

SEA DUCK MANAGEMENT

Given the life history variation and population dynamics presented, how then can sea duck populations be managed? The answer is that one must consider multiple aspects of sea duck population dynamics in conjunction with potential management actions available. For example, elasticity analyses can identify the parameters that have the greatest relative effect on lambda, such as adult survival. However, these same parameters may have little environmental and/or genetic variation and therefore they may not respond to management actions. Given the high elasticity for adult female survival, changes in harvest can be expected to have immediate and large effects on population trends (Gilliland et al. 2009). In fact, both Merkel (2010) and Burnham et al. (2012) report high rates of breeding population increase following a significant decrease in harvest of Common Eiders in Greenland. However, Hario et al. (2009) concluded that changes in survival were not solely responsible for long-term trends in a breeding population of Common Eiders in Sweden. Further, Rönkä et al. (2011) showed positive relationships between productivity and subsequent population size for Common Eiders and Common Mergansers. Stien et al. (2010) demonstrated that removal of avian nest predators increased nest success at one of two Common Eider colonies studied. In the case of nonhunted species, there may be limited options to manage adult female survival. Understanding the retrospective population variance decomposition can help to identify parameters that, at least in the past, have resulted in variation in population dynamics. However, these are typically parameters with low relative influence on population growth. Therefore, much larger management efforts would be necessary to influence population growth via these parameters. Ideally, management should focus on actions that have the potential to influence more than one life history parameter simultaneously, thereby taking advantage of the effects of correlations among the population parameters. For example, lead poisoning has been identified as a source of mortality for incubating Spectacled Eiders on the Yukon–Kuskokwim Delta in Alaska (Flint et al. 1997, Grand et al. 1998). Thus, elimination of lead poisoning would increase annual survival and have slight but positive effects on nesting success as well as duckling and juvenile survival (Flint and Grand 1997, Grand et al. 1998, Flint et al. 2000a). Management actions that both reduce process variance and alter the mean will have the greatest overall effect and increase the likelihood of observing the desired effect (Wisdom et al. 2000). Potential examples are predator control or habitat manipulation designed to eliminate years of low productivity while simultaneously increasing adult survival by reducing predator-related mortality of nesting hens. Such programs would positively influence multiple parameters by reducing variance and increasing the mean.

POPULATION RESPONSE TO MANAGEMENT ACTIONS

Population dynamics as predicted by matrix models assume that a population is at a stable age distribution. That is, given a specific set of life history values that are consistent over time, populations will eventually reach a case where a constant proportion of the population will occur within each age class. Interpretation of sensitivities and elasticities requires the assumption that the population is at this stable distribution and will remain at this stable distribution. However, as soon as one or more of the input parameters change, a corresponding change in lambda will occur and with it a change in the stable age distribution. Thus, as soon as we successfully invoke a management action designed to alter lambda, the previous sensitivities and elasticities no longer apply, nor do sensitivities or elasticities predicted from a new model because the population has not had sufficient time to obtain the new stable age distribution. A period of transition in which populations deviate from a stable age distribution, following some alteration of life history parameters, is termed transient dynamics. During periods of transient dynamics, the strict predictions of population models do not apply (Koons et al. 2006) and the population response to management actions, in terms of changes in lambda, will never be exactly as predicted (de Kroon et al. 2000). Further, since populations exist in stochastic environments, actual life history parameter values are not constant through time but, rather, are constantly varying. Thus, in the most realistic modeling scenario, populations in stochastic environments are rarely at a stable age distribution, but are more likely in a perpetual state of transient dynamics. Thus, expected responses to management actions will rarely follow exactly as the predictions of population models, and the more extreme the management action or effects of a perturbation on lambda, the worse this effect may be (Koons et al. 2006). In spite of these issues, the general outcomes from population models still apply, and the magnitude of these discrepancies is likely less than the measurement error of population size or our ability to measure the true response.

The time scale at which sea duck populations respond to management actions is likely to be fairly long. For example, consider the effect of an increase in productivity for Common Eiders. It will be 3 years before most of the offspring that result from this increase return to breed for the first time and then another 3 years before their offspring return to breed for the first time (Hario and Rintala 2006, Rönkä et al. 2011). Jónsson et al. (2013) found a 3-year lag between climatic parameters that likely increased Common Eider recruitment and population response. Similarly, Rönkä et al. (2011) showed a positive relationship between breeding success and subsequent population size with a time lag corresponding to recruitment age for Common Eiders and goldeneyes. Sæther et al. (2005) demonstrated that observed effects of environmental stochasticity and variation in life history parameters may only be observable at the scale of generation time. Mean generation times for sea ducks, based on our models, varied from 4.1 years in Common Mergansers to 10.7 years in Common Eiders. Accordingly, we can expect relatively long time lags between onset of a management action for long-lived sea ducks and a measurable response. Transient dynamics associated with unstable age distributions may further increase this time lag by causing fluctuating responses to management actions and in extreme cases cause populations to stabilize at unpredicted levels (Koons et al. 2006).

The potential for management actions to influence population dynamics may be asymmetrical. For example, because sea ducks are long lived with low reproductive effort, survival rates tend to be relatively high. Further, there tends to be little environmental or stochastic variation in survival compared to reproductive parameters (Benton and Grant 1996, Iles 2012). As I have demonstrated, sea ducks will tend to show the highest elasticities for adult female survival implying that changing survival would be a logical management action to alter population size. However, the utility of altering survival to influence population size depends on the management objective. If the goal is to increase population size, we would need to increase adult female survival (Merkel 2010, Burnham et al. 2012). Yet, because sea ducks already have relatively high survival, there may not be practical management options to increase survival (de Kroon et al. 2000). Conversely, if the goal is to reduce population size, we have numerous management options for lowering adult female survival. Thus, elasticity analyses can help identify potential management targets, but the utility of those targets may depend on the direction of the management goal.

MANAGEMENT AND DENSITY DEPENDENCE

It is unknown to what degree sea duck populations behave in a density-dependent fashion. Resolving this issue should be a top research priority as potential management actions differ depending on density-dependent effects. For example, if populations are density independent, then targeted management actions at specific demographic parameters are warranted. However, if populations are density dependent, then manipulation of specific demographic parameters may be ineffective. For example, Pehrsson and Nyström (1988) found that increases in nesting success were offset by declines in duckling survival of Long-tailed Ducks. Similarly, Pöysä and Pöysä (2002) increased nesting density of Common Goldeneyes with the addition of nest boxes, but found no net increase in total number of duckling fledged. Thus, it appears that density dependence may have been occurring during brood rearing. Nordstrom et al. (2002) removed American mink from a series of islands in Finland and found that Velvet Scoter populations increased but Common Eider and Goosander populations did not. Similarly, Stien et al. (2010) removed avian predators but only achieved increases in nest success at one of two Common Eider colonies. Thus, scoters were certainly limited by low productivity, but Common Eiders were likely limited during some other portion of their life cycle (Nordstrom et al. 2002). If density dependence exists, then management actions need to be targeted at the density limiting factor. Frequently, management will require habitat manipulation as opposed to direct management of life history parameters. However, if the limiting factor occurs at sea, there are likely few management options available (Flint 2013). Thus, understanding the source and degree of density dependence within sea ducks will dictate the general types of management actions that are likely to be successful.

Understanding the density-dependent response of sea duck populations also would influence harvest regulation and management. If populations are truly density independent, then any human harvest is functionally a source of additive mortality. However, if populations are density dependent, then some portion of harvest mortality is compensatory. Also, in a density-dependent situation, the management goal must be clearly defined. Two obvious options are management for maximum population size and management for maximum sustained yield. The population size objective will be different under these two goals because the maximum sustained yield population size is somewhat less than carrying capacity.

CONCLUSIONS

Numerous populations of sea ducks have declined in recent years (Flint 2013; Chapter 1, this volume). The fact that these negative trends have been observed over decades implies that causative changes in life history parameters have occurred in the past and appear to have been maintained for long periods of time. There are likely no easy management options for altering declining population trends for sea ducks. For most populations, harvest estimates are low, or populations continue to decline in the absence of a known harvest (but see Merkel 2010, Burnham et al. 2012). Accordingly, viable management options to increase annual survival may not exist (Wilson et al. 2007, Coulson 2010). Meanwhile, management actions focused on productivity are constrained by low elasticities of productivity parameters where even intensive management actions are likely to only yield small changes in lambda. Further, these actions will likely have to be maintained for decades before population response may be realized and natural stochastic variation may dampen potential effects of management actions.

Given the results presented here, all sea duck populations tend to show the same patterns of sensitivities and elasticities. Thus, basic models will tend to prescribe the same management action for all species. Perhaps more relevant is what we think we can manage in a population. Can we change mean values or can we manage the variation? If we think we can manage mean values without influencing the variation, which seems unlikely, then prospective analyses should provide a useful tool (Ehrlén and van Groenendael 1998). Conversely, if we believe we have a greater ability to manage variation, perhaps by reducing the frequency of years with reproductive failure, then we will have to target life history characteristics with greater variation, and the retrospective population analyses may provide a more useful tool. In the case of managing variation to obtain a directional goal, we have to acknowledge that in order to simultaneously change the mean and the variance, a combination of approaches will likely be necessary. In that sense, it is not really survival or productivity that should be the sole focus of management actions but, rather, consideration of both components of the life history of the birds.

Last, population models can aid us in interpreting available data for some species. In many cases, estimates of population size are based on surveys of breeding habitats. However, given that most sea duck species do not attempt to breed until >1 year of age and established adult breeders may skip reproduction in some years (Yoccoz et al. 2002, Coulson 2010), some fraction of the population is not present on the breeding grounds (Desholm et al. 2002). The nonbreeding proportion may represent a considerable fraction of the population (>30%), and population objectives based on such surveys need to account for this bias (Coulson 2010).

ACKNOWLEDGMENTS

I thank the USGS Alaska Science Center for support of my time during the development of this manuscript. My early exposure to population modeling by R. Rockwell and B. Grand had a major influence on my thinking on this topic. I also thank D. Esler, S. Iverson, J. Schmutz, C. Taylor, and H. Wilson for providing comments on an earlier draft of this manuscript. Any use of trade names is for descriptive purposes only and does not imply endorsement by the US Government.

LITERATURE CITED

Anderson, B. W., and R. L. Timken. 1972. Sex and age ratios and weights of Common Mergansers. Journal of Wildlife Management 36:1127–1133.

Barbraud, C., and H. Weimerskirch. 2005. Environmental conditions and breeding experience affect costs of reproduction in Blue Petrels. Ecology 86:682–692.

Benton, T. G., and A. Grant. 1996. How to keep fit in the real world: elasticity analyses and selection pressures on life histories in a variable environment. American Naturalist 147:115–139.

Bentzen, R. L., and A. N. Powell. 2012. Population dynamics of King Eiders breeding in northern Alaska. Journal of Wildlife Management 76:1011–1020.

Bilcher, M. E., L. M. Rasmussen, M. K. Sejr, F. R. Merkel, and S. Rysgaard. 2011. Abundance and energy requirements of Eiders (*Somateria* spp.) suggest high predation pressure on macrobenthic fauna in a key wintering habitat in SW Greenland. Polar Biology 34:1105–1116.

Bolen, E. G., and B. R. Chapman. 1981. Estimating winter sex ratios for Buffleheads. Southwestern Naturalist 26:49–52.

Boyd, W. S., B. D. Smith, S. A. Iverson, J. E. Thompson, and S. Schneider. 2009. Apparent survival, natal philopatry, and recruitment of Barrow's Goldeneyes (*Bucephala islandica*) in the Cariboo–Chilcotin region of British Columbia, Canada. Canadian Journal of Zoology 87:337–345.

Burnham, K. K., J. A. Johnson, B. Konkel, and J. L. Burnham. 2012. Nesting Common Eider (*Somateria mollissima*) population quintuples in Northwest Greenland. Arctic 65:456–464.

Burnham, K. P., and G. C. White. 2002. Evaluation of some random-effects methodology applicable to bird ringing data. Journal of Applied Statistics 29:254–264.

Camphuysen, C. J., C. M. Berrevoets, H. J. W. M. Cremers, A. Dekinga, R. Dekker, B. J. Ens, T. M. van der Have et al. 2002. Mass mortality of Common Eiders (*Somateria mollissima*) in the Dutch Wadden Sea, winter 1999/2000: starvation in a commercially exploited wetland of international importance. Biological Conservation 106:303–317.

Caron, C. M., and P. W. C. Paton. 2007. Population trends and habitat use of Harlequin Ducks in Rhode Island. Journal of Field Ornithology 78:254–262.

Caswell, H. 2000. Prospective and retrospective perturbation analyses: their roles in conservation biology. Ecology 81:619–627.

Caswell, H. 2001. Matrix population models, 2nd ed. Sinauer Associates, Sunderland, MA.

Chaulk, K. G., and M. L. Mahoney. 2012. Does spring ice cover influence nest initiation date and clutch size in Common Eiders? Polar Biology 35:645–653.

Chaulk, K. G., G. J. Robertson, and W. A. Montevecchi. 2004. Regional and annual variability in Common Eider nesting ecology in Labrador. Polar Research 23:121–130.

Christensen, T. K. 1999. Effects of cohort and individual variation in duckling body condition on survival and recruitment in the Common Eider *Somateria mollissima*. Journal of Avian Biology 30:203–308.

Cooke, F., G. R. Robertson, C. M. Smith, R. I. Goudie, and W. S. Boyd. 2000. Survival, emigration and winter population structure of Harlequin Ducks. Condor 102:137–144.

Cooke, F., R. F. Rockwell, and D. B. Lank. 1995. The Snow Geese of La Pérouse Bay: natural selection in the wild. Oxford University Press, New York, NY.

Coulson, J. C. 1984. The population dynamics of the Eider Duck *Somateria mollissima* and evidence of considerable non-breeding by adult ducks. Ibis 126:525–543.

Coulson, J. C. 1999. Variation in clutch size of the Common Eider: a study based on 41 breeding seasons on Coquet Island, Northumberland, England. Waterbirds 22:225–238.

Coulson, J. C. 2010. A long-term study of the population dynamics of Common Eiders *Somateria mollissima*: why do several parameters fluctuate markedly? Bird Study 57:1–18.

Coulson, J. C. 2013. Age-related and annual variation in mortality rates of adult female Common Eiders (*Somateria mollissima*). Waterbirds 36:234–239.

Coulson, T., J.-M. Gaillard, and M. Festa-Bianchet. 2005. Decomposing the variation in population growth into contributions from multiple demographic rates. Journal of Animal Ecology 74:789–801.

D'Alba, L., P. Monaghan, and R. G. Nager. 2010. Advances in laying date and increasing population size suggest positive responses to climate change in Common Eiders *Somateria mollissima* in Iceland. Ibis 152:19–28.

de Kroon, H., J. van Groenendael, and J. Ehrlén. 2000. Elasticities: a review of methods and model limitations. Ecology 81:607–618.

Descamps, S., H. G. Gilchrist, J. Bêty, E. I. Buttler, and M. R. Forbes. 2009. Costs of reproduction in a long-lived bird: large clutch size is associated with low survival in the presence of a highly virulent disease. Biology Letters 5:278–281.

Descamps, S., N. G. Yoccoz, J.-M. Gaillard, H. G. Gilchrist, K. E. Erikstad, S. A. Hanssen, B. Cazelles, M. R. Forbes, and J. Bêty. 2010. Detecting population heterogeneity in effects of North Atlantic Oscillations on seabird body condition: get into the rhythm. Oikos 199:1526–1536.

Desholm, M., T. K. Christensen, G. Scheiffarth, M. Hario, Å. Andersson, B. Ens, C. J. Camphuysen et al. 2002. Status of the Baltic/Wadden Sea population of the Common Eider *Somateria m. mollissima*. Wildfowl 53:167–203.

Doak, D. F., K. Gross, and W. F. Morris. 2005. Understanding and predicting the effects of sparse data on demographic analyses. Ecology 86:1154–1163.

Ehrlén, J., and J. van Groenendael. 1998. Direct perturbation analysis for better conservation. Conservation Biology 12:470–474.

Ehrlén, J., J. van Groenendael, and H. de Kroon. 2001. Reliability of elasticity analysis: reply to Mills et al. Conservation Biology 15:278–280.

Erikstad, K. E., and T. Tveraa. 1995. Does the cost of incubation set limits to clutch size in the Common Eider *Somateria mollissima*? Oecologia 103:270–274.

Esler, D. 2000. Applying metapopulation theory to the conservation of migratory birds. Conservation Biology 14:366–372.

Evans, M. R., D. B. Lank, W. S. Boyd, and F. Cooke. 2002. A comparison of the characteristics and fate of Barrow's Goldeneye and Bufflehead nests in nest boxes and natural cavities. Condor 104:610–619.

Flint, P. L. 2013. Changes in size and trends of North American sea duck populations associated with North Pacific oceanic regime shifts. Marine Biology 160:59–65.

Flint, P. L., and J. B. Grand. 1997. Survival of adult female and juvenile Spectacled Eiders during brood rearing. Journal of Wildlife Management 61:218–222.

Flint, P. L., J. B. Grand, T. F. Fondell, and J. A. Morse. 2006a. Population dynamics of Greater Scaup breeding on the Yukon–Kuskokwim Delta, Alaska. Wildlife Monographs 162:1–21.

Flint, P. L., J. B. Grand, J. A. Morse, and T. F. Fondell. 2000a. Late summer survival of adult female and juvenile Spectacled Eiders on the Yukon–Kuskokwim Delta, Alaska. Waterbirds 23:292–297.

Flint, P. L., J. A. Morse, J. B. Grand, and C. L. Moran. 2006b. Correlated growth and survival of Spectacled Eider ducklings: evidence of habitat limitation? Condor 108:901–911.

Flint, P. L., M. R. Petersen, C. P. Dau, J. E. Hines, and J. D. Nichols. 2000b. Annual survival and site fidelity of Steller's Eiders molting along the Alaska Peninsula. Journal of Wildlife Management 64:261–268.

Flint, P. L., M. R. Petersen, and J. B. Grand. 1997. Exposure of Spectacled Eiders and other diving ducks to lead in western Alaska. Canadian Journal of Zoology 75:439–443.

Gilliland, S. G., H. G. Gilchrist, R. F. Rockwell, G. J. Robertson, J.-P. L. Savard, F. Merkel, and A. Mosbech. 2009. Evaluating the sustainability of harvest among northern Common Eiders *Somateria mollissima borealis* in Greenland and Canada. Wildlife Biology 15:24–36.

Gould, W. R., and J. D. Nichols. 1998. Estimation of temporal variability of survival in animal populations. Ecology 79:2531–2538.

Grand, J. B., P. L. Flint, M. R. Petersen, and T. L. Moran. 1998. Effect of lead poisoning on Spectacled Eider survival rates. Journal of Wildlife Management 62:1103–1109.

Grant, A., and T. G. Benton. 2000. Elasticity analysis for density-dependent populations in stochastic environments. Ecology 81:680–693.

Hario, M., M. J. Mazerolle, and P. Saurola. 2009. Survival of female Common Eiders *Somateria m. mollissima* in a declining population of the northern Baltic Sea. Oecologia 159:747–756.

Hario, M., and J. Rintala. 2006. Fledgling production and population trends in Finnish Common Eiders (*Somateria mollissima mollissima*)—evidence of density dependence. Canadian Journal of Zoology 84:1038–1046.

Hartman, G., A. Kölzsch, K. Larsson, M. Norberg, and J. Höglund. 2013. Trends and population dynamics of a Velvet Scoter (*Melanitta fusca*) population: influence of density dependence and winter climate. Journal of Ornithology 154:837–847.

Heppell, S. S., H. Caswell, and L. B. Crowder. 2000. Life histories and elasticity patterns: perturbation analysis for species with minimal demographic data. Ecology 81:654–665.

Iles, D. T. 2012. Drivers of nest success and stochastic population dynamics of the Common Eider (*Somateria mollissima*). Thesis, Utah State University, Logan, UT.

Iverson, S. A., W. S. Boyd, H. M. Regehr, and M. S. Rodway. 2006. Sex and age-specific distributions of sea ducks wintering in the Strait of George, British Columbia: implications for the use of age ratios as in index of recruitment. Technical Report Series, No. 459. Canadian Wildlife Service, Environment Canada, Sackville, NB.

Iverson, S. A., and D. Esler. 2010. Harlequin Duck population injury and recovery dynamics following the 1989 *Exxon Valdez* oil spill. Ecological Applications 20:1993–2006.

Iverson, S. A., B. D. Smith, and F. Cooke. 2004. Age and sex distributions of wintering Surf Scoters: implications for the use of age ratios as an index of recruitment. Condor 106:252–262.

Johnson, D. H., J. D. Nichols, and M. D. Schwartz. 1992. Population dynamics of breeding waterfowl. Pp. 446–485 in B. J. D. Batt, A. D. Afton, M. G. Anderson, C. D. Ankney, D. H. Johnson, J. A. Kadlec, and G. L. Krapu (editors), Ecology and management of breeding waterfowl. University of Minnesota Press, Minneapolis, MN.

Jónsson, J. E., A. Gardarsson, J. A. Gill, U. K. Pétursdóttir, A. Petersen, and T. G. Gunnarsson. 2013. Relationships between long-term demography and weather in a sub-Arctic population of Common Eiders. PLoS One 8:e67093.

Koons, D. N., R. F. Rockwell, and J. B. Grand. 2006. Population momentum: implications for wildlife management. Journal of Wildlife Management 70:19–26.

Krementz, D. G., P. W. Brown, F. P. Kehoe, and C. S. Houston. 1997. Population dynamics of White-winged Scoters. Journal of Wildlife Management 61:222–227.

Lehikoinen, A., T. K. Christensen, M. Ost, M. Kilpi, P. Saurola, and A. Vattulainen. 2008. Large-scale change in the sex ratio of a declining Eider *Somateria mollissima* population. Wildlife Biology 14:288–301.

Lehikoinen, A., M. Kilpi, and M. Ost. 2006. Winter climate affects subsequent breeding success of Common Eiders. Global Change Biology 12:1355–1365.

Lewis, T. L., D. Esler, W. S. Boyd. 2007. Effects of predation by sea ducks on clam abundance in soft-bottom intertidal habitats. Marine Ecology Progress Series 329:131–144.

Mehlum, F. 1991. Breeding population size of the Common Eider *Somateria mollissima* in Kongsfjorden, Svalbard, 1981–1987. Norsk Polarinstitut Skrifter 195:21–29.

Mehlum, F. 2012. Effects of sea ice on breeding numbers and clutch size of a high arctic population of the Common Eider *Somateria mollissima*. Polar Science 6:143–153.

Merkel, F. R. 2010. Evidence of recent population recovery in Common Eiders breeding the western Greenland. Journal of Wildlife Management 74:1869–1874.

Mills, L. S., D. F. Doak, and M. J. Wisdom. 1999. Reliability of conservation actions based on elasticity analysis of matrix models. Conservation Biology 13:815–829.

Milonoff, M., H. Poysa, P. Runko, and V. Ruusila. 2004. Brood rearing costs affect future reproduction in the precocial Common Goldeneye *Bucephala clangula*. Journal of Avian Biology 35:344–351.

Morris, W. F., and D. F. Doak. 2002. Quantitative conservation biology: theory and practice of population viability analyses. Sinauer Associates, Sunderland, MA.

Morris, W. F., and D. F. Doak. 2004. Buffering of life histories against environmental stochasticity: accounting for a spurious correlation between the variabilities of vital rates and their contributions to fitness. American Naturalist 163:579–590.

Murray, D. L., M. G. Anderson, and T. D. Steury. 2010. Temporal shift in density dependence among North American breeding duck populations. Ecology 91:571–581.

Nordstrom, M., J. Hogmander, J. Nummelin, J. Laine, N. Laanetu, and E. Korpimaki. 2002. Variable responses of waterfowl breeding populations to long-term removal of introduced American mink. Ecography 25:385–394.

Oosterhuis, R., and K. Van Dijk. 2002. Effect of food shortage on the reproductive output of Common Eiders *Somateria mollissima* breeding at Griend (Wadden Sea). Atlantic Seabirds 4:29–38.

Ost, M., and B. B. Steele. 2010. Age-specific nest-site preference and success in Eider. Oecologia 162:59–69.

Pearce, J. P., J. A. Reed, and P. L. Flint. 2005. Geographic variation in survival and migratory tendency among North American Common Mergansers. Journal of Field Ornithology 76:109–118.

Pehrsson, O., and K. G. K. Nyström. 1988. Growth and movements of Oldsquaw ducklings in relation to food. Journal of Wildlife Management 52:185–191.

Petersen, M. R., and D. C. Douglas. 2004. Winter ecology of Spectacled Eiders: environmental characteristics and population change. Condor 106:79–94.

Petersen, M. R., and P. L. Flint. 2002. Population structure of Pacific Common Eiders breeding in Alaska. Condor 104:780–787.

Petersen, M. R., B. J. McCaffery, and P. L. Flint. 2003. Postbreeding distribution of Long-tailed Ducks *Clangula hyemalis* from the Yukon–Kuskokwim Delta. Wildfowl 54:129–139.

Pfister, C. A. 1998. Patterns of variance in stage-structured populations: evolutionary predictions and ecological implications. Proceedings of the National Academy of Sciences USA 95:213–218.

Pöysä, H., and S. Pöysä. 2002. Nest-site limitation and density dependence of reproductive output in the Common Goldeneye *Bucephala clangula*: implications for the management of cavity-nesting birds. Journal of Applied Ecology 39:502–510.

Reed, J. A., D. Lacroix, and P. L. Flint. 2007. Depredation of Common Eider nests along the central Beaufort Sea coast: a case where no one wins. Canadian Field-Naturalist 121:308–312.

Robertson, G. J. 2008. Using winter juvenile/adult ratios as indices of recruitment in population models. Waterbirds 31:152–158.

Robertson, G. J., and F. Cooke. 1999. Winter philopatry in migratory waterfowl. Auk 116:20–34.

Robertson, G. J., and H. G. Gilchrist. 1998. Evidence of population declines among Common Eiders breeding in the Belcher Islands, Northwest Territories. Arctic 51:378–385.

Rohwer, F. C. 1988. Inter- and intraspecific relationships between egg size and clutch size in waterfowl. Auk 105:161–176.

Rönkä, M., L. Saari, M. Hario, J. Hänninen, and E. Lehikoinen. 2011. Breeding success and breeding population trends of waterfowl: implications for monitoring. Wildlife Biology 17:225–239.

Rönkä, M. T. H., C. L. V. Saari, E. A. Lehikonen, J. Suomela, and K. Häkkilä. 2005. Environmental changes and population trends of breeding waterfowl in the northern Baltic Sea. Annales Zoologici Fennici 42:587–602.

Roth, J. D. 2003. Variability in marine resources affects arctic fox population dynamics. Journal of Animal Ecology 72:668–676.

Sæther, B. E., and Ø. Bakke. 2000. Avian life history variation and contribution of demographic traits to the population growth rate. Ecology 81:642–653.

Sæther, B. E., R. Lande, S. Engen, H. Weimerskirch, M. Lillegård, R. Altwegg, P. H. Becker et al. 2005. Generation time and temporal scaling of bird population dynamics. Nature 436:99–102.

Samelius, G. 2006. Foraging behaviours and population dynamics of Arctic foxes. Dissertation, University of Saskatchewan, Saskatoon, SK.

Schamber, J. L., P. L. Flint, J. Barry Grand, H. M. Wilson, and J. A. Morse. 2009. Population dynamics of Long-tailed Ducks breeding on the Yukon–Kuskokwim Delta, Alaska. Arctic 62:190–200.

Scribner, K. T., M. R. Petersen, R. L. Fields, S. L. Talbot, J. M. Pearce, and R. K. Chesser. 2001. Sex-biased gene flow in Spectacled Eiders (*Anatidae*): inferences from molecular markers with contrasting modes of inheritance. Evolution 55:2105–2115.

Smith, C. M., R. I. Goudie, and F. Cooke. 2001. Winter age ratios and the assessment of recruitment of Harlequin Ducks. Waterbirds 24:39–44.

Stien, J., N. G. Yoccoz, and R. A. Ims. 2010. Nest predation in declining populations of Common Eiders *Somateria mollissima*: an experimental evaluation of the role of hooded crows *Corvus cornix*. Wildlife Biology 16:123–134.

Swennen, C. 1990. Dispersal and migratory movements of Eiders *Somateria mollissima* breeding in the Netherlands. Ornis Scandinavica 21:17–27.

Swennen, C., P. Duiven, and L. A. F. Reyrink. 1979. Notes on the sex ratio in the Common Eider *Somateria mollissima* (L.). Ardea 67:54–61.

van Tienderen, P. H. 2000. Elasticities and the link between demographic and evolutionary dynamics. Ecology 81:666–679.

Viallefont, A., F. Cooke, and J. D. Lebreton. 1995. Age-specific costs of first-time breeding. Auk 112:67–76.

Weimerskirch, H. 1990. The influence of age and experience on breeding performance of the Antarctic Fulmar, *Fulmarus glacialoides*. Journal of Animal Ecology 59:867–875.

White, G. C., K. P. Burnham, and R. J. Barker. 2008. Evaluation of a Bayesian MCMC random effects inference methodology for capture–mark–recapture data. Pp. 1119–1127 in D. L. Thomson, E. G. Cooch, and M. J. Conroy (editors), Modeling demographic processes in marked populations. Environmental and Ecological Statistics Series (vol. 3), Springer, New York, NY.

Wilson, H. M., P. L. Flint, T. L. Moran, and A. N. Powell. 2007. Survival of breeding Pacific Common Eiders on the Yukon–Kuskokwim Delta, Alaska. Journal of Wildlife Management 71:403–410.

Wilson, H. M., P. L. Flint, A. N. Powell, J. B. Grand, and T. L. Moran. 2012. Population ecology of breeding Pacific Common Eiders on the Yukon–Kuskokwim Delta, Alaska. Wildlife Monographs 182:1–28.

Wisdom, M. J., L. S. Mills, and D. F. Doak. 2000. Life stage simulation analysis: estimating vital-rate effects on population growth for conservation. Ecology 81:628–641.

Wooler, R. D., J. S. Bradley, I. J. Skira, and D. L. Serventy. 1990. Reproductive success of Short-tailed Shearwaters *Puffinus tenuirostris* in relation to their age and breeding experience. Journal of Animal Ecology 59:161–170.

Yoccoz, N. G., K. E. Erikstad, J. O. Bustnes, S. A. Hanssen, and T. Tveraa. 2002. Costs of reproduction in Common Eiders (*Somateria mollissima*): an assessment of relationships between reproductive effort and future survival and reproduction based on observational and experimental studies. Journal of Applied Statistics 29:57–64.

CHAPTER FOUR

Infectious Diseases, Parasites, and Biological Toxins in Sea Ducks*

Tuula E. Hollmén and J. Christian Franson

Abstract. This chapter addresses disease agents in the broad sense, including viruses, bacteria, fungi, protozoan and helminth parasites, and biological toxins. Some of these agents are known to cause mortality in sea ducks, some are thought to be incidental findings, and the significance of others is yet poorly understood. Although the focus of the chapter is on free-living sea ducks, the study of disease in this taxonomic group has been relatively limited and examples from captive sea ducks and other wild waterfowl are used to illustrate the pathogenicity of certain diseases. Much of the early work in sea ducks consisted of anecdotal and descriptive reports of parasites, but it was soon recognized that diseases such as avian cholera, renal coccidiosis, and intestinal infections with acanthocephalans were causes of mortality in wild populations. More recently, adenoviruses, reoviruses, and the newly emergent Wellfleet Bay virus, for example, also have been linked to die-offs of sea ducks. Declining populations of animals are particularly vulnerable to the threats posed by disease and it is important that we improve our understanding of the significance of disease in sea ducks. To conclude, we offer our recommendations for future directions in this field.

Key Words: avian cholera, avian influenza, bacteria, biological toxins, botulism, clinical signs, disease, epidemiology, parasites, pathogen, transmission, viruses, zoonotic.

Diseases caused by infectious organisms, parasites, and biological toxins have received increasing attention in wildlife studies over the recent decades. Advancements in the ecology of disease, diagnostic methods, and theory have helped to increase knowledge about the occurrence, pathology, and population-level effects of diseases in wildlife. Accompanying these developments has been an increased awareness of the emergence of infectious diseases in new hosts and locations and the reemergence of diseases that may have previously been of little concern. Mechanisms for disease emergence include the transmission of infectious agents from reservoir hosts to previously unaffected populations, translocations of wildlife that result in the risk of exposing naïve species or populations, geographic dispersal of infected wildlife, and the evolution of novel strains or organisms (Daszak et al. 2000). Among the emerging and reemerging disease agents, viruses and bacteria are of particular concern. Dobson and Foufopoulos (2001) surveyed emerging pathogens in the late 1990s and listed West Nile virus (WNV), avian

* Hollmén, T. E. and J. C. Franson. 2015. Infectious diseases, parasites, and biological toxins in sea ducks. Pp. 97–123 in J.-P. L. Savard, D. V. Derksen, D. Esler, and J. M. Eadie (editors). Ecology and conservation of North American sea ducks. Studies in Avian Biology (no. 46), CRC Press, Boca Raton, FL.

pox, duck plague, avian cholera, and avian botulism, all of which are discussed in this chapter, as agents of concern for birds. Some of the most important factors modulating the distribution and overall impacts of diseases on wildlife include environmental changes resulting from anthropogenic perturbations. For example, global climate change may affect the distribution of disease vectors as well as host populations (Kovats et al. 2001, Irons et al. 2008). Habitat degradation and fragmentation may alter migratory patterns of wildlife, resulting in spread of disease to new areas, as well as weakened host immune defenses, leading to increased susceptibility to disease (Acevedo-Whitehouse and Duffus 2009, Altizer et al. 2011).

Several life history characteristics of sea ducks may influence their exposure to disease agents as well as the epidemiology of diseases when they occur within populations. As a Northern Hemisphere group that often frequents coastal marine and estuarine environments, sea ducks may encounter certain diseases more or less frequently than cosmopolitan waterfowl and species that prefer freshwater habitats. In North America, some species of sea ducks have suffered substantial mortality from avian botulism in the Great Lakes region, but this disease has historically been associated with dabbling ducks in the Western United States and the Canadian prairies (Rocke and Bollinger 2007). On the other hand, as sea ducks occupy salt water environments, they may be exposed to harmful toxins produced by a variety of marine algae (Landsberg et al. 2007). Food habits can influence exposure to these agents because sea ducks can be exposed by direct consumption of toxins in the aquatic environment or by consuming bivalves or other prey that have accumulated the toxins. Food habits can also influence exposure to those parasites whose life cycle involves an intermediate host.

Many early studies of sea ducks, some dating back to the 1950s, focused on the identification and taxonomy of parasites. Past reports were mainly descriptive and anecdotal, and most did not include information on pathology or mortality. However, parasites that have been associated with mortality in sea ducks include acanthocephalans, intestinal helminths that affect both adults and ducklings, and renal coccidia, protozoa that cause kidney disease, primarily in ducklings. With development of additional diagnostic capabilities for major waterfowl diseases, avian cholera,

avian botulism, and other pathogens were investigated as causes of large-scale die-offs in sea ducks. Avian cholera has caused sporadic disease at nesting and wintering areas since the 1960s but has recently reemerged as a substantial mortality factor in nesting Common Eiders (*Somateria mollissima*) in Northern Canada (Gershman et al. 1964, Descamps et al. 2009). Avian botulism type C has traditionally been responsible for the major botulism die-offs in waterfowl in North America, but botulism type E has recently caused extensive mortality in sea ducks in the Great Lakes region. More recently, research has expanded to include characterization of novel disease agents, application of quantitative techniques to evaluate population-level effects, and epidemiology and ecology of disease in sea ducks. In sea ducks, Wellfleet Bay virus, caused by a novel *Orthomyxovirus*, may be considered an emerging infectious disease that has resulted in mortality of Common Eiders on the East Coast of the United States (Ballard et al. 2012). As population declines have been observed in many of the extant sea duck species, knowledge about potential disease impacts on population dynamics is becoming increasingly important.

This chapter provides a review of information currently available about infectious diseases, parasites, and biological toxins in sea ducks that remains relatively scarce. Comprehensive surveys to determine host and geographic range of diseases have not been conducted, and with a few exceptions, studies on population-level effects of pathogens are still lacking. Information about pathogenicity and host effects is available from captive experiments, clinical studies, or postmortem pathology studies for some disease agents, but few data are available regarding epidemiology and ecology of disease in sea ducks, including transmission and persistence of disease in populations. For some species of sea ducks, no published reports exist for exposure to infectious and parasitic disease agents or biological toxins. Consequently, the depth of coverage of certain disease characteristics, such as host range, pathology, ecology, and effects on the host, will necessarily vary among the disease agents described in this chapter.

This chapter addresses viral, bacterial, and parasitological diseases according to taxonomic group of the disease agent. Biological toxins arising from avian botulism and harmful algae are also included. Some infectious diseases of sea ducks are zoonotic diseases that are capable of infecting humans. Zoonotic diseases such as chlamydiosis could be transferred directly from sea ducks to humans, typically by inhalation of the organism. Other zoonoses require an intermediate vector, as in the case of WNV where an arthropod transfers the disease from bird to human via a blood meal. The information presented in the chapter is supplemented by available reports for other avian species as relevant to further understanding of the disease organism or its potential impacts on the avian host. Recommendations for future study are outlined in the "Conclusions and Future Directions" section.

VIRUSES

Viruses are obligate intracellular agents that utilize the host cell machinery for replication. Viruses are divided into DNA and RNA viruses based on the type of their nucleic acids and further classified into major families, genera, species, and variants. Many viruses hold the capacity for rapid mutation, and virus taxonomy is an actively evolving field. Diagnostic procedures for detection of viruses utilize a suite of laboratory approaches, including cultivation, genetics, visualization and detection of virus particles and infectious units, and serologic assays. Because of their nature as obligate intracellular agents, culture of viruses in a laboratory setting requires the use of live host cells, typically cell cultures or embryonated eggs. The pathogenicity of viruses varies greatly and is often dependent on multiple factors including the virulence of the virus species and variant, host susceptibility, dose, and environmental conditions. Understanding about the occurrence and impact of viruses on sea duck health is increasing. Several major viral pathogens of waterfowl have been recently discovered in sea ducks, novel viruses have been isolated and characterized, and knowledge about effects on individual hosts and populations is developing. The following section is organized by taxonomic families and for each virus group, focusing on key aspects of the pathobiology of the agent and its documented effects on sea ducks.

Duck Plague Virus (Duck Virus Enteritis)

Duck plague is an acute infectious disease of Anseriformes caused by duck herpesvirus 1, a member of the family Herpesviridae. Duck plague primarily affects domestic ducks, although outbreaks have occurred in backyard flocks, nonmigratory park waterfowl, zoological collections, and rarely in free-ranging waterfowl, with mortality reported in at least eight species of sea ducks (Converse and Kidd 2001, Hansen and Gough 2007). Duck plague was first reported from Europe in the early 1920s, although it was misidentified as avian influenza, and has since been confirmed or suspected in Asia, North America, and Brazil (Hansen and Gough 2007). Transmission typically occurs by direct contact with infected waterfowl or from water contaminated with the virus. Sick birds are rarely seen in the field, but clinical signs include bloody discharge from the vent or bill, ulcerative lesions on the inside surface of the mandible, prolapse of the penis in males, weakness, inability to fly, and convulsions (Friend 1999b). Waterfowl that survive the disease may become carriers, and viral shedding has been reported for up to several years (Burgess et al. 1979). Gross lesions of duck plague are associated with damage to the vascular and lymphoid systems, and although findings differ among species, the most common lesions are necrotic, raised plaques along the longitudinal folds of the esophagus, pinpoint areas of necrosis on the liver, hemorrhage on the surface of the heart and other visceral organs, and hemorrhage on the areas of lymphoid tissue in the intestinal tract (Friend 1999b, Hansen and Gough 2007). Carcasses are generally in good body condition because of the acute nature of the disease. A definitive diagnosis requires virus isolation or detection of viral DNA in tissues or swabs.

Reports of duck plague in free-flying sea ducks include isolation of the virus from a Bufflehead (*Bucephala albeola*) that was collected with other species of waterfowl in the vicinity of the first outbreak in domestic ducks in North America, which occurred in New York, United States, in 1967 (Leibovitz 1968). Common Mergansers (*Mergus merganser*) and Common Goldeneyes (*Bucephala clangula*) were found dead during a duck plague epizootic in South Dakota, United States, in 1973 that killed more than 40,000 Mallards (*Anas platyrhynchos*; Proctor et al. 1975, Friend 1999b). Captive Buffleheads, goldeneyes (*Bucephala* spp.), and Common Mergansers died during a duck plague outbreak in a zoological park in Washington, DC, United States, in 1975, and a Barrow's Goldeneye (*Bucephala islandica*) died during a subsequent outbreak at the same park in 2005 (Montali et al. 1976, ProMed Mail 2005). Additional sea ducks reported dying of duck plague in the United States include King Eiders (*Somateria spectabilis*) and Hooded Mergansers (*Lophodytes cucullatus*), although temporal and geographical details were not provided (Converse and Kidd 2001). Duck plague was reported from Hooded Mergansers, Red-breasted Mergansers (*Mergus serrator*), Common Mergansers, and Common Eiders in a zoological collection in England between 1978 and 2003 (Hansen and Gough 2007). Captive Common Eiders died following experimental intramuscular inoculation with duck herpesvirus 1 (van Dorssen and Kunst 1955).

Duck plague could be devastating to local populations of sea ducks, particularly endangered species, and to captive collections. However, reported outbreaks of duck plague are infrequent in free-living waterfowl species, and mortality events have rarely affected sea ducks. In the absence of information regarding the relative susceptibility of sea ducks to duck plague virus, the paucity of reports in wild sea ducks may simply reflect a low frequency of exposure, because the disease has historically been reported primarily in domestic, semidomestic, and captive ducks.

Adenoviruses

Adenoviruses are a group of DNA viruses that affect multiple avian species (Fitzgerald 2007, 2008). Some adenoviruses are known to cause clinical disease and mortality, whereas others have been implicated in disease syndromes of multiple etiologies. Although adenoviruses have significant impacts in domestic poultry, relatively little is known about the occurrence and pathogenicity of this group of viruses in wild avian species, including sea ducks.

Most adenoviruses that infect domestic fowl, ducks, and pigeons are grouped in one of three types (I–III). Aviadenoviruses include multiple species and serotypes and infect domestic poultry. Siadenoviruses cause disease in chickens, turkeys, and pheasants. Atadenoviruses include the egg drop syndrome virus found in chickens and ducks, though are only known to cause illness in chickens. Several recently discovered adenoviruses from wild avian species are not yet classified and may represent other genera.

Among sea ducks, adenoviruses have been associated with die-offs of Long-tailed Ducks

(*Clangula hyemalis*) and Common Eiders. An adenovirus was linked to a mortality event in molting Long-tailed Ducks in the Beaufort Sea off the northern coast of Alaska in 2001 (Hollmén et al. 2003a). Viruses were isolated from intestinal samples of birds that showed lesions of enteritis at postmortem examination. Antibody prevalence was found higher at the location of the die-off as compared to a reference location in the Beaufort Sea with no observed mortality and to another year. The adenovirus isolated from Long-tailed Ducks was not neutralized by reference antisera against the three known adenovirus genera and likely represents a novel adenovirus serotype. In a follow-up experimental inoculation study, the adenovirus isolated from the die-off caused lesions similar to those found in the carcasses providing further evidence of the adenovirus as a cause or contributing factor to the mortality event in Long-tailed Ducks. Pathological lesions observed in the experimental study suggest that the virus found in Long-tailed Ducks may share some characteristics of type I and II avian adenoviruses, and further experimental work showed some pathogenicity of the virus to Mallards (L. F. Skerratt, James Cook University, unpubl. data). Additional genetic characterization has shown evidence of multiple adenovirus genotypes circulating in sea ducks (T. E. Hollmén, Alaska SeaLife Center and University of Alaska, unpubl. data). Isolates from the Long-tailed Ducks were further characterized by sequencing the adenovirus hexon gene, and phylogenetic characterization determined that the majority of viruses from the mortality site were most closely related to a goose adenovirus A type 4 in the *Aviadenovirus* genus (K. L. Edgar, Alaska SeaLife Center, unpubl. data).

In Common Eiders, adenoviruses have been associated with mortality in the Baltic Sea (Hollmén et al. 2003b). Viruses were isolated from intestinal tissues from male eiders found dead during the prebreeding season and were associated with mucosal necrosis and impactions of the small intestine. The carcasses were found in poor body condition with low body mass, marked pectoral muscle atrophy, and absence of subcutaneous fat reserves. The most remarkable lesions were found in the small intestines of affected birds, consisting of 45–60 cm long segments of impactions containing mussel shell fragments and vegetation. The virus was not neutralized by reference antisera against the known adenovirus types and likely represents a novel adenovirus serotype. In a follow-up experimental inoculation study, mallards developed a mild gastrointestinal illness with intestinal distention and chronic inflammatory reactions on the mucosa of the lower intestine and ceca (Hollmén et al. 2003b). The experimental findings lend support to the ability of the eider adenovirus to cause gastrointestinal illness and contribute to mortality in the wild. The large quantities of vegetation found in the impacted intestines led the authors to speculate about dietary shifts from blue mussels (Mytilus edulis) to vegetative matter as a possible contributor to pathology.

Several strains of adenoviruses have been linked to gastrointestinal disease, and in wild Common Eiders, the stress and close proximity of birds during prebreeding courtship may facilitate virus transmission and increased susceptibility among hosts. In the Baltic Sea, where the eider adenovirus was first isolated in 1998, similar mortality events had been described in earlier years in Common Eider males, and it is possible that the virus found in 1998 is linked to periodic die-offs in the region (Hollmén et al. 2003b). The prevalence of the Common Eider adenovirus in other populations of sea ducks is unknown.

Egg drop syndrome virus is another adenovirus of concern in avian species (Adair and Smyth 2008). Serologic evidence of egg drop syndrome virus has been found in Buffleheads and mergansers (Gulka et al. 1984), but it is not currently known whether this group of adenoviruses is capable of causing disease in other sea ducks.

Further studies are needed to better understand the distribution and potential pathogenicity of adenoviruses, especially in light of recent reports linking adenoviruses to mortality in two species of sea ducks, Long-tailed Ducks and Common Eiders. The newly isolated viruses from sea ducks also appear to represent novel types of adenoviruses, and further work is necessary to better characterize these viruses and their potential impact on sea duck health and populations.

Orthoreoviruses

Avian viruses within the genus Orthoreovirus have been detected in a number of avian species, mostly domestic poultry, psittacines, and captive ducks, but also in wild birds (Hollmén and Docherty 2007, Jones 2008). In domestic species, reoviruses cause a large variety of disease syndromes, including arthritis/tenosynovitis, growth retardation, enteritis, pericarditis, hepatitis, splenitis, bursal and lymphoid atrophy, and respiratory disease (Jones 2008). More recently, reoviruses have been linked to mortality in free-ranging birds, including sea ducks (Hollmén and Docherty 2007).

In the late 1990s, a reovirus was associated with large-scale mortality of Common Eider ducklings in Northern Europe (Hollmén et al. 2002). Reoviruses were isolated from bursal tissue of dead ducklings, which were in poor body condition and had hemorrhages in the liver and necrosis of liver and bursal tissues. Additional research, including experimental inoculation in Mallard ducklings, comparative postmortem examinations of ducklings in years of low mortality, and serological surveys of Common Eider females linking high seroprevalences to consequent mortality events in ducklings, provided evidence that reoviruses were responsible for the duckling mortality events. Hollmén et al. (2002) suggested in their conclusion that reoviral infections of young ducklings may cause direct mortality or lead to immunosuppressive effects on the developing bursa of Fabricius and consequent vulnerability to secondary infections or parasite infestations. Reoviruses have been associated with immune system compromise, and some viruses have been shown to be more pathogenic with coinfections by coccidian parasites (Rinehart and Rosenberger 1983; Ruff and Rosenberger 1985a,b; Montgomery et al. 1986). Renal coccidia are commonly found in eider ducklings in the Baltic Sea, and it is possible that coinfections between viruses and coccidia may significantly affect eider duckling health in this population (Hollmén et al. 1996).

Reoviruses also infect sea ducks in North America. Viruses have been isolated from Pacific Common Eiders, Harlequin Ducks (Histrionicus histrionicus), and a Long-tailed Duck in Alaska (T. E. Hollmén, unpubl. data; Skerratt et al. 2005). Additionally, serologic evidence of exposure has been found in Steller's Eiders (Polysticta stelleri) and Spectacled Eiders (Somateria fischeri) (T. E. Hollmén, unpubl. data). However, effects of reoviruses on the health of these sea duck populations are unknown.

In summary, prevalence and effects of reoviruses in sea ducks are largely unknown. Considering the occurrence of reoviruses in several sea duck species and their known capacity to cause a variety of disease syndromes in avian hosts, further study is warranted to better understand potential impacts of this group of viruses on sea duck health.

Infectious Bursal Disease Virus

Infectious bursal disease virus (IBDV) is an RNA virus in the family Birnaviridae that has been associated with bursal disease and immunosuppression in domestic poultry (Eterradossi and Saif 2008). Little is known about the occurrence and pathology of Birnaviruses in sea ducks, but evidence of exposure has been found in two species of eiders on two continents. Hollmén et al. (2000) found positive IBDV titers (≥1:64) in serum samples of Common Eiders from the Baltic Sea and Spectacled Eiders in Alaska. Prevalence of seropositive birds varied from 18% to 96% among species and locations tested. The source of IBDV exposure detected in eiders remained unknown, although Hollmén et al. (2000) speculated that sympatrically nesting gulls in the Baltic Sea may have transmitted virus to Common Eiders after foraging on infected poultry waste at landfills. A similar scenario has been proposed to explain positive antibody titers found in penguins in Antarctica (Gardner et al. 1997). In sea ducks, information about health effects of IBDV is entirely lacking. If the IBDV is capable of causing effects similar to those found in domestic poultry, these viruses may affect duckling survival through bursal infections and immunosuppression, leading to vulnerability to secondary infections and mortality (Hollmén et al. 2000).

Avian Influenza Virus

Influenza viruses are RNA viruses that belong to the family Orthomyxoviridae and have been isolated from a variety of mammalian species as well as domestic and wild birds (Stallknecht et al. 2007). All avian influenza viruses are classified as type A. Influenza type B and type C viruses occur in humans and some other mammals, but have not been isolated from birds (Swayne and Halvorson 2008). Influenza A viruses are categorized into subtypes, or serotypes, based on the characteristics of the surface proteins hemagglutinin (HA) and neuraminidase (NA) of the virus particle. Thus, there are 16 HA subtypes (H1–H16) and 9 NA subtypes (N1–N9) known to occur in wild birds, and many subtype combinations are possible. Avian influenza viruses vary in their ability to cause overt disease and are classified as low pathogenic avian influenza (LPAI) or high pathogenic avian influenza (HPAI) viruses based on virulence in chickens. HPAI virus strains primarily affect poultry, generally causing high morbidity and mortality, and all HPAI virus strains identified to date have been of the H5 or H7 subtypes (World Organization for Animal Health 2009). According to current recommendations by the World Organization for Animal Health (2009), LPAI and HPAI viruses are differentiated based on the results of the intravenous pathogenicity index following inoculation of the virus in chickens and/or deduced amino acid motifs of the HA gene at the cleavage site. Certain H5N1 and H7N9 viruses are zoonotic diseases and have the potential to cause illness or death in humans.

Wild aquatic birds are considered the natural reservoir of influenza A viruses, with most LPAI isolates coming from waterfowl and, to a lesser extent, shorebirds and gulls (Webster et al. 1992, Wallensten 2007). LPAI viruses may be shed in high concentrations in the feces of infected birds, and transmission is primarily by the fecal–oral route through contaminated water (Brown and Stallknecht 2008). Wild birds are believed to be inapparent carriers of LPAI viruses, meaning that they do not exhibit clinical signs and show no or only mild lesions of disease (Cooley et al. 1989, Stallknecht et al. 2007). However, recent studies have raised concern about biological costs, such as effects on body mass and timing of migration, that LPAI infection might have on migrating waterfowl (van Gils et al. 2007, Latorre-Margalef et al. 2009). Reports from sea ducks to date are fewer than for other waterfowl, but both LPAI viruses and the HPAI H5N1 virus have been reported from sea ducks (Stallknecht et al. 2007, Zohari et al. 2008, Hesterberg et al. 2009).

Avian influenza virus was first recognized in Italy in 1878 as a high pathogenic form in poultry. It was subsequently reported in Germany in 1890 and in the United States in 1924 and had been detected in other parts of Europe, Asia, Egypt, and South America by 1930 (Stallknecht et al. 2007, Swayne and Halvorson 2008). In wild birds, the first isolation of an HPAI virus was an H5N3 virus from Common Terns (Sterna hirundo) during a mortality event in South Africa in 1961 (Becker 1966). After the South African die-off, evidence of LPAI virus exposure in wild birds accumulated as investigators began sampling more species. However, it was not until late 2002 that HPAI viruses were again identified as a cause of mortality in wild birds, when an outbreak of HPAI H5N1 virus killed various species of waterfowl in two Hong Kong parks and free-living waterbirds in the vicinity (Ellis et al. 2004, Stallknecht et al. 2007). The HPAI H5N1 virus subsequently spread to Asia, Europe, and Africa (Yee et al. 2009). As the virus spread, HPAI H5N1 infection in wild migratory birds, as evidenced by mortality and detection of virus shedding, was reported in Anseriformes and several other orders of birds (Liu et al. 2005, Bragstad et al. 2007, Kou et al. 2009). Experimental studies with HPAI H5N1 viruses in several species, including waterfowl and gulls, have shown oropharyngeal shedding to be of greater concentration and longer duration than fecal shedding (Brown et al. 2006, 2008).

Clinical signs of HPAI reported in waterfowl and other waterbirds have included weakness and inability to fly, diarrhea, and neurological signs such as tremors and ataxia (Rowan 1962, Ellis et al. 2004, Liu et al. 2005). A variety of gross and microscopic pathologic observations have been reported in wild waterfowl affected with HPAI, including necrosis and hemorrhage in the pancreas, congestion and encephalitis in the brain, focal necrosis in the spleen and liver, and congestion, edema, and hemorrhage in the lungs (Ellis et al. 2004, Liu et al. 2005, Bröjer et al. 2009, Ogawa et al. 2009). Some of these findings have also been noted in experimental studies with HPAI in birds, but the expression of disease varies among and within species in challenge studies. In one study, Tufted Ducks (*Aythya fuligula*) and Eurasian Pochards (*A. ferina*) that developed severe clinical signs died or were euthanized within 4 days of inoculation with HPAI H5N1 virus, whereas mildly affected birds recovered after 7 or 8 days. In contrast, four species of dabbling ducks were clinically unaffected (Keawcharoen et al. 2008). In Wood Ducks (*Aix sponsa*) inoculated with HPAI H5N1 viruses, three individuals developed severe neurological signs and died or were euthanized 7 or 8 days later, one developed clinical signs but recovered, and two others showed no signs (Brown et al. 2006). In the same study, Mallard, Blue-winged Teal (*Anas discors*), Northern Pintail (*A. acuta*), and Redheads (*Aythya americana*) developed no clinical signs (Brown et al. 2006). However, 100% mortality was reported in all individuals of four species of swans, including birds directly inoculated and those acquiring the virus from contact with infected birds (Brown et al. 2008). Susceptibility of swans to HPAI H5N1 virus strains is also reflected in field reports of mortality (Bragstad et al. 2007, Nagy et al. 2007, Weber et al. 2007, Ogawa et al. 2009, Usui et al. 2009).

Clinical signs or lesions are not typically associated with LPAI, and no lesions are specific for HPAI. Thus, a diagnosis of avian influenza in live or dead birds must be confirmed by laboratory testing. Laboratory methods include direct detection of influenza A viral proteins or nucleic acids in samples with technologies such as reverse transcriptase polymerase chain reaction (RT-PCR) or virus isolation in chicken embryos (Swayne and Halvorson 2008). When such methods are used on cloacal or oropharyngeal swab material from birds, active infection and viral shedding can be detected. The presence of antibodies to avian influenza in serum or plasma samples is an indication of current or previous infection with the virus. Additional research is needed to better understand the duration of avian influenza antibody persistence in birds, but antibodies likely persist in waterfowl for at least several months (Hoye et al. 2011, Hénaux et al. 2013). Among sea ducks and other waterfowl sampled in Alaska, the frequency of birds testing positive for avian influenza antibodies in serum was higher than the frequency of birds shedding virus (Wilson et al. 2013). Antibodies to avian influenza viruses were detected in serum samples of 51% of Long-tailed Ducks, 69% of Black Scoters (*Melanitta americana*), and 86% of eiders (Common, Spectacled, and Steller's combined), whereas virus was detected in <5% of individuals of those species (Wilson et al. 2013). In another Alaskan study, Heard et al. (2008) reported plasma antibodies to avian influenza in 42% of Harlequin Ducks and 14% of Barrow's Goldeneyes.

Concern over the emergence and spread of HPAI H5N1 resulted in the development of wild bird surveillance and monitoring programs in many areas of the world (U.S. Interagency Working Group 2006, Gaidet et al. 2007, Pittman et al. 2007, Parmley et al. 2008). Sea ducks have been included in many surveillance programs and are of high priority for monitoring in Alaska because of migratory patterns of several species that provide the potential for them to carry HPAI

H5N1 viruses from East Asia to North America. Phylogenetic analysis of avian influenza viruses isolated from Steller's Eiders on the Alaska Peninsula revealed that 4.9% of the viral genes were of Eurasian origin supporting viral gene flow between continents (Ramey et al. 2011). However, published reports of HPAI H5N1 viruses in sea ducks exist for only two species, Common Merganser and Smew (*Mergellus albellus*), that were found dead or moribund during widespread wild bird mortality in the European Union (EU) in 2006 (Zohari et al. 2008, Hesterberg et al. 2009). Although these two species were not considered to be high-risk, and relatively few birds were tested, 12 of 92 (13%) Common Mergansers and 2 of 7 (29%) Smews were positive for HPAI H5N1 in the EU-wide report (Hesterberg et al. 2009).

The isolation of LPAI subtypes from sea ducks dates back to at least 1978 when H4N1 was recovered from a Bufflehead in Canada (Hinshaw et al. 1980). Influenza virus was also reported from Buffleheads in Ohio and California, United States, in 1986 and 2007–2008, respectively (Slemons et al. 1991, Hill et al. 2010). Parmley et al. (2008) detected avian influenza viruses by RT-PCR in 2 of 18 Common Goldeneye and 8 of 26 Hooded Mergansers in Canada in 2005. Of 2,332 sea ducks from 11 species sampled in Alaska from early 2006 through early 2007, 0.69% were RT-PCR positive for avian influenza, and viruses (subtypes not reported) were isolated from 1% of Steller's Eiders, 0.34% of Spectacled Eiders, 1.4% of King Eiders, 1% of Common Eiders, and 2% of Long-tailed Ducks (Ip et al. 2008). Ramey et al. (2010) reported avian influenza subtype H10N1 from a Harlequin Duck and H10N9 from a King Eider sampled at St. Lawrence Island, Alaska. Subtypes H3N8, H4N7, and H7N7 were described from King and Spectacled Eiders in western Alaska, and several subtypes have been reported in Steller's Eiders from the Alaska Peninsula (Ramey et al. 2011, Reeves et al. 2013).

In Europe, an H2N2 strain was reported from two Long-tailed Ducks and three White-winged Scoters (*Melanitta fusca*) in Germany in 1981 (Sinnecker et al. 1983). Jonassen and Handeland (2007) reported H6N2 from a Black Scoter (*Melanitta nigra*) in Norway in 2005, but samples from Common Goldeneye, Common Merganser, and Red-breasted Merganser were negative. In a study in Northern Europe from 1998 to 2006, samples from 2 of 37 Common Eiders tested positive for influenza A virus by RT-PCR, but three other sea duck species tested were negative (Munster et al. 2007). In another study, influenza viruses were detected in 22.2% of Common Eiders in Norway and 3.4% of Common Eiders in Denmark, where one H5 subtype was identified (Germundsson et al. 2010, Hjulsager et al. 2012).

Less is known about avian influenza viruses in sea ducks than in other waterfowl, but it is reasonable to assume that these viruses circulate naturally within sea duck populations. As with other wild birds, sea ducks are presumed to be inapparent carriers of LPAI viruses, albeit with low viral shedding rates. However, HPAI H5N1 has been detected in individuals of at least two species of sea ducks found dead during mortality events, and it is possible that sea ducks are similar to some diving ducks in being more susceptible to infection than dabbling ducks. Thus, as with other infectious diseases, the potential effects of highly pathogenic forms of avian influenza viruses in sea ducks are of concern, particularly for threatened populations.

Wellfleet Bay Virus

Between 1998 and 2011, estimated losses of more than 6,000 Common Eiders along the coast of Cape Cod in Massachusetts, United States, have been attributed to a novel virus of the family Orthomyxoviridae, tentatively named Wellfleet Bay virus (Ballard et al. 2012, Pello and Olsen 2013). Little is currently known about the epidemiology or pathology associated with this newly discovered virus. To date, the virus has only been found in Common Eiders, and deaths have apparently been restricted to the Cape Cod area. However, 18 of 66 (27%) Common Eiders sampled along the coast of Rhode Island were seropositive to Wellfleet Bay virus (Olsen et al. 2012).

Poxviruses

Avian poxviruses are DNA viruses in the genus *Avipoxvirus*. Avian pox is characterized by proliferative lesions on the skin and typically is not fatal unless the lesions occur around the eyes or on mucous membranes in the oral cavity (van Riper and Forrester 2007). Poxviruses can be transmitted by biting insects as they carry the virus from infected to susceptible birds and by direct contact with infected birds or contaminated objects (van Riper and Forrester 2007). Little is known about poxvirus infections in sea ducks, but viruses belonging to this group have been found in at least three sea duck species in North America, including a Harlequin Duck from Alaska, Common Goldeneye from New York and Saskatchewan, and a Black Scoter from Pennsylvania (Hansen 1999). In the Harlequin Duck, the poxvirus was associated with bumblefoot-type lesions.

West Nile Virus

WNV is an arbovirus (virus transmitted by arthropods) of the family Flaviviridae and is a zoonotic disease. The predominant vectors are ornithophilic mosquitoes of the genus *Culex*, which transfer the virus from animal to animal as they consume blood meals. Birds are the primary hosts of WNV, and the virus has been reported from at least 326 native and exotic avian species in North America, including Barrow's Goldeneye, Common Goldeneye, Common Merganser, Hooded Merganser, Bufflehead, and Smew (Meece et al. 2006, Centers for Disease Control and Prevention 2011). The virus was discovered in 1937 in Uganda, and its historical range extends from Africa to the Middle East, Europe, and Asia (McLean and Ubico 2007). In 1999, WNV was reported in North America, the first documentation of the virus in the Western Hemisphere, and it has since spread throughout the United States and into Canada, Central and South America, and the Caribbean and now occurs on all continents except Antarctica (Artsob et al. 2009, Weissenböck et al. 2010). The virus strain introduced to North America was of greater virulence in birds, and whereas WNV typically has not been associated with bird mortality in most other parts of the world, outbreaks of WNV in North America have been characterized by substantial numbers of avian deaths, particularly in corvids (McLean and Ubico 2007, Brault 2009). Bird species differ considerably in susceptibility to disease when infected by the North American strain of WNV and clinical signs may be absent or may include lethargy, ruffled feathers, unusual posture, inability to hold the head up, and staggering gait (Komar et al. 2003). Gross lesions may include hemorrhage on the brain, enlargement of the spleen, and light patchy areas on the surface of the heart. Diagnostic methods include virus isolation, polymerase chain reaction, and/or immunohistochemistry (Steele et al. 2000).

Although waterfowl have generally been considered relatively resistant to the virus, WNV mortality was diagnosed in Mallard and Bronzewinged (Spectacled) Ducks (*Anas specularis*) in a zoological collection in the 1999 outbreak in New York, United States, and in captive Lesser Scaup (*Aythya affinis*) ducklings in Canada (Steele et al. 2000, Himsworth et al. 2009). In another disease outbreak in captive waterfowl, WNV was detected in one dead Barrow's Goldeneye, and serum antibody against WNV was detected in another Barrow's Goldeneye with clinical signs of exposure (Meece et al. 2006). Common Goldeneye, Common Merganser, Hooded Merganser, Bufflehead, and Smew appear in a US WNV mortality database, but details are unavailable, except that all but the Hooded Mergansers were listed as captive birds (Centers for Disease Control and Prevention 2011).

Little evidence indicates that WNV is an important disease of sea ducks, but the virus has greatly expanded its range in recent years and a mutation associated with increased replication in avian hosts has been identified (Brault 2009). As of 2004, WNV activity among birds in North America was detected as far north as 57.7°, and further northward expansion could result from the effects of global climate change on such factors as changes in vector population distributions and increased duration of the transmission season (Parkinson and Butler 2005, Brault 2009).

BACTERIA

Bacteria are ubiquitous unicellular organisms, ranging from symbiotic to highly pathogenic in their relationship with host organisms. Bacteria are classified based on their shape, structure, and biochemical properties, and their taxonomy is constantly evolving with the advancement of identification and characterization techniques. In this section, we provide an overview of the primary bacterial pathogens reported in sea ducks to date. As is true for potential pathogens in general, information about bacterial diseases in sea ducks is limited to a set of commonly studied agents that are reported from relatively few sea duck populations. However, because bacteria are cosmopolitan and often have been found in environments inhabited by sea ducks, it is likely that bacteria play a key role in sea duck health and survival.

Avian Cholera

Avian cholera (avian pasteurellosis, fowl cholera) is an infectious disease of birds caused by the gram-negative bacterium *Pasteurella multocida*, with natural infections reported in more than 180 species of birds including 9 species of sea ducks (Samuel et al. 2007). The disease was first described in domestic birds in Europe in the late 1700s, mortality occurred in several European countries throughout the 1800s, and was reported in domestic birds in North America in the late 1800s (Samuel et al. 2007, Glisson et al. 2008). The first report of disease consistent with avian cholera in wild waterfowl was from Egyptian Geese (*Alopochen aegyptiaca*) and Spur-winged Geese (*Plectropterus gambensis*) on Lake Nakuru, Kenya, in 1940 (Hudson 1959). In North America, avian cholera was first reported in wild waterfowl in the 1940s, and large-scale die-offs were common from the 1970s until the late 1990s (Samuel et al. 2007). The disease was reported in wild birds in Europe in the 1970s and in Korea in 2000 (Samuel et al. 2007). Three subspecies of *P. multocida* are recognized (*P. m. multocida, P. m. gallicida,* and *P. m. septica*). Further subgroupings or serotypes are based on serologic determination of capsular antigens (A, B, D, E, and F) and somatic antigens (1–16) (Glisson et al. 2008). Among 295 isolates obtained from waterfowl and other wild birds in California, United States, 63% were *P. m. multocida*, 37% were *P. m. gallicida*, and 0.7% were *P. m. septica* (Hirsh et al. 1990). Most *P. multocida* strains in birds are capsular type A (Rhodes and Rimler 1987), and some strains may react with more than one somatic antigen and are identified accordingly by two serotype numbers (e.g., 3,4). At least 9 of the 16 serotypes have been reported in wild birds, with serotypes 1,3 and 3,4 commonly reported in sea ducks (Samuel et al. 2007). Molecular techniques have been used to further differentiate strains of *P. multocida* (Samuel et al. 2007, Blehert et al. 2008).

Wild waterfowl, including sea ducks, are thought to acquire avian cholera primarily by ingestion and inhalation (Botzler 1991). Nasal discharges may contain large amounts of *P. multocida*, and the ingestion of water contaminated with exudates from sick and dead birds is an important mode of transmission of avian cholera in waterfowl (Wobeser 1997). Similarly, when water containing *P. multocida* is aerosolized as the surface is disturbed by large concentrations of birds, the aerosols can serve as a source for inhalation of the organism. An important area of investigation regarding the epizootiology of avian cholera in waterfowl focuses on where the organism survives from year to year, whether in the environment, in birds that carry and shed it, or both. Contaminated wetland environments play an important role in the transmission of avian cholera in waterfowl, but it is unknown whether wetland ecosystems can serve as year-round reservoirs of *P. multocida* (Samuel et al. 2007). However, some species of waterfowl function as reservoirs (Samuel et al. 2007).

Avian cholera epizootics are often characterized by rapid onset with high mortality and have resulted in the deaths of tens of thousands of waterfowl in single events, particularly in North America (Friend 1999a). Avian cholera is an acute disease of waterfowl, and the overwhelming majority of birds are simply found dead, with no opportunity for the observation of clinical signs. For example, nesting female Common Eiders have been found dead on or near their nests (Korschgen et al. 1978). Clinical signs of avian cholera that have been observed in birds are nonspecific and include lethargy, convulsions, erratic flight, mucous discharge from the bill, and matting of the feathers around the vent, eyes, and bill (Friend 1999a). Dead birds are typically in good body condition but may have congested lungs, pinpoint pale foci on the liver, and mucoid material in the intestine (Samuel et al. 2007). A definitive diagnosis of avian cholera requires isolation and identification of *P. multocida*. Although liver samples are generally preferred, blood and a variety of other tissues, including bone marrow from scavenged carcasses, can be used for culture. Selective media have been developed that may increase culture success (Samuel et al. 2007).

Avian cholera has been reported in at least nine species of sea ducks and has caused substantial mortality, particularly in nesting Common Eiders and wintering Long-tailed Ducks. An early report describes the estimated mortality of over 200 Common Eiders on nesting islands off the coast of Maine, United States, in 1963 (Gershman et al. 1964). Korschgen et al. (1978) studied avian cholera in eider nesting areas along coastal Maine from 1970 through 1976, reporting mortality of over 600 birds during five nesting seasons. The authors also isolated the organism from a stagnant pool containing dead eiders, from oropharyngeal swabs of 1 of 357 live nesting eiders, and from organ cultures of 1 of 236 Common Eiders collected during the overwintering period. Additional avian cholera mortality on nesting islands in Maine occurred during several years in the 1980s (Skerratt et al. 2005). Of a total population of about 4,000 nesting females at Île Blanche, Québec, an estimated 1,700 Common Eiders died during avian cholera outbreaks in 1964 and 1966 (Reed and Cousineau 1967). The severity of the outbreaks, at least in part, was attributed to high nesting densities, physiological stress of egg laying and incubation, and the occurrence of stagnant pools in the nesting area. Outbreaks occurred again on Île Blanche in 1976, 1985, and 2002, the latter event resulting in ~7,000 mortalities (87% females) or 19% of the nesting population in the St. Lawrence estuary (Joint Working Group on the Management of the Common Eider 2004). At Southampton Island, Nunavut, Canada, avian cholera outbreaks occurred during 2005–2007, resulting in the mortality of 5%–32% of breeding Common Eiders (Descamps et al. 2009). Descamps et al. (2012) modeled the effects of avian cholera on this population, finding more than one outbreak per decade to be unsustainable for the colony and predicting that more than four outbreaks per decade would drive the colony to extinction within 20 years. In a Danish nesting colony of Common Eiders, avian cholera accounted for a 14% decrease in annual survival of adult females (Tjørnløv et al. 2013).

An estimated 80–100 Common Eider females died of avian cholera, many on or near their nests, on the island of Vlieland, the Netherlands, in 1984 (Swennen and Smit 1991). The authors noted that the age composition of females at the colony was affected, as 5 years later, there was a greater frequency of young breeding females at the die-off site than other nesting areas (Swennen and Smit 1991). According to Christensen et al. (1997), an outbreak of avian cholera in Denmark in 1996 killed at least 900 male and female Common Eiders in late winter and 3,146 females at several breeding colonies, representing 35%–95% of the females within each colony. Subsequent avian cholera outbreaks in Denmark in 2001 and 2003 involved Common Eiders and several other species of waterbirds, including Red-breasted Mergansers (Pedersen et al. 2003). In 1997, a suspected avian cholera die-off killed an estimated 500–1,000 birds, including Common Eiders, on an island off the coast of Gotland, Sweden (Persson 1998). Mortality occurred again the following year, and in May of 1998, *P. multocida* was isolated from several species, including Common Eiders (Persson 1998). Aside from outbreaks in eiders, avian cholera has also killed tens of thousands of wintering sea ducks of other taxa, primarily during three outbreaks in the Chesapeake Bay area of Maryland and Virginia, United States. Locke et al. (1970) reported that during and after an avian cholera die-off in early 1970 involving primarily sea ducks, at least 4,780 carcasses were collected including Long-tailed Duck (70.5%), White-winged Scoter (18.7%), Bufflehead (2.6%), and Common Goldeneye (1.8%). Avian cholera occurred again in the Chesapeake Bay in 1978. Of 31,295 dead birds collected during the outbreak, most were diving ducks, with Long-tailed Ducks making up 80% of the total (Montgomery et al. 1979). In addition to Long-tailed Ducks, *P. multocida* was isolated from tissues of goldeneye, Bufflehead, Surf Scoter (*Melanitta perspicillata*), White-winged Scoter, Black Scoter, and Red-breasted Merganser (Montgomery et al. 1979). Another outbreak in Chesapeake Bay in 1994 killed more than 36,700 birds, with Long-tailed Ducks comprising 86% of the total (Hindman et al. 1997). In the United States, additional cases of avian cholera associated with low level mortality were identified in migrating and wintering sea ducks, including Common Merganser, in seven states (Skerratt et al. 2005). In Europe, carcasses of 44 sea ducks, including Common Goldeneye, White-winged Scoter, Black Scoter, and Red-breasted Merganser, were picked up during a mortality event attributed to avian cholera in January and February 1977 in the Netherlands (Mullié et al. 1979).

As transmission of avian cholera may be facilitated by high bird densities, a few attempts have been made to disinfect small pools and ponds or to modify habitat to prevent the disease from recurring at Common Eider nesting areas (Gershman et al. 1964, Swennen and Smit 1991, Dallaire and Giroux 2005). However, even if avian cholera did not occur in subsequent years, it is not possible to know if such actions prevented disease reemergence. During epizootics, carcass removal and disposal is generally recommended to lessen contamination of the environment, reduce the likelihood of a decoy effect that can attract susceptible birds to the site, and prevent the spread of the disease by scavengers, but the success of carcass removal has not been rigorously tested (Botzler 1991). Since the late 1990s, avian cholera mortality events have declined in North America (Samuel et al. 2007). However, it can be anticipated that sporadic outbreaks of avian cholera, some involving wintering sea ducks, will continue, and this disease has the potential to have severe impacts on nesting populations.

Colibacillosis

The disease caused by avian pathogenic *Escherichia coli* is often referred to as avian colibacillosis. *E. coli* is a gram-negative rod bacterium belonging to the family Enterobacteriaceae (Bopp et al. 2003). Many strains of *E. coli* are part of normal intestinal flora of birds, and the bacterium is ubiquitous in animal feces. However, some strains are known to be pathogenic and may cause a variety of diseases in birds, including intestinal, respiratory, cardiac, and hepatic illness.

Little is known about the prevalence of *E. coli* or occurrence of colibacillosis in sea ducks. In a recent study in Alaska, Steller's Eiders and Harlequin Ducks were found to be positive for *E. coli* carrying virulence attributes associated with avian colibacillosis (Hollmén et al. 2011). In this study, serum biochemistry provided further evidence of pathogenicity in Steller's Eiders. The source of the bacterium found in sea ducks was suspected to be from sewage outfall sites in nearshore foraging habitats. Information about other species and geographic regions is lacking.

Chlamydiosis

Avian chlamydiosis (also known as psittacosis or ornithosis) is a bacterial disease of birds caused by *Chlamydophila psittaci* (previously *Chlamydia psittaci*). The severity of disease varies according to the strain of the organism and host species, and although infections may be common in wild birds, clinical signs are often considered to be mild and mortality rates are typically low. The mode of transmission of chlamydia in wild birds is not well known, but the primary routes are likely the ingestion and inhalation of infectious material when birds come in contact with individuals that are shedding the organism in feces or through nasal and ocular secretions, or with environments contaminated with such discharges (Andersen and Franson 2007). Chlamydiosis is a zoonotic disease that can be transferred from waterfowl or other birds to humans (Wobeser and Brand 1982).

Few clinical signs associated with chlamydiosis have been reported in wild birds, as most infections are thought to be latent or without signs of illness. If susceptible hosts are infected with more virulent strains, expected signs may include ocular and nasal discharge, diarrhea, weakness, ruffled feathers, tremors, abnormal gait, or, in severe cases, respiratory distress and sudden death (Andersen and Franson 2007). Gross lesions include a greatly enlarged spleen or liver and pericarditis, an inflammation and thickening of the pericardial sac surrounding the heart (Franson and Pearson 1995). Diagnostic confirmation of chlamydiosis in wild birds requires demonstration of the organism in tissue samples or smears of feces or exudate, typically by laboratory isolation or visualization with special stains, or detection of chlamydial DNA by molecular techniques (Andersen and Franson 2007). Serological tests for antibodies have limited use in the diagnosis of chlamydiosis in wild birds but have been used in epizootiological studies to detect past infection in waterfowl (Andersen and Franson 2007, Docherty et al. 2012). Among sea ducks, chlamydiosis has been reported in the Common Goldeneye, Black Scoter, and Smew (Burkhart and Page 1971).

Erysipelothrix rhusiopathiae

E. rhusiopathiae is a gram-positive rod bacterium in the genus *Erysipelothrix* (Bille et al. 2003). Transmission of *Erysipelothrix* is thought to be through ingestion or wound infection, and in wild birds, there are typically no specific clinical signs other than acute death (Wolcott 2007). *E. rhusiopathiae* is a zoonotic organism, capable of transmitting between animals and humans and causing illness in humans. *E. rhusiopathiae* was detected in Common Mergansers that were found dead in a large mortality event involving multiple avian species in the Great Salt Lake in the 1970s. The bacterium was determined to be the cause of that die-off, but no further information is available about the occurrence and pathogenicity of this organism in sea ducks.

Mycobacterium avium

Mycobacteria are acid-fast rod bacteria belonging to the family Mycobacteriaceae. The genus *Mycobacterium* includes many well-known pathogens, among them *M. avium*, the most common cause of avian tuberculosis. Avian tuberculosis affects many species of birds and is typically a chronic, slow-developing disease resulting in poor appetite, weight loss, weakness, and lethargy leading to death (Converse 2007). Lesions often consist of light-colored nodules in abdominal viscera if Mycobacteria were ingested or, in lungs and air sacs, if the organism was inhaled. Lesions may be localized in one tissue or broadly disseminated (Converse 2007). Sea ducks of several species are known to be susceptible to tuberculosis as the disease was diagnosed in nearly 50% of tribe Mergini mortalities in a captive wildfowl collection (Cromie et al. 1991). Little is known about the occurrence and manifestations of mycobacterial disease in wild sea ducks, but characteristic lesions were noted in dead Common Eiders examined in Scotland. Additionally, typical acid-fast organisms were seen in lung tissue of a Black Scoter from the German North Sea coast, and *M. avium* was detected in disseminated abscesses in a Long-tailed Duck found dead of traumatic injury on the North Slope of Alaska (Garden 1961, Skerratt et al. 2005, Siebert et al. 2012).

FUNGI

Fungi are single-celled or multicellular organisms that vary in pathogenicity, from opportunistic colonizers to true pathogens. Fungi are ubiquitous in the environment. Multiple genera of fungi are reported as pathogenic, with the genus *Aspergillus* most commonly reported in birds.

Aspergillosis

Aspergillosis is a disease caused by fungal spores of *Aspergillus fumigatus* or *A. niger*. *Aspergillus* spores are ubiquitous in soil and organic matter, and disease is most commonly caused by inhalation of spores from an environmental source. Infection of the respiratory tract is a common form of the disease, but infections also have been found to occur in the skin, bones, eyes, gastrointestinal tract, and central nervous system (Kunkle 2003, Beernaert et al. 2010). Both acute and chronic forms of the disease can occur in birds, and the most common source of exposure in captive held birds is via moldy feed or bedding materials. Impaired immunity may predispose individuals to more severe manifestation of the disease (Beernaert et al. 2010).

A. fumigatus is the most pathogenic species in wild birds and has been found to affect sea ducks. *A. fumigatus* has caused disseminated granulomas in Common Eiders, Buffleheads, and Common Mergansers (Skerratt et al. 2005). In captivity, eiders are thought to be particularly susceptible to aspergillosis on account of multiple reports of infections (Hubben 1958, Hillgarth and Kear 1979, Stetter et al. 1994). Some serologic evidence supports exposure to *Aspergillus* spores in wild sea ducks, including Steller's Eiders and Spectacled Eiders (T. E. Hollmén, unpubl. data). Potential impacts on wild populations of sea ducks are currently unknown.

PARASITES

Parasites are defined as organisms that live in or on another organism by deriving nutrients from its host. Parasites may be unicellular or multicellular organisms and are classified based on their structure and life cycle. Much of the work relating to parasites of wild birds, including sea ducks, has involved taxonomic classification of specimens, and it is likely that the majority of parasites of wild birds have yet to be identified. Available information from wild birds is largely descriptive, and little is known about potential population-level impacts and fitness trade-offs relating to parasitism. Nevertheless, some exciting work in this field has been conducted in sea ducks. In an experimental field study of incubating Common Eiders, Hanssen et al. (2003) found that return rates of unsuccessful females treated with an antiparasitic drug were higher than a reference group of unsuccessful females that were untreated. The authors suggested that drug treatment may have resulted in reduced mortality, leading to a higher rate of return.

This section is organized into two general parts, summarizing information about unicellular protozoan parasites and multicellular metazoan parasites, especially helminths. A vast body of taxonomic literature exists describing helminths and other parasites recovered from specimens of sea ducks. A large portion of this literature consists of descriptive reports, many of which lack information about the pathogenicity of the organism or significance to the host. We include a summary compilation of these reports, arranged by taxonomic class and species of parasite among the various sea duck species, with references to guide the reader to the wealth of source information on that topic in the electronic Appendix (www.crcpress.com/product/isbn/9781482248975). Most species of parasites detected in sea ducks have not been commonly associated with disease, or their pathogenicity has not been further studied.

Protozoan Parasites

Coccidia

Coccidia are unicellular parasites that are known to cause a wide variety of diseases in multiple avian groups, including sea ducks. In sea ducks, coccidian parasites belonging to the genus Eimeria have been most commonly associated with renal disease, but intestinal manifestations of coccidiosis also have been reported.

Renal coccidia have been detected in several species of sea ducks, and infections are generally thought to be most pathogenic to young birds (Wobeser 1997). Franson and Derksen (1981) reported renal coccidia resembling E. somateriae in adult Long-tailed Ducks in Alaska. The parasites were associated with macroscopic or microscopic lesions in the affected kidney tissues, but the authors concluded that the birds were in good body condition and that the renal parasites were likely not causing clinical illness in the infected individuals. Renal coccidiosis has been found in Common Eider ducklings in multiple breeding areas, and the disease has been linked to illness and mortality in ducklings (Christiansen 1952, Waldén 1961, Persson et al. 1974, Mendenhall 1976, Hollmén et al. 1996, Skirnisson 1997). In infected ducklings, kidneys were swollen and mottled with white foci. Microscopic findings included destruction of cellular tissue with large numbers of oocytes, and several authors have suggested renal coccidiosis as a significant source of mortality in young Common Eider ducklings. Renal coccidia also have been reported in other sea duck species, including Common Goldeneyes in Canada and in Red-breasted Mergansers in Florida (Gajadhar et al. 1983, Forrester and Spalding 2003).

Skirnisson (1997) described an outbreak of intestinal and renal coccidiosis, caused by Eimeria sp. and E. somateriae, respectively, in Common Eider ducklings in Iceland in 1993. The author associated the outbreak with poor foraging conditions among malnourished ducklings. Christiansen and Madsen (1948) described outbreaks of intestinal coccidiosis caused by E. bucephalae in young Common Goldeneyes in Denmark, and Skerratt et al. (2005) reported intestinal coccidiosis in Common Eiders and White-winged Scoters in the United States.

Sarcocystis

Sarcocystis sp. are coccidian parasites with an indirect life cycle involving an intermediate host, often avian, and a carnivorous definitive host. In birds, including sea ducks, a characteristic of infection with this parasite is the formation of sarcocysts, elongate spindle-shaped structures that can be visualized in muscle tissue (Greiner 2008). In North America, Sarcocystis rileyi has been documented in White-winged Scoters and goldeneyes (Bucephala spp.), and Sarcocystis sp. was reported in Steller's Eiders and a Black Scoter (Skerratt et al. 2005, Greiner 2008). Kutkiené and Sruoga (2004) found Sarcocystis sp. in Black Scoters, White-winged Scoters, Long-tailed Ducks, Common Mergansers, one Common Goldeneye, and one Common Eider in Lithuania. Health effects of Sarcocystis sp. on the sea duck host are unknown, but many species of birds can have large numbers of sarcocysts in muscle tissue without exhibiting clinical signs of disease (Greiner 2008).

Hemoparasites

Hemoparasites are unicellular vector-transmitted protozoan blood parasites with complex indirect life cycles. Hemoparasites have been widely studied in many avian species and have been reported to occur in sea ducks. Hemoparasites are known to be pathogenic in many avian species, with potential population-level effects. Historically, few blood parasites have been reported in birds from northerly latitudes. However, due to their life cycle involving transmission by vectors, some species have been suspected to exhibit northward range expansion due to temperature increase and redistribution of vectors associated with climate warming. Such ecological changes may expose more northerly species, such as sea ducks, to new parasites. Currently, little information is available about baseline prevalence, health effects on the host, or potential population-level effects of hemoparasites on sea duck populations.

Avian malaria is a mosquito-transmitted disease caused by unicellular protozoan parasites of the genus *Plasmodium*. *Plasmodium* sp. has been reported as a cause of death in Spectacled Eiders in captivity. Although the affected captive eiders were maintained and exposed to infection at latitudes south from their natural range, these findings indicate the potential pathogenicity of *Plasmodium* sp. in eiders. *Plasmodium circumflexum* was found in a Spectacled Eider that was trapped on the Yukon Delta, Alaska, and moved to captivity in Manitoba, where it later died (Savage and McTavish 1951). In free-living sea ducks of North America, *Plasmodium* sp. has been reported in Common Goldeneyes, Buffleheads, Hooded Mergansers, Common Mergansers, Spectacled Eiders, and Common Eiders, but the consequences of infection for these species are unknown (Greiner et al. 1975, DeJong et al. 2001).

Leucocytozoon spp. are transmitted by black flies in the family Simuliidae. *Leucocytozoon simondi* has been identified in multiple species of sea ducks, including Long-tailed Ducks, all three mergansers, Surf Scoters, Black Scoters, Smew, Common Goldeneyes, Harlequin Ducks, Spectacled Eiders, Common Eiders, and Steller's Eiders (Forrester and Greiner 2008, and references therein; T. E. Hollmén, unpubl. data). Identifications have been reported in sea duck samples from United States, Canada, and Russia. The pathogenicity of *Leucocytozoon* spp. in sea ducks is unknown.

Haemoproteus spp. include some of the most widely spread hemoparasites of wild birds, but their pathogenicity and ecology are poorly understood. Little is known about manifestations of *Haemoproteus* infections in sea ducks, but parasites representing this group have been found in goldeneyes, Long-tailed Ducks, Hooded Mergansers, Common Mergansers, Red-breasted Mergansers, Black Scoters, Surf Scoters, and Common Eiders (Greiner et al. 1975; DeJong et al. 2001; T. E. Hollmén, unpubl. data).

Helminths

A large variety of helminths, including cestodes, nematodes, trematodes, and acanthocephalans, have been reported in sea ducks. McDonald (1969) provided a bibliography of helminths in Anatidae, and a large number of reports have followed.

We have compiled reports of helminths and other parasites in sea ducks and provide a list of references in the electronic Appendix (www.crcpress.com/product/isbn/9781482248975). Some helminths have been commonly reported to occur in sea ducks, such as the gizzard worm *Amidostomum* sp. Nevertheless, with the exception of acanthocephalans, little is known about the health effects of most helminth species on the sea duck host and host populations. Therefore, we focus our discussion on acanthocephalans but present a complete list of the occurrence of helminths in sea ducks in the electronic Appendix (www.crcpress.com/product/isbn/9781482248975).

Acanthocephalans

Acanthocephalans, or thorny-headed worms, have been reported in multiple species of sea ducks and have been associated with pathology both by observations during postmortem examination of carcasses and in experimental studies. Acanthocephalans cause intestinal infections, and mortality may be caused by heavy infestation and penetration of the proboscii (mouth parts) of the worms through the intestinal wall, leading to inflammation of the coelomic wall.

Acanthocephalans have been widely reported in Common Eiders and implicated in mortality of both adults and ducklings (Clark et al. 1958, Kulatchkova 1960, Garden et al. 1964, Grenquist 1970, Bishop and Threlfall 1974, Persson et al. 1974, Itamies et al. 1980). In Common Eiders on the Baltic Sea, acanthocephalans have been associated with large-scale mortality events and are considered a significant cause of duckling mortality and a potential factor in eider population declines (Grenquist 1951; Persson et al. 1974; Hario et al. 1992; Hollmén et al. 1996, 1999). Eiders that breed on the archipelago of the Baltic Sea become infected with *Polymorphus minutus* worms by consuming amphipods (*Gammarus* sp.). Amphipods are one of the primary food resources of young ducklings, thus resulting in exposure to infection at a young age. Experimental infections have shown that even relatively low numbers of parasites can cause measurable health effects in ducklings (Hollmén et al. 1999). Therefore, Hollmén et al. (1999) concluded that acanthocephalans may have contributed, alone or with other cofactors, to the low duckling survival and population decline of Common Eiders observed in the Baltic Sea in the 1990s.

Bustnes and Galaktionov (2004) studied trade-offs between energy intake and avoidance of acanthocephalan parasites in Steller's Eiders in northern Norway. The authors hypothesized that eiders in good body condition avoided intermediate hosts of acanthocephalan parasites as prey and thus negative consequences on their body condition. Indeed, they found a negative relationship between body condition and the proportion of prey known to be an intermediate host to acanthocephalans, providing evidence for trade-offs between energy intake and parasite avoidance.

BIOLOGICAL TOXINS

Biological toxins or biotoxins are poisonous substances produced by living cells or organisms. We present information about two types of biotoxins that affect sea ducks.

Avian Botulism

Botulism is caused by neurotoxins produced by the bacterium Clostridium botulinum, of which seven types (A–G) are recognized. The disease is typically caused by the ingestion of the toxins in food items. Botulism type C is the form that has the biggest impact on birds in general, and although it is most common in filter feeding and dabbling waterfowl and probing shorebirds, it has been diagnosed in at least 263 species from 39 families, including eight species of sea ducks (Rocke and Bollinger 2007). Optimal growth of type C strains occurs at relatively warm temperatures of 25°C–40°C, and most outbreaks occur in wetlands in the summer and fall when ambient temperatures are higher (Rocke and Bollinger 2007). Reports of bird mortalities in western North America that may have been caused by avian botulism type C date back to the late 1800s. By 1930, duck sickness had been observed in at least 14 western states in the United States and the Canadian provinces of Saskatchewan and Alberta (Kalmbach and Gunderson 1934). The magnitude of type C botulism mortality is highly variable, but outbreaks involving losses of 50,000 waterbirds have historically been relatively common (Wobeser 1997, Rocke and Bollinger 2007).

Botulinum neurotoxins interfere with transmission of impulses at the neuromuscular junction, leading to clinical signs of paralysis. Affected birds may exhibit difficulty taking flight or landing, leg weakness that may result in attempts to propel themselves by their wings, recumbency, and paralysis of the nictitating membrane, and as the cervical muscles become affected, the head may rest on the ground or in the water, leading to drowning as the immediate cause of death (Wobeser 1997, Rocke and Friend 1999). Typically, no gross or microscopic lesions are observed. A preliminary diagnosis of botulism can be made on the basis of clinical signs but must be confirmed by laboratory testing for the toxin. The most commonly used test is the mouse protection test (Quortrup and Sudheimer 1943). Briefly, a blood sample is collected from a vein of a live bird or from the heart of a bird that recently died. Serum is then inoculated into two groups of mice: one protected by specific antitoxin and a reference group. If the protected mouse survives and the unprotected mouse dies or becomes sick, the sample is considered positive for botulism toxin (Rocke and Bollinger 2007).

Treatment can be successful in 75%–90% of waterfowl with type C botulism (Rocke and Friend 1999). Thus, treatment is a viable option for sea ducks, particularly if endangered species or valuable captive collections are affected. Mildly affected birds may recover if simply given easy access to water and food and protection from inclement weather and predators, whereas birds exhibiting more severe clinical signs can be dosed orally with water and, in cases of type C botulism, can be injected with type C antitoxin (Rocke and Friend 1999). Type E antitoxin is available, but little is known about its effectiveness in birds (Rocke and Bollinger 2007). The use of a toxoid vaccine against type C botulism has shown some success in protecting waterfowl during botulism outbreaks (Martinez and Wobeser 1999, Rocke et al. 2000).

Sea ducks typically make up a small proportion of waterfowl afflicted with type C botulism, but a 1998 outbreak in Alberta, Canada, killed an estimated 100,000–250,000 Buffleheads (Skerratt et al. 2005, Rocke and Bollinger 2007). Early reports of sea ducks dying of probable type C botulism were from the Western United States and included Common Goldeneyes, Buffleheads, and Red-breasted Mergansers in 1932 (Kalmbach and Gunderson 1934). A small number of Black Scoters and Common Eiders were among birds found dead of avian botulism type C in Britain in 1975 (Lloyd et al. 1976). Type C botulism was also confirmed in a wild Hooded Merganser in Wisconsin, United States, in 1984 and in Common Eiders and Barrow's Goldeneyes in a zoological park in Colorado, United States, in the mid-1990s (Cambre and Keimy 1993, Skerratt et al. 2005).

In recent years, botulism type E has posed a greater threat to sea ducks, causing considerable mortality in fish-eating birds in the Great Lakes region of North America. Included in the 31 species of birds known to have been diagnosed with botulism type E are eight sea ducks (Rocke and Bollinger 2007). *C. botulinum* type E prefers temperate, fresh, and brackish water environments. The ability of spores to germinate at lower temperatures gives it an advantage in cold water environments, although outbreaks have also occurred in the hot, saline environment of the Salton Sea in Southern California, United States (Rocke and Bollinger 2007). In Lake Michigan in 1963 and 1964, botulism type E was reportedly responsible for the deaths of at least 12,600 birds, mostly Common Loons (*Gavia immer*) and three species of gulls, although several Long-tailed Ducks, unidentified mergansers, and Common Goldeneyes were also found dead (Fay et al. 1965). Die-offs continued sporadically in Lake Michigan through the early 1980s, mostly in species other than sea ducks, although Long-tailed Ducks and White-winged Scoters were apparently affected in a 1976 mortality event (Brand et al. 1983, 1988). Type E botulism mortality in the Great Lakes shifted east from Lake Michigan in the 1960s through the 1980s to lakes Huron, Erie, and Ontario in the late 1990s and early 2000s. Multiple genetically distinct strains of *C. botulinum* have been involved with mortality events in recent years (Rocke and Bollinger 2007, Hannett et al. 2011). Along with the geographic shift came an increase in the frequency of sea ducks being affected, such as Long-tailed Ducks, Red-breasted Mergansers, Common Mergansers, and Surf Scoters (Carpentier 2000). For example, estimated bird mortalities attributed to botulism type E in the Great Lakes included more than 2,000 Red-breasted Mergansers in 2000 and more than 13,000 Long-tailed Ducks in 2002 (Adams et al. 2009).

Cold marine environments are not favorable to outbreaks of avian botulism, but sea ducks that use fresh water habitats during part of their life cycle and species that come ashore to nest near brackish or fresh water may be at risk. Apart from a massive mortality event in Buffleheads in Alberta,

Canada, sea ducks historically have made up a relatively small proportion of birds dying of botulism type C. In contrast, type E botulism mortality in sea ducks has recently been on the rise in the Great Lakes region of North America. Type C or type E avian botulism have the potential for devastating effects on local populations, endangered species, and captive collections of sea ducks. Skerratt et al. (2005) suggested that the detection of type E botulism in a Common Goldeneye in Alaska may provide a warning signal about the expansion of the disease in North American sea ducks.

Algal Toxins

About 80 marine and 55 species of freshwater toxin-producing microalgae have been identified (Landsberg et al. 2007). Rapid growth of toxic algae can result in harmful algal blooms that may have adverse effects on a variety of aquatic organisms. The major groups of toxins include saxitoxins, brevetoxins, domoic acid, and cyanotoxins (anatoxin, microcystin, nodularin), which are primarily neurotoxic or hepatotoxic. Wildlife may be exposed by consuming contaminated food items or from direct ingestion of contaminated water during foraging or drinking (Landsberg et al. 2007). For example, paralytic shellfish poisoning, caused by saxitoxin, is a well-known public health threat that can also affect birds that consume mollusks. A wide variety of clinical signs can be expected in birds affected by algal toxins, including weakness, inability to fly, abnormal head movements and postures, vomiting, lack of awareness, and convulsions. Gross lesions may include hemorrhage and necrosis in muscle, presumably due to the impaired muscle function resulting from neurologic signs or jaundiced, necrotic livers associated with hepatotoxins (Landsberg et al. 2007). Definitive diagnosis is difficult because of factors such as a lack of characteristic lesions, the fact that lethal doses of algal toxins are not known for individual bird species, and analytical challenges (Landsberg et al. 2007).

The link between harmful algal blooms and avian mortalities dates back to the late nineteenth century. Occasional deaths and health effects associated with saxitoxins, brevetoxins, and microcystins have been reported since the 1940s in sea ducks, including White-winged Scoters, Black Scoters, Common Eiders, Red-breasted Mergansers, and Common Goldeneyes (Landsberg et al. 2007). Nodularin, a cyanobacterial hepatotoxin, has been detected in tissues of mussels (*Mytilus edulis*, *Dreissena polymorpha*), clams (*Macoma balthica*), and Common Eiders in the Baltic Sea (Sipiä et al. 2002, 2006, 2008). Dead Surf Scoters, along with other birds, were found in association with a mortality event in California sea lions (*Zalophus californianus*) that was attributed to domoic acid toxicity (Gulland et al. 1999). Avian vacuolar myelinopathy, a neurological disease affecting birds and suspected to be caused by a cyanotoxin, has also been diagnosed in Buffleheads (Augspurger et al. 2003, Landsberg et al. 2007).

Although harmful algal blooms have not been shown to be a major mortality factor for sea ducks, their full impact may be underestimated because, in addition to direct lethal effects, exposure to the toxins may make birds more susceptible to stressors in the environment (Shumway et al. 2003). Furthermore, the epizootiology of algal blooms in relation to sea ducks may be affected in yet unknown ways as a result of global climate change (Hallegraeff 2010).

CONCLUSIONS AND FUTURE DIRECTIONS

Sea ducks face threats from a wide range of infectious and parasitic diseases. As in other wild birds, the magnitude of impacts in specific species and populations are likely to vary due to the diversity of disease-causing agents, variety of sea duck life histories, and differences in ecological factors. Avian cholera is known to be an important mortality factor in nesting Common Eiders, with a potential to devastate local nesting colonies, and has killed many thousands of wintering sea ducks. An estimated 250,000 Buffleheads died in one botulism type C outbreak, and botulism type E has recently emerged as an important cause of mortality, killing thousands of sea ducks. Reoviruses and acanthocephalans have been associated with large-scale duckling mortality. Some other diseases, such as aspergillosis and avian tuberculosis, are known to be significant problems in captive birds but may cause more limited impacts in wild sea duck populations. Our knowledge about infectious diseases, parasites, and biological toxins in sea ducks is incomplete and we have yet to learn about many existing and emerging diseases that have the potential to cause substantial mortality but so far have not been reported to do so in sea ducks. For instance, duck plague can be highly fatal, but virtually nothing is known about this disease in sea ducks. Myriad additional examples of potential disease agents that may be significant for sea ducks exist but have not been adequately studied. For example, coronaviruses are widely spread among birds and were found present in Northern Pintails in the Beringia region (Muradrasoli et al. 2010). Because of the documented presence in a sympatric waterfowl species, coronaviruses should be considered a potential agent infecting sea ducks.

Most investigations and publications relating to sea duck diseases are still focused on a relatively small number of species and geographic locations. For many infectious and parasitic agents, systematic surveys across taxa and geographic range of particular species have not been conducted, and information about disease occurrence is entirely lacking for some populations and species of sea ducks. Host species–specific and disease agent–specific assay techniques are still lacking for most host–disease systems, and much work remains to be done in the development of sensitive and specific laboratory tools to identify disease agents and measure effects on their host. Furthermore, few controlled experiments have been conducted to date to characterize pathology associated with disease agents and to determine mechanistic cause–effect relationships. As a consequence, much of the current knowledge about potential impacts of disease agents on the sea duck host is derived from other avian species. More detailed understanding of host–disease interactions would also increase our understanding of subclinical and sublethal effects of diseases, which may significantly contribute to overall sea duck population health—a topic with increasing importance under scenarios of environmental change and potential for multiple stressors to influence the health of wild sea duck populations.

Shortcomings aside, many important advances have been made recently that increase understanding of disease in sea ducks. Focused investigations of mortality events have led to more detailed understanding of possible mechanisms of disease impacts on sea duck hosts and populations, and isolation of disease agents from sea ducks has led to the development of species-specific detection tools (Hollmén et al. 2002, 2003a). Furthermore, application of quantitative modeling techniques has facilitated evaluation of population-level impacts of disease (Descamps et al.

2012, Tjørnløv et al. 2013), and interdisciplinary surveillance efforts have led to more comprehensive understanding of disease ecology in sea ducks (Ip et al. 2008; Ramey et al. 2010, 2011).

However, much remains to be learned to address the gaps in understanding of disease in sea ducks and other wild birds (Friend et al. 2001, Sea Duck Joint Venture 2008). Many populations of sea ducks have exhibited precipitous declines, and it is critical to evaluate the role that disease may play in causing or contributing to the observed declines (Smith et al. 2006). In the era of rapid environmental change due to large-scale climatic effects and other anthropogenic influences, the ecology of diseases may be rapidly changing, and additional populations may become vulnerable to disease impacts. Many sea duck populations are distributed at high latitudes and in areas on the receiving end of northbound expanding species, including potential disease vectors (Kovats et al. 2001, Garamszegi 2011, Loiseau et al. 2012). Recent, warmer conditions of the arctic and subarctic may facilitate and support transmission and persistence of diseases and vectors novel to these regions. Local populations, with potentially limited capacity of defense against existing and emerging pathogens, may be exposed to new disease risk. Furthermore, if population declines lead to smaller, isolated sea duck populations, these populations may become more vulnerable to mortality events caused by pathogens. Overall, changing environments and ecological conditions may lead to emergence of new disease threats (Daszak et al. 2000, Dobson and Foufopoulos 2001, Friend et al. 2001, Plowright et al. 2008). Understanding ecological drivers and effects of disease emergence in a time of large-scale environmental change will require interdisciplinary efforts, including fields of ecology, wildlife biology, veterinary medicine, epidemiology, physical environmental sciences, and social sciences.

In summary, we believe that the field of sea duck disease study offers many areas for future research and conclude by offering our views and priorities on some avenues for future direction. Rather than suggesting specific disease agents and host species for future research, we focus on broader perspectives. We arrange our recommendations into three categories of research: host–pathogen interactions and effects on the host, epidemiology of disease and effects on populations, and ecology of disease and predicting future impacts of disease on sea ducks. To better understand host–pathogen interactions, more experimental work needs to be carried out under controlled laboratory conditions and in the field. This work may be conducted in sea ducks, with appropriate surrogates, or in laboratory models. The aims of such studies should include characterizing disease pathogenesis, pathology, and impacts on host physiology, including immune response. To address the need to improve understanding of population-level impacts, many gaps in knowledge about epidemiology of sea duck disease need to be filled. We will need to better understand transmission dynamics of diseases and host resistance, susceptibility, and persistence of disease among sea ducks. To accomplish this goal, an improved understanding of immune responses is necessary. Environmental and species reservoirs of disease should also be investigated. We need to continue development of laboratory-based detection methods that offer the sensitivity and specificity to address the questions at hand. Also, further development of epidemiological and other quantitative modeling tools to estimate population-level impacts are needed. And last but not least, to advance our ability to understand and anticipate potential future impacts of diseases on sea ducks, a greater focus on the ecology of diseases is critically important. As a starting point, we should aim at collection and synthesis of baseline data, especially from populations at risk of being impacted by predicted environmental change or new threats. Collection and analysis of long-term data sets would provide valuable background and comparison for future investigations. We conclude by restating the importance of understanding the ecological factors influencing the distribution, transmission, and impacts of disease in the face of large-scale environmental change and declining sea duck populations. To accomplish our broad objectives, more resources to support long-term interdisciplinary research are needed.

LITERATURE CITED

Acevedo-Whitehouse, K., and A. L. J. Duffus. 2009. Effects of environmental change on wildlife health. Philosophical Transactions of the Royal Society of London B 364:3429–3438.

Adair, B. M., and J. A. Smyth. 2008. Egg drop syndrome. Pp. 266–276 in Y. M. Saif, A. M. Fadly, J. R. Glisson, L. R. McDougald, L. K. Nolan, and D. E. Swayne (editors), Diseases of poultry, 12th ed. Blackwell Publishing, Ames, IA.

Adams, D., K. Roblee, and W. Stone. [online]. 2009. NYS waterbird mortality as a result of type E botulism and nonnative invasive species. P. 3 in Abstract Theme B. Odum Conference, 30 April–1 May 2009, Rensselaerville, New York. New York Invasive Species Research Institute, Cornell University, Ithaca, New York. <http://www.nyisri.org/odumposterB.aspx> (24 November 2013).

Altizer, S., R. Bartel, and B. A. Han. 2011. Animal migration and infectious disease risk. Science 331:296–302.

Andersen, A. A., and J. C. Franson. 2007. Avian chlamydiosis. Pp. 303–316 in N. J. Thomas, D. B. Hunter, and C. T. Atkinson (editors), Infectious diseases of wild birds. Blackwell Publishing, Ames, IA.

Artsob, H., D. J. Gubler, D. A. Enria, M. A. Morales, M. Pupo, M. L. Bunning, and J. P. Dudley. 2009. West Nile virus in the New World: trends in the spread and proliferation of West Nile virus in the Western Hemisphere. Zoonoses and Public Health 56:357–369.

Augspurger, T., J. R. Fischer, N. J. Thomas, L. Sileo, R. E. Brannian, K. J. G. Miller, and T. E. Rocke. 2003. Vacuolar myelinopathy in waterfowl from a North Carolina impoundment. Journal of Wildlife Diseases 39:412–417.

Ballard, J., A. Allison, J. Brown, S. Gibbs, C. Dwyer, J. C. Ellis, S. Courchesne, R. Mickley, and K. Keel. 2012. Diagnostic findings from Common Eider (Somateria mollissima) mortality events in the northeastern United States associated with Wellfleet Bay virus, a novel orthomyxovirus. P. 77 in Proceedings of the United States Animal Health Association, 18–24 October, Greensboro, NC.

Becker, W. B. 1966. The isolation and classification of tern virus: influenza Virus A/Tern/South Africa/1961. Journal of Hygiene 64:309–320.

Beernaert, L. A., F. Pasmans, L. Van Waeyenberghe, F. Haesebrouck, and A. Martel. 2010. Aspergillus infections in birds: a review. Avian Pathology 39:325–321.

Bille, J., J. Rocourt, and B. Swaminathan. 2003. Listeria and Erysipelothrix. Pp. 461–471 in P. R. Murray, E. J. Baron, J. H. Jorgensen, M. A. Pfaller, and R. H. Yolken (editors), Manual of clinical microbiology, 8th ed. ASM Press, Washington, DC.

Bishop, C. A., and W. Threlfall. 1974. Helminth parasites of the Common Eider Duck, Somateria mollissima (L.), in Newfoundland and Labrador. Proceedings of the Helminthological Society of Washington 41:25–35.

Blehert, D. S., K. L. Jefferson, D. M. Heisey, M. D. Samuel, B. M. Berlowski, and D. J. Shadduck. 2008. Using amplified fragment length polymorphism analysis to differentiate isolates of Pasteurella multocida serotype 1. Journal of Wildlife Diseases 44:209–225.

Bopp, A. C., F. W. Brenner, P. I. Fields, J. G. Wells, and N. A. Strockbine. 2003. Escherichia, Shigella and identification. Pp. 654–671 in P. R. Murray, E. J. Baron, J. H. Jorgensen, M. A. Pfaller, and R. H. Yolken (editors), Manual of clinical microbiology, 8th ed. ASM Press, Washington, DC.

Botzler, R. G. 1991. Epizootiology of avian cholera in wildfowl. Journal of Wildlife Diseases 27:367–395.

Bragstad, K., P. H. Jørgensen, K. Handberg, A. S. Hammer, S. Kabell, and A. Fomsgaard. 2007. First introduction of highly pathogenic H5N1 avian influenza A viruses in wild and domestic birds in Denmark, Northern Europe. Virology Journal 4:43.

Brand, C. J., R. M. Duncan, S. P. Garrow, D. Olson, and L. E. Schumann. 1983. Waterbird mortality from botulism type E in Lake Michigan: an update. Journal of Wildlife Diseases 95:269–275.

Brand, C. J., S. M. Schmidt, R. M. Duncan, and T. M. Cooley. 1988. An outbreak of type E botulism among Common Loons (Gavia immer) in Michigan's Upper Peninsula. Journal of Wildlife Diseases 24:471–476.

Brault A. C. 2009. Changing patterns of West Nile virus transmission: altered vector competence and host susceptibility. Veterinary Research 40:43.

Bröjer, C., E. O. Ågren, H. Uhlhorn, K. Bernodt, T. Mörner, D. S. Jansson, R. Mattsson, S. Zohari, P. Thorén, M. Berg, and D. Gavier-Widén. 2009. Pathology of natural highly pathogenic avian influenza H5N1 infection in wild Tufted Ducks (Aythya fuligula). Journal of Veterinary Diagnostic Investigation 21:579–587.

Brown, J. D., and D. E. Stallknecht. 2008. Wild bird surveillance for the influenza virus. Pp. 85–97 in E. Spackman (editor), Methods in molecular biology 436: avian influenza virus. Humana Press, Totowa, NJ.

Brown, J. D., D. E. Stallknecht, J. R. Beck, D. L. Suarez, and D. E. Swayne. 2006. Susceptibility of North American ducks and gulls to H5N1 highly pathogenic avian influenza viruses. Emerging Infectious Diseases 12:1663–1670.

Brown, J. D., D. E. Stallknecht, and D. E. Swayne. 2008. Experimental infection with swans and geese with highly pathogenic avian influenza virus (H5N1) of Asian lineage. Emerging Infectious Diseases 14:136–142.

Burgess, E. C., J. Ossa, and T. M. Yuill. 1979. Duck plague: a carrier state in waterfowl. Avian Diseases 23:940–949.

Burkhart, R. L., and L. A. Page. 1971. Chlamydiosis. Pp. 118–140 in J. W. Davis, R. C. Anderson, L. Karstad, and D. O. Trainer (editors), Infectious and parasitic diseases of wild birds. Iowa State University Press, Ames, IA.

Bustnes, J. O., and K. V. Galaktionov. 2004. Evidence of a state-dependent trade-off between energy intake and parasite avoidance in Steller's Eiders. Canadian Journal of Zoology 82:1566–1571.

Cambre, R. C., and D. Keimy. 1993. Vaccination of zoo birds against avian botulism with mink botulism vaccine. Pp. 335–337 in Proceedings of the American Association of Zoo Veterinarians Conference, 9–14 October 1993, St. Louis, MO.

Carpentier, G. 2000. Avian botulism outbreak along the lower Great Lakes. Ontario Birds 18:84–91.

Centers for Disease Control and Prevention. [online]. 2011. West Nile virus. Vertebrate ecology. <http://www.cdc.gov/ncidod/dvbid/westnile/birdspecies.htm> (15 February 2011).

Christensen, T. K., T. Bregnballe, T. H. Anderson, and H. H. Dietz. 1997. Outbreak of Pasteurellosis among wintering and breeding Common Eiders *Somateria mollissima* in Denmark. Wildlife Biology 3:125–128.

Christiansen, M. 1952. Nyrecoccidiose hos vildtlevend andefugle (Anseriformes). Nordisk Veterinaermedicin 4:1173–1191 (in Danish).

Christiansen, M., and H. Madsen. 1948. *Eimeria bucephalae* n sp. Pathogenic in goldeneye in Denmark. Danish Review Game Biology 1:62–73.

Clark, G. M., D. O'Mera, and J. M. van Weelden. 1958. An epizootic among eider ducks involving an acanthocephalan worm. Journal of Wildlife Management 22:204–205.

Converse, K. A. 2007. Avian tuberculosis. Pp. 289–302 in N. J. Thomas, D. B. Hunter, and C. T. Atkinson (editors), Infectious diseases of wild birds. Blackwell Publishing, Ames, IA.

Converse, K. A., and G. A. Kidd. 2001. Duck plague epizootics in the United States, 1967–1995. Journal of Wildlife Diseases 37:347–357.

Cooley, A. J., H. Van Campen, M. S. Philpott, B. C. Easterday, and V. S. Hinshaw. 1989. Pathological lesions in the lungs of ducks infected with influenza A virus. Veterinary Pathology 26:1–5.

Cromie, R. L., M. J. Brown, D. J. Price, and J. L. Stanford. 1991. Susceptibility of captive wildfowl to avian tuberculosis: the importance of genetic and environmental factors. Tubercle 72:105–109.

Dallaire, A. D., and J. F. Giroux. 2005. Avian cholera in the St. Lawrence estuary in the Common Eider (*Somateria mollissima dresseri*). Canadian Cooperative Wildlife Health Centre Newsletter 11:6–7.

Daszak, P., A. A. Cunningham, and A. D. Hyatt. 2000. Emerging infectious diseases of wildlife—threats to biodiversity and human health. Science 287:443–449.

DeJong, R. L., R. L. Reimink, and H. D. Blankespoor. 2001. Hematozoa of hatch-year Common Mergansers from Michigan. Journal of Wildlife Diseases 37:403–407.

Descamps, S., H. G. Gilchrist, J. Bêty, E. I. Buttler, and M. R. Forbes. 2009. Costs of reproduction in a long-lived bird: large clutch size is associated with low survival in the presence of highly virulent disease. Biology Letters 5:278–281.

Descamps, S., S. Jenouvrier, H. G. Gilchrist, and M. R. Forbes. 2012. Avian cholera, a threat to the viability of an arctic seabird colony? PLoS One 7:e29659.

Dobson, A., and J. Foufopoulos. 2001. Emerging infectious pathogens of wildlife. Philosophical Transactions of the Royal Society of London B 356:1001–1012.

Docherty, D. E., J. C. Franson, R. E. Brannian, R. R. Long, C. A. Radi, D. Kruger, and R. F. Johnson Jr. 2012. Avian botulism and avian chlamydiosis in wild water birds, Benton Lake National Wildlife Refuge, Montana, USA. Journal of Zoo and Wildlife Medicine 43:885–888.

Ellis, T. M., R. B. Bousfield, L. A. Bissett, K. C. Dyrting, G. S. M. Luk, S. T. Tsim, K. Sturm-Ramirez, R. G. Webster, Y. Guan, and J. S. M. Peiris. 2004. Investigation of outbreaks of highly pathogenic H5N1 avian influenza in waterfowl and wild birds in Hong Kong in late 2002. Avian Pathology 33:492–505.

Eterradossi, N., and Y. M. Saif. 2008. Infectious bursal disease. Pp. 185–208 in Y. M. Saif, A. M. Fadly, J. R. Glisson, L. R. McDougald, L. K. Nolan, and D. E. Swayne (editors), Diseases of poultry, 12th ed. Blackwell Publishing, Ames, IA.

Fay, L. D., O. W. Kaufmann, and L. A. Ryel. 1965. Mass mortality of water-birds in Lake Michigan 1963–64. Pp. 36–46 in Proceedings, Eighth Conference on Great Lakes Research 1965, Great Lakes Research Division Publication, No. 13, University of Michigan, Ann Arbor, MI.

Fitzgerald, S. D. 2007. Avian adenoviruses. Pp. 182–193 in N. J. Thomas, D. B. Hunter, and C. T. Atkinson (editors), Infectious diseases of wild birds. Blackwell Publishing, Ames, IA.

Fitzgerald, S. D. 2008. Adenovirus infections, introduction. Pp. 251–266 in Y. M. Saif, A. M. Fadly, J. R. Glisson, L. R. McDougald, L. K. Nolan, and D. E. Swayne (editors), Diseases of poultry, 12th ed. Blackwell Publishing, Ames, IA.

Forrester, D. J., and E. C. Greiner. 2008. Leucocytozoonosis. Pp. 54–107 in C. T. Atkinson, N. J. Thomas, and D. B. Hunter (editors), Parasitic diseases of wild birds. Blackwell, Ames, IA.

Forrester, D. J., and M. G. Spalding. 2003. Parasites and diseases of wild birds in Florida. University Press of Florida, Gainesville, FL.

Franson, J. C., and D. V. Derksen. 1981. Renal coccidiosis in Oldsquaws (*Clangula hyemalis*) from Alaska. Journal of Wildlife Disease 17:237–239.

Franson, J. C., and J. E. Pearson. 1995. Probable epizootic chlamydiosis in wild California (*Larus californicus*) and Ring-billed (*Larus delawarensis*) Gulls in North Dakota. Journal of Wildlife Diseases 31:424–427.

Friend, M. 1999a. Avian cholera. Pp. 75–92 in M. Friend and J. C. Franson (editors), Field manual of wildlife diseases: general field procedures and diseases of birds. U.S. Geological Survey Information and Technology Report 1999–001. U.S. Geological Survey, Washington, DC.

Friend, M. 1999b. Duck plague. Pp. 141–151 in M. Friend and J. C. Franson (editors), Field manual of wildlife diseases: general field procedures and diseases of birds. U.S. Geological Survey Information and Technology Report 1999–001. U.S. Geological Survey, Washington, DC.

Friend, M., R. G. McLean, and F. J. Dein. 2001. Disease emergence in birds: challenges for the twenty-first century. Auk 118:290–303.

Gaidet, N., T. Dodman, A. Caron, G. Balanca, S. Desvaux, F. Goutard, G. Cattoli et al. 2007. Influenza surveillance in wild birds in Eastern Europe, the Middle East, and Africa: preliminary results from an ongoing FAO-led survey. Journal of Wildlife Diseases 43:S22–S28.

Gajadhar, A. A., R. J. Cawthorn, G. A. Wobeser, and P. H. G. Stockdale. 1983. Prevalence of renal coccidia in wild waterfowl in Saskatchewan. Canadian Journal of Zoology 61:2631–2633.

Garamszegi, L. 2011. Climate change increases the risk of malaria in birds. Global Change Biology 17:1751–1759.

Garden, E. A. 1961. Tuberculosis in eiders. Pp. 165–166 in H. Boyd and P. Scott (editors), The twelfth annual report of The Wildfowl Trust. Bailey & Son, Dursley, U.K.

Garden, E. A., C. Rayski, and V. M. Thom. 1964. A parasitic disease in eider ducks. Bird Study 11:280–287.

Gardner, H., K. M. Riddle, S. Brouwer, and L. Gleeson. 1997. Poultry virus infection in Antarctic penguins. Nature 387:245.

Germundsson, A., K. I. Madslien, M. J. Hjortaas, K. Handeland, and C. M. Jonassen. 2010. Prevalence and subtypes of influenza A viruses in wild waterfowl in Norway 2006–2007. Acta Veterinaria Scandinavica 52:28.

Gershman, M., J. F. Witter, H. E. Spencer Jr., and A. Kalvaitis. 1964. Case report: epizootic of fowl cholera in the Common Eider Duck. Journal of Wildlife Management 28:587–589.

Glisson, J. R., C. L. Hofarce, and J. P. Christensen. 2008. Fowl cholera. Pp. 739–758 in Y. M. Saif, A. M. Fadly, J. R. Glisson, L. R. McDougald, L. K. Nolan, and D. E. Swayne (editors), Diseases of poultry, 12th ed. Blackwell Publishing, Ames, IA.

Greiner, E. C. 2008. *Isospora, Atoxoplasma,* and *Sarcocystis.* Pp. 108–119 in C. T. Atkinson, N. J. Thomas, and D. B. Hunter (editors), Parasitic diseases of wild birds. Wiley-Blackwell, Ames, IA.

Greiner, E. C., G. F. Bennett, E. M. White, and R. F. Coombs. 1975. Distribution of the avian hematozoa of North American. Canadian Journal of Zoology 53:1762–1787.

Grenquist, P. 1951. On the fluctuations in the numbers of waterfowl in the Finnish Archipelago. Pp. 494–496 in S. Hörstadius (editor), Proceedings of the Xth International Ornithological Congress, Uppsala, Sweden.

Grenquist, R. 1970. On mortality of the Eider Duck (*Somateria mollissima*) caused by acanthocephalan parasites. Suomen Riista 22:24–34 (in Finnish).

Gulka, C. M., T. H. Piela, V. J. Yates, and C. Bagshaw. 1984. Evidence of exposure of waterfowl and other aquatic birds to the hemagglutinating duck adenovirus identical to EDS-76 virus. Journal of Wildlife Diseases 20:1–5.

Gulland, F. M. D., M. Haulena, M. Lander, L. J. Lowenstine, K. Lefebvre, T. Lipscomb, T. Knowles, C. Scholin, T. Spraker, V. Trainer, and F. Van Dolah. 1999. Domoic acid toxicity in California sea lions (*Zalophus californianus*) stranded along the central California coast. Proceedings of the International Association for Aquatic Animal Medicine 30:111–113.

Hallegraeff, G. M. 2010. Ocean climate change, phytoplankton community responses, and harmful algal blooms: a formidable predictive challenge. Journal of Phycology 46:220–235.

Hannett, G. E., W. B. Stone, S. W. Davis, and D. Wroblewski. 2011. Biodiversity of *Clostridium botulinum* type E associated with a large outbreak of botulism in wildlife from Lake Erie and Lake Ontario. Applied and Environmental Microbiology 77:1061–1068.

Hansen, W. 1999. Avian pox. Pp. 163–169 in M. Friend and J. C. Franson (editors), Field manual of wildlife diseases: general field procedures and diseases of birds. U.S. Geological Survey Information and Technology Report 1999–001. U.S. Geological Survey, Washington, DC.

Hansen, W. R., and R. E. Gough. 2007. Duck plague (duck virus enteritis). Pp. 87–107 in N. J. Thomas, D. B. Hunter, and C. T. Atkinson (editors), Infectious diseases of wild birds. Blackwell Publishing, Ames, IA.

Hanssen, S. A., I. Folstad, K. E. Erikstad, and A. Oksanen. 2003. Costs of parasites in Common Eiders: effects of antiparasite treatment. Oikos 100:105–111.

Hario, M., K. Selin, and T. Soveri. 1992. Is there a parasite-induced deterioration in eider fecundity? Suomen Riista 38:23–33.

Heard, D. J., D. M. Mulcahy, S. A. Iverson, D. J. Rizzolo, E. C. Greiner, J. Hall, H. Ip, and D. Esler. 2008. A blood survey of elements, viral antibodies, and hemoparasites in wintering Harlequin Ducks (*Histrionicus histrionicus*) and Barrow's Goldeneyes (*Bucephala islandica*). Journal of Wildlife Diseases 44:486–493.

Hénaux, V., J. Parmley, C. Soos, and M. D. Samuel. 2013. Estimating transmission of avian influenza in wild birds from incomplete epizootic data: implications for surveillance and disease spread. Journal of Applied Ecology 50:223–231.

Hesterberg, U., K. Harris, D. Stroud, V. Guberti, L. Busani, M. Pittman, V. Piazza, A. Cook, and I. Brown. 2009. Avian influenza surveillance in wild birds in the European Union in 2006. Influenza and Other Respiratory Viruses 3:1–14.

Hill, N. J., J. Y. Takekawa, C. J. Cardona, J. T. Ackerman, A. K. Schultz, K. A. Spragens, and W. M. Boyce. 2010. Waterfowl ecology and avian influenza in California: do host traits inform us about viral occurrence? Avian Diseases 54:426–432.

Hillgarth, N., and J. Kear. 1979. Diseases of seaducks in captivity. Wildfowl 30:135–141.

Himsworth, C. G., K. E. B. Gurney, A. S. Neimanis, G. A. Wobeser, and F. A. Leighton. 2009. An outbreak of West Nile virus infection in captive Lesser Scaup (*Aythya affinis*) ducklings. Avian Diseases 53:129–134.

Hindman, L. J., W. F. Harvey IV, G. R. Costanzo, K. A. Converse, and G. Stein Jr. 1997. Avian cholera in Ospreys: first occurrence and possible mode of transmission. Journal of Field Ornithology 68:503–508.

Hinshaw, V. S., R. G. Webster, and B. Turner. 1980. The perpetuation of orthomyxoviruses and paramyxoviruses in Canadian waterfowl. Canadian Journal of Microbiology 26:622–629.

Hirsh, D. C., D. A. Jessup, K. P. Snipes, T. E. Carpenter, D. W. Hird, and R. H. McCapes. 1990. Characteristics of *Pasteurella multocida* isolated from waterfowl and associated avian species in California. Journal of Wildlife Diseases 26:204–209.

Hjulsager, C. K., S. Ø. Breum, R. Trebbien, K. J. Handberg, O. R. Therkildsen, J. J. Madsen, K. Thorup, J. A. Baroch, T. J. DeLiberto, L. E. Laresen, and P. H. Jøgensen. 2012. Surveillance for avian influenza viruses in wild birds in Denmark and Greenland, 2007–10. Avian Diseases 56:992–998.

Hollmén, T., and D. E. Docherty. 2007. Orthoreovirus. Pp. 177–181 in N. J. Thomas, D. B. Hunter, and C. T. Atkinson (editors), Infectious diseases of wild birds. Blackwell Publishing, Ames, IA.

Hollmén, T., J. C. Franson, D. E. Docherty, M. Kilpi, M. Hario, L. H. Creekmore, and M. Petersen. 2000. Infectious bursal disease virus antibodies in Eider Ducks and Herring Gulls. Condor 102:688–691.

Hollmén, T., J. C. Franson, P. L. Flint, J. B. Grand, R. B. Lanctot, D. E. Docherty, and H. M. Wilson. 2003a. An adenovirus linked to mortality and disease in Long-tailed Ducks in Alaska. Avian Diseases 47:173–179.

Hollmén, T., J. C. Franson, M. Kilpi, D. E. Docherty, W. R. Hansen, and M. Hario. 2002. Isolation and characterization of a reovirus from Common Eiders (*Somateria mollissima*) from Finland. Avian Diseases 46:478–484.

Hollmén, T., J. C. Franson, M. Kilpi, D. E. Docherty, and V. Myllys. 2003b. An adenovirus associated with intestinal impaction and mortality of male Common Eiders (*Somateria mollissima*) in the Baltic Sea. Journal of Wildlife Diseases 39:114–120.

Hollmén, T., M. Hario, and J. T. Lehtonen. 1996. Description of an epizootic in eider ducklings from the Gulf of Finland. Suomen Riista 42:32–39.

Hollmén, T., J. T. Lehtonen, S. Sankari, T. Soveri, and M. Hario. 1999. An experimental study on the effects of polymorphiasis on Common Eider ducklings. Journal of Wildlife Diseases 35:466–473.

Hollmén, T. E., C. DebRoy, P. L. Flint, D. E. Safine, J. Schamber, A. Riddle, and K. Trust. 2011. Molecular typing of *Escherichia coli* strains associated with threatened sea ducks and near-shore marine habitats of southwest Alaska. Environmental Microbiology Reports 3:262–269.

Hoye, B. J., V. J. Munster, H. Nishiura, R. A. M. Fouchier, J. Madsen, and M. Klaassen. 2011. Reconstructing an annual cycle of interaction: natural infection and antibody dynamics to avian influenza along a migratory flyway. Oikos 120:748–755.

Hubben, K. 1958. Case report—*Aspergillus* meningoencephalitis in turkeys and ducks. Avian Diseases 2:110–116.

Hudson, J. R. 1959. Pasteurellosis. Pp. 413–436 in A. Stableforthe and I. Galloway (editors), Infectious diseases of animals (vol. 2). Butterworths Scientific, London, U.K.

Ip, H. S., P. L. Flint, J. C. Franson, R. J. Dusek, D. V. Derksen, R. E. Gill Jr., C. R. Ely et al. 2008. Prevalence of influenza A viruses in wild migratory birds in Alaska: patterns of variation in detection at a crossroads of intercontinental flyways. Virology Journal 5:art71.

Irons, D. B., T. Anker-Nilssen, A. J. Gaston, G. V. Byrd, K. Falk, G. Gilchrist, M. Hario et al. 2008. Fluctuations in circumpolar seabird populations linked to climate oscillations. Global Change Biology 14:1455–1463.

Itamies, J., T. Valtonen, and H. P. Fagerholm. 1980. *Polymorphus minutus* (Acanthocephala) infestation of eiders and its role as a possible cause of death. Annales Zoological Fennica 17:285–289.

Joint Working Group on the Management of the Common Eider. 2004. Québec Management Plan for the Common Eider *Somateria mollissima dresseri*. A special publication of the Joint Working Group on the Management of the Common Eider, Québec, QC.

Jonassen, C. M., and K. Handeland. 2007. Avian influenza virus screening in wild waterfowl in Norway, 2005. Avian Diseases 51:425–428.

Jones, R. C. 2008. Reovirus infections. Pp. 309–328 in Y. M. Saif, A. M. Fadly, J. R. Glisson, L. R. McDougald, L. K. Nolan, and D. E. Swayne (editors), Diseases of poultry, 12th ed. Blackwell Publishing, Ames, IA.

Kalmbach, E. R., and M. F. Gunderson. 1934. Western duck sickness: a form of botulism. U.S. Department of Agriculture Technical Bulletin, No. 411. U.S. Department of Agriculture, Washington, DC.

Keawcharoen, J., D. van Riel, G. van Amerongen, T. Bestebroer, W. E. Beyer, R. van Lavieren, A. D. M. E. Osterhaus, R. A. M. Fouchier, and T. Kuiken. 2008. Wild ducks as long-distance vectors of highly pathogenic avian influenza virus (H5N1). Emerging Infectious Diseases 14:600–607.

Komar, N., S. Langevin, S. Hinten, N. Nemeth, E. Edwards, D. Hettler, B. Davis, R. Bowen, and M. Bunning. 2003. Experimental infection of North American birds with the New York 1999 strain of West Nile virus. Emerging Infectious Diseases 9:311–322.

Korschgen, C. E., H. C. Gibbs, and H. L. Mendall. 1978. Avian cholera in eider ducks in Maine. Journal of Wildlife Diseases 14:254–258.

Kou, Z., Y. Li, Z. Yin, S. Guo, M. Wang, X Gao, P. Li et al. 2009. The survey of H5N1 flu virus in wild birds in 14 provinces of China from 2004 to 2007. PLoS One 4:e6926.

Kovats, R. S., D. H. Campbell-Lendrum, A. J. McMichael, A. Woodward, and J. S. H. Cox. 2001. Early effects of climate change: do they include changes in vector-borne disease? Philosophical Transactions of the Royal Society of London B 356:1057–1068.

Kulatchkova, V. G. 1960. The mortality of the chicks of the Common Eider and its causes. Trudy Kandalaksh. Gos. Zapov. (Transactions of the Kandalaksha State Reserve) 3:91–106.

Kunkle, R. A. 2003. Fungal infections. Pp. 883–902 in Y. M. Saif, H. J. Barnes, J. R. Glisson, A. M. Fadly, L. R. McDougald, and D. E. Swayne (editors), Diseases of poultry, 11th ed. Blackwell Publishing, Ames, IA.

Kutkiené, L., and A. Sruoga. 2004. *Sarcocystis* spp. in birds of the order Anseriformes. Parasitology Research 92:171–172.

Landsberg, J. H., G. A. Vargo, L. J. Flewelling, and F. E. Wiley. 2007. Algal biotoxins. Pp. 431–455 in N. J. Thomas, D. B. Hunter, and C. T. Atkinson (editors), Infectious diseases of wild birds. Blackwell Publishing, Ames, IA.

Latorre-Margalef, N., G. Gunnarsson, V. J. Munster, R. A. M. Fouchier, A. D. M. E. Osterhaus, J. Elmberg, B. Olsen et al. 2009. Effects of influenza A virus infection on migrating mallard ducks. Proceedings of the Royal Society B 276:1029–1036.

Leibovitz, L. 1968. Progress report: duck plague surveillance of American Anserifomes. Bulletin of the Wildlife Disease Association 4:87–91.

Liu, J., H. Xiao, F. Lei, Q. Zhu, K. Qin, X.-W. Zhang, X.-L. Zhang et al. 2005. Highly pathogenic H5N1 influenza virus infection in migratory birds. Science 309:1206.

Lloyd, C. S., G. J. Thomas, J. W. MacDonald, E. D. Borland, K. Strandring, and J. L. Smart. 1976. Wild bird mortality caused by botulism in Britain, 1975. Biological Conservation 10:119–129.

Locke, L. N., V. Stotts, and G. Wolfhard. 1970. An outbreak of fowl cholera in waterfowl in the Chesapeake Bay. Journal of Wildlife Diseases 6:404–407.

Loiseau, C., R. J. Harrigan, A. J. Cornel, S. L. Guers, M. Dodge, T. Marzec, J. S. Carlson, B. Seppi, and R. N. M. Sehgal. 2012. First evidence and predictions of *Plasmodium* transmission in Alaska bird populations. PLoS One 7:e44729.

Martinez, R., and G. Wobeser. 1999. Immunization of ducks for type C botulism. Journal of Wildlife Diseases 35:710–715.

McDonald, M. E. 1969. Catalogue of helminthes of waterfowl (Anatidae). U.S. Fish and Wildlife Service Special Scientific Report-Wildlife 126. U.S. Fish and Wildlife Service, Anchorage, AK.

McLean, R. G., and S. R. Ubico. 2007. Arboviruses in birds. Pp. 17–62 in N. J. Thomas, D. B. Hunter, and C. T. Atkinson (editors), Infectious diseases of wild birds. Blackwell Publishing, Ames, IA.

Meece, J. K., T. A. Kronenwetter-Koepel, M. F. Vandermause, and K. D. Reed. 2006. West Nile virus infection in commercial waterfowl operation, Wisconsin. Emerging Infectious Diseases 12:1451–1453.

Mendenhall, V. 1976. Survival and causes of mortality in eider ducklings on the Ythan Estuary, Aberdeenshire, Scotland. Wildfowl 27:160.

Montali, R. J., M. Bush, and G. A. Greenwell. 1976. An epornitic of duck viral enteritis in a zoological park. Journal of the American Veterinary Medical Association 169:954–958.

Montgomery, R. D., G. Stein Jr., V. D. Stotts, and F. H. Settle. 1979. The 1978 epornitic of avian cholera on the Chesapeake Bay. Avian Diseases 24:966–978.

Montgomery, R. D., P. Villegas, D. L. Dawe, and J. Brown. 1986. A comparison between the effect of an avian reovirus and infectious bursal disease virus on selected aspects of the immune system of the chicken. Avian Diseases 30:298–308.

Mullié, W. C., T. Smit, and L. Moraal. 1979. Vogelcholera (Pasteurellosis) als oorzaak van sterfte onder watervogels in het Deltagebied in 1977. Vogeljaar 27:11–20 (in Dutch).

Munster, V. J., C. Baas, P. Lexmond, J. Waldenström, A. Wallensten, T. Fransson, G. F. Rimmelzwaan et al. 2007. Spatial, temporal, and species variation in prevalence of influenza A viruses in wild migratory birds. PLoS Pathogens 3:e61.

Muradrasoli, S., A. Balint, J. Wahlgren, J. Waldenström, S. Belák, J. Blomberg, and B. Olsen. 2010. Prevalence and phylogeny of coronaviruses in wild birds from the Bering Strait Area (Beringia). PLoS One 5:e13640.

Nagy, A., J. Machova, J. Hornickova, M. Tomci, I. Nagl, B. Horyna, and I. Holko. 2007. Highly pathogenic avian influenza virus subtype H5N1 in Mute Swans in the Czech Republic. Veterinary Microbiology 120:9–16.

Ogawa, S., Y. Yamamoto, M. Yamada, M. Mase, and K. Nakamura. 2009. Pathology of Whooper Swans (Cygnus cygnus) infected with H5N1 avian influenza virus in Akita, Japan, in 2008. Journal of Veterinary Medical Science 71:1377–1380.

Olsen, G. H., S. Gibbs, J. Beuth, J. R. Ballard, and C. Dwyer. 2012. Searching for Wellfleet Bay virus in Common Eiders in Rhode Island. P. 355 in Proceedings of the Association of Avian Veterinarians 33rd Annual Conference. Association of Avian Veterinarians, 11–15 August 2012, Louisville, KY.

Parkinson, A. J., and J. C. Butler. 2005. Potential impacts of climate change on infectious diseases in the Arctic. International Journal of Circumpolar Health 64:478–486.

Parmley, E. J., N. Bastien, T. F. Booth, V. Bowes, P. A. Buck, A. Breault, D. Caswell et al. 2008. Wild bird influenza survey, Canada, 2005. Emerging Infectious Diseases 14:84–87.

Pedersen, K., H. H. Dietz, J. C. Jørgensen, T. K. Christensen, T. Bregnballe, and T. H. Andersen. 2003. Pasteurella multocida from outbreaks of avian cholera in wild and captive birds in Denmark. Journal of Wildlife Diseases 39:808–816.

Pello, S. J., and G. H. Olsen. 2013. Emerging and reemerging diseases of avian wildlife. Veterinary Clinics of North America: Exotic Animal Practice 16:357–381.

Persson, L. 1998. Fågelkolera på ejder och andra Fåglar på Västergarns utholme. Bläcku 24:91–93 (in Swedish).

Persson, L., K. Borg, and H. Falt. 1974. On the occurrence of endoparasites in eider ducks in Sweden. Viltrevy 9:1–24.

Pittman, M., A. Laddomada, R. Freigofas, V. Piazza, A. Brouw, and I. H. Brown. 2007. Surveillance, prevention, and diseases management of avian influenza in the European Union. Journal of Wildlife Diseases 43:S64–S70.

Plowright, R. K., S. H. Sokolow, M. E. Gorman, P. Daszak, and J. E. Foley. 2008. Causal inference in disease ecology: investigating ecological drivers of disease emergence. Frontiers in Ecology and the Environment 6:420–429.

Proctor, S. J., G. L. Pearson, and L. Liebovitz. 1975. Color atlas of wildlife pathology 2. Duck plague in free-flying waterfowl observed during the Lake Andes epizootic. Wildlife Disease Microfische, No. 67. Wildlife Disease Association, Ames, IA.

ProMed Mail. [online]. 2005. Duck viral enteritis—USA (District of Columbia). International Society for Infectious Diseases. Archive No. 20050602.1537. <www.promedmail.org> (20 May 2010).

Quortrup, E. R., and R. L. Sudheimer. 1943. Detection of botulinus toxin in the blood stream of wild ducks. Journal of the American Veterinary Medical Association 102:264–266.

Ramey, A. M., J. M. Pearce, C. R. Ely, L. M. S. Guy, D. B. Irons, D. V. Derksen, and H. S. Ip. 2010. Transmission and reassortment of avian influenza viruses at the Asian-North American interface. Virology 406:352–359.

Ramey, A. M., J. M. Pearce, A. B. Reeves, J. C. Franson, M. R. Petersen, and H. S. Ip. 2011. Evidence for limited exchange of avian influenza viruses between seaducks and dabbling ducks at Alaska Peninsula coastal lagoons. Archives of Virology 156:1813–1821.

Reed, A., and J. G. Cousineau. 1967. Epidemics involving the Common Eider (Somateria mollissima) at Ile Blanche, Quebec. Naturaliste Canadien 94:327–334.

Reeves, A. B., J. M. Pearce, A. M. Ramey, C. R. Ely, J. A. Schmutz, P. L. Flint, D. V. Derksen, H. S. Ip, and K. A. Trust. 2013. Genomic analysis of avian influenza viruses from waterfowl in western Alaska, USA. Journal of Wildlife Diseases 49:600–610.

Rhodes, K. R., and R. B. Rimler. 1987. Capsular groups of Pasteurella multocida isolated from avian hosts. Avian Diseases 31:895–898.

Rinehart, C. L., and J. K. Rosenberger. 1983. Effects of avian reoviruses on the immune responses of chickens. Poultry Science 62:1488–1489.

Rocke, T. E., and T. K. Bollinger. 2007. Avian botulism. Pp. 377–416 in N. J. Thomas, D. B. Hunter, and C. T. Atkinson (editors), Infectious diseases of wild birds. Blackwell Publishing, Ames, IA.

Rocke, T. E., and M. Friend. 1999. Avian botulism. Pp. 271–281 in M. Friend and J. C. Franson (editors), Field manual of wildlife diseases: general field procedures and diseases of birds. U.S. Geological Survey Information and Technology Report 1999–001. U.S. Geological Survey, Washington, DC.

Rocke, T. E., M. D. Samuel, P. K. Swift, and G. S. Yarris. 2000. Efficacy of a type C botulism vaccine in Green-winged Teal. Journal of Wildlife Diseases 36:489–493.

Rowan, M. K. 1962. Mass mortality among European Common Terns in South Africa in April–May 1961. British Birds 55:103–114.

Ruff, M. D., and J. K. Rosenberger. 1985a. Concurrent infections with reoviruses and coccidia in broilers. Avian Diseases 29:465–478.

Ruff, M. D., and J. K. Rosenberger. 1985b. Interaction of low-pathogenicity reoviruses and low levels of infection with several coccidial species. Avian Diseases 29:1057–1065.

Samuel, M. D., R. G. Botzler, and G. A. Wobeser. 2007. Avian cholera. Pp. 239–269 in N. J. Thomas, D. B. Hunter, and C. T. Atkinson (editors), Infectious diseases of wild birds. Blackwell Publishing, Ames, IA.

Savage, A., and W. B. McTavish. 1951. Plasmodium circumflexum in a Manitoba duck. Journal of Parasitology 37:533–534.

Sea Duck Joint Venture. [online]. 2008. Sea duck joint venture strategic plan 2008–2012. U.S. Fish and Wildlife Service, Anchorage, AK. <http://www.seaduckjv.org/sdjv_strategic_plan_2008–12_final_17dec2008.pdf> (4 December 2013).

Shumway, S. E., S. M. Allen, and P. D. Boersma. 2003. Marine birds and harmful algal blooms: sporadic victims of under-reported events? Harmful Algae 2:1–17.

Siebert, U., P. Schwemmer, N. Guse, T. Harder, S. Garthe, E. Prenger-Berninghoff, and P. Wohlsein. 2012. Health status of seabirds and coastal birds found at the German North Sea coast. Acta Veterinaria Scandinavica 54:43.

Sinnecker, R., H. Sinnecker, E. Zilske, and D. Köhler. 1983. Surveillance of pelagic birds for influenza A viruses. Acta Virologica 27:75–79.

Sipiä, V. O., H. T. Kankaanpää, S. Pflugmacher, J. Flinkman, A. Furey, and K. J. James. 2002. Bioaccumulation and detoxification of nodularin in tissues of flounder (Platichthys flesus), mussels (Mytilus edulis, Dreissena polymorpha), and clams (Macoma balthica) from the northern Baltic Sea. Ecotoxicology and Environmental Safety 53:305–311.

Sipiä, V. O., M. Neffling, J. S. Metcalf, S. M. K. Nybom, J. A. O. Meriluoto, and G. A. Codd. 2008. Nodularin in feathers and liver of Eiders (Somateria mollissima) caught from the western Gulf of Finland in June–September 2005. Harmful Algae 7:99–105.

Sipiä, V. O., O. Sjövall, T. Valtonen, D. L. Barnaby, G. A. Codd, J. S. Metcalf, M. Kilpi, O. Mustonen, and J. A. O. Meriluoto. 2006. Analysis of nodularin-R in Eider (Somateria mollissima), roach (Rutilus rutilus L.), and flounder (Platichthys flesus L.), liver and muscle samples from the western Gulf of Finland, northern Baltic Sea. Environmental Toxicology and Chemistry 25:2834–2839.

Skerratt, L. F., J. C. Franson, C. U. Meteyer, and T. E. Hollmén. 2005. Causes of mortality in sea ducks (Mergini) necropsied at the USGS-National Wildlife Health Center. Waterbirds 28:193–207.

Skirnisson, K. 1997. Mortality associated with renal and intestinal coccidiosis in juvenile eiders in Iceland. Parassitologia 39:325–330.

Slemons, R. D., M. C. Shieldcastle, L. D. Heyman, K. E. Bednarik, and D. A. Senne. 1991. Type A influenza viruses in waterfowl in Ohio and implications for domestic turkeys. Avian Diseases 35:165–173.

Smith, K. F., D. F. Sax, and K. D. Lafferty. 2006. Evidence for the role of infectious disease in species extinction and endangerment. Conservation Biology 20:1349–1357.

Stallknecht, D. E., E. Nagy, D. B. Hunter, and R. D. Slemons. 2007. Avian influenza. Pp. 108–130 in N. J. Thomas, D. B. Hunter, and C. T. Atkinson (editors), Infectious diseases of wild birds. Blackwell Publishing, Ames, IA.

Steele, K. E., M. J. Linn, R. J. Schoepp, N. Komar, T. W. Geisbert, R. M. Manduca, P. P. Calle et al. 2000. Pathology of fatal West Nile virus infections in native and exotic birds during the 1999 outbreak in New York City, New York. Veterinary Pathology 37:208–224.

Stetter, M. D., B. J. Mangold, P. P. Calle, B. L. Raphael, J. G. Trupkiewicz, N. Harmati, and R. A. Cook. 1994. Aspergillosis in captive Pacific Eiders (Somateria mollissima). Proceedings of the International Association for Aquatic Animal Medicine 25:110–117.

Swayne, D. E., and D. A. Halvorson. 2008. Influenza. Pp. 153–184 in Y. M. Saif, A. M. Fadly, J. R. Glisson, L. R. McDougald, L. K. Nolan, and D. E. Swayne (editors), Diseases of poultry, 12th ed. Blackwell Publishing, Ames, IA.

Swennen, C., and T. Smit. 1991. Pasteurellosis among breeding Eiders Somateria mollissima in The Netherlands. Wildfowl 42:94–97.

Tjørnløv, R. S., J. Humaidan, and M. Frederiksen. 2013. Impacts of avian cholera on survival of Common Eiders Somateria mollissima in a Danish colony. Bird Study 60:321–326.

U.S. Interagency Working Group. [online]. 2006. An early detection system for highly pathogenic N5N1 avian influenza in wild migratory birds: U.S. interagency strategic plan. <http://www.nwhc.usgs.gov/publications/other/Final_Wild_Bird_Strategic_Plan_0322.pdf> (23 November 2013).

Usui, T., T. Yamaguchi, H. Ito, H. Ozaki, T. Murase, and T. Ito. 2009. Evolutionary genetics of highly pathogenic H5N1 avian influenza viruses isolated from Whooper Swans in northern Japan in 2008. Virus Genes 39:319–323.

van Dorssen, C. A., and H. Kunst. 1955. Over de gevoeligheid van eenden en diverse andere watervogels voor eendenpest. Tijdschrift Voor Diergeneeskunde 80:1286–1295 (in Dutch).

van Gils, J. A., V. J. Munster, R. Radersma, D. Liefhebber, R. A. M. Fouchier, and M. Klaassen. 2007. Hampered foraging and migratory performance in swans infected with low-pathogenic avian influenza A virus. PLoS One 2:e184.

van Riper III, C., and D. J. Forrester. 2007. Avian pox. Pp. 131–176 in N. J. Thomas, D. B. Hunter, and C. T. Atkinson (editors), Infectious diseases of wild birds. Blackwell Publishing, Ames, IA.

Waldén, H. W. 1961. Observations on renal coccidia in Swedish anseriform birds, with notes concerning two new species, Eimeria boschadis and Eimeria christianseni (Sporozoa, Telosporidia). Arkiv for Zoologi 15:97–104.

Wallensten, A. 2007. Influenza virus in wild birds and mammals other than man. Microbial Ecology in Health and Disease 19:122–139.

Weber, S., T. Harder, E. Starick, M. Beer, O. Werner, B. Hoffmann, T. C. Mettenleiter, and E. Mundt. 2007. Molecular analysis of highly pathogenic avian influenza virus of subtype H5N1 isolated from wild birds and mammals in northern Germany. Journal of General Virology 88:554–558.

Webster, R. G., W. J. Bean, O. T. Gorman, T. M. Chambers, and Y. Kawaoka. 1992. Evolution and ecology of influenza A viruses. Microbiological Reviews 56:152–179.

Weissenböck, H., Z. Hubálek, T. Bakonyi, and N. Nowotny. 2010. Zoonotic mosquito-borne flaviviruses: worldwide presence of agents with proven pathogenicity and potential candidates of future emerging diseases. Veterinary Microbiology 140:271–280.

Wilson, H. M., J. S. Hall, P. L. Flint, J. C. Franson, C. R. Ely, J. A. Schmutz, and M. D. Samuel. 2013. High seroprevalence of antibodies to avian influenza viruses among wild waterfowl in Alaska: implications for surveillance. PLoS One 83:e58308.

Wobeser, G. A. 1997. Diseases of wild waterfowl, 2nd ed. Plenum Press, New York, NY.

Wobeser, G. A., and C. J. Brand. 1982. Chlamydiosis in 2 biologists investigating disease occurrences in wild waterfowl. Wildlife Society Bulletin 10:170–172.

Wolcott, M. J. 2007. Erysipelas. Pp. 332–340 in N. J. Thomas, D. B. Hunter, and C. T. Atkinson (editors), Infectious diseases of wild birds. Blackwell Publishing, Ames, IA.

World Organization for Animal Health. 2009. Avian influenza. Chapter 2.3.4. Pp. 436–452 in Manual of diagnostic tests and vaccines for terrestrial animals. World Organization for Animal Health, Paris, France.

Yee, K. S., T. E. Carpenter, and C. J. Cardona. 2009. Epidemiology of H5N1 avian influenza. Comparative Immunology, Microbiology, and Infectious Diseases 32:325–340.

Zohari, S., P. Gyarmati, P. Thorén, G. Czifra, C. Bröjer, S. Belák, and M. Berg. 2008. Genetic characterization of the NS gene indicates co-circulation of two sub-lineages of highly pathogenic avian influenza virus of H5N1 subtype in Northern Europe in 2006. Virus Genes 36:117–125.

CHAPTER FIVE

Breeding Costs, Nutrient Reserves, and Cross-Seasonal Effects*

DEALING WITH DEFICITS IN SEA DUCKS

Ray T. Alisauskas and Jean-Michel Devink

Abstract. We reviewed reproductive life histories and associated nutritional requirements of egg production and incubation for 18 species and subspecies of sea ducks that breed in North America. We also refer to life histories of some European subspecies. We found that basic information for several species remains unavailable for egg composition, egg-laying rates, follicular growth rates and estimates of incubation constancy. Relationships among various life-history traits associated with egg production and incubation by sea ducks revealed that phylogeny and body mass both affect daily and total energetic costs. For example, regression of some life-history traits associated with egg production that were related to body mass across all species or subspecies under consideration showed different patterns when compared between Somatereae (eiders) and Mergeae (non-eiders). Also missing for most species were estimates of the proportion of egg nutrients supplied by endogenous stores. Inferences about nutrient supply to eggs were highly variable, regardless of whether estimation relied on analysis of stable isotopes or the regression of cumulative nutrient production on somatic nutrient reserves. The proportion of egg nutrient supplied by endogenous reserves showed no clear pattern, judging from the lack of relationship to other life history traits. There is a general lack of understanding of whether endogenous nutrient reserves used in reproduction are acquired by sea ducks from wintering habitat, distant staging or those areas proximal to nesting sites. The geographic sourcing of nutrients used in sea duck reproduction should receive additional study because such areas may influence population-level recruitment. Reported incubation constancy in sea ducks ranged from 81% to 99%, and reflected the range of strategies to source nutrients for reproduction. Based on differences in body mass before and after incubation, we estimated that between 8% and 94% of energy requirements during incubation by

* Alisauskas, R. T. and J.-M. Devink. 2015. Breeding Costs, Nutrient Reserves, and Cross-Seasonal Effects: Dealing with Deficits in Sea Ducks. Pp. 125–168 in J.-P. L. Savard, D. V. Derksen, D. Esler, and J. M. Eadie (editors). Ecology and conservation of North American sea ducks. Studies in Avian Biology (no. 46), CRC Press, Boca Raton, FL.

different species or subspecies were met with endogenous reserves. The gradient in reliance on endogenous nutrients during incubation across species was a strong function of body mass ($r^2 = 0.84$). We also discuss the potential interplay of contaminants, nutrition and reproduction, and suggest that determination of nutrient reserve thresholds for breeding are an important research goal.

Key Words: body condition, capital, carryover effects, contaminants, cross-seasonal effects, egg formation, energetic costs, income breeding, incubation, nutrients, reproductive strategies.

Most animals experience seasonal variability in energy costs and nutrient requirements of basic metabolism and reproduction but also seasonal variability in availability of resources required to meet such costs. Migratory species, in particular, face not only temporal but also spatial variation in resource availability in the habitats and landscapes through which they travel over the annual cycle. Resource pulses at a broad continental scale are relatively predictable. Movement through such landscapes between winter and breeding areas is probably highly adapted to these pulses, though annual variation in seasonal resource availability can result even in the same location. An evolved strategy among many migratory species that is particularly well documented in waterfowl is an ability to store energy and macronutrients during periods of high resource availability (Alisauskas and Ankney 1992). Resources may then be used during parts of the annual cycle with reduced food availability or when competing behaviors interfere with the amount of time that animals can feed each day.

The ability to achieve proper levels of energy stores or nutritional condition can influence fitness by affecting the probability of breeding (Bond et al. 2008), or the probability of survival (Blums et al. 2005). If individuals do attempt to breed, then the number of offspring produced per female can be influenced through effects of prebreeding body condition on clutch size (Ankney and MacInnes 1978, Warren et al. 2013), nest success (Kellett and Alisauskas 2000, Bentzen et al. 2008a), and possibly survival of young before fledging (Bustnes et al. 2002). Among breeders, individual variation in body condition of Common Eiders (*Somateria mollissima*) can influence egg size and induce variation in onset of incubation, with consequences for hatch synchrony and survival of ducklings (Hanssen et al. 2002). Egg size, in turn, can affect locomotor and foraging ability of ducklings with clear implications for survival of free-ranging offspring (Anderson and Alisauskas 2001).

Nutritional conditioning in the prenesting phase drives the cumulative effects of these relationships and can influence average annual recruitment (Alisauskas 2002) and therefore demographics at the local or population level. Indeed, Coulson (2013) suggested that survival and recruitment, and thus dynamics of a Common Eider population in England, were influenced by the interplay between food reserves and immunocompetence. Most mortality occurred shortly after hatch and was associated with low recruitment. Both events were linked to a failure to store sufficient nutrient reserves to produce eggs and complete incubation during which food intake is constrained by high nest attendance, resulting in mortality through starvation. However, mass mortality during winter has also been documented in Common Eiders (Camphuysen et al. 2002) and from starvation during spring migration in King Eiders (*Somateria spectabilis*, Barry 1968, Fournier and Hines 1994).

Individual variation in prebreeding nutrient storage schedules can have implications for clutch size or other fitness metrics, through the cost of delay phenomenon. Warren et al. (2013) concluded that prebreeding environmental conditions influenced the ability of temperate ducks to store nutrients, which affected their subsequent clutch size. Annual or lifetime recruitment is also an important metric of fitness relative to other members of the population; life history strategies that improve access to resources or improve the efficiency of storage and subsequent supply of nutrient reserves for reproduction may be highly adaptive. Thus, an understanding of individual, seasonal, and annual variation in food acquisition and storage may provide insights into the evolution of life-history traits and provide guidance in determining the underlying causes of population change.

WHY STUDY CROSS-SEASONAL NUTRITIONAL EFFECTS?

Metrics of population change and their immediate ecological drivers are of direct interest to conservationists and managers. Thus, tests of hypotheses about the influence of resource demand associated with key activities of spring migration and reproduction and interplay with resource supply that is mediated by climate and food availability are important steps toward understanding important sources of ecological variation in recruitment to sea duck populations. Besides predation and disease, climatic variation may be the overarching ecological influence on both survival and recruitment of most organisms, (1) through its direct effects on the production of food and thus annual carrying capacity, (2) through a role in influencing energy budgets and nutritional demands of individuals, (3) by affecting resource access by individual animals, and (4) by indirectly influencing predator–prey dynamics and other trophic interactions.

Cross-seasonal effects have been implied for waterfowl since the hypothesis of Ryder (1970) that constraints associated with nutrient storage and use for egg production were a key factor behind the evolution of clutch size in arctic-nesting Ross's Goose (*Chen rossii*). Such effects are recognized as functional ecological links between events during winter, spring migration, breeding areas, and fall migration (Harrison et al. 2011). Sedinger and Alisauskas (2014) distinguished cross-seasonal effects at a population level from carryover effects at an individual level, but temporally lagged or spatially disjunct ecological influences on survival and recruitment appear to be increasing with attention to other taxa besides waterfowl (Catry et al. 2013). The notion that events during a part of the annual cycle can influence vital rates during a subsequent life stage implies a connection to the ability of animals to garner sufficient nutrients as determined by food availability and individual foraging efficiencies. Given that resource levels, nutrient storage, and vital rates can be connected, identifying the location and timing of nutrient storage may have important conservation applications, including identifying potential risks to sea duck populations from human activities (Dickson and Smith 2013).

Evidence is accumulating that oceanic conditions appear to have a large influence on the population biology of sea ducks by affecting their distribution (Zipkin et al. 2010), with likely consequences for altered feeding opportunities from changes in winter fidelity. Indeed, carryover effects of winter climate have been detected on breeding success of Common Eiders (Lehikoinen et al. 2006). Integrated climate metrics such as the North Atlantic Oscillation seem to have a strong cross-seasonal influence on the prebreeding body condition of some arctic-nesting sea ducks (Descamps et al. 2010). It is unclear whether the mechanism behind such correlations may have more to do with direct impacts on winter feeding or perhaps with delayed spring migration leading to an abbreviated spring feeding schedule (Mehlum 2012). Nevertheless, ecological links between winter conditions and survival or breeding probability were likely behind the connection, illustrated by Flint (2013), between large-scale oceanic regime shifts during particular years in the North Pacific and concurrent shifts in population trajectories of multiple sea duck species. The demographic mechanism driving alteration to population trajectories was not identified, but the cross-seasonal connection between winter conditions and population growth may have been mediated through changes in breeding propensity at least in some years. Nevertheless, Flint (2013) acknowledged that the mechanisms between regime shift and population change in sea ducks are complex, that some species are more likely to be influenced by such regime shifts than others, that the ecological mechanisms likely vary by location, and that both survival and particular elements of recruitment may be involved. The importance of climate and weather on resource abundance and feeding and body condition, respectively, on multiple spatial and temporal scales is probably key to such ultimate large-scale population effects.

Although energy or nutrient limitation can influence survival, in this chapter, we focus on the lagged ecological effects of events during winter and spring migration as they relate to the per-capita recruitment of young at individual (carryover effects) and population levels (cross-seasonal effects; see Sedinger and Alisauskas 2014). We review the energy or nutritional costs associated with the breeding process and the ability of sea ducks to garner sufficient food to meet such costs. Events perhaps as early as late winter leading up to and including the nesting season might be most pertinent to individual or population level recruitment. Compared to annual adult survival, the recruitment process is somewhat more complex ecologically since productivity is a function of multiple reproductive components such as clutch size, brood size, and between-state transition probabilities including breeding propensity, nest and egg success, and offspring survival to fledging. Each of these components of reproduction may be affected differently by prebreeding nutrient schedules. We focus attention toward the role of carryover versus cross-seasonal effects of nutrition on breeding propensity (Reynolds 1972, Alisauskas and Ankney 1994), egg production (Alisauskas and Ankney 1992), and incubation (Afton and Paulus 1992). Residual body reserves of breeding adults after hatching may also influence duckling survival and so may additionally modify recruitment. We also review the interplay between carryover effects and interspecific variation in body size.

ESTIMATING ENERGETIC COSTS

Three relatively distinct periods of the breeding cycle are usually considered when examining the energetic requirements of breeding: clutch formation, incubation, and post-hatch. Each of these periods requires lipids, proteins, minerals, and additional nutrients for tissue production or energy in excess of the minimum requirements for maintenance and survival of adults. Sources of nutrients used during each of these periods are not necessarily the same and some sea duck species may use one strategy for the clutch formation stage while employing an alternate strategy for the incubation period depending on particular breeding conditions and life-history strategies. This chapter will focus on egg laying and incubation, whereas brood rearing is reviewed by Savard and Petersen (Chapter 9, this volume).

Egg formation involves complex biochemical processes that result in the production of the gross nutrients of fat and protein and construction of eggshells from mineral sources. A convenient metric of costs is their expression as energy equivalents (kJ). However, an energetic approach may distract attention from the availability of specific micronutrients that could limit the rate of egg production (Alisauskas

and Ankney 1992), even though the supply of macronutrients might appear to be superfluous with respect to demand. Thus, we view egg production as a question of sufficient daily rates of specific nutrient supply to meet the diverse requirements of egg protein, yolk lipid, and eggshell synthesis, after meeting existing energy requirements. On the other hand, most pertinent to incubation is the sufficient supply of energy, through catabolism of fat or protein reserves remaining after egg production. Importantly, accessibility to water may complicate the relationship between preincubation energy reserves and nest success in some species, such as shown for arctic-nesting Greater Snow Geese (*Chen caerulescens atlanticus*; Lecomte et al. 2009). Thirst and hydration as limiting factors for incubation constancy have not been studied in detail in sea ducks, but Bolduc and Guillemette (2003) reported that incubating Common Eiders rarely feed during incubation breaks but frequently drink immediately after landing on the sea.

COSTS OF EGG PRODUCTION

The ovarian cycle has been characterized as six distinct periods (Halse 1985). During early spring, primary oocytes of sexually mature females transform from the resting follicle stage to developing follicles and increase to between 1 and 5 mm in size due to the deposition of white yolk (or primordial yolk; Halse 1985). Such follicles are considered to be in the stage of pre-rapid follicular growth (pre-RFG). The quantity of nutrients invested in advancing resting follicles to the pre-RFG stage is small relative to later stages. To our knowledge, there is no information available for sea ducks on the relationship between the condition of the female during this stage, which occurs either on the wintering grounds or during migration, and the determination of whether or not a female attempts to reproduce in a given year. This stage is an important aspect to the question of carryover effects of proper nutrition on breeding probability, with cross-seasonal relevance to the role of recruitment in sea duck population dynamics.

RFG is caused by the deposition of true yolk as opposed to white yolk. The timing of RFG in sea ducks is likely during the later phases of spring migration or perhaps after arrival on the breeding grounds, but we are unaware of studies that have reported RFG at spring staging areas. Immediately preceding RFG, the oviduct, which is composed primarily of protein, begins to thicken and enlarge in preparation for ovulation and egg laying. True yolk is composed of about 32% lipid, 23% protein, and 45% water, with a small amount of carbohydrates and minerals, and the composition is generally consistent across ducks (Anatidae; Sotherland and Rahn 1987, Alisauskas and Ankney 1992, Lavers et al. 2006). Rapid growth occurs simultaneously in several follicles, with the time between initiation of each successive follicle largely corresponding to the interval between laying of successive eggs. Follicular growth rates seem highly variable based on the limited information available. Common Eiders require 6 days to produce a fully developed follicle with a dry mass of about 30 g, Common Mergansers require 10 days to produce a follicle of about 20 g dry mass, and Common Goldeneyes require ~8.5 days to produce a follicle of 13 g dry mass (Alisauskas and Ankney 1992; Figures 5.2-5.5). Once ovulated, follicles enter the oviduct where albumen envelopes the yolk before deposition of the shell in the shell gland. This period is energetically demanding as it requires large investment of lipids and proteins into several follicles that were initiated sequentially but with overlapping schedules of nutrient supply (King 1973, Alisauskas and Ankney 1992). Total duration of RFG from the first follicle to begin development until the final egg is laid varies according to the laying intervals and clutch sizes across individuals.

INTERSPECIFIC PATTERNS OF EGG AND CLUTCH PRODUCTION

Egg mass, egg composition, and clutch size govern the total nutritional costs of egg production per nesting attempt. All three determinants of egg production costs show considerable interspecific variation and tend to covary with mean body mass (Table 5.1). In many regards beyond differences in size and mass, life history patterns of sea ducks of subtribe Somatereae (hereafter, eiders) differ markedly from those in subtribe Mergeae (hereafter, non-eiders; Livezey 1995). Nevertheless, interspecific variation in absolute egg mass shows a strong positive relationship with mean adult body mass common to both eiders and non-eiders (Figure 5.1). The average egg size for each species

may have evolved to correspond to the amount of nutrition sufficient for the production of precocial offspring, as well as the asymptotic adult mass that growing ducklings must achieve and the structural constraints of females that produce the eggs. When expressed as a proportion of body mass, however, larger species tend to lay proportionally smaller eggs (Figure 5.2), a pattern shared by both eiders and non-eiders. Although there is a weak apparent trade-off between mean egg size and mean clutch size overall (Figure 5.3), different patterns emerge when tribes are considered separately. For example, the negative relationship remains strong among eiders but shows a weak, nonsignificant positive trend in non-eiders. A priori, such trade-offs predict greater reliance on endogenous nutrients for egg production by eiders, but perhaps not among non-eiders, or at best to a lesser degree.

Larger species also tend to lay the smallest clutches (Figure 5.4). Similar to relationships for egg size, only eiders show a statistically significant inverse relationship driven largely by the small body mass and large clutch size of Steller's Eider (*Polysticta stelleri*). The relationship among non-eiders is opposite to that for eiders although the slope of the relationship among non-eiders is not different statistically from zero. Indeed, Steller's Eider life history in this regard is more similar to the non-eiders than to the *Somateria* species of eiders, even though these two genera are considered a monophyletic sister group to the Mergini (Livezey 1995).

It is useful to distinguish between total costs associated with producing an entire clutch and daily costs of nutrient or energy demand during egg production. The latter provide a daily schedule of demand that may have evolved to be consistent with other aspects of specific life-history traits, such as diet or body size. Maximum total daily cost associated with egg production during a nesting attempt is governed by the energy equivalents of eggs and the overlap in schedules of production between sequential eggs, indexed by rate of RFG, and the egg-laying rate (King 1973). Alisauskas and Ankney (1992) found that duration of RFG (days) among 12 waterfowl species was positively related to mean egg mass following

$$RFG = 1.273E^{0.43}\left(r^2 = 0.71\right)$$

and was significantly more rapid than for other avian species with the same egg masses

TABLE 5.1

Interspecific variation in traits of egg production by North American sea ducks.

Species and Subspecies	Female body mass (g)		Egg mass (g)		Laying frequency (eggs/day)		Clutch Size (eggs)		Clutch mass (g)[a]
	Range	Mean	Range	Mean	Range	Mean	Range	Mean	Mean
Bucephala albeola, Bufflehead, BUFF		337 (Gauthier 1993)		37.4 (Erskine 1971b)	0.5–0.67	0.67 (Gauthier 1993)	6–11	7.8 (Zammuto 1986)	292
Bucephala clangula, Common Goldeneye, COGO	61–66	859 (Eadie et al. 1995)		63.5 (Eadie et al. 1995)	0.5–0.67	0.67 (Eadie et al. 1995)	6–9	7.1 (Mallory et al. 1994)	451
Bucephala islandica, Barrow's Goldeneye, BAGO		843 (Eadie et al. 2000)		67.5 (Eadie et al. 2000)		0.5 (Thompson 1996)	1–12	10.2 (Zammuto 1986)	689
Clangula hyemalis, Long-tailed Duck, LTDU	500–950	814 (Robertson and Savard 2002)		42.7 (Peterson and Ellarson 1979)		1 (Alison 1975)		7.5 (Zammuto 1986)	320
Histrionicus histrionicus, Harlequin Duck, HADU		570 (Robertson and Goudie 1999)		53 (Schönwetter 1960)		0.5 (Robertson and Goudie 1999)	3–9	6 (Robertson and Goudie 1999)	318
Lophodytes cucullatus, Hooded Merganser, HOME		554 (Dugger et al. 2009)		59 (Dugger et al. 2009)	0.5–0.67	0.67 (Dugger et al. 2009)	5–44	10.3 (Mallory et al. 2002)	608
Melanitta fusca, White-winged Scoter, WWSC		1450 (Brown and Fredrickson 1997)		82.4 (Brown 1981)		0.75 (Brown 1981)	6–12	8.8 (Brown and Fredrickson 1997)	725
Melanitta americana, Black Scoter, BLSC		987 (Bordage and Savard 1995)		60.7 (Dement'Ev and Gladkov 1967)	0.5–1	0.75 (Bengtson 1972)	5–10	8.7 (Bengtson 1972)	528
Melanitta perspicillata, Surf Scoter, SUSC	76–79	1025 (Vermeer and Bourne 1984)		77.5 (Savard et al. 1998)		0.67 (Savard et al. 1998)	6–9	7.6 (Morrier et al. 1997)	589

Species								
Mergus merganser, Common Merganser, COME	895–1770	1232 (Cramp and Simmons 1977)		70 (Mallory and Metz 1999)	0.75 (Mallory and Lumsden 1994)	6–17	9.6 (Zammuto 1986)	672
Mergus serrator, Red-breasted Merganser, RBME		998 (Titman 1999)		67.9 (Titman 1999)	0.67 (Curth 1954)	5–24	9.5 (Zammuto 1986)	645
Polysticta stelleri, Steller's Eider, STEI		852 (Fredrickson 2001)		55 (Fredrickson 2001)	1 (Solovieva 1997)	5–8	7.7 (Solovieva 1997)	424
Somateria fischeri, Spectacled Eider, SPEI		1623 (Petersen et al. 2000)		71.2 (Dau 1974)	1 (Dau 1974)	1–11	4.7 (Petersen et al. 2000)	335
Somateria mollissima borealis, Common Eider, COEI-b		1648 (Goudie et al. 2000)		100.9 (Goudie et al. 2000)	1 (Cooch 1965)	3–5	3.5 (Robertson and Gilchrist 1998)	351
Somateria mollissima dresseri, Common Eider, COEI-d		2529 (Guillemette and Ouellet 2005a)		113.1 (Goudie et al. 2000)	1 (Cooch 1965)	3–6	4.0 (Goudie et al. 2000)	448
Somateria mollissima sedentaria, Common Eider, COEI-s		1955 (Goudie et al. 2000)		103.5 (Goudie et al. 2000)	1 (Cooch 1965)	3–7	4.4 (Goudie et al. 2000)	450
Somateria mollissima v-nigrum, Common Eider, COEI-v		2620 (Alisauskas and Ankney 1992)		117 (Alisauskas and Ankney 1992)	1 (Cooch 1965)	3–8	4.6 (Goudie et al. 2000)	532
Somateria spectabilis, King Eider, KIEI		1704 (Scott et al. 2010)	67–71.6	69.3 (Powell and Suydam 2012)	1 (Palmer 1976)	2–8	4.9 (Bentzen and Powell 2012)	340

[a] Clutch mass = average clutch size × average egg mass.

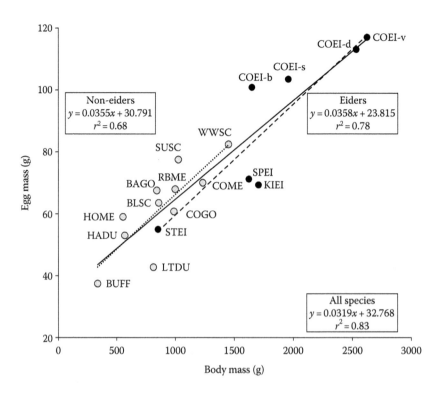

Figure 5.1. Interspecific relationship between mean egg mass and mean female body mass among sea ducks (from Table 5.1). Regressions are shown for all species (solid line), for eider subtribe Somatereae (solid circles, dashed line), and for non-eider subtribe Mergeae (open circles, dotted line).

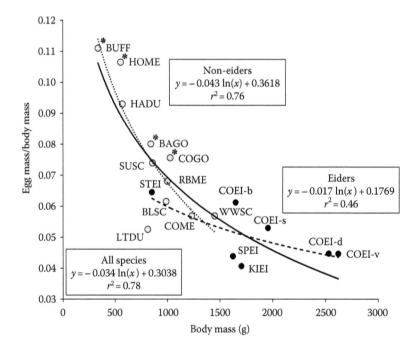

Figure 5.2. Mean egg mass as a proportion of mean female body mass in sea ducks (from Table 5.1). Cavity nesters are shown with a "*". Regressions are shown for all species (solid line), for eider subtribe Somatereae (solid circles, dashed line), and for non-eider subtribe Mergeae (open circles, dotted line).

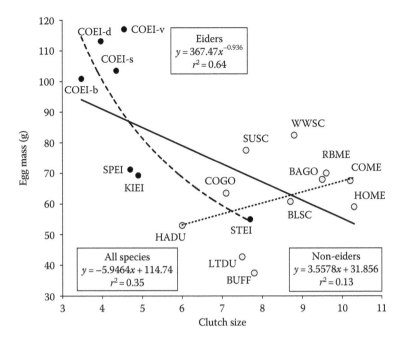

Figure 5.3. Relationship between mean egg mass and interspecific variation in mean clutch size of sea ducks (from Table 5.1). Regressions are shown for all species (solid line), for eider subtribe Somatereae (solid circles, dashed line), and for non-eider subtribe Mergeae (open circles, dotted line).

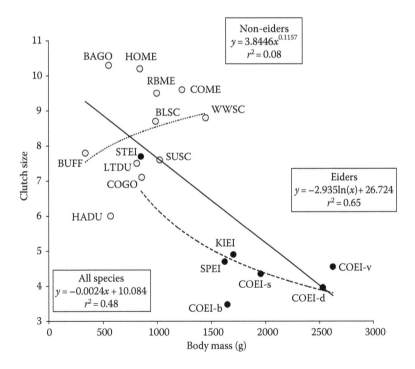

Figure 5.4. Relationship between mean clutch size and interspecific variation in mean body mass in female sea ducks (from Table 5.1). Regressions are shown for all species (solid line), for eider subtribe Somatereae (solid circles, dashed line), and for non-eider subtribe Mergeae (open circles, dotted line).

attesting to the ability of waterfowl to rapidly acquire or mobilize nutrients for egg production.

Egg-laying rates tend to be reported as averages for the entire clutch (Table 5.1). Some estimates of laying rates are likely influenced by the frequency of visits to nests by observers, such that estimation of timing for sequentially laid eggs may be imprecise due to unquantified sampling error. Moreover, there is likely variation among individual females, as there certainly is among species. For example, Thompson (1996) demonstrated that the time between sequential eggs declined as more eggs were laid by female Bufflehead (*Bucephala albeola*) and Barrow's Goldeneye (*B. islandica*). In contrast, Watson et al. (1993) found that laying interval increased between sequential eggs laid later in Common Eider clutches. Such level of detail in the timing of egg laying for other species would increase the accuracy of inferred daily schedules about nutritional requirements of egg production, as has been illustrated for some other waterfowl species (Alisauskas and Ankney 1992, 1994). Longer intervals also suggest that plasticity in egg-laying rates may provide flexibility in tactics of nutrient supply through reductions in peak costs if the energetic demand is spread over a longer laying interval.

In addition to follicular growth rates and laying intervals, final egg size and composition is pertinent to the estimation of nutrient requirements for their production. Unfortunately, our survey about detailed fat, protein, and mineral composition of sea duck eggs revealed information for only seven species, of which only three have been published (Table 5.2). The surprising lack of information identifies a clear need for basic information about egg composition for many species of sea ducks. The seven species represent a range in mean body size from about 337 g to 2 kg in body mass and span temperate to arctic-nesting latitudes (Table 5.2). Despite variability in breeding biology, there seems to be a consistency in proportional composition judging from the narrow range in the proportion of eggs comprised of water (58%–64%), lipids (13%–17%), protein (14%–16%), and minerals (7%–9%). Egg protein is almost equally partitioned into albumen (6%–8%) and yolk protein (6%–9%). These ranges compare favorably with the assumption by Alisauskas and Ankney (1992) that average composition of waterfowl eggs was 15% lipid, 9% albumen, and 6% yolk protein (15% total protein). The resulting range in energy density of eggs was 8.6–10.6 kJ/g of fresh egg mass (Table 5.2), assuming 23.6 kJ/g protein and 39.5 kJ/g lipid (Brody 1945), comparable to a mean for all waterfowl assuming 9.5 kJ/g of fresh egg mass (Alisauskas and Ankney 1992). Nevertheless, published information about composition of eggs is still needed for 11 taxa of sea ducks that breed in North America and additional species that breed elsewhere. Several papers have updated egg-laying rates by sea ducks since the survey conducted by Alisauskas and Ankney (1992). More complete information about follicular growth rates would be useful for different species and subspecies of Common Eiders in the *Somateria mollissima* complex. Such data might provide additional insights about the interplay between limits to the daily supply of specific nutrients to sea duck eggs and the relative contribution of endogenous versus exogenous nutrients.

The larger eiders seem adapted to produce eggs at the highest rate—apparently fixed at 1 egg/day—complementing their rapid rate of follicular growth (see the previous texts). Steller's Eiders are included in this clade (Livezey 1995) but are relatively small compared to the other eiders (Table 5.1, Figure 5.5) and resemble the non-eiders in other life-history and ecological aspects. Conversely, Long-tailed Ducks (*Clangula hyemalis*) are the only species among non-eiders with an egg-laying rate equal to that of eiders. When eiders and non-eiders are considered separately, however, laying rate is unrelated to body mass in either group, suggesting that phylogeny is more relevant to laying rate than metabolism. These inferences are subject to inclusion of additional data from further study of laying rate across taxa, populations, and individuals.

Besides potential limitations to the daily rate of nutrient supply, total nutrient content of the clutch provides a separate and overall measure of nutrient allocation to eggs. Clutch mass is a convenient metric for quantifying absolute production costs for each species, since composition of sea duck eggs is largely proportional (Table 5.2). Clutch mass is unrelated to female body mass among sea ducks in a naïve assessment if subtribe is ignored (Figure 5.6), but stratification by subtribe reveals positive relationships of clutch mass to body mass in each. The slope of this steep relationship is statistically different from zero in non-eiders, but not in eiders. Cavity nesters are not exceptional compared to other non-eiders (Figure 5.6). If comparisons are restricted

TABLE 5.2

Egg composition among sea ducks.

| Species and subspecies | Location | n | Fresh Mass (g) | Mineral | Dry mass (g) | | | | Egg Protein | Egg Water (g) |
					Dry Albumen	Dry Yolk	Yolk Lipid	Yolk Protein		
Bucephala albeola, Bufflehead (Lavers et al. 2006)	BC	123	36.7	3.4	2.7	7.6	5.1	2.5	5.2	23.0
Bucephala clangula, Common Goldeneye (Lavers et al. 2006)	BC, ON, MN		63.9							
Bucephala islandica, Barrow's Goldeneye (Lavers et al. 2006)	BC	226	66.4	6.2	4.9	13.3	8.9	4.3	9.2	42.0
Clangula hyemalis, Long-tailed Duck (R. Alisauskas and D. Kellett. Environment Canada, unpubl. data)	NU	10	39.6	3.0	3.1	8.9	5.7	3.2	6.3	24.6
Histrionicus histrionicus, Harlequin Duck										
Lophodytes cucullatus, Hooded Merganser										
Melanitta fusca, White-winged Scoter										
Melanitta nigra, Black Scoter (D. Esler and P. Flint, U.S. Geological Survey, unpubl. data)	AK	18	76.8	6.7	4.8	17.5	10.4	7.1	11.9	47.8
Melanitta perspicillata, Surf Scoter (S. De La Cruz and J. Takekawa, U.S. Geological Survey, unpubl. data)	NWT	15	62.5	5.9	4.4	12.0				
Mergus merganser, Common Merganser										
Mergus serrator, Red-breasted Merganser										
Polysticta stelleri, Steller's Eider										
Somateria fischeri, Spectacled Eider										
Somateria mollissima borealis, Common Eider										
Somateria mollissima dresseri, Common Eider										
Somateria mollissima sedentaria, Common Eider (Swennen and Van der Meer 1995, mean of 3 years)	Netherlands	118	103.1	8.9	7.2	27.2	17.4	9.8	17.0	59.8
Somateria mollissima v-nigrum, Common Eider										
Somateria spectabilis, King Eider (R. Alisauskas and D. Kellett, Environment Canada, unpubl. data)	NU	10	61.1	4.4	4.2	14.0	8.8	5.2	9.4	38.5

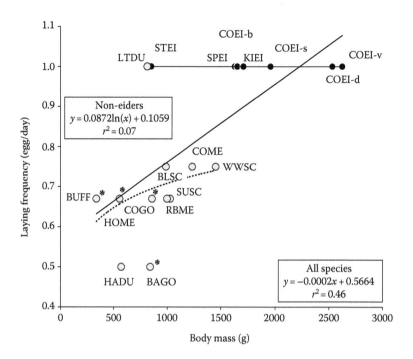

Figure 5.5. Relationship between mean egg-laying rate and interspecific variation in mean body mass of female sea ducks (from Table 5.1). Cavity nesters are shown with a "*". Regressions are shown for all species (solid line), for eider subtribe Somatereae (solid circles, dashed line), and for non-eider subtribe Mergeae (open circles, dotted line).

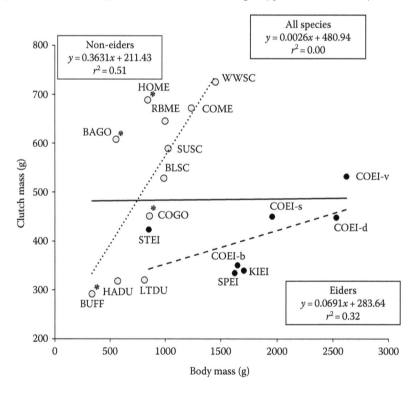

Figure 5.6. Mean clutch mass and mean body mass of female sea ducks. Cavity nesters are shown with a "*". Regressions are shown for all species (solid line), for eider subtribe Somatereae (solid circles, dashed line), and for non-eider subtribe Mergeae (open circles, dotted line).

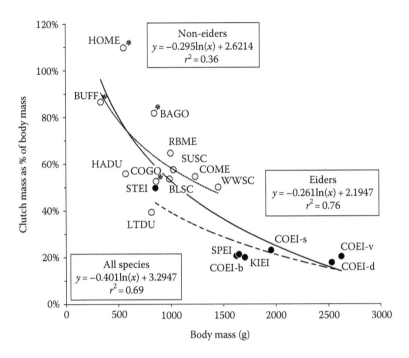

Figure 5.7. Mean clutch mass in relation to mean body mass of female sea ducks (from Table 5.1). Cavity nesters are shown with a "*". Regressions are shown for all species (solid line), for eider subtribe Somatereae (solid circles, dashed line), and for non-eider subtribe Mergeae (open circles, dotted line).

to a range in body mass overlapping both groups (852–1450 g), non-eiders produced about 1.5–2.0 times more egg nutrients than eiders per gram of body mass. However, the smallest three species of non-eiders were more similar to eiders than to other non-eiders (Figure 5.6). Thus, absolute nutrient allocation to eggs was relatively invariant along a metabolic rate scale determined by body mass of eiders but strongly dependent in non-eiders.

An expression of nutrient production as a proportion of body mass rescales absolute costs in a manner consistent with metabolic size and can reveal different patterns that may have pertinence to metabolic constraints. When clutch mass is so expressed, both eiders and non-eiders share a similar declining trajectory in proportional nutrient production with increasing metabolic size (Figure 5.7). Negative relationships in both groups have slopes that are statistically different from zero. When adjusted in this way, however, proportional nutrient production is somewhat greater among non-eiders than among eiders of similar body size. Cavity nesters among non-eiders tend to have the smallest body sizes but the highest

proportional costs of clutch production, relative to ground-nesting non-eiders.

We view the foregoing analyses as exploratory and of possible relevance to constraints on production costs, rather than formal tests of hypotheses about sea duck life-history strategies. Nevertheless, eiders are clearly distinct from other sea ducks in that the largest species invest proportionately far fewer nutrients in clutch production (Figure 5.7), although they show the highest rates of egg production (Figure 5.8). Such a trade-off may be consistent with a greater reliance by the larger-bodied eiders on endogenous reserves, allowing the provision of an uninterrupted supply of nutrients for eggs. Smaller egg mass in proportion to body mass reduces daily rate limits (Figure 5.2); furthermore, a clutch size smaller than the number of days required for RFG by birds that lay 1 egg/day means that maximum daily costs equivalent to 1 egg are not reached by eiders (Alisauskas and Ankney 1992). The lower clutch size and reduced proportional nutrient allocation of eiders may permit sparing of endogenous reserves for use during incubation with high daily constancy.

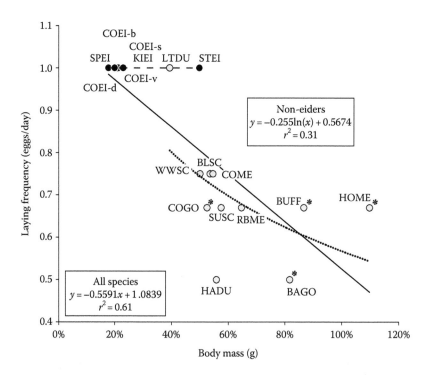

Figure 5.8. Relationship between mean egg-laying frequency and total nutrient allocation expressed as percentage of body mass for female sea ducks. Cavity nesters are shown with a "*". Regressions are shown for all species (solid line), for eider subtribe Somatereae (solid circles, dashed line), and for non-eider subtribe Mergeae (open circles, dotted line).

INCUBATION STRATEGIES

Similar to rates of egg production, considerable interspecific variation in incubation duration exists among sea ducks and ranges from 23 to 33 days. Much of the variation in incubation duration among waterfowl is related to body mass (Afton and Paulus 1992). About 35% of the interspecific variation in incubation duration by sea ducks is explained by an apparent inverse relationship with mean body mass (Figure 5.9). We report species averages but incubation duration also varies among subspecies and populations of Common Eider (Bolduc and Guillemette 2003). Closer inspection of the data in Table 5.3 following stratification shows that the apparent decline in duration with increasing body mass is driven largely by phylogeny instead of body mass, as slopes for eiders and non-eiders were not statistically different from zero (Figure 5.9). Although nonsignificant, about one-third of the variation in incubation duration by eiders is positively related to body mass, suggesting greater nutrient storage potential is consistent with the large body mass of eiders versus non-eiders. Incubation duration may

also be governed by ontogenic constraints of growing embryos. Egg size and body mass are strongly related (Figures 5.1 and 5.2), but a greater share of the variance in incubation duration is related to egg size than to body mass among eiders, and not other species of sea ducks (Figure 5.10). Thus, growth requirements of embryos appear to influence the incubation duration in eiders. Incubation duration is also linked via body reserves to relative nutritional investment in eggs across all species, because there is a strong positive relationship between the duration of incubation and absolute clutch mass (Figure 5.11). Such a pattern is expected because of the need to supply more heat to a larger number of developing embryos. At a common body size of 400 g, non-eiders have a predicted duration of >28 days compared to only 24 days for eiders. The reduced duration in eiders may be related to a greater reliance on reserves permitting higher average rate of nest attendance. Among non-eiders, further differences in duration of incubation are present between cavity nesters (incubation duration ≥30 days) and open nesters (≤30 days, except for Common Merganser *Mergus merganser*, Figure 5.11).

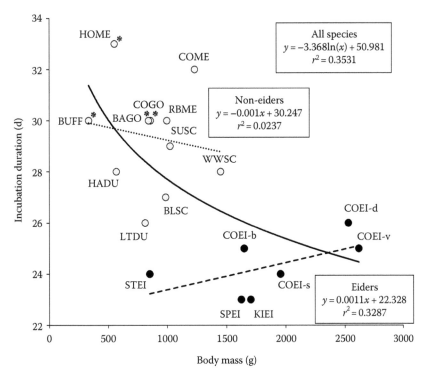

Figure 5.9. Mean incubation duration in relation to interspecific variation in mean body mass of female sea ducks (from Tables 5.1 and 5.3). Cavity nesters are shown with a "*". Regressions are shown for all species (solid line), for eider subtribe Somatereae (solid circles, dashed line), and for non-eider subtribe Mergeae (open circles, dotted line).

Thus, variation in predation risk further complicates patterns of energy constraints among sea ducks. Incubation duration is an even greater function of clutch mass when expressed as a rough metabolic proportion of body mass (Figure 5.12). Although some of this variation was due to phylogenetic differences in body size between eiders and non-eiders, the slope against clutch mass as a proportion of body mass is statistically different from zero only in non-eiders. Differences in response of incubation duration to metabolically scaled clutch size may stem from variation in the extent that nutrient reserves supply energy during incubation, which is predicted to be high in eiders, but highly variable in non-eiders. The difference may also stem from competing demands of heat transfer to eggs versus self-maintenance of incubating hens. Female requirements are highly variable in non-eiders because four species of cavity nesters have different thermal environments from ground nesters.

Less information is available about incubation constancy in sea ducks or the proportion of the 24 h day spent attending nests. Based on data from 11 studies for which there is corresponding data about body mass of females (Table 5.3),

larger species spend a higher proportion of their incubation periods attending eggs (Figure 5.13) indicating a greater reliance on catabolism of stored nutrients that were not used during egg production. Incubation constancy shows a weak negative relationship with incubation duration as body mass increases (Figure 5.14), but this pattern disappears when data are stratified by eider versus non-eider. Again, phylogeny and body size are factors here, as observed for incubation duration, although these inferences may change when data gaps for incubation constancy are addressed for Buffleheads, Barrow's Goldeneye, Harlequin Ducks, Surf Scoters, and Black Scoters (Table 5.3).

Exploratory analysis of incubation patterns illustrate the strong linkages among some coevolved traits such as body mass, clutch mass, incubation constancy, and incubation duration across sea ducks, with implications for endogenous nutrient use. Risks and energy costs of incubation by sea ducks seem to be influenced by egg size in eiders but by clutch mass in non-eiders, further modified by the presence of cavity nesting by some non-eiders presumably associated with reduced predation risk.

TABLE 5.3
Interspecific patterns of incubation by sea ducks.

Species and subspecies	Post-laying body mass (g)	Hatch body mass (g)	Proportion of post-laying mass lost (g)	Mass loss (g/Day)	Incubation constancy	Incubation duration (Days)
Bucephala albeola, Bufflehead, BUFF	—	—	—	—	—	30 (Erskine 1971b)
Bucephala clangula, Common Goldeneye, COGO	751 (Mallory and Weatherhead 1993)	626 (Mallory and Weatherhead 1993)	0.20	4.2	0.81 (Mallory and Weatherhead 1993)	30 (Eadie et al. 1995)
Bucephala islandica, Barrow's Goldeneye, BAGO	—	—	—	—	—	30 (Eadie et al. 2000)
Clangula hyemalis, Long-tailed Duck, LTDU	618 (Kellett et al. 2005)	575 (Kellett et al. 2005)	0.07	1.7	0.84 (Lawson 2006)	26 (Alison 1975)
Histrionicus histrionicus, Harlequin Duck, HADU	—	—	—	—	—	28 (Bengtson 1972)
Lophodytes cucullatus, Hooded Merganser, HOME	539 (Zicus 1997)	506 (Zicus 1997)	0.06	1.0	—	33 (Dugger et al. 2009)
Lophodytes cucullatus, Hooded Merganser, HOME	565 (Mallory et al. 1994)	475 (Mallory et al. 1994)	0.16	2.7	0.85	33 (Dugger et al. 2009)
Melanitta fusca, White-winged Scoter, WWSC	1498 (Dobush 1986)	1154 (Dobush 1986)	0.23	12.3	0.84	28 (Brown 1977)
Melanitta nigra, Black Scoter, BLSC	—	—	—	—	—	27 (Bordage and Savard 1995)
Melanitta perspicillata, Surf Scoter, SUSC	—	—	—	—	—	29 (Savard et al. 1998)
Mergus merganser, Common Merganser, COME	1200 (Mallory and Lumsden 1994)	876 (Mallory and Lumsden 1994)	0.27	10.1	0.88	32 (Mallory and Metz 1999)

Species						
Mergus serrator, Red-breasted Merganser, RBME	—	—	—	—	0.91	30 (Titman 1999)
Polysticta stelleri, Steller's Eider, STEI	—	—	—	—	0.81 (Quakenbush et al. 2004)	24 (Quakenbush et al. 2004)
Somateria fischeri, Spectacled Eider, SPEI	1415 (Grand and Flint 1997)	1047 (Grand and Flint 1997)	0.26	16.0	0.90 (Flint and Grand 1999)	23 (Grand and Flint 1997)
Somateria mollissima borealis, Common Eider, COEI-b	1773 (Parker and Holm 1990)	1368 (Parker and Holm 1990)	0.23	16.2	—	25 (Goudie et al. 2000)
Somateria mollissima dresseri, Common Eider, COEI-d	—	—	—	—	—	26 (Bourget 1973)
Somateria mollissima sedentaria, Common Eider, COEI-s	1,625 (Bottitta et al. 2003)	1,253 (Bottitta et al. 2003)	0.23	15.5	0.99 (Bottitta 1999)	24 (Bottitta 1999)
Somateria mollissima v-nigrum, Common Eider, COEI-v	—	—	—	—	—	25 (Goudie et al. 2000)
Somateria spectabilis, King Eider, KIEI	1619 (Kellett and Alisauskas 2000)	1128 (Kellett and Alisauskas 2000)	0.30	21.3	0.98 (Bentzen et al. 2008a)	23 (Powell and Suydam 2012)

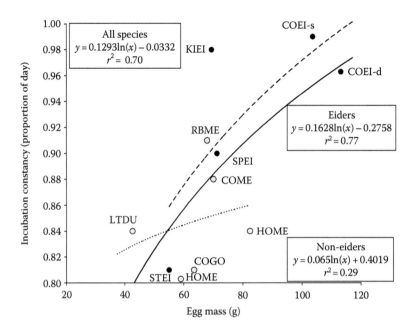

Figure 5.10. Mean incubation duration in relation to interspecific variation in egg size of sea ducks (from Tables 5.1 and 5.3). Regressions are shown for all species (solid line), for eider subtribe Somatereae (solid circles, dashed line), and for non-eider subtribe Mergeae (open circles, dotted line).

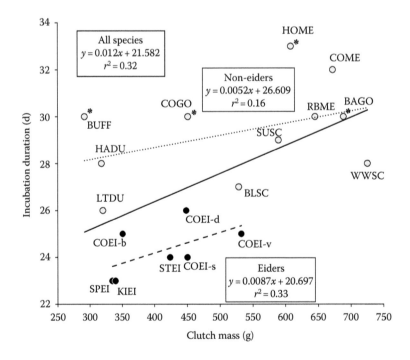

Figure 5.11. Mean incubation duration (days) as a function of mean clutch mass of female sea ducks (from Tables 5.1 and 5.3). Cavity nesters are shown with a "*". Regressions are shown for all species (solid line), for eider subtribe Somatereae (solid circles, dashed line), and for non-eider subtribe Mergeae (open circles, dotted line).

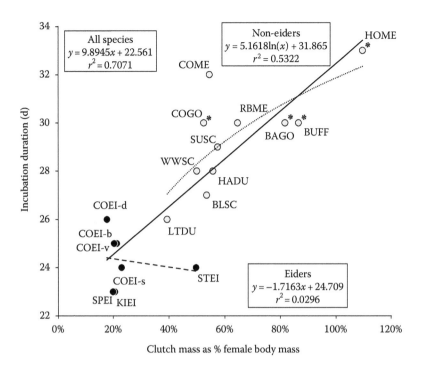

Figure 5.12. Mean incubation duration in relation to interspecific variation in mean clutch mass of sea ducks (from Tables 5.1 and 5.3). Cavity nesters are shown with a "*". Regressions are shown for all species (solid line), for eider subtribe Somatereae (solid circles, dashed line), and for non-eider subtribe Mergeae (open circles, dotted line).

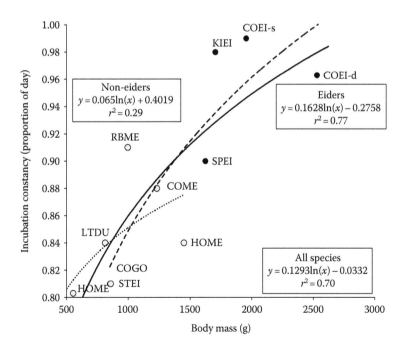

Figure 5.13. Mean incubation constancy (expressed as a proportion of 24 h day) in interspecific variation in mean body mass of female sea ducks (from Tables 5.1 and 5.3). Regressions are shown for all species (solid line), for eider subtribe Somatereae (solid circles, dashed line), and for non-eider subtribe Mergeae (open circles, dotted line).

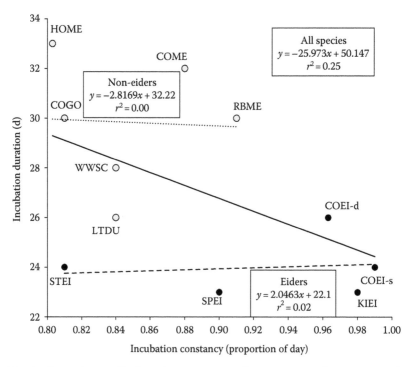

Figure 5.14. Mean incubation duration (days) in relation to interspecific variation in incubation constancy by sea ducks (from Table 5.3). Regressions are shown for all species (solid line), for eider subtribe Somatereae (solid circles, dashed line), and for non-eider subtribe Mergeae (open circles, dotted line).

SOURCES OF ENERGY AND NUTRIENTS

Lipids, proteins, and minerals used in the creation of reproductive tissues ultimately must be derived from nutrients assimilated through an individual's diet. However, the ability of some birds to store nutrients before RFG permits greater flexibility in timing of egg laying, incubation constancy, level of nest defense, and relative nutrient investment into reproduction. The use of exogenous nutrients from diet for clutch production and incubation has been termed *income breeding* (Drent and Daan 1980). The opposite extreme, where an individual depends on endogenous nutrients stored for meeting nutritional requirements or energy requirements, has been characterized as *capital breeding* by Drent and Daan (1980). Few species likely employ either strategy completely, so that the relative use of endogenous nutrients for avian reproduction has been characterized as the capital–income continuum (Meijer and Drent 1999). Determinations about the extent of capital breeding strategies are further complicated by the fact that endogenous nutrients may be acquired during late winter, spring migration, or in foraging habitats adjacent to nesting locations (Klaassen et al. 2006).

The distinction between local and distant capital breeding strategies is important to understanding relative value of habitats where reproductive nutrients are acquired, with implications for areas where habitat conservation might be directed.

Female sea ducks begin incubation toward the end of clutch completion, though incubation in Common Eiders has been reported to begin after the second egg is laid (Swennen et al. 1993). Incubation constancy ranges from 81% to 99% in sea ducks, which leaves little time to acquire nutrients needed for self-maintenance and incubation. Variation in incubation constancy in sea ducks may be related to a trade-off strategy between investing greater endogenous resources for incubation and the benefits of higher nest attendance and defense and investing more resources in larger clutches and taking more frequent breaks during incubation (Swennen et al. 1993). Common and King Eiders have the highest reported incubation constancy of sea ducks at 99% and 98%, respectively, but at the expense of up to 36% of their preincubation body mass during this period (Table 5.3; Korschgen 1977, Parker and Holm 1990, Kellett and Alisauskas 2000, Bolduc and Guillemette 2003, Bentzen et al. 2008b). Spectacled Eiders (*Somateria fischeri*) also have a relatively high incubation constancy rate and have been reported to lose ~26% of their preincubation body mass (Flint and Grand 1999). Species with lower incubation constancies, such as Common Goldeneye (*Bucephala clangula*), with a reported rate of 81%, show a lower energetic investment toward incubation losing ~20% of their body mass (Mallory and Weatherhead 1993). White-winged Scoters (*Melanitta fusca*), with a reported incubation constancy of 84%, lose ~23%–25% of their body mass during incubation with about 36% of that mass being lipids, 16% protein, and 3% mineral (Dobush 1986, Brown and Fredrickson 1987). Thus, incubation may require storage of large quantities of endogenous nutrients, primarily lipids, for supplying energy demands.

Various studies have examined factors that affect intraspecific variation in energetics of incubation, particularly among ground-nesting Common Eiders (D'Alba et al. 2009). Bottitta et al. (2003) experimentally lengthened the incubation period of Common Eider and found that experimental females tended to lose more mass than control groups and had lower incubation constancy during extended incubation periods. The cost of incubation and trade-off of constancy may also impact fitness as female eiders in poor body condition tend to have lower constancy rates and also tend to abandon nests and young at higher rates (Bustnes et al. 2002, Hanssen et al. 2003, Bourgeon et al. 2006). Lower incubation constancy late in incubation has been observed in several sea duck species and may be driven by the requirement to compensate for exhausted nutrient stores with increased food intake (Criscuolo et al. 2002, Bolduc and Guillemette 2003, Bottitta et al. 2003, Bentzen et al. 2008b). However, this pattern has not been reported in all studies. Flint and Grand (1999) reported that incubation constancy in Spectacled Eiders was unrelated to loss of body mass or incubation period. All females lost mass, but variation in rates of mass loss suggested variation in food intake, metabolic rate, or a combination of both factors. Nest site selection may also play a key role in the nest success of sea ducks, as sheltered nests tend to decrease energetic costs and had lower mass loss and increased nest success (Kilpi and Lindstrom 1997, D'Alba et al. 2009). These studies attributed increased nest success to thermal benefits of natural and artificial shelters rather than to any differences in predation. Reduced incubation constancy may also compound thermal stress because of the need to rewarm developing embryos, particularly early in incubation.

Greater stores of somatic nutrients provide benefits for buffering periodic difficulties in meeting costs directly with dietary nutrients, but maintaining larger nutrient stores has a number of potential costs in birds. Disadvantages of maintaining high body mass may include (1) higher metabolic costs incurred to maintain nutrient stores, which may require additional foraging; (2) increased egg predation risk due to increased time devoted to food acquisition; (3) compromised maneuverability of heavier birds, leading to higher predation risk; (4) increased risk of injury from the strain of transporting additional mass; (5) reduced mobility and foraging efficiency associated with such increased mass; (6) possible pathological costs associated with body mass beyond optimal levels of storage; and (7) impingement on future ability to acquire sufficient nutrients that may result in trade-offs between current and future breeding attempts (Witter and Cuthill 1993). Whether cumulative benefits in the degree of nutrient storage are unidirectional or show some optimum at intermediate levels, we suggest that this relationship deserves further attention. Trade-offs between potential costs of storing large amounts of adipose and muscle tissue and benefits stemming from increased nutrient and energy availability might be most pertinent to variation in relative fitness among individuals. At a population level, costs associated with maintenance of larger somatic nutrient and lipid levels might drive well-described patterns of seasonal variation in body mass and nutrient stores of sea ducks (Peterson and Ellarson 1979). Seasonal hyperphagia in sea ducks generally occurs in response to expected periods of high energetic demands during migration and reproduction (Guillemette 2001). The increase in body mass is primarily the result of accumulating large stores of lipids, although additional muscle mass also occurs from use–disuse phenomena associated with increased flight activity and increased power requirements of carrying additional lipid mass (Ankney 1984). In addition to lipid and protein, calcium can be accumulated in the medullary bone and more labile cortical bone before reproduction (Simkiss 1961), as calcium represents the most important mineral in eggshells (Dacke et al. 1993).

Somatic Fat

Lipids serve a variety of functions in vertebrates. Some are stored as fuel while others form essential function in membrane structure. Lipids represent the most efficient fuel that animals can accumulate and deposit in eggs, yielding ~33 kJ/g (Sotherland and Rahn 1987), or approximately 8–10 times more energy on a wet weight basis than either protein or carbohydrates (McWilliams et al. 2004). The most abundant forms of lipids in animals are triglycerides, phospholipids, and wax esters, all of which are normally comprised of one or more fatty acids bonded to other molecules (Williams and Buck 2010). Somatic fat stores in migratory birds are composed of triglycerides containing three fatty acids esterified to a glycerol molecule (McWilliams et al. 2004). There are a large number of fatty acids found in nature, and the structure of a fatty acid determines several key chemical and physical properties. The structure of fatty acids is expressed in shorthand as the number of carbon atoms per chain:number of double carbon bonds. Migratory birds acquire a wide range of lipids in their diverse diets, but palmitic (16:0), oleic (18:1), and linoleic acids (18:2) comprise ~75% of lipids in body fat (Williams and Buck 2010). Nevertheless, diet plays a role in determining the fatty acid composition of somatic lipids in waterfowl (Alisauskas and Ankney 1992), and preferential foraging by prebreeding sea ducks may be associated with requirements for specific fatty acids during reproduction (Anderson et al. 2009) or a need for foods with higher fat content (Pierce and McWilliams 2004). Alisauskas and Ankney (1992) point out that a mismatch between profiles in dietary and tissue fatty acids can impinge upon the efficiency of bioproduction. Deficiencies in dietary supplies of linoleic acid can reduce egg size (Griminger 1986).

Lipid accumulation in birds is a function of energy intake, energy expenditure, and excreted energy (urinary and fecal energy). Increased energy intake through intense hyperphagia in sea ducks has been observed to meet the prebreeding, fattening period (Guillemette 2001, Rigou and Guillemette 2010). Specific prey targeted by hyperphagic sea ducks likely also play a role in their lipid accumulation before migration and reproduction—when there can be a focus on foods high in lipid and energy content (Anderson et al. 2009). The nutrient composition of prey is important to the process of lipid acquisition as diets with a high protein–calorie ratio minimize lipid storage and enhance muscle building. Protein metabolism increases the insulin-like growth factor I, which inhibits fat deposition (McWilliams et al. 2004). Conversely, a low protein–calorie ratio improves the efficiency of fat accumulation in birds (Klasing 1998).

Lipids deposited in somatic tissues may originate from either dietary fatty acids or synthesized fatty acids (Alisauskas and Ankney 1992). Dietary fatty acids absorbed through the gut are transported to the liver where they are converted to lipoproteins, then secreted into blood and transported to adipocytes where they are deaminated and deposited as triglycerides. Alternatively, some dietary fatty acids may be metabolized and converted to different fatty acid structures in the liver prior to deposition in order to accumulate more favorable fatty acids for migration and reproduction. Lipids may also be synthesized in the liver from carbohydrates through *de novo* synthesis (Klasing 1998). For example, Hammer et al. (1998) concluded that fatty acids in somatic fat stores of Redheads (*Aythya americana*) were synthesized from carbohydrates or proteins in their sea grass diet rather than direct incorporation of fatty acids. The specific composition of lipids stored in females is likely a function of the nutrients used for reproduction. Connor et al. (1996) found that the mobilization of fatty acids into plasma was not proportional to their content in adipose tissue, but was influenced by their molecular structure. Alisauskas and Ankney (1992) suggested that there may be metabolic inefficiencies in converting somatic fats to egg lipids, if fatty acid profiles differ between source and product. Fatty acid composition of King Eider eggs showed a remarkably similar profile to other species of wild ducks with different diets, and the three most common fatty acids in egg yolks of these species were 16:0, 16:1, and 18:1 (Speake et al. 2002). This pattern suggests that efficiencies in bioproduction are probably different among species that differ in their diets.

Protein

Protein stores in sea ducks, as in other waterfowl, are typically associated with the larger muscle groups such as the pectoral muscles, legs, gizzard, and the liver. Unlike lipids, protein is found in active muscle tissues that play roles in mobility, foraging, and digestion. Therefore, the concept of stored protein for later reproductive use in birds is somewhat enigmatic. Protein is not typically used as a source of energy in waterfowl, unless under severe nutritional stress (Blem 1990). Changes in protein stores of birds are more likely a function of other seasonal activities, such as increased digestive function (gizzard mass), increased lift force (pectoral muscles), or increased foraging (leg muscles) associated with use and disuse (Ankney 1984). Greater protein mass of prebreeding sea ducks may simply be a function of their increased activities associated with breeding, including increased foraging to acquire nutrients for egg production, increased daily flight costs of migration, increased flight costs per distance travelled due to carrying large stores of lipids, and any developed reproductive organs.

Carbohydrates

Carbohydrates range in structure from simple sugars to complex polysaccharides and are not stored in any significant amount in avian species. Glycogen, the analogue of starch, is stored in the liver and muscle tissues of birds and is available as a short-term energy source, but is not an optimal energy store due to lower mass-specific energy content. In carnivorous birds, such as sea ducks, gluconeogenesis occurs constantly in the liver from amino acids derived from the diet or somatic tissues, which produces a more steady level of glucose than in granivorous species (Pollock 2002). This feature allows sea ducks and other carnivores to feed at longer intervals compared to granivorous species that must forage more frequently. The carnivorous diet of sea ducks also leads to physiological adaptations in the digestive system that result in poor absorptive capacity of ingested carbohydrates, such as glucose (Pollock 2002). As glucose may be synthesized from other nutrients and provides a lower energetic return than lipids per unit mass, the storage of large quantities of carbohydrates in somatic tissues has little benefit. Nevertheless, ash-free, lipid-free dry contents of eggs contain ~5% carbohydrates (Sotherland and Rahn 1987),

and carbohydrates in eggs are important as an energy source in the first 10 days for the developing embryo and are completely synthesized from triglyceride fatty acids thereafter (Barrett et al. 1970).

Calcium

Due to the large amount of calcium required in the production of calcified egg shells of birds (Dhondt and Hochachka 2001), the ability to store this mineral is important to meet nutritional requirements during egg production. Increased estrogen preceding egg formation motivates formation of medullary bone found primarily in the femur and tibiotarsus but also to a lesser extent in the humerus, radius, and ulna (de Matos 2008). Medullary bone formation typically begins about two weeks before egg laying in chickens and can increase skeletal mass by 25% (Taylor and Dacke 1984). Medullary bone can supply 30%–40% of calcium deposited in eggs of chickens consuming a diet containing 2% calcium, though patterns of medullary bone use by sea ducks are not well understood. However, if calcium stored as medullary bone is insufficient to provide more than a small fraction of the total calcium required for eggshell formation, it is more likely that this storage mechanism serves as a short-term reservoir that is replenished during periods of intense feeding between eggs.

Little research has been conducted on calcium dynamics in sea ducks, although studies have noted that King Eiders may consume small bones during egg laying, which is believed to be linked to the calcium demands of eggshell formation (Uspenski 1972 in Krapu and Reinecke 1992). This incidental observation suggests that stores of endogenous calcium are sufficient to meet the needs for egg formation in sea ducks. Similar observations exist for arctic-breeding geese that consume empty clam shells (Flint et al. 1998) or eggshells from previous years' production of goose eggs (Gloutney et al. 2001) before and during, but not after, egg laying. Therefore, even in large sea ducks such as eiders with potentially greater calcium storage capacity, dietary calcium is probably required to meet the demands of egg formation. The calcium budgets of breeding sea ducks merits further study.

TECHNIQUES FOR DETERMINING SOURCES OF REPRODUCTIVE NUTRIENTS

Studies that determine relative contributions of dietary and endogenous nutrients toward avian reproduction should consider all activities associated with both the production and rearing of the young. Earlier studies of nutrient allocation strategies in waterfowl and sea ducks examined only the sources of nutrients invested in egg production. However, it is well known that in some species of sea ducks, incubation is also a costly activity that requires investment by the female, given that males do not participate. Because nutrient investment by breeding female sea ducks far exceeds those by males, this section focuses on techniques used to assess strategies of energy and nutrient supply by breeding females.

Body Composition Analysis

Traditionally, assessing reproductive energetic strategies was done by collecting females at various stages of the breeding cycle, from pre-laying to clutch completion and through incubation (Ankney and MacInnes 1978, Peterson and Ellarson 1979). Females were dissected and gross composition (lipid, protein, and mineral) estimated from various laboratory methods and compared between stages. Specific methods for determining nutrient mass in birds and eggs are described in detail by Alisauskas and Ankney (1992). One approach involves regressing absolute mass of somatic nutrient stores (g) over cumulative investment in reproductive tissues (g) to estimate the slope (Alisauskas and Ankney 1985), which allows inferences about relative allocation to reproductive nutrients (see also Sedinger et al. 1997). If the slope of the relationship between whole somatic protein and cumulative content of clutch protein was zero when females were sampled, then a reasonable inference would be that there was no net allocation of endogenous protein to eggs. Alternatively, if the same relationship for protein or lipid demonstrated a slope approaching negative one with one gram of lipid lost for each gram of clutch lipid, then a largely capital strategy would be inferred. Linear or nonlinear relationships between cumulative allocation to egg nutrients and respective somatic tissues could be considered with a multimodel approach (Burnham and Anderson 2002). Besides quantifying changes in somatic nutrient

stores during clutch formation, body composition analysis can also lead to estimates of nutrient depletion during incubation. As shown in Table 5.3 and elsewhere (Chapter 10, this volume), high incubation constancy (81%–99%) of different species of sea ducks places temporal constraints on feeding ability and the extent to which energy requirements can be met with either endogenous or exogenous nutrients.

There are two key limitations to using body composition analysis for assessing nutrient allocation strategies in sea ducks (DeVink et al. 2011). First, this method is based on an assessment of a generalized or average strategy based on regression parameter estimates for the sample of birds analyzed. Lethal sampling of birds for body composition analysis prevents temporally repeated sampling of individuals so that knowledge about an individual's energetic strategy is not possible (Alisauskas and Ankney 1992). Second, without specific information about the metabolic requirements of the birds and dietary intake, it is impossible to differentiate use of lipids for clutch formation versus metabolic requirements (Meijer and Drent 1999).

Body composition analysis, however, provides a method for estimation of average nutrient composition of females leading up to egg formation. Estimates can be informative for comparing breeders with nonbreeders and for detection of endogenous nutrient thresholds for breeding in a species (Alisauskas and Ankney 1994, Bond et al. 2008). Nutritional determinants of breeding probability can be an important piece of information most relevant to governance of recruitment at a population level. The ability to link breeding likelihood to proper nutritional conditioning may provide proper focus toward assessing landscape change and conservation applications, if the timing and location of nutrient storage is known.

Body composition analysis has been used to assess strategies of nutrient supply in White-winged Scoters (Dobush 1986, DeVink et al. 2011), Bufflehead, and Barrow's Goldeneyes (Thompson 1996, Table 5.4). The massive decline in nutrient reserves during and after egg laying in Common Eiders also seems like compelling evidence for their critical importance during both egg laying and incubation (Korschgen 1977). Dobush (1986) found that White-winged Scoters nesting in Alberta and Saskatchewan did not use fat or protein during egg laying despite storing somatic

TABLE 5.4

Inferred contribution of endogenous macronutrients to egg nutrients, using the regression method of Alisauskas and Ankney (1985).

	Egg somatic nutrient decline (g)/Egg nutrient produced (g)			
	Protein	Lipid	Mineral	Source
Bucephala albeola, BUFF	0–2.08	0.26	0–0.08	Thompson (1996)
Bucephala islandica, BAGO	0–0.11	0.65	0.03	Thompson (1996)
Melanitta fusca, WWSC	0.08	0.06	0.04	Dobush (1986)
Melanitta fusca, WWSC	0.28	0.47	—[a]	DeVink et al. (2011)

[a] Not estimated.

protein on nesting areas before egg synthesis. However, most sea duck species remain unstudied in this regard.

Stable-Isotope Analysis

Stable-isotope analysis has been a relatively recent innovation in the field of avian nutrition during egg formation and has been used with breeding diving and dabbling ducks (Hobson et al. 2004), arctic-nesting geese (Gauthier et al. 2003), and sea ducks (Hobson et al. 2005). The approach relies on geographical or trophic-level variation in ratios of naturally occurring stable isotopes of several elements. Stable-isotope values are expressed in parts per thousand (‰) relative to international standards and are denoted as $\delta^{13}C$, for example, for carbon isotope ratios where

$$\delta^{13}C = \left[\left(\frac{R_{sample}}{R_{standard}} \right) \times 1000 \right],$$

and $R = {}^{13}C/{}^{12}C$ (reviewed in Hobson 1999, 2006, see also Bond and Hobson 2012). In consumers that migrate between areas with isotopically distinct food webs (hereafter, isoscapes), body tissues should initially integrate and reflect the origin of nutrients derived from the previous location, and the duration of the signal should also depend on the turnover rate of the tissue, measured as the difference in isotopic values between ambient food and somatic tissue. Partly because of differences in atomic mass between the heavy and light isotope of the same element, each isotope can have differential rates of transfer during metabolism, such as when somatic tissue is catabolized, ingested

foods are converted to stored tissue, or either of these are converted to egg tissue. The difference is called isotopic discrimination, denoted as $\Delta^{13}C$, using carbon as an example. If the way in which stable-isotope values of macromolecules change or discriminate between dietary or somatic sources and egg macromolecules is known, it is possible to estimate the proportional contributions of each of these sources to egg formation using standard mixing model approaches (Hobson et al. 1997a, b; Hobson 2006). Mixing models are particularly useful for sea ducks that move from marine wintering or staging areas to freshwater breeding environments, as the two ecosystems generally have unique isoscapes. The use of stable isotopes of multiple elements may provide a more robust approach to this assessment to effectively triangulate nutrient sources in multiple axes. Contrary to body composition analysis, this method allows the investigator to determine the nutrient source for eggs produced by each individual in a sampled population to determine the range of strategies used in a population and the relationship to relative body condition (Oppel et al. 2010). Moreover, the method does not require knowledge of either the quantities of specific nutrients invested in clutches or associated efficiencies of biosynthesis, which is required for body composition analysis.

Limitations exist to the application of stable-isotope analysis for understanding strategies of nutrient allocation to eggs by sea ducks. Hobson (2006) noted that discreteness of isoscapes of wintering and breeding habitats used by birds is an important prerequisite for using stable-isotope analysis for egg nutrition by sea ducks. When stable-isotope values of endogenous nutrients and dietary nutrients are not statistically different, it is impossible to differentiate the sources of nutrients in eggs and reproductive tissues. The similarity of signals between endogenous lipids and dietary lipids precluded an understanding of their relative contributions to egg lipid synthesis by King Eiders (Oppel et al. 2010) and White-winged Scoters (DeVink et al. 2011). Hobson (2006) also illustrated how differences in isotopic ratios associated with freshwater compared to marine habitat may become confounded. Prolonged consumption of food by females at or near freshwater breeding sites before RFG begins would dilute the marine signal that they arrive with. The resulting shift in isotopic ratios from values acquired from marine areas to values more similar to local freshwater isoscapes could occur simply due to turnover of endogenous tissue, regardless of whether or not there is net deposition of somatic nutrients. For example, Oppel et al. (2009) provided clear evidence of a shift in $\delta^{13}C$ in red blood cells from a marine signal toward the local freshwater signal by King Eiders for several days after arrival on breeding areas. A confounding mixing of stable isotopes from distant capital stored at marine areas with local capital stores from freshwater breeding areas (sensu Klaassen et al. 2006), could tend to shift inference toward an increased importance of exogenous sources. It was clear that local freshwater diets were incorporated by King Eiders after arrival but it was not known whether food intake between arrival and nesting was sufficient to induce prenesting storage of nutrient reserves from local freshwater foods (Oppel et al. 2009, 2011), as previously demonstrated by Dobush (1986) for White-winged Scoters.

A further challenge is that most studies to date have relied on discrimination values derived from experimental studies using surrogate species. Moreover, there are at least 10 possible pathways between stable-isotope ratios for carbon (C) and nitrogen (N) from endogenous versus exogenous sources to the main egg constituents of albumen, yolk lipid, and yolk protein. Evidence is accumulating that discrimination values vary by each pathway, by specific fatty acids synthesized, across species, possibly among individuals of the same species, and even among different tissues of the same individual animal (Hobson 1995, Polito et al. 2009, Budge et al. 2011, Federer et al. 2012). Such diversity in discrimination values may impinge on the general applicability of using stable-isotope analysis for studies of egg provisioning, without confirming discrimination for the metabolic pathways of interest. This uncertainty may mean that inferences about relative contributions to egg syntheses of nutrients from different sources should be viewed as preliminary and subject to a fuller understanding of the complexities behind variation in discrimination values.

Most researchers that have used stable-isotope analyses to investigate nutrient allocation to sea duck eggs (Table 5.5) have employed discrimination values between diet and egg components determined by Hobson (1995). Recent studies have also begun to identify factors that affect discrimination values specifically in sea duck tissues. For example, variation in fatty acid structure has an associated range in discrimination factors from 0‰ to 4‰ for dietary to somatic lipids in Steller's Eider and Spectacled Eider (Budge et al. 2011). The library of discrimination values from diet to tissue and diet to egg has

TABLE 5.5

Proportion of egg components inferred to have endogenous source of nutrient from using analysis of stable-isotope values (^{13}C and ^{15}N).

	Albumen	Yolk lipid	Yolk protein	Source
Bucephala islandica	0.29	0.05	0.24	Hobson et al. (2005)
Clangula hyemalis	0.01	0.62	0.21	Lawson (2006)
Histrionicus histrionicus	0.01	0.00	0.05	Bond et al. (2007)[b]
Melanitta fusca	—[a]	—[a]	0.77	DeVink et al. (2011)
Somateria mollissima borealis	0.02	0.71	0.14	Sénéchal et al. (2011)[a]
Somateria spectabilis	0.02	0.16	0.03	Lawson (2006)
Somateria spectabilis	0.07	—[a]	0.28	Oppel et al. (2010)

[a] Not estimated.

[b] Red blood cells from winter birds as endpoints for endogenous nutrients.

been growing but a fundamental problem remains: determination of discrimination of C and N from tissue to eggs has yet to be determined for any sea duck species. Instead, discrimination of endogenous sources of C and N from the muscles and fat reserves of breeding females to egg components was assumed according to reasoning by Gauthier et al. (2003). Determination of isotopic discrimination for endogenous fats and protein as they are allocated to egg albumen, egg lipid, and yolk proteins in free-ranging birds may require the use of a study species that is completely anorexic during the rapid ovarian cycle. One possibility is the Emperor Penguin (*Aptenodytes forsteri*), which is known to fast in the wild during egg synthesis and which has successfully been induced to breed (Groscolas et al. 1986). Depending on the distance between isotope values among endogenous body tissues, exogenous foods consumed during egg production, and the egg nutrients, inferences about the relative contributions of endogenous versus exogenous sources can be sensitive to assumed discrimination values. Sensitivity of inferences to assumed discrimination values were considered to be high in a study of diet reconstruction of two species of seabirds (Bond and Diamond 2011). For example, mean estimated percentage of tern diets composed of krill ranged from 0.11 to 0.91, when $\Delta^{13}C$ ranged from $-0.81‰$ to 2.1 ‰ and $\Delta^{15}N$ ranged from 1.4‰ to 3.49‰, which span published values for marine prey consumed by seabirds. The influence of isotopic discrimination found by Bond and Diamond (2011) had clear implications for direction of conservation and management prescriptions toward at least one endangered species of seabird.

Stable-isotope analysis can also be used to identify marine versus freshwater sources of food used for storage not only of fat reserves for egg production but also of fat catabolized during incubation (Schmutz et al. 2006). However, it is less clear whether it can be used for estimating the fraction of incubation energy budgets met with catabolism of body fat. Instead, the problem can be addressed in a general way simply with knowledge about mass dynamics and estimates of metabolic rate.

To date, stable-isotope analysis has been used to examine relative contributions of endogenous versus exogenous nutrients to eggs of five species of sea ducks: White-winged Scoters (DeVink et al. 2011), Barrow's Goldeneyes (Hobson et al. 2005), Long-tailed Ducks (Lawson 2006), Harlequin Ducks (*Histrionicus histrionicus*; Bond et al. 2007),

King Eiders (Lawson 2006, Oppel et al. 2010), and Common Eiders (Sénéchal et al. 2011). Together, these studies suggest great variation in degree of reliance on endogenous nutrients for egg components and egg formation among species (Table 5.5), with no apparent pattern evident that might relate to other aspects of their ecology. Some of the uncertainty may be due to a reliance by some studies on somatic tissue sampled exclusively from birds during winter (Bond et al. 2007, Oppel et al. 2010), or from their winter diets (Lawson 2006), instead of using tissues sampled from egg-laying birds as the endogenous reference (see earlier citations). In most cases, conclusions support a variable although reduced importance for endogenous nutrients during clutch production (Table 5.5) than generally implied by the work of Milne (1976) or Korschgen (1977). The generally higher importance of income instead of capital nutrition during egg production is largely consistent with Hobson's (2006) review of this topic in a wide range of avian taxa, which concluded that most birds rely completely on foods consumed concurrently with egg synthesis based on inferences from stable-isotope analyses.

Our compilation of results also suggest there is considerable variability in estimates of reliance on endogenous nutrients (Table 5.5), which can also vary from the first to the last eggs laid in the same clutch. Some of the variation is undoubtedly due to sampling error, but additional variation may be related to age or experience in breeding, possibly due to constraints faced by younger birds, or may represent flexibility in allocation strategies that optimize nutrient allocation to eggs while retaining excess nutrients for incubation.

Fatty Acid Analysis

The use of fatty acid signatures for determining dietary composition of vertebrate predators is a relatively new technique (Iverson 1993), but has been applied to sea duck feeding ecology (Wang et al. 2010). Lipids in sea ducks and other waterfowl include triglycerides, phospholipids, and wax esters that are composites of various fatty acids (Alisauskas and Ankney 1992, Austin 1993). Proportions of specific fatty acids in tissues may be useful in determining the diet of individual predators given that vertebrates are capable of synthesizing only a small proportion of the fatty acids that may comprise somatic lipids (Iverson et al. 2004). Fatty acid analysis is particularly useful in marine systems, compared to freshwater systems, due to the specific polyunsaturated fatty acids that originate in phytoplankton (Williams and Buck 2010). Analysis of fatty acids may provide a new tool for determining sources of nutrients in eggs of sea ducks (Williams and Buck 2010), but there are some limitations.

The first limitation is that this method is specific to lipids and does not provide information on the sources of protein or minerals to reproductive activities. The second limitation is that fatty acids of ingested food items may be broken down and reconstructed when deposited in adipose tissues and further allocated to eggs in significantly different proportions than are found in adipose tissues, which was the case in purely capital breeding Emperor Penguins (Speake et al. 1999). To employ fatty acid analysis requires that lipid metabolism be predictable and quantifiable, which has been debated (Grahl-Nielsen 2009, Thiemann et al. 2009). In species of sea ducks where egg lipids are deposited using an income strategy, fatty acid analysis may be useful for determining dietary items selected during the egg formation period (Williams and Buck 2010). Given the interspecific range of nutrient allocation strategies, this application would likely require the additional use of another assessment tool such as stable-isotope analysis.

There may be benefits to using multiple methods for assessing nutrient allocation strategies, where possible. One benefit occurs when certain nutrients are not distinguishable using stable isotopes (Oppel et al. 2010), but inferences can be obtained using body composition analysis (DeVink et al. 2011). Combining analyses may also provide the investigator with the ability to make a more precise determination of nutrient strategies during the clutch formation and incubation stages, both of which are important for successful reproduction in sea ducks. The combined use of stable-isotope analysis and fatty acid analysis has also been employed to better determine feeding guilds and diets used to produce eggs in other marine birds (Anderson et al. 2009, Ramirez et al. 2009, Budge et al. 2011). More research is needed in this application to better understand stable-isotope dynamics in fatty acid metabolism.

Blood Metabolites

Concentrations of various metabolites in plasma can provide information about the extent to which energy expenditure exceeds food intake. Increased concentrations of plasma triglycerides occur during transport to adipose tissue, and increases in free fatty acids and glycerol are thought to be associated with lipid catabolism (Jenni-Eiermann and Jenni 1996). Similarly, β-hydroxybutyrate results from lipid catabolism and can index use of fat reserves during fasting, while uric acid concentration has been used to index protein catabolism (Boismenu et al. 1992). Corticosterone can indicate general nutritional condition in birds often interpreted as stress when individuals fall below the required plane of nutrition (Love et al. 2005). Bentzen et al. (2008a) measured concentrations of various plasma metabolites in King Eiders incubating at two study areas. The results were consistent with the notion that most energy requirements were met with fat reserves, but the authors concluded that there was little catabolism of protein based on the absence of an increase in uric acid concentration. However, feeding opportunities are severely curtailed when females show 98% incubation constancy and uric acid may have been elevated throughout incubation; without reference to uric acid levels of King Eiders at known states of protein equilibrium or storage, it might have been difficult to assess whether levels were relatively elevated throughout incubation.

Hollmén et al. (2001) used similar methods and found that dynamics in blood metabolite levels measured in specific populations of Common Eiders varied across years. Moreover, spatial variation in blood metabolite levels were related to population declines of Common Eider in some areas, suggesting that population dynamics were linked to nutrition and possibly disease. In any event, such approaches offer some insights into endogenous and exogenous energy resource use during incubation and perhaps confirm patterns determined with other methods, such as modeling loss of body mass.

ENDOGENOUS NUTRIENT USE DURING INCUBATION

All taxa of sea ducks for which there are available data show a reliance on endogenous nutrients during incubation (Table 5.3), judging from differences in mass before incubation and at hatch. Females lost between 6% and 30% of their preincubation body mass over the course of incubation. When scaled against respective incubation durations, daily rate of loss in body mass ranged from 1.0 to 21.3 g/day. Average mass loss of 37.5 g/day for five Common Eiders from Denmark appears to be the highest rate published (Bolduc and Guillemette 2003), but this population had the heaviest mass of the eight populations considered in their review. Bolduc and Guillemette (2003) indicated that the rate of mass loss across all species and subspecies of sea ducks was strongly related to body mass at the start of incubation, body mass at hatch, and incubation constancy. However, such relationships do not exist intraspecifically, at least for Common Eiders sampled in Denmark (Bolduc and Guillemette 2003), suggesting different tactics or individual abilities in sustaining mass and constancy during successful incubation (Flint and Grand 1999).

To determine the role of endogenous nutrient for meeting energy demand during incubation, the total energy budget of incubation needs to be calculated. We assumed that basal metabolic rate (BMR, kJ/day) and resting metabolic rate (RMR, kJ/day; Blem 1990) characterize energy expenditure when attending nests, and during recesses respectively, during incubation. Aschoff and Pohl (1970) found that BMR (kcal/h) and RMR (kcal/h) scaled to body mass (kg) according to the following respective equations:

$$RMR = 3.79 \times M^{0.73}$$

$$BMR = 3.06 \times M^{0.73}$$

Thus, expressed as kJ/day, these equations become

$$RMR = I \times 24 \times 4.184 \times 3.79 \times M^{0.73}$$

$$BMR = I \times 24 \times 4.184 \times 3.06 \times M^{0.73}$$

since 1 kcal = 4.184 kJ and I represents the median incubation duration in days (Table 5.5).

We calculated daily energy expenditure, DEE, as

$$DEE = P \times BMR + (1 - P) \times RMR$$

where P is incubation constancy (Table 5.5).

To estimate the contribution of endogenous nutrients to satisfy energy demand during incubation, we relied on Dobush's (1986) data for loss of body mass by White-winged Scoters, to estimate fat and protein catabolism. Dobush (1986) studied a population nesting near the southern edge of breeding range of the species in Saskatchewan and found that females largely used exogenous nutrients to produce eggs; however, nesting females lost 344 g of body mass during the 28-day incubation period (Table 5.3). Declines in body mass (g) were the result of catabolism of 125.3 g of body fat and 55.9 g of protein so that each g of body mass meant a corresponding loss of 0.36 g of fat and 0.16 g of protein. Korschgen (1977) reported that body mass of Common Eiders in Maine declined by 540 g during incubation, which represented a decline of 227 g of fat and 86 g of protein. Thus, for Common Eiders, a 1 g decline in body mass during incubation represented 0.42 g decline in fat and 0.16 g of protein, comparing favorably with respective values from Dobush's (1986) study. We used the median, 0.39, between the values for fat loss g/g of body mass decline determined for Common Eiders and White-winged Scoters. If it is assumed that these ratios apply to mass loss in other species, then fat and protein catabolism during incubation could be predicted for each species based on its loss of body mass. Thus, the energy liberated by catabolism of fat was approximately

$$E_{fat} = 0.39 \times L \times 39.53$$

and that from catabolism of protein was

$$E_{protein} = 0.16 \times L \times (23.64 - 5.2)$$

where energy equivalents for fat and protein are 39.53 and 23.64 kJ/g (Brody 1945) except that 5.2 kJ of protein is lost as urine and not available for meeting energy costs of RMR. The energy supplied from both endogenous sources are simply the sum of E_{fat} and $E_{protein}$. In this manner, we can estimate the relative contribution of endogenous energy stores to incubation energy demands as the quotient of the available energy released from catabolism of body stores over predicted DEE (Table 5.6) for at least 11 species or subspecies of sea ducks. Endogenous sources for incubation energy ranged from 9% for Hooded Mergansers (Lophodytes cucullatus) to 116% for King Eiders (Figure 5.15). We did not understand the extent to which different species may fly during incubation breaks and did not attempt to estimate daily flight costs, but we assume them to be small relative to the metabolic rate expended during the rest of the day. Our estimates of >100% energy supplied by endogenous reserves likely is related to our underestimate of DEE, because we ignored

TABLE 5.6

Estimated DEE, total energy expended during incubation (EE), and estimated contributions to incubation energy from catabolism of endogenous reserves in North American sea ducks based on data in Table 5.3.

| Species or subspecies | DEE (kJ/days) | EE during incubation (kJ) | Energy equivalent from catabolism | | | Proportion supplied endogenously |
			Fat catabolism (kJ)	Protein catabolism (kJ)	Total (kJ)	
Bucephala albeola, Bufflehead, BUFF	—	—	—	—	—	—
Bucephala clangula, Common Goldeneye, COGO	228	6,845	1782	474	2255	0.28
Bucephala islandica, Barrow's Goldeneye, BAGO	—	—	—	—	—	—
Clangula hyemalis, Long-tailed Duck, LTDU	213	5,538	612	163	775	0.12
Histrionicus histrionicus, Harlequin Duck, HADU	—	—	—	—	—	—
Lophodytes cucullatus, Hooded Merganser, HOME	196	6,457	509	97	606	0.09
Lophodytes cucullatus, Hooded Merganser, HOME	187	6,166	1394	267	1660	0.27
Melanitta fusca, White-winged Scoter, WWSC	354	9,917	5303	1015	6318	0.64
Melanitta nigra, Black Scoter, BLSC	—	—	—	—	—	—
Melanitta perspicillata, Surf Scoter, SUSC	—	—	—	—	—	—
Mergus merganser, Common Merganser, COME	287	9,182	4995	956	5951	0.65
Mergus serrator, Red-breasted Merganser, RBME	—	—	—	—	—	—
Polysticta stelleri, Steller's Eider, STEI	—	—	—	—	—	—
Somateria fischeri, Spectacled Eider, SPEI	325	7,485	5673	1086	6759	0.90
Somateria mollissima borealis, Common Eider, COEI-b	478	11,960	6244	1195	7439	0.62
Somateria mollissima dresseri, Common Eider, COEI-d	373	9,706	8325	1593	9918	1.02
Somateria mollissima sedentaria, Common Eider, COEI-s	363	8,715	5735	1098	6833	0.78
Somateria mollissima v-nigrum, Common Eider, COEI-v	—	—	—	—	—	—
Somateria spectabilis, King Eider, KIEI	337	7,754	7570	1449	9018	1.16

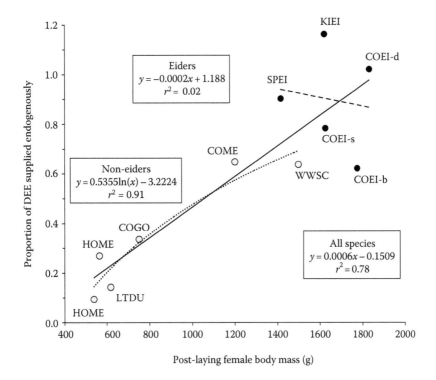

Figure 5.15. Estimated contribution of (endogenous) nutrient reserves to meeting energy expenditure during incubation, as a function of interspecific variation in body mass of female sea ducks (from Table 5.6). Regressions are shown for all species (solid line), for eider subtribe Somatereae (solid circles, dashed line), and for non-eider subtribe Mergeae (open circles, dotted line).

flight costs. Nevertheless, interspecific strategies for meeting energy costs of incubation are highly variable among sea ducks but driven largely by interspecific variation in body mass. While we used data from Korschgen (1977) and Dobush (1986) for changes in body components during incubation, we urge similar determinations for other species to verify whether or not our assumption that loss of fat and protein in other sea ducks was proportionally similar to the mass lost by Common Eiders and White-winged Scoters. We assume that there is also considerable variation within species judging from variable trajectories of mass loss during incubation (Flint and Grand 1999).

Mass dynamics during incubation may provide additional insights. Both King Eiders and Long-tailed Ducks show monotonic declines in mass (Kellett et al. 2005), but mass loss in Common Goldeneyes can be more complicated depending on location, ambient food resources, and individual variation in incubation mass dynamics (Mallory and Weatherhead 1993, Zicus and Riggs 1996). Although such facultative adjustments by different individuals, perhaps of different quality,

age, or experience, are likely, all species of sea ducks show a reliance on endogenous nutrients during incubation, the extent of which increases dramatically in species with large mean body size.

TIMING AND LOCATION OF NUTRIENT STORAGE

We found comparatively little information about mass or body component dynamics over the annual cycle; thus, it remains difficult to characterize strategies for nutrient storage by different species of sea ducks. The challenge is that many species of sea ducks are circumpolar and so cover a wide geographic range not only during breeding but also during winter and can have multiple migration corridors of importance. Hence, schedules of storage for specific nutrients may be as diverse as the geography that certain species may cover as they move from diverse winter to possibly equally diverse breeding habitats. The challenges are compounded by the fact that much of the range of sea ducks, particularly those that breed in the arctic, is difficult to access, thereby posing challenges for sampling birds. However, some species such as Spectacled Eiders or local populations of some sea ducks such as Black Scoters (*Melanitta nigra americana*) that breed in eastern North America are largely closed populations with relatively restricted ranges both during breeding and winter, with relatively narrow and probably traditional migration corridors that show little spatial change over time. Even so, there seems to be a general dearth of information about timing and location of mass gain, let alone storage schedules of specific nutrients.

Most information about the annual cycle of body mass in sea ducks is restricted to papers published in the 1970s. Erskine (1971a) studied body mass of Common Mergansers shot in Nova Scotia. Following a decline in mass from December to March, he noted that the increased body mass in female mergansers during spring was probably a common feature in all ducks. Milne (1976) also noted a large prenesting increase in body mass of adult female Common Eiders from Scotland, most of which was due to increased fat storage sometime before laying in April. Similar to results of Korschgen (1977) for breeding Common Eiders in Maine, Milne (1976) documented how egg laying and incubation were accompanied by a strong decline in fat and protein reserves followed by recovery in somatic protein before molt in July. Peterson and Ellarson (1979) studied body composition of Long-tailed Ducks wintering on Lake Michigan and compared them to data from a breeding ground near northwest Hudson Bay. The age and sex classes showed some parallelism in fat dynamics over time, but adult females exhibited the greatest changes in fat dynamics. Following an increase during fall to midwinter (January) levels of 72–106 g, total body fat declined steadily until April to 31–50 g. Between April and May, intense lipogenesis occurred so that body fat increased by 1.8–3.5 times to achieve 135–142 g of fat. By the time that paired female Long-tailed Ducks were on the breeding grounds, fat had declined from 135 to 81 g before laying, suggesting that 40% of fat stored at Lake Michigan had been used for migration. During egg production, females lost 69% of their prenesting body fat and had only 7 g remaining when with broods. Overall, female Long-tailed Ducks had lost 35% of their premigratory body mass from Lake Michigan. By the time females were with broods, 30% of their premigratory protein and 95% of their premigratory body fat had been used. Similar monthly schedules of somatic tissue storage and catabolism are not available

for other species of sea ducks. It would be useful to compare information about the constancy of body composition of Common Eiders during and between winters in Greenland (Jamieson et al. 2006), with information from spring staging areas used by this population to identify critical locations for spring conditioning when these birds migrate to nesting areas. Long-term data of the type presented by Descamps et al. (2010) are also valuable in that they permit assessing climatic or weather effects on fat storage and thus body mass of prebreeding sea ducks after arrival on breeding areas.

Premigratory or prebreeding storage of nutrient reserves seems to occur in all sea duck species studied to date, suggesting that nutritional conditioning in preparation for the nutritional demands of egg production or incubation may be a common strategy in sea ducks. Dobush (1986) found that female White-winged Scoters stored protein after arrival on their breeding grounds in Saskatchewan and Alberta, but there was no change in body fat or minerals, providing evidence that some sea ducks store local capital. Oppel and Powell (2010) used carbon isotope ratios in red blood cells of King Eiders to estimate arrival time on freshwater breeding areas, based on the idea that time since arrival should be reflected in a mixed signal of local freshwater and final marine staging areas. If so, the marine-based signal of body reserves, characteristic of King Eiders immediately after arrival on freshwater areas, would be diluted by the local freshwater signal with continued feeding. The longer that prenesting mixing of local isotopic signals from freshwater foods dilute the marine signal of new arrivals, the more confounded the signal of endogenous nutrients would become. Thus, there is potential for storage of reserves by King Eiders on nesting areas before breeding, as observed by Dobush (1986) for White-winged Scoters. Importantly, Klaassen et al. (2006) distinguished nutrient reserves used by arctic nesters as either local or distant capital, in addition to exogenous nutrients used to supply breeding requirements in the so-called income fashion. Timing of arrival and body mass of prenesting King Eiders in northern Alaska were unrelated to widely dispersed winter locations of individual birds (Oppel and Powell 2009). An apparent absence of a carryover effect from winter to prebreeding mass suggests that net nutrient storage by King Eiders for breeding probably occurs during spring migration.

Understanding where and when sea ducks become hyperphagic for prebreeding storage would enable conservationists to consider protective legislation for such regions, particularly for species of sea ducks with relatively small geographic ranges or with only a few key wintering or staging habitats typically used for nutrient storage. Past work using radio telemetry studies of sea duck migration and movements has helped identify important winter and staging areas (Bustnes et al. 2010b, Dickson and Smith 2013). In combination with such information, knowledge about foraging and nutritional ecology at important nodes in the annual cycle is needed. Better data would help to properly focus attention on the areas where climatic or meteorological variation may play an important role in governing carryover and cross-seasonal effects between wintering areas and successful spring breeding (Lehikoinen et al. 2006, Descamps et al. 2010).

LIMITATIONS ON STORAGE

Lovvorn and Jones (1994) suggested that the high wing loading of diving ducks due to their relatively small narrow wings was not an adaptation for diving, but rather one that allows high-speed flight given the use of open aquatic environments by these species—a hypothesis that is probably equally relevant to sea ducks. A potential drawback of relatively small, narrow wings is constraints on the amount of adipose tissue that may be accumulated and transported during migration for reproductive use (Meunier 1951). Several authors have proposed that large body size (Pennycuick 1975) and low mass of flight muscle (Marden 1987) that are characteristic of the tribe Mergini might limit the ability to take flight. Thus, somatic tissue mass may be constrained by an individual's ability to fly in an energy-efficient manner between feeding and nesting sites and to avoid predators (Pelletier et al. 2008).

Based on observational flight behavior and wing loading measurements, Meunier (1951) proposed a general wing loading threshold of 2.5 g mass/cm^2 wing area, above which flightlessness would result. This threshold has been supported by observations of flightlessness in steamer ducks (Humphrey and Livezey 1982, Livezey and Humphrey 1986). However, this approach does not account for physiological factors, such as wing beat frequency and flight muscle power, which

can further influence an individual's ability to attain flight. Using experimental tests of various flying vertebrates, Marden (1987) determined that flight can be achieved only if the breast muscle mass is ≥16% of the total body mass. However, these experiments were not conducted on species over 400 g mass, and the relationship of lift to muscle mass declines in larger birds.

Marden (1994) proposed an alternative model to determine flightlessness in birds based on specific lift that estimates a bird's ability to overcome the force of gravity (9.8 N/kg). Guillemette and Ouellet (2005a) tested these models with Common Eiders collected during prebreeding for which their flight aptitude was known, including flightlessness. The authors concluded that measures of wing loading or the flight muscle mass ratio are not accurate predictors of flightlessness. Instead, Marden's (1994) model of takeoff performance best predicted flightlessness in eiders and other species. Female Common Eiders likely have adapted to their periodically high body mass associated with reproduction by having larger relative wing surface area for their structural size and greater relative flight muscle mass compared to males (Ouellet et al. 2008), though there was no measured difference in catalytic capacity of flight muscles to increase their efficiency.

Sea ducks range in body mass from several hundred grams to several kilograms and are generally characterized by relatively small wing size proportional to their body mass, thus requiring a running takeoff to attain flight (Norberg 1990, Lovvorn and Jones 1994). During prebreeding, female eiders undergo hyperphagia and gorge on food to the point that they may become temporarily flightless (Guillemette and Ouellet 2005b, Rigou and Guillemette 2010). During clutch formation, females also must contend with the added mass of reproductive organs, which may further physically interfere with the ability to store adipose tissue. There are unusual instances where nesting sea ducks may not require flight to move from feeding areas to their nest site. For example, some eiders nesting in the Bay of Fundy have been observed swimming and walking to their nesting site following a foraging bout (J.-M. DeVink, Canadian Wildlife Service, unpubl. data). Flightlessness in relation to prebreeding hyperphagia and its effect on limiting nutrient acquisition by sea ducks deserves further attention.

ROLE OF CONTAMINANTS

Due to their primarily carnivorous diet, sea ducks occupy relatively high trophic positions. Many sea ducks forage in coastal marine environments where contaminant concentrations may be elevated from industrialization (Peterson and Ellarson 1976, Harrison and Peak 1995, Gurney et al. 2014), raising the risk of exposure to relatively high concentrations of contaminants in some areas. Many organic contaminants are increasing in northern environments due to anthropogenic inputs and may be increasing in concentrations across the breeding ranges of sea ducks due to long-range transport mechanisms (Fisk et al. 2005). Some heavy metals such as cadmium (Cd), mercury (Hg), and selenium (Se) apparently occur naturally at elevated levels. However, anthropogenic activities may be increasing concentrations from industrial activities such as mining effluents or may cause them to be more bioavailable due to processes such as acidification (Scheuhammer 1991). Moreover, many sea ducks consume filter-feeding bivalves including zebra mussels (*Dreissena polymorpha*), quagga mussels (*D. bugensis*), and clams that can bioaccumulate contaminants to higher concentrations than ambient levels (De Kock and Bowmer 1993). During periods of hyperphagia in contaminated environments, rates of contaminant intake may increase over a relatively short time. Besides potential contaminant effects on fitness of sea ducks outside the breeding season (Wayland et al. 2008), contaminants stored along with nutrient reserves during winter or spring migration can be transported to breeding areas (Gurney et al. 2014). Potential effects of contaminants may include lower body condition of females, which could impact breeding propensity, egg production, or reproductive success (Wayland et al. 2008). Additionally, contaminants stored with somatic tissues may be mobilized when reserves are catabolized for use during egg formation or result in physiological effects in breeding females. Here, we consider such interactions of contaminants with reproduction on females and maternal transfer to eggs.

Trace elements have been quantified in tissues of breeding females and eggs of various sea duck species and locations throughout much of their circumpolar range. Certain trace element concentrations can be relatively high in coastal marine sediments and may be accumulated by sea ducks foraging on benthic prey (Franson et al. 2004;

Gurney et al. 2014; Chapter 6, this volume). Some species breed in freshwater environments where trace element concentrations may be lower and the amount deposited into eggs may be a function of the nutrient allocation strategy of the species, as was hypothesized for selenium in Lesser Scaup (*Aythya affinis*; DeVink et al. 2008). As trace elements may bind to certain amino acids, they are potentially transferred to eggs during formation. If proteins used for synthesis of yolk precursors, such as vitellogenin, are derived from dietary sources, and not somatic tissues, trace elements may not be transferred to eggs. Alternatively, the follicle wall may serve as a protective barrier preventing potential teratogenic contaminants, such as cadmium, from entering the follicle (Sato et al. 1997). Indeed, Sato et al. (1999) observed that chickens injected with cadmium accumulated high concentrations of the element in the follicle walls, where synthesis of cadmium-binding metallothionein may serve to protect the egg from toxic elements. The ability of the follicle membrane to bind other trace elements is unknown but deserves further attention. These two processes are consistent with observations of relatively high concentrations of trace elements in somatic tissues of sea ducks but lower concentrations in their eggs and follicles (Henny et al. 1995; Franson et al. 2000, 2004; Burger et al. 2007; DeVink et al. 2008).

While concentrations of trace elements in eggs have generally been below levels of concern, higher levels in females may have deleterious effects—for instance, in incubating females that are nutritionally stressed. Wayland et al. (2002) observed a positive relationship between cadmium concentrations in the kidneys and stress responses of female Common Eiders captured during incubation, though the same relationship was not detected in prenesting females. Interestingly, the stress response was inversely related to selenium concentrations in prenesting eiders but positively correlated with body condition in other waterfowl (DeVink et al. 2008). During incubation, female sea ducks of several species rely heavily on endogenous nutrients to meet energetic needs (Table 5.6). Mobilization of somatic nutrients may release contaminants, which could have deleterious effects on incubating hens. Blood lead concentrations have been positively correlated with incubation stage of Common Eiders (Franson et al. 2000), where this contaminant inhibited the activity of the enzyme delta-aminolevulinic acid dehydratase (ALAD) required for the synthesis of heme, a component of hemoglobin and other hemoproteins. Other trace elements such as selenium affect biological functions through oxidative stress (Hoffman et al. 1998, Franson et al. 2004) and the dynamics of these, particularly during incubation, deserve further study in breeding sea ducks. On the other hand, Gurney et al. (2014) found that elevated cadmium, lead, and selenium levels were unrelated to body mass or nest initiation dates of White-winged Scoters in Saskatchewan, regardless of whether they wintered on the Atlantic or Pacific coasts.

Many organic contaminants are lipophilic and may accumulate in fat stored by female sea ducks. When bound to lipids, these contaminants have minimal negative effects on hens. However, nutrient dynamics may release these compounds where they may be mobilized to sensitive organs and tissues or deposited into developing eggs. Indeed, Bustnes et al. (2010a) observed that concentrations of organochlorines (OCs), polychlorinated biphenyls (PCBs), and hexachlorobenzene increased 8.2-, 3.6-, and 1.7-fold, respectively, in the blood of Common Eiders late in incubation (day 20) compared to the same birds sampled in early incubation (day 5). Increases are likely due to the heavy reliance of this species on somatic lipids as a source of energy during incubation (Table 5.6). Similar results were found in Common Eiders in Iceland by Olafsdottir et al. (1998), who observed that concentrations of OCs increased by up to 10-fold between winter levels and those of incubating birds. Bustnes et al. (2010a) also found a correlation between certain OCs and clutch size. Females in poorer condition typically lay smaller clutches, so that females with higher OC levels may have been in poorer prebreeding condition, though the cause–effect relationship underlying this correlation is uncertain. PCBs are also known to cause eggshell thinning at higher concentrations through prevention of calcium deposition. However, concentrations of PCBs in eggs of Spectacled Eiders from Alaska were generally low and did not correlate with eggshell thickness (Wang et al. 2005). While concentrations did not reach harmful levels in this population, the results indicate the potential for contaminants to affect incubating females, particularly birds nesting in areas with higher concentrations in the environment. There may be other emerging lipophilic contaminants such as polybrominated diphenyl ethers with possible deleterious effects that have not been adequately studied (Hites 2004).

While most contaminants affect breeding sea ducks through biochemical processes leading to toxicity, physical oiling of plumage may also cause serious problems. Sea ducks are more susceptible to oiling than many other marine avifauna due to their use of near-shore areas, and a high proportion of their lifetime is spent in such habitats (Camphuysen and Heubeck 2001). Thermoregulation by Common Eiders is rapidly affected by oiling and individuals may become hypothermic within 70 min of their plumage contacting surface oil (Jenssen and Ekker 1991). Challenges to thermoregulation from oiling could lead to individuals having lower body mass (Esler et al. 2002), which in turn could lead to a lack of somatic nutrients required for egg production and incubation.

Overall, there is little evidence that contaminants currently have widespread negative effects on reproduction or population dynamics in sea ducks (Wayland et al. 2008) or on the energetic demands of reproduction (Stout et al. 2002, Mallory et al. 2004, Fisk et al. 2005). However, part of this apparent lack of large-scale effect may be due to the absence of research in some remote polar areas that may be used by a large fraction of sea duck populations. Local events have been shown to impact population dynamics for extended periods of time, such as the Exxon Valdez oil spill (Iverson and Esler 2010), and several similar events combined may indeed have widespread effects. Research on contaminants in impacted areas deserves greater attention as we increase our understanding of the effects of contaminants on physiological processes.

INFORMATION NEEDS AND FUTURE RESEARCH DIRECTIONS

Despite good information about egg and clutch size (Table 5.1), and a reasonably good understanding of the trade-offs and correlations in life-history traits among sea ducks (Figures 5.1-5.10), our review has uncovered some important gaps in information about the basic costs of egg production for many species (Table 5.2). Knowledge about the role of nutrient reserves in fueling incubation remains incomplete (Table 5.3). Basic descriptions of the seasonal dynamics of body mass and body composition could shed light on habitats and important bird areas that could prove to be important to the conservation of some populations of sea ducks. Research during the last decade suggests that both endogenous and exogenous sources of nutrients could be important for egg production and incubation, but there have been no advances in understanding the details of how carryover effects influence breeding propensity and associated nutrient thresholds. Much of the reason for the continued absence of basic information about sea duck biology is their use of remote marine or arctic habitats that are difficult to access and can, pose dangerous challenges for researchers. Advances have been made with the use of stable isotopes for inferring the relative contributions of endogenous and exogenous nutrients, but the inferences are based on untested assumptions about discrimination factors pertinent to sea ducks, particularly for endogenous nutrients used for egg synthesis. As well, we urge researchers to consider the role that local versus distant capital may play in addition to exogenous foods (Klaassen et al. 2006) and to use caution when inferring endogenous versus exogenous contributions, particularly when local capital and exogenous nutrients are derived from the same foods.

Last, while our focus has been on carryover nutritional effects on egg production and incubation, the role of nutrition in governing breeding probability among sexually mature sea ducks has received little attention (Bond et al. 2008, Wayland et al. 2008). Identifying the ecological factors that govern breeding probability is a difficult area of study. Determining nutritional thresholds for breeding requires that comparative nutrition of breeders and nonbreeders is understood (Alisauskas and Ankney 1994). Although breeding sea ducks, particularly incubating females, are rather easily captured and sampled, nonbreeders remain highly mobile and elusive to researchers. An important area for future research is the study of cross-seasonal linkages between food abundance at important winter or spring staging areas and subsequent sea duck recruitment over broad scales (Sedinger and Alisauskas 2014). Key to that understanding are determinations of the nutrient thresholds that govern breeding probability (Reynolds 1972, Bond et al. 2008) and the mediating role of climate and weather on whether individual foraging efficiencies carry over to affect whether such thresholds can be met or surpassed.

ACKNOWLEDGMENTS

S. De La Cruz, D. Esler, P. Flint, D. Kellett, and J. Takekawa kindly supplied unpubl. data. We thank R. Bentzen, D. Derksen, K. Hobson, T. Lewis, and J. Schmutz for their thoughtful reviews and helpful comments.

LITERATURE CITED

Afton, A. D., and S. L. Paulus. 1992. Incubation and brood care. Pp. 62–108 in B. D. J. Batt, A. Afton, and M. Anderson (editors), The ecology and management of breeding waterfowl. University of Minnesota Press, Minneapolis, MN.

Alisauskas, R. T. 2002. Arctic climate, spring nutrition, and recruitment in midcontinent Lesser Snow Geese. Journal of Wildlife Management 66:181–193.

Alisauskas, R. T., and C. D. Ankney. 1985. Nutrient reserves and the energetics of reproduction in American Coots. Auk 102:133–144.

Alisauskas, R. T., and C. D. Ankney. 1992. The cost of egg laying and its relation to nutrient reserves in waterfowl. Pp. 30–61 in B. D. J. Batt, A. Afton, and M. Anderson (editors), The ecology and management of breeding waterfowl. University of Minnesota Press, Minneapolis, MN.

Alisauskas, R. T., and C. D. Ankney. 1994. Nutrition of breeding female Ruddy Ducks: the role of nutrient reserves. Condor 96:878–897.

Alisauskas, R. T., and D. K. Kellett. 2014. Age-specific in situ recruitment of female King Eiders estimated with mark-recapture. Auk 131:129–140.

Alison, R. M. 1975. Breeding biology and behavior of the Oldsquaw (Clangula hyemalis L.). Ornithological Monographs 18:1–52.

Anderson, E. M., J. R. Lovvorn, D. Esler, W. S. Boyd, and K. C. Stick. 2009. Using predator distributions, diet, and condition to evaluate seasonal foraging sites: sea ducks and herring spawn. Marine Ecology Progress Series 386:287–302.

Anderson, V. R., and R. T. Alisauskas. 2001. Egg size, body size, locomotion, and feeding performance in captive King Eider ducklings. Condor 103:195–199.

Ankney, C. D. 1984. Nutrient reserve dynamics of breeding and molting Brant. Auk 101:361–370.

Ankney, C. D., and C. D. MacInnes. 1978. Nutrient reserves and reproductive performance of female Lesser Snow Geese. Auk 95:459–471.

Aschoff, J., and H. Pohl. 1970. Rhythmic variations in energy metabolism. Federation Proceedings 29:1541–1552.

Austin, J. E. 1993. Fatty acid composition of fate depots in wintering Canada Geese. Wilson Bulletin 105:339–347.

Barrett, J., C. W. Ward, and D. Fairbairn. 1970. The glyoxylate cycle and the conversion of triglycerides to carbohydrates in developing eggs of *Ascaris lumbricoides*. Comparative Biochemistry and Physiology 35:577–586.

Barry, T. W. 1968. Observations on natural mortality and native use of eider ducks along the Beaufort Sea coast. Canadian Field Naturalist 82:140–144.

Bengtson, S. A. 1972. Reproduction and fluctuations in the size of duck populations at Lake Myvatn, Iceland. Oikos 23:35–58.

Bentzen, R. L. and A. N. Powell. 2012. Population dynamics of King Eiders breeding in northern Alaska. Journal of Wildlife Management 76:1011–1020.

Bentzen, R. L., A. N. Powell, and R. S. Suydam. 2008a. Factors influencing nesting success of King Eiders on northern Alaska's coastal plain. Journal of Wildlife Management 72:1781–1789.

Bentzen, R. L., A. N. Powell, T. D. Williams, and A. S. Kitaysky, 2008b. Characterizing the nutritional strategy of incubating King Eiders *Somateria spectabilis* in northern Alaska. Journal of Avian Biology 39:683–690.

Blem, C. R. 1990. Avian energy storage. Pp. 59–113 in D. M. Power (editor), Current Ornithology (vol. 7), Plenum Press, New York, NY.

Blums, P., J. D. Nichols, M. S. Lindberg, and A. Mednis. 2005. Individual quality, survival variation and patterns of phenotypic selection on body condition and timing of nesting in birds. Oecologia 143:365–376.

Boismenu, C., G. Gauthier, and J. Larochelle. 1992. Physiology of prolonged fasting in Greater Snow Geese (*Chen caerulescens atlantica*). Auk 109:511–521.

Bolduc, F., and M. Guillemette. 2003. Incubation constancy and mass loss in the Common Eider *Somateria mollissima*. Ibis 145:329–332.

Bond, A. L., and A. W. Diamond. 2011. Recent Bayesian stable-isotope mixing models are highly sensitive to variation in discrimination factors. Ecological Applications 21:1017–1023.

Bond, A. L., and K. A. Hobson. 2012. Reporting stable-isotope ratios in ecology: recommended terminology, guidelines and best practices. Waterbirds 35:324–331.

Bond, J. C., D. Esler, and K. A. Hobson. 2007. Isotopic evidence for sources of nutrients allocated to clutch formation by Harlequin Ducks. Condor 109:698–704.

Bond, J. C., D. Esler, and T. D. Williams. 2008. Breeding propensity of female Harlequin Ducks. Journal of Wildlife Management 72:1388–1393.

Bordage, D., and J.-P. L. Savard. 1995. Black Scoter (*Melanitta nigra*). in A. Poole and F. Gill (editors), The Birds of North America, No. 177. The Academy of Natural Sciences, Philadelphia, PA and the American Ornithologists' Union, Washington, DC.

Bottitta, G. E. 1999. Energy constraints on incubating Common Eiders in the Canadian Arctic (East Bay, Southampton Island, Nunavut). Arctic 52:425–437.

Bottitta, G. E., E. Nol, and H. G. Gilchrist. 2003. Effects of experimental manipulation of incubation length on behavior and body mass of Common Eiders in the Canadian arctic. Waterbirds 26:100–107.

Bourgeon, S., F. Criscuolo, F. Bertile, T. Raclot, G. W. Gabrielsen, and S. Massemin. 2006. Effects of clutch sizes and incubation stage on nest desertion in the female Common Eider *Somateria mollissima* nesting in the high Arctic. Polar Biology 29:358–363.

Bourget, A. A. 1973. Relation of eiders and gulls nesting in mixed colonies in Penobscot Bay, Maine. Auk 90:809–820.

Brody, S. 1945. Bioenergetics and growth. Reinhold, New York.

Brown, P. W. 1977. Breeding biology of White-winged Scoter (*Melanitta fusca deglandi*). Thesis, Iowa State University, Ames, IA.

Brown, P. W. 1981. Reproductive ecology of the White-winged Scoter. Dissertation. University of Missouri, Columbia, MO.

Brown, P. W., and L. H. Fredrickson. 1987. Body and organ weights and carcass composition of breeding female White-winged Scoters. Wildfowl 38:103–107.

Brown, P. W., and L. H. Fredrickson. 1997. White-winged Scoter (*Melanitta fusca*). in A. Poole and F. Gill (editors), The Birds of North America, No. 274, The Academy of Natural Sciences, Philadelphia, PA and the American Ornithologists' Union, Washington, DC.

Budge, S. M., S. W. Wang, T. Hollmén, and M. J. Wooller. 2011. Carbon isotopic fractionation in eider adipose tissue varies with fatty acid structure: implications for trophic studies. Journal of Experimental Biology 214:3790–3800.

Burger, J., M. Gochfeld, C. Jeitner, D. Snigaroff, R. Snigaroff, T. Stamm, and C. Volz. 2007. Assessment of metals in down feathers of female Common Eiders and their eggs from the Aleutians: arsenic, cadmium, chromium, lead, manganese, mercury, and selenium. Environmental Monitoring and Assessment 143:247–256.

Burnham, K. P., and D. R. Anderson. 2002. Model selection and multimodel inference: a practical information-theoretic approach, 2nd ed. Springer, New York, NY.

Bustnes, J. O., K. E. Erikstad, and T. H. Bjorn. 2002. Body condition and brood abandonment in Common Eiders breeding in the high Arctic. Waterbirds 25:63–66.

Bustnes, J. O., B. Moe, D. Herzke, S. A. Hanssen, T. Nordstad, K. Sagerup, G. W. Gabrielsen, and K. Borgå. 2010a. Strongly increasing blood concentrations of lipid-soluble organochlorines in high arctic Common Eiders during incubation fast. Chemosphere 79:320–325.

Bustnes, J. O., A. Mosbech, C. Sonne, and G. H. Systad. 2010b. Migration patterns, breeding and moulting locations of King Eiders wintering in north-eastern Norway. Polar Biology 33:1379–1385.

Camphuysen, C. J., C. M. Berrevoets, H. J. W. M. Cremers, A. Dekinga, R. Dekker, B. J. Ens, T. M. van der Have et al. 2002. Mass mortality of Common Eiders (Somateria mollissima) in the Dutch Wadden Sea, winter 1999/2000: starvation in a commercially exploited wetland of international importance. Biological Conservation 106:303–317.

Camphuysen, C. J., and M. Heubeck. 2001. Marine oil pollution and beached bird surveys: the development of a sensitive monitoring instrument. Environmental Pollution 112:443–461.

Catry, P., M. P. Dias, R. A. Phillips, and J. P. Granadeiro. 2013. Carry-over effects from breeding modulate the annual cycle of a long-distance migrant: an experimental demonstration. Ecology 94:1230–1235.

Connor, W. E., D. S. Lin, and C. Colvis. 1996. Differential mobilization of fatty acids from adipose tissue. Journal of Lipid Research 37:290–298.

Cooch, F. G. 1965. The breeding biology and management of the Northern Eider (Somateria mollissima borealis), Cape Dorset Area, Northwest Territories. Canadian Wildlife Service, Wildlife Management Bulletin, Series 2, No. 10. Canadian Wildlife Service, Ottawa, ON.

Coulson, J. C. 2013. Age-related annual variation in mortality rates of adult female Common Eiders (Somateria mollissima). Waterbirds 36:234–239.

Cramp, S., and K. E. L. Simmons. 1977. Handbook of the birds of Europe, the Middle East and North Africa: the birds of the Western Palearctic. Ostrich to ducks (vol. 1). Oxford University Press, Oxford, U.K.

Criscuolo, F., G. W. Gabrielsen, J.-P. Gendner, and Y. Le Maho. 2002. Body mass regulation during incubation in female Common Eiders Somateria mollissima. Journal of Avian Biology 33:83–88.

Curth, P. 1954. Der Mittelsager: Soziologie und Brutbiologie. Neue BrehmBucherei Heft 126. A. Ziemsen Verlag, Leipzig, Germany (in German).

Dacke, C. G., S. Arkle, D. J. Cook, I. M. Wormstone, S. Jones, M. Saidi, and Z. A. Bascal. 1993. Medullary bone and avian calcium regulation. Journal of Experimental Biology 184:63–88.

D'Alba, L., P. Monaghan, and R. G. Nager. 2009. Thermal benefits of nest shelter for incubating female eiders. Journal of Thermal Biology 34:93–99.

Dau, C. P. 1974. Nesting biology of the Spectacled Eider, Somateria fischeri (Brandt) on the Yukon-Kuskokwim Delta, Alaska. M.S. thesis, University of Alaska, Fairbanks, AK.

De Kock, W. C., and C. T. Bowmer. 1993. Bioaccumulation, biological effects, and food chain transfer of contaminants in the zebra mussel (Dreissena polymorpha). Pp. 503–533, 518 in T. F. Nalepa and D. W. Schloesser (editors), Zebra Mussels: biology, impacts, and control. Lewis Publishers, Boca Raton, FL.

de Matos, R. 2008. Calcium metabolism in birds. Veterinary Clinics Exotic Animal Practice 11:59–82.

Dement'ev, G. P., and N. A. Gladkov (editors). 1967. Birds of the Soviet Union (vol. 2). Publishing House Sovetskaya Nuka, Moscow, USSR. (English translation, translated from Russian by the Israel Program for Scientific Translation, Jerusalem Israel, in 1967).

Descamps, S., N. G. Yoccoz, J.-M. Gaillard, H. G. Gilchrist, K. Erikstad, S. A. Hanssen, B. Cazelles, M. R. Forbes, and J. Bety. 2010. Detecting population heterogeneity in effects of North Atlantic Oscillations on seabird body condition: get into the rhythm. Oikos 119:1526–1536.

DeVink, J.-M., R. G. Clark, S. M. Slattery, and M. Wayland. 2008. Is selenium affecting body condition and reproduction in boreal breeding scaup, scoters, and Ring-necked Ducks? Environmental Pollution 152:116–122.

DeVink, J.-M., S. M. Slattery, R. G. Clark, R. T. Alisauskas, and K. A. Hobson. 2011. Combining stable-isotope and body-composition analyses to assess nutrient-allocation strategies in breeding White-winged Scoters (Melanitta fusca). Auk 128:166–174.

Dhondt, A. A., and W. M. Hochachka. 2001. Variations in calcium used by birds during the breeding season. Condor 103:592–598.

Dickson, D. L., and P. A. Smith. 2013. Habitat used by Common and King Eiders in spring in the southeast Beaufort Sea and overlap with resource exploitation. Journal of Wildlife Management 77:777–790.

Dobush, G. R. 1986. The accumulation of nutrient reserves and their contribution to reproductive success in the White-winged Scoter. Thesis, University of Guelph, Guelph, ON.

Drent, R. H., and S. Daan. 1980. The prudent parent: energetic adjustments in avian breeding. Ardea 68:225–252.

Dugger, B. D., K. M. Dugger, and L. H. Fredrickson. 2009. Hooded Merganser (Lophodytes cucullatus). in A. Poole (editor), The Birds of North America, No. 98. The Academy of Natural Sciences, Philadelphia, PA and the American Ornithologists' Union, Washington, DC.

Eadie, J. M., M. L. Mallory, and H. G. Lumsden. 1995. Common Goldeneye (Bucephala clangula). in A. Poole and F. Gill (editors), The Birds of North America, No. 170. The Academy of Natural Sciences, Philadelphia, PA and the American Ornithologists' Union, Washington, DC.

Eadie, J. M., J.-P. L. Savard, and M. L. Mallory. 2000. Barrow's Goldeneye (*Bucephala islandica*). in A. Poole, and F. Gill (editors), The Birds of North America, No. 548. The Academy of Natural Sciences, Philadelphia, PA and the American Ornithologists' Union, Washington, DC.

Erskine, A. J. 1971a. Growth and annual cycles in weights, plumages and reproductive organs of Goosanders in eastern Canada. Ibis 113:42–58.

Erskine, A. J. 1971b. Buffleheads. Canadian Wildlife Service Monograph Series, No. 4. Environment Canada, Ottawa, ON.

Esler, D., T. D. Bowman, K. A. Trust, B. E. Ballachey, T. A. Dean, S. C. Jewett, and C. E. O'Clair. 2002. Harlequin Duck population recovery following the 'Exxon Valdez' oil spill: progress, process and constraints. Marine Ecology Progress Series 241:271–286.

Federer, R. N., T. E. Hollmén, D. Esler, and M. J. Wooller. 2012. Stable carbon and nitrogen isotope discrimination factors for quantifying Spectacled Eider nutrient allocation to egg production. Condor 114:726–732.

Fisk, A. T., C. A. de Wit, M. Wayland, Z. Z. Kuzyk, N. Burgess, R. Letcher, B. Braune et al. 2005. An assessment of the toxicological significance of anthropogenic contaminants in Canadian arctic wildlife. Science of the Total Environment 351:57–93.

Flint, P. L. 2013. Changes in size and trends of North American sea duck populations associated with North Pacific oceanic regime shifts. Marine Biology 160:59–65.

Flint, P. L., and J. B. Grand. 1999. Incubation behavior of Spectacled Eiders on the Yukon-Kuskokwim Delta, Alaska. Condor 101:413–416.

Flint, P. L., C. L. Moran, and J. L. Schamber. 1998. Survival of Common Eider *Somateria mollissima* adult females and ducklings during brood rearing. Wildfowl 49:103–109.

Fournier, M. A., and J. E. Hines. 1994. Effects of starvation on muscle and organ mass of King Eiders *Somateria spectabilis* and the ecological and management implications. Wildfowl 45:188–197.

Franson, J. C., T. E. Hollmén, P. L. Flint, J. B. Grand, and R. B. Lanctot. 2004. Contaminants in molting Long-tailed Ducks and nesting Common Eiders in the Beaufort Sea. Marine Pollution Bulletin 48:504–513.

Franson, J. C., T. E. Hollmén, R. H. Poppenga, M. Hario, M. Kilpi, and M. R. Smith. 2000. Selected trace elements and organochlorines: some findings in blood and eggs of nesting Common Eiders (*Somateria mollissima*) from Finland. Environmental Toxicology and Chemistry 19:1340–1347.

Fredrickson, L. H. 2001. Steller's Eider (*Polysticta stelleri*). in A. Poole and F. Gill (editors), The Birds of North America, No. 571. The Academy of Natural Sciences, Philadelphia, PA and the American Ornithologists' Union, Washington, DC.

Gauthier, G. 1993. Bufflehead (*Bucephala albeola*). in A. Poole and F. Gill (editors), The Birds of North America, No. 67. The Academy of Natural Sciences, Philadelphia, PA and the American Ornithologists' Union, Washington, DC.

Gauthier, G., J. Bêty, and K. A. Hobson. 2003. Are Greater Snow Geese capital breeders? New evidence from a stable-isotope model. Ecology 84:3250–3264.

Gloutney, M. L., R. T. Alisauskas, S. M. Slattery, and A. D. Afton. 2001. Foraging time and dietary intake by nesting Ross' and Lesser Snow Geese. Oecologia 127:78–86.

Goudie, R. I., G. J. Robertson, and A. Reed. 2000. Common Eider (*Somateria mollissima*). in A. Poole and F. Gill (editors), The Birds of North America, No. 546. The Academy of Natural Sciences, Philadelphia, PA and the American Ornithologists' Union, Washington, DC.

Grahl-Nielsen, O. 2009. Exploration of the foraging ecology of marine mammals by way of the fatty acid composition of their blubber. Marine Mammal Science 25:239–242.

Grand, J. B., and P. L. Flint. 1997. Productivity of nesting Spectacled Eiders on the lower Kashunuk River, Alaska. Condor 99:926–932.

Griminger, P. 1986. Lipid metabolism. Pp. 345–358 in P. D. Sturkie (editor), Avian physiology. Springer-Verlag, New York, NY.

Groscolas, R., M. Jallageas, A. Goldsmith, and I. Assenmacher. 1986. The endocrine control of reproduction and molt in male and female Emperor (*Aptenodytes forsteri*) and Adélie (*Pygoscelis adeliae*) Penguins. I. Annual changes in plasma levels of gonadal steroids and luteinizing hormone. General Comparative Endocrinology 62:43–53.

Guillemette, M. 2001. Foraging before spring migration and before breeding in Common Eiders: does hyperphagia occur? Condor 103:633–638.

Guillemette, M., and J.-F. Ouellet. 2005a. Temporary flightlessness in pre-laying Common Eiders *Somateria mollissima*: are females constrained by excessive wing-loading or by minimal flight muscle ratio? Ibis 147:293–300.

Guillemette, M., and J.-F. Ouellet. 2005b. Temporary flightlessness as a potential cost of reproduction in pre-laying Common Eiders *Somateria mollissima*. Ibis 147:301–306.

Gurney, K. E. B., C. J. Wood, R. T. Alisauskas, J.-M. DeVink, and S. M. Slattery. 2014. Identifying carry-over effects of wintering area on reproductive parameters in White-winged Scoters: an isotopic approach. Condor 116:251–264.

Halse, S. A. 1985. Gonadal cycles and levels of luteinizing hormone in wild Spur-winged Geese, *Plectropterus gambensis*. Journal of Zoology 205:335–355.

Hammer, B. T., M. L. Fogel, and T. C. Hoering. 1998. Stable carbon isotope ratios of fatty acids in seagrass and redhead ducks. Chemical Geology 152:29–41.

Hanssen, S. A., H. Engebretsen, and K. E. Erikstad. 2002. Incubation start and egg size in relation to body reserves in the Common Eider. Behavioral Ecology and Sociobiology 52:282–288.

Hanssen, S. A., K. E. Erikstad, V. Johnsen, and O. B. Jan. 2003. Differential investment and costs during avian incubation determined by individual quality: an experimental study of the Common Eider (*Somateria mollissima*). Proceedings of the Royal Society of London B 270:531–537.

Harrison, R. M., and J. D. Peak. 1995. Global disposition of contaminants. Pp. 633–651 in D. J. Hoffman, B. A. Rattner, G. A. Burton Jr., and J. Cairns Jr. (editors), Handbook of ecotoxicology. Lewis Publishers, Boca Raton, FL.

Harrison, X. A., J. D. Blount, R. Inger, D. R. Norris, and S. Bearhop. 2011. Carryover effects as drivers of fitness differences in animals. Journal of Animal Ecology 80:4–18.

Henny, C. J., D. D. Rudis, T. J. Roffe, and E. Robinson-Wilson. 1995. Contaminants and sea ducks in Alaska and the circumpolar region. Environmental Health Perspectives 103S:41–49.

Hites, R. A. 2004. Polybrominated diphenyl ethers in the environment and in people: a meta-analysis of concentrations. Environmental Science and Technology 38:945–956.

Hobson, K. A. 1995. Reconstructing avian diets using stable-carbon and nitrogen isotope analysis of egg components: patterns of isotopic fractionation and turnover. Condor 97:752–762.

Hobson, K. A. 1999. Tracing origins and migration of wildlife using stable isotopes: a review. Oecologia 120:314–326.

Hobson, K. A. 2006. Using stable isotopes to quantitatively track endogenous and exogenous nutrient allocations to eggs of birds that travel to breed. Ardea 94:359–369.

Hobson, K. A., L. Atwell, L. I. Wassenaar, and T. Yerkes. 2004. Estimating endogenous nutrient allocations to reproduction in Redhead ducks: a dual isotope approach using δD and δ¹³C measurements of female and egg tissues. Functional Ecology 18:737–745.

Hobson, K. A., H. L. Gibbs, and M. L. Gloutney. 1997a. Preservation of blood and tissue samples for stable-carbon and stable-nitrogen isotope analysis. Canadian Journal of Zoology 75:1720–1723.

Hobson, K. A., K. D. Hughes, and P. J. Ewins. 1997b. Using stable-isotope analysis to identify endogenous and exogenous sources of nutrients in eggs of migratory birds: applications to Great Lakes contaminants research. Auk 114:467–478.

Hobson, K. A., J. E. Thompson, M. R. Evans, and W. S. Boyd. 2005. Tracing nutrient allocation to reproduction in Barrow's Goldeneye. Journal of Wildlife Management 69:1221–1228.

Hoffman, D. J., H. M. Ohlendorf, C. M. Marn, and G. W. Pendleton. 1998. Association of mercury and selenium with altered glutathione metabolism and oxidative stress in diving ducks from the San Francisco Bay region, USA. Environmental Toxicology and Chemistry 17:167–172.

Hollmén T., J. C. Franson, M. Hario, S. Sankari, M. Kilpi, and K. Lindström. 2001. Use of serum biochemistry to evaluate nutritional status and health of incubating Common Eiders (*Somateria mollissima*) in Finland. Physiological and Biochemical Zoology 74:333–342.

Humphrey, P. S., and B. C. Livezey. 1982. Flightlessness in flying steamer-ducks. Auk 99:368–372.

Iverson, S. J. 1993. Milk secretion in marine mammals in relation to foraging: can milk fatty acids predict diet? Symposium of the Zoological Society of London 66:263–291.

Iverson, S. J., and D. Esler. 2010. Harlequin Duck population injury and recovery dynamics following the 1989 Exxon Valdez oil spill. Ecological Applications 20:1993–2006.

Iverson, S. J., C. Field, W. D. Bowen, and W. Blanchard. 2004. Quantitative fatty acid signature analysis: a new method for estimation predator diets. Ecological Monographs 74:211–235.

Jamieson, S. E., H. G. Gilchrist, F. R. Merkel, K. Falk and A. W. Diamond. 2006. An evaluation of methods used to estimate carcass composition of Common Eiders *Somateria mollissima*. Wildlife Biology 12:219–226.

Jenni-Eiermann, S., and L. Jenni. 1996. Metabolic differences between the postbreeding, moulting and migratory periods in feeding and fasting passerine birds. Functional Ecology 10:62–72.

Jenssen, B. M., and M. Ekker. 1991. Dose dependent effects of plumage-oiling on thermoregulation of Common Eiders *Somateria mollissima* residing in water. Polar Research 10:579–584.

Kellett, D. K., and R. T. Alisauskas. 2000. Body-mass dynamics of King Eiders during incubation. Auk 117:812–817.

Kellett, D. K., R. T. Alisauskas, K. R. Mehl, K. L. Drake, J. J. Traylor, and S. L. Lawson. 2005. Body mass of Long-tailed Ducks (Clangula hyemalis) during incubation. Auk 122:313–318.

Kilpi, M., and L. Lindstrom. 1997. Habitat-specific clutch size and cost of incubation in Common Eiders, Somateria mollissima. Oecologia 111:297–301.

King, J. R. 1973. Energetics of reproduction in birds. Pp. 78–120 in D. S. Farner (editor), Breeding biology of birds. National Academy of Sciences, Washington, DC.

Klaassen, M., K. F. Abraham, R. L. Jefferies, and M. Vrtiska. 2006. Factors affecting the site of investment, and the reliance on savings for arctic breeders: the capital–income dichotomy revisited. Ardea 94:371–384.

Klasing, K. C. 1998. Comparative avian nutrition. CAB International, New York, NY.

Korschgen, C. E. 1977. Breeding stress of female eiders in Maine. Journal of Wildlife Management 41:360–373.

Krapu, G. L., and K. J. Reinecke. 1992. Foraging ecology and nutrition. Pp. 30–61 in B. D. J. Batt, A. Afton, and M. Anderson (editors), The ecology and management of breeding waterfowl. University of Minnesota Press, Minneapolis, MN.

Lavers, J. L., J. E. Thompson, C. A. Paszkowski, and C. D. Ankney. 2006. Variation in size and composition of Bufflehead (Bucephala albeola) and Barrow's Goldeneye (Bucephala islandica) eggs. Wilson Journal of Ornithology 118:173–177.

Lawson, S. L. 2006. Comparative reproductive strategies between Long-tailed Ducks and King Eiders at Karrak Lake, Nunavut: use of energy resources during the nesting season. Thesis, University of Saskatchewan, Saskatoon, SK.

Lecomte, N., G. Gauthier, and J.-F. Giroux. 2009. A link between water availability and nesting success mediated by predator–prey interactions in the Arctic. Ecology 90:465–475.

Lehikoinen, A., M. Kilpi, and M. Öst. 2006. Winter climate affects subsequent breeding success of Common Eiders. Global Change Biology 12:1355–1365.

Livezey, B. C. 1995. Phylogeny and evolutionary ecology of modern seaducks (Anatidae: Mergini). Condor 97:233–255.

Livezey, B. C., and P. S. Humphrey. 1986. Flightlessness in steamer-ducks (Anatidae: Tachyeres): its morphological bases and a probable evolution. Evolution 40:540–558.

Love, O. P., E. H. Chin, K. E. Wynne-Edwards, and T. D. Williams. 2005. Stress hormones: a link between maternal condition and sex-biased reproductive investment. American Naturalist 166:751–766.

Lovvorn, J. R., and D. R. Jones. 1994. Biomechanical conflicts between adaptations for diving and aerial flight in estuarine birds. Estuaries 17:62–75.

Mallory, M. L., B. M. Braune, M. Wayland, H. G. Gilchrist, and D. L. Dickson. 2004. Contaminants in Common Eiders (Somateria mollissima) of the Canadian Arctic. Environmental Reviews 12:197–218.

Mallory, M. L., and H. G. Lumsden. 1994. Notes on egg laying and incubation in the Common Merganser. Wilson Bulletin 106:757–759.

Mallory, M. L., D. K. McNicol, and P. J. Weatherhead. 1994. Habitat quality and reproductive effort of Common Goldeneyes nesting near Sudbury, Canada. Journal of Wildlife Management 58:552–560.

Mallory, M., and K. Metz. 1999. Common Merganser (Mergus merganser). in A. Poole and F. Gill (editors), The Birds of North America, No. 442. The Academy of Natural Sciences, Philadelphia, PA and the American Ornithologists' Union, Washington, DC.

Mallory, M. L., A. Taverner, B. Bower, and D. Crook. 2002. Wood Duck and Hooded Merganser breeding success in nest boxes in Ontario. Wildlife Society Bulletin 30:310–316.

Mallory, M. L., and P. J. Weatherhead. 1993. Incubation rhythms and mass loss of Common Goldeneyes. Condor 95:849–859.

Marden, J. H. 1987. Maximum lift production during takeoff in flying animals. Journal of Experimental Biology 130:235–258.

Marden, J. H. 1994. From damselflies to pterosaurs: how burst and sustainable flight performance scale with size. American Journal of Physiology 266:R1077–R1084.

McWilliams, S. R., C. Guglielmo, B. Pierce, and M. Klaassen. 2004. Flying, fasting, and feeding in birds during migration: a nutritional and physiological ecology perspective. Journal of Avian Biology 35:377–393.

Mehlum, F. 2012. Effects of sea ice on breeding numbers and clutch size of a high arctic population of the Common Eider Somateria mollissima. Polar Science 6:143–153.

Meijer, T., and R. Drent. 1999. Re-examination of the capital and income dichotomy in breeding birds. Ibis 141:399–141.

Meunier, K. 1951. Korrelation und Umkonstruktion in den Grössenbeziehungen zwischen Vogelflugel und Vogelkörper. Biologia Generalis 19:403–443.

Milne, H. 1976. Body weights and carcass composition of the Common Eider. Wildfowl 27:115–122.

Morrier, A., L. Lesage, A. Reed, and J.-P. L. Savard. 1997. Étude sur l'écologie de la Macreuse à Front Blanc au lac Malbaie, Réserve des Laurentides-1994–1995. Technical. Report Series, No. 301, Canadian Wildlife Service, QC.

Norberg, U. M. 1990. Vertebrate flight. Springer-Verlag, Berlin, Germany.

Olafsdottir, K., K. Skirnisson, G. Gylfadottir, and T. Johannesson. 1998. Seasonal fluctuations of organochlorine levels in the Common Eider (*Somateria mollissima*) in Iceland. Environmental Pollution 103:153–158.

Oppel, S., and A. N. Powell. 2009. Does winter region affect spring arrival time and body mass of King Eiders in northern Alaska? Polar Biology 32:1203–1209.

Oppel, S., and A. N. Powell. 2010. Carbon isotope turnover in blood as a measure of arrival time in migratory birds using isotopically distinct environments. Journal of Ornithology 151:123–131.

Oppel, S., A. N. Powell, and M. G. Butler. 2011. King Eider foraging effort during the pre-breeding period in Alaska. Condor 113:52–60.

Oppel, S., A. N. Powell, and D. M. O'Brien. 2009. Using eggshell membranes as a non-invasive tool to investigate the source of nutrients in avian eggs. Journal of Ornithology 150:109–115.

Oppel, S., A. N. Powell, and D. M. O'Brien. 2010. King Eiders use an income strategy for egg production: a case study for incorporating individual dietary variation into nutrient allocation research. Oecologia 164:1–12.

Ouellet, J.-F., M. Guillemette, and P. U. Blier. 2008. Morphological and physiological aspects of the takeoff aptitudes of female Common Eiders (*Somateria mollissima*) during the pre-laying period. Canadian Journal of Zoology 86:462–469.

Palmer, R. S. 1976. Handbook of North American birds (vol. 3), Waterfowl (Part 2). Yale University Press, New Haven, CT.

Parker, H., and H. Holm. 1990. Patterns of nutrient and energy expenditure in female Common Eiders nesting in the high arctic. Auk 107:660–668.

Pelletier, D., M. Guillemette, J.-M. Grandbois, and P. J. Butler. 2008. To fly or not to fly: high flight costs in a large sea duck do not imply an expensive lifestyle. Proceedings of the Royal Society B 275:2117–2124.

Pennycuick, C. J. 1975. Mechanics of flight. Pp. 1–75 in D. S. Farner and J. R. King (editors), Avian biology (vol. 5). Academic Press, New York, NY.

Petersen, M. R., J. B. Grand, and C. P. Dau. 2000. Spectacled Eider (*Somateria fischeri*). in A. Poole and F. Gill (editors), The Birds of North America, No. 547. The Academy of Natural Sciences, Philadelphia, PA and the American Ornithologists' Union, Washington, DC.

Peterson, S. R., and R. S. Ellarson. 1976. Total mercury residues in livers and eggs of Oldsquaws. Journal of Wildlife Management 40:704–709.

Peterson, S. R., and R. S. Ellarson. 1979. Changes in Oldsquaw *Clangula hyemalis* carcass weight. Wilson Bulletin 91:288–300.

Pierce, B. J., and S. R. McWilliams. 2004. Diet quality and food limitation affect the dynamics of body composition and digestive organs in a migratory songbird (*Zonotrichia albicollis*). Physiological and Biochemical Zoology 77:471–483.

Polito, M. J., S. Fisher, C. R. Tobias, and S. D. Emslie. 2009. Tissue-specific isotopic discrimination factors in Gentoo Penguin (*Pygoscelis papua*) egg components: implication for dietary reconstruction using stable isotopes. Journal of Experimental Marine Biology and Ecology 372:106–112.

Pollock, C. 2002. Carbohydrate regulation in avian species. Seminars in Avian and Exotic Pet Medicine 11:57–64.

Powell, A. N., and R. S. Suydam. 2012. King Eider (*Somateria spectabilis*). in A. Poole and F. Gill (editors), The Birds of North America, No. 547. The Academy of Natural Sciences, Philadelphia, PA and the American Ornithologists' Union, Washington, DC.

Quakenbush, L., R. Suydam, T. Obritschkewitsch, and M. Deering. 2004. Breeding biology of Steller's Eiders (*Polysticta stelleri*) near Barrow, Alaska, 1991–99. Arctic 57:166–182.

Ramirez, R., L. Jover, C. Sanpera, X. Ruiz, E. Piqué, and R. Guitart. 2009. Combined measurements of egg fatty acids and stable isotopes as indicators of feeding ecology in lake-dwelling birds. Freshwater Biology 54:1832–1842.

Reynolds, C. M. 1972. Mute swan weights in relation to breeding. Wildfowl 23:111–118.

Rigou, Y., and M. Guillemette. 2010. Foraging effort and pre-laying strategy in breeding Common Eiders. Waterbirds 33:314–322.

Robertson, G. J., and H. G. Gilchrist. 1998. Evidence of population declines among Common Eiders breeding in the Belcher Islands, Northwest Territories. Arctic 51:378–385.

Robertson, G. J., and R. I. Goudie. 1999. Harlequin Duck (*Histrionicus histrionicus*). in A. Poole and F. Gill (editors), The Birds of North America, No. 466. The Academy of Natural Sciences, Philadelphia, PA and the American Ornithologists' Union, Washington, DC.

Robertson, G. J., and J.-P. L. Savard. 2002. Long-tailed Duck (*Clangula hyemalis*). in A. Poole and F. Gill (editors), The Birds of North America, No. 651. The Academy of Natural Sciences, Philadelphia, PA and the American Ornithologists' Union, Washington, DC.

Ryder, J. P. 1970. A possible factor in the evolution of clutch size in Ross's Goose. Wilson Bulletin 82:5–13.

Sato, S., S. Okabe, T. Emoto, M. Kurasaki, and Y. Kojima. 1997. Restriction of cadmium transfer to eggs from laying hens exposed to cadmium. Journal of Toxicology and Environmental Health 51:15–22.

Sato, S., S. Okabe, T. Emoto, M. Kurasaki, and Y. Kojima. 1999. Restriction of cadmium transfer to egg from laying hen exposed to cadmium: involvement of metallothionein in the ovaries. Pp. 315–320 in C. Klaassen (editor), Metallothionein IV. Birkhauser Verlag Ag, Basel, Switzerland.

Savard, J.-P. L., D. Bordage, and A. Reed. 1998. Surf Scoter (Melanitta perspicillata). in A. Poole and F. Gill (editors), The Birds of North America, No. 363. The Academy of Natural Sciences, Philadelphia, PA and the American Ornithologists' Union, Washington, DC.

Scheuhammer, A. M. 1991. Effects of acidification on the availability of toxic metals and calcium to wild birds and mammals. Environmental Pollution 71:329–375.

Schmutz, J. A., K. A. Hobson, and J. A. Morse. 2006. An isotopic assessment of protein from diet and endogenous stores: effects on egg production and incubation behaviour of geese. Ardea 94:385–397.

Schönwetter, M. 1960. Handbuch der Oologie. in W. Meise (editor), Lieferung 1. Akademie Verlag, Berlin, Germany (in German).

Scott, C. A., J. A. K. Mazet, and A. N. Powell. 2010. Health evaluation of western arctic King Eiders (Somateria spectabilis). Journal of Wildlife Diseases 46:1290–1294.

Sedinger, J. S., and R. T. Alisauskas. 2014. Cross-seasonal effect and the dynamics of waterfowl populations. Wildfowl 4(Special Issue):277–304.

Sedinger, J. S., C. D. Ankney, and R. T. Alisauskas. 1997. Refined methods for assessment of nutrient reserve use and regulation of clutch size. Condor 99:836–840.

Sénéchal, E., J. Bêty, H. G. Gilchrist, K. A. Hobson, and S. E. Jamieson. 2011. Do purely capital layers exist among flying birds? Evidence of exogenous contribution to arctic-nesting Common Eider eggs. Oecologia 165:593–604.

Simkiss, K. 1961. Calcium metabolism and avian reproduction. Biological Reviews 36:321–367.

Solovieva, D. 1997. Timing habitat use and breeding biology of Steller's Eider in the Lena Delta, Russia. Pp. 35–39 in Proceedings from Steller's Eider Workshop. Seaduck Specialist Group Bulletin, No. 7. Wetlands International, Kalø, Denmark.

Sotherland, P. R., and H. Rahn. 1987. On the composition of bird eggs. Condor 89:48–65.

Speake, B. K., F. Decrock, P. F. Surai, and R. Groscolas. 1999. Fatty acid composition of the adipose tissue and yolk lipids of a bird with a marine-based diet, the Emperor Penguin (Aptenodytes forsteri). Lipids 34:283–290.

Speake, B. K., P. F. Surai, and G. R. Bortolotti. 2002. Fatty acid profiles of yolk lipids of five species of wild ducks (Anatidae) differing in dietary preference. Journal of Zoology 257:533–538.

Stout, J. H., K. A. Trust, J. F. Cochrane, R. S. Suydam, and L. T. Quakenbush. 2002. Environmental contaminants in four eider species from Alaska and arctic Russia. Environmental Pollution 119:215–226.

Swennen, C., J. C. H. Ursem, and P. Duiven. 1993. Determinate laying and egg attendance on Common Eiders. Ornis Scandinavica 24:48–52.

Swennen, C., and J. Van der Meer. 1995. Composition of eggs of Common Eiders. Canadian Journal of Zoology 73:584–588.

Taylor, T. G., and C. G. Dacke. 1984. Calcium metabolism and its regulation. Pp. 125–170 in F. M. Freeman (editor), Physiology and biochemistry of the domestic fowl (vol. 5). Academic Press, London, U.K.

Thiemann, G. W., S. J. Iverson, and I. Stirling. 2009. Using fatty acids to study marine mammal foraging: the evidence from an extensive and growing literature. Marine Mammal Science 25:243–249.

Thompson, J. E. 1996. Comparative reproductive ecology of female Buffleheads (Bucephala albeola) and Barrow's Goldeneyes (Bucephala islandica). Dissertation, University of Western Ontario, London, ON.

Titman, R. D. 1999. Red-breasted Merganser (Mergus serrator). in A. Poole and F. Gill (editors), The Birds of North America, No. 443. The Academy of Natural Sciences, Philadelphia, PA and the American Ornithologists' Union, Washington, DC.

Uspenski, S. M. 1972. Die Eiderenten. A. Ziemsen Verlag. Wittenberg, Lutherstadt, Germany (in German).

Vermeer, K., and N. Bourne. 1984. The White-winged Scoter diet in British Columbia waters: resource partitioning with other scoters. Pp. 30–38 in D. N. Nettleship, G. A. Sanger, and P. F. Springer (editors), Marine birds: their feeding ecology and commercial fisheries relationships. Canadian Wildlife Service Special Publication, Ottawa, ON.

Wang, D., K. Huelck, S. Atkinson, and Q. X. Li. 2005. Polychlorinated biphenyls in eggs of Spectacled Eiders (Somateria fischeri) from the Yukon-Kuskokwim Delta, Alaska. Bulletin of Environmental Contamination and Toxicology 75:760–767.

Wang, S. W., T. E. Hollmén, and S. J. Iverson. 2010. Validating quantitative fatty acid signature analysis to estimate diets of Spectacled and Steller's Eiders (*Somateria fischeri* and *Polysticta stelleri*). Journal of Comparative Physiology B 180:125–139.

Warren, J. M., K. A. Cutting, and D. A. Koons. 2013. Body condition dynamics and the cost-of-delay hypothesis in a temperate-breeding duck. Journal of Avian Biology 44:1–8.

Watson, M. D., G. J. Robertson, and F. Cooke. 1993. Egg-laying time and laying interval in the Common Eider. Condor 95:869–878.

Wayland, M., K. L. Drake, R. T. Alisauskas, D. K. Kellett, J. Traylor, C. Swoboda, and K. Mehl. 2008. Survival rates and blood metal concentrations in two species of free-ranging North American sea ducks. Environmental Toxicology and Chemistry 27:698–704.

Wayland, M., H. G. Gilchrist, T. Marchant, J. Keating, and J. E. Smits. 2002. Immune function, stress response, and body condition in arctic-breeding Common Eiders in relation to cadmium, mercury, and selenium concentrations. Environmental Research 90:47–60.

Williams, C. T., and C. L. Buck. 2010. Using fatty acids as dietary tracers in seabird trophic ecology: theory, applications and limitations. Journal of Ornithology 151:531–543.

Witter, M. S., and I. C. Cuthill. 1993. The ecological costs of avian fat storage. Philosophical Transactions of the Royal Society of London B 340:73–92.

Zammuto, R. 1986. Life histories of birds: clutch size, longevity, and body mass among North American game birds. Canadian Journal of Zoology 64:2739–2749.

Zicus, M. C. 1997. Female Hooded Merganser body mass during nesting. Condor 99:220–224.

Zicus, M. C., and M. R. Riggs. 1996. Change in body mass of female Common Goldeneyes during nesting and brood rearing. Wilson Bulletin 108:61–71.

Zipkin, E. F., B. Gardner, A. T. Gilbert, A. F. O'Connell, J. A. Royle, and E. D. Silverman. 2010. Distribution patterns of wintering sea ducks in relation to the North Atlantic Oscillation and local environmental characteristics. Oecologia 163:893–902.

CHAPTER SIX

Contaminants in Sea Ducks*

METALS, TRACE ELEMENTS, PETROLEUM, ORGANIC POLLUTANTS, AND RADIATION

J. Christian Franson

Abstract. Exposure to lead and petroleum has caused deaths of sea ducks, but relatively few contaminants have been shown to cause mortality or be associated with population level effects. This chapter focuses primarily on field reports of contaminant concentrations in tissues of sea ducks in North America and Europe and results of some pertinent experimental studies. Much of the available interpretive data for contaminants in waterfowl come from studies of freshwater species. Limits of available data present a challenge for managers interested in sea ducks because field reports have shown that marine birds may carry greater burdens of some pollutants than freshwater species, particularly metals. It is important, then, to distinguish poisoning due to a particular contaminant as a cause of death in sea ducks versus simple exposure based solely on tissue residues. A comprehensive approach that incorporates information on field circumstances, any observed clinical signs and lesions, and tissue residues is recommended when evaluating contaminant concentrations in sea ducks.

Key Words: biomonitoring, cadmium, carbamate, ecotoxicology, lead, mercury, oil spill, organic pollutant, organophosphate, radiation, risk assessment, selenium.

When released into the environment, many pollutants are globally distributed by a complex system that includes atmospheric, freshwater, and marine transport, with exchange between various environmental compartments and biota (Harrison et al. 2003). As a consequence of this environmental cycling and transport, including the phenomenon of arctic haze, certain contaminants released in both hemispheres may reach remote northerly sea duck habitats, where some chemicals are slow to degrade and may biomagnify in food chains (Shaw 1995, Bard 1999, Muir et al. 1999, Blais 2005). Local and regional sources and pathways also contribute to the overall contaminant burden in the arctic, as well as subarctic and temperate regions used by sea ducks. For example, contamination may result from the release of metals from smelters and industrial areas; persistent organic pollutants (POPs) from combustion, inappropriate

* Franson, J. C. 2015. Contaminants in Sea Ducks: Metals, Trace Elements, Petroleum, Organic Pollutants, and Radiation. Pp. 169–240 in J.-P. L. Savard, D. V. Derksen, D. Esler, and J. M. Eadie (editors). Ecology and conservation of North American sea ducks. Studies in Avian Biology (no. 46), CRC Press, Boca Raton, FL.

disposal, or point sources such as military sites; petroleum hydrocarbons from natural seeps, exploration activities, spills, and shipping; and residual radiation from early nuclear weapon testing and more recent inputs from reactor accidents (Arctic Monitoring and Assessment Programme 1998). Climate change is likely to have significant consequences for the ecotoxicology of many contaminants, especially in arctic and subarctic regions (Schiedek et al. 2007). For instance, rising temperatures may increase the uptake and the excretion of pollutants, enhance their toxicity or transform some to more bioactive metabolites, alter homeostasis and physiological responses, increase volatilization of POPs and pesticides to the atmosphere, change food webs by altering nutrient cycles or critical habitat, and result in the release of some contaminants from melting permafrost (Macdonald et al. 2005, Noyes et al. 2009, Rydberg et al. 2010). Fundamental changes in ecosystems as a result of climate change may additionally cause shifts in ranges of disease vectors that, when coupled with contaminant-induced immunosuppression, may increase the susceptibility of wildlife to infectious agents (Noyes et al. 2009). All of these issues may be particularly important for populations living at the edge of their physiological tolerance where the capacity for acclimation is limited (Noyes et al. 2009).

The exposure of sea ducks and other wildlife to environmental pollutants depends on a number of factors including the physiochemical nature of individual contaminants and their mixtures, the extent of habitat pollution, environmental conditions, and natural history characteristics of the species as they influence physical contact with the environment and the contaminants therein (Golden and Rattner 2003, Rattner and Heath 2003, Smith et al. 2007). Ingestion is the principal route of exposure to many contaminants for sea ducks and other birds. Other pathways of animal exposure include skin contact, inhalation, and maternal transfer. Primary exposure is the direct consumption of the contaminant, such as the ingestion of lead shotgun pellets while ducks are feeding or selecting grit, or ingestion of petroleum compounds when preening. Secondary exposure refers to the ingestion of a contaminant present in tissues of prey or other food items. Exposure may result from accumulation in a living organism or when prey have died as a result of exposure to the contaminant.

Accumulation in biota is described by bioconcentration, the accumulation of contaminants by an organism directly from the environment; biomagnification, the increase in concentrations with increasing trophic level; and bioaccumulation, the total accumulation from the environment and foods. Biomagnification is characteristic of many organic compounds, such as polychlorinated biphenyls (PCBs) and chlorinated pesticides and their breakdown products, as well as some metals, such as mercury (Ricca et al. 2008). Not only does the extent of pollution in the environment affect the bioaccumulation and exposure of animals to contaminants, but the physical characteristics of the environment can influence their toxicity. For example, the toxicity of some pollutants is enhanced by cold temperatures, putting sea ducks living in low ambient temperatures at greater risk for adverse effects (Rattner and Heath 2003). In estuarine and marine environments, sea ducks depend on the salt gland to eliminate excess salt and maintain water balance. Alterations of salt gland function and osmoregulation have been reported in birds experimentally exposed to some contaminants (Eastin et al. 1982, Bennett et al. 2000, Hughes et al. 2003, Rattner and Heath 2003).

The aim of this chapter is to address the presence and effects of contaminants in sea ducks with regard to sources, exposure, toxicity, and commonly used biomarkers. After a brief discussion on natural history factors as they relate to contaminant exposure, the focus of my chapter is on metals and trace elements, POPs, organophosphorus and carbamate compounds, petroleum, and radiation. Interactions among various contaminants, and between contaminants and many dietary constituents, can affect their toxicity in ways that are synergistic, additive, or antagonistic. Major interactions are mentioned briefly, but citations are included that refer readers to sources of further information on this complex topic. Because toxicity data specific for sea ducks are lacking for most contaminants, frequent references are made to interpretations and criteria widely accepted for waterfowl or birds in general. Studies reporting tissue concentrations from sea ducks sampled or collected in the field are cited for many contaminants to provide a representative, but not necessarily exhaustive, summary of the residue literature for sea ducks. Residue concentrations from the literature reported as parts per million (ppm) or parts per billion (ppb) were converted to the International System of Units (SI units) and expressed as mg/kg. To provide consistency among estimates, concentrations reported in µg/g were converted to mg/kg (1 µg/g = 1 mg/kg). Residues in blood reported on a volumetric basis, such as µg/dL (100 µg/dL = ~1 mg/kg), µg/mL (1 µg/mL = ~1 mg/kg), or ng/mL (1000 ng/mL = ~1 mg/kg), were not converted. Authors vary according to their preference to express residues on wet weight (ww) or dry weight (dw) basis, and residues in this chapter are expressed following the original publication. In addition, fresh weight (fw) is sometimes used and is considered to be equivalent to wet weight. Reporting on a dry weight basis is often recommended to control for variation in moisture content (Adrian and Stevens 1979). If moisture data are provided, conversion between wet weight and dry weight may be calculated with the following formula:

Dry weight concentration

$$= \frac{\text{wet weight concentration}}{1 - \text{percent moisture}},$$

with percent moisture expressed as a decimal. Otherwise, moisture data from tissues of similar birds reported elsewhere or from a reference work may be used for a rough approximation. As a case in point, using the moisture content of Mallard (*Anas platyrhynchos*) tissues reported by Scanlon (1982), 1 mg/kg ww equals ~4.6 mg/kg dw for blood, 3.1 mg/kg dw for liver, 4.3 mg/kg dw for kidney, and 1.2 mg/kg dw for bone. Contaminant concentrations are sometimes expressed on a lipid weight (lw) basis, particularly in eggs. If lipid content is provided as a percentage of fresh egg mass, conversion between wet weight and lipid weight may be accomplished with the following formula:

Wet weight concentration

$$= \frac{\text{lipid weight concentration}}{1 / \text{percent lipid}},$$

with percent lipid expressed as a decimal.

NATURAL HISTORY AND CONTAMINANTS

Natural history characteristics play an important role in the exposure to and effects of contaminants on populations and species. Ingestion is a primary route of exposure, and diet is a significant factor in determining the amount and types of pollutants an animal is exposed to. Molluscs, for example, the preferred food items of many sea ducks during parts of their life cycle, concentrate both metals and organic contaminants (Boening 1999, Wang et al. 2008). Mussels themselves may be negatively impacted by contaminants (Aarab et al. 2008, Gagné et al. 2008), thereby potentially affecting food availability for sea ducks and other animals that prey on them. Food habits and other natural history characteristics are addressed by Golden and Rattner (2003) in their description of a contaminant vulnerability index, part of a proposed system of ranking terrestrial vertebrate species for biomonitoring and risk assessment studies. Their vulnerability index ranks the susceptibility of populations to the effects of a contaminant or group of contaminants and consists of three components: exposure potential, sensitivity, and resilience of a population. The importance of various natural history characteristics in relation to the significance of contaminant exposure varies within and among contaminant groups such as POPs, metals and trace elements, petroleum, or agricultural chemicals. For POPs and mercury, examples of biomagnifying contaminants, Golden and Rattner (2003) defined exposure potential in terms of six factors: (1) dietary preference; (2) longevity; (3) use of agricultural, industrial, or urbanized areas; (4) residency; (5) social structure; and (6) range. A dietary preference for fish makes an animal most vulnerable for the exposure potential component of the vulnerability index for POPs and mercury, as fish tend to bioconcentrate and bioaccumulate these contaminants (Golden and Rattner 2003). Thus, if the remaining five natural history components were similar among species for a particular biomagnifying contaminant or group, it would be expected that mergansers, being piscivorous, would have a greater exposure potential than sea ducks consuming invertebrates. Accordingly, in a study of waterfowl harvested in Canada, Braune and Malone (2006a) reported that mergansers generally had higher levels of mercury and organochlorines than other sea ducks. However, if in addition to differences in dietary preference there are differences in one or more of the other natural history characteristics among species, dietary preference may not be the main predictor of exposure potential for biomagnifying contaminants. Among sea ducks other than mergansers in the Canadian study, Long-tailed Ducks (*Clangula hyemalis*) had the highest levels of organochlorines. The authors suggested that those findings might have been related, in part, to the fact that large numbers of Long-tailed Ducks overwinter in the Great Lakes–St. Lawrence River system, an area known to be contaminated with organochlorines (Braune and Malone 2006a). Fish comprise little of the diet of Long-tailed Ducks wintering in the Great Lakes (Schummer et al. 2008), and this may be a situation where the use of an organochlorine-polluted environment by Long-tailed Ducks outweighed the piscivorous dietary preference of mergansers. These findings suggest that a complex interaction exists among natural history characteristics of different species and the biomagnifying contaminants to which they are exposed.

The two remaining components of the vulnerability index model of Golden and Rattner (2003) are sensitivity and population resilience. Determinants of sensitivity vary somewhat among contaminants. Thus, the authors proposed that sensitivity to POPs is based on feeding specialization and the ability to metabolize and clear these contaminants. Generalized feeders are least vulnerable, while specialized feeders are most vulnerable. Fish-eating birds generally have a low ability to clear POPs and therefore are more vulnerable than nonpiscivorous animals (Golden and Rattner 2003). A variety of additional considerations may affect sensitivity. For instance, species with less frequent molts have fewer opportunities to eliminate mercury, as up to 90% of the body burden is deposited in feathers (Burger 1993). Last, Golden and Rattner (2003) suggest that factors important to population resilience following a harmful contaminant exposure may be defined in terms of abundance, distribution, reproductive potential, and age at first breeding. Using POPs as a case in point, a species of sea duck would be most vulnerable with regard to sensitivity and population resilience if they are a specialized fish eater, rare in abundance but endemic within the area of exposure, produce few offspring per year, and have delayed age at maturity. Thus, vulnerability of sea ducks to contaminants varies according to differing natural history characteristics among species as they relate to exposure potential, sensitivity, and population resilience among different contaminants and contaminant groups. A comprehensive assessment of these interactions is an important part of evaluating the effects of a contaminant or contaminant group on a population or species of

animals, including sea ducks. Additionally, body condition and normal seasonal changes in fat reserves and physiology associated with such factors as migration, nesting, and molt may affect contaminant residues in tissues (Stewart et al. 1994, Rattner and Jehl 1997, Debacker et al. 2000, Wayland et al. 2005).

METALS AND TRACE ELEMENTS

Lead

Primary sources of lead available to waterfowl, including sea ducks, are spent lead shotgun pellets, lead fishing tackle, and lead in sediments and food items as a result of releases into the environment from mining and smelting wastes, industrial effluents, municipal wastewater, and other anthropogenic sources (Sanderson and Bellrose 1986, Sileo et al. 2001, Scheuhammer et al. 2003, Sweeney and Sañudo-Wilhelmy 2004, Cave et al. 2005, Wilson et al. 2005). Without ingested particulate lead, or detailed information on recent habitat use and movements that may focus an investigation, it is often extremely difficult or impossible to identify the source of exposure. However, the determination of stable lead isotope ratios in bird tissues is a tool being used to assist in distinguishing among sources (Scheuhammer and Templeton 1998, Meharg et al. 2002, Matz and Flint 2009).

Regulations restricting the use of lead ammunition have been instituted in at least 29 countries, and studies in the United States and Canada have demonstrated reduced levels of lead exposure in waterfowl following bans of lead shot for waterfowl hunting (Anderson et al. 2000, Samuel and Bowers 2000, Stevenson et al. 2005, Avery and Watson 2009). However, lead shotgun pellets are still widely used around the world, and even where the use of lead shot has been prohibited, residual lead pellets may remain available in some wetland environments for some time after the regulations have gone into effect (Rocke et al. 1997, Flint and Schamber 2010). Restrictions on lead fishing weights, particularly smaller sizes considered more likely to be ingested by waterbirds, have been implemented on certain federal lands and by some state agencies in the United States, as well as in several other countries, including Great Britain, Canada, and Denmark (Rattner et al. 2008).

Lead is a highly toxic metal that affects all body systems and produces a variety of adverse biochemical, neurological, immunosuppressive, and renal effects (Trust et al. 1990, Rocke and Samuel 1991, Eisler 2000a). Waterfowl and other birds that use grit for mechanical breakdown of food may select lead particles and be at greater risk than other species for ingesting lead shotgun pellets and fishing tackle. Furthermore, the mechanical grinding action of the gizzard facilitates the gastric dissolution of lead by stomach acids, resulting in the absorption of lead salts by the blood and their subsequent distribution to soft tissues and bone. Diet is the primary modifier of lead absorption and its toxic effects in many species, including waterfowl (Eisler 2000a, Hamilton and Hoffman 2003). Both the composition and the physical form of the diet relate specifically to the toxicity of ingested lead pellets. In experimental studies with Mallards, nutritionally adequate diets high in protein and calcium mitigated the effects of lead shot poisoning, and ducks fed items of small size as well as the succulent parts of aquatic plants were less affected by lead toxicity than those fed larger food items of greater hardness (Jordan and Bellrose 1951, Sanderson and Bellrose 1986). Trace elements and other dietary factors known to influence the uptake and toxicity of lead include zinc, iron, cadmium, mercury, copper, magnesium, vitamin E, vitamin D, ascorbic acid, thiamin, and phosphorus (Eisler 2000a, Pattee and Pain 2003). The interpretation of the interactions of these dietary constituents with lead is extremely complex because both deficiencies and excesses of some of the required nutrients enhance lead absorption and toxicity (Eisler 2000a, Pattee and Pain 2003). Muscle wasting and loss of fat reserves are the most consistent observations associated with lead poisoning across avian taxa (Beyer et al. 1988, Locke and Thomas 1996). Clinical signs observed in lead-poisoned waterfowl following chronic exposure include anorexia, emaciation, lethargy, wing droop, ataxia, green diarrhea and staining around the vent, and neurological signs such as leg paralysis or convulsions (Wobeser 1997). Other gross lesions include impactions of the esophagus or proventriculus, distended gallbladder, dark discolored gizzard lining, light areas (evidence of necrosis) in heart or gizzard muscle, pale tissues, and atrophied internal organs (Wobeser 1997). In cases where birds die rapidly following acute exposure to lead, few of these lesions may be observed. Particulate lead may or may not be present in the stomach or gizzard. The occurrence of microscopic acid-fast intranuclear inclusion bodies in kidney tubules is highly suggestive of lead poisoning, although they are not present in all birds that die of lead toxicosis (Locke et al. 1967, Beyer et al. 1988, Ochiai et al. 1993).

Lead disrupts hematopoiesis, often resulting in anemia, by inhibiting delta-aminolevulinic acid dehydratase (ALAD) and heme synthetase, two enzymes that are involved in hemoglobin synthesis (Eisler 2000a). Inhibition of heme synthetase causes an accumulation of protoporphyrin in the blood. These biochemical effects of lead form the basis for the two biomarkers that have been commonly employed for lead exposure in birds: measurements of ALAD activity and the concentration of protoporphyrin in the blood (Dieter et al. 1976, Roscoe et al. 1979). The ALAD assay is sensitive, as enzyme activity is reduced quickly after lead ingestion and by low concentrations of lead (5 μg/dL) in the blood, but the relationship between ALAD and blood lead is nonlinear (Pain and Rattner 1988, Pain 1989, Franson et al. 2002b). This assay has been applied as a biomarker of exposure in sea ducks, including Common Eiders (*Somateria mollissima*) and Long-tailed Ducks (Franson et al. 2000a, b, 2004). The protoporphyrin assay is less sensitive, and changes in protoporphyrin concentrations in blood lag somewhat behind increases in blood lead (Roscoe et al. 1979, Franson et al. 1986).

Tissues often used for monitoring lead exposure in birds include liver, kidney, blood, bone, and in some cases feathers. In live birds, measurement of the lead concentration in the blood provides direct evidence of exposure. If it is unknown when a bird was exposed to lead, it is useful to analyze sequential samples of blood, as increasing lead concentrations indicate recent or continuous exposure, whereas decreasing concentrations suggest that exposure occurred in the past and is not recurrent. A portable analyzer has been used to measure blood lead concentrations in birds in the field, including sea ducks (Brown et al. 2006). Liver and kidney are the tissues most commonly used to assess recent lead exposure in dead birds (Franson and Pain 2011). Bone undergoes rapid uptake of lead but slow release and is generally considered a reflection of lifetime lead exposure in individuals, although bone lead concentrations have also been used to determine geographical patterns of lead poisoning in populations (Stendell et al. 1979, Scheuhammer and Dickson 1996). The use of feathers for monitoring metals, including lead, in live or dead birds is based on the premise that metals are deposited during the period of feather growth, reflecting circulating blood concentrations at the time of feather formation. However, a variety of factors, including feather type, timing of molt and exposure, and variations among feather parts can affect the interpretation of metals in feathers (Burger 1993). Lead concentrations in feathers of juvenile birds have been shown to be an indicator of dietary exposure at the time of feather formation and may be correlated with lead in some tissues (Golden et al. 2003). Conversely, lead from the atmosphere may be deposited on feathers, calling into question the utility of using feathers for monitoring lead absorption (Pain et al. 2005). In fact, exogenous contamination is a factor to consider in the interpretation of many heavy metals in feathers (Jaspers et al. 2004). Although some lead is transferred to eggs, a laboratory experiment with Japanese Quail (*Coturnix coturnix*) showed that concentrations were much lower than in the diet, and studies with Common Eiders failed to detect correlations between lead concentrations in eggs and concentrations in feathers or blood of laying females (Leonzio and Massi 1989, Grand et al. 2002, Burger et al. 2008). Thus, blood is the preferred tissue for nonlethal sampling of lead exposure. In waterfowl, blood lead concentrations of ≥20 µg/dL (~0.2 mg/kg ww) are generally considered evidence of lead exposure and subclinical poisoning, whereas concentrations >100 µg/dL (>1.0 mg/kg ww) may be associated with severe clinical poisoning (Franson and Pain 2011). Similarly, <2 mg/kg ww lead in liver or kidney are considered normal background concentrations, and ≥6 mg/kg ww lead in liver or kidney, accompanied by lesions of lead poisoning, are indicative of lead poisoning as the cause of death (Franson and Pain 2011). If evidence of acute lead exposure exists, a lead concentration of >20 mg/kg dw in bone is consistent with poisoning (Franson and Pain 2011). In dead birds, lead exposure based on tissue residues alone does not necessarily equate with lead poisoning as a definitive cause of death, which requires an examination of the carcass for pathologic lesions consistent with lead poisoning (Wobeser 1997). If a necropsy was not performed, a more conservative approach

is recommended. Beyer et al. (1998) reported that in 95% of waterfowl diagnosed as lead poisoned, liver lead concentrations were 38 mg/kg dw (about 10 mg/kg ww) or greater, whereas <1% of birds dying of other causes had liver lead concentrations of that magnitude. The authors concluded that lead concentrations of 38 mg/kg dw or greater in liver could be used to identify lead-poisoned waterfowl in the absence of pathological observations (Beyer et al. 1998). Tissues of sea ducks shot with lead shot may be contaminated with lead, whether or not pellets or fragments are visible, thereby complicating interpretations (Frank 1986a, Scheuhammer et al. 1998a).

The prevalence of lead shotgun pellets in gizzards of sea ducks has been studied in a variety of surveys, many differentiating between ingested and shot-in pellets but without accompanying lead residues in tissues or, if lead analysis was done, the tissue concentrations were below those consistent with a diagnosis of lead poisoning as the cause of death. A food habit study from the 1930s reported lead shot in gizzards of about 1% of Barrow's Goldeneyes (*Bucephala islandica*), Harlequin Ducks (*Histrionicus histrionicus*), Common Eiders, and Black Scoters (*Melanitta nigra*), 1.7% of Surf Scoters (*Melanitta perspicillata*), and 2.5% of White-winged Scoters (*Melanitta fusca*), but the authors did not indicate how the birds were collected and if some of the pellets may have been shot in (Shillinger and Cottam 1937). However, Bellrose (1959) reported the presence of ingested lead shot in gizzards of 3.5% of Common Goldeneyes (*Bucephala clangula*), 0.69% of Buffleheads (*Bucephala albeola*), and 1.46% of mergansers (*Mergus* spp.) examined in North America prior to 1953. Danell et al. (1977) found ingested lead pellets in 0%–33% of Common Goldeneyes examined from five locations in Sweden, with an overall prevalence of about 10%. Ingested lead shot were found in 32% and 17% of Common Goldeneyes in Finland and Ireland, respectively (Danell 1980, Butler 1990). Eleven percent of Long-tailed Ducks collected for a food habit study in northern Alaska during 1979 and 1980 had lead shot in their gizzards (Taylor 1986). Ingested lead shot were found in 11.8% of Common Goldeneyes and 1.7% of Buffleheads sampled in the Mississippi Flyway of the United States in the late 1970s (Anderson et al. 1987). Scanlon et al. (1980) reported ingested shot in 1 of 31 Buffleheads, but in none of 31 Long-tailed

Ducks and 53 White-winged Scoters examined from Maryland. The Bufflehead with ingested shot had 10.7 mg/kg dw lead in its liver, whereas the birds without ingested shot had mean liver lead concentrations of 5.4–6.4 mg/kg dw (Scanlon et al. 1980). Mudge (1983) examined gizzards of two Common Eiders, one White-winged Scoter and 18 Common Goldeneyes, finding lead pellets in two of the goldeneyes. The liver lead concentration of one of the Common Goldeneyes with ingested lead shot was 9.2 mg/kg dw, but the other was not tested, and the mean lead concentration in livers of 13 goldeneyes without ingested pellets was 2.4 mg/kg dw (Mudge 1983). Lead shot were found in gizzards of 0.4% of Buffleheads and 5.1% of Long-tailed Ducks, but none in Common Goldeneyes, collected in 2002–2004 at Lake Ontario, Canada (Schummer et al. 2011). The authors of that study attributed the greater frequency of lead shot ingestion in Long-tailed Ducks compared with the other species to differences in habitat use and diets. Ingested lead shotgun pellets have been most frequently associated with lead exposure and poisoning in sea ducks, but ingested lead fishing weights have been reported in at least two Common Mergansers (*Mergus merganser*), a Red-breasted Merganser (*M. serrator*), a White-winged Scoter, and a Long-tailed Duck (Scheuhammer et al. 2003, Skerratt et al. 2005, Schummer et al. 2011).

Reports of sea ducks confirmed or suspected to have died of lead poisoning have been published for at least five species in five countries (Table 6.1). Liver lead concentrations in lead poisoned sea ducks have been as high as 250 mg/kg dw in Common Goldeneyes and 38 and 52 mg/kg ww in Spectacled Eiders (*Somateria fischeri*) and Common Eiders, respectively (Table 6.1). After lead poisoning was documented in Spectacled Eiders found moribund and dead on the Yukon–Kuskokwim Delta, studies with live birds during the early to mid-1990s found that 11.6% of Spectacled Eiders on their nesting grounds had radiographic evidence of ingested shot (Franson et al. 1995b, Flint et al. 1997). Although it was not possible to distinguish between lead and steel or other nontoxic shot on radiographs, 82% of the birds with paired blood samples and positive radiographs had blood lead concentrations ≥0.2 mg/kg ww, providing strong evidence that the ingested pellets were lead (Flint et al. 1997). The probability of lead exposure of adult female Spectacled Eiders increased with time on the breeding grounds and during the brood-rearing period (Flint et al. 1997). Further studies indicated low survival of adult female Spectacled Eiders during brood rearing, showed that females exposed to lead prior to hatch had lower overwinter survival rates than those not exposed to lead, and identified lead exposure as a contributor to juvenile mortality, suggesting that lead exposure and poisoning may impede recovery of local Spectacled Eider populations (Flint and Grand 1997, Grand et al. 1998, Flint et al. 2000). On the Yukon–Kuskokwim Delta, Common Eiders have typically exhibited lower levels of lead exposure than Spectacled Eiders, perhaps in large part because Common Eiders fast during incubation, whereas Spectacled Eiders feed in small ponds where lead shot may persist for years (Flint et al. 1997, Flint and Schamber 2010). Therefore, blood lead concentrations of ≥0.2 mg/kg ww were found in <5% of nesting Common Eiders, compared with about 25% of Spectacled Eiders sampled at hatch (Flint et al. 1997, Wilson et al. 2007). In another study on the Yukon–Kuskokwim Delta, female Common Eiders and Spectacled Eiders sampled at hatch had mean blood lead concentrations of 0.14 and 2.02 mg/kg ww, respectively (Grand et al. 2002). On the other hand, in northern Alaska, Spectacled Eiders displayed less evidence of lead exposure, as only 3.2% had blood lead concentrations ≥0.2 mg/kg ww, suggesting that the relative availability of lead there is lower than on the Yukon–Kuskokwim Delta (Wilson et al. 2004). Additional field surveys of lead concentrations in tissues have been reported for several species of sea ducks (Table 6.2). Often in these studies, gizzards were not examined or no comment was made on the presence or absence of ingested lead.

Lead poisoning has been documented as the cause of death for several species of sea ducks, and additional reports detail the frequency of ingested lead pellets along with the prevalence of elevated concentrations of lead in tissues. When evaluating the overall significance of lead exposure, one should consider not only acute mortality but also the adverse health effects of chronic sublethal exposures that are associated with adverse effects that may impact survival.

TABLE 6.1
Reports of sea ducks confirmed or suspected to have died of lead poisoning.

Species and location	No. lead poisoned	No. with ingested lead pellets	Lead concentration	Source
Spectacled Eider (*S. fischeri*)				
United States (Alaska)	4	1	26–38 ww, 8.5 ww (blood, 1 bird)	Franson et al. (1995b)
	1[a]	1	>12 ww	Skerratt et al. (2005)
Common Eider (*S. mollissima*)				
Denmark	1	1	13[b], 100[b] (kidney)	Clausen and Wolstrup (1979)
	3	1	>7 ww (liver and kidney)	Karlog et al. (1983)
Finland	4	0	47.9–81.7 dw	Hollmén et al. (1998)
	2	1	34–98 dw, 52–695 dw (kidney)	Franson et al. (2000b)
United States (Alaska)	1	1	52 ww	Franson et al. (1995b)
Black Scoter (*M. nigra*)				
The Netherlands	1	1	NR[c]	Smit et al. (1988)
Common Goldeneye (*B. clangula*)				
Denmark	2	2	23–53[b], 14[b] (kidney, 1 bird)	Clausen and Wolstrup (1979)
The Netherlands	1	1	NR	Smit et al. (1988)
Germany	1	1	NR	Borkenhagen (1979)
Norway	2	NR	>5 ww	Holt and Frøslie (1989)
United States (Illinois)	4	4	62–65[d] ww, 61–95 ww (kidney), 66–98 dw (bone)	Anderson (1975)
United States (Idaho)	1	0	38 ww, 12.3 ww (blood clot)	Blus et al. (1995)
United States (state NR)	3	NR	96–250 dw	Beyer et al. (1998)
	8[e]	NR	>12 ww	Skerratt et al. (2005)
Common Merganser (*M. merganser*)				
United States (1 Idaho, 1 NR)	2	1[f]	>12 ww	Skerratt et al. (2005)

Lead concentrations in liver tissue, unless otherwise specified, expressed as mg/kg (ppm) ww (wet weight or fresh weight) or dw (dry weight).

[a] Exclusive of those reported by Franson et al. (1995b).
[b] Wet weight presumed but not stated.
[c] Not reported.
[d] Tissue results are from two of four birds.
[e] Exclusive of those reported by Beyer et al. (1998).
[f] Ingested lead fishing weight.

Lead exposure and poisoning are of particular concern for endangered or threatened species and for small, isolated populations. Although ingested lead shotgun pellets have accounted for most reports of lead exposure and poisoning in sea ducks, ingested lead fishing weights have been reported as well. Lead persists in the environment for years, but studies in North America with several species of waterfowl have shown that the implementation of nontoxic shot regulations has reduced lead exposure and estimated mortality.

TABLE 6.2

Concentrations of Lead, Selenium, Mercury, Cadmium, Copper, Chromium, Nickel, and Zinc in tissues of sea ducks from representative field studies.

Species and location	Tissue	Lead	Selenium	Mercury	Cadmium	Copper	Chromium	Nickel	Zinc	Source
Steller's Eider (P. stelleri)										
United States (Alaska), Russia	Liver	ND–26 dw	17.6–25.6[b] dw	2.04–4.27[b] dw	5.93–8.25[b] dw	28.8–33[b] dw	—[c]	—	106–129[b] dw	Stout et al. (2002)
	Kidney	ND–93 dw	9.11–12.80[b] dw	0.96–1.69[b] dw	17.9–39.8[b] dw	25.5–26.2[b] dw	0.14–0.55[b] dw	—	87.2–116[b] dw	Stout et al. (2002)
United States (Alaska)	Liver	—	14 dw (1 bird)	1.1 dw (1 bird)	13.3 dw (1 bird)	26 dw (1 bird)	—	—	—	Henny et al. (1995)
	Kidney	—	—	—	97 dw (1 bird)	—	—	—	—	Henny et al. (1995)
	Blood	ND–59 µg/dL	—	—	—	—	—	—	—	Brown et al. (2006)
Spectacled Eider (S. fischeri)										
United States (Alaska), Russia	Liver	ND–174 dw	43.5–124[b] dw	1.18–1.31[b] dw	18.2–37.0[b] dw	455–485[b] dw	—	—	158–170[b] dw	Stout et al. (2002)
	Kidney	ND–110 dw	25.7–67.6[b] dw	0.57–0.65[b] dw	41.0–99.8[b] dw	62.6–71.1[b] dw	0.36–0.56[b] dw	—	134–145[b] dw	Stout et al. (2002)
United States (Alaska)	Liver	—	35–77 dw	0.4–1.1 dw	5.2–30.2 dw	12–345 dw	—	—	95–146 dw	Henny et al. (1995)
	Liver	ND	124.41 dw	1.13 dw	33.8 dw	558.6 dw	ND	0.616 dw (1 bird)	157.6 dw	Trust et al. (2000b)
	Liver	0.05–0.08[b] dw	171.79–235.59[b] dw	1.18–1.42[b] dw	34.58–36.20[b] dw	238.68–733.59[b] dw	—	—	129.99–138.12[b] dw	Lovvorn et al. (2013)
	Kidney	—	—	—	5.8–113 dw	—	—	—	—	Henny et al. (1995)
	Kidney	ND–3.85 dw	68.86 dw	0.73 dw	95.55 dw	67.5 dw	ND–0.637 dw	0.507 dw (1 bird)	129.6 dw	Trust et al. (2000b)

(Continued)

TABLE 6.2 (Continued)

Concentrations of Lead, Selenium, Mercury, Cadmium, Copper, Chromium, Nickel, and Zinc in tissues of sea ducks from representative field studies.

Species and location	Tissue	Lead	Selenium	Mercury	Cadmium	Copper	Chromium	Nickel	Zinc	Source
Spectacled Eider (S. fischeri)										
	Kidney	0.18–0.29[b] dw	96.17–97.37[b] dw	0.71–0.72[b] dw	163.42–201.81[b] dw	69.02–76.83[b] dw	—	—	131.28–137.68[b] dw	Lovvorn et al. (2013)
	Muscle	0.04–0.06[b] dw	22.63–24.40[b] dw	0.17–0.21[b] dw	2.05–2.66[b] dw	15.46–16.14[b] dw	—	—	41.28–43.49[b] dw	Lovvorn et al. (2013)
	Blood	≥0.2 ww in 6.6%–35.8%	—	—	—	—	—	—	—	Flint et al. (1997)
	Blood	0.14–2.02[b] ww	1.96–19.26[b] ww	0.14–0.15[b] ww	0.1–0.27[b] ww	—	—	—	—	Grand et al. (2002)
	Blood	ND–4.3 ww	1.15–14.7[b] ww	0.07–0.22[b] ww	0.02–0.05[b] ww	—	—	—	—	Wilson et al. (2004)
	Eggs	0.09–0.14[b] ww	1.20–1.42[b] ww	0.06 ww	ND	—	—	—	—	Grand et al. (2002)
King Eider (S. spectabilis)										
Russia	Liver	—	—	5.75 dw (1 bird)	25.5 dw (1 bird)	696 dw (1 bird)	—	ND	201 dw (1 bird)	Kim et al. (1996a)
	Liver	—	8.75–13.1[b] dw	0.72–3.03[b] dw	3.61–5.65[b] dw	50.5–446[b] dw	—	—	145–164[b] dw	Savinov et al. (2003)
	Kidney	—	—	3.92 dw (1 bird)	124 dw (1 bird)	62.6 dw (1 bird)	—	ND	181 dw (1 bird)	Kim et al. (1996a)
	Muscle	—	—	0.29 dw (1 bird)	0.66 dw (1 bird)	19 dw (1 bird)	—	ND	32 dw (1 bird)	Kim et al. (1996a)
	Muscle	—	1.97–3.74[b] dw	0.21–0.59[b] dw	0.07–0.47[b] dw	16.6–34.6[b] dw	—	—	45.7–69.3[b] dw	Savinov et al. (2003)
	Feathers	—	—	0.57 dw (1 bird)	ND	86.5 dw (1 bird)	—	ND	133 dw (1 bird)	Kim et al. (1996a)

Greenland	Liver	—	6.34 ww	0.44 ww	0.448–4.52[b] ww	—	—	—	—	—	Dietz et al. (1996)
	Liver	—	9.56 ww	0.515 ww	4.58 ww	—	—	—	—	—	Johansen et al. (2003)
	Kidney	—	6.66 ww	0.276 ww	0.916–18.3[b] ww	—	—	—	—	—	Dietz et al. (1996)
	Muscle	—	0.539 ww	0.109 ww	0.14–0.316[b] ww	—	—	—	—	—	Dietz et al. (1996)
	Muscle	—	0.81 ww	0.113 ww	0.486 ww	—	—	—	—	—	Johansen et al. (2003)
Canada (Arctic)	Liver	—	18.7–35.6[b] dw	1.7–3.8[b] dw	23.3–33.5[b] dw	86.7–224.5[b] dw	—	—	135.3–188.2[b] dw	—	Wayland et al. (2001b)
	Kidney	—	—	—	111.7–173.7[b] dw	—	—	—	—	—	Wayland et al. (2001b)
Canada (northern)	Liver	—	2.6–20 ww	0.44–1.3 ww	—	—	—	—	—	—	Braune and Malone (2006b)
Canada (NU)	Blood	0.011–0.021[b] ww	3.6–4.9[b] ww	0.13–0.18[b] ww	5.9–12.1[b] ng/mL	—	—	—	—	—	Wayland et al. (2008a,b)
United States (Alaska)	Liver	>7 dw (3 birds)	27.6–34.5[b] dw	1.60–2.49[b] dw	14.4–24.6[b] dw	104–408[b] dw	—	—	133–152[b] dw	—	Stout et al. (2002)
	Kidney	—	22.1–23.1[b] dw	0.98–1.3[b] dw	53.1–78[b] dw	55.4–63.6[b] dw	0.59–0.92[b] dw	—	115–136[b] dw	—	Stout et al. (2002)
	Blood	0.1–0.13[b] ww	8.68–10.18[b] ww	0.22–0.31[b] ww	0.03–0.97[b] ww	—	—	—	—	—	Wilson et al. (2004)
Common Eider (S. mollissima)											
Scotland	Liver	—	—	2.14 dw (1 bird)	—	—	—	—	—	—	Jones et al. (1972)
	Liver	—	—	24.6–48 dw	—	—	—	—	—	—	Dale et al. (1973)
	Kidney	—	—	1.82 dw (1 bird)	—	—	—	—	—	—	Jones et al. (1972)
Denmark	Liver	<2.9 to >7[d] ww	—	—	13 dw	333 dw	—	—	—	—	Karlog et al. (1983)

(Continued)

TABLE 6.2 (Continued)

Concentrations of Lead, Selenium, Mercury, Cadmium, Copper, Chromium, Nickel, and Zinc in tissues of sea ducks from representative field studies.

Species and location	Tissue	Lead	Selenium	Mercury	Cadmium	Copper	Chromium	Nickel	Zinc	Source
Common Eider (S. mollissima)										
	Kidney	<2.9 to >7[d] ww	—	1 dw	38 dw	—	—	—	—	Karlog et al. (1983)
Greenland	Liver	—	—	0.48–1.19[b] ww	1.35–5.17[b] ww	—	—	—	43.92 ww	Nielsen and Dietz (1989)
	Liver	0.048 ww	6.09–6.37[b] ww	0.046–0.784[b] ww	0.141–3.12[b] ww	—	—	—	—	Dietz et al. (1996)
	Liver	0.06 ww	7.1 ww	0.891 ww	2.99 ww	—	—	—	—	Johansen et al. (2003)
	Kidney	—	—	0.24–0.30[b] ww	5.2–20.1[b] ww	—	—	—	33.55 ww	Nielsen and Dietz (1989)
	Kidney	0.082 ww	5.06–6.06[b] ww	0.105–0.302[b] ww	0.852–13.1[b] ww	—	—	—	—	Dietz et al. (1996)
	Muscle	—	—	0.11–0.2[b] ww	0.02–0.24[b] ww	—	—	—	11.91 ww	Nielsen and Dietz (1989)
	Muscle	ND	0.63–0.907[b] ww	0.016–0.166[b] ww	0.022–0.122[b] ww	—	—	—	—	Dietz et al. (1996)
	Muscle	ND	1.1 ww	0.151 ww	0.163	—	—	—	—	Johansen et al. (2003)
Svalbard	Liver	ND	8.9 ww	1.0 ww	4.3 ww	270 ww	—	—	50 ww	Norheim (1987)
	Liver	—	2.8–25.8 ww	—	4.5–16.4 ww	20–1050 ww	—	—	47–150 ww	Norheim and Borch-Iohnsen (1990)
	Liver	—	9.21 dw	1.8 dw	17.2 dw	228 dw	—	—	281 dw	Savinov et al. (2003)
	Kidney	ND	—	—	14 ww	13 ww	—	—	33 ww	Norheim (1987)
	Muscle	—	3.17 dw	0.4 dw	0.43 dw	19.7 dw	—	—	36.4 dw	Savinov et al. (2003)
The Netherlands	Liver	0.7 dw (median)	—	—	6.5 dw (median)	—	—	—	—	Hontelez et al. (1992)

Location	Tissue									Reference
	Kidney	0.7 dw (median)	—	—	15.3 dw (median)	—	—	—	—	Hontelez et al. (1992)
	Bone	3.0 dw (median)	—	—	—	—	—	—	—	Hontelez et al. (1992)
Norway	Liver	—	—	0.6 ww	13 dw	367 dw	1 dw	1 dw	204 dw	Lande (1977)
	Kidney	—	—	—	25 dw	43 dw	1 dw	2 dw	117 dw	Lande (1977)
	Muscle	—	—	0.2 ww	2 dw	13 dw	1 dw	2 dw	33 dw	Lande (1977)
	Eggs	—	—	0.2 ww	1 dw	4 dw	1 dw	1 dw	56 dw	Lande (1977)
	Eggs	—	—	0.06 ww	—	—	—	—	—	Barrett et al. (1996)
Finland	Liver	ND–81.7[d] dw	15.3–47.0[b] dw	ND–9.34	15.9–27.5[b] dw	137–1540[b] dw	ND–4.24 dw	—	147–443[b] dw	Hollmén et al. (1998)
	Liver	ND–98[d] dw	ND–53.7 dw	7.5 dw (1 bird)	ND–35.3 dw	43.0–1381[b] dw	ND	—	137–272[b] dw	Franson et al. (2000a)
	Kidney	ND–695[d] dw	ND–14.7	ND	10.8–135 dw	44.1–127[b] dw	ND	—	127–274[b] dw	Franson et al. (2000a)
	Blood	0.37 ww	1.98 ww	ND–0.22 ww	—	—	—	—	—	Hollmén et al. (1998)
	Blood	0.01–14.2 ww	1.26–2.86[b] ww	ND–0.31 ww	—	—	—	—	—	Franson et al. (2000a)
	Eggs	—	0.42–0.94 ww	0.04–0.46 ww	—	—	—	—	—	Franson et al. (2000a)
Sweden	Liver	—	—	—	2.2 ww (median)	—	—	—	—	Frank (1986b)
	Kidney	—	—	—	5.1 ww (median)	—	—	—	—	Frank (1986b)
Canada (eastern)	Eggs	—	—	0.03–0.08[b] ww	—	—	—	—	—	Pearce et al. (1979)
Canada (NB)	Liver	—	—	0.987 ww	—	—	—	—	—	Braune (1987)
	Kidney	—	—	0.358 ww	—	—	—	—	—	Braune (1987)
	Muscle	—	—	0.153 ww	—	—	—	—	—	Braune (1987)

(Continued)

TABLE 6.2 (Continued)

Concentrations of Lead, Selenium, Mercury, Cadmium, Copper, Chromium, Nickel, and Zinc in tissues of sea ducks from representative field studies.

Species and location	Tissue	Lead	Selenium	Mercury	Cadmium	Copper	Chromium	Nickel	Zinc	Source
Common Eider (S. mollissima)										
	Blood	—	—	0.051 ww	—	—	—	—	—	Bond and Diamond (2009)
	Eggs	—	—	0.322–0.42[b] ww	—	—	—	—	—	Bond and Diamond (2009)
	Feathers	—	—	1.539 ww	—	—	—	—	—	Bond and Diamond (2009)
Canada (Arctic)	Liver	—	10.2–29.1[b] dw	1.3–1.6[b] dw	10.5–15.8[b] dw	82–123.5[b] dw	—	—	120.5–152.4[b] dw	Wayland et al. (2001b)
	Kidney	—	—	—	68.1–117.2[b] dw	—	—	—	—	Wayland et al. (2001b)
Canada (NU)	Liver	—	18.5–20.1[b] dw	1.8–3.9[b] dw	—	—	—	—	—	Wayland et al. (2001a)
	Liver	—	14.1–32.1[b] dw	2.6–3.7[b] dw	—	—	—	—	—	Wayland et al. (2002)
	Liver	—	16.2 dw	3.3 dw	—	—	—	—	—	Wayland et al. (2003)
	Liver	ND–7.08[b] dw	28.6–38.5[b] dw	1.2–2.05[b] dw	10.2–15.0[b] dw	46.9–2051[b] dw	ND	ND	127.2–168.4[b] dw	Mallory et al. (2004a, b)
	Liver	—	16.6 dw	2.0–3.5[b] dw	—	—	—	—	—	Wayland et al. (2005)
	Kidney	—	—	—	67–168.6[b] dw	—	—	—	—	Wayland et al. (2001a)
	Kidney	—	—	—	81.2–162.6[b] dw	—	—	—	—	Wayland et al. (2002)
	Kidney	—	—	—	164.6 dw	—	—	—	—	Wayland et al. (2003)
	Kidney	—	—	—	74–164[b] dw	—	—	—	—	Wayland et al. (2005)
	Muscle	ND	3.29–6.28[b] dw	0.37–0.5[b] dw	0.4–0.62[b] dw	23.7–23.8[b] dw	ND	—	42.8–60.6[b] dw	Mallory et al. (2004a, b)

Location	Tissue									References
	Blood	3.51–4.68 μg/dL	0.23 μg/dL	0.27–0.66 μg/dL	—	—	—	—	—	Wayland et al. (2001a)
Canada (NU), Greenland	Bone	0.66–1.09b dw	—	—	—	—	—	—	—	Ethier et al. (2007)
United States (Maine)	Blood	0.11 ww	—	—	—	—	—	—	—	Goodale et al. (2008)
	Eggs	0.12 ww	—	—	—	—	—	—	—	Mierzykowski et al. (2005)
	Eggs	0.14 ww	—	—	—	—	—	—	—	Goodale et al. (2008)
United States (Alaska)	Liver	>7 dw (1 bird)	7.85–9.29b dw	1.59–1.94b dw	16.1–19.8b dw	79.9–607b dw	—	—	127–139b dw	Stout et al. (2002)
	Kidney	7.34–10.1 dw	0.85–0.96 dw	77.6–88.1 dw	42.6–42.9b dw	0.11–0.17b dw	—	—	133–142b dw	Stout et al. (2002)
	Blood	≥0.2 ww in 4%	—	—	—	—	—	—	—	Flint et al. (1997)
	Blood	0.14b ww	7.29 ww	0.15 ww	ND	—	—	—	—	Grand et al. (2002)
	Blood	0.25 dw	36.1 dw	0.73–0.87b dw	ND–0.08 dw	1.96 dw	0.91–0.93 dw	ND	26.2 dw	Franson et al. (2004)
	Blood	ND–7.0 ww	5.77–11.4 ww	—	—	—	—	—	—	Wilson et al. (2007)
	Eggs	0.211 ww	0.431 ww	0.076 ww	—	—	—	—	—	Burger and Gochfeld (2007)
	Eggs	0.254–0.332b dw	0.992–2.1b dw	0.413–0.466b dw	0.001–0.002b dw	—	0.248–0.497b dw	—	—	Burger et al. (2008)
	Eggs	ND–1.73 dw	2.28 dw	0.57 dw	ND–0.07 dw	5.5 dw	0.26 dw	0.17–0.26b dw	50.3 dw	Franson et al. (2004)
	Feathers	0.489–0.642b dw	1.60–1.64b dw	0.886–1.24b dw	0.025–0.028b dw	—	1.74–1.79b dw	—	—	Burger et al. (2008)
	Feathers	0.992 dw	0.878 dw	0.840 dw	0.080 dw	—	0.172 dw	—	—	Burger and Gochfeld (2009)

TABLE 6.2 (Continued)

Concentrations of Lead, Selenium, Mercury, Cadmium, Copper, Chromium, Nickel, and Zinc in tissues of sea ducks from representative field studies.

Species and location	Tissue	Lead	Selenium	Mercury	Cadmium	Copper	Chromium	Nickel	Zinc	Source
Harlequin Duck (H. histrionicus)										
United States (Alaska)	Blood	0.026 ww	5.5 ww	0.82 ww	—	0.46 ww	0.25 ww	0.007 ww	5.01 ww	Heard et al. (2008)
Surf Scoter (M. perspicillata)										
Canada (BC)	Liver	0.14–0.24[b] ww	—	0.93–2.12[b] ww	—	10.41–10.96[b] ww	—	—	31.02–35.92[b] ww	Vermeer and Peakall (1979)
	Liver	—	35.2–38.9[b] dw	1.30–1.42[b] dw	—	41.5–43.7[b]dw	—	—	—	Elliott et al. (2007)
	Kidney	—	—	—	4.8–37.9[b] dw	—	—	—	103–129[b] dw	Elliott et al. (2007)
Canada (Pacific Northwest)	Liver	—	—	—	6.0–77.7[b] ww	31.6–84.1[b] ww	—	—	111.7–139.0[b] ww	Barjaktarovic et al. (2002)
	Kidney	—	—	—	13.3–178.7[b] ww	19.9–33.9[b] ww	—	—	93.8–151[b] ww	Barjaktarovic et al. (2002)
Canada (northern)	Liver	—	3.7–17 ww	0.23–3.8 ww	—	—	—	—	—	Braune and Malone (2006b)
United States (California)	Liver	ND–2.7 dw	34.4 dw	12.5 dw	—	49.8 dw	ND–1.5 dw	ND–0.54 dw	131 dw	Ohlendorf et al. (1986)
	Liver	0.178–0.966[b] dw	—	—	4.87–9.76[b] ww	29.3–58.3[b] ww	—	—	103–133[b] dw	Ohlendorf et al. (1991)
	Liver	—	20–119[b] dw	10–19[b] dw	6–8[b] dw	—	—	—	—	Hoffman et al. (1998)
	Liver	—	—	3.11 dw	—	—	—	—	—	Eagles-Smith et al. (2009a)
	Kidney	—	—	—	24.6 dw	—	—	—	—	Ohlendorf et al. (1986)

Location	Tissue									Reference
United States (Oregon, Washington)	Liver	ND–9.6 dw	4.9–43.4[b] dw	0.41–4.19[b] dw	0.33–12.9[b] dw	28.6–119.2[b] dw	ND–0.72 dw	ND–0.32 dw	87.5–155.2[b] dw	Henny et al. (1991)
	Kidney	—	—	—	0.92–54.8[b] dw	—	—	—	—	Henny et al. (1991)
United States (Alaska)	Liver	—	14 dw (1 bird)	1.3–7.2 dw	5–15 dw	75–110 dw	—	—	193–200 dw	Henny et al. (1995)
	Kidney	—	—	—	16.5 dw (1 bird)	—	—	—	—	Henny et al. (1995)
White-winged Scoter (M. fusca)										
Scotland	Liver	—	—	5.4 dw (1 bird)	—	—	—	—	—	Dale et al. (1973)
Canada (Pacific Northwest)	Liver	—	—	—	10.4–13[b] ww	47.8–93[b] ww	—	—	77.8–94[b] ww	Barjaktarovic et al. (2002)
	Kidney	—	—	—	12.6–75.1[b] ww	31.3–48.4[b] ww	—	—	97.4–106[b] ww	Barjaktarovic et al. (2002)
Canada (NT)	Liver	—	32.6 dw	1.0 dw	—	—	—	—	—	DeVink et al. (2008)
Canada (SK)	Blood	0.04 ww	3.9 ww	0.19 ww	4.6 ng/mL	—	—	—	—	Wayland et al. (2008b)
United States (New York)	Liver	ND	—	2.0–10 ww	2.0 ww (1 bird)	—	—	—	—	Baker et al. (1976)
	Muscle	14 ww 1 bird	—	0.2 ww (1 bird)	ND	—	—	—	—	Baker et al. (1976)
United States (Maryland)	Liver	5.7 dw	—	—	—	—	—	—	—	Scanlon et al. (1980)
United States (Chesapeake Bay)	Liver	5.4 dw	—	—	3.1 dw	39.2 dw	—	—	159 dw	Di Giulio and Scanlon (1984)
United States (Alaska)	Liver	0.2–0.9 dw	12–85 dw	0.28–12 dw	2.8–95.3 dw	20–172 dw	0.2–0.98 dw	0.05–14.8 dw	63–371 dw	Henny et al. (1995)
	Kidney	—	—	—	7.1–375 dw	—	—	—	—	Henny et al. (1995)

(Continued)

TABLE 6.2 (Continued)

Concentrations of Lead, Selenium, Mercury, Cadmium, Copper, Chromium, Nickel, and Zinc in tissues of sea ducks from representative field studies.

Species and location	Tissue	Lead	Selenium	Mercury	Cadmium	Copper	Chromium	Nickel	Zinc	Source
Black Scoter (M. nigra)										
United States (Alaska)	Liver	—	24–32 dw	2.5–3.2 dw	30.3–33.4 dw	165–245 dw	—	—	199–287 dw	Henny et al. (1995)
	Kidney	—	—	—	59.3–68.7 dw	—	—	—	—	Henny et al. (1995)
	Muscle	—	—	0.135 ww	—	—	—	—	—	Rothshchild and Duffy (2005)
	Brain	—	—	0.105 ww	—	—	—	—	—	Rothshchild and Duffy (2005)
	Blood	ND–20 μg/dL	—	—	—	—	—	—	—	Brown et al. (2006)
Long-tailed Duck (C. hyemalis)										
Russia	Liver	—	—	27.1–31.4[b] dw	12.7 dw	30.0 dw	—	ND	106 dw	Kim et al. (1996a,b)
	Kidney	—	—	3.69–6.0[b] dw	79.5 dw	23.9 dw	—	ND	115 dw	Kim et al. (1996a,b)
	Muscle	—	—	1.14–1.5[b] dw	0.42 dw	20.8 dw	—	ND	32.1 dw	Kim et al. (1996a,b)
	Feathers	—	—	0.7–1.96[b] dw	0.62 dw	24.4 dw	—	ND	86.1 dw	Kim et al. (1996a,b)
Poland	Liver	0.13–0.14[b] ww	—	—	0.59–0.76[b] ww	1.3–1.5[b] ww	—	ND–0.047 ww	14–15[b] ww	Szefer and Falandysz (1983)
	Muscle	0.028–0.13[b] ww	—	—	0.005–0.02[b] ww	0.59–0.96 ww	—	ND–0.064 ww	3.6–13[b] ww	Szefer and Falandysz (1983)
	Brain	ND–7.95 ww	—	—	ND–0.123 ww	1.65–4.82 ww	—	—	7.7–15.1 ww	Kalisińska and Szuberla (1996)
	Feathers	1.2–5.2[b] dw	—	—	0.046–0.1[b] dw	6.3 dw	—	0.21–0.44[b] dw	110 dw	Szefer and Falandysz (1983)

Location	Tissue									Reference
Sweden	Liver	—	—	—	—	4.1 ww (median)	—	—	—	Frank (1986b)
	Kidney	—	—	—	—	9.7 ww (median)	—	—	—	Frank (1986b)
Canada (ON)	Liver	—	—	22.69 dw	—	—	—	—	—	Schummer et al. (2010)
Canada (Arctic)	Liver	—	ND–0.34 ww	5.4–27[b] dw	0.9–8.0[b] dw	—	16–40[b] dw	ND	22–65 ww	Braune et al. (2005a)
	Kidney	—	—	—	—	21–129[b] dw	—	ND	—	Braune et al. (2005a)
Canada (Hudson Bay)	Liver	—	—	—	0.29–1.30[b] ww	—	—	—	—	Peterson and Ellarson (1976)
	Eggs	—	—	—	0.20 ww	—	—	—	—	Peterson and Ellarson (1976)
United States (Lake Michigan)	Liver	—	—	—	0.68–0.80[b] ww	—	—	—	—	Peterson and Ellarson (1976)
United States (Maryland)	Liver	—	6.4 dw	—	—	—	—	—	—	Scanlon et al. (1980)
United States (Chesapeake Bay)	Liver	—	7.1 dw	—	—	5.47 dw	20.1 dw	—	158 dw	Di Giulio and Scanlon (1984)
	Liver	—	ND–48 dw	11.6–18.2[b] dw	0.347–2.01[b] dw	2.21–4.57[b] dw	—	—	—	Mashima et al. (1998)
	Kidney	—	—	—	—	9.02–25 dw	—	—	—	Mashima et al. (1998)
United States (Alaska)	Muscle	—	—	—	0.151 ww	—	—	—	—	Rothschild and Duffy (2005)
	Brain	—	—	—	0.105 ww	—	—	—	—	Rothschild and Duffy (2005)
	Blood	—	≥0.2 ww in 21%	—	—	—	—	—	—	Flint et al. (1997)
	Blood	0.07 dw	48.8 dw	0.57–0.67[b] dw	ND–0.03 dw	1.38 dw	0.51–0.74[b] dw	ND	20.4 dw	Franson et al. (2004)

(Continued)

TABLE 6.2 (*Continued*)

Concentrations of Lead, Selenium, Mercury, Cadmium, Copper, Chromium, Nickel, and Zinc in tissues of sea ducks from representative field studies.

Species and location	Tissue	Lead	Selenium	Mercury	Cadmium	Copper	Chromium	Nickel	Zinc	Source
Bufflehead (*B. albeola*)										
Canada (MB)	Liver	—	—	0.36–3.8 ww	—	—	—	—	—	Driver and Derksen (1980)
	Muscle	—	—	0.1–0.84 ww	—	—	—	—	—	Driver and Derksen (1980)
Canada (northern)	Liver	—	1.6–4.0 ww	0.35–1.4 ww	—	—	—	—	—	Braune and Malone (2006b)
Canada (ON)	Liver	—	12.36 dw	—	—	—	—	—	—	Schummer et al. (2010)
United States (New York)	Liver	ND	—	0.099–0.11 ww	ND	—	—	—	—	Baker et al. (1976)
	Muscle	ND	—	0.083–0.09 ww	2 ww (1 bird)	—	—	—	—	Baker et al. (1976)
United States (Maryland)	Liver	5.4 dw	—	—	—	—	—	—	—	Scanlon et al. (1980)
United States (Maryland)	Liver	10.7[e] dw (1 bird)	—	—	—	—	—	—	—	Scanlon et al. (1980)
United States (Chesapeake Bay)	Liver	5.1[f] dw	—	—	1.24 dw	36.7 dw	—	—	116 dw	Di Giulio and Scanlon (1984)
	Kidney	36.4[f] dw	—	—	5.51 dw	16.2 dw	—	—	83 dw	Di Giulio and Scanlon (1984)
	Bone	12.4[f] dw	—	—	—	—	—	—	175 dw	Di Giulio and Scanlon (1984)
United States (Michigan)	Liver	—	32.1 dw	1.07 dw	2.6 dw	—	0.58 dw	ND	—	Custer and Custer (2000)
United States (Nevada)	Liver	—	—	2.63 ww	—	—	—	—	—	Gerstenberger (2004)
	Muscle	—	—	0.91 ww (1 bird)	—	—	—	—	—	Gerstenberger (2004)

United States (Alaska)	Muscle	—	—	0.078 ww	—	—	—	—	Rothschild and Duffy (2005)
	Brain	—	—	0.049 ww	—	—	—	—	Rothschild and Duffy (2005)
Common Goldeneye (*B. clangula*)									
Sweden (ducklings)	Liver	0.26–0.59[b] ww	1.5–2.33[b] ww	0.524–0.911[b] ww	0.079–0.22[b] ww	10–12[b] ww	—	30–34[b] ww	Eriksson et al. (1989)
Finland	Liver	—	—	0.38–2.15[b] ww	—	—	—	—	Särkkä et al. (1978)
	Muscle	—	—	0.16–0.24[b] ww	—	—	—	—	Särkkä et al. (1978)
United Kingdom	Liver	2.4 dw	—	—	—	—	—	—	Mudge (1983)
	Liver	9.2[c] dw (1 bird)	—	—	—	—	—	—	Mudge (1983)
	Bone	ND–16.9 dw	—	—	—	—	—	—	Mudge (1983)
Canada (eastern)	Liver	—	—	0.63 ww (1 bird)	—	—	—	—	Fimreite et al. (1971)
Canada (ON)	Liver	—	—	4.23 ww	—	—	—	—	Fimreite (1974)
	Liver	—	—	2.43–8.23 ww	—	—	—	—	Annett et al. (1975)
	Liver	—	12.03 dw	—	—	—	—	—	Schummer et al. (2010)
	Muscle	—	—	7.8 ww	—	—	—	—	Vermeer et al. (1973)
	Muscle	—	—	1.22 ww	—	—	—	—	Fimreite (1974)
	Muscle	—	—	0.62–2.6 ww	—	—	—	—	Annett et al. (1975)

(Continued)

TABLE 6.2 (Continued)

Concentrations of Lead, Selenium, Mercury, Cadmium, Copper, Chromium, Nickel, and Zinc in tissues of sea ducks from representative field studies.

Species and location	Tissue	Lead	Selenium	Mercury	Cadmium	Copper	Chromium	Nickel	Zinc	Source
Common Goldeneye (B. clangula)										
Canada (MB)	Liver	—	—	0.58–3.6 ww	—	—	—	—	—	Driver and Derksen (1980)
	Muscle	—	—	0.12–0.72 ww	—	—	—	—	—	Driver and Derksen (1980)
Canada (northern)	Liver	—	2.2–3.9 ww	0.64–1.3 ww	—	—	—	—	—	Braune and Malone (2006b)
Canada (ON), ducklings	Liver	—	—	—	—	—	—	0.45–0.79[b] ww	—	Outridge and Scheuhammer (1993b)
	Kidney	—	—	—	—	—	—	0.12–0.34[b] ww	—	Outridge and Scheuhammer (1993b)
United States (Minnesota)	Eggs	—	—	0.11 ww	—	—	—	—	—	Zicus et al. (1988)
United States (Idaho)	Kidney	—	—	—	0.19 ww (1 bird)	—	—	—	—	Blus et al. (1995)
United States (Michigan)	Liver	—	36.2 dw	1.03 dw	4.25 dw	—	0.66 dw	ND	—	Custer and Custer (2000)
United States (Nevada)	Liver	—	—	0.42 ww	—	—	—	—	—	Gerstenberger (2004)
	Muscle	—	—	0.40 ww	—	—	—	—	—	Gerstenberger (2004)
United States, Canada (northeast)	Liver	—	—	1.5 ww	—	—	—	—	—	Evers et al. (2005)
	Muscle	—	—	0.33 ww	—	—	—	—	—	Evers et al. (2005)
	Blood	—	—	0.21 ww	—	—	—	—	—	Evers et al. (2005)
	Eggs	—	—	0.33 ww	—	—	—	—	—	Evers et al. (2005)
	Feathers	—	—	2.8 ww	—	—	—	—	—	Evers et al. (2005)

Location	Tissue									Reference
United States (Alaska)	Muscle	—	—	0.039 ww (1 bird)	—	—	—	—	—	Rothschild and Duffy (2005)
	Brain	—	—	ND	—	—	—	—	—	Rothschild and Duffy (2005)
United States (Utah)	Liver	0.17–0.39[b] ww	2.7–6.77[b] ww	3.1–14.6[b] ww	0.19–0.34[b] ww	—	11.2–13.9[b] ww	—	39.9–48.7[b] ww	Vest et al. (2009)
Barrow's Goldeneye (B. islandica)										
Canada (northern)	Liver	—	3.0–9.7 ww	0.12–1.3 ww	—	—	—	—	—	Braune and Malone (2006b)
Canada (eastern)	Liver	ND–0.18[b] dw	6.99–36.9[b] dw	2.38–4.44[b] dw	ND–5.06[b] dw	—	—	—	—	Ouellet et al. (2012)
Canada (eastern)	Kidney	0.19–0.49[b] dw	8.88–44.0[b] dw	1.36–2.76[b] dw	ND–47.4[b] dw	—	—	—	—	Ouellet et al. (2012)
Canada (eastern)	Bone	1.27–1.52[b] dw	—	—	—	—	—	—	—	Ouellet et al. (2012)
United States (Alaska)	Liver	ND–3.0 dw	—	—	4.7–7.6[b] dw	0.5–2.9[b] dw	31.1–51.2[b] dw	—	126–154[b] dw	Franson et al. (1995a)
	Kidney	ND–4.9 dw	—	—	13.6–55.7[b] dw	2.5–6.0[b] dw	14.9–21.1[b] dw	—	89.3–125[b] dw	Franson et al. (1995a)
	Muscle	ND–3.0 dw	—	—	ND–0.9 dw	2.3–6.3[b] dw	8.6–22.3[b] dw	—	21.0–39.6[b] dw	Franson et al. (1995a)
	Blood	0.021 ww	9.8 ww	0.98 ww	—	0.25 ww	0.38 ww	0.006 ww	4.14 ww	Heard et al. (2008)
Hooded Merganser (L. cucullatus)										
Canada (ON)	Liver	—	—	1.62–4.5 ww	—	—	—	—	—	Annett et al. (1975)
	Muscle	—	—	0.65–2.94 ww	—	—	—	—	—	Annett et al. (1975)
	Muscle	—	—	12.31 ww	—	—	—	—	—	Vermeer et al. (1973)

(Continued)

TABLE 6.2 (*Continued*)

Concentrations of Lead, Selenium, Mercury, Cadmium, Copper, Chromium, Nickel, and Zinc in tissues of sea ducks from representative field studies.

Species and location	Tissue	Lead	Selenium	Mercury	Cadmium	Copper	Chromium	Nickel	Zinc	Source
Canada (ON), ducklings	Liver	—	—	—	—	—	—	0.07–0.14[b] ww	—	Outridge and Scheuhammer (1993b)
	Kidney	—	—	—	—	—	—	0.99–1.24[b] ww	—	Outridge and Scheuhammer (1993b)
Canada (MB)	Liver	—	—	0.87 ww (1 bird)	—	—	—	—	—	Driver and Derksen (1980)
	Muscle	—	—	0.26 ww (1 bird)	—	—	—	—	—	Driver and Derksen (1980)
	Feathers	—	—	0.26 ww (1 bird)	—	—	—	—	—	Driver and Derksen (1980)
United States (Wisconsin)	Eggs	—	—	0.96 ww	—	—	—	—	—	Faber and Hickey (1973)
United States (14 states)	Eggs	—	—	0.16–1.49[b] ww	—	—	—	—	—	White and Cromartie (1977)
United States (Minnesota)	Eggs	—	—	0.45 ww	—	—	—	—	—	Zicus et al. (1988)
United States, Canada (northeast)	Liver	—	—	4.7 ww	—	—	—	—	—	Evers et al. (2005)
	Muscle	—	—	0.96 ww	—	—	—	—	—	Evers et al. (2005)
	Blood	—	—	0.68–0.88[b] ww	—	—	—	—	—	Evers et al. (2005)
	Eggs	—	—	0.64 ww	—	—	—	—	—	Evers et al. (2005)
	Feathers	—	—	10.4 ww	—	—	—	—	—	Evers et al. (2005)

Common Merganser (*M. merganser*)									
Poland	Liver	—	—	9.05–17.37[b] dw	—	—	—	—	—	Kalisińska et al. (2010)
	Liver	—	2.57–3.32[b] dw	9.37–18.8[b] dw	—	—	—	—	—	Kalisińska et al. (2014)
	Kidney	—	—	6.05–11.72[b] dw	—	—	—	—	—	Kalisińska et al. (2010)
	Kidney	—	2.02–2.45[b] dw	7.27–11.6[b] dw	—	—	—	—	—	Kalisińska et al. (2014)
	Muscle	—	—	1.89–2.56[b] dw	—	—	—	—	—	Kalisińska et al. (2010)
	Brain	—	—	0.98–1.54[b] dw	—	—	—	—	—	Kalisińska et al. (2010)
Sweden	Liver	—	—	—	0.19 ww (median)	—	—	—	—	Frank (1986b)
	Kidney	—	—	—	0.87 ww (median)	—	—	—	—	Frank (1986b)
Canada (ON)	Liver	—	—	26.2 ww	—	—	—	—	—	Fimreite (1974)
	Liver	—	—	1.5–98.56 ww	—	—	—	—	—	Annett et al. (1975)
	Muscle	—	—	6.79 ww	—	—	—	—	—	Vermeer et al. (1973)
	Muscle	—	—	6.12 ww	—	—	—	—	—	Fimreite (1974)
	Muscle	—	—	0.51–8.36 ww	—	—	—	—	—	Annett et al. (1975)
Canada (ON), ducklings	Liver	—	—	—	—	—	—	0.12 ww	—	Outridge and Scheuhammer (1993b)
	Kidney	—	—	—	—	—	—	ND–0.5 ww	—	Outridge and Scheuhammer (1993b)

(Continued)

TABLE 6.2 (Continued)

Concentrations of Lead, Selenium, Mercury, Cadmium, Copper, Chromium, Nickel, and Zinc in tissues of sea ducks from representative field studies.

Species and location	Tissue	Lead	Selenium	Mercury	Cadmium	Copper	Chromium	Nickel	Zinc	Source
Common Merganser (M. merganser)										
Canada (eastern)	Liver	—	9.7 dw	15 dw	—	—	—	—	—	Scheuhammer et al. (1998b)
	Kidney	—	8.5 dw	11 dw	—	—	—	—	—	Scheuhammer et al. (1998b)
	Muscle	—	1.8 dw	3.0 dw	—	—	—	—	—	Scheuhammer et al. (1998b)
United States (Michigan, Wisconsin)	Eggs	—	—	0.57 ww	—	—	—	—	—	Faber and Hickey (1973)
	Eggs	—	—	0.52–0.56[b] ww	—	—	—	—	—	White and Cromartie (1977)
United States (Lake Michigan)	Eggs	—	—	0.58 ww	—	—	—	—	—	Haseltine et al. (1981)
United States (Alaska)	Liver	ND	—	—	0.4–9.9 dw	14.8–39.6 dw	ND–1.3 dw	—	74.1–119 dw	Franson et al. (1995a)
	Kidney	ND–12.7 dw	—	—	1.5–104 dw	1.6–22.6 dw	0.6–11.1 dw	—	65.9–116 dw	Franson et al. (1995a)
	Muscle	ND–1.5 dw	—	—	ND	1.0–23.0 dw	0.6–4.5 dw	—	28.9–193 dw	Franson et al. (1995a)
United States (Nevada)	Liver	—	—	2.61 ww	—	—	—	—	—	Gerstenberger (2004)
	Muscle	—	—	0.22 ww	—	—	—	—	—	Gerstenberger (2004)
United States, Canada (northeast)	Muscle	—	—	1.71 ww	—	—	—	—	—	Evers et al. (2005)
	Blood	—	—	0.6–1.57[b] ww	—	—	—	—	—	Evers et al. (2005)
	Eggs	—	—	1.43 ww	—	—	—	—	—	Evers et al. (2005)
	Feathers	—	—	8.0–8.8 ww	—	—	—	—	—	Evers et al. (2005)

Red-breasted Merganser (M. serrator)

Location	Tissue								Reference
Scotland	Liver	—	—	122 dw (1 bird)	—	—	—	—	Dale et al. (1973)
Sweden	Liver	—	—	—	0.54 ww (median)	—	—	—	Frank (1986b)
	Kidney	—	—	—	5.1 ww (median)	—	—	—	Frank (1986b)
Finland	Liver	—	—	2.78–23.23[b] ww	—	—	—	—	Särkkä et al. (1978)
	Muscle	—	—	1.32–5.44[b] ww	—	—	—	—	Särkkä et al. (1978)
Canada (NB)	Liver	—	—	0.81 ww	—	—	—	—	Fimreite et al. (1971)
United States (Lake Michigan)	Eggs	—	—	1.59 ww	—	—	—	—	Faber and Hickey (1973)
	Eggs	0.93 ww	0.61–0.74[b] ww	0.51–0.52[b] ww	ND	0.75 ww	ND–0.24 ww	15 ww	Haseltine et al. (1981)
United States (Wisconsin)	Eggs	—	—	0.56 ww	—	—	—	—	White and Cromartie (1977)

Mean, minimum–maximum, or as otherwise specified, expressed as mg/kg (ppm) ww (wet weight or fresh weight) or dw (dry weight). Some blood values are reported as μg/dL (100 μg/dL = ~1 mg/kg ww) or ng/mL (1000 ng/mL = ~1 mg/kg ww).

[a] ND = not detected.

[b] Minimum–maximum of 2 or more means, a single value and 1 mean, or mean of detectable concentrations only.

[c] Not analyzed or results not reported.

[d] See Table 6.1 for lead-poisoned sea ducks reported in this study.

[e] Ingested lead shot present in gizzard.

[f] Ingested lead shot found in 1 of 37 birds, but tissue lead concentrations were not identified for that bird.

Selenium

The trace element selenium is a component of the earth's crust that is redistributed naturally in the environment through volcanic activity, terrestrial weathering of rocks and soils, wildfires, and volatilization from plants and water bodies (Maher et al. 2010). It also becomes available through a variety of anthropogenic activities and inputs, such as coal mining; fossil fuel combustion; gold, silver, and nickel mining; metal smelting; agricultural irrigation; and municipal, industrial, and oil refinery wastes (Lemly 2004, Maher et al. 2010). Future anthropogenic sources of selenium will continue to be influenced by industrial development, and climate change will affect atmospheric transport of selenium aerosols (Maher et al. 2010). Globally, selenium cycles through terrestrial, atmospheric, and aquatic compartments, but marine systems comprise the largest part of the cycle with volatilization and recycling through biota representing the main pathways (Haygarth 1994). Selenium is biologically essential for animal nutrition, but the margin between essential and toxic levels for most animals is narrow (Ohlendorf 2003). Selenium exists in a variety of chemical and physical forms, or *species*, that cycle through aquatic ecosystems (Maher et al. 2010). Elemental selenium is practically insoluble in water and, although inorganic selenium compounds are toxic, the organic selenides present the greatest hazard to birds (Ohlendorf and Heinz 2011). Selenomethionine, an organic selenide likely to be ingested by aquatic birds, is highly toxic and results in bioaccumulation of selenium in eggs that may cause reduced hatchability and deformities of embryos (Spallholz and Hoffman 2002, Janz et al. 2010, Ohlendorf and Heinz 2011).

Selenium toxicity has long been recognized in domestic animals, but only more recently in fish and wildlife. Studies in the Central Valley of California in the early to mid-1980s identified selenium in agricultural drain water as the cause of a disease syndrome in aquatic birds, where high selenium concentrations occurred in the tissues of adults and juveniles and high levels in eggs were associated with embryonic deaths and developmental abnormalities (Ohlendorf 2002). A series of experimental investigations followed, in which Mallards were commonly used as a model for waterfowl, confirming the health and reproductive effects observed in the field were attributable to selenium toxicity (Ohlendorf 2002). Adverse effects of selenium exposure demonstrated in field and laboratory studies, many with freshwater birds, include changes in plasma enzyme activities, teratogenesis, embryonic death and reduced hatchability, reduced growth of young, decreased survival and body mass in adults, impaired immune response, and histopathologic lesions (Hoffman 2002). Altered glutathione metabolism is associated with many of these adverse effects, and measurements of oxidative stress, such as increased glutathione peroxidase activity, increased ratio of oxidized to reduced glutathione, and increased thiobarbituric acid reactive substances, have been used as biomarkers of selenium toxicity (Hoffman 2002). Based on laboratory studies with Mallards, key lesions of selenium poisoning include emaciation (25%–50% loss of normal body mass) with severe atrophy of fat and breast muscle, splenic atrophy, absence of thymus, liver necrosis or enlarged nodular liver, pale pancreas, loss of nails, and bilateral loss of feathers of the head and neck (Albers et al. 1996, O'Toole and Raisbeck 1997). Many of the same gross lesions were found in an experimental study of selenium exposure in Common Eiders, as well as microscopic lesions consisting of severe depletion of lymphoid organs, hepatic lipidosis, and necrosis of feather pulp and feather epithelium (Franson et al. 2007).

Selenium concentrations in liver, kidney, muscle, blood, eggs, and feathers have been measured in field surveys of birds, including sea ducks (Table 6.2), but selenium levels in liver and eggs are typically used to assess the risk of adverse effects (Ohlendorf and Heinz 2011). When considering selenium concentrations in tissues in relation to possible toxic effects, it should be noted that the interactions of selenium with other trace elements, metals, and nutrients are many and complex and are affected by a number of factors. The best known of these interactions is with mercury and is characterized by mutual antagonism and a protective effect of selenium against mercury toxicity and vice versa (Cuvin-Aralar and Furness 1991, Khan and Wang 2009). Thus, an evaluation of selenium residues in tissues is ideally done in conjunction with mercury residues unless it is already known that mercury levels are low. Studies have been conducted on interactions of selenium with many nutritional factors and trace elements, such as arsenic, boron, sulfur, and dietary protein. As with some other contaminants, interactions may be synergistic or antagonistic (Heinz and Hoffman 1998, Hamilton and Hoffman 2003, Ohlendorf 2003, Ohlendorf and Heinz 2011).

In freshwater and terrestrial birds, liver selenium concentrations of <10 mg/kg dw represent a low probability of adverse effects, but levels >20 mg/kg dw are considered evidence of selenium toxicosis when accompanied by characteristic lesions (Ohlendorf and Heinz 2011). Some species of marine birds may have considerably higher selenium thresholds, with liver concentrations of 20–75 mg/kg dw presenting merely a low probability of adverse effects (Ohlendorf and Heinz 2011). Even greater concentrations have been found in some sea ducks without reported detrimental effects. For example, maximum liver selenium concentrations of 171–489 mg/kg dw have been reported in individual male Spectacled Eiders (Trust et al. 2000b, Stout et al. 2002, Lovvorn et al. 2013).

Some metals, including selenium, are more or less readily transferred from females to eggs, but the degree to which concentrations in eggs reflect dietary exposure at the time of formation varies among species according to the extent of exogenous versus endogenous resources contributed to eggs (Hobson et al. 2005, Bond et al. 2007). Because selenium is teratogenic and embryotoxic, the concentration of selenium in the egg provides a sensitive measure for evaluating hazards to birds (Ohlendorf and Heinz 2011). Although mean background selenium concentrations in eggs of freshwater and terrestrial species are generally <3 mg/kg dw, levels in some marine birds may be higher (Ohlendorf and Heinz 2011). Considerable variation exists among species regarding the levels of selenium in eggs that are associated with adverse effects. For example, the estimated selenium concentrations in eggs expected to cause a 10% incidence of teratogenic effects in Mallards (*sensitive* species), Black-necked Stilts (*Himantopus mexicanus*) (*average* species), and American Avocets (*Recurvirostra americana*) (*tolerant* species) are 23 mg/kg dw, 37 mg/kg dw, and 74 mg/kg dw, respectively (Ohlendorf and Heinz 2011). However, in sensitive species, even when selenium concentrations are <8 mg/kg dw in eggs, there is a low probability of reproductive impairment characterized by reduced hatchability, and the probability is elevated in sensitive and moderately sensitive

species when selenium in eggs is >12 mg/kg dw (Ohlendorf and Heinz 2011).

The background selenium concentration in the blood of freshwater birds is normally <0.4 mg/kg ww, yet marine birds sampled in unpolluted areas often have higher concentrations (Ohlendorf and Heinz 2011). The relationship between selenium concentrations in the blood and possible toxicity is not as well understood as it for concentrations in liver and eggs, but when experimental Mallards began to die of selenium poisoning, the mean concentrations of selenium in the blood of survivors were about 5–14 mg/kg ww (Heinz and Fitzgerald 1993). Selenium in feathers should be used with caution for biomonitoring. Selenium may have been deposited in feathers at the time and location they were formed, concentrations may increase over time in the same feather and vary within and among feather types, some of the selenium content of feathers may be the result of external contamination, and concentrations may leach out during washing (Burger 1993, Ohlendorf and Heinz 2011). Thus, unless perhaps sampled when growing, feathers may not be particularly useful for evaluating potential selenium toxicity in birds (Ohlendorf and Heinz 2011). Burger (1993) summarized 42 field studies with a variety of bird species, reporting an average selenium concentration in feathers of 6.0 mg/kg dw. The mean concentration in feathers of Common Eiders nesting on the Aleutian Islands was 1.6 mg/kg (Burger et al. 2008).

Sea ducks, in particular, may accumulate greater concentrations of selenium in their tissues than freshwater birds. An experimental study demonstrated that, when exposed to similar dietary selenium concentrations, Common Eiders had several times more selenium in their livers than Mallards (Franson et al. 2007). Mallards fed 60 and 80 mg/kg selenium had mean selenium concentrations in liver tissue of 60.6 mg/kg ww and 99 mg/kg dw, respectively (Albers et al. 1996, O'Toole and Raisbeck 1997). However, Common Eiders on dietary selenium concentrations of 60 and 80 mg/kg had hepatic selenium concentrations of 735 mg/kg dw (or 190 mg/kg ww) and 1252 mg/kg dw (or 343 mg/kg ww), respectively (Franson et al. 2007). Common Eiders fed 20 mg/kg selenium had 351 mg/kg dw (or 101 mg/kg ww) selenium in liver tissue, compared with 49 mg/kg dw and 29.6 mg/kg ww in Mallards fed 20 and 25 ppm, respectively

(Albers et al. 1996, O'Toole and Raisbeck 1997, Franson et al. 2007). The mean concentration of selenium in the blood of eiders fed 60 mg/kg selenium reached 17.3 mg/kg ww, which compares with about 16 mg/kg ww in Mallards fed 60 mg/kg selenium for approximately the same time (O'Toole and Raisbeck 1997). Accordingly, although the selenium concentrations in the livers of eiders were much higher than in livers of the Mallards in these experiments, concentrations in the blood of the two species were similar.

Wintering sea ducks feeding at high trophic levels in marine environments where selenium levels are greater may arrive on breeding grounds with higher selenium concentrations in their tissues (DeVink et al. 2007). Lovvorn et al. (2013) reported mean liver selenium concentrations of 171.8 mg/kg dw in female Spectacled Eiders and 235.6 mg/kg dw in males collected on their wintering grounds in the Bering Sea. The authors suggested that certain characteristics of the eiders, such as sustained high food intake and an inherent propensity to accumulate trace elements, combined with atmospheric and oceanographic processes that concentrate trace elements in the benthic food web of the wintering area, contributed to the high selenium levels (Lovvorn et al. 2013).

Studies of the dynamics of selenium in the blood of sea ducks have shown that concentrations decline when birds are inland during the breeding season, supporting the hypothesis that the major source of selenium exposure comes from the marine environment. Daily rates of decline in blood selenium concentrations in sea ducks on breeding grounds have been reported to be 1.9% and 2.3% in Spectacled Eiders and 0.96% in Common Eiders (Grand et al. 2002; Wilson et al. 2004, 2007). Consequently, mean selenium concentrations in the blood of female Spectacled Eiders declined from 12.8 mg/kg ww during incubation to 4.32 mg/kg ww during the brood-rearing period, and blood selenium concentrations in ducklings averaged 1.96 mg/kg ww, suggesting that local foods were not a significant source of selenium (Grand et al. 2002). In another study, Wilson et al. (2004) reported mean selenium concentrations of 14.7 and 6.1 mg/kg ww in the blood of prebreeding male and female Spectacled Eiders, respectively, whereas concentrations were 4.0 mg/kg ww in nesting females, 2.9 mg/kg ww in females during the brood-rearing period, and 1.1 mg/kg ww in ducklings.

These results lead to three major findings: selenium levels in marine birds decline during the breeding season after they leave the marine environment, selenium levels are lower among birds of the same species feeding in freshwater environments compared with marine habitats, and concentrations of selenium are lower among young compared to adults, even when food habits are similar (Goede et al. 1989; Goede 1993; Franson et al. 1999, 2002a). Birds returning to a marine environment after the breeding season reacquire selenium, such as molting male Long-tailed Ducks in the Beaufort Sea off the coast of Alaska that had a mean selenium level in blood of 48.8 mg/kg dw (Franson et al. 2004).

High levels of selenium are typically found in blood of some sea ducks, but concentrations reported in eggs have been low. In Alaska, blood selenium concentrations in female Spectacled Eiders laying all viable eggs and birds laying one or more nonviable eggs were 13.3 and 10.4 mg/kg ww, respectively, but selenium concentrations in both viable and nonviable eggs were low (1.42 and 1.20 mg/kg ww, respectively) and were not significantly different (Grand et al. 2002). In another study, female Common Eiders had a mean selenium concentration of 36.1 mg/kg dw in their blood, but 2.28 mg/kg dw in eggs, well below the concentration of 12 mg/kg dw suggested as a threshold for an elevated probability of reproductive impairment (Franson et al. 2004, Ohlendorf and Heinz 2011). Although field studies have not shown teratogenic effects, differences in concentrations between viable and nonviable eggs, or adverse effects of selenium in blood of sea ducks, Wilson et al. (2007) found a positive relationship between the probability of at least one nonviable egg in Common Eider nests and the selenium concentration in the blood of the female. The authors reported that the odds of at least one nonviable egg occurring in a clutch increased by 1.15 times for each increase of 1 mg/kg ww in blood selenium (Wilson et al. 2007).

Selenium is a required trace element, but excess exposure may cause poisoning in wild birds consuming food items with high levels of selenium. Moreover, mortality and reproductive impairment have been reported in areas where selenium was leached from soils into irrigation drain water. Laboratory studies of selenium exposure have also demonstrated mortality and severe adverse reproductive effects in freshwater birds. Background concentrations of selenium in tissues of marine birds, including sea ducks, are often much greater than concentrations in freshwater species, without apparent adverse effects. Similarly, an experimental study of selenium exposure with Common Eiders demonstrated they developed many of the same clinical signs and lesions as in studies with Mallards, but eiders had much higher selenium concentrations in tissues. Field studies with Common Eiders have shown that although females may have relatively high selenium levels in their blood, little is transferred to eggs, the critical indicator of reproductive effects. Nonetheless, one study with Common Eiders found that the probability of a nest containing at least one nonviable egg was positively related to blood selenium concentrations in females (Wilson et al. 2007).

Mercury

Mercury is a nonessential, toxic heavy metal that occurs in soils and sediments where its ore, cinnabar, has been mined for centuries. Mercury can enter the environment naturally as a gas from degassing from mercury mineral deposits, volcanic activity, contaminated aquatic and terrestrial systems, and forest fires or in solution or a particulate form (Eisler 2000a, Wang et al. 2004). Important sources of anthropogenic mercury emissions to the atmosphere include combustion of fossil fuels, metal production, cement production, waste incineration, and the use of mercury in various products (Pacyna et al. 2010). Erosion of soil contaminated with mercury, urban sewage discharge, and mining activities for precious metals and lead contribute to mercury contamination in aquatic systems (Wang et al. 2004). Historical sources of mercury pollution include the use of pesticides and fungicides containing mercury, contamination from chloralkali plants, and discharges associated with the manufacture of pulp and paper (Wang et al. 2004). The global cycle of mercury involves its movement among atmospheric, terrestrial, aquatic, and biotic compartments. Of particular relevance to sea ducks is the transport of mercury to arctic ecosystems through the atmosphere, ocean currents, and rivers, and anthropogenic emissions have greatly increased the mercury burden of arctic biota, making the arctic an important global sink for mercury (Leitch et al. 2007, Poissant et al. 2008, Dietz et al. 2009). In two species of seabirds that overwinter in northern waters, for example,

mercury concentrations were about twice as high in 2003 than in 1975 (Braune 2007).

Ingested inorganic mercury is poorly absorbed, accumulating in the kidney where it binds to metallothioneins, proteins of low molecular weight that bind heavy metal ions, offering some protection against the toxicity of mercury and cadmium, and regulating metabolism of the nutritionally essential copper and zinc in response to dietary and physiological changes (Dunn et al. 1987, Scheuhammer 1987). Adverse reproductive effects of inorganic mercury in birds include delayed gonadal development, reduced egg fertility, and depressed growth (Scheuhammer 1987). Methylation, the conversion of inorganic mercury to its more toxic form, methylmercury, occurs primarily by microbial activity in sediments and wetlands (Wiener et al. 2003). Methylmercury is thought to be the most bioavailable and toxic form of mercury to wildlife, accumulates and biomagnifies in food chains, and accounts for a greater percentage of total mercury at higher trophic levels (Eisler 2000a, Shore et al. 2011). In birds, consumption of fish is the primary route of methylmercury exposure, and the risk of such exposure is increased in acidified environments (Scheuhammer 1991, Wiener et al. 2003). Intestinal absorption of methylmercury is nearly 100%, it is more slowly metabolized and has a longer biological half-life than other organic forms, and it readily moves across the blood–brain barrier, accounting for central nervous system dysfunction and pathology (Scheuhammer 1987). For birds, major excretory routes for methylmercury include feathers, feces, and, in breeding females, production of eggs (Wiener et al. 2003). Exposure of birds to methylmercury may cause direct mortality, teratogenic effects on embryos, and sublethal effects that include altered behavior and impaired reproduction resulting from fewer eggs laid, reduced hatching success, and greater early mortality in hatchlings (Eisler 2000a, Shore et al. 2011). Clinical signs of methylmercury poisoning in live animals are characterized by loss of appetite, emaciation, and central nervous system signs, including loss of muscle coordination, paralysis, and tremors, while pathological changes in animals dying of mercury poisoning include microscopic lesions in the peripheral and central nervous systems, liver, and kidney (Heinz 1996). Adverse immunological effects have occurred in birds exposed to mercury, and there is evidence suggesting that there may be a greater potential for infection by disease organisms in birds with elevated mercury levels in tissues (Kenow et al. 2007, Scheuhammer et al. 2007). The evaluation of mercury concentrations in tissues is complicated by several factors, including the observation that some marine birds may normally accumulate higher concentrations of mercury than those shown to cause adverse effects in other avian species, the variability of mercury distribution in tissues among species and life stages, body condition, the ratio of inorganic to methylmercury in the tissue, and the potential protective effect of selenium on mercury toxicity (Scheuhammer 1987, Ohlendorf 1993, Eisler 2000a, Wayland et al. 2005, Eagles-Smith et al. 2008, Scheuhammer et al. 2008).

The liver and kidney have often been used to monitor mercury exposure in birds. If exposure is primarily to inorganic mercury, the total mercury concentration is typically higher in kidney than in liver, but if exposure is primarily to methylmercury, the kidney to liver ratio of total mercury will be close to or even <1.0 (Scheuhammer 1987). Evidence exists that some aquatic birds can demethylate mercury, converting methylmercury to the less toxic inorganic form, and that demethylation rates differ among taxa (Thompson and Furness 1989, Kim et al. 1996b, Eagles-Smith et al. 2009b). The amount of organic mercury in livers of 12 species of seabirds, expressed as a percentage of total mercury, ranged from 2.6% in Wandering Albatrosses (*Diomedea exulans*) to 92.6% in Little Shearwaters (*Puffinus assimilis*, Thompson and Furness 1989). In livers of Long-tailed Ducks, organic mercury accounted for 28% of total mercury (Kim et al. 1996b). As total mercury increases in the liver, an increasingly greater proportion occurs as inorganic mercury, which accumulates in association with selenium, forming nontoxic complexes (Scheuhammer et al. 2007, Shore et al. 2011). In adults of four waterbird species, demethylation occurred above a threshold of 8.5 mg/kg dw total mercury in the liver, and selenium concentrations were positively correlated with inorganic mercury levels above, but not below, the demethylation threshold (Eagles-Smith et al. 2009b). Because of demethylation and the association of the resulting inorganic mercury with selenium in inert complexes, it is useful to analyze tissues for organic mercury, total mercury, and

selenium when evaluating toxicity of mercury or selenium (Scheuhammer et al. 1998b). Body condition has an impact on tissue concentrations of mercury, as well as other metals, particularly in the liver. Greater hepatic mercury concentrations were found in lighter, nesting Common Eiders than in heavier, prenesting birds, but there was no difference in total mercury content in liver between the two groups (Wayland et al. 2005). The authors attributed the increased mercury levels in nesting birds to a decline in liver mass as fat reserves were depleted (Wayland et al. 2005). Thus, the total mercury content in the liver may have been relatively stable, but the concentration increased as liver mass decreased. Body and liver mass were also found to decrease with increasing hepatic mercury concentrations in Surf Scoters (Hoffman et al. 1998).

Feathers have also been used to monitor for mercury exposure in birds because they may be an important excretory route for mercury in some species. Following molt, newly grown feathers have been reported to contain as much as 93% of the body burden of mercury in adults (Braune and Gaskin 1987). Mercury in feathers consists of nearly 100% methylmercury, and concentrations vary among feather types, between feathers of the same type, and, to some extent, within individual feathers (Furness et al. 1986, Thompson and Furness 1989). Variability of mercury levels among different types of feathers and between feathers of the same type has been suggested to be associated with sequence of feather molt and a redistribution of methylmercury from body tissues into the feathers. Reports indicate that, in general, replacement feathers that grow early have greater mercury concentrations than those that grow later, and these findings have been attributed to the transfer of accumulated mercury from body tissues to the growing feathers (Furness et al. 1986, Honda et al. 1986, Braune and Gaskin 1987). Species with slow molting patterns may rely more on demethylation to detoxify methylmercury than on excretion through the feathers (Kim et al. 1996b). When interpreting mercury concentrations in feathers, it is also important to consider the contribution of methylmercury in the diet at the time when new feathers were growing (Burger 1993). Eggs are useful for monitoring mercury exposure because deposition of mercury into eggs is an important route of elimination for reproducing females, and concentrations in eggs provide a means to evaluate the effects of mercury on embryos and hatchlings, which contribute to the adverse effects of mercury on reproduction (Wiener et al. 2003). There is a relationship between dietary mercury and concentrations in eggs at the time they are formed, and studies with Mallards have shown that it may take no more than a week of feeding in a contaminated area for harmful concentrations to be excreted into eggs (Heinz et al. 2009a).

Shore et al. (2011) provided estimates of tissue mercury concentrations associated with death in nonmarine birds as follows (in mg/kg ww): brain, >15 (adults) and >3 (developing young); liver, >20; and kidney, >40. Mercury residues (in mg/kg ww) of >2 in liver, >0.6 in eggs, and >5 in feathers may be associated with impaired reproduction in nonmarine birds (Eisler 2000a, Shore et al. 2011). However, there is considerable variability among species in the sensitivity of embryos to mercury exposure, and although selenium may have a protective effect against mercury poisoning in adults, studies with Mallards have shown that selenium and mercury in combination can have more adverse reproductive effects than either metal alone (Heinz and Hoffman 1998, Hamilton and Hoffman 2003, Heinz et al. 2009b). Because marine birds seem to tolerate high levels of mercury in tissues, Heinz (1996) stressed that the signs and histopathological effects are important considerations in evaluating the effects of mercury in these species. Mercury concentrations in tissues of sea ducks have been reported in many field studies (Table 6.2).

Early reports of mercury poisoning included cases in Scandinavia where terrestrial birds consumed seed treated with organic mercury fungicides. More recently, mercury contamination has been associated with aquatic environments as a result of long-range atmospheric transport, and high concentrations in wildlife have been found in remote locations (Eisler 2000a). Among sea ducks, fish-eating species are generally at greater risk than others for mercury poisoning because of the increase in mercury concentrations through the food chain. As with some other metals, mercury levels in sea ducks from marine environments may be expected to be higher than those in freshwater species. Interpretation is not straightforward because tissue concentrations indicative of toxicity have not received as much study in sea ducks as in inland birds.

Cadmium

Cadmium is a biologically nonessential metal found in the earth's crust in association with zinc ores and is used in electroplating, pigment production, and the production of plastic stabilizers and batteries (Eisler 2000a). Anthropogenic sources of cadmium include dust and wastewater from smelters, burning of fossil fuels and cadmium-containing materials, fertilizers, and municipal wastewater and sludge, with the highest environmental concentrations being found near smelters and industrialized areas (Eisler 2000a). Uptake of cadmium is primarily through the respiratory route and, to a lesser extent, in the intestine where greater proportions are absorbed at higher doses (Wayland and Scheuhammer 2011). Cadmium concentrations increase with age, even in free-ranging wildlife exposed to low levels of dietary cadmium (Scheuhammer 1987). Several studies of cadmium levels in liver and kidney tissue of eiders have demonstrated this characteristic, as Common Eiders and King Eiders (*Somateria spectabilis*) >1 year of age had considerably greater cadmium than eiders <1 year old, and, among eider ducklings, cadmium concentrations increased between 3 and 10 weeks of age (Karlog et al. 1983, Frank 1986b, Nielsen and Dietz 1989, Dietz et al. 1996, Hollmén et al. 1998, Franson et al. 2000a). Birds also tend to accumulate higher cadmium concentrations in kidney tissue than in the liver. In a study combining results from a variety of avian species, cadmium concentrations in kidneys were twice those of liver, but those two tissues accounted for 92% of the cadmium measured in kidney, liver, brain, bone, and blood (García-Fernández et al. 1996). In sea ducks, as in other birds, cadmium concentrations in muscle and blood are typically lower than levels in kidney or liver (Lande 1977, Franson et al. 1995a, Grand et al. 2002, Wilson et al. 2004, Wayland et al. 2008a). Thus, Wayland et al. (2001a) reported mean cadmium concentrations of 0.27–0.66 µg/dL (~0.0027–0.0066 mg/kg ww) and 67–169 mg/kg dw in blood and kidney, respectively, of Common Eiders from Canada. Marine biota generally have higher cadmium residues than freshwater or terrestrial organisms, as total cadmium levels tend to be higher in seawater, and aquatic invertebrates may accumulate cadmium concentrations that are considerably greater than the surrounding aqueous environment (Scheuhammer 1987, Eisler

2000a). Pelagic seabirds have higher concentrations of cadmium in tissues than coastal marine birds and terrestrial birds have the lowest levels (Wayland and Scheuhammer 2011).

Cadmium is another of the metals that induces the synthesis of the protein metallothionein, and most of the cadmium in bird tissues occurs in the liver and kidneys bound to metallothionein (Wayland and Scheuhammer 2011). Studies of Surf Scoters and a number of free-living seabird species have shown correlations between renal cadmium and metallothionein concentrations (Elliott et al. 1992, Elliott and Scheuhammer 1997, Barjaktarovic et al. 2002, Braune and Scheuhammer 2008). However, in a study of Spectacled Eiders, renal cadmium and metallothionein were not correlated, and it was theorized that metallothionein induction was not as efficient in this species as in other marine birds or that the eiders were utilizing energy to overcome stressors at the expense of metallothionein synthesis (Trust et al. 2000b). The authors reported that the combination of copper, cadmium, and zinc (all metallothionein-inducing metals) provided a better model for metallothionein concentrations in eiders than cadmium alone.

A variety of adverse effects of cadmium exposure in birds have been reported, including intestinal damage with altered nutrient uptake, osmoregulatory imbalance, growth suppression, anemia, altered behavior, damage to ovaries and testes, suppression of egg production, and endocrine disruption (Wayland and Scheuhammer 2011). Of particular interest regarding sea ducks are potential effects on osmoregulation. Bennett et al. (2000) evaluated the effect of cadmium ingestion on salt gland and kidney function in Pekin Ducks as a model for marine ducks. The effects of cadmium included enlarged salt glands and kidneys, renal tubular damage, and lower glomerular filtration rate. Pekin Ducks are less salt tolerant than sea ducks and were exposed to high cadmium concentrations, and the authors concluded that further studies are needed to determine if the effects on sea ducks would be similar. Studies with Common Eiders have shown that spleen mass was negatively correlated with renal cadmium concentrations, suggesting a possible effect on the immune system (Wayland et al. 2001b). Renal cadmium concentrations were greater in lighter, nesting Common Eiders than in heavier, prenesting females, but there was no difference in the total kidney cadmium

content (Wayland et al. 2005). Cadmium may exert adverse effects on a variety of systems, but the primary target of chronic cadmium toxicity is the kidney, where elevated concentrations may result in degeneration and renal tubular cell necrosis (Cain et al. 1983, Scheuhammer 1987). A mean cadmium concentration of 132 mg/kg ww in kidney tissue was associated with renal pathology in experimental Wood Ducks (*Aix sponsa*) fed 100 mg/kg cadmium from 1 week to 3 months of age (Mayack et al. 1981). White et al. (1978) fed Mallards up to 200 mg/kg cadmium in feed for 3 months and, although there was no mortality or weight loss, the frequency and severity of kidney lesions increased with dosage and time, and cadmium residues in kidneys of birds receiving the highest dose were 134 and 77 mg/kg ww after 60 and 90 days, respectively.

When monitoring for low-level cadmium exposure in situations where renal toxicity is unlikely, kidney is the tissue of choice because concentrations in that tissue will typically be greater than in the liver (Scheuhammer 1987). However, cadmium concentrations in the kidney can be expected to fall in relation to necrosis of the tubular epithelium in the kidney caused by cadmium exposure (White et al. 1978). Thus, when cadmium exposure is high, concentrations in liver may be greater than kidney (Mayack et al. 1981, Scheuhammer 1988). Little cadmium is transferred to eggs, and eggs are not considered useful for effective biomonitoring (Scheuhammer 1987, Ohlendorf 1993). Wayland and Scheuhammer (2011) reviewed several studies of the use of feathers for assessing cadmium exposure in wild birds, some of which showed that concentrations in feathers were correlated with those in kidney or liver, whereas some did not. Conflicting findings and reports of potential cadmium contamination of feathers from uropygial gland secretions and from external sources led to the conclusion that uncertainty exists regarding the accuracy of using feathers to evaluate cadmium exposure in birds. Wayland and Scheuhammer (2011) calculated that a cadmium concentration of about 65 mg/kg ww in kidneys of birds would result in a 50% probability of altered metabolism or tissue damage and suggested that a more liberal level above which adverse effects are likely is 100 mg/kg ww in kidney tissue. In the liver, the threshold concentration for negative effects may be in the range of 45–70 mg/kg ww (Wayland and Scheuhammer 2011). Cadmium exposure in sea ducks has been demonstrated by elevated tissue concentrations, with some values in individual Common Eiders and Spectacled Eiders approaching the equivalent of 100 mg/kg ww in kidney tissue, but with little evidence of adverse health effects (Wayland et al. 2001a, 2002, 2003; Lovvorn et al. 2013). Using a modeling approach, Bendell (2011) estimated the amount of cadmium ingested by Surf Scoters that consumed mussels, concluding that in some areas of the Pacific Northwest, sea ducks could be exposed to toxicologically significant cadmium levels when feeding on mussels in aquaculture situations.

Other Metals

A variety of metals in addition to lead, selenium, mercury, and cadmium have been measured in tissues of wild birds, including sea ducks, but little information is available on the toxic effects of many of these metals. This section focuses on chromium, nickel, and zinc, three metals that tend to be elevated when samples are taken from animals captured at or near point sources of contamination. Also included is copper, which has been reported to be stored at high concentrations in tissues of some sea ducks.

Chromium is naturally mobilized by weathering processes, but anthropogenic inputs to the environment are much greater and include sources related to the production of stainless steel, chrome-plated metals, electroplated materials, and paint pigments, as well as chromium disposal in sewage and solid wastes containing consumer products (Eisler 2000a). Chemical forms such as hexavalent chromium (Cr^{+6}) are toxic, whereas trivalent chromium (Cr^{+3}) is an essential nutrient for some species, although it may also cause adverse health effects at high exposures (Eisler 2000a). Wildlife are exposed to chromium primarily through ingestion, and trivalent chromium accumulates mainly in the liver, whereas hexavalent chromium has a wider organ distribution, including spleen, kidneys, lungs, and bone (Outridge and Scheuhammer 1993a). Experimental studies in laboratory animals have shown that chromium and cadmium can be antagonistic, each counteracting some of the adverse effects of the other (Eisler 2000a). With the exception of one study that examined behavioral effects, experimental data on chromium toxicity in avian wildlife are lacking. Heinz and Haseltine (1981) fed adult

American Black Ducks (*Anas rubripes*) diets with up to 200 mg/kg ww added chromium, as chromium potassium sulfate, and tested ducklings from those adults for avoidance behavior, reporting no significant effects. Chromium concentrations in wildlife tissues of ≥4 mg/kg dw have been viewed as evidence of probable chromium contamination (Eisler 2000a). However, chromium concentrations reported in field studies are highly variable, and because of the paucity of toxicological data in wildlife, the significance of elevated chromium levels is unclear (Outridge and Scheuhammer 1993a). A summary of several field reports revealed that chromium concentrations in tissues of birds ranged from about 0.1 to 15 mg/kg dw in uncontaminated environments and from 1 to 700 mg/kg dw in contaminated areas, depending on the species and tissue analyzed (Outridge and Scheuhammer 1993a). Chromium concentrations have been reported to be highest at low trophic levels in both marine and terrestrial food chains (Outridge and Scheuhammer 1993a, Eisler 2000a). Chromium concentrations reported in sea ducks have generally been low (Table 6.2) and have not been shown to be associated with adverse effects.

Nickel, an essential trace metal for at least some vertebrate species, has a variety of metallurgical and chemical uses and is widespread in the environment both from natural sources, such as dust from weathering of rocks and soils, volcanic emissions, and forest fires, and anthropogenic releases including mining, smelting, refining, fossil fuel combustion, and waste incineration (Eisler 2000a). An experimental study with Mallard ducklings showed lower bone density in females fed 800 mg nickel per kg of diet and tremors, paresis, ataxia, reduced body weight, and death in ducklings fed 1,200 mg/kg nickel (Cain and Pafford 1981). Livers and kidneys of control birds that received no added dietary nickel had <1 mg/kg ww nickel, whereas ducklings that died had nickel concentrations of up to 23 mg/kg ww in liver and 74 mg/kg ww in kidney (Cain and Pafford 1981). Nickel residues in tissues of birds from uncontaminated environments are usually <5 mg/kg dw, and tissue concentrations in polluted areas have ranged up to 80 mg/kg dw (Outridge and Scheuhammer 1993b). Although levels in tissues were not reliable indicators of toxicity because adverse effects occurred in the absence of high concentrations, Outridge and Scheuhammer (1993b) suggested that when nickel residues are >10 mg/kg dw in kidney or >3 mg/kg dw in liver, toxicosis should be suspected. Detectable levels of nickel have been reported in field collections of tissues from at least 10 species of sea ducks (Table 6.2).

Zinc is an essential trace element but also has been identified as a toxin in animals as a result of the ingestion of zinc-containing objects and exposure to emissions from zinc mines and smelters (Eisler 2000a). Experimental studies of zinc poisoning in Mallard ducks, from zinc added to feed and dosing with zinc shotgun pellets, were characterized by high mortality and maximum mean zinc concentrations of up to 483, 519, and 2672 mg/kg ww in liver, kidney, and pancreas, respectively (Gasaway and Buss 1972, Levengood et al. 1999). Zinc concentrations in liver, kidney, and pancreas of controls were 54–63, 26–27, and 77–100 mg/kg ww, respectively (Gasaway and Buss 1972, Levengood et al. 1999). The maximum concentration of zinc in plasma of poisoned Mallards was 16 mg/kg ww, compared with a maximum of 3.5 mg/kg ww in controls (Levengood et al. 1999). A captive King Eider diagnosed with zinc poisoning from ingesting two pennies had a blood serum zinc concentration of 26 mg/kg (ww presumed but not stated), which dropped to 4.1 mg/kg 14 days after removal of the coins (Culver 2007). In another captive bird collection, a Common Merganser that survived zinc poisoning had a serum zinc concentration of 16.6 mg/L ww 2 days after removal of a penny from its ventriculus, and four Barrow's Goldeneyes that died of zinc poisoning from ingested pennies had hepatic zinc concentrations of 242–548 mg/kg ww (Zdziarski et al. 1994). In wild waterfowl, zinc toxicosis has been reported in Canada Geese (*Branta canadensis*), Mallards, and a Trumpeter Swan (*Cygnus buccinator*) from a lead and zinc mining area in south central United States, with zinc concentrations as high as 2900 mg/kg dw in liver, 970 mg/kg dw in kidney, and 3200 mg/kg ww in pancreas (Sileo et al. 2003, Beyer et al. 2004, Carpenter et al. 2004). Zinc concentrations reported in tissues of wild sea ducks have been considerably lower than concentrations associated with toxicity in waterfowl (Table 6.2).

Copper is an essential trace element, with concentrations in tissues of wild birds varying considerably within and among species and species groups. Among species, differences in copper levels in tissues are not necessarily the result of variations in dietary intake, but may be related to differences in excretion (Davis and Mertz 1987). Copper is widely used in industrial and electronic applications and copper sulfate is incorporated in a variety of agricultural products, including fungicides, algicides, and nutritional supplements (Eisler 2000a). Copper-based paints, used to kill bottom fouling organisms on boat hulls, are an alternative to paints containing tributyltin (TBT) and are a potential source of dissolved copper in marine and estuarine environments (Turner 2010). Elevated copper concentrations in the environment are often the result of anthropogenic inputs near copper smelters and mines, areas receiving municipal and industrial wastes, and marinas, as well as in soils and sediments after long-term applications of copper-based fungicides and algicides (Eisler 2000a). High copper concentrations in livers, typically accompanied by considerably lower levels in kidney tissue, have been reported in several species of sea ducks, particularly eiders (Table 6.2). In Spectacled Eiders, Trust et al. (2000b) found mean copper concentrations of 558.6 mg/kg in liver and 67.5 mg/kg in kidney. However, metallothionein concentrations were lower than expected, and the authors suggested that metallothionein induction in Spectacled Eiders may be less efficient than in other marine birds or that energy resources must be used to overcome other stressors with little left for metallothionein synthesis. Additionally, evidence exists that in some eiders, males accumulate more copper than females. For example, mean concentrations of copper in livers of male Common Eiders and King Eiders collected in Alaska were 607 and 408 mg/kg dw, respectively, compared with 79.9 and 104 mg/kg dw, respectively, in females of the same species (Stout et al. 2002). Common Eider males in the Baltic Sea had mean hepatic copper concentrations ranging from 604 to 1381 mg/kg dw, whereas the mean concentration in females was 43 mg/kg dw (Franson et al. 2000a). Norheim and Borch-Iohnsen (1990) suggested that the large range of hepatic copper within Common Eiders may be the result of differences in the relative intake of food items, such as mussels, snails, and crustaceans that have high copper levels in their blood pigments. However, the variability in copper levels among species and species groups may not simply be the result of dietary intake, but may also be associated with differences in excretion (Davis and Mertz 1987). In any case, although high copper concentrations have been reported in tissues of some birds, little information is available on its toxicity in avian wildlife, making interpretation of tissue values difficult (Eisler 2000a).

Sea ducks are exposed to many heavy metals from natural and anthropogenic sources, and it is frequently difficult to evaluate the significance of exposure and tissue concentrations in relation to adverse effects. The Arctic Monitoring and Assessment Programme (2005) used three types of parameters to address the biological effects of heavy metals: (1) observed effects in individuals, populations, or communities, (2) physiological responses as indicators of effects, and (3) tissue concentrations or biomarkers as indicators of exposure. Except for lead, published reports of adverse effects of heavy metals on individuals or populations of wild sea ducks are rare, and, although physiological responses and biomarkers have been measured in a number of studies, evaluations of the results are often hindered by the lack of experimental interpretive data specific to sea ducks. Similarly, tissue concentrations may be difficult to interpret because toxicological criteria are based primarily on studies of freshwater birds, and reports indicate that sea ducks may tolerate higher levels of some heavy metals. Factors complicating the interpretation of metal residues in tissues include differences in exposure regimes between experimental studies and natural situations and the interactions among some metals that may affect their toxicity. All metals can be toxic at high enough concentrations, and where possible, field evidence of exposure should be linked to concentrations in tissues and any measurable effects in an effort to better understand the ecotoxicology of metals (Arctic Monitoring and Assessment Programme 2005).

PETROLEUM

Petroleum consists of crude oils, which vary considerably in chemical composition, viscosity, and other physical characteristics, and a variety of refined oil products (Albers 2003). Exposure of wildlife to petroleum results from natural seeps and anthropogenic sources including oil spills from the rupture of pipelines and tanker accidents at sea, discharges from ship ballast, oil and gas exploration and development, production and refining activities, and municipal and industrial waste discharges (Arctic Monitoring and Assessment Programme 1998). Spilled oil may be retained along coasts, potentially resulting in the accumulation of oil on beaches, incorporation of petroleum compounds in sediments, rerelease of buried oil from those sediments into surface waters, floating oil slicks, and entrapment of oil under ice (Arctic Monitoring and Assessment Programme 1998). Climate change will likely have an impact on the potential threat of oil exposure to wildlife, including sea ducks. Projected retreat of sea ice in arctic areas may improve access to ports and onshore areas, result in longer shipping seasons with more oil tanker traffic, increase coastal erosion, and degrade permafrost that could harm existing facilities and cause the release of contained waste materials (Arctic Monitoring and Assessment Programme 2007). When crude oil and petroleum products are released into the environment, they undergo physical and chemical alterations called weathering, and the dispersal of oil from spills or other discharges into water bodies is influenced by many factors including currents and tides, wind, interactions with ice, and the chemical and physical properties of the oil itself. Bioremediation, the addition of materials to the environment to accelerate the degradation of petroleum and some other contaminants, is a tool that has been used in the field after a number of oil spills (Swannell et al. 1996). Nutrients are added to stimulate the natural microbiological degradation process and, in some cases, the area may be seeded with microbes to augment those occurring naturally (Swannell et al. 1996). The success of bioremediation is affected by a number of factors, particularly in cold environments, and although it has been shown to enhance biodegradation of petroleum on shorelines, there is little evidence that bioremediation is effective at sea (Swannell et al. 1996, Si-Zhong et al. 2009). Oil dispersants, detergent-like compounds containing surfactants that remove oil from the water surface and move it down into the water column as droplets, are used in an attempt to prevent oil from reaching shorelines, to reduce the impact of oil on birds and mammals at the water surface, and to promote the biodegradation of oil in the water column (Fingas 2011). Although the dispersants used today are less toxic than oil, sea ducks and other diving birds can be exposed to the oil droplets created by dispersants in the water column while foraging for food (Peakall et al. 1987, Fingas 2011).

Routes of petroleum exposure and adverse effects for sea ducks and other birds include inhalation of volatile fractions, external contamination of feathers, eggshell oiling, ingestion of oil from preening or on food items, habitat alterations, and potential changes in the prey base. External oiling is a major cause of acute mortality in birds associated with oil spills. Feather surfaces repel water because of interlocking structural elements and the high surface tension of water, but lower surface tension oils are absorbed by feathers causing the plumage to become matted, compromising waterproofing, insulation, and buoyancy, often leading to death by hypothermia, drowning, or starvation (Leighton 1993, Jenssen 1994, Jessup and Leighton 1996). Preening spreads oil throughout the plumage, enhancing its effect, and heat loss from birds is greater in water than in air of the same temperature (Jenssen 1994). Jenssen and Ekker (1991a) reported dose-related and time-related effects of plumage oiling on thermoregulation in Common Eiders in an experimental study simulating an encounter with an oil slick. When 70 mL of crude oil was introduced to a chamber containing 25 L of seawater, eiders became hypothermic within 70 min, and the authors concluded that the thermoregulatory effects of oiling were more severe after the birds had time to preen the oil into a greater part of their plumage. Another study found that oil-dispersant mixtures had greater thermoregulatory effects than crude oil alone, and Common Eiders were more susceptible to adverse effects caused by the mixtures than Mallards (Jenssen and Ekker 1991b). The authors suggested that species having an air-filled plumage with high insulating characteristics are probably more vulnerable to the thermoregulatory effects of oiling than species with less insulation because the air-filled plumage is more subject to collapse and matting than compact plumage.

The oil consumed during feather preening by oiled birds, and from contaminated food or water, can have a variety of effects that Leighton (1993) categorized into three major groups: physiological stress responses, reduced reproduction, and destruction of red blood cells. Specifically, ingested oil may cause gastrointestinal irritation, pneumonia, dehydration, impaired osmoregulation, immune suppression, anemia, alterations in reproductive hormones and egg structure, impaired growth, and abnormal behavior (Leighton 1993, Albers 2003). Hemolytic anemia has been specifically documented in White-winged Scoters exposed to fuel oil (Yamato et al. 1996). Oil is also extremely embryotoxic, and small amounts of surface contamination transferred to eggs from oiled plumage during incubation or from nest material can result in death of embryos, particularly during the first half of incubation (Leighton 1993). Albers and Szaro (1978) applied 5 or 20 µL of No. 2 fuel oil to Common Eider eggs in the field when embryos were <17 days old and examined the eggs 7 days later. They reported that embryonic mortality in the 20 µL group was 20% greater than controls, but that 5 µL of fuel oil did not significantly affect mortality. Many of the toxic effects of petroleum exposure are attributed to polycyclic aromatic hydrocarbons (PAHs) and measurements of 7-ethoxyresorufin-O-deethylase (EROD) to assess induction of cytochrome P450 mixed-function oxygenase systems in the liver have been used as biomarkers of these and other toxins in birds, including sea ducks (Trust et al. 2000a,b; Albers 2006; Harris et al. 2007; Miles et al. 2007). Some PAHs that were shown in experimental studies to be toxic for Common Eider embryos have been found in Common Eider eggs collected in the field, but at low concentrations. Thus, a mixture of 18 PAHs injected into Common Eider eggs at concentrations of 0.2 and 2.0 mg/kg of egg caused embryo mortality of 18% and 94%, respectively (Brunström et al. 1990). At the 0.2 mg/kg dosage, each of the 18 PAHs in the mixture was injected into the eggs at concentrations of about 9–13 ng/g ww, but Common Eider eggs from the Beaufort Sea had detectable levels of only 11 of the 18 PAHs, with maximum concentrations of ≤1 ng/g ww (Brunström et al. 1990, Franson et al. 2004). Benzo[k]fluoranthene injected at 0.2 mg/kg level caused 44% mortality in Common Eider embryos, but this compound was not detected in Common Eider eggs in the field (Brunström et al. 1990, Franson et al. 2004). The maximum concentrations of phenanthrene, fluoranthene, and pyrene found in Common Eider eggs from the Beaufort Sea were 1.1, 0.97, and 0.54 ng/g ww, respectively (Franson et al. 2004). Brunström et al. (1991) observed no mortality in chicken (Gallus domesticus) embryos when those three PAHs were each injected at a concentration of 300 ng of compound per gram of egg. Although PAH concentrations reported in field collections of Common Eider eggs in Alaska have been low, some evidence exists of an association between PAHs and productivity elsewhere in the wild. Bustnes (2013) studied a Common Eider colony in a Norwegian fjord that received high inputs of PAHs for decades and reported that, after the pollution level declined, reproductive output improved considerably.

Worldwide, the estimated input of oil in metric tonnes into the marine environment from ships and other sea-based activities was >300,000 per year between 1968 and 1977, >440,000 per year between 1978 and 1987, and >1.2 million per year between 1988 and 1997 (Joint Group of Experts on the Scientific Aspects of Marine Environmental Protection 2007). Large oil tanker spills and oil well blowouts are highly visible events that receive a large amount of attention and can have catastrophic effects on local and more widespread environments, depending on the amount spilled and its movement and redistribution. Vermeer and Vermeer (1975) summarized information on bird mortality associated with 16 oil tanker accidents and 20–tanker discharges between 1937 and 1972, many of which involved sea ducks. Early large-scale releases of oil into the environment include leakage resulting from the break up of two tankers during a storm near the coast of Massachusetts in 1952 (Burnett and Snyder 1954). Common Eiders were heavily affected by that spill, with wintering population estimates in the area dropping from 500,000 the previous year to 150,000 after the spill (Burnett and Snyder 1954). During five oil spill events of generally unknown sources in Denmark between 1968 and 1971, the minimum estimated number of Common Eiders, Black Scoters, and White-winged Scoters lost totaled more than 40,000 birds (Joensen 1972). Oil tanker and barge accidents along the Atlantic coast of Canada in 1970 killed an estimated 12,000 birds, including Long-tailed Ducks, Red-breasted Mergansers, and Common Eiders (Brown et al. 1973). The collision of 2 tankers near San Francisco Bay in 1971 resulted in an estimated mortality of 20,000 birds, with Surf Scoters and White-winged Scoters among the species experiencing the heaviest losses (Smail et al. 1972). The 1996 Sea Empress oil tanker spill in southwest Wales, United Kingdom, impacted an important wintering ground for the Black Scoter (Banks et al. 2008). Although peak winter counts of Black Scoters were substantially lower during the two winters following the spill, after 10 years, counts were similar to those recorded before the spill, and it was suggested that the return of scoters to previously contaminated feeding areas implied that the ecosystem had regenerated sufficiently to support its top predator (Banks et al. 2008). Oiled

King Eiders and Long-tailed Ducks were found after the freighter Citrus collided with another vessel, spilling oil near St. Paul Island, Alaska, in 1996 (Flint et al. 1999). Castège et al. (2004) reported a 20% decrease in Black Scoter populations in the Bay of Biscay, France, two years after the oil spill from the tanker Erika in 1999. Carcasses of several species of sea ducks were collected at Unalaska Island, Alaska, in association with the breakup and oil spill from the freighter Selendang Ayu in 2004 (Byrd and Daniel 2008). Based on EROD activity in liver biopsies, Flint et al. (2012) found evidence that Harlequin Ducks were exposed to hydrocarbons more than 3 years after the Selendang Ayu spill. Large-scale oil well blowouts occurred off the coast of Santa Barbara, California (1969), in the Ekofisk oil field in the North Sea (1977), and Ixtoc 1 in the Gulf of Mexico (1979). Recently, individuals of at least three sea duck species (Bufflehead, Red-breasted Merganser, and Surf Scoter) were picked up in the autumn of 2010 in association with the Deepwater Horizon oil well blowout in the Gulf of Mexico earlier that year (U.S. Fish and Wildlife Service 2011).

Effects on biota of the 1989 Exxon Valdez oil spill in Prince William Sound, Alaska, have been extensively studied. A variety of approaches have been used to evaluate consequences of the spill on sea ducks, including winter and summer population surveys, survival studies, measurements of physiological biomarkers, and risk assessments based on PAH concentrations in prey and environmental samples. Overall findings associated with the spill included initial negative effects on numbers of several species of sea ducks, followed by lingering consequences demonstrated by the measurement of biomarkers. Perhaps because of the varying methodologies used and endpoints assessed, conclusions of some research differ with regard to the persistence of impacts on sea ducks, particularly Harlequin Ducks, two decades after the oil spill. Thus, Esler et al. (2010) reported that hepatic EROD activity was higher in Harlequin Ducks from oiled areas than unoiled areas during 2005–2009, interpreting those results to indicate that harlequins continued to be exposed to Exxon Valdez oil up to 20 years after the spill. Furthermore, although studies of female survival suggested that direct effects on Harlequin Ducks had largely abated 11–14 years after the spill, a 24-year timeline to full recovery of Harlequin

Duck populations has been projected, based on the most likely combinations of variables (Esler and Iverson 2010, Iverson and Esler 2010). On the other hand, the authors of risk assessments incorporating model elements of PAH concentrations in prey species, sediments, and seawater to project exposures concluded that toxicological risks to Harlequin Ducks from residual Exxon Valdez oil were essentially nonexistent 20 years after the spill (Neff et al. 2011, Harwell et al. 2012). Esler et al. (2002) and Wiens et al. (2010) review many of the issues surrounding the Exxon Valdez oil spill in relation to Harlequin Duck populations.

Acute mortality estimates after the Exxon Valdez oil spill were from 100,000 to 300,000 birds killed, with sea ducks accounting for nearly 25% of carcasses retrieved in Prince William Sound and 5.3% of those collected within the broader area affected by the spill (Piatt et al. 1990). Day et al. (1997) studied habitat use by marine birds in Prince William Sound during summer surveys for 2.5 years following the oil spill, relating species abundance to oiling gradient. The authors reported Barrow's Goldeneyes and Buffleheads were negatively impacted in their use of oil-affected habitats and did not recover by the end of the study in 1991. By contrast, Harlequin Ducks, Black Scoters, Common Mergansers, and Red-breasted Mergansers exhibited initial negative effects on abundance but recovered by the end of the study (Day et al. 1997). When pre- and postspill summer bird densities in Prince William Sound were compared, negative effects on Common and Barrow's Goldeneyes and Common and Red-breasted Mergansers were reported through 1998, whereas Harlequin Ducks exhibited negative effects in 1990 and 1991 only (Irons et al. 2000). Wiens et al. (2004) reported summer surveys indicated no evidence of an oiling impact on habitat occupancy by Harlequin Ducks and Barrow's Goldeneyes, results for Common Goldeneyes were inconsistent with regard to oiling effects, and Common Mergansers exhibited initial negative effects followed by recovery. However, winter radiotelemetry studies and surveys during the mid-1990s indicated that survival and densities of Harlequin Ducks in oiled areas were lower than in unoiled areas, and populations had not fully recovered by 1998, whereas populations of Barrow's Goldeneyes had recovered by that time (Esler et al. 2000a–c, 2002). Using winter and summer surveys, Lance

et al. (2001) reported that Harlequin Ducks and Buffleheads showed weak evidence of recovery by 1998, although neither showed positive trends in both winter and summer, but mergansers, scoters, and goldeneyes displayed evidence of continuing effects. Field sampling from 1996 to 1998 showed EROD activity in livers of Harlequin Ducks and Barrow's Goldeneyes was greater in birds from oiled areas than in those from unoiled areas, a finding attributed to exposure to residual oil (Trust et al. 2000a). By 2009, there was no difference in EROD activity between oiled and unoiled areas for Barrow's Goldeneyes but, as mentioned earlier, elevated EROD activity persisted in Harlequin Ducks that year (Esler et al. 2010, 2011). The possibility of long-term oil exposure of sea ducks that prey on mussels is supported by the findings of Carls et al. (2001), who monitored the persistence of PAHs in mussel beds in Prince William Sound and the Gulf of Alaska. The authors reported that in 1995, mean hydrocarbon concentrations were twice the background concentrations in mussels from 18 of 31 sites, concluding that significant contamination may persist at some locations for several decades. In a recent update, the Exxon Valdez Oil Spill Trustee Council (2010) lists Harlequin Ducks and Barrow's Goldeneyes as recovering.

Many small oil spills and various discharges occur annually, accounting for substantial petroleum contamination. For example, hundreds of oil spills are registered each year along shipping routes from the southwest Baltic Sea to the Gulf of Finland, and the number of oiled Long-tailed Ducks counted on the southern coast of Gotland, Sweden, during the winters of 1996 through early 2004 was as high as 35,000 per year (Larsson and Tydén 2005). When oil does not make landfall, bird mortalities may not be detected unless oiled birds swim to shore or carcasses drift onto beaches (Fowler and Flint 1997, Flint and Fowler 1998). In both large and small oil spills, modeling of bird mortality can be an important tool in estimating losses (Van Pelt and Piatt 1995, Flint et al. 1999, French-McCay 2004, Wiese and Robertson 2004, Byrd et al. 2009).

Oil spills and leaks from development and transport of petroleum resources are a significant threat to sea ducks, both in aquatic habitats used during migration and wintering and in terrestrial nesting areas that may be in close proximity to oil fields. Considerations in the overall assessment of the effects of petroleum exposure on sea ducks include the obvious mortality caused by external oiling, as well as physiological effects, reduced reproductive capacity, and alterations to habitat and the availability and quality of food items. Oil spills and pollution in arctic and subarctic regions can pose additional hazards because of the effects of cold temperatures on the physical state of petroleum, bioremediation, and ecosystem recovery. A review of findings related to the Exxon Valdez spill led Peterson et al. (2003) to call for broad-based studies of oil ecotoxicology, expanding the scope beyond estimates of acute toxicity to include assessments of delayed, chronic, and indirect effects over longer time periods.

PERSISTENT ORGANIC POLLUTANTS

Common characteristics of most POPs include low solubility in water (hydrophobicity), an affinity for accumulation in fat (lipophilicity), biomagnification, and resistance to biodegradation (Arctic Monitoring and Assessment Programme 1998). The global distribution of POPs has led international organizations to address issues related to environmental contamination with these chemicals. Compounds often referred to as legacy POPs, such as dichlorodiphenyltrichloroethane (DDT), dioxins, hexachlorobenzene (HCB), and PCBs, have been banned or greatly restricted in many countries, yet are still found in the environment because of their persistent nature (Scheringer 2009). Although some classes of POPs have declined in the arctic in the last 20–30 years, parts of arctic systems function as sinks for persistent contaminants (Macdonald et al. 2000, Rigét et al. 2010). Thus, concentrations of many POPs in the arctic are of a magnitude generally not attributable to known sources in arctic environments, but rather they are thought to arrive after evaporation in lower latitudes and northerly long-range atmospheric transport, followed by deposition onto land and water surfaces through a process of cold condensation (Wania and Mackay 1993, Arctic Monitoring and Assessment Programme 2004). Organic compounds condense at different temperatures depending on their volatility, and those compounds with a low vapor pressure and relatively high water solubility may preferentially accumulate and become enriched in polar regions (Wania and Mackay 1993, Wania 2003). In addition, POPs may be transported north via pelagic organisms and migratory birds that migrate through various climate zones and may introduce these lipophilic pollutants to arctic food webs characterized by animals with thick layers of fat, where they tend to bioaccumulate and biomagnify at higher trophic levels (Borgå et al. 2001). Other sources of POPs may include emissions from military bases, harbors, or landfills that may be important locally or regionally but are minor contributors from a circumpolar perspective (Arctic Monitoring and Assessment Programme 2004). Some less persistent chlorinated pesticides and others, including endosulfan and organotins, are of concern because of their toxicity and presence in remote environments (Arctic Monitoring and Assessment Programme 2004).

Many of the organochlorine pesticides became available for use in the 1940s and 1950s. They are generally neurotoxic and, although acute mortality may occur within hours of high-level exposure, death more typically occurs following an accumulation of residues over weeks or months (Blus 2003). Signs of organochlorine toxicity often include ataxia, abnormal posture, tremors, spasms, and convulsions. Birds that die after chronic exposure will usually be emaciated, whereas those dying acutely may exhibit no lesions. Brain tissue is typically analyzed for chemical residues in diagnostic evaluations of organochlorine poisoning, although concentrations of particular compounds associated with mortality may vary among species (Blus 2011, Elliott and Bishop 2011). Body condition is another important factor to consider, because as birds lose weight, the organochlorines in fat reserves are mobilized and redistributed to other lipid-rich tissues, particularly the brain (Blus et al. 1996). Bustnes et al. (2010) reported that increases of certain organochlorines in the blood of Common Eiders were associated with poor body condition late in the incubation period. Olafsdottir et al. (1998) studied seasonal fluctuations of organochlorine levels in breast muscle and liver of Common Eiders in Iceland, noting that concentrations increased in females between February and June. The authors attributed this finding to the relocation of organochlorines from diminishing adipose tissue, as nesting females lost weight, to lipids in other tissues in the body. Animals are sometimes exposed to many organochlorines simultaneously, residues of most are positively correlated in tissues, and the various chemicals may be additive, synergistic, or even antagonistic (Blus 2003). Stickel et al. (1970) developed an additive scale for *toxic equivalents* of DDT and its principal metabolites, dichlorodiphenyldichloroethane (DDD) and dichlorodiphenyldichloroethylene (DDE). The authors considered 1 mg/kg DDT equal to 5 mg/kg of DDD or 15 mg/kg of DDE in avian brain tissue, with lethal toxicity occurring at 10 mg/kg DDT or greater (Stickel et al. 1970). Recently, Elliott and Bishop (2011) proposed an expanded scheme for toxic equivalents that includes additional organochlorines and is based on the toxicity of dieldrin. Interactions among organochlorines may also result in enhanced residues of one or more compounds. For instance, concentrations of chlordane in brains of birds exposed to chlordane followed by endrin were greater than in birds exposed to chlordane alone (Ludke 1976).

In the absence of direct mortality in birds, well-known sublethal effects of chlorinated pesticide exposure are associated with reproductive impairment and include failure to breed, eggshell thinning, and a decrease in hatchability and survival of young, most of which have been attributed to DDE (Blus et al. 1996, Blus 2011). Organochlorine residues in eggs have frequently been used for monitoring trends of exposure and evaluating the potential for adverse reproductive effects, but care must be taken when interpreting findings in relation to toxic effects because of the wide variation in sensitivity among species (Blus 2003). The induction of liver microsomal enzyme activity is a preparatory step for the metabolism, detoxification, and elimination of contaminants, including organochlorines, and measurement of these enzymes has been used as a biomarker for exposure (Blus et al. 1996). Exposure to organochlorine pesticides may result in a variety of other physiological responses including effects on estrogenic activity, adrenal function, immune system function, and others (Blus 2003).

Uses of legacy POPs are now restricted to narrowly prescribed purposes within a framework of registration for specific exemptions (United Nations Environmental Programme 2001). The single most widely recognized of the chlorinated hydrocarbon pesticides is DDT, which has been banned in many countries, but it is still used against mosquitoes to control malaria in some parts of the world (van den Berg 2009, World Health Organization 2011). The adverse effects of DDT and its primary metabolite DDE on survival and reproductive success have been demonstrated in a variety of avian species, and decreasing environmental levels have been important in the recovery of bird populations (Arctic Monitoring and Assessment Programme 2009). Of the remaining legacy POPs, chlordane has been used to control termites and as a broad spectrum agricultural pesticide, toxaphene and mirex as insecticides, and HCB as a fungicide. Polychlorinated dibenzo-p-dioxins (PCDDs) and polychlorinated dibenzofurans (PCDFs) have no commercial use and are released into the environment as by-products in several chemical manufacturing processes (Harris and Elliott 2011). Of the 209 possible PCB congeners, or related compounds within the PCB group, about 100–150 were incorporated into formulations that are now distributed throughout the environment (Rice et al. 2003). PCBs were used extensively as insulating or cooling agents in electrical transformers and capacitors, as well as in paints, carbonless copy paper, cutting oils, sealants, and pesticide extenders (Eisler 2000b). Similar to other legacy POPs, PCBs bioaccumulate and biomagnify in food chains. PCBs have been shown to cause direct mortality, changes in plumage color, embryonic defects, reproductive failure, immunotoxicity, liver damage, tumors, and emaciation, and a variety of tissues have been used to monitor for PCBs (Eisler 2000b, Harris and Elliott 2011). Because the various PCBs, PCDDs, and PCDFs differ in their toxicity, a system of toxicity equivalency was developed whereby the relative toxicity of a congener is compared to the dioxin 2,3,7,8-TCDD, the most toxic compound of the group (Rice et al. 2003). Similarly, because the risk of cumulative exposure to several PCBs, PCDDs, and PCDFs is often greater than that posed by individual exposure, each congener's toxicity equivalent concentration (actual concentration × toxicity equivalency factor [TEF]) can be summed to derive a total value of exposure (Rice et al. 2003). The sensitivity to dioxin-like contaminants also varies greatly among species of birds, and genetic sequencing has identified characteristics that provide a molecular understanding of dioxin susceptibility (Karchner et al. 2006). Common Eiders are predicted to be relatively insensitive compared to birds of other Orders, such as Galliformes and Passeriformes, which are predicted to have intermediate sensitivity to dioxin-like toxicity (Harris and Elliott 2011). Tissue concentrations of PCBs,

PCDDs, and PCDFs reported as lowest effect levels in hatchlings and threshold criteria for survival and reproduction in birds are summarized by Harris and Elliott (2011).

Legacy POPs have been measured in a variety of sea ducks in many countries (Table 6.3). Several of these chemicals have exhibited declines over the years. In eggs of Red-breasted Mergansers sampled from nesting areas in Lake Michigan, several organochlorines declined substantially between sampling periods of 1977–1978, 1990, and 2002 (Haseltine et al. 1981, Heinz and Stromborg 2009). For example, DDE declined by 66% between 1977–1978 and 1990 and an additional 36% between 1990 and 2002, whereas dieldrin concentrations declined by 16% between 1977–1978 and 1990 and an additional 96% between 1990 and 2002 (Heinz and Stromborg 2009). However, although PCB concentrations in eggs of Red-breasted Mergansers declined between 1977–1978 and 1990, the relative potency of the mixtures, based on toxic equivalency factors, remained similar (Williams et al. 1995). In Common Merganser eggs sampled in Denmark during 1973–1976, Hansen and Kraul (1981) found no significant correlation between shell thickness and concentrations of dieldrin, DDE and PCBs, although shell thickness was 22% less than that of eggs sampled prior to the DDT era. Trust et al. (2000b) analyzed brains of 20 Spectacled Eiders from Alaska for organochlorine compounds, reporting detectable levels of HCB (0.062 mg/kg dw) and beta-hexachlorocyclohexane (0.053 mg/kg dw) in one bird and DDE in two birds (0.062 and 0.10 mg/kg dw). O'Keefe et al. (2006) reported PCDD concentrations of 190–2200 pg/g lw in fat and 190–2400 pg/g lw in liver and PCDF concentrations of 122–2638 pg/g lw in fat and 55–3910 pg/g lw in liver of Common Mergansers from New York. No organochlorines in appreciable concentrations were found in liver and kidney tissues of 44 eiders (Steller's Eider [*Polysticta stelleri*], Spectacled Eider, King Eider, Common Eider) collected in Alaska and arctic Russia (Stout et al. 2002).

Other pesticides of concern for sea ducks with certain characteristics of POPs, and introduced many years ago, include organotin compounds. Although inorganic tin poisoning has been reported in mammals, inorganic tin and its salts are not considered particularly toxic because of their poor absorption, low solubility, and rapid tissue turnover (Eisler 2000a). However, of 260 organotins known, 36 are listed as toxic, with the triorganotin compounds being the most harmful (Eisler 2000a). TBT is highly toxic to aquatic life, more so than its derivatives monobutyltin (MBT) and dibutyltin (DBT) (Eisler 2000a, Hoch 2001). Organotins have a variety of applications, including their use as heat and light stabilizers for polyvinyl chloride plastics, wood preservatives, and agricultural pesticides to control certain plant diseases and insect pests (Hoch 2001). Of particular relevance to marine ecosystems is the use of organotins, primarily TBT, in antifouling paints applied to boat hulls to control organisms such as barnacles, mussels, and other invertebrates that create roughness and reduce speed (Hoch 2001). The toxicity and environmental impacts of organotins as antifoulants led to their restriction in many countries in the late 1980s. An international treaty went into effect in 2008 requiring signatories to prohibit and/or restrict the use of harmful organotins in antifouling paints on their ships and ships that enter their ports (Sonak et al. 2009).

TBT is moderately lipophilic, bioaccumulates, and has been shown to cause developmental and reproductive effects in invertebrates and fish (Hoch 2001, Arctic Monitoring and Assessment Programme 2004, Antizar-Ladislao 2008). Less is known about the toxicity of organotins in birds, but TBT oxide at 60 and 150 mg/kg of diet resulted in decreased hatchability and increased embryonic mortality in Japanese Quail, although no overt effects were noted in parent quail (Coenen et al. 1992). Dietary concentrations of 50 mg tin as trimethyltin chloride per kg food caused tremors, ataxia, and lethargy in Mallard ducklings, 100% mortality within 5 days, and histopathologic lesions in the brain and spinal cord (Fleming et al. 1991). Organotins are typically measured in soft tissues, such as liver and kidney, of birds but higher concentrations have been reported in feathers, suggesting a natural excretion mechanism (Guruge et al. 1996). Analytical results in tissues are typically reported as the amount of organotin compound per gram of tissue or the amount of tin (Sn) per gram of tissue.

Kannan and Falandysz (1997) reported 280 and 4600 ng/g ww total butyltins (sum of MBT, DBT, and TBT) in livers of two Long-tailed Ducks from the southern Baltic Sea. Total butyltins in livers of Common Eiders sampled in Denmark ranged from 12 to 202 ng/g ww and a Black

TABLE 6.3

Concentrations of PCBs, HCB, ΣDDT (DDT + DDE + DDD), DDE, dieldrin, and chlordane compounds in tissues of sea ducks from representative field studies.

Species and location	Tissue	PCBs	HCB	ΣDDT	DDE	Dieldrin	ΣChlordane	Source
Steller's Eider (P. stelleri)								
United States (Alaska)	Plasma	5–22[a] ww	—[b]	—	—	—	—	Miles et al. (2007)
Spectacled Eider (S. fischeri)								
United States (Alaska)	Eggs	6.60–44.9 ww	—				—	Wang et al. (2005)
King Eider (S. spectabilis)								
Greenland	Fat	3,300 dw	—	—	1,733 dw	—	—	Braestrup et al. (1974)
	Liver	27.3 ww	—	5.7 ww	—	—	16.0 ww	Johansen et al. (2003)
	Muscle	14.5 ww	—	3.89 ww	—	—	4.98 ww	Johansen et al. (2003)
	Liver	—	56 lw	—	—	—	—	Vorkamp et al. (2004)
	Muscle	—	62 lw	—	—	—	—	Vorkamp et al. (2004)
Canada	Muscle	3–31 ww	1–17[c] ww	<1–18 ww	—	<1–2 ww	1–12 ww	Braune and Malone (2006a)
Common Eider (S. mollissima)								
Russia, Norway	Liver	0.8–54.3 ww	0.3–7.3 ww	0.5–18.2 ww	—	—	0.3–6.9 ww	Savinova et al. (1995)
	Muscle	4.5 ww	4.6 ww	2.7 ww	—	—	6.9 ww	Savinova et al. (1995)
	Brain	1.8–10.0 ww	0.3–2.2 ww	0.5–2.8 ww	—	—	0.1–0.8 ww	Savinova et al. (1995)
	Fat	105.1–1788.1 ww	3.2–128.6 ww	69.8–513.6 ww	—	—	4.1–88.6 ww	Savinova et al. (1995)
Poland	Fat	12,000 lw	200 lw	2,500 lw	—	—	—	Falandysz and Szefer (1982)
Norway	Eggs	80 ww	ND[d]	30 ww	20 ww	—	ND	Barrett et al. (1996)
	Eggs	109–880 lw	7.1–26.4 lw	47.4–220 lw	36.3–180 lw	—	9.3–74.7 lw	Herzke et al. (2009)
Sweden	Eggs	108.4 ww	—	20.1 ww	—	—	—	Carlsson et al. (2011)
Svalbard	Liver	ND	2 ww	—	11 ww	—	—	Norheim and Kjos-Hanssen (1984)
	Fat	3,100 ww	130 ww	—	680 ww	—	—	Norheim and Kjos-Hanssen (1984)
	Liver	10–150 ww	—	—	—	—	—	Mehlum and Daelemans (1995)
	Yolk sac	262 lw	50.1 lw	—	78.4 lw	—	16.0[e] lw	Murvoll et al. (2007)

Spitsbergen	Muscle	2,600 lw	—	—	980 lw	—	—	Andersson et al. (1988)
Greenland	Fat	2,000 dw (1 bird)	—	—	800 dw (1 bird)	—	—	Braestrup et al. (1974)
	Liver	25.5 ww	—	5.78 ww	—	—	12.8 ww	Johansen et al. (2003)
	Muscle	23.0 ww	—	3.84 ww	—	—	4.57 ww	Johansen et al. (2003)
	Liver	71 lw	—	—	—	—	—	Vorkamp et al. (2004)
	Muscle	50 lw	—	—	—	—	—	Vorkamp et al. (2004)
Iceland	Eggs	560 lw	34 lw	—	170 lw	—	22[f] lw	Jörundsdóttir et al. (2010)
Finland	Eggs	—	ND	8.9–60.5[g] ww	8.9–60.5 ww	ND	ND	Franson et al. (2000b)
Canada (eastern)	Eggs	520–4,670[a] ww	—	—	290–590[a] ww	290–590[a] ww	10–20[a] ww	Pearce et al. (1979)
Canada (Nunavut)	Liver	240 ww	—	3 ww	—	—	7 ww	Muir et al. (1992)
	Liver	2.37–9.28[a] ww	1.46–4.26[a,c] ww	1.21–4.99[a] ww	—	—	3.12–12.68[a] ww	Mallory et al. (2004a)
	Muscle	1.74–3.29[a] ww	1.08–1.61[a,c] ww	1.65–2.17[a] ww	—	—	1.28–1.34[a] ww	Mallory et al. (2004a)
Canada	Muscle	ND–97 ww	ND–8[c] ww	ND–26 ww	—	ND–4 ww	ND–13 ww	Braune and Malone (2006a)
United States (Maine)	Eggs	1,600 ww	—	—	230 ww	—	—	Szaro et al. (1979)
United States (Alaska)	Eggs	15.12 ww	7.47 ww	8.02[g] ww	8.02 ww	2.55 ww	4.8–20.3	Franson et al. (2004)
Harlequin Duck (*H. histrionicus*)								
Greenland	Fat	3,540 dw	—	—	1,220 dw	—	—	Braestrup et al. (1974)
United States (Alaska)	Plasma	7–19 ww	—	—	—	—	—	Miles et al. (2007)
Surf Scoter (*M. perspicillata*)								
Canada (BC)	Liver	159[h] ww	1.3[h] ww	—	19.7[h] ww	—	1.6[h] ww	Elliott and Martin (1998)
	Muscle	14–110[h] ww	1.6–5.3[h] ww	—	16–30[h] ww	—	0.2–0.5[h] ww	Elliott and Martin (1998)
Canada	Muscle	ND–353 ww	<1–12[c] ww	ND–127 ww	—	ND–3 ww	ND–10 ww	Braune and Malone (2006a)
Canada (BC)	Liver	2–33[a] ww	ND–2 ww	2–18[a] ww	2–18[a] ww	—	ND–2 ww	Wilson et al. (2010)

(Continued)

TABLE 6.3 (Continued)

Concentrations of PCBs, HCB, ΣDDT (DDT + DDE + DDD), DDE, dieldrin, and chlordane compounds in tissues of sea ducks from representative field studies.

Species and location	Tissue	PCBs	HCB	ΣDDT	DDE	Dieldrin	ΣChlordane	Source
Surf Scoter								
United States (New York; 1 bird)	Liver	690 ww	—	—	ND	—	—	Kim et al. (1984)
	Muscle	710 ww	—	—	ND	—	—	Kim et al. (1984)
	Fat	14,000 ww	—	—	ND	—	—	Kim et al. (1984)
United States (California)	Carcass	947–2,770[a] ww	—	—	298–2,390[a] ww	110–380 ww	—	Ohlendorf et al. (1991)
White-winged Scoter (M. fusca)								
Poland	Fat	19,000–22,000[a] lw	450–530[a] lw	3,200–5,200[a] lw	—	—	—	Falandysz and Szefer (1982)
Canada (BC)	Muscle	7.61[b] ww	1.4[b] ww	—	8.1[b] ww	<0.3[b] ww	—	Elliott and Martin (1998)
Canada	Muscle	ND–395 ww	<1–8[c] ww	ND–46 ww	—	ND–5 ww	ND–10 ww	Braune and Malone (2006a)
United States (New York)	Liver	470–830 ww	—	—	ND–100 ww	—	—	Kim et al. (1984)
	Muscle	220–940 ww	—	—	90–140 ww	—	—	Kim et al. (1984)
	Brain	260 ww	—	—	ND	—	—	Kim et al. (1984)
	Fat	5,400–11,000 ww	—	—	ND–170 ww	—	—	Kim et al. (1984)
Black Scoter (M. nigra)								
Poland	Fat	9,800–15,000[a] lw	300–380[a] lw	1,800–3,100[a] lw	—	—	—	Falandysz and Szefer (1982)
Canada	Muscle	1–144 ww	<1–3[c] ww	<1–69 ww	—	ND–5	ND–7 ww	Braune and Malone (2006a)
Long-tailed Duck (C. hyemalis)								
Poland	Fat	13,000–18,000[a] lw	700–990[a] lw	2,300–5,400[a] lw	—	—	—	Falandysz and Szefer (1982)
Greenland	Fat	4,333 dw	—	—	1,033 dw	—	—	Braestrup et al. (1974)
Canada	Muscle	3–1,080 ww	1–36[c] ww	<1–433 ww	—	ND–120 ww	1–102 ww	Braune and Malone (2006a)

Location	Tissue						Reference
Canada (Hudson Bay)	Eggs	48,000 ww	—	7,600 ww	—	—	Muir et al. (1992)
	Carcass	18,000–25,000[a] ww	6400–6500[a] ww	—	—	—	Muir et al. (1992)
Canada (NT)	Carcass	6,880 ww	—	1,090 ww	510 ww	—	Johnstone et al. (1996)
Lake Michigan (United States)	Carcass	4,000–107,000[a] ww	2,000–42,000[a] ww	—	—	—	Peterson and Ellarson (1978)
Bufflehead (B. albeola)							
Canada (BC)	Liver	315[h] ww	5.4[h] ww	2.7[h] ww	1.1[h] ww	—	Elliott and Martin (1998)
	Muscle	3.54–239[h] ww	1.2–3.2[h] ww	2.0–6.8[h] ww	<0.4[h] ww	—	Elliott and Martin (1998)
Canada	Muscle	ND–48 ww	ND–6[c] ww	ND–8[h] ww	ND–8 ww	ND–18 ww	Braune and Malone (2006a)
United States (New York)	Liver	70–80 ww	—	ND–10 ww	—	—	Kim et al. (1984)
	Muscle	50–580 ww	—	ND–130 ww	—	—	Kim et al. (1984)
	Fat	240–53,000 ww	—	ND	—	—	Kim et al. (1984)
	Muscle	150 ww	20 ww	—	—	—	Foley (1992)
	Fat	6,500 ww	64 ww	1,350 ww	220 ww	291 ww	Foley (1992)
United States (Great Lakes)	Liver	693–3,722[a] ww	30–37[a] ww	—	—	—	Mazak et al. (1997)
Common Goldeneye (B. clangula)							
Poland	Fat	5,500–13,000[a] lw	200–310[a] lw	1,400–2,900[a] lw	—	—	Falandysz and Szefer (1982)
Finland	Liver	310–340[a] ww	—	120–230[a] ww	—	—	Särkkaä et al. (1978)
	Muscle	150–220[a] ww	—	140–160[a] ww	—	—	Särkkaä et al. (1978)
	Eggs	1,350 ww	45 ww	750 ww	—	—	Paasivirta et al. (1981)
Canada (BC)	Muscle	18.1–119[h] ww	1.6–1.8[h] ww	1.8–11[h] ww	<0.4[h] ww	—	Elliott and Martin (1998)
Canada	Muscle	ND–256 ww	<1–114 ww	—	ND–9 ww	ND–22 ww	Braune and Malone (2006a)

(Continued)

TABLE 6.3 (Continued)

Concentrations of PCBs, HCB, ΣDDT (DDT + DDE + DDD), DDE, dieldrin, and chlordane compounds in tissues of sea ducks from representative field studies.

Species and location	Tissue	PCBs	HCB	ΣDDT	DDE	Dieldrin	ΣChlordane	Source
Common Goldeneye								
United States (New York; 1 bird)	Liver	410 ww	—	—	ND	—	—	Kim et al. (1984)
	Brain	330 ww	—	—	ND	—	—	Kim et al. (1984)
	Fat	1,200 ww	—	—	ND	—	—	Kim et al. (1984)
United States (Michigan)	Carcass	7,600 ww	1,700 ww	600 ww	480 ww	—	94 ww	Smith et al. (1985)
United States (New York)	Fat	2,900–34,600[a] ww	40–150[a] ww	510–4,830[a] ww	480–4,600[a] ww	100–690[a] ww	40–250[a,c] ww	Foley and Batcheller (1988)
United States (Minnesota)	Eggs	1,520 ww	—	—	520 ww	—	—	Zicus et al. (1988)
Barrow's Goldeneye (B. islandica)								
Canada (BC)	Liver	138[h] ww	0.6[h] ww	—	3.5[h] ww	3.0[h] ww	—	Elliott and Martin (1998)
	Muscle	57.2[h] ww	2.5[h] ww	—	2.5[h] ww	<0.3[h] ww	—	Elliott and Martin (1998)
Canada	Muscle	2–72 ww	<1–17[c] ww	2–28 ww	—	ND–4 ww	<1–5 ww	Braune and Malone (2006a)
Canada (eastern)	Liver	34.8–236.4[a] ww	2.0–3.7[a] ww	9.9–65.9[a] ww	9.2–57.1[a] ww	1.7–5.4[a] ww	2.4–4.9[c] ww	Ouellet et al. (2012)
Hooded Merganser (L. cucullatus)								
Canada (BC)	Muscle	78.4[h] ww	3.8[h] ww	—	6.2[h] ww	<0.3[h] ww	—	Elliott and Martin (1998)
Canada (QC)	Eggs	425[h] ww	—	306[g,h] ww	306[h] ww	21.3[h] ww	—	Champoux (1996)
Canada	Muscle	2–621 ww	<1–4[c] ww	6–2,610 ww	—	ND–5 ww	ND–50	Braune and Malone (2006a)
United States (14 states)	Eggs	440–4,910[a] ww	—	—	340–13,200[a] ww	ND–1,340 ww	—	White and Cromartie (1977)
United States (Wisconsin)	Eggs	17,600 lw	—	—	14,600 lw	250 lw	—	Faber and Hickey (1973)
	Eggs	920 ww	5 ww	254 ww	229 ww	15 ww	26 ww	Custer et al. (2002)

Location	Tissue							Reference
United States (New York; 1 bird)	Liver	1,300 ww	—	—	230 ww	—	—	Kim et al. (1984)
	Muscle	1,600 ww	—	—	400 ww	—	—	Kim et al. (1984)
	Fat	3,900 ww	—	—	1,100 ww	—	—	Kim et al. (1984)
United States (Minnesota)	Eggs	660 ww	—	—	620 ww	—	—	Zicus et al. (1988)
Common Merganser (*M. merganser*)								
Poland (1 bird)	Liver	2,800 ww	140 ww	1,600 ww	—	—	—	Dubrawski and Falandyz (1980)
	Muscle	1,900 ww	90 ww	1,500 ww	—	—	—	Dubrawski and Falandyz (1980)
	Fat	62,000 ww	2,700 ww	47,000 ww	—	—	—	Dubrawski and Falandyz (1980)
	Fat	79,000 lw	450 lw	9,700 lw	—	—	—	Falandysz and Szefer (1984)
Denmark	Eggs	13,500 ww	—	—	3,900 ww	120 ww	—	Hansen and Kraul (1981)
Canada (BC)	Liver	78–542[h] ww	14[h] ww	—	110–229[h] ww	3.3–4.4[h] ww	—	Elliott and Martin (1998)
	Muscle	38.3–337[h] ww	7.4–9.8[h] ww	—	44–140[h] ww	0.8–1.6[h] ww	—	Elliott and Martin (1998)
Canada (QC)	Eggs	12,983[h] ww	—	1,771[h] ww	1,720[h] ww	572[h] ww	—	Champoux (1996)
Canada	Muscle	23–2,440 ww	<1–27[c] ww	5–834 ww	—	ND–172 ww	<1–116 ww	Braune and Malone (2006a)
United States (Michigan, Wisconsin)	Eggs	260,200 lw	—	—	110,600 lw	7,400 lw	—	Faber and Hickey (1973)
United States (New York)	Liver	770–2,900 ww	—	—	60–770	—	—	Kim et al. (1984)
	Liver	81–14,200 ww	—	—	—	—	—	O'Keefe et al. (2006)
	Muscle	610–20,000 ww	—	—	ND–2,000 ww	—	—	Kim et al. (1984)
	Brain	300–2,300 ww	—	—	ND–200 ww	—	—	Kim et al. (1984)
	Fat	1,700–9,800 ww	—	—	180–2,000 ww	—	—	Kim et al. (1984)
United States (Wisconsin)	Eggs	79,430 ww	ND	—	24,440 ww	640 ww	820 ww	White and Cromartie (1977)

TABLE 6.3 (Continued)

Concentrations of PCBs, HCB, ΣDDT (DDT + DDE + DDD), DDE, dieldrin, and chlordane compounds in tissues of sea ducks from representative field studies.

Species and location	Tissue	PCBs	HCB	ΣDDT	DDE	Dieldrin	ΣChlordane	Source
Red-breasted Merganser (M. serrator)								
Poland	Fat	4,300–320,000 lw	89–970 lw	820–38,000 lw	680–38,000 lw	—	—	Falandysz and Szefer (1984)
Finland	Liver	1,810–1,890ᵃ ww		1,770–3,160ᵃ ww				Särkkää et al. (1978)
	Muscle	180–1,930ᵃ ww		80–2,730ᵃ ww				Särkkää et al. (1978)
Canada	Muscle	6–1,700 ww	<1–12ᶜ ww	2–951 ww	—	ND–48 ww	ND–37 ww	Braune and Malone (2006a)
United States (Lake Michigan)	Eggs	489,000 lw	—	—	257,100 lw	4,460 lw	—	Faber and Hickey (1973)
United States (New York)	Liver	740–1,700 ww	—	—	ND–130 ww	—	—	Kim et al. (1984)
	Muscle	710–740 ww	—	—	ND	—	—	Kim et al. (1984)
	Brain	110–430 ww	—	—	ND	—	—	Kim et al. (1984)
	Fat	1,200–26,000 ww	—	—	170–2,600 ww	—	—	Kim et al. (1984)
United States (Wisconsin)	Eggs	44,670 ww	110 ww	—	15,730 ww	1,000 ww	570 ww	White and Cromartie (1977)

Mean, minimum–maximum, or as otherwise specified, expressed as ng/g (ppb) ww (wet weight or fresh weight), lw (lipid weight), or dw (dry weight).

ᵃ Minimum–maximum of 2 or more means or a single value and 1 mean.

ᵇ Not analyzed or results not reported.

ᶜ Includes HCB, pentachlorobenzene, and tetrachlorobenzene.

ᵈ ND = not detected.

ᵉ Oxychlordane only.

ᶠ Trans-nonachlor only.

ᵍ Consists of DDE only; DDT and DDD below detection limits.

ʰ Values from pooled samples.

Scoter had 20 ng/g ww in its liver (Strand and Jacobsen 2005). Total butyltin concentrations in livers of Surf Scoters, Common Mergansers, Long-tailed Ducks, and Harlequin Ducks from coastal British Columbia were 28–1100, 229–381, 654, and 241 ng/g ww, respectively (Kannan et al. 1998). In a later study in British Columbia, mean total butyltins were 46.7 and 34.8 ng Sn/g ww, or ~102 and 72 ng organotin/g ww (Meador 2011), in female and male Surf Scoters, respectively (Elliott et al. 2007). In the studies of sea ducks where butyltins were measured separately, concentrations of the more toxic TBT were considerably lower, or below detection limits, in relation to MBT and DBT (Kannan et al. 1998, Strand and Jacobson 2005, Elliott et al. 2007).

Organotins break down more slowly in sediments than water, and contaminated sediments may serve as a long-term reservoir of these compounds, particularly in or near ports, harbors, shipping channels, and other areas where organotin-based antifouling coatings have been used (Hoch 2001, Antizar-Ladislao 2008). Some sea ducks may winter in the vicinity of such contaminated areas and could be exposed to elevated levels of TBT in their diet because blue mussels (*Mytilus edulis*), for example, tend to accumulate greater concentrations of TBT than the less toxic MBT and DMT (Hong et al. 2002, Rüdel et al. 2003). Maximum concentrations of TBT (ww basis) in blue mussels, for example, reported in a number of studies included 153 ng/g in British Columbia, 210 ng/g in the North Sea, 340 ng/g in the Baltic Sea, and 1200 ng/g in coastal Korea (Hong et al. 2002, Rüdel et al. 2003). These concentrations (0.153–1.2 mg/kg) are considerably less than the level of 60 mg TBT per kg of food associated with adverse effects in Japanese Quail (Coenen et al. 1992). However, sea ducks could also be impacted by reduced populations of prey, because TBT can cause impaired reproduction and direct mortality in molluscs (Meador 2011).

Criteria have been established for the identification and listing of new and emerging POPs based on persistence, bioaccumulation, potential for long-range environmental transport, and adverse effects (United Nations Environmental Programme 2001). Similar to legacy POPs, some of these are present in arctic environments where they have never been used, for example, polybrominated diphenyl ethers (PBDEs) and perfluorinated organic compounds (PFCs), the most widely known of which are perfluorooctane sulfonate (PFOS) and perfluorooctanoic acid (PFOA, Arctic Monitoring and Assessment Programme 2009). PBDEs are components of brominated fire retardants used in computer casings and monitors, television sets, other electronic equipment, electrical cables, switches, building materials, and textiles and enter the environment when these products are incinerated or deposited in landfills (Darnerud et al. 2001). PBDEs are widely transported and distributed to remote areas of the environment, are lipophilic, and have been shown to bioaccumulate, and at least some congeners tend to biomagnify in food chains (Darnerud et al. 2001, de Wit 2002, de Wit et al. 2010). Field reports from a variety of biota generally indicate sharp increases in PBDE burdens since the 1970s and 1980s, but with evidence of leveling off in at least some species during the late 1990s and early 2000s (Ikonomou et al. 2002, Braune et al. 2005b, Elliott et al. 2005, Law et al. 2006, Helgason et al. 2009). Experimental studies with birds have reported evidence of effects of PBDEs on the immune system, thyroxine and retinol concentrations, glutathione metabolism and oxidative stress, and reproductive behaviors (Fernie et al. 2005a,b, 2008). Other adverse effects of PBDEs varied among avian species and included decreased pipping of eggs and hatching success, induction of EROD, and histological changes in the bursa of Fabricius (McKernan et al. 2009). Furthermore, in vitro testing has shown that certain brominated fire retardants have endocrine disrupting potencies (Hamers et al. 2006). Few published reports exist for PBDEs in sea ducks, but studies with Common Eiders show relatively low levels as compared with seabirds. In a study at Svalbard in the Norwegian arctic, Murvoll et al. (2007) reported total PBDE concentrations of 2 ng/g lw in yolk sacs of hatchling Common Eiders but 90 ng/g lw in yolk sacs of Brünnich's Guillemot (*Uria lomvia*). At a remote site in Norway, PBDE concentrations in eggs of Common Eider and European Shag (*Phalacrocorax aristotelis*) were 5.5–7.2 and 90–97 ng/g lw, respectively (Herzke et al. 2009). In the same study, Common Eider eggs collected near Trondheim, Norway, showed more evidence of PBDE exposure with total PBDE concentrations of 27.4–48.2 ng/g lw. Carlsson et al. (2011) reported total PBDE concentrations of 0.6 ng/g ww in eggs of Common Eiders in Sweden. The concentrations reported in these sea duck eggs are much lower than the lowest level (1,800 ng/g ww or 32,000 ng/g lw) of PBDEs found to impair pipping and hatching success in American Kestrels (*Falco sparverius*), and Mallards subjected to the same exposure exhibited no effects (McKernan et al. 2009). In a Canadian arctic study, White-winged Scoters had 71 ng/g lw of total PBDE concentrations in their livers, compared with 20 ng/g lw in livers of Common Eiders (Kelly et al. 2008). Liver PBDE concentrations were 2.34–19.25 ng/g ww in Surf Scoters sampled in British Columbia, Canada, and 4.02 ng/g ww in Barrow's Goldeneyes from the St. Lawrence ecosystem in eastern Canada (Wilson et al. 2010, Ouellet et al. 2012).

PFCs have been used in various industrial and commercial applications such as lubricants, refrigerants, adhesives, fire fighting foams, and paints; two of the most widely known of this class of chemicals, PFOS and PFOA, have been found in water and biota in many parts of the world as a result of long-range global transport (Simcik 2005). PFOS is often found in wildlife tissues, but PFOA is detected only occasionally and at lower concentrations (Martin et al. 2004). Experimental studies with mammals indicate that PFCs act as endocrine disrupters by altering testosterone and estradiol levels; PFOS causes liver enlargement, reduced serum cholesterol, and reproductive toxicity; and both PFOS and PFOA promote liver tumors (Arctic Monitoring and Assessment Programme 2004, Jensen and Leffers 2008). Little information is available on PFC toxicity in birds, although Newsted et al. (2005) reported that, in Mallards, the lowest observed adverse effect level (LOAEL) in an acute toxicity study was 48 μg/mL in serum and 30 μg/g in liver. The authors also reported that the no observed adverse effect level (NOAEL) in a reproductive study of Mallards was 53 μg/mL in egg yolk (the LOAEL was not provided). In field collections, concentrations of PFOS in blood of White-winged Scoters, Common Eiders, and Long-tailed Ducks from the Baltic Sea were in the pg/mL range, far less than the LOAEL in Mallards, and PFOA concentrations in blood of the same three species were <1 pg/mL (Newsted et al. 2005, Falandysz et al. 2007). Common Eider eggs collected near Trondheim, Norway, had PFOS and PFOA concentrations of 29 and 3.2 ng/g ww, respectively, whereas eggs from remote islands off the Norwegian coast had 15 ng/g ww PFOS and 1.4 ng/g ww PFOA

(Herzke et al. 2009). Again, these concentrations are far less than the NOAEL calculated for Mallards (Newsted et al. 2005). In Common Eider eggs collected in Sweden, total PFCs were 91.1 ng/g ww (Carlsson et al. 2011). Considering the sum of POPs and PFCs, the relative contribution of PFCs was higher in Common Eider eggs than European Shag eggs, and the authors attributed these findings to the two species feeding at different trophic levels, different properties of lipophilic POPs, and different breeding strategies (Herzke et al. 2009). In liver tissues from 10 species of waterfowl sampled in New York, including Hooded Merganser (*Lophodytes cucullatus*), Surf Scoter, Bufflehead, Common Goldeneye, and Common Merganser, PFOS was the most abundant PFC and occurred in all samples, whereas PFOA was not found in any bird livers (Sinclair et al. 2006). Birds categorized as piscivorous had greater PFOS levels than nonpiscivorous birds, and the highest hepatic concentrations were 635 and 441 ng/g ww in Buffleheads and Common Mergansers, respectively, considerably lower than the LOAEL reported for Mallards (Newsted et al. 2005, Sinclair et al. 2006).

Because of long-range transport and long half-lives, many POPs that have not been used in years are still measurable in the environment and animals in remote areas. In addition, certain organochlorine pesticides, such as DDT, continue to be used for disease vector control in some countries. Overall, however, there is evidence for a decline of legacy organochlorines as reported, for example, in marine biota in arctic Canada between the 1970s and 1990s (Braune et al. 2005b). Although evidence of exposure to legacy POPs in sea ducks as a group has generally been low, these chemicals still should be considered in contaminant evaluations because of the potential for point source emissions and the fact that piscivorous species, in particular, are at risk because of food chain biomagnification. While legacy POPs decline slowly because of global cycling and long half-lives, recently emitted new types of persistent chemicals such as PBDEs may increase in step with emissions and the question remains as to the extent that temporal trends in POPs measured at various locations apply throughout the circumpolar arctic (Arctic Monitoring and Assessment Programme 2004). The potential impact of these emerging POPs on sea ducks is unclear.

ORGANOPHOSPHORUS AND CARBAMATE PESTICIDES

Organophosphorus and carbamate pesticides, also referred to as anticholinesterase or cholinesterase-inhibiting pesticides, are used to control a variety of agricultural pests and to prevent livestock and human diseases. This group of chemicals includes some 200 organophosphorus compounds, such as diazinon, famphur, fenthion, malathion, parathion, and 50 carbamates, including carbaryl, carbofuran, methiocarb, and methomyl (Hill 2003). Although highly toxic, these two classes of compounds are shorter lived in the environment than the more persistent chlorinated hydrocarbons and do not accumulate in food chains (Eisler 2000b, Hill 2003). Although certain uses of organophosphorus and carbamate pesticides have been prohibited or restricted in some countries because of adverse effects in wildlife, many are still considered to be in the category of current use rather than legacy pesticides. In general, wildlife are exposed to these pesticides through ingestion of contaminated water, soil, foliage, invertebrates and other food items, treated seeds, pesticide granules, and less commonly from dermal exposure and inhalation (Hill 2003).

Organophosphorus and carbamate compounds exert their toxic effects through inhibition of the enzyme cholinesterase, leading to an accumulation of acetylcholine in the nervous system and resulting in overstimulation of neurons, eventual paralysis of respiratory muscles, and acute mortality from suffocation (Fairbrother 1996). Thus, measurement of cholinesterase activity in the brain of dead animals or in blood of live animals is a sensitive biomarker of organophosphorus and carbamate pesticide exposure, and a definitive diagnosis of poisoning is characterized by cholinesterase inhibition of >50% and supported by the finding of pesticide residues in gut contents (Fairbrother 1996). A technique referred to as thermal reactivation is used to distinguish between organophosphorus and carbamate compounds. When the cholinesterase activity in the initial test is inhibited compared with controls, the brain homogenate is heated and retested. If a carbamate was involved, the cholinesterase activity characteristically increases, whereas it remains inhibited if an organophosphate was the cause of poisoning (Smith et al. 1995). Anticholinesterase compounds have been determined to be the cause of many avian mortality events, and experimental studies with nonfatal doses have shown a variety of behavioral effects, including alterations of general activity and alertness, aggression, foraging, navigation, and reproduction (Hill 2003, Fleischli et al. 2004). Of relevance to sea ducks is the potential for effects of organophosphorus compounds on salt gland function and, hence, osmoregulation. Herin et al. (1978) reported an inverse relationship between chlorpyrifos and salt gland function in Mallard ducklings. In Black Ducks, Eastin et al. (1982) reported fenthion altered enzyme activity in salt gland, but Rattner et al. (1983) concluded that exposure to environmentally realistic levels of organophosphorus pesticides did not affect osmoregulatory function in Black Ducks. Because of their relatively short persistence in the environment, the chronic toxicity of organophosphorus and carbamate pesticides has not been extensively studied in wildlife, and although these pesticides have not caused the kind of population effects attributed to organochlorines, questions remain regarding potential long-term hazards of repeated exposure, particularly at different stages in the reproductive cycle (Hill 2003).

Organophosphorus and carbamate pesticides are frequently used in terrestrial landscapes, and the risk of exposure to sea ducks in marine environments is low. However, sea ducks could potentially encounter these pesticides as a result of runoff, drift, or from applications to estuarine wetlands in some parts of the world for mosquito or other pest control. Furthermore, organophosphorus pesticides such as chlorpyrifos, diazinon, terbufos, and methyl parathion have been found in biota and environmental samples from subarctic and arctic Alaska and Canada (Arctic Monitoring and Assessment Programme 2009). Although little information is available on these compounds affecting sea ducks, one report lists a Bufflehead found dead during a mortality event attributed to anticholinesterase pesticides (Fleischli et al. 2004).

RADIATION

Radioactivity is the property of isotopes of elements or radionuclides that spontaneously emit radiation by decay of their unstable nuclei (Meyers-Schöne and Talmage 2003). As they decay, radionuclides release energy from the emission of alpha, beta, or neutron particles, or electromagnetic radiation as gamma or x-rays, collectively referred to as ionizing radiation because of an ability to produce charged particles in matter through which they pass (Sample and Irvine 2011). Adverse effects of ionizing radiation include cell death, organ and tissue damage, reduced growth, altered behavior, increased frequency of tumors and gene mutations, and lowered reproduction. In most species tested, including birds, there is a direct relationship between dose and mortality at high single exposures to radiation (Eisler 2000c, Sample and Irvine 2011). Ionizing radiation of particular importance in avian radioecology includes alpha particles that can cause great biological damage but have low penetrating power, beta particles that have little mass but greater penetrating power, and gamma rays of high energy and great ability to penetrate biological materials (Brisbin 1991).

Natural background sources of ionizing radiation include sunlight, cosmic rays, and naturally occurring radionuclides. Anthropogenic sources of radiation include radionuclides released from uranium mining, nuclear fuel production, operation of nuclear reactors, nuclear accidents, waste storage sites, and nuclear weapon testing (Eisler 2000c, Meyers-Schöne and Talmage 2003, Sample and Irvine 2011). Radioactive materials may move through the environment in the atmosphere depending on the magnitude and direction of wind currents; through aquatic systems according to a variety of factors including water depth, temperature, tides, and groundwater characteristics; and by biological systems as animals disperse accumulated radionuclides along their migratory pathways or within their home ranges (Eisler 2000c). The main sources of radioactive contamination in arctic regions are global fallout from atmospheric nuclear weapon tests between 1945 and 1980, liquid discharges from nuclear reprocessing plants, and fallout following the Chernobyl accident in 1986 (Strand et al. 2002). Waterfowl apparently eliminate accumulated radionuclides relatively rapidly. In an experiment where Mallards were transferred from liquid radioactive waste ponds to an uncontaminated environment, the estimated half-life of several radionuclides ranged from 10 to 86 days (Halford et al. 1983). Radionuclides in soft tissues were highest in gut, followed by liver and muscle (Halford et al. 1983).

Studies in which sea duck tissues were analyzed for radionuclides have generally reported low or nondetectable activity. Trust et al. (2000b) analyzed muscle tissue of Spectacled Eiders from St. Lawrence Island for ^{137}Cs, but it was below the minimum detectable concentration in all samples. Burger and Gochfeld (2007) measured ^{137}Cs, ^{129}I, ^{60}Co, and ^{152}Eu in muscle of Common Eiders from Amchitka and Kiska Islands in the Aleutians, finding that all were below the minimum detectable concentration. A Red-breasted Merganser collected on the coast of England, in an area thought to be affected by radionuclides, had detectable activities of ^{137}Cs and ^{134}Cs in muscle (144 and 8 becquerels (Bq)/kg ww, respectively) and liver (251 and 13 Bq/kg ww, respectively), whereas $^{239/240}$Pu and ^{238}Pu were near or below the lower limit of detection (<0.01–0.02 Bq/kg, ww) (Lowe 1991). The ^{137}Cs concentrations in muscle tissues of two Common Goldeneyes collected in Ireland after the Chernobyl accident were 9 and 16 Bq/kg (ww presumed but not stated, Pearce 1995). A study in the Baltic Sea found that waterbird species that consumed a varied diet had higher levels of ^{210}Po than birds that ate mainly fish, and concentrations of ^{210}Po in liver of Long-tailed Ducks, White-winged Scoters, and Common Eiders were 19, 5, and 3 Bq/kg ww, respectively (Skwarzec and Fabisiak 2007). The authors also conducted a study showing that >63% of ^{210}Po in feathers was adsorbed, presumably from the atmosphere (Skwarzec and Fabisiak 2007). Concentrations of ^{137}Cs, ^{90}Sr, and $^{239/240}$Pu in muscle of Common Eiders from the Barents Sea were <0.5 Bq/kg dw (Matishov et al. 2003). In comparison, the mean ^{137}Cs concentration in livers from American Coots (Fulica americana) collected in the 1970s from a wastewater pond at a nuclear facility in Washington State, United States, was 440 picocurie/g dw, or about 16,280 Bq/kg dw (Cadwell et al. 1979).

According to the Arctic Monitoring and Assessment Programme (2010), progress has been made in the reduction of threats and risks of radiation exposure in the arctic, including control of planned discharges, mitigation of accident risks, and the development of systems to manage nuclear legacy sites. Although declining in recent years, of continuing potential concern for sea ducks are discharges from nuclear reprocessing plants into marine environments, and thereby sediments, particularly because climate change may influence the remobilization of existing radioactive materials from the seabed to the water column (Arctic Monitoring and Assessment Programme 2010). In recent years, efforts have been devoted to the development of tools for assessing risk of exposure to radiation in animals, some of which have been summarized by Sample and Irvine (2011).

SUMMARY

Except for lead and oil, few contaminants have been diagnosed as the cause of death for sea ducks or shown to be associated with population level effects. However, the impacts of sublethal contaminant exposure that could make sea ducks more subject to predation, prevent adequate foraging, impair reproduction, or cause immunosuppression rendering them more susceptible to infectious diseases are not well understood. Thus, a consideration of contaminants should be part of any thorough evaluation of population health. The interpretation of tissue residues of many contaminants can be a challenge in any avian species, but with sea ducks, this is complicated further by the fact that most experimental laboratory studies on which interpretive recommendations are based have been carried out in freshwater species. Field studies, however, have shown that marine birds carry greater burdens of some pollutants than freshwater species, particularly metals. For example, tissue concentrations of selenium and mercury that may be considered toxic in wild freshwater birds may result in no apparent adverse effects in sea ducks. In any case, comparisons of tissue residues from experimental studies with field data should be done with caution because wild birds with potentially marginal diets and exposed to natural environmental conditions may be more susceptible to contaminant poisoning than birds in controlled situations. Even when interpretive guidelines for a particular contaminant, such as lead, are available, poisoning as a cause of death should be distinguished from exposure based solely on tissue residues. When investigating possible contaminant mortality, a thorough evaluation of field circumstances, any observed clinical signs, gross lesions and pathological findings, tissue residues, and other laboratory testing to rule out potential infectious or parasitic diseases and other conditions is ideal because few signs and lesions are specific to contaminants.

Much of the published literature on contaminants in sea ducks consists of surveys that evaluate tissue residues in relation to known contaminant issues, studies that address hypotheses regarding the effects of contaminants, and reports that provide what are considered to be normal background concentrations at specific geographic locations. Such studies are particularly useful in building a knowledge base for the effects of contaminants in sea ducks when they relate tissue residues to effects on specific physiological or other parameters, known sources and magnitude of exposure, and temporal changes in contaminants in the broader environment. Studies of a variety of species, including some sea ducks, have shown that levels of certain long-lived contaminants have declined in the environment, probably in large part because of restrictions on their use and replacement with shorter-lived or less toxic chemicals. However, new and emerging chemicals with the potential to be problematic continue to be identified as they appear in biota far removed from their sources of emission. Last, global climate change will influence the ecotoxicology of pollutants and alter natural systems in ways that may add new stressors impacting the effects of contaminants in sea ducks and other wildlife.

LITERATURE CITED

Aarab, N., D. M. Pampanin, A. Nævdal, K. B. Øysæd, L. Gastaldi, and R. K. Bechmann. 2008. Histopathology alterations and histochemistry measurements in mussel, Mytilus edulis collected offshore from an aluminum smelter industry (Norway). Marine Pollution Bulletin 57:569–574.

Adrian, W. J., and M. L. Stevens. 1979. Wet versus dry weights for heavy metal toxicity determinations in duck liver. Journal of Wildlife Diseases 15:125–126.

Albers, P. H. 2003. Petroleum and individual polycyclic aromatic hydrocarbons. Pp. 341–371 in D. J. Hoffman, B. A. Rattner, G. A. Burton Jr., and J. Cairns Jr. (editors), Handbook of ecotoxicology, 2nd ed. Lewis Publishers, Boca Raton, FL.

Albers, P. H. 2006. Birds and polycyclic aromatic hydrocarbons. Avian and Poultry Biology Reviews 17:125–140.

Albers, P. H., D. E. Green, and C. J. Sanderson. 1996. Diagnostic criteria for selenium toxicosis in aquatic birds: dietary exposure, tissue concentrations, and macroscopic effects. Journal of Wildlife Diseases 32:468–485.

Albers, P. H., and R. C. Szaro. 1978. Effects of No. 2 fuel oil on Common Eider eggs. Marine Pollution Bulletin 9:138–139.

Anderson, W. L. 1975. Lead poisoning in waterfowl at Rice Lake, Illinois. Journal of Wildlife Management 39:264–270.

Anderson, W. L., S. P. Havera, and R. A. Montgomery. 1987. Incidence of ingested shot in waterfowl in the Mississippi flyway, 1977–1979. Wildlife Society Bulletin 15:181–188.

Anderson, W. L., S. P. Havera, and B. W. Zercher. 2000. Ingestion of lead and nontoxic shotgun pellets by ducks in the Mississippi flyway. Journal of Wildlife Management 64:848–857.

Andersson, Ö., C. Linder, M. Olsson, L. Reutergårdh, U. Uvemo, and U. Wideqvist. 1988. Spatial differences and temporal trends of organochlorine compounds in biota from the Northwestern Hemisphere. Archives of Environmental Contamination and Toxicology 17:755–765.

Annett, C. S., F. M. D'Itri, J. R. Ford, and H. H. Prince. 1975. Mercury in fish and waterfowl from Ball Lake, Ontario. Journal of Environmental Quality 4:219–222.

Antizar-Ladislao, B. 2008. Environmental levels, toxicity and human exposure to tributyltin (TBT)-contaminated marine environment. A review. Environmental International 34:292–308.

Arctic Monitoring and Assessment Programme. 1998. AMAP assessment report: arctic pollution issues. Arctic Monitoring and Assessment Programme (AMAP), Oslo, Norway.

Arctic Monitoring and Assessment Programme. 2004. AMAP assessment 2002: persistent organic pollutants (POPs) in the arctic. Arctic Monitoring and Assessment Programme (AMAP), Oslo, Norway.

Arctic Monitoring and Assessment Programme. 2005. AMAP assessment 2002: heavy metals in the arctic. Arctic Monitoring and Assessment Programme (AMAP), Oslo, Norway.

Arctic Monitoring and Assessment Programme. 2007. Arctic oil and gas 2007. Arctic Monitoring and Assessment Programme (AMAP), Oslo, Norway.

Arctic Monitoring and Assessment Programme. 2009. Arctic pollution 2009. Arctic Monitoring and Assessment Programme (AMAP), Oslo, Norway.

Arctic Monitoring and Assessment Programme. 2010. AMAP assessment 2009: radioactivity in the arctic. Arctic Monitoring and Assessment Programme (AMAP), Oslo, Norway.

Avery, D., and R. T. Watson. 2009. Regulation of lead-based ammunition around the world. Pp. 161–168 in R. T. Watson, M. Fuller, M. Pokras, and W. G. Hunt (editors), Ingestion of lead from spent ammunition: implications for wildlife and humans. The Peregrine Fund, Boise, ID.

Baker, F. D., C. F. Tumasonis, W. B. Stone, and B. Bush. 1976. Levels of PCB and trace metals in waterfowl in New York state. New York Fish and Game Journal 23:82–91.

Banks, A. N., W. G. Sanderson, B. Hughes, P. A. Cranswick, L. E. Smith, S. Whitehead, A. J. Musgrove, B. Haycock, and N. P. Fairney. 2008. The Sea Empress oil spill (Wales, UK): effects on Common Scoter Melanitta nigra in Carmarthen Bay and status ten years later. Marine Pollution Bulletin 56:895–902.

Bard, S. M. 1999. Global transport of anthropogenic contaminants and the consequences for the arctic marine ecosystem. Marine Pollution Bulletin 38:356–379.

Barjaktarovic, L., J. E. Elliott, and A. M. Scheuhammer. 2002. Metal and metallothionein concentrations in scoter (Melanitta spp.) from the Pacific Northwest of Canada, 1989–1994. Archives of Environmental Contamination and Toxicology 43:486–491.

Barrett, R. T., J. U. Skaare, and G. W. Gabrielsen. 1996. Recent changes in levels of persistent organochlorines and mercury in eggs of seabirds from the Barents Sea. Environmental Pollution 92:13–18.

Bellrose, F. C. 1959. Lead poisoning as a mortality factor in waterfowl populations. Illinois Natural History Survey Bulletin 27:235–288.

Bendell, L. I. 2011. Sea ducks and aquaculture: the cadmium connection. Ecotoxicology 20:474–478.

Bennett, D. C., M. R. Hughes, J. E. Elliott, A. M. Scheuhammer, and J. E. Smits. 2000. Effect of cadmium on Pekin Duck total body water, water flux, renal filtration, and salt gland function. Journal of Toxicology and Environmental Health 59:43–56.

Beyer, W. N., J. Dalgarn, S. Dudding, J. B. French, R. Mateo, J. Miesner, L. Sileo, and J. Spann. 2004. Zinc and lead poisoning in wild birds in the tri-state mining district (Oklahoma, Kansas, and Missouri). Archives of Environmental Contamination and Toxicology 48:108–117.

Beyer, W. N., J. C. Franson, L. N. Locke, R. K. Stroud, and L. Sileo. 1998. Retrospective study of the diagnostic criteria in a lead-poisoning survey of waterfowl. Archives of Environmental Contamination and Toxicology 35:506–512.

Beyer, W. N., J. W. Spann, L. Sileo, and J. C. Franson. 1988. Lead poisoning in six captive avian species. Archives of Environmental Contamination and Toxicology 17:121–130.

Blais, J. M. 2005. Biogeochemistry of persistent bioaccumulative toxicants: processes affecting the transport of contaminants to remote areas. Canadian Journal of Fisheries and Aquatic Sciences 62:236–243.

Blus, L. J. 2003. Organochlorine pesticides. Pp. 313–339 in D. J. Hoffman, B. A. Rattner, G. A. Burton Jr., and J. Cairns Jr. (editors), Handbook of ecotoxicology, 2nd ed. Lewis Publishers, Boca Raton, FL.

Blus, L. J. 2011. DDT, DDD, and DDE in birds. Pp. 425–444 in W. N. Beyer and J. P. Meador (editors), Environmental contaminants in biota: interpreting tissue concentrations, 2nd ed. CRC Press, Boca Raton, FL.

Blus, L. J, C. J. Henny, D. J. Hoffman, and R. A. Grove. 1995. Accumulation in and effects of lead and cadmium on waterfowl and passerines in northern Idaho. Environmental Pollution 89:311–318.

Blus, L. J., S. N. Wiemeyer, and C. J. Henny. 1996. Organochlorine pesticides. Pp. 61–70 in A. Fairbrother, L. N. Locke, and G. L. Hoff (editors), Noninfectious diseases of wildlife, 2nd ed. Iowa State University Press, Ames, IA.

Boening, D. W. 1999. An evaluation of bivalves as biomonitors of heavy metals pollution in marine waters. Environmental Monitoring and Assessment 55:459–470.

Bond, A. L., and A. W. Diamond. 2009. Mercury concentrations in seabird tissues from Machias Seal Island, New Brunswick, Canada. Science of the Total Environment 407:4340–4347.

Bond, J. C., D. Esler, and K. A. Hobson. 2007. Isotopic evidence for sources of nutrients allocated to clutch formation by Harlequin Ducks. Condor 109:698–704.

Borgå, K., G. W. Gabrielsen, and J. U. Skaare. 2001. Biomagnification of organochlorines along a Barents Sea food chain. Environmental Pollution 113:187–198.

Borkenhagen, V. P. 1979. Lead-poisoning of waterfowl. Zeitschrift für Jagdwissenschaft 25:178–179.

Braune, B. M. 1987. Comparison of total mercury levels in relation to diet and molt for nine species of marine birds. Archives of Environmental Contamination and Toxicology 16:217–224.

Braune, B. M. 2007. Temporal trends of organochlorines and mercury in seabird eggs from the Canadian Arctic, 1975–2003. Environmental Pollution 148:599–613.

Braune, B. M., and D. E. Gaskin. 1987. Mercury levels in Bonaparte's Gulls (Larus philadelphia) during autumn molt in the Quoddy Region, New Brunswick, Canada. Archives of Environmental Contamination and Toxicology 16:539–549.

Braune, B. M., K. A. Hobson, and B. J. Malone. 2005a. Regional differences in collagen stable isotope and tissue trace element profiles in populations of Long-tailed Duck breeding in the Canadian Arctic. Science of the Total Environment 346:156–168.

Braune, B. M., and B. J. Malone. 2006a. Organochlorines and mercury in waterfowl harvested in Canada. Environmental Monitoring and Assessment 114:331–359.

Braune, B. M., and B. J. Malone. 2006b. Mercury and selenium in livers of waterfowl harvested in northern Canada. Archives of Environmental Contamination and Toxicology 50:284–289.

Braune, B. M., P. M. Outridge, A. T. Fisk, D. C. G. Muir, P. A. Helm, K. Hobbs, P. F. Hoekstra et al. 2005b. Persistent organic pollutants and mercury in marine biota of the Canadian Arctic: an overview of spatial and temporal trends. Science of the Total Environment 351–352:4–56.

Braune, B. M., and A. M. Scheuhammer. 2008. Trace element and metallothionein concentrations in seabirds from the Canadian Arctic. Environmental Toxicology and Chemistry 27:645–651.

Brisbin Jr., I. L., 1991. Avian radioecology. Pp. 69–140 in D. M. Power (editor), Current ornithology (vol. 8). Plenum Press, New York.

Brown, C. S., J. Luebbert, D. Mulcahy, J. Schamber, and D. H. Rosenberg. 2006. Blood lead levels of wild Steller's Eiders (Polysticta stelleri) and Black Scoters (Melanitta nigra) in Alaska using a portable blood lead analyzer. Journal of Zoo and Wildlife Medicine 37:361–365.

Brown, R. G. B., D. I. Gillespie, A. R. Lock, A. Pearce, and G. H. Watson. 1973. Bird mortality from oil slicks off eastern Canada, February–April 1970. Canadian Field-Naturalist 87:225–234.

Brunström, B., D. Broman, and C. Näf. 1990. Embryotoxicity of polycyclic aromatic hydrocarbons (PAHs) in three domestic avian species, and of PAHs and coplanar polychlorinated biphenyls (PCBs) in the Common Eider. 1990. Environmental Pollution 67:133–143.

Brunström, B., D. Broman, and C. Näf. 1991. Toxicity and EROD-inducing potency of 24 polycyclic aromatic hydrocarbons (PAHs) in chick embryos. Archives of Toxicology 65:485–489.

Burger, J. 1993. Metals in avian feathers: bioindicators of environmental pollution. Reviews in Environmental Toxicology 5:203–311.

Burger, J., and M. Gochfeld. 2007. Metals and radionuclides in birds and eggs from Amchitka and Kiska Islands in the Bering Sea/Pacific Ocean ecosystem. Environmental Monitoring and Assessment 127:105–117.

Burger, J., and M. Gochfeld. 2009. Mercury and other metals in feathers of Common Eider (*Somateria mollissima*) and Tufted Puffin (*Fratercula cirrhata*) from the Aleutian Chain of Alaska. Archives of Environmental Contamination and Toxicology 56:596–606.

Burger, J., M. Gochfeld, C. Jeitner, D. Snigaroff, R. Snigaroff, T. Stamm, and C. Volz. 2008. Assessment of metals in down feathers of female Common Eiders and their eggs from the Aleutians: arsenic, cadmium, chromium, lead, manganese, mercury, and selenium. Environmental Monitoring Assessment 143:247–256.

Burnett, F. L., and D. E. Snyder. 1954. Blue crab as starvation food of oiled American Eiders. Auk 71:315–316.

Bustnes, J. O. 2013. Reproductive recovery of a Common Eider *Somateria mollissima* population following reductions in discharges of polycyclic aromatic hydrocarbons (PAHs). Bulletin of Environmental Contamination and Toxicology 91:202–207.

Bustnes, J. O., B. Moe, D. Herzke, S. A. Hanssen, T. Nordstad, K. Sagerup, G. W. Gabrielsen, and K. Borgå. 2010. Strongly increasing blood concentrations of lipid-soluble organochlorines in high arctic Common Eiders during incubation fast. Chemosphere 79:320–325.

Butler, D. 1990. The incidence of lead shot ingestion by waterfowl in Ireland. Irish Naturalists' Journal 23:309–313.

Byrd, G. V., and L. Daniel. 2008. Bird species found oiled, December 2004–January 2005, at Unalaska Island following the M/V *Selendang Ayu* oil spill. Preassessment Data Report #9. U.S. Fish and Wildlife Service, Alaska Maritime National Wildlife Refuge, Homer, AK.

Byrd, G. V., J. H. Reynolds, and P. L. Flint. 2009. Persistence rates and detection probabilities of bird carcasses on beaches of Unalaska Island, Alaska, following the wreck of the M/V *Selendang Ayu*. Marine Ornithology 37:197–204.

Cadwell, L. L., R. G. Schreckhise, and R. E. Fitzner. 1979. Cesium-137 in Coots (*Fulica americana*) on Hanford waste ponds: contribution to population dose and offsite transport estimates. Pp. 485–491 in J. E. Watson Jr. (editor), Low-level radioactive waste management: proceedings of Health Physics Society twelfth midyear topical symposium, 11–15 February, 1979, Williamsburg, VA. U.S. Environmental Protection Agency, Washington, DC.

Cain, B. W., and E. A. Pafford. 1981. Effects of dietary nickel on survival and growth of Mallard ducklings. Archives of Environmental Contamination and Toxicology 10:737–745.

Cain, B. W., L. Sileo, J. C. Franson, and J. Moore. 1983. Effects of dietary cadmium on Mallard ducklings. Environmental Research 32:286–297.

Carls, M. G., M. M. Babcock, P. M. Harris, G. V. Irvine, J. A. Cusick, and S. D. Rice. 2001. Persistence of oiling in mussel beds after the *Exxon Valdez* oil spill. Marine Environmental Research 51:167–190.

Carlsson, P., D. Herzke, M. Wedborg, and G. W. Gabrielsen. 2011. Environmental pollutants in the Swedish marine ecosystem, with special emphasis on polybrominated diphenyl ethers (PBDE). Chemosphere 82:1286–1292.

Carpenter, J. W., G. A. Andrews, and W. N. Beyer. 2004. Zinc toxicosis in a free-flying Trumpeter Swan (*Cygnus buccinator*). Journal of Wildlife Diseases 40:769–774.

Castège, I., G. Hémery, N. Roux, J. d'Elbée, Y. Lalanne, F. D'Amico, and C. Mouchès. 2004. Changes in abundance and at-sea distribution of seabirds in the Bay of Biscay prior to, and following the "*Erika*" oil spill. Aquatic Living Resources 17:361–367.

Cave, R. R., J. E. Andrews, T. Jickells, and E. G. Coombes. 2005. A review of sediment contamination by trace metals in the Humber catchment and estuary, and the implications for future estuary water quality. Estuarine, Coastal and Shelf Science 62:547–557.

Champoux, L. 1996. PCBs, dioxins and furans in Hooded Merganser (*Lophodytes cucullatus*), Common Merganser (*Mergus merganser*) and mink (*Mustela vison*) collected along the St. Maurice River near La Tuque, Quebec. Environmental Pollution 92:147–153.

Clausen, B., and C. Wolstrup. 1979. Lead poisoning in game from Denmark. Danish Review of Game Biology 11:1–22.

Coenen, T. M. M., A. Brouwer, I. C. Enninga, and J. H. Koeman. 1992. Subchronic toxicity and reproduction effects of tri-n-butyltin oxide in Japanese Quail. Archives of Environmental Contamination and Toxicology 23:457–463.

Culver, S. 2007. Zinc toxicosis in a King Eider. Veterinary Technician 28:634–637.

Custer, C. M., and T. W. Custer. 2000. Organochlorine and trace element contamination in wintering and migrating diving ducks in the southern Great Lakes, USA, since the zebra mussel invasion. Environmental Toxicology and Chemistry 19:2821–2829.

Custer, T. W., C. M. Custer, and R. K. Hines. 2002. Dioxins and congener-specific polychlorinated biphenyls in three avian species from the Wisconsin River, Wisconsin. Environmental Pollution 119:323–332.

Cuvin-Aralar, M. L. A., and R. W. Furness. 1991. Mercury and selenium interaction: a review. Ecotoxicology and Environmental Safety 21:348–364.

Dale, I. M., M. S. Baxter, J. A. Bogan, and W. R. P. Bourne. 1973. Mercury in seabirds. Marine Pollution Bulletin 4:77–79.

Danell, K. 1980. Konsekvenser av blyhagelanvändning för vattenfåglar och deras miljö (The detrimental effects of using lead-shot pellets for shooting waterfowl). Sveriges Lantbruksuniversitet, Institutionen för Viltekologi Rapport 4, Uppsala, Sweden (in Swedish with English summary).

Danell, K., Å. Andersson, and V. Marcström. 1977. Lead shot pellets dispersed by hunters—ingested by ducks. Ambio 6:235–237.

Darnerud, P. O., G. S. Eriksen, T. Jóhannesson, P. B. Larsen, and M. Viluksela. 2001. Polybrominated diphenyl ethers: occurrence, dietary exposure, and toxicology. Environmental Health Perspectives 109:49–68.

Davis G. K., and W. Mertz. 1987. Copper. Pp. 301–364 in W. Mertz (editor), Trace elements in human and animal nutrition, 5th ed. (vol. 1). Academic Press, San Diego, CA.

Day, R. H., S. M. Murphy, J. A. Wiens, G. D. Hayward, E. J. Harner, and L. N. Smith. 1997. Effects of the Exxon Valdez oil spill on habitat use by birds in Prince William Sound, Alaska. Ecological Applications 7:593–613.

Debacker, V., T. Jauniaux, F. Coignoul, and J.-M. Bouquegneau. 2000. Heavy metals contamination and body condition of wintering Guillemots (Uria aalge) at the Belgian coast from 1993 to 1998. Environmental Research (Section A) 84:310–317.

DeVink, J. A., R. G. Clark, S. M. Slattery, and A. M. Scheuhammer. 2007. Cross-seasonal association between winter trophic status and breeding ground selenium levels in boreal White-winged Scoters. Avian Conservation and Ecology 3:art3.

DeVink, J.-M. A., R. G. Clark, S. M. Slattery, and M. Wayland. 2008. Is selenium affecting body condition and reproduction in boreal breeding scaup, scoters, and Ring-necked Ducks? Environmental Pollution 152:116–122.

de Wit, C. A. 2002. An overview of brominated flame retardants in the environment. Chemosphere 46:583–624.

de Wit, C. A., D. Herzke, and K. Vorkamp. 2010. Brominated flame retardants in the Arctic environment—trends and new candidates. Science of the Total Environment 408:2885–2918.

Dieter, M. P., M. C. Perry, and B. M. Mulhern. 1976. Lead and PCB's in Canvasback ducks: relationship between enzyme levels and residues in blood. Archives of Environmental Contamination and Toxicology 5:1–13.

Dietz, R., P. M. Outridge, and K. A. Hobson. 2009. Anthropogenic contributions to mercury levels in present-day Arctic animals—a review. Science of the Total Environment 407:6120–6131.

Dietz, R., F. Riget, and P. Johansen. 1996. Lead, cadmium, mercury and selenium in Greenland marine animals. Science of the Total Environment 186:67–93.

Di Giulio, R. T., and P. F. Scanlon. 1984. Heavy metals in tissues of waterfowl from the Chesapeake Bay, USA. Environmental Pollution (Series A) 35:29–48.

Driver, E. A., and A. J. Derksen. 1980. Mercury levels in waterfowl from Manitoba, Canada, 1971–1972. Pesticides Monitoring Journal 14:95–101.

Dubrawski, R., and J. Falandysz. 1980. Chlorinated hydrocarbons in fish-eating birds from the Gdańsk Bay, Baltic Sea. Marine Pollution Bulletin 11:15–18.

Dunn, M. A., T. L. Blalock, and R. J. Cousins. 1987. Metallothionein (42525A). Proceedings of the Society for Experimental Biology and Medicine 185:107–119.

Eagles-Smith, C. A., J. T. Ackerman, T. L. Adelsbach, J. Y. Takekawa, A. K. Miles, and R. A. Keister. 2008. Mercury correlations among six tissues for four waterbird species breeding in San Francisco Bay, California, USA. Environmental Toxicology and Chemistry 27:2136–2153.

Eagles-Smith, C. A., J. T. Ackerman, S. E. W. De La Cruz, and J. Y. Takekawa. 2009a. Mercury bioaccumulation and risk to three waterbird foraging guilds is influenced by foraging ecology and breeding stage. Environmental Pollution 157:1993–2002.

Eagles-Smith, C. A., J. T. Ackerman, J. Yee, and T. L. Adelsbach. 2009b. Mercury demethylation in waterbird livers: dose–response thresholds and differences among species. Environmental Toxicology and Chemistry 28:568–577.

Eastin Jr., W. C., W. J. Fleming, and H. C. Murray. 1982. Organophosphate inhibition of avian salt gland Na, K-ATPase activity. Comparative Biochemistry and Physiology 73C:101–107.

Eisler, R. 2000a. Handbook of chemical risk assessment. Metals (vol. 1), Lewis Publishers, Boca Raton, FL.

Eisler, R. 2000b. Handbook of chemical risk assessment. Organics (vol. 2), Lewis Publishers, Boca Raton, FL.

Eisler, R. 2000c. Handbook of chemical risk assessment. Metalloids, radiation, cumulative index to chemicals and species (vol. 3), Lewis Publishers, Boca Raton, FL.

Elliott, J. E., and C. A. Bishop. 2011. Cyclodiene and other organochlorine pesticides in birds. Pp. 447–475 in W. N. Beyer and J. P. Meador (editors), Environmental contaminants in biota: interpreting tissue concentrations, 2nd ed. CRC Press, Boca Raton, FL.

Elliott, J. E., M. L. Harris, L. K. Wilson, B. D. Smith, S. P. Batchelor, and J. Maguire. 2007. Butyltins, trace metals and morphological variables in Surf Scoter (Melanitta perspicillata) wintering on the south coast of British Columbia, Canada. Environmental Pollution 149:114–124.

Elliott, J. E., and P. A. Martin. 1998. Chlorinated hydrocarbon contaminants in grebes and seaducks wintering on the coast of British Columbia, Canada: 1988–1993. Environmental Monitoring and Assessment 53:337–362.

Elliott, J. E., and A. M. Scheuhammer. 1997. Heavy metal and metallothionein concentrations in seabirds from the Pacific coast of Canada. Marine Pollution Bulletin 34:794–801.

Elliott, J. E., A. M. Scheuhammer, F. A. Leighton, and P. A. Pearce. 1992. Heavy metal and metallothionein concentrations in Atlantic Canadian seabirds. Archives of Environmental Contamination and Toxicology 22:63–73.

Elliott, J. E., L. K. Wilson, and B. Wakeford. 2005. Polybrominated diphenyl ether trends in eggs of marine and freshwater birds from British Columbia, Canada, 1979–2002. Environmental Science and Technology 39:5584–5591.

Eriksson, M. O. G., L. Henrikson, and H. G. Oscarson. 1989. Metal contents in liver tissues of non-fledged Goldeneye, *Bucephala clangula*, ducklings: a comparison between samples from acidic, circumneutral, and limed lakes in south Sweden. Archives of Environmental Contamination and Toxicology 18:255–260.

Esler, D., B. E. Ballachey, K. A. Trust, S. A. Iverson, J. A. Reed, A. K. Miles, J. D. Henderson et al. 2011. Cytochrome P4501A biomarker indication of the timeline of chronic exposure of Barrow's Goldeneyes to residual *Exxon Valdez* oil. Marine Pollution Bulletin 62:609–614.

Esler, D., T. D. Bowman, T. A. Dean, C. E. O'Clair, S. C. Jewett, and L. L. McDonald. 2000a. Correlates of Harlequin Duck densities during winter in Prince William Sound, Alaska. Condor 102:920–926.

Esler, D., T. D. Bowman, C. E. O'Clair, T. A. Dean, and L. L. McDonald. 2000b. Densities of Barrow's Goldeneyes during winter in Prince William Sound, Alaska in relation to habitat, food and history of oil contamination. Waterbirds 23:423–429.

Esler, D., T. D. Bowman, K. A. Trust, B. E. Ballachey, T. A. Dean, S. C. Jewett, and C. E. O'Clair. 2002. Harlequin Duck population recovery following the 'Exxon Valdez' oil spill: progress, process and constraints. Marine Ecology Progress Series 241:271–286.

Esler, D., and S. A. Iverson. 2010. Female Harlequin Duck winter survival 11 to 14 years after the *Exxon Valdez* oil spill. Journal of Wildlife Management 74:471–478.

Esler, D., J. A. Schmutz, R. L. Jarvis, and D. M. Mulcahy. 2000c. Winter survival of adult female Harlequin Ducks in relation to history of contamination by the *Exxon Valdez* oil spill. Journal of Wildlife Management 64:839–847.

Esler, D., K. A. Trust, B. E. Ballachey, S. A. Iverson, T. L. Lewis, D. J. Rizzolo, D. M. Mulcahy et al. 2010. Cytochrome P4501A biomarker indication of oil exposure in Harlequin Ducks up to 20 years after the *Exxon Valdez* oil spill. Environmental Toxicology and Chemistry 29:1138–1145.

Ethier, A. L. M., B. M. Braune, A. M. Scheuhammer, and D. E. Bond. 2007. Comparison of lead residues among avian bones. Environmental Pollution 145:915–919.

Evers, D. C., N. M. Burgess, L. Champoux, B. Hoskins, A. Major, W. M. Goodale, R. J. Taylor, R. Poppenga, and T. Daigle. 2005. Patterns and interpretation of mercury exposure in freshwater avian communities in northeastern North America. Ecotoxicology 14:193–221.

Exxon Valdez Oil Spill Trustee Council. [online]. 2010. Exxon Valdez oil spill restoration plan: 2010 update injured resources and services. <http//www.evostc.state.ak.us> (13 July 2012).

Faber, R. A., and J. J. Hickey. 1973. Eggshell thinning, chlorinated hydrocarbons, and mercury in inland aquatic bird eggs, 1969 and 1970. Pesticides Monitoring Journal 7:27–36.

Fairbrother, A. 1996. Cholinesterase-inhibiting pesticides. Pp. 52–60 in A. Fairbrother, L. N. Locke, and G. L. Hoff (editors), Noninfectious diseases of wildlife, 2nd ed. Iowa State University Press, Ames, IA.

Falandysz, J., and P. Szefer. 1982. Chlorinated hydrocarbons in diving ducks wintering in Gdańsk Bay, Baltic Sea. Science of the Total Environment 24:119–127.

Falandysz, J., and P. Szefer. 1984. Chlorinated hydrocarbons in fish-eating birds wintering in the Gdańsk Bay, 1981–82 and 1982–83. Marine Pollution Bulletin 15:298–301.

Falandysz, J., S. Taniyasu, N. Yamashita, P. Rostkowski, K. Zalewski, and K. Kannan. 2007. Perfluorinated compounds in some terrestrial and aquatic wildlife species from Poland. Journal of Environmental Science and Health Part A 42:715–719.

Fernie, K. J., G. Mayne, J. L. Shutt, C. Pekarik, K. A. Grasman, R. J. Letcher, and K. Drouillard. 2005a. Evidence of immunomodulation in nestling American Kestrels (*Falco sparverius*) exposed to environmentally relevant PBDEs. Environmental Pollution 138:485–493.

Fernie, K. J., J. L. Shutt, R. J. Letcher, J. I. Ritchie, K. Sullivan, and D. M. Bird. 2008. Changes in reproductive courtship behaviors of adult American Kestrels (*Falco sparverius*) exposed to environmentally relevant levels of the polybrominated diphenyl ether mixture, DE-71. Toxicological Sciences 102:171–178.

Fernie, K. J., J. L. Shutt, G. Mayne, D. Hoffman, R. J. Letcher, K. Drouillard, and I. J. Ritchie. 2005b. Exposure to polybrominated diphenyl ethers (PBDEs): changes in thyroid, vitamin A, glutathione homeostasis, and oxidative stress in American Kestrels (*Falco sparverius*). Toxicological Sciences 88:375–383.

Fimreite, N. 1974. Mercury contamination of aquatic birds in northwestern Ontario. Journal of Wildlife Management 38:120–131.

Fimreite, N., W. N. Holsworth, J. A. Keith, P. A. Pearce, and I. M. Gruchy. 1971. Mercury in fish and fish-eating birds near sites of industrial contamination in Canada. Canadian Field-Naturalist 85:211–220.

Fingas, M. 2011. Oil science and technology. Gulf Professional Publishing, Houston, TX.

Fleischli, M. A., J. C. Franson, N. J. Thomas, D. L. Finley, and W. Riley Jr. 2004. Avian mortality events in the United States caused by anticholinesterase pesticides: a retrospective summary of National Wildlife Health Center Records from 1980 to 2000. Archives of Environmental Contamination and Toxicology 46:542–550.

Fleming, W. J., E. F. Hill, J. J. Momot, and V. F. Pang. 1991. Toxicity of trimethyltin and triethyltin to Mallard ducklings. Environmental Toxicology and Chemistry 10:255–260.

Flint, P. L., and A. C. Fowler. 1998. A drift experiment to assess the influence of wind on recovery of oiled seabirds on St. Paul Island, Alaska. Marine Pollution Bulletin 36:165–166.

Flint, P. L., A. C. Fowler, and R. F. Rockwell. 1999. Modeling bird mortality associated with the M/V Citrus oil spill off St. Paul Island, Alaska. Ecological Modeling 117:261–267.

Flint, P. L., and J. B. Grand. 1997. Survival of Spectacled Eider adult females and ducklings during brood rearing. Journal of Wildlife Management 61:217–221.

Flint, P. L., J. B. Grand, J. A. Morse, and T. F. Fondell. 2000. Late summer survival of adult female and juvenile Spectacled Eiders on the Yukon–Kuskokwim Delta, Alaska. Waterbirds 23:292–297.

Flint, P. L., M. R. Petersen, and J. B. Grand. 1997. Exposure of Spectacled Eiders and other diving ducks to lead in western Alaska. Canadian Journal of Zoology 75:439–443.

Flint, P. L., and J. L. Schamber. 2010. Long-term persistence of spent lead shot in tundra wetlands. Journal of Wildlife Management 74:148–151.

Flint, P. L., J. L. Schamber, K. A. Trust, A. K. Miles, J. D. Henderson, and B. W. Wilson. 2012. Chronic hydrocarbon exposure of Harlequin Ducks in areas affected by the *Selendang Ayu* oil spill at Unalaska Island, Alaska. Environmental Toxicology and Chemistry 31:2828–2831.

Foley, R. E. 1992. Organochlorine residues in New York waterfowl harvested by hunters in 1983–1984. Environmental Monitoring and Assessment 21:37–48.

Foley, R. E., and G. R. Batcheller. 1988. Organochlorine contaminants in Common Goldeneye wintering on the Niagara River. Journal of Wildlife Management 52:441–445.

Fowler, A. C., and P. L. Flint. 1997. Persistence rates and detection probabilities of oiled King Eider carcasses on St. Paul Island, Alaska. Marine Pollution Bulletin 34:522–526.

Frank, A. 1986a. Lead fragments in tissues from wild birds: a cause of misleading analytical results. Science of the Total Environment 54:275–281.

Frank, A. 1986b. In search of biomonitors for cadmium: cadmium content of wild Swedish fauna during 1973–1976. Science of the Total Environment 57:57–65.

Franson, J. C., G. M. Haramis, M. C. Perry, and J. F. Moore. 1986. Blood protoporphyrin for detecting lead exposure in Canvasbacks. Pp. 32–37 *in* J. S. Feierabend, and A. B. Russell (editors), Lead poisoning in waterfowl—a Workshop. National Wildlife Federation, Washington, DC.

Franson, J. C., D. J. Hoffman, and J. A. Schmutz. 2002a. Blood selenium concentrations and enzyme activities related to glutathione metabolism in wild Emperor Geese. Environmental Toxicology and Chemistry 21:2179–2184.

Franson, J. C., D. J. Hoffman, A. Wells-Berlin, M. C. Perry, V. Shearn-Bochsler, D. L. Finley, P. L. Flint, and T. Hollmén. 2007. Effects of dietary selenium on tissue concentrations, pathology, oxidative stress, and immune function in Common Eiders (*Somateria mollissima*). Journal of Toxicology and Environmental Health 70:861–874.

Franson, J. C., T. Hollmén, M. Hario, M. Kilpi, and D. L. Finley. 2002b. Lead and delta-aminolevulinic acid dehydratase in blood of Common Eiders (*Somateria mollissima*) from the Finnish archipelago. Ornis Fennica 79:87–91.

Franson, J. C., T. Hollmén, R. H. Poppenga, M. Hario, and M. Kilpi. 2000a. Metals and trace elements in tissues of Common Eiders (*Somateria mollissima*) from the Finnish archipelago. Ornis Fennica 77:57–63.

Franson, J. C., T. Hollmén, R. H. Poppenga, M. Hario, M. Kilpi, and M. R. Smith. 2000b. Selected trace elements and organochlorines: some findings in blood and eggs of nesting Common Eiders (*Somateria mollissima*) from Finland. Environmental Toxicology and Chemistry 19:1340–1347.

Franson, J. C., T. E. Hollmén, P. L. Flint, J. B. Grand, and R. B. Lanctot. 2004. Contaminants in molting Long-tailed Ducks and nesting Common Eiders in the Beaufort Sea. Marine Pollution Bulletin 48:504–513.

Franson, J. C., P. S. Koehl, D. V. Derksen, T. C. Rothe, C. M. Bunck, and J. F. Moore. 1995a. Heavy metals in seaducks and mussels from Misty Fjords National Monument in Southeast Alaska. Environmental Monitoring and Assessment 36:149–167.

Franson, J. C., and D. J. Pain. 2011. Lead in birds. Pp. 563–593 in W. N. Beyer and J. P. Meador (editors), Environmental contaminants in biota: interpreting tissue concentrations, 2nd ed. CRC Press, Boca Raton, FL.

Franson, J. C., M. R. Petersen, C. U. Meteyer, and M. R. Smith. 1995b. Lead poisoning of Spectacled Eiders (Somateria fischeri) and of a Common Eider (Somateria mollissima) in Alaska. Journal of Wildlife Diseases 31:268–271.

Franson, J. C., J. A. Schmutz, L. H. Creekmore, and A. C. Fowler. 1999. Concentrations of selenium, mercury, and lead in blood of Emperor Geese in western Alaska. Environmental Toxicology and Chemistry 18:965–969.

French-McCay, D. P. 2004. Oil spill modeling: development and validation. Environmental Toxicology and Chemistry 23:2441–2456.

Furness, R. W., S. J. Muirhead, and M. Woodburn. 1986. Using bird feathers to measure mercury in the environment: relationships between mercury content and molt. Marine Pollution Bulletin 17:27–30.

Gagné, F., T. Burgeot, J. Hellou, S. St-Jean, É. Farcy, and C. Blaise. 2008. Spatial variations in biomarkers of Mytilus edulis mussels at four polluted regions spanning the Northern Hemisphere. Environmental Research 107:201–217.

García-Fernández, A. J., J. A. Sanchez-Garcia, M. Gomez-Zapata, and A. Luna. 1996. Distribution of cadmium in blood and tissues of wild birds. Archives of Environmental Contamination and Toxicology 30:252–258.

Gasaway, W. C., and I. O. Buss. 1972. Zinc toxicity in the Mallard duck. Journal of Wildlife Management 36:1107–1117.

Gerstenberger, S. L. 2004. Mercury concentrations in migratory waterfowl harvested from southern Nevada Wildlife Management Areas, USA. Environmental Toxicology 19:35–44.

Goede, A. A. 1993. Selenium in eggs and parental blood of a Dutch marine wader. Archives of Environmental Contamination and Toxicology 25:79–84.

Goede, A. A., T. Nygard, M. de Bruin, and E. Steinnes. 1989. Selenium, mercury, arsenic and cadmium in the lifecycle of the Dunlin, Calidris alpina, a migrant wader. Science of the Total Environment 78:205–218.

Golden, N. H., and B. A. Rattner. 2003. Ranking terrestrial vertebrate species for utility in biomonitoring and vulnerability to environmental contaminants. Reviews in Environmental Contamination and Toxicology 176:67–136.

Golden, N. H., B. A. Rattner, J. B. Cohen, D. J. Hoffman, E. Russek-Cohen, and M. A. Ottinger. 2003. Lead accumulation in feathers of nestling Black-crowned Night Herons (Nycticorax nycticorax) experimentally treated in the field. Environmental Toxicology and Chemistry 22:1517–1524.

Goodale, M. W., D. C. Evers, S. E. Mierzykowski, A. L. Bond, N. M. Burgess, C. I. Otorowski, L. J. Welch et al. 2008. Marine foraging birds as bioindicators of mercury in the Gulf of Maine. EcoHealth 5:409–425.

Grand, J. B., P. L. Flint, M. R. Petersen, and C. L. Moran. 1998. Effect of lead poisoning on Spectacled Eider survival rates. Journal of Wildlife Management 62:1103–1109.

Grand, J. B., J. C. Franson, P. L. Flint, and M. R. Petersen. 2002. Concentrations of trace elements in eggs and blood of Spectacled and Common Eiders on the Yukon–Kuskokwim Delta, Alaska, USA. Environmental Toxicology and Chemistry 21:1673–1678.

Guruge, K. S., S. Tanabe, H. Iwata, R. Taksukawa, and S. Yamagishi. 1996. Distribution, biomagnification, and elimination of butyltin compound residues in Common Cormorants (Phalacrocorax carbo) from Lake Biwa, Japan. Archives of Environmental Contamination and Toxicology 31:210–217.

Halford, D. K., O. D. Markham, and G. C. White. 1983. Biological elimination rates of radioisotopes by Mallards contaminated in a liquid radioactive waste disposal area. Health Physics 45:745–756.

Hamers, T., J. H. Kamstra, E. Sonneveld, A. J. Murk, M. H. A. Kester, P. L. Andersson, J. Legler, and A. Brouwer. 2006. In vitro profiling of the endocrine-disrupting potency of brominated flame retardants. Toxicological Sciences 92:157–173.

Hamilton, S. J., and D. J. Hoffman. 2003. Trace element and nutrition interactions in fish and wildlife. Pp. 1197–1235 in D. J. Hoffman, B. A. Rattner, G. A. Burton Jr., and J. Cairns Jr. (editors), Handbook of ecotoxicology, 2nd ed. Lewis Publishers, Boca Raton, FL.

Hansen, S. G., and I. Kraul. 1981. Shell thickness and residues of dieldrin, DDE, and PCB in eggs of Danish Goosanders Mergus merganser. Ornis Scandinavica 12:160–165.

Harris, M. L., and J. E. Elliott. 2011. Effects of polychlorinated biphenyls, dibenzo-p-dioxins and dibenzofurans, and polybrominated diphenyl ethers in wild birds. Pp. 477–528 in W. N. Beyer and J. P. Meador (editors), Environmental contaminants in biota: interpreting tissue concentrations, 2nd ed. CRC Press, Boca Raton, FL.

Harris, M. L., L. K. Wilson, S. Trudeau, and J. E. Elliott. 2007. Vitamin A and contaminant concentrations in Surf Scoters (Melanitta perspicillata) wintering on the Pacific coast of British Columbia, Canada. Science of the Total Environment 378:366–375.

Harrison, R. M., S. Harrad, and J. Lead. 2003. Global disposition of contaminants. Pp. 855–875 in D. J. Hoffman, B. A. Rattner, G. A. Burton Jr., and J. Cairns Jr. (editors), Handbook of ecotoxicology, 2nd ed. Lewis Publishers, Boca Raton, FL.

Harwell, M. A., J. H. Gentile, K. R. Parker, S. M. Murphy, R. H. Day, A. E. Bence, J. M. Neff, and J. A. Wiens. 2012. Quantitative assessment of current risks to Harlequin Ducks in Prince William Sound, Alaska, from the Exxon Valdez oil spill. Human and Ecological Risk Assessment 18:261–328.

Haseltine, S. D., G. H. Heinz, W. L. Reichel, and J. F. Moore. 1981. Organochlorine and metal residues in eggs of waterfowl nesting on islands in Lake Michigan off Door County, Wisconsin, 1977–78. Pesticides Monitoring Journal 15:90–97.

Haygarth, P. M. 1994. Global importance and global cycling of selenium. Pp. 1–27 in W. T. Frankenberger Jr. and S. Benson (editors), Selenium in the environment. Marcel Dekker, New York, NY.

Heard, D. J., D. M. Mulcahy, S. A. Iverson, D. J. Rizzolo, E. C. Greiner, J. Hall, H. Ip, and D. Esler. 2008. A blood survey of elements, viral antibodies, and hemoparasites in wintering Harlequin Ducks (Histrionicus histrionicus) and Barrow's Goldeneyes (Bucephala islandica). Journal of Wildlife Diseases 44:486–493.

Heinz, G. H. 1996. Mercury poisoning in wildlife. Pp. 118–127 in A. Fairbrother, L. N. Locke, and G. L. Hoff (editors), Noninfectious diseases of wildlife, 2nd ed. Iowa State University Press, Ames, IA.

Heinz, G. H., and M. A. Fitzgerald. 1993. Overwinter survival of Mallards fed selenium. Archives of Environmental Contamination and Toxicology 25:90–94.

Heinz, G. H., and S. D. Haseltine. 1981. Avoidance behavior of young Black Ducks treated with chromium. Toxicology Letters 8:307–310.

Heinz, G. H., and D. J. Hoffman. 1998. Methylmercury chloride and selenomethionine interactions on health and reproduction in Mallards. Environmental Toxicology and Chemistry 17:139–145.

Heinz, G. H., D. J. Hoffman, J. D. Klimstra, and K. R. Stebbins. 2009a. Rapid increases in mercury concentrations in the eggs of Mallards fed methylmercury. Environmental Toxicology and Chemistry 28:1979–1981.

Heinz, G. H., D. J. Hoffman, J. D. Klimstra, K. R. Stebbins, S. L. Kondrad, and C. A. Erwin. 2009b. Species differences in the sensitivity of avian embryos to methylmercury. Archives of Environmental Contamination and Toxicology 56:129–138.

Heinz, G. H., and K. L. Stromborg. 2009. Further declines in organochlorines in eggs of Red-breasted Mergansers from Lake Michigan, 1977–1978 versus 1990 versus 2002. Environmental Monitoring and Assessment 159:163–168.

Helgason, L. B., A. Polder, S. Føreid, K. Bæk, E. Lie, G. W. Gabrielsen, R. T. Barrett, and J. U. Skaare. 2009. Levels and temporal trends (1983–2003) of polybrominated diphenyl ethers and hexabromocyclododecanes in seabird eggs from north Norway. Environmental Toxicology and Chemistry 28:1096–1103.

Henny, C. J., L. J. Blus, R. A. Grove, and S. P. Thompson. 1991. Accumulation of trace elements and organochlorines by Surf Scoters wintering in the Pacific Northwest. Northwestern Naturalist 72:43–60.

Henny, C. J., D. D. Rudis, T. J. Roffe, and E. Robinson-Wilson. 1995. Contaminants and sea ducks in Alaska and the circumpolar region. Environmental Health Perspectives 103:41–49.

Herin, R. A., J. E. Suggs, E. M. Lores, L. T. Heiderscheit, J. D. Farmer, and D. Prather. 1978. Correlation of salt gland function with levels of chlorpyrifos in the feed of Mallard ducklings. Pesticide Biochemistry and Physiology 9:157–164.

Herzke, D., T. Nygård, U. Berger, S. Huber, and N. Røv. 2009. Perfluorinated and other persistent halogenated organic compounds in European Shag (Phalacrocorax aristotelis) and Common Eider (Somateria mollissima) from Norway: A suburban to remote pollutant gradient. Science of the Total Environment 408:340–348.

Hill, E. F. 2003. Wildlife toxicology of organophosphorus and carbamate pesticides. Pp. 281–312 in D. J. Hoffman, B. A. Rattner, G. A. Burton Jr., and J. Cairns Jr. (editors), Handbook of ecotoxicology, 2nd ed. Lewis Publishers, Boca Raton, FL.

Hobson, K. A., J. E. Thompson, M. R. Evans, and S. Boyd. 2005. Tracing nutrient allocation to reproduction in Barrow's Goldeneye. Journal of Wildlife Management 69:1221–1228.

Hoch, M. 2001. Organotin compounds in the environment—an overview. Applied Geochemistry 16:719–743.

Hoffman, D. J. 2002. Role of selenium toxicity and oxidative stress in aquatic birds. Aquatic Toxicology 57:11–26.

Hoffman, D. J., H. M. Ohlendorf, C. M. Marn, and G. W. Pendleton. 1998. Association of mercury and selenium with altered glutathione metabolism and oxidative stress in diving ducks from the San Francisco Bay Region, USA. Environmental Toxicology and Chemistry 17:167–172.

Hollmén, T., J. C. Franson, R. H. Poppenga, M. Hario, and M. Kilpi. 1998. Lead poisoning and trace elements in Common Eiders Somateria mollissima from Finland. Wildlife Biology 4:193–203.

Holt, G., and A. Frøslie. 1989. Blyforgiftning hos andefugler. Norsk Veterinærtidsskrift 101:759–765 (in Norwegian with English summary).

Honda, K., T. Nasu, and R. Tatsukawa. 1986. Seasonal changes in mercury accumulation in the Black-eared Kite, *Milvus migrans lineatus*. Environmental Pollution 42:325–334.

Hong, H. K., S. Takahashi, B. Y. Min, and S. Tanabe. 2002. Butyltin residues in blue mussels (*Mytilus edulis*) and arkshells (*Scapharca broughtonii*) collected from Korean coastal waters. Environmental Pollution 117:475–486.

Hontelez, L. C. M. P., H. M. van den Dungen, and A. J. Baars. 1992. Lead and cadmium in birds in the Netherlands: a preliminary survey. Archives of Environmental Contamination and Toxicology 23:453–456.

Hughes, M. R., D. C. Bennett, D. A. Gray, P. J. Sharp, A. M. Scheuhammer, and J. E. Elliott. 2003. Effects of cadmium ingestion on plasma and osmoregulatory hormone concentrations in male and female Pekin Ducks. Journal of Toxicology and Environmental Health 66:565–579.

Ikonomou, M. G., S. Rayne, and R. F. Addison. 2002. Exponential increases of the brominated flame retardants, polybrominated diphenyl ethers, in the Canadian arctic from 1981 to 2000. Environmental Science and Technology 36:1886–1892.

Irons, D. B., S. J. Kendall, W. P. Erickson, L. L. McDonald, and B. K. Lance. 2000. Nine years after the Exxon *Valdez* oil spill: effects on marine bird populations in Prince William Sound, Alaska. Condor 102:723–737.

Iverson, S. A., and D. Esler. 2010. Harlequin Duck population injury and recovery dynamics following the 1989 Exxon *Valdez* oil spill. Ecological Applications 20:1993–2006.

Janz, D. M., D. K. DeForest, M. L. Brooks, P. M. Chapman, G. Gilrohn, D. Hoff, W. A. Hopkins et al. 2010. Selenium toxicity to aquatic organisms. Pp. 141–231 in P. M. Chapman, W. J. Adams, M. L. Brooks, C. G. Delos, S. N. Luoma, W. A. Maher, H. M. Ohlendorf, T. S. Presser, and D. P. Shaw (editors), Ecological assessment of selenium in the aquatic environment. CRC Press, Boca Raton, FL.

Jaspers, V., T. Dauwe, R. Pinxten, L. Bervoets, R. Blust, and M. Eens. 2004. The importance of exogenous contamination on heavy metal levels in bird feathers. A field experiment with free-living Great Tits, *Parus major*. Journal of Environmental Monitoring 6:356–360.

Jensen, A. A., and H. Leffers. 2008. Emerging endocrine disrupters: perfluoroalkylated substances. International Journal of Andrology 31:161–169.

Jenssen, B. M. 1994. Review article: effects of oil pollution, chemically treated oil, and cleaning on the thermal balance of birds. Environmental Pollution 86:207–215.

Jenssen, B. M., and M. Ekker. 1991a. Dose dependent effects of plumage-oiling on thermoregulation of Common Eiders *Somateria mollissima* residing in water. Polar Research 10:579–584.

Jenssen, B. M., and M. Ekker. 1991b. Effects of plumage contamination with crude oil dispersant mixtures on thermoregulation in Common Eiders and Mallards. Archives of Environmental Contamination and Toxicology 20:398–403.

Jessup, D. A., and F. A. Leighton. 1996. Oil pollution and petroleum toxicity to wildlife. Pp. 141–156 in A. Fairbrother, L. N. Locke, and G. L. Hoff (editors), Noninfectious diseases of wildlife, 2nd ed. Iowa State University Press, Ames, IA.

Joensen, A. H. 1972. Studies on oil pollution and sea birds in Denmark 1968–1971. Danish Review of Game Biology 6:1–32.

Johansen, P., D. Muir, G. Asmund, and R. Dietz. 2003. Contaminants in subsistence animals in Greenland. Pp. 21–51 in B. Deutch and J. C. Hansen (editors), AMAP Greenland and the Faroe Islands 1997–2001. Human health (vol. 1), Danish Ministry of Environment, Copenhagen, Denmark.

Johnstone, R. M., G. S. Court, A. C. Fesser, D. M. Bradley, L. W. Oliphant, and J. D. MacNeil. 1996. Long-term trends and sources of organochlorine contamination in Canadian tundra Peregrine Falcons, *Falco peregrinus tundrius*. Environmental Pollution 93:109–120.

Joint Group of Experts on the Scientific Aspects of Marine Environmental Protection. 2007. Estimates of oil entering the marine environment from sea-based activities. Reports and Studies GESAMP, No. 75.

Jones, A. M., Y. Jones, and W. D. P. Stewart. 1972. Mercury in marine organisms of the Tay region. Nature 238:164–165.

Jordan, J. S., and F. C. Bellrose. 1951. Lead poisoning in wild waterfowl. Illinois Natural History Survey Biological Notes, No. 26. Illinois Natural History Survey, Urbana, IL.

Jörundsdóttir, H., K. Löfstrand, J. Svavarsson, A. Bignert, and Å. Bergman. 2010. Organochlorine compounds and their metabolites in seven Icelandic seabird species—a comparative study. Environmental Science and Technology 44:3252–3259.

Kalisińska, E., H. Budis, J. Podlasińska, N. Łanocha, and K. M. Kavetska. 2010. Body condition and mercury concentration in apparently healthy Goosander (*Mergus merganser*) wintering in the Odra estuary, Poland. Ecotoxicology 19:1382–1399.

Kalisińska, E., J. Gorecki, A. Okonska, B. Pilarczyk, A. Tomza-Marciniak, H. Budis, N. Łanocha, D. I. Kosik-Bogacka, K. M. Kavetska, M. Macherzynski, and J. Golas. 2014. Hepatic and nephric mercury and selenium concentrations in Common Mergansers, *Mergus merganser*, from Baltic region, Europe. Environmental Toxicology and Chemistry 33:421–430.

Kalisińska, E., and U. Szuberla. 1996. Heavy metals in the brain of Long-tailed Duck (*Clangula hyemalis*) wintering in the Pomeranian Bay, Poland. Biological Trace Element Research 55:191–197.

Kannan, K., and J. Falandysz. 1997. Butyltin residues in sediment, fish, fish-eating birds, harbour porpoise and human tissues from the Polish coast of the Baltic Sea. Marine Pollution Bulletin 34:203–207.

Kannan, K., K. Senthilkumar, J. E. Elliott, L. A. Feyk, and J. P. Giesy. 1998. Occurrence of butyltin compounds in tissues of water birds and seaducks from the United States and Canada. Archives of Environmental Contamination and Toxicology 35:64–69.

Karchner, S. I., D. G. Franks, S. W. Kennedy, and M. E. Hahn. 2006. The molecular basis for differential dioxin sensitivity in birds: role of the aryl hydrocarbon receptor. Proceedings of the National Academy of Sciences 103:6252–6257.

Karlog, O., K. Elvestad, and B. Clausen. 1983. Heavy metals (cadmium, copper, lead and mercury) in Common Eiders (*Somateria mollissima*) from Denmark. Nordisk Veterinär-Medicin 35:448–451.

Kelly, B. C., M. G. Ikonomou, J. D. Blair, and F. A. P. C. Gobas. 2008. Bioaccumulation behavior of polybrominated diphenyl ethers (PBDEs) in a Canadian Arctic marine food web. Science of the Total Environment 401:60–72.

Kenow, K. P., K. A. Grasman, R. K. Hines, M. W. Meyer, A. Gendron-Fitzpatrick, M. G. Spalding, and B. R. Gray. 2007. Effects of methylmercury exposure on the immune function of juvenile Common Loons (*Gavia immer*). Environmental Toxicology and Chemistry 26:1460–1469.

Khan, M. A. K., and F. Wang. 2009. Mercury-selenium compounds and their toxicological significance: toward a molecular understanding of the mercury-selenium antagonism. Environmental Toxicology and Chemistry 28:1567–1577.

Kim, E. Y., H. Ichihashi, K. Saeki, G. Atrashkevich, S. Tanabe, and R. Tatsukawa. 1996a. Metal accumulation in tissues of seabirds from Chaun, northeast Siberia, Russia. Environmental Pollution 92:247–252.

Kim, E. Y., T. Murakami, K. Saeki, and R. Tatsukawa. 1996b. Mercury levels and its chemical form in tissues and organs of seabirds. Archives of Environmental Contamination and Toxicology 30:259–266.

Kim, K. S., M. J. Pastel, J. S. Kim, and W. B. Stone. 1984. Levels of polychlorinated biphenyls, DDE, and mirex in waterfowl collected in New York State, 1979–1980. Archives of Environmental Contamination and Toxicology 13:373–381.

Lance, B. K., D. B. Irons, S. J. Kendall, and L. L. McDonald. 2001. An evaluation of marine bird population trends following the Exxon Valdez oil spill, Prince William Sound, Alaska. Marine Pollution Bulletin 42:298–309.

Lande, E. 1977. Heavy metal pollution in Trondheimsfjorden, Norway, and the recorded effects on the fauna and flora. Environmental Pollution 12:187–198.

Larsson, K., and L. Tydén. 2005. Effekter av oljeutsläpp på övervintrande alfågel *Clangula hyemalis* vid Hoburgs bank i centrala Östersjön mellan 1996/97 och 2003/04. Ornis Svecica 15:161–171 (in Swedish with English summary).

Law, R. J., C. R. Allchin, J. de Boer, A. Covaci, D. Herzke, P. Lepom, S. Morris, J. Tronczynski, and C. A. de Wit. 2006. Levels and trends of brominated flame retardants in the European environment. Chemosphere 64:187–208.

Leighton, F. A. 1993. The toxicity of petroleum oils to birds. Environmental Reviews 1:92–103.

Leitch, D. R., J. Carrie, D. Lean, R. W. Macdonald, G. A. Stern, and F. Wang. 2007. The delivery of mercury to the Beaufort Sea of the Arctic Ocean by the Mackenzie River. Science of the Total Environment 373:178–195.

Lemly, A. D. 2004. Aquatic selenium pollution is a global environmental safety issue. Ecotoxicology and Environmental Safety 59:44–56.

Leonzio, C., and A. Massi. 1989. Metal biomonitoring in bird eggs: a critical experiment. Bulletin of Environmental Contamination and Toxicology 43:402–406.

Levengood, J. M., G. C. Sanderson, W. L. Anderson, G. L. Foley, L. M. Skowron, P. W. Brown, and J. W. Seets. 1999. Acute toxicity of ingested zinc shot to game-farm Mallards. Illinois Natural History Survey Bulletin 36:1–36.

Locke, L. N., G. E. Bagley, and L. T. Young. 1967. The ineffectiveness of acid-fast inclusions in diagnosis of lead poisoning in Canada Geese. Bulletin of the Wildlife Disease Association 3:176.

Locke, L. N., and N. J. Thomas. 1996. Lead poisoning of waterfowl and raptors. Pp. 108–117 in A. Fairbrother, L. N. Locke, and G. L. Hoff (editors), Noninfectious diseases of wildlife, 2nd ed. Iowa State University Press, Ames, IA.

Lovvorn, J. R., M. F. Raisbeck, L. W. Cooper, G. A. Cutter, M. W. Miller, M. L. Brooks, J. M. Grebmeier, A. C. Matz, and C. M. Schaefer. 2013. Wintering eiders acquire exceptional Se and Cd burdens in the Bering Sea: physiological and oceanographic factors. Marine Ecology Progress Series 489:245–261.

Lowe, V. P. W. 1991. Radionuclides and the birds at Ravenglass. Environmental Pollution 70:1–26.

Ludke, J. L. 1976. Organochlorine pesticide residues associated with mortality: additivity of chlordane and endrin. Bulletin of Environmental Contamination and Toxicology 16:253–260.

Macdonald, R. W., L. A. Barrie, T. F. Bidleman, M. L. Diamond, D. J. Gregor, R. G. Semkin, W. M. J. Strachan et al.. 2000. Contaminants in the Canadian Arctic: 5 years of progress in understanding sources, occurrence and pathways. Science of the Total Environment 254:93–234.

Macdonald, R. W., T. Harner, and J. Fyfe. 2005. Recent climate change in the Arctic and its impact on contaminant pathways and interpretation of temporal trend data. Science of the Total Environment 342:5–86.

Maher, W., A. Roach, M. Doblin, T. Fan, S. Foster, R. Garrett, G. Möller, L. Oram, and D. Wallschläger. 2010. Environmental sources, speciation, and partitioning of selenium. Pp. 47–92 in P. M. Chapman, W. J. Adams, M. L. Brooks, C. G. Delos, S. N. Luoma, W. A. Maher, H. M. Ohlendorf, T. S. Presser, and D. P. Shaw (editors), Ecological assessment of selenium in the aquatic environment. CRC Press, Boca Raton, FL.

Mallory, M. L., B. M. Braune, M. Wayland, H. G. Gilchrist, and D. L. Dickson. 2004a. Contaminants in Common Eiders (Somateria mollissima) of the Canadian arctic. Environmental Reviews 12:197–218.

Mallory, M. L., M. Wayland, B. M. Braune, and K. G. Drouillard. 2004b. Trace elements in marine birds, arctic hare and ringed seals breeding near Qikiqtarjuaq, Nunavut, Canada. Marine Pollution Bulletin 49:119–141.

Martin, J. W., M. M. Smithwick, B. M. Braune, P. F. Hoekstra, D. C. G. Muir, and S. A. Mabury. 2004. Identification of long-chain perfluorinated acids in biota from the Canadian Arctic. Environmental Science and Technology 38:373–380.

Mashima, T. Y., W. J. Fleming, and M. K. Stoskopf. 1998. Metal concentrations in Oldsquaw (Clangula hyemalis) during an outbreak of avian cholera, Chesapeake Bay, 1994. Ecotoxicology 7:107–111.

Matishov, D. G., G. G. Matishov, and N. V. Lebedeva. 2003. Content of artificial radionuclides in the birds of the Barents Sea and the Sea of Azov. Doklady Biological Sciences 389:157–159.

Matz, A., and P. Flint. 2009. Lead isotopes indicate lead shot exposure in Alaska-breeding waterfowl. P. 174 in R. T. Watson, M. Fuller, M. Pokras, and W. G. Hunt (editors), Ingestion of lead from spent ammunition: implications for wildlife and humans. The Peregrine Fund, Boise, ID.

Mayack, L. A., P. B. Bush, O. J. Fletcher, R. K. Page, and T. T. Fendley. 1981. Tissue residues of dietary cadmium in Wood Ducks. Archives of Environmental Contamination and Toxicology 10:637–645.

Mazak, E. J., H. J. MacIsaac, M. R. Servos, and R. Hesslein. 1997. Influence of feeding habits on organochlorine contaminant accumulation in waterfowl on the Great Lakes. Ecological Applications 7:1133–1143.

McKernan, M. A., B. A. Rattner, R. C. Hale, and M. A. Ottinger. 2009. Toxicity of polybrominated diphenyl ethers (DE-71) in chicken (Gallus gallus), Mallard (Anas platyrhynchos), and American Kestrel (Falco sparverius) embryos and hatchlings. Environmental Toxicology and Chemistry 28:1007–1017.

Meador, J. P. 2011. Organotins in aquatic biota: occurrence in tissue and toxicological significance. Pp. 255–284 in W. N. Beyer, and J. P. Meador (editors), Environmental contaminants in biota: interpreting tissue concentrations, 2nd ed. CRC Press, Boca Raton, FL.

Meharg, A. A., D. J. Pain, R. M. Ellam, R. Baos, V. Olive, A. Joyson, N. Powell, A. J. Green, and F. Hiraldo. 2002. Isotopic identification of the sources of lead contamination for White Storks (Ciconia ciconia) in a marshland ecosystem (Doñana, S.W. Spain). Science of the Total Environment 300:81–86.

Mehlum, F., and F. F. Daelemans. 1995. PCBs in Arctic birds from the Svalbard region. Science of the Total Environment 160/161:441–446.

Meyers-Schöne, L., and S. S. Talmage. 2003. Nuclear and thermal. Pp. 615–643 in D. J. Hoffman, B. A. Rattner, G. A. Burton Jr., and J. Cairns Jr. (editors), Handbook of ecotoxicology, 2nd ed. Lewis Publishers, Boca Raton, FL.

Mierzykowski S. E., L. J. Welch, W. Goodale, D. C. Evers, C. S. Hall, S. W. Kress, and R. B. Allen. 2005. Mercury in bird eggs from coastal Maine. U.S. Fish and Wildlife Service Special Project Report FY05-MEFO-1-EC. U.S. Fish and Wildlife Service, Old Town, ME.

Miles, A. K., P. L. Flint, K. A. Trust, M. A. Ricca, S. E. Spring, D. E. Arrieta, T. Hollmen, and B. W. Wilson. 2007. Polycyclic aromatic hydrocarbon exposure in Steller's Eiders (Polysticta stelleri) and Harlequin Ducks (Histronicus histronicus) in the eastern Aleutian Islands, Alaska, USA. Environmental Toxicology and Chemistry 26:2694–2703.

Mudge, G. P. 1983. The incidence and significance of ingested lead pellet poisoning in British wildfowl. Biological Conservation 27:333–372.

Muir, D., B. Braune, B. DeMarch, R. Norstrom, R. Wagemann, L. Lockhart, B. Hargrave, D. Bright, R. Addison, J. Payne, and K. Reimer. 1999. Spatial and temporal trends and effects of contaminants in the Canadian Arctic marine ecosystem: a review. Science of the Total Environment 230:83–144.

Muir, D. C. G., R. Wagemann, B. T. Hargrave, D. J. Thomas, D. B. Peakall, and R. J. Norstrom. 1992. Arctic marine ecosystem contamination. Science of the Total Environment 122:75–134.

Murvoll, K. M., J. U. Skaare, H. Jensen, and B. M. Jenssen. 2007. Associations between persistent organic pollutants and vitamin status in Brünnich's Guillemot and Common Eider hatchlings. Science of the Total Environment 381:134–145.

Neff, J. M., D. S. Page, and P. D. Boehm. 2011. Exposure of sea otters and Harlequin Ducks in Prince William Sound, Alaska, USA, to shoreline oil residues 20 years after the Exxon Valdez oil spill. Environmental Toxicology and Chemistry 30:659–672.

Newsted, J. L., P. D. Jones, K. Coady, and J. P. Giesy. 2005. Avian toxicity reference values for perfluorooctane sulfonate. Environmental Science and Technology 39:9357–9362.

Nielsen, C. O., and R. Dietz. 1989. Heavy metals in Greenland seabirds. Meddelelser om Grønland, Bioscience 29:1–26.

Norheim, G. 1987. Levels and interactions of heavy metals in sea birds from Svalbard and the Antarctic. Environmental Pollution 47:83–94.

Norheim, G., and B. Borch-Iohnsen. 1990. Chemical and morphological studies of liver from Eider (Somateria mollissima) in Svalbard with special reference to the distribution of copper. Journal of Comparative Pathology 102:457–466.

Norheim, G., and B. Kjos-Hanssen. 1984. Persistent chlorinated hydrocarbons and mercury in birds caught off the west coast of Spitsbergen. Environmental Pollution 33:143–152.

Noyes, P. D., M. K. McElwee, H. D. Miller, B. W. Clark, L. A. Van Tiem, K. C. Walcott, K. N. Erwin, and E. D. Levin. 2009. The toxicology of climate change: environmental contaminants in a warming world. Environment International 35:971–986.

Ochiai, K., K. Jin, M. Goryo, T. Tsuzuki, and C. Itakura. 1993. Pathomorphologic findings of lead poisoning in White-fronted Geese (Anser albifrons). Veterinary Pathology 30:522–528.

Ohlendorf, H. M. 1993. Marine birds and trace elements in the temperate North Pacific. Pp. 232–240 in K. Vermeer, K. T. Briggs, K. H. Morgan, and D. Siegel-Causey (editors), The status, ecology, and conservation of marine birds of the North Pacific. Canadian Wildlife Service Special Publication, Ottawa, ON.

Ohlendorf, H. M. 2002. The birds of Kesterson Reservoir: a historical perspective. Aquatic Toxicology 57:1–10.

Ohlendorf, H. M. 2003. Ecotoxicology of selenium. Pp. 465–500 in D. J. Hoffman, B. A. Rattner, G. A. Burton Jr., and J. Cairns Jr. (editors), Handbook of ecotoxicology, 2nd ed. Lewis Publishers, Boca Raton, FL.

Ohlendorf, H. M., and G. H. Heinz. 2011. Selenium in birds. Pp. 669–701 in W. N. Beyer, and J. P. Meador (editors), Environmental contaminants in biota: interpreting tissue concentrations, 2nd ed. CRC Press, Boca Raton, FL.

Ohlendorf, H. M., R. W. Lowe, P. R. Kelly, and T. E. Harvey. 1986. Selenium and heavy metals in San Francisco Bay diving ducks. Journal of Wildlife Management 50:64–70.

Ohlendorf, H. M., K. C. Marois, R. W. Lowe, T. E. Harvey, and P. R. Kelly. 1991. Trace elements and organochlorines in Surf Scoters from San Francisco Bay, 1985. Environmental Monitoring and Assessment 18:105–122.

O'Keefe, P. W., W. C. Clayton, S. Conner, B. Bush, and C. Hong. 2006. Organic pollutants in wild ducks from New York State: I. Interspecies differences in concentrations and congener profiles of PCBs and PCDDs/ PCDFs. Science of the Total Environment 361:111–113.

Olafsdottir, K., K. Skirnisson, G. Gylfadottir, and T. Johannesson. 1998. Seasonal fluctuations of organochlorine levels in the Common Eider (Somateria mollissima) in Iceland. Environmental Pollution 103:153–158.

O'Toole, D., and M. F. Raisbeck. 1997. Experimentally induced selenosis of adult Mallard ducks: clinical signs, lesions, and toxicology. Veterinary Pathology 34:330–340.

Ouellet, J., L. Champoux, and M. Robert. 2012. Metals, trace elements, polychlorinated biphenyls, organochlorine pesticides, and brominated flame retardants in tissues of Barrow's Goldeneyes (Bucephala islandica) wintering in the St. Lawrence marine ecosystem, eastern Canada. Archives of Environmental Contamination and Toxicology 63:429–436.

Outridge, P. M., and A. M. Scheuhammer. 1993a. Bioaccumulation and toxicology of chromium: implications for wildlife. Reviews of Environmental Contamination and Toxicology 130:31–77.

Outridge, P. M., and A. M. Scheuhammer. 1993b. Bioaccumulation and toxicology of nickel: implications for wild mammals and birds. Environmental Reviews 1:172–197.

Paasivirta, J., J. Särkkä, J. Pellinen, and T. Humppi. 1981. Biocides in eggs of aquatic birds. Completion of a food chain enrichment study for DDT, PCB and Hg. Chemosphere 10:787–794.

Pacyna, E. G., J. M. Pacyna, K. Sundseth, J. Munthe, K. Kindbom, S. Wilson, F. Steenhuisen, and P. Maxson. 2010. Global emission of mercury to the atmosphere from anthropogenic sources in 2005 and projections to 2020. Atmospheric Environment 44:2487–2499.

Pain, D. J. 1989. Haematological parameters as predictors of blood lead and indicators of lead poisoning in the Black Duck (Anas rubripes). Environmental Pollution 60:67–81.

Pain, D. J., A. A. Meharg, M. Ferrer, M. Taggart, and V. Penteriani. 2005. Lead concentrations in bones and feathers of the globally threatened Spanish Imperial Eagle. Biological Conservation 121:603–610.

Pain, D. J., and B. A. Rattner. 1988. Mortality and hematology associated with the ingestion of one number four lead shot in Black Ducks, *Anas rubripes*. Bulletin of Environmental Contamination and Toxicology 40:159–164.

Pattee, O. H., and D. J. Pain. 2003. Lead in the environment. Pp. 373–408 *in* D. J. Hoffman, B. A. Rattner, G. A. Burton Jr., and J. Cairns Jr. (editors), Handbook of ecotoxicology, 2nd ed. Lewis Publishers, Boca Raton, FL.

Peakall, D. M., P. G. Wells, and D. Mackay. 1987. A hazard assessment of chemically dispersed oil spills and seabirds. Marine Environmental Research 22:91–106.

Pearce, J. 1995. Radiocesium in migratory bird species in Northern Ireland following the Chernobyl accident. Bulletin of Environmental Contamination and Toxicology 54:805–811.

Pearce, P. A., D. B. Peakall, and L. M. Reynolds. 1979. Shell thinning and residues of organochlorines and mercury in seabird eggs, eastern Canada, 1970–76. Pesticides Monitoring Journal 13:61–68.

Peterson, C. H., S. D. Rice, J. W. Short, D. Esler, J. L. Bodkin, B. E. Ballachey, and D. B. Irons. 2003. Long-term ecosystem response to the Exxon Valdez oil spill. Science 302:2082–2086.

Peterson, S. R., and R. S. Ellarson. 1976. Total mercury residues in livers and eggs of Oldsquaws. Journal of Wildlife Management 40:704–709.

Peterson, S. R., and R. S. Ellarson. 1978. p,p'-DDE, polychlorinated biphenyls, and endrin in Oldsquaws in North America, 1969–1973. Pesticides Monitoring Journal 11:170–181.

Piatt, J. F., C. J. Lensink, W. Butler, M. Kendziorek, and D. R. Nysewander. 1990. Immediate impact of the 'Exxon Valdez' oil spill on marine birds. Auk 107:387–397.

Poissant, L., H. H. Zhang, J. Canário, and P. Constant. 2008. Critical review of mercury fates and contamination in the arctic tundra ecosystem. Science of the Total Environment 400:173–211.

Rattner, B. A., W. J. Fleming, and H. C. Murray. 1983. Osmoregulatory function in ducks following ingestion of the organophosphorus insecticide fenthion. Pesticide Biochemistry and Physiology 20:246–255.

Rattner, B. A., J. C. Franson, S. R. Sheffield, C. I. Goddard, N. J. Leonard, D. Stang, and P. J. Wingate. 2008. Sources and implications of lead ammunition and fishing tackle on natural resources. The Wildlife Society Technical Review 08-01. The Wildlife Society and the American Fisheries Society, Bethesda, MD.

Rattner, B. A., and A. G. Heath. 2003. Environmental factors affecting contaminant toxicity in aquatic and terrestrial vertebrates. Pp. 679–699 in

D. J. Hoffman, B. A. Rattner, G. A. Burton Jr., and J. Cairns Jr. (editors), Handbook of ecotoxicology, 2nd ed. Lewis Publishers, Boca Raton, FL.

Rattner, B. A., and J. R. Jehl Jr. 1997. Dramatic fluctuations in liver mass and metal content of Eared Grebes (*Podiceps nigricollis*) during autumnal migration. Bulletin of Environmental Contamination and Toxicology 59:337–343.

Ricca, M. A., A. K. Miles, and R. G. Anthony. 2008. Sources of organochlorine contaminants and mercury in seabirds from the Aleutian archipelago of Alaska: inferences from spatial and trophic variation. Science of the Total Environment 406:308–323.

Rice, C. P., P. W. O'Keefe, and T. J. Kubiak. 2003. Sources, pathways, and effects of PCBs, dioxins, and dibenzofurans. Pp. 501–573 in D. J. Hoffman, B. A. Rattner, G. A. Burton Jr., and J. Cairns Jr. (editors), Handbook of ecotoxicology, 2nd ed. Lewis Publishers, Boca Raton, FL.

Rigét, F., A. Bignert, B. Braune, J. Stow, and S. Wilson. 2010. Temporal trends of legacy POPs in Arctic biota, an update. Science of the Total Environment 408:2874–2884.

Rocke, T. E., C. J. Brand, and J. G. Mensik. 1997. Site-specific lead exposure from lead pellet ingestion in sentinel Mallards. Journal of Wildlife Management 61:228–234.

Rocke, T. E., and M. D. Samuel. 1991. Effects of lead shot ingestion on selected cells of the Mallard immune system. Journal of Wildlife Diseases 27:1–9.

Roscoe, D. E., S. W. Nielsen, A. A. Lamola, and D. Zuckerman. 1979. A simple, quantitative test for erythrocytic protoporphyrin in lead-poisoned ducks. Journal of Wildlife Diseases 15:127–136.

Rothschild, R. F. N., and L. K. Duffy. 2005. Mercury concentrations in muscle, brain, and bone of Western Alaskan waterfowl. Science of the Total Environment 349:277–283.

Rüdel, H., P. Lepper, J. Steinhanses, and C. Schröter-Kermani. 2003. Retrospective monitoring of organotin compounds in marine biota from 1985 to 1999: results from the German Environmental Specimen Bank. Environmental Science and Technology 37:1731–1738.

Rydberg, J., J. Klaminder, P. Rosén, and R. Bindler. 2010. Climate driven release of carbon and mercury from permafrost mires increases mercury loading to sub-arctic lakes. Science of the Total Environment 408:4778–4783.

Sample, B. E., and C. Irvine. 2011. Radionuclides in biota. Pp. 703–732 in W. N. Beyer and J. P. Meador (editors), Environmental contaminants in biota: interpreting tissue concentrations, 2nd ed. CRC Press, Boca Raton, FL.

Samuel, M. D., and E. F. Bowers. 2000. Lead exposure of American Black Ducks after implementation of non-toxic shot. Journal of Wildlife Management 64:947–953.

Sanderson, G. C., and F. C. Bellrose. 1986. A review of the problem of lead poisoning in waterfowl. Illinois Natural History Survey Special Publication 4. Illinois Natural History Survey, Champaign, IL.

Särkkä, J., M. Hattula, J. Janatuinen, J. Paasivirta, and R. Palokangas. 1978. Chlorinated hydrocarbons and mercury in birds of Lake Päijänne, Finland—1972–74. Pesticides Monitoring Journal 12:26–35.

Savinov, V. M., G. W. Gabrielsen, and T. N. Savinova. 2003. Cadmium, zinc, copper, arsenic, selenium and mercury in seabirds from the Barents Sea: levels, inter-specific and geographical differences. Science of the Total Environment 306:133–158.

Savinova, T. N., A. Polder, G. W. Gabrielsen, and J. U. Skaare. 1995. Chlorinated hydrocarbons in seabirds from the Barents Sea area. Science of the Total Environment 160/161:497–504.

Scanlon, P. F. 1982. Wet and dry weight relationships of Mallard (Anas platyrhynchos) tissues. Bulletin of Environmental Contamination and Toxicology 29:615–617.

Scanlon, P. F., V. D. Stotts, R. G. Oderwald, T. J. Dietrick, and R. J. Kendall. 1980. Lead concentrations in livers of Maryland waterfowl with and without ingested lead shot present in gizzards. Bulletin of Environmental Contamination and Toxicology 25:855–860.

Scheringer, M. 2009. Long-range transport of organic chemicals in the environment. Environmental Toxicology and Chemistry 28:677–690.

Scheuhammer, A. M. 1987. The chronic toxicity of aluminum, cadmium, mercury, and lead in birds: a review. Environmental Pollution 46:263–295.

Scheuhammer, A. M. 1988. The dose-dependent deposition of cadmium into organs of Japanese Quail following oral administration. Toxicology and Applied Pharmacology 95:153–161.

Scheuhammer, A. M. 1991. Effects of acidification on the availability of toxic metals and calcium to wild birds and mammals. Environmental Pollution 71:329–375.

Scheuhammer, A. M., N. Basu, N. M. Burgess, J. E. Elliott, G. D. Campbell, M. Wayland, L. Champoux, and J. Rodrigue. 2008. Relationships among mercury, selenium, and neurochemical parameters in Common Loons (Gavia immer) and Bald Eagles (Haliaeetus leucocephalus). Ecotoxicology 17:93–101.

Scheuhammer, A. M., and K. M. Dickson. 1996. Patterns of environmental lead exposure in waterfowl in Eastern Canada. Ambio 25:14–20.

Scheuhammer, A. M., M. W. Meyer, M. B. Sandheinrich, and M. W. Murray. 2007. Effects of environmental methylmercury on the health of wild birds, mammals, and fish. Ambio 36:12–18.

Scheuhammer, A. M., S. L. Money, D. A. Kirk, and G. Donaldson. 2003. Lead fishing sinkers and jigs in Canada: review of their use patterns and toxic impacts on wildlife. Occasional Paper, No. 108. Canadian Wildlife Service, Ottawa, ON.

Scheuhammer, A. M., J. A. Perrault, E. Routhier, B. M. Braune, and G. D. Campbell. 1998a. Elevated lead concentrations in edible portions of game birds harvested with lead shot. Environmental Pollution 102:251–257.

Scheuhammer, A. M., and D. M. Templeton. 1998. Use of stable isotope ratios to distinguish sources of lead exposure in wild birds. Ecotoxicology 7:37–42.

Scheuhammer, A. M., A. H. K. Wong, and D. Bond. 1998b. Mercury and selenium accumulation in Common Loons (Gavia immer) and Common Mergansers (Mergus merganser) from eastern Canada. Environmental Toxicology and Chemistry 17:197–201.

Schiedek, D., B. Sundelin, J. W. Readman, and R. W. Macdonald. 2007. Interactions between climate change and contaminants. Marine Pollution Bulletin 54:1845–1856.

Schummer, M. L., S. S. Badzinski, S. A. Petrie, Y.-W. Chen, and N. Belzile. 2010. Selenium accumulation in sea ducks wintering at Lake Ontario. Archives of Environmental Contamination and Toxicology 58:854–862.

Schummer, M. L., I. Fife, S. A. Petrie, and S. S. Badzinski. 2011. Artifact ingestion in sea ducks wintering at northeastern Lake Ontario. Waterbirds 34:51–58.

Schummer, M. L., S. A. Petrie, and R. C. Bailey. 2008. Dietary overlap of sympatric diving ducks during winter on northeastern Lake Ontario. Auk 125:425–433.

Shaw, G. E. 1995. The arctic haze phenomenon. Bulletin of the American Meteorological Society 76:2403–2413.

Shillinger, J. E., and C. C. Cottam. 1937. The importance of lead poisoning in waterfowl. Transactions of the Second North American Wildlife Conference 2:398–403.

Shore, R. F., M. G. Pereira, L. A. Walker, and D. R. Thompson. 2011. Mercury in nonmarine birds and mammals. Pp. 609–624 in W. N. Beyer and J. P. Meador (editors), Environmental contaminants in biota: interpreting tissue concentrations, 2nd ed. CRC Press, Boca Raton, FL.

Sileo, L., W. N. Beyer, and R. Mateo. 2003. Pancreatitis in wild zinc-poisoned waterfowl. Avian Pathology 32:655–660.

Sileo, L., L. H. Creekmore, D. J. Audet, M. R. Snyder, C. U. Meteyer, J. C. Franson, L. N. Locke, M. R. Smith, and D. L. Finley. 2001. Lead poisoning of waterfowl by contaminated sediment in the Coeur d'Alene River. Archives of Environmental Contamination and Toxicology 41:364–368.

Simcik, M. F. 2005. Global transport and fate of perfluorochemicals. Journal of Environmental Monitoring 7:759–763.

Sinclair, E., D. T. Mayack, K. Roblee, N. Yamashita, and K. Kannan. 2006. Occurrence of perfluoroalkyl surfactants in water, fish, and birds from New York State. Archives of Environmental Contamination and Toxicology 50:398–410.

Si-Zhong, Y., J. Hui-Jun, W. Zhi, H. Rui-Xia, J. Yan-Jun, L. Xiu-Mei, and Y. Shao-Peng. 2009. Bioremediation of oil spills in cold environments: a review. Pedosphere 19:371–381.

Skerratt, L. F., J. C. Franson, C. U. Meteyer, and T. E. Hollmén. 2005. Causes of mortality in sea ducks (Mergini) necropsied at the USGS-National Wildlife Health Center. Waterbirds 28:193–207.

Skwarzec, B., and J. Fabisiak. 2007. Bioaccumulation of polonium ^{210}Po in marine birds. Journal of Environmental Radioactivity 93:119–126.

Smail, J., D. G. Ainley, and H. Strong. 1972. Notes on birds killed in the 1971 San Francisco oil spill. California Birds 3:25–32.

Smit, T., T. Bakhuizen, and L. G. Moraal. 1988. Metallic lead as a source of lead intoxication in the Netherlands. Limosa 61:175–178 (in Dutch).

Smith, M. R., N. J. Thomas, and C. Hulse. 1995. Application of brain cholinesterase reactivation to differentiate between organophosphorus and carbamate pesticide exposure in wild birds. Journal of Wildlife Diseases 31:263–267.

Smith, P. N., G. P. Cobb, C. Godard-Codding, D. Hoff, S. T. McMurry, T. R. Rainwater, and K. D. Reynolds. 2007. Contaminant exposure in terrestrial vertebrates. Environmental Pollution 150:41–64.

Smith, V. E., J. M. Spurr, J. C. Filkins, and J. J. Jones. 1985. Organochlorine contaminants of wintering ducks foraging on Detroit River sediments. Journal of Great Lakes Research 11:231–246.

Sonak, S., P. Pangam, A. Giriyan, and K. Hawaldar. 2009. Implications of the ban on organotins for protection of global coastal and marine ecology. Journal of Environmental Management 90:S96–S108.

Spallholz, J. E., and D. J. Hoffman. 2002. Selenium toxicity: cause and effects in aquatic birds. Aquatic Toxicology 57:27–37.

Stendell, R. C., R. I. Smith, K. P. Burnham, and R. E. Christensen. 1979. Exposure of waterfowl to lead: a nationwide survey of residues in wing bones of seven species, 1972–73. U.S. Fish and Wildlife Service Special Scientific Report—Wildlife, No. 223. U.S. Fish and Wildlife Service, Washington, DC.

Stevenson, A. L., A. M. Scheuhammer, and H. M. Chan. 2005. Effects of nontoxic shot regulations on lead accumulation in ducks and American Woodcock in Canada. Archives of Environmental Contamination and Toxicology 48:405–413.

Stewart, F. M., D. R. Thompson, R. W. Furness, and N. Harrison. 1994. Seasonal variation in heavy metal levels in tissues of Common Guillemots, *Uria aalge* from northwest Scotland. Archives of Environmental Contamination and Toxicology 27:168–175.

Stickel, W. H., L. F. Stickel, and F. B. Coon. 1970. DDE and DDD residues correlated with mortality of experimental birds. Pp. 287–294 in W. P. Deichmann (editor), Pesticides Symposia, Inter-American Conference on Toxicology and Occupational Medicine. Helios & Associates, Miami, FL.

Stout, J. H., K. A. Trust, J. F. Cochrane, R. S. Suydam, and L. T. Quakenbush. 2002. Environmental contaminants in four eider species from Alaska and arctic Russia. Environmental Pollution 119:215–226.

Strand, J., and J. A. Jacobsen. 2005. Accumulation and trophic transfer of organotins in a marine food web from the Danish coastal waters. Science of the Total Environment 350:72–85.

Strand, P., B. J. Howard, A. Aarkrog, M. Balonov, Y. Tsaturov, J. M. Bewers, A. Salo, M. Sickel, R. Bergman, and K. Rissanen. 2002. Radioactive contamination in the Arctic—sources, dose assessment and potential risks. Journal of Environmental Radioactivity 60:5–21.

Swannell, R. P. J., K. Lee, and M. McDonagh. 1996. Field evaluations of marine oil spill bioremediation. Microbiological Reviews 60:342–365.

Sweeney, A., and S. A. Sañudo-Wilhelmy. 2004. Dissolved metal contamination in the East River-Long Island sound system: potential biological effects. Marine Pollution Bulletin 48:663–670.

Szaro, R. C., N. C. Coon, and E. Kolbe. 1979. Pesticide and PCB of Common Eider, Herring Gull and Great Black-backed Gull eggs. Bulletin of Environmental Contamination and Toxicology 22:394–399.

Szefer, P., and J. Falandysz. 1983. Investigations of trace metals in Long-tailed Duck (*Clangula hyemalis* L.) from the Gdańsk Bay. Science of the Total Environment 29:269–276.

Taylor, E. J. 1986. Foods and foraging ecology of Oldsquaws (*Clangula hyemalis* L.) on the Arctic Coastal Plain of Alaska. Thesis, University of Alaska, Fairbanks, AK.

Thompson, D. R., and R. W. Furness. 1989. The chemical form of mercury stored in South Atlantic seabirds. Environmental Pollution 60:305–317.

Trust, K. A., D. Esler, B. R. Woodin, and J. J. Stegeman. 2000a. Cytochrome P450 1A induction in sea ducks inhabiting nearshore areas of Prince William Sound, Alaska. Marine Pollution Bulletin 40:397–403.

Trust, K. A., M. W. Miller, J. K. Ringelman, and I. M. Orme. 1990. Effects of ingested lead on antibody production in Mallards (Anas platyrhynchos). Journal of Wildlife Diseases 26:316–322.

Trust, K. A., K. T. Rummel, A. M. Scheuhammer, I. L. Brisbin Jr., and M. J. Hooper. 2000b. Contaminant exposure and biomarker responses in Spectacled Eiders (Somateria fischeri) from St. Lawrence Island, Alaska. Archives of Environmental Contamination and Toxicology 38:107–113.

Turner, A. 2010. Marine pollution from antifouling paint particles. Marine Pollution Bulletin 60:159–171.

United Nations Environment Programme. 2001. Final Act of the Conference of Plenipotentiaries on the Stockholm Convention on Persistent Organic Pollutants. UNEP, Geneva, Switzerland.

United States Fish and Wildlife Service. [online]. 2011. FWS Deepwater Horizon oil spill response: bird impact data and consolidated wildlife reports. <http://www.fws.gov/home/dhoilspill/collectionreports.html> (2 December 2011).

van den Berg, H. 2009. Global status of DDT and its alternatives for use in vector control to prevent disease. Environmental Health Perspectives 117:1656–1663.

Van Pelt, T. I., and J. F. Piatt. 1995. Deposition and persistence of beachcast seabird carcasses. Marine Pollution Bulletin 30:794–802.

Vermeer, K., F. A. J. Armstrong, and D. R. M. Hatch. 1973. Mercury in aquatic birds at Clay Lake, western Ontario. Journal of Wildlife Management 37:58–61.

Vermeer, K., and D. B. Peakall. 1979. Trace metals in seaducks of the Fraser River Delta intertidal area, British Columbia. Marine Pollution Bulletin 10:189–193.

Vermeer, K., and R. Vermeer. 1975. Oil threat to birds on the Canadian west coast. Canadian Field-Naturalist 89:278–298.

Vest, J. L., M. R. Conover, C. Perschon, J. Luft, and J. O. Hall. 2009. Trace element concentrations in wintering waterfowl from the Great Salt Lake, Utah. Archives of Environmental Contamination and Toxicology 56:302–316.

Vorkamp, K., F. Riget, M. Glasius, M. Pécseli, M. Lebeuf, and D. Muir. 2004. Chlorobenzenes, chlorinated pesticides, coplanar chlorobiphenyls and other organochlorine compounds in Greenland biota. Science of the Total Environment 331:157–175.

Wang, D., K. Huelck, S. Atkinson, and Q. X. Li. 2005. Polychlorinated biphenyls in eggs of Spectacled Eiders (Somateria fischeri) from the Yukon–Kuskokwim Delta, Alaska. Bulletin of Environmental Contamination and Toxicology 75:760–767.

Wang, Q., D. Kim, D. D. Dionysiou, G. A. Sorial, and D. Timberlake. 2004. Sources and remediation for mercury contamination in aquatic systems—a literature review. Environmental Pollution 131:323–336.

Wang, Y., T. Wang, A. Li, J. Fu, P. Wang, Q. Zhang, and G. Jiang. 2008. Selection of bioindicators of polybrominated diphenyl ethers, polychlorinated biphenyls, and organochlorine pesticides in mollusks in the Chinese Bohai Sea. Environmental Science and Technology 42:7159–7165.

Wania, F. 2003. Assessing the potential of persistent organic chemicals for long-range transport and accumulation in polar regions. Environmental Science and Technology 37:1344–1351.

Wania, F., and D. Mackay. 1993. Global fractionation and cold condensation of low volatility organochlorine compounds in polar regions. Ambio 22:10–18.

Wayland, M., R. T. Alisauskas, D. K. Kellett, and K. R. Mehl. 2008a. Trace element concentrations in blood of nesting King Eiders in the Canadian arctic. Archives of Environmental Contamination and Toxicology 55:683–690.

Wayland, M., K. L. Drake, R. T. Alisauskas, D. K. Kellett, J. Traylor, C. Swoboda, and K. Mehl. 2008b. Survival rates and blood metal concentrations in two species of free-ranging North American sea ducks. Environmental Toxicology and Chemistry 27:698–704.

Wayland, M., A. J. Garcia-Fernandez, E. Neugebauer, and H. G. Gilcrest. 2001a. Concentrations of cadmium, mercury and selenium in blood, liver and kidney of Common Eider ducks from the Canadian arctic. Environmental Monitoring and Assessment 71:255–267.

Wayland, M., H. G. Gilchrist, D. L. Dickson, T. Bollinger, C. James, R. A. Carreno, and J. Keating. 2001b. Trace elements in King Eiders and Common Eiders in the Canadian arctic. Archives of Environmental Contamination and Toxicology 41:491–500.

Wayland, M., H. G. Gilchrist, T. Marchant, J. Keating, and J. E. Smits. 2002. Immune function, stress response, and body condition in arctic-breeding Common Eiders in relation to cadmium, mercury, and selenium concentrations. Environmental Research Section A 90:47–60.

Wayland, M., H. G. Gilchrist, and E. Neugebauer. 2005. Concentrations of cadmium, mercury and selenium in Common Eider ducks in the eastern Canadian arctic: influence of reproductive stage. Science of the Total Environment 351/352:323–332.

Wayland, M., and A. M. Scheuhammer. 2011. Cadmium in birds. Pp. 645–666 in W. N. Beyer and J. P. Meador (editors), Environmental contaminants in biota: interpreting tissue concentrations, 2nd ed. CRC Press, Boca Raton, FL.

Wayland, M., J. E. G. Smits, H. G. Gilchrist, T. Marchant, and J. Keating. 2003. Biomarker responses in nesting, Common Eiders in the Canadian arctic in relation to tissue cadmium, mercury and selenium concentrations. Ecotoxicology 12:225–237.

White, D. H., and E. Cromartie. 1977. Residues of environmental pollutants and shell thinning in merganser eggs. Wilson Bulletin 89:532–542.

White, D. H., M. T. Finley, and J. F. Ferrell. 1978. Histopathologic effects of dietary cadmium on kidneys and testes of Mallard ducks. Journal of Toxicology and Environmental Health 4:551–558.

Wiener, J. G., D. P. Krabbenhoft, G. H. Heinz, and A. M. Scheuhammer. 2003. Ecotoxicology of mercury. Pp. 409–463 in D. J. Hoffman, B. A. Rattner, G. A. Burton Jr., and J. Cairns Jr. (editors), Handbook of ecotoxicology, 2nd ed. Lewis Publishers, Boca Raton, FL.

Wiens, J. A., R. H. Day, S. M. Murphy, and M. A. Fraker. 2010. Assessing cause–effect relationships in environmental accidents: Harlequin Ducks and the Exxon Valdez oil spill. Pp. 131–189 in C. F. Thompson (editor), Current ornithology (vol. 17). Springer, New York, NY.

Wiens, J. A., R. H. Day, S. M. Murphy, and K. R. Parker. 2004. Changing habitat and habitat use by birds after the Exxon Valdez oil spill. Ecological Applications 14:1806–1825.

Wiese, F. K., and G. J. Robertson. 2004. Assessing seabird mortality from chronic oil discharges at sea. Journal of Wildlife Management 68:627–638.

Williams, L. L., J. P. Giesy, D. A. Verbrugge, S. Jurzysta, G. Heinz, and K. Stromborg. 1995. Polychlorinated biphenyls and 2,3,7,8-tetrachlorodibenzo-p-dioxin equivalents in eggs of Red-breasted Mergansers near Green Bay, Wisconsin USA, in 1977–78 and 1990. Archives of Environmental Contamination and Toxicology 29:52–60.

Wilson, B., B. Lang, and F. B. Pyatt. 2005. The dispersion of heavy metals in the vicinity of Britannia Mine, British Columbia, Canada. Ecotoxicology and Environmental Safety 60:269–276.

Wilson, H. M., P. L. Flint, and A. N. Powell. 2007. Coupling contaminants with demography: effects of lead and selenium in Pacific Common Eiders. Environmental Toxicology and Chemistry 26:1410–1417.

Wilson, H. M., M. R. Petersen, and D. Troy. 2004. Concentrations of metals and trace elements in blood of spectacled and King Eiders in northern Alaska, USA. Environmental Toxicology and Chemistry 23:408–414.

Wilson, L. K., M. L. Harris, S. Trudeau, M. G. Ikonomou, and J. E. Elliott. 2010. Properties of blood, porphyrins, and exposure to legacy and emerging persistent organic pollutants in Surf Scoters (Melanitta perspicillata) overwintering on the south coast of British Columbia, Canada. Archives of Environmental Contamination and Toxicology 59:322–333.

Wobeser, G. A. 1997. Diseases of wild waterfowl, 2nd ed. Plenum Press, New York, NY.

World Health Organization. 2011. The use of DDT in malaria vector control: WHO position statement. Global Malaria Programme, World Health Organization, Geneva, Switzerland.

Yamato, O., I. Goto, and Y. Maede. 1996. Hemolytic anemia in wild seaducks caused by marine oil pollution. Journal of Wildlife Diseases 32:381–384.

Zdziarski, J. M., M. Mattix, R. M. Bush, and R. J. Montali. 1994. Zinc toxicosis in diving ducks. Journal of Zoo and Wildlife Medicine 25:438–445.

Zicus, M. C., M. A. Briggs, and R. M. Pace III. 1988. DDE, PCB, and mercury residues in Minnesota Common Goldeneye and Hooded Merganser eggs, 1981. Canadian Journal of Zoology 66:1871–1876.

CHAPTER SEVEN

Foraging Behavior, Ecology, and Energetics of Sea Ducks*

Ramūnas Žydelis and Samantha E. Richman

Abstract. Sea ducks spend the majority of their life in a cold, marine environment where they must dive, often to great depths, consuming enough food to maintain energy balance. The food consumed is hard shelled, cold, and of low energetic value; yet sea ducks excel when faced with many energetic and thermoregulatory challenges, especially for a relatively small-bodied marine endotherm. The energy cost of thermoregulation and intensive work required to collect daily rations are especially high for sea ducks at high latitudes where they must cope with limited daylight for foraging under harsh winter conditions. To offset these high energetic demands, sea ducks must optimize their decisions about habitat choice, foraging behavior, and prey selection. Foraging behavior strives to maximize gross energy gain and minimize costs to reach energy balance, sometimes under the most extreme of conditions such as the polynyas of the Belcher Islands and St. Lawrence Island. Studying sea ducks under these conditions is equally challenging but through the combined research effort of agencies, organizations, academics, and individuals, we have made great strides in gaining information on the foraging ecology of sea ducks. In this chapter, we review the general aspects of sea duck foraging ecology and diving behavior, foraging energetics, and modeling of energy balance. Each of these topics is discussed in detail, but it is not our intention to review all of the literature available, but rather to highlight the particular discoveries and developments that have greatly increased our understanding of the foraging ecology of sea ducks.

Key Words: allometry, diet composition, diving behavior, ecophysiology, energetics, foraging ecology, individual-based models, metabolic rate, thermoregulation.

S ea ducks in the Tribe Mergini are divers, capable of reaching great depths where they feed on a variety of macroinvertebrates, including molluscs, echinoderms, crustaceans, and fishes. Fish spawning events are important to the diet of sea ducks at certain times of the year but bivalves usually dominate the diet, with the exception of mergansers, which are generally fish specialists. In this chapter, we review general aspects of sea duck foraging ecology and diving behavior, elements of energy gain and cost, and modeling of energy balance and habitat choice.

Formation of the Sea Duck Joint Venture marked the beginning of a period of coordinated investigations of sea duck ecology, which yielded significant progress in attaining new

* Žydelis, R. and S. E. Richman. 2015. Foraging behavior, ecology, and energetics of sea ducks. Pp. 241–265 in J.-P. L. Savard, D. V. Derksen, D. Esler, and J. M. Eadie (editors). Ecology and conservation of North American sea ducks. Studies in Avian Biology (no. 46), CRC Press, Boca Raton, FL.

knowledge about this group of waterbirds. With the development of remote sensing technology, we now have a better understanding of the major habitats or regions used by sea ducks throughout the annual cycle, and ground-based telemetry has provided more localized information on foraging effort, movement patterns, and habitat selection. These studies, in combination with benthic sampling, diet analysis, and data collected from captive studies, have provided information needed for models to estimate energetic carrying capacity or habitat suitability models. Individual-based models require similar types of data but attempt to answer questions at different scales from the individual to the population.

FORAGING ECOLOGY AND DIVING BEHAVIOR

Much of our understanding of animal foraging ecology is based on fundamentals of optimal foraging theory. The theory postulates that organisms forage in such a way as to maximize their net energy intake per unit time (MacArthur and Pianka 1966). The mechanistic explanation of foraging behavior and basic principles of optimal foraging theory help to explain sea duck habitat or patch selection (Kirk et al. 2008), diet choice (Beauchamp et al. 1992), and prey size selection (Bustnes 1998). However, optimal foraging theory also includes constraints such as habitat and diet specialization, digestive processing, social interactions, and anti-predator behavior (Pyke 1984), and therefore some authors dismissed this theory altogether (Pierce and Ollason 1987). Also, sea ducks may not always appear to forage optimally relative to their environment or the spatial/temporal scales of the analyses. Complexity and variability of a natural environment is usually much greater than that of experimental setups, and simple heuristic models of foraging behavior may not immediately explain bird foraging decisions observed in the wild (Heath et al. 2007). Despite debates surrounding the optimal foraging theory (reviewed by Stephens et al. 2007), we maintain that fundamentals of this theory are useful when trying to understand sea duck foraging decisions. Although this chapter does not necessarily aim to review sea duck foraging ecology in the light of optimal foraging theory, we will see that postulates of this theory repeatedly emerge.

Dive Depth and Duration

During the nonbreeding season, sea ducks typically dive for food although occasional observations suggest that in certain habitats, they can forage by head dipping or upending like dabbling ducks (Petersen 1980, Fox and Mitchell 1997, Guillemette 1998, Systad and Bustnes 2001) or even walking on reefs during low tide and eating exposed mussels (Nehls 1995). Harlequin Ducks (*Histrionicus histrionicus*), Buffleheads (*Bucephala albeola*), and Hooded Mergansers (*Lophodytes cucullatus*) typically dive to less than 5 m and use coastal habitats, while most other species aggregate in nearshore waters at depths down to 10–12 m. Long-tailed Ducks (*Clangula hyemalis*), Common Eiders (*Somateria mollissima*), and Scoters (*Melanitta* spp.) can forage over a broader depth range to 20–25 m or deeper. The deepest diving sea ducks are King Eiders (*Somateria spectabilis*) regularly foraging at 30–35 m (Bustnes and Lønne 1997, Mosbech et al. 2006) and Spectacled Eiders (*Somateria fischeri*), which use habitats 40–70 m in depth (Lovvorn et al. 2003; Figure 7.1). There are anecdotal records of Long-tailed Ducks being caught in fishing nets set at 60 m and deeper in the Great Lakes (Schorger 1947). Sea ducks can also forage pelagically without reaching the bottom, such as the Long-tailed Ducks and scoters that feed on pelagic gammarids off the coast of Southern New England near Nantucket (White et al. 2009, White 2013). Fish-eating mergansers also can forage in the water column without reaching the bottom. As a group, sea ducks utilize a rather broad range of depths, which offers them access to ample habitats within shallow seas of the Northern Hemisphere.

For most diving birds, dive time is linearly related to depths at which they are foraging: shallow diving species spend just 10–20 s underwater per dive, whereas for deeper diving species, it usually takes 30–50 s for a single dive (Figure 7.2). It was estimated that the deepest diving Spectacled Eiders in the Bering Sea, where water depths are well over 40 m, must spend about 3 min underwater in a single dive (Lovvorn et al. 2009). While diving capacity is generally linearly related to body mass in air-breathing vertebrates, no such significant relationship was found among ducks (Schreer and Kovacs 1997). It should be recognized, however, that the relationship between foraging depth and dive duration is not always simple and depends on specific habitats used at different times of the year, the type of prey consumed, and the species of sea duck.

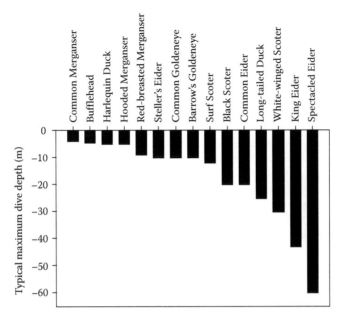

Figure 7.1. Typical maximum diving depths of sea duck species reported in the literature.

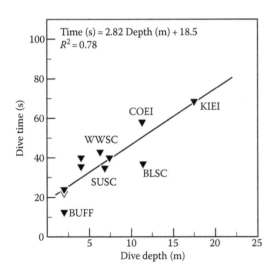

Figure 7.2. Linear relationship between diving depth and average time spent underwater by different sea duck species (time and depth values taken from Gauthier 1993, Guillemette 1998, Kaiser et al. 2006, Heath et al. 2007, Lewis et al. 2007b, Powell and Suydam 2012, Fehmarnbelt Fixed Link Bird Services 2013a). BLSC, Black Scoter; BUFF, Bufflehead; COEI, Common Eider; KIEI, King Eider; LTDU, Long-tailed Duck; SUSC, Surf Scoter; and WWSC, White-winged Scoter.

Foraging Effort

Foraging behavior of sea ducks can be categorized into a dive going underwater and a short pause between consecutive dives, termed a dive cycle. Consecutive dive cycles are often followed by longer breaks (~30 min) at the surface before resuming diving; we define a series of dives separated by relatively short surface intervals as a foraging bout. During the breaks separating foraging bouts, birds usually remain at the surface or engage in other behaviors such as preening and resting. Typically, during individual dives, multiple small prey items are swallowed whole underwater, and through the foraging bout, birds fill their esophagus and proventriculus with prey. Large prey items are often brought to the surface to manipulate and handle before swallowing while the bird is replenishing oxygen stores at the same time (Lewis et al. 2008). Foraging bouts consist of 1 to >50 dives in series, depending on prey availability or accessibility and search time.

Field observations and captive studies demonstrate that sea ducks forage in bouts with multiple dives in series followed by a period of time when the food is processed and passed through the gizzard. Food processing time depends on the amount of material that has to be passed through the gizzard, which ultimately depends on the prey species and size. No studies have specifically addressed this question of how long food is processed in the gizzard before the proventriculus is emptied allowing for another foraging bout to resume. For Common Eiders fed clams, passage rate through the digestive system (mouth to anus) was estimated to take 5 h (Richman and Lovvorn 2003), but that would include multiple foraging bouts under natural conditions. Observational studies indicate that foraging bouts and pauses between bouts are highly variable and depend on a number of biotic (prey density) and abiotic (water depth) factors, habitat, species, and prey type (Beauchamp 1992, Guillemette et al. 1992, Guillemette 1998, Goudie 1999). The amount of time that birds can spend actively foraging is important, especially when species are restricted in time available for foraging due to short day length or tidal currents. Some studies have reported that sea ducks can forage without obvious longer breaks separating foraging bouts, such as Common Eiders and Long-tailed Ducks observed in the southern Baltic Sea (Fehmarnbelt Fixed Link Bird Services 2013a). Foraging without breaks could be explained by birds' feeding on small, thin-shelled bivalves or other soft prey that can pass the gizzard quickly without a need for a digestion break (Fehmarnbelt Fixed Link Bird Services 2013a).

Most of the benthic-feeding sea ducks with large body size tend to feed for lower fractions of time during a day than small species (Figure 7.3). Despite the fact that larger species require more food in absolute terms, smaller ducks have higher mass-specific energy requirements compared to larger species (Ouellet et al. 2013). Thus, smaller species must devote more time to foraging throughout the day to maintain energy balance. Foraging activities occupy the major part of sea duck daytime activity budgets, being second only to resting (Table 7.1).

Within species, time spent foraging can vary substantially depending on many factors, both exogenous and endogenous. Foraging time of wintering Surf Scoters (*Melanitta perspicillata*) was found to increase with decreasing latitude. In Baja California, scoters spend twice as much time foraging as in Alaska and exhibit an intermediate foraging time in British Columbia (VanStratt 2011). The author suggested that such differences are determined by foraging opportunities, which apparently decline towards the southern

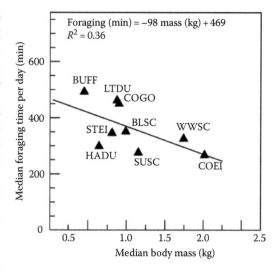

Figure 7.3. Median time spent foraging (minutes/day) during winter months, November–March, as a function of median body mass (values from Nilsson 1970, Goudie 1999, Systad and Bustnes 2001, Schummer et al. 2012, Fehmarnbelt Fixed Link Bird Services 2013a). BLSC, Black Scoter; BUFF, Bufflehead; COEI, Common Eider; COGO, Common Goldeneye; HADU, Harlequin Duck; LTDU, Long-tailed Duck; SUSC, Surf Scoter; STEI, Steller's Eider; and WWSC, White-winged Scoter.

periphery of the distribution range. Surf Scoters in Baja California had a diet of infaunal invertebrates such as ghost shrimp (*Callianassa* spp.) compared to a mostly bivalve diet further north, and in Baja worked hard to collect their daily ration (VanStratt 2011). The northern distributional limit of wintering sea ducks can be determined by daylight. Being primarily diurnal foragers, sea ducks are constrained by short day length at high latitudes, and to maintain energy intake, they increase foraging intensity (Guillemette 1998, Systad et al. 2000). Ambient temperature also has an effect; in colder environments, sea ducks tend to feed more, probably to compensate for increased thermoregulatory demands (Nilsson 1970, Goudie and Ankney 1986, Schummer et al. 2012). However, birds can also increase foraging effort when trying to accumulate fat deposits before long-distance migrations and (or) breeding, a phenomenon known as hyperphagia (Guillemette 2001, Hario and Öst 2002, Rigou and Guillemette 2010, Oppel et al. 2011).

Prey abundance and caloric value also have direct effects on foraging effort. Scoters spend less time diving for the abundant and nutritious eggs of Pacific herring (*Clupea pallasii*) than when

TABLE 7.1
Several examples of sea duck activity budgets shown as percent of daytime spent per certain activity during wintering period.

Species	Location	Month	Feeding	Resting	Locomotion	Comfort	Social	Sources
			\<center\>Activity type\</center\>					
Harlequin Duck	Haida Gwaii, BC	Oct–Feb	51–68	22	10	5	3	Goudie (1999)
Black Scoter	Haida Gwaii, BC	Oct–Feb	53–76	18	8	3	1	Goudie (1999)
Steller's Eider	Izembek, AK	Sep–Mar	70	5	8	15	1.7	Laubhan and Metzner (1999)
	Cold Bay, AK	Sep–Mar	80	1	14	4	1.4	Laubhan and Metzner (1999)
Surf Scoter	Haida Gwaii, BC	Oct–Feb	40–70	21	12	5	2	Goudie (1999)
White-winged Scoter	Haida Gwaii, BC	Oct–Feb	54–75	23	6	6	1	Goudie (1999)
	Puget Sound, WA	Nov–Feb	23–30	11–23	34–44	10–18		Anderson and Lovvorn (2011)

Feeding activity represents foraging bouts, that is, dive and short pauses between consecutive dives. See sources for specific details.

foraging on bivalves (Lewis et al. 2007b). Foraging time increases as prey is depleted (Kirk et al. 2007, Anderson and Lovvorn 2011, VanStratt 2011, Bustnes et al. 2013) or birds completely abandon a depleted area (Kirk et al. 2008, Lovvorn et al. 2013). However, duck responses to depleted food resources vary by habitats and prey type. For example, Surf Scoters in British Columbia were found spending similar amounts of time foraging in three habitat types differing by bivalve prey species, the nutritional content, quality, and quantity, which suggests that despite variations in prey landscapes, Surf Scoters are able to maintain similar intake rates by redistribution and habitat selection (Kirk et al. 2007). Several response mechanisms to changing prey abundance have been observed. For example, Common and King Eiders wintering in northern Norway increased their foraging effort, redistributed themselves within habitats, and dispersed into smaller flocks as food resources declined over winter (Bustnes et al. 2013). These examples imply high adaptability of sea ducks to varying environmental conditions and an ability to employ multiple foraging strategies when responding to changing prey availability or moving between habitat types.

A rather interesting phenomenon is synchronous group foraging of some sea duck species. Sea ducks are often gregarious during nonbreeding periods, rafting together in large groups. When birds are in groups, synchronous foraging occurs when the entire flock dives simultaneously, and

has been documented in Surf Scoters, Barrow's Goldeneyes (*Bucephala islandica*), Common Eiders, and Steller's Eiders (*Polysticta stelleri*, Schenkeveld and Ydenberg 1985, Beauchamp 1992, Guillemette et al. 1993, Laubhan and Metzner 1999). Benefits of synchronous foraging are not fully understood; however, synchrony might help reduce kleptoparasitism by conspecifics and gulls (Schenkeveld and Ydenberg 1985) and maintain flock cohesion during dives (Beauchamp 1992).

Nocturnal Foraging

Sea ducks are thought to be visual foragers and to limit their foraging activity to daylight hours (Owen 1990, McNeil et al. 1992). However, to cope with reduced daylight at high latitudes, some species may feed well after sunset (Systad et al. 2000, Systad and Bustnes 2001). Radio telemetry revealed that Surf Scoters and White-winged Scoters (*Melanitta fusca*) essentially do not dive at night in Baynes Sound, British Columbia (Lewis et al. 2005). Similar results were reported for Harlequin Ducks in Prince William Sound, Alaska (Rizzolo et al. 2005), and for Common Eiders and Long-tailed Ducks in the southern Baltic Sea (Fehmarnbelt Fixed Link Bird Service 2013a). No nocturnal foraging was observed among Common Eiders wintering in the Gulf of St. Lawrence (Guillemette 1998) or in the polynya near the Belcher Islands of Hudson Bay (Heath and Gilchrist 2010). However, under certain conditions or in particular habitats, nocturnal foraging of sea ducks has been observed for several species. Systad and Bustnes (2001) reported nocturnal foraging of Steller's Eiders in northern Norway (70°N) in midwinter and explained this behavior as an adaptation to the short period of daylight and the cold environment. It is also known that molting Common Eiders regularly forage at night in tidal habitats of the Wadden Sea (Nehls 1995) and nocturnal diving was recorded during 10% of the time Common Eiders spent foraging in Denmark (Pelletier et al. 2007). Common Eiders wintering in southwest Greenland use two foraging strategies. They are diurnal foragers in an outer coastal habitat and primarily feed at night in a fjord habitat; the latter strategy was explained as predator avoidance behavior (Merkel and Mosbech 2008). Surf Scoters also forage during day and night in the southern periphery of their wintering habitat of Baja California (VanStratt 2011). In summary, at least some species of sea ducks are capable of feeding in the dark, although diurnal foraging clearly prevails.

Diet Composition

Among the sea ducks, Red-breasted Mergansers (*Mergus serrator*) and Common Mergansers (*M. merganser*) are obligatory piscivores that forage on small (10–15 cm long) and medium-sized fish (10–30 cm), respectively. Prey composition varies with water body and geographic location and typically represents the most abundant fish species of suitable size (Mallory and Metz 1999, Titman 1999). Hooded Mergansers have a more diverse diet where fish comprise about one half of their food, while aquatic insects, crustaceans, crayfish, and molluscs are eaten regularly (Dugger et al. 2009). Anderson et al. (2008) reviewed the diets of sea ducks that forage on diverse taxa, with small sea ducks typically consuming a mix of bivalves, gastropods, crustaceans, and fish, while diets of scoters and eiders consist of mostly bivalves and, to a lesser extent, crustaceans and echinoderms. Some species, such as White-winged Scoter and Spectacled Eider, specialize in foraging on clams and other infaunal prey, while goldeneyes, Harlequin Ducks, and Common Eiders are epifaunal specialists and feed predominately on prey living on the surface of bottom substrate or on submerged vegetation. Long-tailed Duck, Surf Scoter, and Black Scoter (*Mergus americana*) are considered generalist

foragers because they readily feed on either infauna or epifauna, depending on habitat. While these generalizations reflect predominant food choices of sea ducks, there are exceptions. For example, chitons were the major food item of Common Eiders in Iceland (Kristjánsson et al. 2013), sea urchins and crabs comprise a substantial proportion of Common Eider diet in the Gulf of St. Lawrence (Guillemette et al. 1992), and fish are eaten by Common Eiders in the Netherlands (Leopold et al. 2012). Sea duck diets were summarized and conceptualized by Ouellet et al. (2013) who suggested that smaller species use a risk-prone foraging strategy and search for higher-quality, less-abundant prey in part because of their higher mass-specific metabolic rate. In contrast, larger species typically feed using a risk-averse strategy to maximize intake of lower quality but predictably abundant prey. Due to body mass and time constrains, diurnally feeding, small-sized sea ducks would not be able to achieve positive or neutral energy balance if they foraged on voluminous prey of low caloric value, as size of a single meal and a number of meals per day is limited by the birds' morphology and day length (Ouellet et al. 2013). This hypothesis is further supported by activity budgets of sea ducks; the proportion of time spent foraging is inversely related to species size (Figure 7.3).

Sea ducks readily switch their diets to ephemeral but abundant foods. Many species, including Harlequin Duck, Long-tailed Duck, White-winged Scoter, Surf Scoter, and Steller's Eider, turn to feeding on fish eggs when available during mass spawning events. Such phenomena have been documented in British Columbia and Alaska (Rodway 2003, Lewis et al. 2007a, Anderson et al. 2009, Lok et al. 2012), northern Norway (Bustnes and Systad 2001), and the Baltic Sea (Stempniewicz 1995, Žydelis and Esler 2005). Large aggregations of Surf Scoters, along with a few other duck species, have also been recorded at a site where mass spawning of marine polychaetes took place (Lacroix et al. 2005). Another type of opportunistic foraging behavior has been observed in Puget Sound where Surf Scoters and White-winged Scoters were diving next to feeding gray whale (*Eschrichtius robustus*), which presumably exposed benthic invertebrates by disturbing large amounts of bottom sediments (Anderson and Lovvorn 2008). Similarly, Long-tailed Ducks were regularly observed diving in the wake of the Nantucket Island ferry where

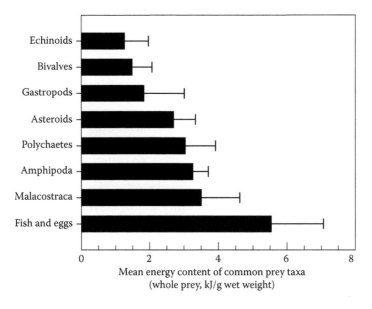

Figure 7.4. Average energetic values (±SD) of principal sea duck prey expressed as kilojoules per gram of fresh mass of the whole organism. Prey energetic values adapted from Ouellet et al. (2013) and Rumohr et al. (1987).

they presumably forage on prey dislodged by the ship propeller (Perry 2012).

Most sea duck foods are of low caloric density with high inorganic content such as sea urchins, molluscs, and crustaceans, while polychaetes, fish, and fish eggs are of high energy density and easily digestible (Figure 7.4). However, energetic value of prey varies substantially by size, season, and stage of reproductive cycle. For example, variation in energy contents of filter-feeding bivalves may range up to 100% throughout the year depending on reproductive stage, phytoplankton availability, and ambient temperature (Bayne and Worrall 1980, Kautsky 1982, Waldeck and Larsson 2013). The flesh-to-shell ratio depends on bivalve size and habitat (Bustnes and Erikstad 1990, Bustnes 1998, Kirk et al. 2007). Therefore, variation in prey energy between locations should be considered whenever possible to most accurately assess the energy intake of sea ducks.

Methods for Studying Sea Duck Diets

Analysis of esophagus and gizzard contents of dead birds is an important method to examine prey species, size preference, and overall diet composition of sea ducks. However, gut content analysis has limitations as (1) soft-bodied prey are often underrepresented because of rapid digestion in the foregut, (2) the esophagus and gizzard

typically contain food from only a single meal and may not be representative of an individual bird's diet, and (3) the sample size is usually limited.

Some information about sea duck diet composition could be collected using focal observations of foraging birds and recording food items brought to the water surface (Lewis et al. 2008), or analyzing remnants of hard prey in bird droppings (Lewis et al. 2007b). These methods can be representative of diets dominated by large bivalve prey, which sea ducks usually handle at the surface (Ydenberg 1988, de Leeuw and van Eerden 1992). However, small or soft-bodied prey would be missed in analyses despite often comprising an important proportion of the diet (Richman and Lovvorn 2004, Anderson et al. 2008).

The challenge in comparing diet studies of species that are often located far from shore is to collect ducks under similar or consistent protocols (Anderson et al. 2008). Authors reporting results of direct diet examination use a range of metrics, including prey counts, frequency of occurrence, percent wet mass, dry mass, ash-free dry mass, and volume, with little consistency, but chosen metrics have important consequences to interpretation. When possible, diet composition should be reported in several standard metrics such as percent composition of prey by mass and percent frequency of occurrence, as well as an index of the size of prey. Multiple metrics are especially important when only biomass (g) is reported. For example, diet items such as clams or mussels can span a large range in sizes that are consumed by ducks; however, shell mass increases exponentially with increasing length, but to a lesser extent for meat content. Thus, a single, individual mussel >80 mm in length can have similar biomass as 30 smaller clams, but because the smaller clams can be easily ingested while underwater, only the smaller clams will be consumed. Also, the total or mass-specific energy content of mussels increases with increasing length to an optimum, before energy content relative to shell fraction declines. For many of the large sea ducks like Common Eider and White-winged Scoter, we unexpectedly find dietary preferences for smaller prey items than we would expect for their body size. Video observations have shown eiders consistently digging between larger prey clumps or near the surface of the sediment in search of smaller mussels (S. E. Richman, pers. obs.). Thus, it is important to report the size composition of ingested prey because energy content can vary with prey size.

Estimating Diet with Biochemical Markers

An alternative to direct diet analysis is the application of stable isotope and fatty acid analyses to describe animal diets and trace migratory origins. Stable isotopes of carbon, hydrogen, oxygen, and even sulfur have revolutionized the way we view animal diets. When differences exist between diets or habitats, biochemical markers can determine how nutrients are allocated from resources to blood, feathers, claws, eggs, and other tissues. For example, a recent major development in the way we understand contributions of endogenous nutrients from the body versus exogenous nutrients from the diet to egg formation by laying female birds involves diets that differ in source nutrients (reviewed by Hobson 2006; Schmutz et al. 2006; Bond et al. 2007; Chapter 5, this volume). This approach has been largely successful in quantifying nutrient allocations but relies on an untested assumption that the isotopic link between endogenous body tissues to eggs laid under a strictly capital strategy resembles that of a carnivorous bird under a purely income strategy (Gauthier et al. 2003). The true capital discrimination factor remains elusive because it is generally difficult to get birds to lay in captivity without food.

To obtain more definitive results from this promising method, we first need to develop species-specific calibration coefficients and discrimination factors on known, homogenized diets. Combination of fatty acid analysis and compound-specific stable isotope analysis (CSIA) would allow quantifying consumer diets and tracing animal diet sources. Using fatty acids from consumers and a comprehensive diet library combined with calibration coefficients that account for consumer metabolism, it is possible to estimate the proportion of diet items in the consumer (quantitative fatty acid signature analysis; Iverson et al. 2004, Wang et al. 2010). Similarly, CSIA focuses on tracking the isotopic signature of specific fatty acids that can only arise from diet. Thus, these dietary fatty acids in an animal should reflect that of the diet source (Budge et al. 2008, 2011). Although these methods have been developed for some of the marine food webs, we could begin acquiring common prey items for sea ducks to contribute to a fatty acid library. At that time, we may be able to apply these methods to the field.

FORAGING ENERGETICS

An approach to foraging energetics is to determine the components of an animal's energy budget that are considered important, such as activity in flight, diving, or resting, and then test those values to see if they are sensitive to a change. Energetic modeling can be approached in a number of ways, but all methods have the same underlying theme of trying to determine gross energy in minus energy costs. The amount of effort an animal must expend to acquire enough resources to maintain a positive energy balance can be broken down into measureable components. Energetics can be viewed as a conceptual flow chart analogous to a cost–benefit relationship in economics, where an animal must balance energy cost with energy gain, and is the basis for the our approach for energy balance (Figure 7.5, adapted from Lovvorn et al. 2009).

For a sea duck foraging in a given habitat, prey is often distributed in dense patches that must be found during the search time. For our purposes, we assume that a bird has found a prey patch and the amount of food that is consumed by a duck during a dive is determined by the prey type (taxa), density of prey, and the size structure of prey. To maintain energy balance, an individual will expend energy to find, consume, and process food,

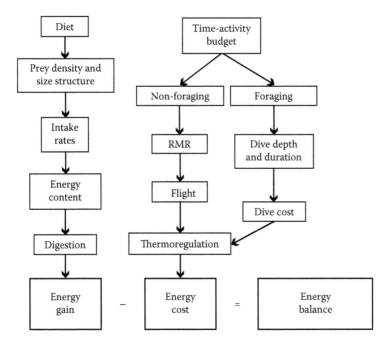

Figure 7.5. Flow chart of the energy balance approach. (Adapted from Lovvorn et al. 2009.)

in addition to maintenance metabolism, different activities such as diving, flying, or swimming, growth, and production of waste products. Here, we will use this energetic approach to evaluate the elements of energy gain and cost for sea ducks. Elements of energy gain incorporate the availability and accessibility of prey and the rate of ingestion and digestion. Elements of energy cost, on the other hand, include the amount of time an animal spends in a particular activity multiplied by the cost of that activity and additional costs of thermoregulation. Here, we briefly discuss individual components on energy gain and cost and describe how these factors allow sea ducks to be successful in the marine environment. Studies of captive birds and field experiments have improved our understanding of how sea ducks allocate energy for daily or seasonal activities, which will influence when and where they migrate or reproduce, as well as habitat selection and ultimately survival.

ELEMENTS OF ENERGY GAIN

The amount of energy gained by a predator varies widely between different prey types such as mussels versus herring spawn and often changes considerably by prey size, season, and location. The blue mussel (*Mytilus edulis*) is a common bivalve prey for many sea ducks but tends to be smaller and thinner shelled at aquaculture sites than populations of the same species in intertidal areas (Kirk et al. 2007) and has a different shell size and composition when growing in warmer and often brackish waters of the Baltic (Kautsky et al. 1990). These different characteristics of primary prey such as mussels or clams, as well as overall species composition, abundance, and size structure of benthic prey communities, all have important consequences to the amount of food consumed (Zwarts and Wanink 1993; Richman and Lovvorn 2003, 2009; Merkel et al. 2007; Lovvorn et al. 2009). Thus, gross energy gain depends on all of these factors and especially the availability and accessibility of prey within a habitat.

For clarity, our discussions of energy budgets will use calories (cal), joules (J), and watts (W) as principal units because they are easily interconverted (1 cal = 4.184 J or 1 J = 0.2389 cal, 1 W = 1 J/s). Note that W and W/kg are not the same measures and the mass of the animal should always be reported so that energy costs can be compared among species and studies.

Energy Intake

Gross energy intake was presented by Richman and Lovvorn (2008) in a comparison of foraging profitability (energy intake minus cost) and threshold prey densities (the density at which energy balance becomes positive), for captive scaup (*Aythya* spp.) and White-winged Scoters foraging in a dive tank using the following equation:

$$\text{Energy intake (J/dive)} = \text{Intake}_{\text{size,depth}} \times \text{AE}_N \times \text{Energy} \times \text{Time}_{\text{bottom}}$$

where energy intake is calculated for individual dives (J/dive). It is important to recognize the unit of choice when estimating energy intake where intake rates are based on functional response curves generated for White-winged Scoters feeding on clams (18–24 and 24–30 mm in length) at a burial depth of 4 or 6 cm in sand (Richman and Lovvorn 2003, 2008). Diving ducks typically show a type II functional response where intake rates are an increasing function with increasing prey density (number of individual prey items/m²) up to a maximum when a bird can only eat so many per second (Richman and Lovvorn 2003). Intake rates are the number of prey items consumed per second while the bird is at the bottom. Energy intake for a particular prey item is then adjusted for the proportional assimilation efficiency (AE$_N$), which is the amount of ingested energy absorbed by the gut corrected for nitrogen retention (~70% for eiders fed clams, Richman and Lovvorn 2003). For White-winged Scoters in a large dive tank, intake rates for prey of differing size and depth in the substrate influenced the amount of energy gained while foraging at the bottom with larger prey and deeper burial depths resulting in lower intake rates (Richman and Lovvorn 2003, 2004). Perhaps the handling associated with larger prey as well as the extra time needed to excavate larger clams from sediments explains the reduced intake rates.

For maintaining positive energy intake, the sum of foraging time and digestive processing may not exceed the total time available for foraging. Because sea ducks consume prey whole, preference for smaller bivalves than expected has been explained by differential availability, handling times, effects of meat/shell ratio on nutrient gain relative to passage rates, or as a means of avoiding prey that are too large to be swallowed (Bustnes and Erikstad 1990). Resistance of shells to crushing in the gizzard may also affect selection of species and sizes of bivalve prey (Navarro et al. 1989). Further, shell lengths that are most

often consumed by sea ducks are 10–30 mm, which may be optimal in terms of the ratio of proportion of meat to the cost of crushing the shell (Bustnes and Erikstad 1990, Nehls 1995, Hamilton et al. 1999).

The time required to process food in the gut can be longer than the time needed to find, handle, and ingest food (Jeschke et al. 2002). Further, passage times through the gut vary widely among prey types. Soft-bodied prey may travel more quickly through the digestive system than mussels that must be crushed in the gizzard before moving into the small intestine. Differences among prey in retention time in the gut can therefore affect acquisition of nutrients and energy (Guillemette 1994, 1998). To complicate matters further, waterfowl show great phenotypic flexibility and rapid adjustments in gut morphology, altering gut capacity, volume, rates of uptake, and absorption (Goudie and Ankney 1986, McWilliams and Karasov 2001).

Body Reserves

If energy intake exceeds energy demands, birds can store energy as body reserves for use during periods when energy expenditure exceeds nutritional intake. Of four basic body components, protein, lipid, mineral, and water, lipid is the main energy reserve because it has minimal structural functions and can be catabolized (Schamber et al. 2009). The energy content of lipid at 39.3 kJ/g is more than twice that of protein at 17.8 kJ/g (Schmidt-Nielsen 1975). Birds use lipid reserves as a buffer in uncertain foraging and environmental conditions, as an energy source to fuel long-distance migrations, and as extra energy for egg production and fasting during the nesting period.

Lipid reserves are an excellent source of energy for birds that must dive and fly. Both lipid and protein yield a lot of energy; however, proteinaceous tissue is more than 70% water and therefore heavier than lipids. The low water content of lipid makes it an ideal storage form although it is a common misconception that having higher lipid reserves for diving birds will influence dive costs due to increase costs of buoyancy. While lipid and adipose tissues are far more buoyant than proteinaceous tissue like muscle, subcutaneous fat provides insulation that does not compress with depth underwater unlike the air layer in the plumage, which is the primary source of insulation for sea ducks. Since the buoyancy of air is nine times greater than that of lipid, air spaces in the respiratory system and plumage are far more influential to diving birds. Birds can make small adjustments of air spaces and easily compensate for any buoyancy changes due to changes in lipid stores (J. R. Lovvorn, Southern Illinois University, pers. comm.).

Body reserves of shorebirds on average consisted of 85% lipid, 10% water, and 5% protein (Kersten and Piersma 1987), and similar levels might be expected for sea ducks. These authors also reported that for 1 g increase in body mass, a bird needs 45.7 kJ of additional energy intake at the energy deposition efficiency of 88%. Body mass is often used to assess bird body condition, especially when mass is used together with additional body size metrics (Schamber et al. 2009). Dynamics of sea duck body condition throughout the annual cycle reflect strategies of nutrient balance and reveal potentially critical periods for birds. Body mass in waterfowl typically is low following the breeding season and through molt but increases in the autumn, declines throughout the winter, and increases again prior to spring migration (Hepp et al. 1986, Kestenholz 1994). Sea ducks follow a similar trend in body mass dynamics (Petterson and Ellarson 1979, Anderson and Lovvorn 2011, Schummer et al. 2012, Palm et al. 2013). Maintaining high body mass in early winter could be viewed as an adaptation when the uncertainty of sufficient food intake during the shortest days of the year and possible harsh winter weather conditions may pose constraints on energy intake. The uncertainty of limited food resources or foraging time presumably relaxes as days get longer with progressing winter and subsequently birds do not need to invest in carrying extra reserves. However, body mass could also decline as a result of not meeting daily nutritional demands due to depletion of food (Anderson and Lovvorn 2011) or severe environmental conditions (Schummer et al. 2012). Wintering White-winged Scoters were found to maintain highest body mass at an exposed offshore site where birds faced unpredictable food resources and increased thermoregulatory costs compared to sheltered nearshore sites in British Columbia and Washington State (Palm et al. 2013). Interestingly, this study showed that despite varying scoter body mass at different sites, levels of plasma metabolites varied little indicating that White-winged

Scoters at all sites maintained physiological homeostasis (Palm et al. 2013).

ELEMENTS OF ENERGY COST

Sea ducks face many challenges to living in a marine environment where maintaining body temperature in cold water is energetically expensive. Nevertheless, Long-tailed Ducks, Steller's Eiders, and goldeneyes are small-bodied sea ducks that inhabit cold water in polar and subpolar regions alongside large-bodied ducks like Common Eiders and Scoters. In these habitats, energy expenditure is high for maintenance metabolism, foraging and nonforaging activities (including flight and diving), and thermoregulation.

Maintenance Energy Costs

Basal metabolic rate (BMR) is the measure of the lowest rate of energy expenditure for individuals that are sexually and physically mature, postabsorptive, at rest, and within a thermoneutral environment (Kleiber 1932). We often use BMR as a common benchmark for comparing species using allometric relationships with body mass on a log scale. However, sea ducks rarely satisfy the conditions for basal metabolism and are always well above those predicted from standard allometric equations (Ellis and Gabrielsen 2002). A more practical measure of metabolism for sea ducks, which includes thermoregulatory costs, is resting metabolic rate (RMR). We can then measure activity-specific metabolic rates such as the costs of swimming, preening, or diving. Daily energy expenditure (DEE) is estimated as the duration of different activities multiplied by their respective costs and summed for 24 h per day. While we often focus on activities of high apparent cost like diving or flying, these activities may occupy a relatively small fraction of the day (Table 7.1). While at sea, sea ducks can spend 40%–80% of their day floating on the water surface, making resting costs the largest component of their DEE. Using a spatially explicit simulation model, Lovvorn et al. (2009) showed that for Spectacled Eiders in the Bering Sea, diving and flying were a minor part of the overall energy budget.

Metabolic rates for a few species of sea ducks have been measured in both air and for birds floating on water at varying temperatures. While differences in experimental setup and behavior

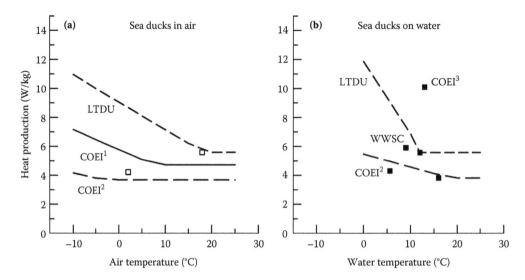

Figure 7.6. RMR for sea ducks in air (a) and floating on water (b) at varying temperatures. Regression lines for single studies in which data were collected over a range of temperatures are given in Table 7.2 (Richman and Lovvorn 2011). COEI, Common Eider ([1]Gabrielsen et al. 1991, [2]Jenssen et al. 1989, [3]Hawkins et al. 2000), WWSC, White-winged Scoter, and LTDU, Long-tailed Duck.

state of the birds make comparisons a challenge, Richman and Lovvorn (2011) found that sea ducks were far more resistant to heat loss than other species of sea birds (Figure 7.6, Table 7.2). In fact, for Common Eiders, there was little difference in RMR between air and water at any temperature. For much smaller Long-tailed Ducks, metabolic rates in air and water were similar above 10°C but increased faster in water at lower temperatures and rose more rapidly than for other larger species of sea ducks (Richman and Lovvorn 2011). Common Eiders showed no clear lower critical temperature in air or on water (Jenssen et al. 1989), perhaps because of their large body size and thick plumage air layer.

A major challenge, however, is the variation and inconsistencies in measures of metabolic rates that may have important implications for energetics models. For example, the metabolic rate of Common Eiders while floating on water at 16°C was measured as 3.83 W/kg by Jenssen et al. (1989), but 10.1 W/kg at similar temperatures (14°C–19°C) by Hawkins et al. (2000). If we were to construct a time–energy budget for a Common Eider floating on water at 15°C for a conservative estimate of 17 h a day (Systad et al. 2000), the total daily cost of resting on the water surface would be ~457 kJ if we used 3.83 W/kg versus 1,100 kJ/day if we used 10.1 W/kg. The differences in the estimates would

more than double the amount of food required per day, an anomaly that would increase if the trend were extrapolated to lower water temperatures (Richman and Lovvorn 2011). For sea ducks that spend far more time resting on water than actively diving, valid estimates of RMR are critical to models of energy balance used to estimate of the amount and quality of habitat they need (Lovvorn et al. 2009).

Activity Costs

Cost of Flight

Nonbreeding sea ducks spend small portions of their day engaged in social activities, preening, or flying, except for migration periods. Using implantable data loggers, Pelletier et al. (2007, 2008) found that nonmigrating Common Eiders fly on average only about 10 min/day. No similar information is available for other sea duck species, but they probably do not spend more than 1%–2% of their time in flight. Although the proportion of time spent flying for sea ducks may be low, the relatively high energy cost of flight means that it cannot be ignored when estimating energy expenditures.

The energy costs of flying can be estimated using freely available software (Program Flight, Pennycuick 2008). The only input parameters

TABLE 7.2
Resting metabolic rate in air (RMR$_{air}$) and on water (RMR$_{water}$) and mean body mass.

Species	N	Mass (kg)	RMR (W)	T_{obs} (°C)	T_{lc} (°C)	Regression (Below T_{lc})	RQ	Sources
In Air								
Common Eider	12	1.66 ± 0.25	4.75 ± 0.39	>7	7	$5.81T_A - 0.14T_A$	0.70	Gabrielsen et al. (1991)
	6	1.79 ± 0.13	4.22 ± 0.35	>2	2		0.77	Hawkins et al. (2000)
	7	1.66 ± 0.17	3.68 ± 0.48	>1.5	1.5	$3.46T_A - 0.07T_A$	0.71	Jenssen et al. (1989)
Long-tailed Duck	5	0.49 ± 0.03	5.6 ± 0.32	>18	18	$9.06T_A - 1.9T_A$	0.71	Jenssen and Ekker (1989)
On Water								
Common Eider	7	1.79 ± 0.13	10.10 ± 2.027	13.7–19			0.77	Hawkins et al. (2000)
	10	1.80 ± 0.09	4.3 ± 0.3	5.6			0.85	Jenssen and Ekker (1991)
	5	1.95 ± 0.09	3.83 ± 0.24	16–25	16	$5.48T_w - 0.09T_w$	0.71	Jenssen et al. (1989)
White-winged Scoter	5	1.09	5.91	9			0.80	Richman and Lovvorn (2008)
Long-tailed Duck	5	0.49 ± 0.03	5.59 ± 0.56	>12		$11.87T_w - 0.50T_w$	0.70	Jenssen et al. (1989)

Measures of resting metabolic rate (RMR), the observed temperature (T_{obs}) in air (T_A) or while floating on water (T_w), and lower critical temperature (T_{lc}). In each study, the respiratory quotient (RQ) was calculated directly from the ration of CO_2 production to oxygen consumption (from Richman and Lovvorn 2011).

required for this mechanical flight model are body mass, wing span, and wing area, and for many bird species, these morphometric measures are available in Alerstam et al. (2007). Sea ducks have a relatively small wing area and consequently high wing loading (body mass divided by wing area), and as a result, they are fast fliers (Lovvorn and Jones 1994, Alerstam et al. 2007). Fast flight, however, is energetically expensive, ranging from 67 to 99 W/kg for the assessed species (Figure 7.7). Flight costs do not correlate with bird mass, but it seems that mergansers, which pursue their prey underwater, have the lowest flight costs compared to ducks that forage on sessile benthic prey. Lovvorn and Jones (1994) tested the hypothesis that high wing loading in foot-propelled divers could be an adaptation to diving; however, they concluded that wing morphology is probably the result of relaxed competing demands for flight maneuverability and takeoff ability in open water environments.

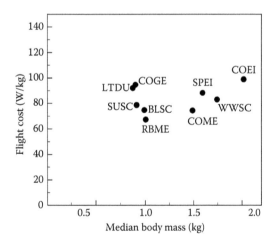

Figure 7.7. Flight costs of several sea duck species estimated using Flight software (Pennycuick 2008); parameters for calculations taken from Alerstam et al. (2007). BLSC, Black Scoter; COEI, Common Eider; COGO, Common Goldeneye; COME, Common Merganser; LTDU, Long-tailed Duck; RBME, Red-breasted Merganser; SUSC, Surf Scoter; SPEI, Spectacled Eider; and WWSC, White-winged Scoter.

Cost of Foraging

For diving birds, the energy costs of locating, pursuing, and capturing prey are complicated because they are obligated to return to the surface to breathe. Because sea ducks dive for their food, they must overcome the hydrodynamic drag associated with propelling the body through the water and the force of buoyancy while submerged (Lovvorn and Jones 1991, Lovvorn et al. 1991). As the duck descends in the water column, air in the respiratory system and plumage is compressed by hydrostatic pressure with increasing depth underwater, which then reduces the costs of countering the force of buoyancy. The drag force acting on a body increases rapidly with increasing speed (Hoerner 1958). These forces act in concert with other physiological limitations as a result of breath holding while exercising. Biomechanical models use measures of plumage volume, compression, and drag from live animals and specimens in tow tanks in concert with measures of oxygen consumption (mL O_2/s) to calculate aerobic efficiency during dives (Lovvorn et al. 1991b, 2004). We can then estimate costs of diving as a function of different depths. Interestingly, the costs of diving are not exceptionally higher than other activities such as flying or swimming (Lovvorn et al. 2009). Deep diving sea ducks often reach depths of over 10 m, where buoyancy becomes negligible and costs of underwater swimming are often lower than surface swimming and much lower than flying. In addition, thermal substitution may further offset energy costs for wing-propelled diving as waste heat generated from two large muscle masses in the breast and legs may also benefit the energy budget by reducing the costs of thermoregulation in cold water.

Diving is an essential part of foraging activity for sea ducks, but measuring the cost of diving for free-ranging birds is often complicated and challenging. Methods such as attaching data microloggers to birds or employing the doubly-labeled water technique require recapture of the same bird to measure metabolism, and cannot provide information on the cost of specific behaviors alone. Direct measures of metabolism using the dive-hole technique developed for penguins and seals in the Antarctic are not possible in most areas where sea ducks winter (Kooyman 1965). While there is great promise in the use of heart-rate data loggers to obtain activity-specific metabolic rates (Butler et al. 1995), calibration for multiple species of differing body size is needed before this method can be used more broadly.

Observations of underwater swim mechanics further reveal details that allow sea ducks to minimize costs. With captive White-winged Scoters diving to 2 m in a tank, Richman and Lovvorn (2008) found that during descent birds used either foot propulsion alone or foot and wing propulsion simultaneously. Only foot strokes were used to resist the upward force of buoyancy while at the bottom, with the wings partially opened lateral to the body, which may increase drag to oppose buoyancy. Heath et al. (2006) described the stroke patterns of Common Eiders diving in the Canadian Arctic (Figure 7.8). Coordination of foot strokes and wing strokes may reduce overall drag by allowing more constant instantaneous speeds throughout the stroke cycle. During the upstroke, the wings had a fairly low angle of attack, whereas during the downstroke, and especially the transition between upstroke and downstroke, the angle of attack was much greater with presumably much higher drag. The recovery stroke, or retraction of the feet with the webbing closed, occurred during the upstroke of the wing, and the power stroke of the feet occurred during the transition of the wing stroke from upstroke to downstroke. This timing may avoid higher drag incurred by more unsteady thrust at the same mean speed, thereby reducing the energy cost of descent (Lovvorn 2001, Heath et al. 2006). Richman and Lovvorn (2008) further partitioned the energy cost of descent for White-winged Scoters, showing that the use of wings in addition to feet reduced the energy cost of descent by ~34%, which is an important advantage.

Thermoregulation

For shallow diving sea ducks, increased work rates to counter buoyancy near the surface may generate excess heat from exercising muscles, which may offset increased thermoregulatory demands of diving in cold water. However, optimal work rates during diving may change with decreasing water temperature. As thermostatic demands increase at lower water temperatures, heat loss may exceed metabolic heat production from exercise, and shivering begins (Lovvorn 2007).

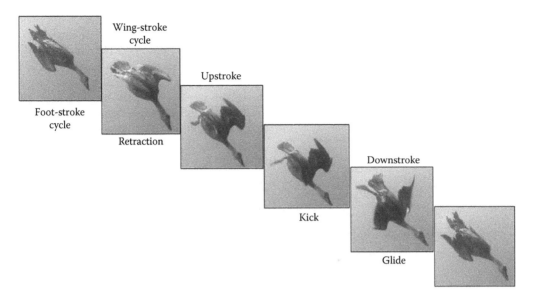

Figure 7.8. Wing- and foot-stroke cycles of the Common Eider diving at the Belcher Islands, Nunavut. (Photos from Heath et al. 2006.)

During exercise, threshold temperatures may set limits where heat loss exceeds metabolic heat production, which will affect an animal's foraging profitability as energy gain minus cost.

When discussing maintenance costs in the previous sections, we showed that thermoregulation can constitute a substantial part of all energy costs. The high thermostatic demands for diving birds can be met by (a) increasing metabolic heat production, (b) regional changes in blood flow and vasomotion, (c) allowing body temperature to decline, or (d) substitution from the heat generated by exercising muscles or digestion. There are both advantages and disadvantages to increasing metabolic heat production as it depletes oxygen stores quickly and requires higher energy intake.

Another important mechanism for thermoregulatory control in changes to blood flow, temporal variation in pulsatile blood flow to the legs, and overall reduction in body temperature during diving is to reduce heat loss and conserve oxygen by metabolic depression (Butler et al. 1995, Culik et al. 1996, Bevan et al. 1997, Handrich et al. 1997, de Leeuw et al. 1998); however, this is not a universal strategy. Some birds increase their body temperature during diving (Stephenson 1994), while others remain unchanged (Gallivan and Rolan 1979; Kaseloo and Lovvorn 2003, 2005). Variation in body temperature suggests that species of differing body mass and thermal inertia may use different strategies for controlling heat loss and may use them in combination with other means of heat conservation.

Daily Energy Requirements and Allometry

Depending on a number of factors, we can calculate an animal's daily energy requirements (DER), which will guide estimates of the amount of food and thus amount of foraging habitat required. These calculations may appear straightforward, but their accuracy varies tremendously depending on what variables or parameter estimates are used in the calculations. The total energy requirement of an animal will change with environmental temperature throughout the day depending on the thermoregulatory capacity of the animal and will further increase with activity and during periods of growth, storage, and reproduction.

A convenient but perhaps oversimplified approach is to use allometric relationships between RMR and body mass to predict food requirements. One of the more common equations used is based on Miller and Eadie's (2006) analysis for waterfowl. However, data for diving ducks and particularly sea ducks are better described by allometric equations developed for seabirds (Ellis and Gabrielsen 2002). Richman and Lovvorn (2011) plotted RMR (W) in air and on water for diving ducks and found a linear relationship described by RMR_{air}

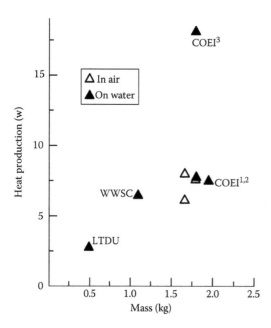

Figure 7.9. RMR (W) for sea ducks measured by open-flow respirometry in air and on water in relation to body mass (kg) (data from Table 7.2, Richman and Lovvorn 2011). COEI, Common Eider ([1]Gabrielsen et al. 1991, [2]Jenssen et al. 1989, [3]Hawkins et al. 2000); WWSC, White-winged Scoter; and LTDU, Long-tailed Duck.

(W) = 3.83 mass (kg) + 0.67 (R^2 = 0.94, P < 0.01) and RMR_{water} (W) = 2.93 mass (kg) + 2.37 (R^2 = 0.64, P < 0.01, excluding Hawkins et al. 2000). For sea ducks, RMR (W) as a function of body mass is linear over a small range of body mass (Figure 7.9), but additional data is needed on metabolic rates of more sea duck species.

When converting RMR to DER, we must make several assumptions to correctly predict energy costs. Some models use detailed thermoregulatory costs and include balanced heat loss from the bird to the environment, which incorporate measures of ambient temperature and wind speed (McKinney and McWilliams 2005). This same model also incorporated estimates of the energy costs of specific daily activities of wintering waterfowl (Tables 7.1 and 7.2).

Threshold Prey Densities

The complement of different-sized predators that depend on a habitat must be able to meet their energy requirements at different levels of prey abundance. For example, when evaluating a habitat, it is important to consider the fraction of prey of different sizes or accessibility that are effectively available to each predator (Werner 1979, Zwarts and Wanink 1984, Persson 1985, Dickman 1988). This principle is especially important when species of differing body size coexist in the same habitat. For example, larger species can search a larger area per unit time, eat greater amounts and broader size ranges of foods, and may be able to withstand prey depletion better than smaller predators restricted to smaller food items and thus a lower fraction of total food biomass (Schoener 1974, Gerritsen and Kou 1985, Goudie and Ankney 1986).

Measures of foraging profitability (energy gain minus cost) between different-sized predators can provide insight into the threshold prey density at which energy balance will switch from positive to negative. Richman and Lovvorn (2009) combined experimental measurements of dive costs and intake rates for small-bodied Lesser Scaup (*Aythya affinis*, ~600 g) and large-bodied White-winged Scoters (~1.2 kg). Using the energy intake equation (see the "Energy Intake" section), the authors incorporated seasonal variations in energy content, assimilation efficiency, size structure, and burial depth of prey and subtracted the costs of diving to assess the threshold prey densities for scaup versus scoters under different prey regimes. The authors showed that prey size and burial depth were important for determining threshold prey densities. In addition, the energy savings of using wing propulsion in addition to foot propulsion by scoters further reduced threshold prey densities by >10% (Richman and Lovvorn 2009). These variables would all have significant effects when scaled up to estimate the carrying capacity of a habitat for a combination of coexisting predators. Such analyses can provide information to estimate the carrying capacity of a particular habitat area for different species or, conversely, to estimate the habitat area required to support given populations of those species (Lovvorn et al. 2013). While some analyses have used allometric estimates of population density or area required for animals of a given body size (Silva and Downing 1994, Gaston and Blackburn 1996), others have estimated (also allometrically) the energy required by the predator population, and then compared that value to either the total biomass of food, or the biomass above some level of maximum profitable depletion (Korschgen et al. 1988, Michot 1997, Goss-Custard et al. 2003, Durell et al. 2006, Miller and Eadie 2006, Laursen et al. 2010). The approach of comparing estimated energy requirements of mollusc-eating predators to the total standing stock of available food might lead to oversimplified understanding of complex natural ecosystems and mislead environmental management decisions. For example, shellfish management policy based on allometric estimates of bird energy demands has led to overexploitation of bivalve resources in the Netherlands, which resulted in food-shortages for mollusc-eating birds and mass mortality of Common Eiders (Camphuysen et al. 2002, Ens et al. 2004). For sea ducks, it appears that the threshold of maximum depletion or giving up density varies substantially with the size and associated depth distribution of their prey resource so that much less of the total standing stock of prey is profitably available to them.

We often assume that total prey density or biomass is a good measure of available foods. However, in conservation efforts, we urge consideration of the size and accessibility of prey in its value to all predators relying on these resources (Lovvorn et al. 2013). Not only must we measure the total amount of prey available (numerically as well as biomass), but we must also provide information of the size structure of that prey community. Not all prey is accessible or ingestible to all species of sea ducks, and total prey biomass is not equal to the functionally available prey biomass. It is important to consider these aspects on prey communities and predators, along with competitive interactions with other species; all of which can have major effects on our estimates of the numbers of different bird species an area of habitat can support.

Individual-Based Modeling

Sea duck habitat use, foraging ecology, and energetics are complex, and it can be challenging to synthesize the many component processes to understand their function and importance in population dynamics. Individual-based modeling (IBM) integrates multiple processes through simulation of actions and interactions of individuals (Goss-Custard et al. 1995a,b; Stillman et al. 2000; Grimm and Railsback 2005). IBMs are becoming increasingly popular in ecological research and have been used to answer a variety of questions about animal behavior, habitat use, bioenergetics, movements, and population dynamics. The key difference between IBMs and other modeling techniques is that the characteristics of each individual animal are tracked through simulation time steps in an IBM, whereas characteristics of the population are averaged in other models. Conceptual frameworks of IBMs can be variable, but standard protocols have been proposed (Grimm et al. 2006). Several modeling platforms exist for creating IBMs, for example, NetLogo (Wilensky 1999) and MORPH (Stillman 2008), and some investigators choose to create their own IBM frameworks. There are only a few cases where IBMs have been applied to study ducks, but more examples are available from other taxonomic bird groups with similar ecology, such as shorebirds and geese (Duriez et al. 2009, Stillman and Goss-Custard 2010).

To answer questions about winter habitat requirements of Spectacled Eiders in light of changing food resources and climate change, Lovvorn et al. (2009) constructed an IBM integrating benthic prey dynamics over three decades and remotely sensed ice data into simulation models of bird energy balance. Simulation results showed that the area and distribution of the habitats of species foraging on sessile, benthic prey were dynamic on a decadal scale, and thus conservation efforts focused on managing fixed areas might be inadequate to protect such species. The modeling analysis and later applications revealed that loss of sea ice due to climate change could severely affect daily energy balance of Spectacled Eiders, which save energy by resting on ice instead of floating on water (Lovvorn et al. 2014).

Foraging ecology of Common Eiders wintering in the sea ice in Hudson Bay was modeled by Heath et al. (2010) by developing a dynamic state variable model, which could be considered a type of IBM. The authors built a set of behavioral-energetic models and through repeated simulations assessed a suit of environmental, behavioral, and physiological factors that influenced the energy budgets of wintering eiders affected by their foraging decisions (Heath and Gilchrist 2010, Heath et al. 2010). Heath et al. (2010) found that due to the strength and periodicity of currents, Common Eiders rested during the seemingly most profitable foraging period in slack currents but maximized their long-term energy intake by foraging in medium–strong current and digesting during slack currents.

Sea duck IBMs have also employed the MORPH platform (Stillman 2008). Kaiser et al. (2002) parameterized an IBM for Common Scoters wintering in Liverpool Bay, United Kingdom, and simulated environmental impact scenarios to predict displacement of scoters from benthic-feeding areas due to offshore wind farms. Similarly, for Common Eiders wintering in the southern Baltic, a MORPH-based IBM was constructed to measure habitat carrying capacity under the baseline conditions and several impact scenarios of a fixed link construction (18 km long bridge or tunnel) between Denmark and Germany (Fehmarnbelt Fixed Link Bird Services 2013a,b). IBMs provide a useful way to integrate multiple ecological processes and analyze ecosystem-level questions, and these models will be more commonly used in sea duck research in the future.

FORAGING HABITAT REQUIREMENTS

Sea ducks use habitats with high prey density and environmental characteristics that minimize the energy costs of foraging. Common Eiders wintering in the Gulf of St. Lawrence aggregate over shallow patches with the highest prey density (Guillemette et al. 1993, Guillemette and Himmelman 1996). Where marine mussel farming is prevalent, sea ducks are attracted to the supplemental food source along migration routes or on wintering areas throughout the Pacific and Atlantic coasts of Canada, United States, and Europe (Ross and Furness 2000, Dionne et al. 2006, Kirk et al. 2007, Richman et al. 2012, Varennes et al. 2013). The phenomenon is similar to the attraction to waste corn as supplemental food for migrating waterfowl.

The distribution of Black Scoters wintering in Liverpool Bay, Irish Sea, UK, coincided with sites that had high abundance and biomass of bivalve prey; however, the scoters' distribution was also shaped by the shipping lanes, which scoters avoid (Kaiser et al. 2006). King Eiders and Common Eiders aggregate on habitat patches with the highest food densities in northern Norway and redistribute themselves over winter as prey is depleted (Bustnes et al. 2013). The use of habitats with high availability of suitable foods has been reported for other sea duck species in different regions (Žydelis and Ruškytė 2005; Kirk et al. 2007, 2008; Anderson and Lovvorn 2011; Fehmarnbelt Fixed Link Bird Services 2013a).

Furthermore, when different types of suitable prey are available within an area, sea ducks clearly prefer habitat patches with higher-quality prey. Loring et al. (2013) found that higher densities of White-winged and Black Scoters wintering in Narragansett Bay, Rhode Island, are associated with mainly sandy-bottom infauna compared to mixed sediments with both infauna and epifauna, despite mixed sediments supporting higher total biomass of benthic organisms. The authors suggested that this pattern occurs because mixed bottoms contain firm sediments that are harder to penetrate and access infauna, which are the main prey for scoters in the area. In contrast, sand-substrate patches provide high-quality feeding habitat with more easily accessible prey. Similarly, Surf Scoters in Puget Sound use habitats where small mussels are abundant early in the wintering season and switch to sea grass habitats and a diet of crustaceans and gastropods in late winter, after mussels are depleted (Anderson and Lovvorn 2012).

Compelling evidence indicates that prey availability is often the most important factor determining sea duck habitat use. Depending on species and locality, other habitat features may also play an important role. Simultaneous consideration of multiple factors allows understanding of the relative importance of food versus other habitat characteristics. Distribution of Surf Scoters in San Francisco Bay was primarily influenced by the availability of herring roe, but other factors such as presence of eelgrass (*Zostera marina*), water depth, and salinity were also highly ranked in models explaining bird distribution (De La Cruz et al. 2014). In contrast, habitat use by Harlequin Ducks and Barrow's Goldeneyes in Prince William Sound, Alaska, was determined by landscape elements such as proximity to streams and shelter from wind and waves. Based on model selection, these variables were more important than prey biomass, suggesting that available food in that area likely well exceeds predation demands and that birds choose habitats that incur lower energy costs (Esler et al. 2000a,b). Sea duck habitat use may be viewed as a hierarchical phenomenon where birds distribute themselves according to general large-scale characteristics defining where species live (Johnson 1980), and then select patches offering the most available and accessible prey for harvest, balanced with the energy cost of obtaining that food.

CONCLUSIONS

Habitat-based conservation planning for sea ducks in the United States and Canada is guided by a philosophy outlined in the North American Waterfowl Management Plan (NAWMP) and implemented by regional Joint Ventures. Currently, the conservation plans developed by all NAWMP habitat Joint Ventures focused on migrating and wintering waterfowl are based on the premise that food is limiting; thus, increasing food abundance or habitats will result in improved demographic or physiological performance. To estimate habitat needs, we use bioenergetics models as a tool; however, model structure and inputs vary. Information needed to build bioenergetics models falls into two broad categories: energy availability on the landscape and energy needed by waterfowl species of interest. Over the past 20 years, we have seen considerable research directed at improving our understanding of both energy availability and energy requirements for sea ducks, but more information is needed to improve our science-based approach for achieving population objectives and habitat conservation.

ACKNOWLEDGMENTS

We are grateful to the editorial team for guidance, understanding, and support when writing this chapter, especially D. Derksen and V. Skean. Particularly instrumental were detailed reviews by M. Guillemain, S. R. McWilliams, and J. R. Lovvorn.

LITERATURE CITED

Alerstam, T., M. Rosén, J. Bäckman, P. G. P. Ericson, and O. Hellgren. 2007. Flight speeds among bird species: allometric and phylogenetic effects. PLoS Biology 5:e197.

Anderson, E. M., and J. R. Lovvorn. 2008. Gray whales may increase feeding opportunities for avian benthivores. Marine Ecology Progress Series 360:291–296.

Anderson, E. M., and J. R. Lovvorn. 2011. Contrasts in energy status and marine foraging strategies of White-winged Scoters (Melanitta fusca) and Surf Scoters (M. perspicillata). Auk 128:248–257.

Anderson, E. M., and J. R. Lovvorn. 2012. Seasonal dynamics of prey size mediate complementary functions of mussel beds and seagrass habitats for an avian predator. Marine Ecology Progress Series 467:219–232.

Anderson, E. M., J. R. Lovvorn, D. Esler, W. S. Boyd, and K. T. Stick. 2009. Using predator distributions, diet, and condition to evaluate seasonal foraging sites: sea ducks and herring spawn. Marine Ecology Progress Series 386:287–302.

Anderson, E. M., J. R. Lovvorn, and M. T. Wilson. 2008. Reevaluating marine diets of Surf and White-winged Scoters: interspecific differences and the importance of soft-bodied prey. Condor 110:285–295.

Bayne, B. L., and C. M. Worrall. 1980. Growth and production of mussels Mytilus edulis from two populations. Marine Ecology Progress Series 3:317–328.

Beauchamp, G. 1992. Diving behavior in Surf Scoters and Barrow's Goldeneyes. Auk 109:819–827.

Beauchamp, G., M. Guillemette, and R. Ydenberg. 1992. Prey selection while diving by Common Eiders, Somateria mollissima. Animal Behaviour 44:417–426.

Bevan, R. M., I. L. Boyd, P. J. Butler, K. Reid, A. J. Woakes, and J. P. Croxall. 1997. Heart rates and abdominal temperatures of free-ranging South Georgian Shags, Phalacrocorax georgianus. Journal of Experimental Biology 2000:661–675.

Bond, J. C., D. Esler, and K. A. Hobson. 2007. Isotopic evidence for sources of nutrients allocated to clutch formation by Harlequin Ducks. Condor 109:698–704.

Budge, S. M., S. W. Wang, T. E. Hollmén, and M. J. Wooller. 2011. Carbon isotopic fractionation in eider adipose tissue varies with fatty acid structure: implications for trophic studies. Journal of Experimental Biology 214:3790–3800.

Budge, S. M., M. J. Wooller, A. M. Springer, S. J. Iverson, C. P. McRoy, and G. J. Divoky. 2008. Tracing carbon flow in an arctic marine food web using fatty acid-stable isotope analysis. Oecologia 157:117–129.

Bustnes, J. O. 1998. Selection of blue mussels, Mytilus edulis, by Common Eiders, Somateria mollissima, by size in relation to shell content. Canadian Journal of Zoology 76:1787–1790.

Bustnes, J. O., and K. E. Erikstad. 1990. Size selection of common mussels, Mytilus edulis, by Common Eiders, Somateria mollissima: energy maximization or shell weight minimization? Canadian Journal of Zoology 68:2280–2283.

Bustnes, J. O., and O. J. Lonne. 1997. Habitat partitioning among sympatric wintering Common Eiders Somateria mollissima and King Eiders Somateria spectabilis. Ibis 139:549–554.

Bustnes, J. O., and G. H. Systad. 2001. Comparative feeding ecology of Steller's Eider and Long-tailed Ducks in winter. Waterbirds 24:407–412.

Bustnes, J. O., G. H. Systad, and R. C. Ydenberg. 2013. Changing distribution of flocking sea ducks as non-regenerating food sources are depleted. Marine Ecology Progress Series 484:249–257.

Butler, P. J., R. M. Bevan, A. J. Woakes, J. P. Croxall, and I. L. Boyd. 1995. The use of data loggers to determine the energetics and physiology of aquatic birds and mammals. Brazilian Journal of Medical and Biological Research 28:1307–1317.

Camphuysen, C. J., C. M. Berrevoets, H. J. W. M. Cremers, A. Dekinga, R. Dekker, B. J. Ens, T. M. van der Have, R. K. H. Kats, T. Kuiken, M. F. Leopold, J. van der Meer, and T. Piersma. 2002. Mass mortality of Common Eiders (*Somateria mollissima*) in the Dutch Wadden Sea, winter 1999/2000: starvation in a commercially exploited wetland of international importance. Biological Conservation 106:303–317.

Culik, B. M., K. Putz, R. P. Wilson, D. Allers, J. Lage, C. A. Bost, and Y. Le Maho. 1996. Diving energetics in King Penguins (*Aptenodytes patagonicus*). Journal of Experimental Biology 199:973–983.

De La Cruz, S. E. W., J. M. Eadie, A. K. Miles, J. Yee, K. A. Spragens, E. C. Palm, and J. Y. Takekawa. 2014. Resource selection and space use by sea ducks during the non-breeding season: implications for habitat conservation planning in urbanized estuaries. Biological Conservation 169:68–78.

de Leeuw, J. J., P. J. Butler, A. J. Woakes, and F. Zegwaard. 1998. Body cooling and its energetic implications for feeding and diving of Tufted Ducks. Physiological Zoology 71:720–730.

de Leeuw, J. J., and M. R. Van Eerden. 1992. Size selection in diving Tufted Ducks *Aythya fuligula* explained by differential handling of small and large mussels *Dreissena polymorpha*. Ardea 80:353–362.

Dickman, C. R. 1988. Body size, prey size, and community structure in insectivorous mammals. Ecology 69:569–580.

Dionne, M., J.-S. Lauzon-Guay, D. J. Hamilton, and M. A. Barbeau. 2006. Protective socking material for cultivated mussels: a potential non-disruptive deterrent to reduce losses to diving ducks. Aquaculture International 14:595–615.

Dugger, B. D., K. M. Dugger, and L. H. Fredrickson. 2009. Hooded Merganser (*Lophodytes cucullatus*). in A. Poole (editor), The Birds of North America, No. 98. The Academy of Natural Sciences, Philadelphia, PA.

Durell, S. E. A. l. V. d., R. A. Stillman, R. W. G. Caldow, S. McGorty, A. D. West, and J. Humphreys. 2006. Modelling the effect of environment change on shorebirds: a case study on Poole Harbour, UK. Biological Conservation 131:459–473.

Duriez, O., S. Bauer, A. Destin., J. Madsen, B. A. Nolet, R. A. Stillman, and M. Klaassen. 2009. What decision rules might Pink-footed Geese use to depart on migration? An individual-based model. Behavioral Ecology 20:560–569.

Ellis, H. I., and G. W. Gabrielsen. 2002. Energetics of free-ranging seabirds. Pp. 359–407 in E. A. Schreiber and J. Burger (editors), Biology of Marine Birds. CRC, Boca Raton, FL.

Ens, B. J., A. C. Small, and J. de Vlas. 2004. The effects of shellfish fishery on the ecosystems of the Dutch Wadden Sea and Oosterschelde. Alterra-rapport 1011, Alterra, Wageningen, Netherlands.

Esler, D., T. D. Bowman, T. A. Dean, C. E. O'Clair, S. C. Jewett, and L. L. McDonald. 2000a. Correlates of Harlequin Duck densities during winter in Prince William Sound, Alaska. Condor 102:920–926.

Esler, D., T. D. Bowman, C. E. O'Clair, T. A. Dean, and L. L. McDonald. 2000b. Densities of Barrow's Goldeneyes in Prince William Sound, Alaska in relation to food, habitat and history of oil contamination. Waterbirds 23:423–429.

Fehmarnbelt Fixed Link Bird Services. [online]. 2013a. Fehmarnbelt fixed link EIA. Bird investigations in Fehmarnbelt—baseline (vol. II). Waterbirds in Fehmarnbelt. Report No. E3TR0011. <http://vvmdocumentation.femern.com/> (1 April 2014).

Fehmarnbelt Fixed Link Bird Services. [online]. 2013b. Fehmarnbelt fixed link EIA. Fauna and Flora—impact assessment—Birds of the Fehmarnbelt area. Report No. E3TR0015. <http://vvmdocumentation. femern.com/> (1 April 2014).

Fox, A. D., and C. Mitchell. 1997. Spring habitat use and feeding behavior of Steller's Eider *Polysticta stelleri* in Varangerfjord, northern Norway. Ibis 139:542–548.

Gabrielsen, G. W., F. Mehlum, H. E. Karlsen, O. Andresen, and H. Parker. 1991. Energy cost during incubation and thermoregulation in the female Common Eider *Somateria mollissima*. Norsk Polarinstitutt Skrifter 195:51–62.

Gallivan, G. J., and K., Roland. 1979. Temperature regulation in freely diving harp seals (*Phoca groenlandica*). Canadian Journal of Zoology 57:2256–2263.

Gaston, K. J., and T. M. Blackburn. 1996. Conservation implications of geographic range size-body size relationships. Conservation Biology 10:638–646.

Gauthier, G. 1993. Bufflehead (*Bucephala albeola*). in A. Poole and F. Gill (editors), The Birds of North America, No. 67. The Academy of Natural Sciences, Philadelphia, PA.

Gauthier, G., J. Betty, and K. A. Hobson. 2003. Are Greater Snow Geese capital breeders? New evidence from a stable-isotope model. Ecology 84:3250–3264.

Gerritsen, J., and J.-I. Kou. 1985. Food limitation and body size. Archiv fur Hydrobiologie Beiheft Ergebnisse der Limnologie 21:173–184.

Goss-Custard, J. D., R. W. G. Caldow, R. T. Clarke, S. E. A. l. V. d. Durell, and W. J. Sutherland. 1995a. Deriving population parameters from individual variations in foraging behaviour. 1. Empirical game-theory distribution model of Oystercatchers *Haematopus ostralegus* feeding on mussels *Mytilus edulis*. Journal of Animal Ecology 64:265–276.

Goss-Custard, J. D., R. W. G. Caldow, R. T. Clarke, and A. D. West. 1995b. Deriving population parameters from individual variations in foraging behavior. 2. Model tests and population parameters. Journal of Animal Ecology 64:277–289.

Goss-Custard, J. D., R. A. Stillman, R. W. G. Caldow, A. D. West, and M. Guillemain. 2003. Carrying capacity in overwintering birds: when are spatial models needed? Journal of Animal Ecology 40:176–187.

Goudie, R. I. 1999. Behaviour of Harlequin Ducks and three species of scoters wintering in the Queen Charlotte Islands, British Columbia. in R. I. Goudie, M. R. Petersen, and G. J. Robertson (editors), Behaviour and ecology of sea ducks. Occasional Paper, No. 100. Canadian Wildlife Service, Environment Canada, Ottawa, ON.

Goudie, R. I., and C. D. Ankney. 1986. Body size, activity budgets, and diets of sea ducks wintering in Newfoundland. Ecology 67:1475–1482.

Grimm, V., U. Berger, F. Bastiansen, S. Eliassen, V. Ginot, J. Giske, J. Goss-Custard et al. 2006. A standard protocol for describing individual-based and agent-based models. Ecological Modelling 198:115–126.

Grimm, V., and S. F. Railsback. 2005. Individual-based modeling and ecology. Princeton University Press, Princeton, NJ.

Guillemette, M. 1994. Digestive-rate constraint in wintering Common Eiders (*Somateria mollissima*): implications for flying capabilities. Auk 111:900–909.

Guillemette, M. 1998. The effect of time and digestion constraints in Common Eiders while feeding and diving over blue mussel beds. Functional Ecology 12:123–131.

Guillemette, M. 2001. Foraging before spring migration and before breeding in Common Eiders: does hyperphagia occur? Condor 103:633–638.

Guillemette, M., and J. H. Himmelman. 1996. Distribution of wintering Common Eiders over mussel beds: does the ideal free distribution apply? Oikos 76:435–442.

Guillemette, M., J. H. Himmelman, C. Barette, and A. Reed. 1993. Habitat selection by Common Eiders in winter and its interaction with flock size. Canadian Journal of Zoology 71:1259–1266.

Guillemette, M., R. C. Ydenberg, and J. H. Himmelman. 1992. The role of energy intake rate in prey and habitat selection of Common Eiders *Somateria mollissima* in winter: a risk-sensitive interpretation. Journal of Animal Ecology 61:599–610.

Hamilton, D. J., T. D. Nudds, and J. Neate. 1999. Size-selective predation of blue mussels (*Mytilus edulis*) by Common Eiders (*Somateria mollissima*) under controlled field conditions. Auk 116:403–416.

Handrich, Y., R. M. Bevan, J.-B. Charrassin, P. J. Butler, K. Ptz, A. J. Woakes, J. Lage, and Y. Le Maho. 1997. Hypothermia in foraging King Penguins. Nature 388:64–67.

Hario, M., and M. Öst. 2002. Does heavy investment in foraging implicate low food acquisition for female Common Eider *Somateria mollissima*? Ornis Fennica 79:111–120.

Hawkins, P. A. J., P. J. Butler, A. J. Woakes, and J. R. Speakman. 2000. Estimation of the rate of oxygen consumption of the Common Eider duck (*Somateria mollissima*), with some measurements of heart rate during voluntary dives. Journal of Experimental Biology 203:2819–2832.

Heath, J. P., and H. G. Gilchrist. 2010. When foraging becomes unprofitable: energetics of diving in tidal currents by Common Eiders wintering in the Arctic. Marine Ecology Progress Series 403:279–290.

Heath, J. P., H. G. Gilchrist, and R. C. Ydenberg. 2006. Regulation of stroke pattern and swim speed across a range of current velocities: diving by Common Eiders wintering in the Canadian Arctic. Journal of Experimental Biology 209:3974–3983.

Heath, J. P., H. G. Gilchrist, and R. C. Ydenberg. 2007. Can dive cycle models predict patterns of foraging behavior? Diving by Common Eiders in an Arctic polynya. Animal Behaviour 73:877–884.

Heath, J. P., H. G. Gilchrist, and R. C. Ydenberg. 2010. Interactions between rate processes with different timescales explain counterintuitive foraging patterns of arctic wintering eiders. Proceedings of the Royal Society B 277:3179–3186.

Hepp, G. R., R. J. Blohm, R. E. Reynolds, J. E. Hines, and J. D. Nichols. 1986. Physiological condition of autumn-banded Mallards and its relationship to hunting vulnerability. Journal of Wildlife Management 50:177–183.

Hobson, K. A. 2006. Using stable isotopes to quantitatively track endogenous and exogenous nutrient allocations to eggs of birds that travel to breed. Ardea 94:359–369.

Hoerner, S. F. 1958. Fluid dynamic drag: practical information on aerodynamic and hydrodynamic resistance. Hoerner Fluid Dynamics, Bricktown, NJ.

Iverson, S. J., C. Field, W. Don Bowen, and W. Blanchard. 2004. Quantitative fatty acid signature analysis: a new method of estimating predator diets. Ecological Monographs 74:211–235.

Jenssen, B. M., and M. Ekker. 1989. Thermoregulatory adaptations to cold in winter-acclimatized Long-tailed Ducks (*Clangula hyemalis*). Pp. 147–152 in C. Bech and R. E. Reinertsen (editors), Physiology of Cold Adaptation in Birds. Plenum, New York.

Jenssen, B. M., and M. Ekker. 1991. Effects of plumage contamination with crude oil dispersant mixtures on thermoregulation in Common Eiders and Mallards. Archives of Environmental Contamination and Toxicology 20:398–403.

Jenssen, B. M., M. Ekker, and C. Bech. 1989. Thermoregulation in winter-acclimatized Common Eiders (*Somateria mollissima*) in air and water. Canadian Journal of Zoology 67:669–673.

Jeschke, J. M., M. Kopp, and R. Tollrian. 2002. Predator functional responses: discriminating between handling and digesting prey. Ecological Monographs 72:95–112.

Johnson, D. H. 1980. The comparison of usage and availability measurements for evaluating resource preference. Ecology 61:65–71.

Kaiser, M. J., A. J. Elliott, M. Galanidi, E. I. S. Rees, R. W. G. Caldow, R. A. Stillman, W. J. Sutherland, and D. A. Showler. 2002. Predicting the displacement of Common Scoter *Melanitta nigra* from benthic feeding areas due to offshore windfarms. University of Wales Bangor Report to COWRIE, Bangor, U.K.

Kaiser, M. J., M. Galanidi, D. A. Showler, A. J. Elliott, R. W. G. Caldow, E. I. S. Rees, R. A. Stillman, and W. J. Sutherland. 2006. Distribution and behaviour of Common Scoter *Melanitta nigra* relative to prey resources and environmental parameters. Ibis 148:110–128.

Kaseloo, P. A., and J. R. Lovvorn. 2003. Heat increment of feeding and thermal substitution in Mallard ducks feeding voluntarily on grain. Journal of Comparative Physiology B 173:207–213.

Kaseloo, P. A., and J. R. Lovvorn. 2005. Effects of surface activity patterns and dive depth on thermal substitution in fasted and fed Lesser Scaup ducks. Canadian Journal of Zoology 83:301–311.

Kautsky, N. 1982. Quantitative studies on gonad cycle, fecundity, reproductive output and recruitment in a Baltic *Mytilus edulis* population. Marine Biology 68:143–160.

Kautsky, N., K. Johannesson, and M. Tedengren. 1990. Genotypic and phenotypic differences between Baltic and North Sea populations of *Mytilus edulis* evaluated through reciprocal transplantations. I. Growth and morphology. Marine Ecology Progress Series 59:203–210.

Kersten, M., and T. Piersma. 1987. High levels of energy expenditure in shorebirds; metabolic adaptations to an energetically expensive way of life. Ardea 75:175–187.

Kestenholz, M. 1994. Body mass dynamics of wintering Tufted Duck *Aythya fuligula* and Pochard *A. ferina* in Switzerland. Wildfowl 45:147–158.

Kirk, M., D. Esler, and W. S. Boyd. 2007. Morphology and density of mussels on natural and aquaculture structure habitats: implication for sea duck predators. Marine Ecology Progress Series 346:179–187.

Kirk, M., D. Esler, S. A. Iverson, and W. S. Boyd. 2008. Movements of wintering Surf Scoters: predator responses to different prey landscapes. Oecologia 155:859–867.

Kleiber, M. 1932. Body size and metabolism. Hilgardia 6:315–353.

Kooyman, G. L. 1965. Techniques used in measuring diving capacities of Weddell Seals. Polar Record 12:391–394.

Korschgen, C. E., L. S. George, and W. L. Green. 1988. Feeding ecology of Canvasbacks staging on Pool 7 of the upper Mississippi River. Pp. 237–249 in M. W. Weller (editor), Waterfowl in winter. University of Minnesota Press, Minneapolis, MN.

Kristjánsson, T. Ö., J. E. Jónsson, and J. Svavarsson. 2013. Spring diet of Common Eiders (*Somateria mollissima*) in Breiðafjörður, West Iceland, indicates non-bivalve preferences. Polar Biology 36:51–59.

Lacroix, D. L., S. Boyd, D. Esler, M. Kirk, T. Lewis, and S. Lipovsky. 2005. Surf Scoters *Melanitta perspicillata* aggregate in association with ephemerally abundant polychaetes. Marine Ornithology 33:61–63.

Laubhan, M. K., and K. A. Metzner. 1999. Distribution and diurnal behavior of Steller's Eiders wintering on the Alaska Peninsula. Condor 101:694–698.

Laursen, K., P. S. Kristensen, and P. Clausen. 2010. Assessment of blue mussel *Mytilus edulis* fisheries on waterbird shellfish-predator management in the Danish Wadden Sea. Ambio 39:476–485.

Leopold, M. F., A. Cervencl, and F. Müller. 2012. Eidereend *Somateria mollissima* eet vis (Fish-eating eider). Sula 45:41–44.

Lewis, T. L., D. Esler, and W. S. Boyd. 2007a. Effects of predation by sea ducks on clam abundance in soft-bottom intertidal habitats. Marine Ecology Progress Series 329:131–144.

Lewis, T. L., D. Esler, and W. S. Boyd. 2007b. Foraging behaviors of Surf Scoters and White-winged Scoters during spawning of Pacific herring. Condor 109:216–222.

Lewis, T. L., D. Esler, and W. S. Boyd. 2008. Foraging behavior of Surf Scoters (*Melanitta perspicillata*) and White-winged Scoters (*M. fusca*) in relation to clam density: inferring food availability and habitat quality. Auk 125:149–157.

Lewis, T. L., D. Esler, W. S. Boyd, and R. Žydelis. 2005. Nocturnal foraging behavior of wintering Surf Scoters and White-winged Scoters. Condor 107:637–647.

Lok, E. K., D. Esler, J. Y. Takekawa, S. De La Cruz, W. S. Boyd, D. R. Nysewander, J. R. Evenson, and D. H. Ward. 2012. Spatiotemporal associations between Pacific herring and Surf Scoter spring migration: evaluating a 'silver wave' hypothesis. Marine Ecology Progress Series 457:139–150.

Loring, P. H., P. W. P. Paton, S. R. McWilliams, R. A. McKinney, and C. A. Oviatt. 2013. Densities of wintering scoters in relation to benthic prey assemblages in a North Atlantic estuary. Waterbirds 36:144–155.

Lovvorn, J. R. 2001. Upstroke thrust, drag effects, and stroke-glide cycles in wing-propelled swimming by birds. American Zoologist 41:154–165.

Lovvorn, J. R. 2007. Thermal substitution and aerobic efficiency: measuring and predicting effects of heat balance on endotherm diving energetics. Philosophical Transactions of the Royal Society of London B 362:2079–2093.

Lovvorn, J. R., E. M. Anderson, A. R. Rocha, W. W. Larned, J. M. Grebmeier, L. W. Cooper, J. M. Kolts, and C. A. North. 2014. Variable wind, pack ice, and prey dispersion affect the long-term adequacy of protected areas for an Arctic sea duck. Ecological Applications 24:396–412.

Lovvorn, J. R., S. E. De La Cruz, J. W. Takekawa, L. E. Shaskey, and S. E. Richman. 2013. Niche overlap, threshold food densities, and limits to prey depletion for a diving duck assemblage in an estuarine bay. Marine Ecology Progress Series 476:251–268.

Lovvorn, J. R., J. M. Grebmeier, L. W. Cooper, J. K. Bump, and S. E. Richman. 2009. Modelling marine protected area for threatened eiders in a climatically changing Bering Sea. Ecological Applications 19:1596–1613.

Lovvorn, J. R., and D. R. Jones. 1991. Body mass, volume, and buoyancy of some aquatic birds, and their relation to locomotor strategies. Canadian Journal of Zoology 69:2888–2892.

Lovvorn, J. R., and D. R. Jones. 1994. Biomechanical conflicts between diving and aerial flight in estuarine birds. Estuaries 17:62–75.

Lovvorn, J. R., D. R. Jones, and R. W. Blake. 1991. Mechanics of underwater locomotion in diving ducks: drag, buoyancy, and acceleration in a size gradient of species. Journal of Experimental Biology 159:89–108.

Lovvorn, J. R., S. E. Richman, J. M. Grebmeier, and L. W. Cooper. 2003. Diet and body condition of Spectacled Eiders wintering in pack ice of the Bering Sea. Polar Biology 26:259–267

Lovvorn, J. R., Y. Watanuki, A. Kato, Y. Naito, and G. A. Liggins. 2004. Stroke patterns and regulation of swim speed and energy cost in free-ranging Brünnich's Guillemots. Journal of Experimental Biology 207:4679–4695.

MacArthur, R. H., and E. R. Pianka. 1966. On the optimal use of a patchy environment. American Naturalist 100:603–609.

Mallory, M., and K. Metz. 1999. Common Merganser (Mergus merganser). in A. Poole and F. Gill (editors), The Birds of North America, No. 442. The Academy of Natural Sciences, Philadelphia, PA.

McKinney, R. A., and S. R. McWilliams. 2005. A new model to estimate daily energy expenditure for wintering waterfowl. Wilson Bulletin 117:44–55.

McNeil, R., P. Drapeau, and J. D. Goss-Custard. 1992. The occurrence and adaptive significance of nocturnal habits in waterfowl. Biological Reviews 67:381–419.

McWilliams, S. R., and W. H. Karasov. 2001. Phenotypic flexibility in digestive system structure and function in migratory birds and its ecological significance. Comparative Biochemistry and Physiology A 128:579–593.

Merkel, F. R., S. E. Jamieson, K. Falk, and A. Mosbech. 2007. The diet of Common Eiders wintering in Nuuk, Southwest Greenland. Polar Biology 30:227–234.

Merkel, F. R., and A. Mosbech. 2008. Diurnal and nocturnal feeding strategies in Common Eiders. Waterbirds 31:580–586.

Michot, T. C. 1997. Carrying capacity of seagrass beds predicted for Redheads wintering in Chandeleur Sound, Louisiana, USA. Pp. 93–102 in J. Goss-Custard, R. Rufino, and A. Luis (editors), Effect of habitat loss and change on waterbirds. Wetlands International Publication No. 42. The Stationery Office, London, U.K.

Miller, M. R., and J. M. Eadie. 2006. The allometric relationship between resting metabolic rate and body mass in wild waterfowl (Anatidae) and an application to estimation of winter habitat requirements. Condor 108:166–177.

Mosbech, A., R. S. Danø, F. Merkel, C. Sonne, G. Gilchrist, and A. Flagstad. 2006. Use of satellite telemetry to locate key habitats for King Eiders Somateria spectabilis in West Greenland. Pp. 769–776 in G. C. Boere, C. A. Galbraith, and D. A. Stroud (editors), Waterbirds around the world. The Stationery Office, Edinburgh, U.K.

Navarro, R. A., C. R. Velasquez, and R. P. Schlatter. 1989. Diet of the surfbird in southern Chile. Wilson Bulletin 101:137–141.

Nehls, G. 1995. Strategien der ernährung und ihre bedeutung für energiehaushalt und ökologie der Eiderente (Somateria mollissima (L., 1758)). Dissertation, Kiel University, Kiel, Germany (in German).

Nilsson, L. 1970. Food-seeking activity of south Swedish diving ducks in the non-breeding season. Oikos 21:145–154.

Ouellet, J.-F., C. Vanpe, and M. Guillemette. 2013. The body size-dependent diet composition of North American sea ducks in winter. PLoS One 8:e65667.

Oppel, S., A. N. Powell, and M. G. Butler. 2011. King Eider foraging effort during the pre-breeding period in Alaska. Condor 113:52.60.

Owen, M. 1990. Nocturnal feeding in waterfowl. Acta XXth Congressus Internationalis Ornithologici 2:1105–1112.

Palm, E. C., D. Esler. E. M. Anderson, T. D. Williams, and M. T. Wilson. 2013. Variation in physiology and energy management of wintering White-winged Scoters in relation to local habitat conditions. Condor 115:750–761.

Pelletier, D., M. Guillemette, J. M. Grandbois, and P. J. Butler. 2007. It is time to move: linking flight and foraging behavior in a diving bird. Biology Letters 3:357–359.

Pelletier, D., M. Guillemette, J.-M. Grandbois, and P. J. Butler. 2008. To fly or not to fly: high costs in a large sea duck do not imply an expensive lifestyle. Proceedings of the Royal Society B 275:2117–2124.

Pennycuick, C. J. 2008. Modelling the flying bird. Elsevier, Burlington, MA.

Perry, M. C. 2012. Foraging behavior of Long-tailed Ducks in a ferry wake. Northeastern Naturalist 19:135–139.

Persson, L. 1985. Asymmetric competition: are larger animals competitively superior? American Naturalist 126:261–266.

Petersen, M. 1980. Observations of wing-feather moult and summer feeding ecology of Steller's Eiders at Nelson Lagoon. Alaska. Wildfowl 31:99–106.

Peterson, S. R., and R. S. Ellarson. 1979. Changes in Oldsquaw weight. Wilson Bulletin 91:288–300.

Pierce, G. J., and J. G. Ollason. 1987. Eight reasons why optimal foraging theory is a complete waste of time. Oikos 49:111–117.

Powell, A. N., and R. S. Suydam. 2012. King Eider (Somateria spectabilis). in A. Poole and F. Gill (editors), The Birds of North America, No. 547. The Academy of Natural Sciences, Philadelphia, PA.

Pyke, G. H. 1984. Optimal foraging theory: a critical review. Annual Review of Ecology and Systematics 15:523–575.

Richman, S. E., and J. R. Lovvorn. 2003. Effects of clam species dominance on nutrient and energy acquisition by Spectacled Eiders in the Bering Sea. Marine Ecology Progress Series 261:283–297.

Richman, S. E., and J. R. Lovvorn. 2004. Relative foraging value to Lesser Scaup ducks of native and exotic clams from San Francisco Bay. Ecological Applications 14:1217–1231.

Richman, S. E., and J. R., Lovvorn. 2008. Cost of diving by wing and foot propulsion in a sea duck, the White-winged Scoter. Journal of Comparative Physiology B 178:321–332.

Richman, S. E., and J. R. Lovvorn. 2009. Predator size, prey size and threshold food densities of diving ducks: does a common prey base support fewer large animals? Journal of Animal Ecology 78:1033–1042.

Richman, S. E., and J. R. Lovvorn. 2011. Effects of air and water temperatures on resting metabolism of auklets and other diving birds. Physiological and Biochemical Zoology 84:316–332.

Richman, S. E., E. Varennes, J. Bonadelli, and M. Guillemette. 2012. Sea duck predation on mussel farms: growing conflict. Proceedings of Aquaculture Canada, Quebec City, QC.

Rigou, Y., and M. Guillemette. 2010. Foraging effort and pre-laying strategy in breeding Common Eiders. Waterbirds 33:314–322.

Rizzolo, D. J., D. Esler, D. D. Roby, and R. L. Jarvis. 2005. Do wintering Harlequin Ducks forage nocturnally at high latitudes? Condor 107:173–177.

Rodway, M. S. 2003. Timing of pairing in Harlequin Ducks: interaction of spacing behaviour, time budgets, and the influx of herring spawn. Doctoral dissertation, Simon Fraser University, Burnaby, BC.

Ross, B. P., and R. W. Furness. 2000. Minimizing the impact of eider ducks on mussel farming. Ornithology Group, University of Glasgow, Glasgow, Scotland.

Rumohr, H., T. Brey, and S. Ankar. 1987. A compilation of biometric conversion factors for benthic invertebrates of the Baltic Sea. Baltic Marine Biologists Publication 9:1–56.

Schamber, J. L., D. Esler, and P. L. Flint. 2009. Evaluating the validity of using unverified indices of body condition. Journal of Avian Biology 40:49–56.

Schenkeveld, L. E., and R. C. Ydenberg. 1985. Synchronous diving by Surf Scoter flocks. Canadian Journal of Zoology 63:2516–2519.

Schmidt-Nielsen, K. 1975. Animal physiology. Adaptation and environment. Cambridge University Press, London, U.K.

Schmutz, J. A., K. A. Hobson, and J. A. Morse. 2006. An isotopic assessment of protein from diet and endogenous stores: effect on egg production and incubation behaviour of geese. Ardea 94:385–397.

Schoener, T. W. 1974. Resource partitioning in ecological communities. Science 185:27–39.

Schorger, A. W. 1947. The deep diving of the Loon and Old-squaw and its mechanism. Wilson Bulletin 59:151–159.

Schreer, J. F., and K. M. Kovacs. 1997. Allometry of diving capacity in air-breathing vertebrates. Canadian Journal of Zoology 75:339–358.

Schummer, M. L., S. A. Petrie, A. M. Bailey, and S. S. Badzinski. 2012. Factors affecting lipid reserves and foraging activity of Buffleheads, Common Goldeneyes, and Long-tailed Ducks during winter at Lake Ontario. Condor 114:62–74.

Silva, M., and J. A. Downing. 1994. Allometric scaling of minimal mammal densities. Conservation Biology 8:732–743.

Stempniewicz, L. 1995. Feeding ecology of the Long-tailed Duck *Clangula hyemalis* wintering in the Gulf of Gdánsk (southern Baltic Sea). Ornis Svecica 5:133–142.

Stephens, D. W., J. S. Brown, and R. C. Ydenberg. 2007. Foraging behavior and ecology. Chicago University Press, Chicago, IL.

Stephenson, R. 1994. Diving energetics in Lesser Scaup (*Aythya affinis*, Eyton). Journal of Experimental Biology 190:155–178.

Stillman, R. A. 2008. MORPH—an individual-based model to predict the effect of environmental change on foraging animal populations. Ecological Modelling 216:265–276.

Stillman, R. A., and J. D. Goss-Custard. 2010. Individual-based ecology of coastal birds. Biological Reviews 85:413–434.

Stillman, R. A., J. D. Goss-Custard, A. D. West, S. E. A. l. V. d. Durell, R. W. G. Caldow, S. McGrorty, and R. T. Clarke. 2000. Predicting mortality in novel environments: tests and sensitivity of a behaviour-based model. Journal of Applied Ecology 37:564–588.

Systad, G. H., and J. O. Bustnes. 2001. Coping with darkness and low temperatures: foraging strategies in Steller's Eiders, *Polysticta stelleri*, wintering at high latitudes. Canadian Journal of Zoology 79:402–406

Systad, G. H., J. O. Bustnes, and K. E. Erikstad. 2000. Behavioral responses to decreasing day length in wintering sea ducks. Auk 117:33–40.

Titman, R. D. 1999. Red-breasted Merganser (*Mergus serrator*). in A. Poole and F. Gill (editors), The Birds of North America, No. 443. The Academy of Natural Sciences, Philadelphia, PA.

VanStratt, C. S. 2011. Foraging effort by Surf Scoter at the peripheries of their wintering distribution: do foraging conditions influence their range? Thesis. Simon Fraser University, Burnaby, BC.

Varennes, É., S. A. Hanssen, J. Bonardelli, and M. Guillemette. 2013. Sea duck predation in mussel farms: the best nets for excluding Common Eiders safely and efficiently. Aquaculture Environment Interactions 4:31–39.

Waldeck, P., and K. Larsson. 2013. Effects of winter water temperature on mass loss in Baltic blue mussels: implications for foraging sea ducks. Journal of Experimental Marine Biology and Ecology 444:24–30.

Wang, S. W., T. E. Hollmén, and S. J. Iverson. 2010. Validating quantitative fatty acid signature analysis to estimate diets of Spectacled and Steller's Eiders (*Somateria fischeri* and *Polysticta stelleri*). Journal of Comparative Physiology B—Biochemical, Systemic, and Environmental Physiology 180:125–139.

Werner, E. E. 1979. Niche partitioning by food size in fish communities. Pp. 311–322 in H. Clepper (editor), Predator-prey systems in fisheries management. Sport Fishing Institute, Washington, DC.

White, T. P. 2013. Spatial ecology of Long-tailed Ducks and White-winged Scoters wintering on Nantucket Shoals, Massachusetts. Ph.D. thesis, City University of New York, New York, NY.

White, T. P., R. R. Veit, and M. C. Perry. 2009. Feeding ecology of Long-tailed Ducks *Clangula hyemalis* wintering on the Nantucket Shoals. Waterbirds 32:293–299.

Wilensky, U. [online]. 1999. NetLogo. Center for connected learning and computer-based modeling. Northwestern University, Evanston, IL. <http://ccl.northwestern.edu/netlogo/> (1 April 2014).

Ydenberg, R. C. 1988. Foraging by diving birds. Proceedings of the XIXth International Ornithological Congress 9:1832–1842.

Zwarts, L., and J. H. Wanink. 1984. How Oystercatchers and Curlews successively deplete clams. Pp. 69–83 in P. R. Evans, J. D. Goss-Custard, and W. G. Hale (editors), Coastal waders and wildfowl in winter. Cambridge University Press, Cambridge, U.K.

Zwarts, L., and J. H. Wanink. 1993. How the food supply harvestable by waders in the Wadden Sea depends on the variation in energy density, body weight, biomass, burying depth and behaviour of tidal-flat invertebrates. Netherlands Journal of Sea Research 31:441–476.

Žydelis, R., and D. Esler. 2005. Response of wintering Steller's Eiders to herring spawn. Waterbirds 28:344–350.

Žydelis, R., and D. Rušktyė. 2005. Winter foraging of Long-tailed Ducks (*Clangula hyemalis*) exploiting different benthic communities in the Baltic Sea. Wilson Bulletin 117:133–141.

CHAPTER EIGHT

Variation in Migration Strategies of North American Sea Ducks*

Margaret R. Petersen and Jean-Pierre L. Savard

Abstract. Migration exerts strong effects on population dynamics, so consideration of migration as a driver of population change is an important area of inquiry. Sea ducks (Mergini) exemplify the wide range in types of migration strategies, which become more variable with the addition of a third migration to distinct molting areas. We discuss the three migrations, summer, fall, and molt, and emphasize similarities and differences within and among species. For each migration, we focus on timing, routes and stopover sites, nutrient reserve acquisition, stopover behavior, flight behavior, interannual constancy at stopover sites, and variation among sexes and ages. Last, we describe individual variation of annual flight paths, discuss inter- and intra-annual fidelity at stopover sites, examine the role of the environment on migration paths, and evaluate variability and limitations of speed and duration of migration.

Key Words: flight speed, migration route, molt migration, nutrient reserves, pairing, phenology, site fidelity, stopover ecology, timing of migration.

Migration represents a historical and evolutionary trade-off between remaining at a single location with reduced survival and reproduction and incurring the cost of migration to move elsewhere, ultimately resulting in increased reproductive potential and survival. Migration has energetic costs, as well as risks of predation, variable stopover habitat availability, and unpredictable weather that can decrease survival and add to costs of moving to a new site. As quality, quantity, and availability of forage on breeding grounds decline as the year progresses, birds often move to wintering areas that have milder or more stable climates where they have more consistent access to foods and can better survive winter. By leaving the wintering area in spring, individuals can take advantage of the annual flush of resources associated with seasonally variable climates of northern breeding areas (Alexander 1998, Alerstam et al. 2003, Newton 2008).

WHAT IS A MIGRATION STRATEGY?

A migration strategy is the tactic of how and why an individual moves between breeding, molting, and wintering areas. To optimize its survival and reproductive output, a migrating bird may arrive

---

* Petersen, M. R. and J.-P. L. Savard. 2015. Variation in migration strategies of North American sea ducks. Pp. 267–304 in J.-P. L. Savard, D. V. Derksen, D. Esler, and J. M. Eadie (editors). Ecology and conservation of North American sea ducks. Studies in Avian Biology (no. 46), CRC Press, Boca Raton, FL.

quickly having used most of its resources, arrive in the best condition possible, and minimize predation while migrating (Alerstam and Lindström 1990, Newton 2008). The strategy used by an individual takes these needs into account, and tactics may change throughout migration (Karlsson et al. 2012). Selection pressures favor migrants that arrive within the most favorable time period that results in sufficient reserves needed to successfully survive and reproduce (reviewed by Rohwer 1992, Farmer and Wines 1998, Prop et al. 2003).

Migration includes both flight and use of stopover sites. When moving from point to point, an individual may stop at stopover sites for a relatively short period (hours to a few days) to rest before continuing or for longer periods of time to feed and replenish energy needed for continuation of migration (Alerstam and Hedenström 1998, Alerstam 2003). Appropriate habitats are not necessarily evenly nor continuously distributed along the most direct route, some stopover sites have limited resources, and predators actively hunt migrants at some sites (Alerstam 2001, Lind and Cresswell 2006). As a result, some species avoid flying across large expanses of water, others primarily follow coastlines or other physical features such as valleys or rivers, and some avoid tall mountains. Alternatively, when it is advantageous to migrate across such barriers, birds may gain extensive reserves prior to leaving and take advantage of favorable winds to improve their chances of a safe arrival (Alerstam and Lindström 1990, Hedenström and Alerstam 1998, Hedenström 2008, Newton 2008, Gill et al. 2009). By flying with a tailwind, a bird can arrive to its destination quicker and expend less energy than if it flew into a head-wind or in calm conditions (Liechti and Bruderer 1998, Liechti 2006).

The relatively slow fuel deposition rate of ducks, geese, swans, and larger birds in comparison to shorebirds, songbirds, and small species can increase the duration of migration and limit the distance of migration (Lindström 1991, 2003; Hedenström and Alerstam 1998; Alerstam 2003). Migration is fraught with potential difficulties, and the response of individuals to these challenges varies and defines their migration strategy.

DOES WATERFOWL MIGRATION DIFFER?

Compared with other birds, ducks, geese, and swans have unusual migration patterns. Waterfowl lose all of their flight feathers in a simultaneous rather than a sequential molt and often migrate to molting locations outside of breeding and wintering areas. In general, waterfowl undergo three migrations annually: from wintering to breeding areas (spring migration), from breeding to molting areas (molt migration), and from molting to wintering areas (fall migration). Fall migration occurs only in species that molt in areas separate from wintering areas; if their wintering area is the same as their molting area, individuals are considered to have a molt migration but not a fall migration (Bellrose 1980, Oppel et al. 2008). For populations that molt at or near breeding areas (most geese and swans, some sea ducks), fall migration is initiated directly from the breeding grounds (Gauthier 1993, Eadie et al. 2000).

Syntheses by Bellrose (1980) and Hochbaum (1955) provided descriptions of waterfowl migration in North America. Bellrose (1980) advanced management and studies of waterfowl in North America when he identified and described the migration corridors or flyways used by ducks, geese, and swans when moving from breeding to wintering areas and return and examined variability of migration strategies among waterfowl. As with other groups of birds (Evans and Davidson 1990, Farmer and Wiens 1999, Newton 2008), migration strategies of waterfowl can vary among seasons, individuals, ages, sexes, and breeding populations.

WHY STUDY SEA DUCK MIGRATION?

Many species of sea ducks in North America have undergone substantial declines over the past decades (Petersen and Hogan 1996), and little is known about their ecology, including migration strategies. Events during one part of the life cycle can have carryover effects in subsequent annual cycle stages (Marra et al. 1998, Sillett et al. 2000, Black et al. 2007, Norris and Marra 2007). For example, several species of sea ducks are capital or partially capital breeders (Hobson et al. 2005, Sénéchal et al. 2011) and depend in part on resources gained or maintained at stopover sites for successful reproduction (Drent and Daan 1980, Klaassen 2003). Catastrophic events such as early ice storms, refreezing of stopover sites, or delayed thaw that occur during migration may ultimately result in mortality or nonbreeding with subsequent effects on sea duck populations (Myers 1958, Barry 1968, Fournier and Hines 1994, Robertson and Gilchrist 1998).

In other taxa, migration is known to exert strong effects on population dynamics, so consideration of migration as a driver of population change in sea ducks is an important area of inquiry.

Another critical consideration for sea duck conservation is the degree of connectivity among breeding, molting, and wintering areas (Webster et al. 2002, Webster and Marra 2005, Jefferies and Drent 2006, Marra et al. 2006, Norris et al. 2006). Migration is the mechanism of this interconnection (Owen and Black 1990, Drent et al. 2007). As emphasized by Boulet and Norris (2006), Pearce and Talbot (2006), and Pearce et al. (2008), the use of multiple techniques including banding, mark–recapture, behavior, isotopic, and genetic analysis is necessary to improve our understanding of connectivity. Ultimately, population structure of sea ducks depends on an understanding of migration strategies (Chapter 2, this volume).

SEASONAL MOVEMENTS

Spring Migration

Timing of Spring Migration

In coastal communities of arctic Canada and Alaska, the appearance of the first flocks of eiders is a harbinger of spring. In more inland waters, goldeneyes and Buffleheads (*Bucephala albeola*) are the first sea ducks to arrive as ice melts on rivers, lakes, and streams of the boreal forest regions throughout North America (Erskine 1971; Eadie et al. 1995, 2000). A common characteristic of many species of sea ducks is that breeding adults arrive to nesting areas early in spring (Table 8.1). The earliest arriving species are often found in the meltwater on lakes and ponds and are the first to use small open areas (Gauthier 1993, Eadie et al. 1995). Although birds may appear as soon as open water occurs, several species arrive chronologically later than other waterbirds at the same location. Individuals that nest late often arrive when ice in lakes and ponds has disappeared and as snow-free breeding habitats become available.

Many sea ducks nest in areas that are only available seasonally and for short periods of time (Palmer 1976; Bellrose 1980; Gauthier 1993; Dugger et al. 1994; Bordage and Savard 1995; Eadie et al. 1995, 2000; Brown and Fredrickson 1997; Savard et al. 1998; Mallory and Metz 1999;

Robertson and Goudie 1999; Titman 1999; Goudie et al. 2000; Petersen et al. 2000; Suydam 2000; Fredrickson 2001; Robertson and Savard 2002; Powell and Suydam 2012). Because of the short breeding season (3–4 months for some species) at high latitude, birds must arrive within a narrow window of time to successfully lay eggs, incubate clutches, and raise young to fledging. The costs of delayed reproduction are illustrated by observations of flightless King Eider (*Somateria spectabilis*) ducklings caught in the ice as ponds freeze in the fall (Barry 1968) or young found in small, open coastal water areas with rapidly forming ice (McLaren and McLaren 1982). Therefore, timing of spring migration is critical for arriving on breeding areas at the optimal date.

Timing of spring migration among sea ducks to the boreal forest is, in part, related to nest site selection. Species that nest predominantly in tree cavities such as goldeneyes, Bufflehead, and some mergansers are the earliest arriving waterbirds in the boreal forest (Table 8.1; Savard 1980). Cavity nesters can have intense competition for breeding territories and nest sites, leading to selection pressure for earlier arrival (Zicus and Hennes 1989, Dugger et al. 1994, Eadie et al. 1995, Mallory and Metz 1999). Scoters nest on the ground, are not territorial, and likely do not compete for nest sites (Eadie and Gauthier 1985). Ground-nesting species must wait for potential nest sites to become snow-free before nesting. Indeed, in northwestern Canada, the distribution of Surf Scoters (*Melanitta perspicillata*) is linked to the snow line at the time of their arrival on the breeding areas (Takekawa et al. 2011).

Selection favors migrants that optimize their timing of arrival and have adequate reserves for reproduction (reviewed by Rohwer 1992, Farmer and Wiens 1998, Prop et al. 2003, Bêty et al. 2004). Body mass may be related to the timing of spring migration. De La Cruz et al. (2009) found that heavier Surf Scoters left wintering areas earlier, spent more time at stopover sites, and arrived on the breeding grounds slightly earlier than lighter birds. Eiders that arrived on nesting colonies late or with insufficient reserves initiated nests later and were more likely to abandon their nests or young (Korschgen 1977, Bustnes and Erikstad 1991, Öst 1999, Bustnes et al. 2002).

Information on chronology of migration of marked individuals suggests that timing of spring migration varies within a species. In some

TABLE 8.1

General timing and route of migration and habitats of North American sea ducks.

Species	Breeding habitat	Nest type[a]	Spring migration Arrival time	Spring migration Route[b]	Molting habitat	Fall migration Departure time[c]	Fall migration Route[d]	Winter habitat[e]	Sources[f]
Steller's Eider	Arctic, subarctic	Open	Later spring	Likely coastal, overland to coast	Coastal subarctic	M—none, variable / F—none, variable / HY—early	None or coastal, overland to coast	Coastal subarctic	1–4
Spectacled Eider	Arctic, subarctic	Open	Early spring	Unknown	Coastal arctic, subarctic	M—late / F—early / HY—early	Coastal, marine waters	Open ocean, ice edge	5,6
King Eider	Arctic	Open	Early spring	Coastal then inland to nearby wetlands	Coastal arctic, subarctic	M—none, variable / F—none, variable / HY—early	Coastal, marine waters	Coastal arctic, subarctic, ice edge	7,8
Common Eider	Arctic, subarctic	Open	Early spring, resident	Coastal to islands and wetlands	Coastal arctic, subarctic	M—late, none / F—early, late, none / HY—unknown	Coastal, marine waters, overland to coast	Coastal arctic, subarctic, ice edge	7,9,10
Harlequin Duck	Rivers, streams, coast, inland	Open	Early spring	Coastal then inland	Coastal subarctic	M—none / F—none / HY—unknown	None or coastal	Coastal subarctic, N Am	11
Surf Scoter	Subarctic, boreal forest	Open	Later spring	Coastal then broad front overland	Coastal arctic, subarctic, boreal forest	M—late / F—late / HY—late	Overland, coastal	Coastal N Am	12–14
White-winged Scoter	Subarctic, boreal forest	Open	Later spring(?)	Coastal then broad front overland	Inland boreal forest, coastal, subarctic, boreal forest	M—unknown / F—unknown / HY—unknown	Overland, coastal	Coastal N Am	15,16
Black Scoter	Subarctic	Open	Later spring(?)	Coastal then broad front overland	Coastal subarctic	M—late / F—late / HY—unknown	Coastal, marine waters, overland	Coastal subarctic, coastal N Am	17
Long-tailed Duck	Arctic, subarctic	Open	Mid-spring, early spring	Coastal, overland	Coastal and inland arctic, subarctic, boreal forest	M—late / F—early / HY—early	Coastal, marine waters, overland	Coastal arctic, subarctic, large inland lakes, ice edge	7,18–22

Species	Winter habitat[e]	Nest type[a]	Spring migration arrival[d]	Spring migration route[b]	Fall migration habitat	Fall departure time[c]	Fall migration route[b]	General wintering area	Ref[f]
Bufflehead	Boreal forest	Hole	Early spring	Coastal(?) then broad front overland(?)	Inland boreal forest	M—late F—late HY—unknown	Overland, coastal	Inland and coastal N Am	23, 24, 25
Common Goldeneye	Boreal forest	Hole	Early spring	Broad front overland	Inland and coastal boreal forest	M—late F—late HY—late	Overland, coastal	Coastal and inland N Am, S of boreal forest	26
Barrow's Goldeneye	Boreal forest	Hole	Early spring	Broad front overland	Inland and coastal boreal forest	M—late F—late HY—unknown	Overland, coastal	Primarily coastal forest	27–29
Hooded Merganser	Forested wetlands	Hole	Early spring, resident	Broad front overland(?)	Unknown	M—late? F—late? HY—unknown	Overland(?)	Coastal and inland N Am to Mexico	30
Common Merganser	Boreal forest	Hole	Early spring, resident	Coastal(?) then broad front overland	Inland and coastal boreal forest?	M—late F—late HY—early	Overland, coastal	Inland and coastal forest and middle N Am	31
Red-breasted Merganser	Arctic, subarctic, boreal forest	Open	Latest spring	Coastal, overland, large lakes	Coastal, inland arctic, subarctic, boreal forest	M—unknown F—unknown HY—early	Coastal	Coastal N Am	32

Spring migration arrival time (breeding adults only) is in relationship to other sea ducks and waterfowl in the same area, and fall migration includes departure time (early—at fledging; in relationship to nesting chronology). Question marks (?) indicate knowledge gaps.

[a] Nest type: open—includes cracks in rocks, beneath heavy brush, under overhangs, and more open areas. Hole—a cavity within a tree or other structure including man-made nest boxes.

[b] The general route of migration is for the bulk of the population. Exceptions occur for all species. All individuals migrating between continents fly over marine waters.

[c] Fall departure time: early, departs soon after regaining flight, and late, remains several weeks or until ice forms in molting/breeding area. Fall migration of males and females begins when the bird leaves the molting area. HY bird's fall migration begins as it leaves the nesting area.

[d] Fall migration route: all birds molting in the boreal forest must migrate overland in fall to reach coastal wintering areas; marine waters exceed 20 m in depth; and distance off shore >100 km, otherwise coastal.

[e] Winter habitat: general wintering habitat used by most birds; arctic, ice edge; subarctic, waters offshore of subarctic; coastal forest, more southern waters offshore of historically forested regions; N Am, ice-free inland waters; coastal N Am, all coastal waters to northern Mexico.

[f] (1) Petersen (1981), (2) Fredrickson (2001), (3) Petersen et al. (2006), (4) D. Rosenberg (pers. comm.), (5) Petersen et al. (1999), (6) Petersen et al. (2000), (7) Portenko (1972), (8) Oppel et al. (2008), (9) Goudie et al. (2000), (10) Mosbech et al. (2006a), (11) Brodeur et al. (2002), (12) Savard et al. (1998), (13) Morrier et al. (2008), (14) De La Cruz et al. (2009), (15) Brown and Fredrickson (1997), (16) J Evenson, Washington Department of Fish and Game (pers. comm.), (17) Bordage and Savard (1995), (18) Alison (1975), (19) Johnson (1985), (20) Robertson and Savard (2002), (21) Mallory et al. (2006), (22) M. R. Petersen (unpubl. data), (23) Gauthier (1993), (24) Erskine (1972), (25) Bellrose (1980), (26) Eadie et al. (1995), (27) Eadie et al. (2000), (28) Chaulk and Turner (2007), (29) Robert et al. (2002), (30) Dugger et al. (1994), (31) Mallory and Metz (1999), (32) Titman (1999).

species, there is a relationship between dates of initiation of spring migration and the distance that birds migrate. King Eiders that wintered at lower latitudes had a tendency to begin spring migration earlier than those in more northern latitudes (Oppel et al. 2008). In contrast, Common Eiders (*Somateria mollissima*) wintering in Greenland that migrated the greatest distances to breeding areas initiated migration later than birds that moved only a short distance (Mosbech et al. 2006a). Petersen (2009) found no correlation between dates of the start of migration and the distances that Common Eiders migrated in spring from coastal Russia and Alaska (Bering Sea) to northern Alaska (Beaufort Sea). In the eastern Pacific, there is evidence that Surf Scoters left southern wintering areas slightly earlier than those from more northern areas (De La Cruz et al. 2009). There was less variation in the dates of arrival to breeding areas than in dates Common and King Eiders and Surf Scoters left wintering areas (Oppel et al. 2008, De La Cruz et al. 2009, Petersen 2009, Takekawa et al. 2011); this suggests that migration timing was adjusted to achieve optimal arrival timing on breeding areas. In addition, Oppel and Powell (2009a) found that the choice of wintering grounds by King Eiders in the Bering Sea and nearby waters had no effect on body mass upon arrival to their breeding grounds.

Spring Migration Routes and Stopover Sites

Spring migration routes and stopover sites for several species of sea ducks are dictated, in part, by the availability of open water especially as birds get closer to their breeding areas (Woodby and Divoky 1982, Alexander et al. 1997, Dickson and Gilchrist 2002, Goudie and Gilliland 2008). King Eiders begin migration to northern Russia, Alaska, and Canada from scattered locations throughout coastal regions in the Bering Sea and North Pacific Ocean and congregate at stopover locations along the coast of western Alaska (Figure 8.1; Oppel et al. 2008, 2009). Migration paths of King Eiders diverge as birds reach the Chukchi Sea; individuals nesting in North America move east and migrate past Point Barrow (Woodby and Divoky 1982; Suydam et al. 1997, 2000; Powell and Suydam 2012), fly over sea ice (Richardson and Johnson 1981) to open waters along the coasts of Alaska and western Canada (Powell and Suydam 2012),

and then disperse to nesting areas (Alexander et al. 1997). Similarly, some Long-tailed Ducks (*Clangula hyemalis*) migrate long distances in spring along the coast, overland, and over sea ice (Petersen et al. 2003; Mallory et al. 2006; M. R. Petersen, U.S. Geological Survey, unpubl. data); stage in open leads in sea ice, often with King and Common Eiders; and then disperse to nesting areas (Portenko 1972, McLaren and McLaren 1982, Alexander et al. 1997). Long-tailed Ducks and Red-breasted Mergansers (*Mergus serrator*) in west Greenland were reported moving east over inland glaciers (regions >1,500 m asl) as well as north along the coast (Alerstam et al. 1986). Harlequin Ducks (*Histrionicus histrionicus*) may migrate along the coast to stopover sites, which are often wintering areas of other individuals (Robertson et al. 1999) and then move inland and up watersheds to breeding areas. Similarly, scoters wintering in marine coastal waters migrate up the coast and use coastal stopover locations and then move inland and disperse to potential breeding areas (De La Cruz et al. 2009, Lok et al. 2011). In eastern North America, satellite telemetry studies did not detect important spring staging areas away from the coast for Barrow's Goldeneyes (*Bucephala islandica*). Stops in inland areas were always short in duration with all significant staging occurring in coastal brackish or salt waters (Robert et al. 2002, Savard and Robert 2013).

Within a species, the distance individuals migrate can vary substantially, thereby resulting in wide variation in initiation and completion dates of spring migration, the use of stopover sites, and migration routes. For example, some Common Eiders moved several 1,000 km in spring, whereas others moved <5 km or never left the bay or part of the coast where they nested (Petersen and Flint 2002; Mosbech et al. 2006a; Petersen 2009; M. R. Petersen, unpubl. data).

Nutrient Reserve Acquisition and Spring Migration

The condition of birds prior to beginning migration likely influences their use of stopover sites (Alerstam and Hedenström 1998) and, for sea ducks, varies with species. For example, adult and subadult Long-tailed Ducks wintering on the Great Lakes substantially increased in weight and lipid content in April and May prior to migration; mass of adult male Long-tailed Ducks increased

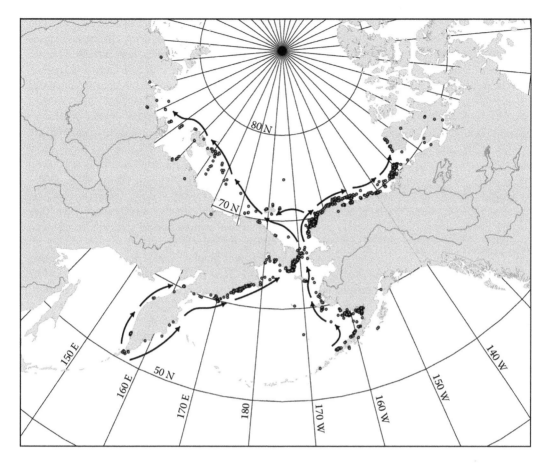

Figure 8.1. Spring migration routes of males and females of the Pacific population of King Eiders (Oppel and Powell 2009b). Birds were marked in arctic Canada and Alaska. When leaving the Bering Sea, females migrated to the east; males migrated both east and west. Lines represent general migration direction and not the actual path. Each dot represents the single best location in the duty cycle. Since the transmitter on/off times varied, the duration a bird was present on a location is unknown. Locations are only included to show a general route of migration.

by 13% and adult females 24% (Peterson and Ellarson 1979). Harlequin Ducks in southern British Columbia, Canada, increased their mass by 7% just before spring migration (Bond and Esler 2006, Esler and Bond 2010) and may have used 2%–3% of their mass during migration (Bond and Esler 2006); however, their use of stopover sites during spring migration was unknown. Surf Scoters and White-winged Scoters (*Melanitta fusca*) also increased mass during late winter and early spring in southeast Alaska, and Surf Scoters continued to gain mass during migration (Anderson et al. 2009).

There is little evidence that hyperphagia occurs in Common Eiders in spring before migration (Guillemette 2001, Merkel et al. 2006). Common Eiders, in comparison to Long-tailed and Harlequin Ducks, are relatively large sea ducks. Accumulation of reserves results in high wing loading, and increased wing loading restricts migration distance due to increased energy needed for flight (Alerstam and Lindström 1990, Hedenström and Alerstam 1998, Åkesson and Hedenström 2007). It may thus be more efficient for a sea duck such as the Common Eider to not accumulate reserves prior to initiating migration (Guillemette 2001, Rigou and Guillemette 2010). Thus, Common Eiders likely use stopover sites to rest, replace reserves lost during migration, and stage until environmental conditions improve. Common Eiders gain the majority of the reserves they use for reproduction in marine habitats at the end of spring migration (local capital; Klaassen et al. 2006) and before they arrive on their nesting colony (Korschgen 1977, Parker and Holm 1990, Christensen 2000, Sénéchal et al. 2011).

Nutrient reserves used during reproduction by Spectacled Eiders (*Somateria fischeri*) and King Eiders likely accumulate at stopover sites during spring migration (Kellett and Alisauskas 2000, Klaassen et al. 2006, Lawson 2006, Oppel et al. 2010) that further emphasizes the importance of stopover sites. It is likely, however, that some sea ducks such as Harlequin Ducks, Long-tailed Duck, Barrow's Goldeneye, and White-winged Scoters depend on stopover sites for rest and replenishment of reserves to continue migration and depend much less on resources acquired at stopover sites during migration to meet the requirements of reproduction (Brown and Fredrickson 1987, Hobson et al. 2005, Arzel et al. 2006, Lawson 2006, Bond et al. 2007b)

Spring Stopover Behavior

Courtship behavior is often seen during spring migration while ducks are at stopover locations and is especially common in species in which one sex winters farther north than the other. Courting scoters, Buffleheads, goldeneyes, and mergansers are often seen at stopover sites during migration and on the breeding grounds until early nesting (Erskine 1971, 1972; Gammonley and Heitmeyer 1990; Eadie et al. 1995, 2000; Mallory and Metz 1999; Titman 1999; O'Connor 2008). Unpaired males and females as well as subadults tend to form separate flocks or occur on the periphery of flocks of paired birds (Kahlert et al. 1998).

Feeding is one of the primary activities at spring stopover areas for some species. As with other waterfowl (Nolet and Drent 1998), the amount of time sea ducks forage during spring migration is likely modulated in response to the availability and quality of preferred forage. Buffleheads at the Klamath Basin stopover site during spring fed >60% of the day, rested <10%, and flew, swam, and engaged in courtship activities the remainder of the time (Gammonley and Heitmeyer 1990). Male Surf Scoters at a spring stopover site on the St. Lawrence River primarily foraged (45%–49%) throughout the day (daylight hours only), and agnostic behavior and courting occurred less frequently (3%–6%; O'Connor 2008). Savard et al. (1998) reported that Surf Scoters increased their foraging activity late in migration (18%–41% in May to 59%–76% in June); presumably, the increased foraging was to build reserves immediately prior to inland migration to breeding areas (Anderson et al. 2009). Common Eiders did not increase their foraging activity during spring migration (Rigou and Guillemette 2010).

In spring, concentrations of scoters, Harlequin Ducks, Common Goldeneyes (*Bucephala clangula*), Buffleheads, Long-tailed Ducks, and Red-breasted Mergansers occurred at Pacific herring (*Clupea pallasi*) spawn locations in British Columbia and Alaska (Sullivan et al. 2002). Rodway (2006, 2007) showed that Harlequin Ducks foraging on energy-rich herring spawn in spring spent less time foraging and devoted more time to courtship and other social behaviors than did Harlequin Ducks in areas with no herring spawn. Similarly, Lewis et al. (2007) found that foraging efforts of Surf and White-winged Scoters dropped significantly when they shifted diet composition in spring from clams to herring spawn.

Fox and Mitchell (1997) found that at a spring staging area in Varangerfjord, Norway, before migration in May, Steller's Eiders (*Polysticta stelleri*) daytime (06:00–21:00 h) activity primarily consisted of feeding (52%), and at night (21:00–06:00 h), Steller's Eiders fed less (37.8%), although there was 24 h of daylight at that time. At the same time, Common Eiders fed much less (only 15% during the day and 11% at night) than Steller's Eiders (Fox and Mitchell 1997).

Flight Behavior during Spring Migration

Sea ducks are generally considered to fly low and fast and flight speed (ground speed minus the effects of winds) and elevation vary with weather and physical characteristics of the landscape. Within a species, flight speeds during migration do not vary among spring, molt, and fall migrations (Table 8.2). Ground speed of migrating birds varies substantially depending on wind speed and direction. Bergman and Donner (1964) found that Common Scoters (*Melanitta nigra*) and Long-tailed Ducks flew about 10% faster when flying high. Long-tailed Ducks, Common Scoters, and Common Eiders increased their elevation when they flew overland, during tailwinds, at night, or in days with poor visibility (Bergman and Donner 1964, Alerstam et al. 1974).

There is high variability within and among species in numbers of eiders in migrating flocks. In spring, Alerstam et al. (1974) reported flock sizes of Common Eiders to over 28,000, but most were <100 birds; Common Eiders were frequently seen to circle over bays, and smaller flocks coalesced into bigger flocks before going overland; flocks flying overwater tended to be smaller. Alerstam et al.

TABLE 8.2
Flight and ground speeds of sea ducks during spring, molt, and fall migrations.

Species	Migration	Flight speed (km h⁻¹)	Ground speed (km h⁻¹)	Location	Sources
Common Eider	Spring	70.0 ± 5.2[b] (n = 354)		S Scandinavia	Rydén and Källander (1964)[d]
Common Scoter	Spring	84 ± 5		Southern Finland	Bergman and Donner (1964)[e]
Common Eider	Spring	74 ± 10[b] (n = 315)	70 ± 5[b] (n = 354)	S Scandinavia	Alerstam et al. (1974)[e]
Long-tailed Duck	Spring	74 ± 5		Southern Finland	Bergman and Donner (1964)[e]
Common Scoter	Spring	79.0 (n = 5)		S Scandinavia	Rydén and Källander (1964)[d]
Common Merganser	Spring	70.4 (n = 6)		S Scandinavia	Rydén and Källander (1964)[d]
Common Goldeneye	Spring	71.2 (n = 4)		S Scandinavia	Rydén and Källander (1964)[d]
Red-breasted Merganser	Spring	72.4 (n = 29)		S Scandinavia	Rydén and Källander (1964)[d]
Steller's Eider	Molt	80 (n = 1)	118 (n = 1)	E Siberian Sea	Alerstam and Gudmundsson (1999)[e]
Eider spp.[a]	Molt		84 ± 13[c] (n = 819)	Barrow, AK	Day et al. (2004)[e]
Long-tailed Duck	Molt	80 (n = 1)	111 (n = 1)	Siberian coast	Alerstam and Gudmundsson (1999)[e]
Common Scoter	Molt	80.9–81.3		Estonia	Jacoby and Jögi (1972)[e]
Barrow's Goldeneye	Molt?		83 (n = 1)	Québec, Canada	Robert et al. (2002)[f]
Long-tailed Duck	Fall		86 (n = 1)	Canada	Mallory et al. (2006)[f]
Long-tailed Duck	All migration		50.2 ± 16.8[b] (n = 3)	Canada	Mallory et al. (2006)[f]

Ground speed can vary substantially depending on wind speed and direction; flight speed is ground speed minus the effects of winds.

[a] King and Common Eiders combined.
[b] SD.
[c] SE.
[d] Speeds calculated from visual observations.
[e] Speeds calculated by radar.
[f] Speeds calculated from satellite telemetry locations (continuous locations within a transmission period).

(1974) also reported an increase in mean flock size as the daily total of migrants increased. King Eiders may form flocks of ≥10,000 birds during migration in spring (Woodby and Divoky 1982, Powell and Suydam 2012). Woodby and Divoky (1982) estimated 500,000 eiders (almost exclusively King Eiders) migrated past Barrow in a 24 h period. Large concentrations of Long-tailed Ducks stage in open leads in sea ice (Portenko 1972, Alexander et al. 1997), but do not necessarily migrate en masse (Richardson and Johnson 1981). Flock sizes of migrating Spectacled Eiders in spring were small (median of six and mode of two individuals) at Cape Romanzof in western Alaska (McCaffery et al. 1999). Many Harlequin Ducks migrate in spring along the coast for short distances as pairs or in small flocks (Robertson et al. 1997, 1999).

Goldeneyes, Buffleheads, and mergansers often migrate in spring during night or day, as singles, pairs, or small flocks (Erskine 1972, Titman 1999) are often dispersed over broad, sparsely inhabited areas (Erskine 1971, Bellrose 1980, Eadie et al. 2000, De La Cruz et al. 2009). In the boreal forest, waters used by migrating sea ducks in spring become available as ice melts; open water becomes available first in the more southern rivers, lakes, and ponds; thus, potential stopover sites are spread over this wide area and may vary among years (De La Cruz et al. 2009).

In southern Scandinavia, Alerstam et al. (1974) used radar tracking combined with visual observations of Common Eiders to determine characteristics of flight during spring migration. Migrating Common Eiders often crossed expanses of land for at least 200 km and migrated above visible sight range and low overwater (some below 30 m, which was below radar detection). Alerstam et al. (1974) also reported Common Eiders following the coastline as well as moving in flight patterns that appeared independent of landforms and found that Common Eiders migrated primarily overwater during day and primarily overland at night. Similarly, in eastern North America, Common Eiders and Surf and White-winged Scoters migrated overland at night (Brown and Fredrickson 1997, Savard et al. 1998), and scoters also migrated during the day along the Atlantic coast (Bond et al. 2007a).

Interannual Constancy at Spring Stopover Sites

Stopover locations are often used in spring of each year by sea ducks (Bellrose 1980, Bordage and Savard 1995, Savard et al. 1998, Mallory and Metz 1999, Titman 1999, Robertson and Savard 2002, Oppel et al. 2009, Petersen 2009, Schamber et al. 2010, Lok et al. 2011). Traditional use of stopover locations can occur if movement to and use of alternate areas requires more time and energy, and the resource quality, quantity, and availability at the traditional site are predictable and adequate during the staging period and results in the same or increased survival.

Some stopover locations consistently have high-quality resources among years, whereas others vary from year to year. For example, Pacific herring roe is an important food for many species of sea ducks during spring migration in the Strait of Georgia, British Columbia, and Southeast Alaska (Vermeer 1981, Vermeer et al. 1997, Sullivan et al. 2002, Rodway et al. 2003, Lewis et al. 2007, Lok et al. 2008). The location, timing, and egg biomass of these spawning events, however, can vary among years, and birds adjust their stopover

locations accordingly to take advantage of this abundant and nutritious resource or rely on alternate foods during spring (Bond and Esler 2006, Lok et al. 2008, Anderson et al. 2009).

Fidelity to breeding sites by females and wintering sites by individuals of both sexes can influence the constancy of migration routes and use of spring stopover sites by individuals. Many sea ducks form pairs during winter (Spurr and Milne 1976; Gowans et al. 1997; Robertson et al. 1999; Rodway 2006, 2007), and each male follows the female to her traditional nesting area. Thus, if females from different nesting areas share a common wintering area, males can potentially pair with females from any breeding area and may be found in one series of stopover locations 1 year and another series the next (Phillips and Powell 2006; M. R. Petersen, unpubl. data). Some male Harlequin Ducks, Buffleheads, and Barrow's Goldeneyes pair with the same female in successive years (Alison 1975, Kuchel 1977, Dzinbal 1982, Gauthier 1987, Gowans et al. 1997), and pairs migrate together in spring to the same breeding area used in previous years (Savard 1985, Smith et al. 2000). This is likely to occur most often in species of sea ducks in which males and females have high fidelity to wintering areas (Anderson et al. 1992; Breault and Savard 1999; Robertson and Cooke 1999; Robertson et al. 1999, 2000; Iverson et al. 2004) and form pairs early in winter (Spurr and Milne 1976, Robertson et al. 1999, Smith et al. 2000, Rodway 2007, Newton 2008). As a result, assuming females use similar routes and stopover locations each year, males will use the same spring migration strategies in successive years. However, the proportion of individuals within a species that use the same stopover sites and spring migration strategy in multiple years is known for only a few individuals. De La Cruz et al. (2009) found that female Surf Scoters used the same migration route they used the previous spring, although the small sample of six birds was not adequate to determine if females repeated their use of spring stopover sites.

Variation in Spring Migration among Sexes and Ages

Adult male and female sea ducks may either: (1) migrate together from wintering areas to the breeding grounds as pairs and thus have similar timing and migration paths, (2) initiate spring migration separately either from different locations or at different times (Kahlert et al. 1998, Coupe and Cooke 1999, Mallory and Metz 1999), or (3) migrate independently until they reach the breeding grounds (Erskine 1971, 1972; Gauthier 1993; Mallory and Metz 1999) and thus have the potential for different timing and use different paths. Most male and female White-winged Scoters migrated together in flocks with equal sex ratios, yet some delayed pair formation until they arrived on the breeding grounds (Brown and Fredrickson 1989). Although most males and females migrated together, in some years, flocks of adult male King Eiders migrated before pairs or females (Woodby and Divoky 1982, Abraham and Finney 1986, Suydam et al. 1997), and some adult male Buffleheads and Common Mergansers (*Mergus merganser*) were the first to arrive on the breeding grounds (Erskine 1971, 1972; Gauthier 1993; Mallory and Metz 1999). Reproduction by early arriving yet unpaired males is predicated on their ability to compete for and acquire a mate before nesting begins or on extrapair copulations (Oring and Sayler 1992, Hario and Hollmén 2004, Steel et al. 2007). Therefore, a spring migration strategy of early arrival by unpaired adult males may be advantageous.

For most species, similarities and differences in migration strategies between sexes are inferred from observations of birds during migration or as they arrive on nesting areas. Among eiders, flocks of only males were observed during spring migration and flocks of eiders that included females were sometimes slightly skewed toward males (Suydam et al. 1997, 2000; McCaffery et al. 1999; Powell and Suydam 2012); flocks of only females were not reported during spring migration. Most female Harlequin Ducks, Long-tailed Ducks, Surf Scoters, goldeneyes, and Buffleheads arrive on the breeding grounds as paired birds (Erskine 1971, Alison 1975, Palmer 1976, Savard et al. 1998, Eadie et al. 2000, Goudie and Gilliland 2008), as do most males (Erskine 1971, 1972; Dau 1974; Eadie et al. 1995, 2000; Petersen et al. 2000). However, male Common Goldeneyes arrived at the St. Lawrence Estuary stopover location in early spring (75.2% males) before females; by late spring, the proportion of adult males decreased as adults left the stopover site (Bourget et al. 2007), thus likely leaving only subadults and nonbreeding adult males. At this same stopover site, sex ratios of early arriving Barrow's Goldeneyes tended to be equal (47.5% male, Bourget et al. 2007).

Migration of subadult sea ducks in spring is poorly described and is inferred from observations of birds on breeding areas, young birds remaining on wintering areas during summer, or flocks of subadults at stopover areas (Palmer 1976). Some subadults migrate to breeding areas and arrive later than adults (Eadie et al. 1995, 2000; Gardarsson 2008; Gardarsson and Einarsson 2008; Oppel and Powell 2010). Some subadult Long-tailed Ducks, White-winged Scoters, Harlequin Ducks, Buffleheads, and Common and Barrow's Goldeneyes migrate to the breeding grounds in spring (Erskine 1971; Palmer 1976; Gauthier 1993; Eadie et al. 1995, 2000; Robertson and Goudie 1999), but migration to the breeding grounds is less common for subadult male King Eiders and Common Mergansers (Pearce and Petersen 2009, Oppel and Powell 2010) and female Harlequin Ducks (Bengtson 1972).

Subadult female sea ducks that nest in cavities migrate to breeding areas late in spring or after nesting of adults and explore potential nest cavities yet rarely nest (Eadie and Gauthier 1985, Zicus and Hennes 1989, Dugger et al. 1994, Eadie et al. 1995, Mallory and Metz 1999). Exploration of nest cavities by nonbreeding birds may increase the probability of occupying a site when they return the next year (Eadie and Gauthier 1985, but see Boyd et al. 2009). Subadult females of most ground-nesting sea ducks have been reported at nesting areas in spring or summer (Robertson and Goudie 1999, Oppel and Powell 2010), although the proportion of subadult females that migrate to the breeding area is known only for King Eiders (88%, 8 of 9; Oppel and Powell 2010). There are no reports of distinctive behavior of searching for nest sites by birds nesting on the ground as described for ducks that nest in cavities. Species that typically nest on the ground, however, may benefit by familiarization with the breeding area. However, subadults of all species may reduce their annual energetic cost of migration if they do not go to the breeding area, but instead move from wintering areas directly to molting areas.

Molt Migration

Timing of Molt Migration

Typically, male sea ducks leave their breeding grounds before adult females (Frimer 1994a, Petersen et al. 1999, Phillips et al. 2006, Oppel et al. 2008). Male sea ducks do not participate in incubation and brood rearing and migrate from breeding grounds to staging and molting areas after incubation begins. Initiation of molt migration of male Spectacled Eiders was later from the most northern breeding area (late June) compared to the most southern locations (early June), although arrival dates to molting areas did not differ (Petersen et al. 1999). For several species (Steller's Eiders, King Eiders, Harlequin Ducks, and Barrow's Goldeneyes), a wide range of arrival dates to molting areas within and among years has been reported (Petersen 1981, Robertson et al. 1999, Robert et al. 2002, Philips et al. 2006) and may be attributed to variability in timing of nesting (Robertson et al. 1997). Phillips et al. (2006) found that in years that male King Eiders left the breeding grounds earlier than normal, they arrived earlier than usual to molting locations. Evidence from Harlequin Ducks, however, suggested that although timing of spring (normal vs. late) varied among years, dates of arrival by males to molting grounds were similar (Robertson et al. 1997).

Male Spectacled Eiders arrived at molting regions 3–6 weeks after leaving breeding areas (Petersen et al. 1999); mean arrival to molting areas of male King Eiders averaged 4 weeks after the start of molt migration (Oppel et al. 2008). Molt migration of male Barrow's Goldeneyes in eastern North America, however, lasted only a few days without well-defined stopover areas; males departed from breeding areas and flew directly to molting sites 1000 km to the north (Robert et al. 2002). Similarly, some male Harlequin Ducks from eastern North America staged only a few days along coastal Labrador before migrating to molting areas in Greenland; however, other male Harlequin Ducks remained up to a month along the coast of Canada before moving to coastal Greenland (Brodeur et al. 2002, Chubbs et al. 2008).

Adult female sea ducks tend to begin molt migration later than adult males (Thomson and Person 1963; Johnson 1971; Frimer 1994a; Suydam et al. 1997, 2000; Petersen et al. 1999; Phillips et al. 2006; Powell and Suydam 2012). Females initiate molt migration over several weeks, and the date they begin migration is a function of the outcome of their breeding attempts. For example, non- or failed breeding female King Eiders migrated before adult females and their young (Portenko

1972, Abraham and Finney 1986, Oppel et al. 2009), as did adult female Spectacled Eiders and Surf Scoters that did not successfully reproduce (Petersen et al. 1999, Savard et al. 2007). Few female Common Eiders in eastern Canada molted more than 100 km from their breeding areas; females that molted at distant areas initiated migration in mid-August (Mosbech et al. 2006a). Farther south, however, Savard et al. (2011) reported that female Common Eiders that molted within 50 km of their nesting colony arrived at molting areas in early August (about August 9; range [July 28–August 16]). Initiation of molt migration of adult male and female Steller's Eiders from nesting regions in Norway was similar, although males arrived to their molting areas before females (Petersen et al. 2006).

The later molt migration by successfully nesting females may influence their selection of molt sites. Successfully breeding females begin molt migration later because of the longer periods of time required to incubate clutches and raise ducklings, which would cause them to arrive too late to complete wing molt at some northern locations (Portenko 1972, Hohman et al. 1992). There is evidence that female King Eiders that nest in northern Alaska molt at lower latitudes than males (Oppel et al. 2008). Similarly, adult female Spectacled Eiders that nested in northern Alaska did not migrate to the most northern molting areas used by males (Petersen et al. 1999). By migrating to more southerly areas, female sea ducks may avoid being caught in ice as it forms in late fall and have an increased amount of daylight available to forage (Chapter 9, this volume).

Molt Migration Routes and Stopover Sites

In his synthesis of migration to molting grounds by waterfowl, Salomonsen (1968) identified four types or strategies of molt migration based on locations of molting sites in relationship to breeding and wintering areas. Type A is defined as a situation in which there is no migration and birds become flightless on or near their breeding area (Figure 8.2a), Type B is characterized as birds migrating to molting sites along spring and fall migration routes (Figure 8.2b), and Type C is represented by birds migrating to wintering areas to molt (Figure 8.2c). Type D is generally characterized as a true molt migration and is typified by birds flying a different direction from the

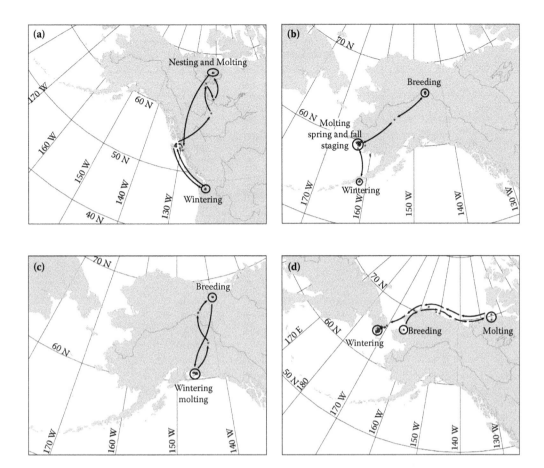

Figure 8.2. Molt migration types following Salomonsen (1968): (a) Type A, remains on the breeding grounds (female White-winged Scoter; J. Evenson, unpubl. data); (b) Type B, migration to molting sites along spring or fall migration paths (female White-winged Scoter; D. Rosenberg, unpubl. data); (c) Type C, molt migration to molting sites on the wintering area (female White-winged Scoter; D. Rosenberg, unpubl. data); and (d) Type D, migration to separate, distinct molting areas (male Common Eider; M. R. Petersen, unpubl. data). Lines represent general migration direction and not the actual path. Each dot represents the single best location in the duty cycle. Since the transmitter on/off times varied, the duration a bird was present on a location is unknown. Locations are only included to show a general route of migration.

expected fall migration route (Figure 8.2d). In Type D, there is separation or variation of routes or molting areas associated with age or sex, or there are geographically distinct regions with perhaps large numbers or densities of birds not associated with wintering, staging, or breeding areas. Salomonsen (1968) further characterized molt migration by the number of birds in molting areas: Type I is defined as single birds or in small groups, Type II as birds in small flocks of hundreds, and Type III as birds in aggregations of thousands to tens of thousands. All of these strategies are used by sea ducks (Table 8.3).

Sea ducks, like all waterfowl, select molting areas with specific characteristics, including high food availability, low predation risk, and predictable open water (Chapter 9, this volume). Molting sites meeting those criteria may occur in any direction from the breeding areas, and thus, the direction of molt migration may be variable (Figure 8.3a) and reflect the importance of molting site characteristics. For example, Spectacled Eiders from the two most northern breeding populations (arctic Russia and northern Alaska) migrated to molting areas along the coast to the south, east, or west of their nesting areas, whereas Spectacled Eiders that nested in western Alaska move both north and west of their breeding areas (Figure 8.3b; Petersen et al. 1999). Some male Harlequin Ducks breeding in eastern Canada fly

TABLE 8.3

Identification of characteristics of molt migration of North American sea ducks in relationship to age and sex, the timing of migration, and migration route.

Species	Age and sex[b]	Initiation	Route[c]	Type	Sources[e]
Steller's Eider	Ad M	July	Nearshore,	BI–III, CII–III	1–6
	Ad F	August–September	offshore,	BI–III, CII–III	
	Failed F	July–August	overland	BI–III, CII–III	
	Subad	April–June		BI–III, CII–III	
Spectacled	Ad M	June–July	Nearshore,	DII–III	7,8
Eider	Ad F	August	offshore,	DII–III	
	Failed F	July–August	overland	DII–III	
	Subad	Unknown		Unknown	
King Eider	Ad M	Early incubation,	Nearshore,	CI–III, DI–III	9–19
	Ad F	June–July	offshore	AI, CI–III, DI–III	
	Failed F	Unknown		CI–III, DI–III	
	Subad	July–August		AI, CIII	
		Unknown			
Common	Ad M	Early–mid-incubation	Nearshore,	DI–III	10–12,17,20–24
Eider[a]	Ad F	August–September	offshore,	AI–III	
	Failed F	July–August	overland	AI–III	
	Subad	June–July		DI–III, CI–II	
Harlequin	Ad M	Early incubation,	Coastal, overland,	BI–II, CI–II	11,25–29
Duck	Ad F	May–July	offshore	BI–II, CI–II	
	Failed F	August–October		BI–II, CI–II	
	Subad	July–August		BI–II, CI–II	
		April–June			
Surf Scoter	Ad M	Mid- to late June, early	Coastal, offshore,	CI, DII–III	11,13,29–35
	Ad F	July	overland	CI–III, DI–III	
	Failed F	Mid- to late August		B, CI–III, DI–III	
	Subad	Mid- to late July		BI–III, CI–III, DI–III	
		Late February–April			
White-	Ad M	June	Coastal, overland	A, C, DI–II	11,21,29,36–38
winged	Ad F	August		A, C, DI–III	
Scoter	Failed F	June–July		A, C, DI–II	
	Subad	March–June		BI–II, DI–II	
Black Scoter	Ad M	June	Coastal, overland,	DII–III	11,29,39–45
	Ad F	Unknown	offshore	DII–III	
	Failed F	July		DII–III	
	Subad	April		BI–III, DII–III	
Long-tailed	Ad M	June–July	Nearshore,	C(?), DI–III	11,12,31,32,46–51
Duck	Ad F	Unknown	offshore,	AI–II, CI–II, DI–III	
	Failed F	July–August	overland	AI–II, CI–II, DI–III	
	Subad	Unknown		Unknown	
Common	Ad M	Early June to mid-July	Overland	DI–III	11,32,52–54
Goldeneye	Ad F	Mid-July to August		AI	
	Failed F	Unknown		Unknown	
	Subad	Later than adult males		DI–III	
Barrow's	Ad M	Late May–June	Coastal, overland	DI–II	11,55,56
Goldeneye	Ad F	Unknown		AI	
	Failed F	July–August		AI	
	Subad	Late May		BI	

(Continued)

TABLE 8.3 (*Continued*)

Identification of characteristics of molt migration of North American sea ducks in relationship to age and sex, the timing of migration, and migration route.

Species	Age and sex[b]	Initiation	Route[c]	Type	Sources[e]
Bufflehead	Ad M	June–July	Overland	AI–II, DI–II	11,46,57,58
	Ad F	Unknown		AI–II, DI–II	
	Failed F	Unknown		AI–II, DI–II	
	Subad	Unknown		Unknown	
Hooded Merganser	Ad M	Early		AI(?)	11,59
	Ad F	Unknown		AI(?)	
	Failed F	Unknown		Unknown	
	Subad	Unknown		Unknown	
Common Merganser	Ad M	Mid-June to mid-July	Overland, coastal	AI, CI, DI–III	11,60–62
	Ad F	Unknown		AI, CI, DI	
	Failed F	Unknown		AI, CI, DI	
	Subad	Mid-June to mid-July		AI, DI–III	
Red-breasted Merganser[a]	Ad M	Late June–early July	Coastal,	Resident, DI–III	11,21,63,64
	Ad F	Unknown	nearshore,	AI, DI	
	Failed F	Unknown	unknown	Unknown	
	Subad	June		DI–III	

The molt migration type is identified following Salomonsen (1968) and represents the current knowledge for the bulk of each species. Question marks (?) indicate knowledge gaps.

[a] Some populations are resident, hence no migration.

[b] Ad M, adult male, breeding male; Ad F, adult female, breeding female that successfully raised young; Failed F, failed female, female that migrated to the breeding grounds and failed to hatch eggs or produce young; Subad, subadult, prenesting aged and nonbreeding bird that did not migrate to the breeding grounds.

[c] Coastal, follows the coastline at an unknown distance from shore; nearshore, within sight from land; offshore, >5 km of shore and crosses water bodies; overland, crosses or migrates over land masses.

[d] Follows Salomonsen (1968).

A: Breeding territory or immediate area.
B: Ordinary autumn migration, subadult during spring migration.
C: Winter quarters.
D: Special area, true molt migration.
  I: Singly or small groups.
  II: Small flocks, hundreds.
  III: Aggregates, >1000.

[e] (1) Petersen (1980), (2) Solovieva (1997), (3) Dau et al. (2000), (4) Flint et al. (2000), (5) U.S. Fish and Wildlife Service (2002), (6) Petersen et al. (2006), (7) Petersen et al. (1999), (8) D. Troy (unpubl. report), (9) Thomson and Pearson (1963), (10) Flock (1973), (11) Palmer (1976), (12) Portenko (1972), (13) Johnson and Richardson (1982), (14) Frimer (1994a), (15) Frimer (1995), (16) Alerstam and Gudmundsson (1999), (17) Day et al. (2004), (18) Phillips et al. (2006), (19) Phillips and Powell (2006), (20) Gauthier et al. (1976), (21) Joensen (1973), (22) Goudie et al. (2000), (23) Mosbech et al. (2006a), (24) M. R. Petersen (unpubl. data), (25) Robertson et al. (1997), (26) Robertson and Goudie (1999), (27) Brodeur et al. (2002), (28) Brodeur et al. (2008), (29) D. Rosenberg (pers. comm.), (30) Andersson (1973), (31) Vermeer and Anweiler (1975), (32) Salter et al. (1980), (33) Savard et al. (1998), (34) Savard et al. (2007), (35) J. Evenson (pers. comm.), (36) Brown and Brown (1981), (37) Brown and Fredrickson (1989), (38) Brown and Fredrickson (1997), (39) Jacoby and Jogi (1972), (40) Kumari (1980), (41) Zhalkyavichus (1982), (42) Herter et al. (1989), (43) Bordage and Savard (1995), (44) Einarsson and Gardarsson (2004), (45) J. Schamber, U.S. Geological Survey (pers. comm.), (46) King (1973), (47) Alison (1975), (48) Derksen et al. (1981), (49) Richardson and Johnson (1981), (50) Petersen et al. (2003), (51) Mallory et al. (2006), (52) Jepsen (1973), (53) Kumari (1980), (54) Eadie et al. (1995), (55) Eadie et al. (2000), (56) Robert et al. (2002), (57) Erskine (1971), (58) Gauthier (1993), (59) Dugger et al. (1994), (60) Mallory and Metz (1999), (61) Pearce and Petersen (2009), (62) Pearce et al. (2009), (63) Titman (1999), (64) Dickson and Gilchrist (2002).

east to the southwest coast of Greenland to molt while others migrate northeast to coastal Labrador (Figure 8.3c; Brodeur et al. 2002, Chubbs et al. 2008, Robert et al. 2008). Herter et al. (1989) found scoters (primarily White-winged Scoters) migrating up the coast past Cape Peirce during their molt migration in the general direction of spring migration by northerly breeding scoters. Pearce et al. (2009) found that some male Common Mergansers molting in south central Alaska had genetic affinities of breeding areas outside of Alaska. Subsequently, four males banded on Kodiak Island during molt were recovered in winter in Washington, Oregon, and California; each of these four birds had genetic affinities to these regions confirming that they had migrated

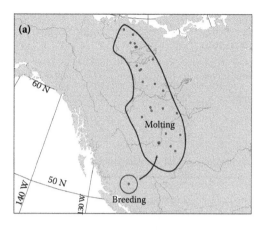

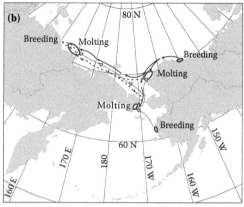

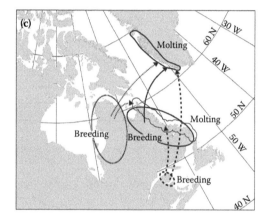

Figure 8.3. Examples of orientation of breeding grounds to molting areas. (a) Male Barrow's Goldeneyes molting on freshwater lakes (W. S. Boyd, unpubl. data), (b) male Spectacled Eiders molting in coastal waters north, east, and south of breeding areas (Petersen et al. 1995, 1999), and (c) male Harlequin Ducks from breeding areas in eastern Canada and molting in coastal waters (Brodeur et al. 2002, Chubbs et al. 2008, Robert et al. 2008, Thomas et al. 2008). Polygons represent general breeding and molting areas. See the papers in *Waterbirds* 31 (special publication 2, 2008) and Robertson and Goudie (1999) for further details on breeding, molting, and wintering areas of Harlequin Ducks. Lines represent general migration direction and not the actual path.

to Alaska for molt and then returned south, outside of Alaska, for winter (Pearce et al. 2009).

All species of sea ducks in North America have multiple molting areas, and molting aggregations typically consist of birds from multiple breeding areas; thus, molt migration strategies may vary substantially among individuals. In Barrow's Goldeneyes, multiple birds from a single breeding area may migrate to the same molt location, but others from that breeding area migrate elsewhere (Savard and Robert 2013; W. S. Boyd, pers. comm.). Some female Common Eiders breeding on the same island in the Gulf of St. Lawrence did not migrate and remained within 80 km of their breeding colony, whereas others migrated 1,000 km to molting areas in Maine

(J.-P. L. Savard, unpubl. data). Similar patterns have been found in other Common Eiders (Savard et al. 2011; M. R. Petersen, unpubl. data), King Eiders (Phillips et al. 2006, Oppel et al. 2008), Spectacled Eiders (Petersen et al. 1999), Steller's Eiders (Dau et al. 2000, Petersen et al. 2006), Longtailed Ducks (Petersen et al. 2003), and Harlequin Ducks (Brodeur et al. 2002, Robert et al. 2008). The exception, based on satellite telemetry data, is female Spectacled Eiders that nest on the Yukon–Kuskokwim Delta. These females only migrated to a single molting area in eastern Norton Sound; no male or female Spectacled Eiders from other areas are known to go there (Petersen et al. 1999), although some King Eiders have been found in western Norton Sound near their molting area.

Using satellite telemetry, D. Troy (unpubl. report) monitored molt migration of Spectacled Eiders along 400 km of the western Beaufort Sea coast and, based on daily locations of individuals in relation to ice distribution, concluded that the lack of open water had a substantial impact on the migration route and use of stopover locations. During the early molt migration period, no nearshore water was available, and males flew overland directly to the Chukchi Sea; when open water in the Beaufort Sea became available later during the migration period, male and failed breeding female Spectacled Eiders migrated along the coast and stopped at coastal locations (D. Troy, unpubl. report).

Stopover Behavior during Molt Migration

Little information is available on the activity patterns of sea ducks at stopover sites during molt migration. As birds begin molt migration, males often form flocks on or adjacent to breeding areas (Bengtson 1966, Erskine 1971, Eadie et al. 2000).

Flight Behavior during Molt Migration

Without handling birds to evaluate their molt and plumage, it is sometimes difficult to determine if sea ducks migrating past a point after the breeding season are on a molt or a fall migration. Based on timing of molt migration and the molting locations of marked King Eiders (all but one were 951–4,727 km from their breeding areas; Phillips et al. 2006, Oppel et al. 2008), we assumed that most King Eiders migrating past Point Barrow from July to September were on a molt migration. The first hatching-year (HY) King Eiders were found in hunter bags after the last few days in August (Johnson 1971, Timson 1976), after which we considered the beginning of fall migration. The status of migrating Long-tailed Ducks at Point Barrow was unknown but likely fall migrants (Johnson and Richardson 1982); however, we include them here for comparison with eiders. King and Common Eiders are often referred to as eiders because of the difficulty of differentiating females and young of the year by species or age and the difficulty in identifying males to species in flight when they are past moderate distances from observers (Thompson and Person 1963, Johnson 1971, Suydam et al. 1997).

Flock sizes of male Long-tailed Ducks, eiders, and scoters during the molt migration along the Bering Sea Coast were <300 birds (Johnson and Richardson 1982). Molt migration and early fall migration observations have been conducted at Point Barrow over many decades. Flock sizes of Long-tailed Ducks were generally <100, and flocks with as many as 2,300 individuals were reported (Thompson and Person 1963, Timson 1976). Flock sizes of eiders flying past Point Barrow varied among years from a mean of 38 individuals to means of 105–110 eiders (or <100) and maximum flock sizes up to 1,100 individuals (Thompson and Person 1963, Johnson 1971, Timson 1976, Day et al. 2004). At Point Barrow, more eiders and Long-tailed Ducks migrated in the evening or at night (Thompson and Person 1963, Johnson 1971, Timson 1976, Day et al. 2004). Similarly, Zalakevicius and Jacoby (1992) and Roed (1971) working in Estonia and south Jutland and south Sweden, respectively, concluded that molt migration of Common Scoters occurred in the evening and at night or only in the evening depending on the location. The total birds per day moving past Point Barrow was greater during good visibility and with tailwinds (Thompson and Person 1963, Johnson 1971, Timson 1976, Day et al. 2004), which was generally found by Roed (1971) for Common Eiders in Jutland and Sweden. Ground speeds, rates of migration, and altitudes of eiders at Point Barrow were most influenced by wind speed and direction (Table 8.2); migration rates were higher with good visibility and strong winds, with weak headwinds, and with strong tailwinds and crosswinds; and altitudes were lower with headwinds than tailwinds (Day et al. 2004), which is similar to that found by Jacoby and Jogi (1972) in Estonia.

In general, eiders migrated at elevations below 4,500 m (Thompson and Person 1963, Day et al. 2004); Brown and Fredrickson (1997) reported that White-winged Scoters flew so low that their wingtips occasionally hit the water during migration. Flock (1973) found that eiders (predominantly King Eiders) migrated low overland as well as higher (as high as 305 m). Herter et al. (1989) reported that during the molt migration, most scoters migrating past Cape Peirce were flying 115 m or less; in the eastern Baltic, Common Scoters flew at altitudes of 1,000–4,500 m overland with tailwinds and 100–600 m overwater (Zalakevicius and Jacoby 1992).

Interannual Constancy at Molt Migration Stopover Sites

Flocks of sea ducks have been found at traditional staging areas during molt migration along coastal North America as well as at inland sites (Palmer 1976, Bellrose 1980, Petersen et al. 1999, Oppel et al. 2009). Over a 10-year period, Oppel et al. (2009) found that male, female, and subadult King Eiders migrating south from northern Alaska and the Northwest Territories, Canada, used stopover sites each year (1997–2007) in the eastern Chukchi Sea during molt migration. Dates of arrival and departure, lengths of stay, and characteristics (distance to shore and depth of water) of locations used by individuals varied by age, sex, and marking location (Oppel et al. 2009).

General migration routes were determined for individuals of only three species, King and Common Eider, and Barrow's Goldeneye, between breeding and molting areas for 2 or more years. Male King Eiders had relatively high fidelity to molting areas, although some males did not return to their previous year's molt site (Phillips and Powell 2006, Knoche et al. 2007) and had low fidelity to breeding areas (Oppel et al. 2008); thus, their complete migration routes changed among years (Phillips et al. 2006). However, portions of the migration route of King Eiders from different breeding areas overlapped in time and space at a common stopover location (eastern Chukchi Sea; Oppel et al. 2009). Phillips et al. (2006) hypothesized that some individuals have high fidelity to a molting area and others follow flocks departing from common stopover sites to different molt sites each year (low fidelity). In contrast, adult female Common Eiders had high fidelity to both breeding and molting areas; however, most migrated less than 100 km and used no stopover locations (M. R. Petersen, unpubl. data). Male Barrow's Goldeneyes tended to be faithful to breeding and molting areas, and general migration routes and some stopover sites could be determined (Savard and Robert 2013; W. S. Boyd, pers. comm.). The exact daily molt migration route and stopover sites and the duration an individual remained at those locations from first migration until several years into adulthood have not been determined for any sea duck.

Variation in Molt Migration among Sexes and Age

The primary migration strategy of many adult female sea ducks is to migrate to coastal marine waters to molt their flight feathers (Petersen et al. 1999, 2006; Mosbech et al. 2006a,b; Phillips et al. 2006; Oppel et al. 2008). Adult female sea ducks that successfully hatch eggs either molt on the breeding grounds at the same time ducklings are present, with no molt migration (Little and Furness 1985, Abraham and Finney 1986), abandon the young before they can fly to molt elsewhere on the breeding area (Eadie et al. 1995, 2000, Savard et al. 2007), or migrate to a distant molt location when the young fledge (Robertson et al. 1997, Petersen et al. 1999). The primary molt migration strategy of most adult goldeneyes, Buffleheads, and mergansers breeding in the boreal forest is to not migrate or migrate to freshwater lakes or wetlands within the boreal forest (Erskine 1971, 1972; Bellrose 1980, Little and Furness 1985, Brown and Fredrickson 1997, Eadie et al. 2000, Hogan et al. 2011); in addition, some male Barrow's Goldeneyes also breeding in the boreal forest migrate to coastal areas before initiating molt (Robert et al. 2002). Similarly, adult female scoters may migrate to molting areas in the boreal forest or to coastal waters. Females within a sea duck species may use a wide variety of migration strategies (Table 8.3; Salomonsen 1968). For example, female Long-tailed Ducks may (1) migrate to large, saltwater lagoons and molt in flocks of hundreds to tens of thousands of birds (Vermeer and Anweiler 1975, Johnson 1985, Petersen et al. 2003), (2) not migrate and molt as dispersed individuals on the nesting area (Portenko 1972, Petersen et al. 2003), or (3) migrate to freshwater lakes and molt in flocks of hundreds to tens of thousands of birds (King 1963, 1973; Derksen et al. 1981).

Females in their second or third year rarely nest, and only some sea ducks in their fourth year nest, yet some of these subadults migrate to nesting areas and, after a brief time, continue to their molting area (Eadie et al. 1995, Pearce and Petersen 2009, Oppel and Powell 2010). Molt migration of subadult sea ducks does not necessarily originate from breeding areas. Subadults and nonbreeding adults of some species migrate directly from wintering grounds to molting areas during spring or summer (Joensen 1973; Robert et al. 2002; Pearce and Petersen 2009; J.-P. L. Savard, unpubl. data) or may undergo molt on wintering or spring stopover areas (Petersen 1980, 1981; Frimer 1995; Robertson and Goudie 1999; Robertson and Savard 2002).

Some subadult sea ducks migrate to coastal marine or estuarine waters where they molt (Salomonsen 1968; Palmer 1976; Petersen 1980, 1981; Robertson and Goudie 1999; Eadie et al. 2000; Oppel and Powell 2010); however, the complete molt migration of individual subadults is described only for King Eiders and Common Mergansers (Pearce and Petersen 2009, Oppel and Powell 2010) and varies by sex. One second-year male Common Merganser migrated during spring and summer directly from its wintering area to its likely molting area, whereas the second-year female moved from her wintering area to the region where she was hatched and then to a nearby coastal area to molt. Oppel and Powell (2010) found that of nine marked subadult female King Eiders, one female migrated to a molting area from its wintering area and the remaining eight began their molt migration from the breeding area; three subadult male King Eiders migrated from their wintering areas to their molting area without going to any breeding areas.

Fall Migration

Timing of Fall Migration

The fall migration of many sea ducks occurs late in the year (Table 8.1). Palmer (1976), Bellrose (1980), and others suggested that fall migration is initiated when ice begins to form at nesting or molting areas. For example, Steller's Eiders in Izembek Lagoon, Alaska, leave this molting area when the lagoon begins to freeze (Laubhan and Metzner 1999), and Long-tailed Ducks continue fall migration as leads in the ice freeze (Portenko 1972). Some sea ducks, particularly male mergansers and goldeneyes, may remain on molting or stopover areas until ice completely covers the open water late into winter (Erskine 1972, Bellrose 1980, Eadie et al. 1995, Bourget et al. 2007). For the more southern populations of Common Eiders and other sea ducks, however, fall migration from molting to wintering areas occurs even though little or no ice is present (Bellrose 1980; M. R. Petersen, unpubl. data). In contrast, many

Common and Spectacled Eiders and some King Eiders did not continue fall migration despite the continuing loss of open water and severe weather but remained in wintering areas as long as some open water was available (Abraham and Finney 1986, Guillemette et al. 1993, Robertson and Gilchrist 1998, Goudie et al. 2000, Mosbech et al. 2006b, Petersen et al. 2012).

Timing of fall migration by individuals from their respective molting areas to their wintering areas has been documented for several species of sea ducks (Petersen et al. 1999, 2006; Mosbech et al. 2006a,b; Phillips et al. 2006; Oppel et al. 2008; Robert et al. 2008; Pearce and Petersen 2009). Petersen et al. (2006) found no difference between male and female Steller's Eiders from the Atlantic population in departure dates from molting areas and dates of arrival to wintering areas. However, departure dates of Spectacled Eiders from molting areas and arrival to wintering areas differed among males and females (Petersen et al. 1999). Similarly, Phillips et al. (2006) found that male King Eiders began fall migration before females. Oppel et al. (2008), however, reported no difference in dates of departure from molting areas between male and female King Eiders and mean dates of arrival of these males and females to wintering areas. Bellrose (1980), Savard (1985), and Eadie et al. (2000) reported that adult male goldeneyes arrived at wintering areas before females and young.

Fall Migration Routes and Stopover Habitats

Some species demonstrate flexibility in their fall migration routes. Some Common Eiders that molted and nested in the St. Lawrence estuary migrated overland (640 km) in fall to Maine, whereas others used a coastal route (2,250 km) following the coast of the Gaspé Peninsula to reach the same destination (Reed 1975; Gauthier et al. 1976; J.-P. L. Savard, unpubl. data). At a broader scale, King Eiders from a central Arctic breeding population have a diverse strategy; some adult breeding females wintered in the east (waters off Greenland and northeastern United States and Canada) while others flew to the west (North Pacific and Bering Sea waters, Mehl et al. 2004).

Stopover patterns also vary among and within species. During fall migration, male Harlequin

Ducks flew directly to their wintering areas with no important stopover areas in between (Robert et al. 2008). Brodeur et al. (2002) and Robert et al. (2008) found that Harlequin Ducks in eastern North America staged at a few coastal sites en route to their wintering areas, whereas Harlequin Ducks in western North America generally molted and wintered in the same area (Robertson et al. 1997, 1999; Iverson and Esler 2006) and thus did not have a fall migration. In contrast, fall migration of some Common Eiders may last several weeks, and birds may use several stopover locations (Mosbech et al. 2006a, Savard et al. 2011). Unlike some shorebirds (Gill et al. 2009), no sea ducks fly for >10,000 km without using stopover locations.

Stopover Behavior during Fall Migration

Bourget et al. (2007) conducted a study of Common and Barrow's Goldeneyes in the St. Lawrence Estuary during fall migration. During fall staging, 31% of the flocks of Barrow's Goldeneyes and 13% of flocks of Common Goldeneyes had >21 individuals. Although goldeneye flocks were small, flocks of Barrow's Goldeneyes were larger in fall (23.1 ± 3.5) than were flocks of Common Goldeneyes (11.3 ± 1.0).

Flight Behavior during Fall Migration

Bergman (1978) evaluated autumn migration of waterfowl between the White Sea and the Baltic region. He found that Long-tailed Ducks migrated during daylight hours; although some ducks flew with a strong tailwind, the bulk of migration occurred with milder tailwinds or in calm conditions. Similarly, Erskine (1972) found that more Buffleheads migrated with tailwinds. When flying overland, the altitude of flocks of Long-tailed Ducks varied between 500 and 2500 m, and birds flew at lower altitudes when overwater (Bergman 1978). At the St. Lawrence Estuary, Gauthier et al. (1976) observed that Common Eiders that migrated low overwater preferred slow headwinds (12–23 km h^{-1}), but avoided headwinds above 24 km h^{-1}. Gauthier et al. (1976) also described the crepuscular behavior of Common Eiders before undertaking an overland night flight: birds took off from water and slowly gained altitude by flying in circles over the shoreline; when they reached a few hundred m high, they headed overland. Passage of a cold front was invariably followed by a pulse in migratory activity; Common Eiders avoided cloudy days to initiate their overland migration, which suggests they may need clear sky to navigate at night (Gauthier et al. 1976).

During fall migration, goldeneyes, Buffleheads, and mergansers migrated at night or day and dispersed over broad, sparsely inhabited areas (Erskine 1971, 1972; Bellrose 1980; Dugger et al. 1994; Eadie et al. 1995, 2000; Mallory and Metz 1999; Titman 1999). During fall migration from boreal forest molting areas, flocks of hundreds to thousands of goldeneyes, Buffleheads, and mergansers have been reported in marine waters and larger lakes (Erskine 1971, Savard 1990, Titman 1999). During fall migration from the St. Lawrence Estuary, most flocks (78%) of Common Eiders comprised fewer than 300 birds with a modal size of about 50 birds; however, a few flocks of >2000 birds were observed (Gauthier et al. 1976).

Interannual Constancy at Fall Stopover Sites

In North America, traditional stopover sites during fall migration have been reported primarily along the coasts, at James Bay, at the Great Lakes, and at large lakes and reservoirs (Bellrose 1980, Ross 1983, Savard et al. 1998, Mallory and Metz 1999, Titman 1999, Robertson and Savard 2002). Several sea duck species that undergo a fall migration have high fidelity to both wintering and molting areas, suggesting that fall migration strategies and use of fall stopover sites might be similar across years. Despite having high fidelity to molting areas at Cape Espenberg (Seward Peninsula, AK) and wintering sites in the Bering Sea (Petersen et al. 2012; M. R. Petersen, unpubl. data), individual adult female Common Eiders either did not use stopover sites or changed stopover locations between two fall migration periods. In contrast, based on long-term data from birds marked with satellite transmitters, W. S. Boyd (pers. comm.) determined that individual Barrow's Goldeneye used a fall stopover location in multiple years.

Variation in Fall Migration among Sexes and Age Groups

Timing and distance of fall migration may vary between sexes (Table 8.1). There is some indication that male Black Scoters (*Melanitta americana*) migrated in fall before females (Bordage and Savard 1995), as did some male Spectacled Eiders (Petersen et al. 1999), King Eiders (Phillips et al. 2006), Common Eiders (Gauthier et al. 1976), Steller's Eiders (Petersen 1980, 1981), and Barrow's Goldeneyes (Savard 1985, Eadie et al. 2000, Savard and Robert 2013). Females of some species (King Eider, Oppel et al. 2008; Common Eider, Goudie et al. 2000; Red-breasted Merganser, Titman 1999; Common Mergansers, Erskine 1972), however, arrived to wintering areas before males (Palmer 1976, Bellrose 1980).

Some HY birds begin their first fall migration soon after fledging (Titman 1999) and others not until ice begins to form on lakes and ponds. Some HY birds of some sea ducks such as Harlequin Ducks migrate in fall at the same time or with adult females (Cooke et al. 2000, Goudie et al. 2000, Regehr et al. 2001), and others, such as HY Common Goldeneyes, migrate before adults (Eadie et al. 1995) yet others after adult females that successfully raised young. However, little is known about the first fall migration of HY sea ducks. One study of Common Mergansers on their first fall migration showed that eight young from two broods marked on the same river dispersed widely before arriving to southern wintering sites rather than taking the more direct route used by the adult female (Pearce and Petersen 2009).

With the exception of King Eiders, little information is available on fall migration and survival of HY sea ducks. Oppel and Powell (2010) found that survival of HY male and female King Eiders during their first fall migration (September to November) was lower than winter (females) or similar to winter survival (males) of adult King Eiders; annual survival of HY King Eiders (0.67) was 1.4 times lower than that of adults (0.94). This is similar to the results of studies of geese; migration of goslings is an especially stressful activity with higher mortality of younger, smaller goslings (Schmutz 1993, Menu et al. 2005).

ANNUAL FLIGHT PATHS

Studies of complete migration movements during spring, molt, and autumn of individual sea ducks have provided insight into similarities and differences in migration strategies within and among species. Migration patterns can be categorized into several general types. As one type, Palmer (1976) and others showed that individuals of many species migrate thousands of km farther than the shortest distance from breeding, molting, and wintering grounds in an elliptical pattern (loop migration; Dingle and Drake 2007, Newton 2008). For example, Surf Scoters that winter along the Pacific coast migrate through the mountain ranges and valleys of British Columbia, the Yukon Territories, and Alaska during spring migration to nesting areas in the boreal forests of western Canada (De La Cruz et al. 2009) and Alaska (D. Rosenberg, pers. comm.); some Surf Scoters then leave breeding areas and fly north to molting areas along the Beaufort Sea coast and the west coast of Alaska; then, during fall migration, they fly southeast to coastal wintering areas (Figure 8.4a; Palmer 1976, Savard et al. 1998). An elliptical pattern is most common among males, which tend to have more distinct and longer molt migrations (Palmer 1976). A second type, a general pattern among females, is to take a direct route when returning to the coast or wintering area (Figure 8.4b; Palmer 1976, Brown and Fredrickson 1997). Some sea ducks use a combination of types; they may migrate along the coast until reaching a point in which they migrate overland to their breeding area, circle back to the coast, and then follow the coast to wintering areas along the same track they used in spring (Figure 8.4c). Another strategy is used by species such as eiders and Long-tailed Ducks that winter and nest in coastal areas; they use the same coastal migration path from breeding to wintering and return with no major overland flights (round trip, Dingle and Drake 2007; leading line, Alerstam 1990; Figure 8.4d). Within a species, individuals may substantially differ in their annual migration paths. For example, some Long-tailed Ducks may use a loop migration and others a round trip or leading-line migration along the coast and yet others directly overwater (M. R. Petersen, unpubl. data).

The use of coastlines, rivers, and other physical features of the environment as migration corridors is common in all sea duck species. However, sea ducks also move across land, fly across expanses of water or ice, or migrate at night when points of reference are obscured (Johnson 1971, Johnson and Richardson 1982, Suydam et al. 2000, Powell and Suydam 2012). Eiders, Long-tailed Ducks, and Harlequin Ducks move from one continent to another and annually fly hundreds of km over

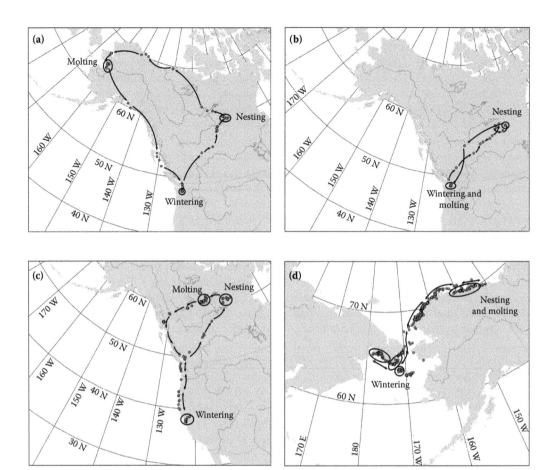

Figure 8.4. General migration strategies of sea ducks: (a) elliptical (*loop*, Dingle and Drake 2007; male Surf Scoter, J. Evenson, unpubl. data), (b) out and back (*inland*, female Surf Scoter: J. Evenson, unpubl. data), (c) combination of elliptical and out and back (*coastal*, female Surf Scoter: J. Evenson, unpubl. data), and (d) out and back (*coastal and overwater*, female Common Eiders: Petersen and Flint 2002; M. R. Petersen, unpubl. data). Lines represent general migration direction and not the actual path. Nesting and molting could not be confirmed by direct observation of nests or flightless birds. A nesting location is thus defined as an area within the general breeding range in which an individual remained for a 10-day period during June–early July (Petersen et al. 2006). A molting area is defined as a location in which the bird remained stationary for 3 or more weeks during mid-August to early October. Each dot represents a single location within a duty cycle. Since the transmitter on/off times varied from 1 to 4 days, the duration a bird was present on a location is unknown. Locations are only included to show a general route of migration.

marine waters out of sight of land. Characteristics of migration, such as flock size, flight path, flight speed and altitude, and duration of movements contribute to our overall view of migration strategies (Tables 8.1, 8.2, and 8.4).

Inter- and Intra-Annual Stopover Site Fidelity

Stopover patterns of sea ducks are limited by the sites that are predictably available (Alerstam and Högstedt 1980, Alerstam and Hedenström 1998). Intra-annual reuse of stopover sites may reflect

optimal characteristics of these systems during much of the year. For example, some locations within the St. Lawrence Estuary are used consistently each year by scoters during both spring and fall migrations (Savard and Falardeau 1997, Savard et al. 1999, Rail and Savard 2003). Variability in use or the presence of various concentrations of scoters during spring and fall may be due to within and among year differences, for example, in forage availability (Bordage and Savard 1995, Larson and Guillemette 2000), breeding phenology (Jones 1965), weather, and ice conditions

TABLE 8.4

Speed and duration of migration by sea ducks.

Species	Age	Sex	Spring migration	Molt migration	Fall migration	Sources[b]
Steller's Eider	Ad	Both	75 km day^{-1} (24–561 km day^{-1}) 24 days (1–51 days), (n = 13)	25 km day^{-1} (9–325 km day^{-1}) 31 days (1–41 days), (n = 13)	874 km day^{-1} (358–1132 km day^{-1}) 4 days (0–9 days), (n = 9)	1,2
Spectacled Eider	Ad	M		20 days (1–54 days), (n = 26)	10 days (4–26 days), (n = 8)	
		F		6 days (1–20 days), (n = 30)	6 days (1–9 days), (n = 12)	2,3
King Eider	Ad	M	61 ± 18 km day^{-1} (n = 26)	77 ± 35 km day^{-1}, (n = 56)[a]		4
		F	46 ± 27 km day^{-1} (n = 16)	122 ± 82 km day^{-1}, (n = 33)[a]		
		Both	62 ± 24 days; (9–10 days), (n = 42)		50 ± 39 km day^{-1}, (11–218 km day^{-1}); (3–105 days), (n = 47)	
Common Eider, S. m. borealis	Ad	Both	57 ± 30 km day^{-1}; 27 ± 10 days (n = 13)	142 ± 67 km day^{-1}; 4 ± 4 days (n = 16)	190 ± 147 km day^{-1}; 16 ± 19 days (n = 21)	5
Common Eider, S. m. v-nigrum	Ad	F	56 ± 29 km day^{-1}; 35 ± 12 days (n = 26); 230 ± 140 km day^{-1}; 9 ± 4 days (n = 8)			2,6
Surf Scoter	Ad	Both	1.5–90.5 days, (n = 74)			7
Barrow's Goldeneye	Ad	M	6 ± 5 day (1–19 days), (n = 13)	19 ± 13 days (4–47 days), (n = 12) 986 ± 178 km day^{-1}, (n = 11)	24 ± 18 days (3–46 days) (n = 4)	8
Common Merganser	HY	Both			15 ± 10 km day^{-1}; 57 ± 17 days (n = 8)	2,9
	Second year	Both		39 ± 2 km day^{-1} (1–61 days), (n = 2)		
	Ad	F	117 km day^{-1}, 1 day, (n = 1)			

NOTES: Speeds of migration are approximate; the duration of premigratory fattening is unknown for sea ducks and excluded in speed calculations. Spring migration is defined as the day an individual leaves its wintering grounds until it arrives to its presumptive breeding grounds; molt migration is from the day it leaves the breeding area until it arrives at its presumptive molting area; and fall migration is from the day it leaves the molting area until its arrival to the wintering area. The duration of migration for birds that do not migrate to a breeding area is calculated based on the day the bird leaves the wintering area until its arrival to the molting area (molt migration). All data are estimated from individuals marked with satellite transmitters. Location data are often not transmitted on a daily basis; thus, migration duration is approximate and its calculation can vary among studies. Data are presented in mean ± SD or median and (minimum–maximum). Both includes M and F.

[a] Differences between males and females are likely a result of bias resulting from locations of nesting sites; differences in speed are unlikely biologically significant (Oppel et al. 2008).

[b] (1) Petersen et al. (2006), (2) M. R. Petersen (unpubl. data), (3) Petersen et al. (1999), (4) Oppel et al. (2008), (5) Mosbech et al. (2006a), (6) Petersen (2009), (7) De La Cruz et al. (2009), (8) Robert et al. (2002), (9) Pearce and Petersen (2009).

(Prach et al. 1981, McLaren and McLaren 1982, Alexander et al. 1997).

Movement patterns of each species are, in part, a result of glacial patterns and life history characteristics (Rand 1948, Ploeger 1968, Alerstam et al. 1986, Newton 2008). Different origins and levels of mixing at breeding and wintering areas likely influence many of the migration patterns seen today (Sutherland 1996, 1998). Migrants share several characteristics, but migration strategies are believed to have developed independently in each species and population (Salomonsen 1955, Lack 1968, Helbig 2003) and, as a result, may differ in response to changes in their environment. The amount of change in migration timing and routes is variable and is an ongoing, dynamic process in response to physical and biological changes in the environment (Tøttrup et al. 2008, Both 2010). The annual use of apparently less optimal migration patterns by different subspecies and breeding populations within a species of sea ducks such as Harlequin Ducks, or Steller's Eiders may be due to variation in the origin and evolution of these migration strategies occurring as a result of isolation during previous glacial periods (Sutherland 1998, Brodeur et al. 2002, Petersen et al. 2006, Newton 2008). The more recent, ongoing development or changes in migration strategies from resident to migrant, or migrant to resident, is partly dependent on genetic variability within a species (Pulido 2007).

The degree of fidelity of individuals to annual migration routes and stopover sites throughout 2 or more years is known for only Barrow's Goldeneye (W. S. Boyd, pers. comm.) and Common Eider (Petersen et al. 2012; M. R. Petersen, unpubl. data). An individual that uses the same migration route (or parts of it) during different migrations within a year may reuse stopover sites (Figure 8.5). This strategy is restricted to individuals that use round trip or leading line types of migration in at least part of their migration (Figures 8.4c–d and 8.5). Adult female Common Eiders nesting at Cape Espenberg in western Alaska have high fidelity to breeding and wintering areas (Petersen et al. 2012) and reuse the same migration route during fall. However, only 16% of these individuals reused stopover sites during both spring and fall (M. R. Petersen, unpubl. data). The ability to determine frequency and duration of use of stopover sites by sea ducks and changes in relationship to resources is complicated by remote locations of many stopover sites and limitations of transmitters that prevent locating individuals daily through all migration periods.

Individual Variation

There is substantial individual variation in the use of stopover sites by waterfowl (Van Eerden et al. 1991, Nolet and Drent 1998, Drent et al. 2003, Eichhorn et al. 2006, Hübner 2006). Migration

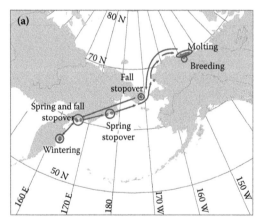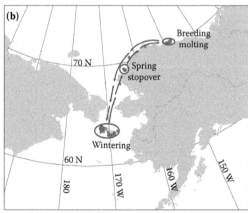

Figure 8.5. Use of spring and fall stopover areas during migration: (a) a single adult female Long-tailed Duck (M. R. Petersen, unpubl. data) and (b) a single adult female Common Eider (M. R. Petersen, unpubl. data). Lines represent general migration direction and not the actual path. Each dot represents a single best location within a duty cycle. Since the transmitter on/off times varied from 1 to 8 days, the duration a bird was present on a location is unknown. Locations are only included to show a general route of migration.

strategies of individuals vary; for instance, Common Eiders from single breeding populations spent variable lengths of time at common stopover sites (Mosbech et al. 2006a, Petersen 2009). In her study of Common Eiders, Petersen (2009) found that some individuals from a single breeding population remained at a spring stopover site for only 1 day (6–30 h), whereas others remained more than a month. Lok et al. (2011) also found individual variation of the length of stay of Surf Scoters that used stopover sites in southeast Alaska. Similar variability was noted for several other sea ducks including Long-tailed Ducks, King Eiders, and Common Eiders during fall migration (Oppel et al. 2008; M. R. Petersen, unpubl. data). How different physiological and environmental factors influences the timing of spring, molt, and fall migration and the use of stopover locations among individuals within and among years remains obscure for sea ducks.

Repeatability of complete migration paths of marked individuals of several species (Berthold et al. 2004, Meyburg et al. 2004, Alerstam et al. 2006, Bobek et al. 2008, Strandberg et al. 2008, Qian et al. 2009, Vardanis et al. 2011) suggests that, although birds may use the same migration corridor among years and among seasons, the reuse of stopover sites by individuals is not consistent. For example, Vardanis et al. (2011) found that the migration route and stopover selection of Marsh Harriers (Circus aeruginosus) varied among years and concluded that flexibility was likely in response to variability in annual weather and habitat conditions. Vardanis et al. (2011), however, found consistency in use by individual Marsh Harriers among years in portions of their migratory routes and stopover sites. Year-to-year consistency in timing of migration of Marsh Harriers and geese occurred more in spring than in fall; there was, however, wide variability in timing of migration among individuals (Bêty et al. 2004, Vardanis et al. 2011).

Variation of migration strategies among years within a species may be affected by the previous year's strategy. For example, Madsen (2001) reported five different spring migration strategies of individual Pink-footed Geese (Anser brachyrhynchus) when migrating from west Jutland, where birds congregate before beginning spring migration, to Svalbard, the breeding area. Pink-footed Geese leaving on the final leg of their spring migration in poor condition changed strategies the next

year, while birds in good condition tended to not change strategies. In only 2 of 6 years, however, did final body condition based on an abdominal profile index differ among birds using different migration strategies.

MIGRATION AND ENVIRONMENT

The most important environmental and physiological factors influencing migration strategies of sea ducks can vary within a migration period. Bauer et al. (2008) showed that, in Pink-footed Geese, the importance of environmental conditions (day length and temperature and their interaction) during spring migration varied along the migration route. Duriez et al. (2009) evaluated several environmental (date, temperature, plant phenology, fixed duration of stay, and plant phenology/date) and physiological (body stores) parameters that likely influenced spring migration of Pink-footed Geese. They found that external cues (plant phenology and date) closely predicted when geese migrated and used stopover locations early in migration, yet factors (date and internal clock) changed in importance the closer birds approached breeding areas. Lok et al. (2012) found that many Surf Scoters migrated in spring in concert with the phenology of spawning of Pacific herring (silver wave hypothesis) and also concluded that there were additional unknown factors that also influenced the timing and use of stopover sites during migration. The interrelationships of environmental and physiological factors that influence the movements of sea ducks remain to be determined.

The stability of migration habitats can influence movement patterns in waterfowl. For instance, Miller et al. (2005) marked Northern Pintail (Anas acuta) over several years and followed individuals throughout a single spring. They found that during spring migration, Northern Pintail used different migration strategies (timing and routes) when moving from the Central Valley of California to the more northern breeding areas relative to extremely cold temperatures and dry conditions at the primary staging area or a change in habitats among years to wet conditions in the Prairie Pothole Region of North America. In contrast, Hupp et al. (2011) did not find annual variation in distribution of breeding areas by Northern Pintails wintering in Japan. Hupp et al. (2011) attributed the difference to the lack of a broad extent of ephemeral wetlands like that found in North America and a tendency of annual consistency in wetlands in the more northern habitats. Highly variable annual, broadscale migration tactics like that of North American Northern Pintails have not been reported for sea ducks. Migration habitats of most sea ducks tend to be predictable among years and, as with Northern Pintails in Japan (Hupp et al. 2011), sea ducks migrate annually to more stable northern habitats (Chapter 13, this volume).

SPEED AND DURATION OF MIGRATION

The overall duration of migration is the total time between stopover sites, the total time for fuelling for flight (including resting and other activities), and flight power (Hedenström and Alerstam 1998, Alerstam 2003, Hedenström 2008). The rate or speed of migration is often expressed as distance/day, that is, the distance of migration divided by the time in migration. The distance based on a predetermined speed a bird can migrate within its annual cycle is restricted by the total amount of time needed for reproduction and molt minus 365 days (Alerstam and Lindström 1990, Hedenström and Alerstam 1998, Alerstam 2003, Hedenström 2008). Larger birds, such as swans, migrate at slower rates than smaller birds, such as terns; thus, to meet annual time constraints, larger birds such as Common Eiders migrate shorter distances between breeding and wintering areas than smaller ducks. Because sea ducks that nest in northern areas have a short period in which they can reproduce, the speed of migration would be expected to result in the bird's arrival at the optimum time for breeding and in sufficient condition to maximize reproduction (Sutherland 1996). Thus, the selection of the migration route must take into account the need to stop and rest, to replenish reserves to continue the flight, and gain reserves for reproduction, as well as the total distance and thus total amount of time of travel (Farmer and Wiens 1998, 1999).

There is substantial variation in sea ducks in the speed and duration of migration within and between species, sex, and time of year (Table 8.4). There is no consistency among the relative speeds of migration during the three migrations or among species. Adult male and female Common Eiders wintering in Greenland migrated at similar speeds (Mosbech et al. 2006a). Molt migration of

King Eiders is quicker than spring migration, and fall migration is slower than any other migration (Oppel et al. 2008, but see Phillips et al. 2006); molt migration occurs over a shorter period of time in female Spectacled Eiders than males (Petersen et al. 1999; M. R. Petersen, unpubl. data). Fall migration, however, is often less variable among sexes; there was no substantial difference in the speed of fall migration between male and female King Eiders (Table 8.4).

The duration of migration is extremely variable among individuals. Spring migration can vary from less than a week to more than a month among individuals from the same wintering area (Robert et al. 2002, Mosbech et al. 2006b, De La Cruz et al. 2009, Petersen 2009). For some species, the duration of migration varies with the distance birds migrate. Male King Eiders that wintered the farthest south took longer to reach their breeding grounds than females and migrated at a faster rate (Table 8.4; Oppel et al. 2008). In contrast, Common Eiders in Greenland that wintered the farthest from their breeding grounds migrated the latest, although the duration of spring migration varied little with distance; as a result, birds that migrated the greater distance migrated at a faster rate (Mosbech et al. 2006b).

Some of the variation in duration of molt migration among sexes has been attributed to males leaving the females soon after incubation is initiated (Table 8.4). Female sea ducks remain longer on the breeding grounds than males before beginning molt migration or do not migrate (Bellrose 1980; Gauthier 1993; Dugger et al. 1994; Bordage and Savard 1995; Eadie et al. 1995, 2000; Brown and Fredrickson 1997; Savard et al. 1998; Mallory and Metz 1999; Robertson and Goudie 1999; Titman 1999; Goudie et al. 2000; Petersen et al. 2000; Suydam 2000; Fredrickson 2001; Robertson and Savard 2002; Powell and Suydam 2012). As a result, for example, successfully nesting female King Eiders have a shorter period of time in which they must complete molt migration and remige molt before ice forms and daylight for foraging is substantially reduced (Salomonsen 1968, Frimer 1994b); thus, they migrate to molting areas at a faster rate (Table 8.4; Oppel et al. 2008). However, based on satellite telemetry data (Table 8.4), there was a significant difference in duration between male and female Spectacled Eiders only during molt migration. The lack of differences in migration duration found between sexes of sea ducks could be a result of small sample sizes (number of species, individuals within a species) and samples containing no or only a few successfully nesting females.

CONCLUSIONS

Uniqueness of the Tribe Mergini

Migration strategies of ducks are highly variable, yet some characteristics result in broad differences among the three tribes: Anatini (dabbling ducks), Aythyini (diving ducks), and Mergini (sea ducks). In general, dabbling and diving ducks reach sexual maturity in their second year, while sea ducks typically do not breed for the first time until their third or fourth summer. As a result of delayed maturity, there are more subadults, and spring, molt, and fall migration strategies may differ among age groups. Sea ducks may winter at high latitudes relatively close to their northern nesting areas. The tolerance of sea ducks to severe winters at northern latitudes may have reduced the energetic demands of migration greater than if they wintered at more southern latitudes. Associated with reduced demands of migration is development of high wing loading with large body size in proportion to small wings so that, although they are less efficient flyers, in combination with other adaptations, sea ducks became more efficient in foraging on benthic invertebrates in marine waters. Because of their ability to exploit prey in these environments, they can remain longer in winter in open waters of more northern latitudes. There are other physical and physiological changes that contribute to cold tolerance and are not necessarily attributed to migration, but allow sea ducks to exploit northern latitudes.

Information Gaps

We have little knowledge of the interrelationships of migration strategies to vital rates of individual sea ducks and ultimately changes in the status of each species. An understanding of migration, particularly in regard to the use of stopover sites, is vitally important to the conservation and management of sea ducks. There are basic information needs that are common to all or most sea ducks. Research needs include the following: (1) determination of key stopover sites; (2) identification of breeding populations (species, subspecies,

breeding groups) using various stopover sites and timing of arrival and length of stay of individuals within each group; (3) an understanding of how and why sea ducks use stopover sites (resting, foraging, courtship) throughout migration; (4) knowledge of how and why migration patterns vary (one group may use several stopover sites and another group none); (5) an understanding of how migration patterns influence connectivity of breeding, molting, and wintering sites; (6) knowledge of how migration in one period can influence the migration of another and ultimately reproduction and survival; (7) description and characterization of dispersal and first- and later-year migration patterns of HY and subadults; and (8) insight into how migration patterns change in regard to environmental variation, physiological condition, and age of individuals.

It is unknown if an individual has high fidelity to a particular wintering, breeding, molting, and stopover sites throughout its life and thus has similar migration patterns among years or if the probability of returning to an area is variable and dependent on factors such as annual changes in physiology, reproduction, and environmental conditions. Additional studies that examine the movement patterns of individuals over several years are needed to better understand site fidelity in sea ducks as it may relate to variability in migration strategies.

There is substantial information on timing of migration and migration routes of some species and populations of sea ducks (Bellrose 1980, Petersen et al. 1999, Brodeur et al. 2002, Robert et al. 2002, Mosbech et al. 2006a,b, Phillips et al. 2006, De La Cruz 2009) and only some on the habitats selected and activity of sea ducks at stopover locations (Gammonley and Heitmeyer 1990, Arzel et al. 2006, O'Connor 2008, Lok et al. 2011; Chapter 13, this volume). Some individuals of all species of sea ducks use stopover locations during one or more migration periods (Bellrose 1980; Gauthier 1993; Dugger et al. 1994; Bordage and Savard 1995; Eadie et al. 1995, 2000; Brown and Fredrickson 1997; Savard et al. 1998; Mallory and Metz 1999; Robertson and Goudie 1999; Titman 1999; Goudie et al. 2000; Petersen et al. 2000; Suydam 2000; Fredrickson 2001; Robertson and Savard 2002; Powell and Suydam 2012). The use of stopover locations likely varies with a bird's location in relationship to the goal, season, age and sex of individuals, weather conditions, physiological condition, previous use, and quality of stopover locations (Alerstam and Hedenström 1998; Farmer and Wiens 1998, 1999; Houston 1998; Mosbech et al. 2006a; Newton 2008; Schaub et al. 2008; Vardanis et al. 2011).

Changes in the Environment

The migration patterns of sea ducks must be flexible with variability expressed within and among individuals, and this flexibility will contribute toward adaptations in response to future changes in the environment (although see Lok et al. 2013). Murphy-Klassen et al. (2005) found that arrival dates of Buffleheads, Common Goldeneyes, Hooded Mergansers (Lophodytes cucullatus), and Red-breasted Mergansers were earlier in spring and concluded that the long-term trend of increasing temperature has contributed to earlier migration dates over the 63-year period of their study (but see Drever et al. 2012). As ice forms later and melts earlier as northern temperatures increase (Douglas 2010), birds are likely to remain at northern fall stopover sites longer than with early ice up. In contrast, the resulting increase in the concentration of ice and its persistence in polynya during winter and spring in Baffin Bay, Davis Strait, coastal West Greenland, and Lancaster Sound may result in decreased availability of critical shallower waters used by King and Common Eiders and their resulting movement to less preferred (deeper water, different prey) foraging areas (Heide-Jørgensen and Laidre 2004).

Over the last decades, the winter distribution and timing of migration, thus migration patterns, have changed in several species of sea ducks in response to invasive species and man's activities. Wormington and Leach (1992) found that corresponding with the arrival of the introduced zebra mussel (Dreissena polymorpha), some diving (primarily Aythya) and sea ducks (primarily Melanitta and Bucephala) remained at fall stopover sites later into fall and early winter. Similarly, the timing of fall migration has changed substantially for segments of various migrant populations in response to construction of water bodies that retain open water; in addition, warm water effluent from generating plants has changed the winter distribution (birds wintering farther north) and thus fall migration characteristics of such species as the Common Merganser (Bellrose 1980).

ACKNOWLEDGMENTS

We thank D. V. Derksen, J. Geiselman, and J. M. Pearce, Alaska Science Center, U.S. Geological Survey, and R. Elliot, Science and Technology, Environment Canada, for the opportunity to prepare this chapter. We especially thank W. S. Boyd, S. E. De La Cruz, J. R. Evenson, D. R. Nysewander, A. N. Powell, and D. H. Rosenberg for sharing their satellite telemetry data and J. Terenzi for the preparation of the figures. We thank W. S. Boyd, R. E. Gill Jr., J. W. Hupp, S. W. De La Cruz, S. Oppel, and D. Esler for reviewing this manuscript. The use of trade names is for descriptive purposes only and does not imply endorsement by the US Government.

LITERATURE CITED

Abraham, K. F., and G. H. Finney. 1986. Eiders of the eastern Canadian arctic. Pp. 55–73 in A. Reed (editor), Eider ducks in Canada. Canadian Wildlife Service Report Series, No. 47. Canadian Wildlife Service, Ottawa, ON.

Åkesson, S., and A. Hedenström. 2007. How migrants get there: migratory performance and orientation. BioScience 57:123–133.

Alerstam, T. 1990. Ecological causes and consequences of bird orientation. Experientia 46:405–415.

Alerstam, T. 2001. Detours in bird migration. Journal of Theoretical Biology 209:319–331.

Alerstam, T. 2003. Bird migration speed. Pp. 253–267 in P. Berthold, E. Gwinner, and E. Sonnenschein (editors), Avian migration. Springer-Verlag, Berlin, Germany.

Alerstam, T., C.-A. Bauer, and G. Roos. 1974. Spring migration of Eiders Somateria mollissima in southern Scandinavia. Ibis 116:194–210.

Alerstam, T., and G. A. Gudmundsson. 1999. Migration patterns of tundra birds: tracking radar observations along the Northeast Passage. Arctic 52:346–371.

Alerstam, T., M. Hake, and N. Kjellén. 2006. Temporal and spatial patterns of repeated migratory journeys by Ospreys. Animal Behaviour 71:555–566.

Alerstam, T., and A. Hedenström. 1998. The development of bird migration theory. Journal of Avian Biology 29:343–369.

Alerstam, T., A. Hedenström, and S. Åkesson. 2003. Long-distance migration: evolution and determinants. Oikos 103:247–260.

Alerstam, T., C. Hjort, G. Högstedt, P. E. Jönsson, J. Karlsson, and B. Larsson. 1986. Spring migration of birds across the Greenland Inland ice. Meddelelser om Grønland, Bioscience 21:3–38.

Alerstam, T., and G. Högstedt. 1980. Spring predict-ability and leap-frog migration. Ornis Scandinavica 11:196–200.

Alerstam, T., and Å. Lindström. 1990. Optimal bird migration: the relative importance of time, energy, and safety. Pp. 331–351 in E. Gwinner (editor), Bird migration. Springer-Verlag, Berlin, Germany.

Alexander, R. M. 1998. When is migration worth-while for animals that walk, swim or fly? Journal of Avian Biology 29:387–394.

Alexander, S. A., D. L. Dickson, and S. E. Westover. 1997. Spring migration of eiders and other water-birds in offshore areas of the western Arctic. Pp. 6–20 in D. L. Dickson (editor), King and Common Eiders of the western Canadian Arctic. Canadian Wildlife Service, Occasional Paper, No. 94. Canadian Wildlife Service, Ottawa, ON.

Alison, R. M. 1975. Breeding biology and behavior of the Oldsquaw (Clangula hyemalis L.). Ornithological Monographs 18:1–52.

Anderson, E. M., J. R. Lovvorn, D. Esler, W. S. Boyd, and K. C. Stick. 2009. Using predator distributions, diet, and condition to evaluate seasonal foraging sites: sea ducks and herring spawn. Marine Ecology Progress Series 386:287–302.

Anderson, M. G., J. M. Rhymer, and F. C. Rohwer. 1992. Philopatry, dispersal, and genetic structure of waterfowl populations. Pp. 365–395 in B. D. J. Batt, A. D. Afton, M. G. Anderson, C. D. Ankney, D. H. Johnson, J. A. Kadlec, and G. L. Krapu (editors), Ecology and management of breeding waterfowl. University of Minnesota Press, Minneapolis, MN.

Andersson, M. 1973. Birds of Nuvagapak Point, north-eastern Alaska. Arctic 26:186–197.

Arzel, C., J. Elmberg, and M. Guillemain. 2006. Ecology of spring-migrating Anatidae: a review. Journal of Ornithology 147:167–184.

Barry, T. W. 1968. Observations of natural mortality and native use of eider ducks along the Beaufort Sea coast. Canadian Field-Naturalist 82:140–144.

Bauer, S., P. Gienapp, and J. Madsen. 2008. The rel-evance of environmental conditions for departure decision changes en route in migrating geese. Ecology 89:1953–1960.

Bellrose, F. C. 1980. Ducks, geese and swans of North America, 3rd ed. Wildlife Management Institute, Stackpole Books, Harrisburg, PA.

Bengtson, S.-A. 1966. Field studies on the Harlequin Duck in Iceland. Wildfowl 17:79–94.

Bengtson, S.-A. 1972. Breeding ecology of the Harlequin Duck Histrionicus histrionicus (L.) in Iceland. Ornis Scandinavica 3:1–19.

Bergman, G. 1978. Effects of wind conditions on the autumn migration of waterfowl between the White Sea area and the Baltic region. Oikos 30:393–397.

Bergman, G., and K. O. Donner. 1964. An analysis of the spring migration of the Common Scoter and the Long-tailed Duck in southern Finland. Acta Zoologica Fennica 105:1–59.

Berthold, P., M. Kaatz, and U. Querner. 2004. Long-term satellite tracking of White Stork (Ciconia ciconia) migration: constancy versus variability. Journal of Ornithology 145:356–359.

Bêty, J., J.-F. Giroux, and G. Gauthier. 2004. Individual variation in timing of migration: causes and repro-ductive consequences in Greater Snow Geese (Anser caerulescens atlanticus). Behavioural Ecology and Sociobiology 57:1–8.

Black, J. M., J. Prop, and K. Larsson. 2007. Wild goose dilemmas: population consequences of indi-vidual decisions in Barnacle Geese. Branta Press, Groningen, Netherlands.

Bobek, M., R. Hampl, L. Peške, F. Pojer, J. Šimek, and S. Bureš. 2008. African odyssey project—satellite tracking of Black Storks Ciconia nigra breeding at a migra-tory divide. Journal of Avian Biology 39:500–506.

Bond, A. L., P. W. Hicklin, and M. Evans. 2007a. Daytime spring migrations of scoters (Melanitta spp.) in the Bay of Fundy. Waterbirds 30:566–572.

Bond, J. C., and D. Esler. 2006. Nutrient acquisition by female Harlequin Ducks prior to spring migration and reproduction: evidence for body mass optimi-zation. Canadian Journal of Zoology 84:1223–1229.

Bond, J. C., D. Esler, and K. A. Hobson. 2007b. Isotopic evidence for sources of nutrients allocated to clutch formation by Harlequin Ducks. Condor 109:698–704.

Bordage, D., and J.-P. L. Savard. 1995. Black Scoter (Melanitta nigra). in A. Poole and F. Gill (editors), The Birds of North America, No. 177. The Academy of Natural Sciences, Philadelphia, PA.

Both, C. 2010. Flexibility of timing of avian migra-tion to climate change masked by environmental constraints en route. Current Biology 20:243–248.

Boulet, M., and D. R. Norris. 2006. Past and pres-ent of migratory connectivity. Ornithological Monographs 61:1–13.

Bourget, D., J.-P. L. Savard, and M. Guillemette. 2007. Distribution, diet and dive behavior of Barrow's and Common Goldeneyes during spring and autumn in the St. Lawrence Estuary. Waterbirds 30:230–240.

Boyd, W. S., B. D. Smith, S. A. Iverson, M. R. Evans, J. E. Thompson, and S. Schneider. 2009. Apparent survival, natal philopatry, and recruitment of Barrow's Goldeneyes (Bucephala islandica) in the Cariboo-Chilcotin region of British Columbia, Canada. Canadian Journal of Zoology 87:337–345.

Breault, A. M., and J.-P. L. Savard. 1999. Philopatry of Harlequin Ducks moulting in southern British Columbia. Pp. 41–44 in R. I. Goudie, M. R. Petersen, and G. J. Roberts (editors), Behaviour and ecology of sea ducks. Canadian Wildlife Service, Occasional Paper, No. 100. Canadian Wildlife Service, Ottawa, ON.

Brodeur, S., J.-P. L. Savard, M. Robert, A. Bourget, G. Fitzgerald, and R. D. Titman. 2008. Abundance and movements of Harlequin Ducks breeding on rivers of the Gaspé Peninsula, Québec. Waterbirds 31 (Special Publication 2):122–129.

Brodeur, S., J.-P. L. Savard, M. Robert, P. Laporte, P. Lamothe, R. D. Titman, S. Marchand, S. Gilliland, and G. Fitzgerald. 2002. Harlequin Duck *Histrionicus histrionicus* population structure in eastern Nearctic. Journal of Avian Biology 33:127–137.

Brown, P. W., and M. A. Brown. 1981. Nesting biology of the White-winged Scoter. Journal of Wildlife Management 45:38–45.

Brown, P. W., and L. H. Fredrickson. 1987. Time budget and incubation behavior of breeding White-winged Scoters. Wilson Bulletin 99:50–55.

Brown, P. W., and L. H. Fredrickson. 1989. White-winged Scoter, *Melanitta fusca*, populations and nesting on Redberry Lake, Saskatchewan. Canadian Field-Naturalist 103:240–247.

Brown, P. W., and L. H. Fredrickson. 1997. White-winged Scoter (*Melanitta fusca*). in A. Poole and F. Gill (editors), The Birds of North America, No. 274. The Academy of Natural Sciences, Philadelphia, PA.

Bustnes, J. O., and K. E. Erikstad. 1991. Parental care in the Common Eider (*Somateria mollissima*): factors affecting abandonment and adoption of young. Canadian Journal of Zoology 69:1538–1545.

Bustnes, J. O., K. E. Erikstad, and T. H. Bjørn. 2002. Body condition and brood abandonment in Common Eiders breeding in the high arctic. Waterbirds 25:63–66.

Chaulk, K. G., and B. Turner. 2007. The timing of waterfowl arrival and dispersion during spring migration in Labrador. Northeastern Naturalist 14:375–386.

Christensen, T. K. 2000. Female pre-nesting foraging and male vigilance in Common Eider *Somateria mollissima*. Bird Study 47:311–319.

Chubbs, T. E., P. G. Trimper, G. W. Humphries, P. W. Thomas, L. T. Elson, and D. K. Laing. 2008. Tracking seasonal movements of adult male Harlequin Ducks from central Labrador using satellite telemetry. Waterbirds 31 (Special Publication 2):173–182.

Cooke, F., G. J. Robertson, and C. M. Smith. 2000. Survival, emigration, and winter population structure of Harlequin Ducks. Condor 102:137–144.

Coupe, M., and F. Cooke. 1999. Factors affecting the pairing chronologies of three species of mergansers in southwest British Columbia. Waterbirds 22:452–458.

Dau, C. P. 1974. Nesting biology of the Spectacled Eider *Somateria fischeri* (Brandt) on the Yukon-Kuskokwim Delta, Alaska. Thesis, University of Alaska, Fairbanks, AK.

Dau, C. P., P. L. Flint, and M. R. Petersen. 2000. Distribution of recoveries of Steller's Eiders banded on the lower Alaska Peninsula, Alaska. Journal of Field Ornithology 71:541–548.

Day, R. H., J. R. Rose, A. K. Prichard, R. J. Blaha, and B. A. Cooper. 2004. Environmental effects on the fall migration of eiders at Barrow, Alaska. Marine Ornithology 32:13–24.

De La Cruz, S. E. W., J. Y. Takekawa, M. T. Wilson, D. R. Nysewander, J. R. Evenson, D. Esler, W. S. Boyd, and D. H. Ward. 2009. Spring migration routes and chronology of Surf Scoters (*Melanitta perspicillata*): a synthesis of Pacific Coast studies. Canadian Journal of Zoology 87:1069–1086.

Derksen, D. V., T. C. Rothe, and W. D. Eldridge. 1981. Use of wetland habitats by birds in the National Petroleum Reserve-Alaska. U.S. Department of Interior, Fish and Wildlife Service Resource Publication, No. 141, U.S. Department of Interior, Washington, DC.

Dickson, D. L., and H. G. Gilchrist. 2002. Status of marine birds of the southeastern Beaufort Sea. Arctic 55 (Suppl. 1):46–58.

Dingle, H., and V. A. Drake. 2007. What is migration? BioScience 57:113–121.

Douglas, D. C. 2010. Arctic sea ice decline: projected changes in timing and extent of sea ice in the Bering and Chukchi Seas. U.S. Geological Survey Open-file Report 2010–1176. U.S. Geological Survey, Reston, VA.

Drent, R., C. Both, M. Green, J. Madsen, and T. Piersma. 2003. Pay-offs and penalties of competing migratory schedules. Oikos 103:274–292.

Drent, R. H., and S. Daan. 1980. The prudent parent: energetic adjustments in avian breeding. Ardea 68:225–252.

Drent, R. H., G. Eichhorn, A. Flagstad, A. J. Van der Graff, K. E. Litvin, and J. Stahl. 2007. Migratory connectivity in Arctic geese: spring stopovers are the weak links in meeting targets for breeding. Journal of Ornithology 148 (Suppl. 2):S501–S514.

Drever, M. C., R. G. Clark, C. Derksen, S. M. Slattery, P. Toose, and T. D. Nudds. 2012. Population vulnerability to climate change linked to timing of breeding in boreal ducks. Global Change Biology 18:480–492.

Dugger, B. D., K. M. Dugger, and L. H. Fredrickson. 1994. Hooded Merganser (*Lophodytes cucullatus*). in A. Poole and F. Gill (editors), The Birds of North America, No. 98. The Academy of National Sciences, Philadelphia, PA.

Duriez, O., S. Bauer, A. Destin, J. Madsen, B. A. Nolet, R. A. Stillman, and M. Klaassen. 2009. What decision rules might Pink-footed Geese use to depart on migration? An individual-based model. Behavioral Ecology 20:560–569.

Dzinbal, K. A. 1982. Ecology of Harlequin Ducks in Prince William Sound, Alaska, during summer. Thesis, Oregon State University, Corvallis, OR.

Eadie, J. M., and G. Gauthier. 1985. Prospecting for nest sites by cavity-nesting ducks of the genus *Bucephala*. Condor 87:528–534.

Eadie, J. M., M. L. Mallory, and H. G. Lumsden. 1995. Common Goldeneye (*Bucephala clangula*). in A. Poole and F. Gill (editors), The Birds of North America, No. 170. The Academy of Natural Sciences, Philadelphia, PA.

Eadie, J. M., J.-P. L. Savard, and M. L. Mallory. 2000. Barrow's Goldeneye (*Bucephala islandica*). in A. Poole and F. Gill (editors), The Birds of North America, No. 548. The Academy of Natural Sciences, Philadelphia, PA.

Eichhorn, G., V. Afanasyev, R. H. Drent, and H. P. van der Jeugd. 2006. Spring stopover routines in Russian Barnacle Geese *Branta leucopsis* tracked by resightings and geolocation. Ardea 94:667–678.

Einarsson, Á., and A. Gardarsson. 2004. Moulting diving ducks and their food supply. Aquatic Ecology 38:297–307.

Erskine, A. J. 1971. Buffleheads. Canadian Wildlife Service, Monograph Series, No. 4. Environment Canada, Ottawa, ON.

Erskine, A. J. 1972. Populations, movements and seasonal distribution of mergansers. Canadian Wildlife Service Report Series, No. 17. Canadian Wildlife Service, Ottawa, ON.

Esler, D., and J. C. Bond. 2010. Cross-seasonal dynamics in body mass of male Harlequin Ducks: a strategy for meeting costs of reproduction. Canadian Journal of Zoology 88:224–230.

Evans, P. R., and N. C. Davidson. 1990. Migration strategies and tactics of waders breeding in arctic and north temperate latitudes. Pp. 387–398 in E. Gwinner (editor), Bird migration. Springer-Verlag, Berlin, Germany.

Farmer, A. H., and J. A. Wiens. 1998. Optimal migration schedules depend on the landscape and the physical environment: a dynamic modeling view. Journal of Avian Biology 29:405–415.

Farmer, A. H., and J. A. Wiens. 1999. Models and reality: time-energy trade-offs in Pectoral Sandpiper (*Calidris melanotos*) migration. Ecology 80:2566–2580.

Flint, P. L., M. R. Petersen, C. P. Dau, J. E. Hines, and J. D. Nichols. 2000. Annual survival and site fidelity of Steller's Eiders molting along the Alaska Peninsula. Journal of Wildlife Management 64:261–268.

Flock, W. L. 1973. Radar observations of bird movements along the Arctic Coast of Alaska. Wilson Bulletin 85:259–275.

Fournier, M. A., and J. E. Hines. 1994. Effects of starvation on muscle and organ mass of King Eiders *Somateria spectabilis* and the ecological and management implications. Wildfowl 45:188–197.

Fox, A. D., and C. Mitchell. 1997. Spring habitat use and feeding behaviour of Steller's Eider *Polysticta stelleri* in Varangerfjord, northern Norway. Ibis 139:542–548.

Fredrickson, L. H. 2001. Steller's Eider (*Polysticta stelleri*). in A. Poole and F. Gill (editors), The Birds of North America, No. 571. The Academy of Natural Sciences, Philadelphia, PA.

Frimer, O. 1994a. Autumn arrival and moult in King Eiders (*Somateria spectabilis*) at Disko, West Greenland. Arctic 47:137–141.

Frimer, O. 1994b. The behaviour of moulting King Eiders *Somateria spectabilis*. Wildfowl 45:176–187.

Frimer, O. 1995. Adaptations by the King Eider *Somateria spectabilis* to its moulting habitat: review of a study at Disco, West Greenland. Dansk Ornithologisk Forenenigs Tidsskrift 89:135–142.

Gammonley, J. H., and M. E. Heitmeyer. 1990. Behavior, body condition, and foods of Buffleheads and Lesser Scaups during spring migration through the Klamath Basin, California. Wilson Bulletin 102:672–683.

Gardarsson, A. 2008. Harlequin Ducks in Iceland. Waterbirds 31 (Special Publication 2):8–14.

Gardarsson, A., and Á. Einarsson. 2008. Relationships among food, reproductive success and density of Harlequin Ducks on the River Laxá at Myvatn, Iceland (1975–2002). Waterbirds 31 (Special Publication 2):84–91.

Gauthier, G. 1987. Further evidence of long-term pair bonds in ducks of the genus *Bucephala*. Auk 104:521–522.

Gauthier, G. 1993. Bufflehead (*Bucephala albeola*). in A. Poole and F. Gill (editors), The Birds of North America, No. 67. The Academy of Natural Sciences, Philadelphia, PA.

Gauthier, J., J. Bédard, and A. Reed. 1976. Overland migration by Common Eiders of the St. Lawrence estuary. Wilson Bulletin 88:333–344.

Gill Jr., R. E., T. L. Tibbits, D. C. Douglas, C. M. Handel, D. M. Mulcahy, J. C. Gottschalck, N. Warnock, B. J. McCaffery, P. F. Battley, and T. Piersma. 2009. Extreme endurance flights by landbirds crossing the Pacific Ocean: ecological corridor rather than barrier? Proceedings of the Royal Society B 276:447–457.

Goudie, R. I., and S. G. Gilliland. 2008. Aspects of distribution and ecology of Harlequin Ducks on the Torrent River, Newfoundland. Waterbirds 31 (Special Publication 2):92–103.

Goudie, R. I., G. J. Robertson, and A. Reed. 2000. Common Eider (*Somateria mollissima*). *in* A. Poole and F. Gill (editors), The Birds of North America, No. 546. The Academy of Natural Sciences, Philadelphia, PA.

Gowans, B., G. J. Robertson, and F. Cooke. 1997. Behaviour and chronology of pair formation by Harlequin Ducks *Histrionicus histrionicus*. Wildfowl 48:135–146.

Guillemette, M. 2001. Foraging before spring migration and before breeding in Common Eiders: does hyperphagia occur? Condor 103:633–638.

Guillemette, M., J. H. Himmelman, C. Barette, and A. Reed. 1993. Habitat selection by Common Eiders in winter and its interaction with flock size. Canadian Journal of Zoology 71:1259–1266.

Hario, M., and T. E. Hollmén. 2004. The role of mate-guarding in pre-laying Common Eiders (*Somateria m. mollissima*) in the northern Baltic Sea. Ornis Fennica 81:119–127.

Hedenström, A. 2008. Adaptations to migration in birds: behavioural strategies, morphology and scaling effects. Philosophical Transactions of the Royal Society B 363:287–299.

Hedenström, A., and T. Alerstam. 1998. How fast can birds migrate? Journal of Avian Biology 29:424–432.

Heide-Jørgensen, M. P., and K. L. Laidre. 2004. Declining extent of open-water refugia for top predators in Baffin Bay and adjacent waters. Ambio 33:487–494.

Helbig, A. J. 2003. Evolution of bird migration: a phylogenetic and biogeographic perspective. Pp. 3–20 *in* P. Berthold, E. Gwinner, and E. Sonnenschein (editors), Avian migration. Springer-Verlag, Berlin, Germany.

Herter, D. R., S. M. Johnston, and A. P. Woodman. 1989. Molt migration of scoters at Cape Peirce, Alaska. Arctic 42:248–252.

Hobson, K. A., J. E. Thompson, M. R. Evan, and S. Boyd. 2005. Tracing nutrient allocation to reproduction in Barrow's Goldeneye. Journal of Wildlife Management 69:1221–1228.

Hochbaum, H. A. 1955. Travels and traditions of waterfowl. University of Minnesota Press, Minneapolis, MN.

Hogan, D., J. E. Thompson, D. Esler, and W. S. Boyd. 2011. Discovery of important postbreeding sites for Barrow's Goldeneye in the Boreal Transition Zone of Alberta. Waterbirds 34:261–268.

Hohman, W. L., C. D. Ankney, and D. H. Gordon. 1992. Ecology and management of postbreeding waterfowl. Pp. 128–189 *in* B. D. J. Batt, A. D. Afton, M. G. Anderson, C. D. Ankney, D. H. Johnson, J. A. Kadlec, and G. L. Krapu (editors), Ecology and management of breeding waterfowl. University of Minnesota Press, Minneapolis, MN.

Houston, A. I. 1998. Models of optimal avian migration: state, time and predation. Journal of Avian Biology 29:395–404.

Hübner, C. E. 2006. The importance of pre-breeding areas for the arctic Barnacle Goose *Branta leucopsis*. Ardea 94:701–713.

Hupp, J. W., N. Yamaguchi, P. L. Flint, J. M. Pearce, K. Tokita, T. Shimada, A. M. Ramey, S. Kharitonov, and H. Higuchi. 2011. Variation in spring migration routes and breeding distribution of Northern Pintails *Anas acuta* that winter in Japan. Journal of Avian Biology 42:289–300.

Iverson, S. A., and D. Esler. 2006. Site fidelity and demographic implications of winter movements by a migratory bird, the Harlequin Duck *Histrionicus histrionicus*. Journal of Avian Biology 37:219–228.

Iverson, S. A., D. Esler, and D. J. Rizzolo. 2004. Winter philopatry of Harlequin Ducks in Prince William Sound, Alaska. Condor 106:711–715.

Jacoby, V., and A. Jögi. 1972. The moult migration of the Common Scoter in light of radar and visual observation data. Communications of the Baltic Commission for Study of Bird Migration 7:118–139 (in Russian with English summary).

Jefferies, R. L., and R. H. Drent. 2006. Arctic geese, migratory connectivity and agricultural change: calling the sorcerer's apprentice to order. Ardea 94:537–554.

Jepsen, P. U. 1973. Studies of the moult migration and wing-feather moult of the goldeneye (*Bucephala clangula*) in Denmark. Danish Review of Game Biology 8:1–23.

Joensen, A. H. 1973. Moult migration and wing-feather moult of seaducks in Denmark. Danish Review of Game Biology 8:1–42.

Johnson, L. L. 1971. The migration, harvest, and importance of waterfowl at Barrow, Alaska. Thesis, University of Alaska, Fairbanks, AK.

Johnson, S. R. 1985. Adaptations of the Long-tailed Duck (*Clangula hyemalis* L.) during the period of molt in arctic Alaska. Acta XVIII Congressus of Internationalis Ornithologica 18:530–540.

Johnson, S. R., and W. J. Richardson. 1982. Waterbird migration near the Yukon and Alaskan Coast of the Beaufort Sea: II: molt migration of seaducks in summer. Arctic 35:291–301.

Jones Jr., R. D., 1965. Returns from Steller's Eiders banded in Izembek Bay, Alaska. Wildfowl 16:83–85.

Kahlert, J., M. Coupe, and F. Cooke. 1998. Winter segregation and timing of pair formation in Red-breasted Merganser *Mergus serrator*. Wildfowl 49:161–172.

Karlsson, H., C. Nilsson, J. Bäckman, and T. Alerstam. 2012. Nocturnal passerine migrants fly faster in spring than in autumn: a test of the time minimization hypothesis. Animal Behaviour 83:87–93.

Kellett, D. K., and R. T. Alisauskas. 2000. Body-mass dynamics of King Eiders during incubation. Auk 117:812–817.

King, J. G. 1963. Duck banding in arctic Alaska. Journal of Wildlife Management 27:356–362.

King, J. G. 1973. A cosmopolitan duck moulting resort; Takslesluk Lake Alaska. Wildfowl 24:103–109.

Klaassen, M. 2003. Relationships between migration and breeding strategies in Arctic breeding birds. Pp. 237–249 in P. Berthold, E. Gwinner, and E. Sonnenschein (editors), Avian migration. Springer-Verlag, Berlin, Germany.

Klaassen, M., K. F. Abraham, R. L. Jefferies, and M. Vrtiska. 2006. Factors affecting the site of investment, and the reliance on savings for arctic breeders: the capital–income dichotomy revisited. Ardea 94:371–384.

Knoche, M. J., A. N. Powell, L. T. Quakenbush, M. J. Wooller, and L. M. Phillips. 2007. Further evidence for site fidelity to wing molt locations by King Eiders: integrating stable isotope analyses and satellite telemetry. Waterbirds 30:52–57.

Korschgen, C. E. 1977. Breeding stress of female eiders in Maine. Journal of Wildlife Management 41:360–373.

Kuchel, C. R. 1977. Some aspects of the behavior and ecology of Harlequin Ducks breeding in Glacier National Park, Montana. Thesis, University of Montana, Missoula, MT.

Kumari, E. 1980. Moult and moult migration of waterfowl in the Baltic basin. Acta Ornithologica 17:37–44.

Lack, D. 1968. Bird migration and natural selection. Oikos 19:1–9.

Larsen, J. K., and M. Guillemette. 2000. Influence of annual variation in food supply on abundance of wintering Common Eiders Somateria mollissima. Marine Ecology Progress Series 201:301–309.

Laubhan, M. K., and K. A. Metzner. 1999. Distribution and diurnal behavior of Steller's Eiders wintering on the Alaska Peninsula. Condor 101:694–698.

Lawson, S. L. 2006. Comparative reproductive strategies between Long-tailed Ducks and King Eiders at Karrak Lake, Nunavut: use of energy resources during the nesting season. Thesis, University of Saskatchewan, Saskatoon, SK.

Lewis, T. L., D. Esler, and W. S. Boyd. 2007. Foraging behaviors of Surf Scoters and White-winged Scoters during spawning of Pacific herring. Condor 109:216–222.

Liechti, F. 2006. Birds: blowin' by the wind? Journal of Ornithology 147:202–211.

Liechti, F., and B. Bruderer. 1998. The relevance of wind for optimal migration theory. Journal of Avian Biology 29:561–568.

Lind, J., and W. Cresswell. 2006. Anti-predation behaviour during bird migration: the benefit of studying multiple behavioural dimensions. Journal of Ornithology 147:310–316.

Lindström, Å. 1991. Maximum fat deposition rates in migrating birds. Ornis Scandinavica 22:12–19.

Lindström, Å. 2003. Fuel deposition rates in migrating birds: causes, constraints and consequences. Pp. 307–320 in P. Berthold, E. Gwinner, and E. Sonnenschein (editors), Avian migration. Springer-Verlag, Berlin, Germany.

Little, B., and R. W. Furness. 1985. Long-distance moult migration by British Goosanders Mergus merganser. Ringing and Migration 6:77–82.

Lok, E. K., D. Esler, J. Y. Takekawa, S. W. De La Cruz, W. S. Boyd, D. R. Nysewander, J. R. Evenson, and D. H. Ward. 2011. Stopover habitats of spring migrating Surf Scoters in southeast Alaska. Journal of Wildlife Management 75:92–100.

Lok, E. K., D. Esler, J. Y. Takekawa, S. W. De La Cruz, W. S. Boyd, D. R. Nysewander, J. R. Evenson, and D. H. Ward. 2012. Spatiotemporal associations between Pacific herring spawn and Surf Scoter spring migration: evaluating a 'silver wave' hypothesis. Marine Ecology Progress Series 457:139–150.

Lok, E. K., M. Kirk, D. Esler, and W. S. Boyd. 2008. Movements of pre-migratory Surf and White-winged Scoters in response to Pacific herring spawn. Waterbirds 31:385–393.

Lok, T., O. Overdijk, and T. Piersma. 2013. Migration tendency delays distributional response to differential survival prospects along a flyway. American Naturalist 181:520–531.

Madsen, J. 2001. Spring migration strategies in Pink-footed Geese Anser brachyrhynchus and consequences for spring fattening and fecundity. Ardea 89 (special issue):43–55.

Mallory, M. L., J. Akearok, N. R. North, D. V. Weseloh, and S. Lair. 2006. Movements of Long-tailed Ducks wintering on Lake Ontario to breeding areas in Nunavut, Canada. Wilson Journal of Ornithology 118:494–501.

Mallory, M., and K. Metz. 1999. Common Merganser (Mergus merganser). in A. Poole and F. Gill (editors), The Birds of North America, No. 442. The Academy of Natural Sciences, Philadelphia, PA.

Marra, P. P., K. A. Hobson, and R. T. Holmes. 1998. Linking winter and summer events in a migratory bird by using stable-carbon isotopes. Science 282:1884–1886.

Marra, P. P., D. R. Norris, S. M. Haig, M. Webster, and J. A. Royle. 2006. Migratory connectivity. Pp. 157–183 in K. Crooks and M. Sanjayan (editors), Connectivity conservation. Cambridge University Press, New York, NY.

McCaffery, B. J., M. L. Wege, and C. A. Nicolai. 1999. Spring migration of Spectacled Eiders at Cape Romanzof, Alaska. Western Birds 30:167–173.

McLaren, P. L., and M. A. McLaren. 1982. Waterfowl populations in eastern Lancaster Sound and western Baffin Bay. Arctic 35:149–157.

Mehl, K. R., R. T. Alisauskas, K. A. Hobson, and D. K. Kellett. 2004. To winter east or west? Heterogeneity in winter philopatry in a central-arctic population of King Eiders. Condor 106:241–251.

Menu, S., G. Gauthier, and A. Reed. 2005. Survival of young Greater Snow Geese (*Chen caerulescens atlantica*) during fall migration. Auk 122:479–496.

Merkel, F. R., A. Mosbech, C. Sonne, A. Flagstad, K. Falk, and S. E. Jamieson. 2006. Local movements, home ranges and body condition of Common Eiders *Somateria mollissima* wintering in southwest Greenland. Ardea 94:639–650.

Meyburg, B.-U., C. Meyburg, T. Bělka, O. Šreibr, and J. Vrana. 2004. Migration, wintering and breeding of a Lesser Spotted Eagle (*Aquila pomarina*) from Slovakia tracked by satellite. Journal of Ornithology 145:1–7.

Miller, M. R., J. Y. Takekawa, J. P. Fleskes, D. L. Orthmeyer, M. L. Casazza, and W. M. Perry. 2005. Spring migration of Northern Pintails from California's Central Valley wintering area tracked with satellite telemetry: routes, timing, and destinations. Canadian Journal of Zoology 83:1314–1332.

Mosbech, A., R. S. Danø, F. Merkel, C. Sonne, G. Gilchrist, and A. Flagstad. 2006b. Use of satellite telemetry to locate key habitats for King Eiders *Somateria spectabilis* in West Greenland. Pp. 769–776 in G. C. Boere, C. A. Galbraith, and D. A. Stroud (editors), Waterbirds around the world. The Stationary Office, Edinburgh, U.K.

Mosbech, A., G. Gilchrist, F. Merkel, C. Sonne, A. Flagstand, and H. Nyegaard. 2006a. Year-round movements of northern Common Eiders *Somateria mollissima borealis* breeding in Arctic Canada and West Greenland followed by satellite telemetry. Ardea 94:651–665.

Morrier, A., L. Lesage, A. Reed, and J.-P. L. Savard. 2008. Étude sur l'écologie de la Macreuse à Front Blanc au Lac Malbaie, Réserve des laurentides, 1994–1995. Série de rapports techniques, No. 301. Région du Québec, Service canadien de la faune, Section Conservation des populations, Environnement Canada, QC (in French).

Murphy-Klassen, H. M., T. J. Underwood, S. G. Sealy, and A. A. Czyrnyj. 2005. Long-term trends in spring arrival dates of migrant birds at Delta Marsh, Manitoba, in relation to climate change. Auk 122:1130–1148.

Myers, M. T. 1958. Preliminary studies of the behavior, migration and distributional ecology of eider ducks in Northern Alaska, 1958. Interim Progress Report. Contracts AINA-37 and ORN-224. Arctic Institute of North America, Calgary, AB.

Newton, I. 2008. The migration ecology of birds. Elsevier, Oxford, U.K.

Nolet, B. A., and R. H. Drent. 1998. Bewick's Swans refuelling on pondweed tubers in the Dvina Bay (White Sea) during their spring migration: first come, first served. Journal of Avian Biology 29:574–584.

Norris, D. R., and P. P. Marra. 2007. Seasonal interactions, habitat quality, and population dynamics in migratory birds. Condor 109:535–547.

Norris, D. R., M. B. Wunder, and M. Boulet. 2006. Perspectives on migratory connectivity. Ornithological Monographs 61:79–88.

O'Connor, M. 2008. Surf Scoter (*Melanitta perspicillata*) ecology on spring staging grounds and during the flightless period. Thesis, McGill University, Montreal, QC.

Oppel, S., D. L. Dickson, and A. N. Powell. 2009. International importance of the eastern Chukchi Sea as a staging area for migrating King Eiders. Polar Biology 32:775–783.

Oppel, S., and A. N. Powell. 2009a. Does winter region affect spring arrival time and body mass of King Eiders in northern Alaska? Polar Biology 32:1203–1209.

Oppel, S., and A. N. Powell. 2009b [online]. Satellite telemetry of King Eiders from northern Alaska 2002–2009. OBIS-SEAMAP. <http://seamap.env.duke.edu/datasets/detail/487> (27 January 2010).

Oppel, S., and A. N. Powell. 2010. Age-specific survival estimates of King Eiders derived from satellite telemetry. Condor 112:323–330.

Oppel, S., A. N. Powell, and D. L. Dickson. 2008. Timing and distance of King Eider migration and winter movements. Condor 110:296–305.

Oppel, S., A. N. Powell, and D. M. O'Brien. 2010. King Eiders use an income strategy for egg production: a case study for incorporating individual dietary variation into nutrient allocation research. Oecologia 164:1–12.

Oring, L. W., and R. D. Sayler. 1992. The mating systems of waterfowl. Pp. 190–213 in B. D. J. Batt, A. D. Afton, M. G. Anderson, C. D. Ankney, D. H. Johnson, J. A. Kadlec, and G. L. Krapu (editors), Ecology and management of breeding waterfowl. University of Minnesota Press, Minneapolis, MN.

Öst, M. 1999. Within-season and between-year variation in the structure of Common Eider broods. Condor 101:598–606.

Owen, M., and J. M. Black. 1990. Waterfowl ecology. Chapman & Hall, New York, NY.

Palmer, R. S. (editor). 1976. Handbook of North American birds (vol. 3), Waterfowl (Part 2). Yale University Press, New Haven, CT.

Parker, H., and H. Holm. 1990. Patterns of nutrient and energy expenditure in female Common Eiders nesting in the high arctic. Auk 107:660–668.

Pearce, J. M., P. Blums, and M.S. Linberg. 2008. Site fidelity is an inconsistent determinant of population structure in the Hooded Merganser (*Lophodytes cucullatus*): evidence from genetic, mark-recapture, and comparative data. Auk 125:711–722.

Pearce, J. M., and M. R. Petersen. 2009. Post-fledging movements of juvenile Common Mergansers (*Mergus merganser*) in Alaska as inferred by satellite telemetry. Waterbirds 32:133–137.

Pearce, J. M., and S. L. Talbot. 2006. Demography, genetics, and the value of mixed messages. Condor 108:474–479.

Pearce, J. M., D. Zwiefelhofer, and N. Maryanski. 2009. Mechanisms of population heterogeneity among molting Common Mergansers on Kodiak Island, Alaska: implications for genetic assessments of migratory connectivity. Condor 111:283–293.

Petersen, M. R. 1980. Observations of wing-feather moult and summer feeding ecology of Steller's Eiders at Nelson Lagoon, Alaska. Wildfowl 31:99–106.

Petersen, M. R. 1981. Populations, feeding ecology and molt of Steller's Eiders. Condor 83:256–262.

Petersen, M. R. 2009. Multiple spring migration strategies in a population of Pacific Common Eiders. Condor 111:59–70.

Petersen, M. R., J. O. Bustnes, and G. H. Systad. 2006. Breeding and moulting locations and migration patterns of the Atlantic population of Steller's Eiders *Polysticta stelleri* as determined from satellite telemetry. Journal of Avian Biology 37:58–68.

Petersen, M. R., D. C. Douglas, and D. M. Mulcahy. 1995. Use of implanted satellite transmitters to locate Spectacled Eiders at-sea. Condor 97:276–278.

Petersen, M. R., D. C. Douglas, H. M. Wilson, and S. E. McCloskey. 2012. Effects of sea ice on winter site fidelity of Pacific Common Eiders (*Somateria mollissima v-nigrum*). Auk 129:399–408.

Petersen, M. R., and P. L. Flint. 2002. Population structure of Pacific Common Eiders breeding in Alaska. Condor 104:780–787.

Petersen, M. R., J. B. Grand, and C. P. Dau. 2000. Spectacled Eider (*Somateria fischeri*). in A. Poole and F. Gill (editors), The Birds of North America, No. 547. The Academy of Natural Sciences, Philadelphia, PA.

Petersen, M. R., and M. E. Hogan. 1996. Seaducks: a time for action. International Waterfowl Symposium 7:62–67.

Petersen, M. R., W. W. Larned, and D. C. Douglas. 1999. At-sea distribution of Spectacled Eiders: a 120-year-old mystery resolved. Auk 116:1009–1020.

Petersen, M. R., B. J. McCaffery, and P. L. Flint. 2003. Post-breeding distribution of Long-tailed Ducks *Clangula hyemalis* from the Yukon-Kuskokwim Delta, Alaska. Wildfowl 54:103–113.

Peterson, S. R., and R. S. Ellarson. 1979. Changes in Oldsquaw carcass weight. Wilson Bulletin 91:288–300.

Phillips, L. M., and A. N. Powell. 2006. Evidence for wing molt and breeding site fidelity in King Eiders. Waterbirds 29:148–153.

Phillips, L. M., A. N. Powell, and E. A. Rexstad. 2006. Large-scale movements and habitat characteristics of King Eiders throughout the nonbreeding period. Condor 108:887–900.

Ploeger, P. L. 1968. Geographic differentiation in Arctic Anatidae as a result of isolation during the last glacial. Ardea 56:1–159.

Portenko, L. A. 1972. Birds of the Chukchi Peninsula and Wrangel Island (vol. 1). Nauka Publishers, Leningrad, Russia (English translation published by Amerind Publishing, New Delhi, India, 1981).

Powell, A. N., and, R. S. Suydam. 2012 [online]. King Eider (*Somateria spectabilis*). in F. Gill (editor), The Birds of North America. Cornell Lab of Ornithology, Ithaca, New York. <http:bna.birds.cornell.edu/review/species/491> (28 January 2014).

Prach, R. W., H. Boyd, and F. G. Cooch. 1981. Polynyas and seaducks. Pp. 67–70 in I. Sterling and H. Cloator (editors), Canadian Wildlife Service Occasional Paper, No. 45. Canadian Wildlife Service, Ottawa, ON.

Prop, J., J. M. Black, and P. Shimmings. 2003. Travel schedules to the high arctic: Barnacle Geese trade-off the timing of migration with accumulation of fat deposits. Oikos 103:403–414.

Pulido, F. 2007. The genetics and evolution of avian migration. BioScience 57:165–174.

Qian, F., H. Wu, L. Gao, H. Zhang, F. Li, X. Zhong, X. Yang, and G. Zheng. 2009. Migration routes and stopover sites of Black-necked Cranes determined by satellite tracking. Journal of Field Ornithology 80:19–26.

Rail, J.-F., and J.-P. L. Savard. 2003. Identification des aires de mue et de repos au printemps des macreuses (*Melanitta* sp.) et de l'Eider à Duvet (*Somateria mollissima*) dans l'estuaire et le golfe du Saint-Laurent. Technical Report Series, No. 408. Canadian Wildlife Service, Quebec Region, Environment Canada, Sainte-Foy, QC (in French).

Rand, A. L. 1948. Glaciation, an isolating factor in speciation. Evolution 2:314–321.

Reed, A. 1975. Migration, homing, and mortality of breeding female Eiders *Somateria mollissima dresseri* of the St. Lawrence estuary, Quebec. Ornis Scandinavica 6:41–47.

Regehr, H. M., C. M. Smith, B. Arquilla, and F. Cooke. 2001. Post-fledgling broods of migratory Harlequin Ducks accompany females to wintering areas. Condor 103:408–412.

Richardson, W. J., and S. R. Johnson. 1981. Waterbird migration near the Yukon and Alaskan coast of the Beaufort Sea: I. timing, routes and numbers in spring. Arctic 34:108–121.

Rigou, Y., and M. Guillemette. 2010. Foraging effort and pre-laying strategy in breeding Common Eiders. Waterbirds 33:314–322.

Robert, M., R. Benoit, and J.-P. L. Savard. 2002. Relationship among breeding, molting, and wintering areas of male Barrow's Goldeneyes (*Bucephala islandica*) in eastern North America. Auk 119:676–684.

Robert, M., G. H. Mittelhauser, B. Jobin, G. Fitzgerald, and P. Lamothe. 2008. New insights on Harlequin Duck population structure in eastern North America as revealed by satellite telemetry. Waterbirds 31 (Special Publication 2):159–172.

Robertson, G. J., and F. Cooke. 1999. Winter philopatry in migratory waterfowl. Auk 116:20–34.

Robertson, G. J., F. Cooke, R. I. Goudie, and W. S. Boyd. 1997. The timing of arrival and moult chronology of Harlequin Ducks *Histrionicus histrionicus*. Wildfowl 48:147–155.

Robertson, G. J., F. Cooke, R. I. Goudie, and W. S. Boyd. 1999. Within-year fidelity of Harlequin Ducks to a moulting and wintering area. Pp. 45–51 in R. I. Goudie, M. R. Petersen, and G. J. Robertson (editors), Behaviour and ecology of sea ducks. Canadian Wildlife Service, Occasional Paper, No. 100. Canadian Wildlife Service, Ottawa, ON.

Robertson, G. J., F. Cooke, R. I. Goudie, and W. S. Boyd. 2000. Spacing patterns, mating systems, and winter philopatry in Harlequin Ducks. Auk 117:299–307.

Robertson, G. J., and H. G. Gilchrist. 1998. Evidence of population declines among Common Eiders breeding in the Belcher Islands, Northwest Territories. Arctic 51:378–385.

Robertson, G. J., and R. I. Goudie. 1999. Harlequin Duck (*Histrionicus histrionicus*). in A. Poole and F. Gill (editors), The Birds of North America, No. 466. The Academy of Natural Sciences, Philadelphia, PA.

Robertson, G. J., and J.-P. L. Savard. 2002. Long-tailed Duck (*Clangula hyemalis*). in A. Poole and F. Gill (editors), The Birds of North America, No. 651. The Academy of Natural Sciences, Philadelphia, PA.

Rodway, M. S. 2006. Have winter spacing patterns of Harlequin Ducks been partially shaped by sexual selection? Waterbirds 29:415–426.

Rodway, M. S. 2007. Timing of pairing in waterfowl II: testing the hypotheses with Harlequin Ducks. Waterbirds 30:506–520.

Rodway, M. S., H. M. Regehr, J. Ashley, P.V. Clarkson, I. R. Goudie, D. E. Hay, C. M. Smith, and K. G. Wright. 2003. Aggregative response of Harlequin Ducks to herring spawning in the Strait of Georgia, British Columbia. Canadian Journal of Zoology 81:504–514.

Roed, A. U. 1971. Fældingstræk over Sønderjylland og ved Kalmarsund. Flora og Fauna 77:45–51 (in Danish with English summary).

Rohwer, F. C. 1992. The evolution of reproductive patterns in waterfowl. Pp. 486–539 in B. D. J. Batt, A. D. Afton, M. G. Anderson, C. D. Ankney, D. H. Johnson, J. A. Kadlec, and G. L. Krapu (editors), Ecology and management of breeding waterfowl. University of Minnesota Press, Minneapolis, MN.

Ross, R. K. 1983. An estimate of the Black Scoter, *Melanitta nigra*, population moulting in James and Hudson bays. Canadian Field-Naturalist 97:147–150.

Rydén, O., and H. Källander. 1964. Beräkning av ejderns (*Somateria mollissima*) sträckhastighet. Vår Fågelvärld 23:151–158 (in Swedish).

Salomonsen, F. 1955. The evolutionary significance of bird-migration. Danske Biologiske Meddelelser 22:1–62.

Salomonsen, F. 1968. The moult migration. Wildfowl 19:5–24.

Salter, R. E., M. A. Gollop, S. R. Johnson, W. R. Koski, and C. E. Tull. 1980. Distribution and abundance of birds on the Arctic Coastal Plain of northern Yukon and adjacent Northwest Territories, 1971–1976. Canadian Field-Naturalist 94:219–238.

Savard, J.-P. L. 1980. Some observations on spring migration of waterfowl in Central British Columbia. Regional Report, Canadian Wildlife Service, Pacific and Yukon Region, Delta, BC.

Savard, J.-P. L. 1985. Evidence of long-term pair bonds in Barrow's Goldeneye *Bucephala islandica*. Auk 102:389–391.

Savard, J.-P. L. 1990. Population de sauvagine hivernant dans l'estuaire du Saint-Laurent: écologie, distribution et abondance. Canadian Wildlife Service Technical Report Series, No. 89. Quebec Region, QC (in French).

Savard, J.-P. L., J. Bédard, and A. Nadeau. 1999. Spring and early summer distribution of scoters and eiders in the St. Lawrence estuary. Pp. 60–65 in R. I. Goudie, M. R. Petersen, and G. J. Robertson (editors), Behaviour and ecology of sea ducks. Canadian Wildlife Service, Occasional Paper, No. 100. Canadian Wildlife Service, Ottawa, ON.

Savard, J.-P. L., D. Bordage, and A. Reed. 1998. Surf Scoter (*Melanitta perspicillata*). in A. Poole and F. Gill (editors), The Birds of North America, No. 363. The Academy of Natural Sciences, Philadelphia, PA.

Savard, J.-P. L., and G. Falardeau. 1997. Inventaires aériens hivernaux, printaniers et estivaux dans les estuaires moyen et marin du Saint-Laurent (hiver 1994, été 1994, printemps 1995). Technical Report Series, No. 282. Canadian Wildlife Service, Quebec Region, QC (in French).

Savard, J.-P. L., L. Lesage, S. C. Gilliland, H. G. Gilchrist, and J.-F. Giroux. 2011. Molting, staging and wintering locations of Common Eiders breeding in the Gyrfalcon Archipelago, Ungava Bay. Arctic 64:197–206.

Savard, J.-P. L., A. Reed, and L. Lesage. 2007. Chronology of breeding and molt migration in Surf Scoters (*Melanitta perspicillata*). Waterbirds 30:223–229.

Savard, J.-P. L., and M. Robert. 2013. Relationships among breeding, molting and wintering areas of adult female Barrow's Goldeneyes (*Bucephala islandica*) in eastern North America. Waterbirds 36:34–42.

Schamber, J. L., P. L. Flint, and A. N. Powell. 2010. Patterns of use and distribution of King Eiders and Black Scoters during the annual cycle in northeastern Bristol Bay, Alaska. Marine Biology 157:2169–2176.

Schaub, M., L. Jenni, and F. Bairlein. 2008. Fuel stores, fuel accumulation, and the decision to depart from a migration stopover site. Behavioural Ecology 19:657–666.

Schmutz, J. A. 1993. Survival and pre-fledging body mass in juvenile Emperor Geese. Condor 95:222–225.

Sénéchal, É., J. Bêty, H. G. Gilchrist, K. A. Hobson, and S. E. Jamieson. 2011. Do purely capital layers exist among flying birds? Evidence of exogenous contribution to arctic-nesting Common Eider eggs. Oecologia 165:593–604.

Sillett, T. S., R. T. Holmes, and T. W. Sherry. 2000. Impacts of a global climate cycle on population dynamics of a migratory songbird. Science 288:2040–2042.

Smith, C. M., F. Cooke, G. J. Robertson, R. I. Goudie, and W. S. Boyd. 2000. Long-term pair bonds in Harlequin Ducks. Condor 102:201–205.

Solovieva, D. 1997. Timing, habitat use and breeding biology of Steller's Eider in the Lena Delta, Russia. Wetlands International Seaduck Specialist Group Bulletin 7:35–39.

Spurr, E. B., and H. Milne. 1976. Adaptive significance of autumn pair formation in the Common Eider *Somateria mollissima* (L.). Ornis Scandinavica 7:85–89.

Steele, B. B., A. Lehikoinen, M. Öst, and M. Kilpi. 2007. The cost of mate guarding in the Common Eider. Ornis Fennica 84:49–56.

Strandberg, R., R. H. G. Klaassen, M. Hake, P. Olofsson, K. Thorup, and T. Alerstam. 2008. Complex timing of Marsh Harrier *Circus aeruginosus* migration due to pre- and post-migratory movements. Ardea 96:159–171.

Sullivan, T. M., R. W. Butler, and W. S. Boyd. 2002. Seasonal distribution of waterbirds in relation to spawning Pacific herring, *Clupea pallasi*, in the Strait of Georgia, British Columbia. Canadian Field-Naturalist 116:366–370.

Sutherland, W. J. 1996. From individual behaviour to population ecology. Oxford University Press, Oxford, U.K.

Sutherland, W. J. 1998. Evidence for flexibility and constraint in migration systems. Journal of Avian Biology 29:441–446.

Suydam, R. S. 2000. King Eider (*Somateria spectabilis*). in A. Poole and F. Gill (editors), The Birds of North America, No. 491. The Academy of Natural Sciences, Philadelphia, PA.

Suydam, R. S., L. T. Quakenbush, D. L. Dickson, and T. Obritschkewitsch. 2000. Migration of King, *Somateria spectabilis*, and Common, *S. mollissima v-nigra*, Eiders past Point Barrow, Alaska, during spring and summer/fall 1996. Canadian Field-Naturalist 114:444–452.

Suydam, R., L. Quakenbush, M. Johnson, J. C. George, and J. Young. 1997. Migration of King and Common Eiders past Point Barrow, Alaska, in spring 1987, spring 1994, and fall 1994. Pp. 21–28 in D. L. Dickson (editor), King and Common Eiders of the western Canadian Arctic. Occasional Paper, No. 94. Canadian Wildlife Service, Ottawa, ON.

Takekawa, J. Y., S. W. De La Cruz, M. T. Wilson, E. C. Palm, J. Yee, D. R. Nysewander, J. R. Evenson, J. M. Eadie, D. Esler, W. S. Boyd, and D. H. Ward. 2011. Breeding distribution and ecology of Pacific Coast Surf Scoters. Pp. 41–64 in J. B. Wells (editor), Boreal birds of North America: a hemispheric view of their conservation links and significance. Studies in Avian Biology, No. 41, University of California Press, Berkeley, CA.

Thomas, P. W., G. H. Mittelhauser, T. E. Chubbs, P. G. Trimper, R. I. Goudie, G. J. Robertson, S. Brodeur, M. Robert, S. G. Gilliland, and J.-P. L. Savard. 2008. Movements of Harlequin Ducks in eastern North America. Waterbirds 31 (Special Publication 2):188–193.

Thompson, D. Q., and R. A. Person. 1963. The eider pass at Point Barrow, Alaska. Journal of Wildlife Management 27:348–356.

Timson, R. S. 1976. Late summer migration at Barrow, Alaska. Pp. 364–399 in Environmental assessment of the Alaskan Continental Shelf, principal investigators' reports, April–June 1976 (vol. 1). National Oceanic and Atmospheric Administration, Boulder, CO.

Titman, R. D. 1999. Red-breasted Merganser (*Mergus serrator*). in A. Poole and F. Gill, (editors), The Birds of North America, No. 443. The Academy of Natural Sciences, Philadelphia, PA.

Tøttrup, A. P., K. Thorup, K. Rainio, R. Yosef, E. Lehikoinen, and C. Rahbek. 2008. Avian migrants adjust migration in response to environmental conditions en route. Biology Letters 4:685–688.

U.S. Fish and Wildlife Service. 2002. Steller's Eider recovery plan. Fairbanks, AK.

Van Eerden, M. R., M. Zijlstra, and M. J. J. E. Loonen. 1991. Individual patterns of staging during autumn migration in relation to body condition in Greylag Geese *Anser anser* in the Netherlands. Ardea 79:261–264.

Vardanis, Y., R. H. G. Klaassen, R. Strandberg, and T. Alerstam. 2011. Individuality in bird migration: routes and timing. Biology Letters 7:502–505.

Vermeer, K. 1981. Food and populations of Surf Scoters in British Columbia. Wildfowl 32:107–116.

Vermeer, K., and G. G. Anweiler. 1975. Oil threat to aquatic birds along the Yukon coast. Wilson Bulletin 87:467–480.

Vermeer, K., M. Bentley, K. H. Morgan, and G. E. J. Smith. 1997. Association of feeding flocks of Brant and sea ducks with herring spawn at Skidegate Inlet. Pp. 102–107 in K. Vermeer and K. H. Morgan (editors), The ecology, status, and conservation of marine and shoreline birds of the Queen Charlotte Islands. Canadian Wildlife Service Occasional Paper, No. 93. Canadian Wildlife Service, Ottawa, ON.

Webster, M. S., and P. P. Marra. 2005. The importance of understanding migratory connectivity and seasonal interactions. Pp. 199–209 in R. Greenberg and P. P. Marra (editors), Birds of two worlds. Johns Hopkins University Press, Baltimore, MD.

Webster, M. S., P. P. Marra, S. M. Haig, S. Bensch, and R. T. Holmes. 2002. Links between worlds: unraveling migratory connectivity. Trends in Ecology and Evolution 17:76–83.

Woodby, D. A., and G. J. Divoky. 1982. Spring migration of eiders and other waterbirds at Point Barrow, Alaska. Arctic 35:403–410.

Wormington, A., and J. H. Leach. 1992. Concentrations of migrant diving ducks at Point Pelee National Park, Ontario, in response to invasion of zebra mussels, *Dreissena polymorpha*. Canadian Field-Naturalist 106:376–380.

Zalakevicius, M., and V. Jacoby. 1992. Radar and visual observations of the moult migration of Common Scoter within eastern Baltic. IWRB Seaduck Bulletin 1:51.

Zhalakyavichus, M. 1982. Radar observations of migration for moulting by Common Scoter in the Lithuanian SSR. Pp. 274–284 in V. M. Gavriolv and R. L. Potapov (editors), Ornithological studies in the USSR, collection of papers (vol. 2). Zoological Institute, USSR Academy of Sciences, Moscow, USSR.

Zicus, M. C., and S. K. Hennes. 1989. Nest prospecting by Common Goldeneyes. Condor 91:807–812.

Figure A.1. (a) Barrow's Goldeneyes, (b) Common Goldeneyes, and (c) Buffleheads are cavity nesters that exhibit strong intra- and inter-specific territorial behavior, and differ in their distribution in North America. (d) Harlequin Ducks are river-breeding specialists that molt and forage along rocky, marine shores in late summer and winter. Photographs by (a) Gary Kramer, (b) Dirk Derksen, (c) Ian Routley, and (d) Jeff Coats.

Figure A.2. All four eider species winter and molt in marine waters. (a) King, (b) Spectacled, and (c) Steller's Eiders breed on freshwater lakes; (d) the Common Eider is the only true colonial sea duck. Photographs by (a) Ken Wright, (b) Ted Swem, (c) Gary Kramer, and (d) Jeff Coats.

Figure A.3. Most scoters breed on freshwater lakes and molt and winter in salt waters where they form large flocks feeding mostly on clams and mussels. (a,b) The Surf Scoter is endemic to North America, (c) the Black Scoter is now recognized as a full species distinct from the Common Scoter of Europe, and (d) the White-winged Scoter occurs in both Europe and North America. Photographs by (a) Brian Uher-Koch, (b) Milo Burcham, (c) Ryan Askren, and (d) Gary Kramer.

Figure A.4. Mergansers differ by their habitat preferences: (a) the Red-breasted Merganser is associated with salt water, whereas (b) Hooded and (c) Common Mergansers exploit fresh waters for breeding. All mergansers forage primarily on fishes. (d) The Long-tailed Duck breeds in northern ponds and coastal regions and is pelagic throughout the winter. It is the only sea duck with distinct winter and breeding plumages. Photographs by (a,b) Gary Kramer, (c) Ian Routley, and (d) Ryan Askren.

CHAPTER NINE

Remigial Molt of Sea Ducks*

Jean-Pierre L. Savard and Margaret R. Petersen

Abstract. Molt is a dynamic process occurring throughout much of the year in waterfowl. The molt of flight feathers by waterfowl, especially sea ducks, however, occurs over a compressed period of time and in specific areas used each year. We provide an overview of the flight feather molt of sea ducks. We focus on the need to molt and why, the timing and duration of flight feather molt, and the duration birds remain at molting areas; energetics of molt and strategies for managing energetic needs; molt migration; food resources and foraging behavior; predation risks; temporal constraints and competition; response to disturbance; and molt habitats and seasonal differences in habitat used by sea ducks. We conclude by presenting and discussing data gaps and emphasize the continuing need for a holistic approach to sea duck management and international cooperation among countries.

Key Words: delayed maturity, dichromatism, eclipse plumage, energetics, feather molt, flightlessness, food resources, molt migration, synchronous molt.

Sea ducks lose and regrow feathers throughout much of the year. Once grown, feathers are not active tissues and do not self-repair, other than through preening. Therefore, feathers wear down over time due to abrasion, UV light, parasites, and other factors and need to be replaced regularly. Feathers serve many important functions, including concealment, thermoregulation, communication, and locomotion (McGraw et al. 2003, Tickell 2003). Our use of the term molt refers to the loss and regrowth of body and flight feathers. Molting areas or molt sites refer to discrete areas used by birds during the flightless period. The process of molting only contour feathers is referred to as body molt and replacement of major flight feathers such as the primaries and secondaries as remigial molt.

The nutritional demands of molt, although slight during most periods, can affect several aspects of waterfowl life histories, including behavior, migration, and energetic requirements. A general separation of molting, breeding, and migration activities help mitigate molt costs (Payne 1972, Kjellen 1994). Thus, an understanding of molt is necessary to fully interpret the behavioral, migratory, energetic, and general life history attributes of sea ducks. Most of the discussion in this chapter is focused on remigial molt when the nutritional and energetic costs for molt are thought to be at their annual maxima due to overlap between body plumage and remigial molt (Howell et al. 2003a) and when birds are flightless. We provide an overview of sea ducks' need

* Savard, J.-P. L. and M. R. Petersen. 2015. Remigial molt of sea ducks. Pp. 305–335 in J.-P. L. Savard, D. V. Derksen, D. Esler, and J. M. Eadie (editors). Ecology and conservation of North American sea ducks. Studies in Avian Biology (no. 46), CRC Press, Boca Raton, FL.

to molt, why they molt, timing and duration of remigial molt, length of stay at molting areas, energetics of molt, strategies for managing energy needs, molt migration, food resources, foraging behavior, predation risks, temporal constraints, competition, response to disturbance, molt habitats, seasonal differences in habitats used, data gaps, and international concerns.

The period of simultaneous remigial molt, typical in waterfowl, lasts from 3 to 5 weeks and is often referred to as wing molt. Remigial molt represents the culmination of molt into eclipse plumage and is usually completed before molt back into breeding plumage has been initiated. All sea ducks molt their remiges only once a year. The remigial molt period of sea ducks is best considered as a distinct annual cycle stage, as their habitat choice, locations, behavior, and physiology when flightless are often markedly different from other times of year. In turn, behavior and habitat use during molt has important implications for sea duck conservation, in recognition that natural variation or human impacts during remigial molt could have population-level effects.

PLUMAGES

Geese, swans, and whistling ducks have only one body molt per year, but most ducks have two (Hohman et al. 1992). Tropical waterfowl often have only one molt per year, whereas temperate zone ducks have two and hybrids between tropical and temperate zone waterfowl had two molts per year indicating a strong genetic component (Welty 1962). In sea ducks, body feathers are replaced at least twice a year: first, when the male colorful nuptial plumage used in courtship is replaced by a cryptic (eclipse) one (usually in late June, early July), at which point flight feathers also are simultaneously shed, and second, when the nuptial plumage is regrown (Palmer 1976, Hohman et al. 1992, Pyle 2005). In females, part of the nuptial plumage (head and neck) is replaced by the eclipse plumage in the spring and the remaining feathers following incubation but before remigial molt. In Long-tailed Ducks, the male has a molting pattern similar to females in that their bright white winter plumage is replaced partially in the head, neck, and scapulars early in the spring (April) by the dark eclipse plumage that gives the impression of the bird having three distinct plumages: a winter plumage, a nuptial plumage that combines part of the winter

and eclipse plumages, and an eclipse plumage (Robertson and Savard 2002). The species apparently has a supplemental plumage in early fall, molting from its eclipse plumage to this transitional plumage before molting in late fall into its winter plumage (Palmer 1976). Complex patterns have generated some confusion as to the exact nature and role of each of these plumages and their overall molting sequence (Stresemann 1948, Salomonsen 1949, Palmer 1976, Howell et al. 2003a). Part of the confusion relates to the asynchronous molt of various feather tracts (Palmer 1972, Howell et al. 2003b). Scoters, like Long-tailed Ducks, also have a spring molt on the head and neck feathers, but as regrown feathers are the same color as the winter feathers, it does not result in a different obvious plumage as in Long-tailed Ducks (Dwight 1914). Hochbaum (1944) suggested that the main role of the cryptic eclipse plumage is to conceal males from predators during the flightless period, whereas Bailey (1981) argued that the darker cryptic plumage had significant thermal advantages (Walsberg et al. 1978). Further work on characterizing and quantifying molt in sea ducks is needed to clarify molting patterns and understand the role of various plumages (Howell 2010).

There is much debate in the literature about the names used to represent various plumages (Howell et al. 2003b, Thompson 2004, Willoughby 2004, Hawkins 2011), with Palmer (1976; see also Humphrey and Parkes 1959) using opposite definitions to those of Pyle (2005). In Palmer (1976) and others, the nuptial plumage is called alternate and the cryptic plumage basic. However, Pyle (2005) proposed to reverse the name of these plumages, considering the nuptial (breeding) plumage as the basic one and the cryptic one (eclipse) as the alternate. The debate has yet to be fully resolved and centers around which plumage should be considered as the original (basic), ancestral one. To minimize confusion, we use the terms nuptial for the bright breeding plumage primarily seen in males and eclipse for the dark cryptic molting plumage. The nuptial plumage corresponds to the alternate plumage of Palmer (1976) and the basic plumage of Pyle (2005), whereas the eclipse plumage follows the basic plumage of Palmer (1976) and the alternate plumage of Pyle (2005).

Most sea ducks do not reach sexual maturity until 2–3 years old, and delayed maturity affects plumage development. Indeed, subadult birds do not have the definitive nuptial plumage of adults and can be identified in their first and even second years (Delacour 1959; Palmer 1976; Smith et al. 1998; Iverson et al. 2003, 2004; Kear 2005; Pyle 2005). These plumages are often called first plumages as opposed to definite plumages (Pyle 2005).

Here, we use the term juvenile to differentiate birds that hatched the previous summer from all other birds, subadult to identify unique plumages of birds after their first year as juvenile but before their first breeding season, and adult for individuals with full breeding plumage. Age of maturity and timing of plumage changes differ among species: juvenile birds generally include individuals in their first year (hatching) or early second year (hatch-year [HY]), subadult birds are generally in their second or third year (SY or TY), whereas adults are 3 years or older (after-second-year [ASY]). Some species can first attain definitive (adult) plumage in their second year, but more typically during their third year in sea ducks. Molt cycles are more complex in sea ducks because of delayed maturity: subadult plumages can persist longer than in many other species of ducks yielding greater plumage diversity within species. Here, we focus largely on adult birds and refer to subadult and juvenile birds when plumages differ from adults.

All North American sea ducks are dichromatic with males of most sea duck species harboring a brighter, more conspicuous plumage than females that likely evolved in part through sexual selection for mates (Lack 1974). The more cryptic and dull plumage of females likely results from advantages of concealment during incubation and/or brood-rearing stages of reproduction. Cryptic coloration also may indicate absence of sexual selective forces on female plumage, as males are the sex competing for mates. Competition between males is particularly strong in sea ducks as males tend to outnumber adult females (Sargeant and Raveling 1992, Donald 2007) and males are usually larger bodied than females. However, Kimball and Ligon (1999) suggested that originally both sexes were brightly colored and that selection occurred on the female for cryptic plumage. Waterfowl exhibit estrogen-dependent plumage dichromatism, an ancestral condition in birds where females deprived of estrogen develop male plumage. Thus, if the ancestral plumage was the bright plumage, selection may have been for the cryptic plumage of the females (Voitkevich 1966, Kimball and Ligon 1999). Clearly, the origin of plumages in waterfowl has yet to be fully resolved.

WHY BECOME FLIGHTLESS?

Most species of birds molt their flight feathers sequentially to maintain an ability to fly during feather replacement (Gin and Melville 1983). However, waterfowl diverge from that ancestral pattern by molting all flight feathers simultaneously (King 1974; Figure 9.1). This peculiar molting adaptation may be related to their association with aquatic habitats, which allows them to find sufficient food and escape predation when flightless (Hohman et al. 1992), and to their high wing loading (Guillemette and Ouellet 2005), which may compromise flying ability with the loss of

Figure 9.1. Remigial molt progression of a male Surf Scoter (no. 630). (a) 23 July 2008, (b) 30 July 2008, (c) 6 August 2008. (Photos by J.-P. L. Savard.)

only a few feathers (Welty 1962). Allometric relationships between feather length, feather growth rate, and body mass favor simultaneous molt in large birds or in birds with high wing loading (Rohwer et al. 2009). Feather growth rates do not differ between species with simultaneous and sequential wing molt so that simultaneous molting is quite time efficient as all the flight feathers are replaced in the time it will take to grow the longest primary (Rohwer et al. 2009). Also, flightlessness during remigial molt is associated with significant energy savings that can be redirected to feather growth (Guillemette et al. 2007).

TIMING OF REMIGIAL MOLT

Intraspecific Variation

The timing of molt is under hormonal control (Payne 1972, Bluhm 1988) and thus likely to vary among individuals. Male White-winged Scoters, Surf Scoters, Long-tailed Ducks, and Harlequin Ducks undertake their remigial molt before females, even in captivity when both sexes are maintained under similar diets and environmental conditions (J.-P. L. Savard, Environment Canada, unpubl. data). Captive birds of the same species and sex can show significant individual variation in timing of molt (J.-P. L. Savard, unpubl. data). In Northern Gadwall (*Anas strepera*), Oring (1966) found that nonbreeding males that were never active in courtship molted earlier than breeding males and that nonbreeding males active in courtship molted the latest. The authors attributed the difference to differences in gonadal development that affected timing of molt initiation, with individuals molting only after gonads had regressed (Payne 1972). Owens and King (1979) found that heavier male Mallards molted earlier than lighter ones.

Males begin to molt into their nuptial plumage during or soon after remigial growth is complete. For species in which some males and females molt remiges on their wintering areas, such as some Harlequin Ducks, acquisition of full breeding plumage immediately after remigial molt may facilitate pair formation in early winter (Gowans et al. 1997; Robertson et al. 1997, 1998; Rodway 2007). Red-breasted Mergansers form pairs in late winter (February), but do not complete molt to their nuptial plumage until late November or early December (Titman 1999), 2–3 months after completing remigial molt. Interestingly, the species is a late breeder compared to most other sea

ducks (Titman 1999). For Harlequin Ducks molting on their wintering areas, Cooke et al. (1997) estimated 87 days as the period from initiation of the molt into eclipse plumage and the return into nuptial plumage: molt from nuptial to eclipse (20 days), period in eclipse plumage (52 days), and molt from eclipse to nuptial plumage (15 days).

In all sea ducks as well as in most other ducks, males undertake remigial molt migration before females (Bellrose 1980, Hohman et al. 1992). Intraspecific variation in timing of remigial molt according to age and sex has been documented in several species of sea ducks and is likely the norm for all species. In Steller's Eider, subadults begin their remigial molt first, followed by adult males, unsuccessful females, and successful females (Petersen 1980, 1981; Metzner 1993). A similar pattern occurs in other species, including Surf Scoters (Savard et al. 2007, Dickson et al. 2012), Common and White-winged Scoters (Joensen 1973, Dickson et al. 2012), eiders (Thompson and Pearson 1963; Timson 1976; Frimer 1994a, 1995a; Suydam et al. 1997, 2000; Petersen et al. 1999; Dickson 2012; D. Troy, BP Exploration, Alaska, unpubl. report), Common Goldeneyes (Jepsen 1973), Barrow's Goldeneyes (Hogan 2012), and Harlequin Ducks (Robertson et al. 1997). Differences in timing between age groups can vary from nearly a month as in Steller's Eiders (Petersen 1981) to only a few days as in Red-breasted Mergansers (Craik et al. 2009). Thus, during the seasonal period at a given molting site, a variable proportion of birds can be flightless at the same time. Most individuals would complete their remigial molt in 4–6 weeks, but a major molting site could have flightless birds present for up to 4 months.

Adult males and subadults of most species molt their flight feathers before adult females, and flocks of flightless birds are often dominated by one sex or a given age class (i.e., adult males, subadult and nonbreeding birds, adult females). It is also likely that unpaired adult males that migrated directly to molting areas undergo remigial molt before paired males that accompanied females to breeding areas. Variation in reproductive status could explain the large range observed in timing of remigial molt in adult male Barrow's Goldeneyes and Surf and White-winged Scoters (Dickson et al. 2012, Hogan et al. 2013b).

In many species of ducks, the female typically becomes flightless after her young have fledged (Hochbaum 1955, Hohman et al. 1992). Breeding

TABLE 9.1
Relationship between parental care and subsequent molt locations of adult females.

Species	Brood size during brood rearing[a]	Duration of female bond with young[b]	Female molt location	Sources[c]
Steller's Eider	Brood	Fledging	Molt migration >50 km	1,2
Spectacled Eider	Brood	Fledging	Molt migration >50 km	1–5
King Eider	Brood, crèche	Fledging (?), brood (?), prefledging (?), postfledging	Breeding grounds; molt migration >50 km	6–8
Common Eider	Brood, crèche	Prefledging, 3 days to ~7 weeks	Brood-rearing area (<50 km)	9–12
Harlequin Duck	Brood	Variable, postfledging	>50 km from nest	13
Surf Scoter	Brood, amalgamation	Prefledging	>50 km from nest	14
White-winged Scoter	Brood, amalgamation	Prefledging ~1–3 weeks	>50 km or brood-rearing region	15,16
Black Scoter	Brood, amalgamation	Prefledging	>50 km	17
Long-tailed Duck	Brood, crèche, or amalgamation	Prefledging >2 weeks	>50 km or brood-rearing region	18–20
Bufflehead	Brood, amalgamation	Prefledging ~5–6 weeks	Unknown distance, brood-rearing region	21–23
Common Goldeneye	Brood, amalgamation	Prefledging ~5–6 weeks, or ~4 to ~35 days	Unknown distance, brood-rearing region	24–26
Barrow's Goldeneye	Brood, amalgamation	Prefledging, 5–6 weeks	>50 km or brood-rearing region; some >500 km	27,28
Hooded Merganser	Brood	Prefledging, 5–6 weeks(?)	Unknown	29,30
Common Merganser	Brood, amalgamation	Brood, prefledging, 3–50 days	Brood-rearing region or distant	31–33
Red-breasted Merganser	Brood, amalgamation	Brood, prefledging	With brood or brood-rearing region	34,35

[a] Brood refers to a single breeding female with her young; crèche is one or more females with mixed broods; amalgamation is similar to crèche but generally occurs in high-density nesting areas and is likely a result of female interactions (Erskine 1971; Munro and Bédard 1977; Savard 1987; Gauthier 1993; Eadie et al. 1995, 2000; Savard et al. 1998). Question marks (?) indicate knowledge gaps.

[b] Prefledging is when young cannot fly (1 day to <flight); postfledging equates to the female leaving when or after her brood fledges and she molts elsewhere; brood means female was flightless with her brood.

[c] (1) Solovieva (1997), (2) Fredrickson (2001), (3) Dau (1974), (4) Petersen et al. (1999), (5) Flint et al. (2000a), (6) Palmer (1976), (7) Abraham and Finney (1986), (8) Suydam (2000), (9) Bustnes and Erikstad (1991), (10) Öst (1999), (11) Goudie et al. (2000), (12) Canadian Wildlife Service (unpubl. data), (13) Robertson and Goudie (1999), (14) Savard et al. (1998), (15) Brown and Brown (1981), (16) Brown and Fredrickson (1989), (17) Bordage and Savard (1995), (18) Alison (1976), (19) Petersen et al. (2003), (20) Portenko (1981), (21) Erskine (1961), (22) Erskine (1971), (23) Gauthier (1993), (24) Pöysä (1992), (25) Eadie et al. (1995), (26) Pöysä et al. (1997), (27) Eadie et al. (2000), (28) J.-P. L. Savard (unpubl. data), (29) Beard (1964), (30) Dugger et al. (1994), (31) Erskine (1972), (32) Little and Furness (1985), (33) Mallory and Metz (1999), (34) Titman (1999), (35) Craik (2009).

female sea ducks can molt after their young fledge, leave before young fledge, or molt while raising young (Table 9.1), for example, Spectacled Eiders that bred successfully will depart on a molt migration after their young fledge (Petersen et al. 1999). Many Common Eider females, however, abandon their broods soon after hatch when crèches form (Munro and Bédard 1977). Similarly, female

Surf Scoters (Savard et al. 1998, 2007), Barrow's Goldeneyes (Eadie et al. 2000), and Common Goldeneyes (Pöysä 1992) leave their broods before they are fledged. Some species have a mix of strategies; flightless adult female Common Mergansers (*Mergus merganser*) have been captured with broods (J. M. Pearce, U.S. Geological Survey, pers. comm.), yet many females abandon broods before fledging

(Table 9.1). Proximal factors leading a female to abandon its brood or undergo remigial molt with her young are not known but likely involve variation in hormonal levels or body condition among females (Bustnes and Erikstad 1991, Öst 1999).

Interspecific Variation

There are differences in timing of remigial molt among species. In captivity, Surf Scoters initiated remigial molt before White-winged Scoters (~3 weeks), Long-tailed Ducks (~3 weeks), and Harlequin Ducks (~4 weeks) (Table 9.2, J.-P. L. Savard, unpubl. data). Dickson et al. (2012) also reported that Surf Scoters (n = 2,099) molting along the Pacific coast tended to molt earlier (ASY males, 4.5 days, n = 90; ASY females, 1.3 days, n = 79) than White-winged Scoters (n = 169), but their data were highly variable. Furthermore, the authors could not distinguish paired and unpaired males, or successful and unsuccessful breeding females, which complicates comparisons between species as the proportion of different stage classes in the sample would greatly affect estimates. However, their data for second-year males of both species in the same year within the same molting site support earlier timing of molt by about a week in Surf Scoters. For each scoter species, timing of molt varied with cohort, site (Alaska vs. British Columbia), and year (Dickson et al. 2012). Data from the St. Lawrence Estuary and the Labrador coast also suggest earlier molting of Surf Scoters than White-winged Scoters (J.-P. L. Savard and S. G. Gilliland, unpubl. data). Given that timing of remigial molt varies with age and sex within a species and because of the wide range in initiation of remigial molt within a given cohort (1.5–2.5 months; Dickson et al. 2012), it is difficult to compare the timing of remigial molt among species in the wild, and few studies have done so to date. In general, it is likely that at least for breeding females, differences between species likely parallel breeding phenology.

LENGTH OF THE REMIGIAL MOLTING PERIOD

Molt dynamics of sea ducks contain many knowledge gaps, especially for subadults and females, which warrant caution in interpretation of existing data. In general, within a species or among

closely related species, birds that weigh less regain flight sooner than heavier birds (Dean 1978, Ankney 1979, Geldenhuys 1983); thus, one would expect the flightless period to be shorter in small-bodied species such as Buffleheads (*Bucephala albeola*) than the larger species such as Common Eiders. Indeed, Guillemette et al. (2007) found an average flightless period of 36 ± 8 days for 10 female Common Eiders, whereas estimates for Buffleheads are around 29–30 days (A. Breault, unpubl. data). However, bird condition positively affected feather growth in Barrow's Goldeneyes (van de Wetering and Cooke 2000) and other waterfowl (Pehrsson 1987, Hohman and Crawford 1995). Differences in length of the flightless period between small and large species are minimized because remiges of larger birds grow more quickly than those of smaller species (Table 9.2). Physiological limits set the speed at which a feather can grow (Rohwer et al. 2009) with molt rates ranging from 2.7 mm/day for a sparrow to 9 mm/day for a swan (Bridge 2011). In sea ducks, primary feather growth rates range from 3.5 mm/day in Harlequin Ducks to 4.5 mm/day in White-winged Scoters (Table 9.2).

In a study of male Barrow's Goldeneyes, the growth rate of primaries remained constant during most of the remigial molt, and the average daily percent change was 2.6% of final feather length (range 2.4%–3.0%; van de Wetering and Cooke 2000), which is similar to the rate reported in other waterfowl (2%–3%; Hohman et al. 1992). Faster growth in heavier birds and earlier flight capacity in lighter ones may result in similar flightless periods in both sexes. However, data on captive sea ducks indicate that remigial growth is not constant but increases quickly in the first week following loss of old feathers, remains high for at least 2 weeks, and slowly decreases in the last week (J.-P. L. Savard, unpubl. data). The decrease in growth rate occurs when primaries are about 70%–80% of their final length, which corresponds to the period when flight capacities are reacquired in waterfowl (Hohman et al. 1992, Brown and Saunders 1998). Length of primary feathers when flight resumes is estimated to be 77%–80% in Barrow's Goldeneyes (Hogan et al. 2013b), 83%–98% in Surf Scoters, and 88%–99% in White-winged Scoters (Dickson et al. 2012). However, Dickson et al. based their estimates on the length of primaries of captured birds, assuming that captured birds could not fly. The assumption may not be true as flying birds are sometimes captured in submerged nets during capture drives (S. G. Gilliland, pers. comm.), that would tend to overestimate the length at which flight capacities are regained. However, bird escape speed by diving or skipping over water likely increases as remiges grow.

Dickson et al. (2012) estimated a period of 7.3 ± 0.04 days in scoters between loss of old feathers and initiation of regrowth, and Hogan (2012) and Hogan et al. (2013b) estimated this period at 6.5 ± 1.2 days in Barrow's Goldeneyes. Inclusion of an early flightless period of 6 days yields a total flightless period ranging between 42.9 and 49.3 days in White-winged Scoters (Table 9.2). In captivity, 80% of primary growth was achieved in about the same length of time for male and female Harlequin Ducks, Surf Scoters, White-winged Scoters, and Long-tailed Ducks (Table 9.2).

The length of the flightless period varies between species and possibly also between molting cohorts and molting habitats. Usually, larger species have a longer flightless period. Estimates of the flightless period of captive birds range from 4.6 weeks in Long-tailed Ducks to 7.4 weeks in White-winged Scoters (Table 9.2). Hogan (2012) indicated that the flightless period tended to be longer in diving than dabbling waterfowl of similar body size, possibly because dabblers are more vulnerable to predation by terrestrial predators as they often hide in shoreline emergent vegetation and generally have lower wing loading.

LENGTH OF STAY AT MOLTING AREAS

The length of time birds remain on areas where they molt remiges varies substantially. Adult males, and some females, of many species initiate their molt into eclipse plumage during the molt migration (Thompson and Pearson 1963, Frimer 1994a). Breeding females that may suspend their molt into eclipse plumage during incubation resume it during their molt migration. Red-breasted Mergansers arrive at their molting location 2–3 weeks before the flightless period (Craik et al. 2009). Variation in timing of remigial molt may be related to the progress made in the molt into eclipse plumage as they arrive on molting areas. Males molt into eclipse plumage before losing their flight feathers during the month or so between leaving the female and becoming flightless. Birds such as Barrow's Goldeneyes and Harlequin Ducks that

arrive early on the molting areas complete their molt into eclipse plumage there (but see Chubbs et al. 2008). Most species molt from their eclipse to nuptial plumage soon after remigial feather growth is completed or even during the late stage of remigial growth. Some sea ducks lose their rectrices while flightless (Robertson et al. 1997; Craik et al. 2009; M. R. Petersen, unpubl. data), and the growth of tail feathers is completed by the time remige growth is complete, while others molt them a few weeks later (Langlois 2006).

The length of time an individual remains on its remigial molting area is, in part, influenced by the formation of ice and the availability and quality of forage (Salomonsen 1968). Many sea ducks remain on molting areas long after having regained flight capabilities, replacing most of their eclipse feathers with a nuptial plumage (Brodeur et al. 2002; Howell 2002; Robert et al. 2002, 2008; Mosbech et al. 2006b; Savard et al. 2011; Hogan 2012; Savard and Robert 2013). In more northern areas, departure from molting areas may coincide with ice formation, but this is not the case in southern areas where ice is not a factor, which suggests that other factors such as day length (Payne 1972) or degree of completion of the body molt may be important to initiate fall migration. Additional research is needed to determine the environmental factors triggering departure from molting sites.

ENERGETICS OF MOLT

Relative costs of feather replacement are inversely related to body size (Hohman et al. 1992, Rohwer et al. 2009) such that larger species take longer to regrow their feathers. Like most production processes, molt results in additional energetic and nutritional requirements. Molt also reduces insulation, impairs diving and flight capacities (Hohman et al. 1992, Murphy 1996), and requires physiological and behavioral adjustments (King 1981). Eclipse plumages are not as dense as nuptial plumages and thus have quite lower thermal properties. In mallards, dry mass of the eclipse plumage is <60% of the nuptial plumage, and wing plumage comprises about 40% of eclipse plumage mass (Hohman et al. 1992). The difference in plumage density particularly affects sea ducks as they feed by diving underwater and feathers are known to be poor insulators in water (Turcek 1966). Insulation may explain in part why remigial molt occurs during the warmest period

TABLE 9.2

Average growth rates of primaries and estimated flightless period of some sea ducks.

Species	Sex	Body Mass[a] (g)	Mean primary growth rate (mm day^{-1}) (n)	Ninth primary growth period (days)	Preemergent period (days) (n)	Proportion of ninth primary for flight (%)	Estimated flightless period (days)	Status
White-winged Scoter	Male		4.5 ± 0.2 SD (8)[b]	36.0[b]		80	42[c]	Captive
White-winged Scoter	Male	1588–1722		42.4[d]		88–99[d]	44.4–49.3[c]	Wild
White-winged Scoter	Female		4.4 ± 0.3 SD (7)[b]	31.0[b]		80	37[c]	Captive
White-winged Scoter	Female	1179–1732		39.8[d]		89–97[d]	42.9–45.8[d]	Wild
White-winged Scoter	Combined		3.8 ± 0.4 SE (12)[d]			80		Captive
White-winged Scoter	Combined		4.7 ± 0.1 SE (8)[d]		7.3 ± 0.04 SE (30)[d]			Wild
Surf Scoter	Male		4.11 ± 1.18 SD (4)[b]	32.7[b]		80	38.7[c]	Captive
Surf Scoter	Male	1148–1153		40.8[d]		86–98[d]	42.4–47.5[d]	Wild
Surf Scoter	Female		4.02 ± 0.16 SD (3)[b]	31.5[b]		80	37.5[c]	Captive
Surf Scoter	Female	1047–1025		38.2[d]		83–93[d]	39.2–42.4[d]	Wild
Surf Scoter	Combined		3.8 ± 0.2 SE (9)[c]		7.3 ± 0.04 SE[c]			Wild
Barrow's Goldeneye	Male	1090–1160	4.04 ± 0.05 SE (22)[e] 3.94 ± 0.13 SE[f]	39 ± 0.5 SE (38)[f]		80[e] 77[f]	37[c], (33 – 40)[g] 36.5 ± 0.5[f]	Wild
Barrow's Goldeneye	Female	730–850		36 ± 0.7 SE[f]		77[f]	34.5 ± 0.8[f]	Wild

Barrow's Goldeneye	Combined		3.9 ± 0.1 SE (38)[f]		6.5 ± 1.2 SE[f]		Wild
Long-tailed Duck	Male		4.53 ± 0.35 SD (3)[b]	27.6[b]		25.8[g]	Captive
Long-tailed Duck	Male	862–1070					Wild
Long-tailed Duck	Female		4.31 ± 0.16 SD (5)[b]	26.3[b]		25.9[g]	Captive
Long-tailed Duck	Female	730–734					Wild
Harlequin Duck	Male		3.69 ± 0.12 SD (13)[b]	30.6[b]		27.9[g]	Captive
Harlequin Duck	Male	615–751	3.5 ± 0.06 SE (68)[h]	37[h]		32.4[g]	Wild
Harlequin Duck	Female		3.67 ± 0.17 (6)[b]	29.4[b]		27.1[g]	Captive
Harlequin Duck	Female	561–642	3.5 ± 0.06 SE (68)[h]	37[h]		32.4[g]	Wild

[a] Winter mass of wild birds: White-winged Scoter (Brown and Fredrickson 1997), Surf Scoter (Savard et al. 1998), Barrow's Goldeneye (Eadie et al. 2000), Long-tailed Duck (Robertson and Savard 2002), Harlequin Duck (Robertson and Goudie 1999).

[b] J.-P. L. Savard (unpubl. data) from captive birds. The captive population was located in Patuxent, Washington, DC (M. Perry, pers. comm.).

[c] Calculated following Dickson et al. (2012).

[d] Dickson et al. (2012).

[e] van de Wetering and Cooke (2000).

[f] Hogan et al. (2013b).

[g] Estimated flightless period = [ninth primary growth period (days) × proportion of ninth primary for flight (80%)] + (6 days preemergence time).

[h] Iverson and Esler (2007).

of the year (July–August), which may help compensate for reduced insulative capacities (Hohman et al. 1992). Many sea ducks undertake extensive molt migrations to reach secluded, productive, and predator-reduced areas (Salomonsen 1968). In these molting areas, they can allocate more energy to the molting process.

Feathers are mostly made of proteins (~90%) and plumage accounts for 4%–12% of body mass (Murphy 1996). In aquatic birds, plumage usually accounts for <6.3% of body mass (Hohman et al. 1992). The energy content of plumage has been estimated at 22 kJ/g dry mass (Murphy and King 1982) and does not vary much among species (Hohman et al. 1992). The energetic cost of replacing flight feathers is substantial in female Common Eiders (139 kJ/day) as their daily and resting metabolic rates increase by 9% and 12%, respectively, during remigial molt (Guillemette et al. 2007). However, flightlessness results in reductions of 6% in daily metabolic rate and 14% in resting metabolic rate relative to periods outside of the remigial molt stage. Reduced foraging efforts, associated with rich molting areas, further reduce energy expenditure. Such energy saving may have contributed to the evolution of temporary flightlessness in birds (Guillemette et al. 2007). Flight and foraging are energetically demanding activities in sea ducks.

STRATEGY OF ENERGY MANAGEMENT DURING REMIGIAL MOLT

Behavior and Body Mass Dynamics

Several studies of geese, dabbling ducks, and diving ducks suggest that the flightless period is not nutritionally stressful (Ankney 1979, Hohman 1993, Thompson and Drobney 1996, Fox and Kahlert 2005) and that mass loss during the flightless period may be an adaptation to regain flight capabilities (Brown and Saunders 1998). In some cases, it may not be necessary to carry somatic reserves because of food abundance on the molting areas. Patterns in body mass dynamics of molting sea ducks are highly variable: Harlequin Ducks (Histrionicus histrionicus) lost mass during the flightless period but spent most of their time resting (Adams et al. 2000); Barrow's Goldeneyes (Bucephala islandica) lost mass in the Yukon Territories (van de Wetering and Cooke 2000), but gained mass in Alberta (Hogan 2012, Hogan et al. 2013a);

Common Scoters (Melanitta nigra) and most Steller's Eiders did not lose mass (Petersen 1981, Fox et al. 2008); and Surf Scoters (M. perspicillata) and White-winged Scoters (M. fusca) gain mass during the flightless period (Dickson 2011). These observations support the conclusion that sea ducks are not energetically stressed during remigial molt, likely because of the high productivity of most molting areas and that the strategy of losing weight during the flightless period to regain flight capacities quickly is less important for sea ducks than for dabbling ducks and geese.

Waterfowl meet energetic requirements of molt through their diet and stored reserves, in varying proportions by species and even individuals. In general, sea ducks molt in productive sites and with a strategy of minimizing energy expenditures; in captivity, Common Eiders met the energetic costs of molt by feeding less but more efficiently and resting more (Portugal et al. 2010). As the growth rate of feather shafts of primaries is constant throughout day and night, a mix of endogenous and exogenous nutrient resources was used: pools of nutrients in liver and muscle tissues serve as a buffer between feeding bouts (Hohman et al. 1992). Molting Long-tailed Ducks frequented productive, sheltered lagoons and met their energy requirements through foraging (Howell 2002). However, Johnson (1985) reported that Long-tailed Ducks were fattest just before feather replacement suggesting that they may also use some endogenous resources. Fox et al. (2014) suggest that the strategy used during remigial molt is highly variable within and among species and is influenced by individual condition just prior to remigial molt as well as environmental and habitat conditions. Such variability has been shown in Brant Geese (Branta bernicla, Fondell et al. 2013).

Organ and Muscle Mass Dynamics

Rapid changes of muscle and body organs in response to different seasonal metabolic requirements have been well documented (Gaunt et al. 1990, Jehl 1997, Piersma and Lindström 1997, Piersma and Gill 1998). In geese, wing muscles atrophy and leg muscles hypertrophy during the flightless period (Ankney 1984, Fox and Kahlert 2005). Hohman et al. (1992) suggested that compensatory changes in breast and leg muscles during remigial molt appeared to be ubiquitous in waterfowl. Fox et al. (2008) confirmed an increase in breast muscle mass of Common Scoters from mid to the end of remigial molt but concluded that there were no change in leg muscle mass based on a non-significant test (P = 0.07). The authors' interpretation may be too conservative because their data suggest a decrease in leg muscle mass of Common Scoters during the latter part of remigial molt—the results suggest that pectoral muscle mass decreased and leg muscle mass increased at the onset of remigial molt. Howell (2002) found a similar dynamic between pectoral and leg muscles in Long-tailed Ducks during remigial molt, and this pattern also occurs in several species of dabbling ducks (Young and Boag 1982, DuBowy 1985). However, other species of sea ducks need to be studied to confirm whether the pattern described earlier is typical or not in sea ducks. Fox et al. (2008) also found a decrease in gizzard mass (P = 0.02) and in liver mass (P = 0.08), which has been observed in a few other waterfowl (DuBowy 1985, Thompson and Drobney 1996). Reduction in the size of organs and muscles also contributes to reduction in energy expenditure as there is a cost associated with maintaining large organs and muscles (Ferrell 1988).

MOLT MIGRATION

Migration

Molt migration, the movement to areas where birds become flightless, is common among waterfowl and occurs in at least nine other families of birds (Jehl 1990). Molt migration in sea ducks can be discrete or highly spectacular depending on species, sex, or age groups. Coastal molt migrations are the most spectacular and often involve hundreds to thousands of birds (Salomonsen 1968, Joensen 1973, Johnson and Richardson 1982). Migrations overland are more discrete, usually occur at night, and involve individuals or small groups. Little staging occurs during nocturnal migration overland with most birds flying directly and rapidly to their molting location (Brodeur et al. 2002, Robert et al. 2002, Savard et al. 2007, Savard and Robert 2013). Molt migration can have implications for understanding connectivity and structure of sea duck populations (Chapter 8, this volume).

Relationships of Molting Areas to Breeding and Wintering Locations

Little is known about the process of molt site selection by sea ducks. The small population of

Barrow's Goldeneyes in eastern North America molts throughout a large area in a variety of habitats and locations ranging from coastal estuaries to inland lakes (Benoit et al. 2001, Robert et al. 2002). There is no apparent connectivity of Barrow's Goldeneyes among breeding, molting, or wintering areas at a landscape scale; rather, individuals at a molting site may come from different breeding locations and use different winter areas. Indeed, several species have shown little connectivity between breeding and molting areas. Examples include Barrow's Goldeneye in eastern North America (Robert et al. 2002, Savard and Robert 2013), Steller's Eider (Figure 9.2a,b, Dau et al. 2000, Petersen et al. 2006), Common Mergansers (Pearce et al. 2009), Surf Scoters (J.-P. L. Savard, unpubl. data), Long-tailed Ducks (Petersen et al. 2003), and Harlequin Ducks (Brodeur et al. 2002). In contrast, there is some connectivity of adult female Common Eiders breeding in Alaska (Petersen and Flint 2002, Petersen et al. 2012). Barrow's Goldeneyes that wintered in Alaska molted in Alaska and in the Yukon Territories, whereas Barrow's Goldeneyes that wintered in southern British Columbia, Washington state, and Oregon molted in Alberta and Northwest Territories (Figure 9.3, D. Esler, U.S. Geological Survey and W. S. Boyd, Canadian Wildlife Service, unpubl. data), indicating a large-scale population structure. Pearce et al. (2014) showed that the structuring likely originated during the last glacial period and was associated with genetic structure as well. Also, in eastern North America, Harlequin Ducks that bred in northern Quebec wintered and molted in Greenland. Birds breeding in southern Quebec wintered in Maine, United States, but molted either along the eastern North American coast or in Greenland (Brodeur et al. 2002, Robert et al. 2008); however, the north–south demarcation between breeding subpopulations has yet to be determined.

Structure among breeding, molting, and wintering sites is likely, but the scale may vary among species and possibly between sexes and has yet to be determined for most species. Selection of a molting location could have a genetic component. Satellite telemetry studies indicate that sea ducks do not necessarily migrate to the closest molting site but can move thousands of kilometers when suitable molting habitat is available within a few hundred kilometers (Petersen et al. 1999, Brodeur et al. 2002, Robert et al. 2008, Savard

Figure 9.2. (a) Flock at rest, and (b) synchronous diving escape behavior of molting Steller's Eiders at Nelson Lagoon, Alaska. (Photo by M. R. Petersen.)

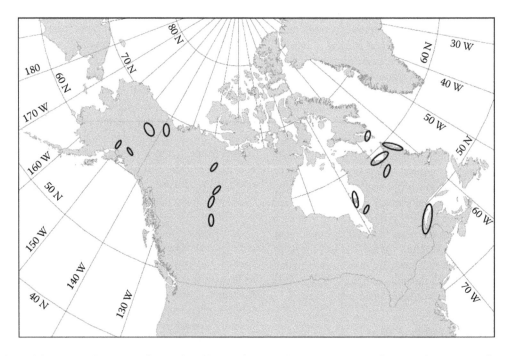

Figure 9.3. Major molting areas of Barrow's Goldeneyes from populations wintering on the east and west coasts of North America (King 1963, 1973; van de Wetering 1997; Robert et al. 2002; Hogan et al. 2011).

and Robert 2013). Patterns of glacier retreat and the location of glacial refugia during the last glaciation may have constrained selection of molting locations by sea ducks and partly shaped current relationships among breeding, wintering, and molting locations (Ploeger 1968, Pielou 1991).

Movements and Fidelity

Sea ducks are quite sedentary when flightless but may occasionally swim several kilometers to explore new feeding sites or in response to disturbance (O'Connor 2008). Surf Scoters used home ranges averaging 25 km^2 (range = 15–34 km^2) with an average core area of 3 km^2 (based on 50% utilization distribution, O'Connor 2008). However, movements of 10 km occurred occasionally. Similar movements (<15 km) have been reported for flightless Long-tailed Ducks (Flint et al. 2004). In Ungava Bay, Canada, the home range of molting female Common Eiders averaged 48 km^2 (range = 19–102 km^2; n = 7) with an average core area (50%) of 5.6 km^2 (range = 0.8–13.1, Savard et al. 2011).

Repeated use of molting areas from year to year is well documented in sea ducks and other waterfowl. Site fidelity suggests that these habitats have predictable resources (Salomonsen 1968, Savard 1988, Hohman et al. 1992). Individual-level site fidelity has been documented in Harlequin Ducks (Cooke et al. 1997, Breault and Savard 1999), King Eiders (Kellett 1999, Phillips and Powell 2006, Dickson 2012), Steller's Eiders (Flint et al. 2000b), Buffleheads (Erskine 1961), Barrow's Goldeneyes (Savard and Robert 2013), and Common Eiders (J.-P. L. Savard, unpubl. data; M. R. Petersen, unpubl. data). However, changes in molting location between years by individuals have been documented in several species including Barrow's Goldeneyes (Savard and Robert 2013), Harlequin Ducks (Brodeur et al. 2008b), and Surf and White-winged Scoters (J.-P. L. Savard, unpubl. data), but their prevalence, cause, and frequency are unknown. Change in age, pairing status, reproductive success, and bird condition may induce a sea duck to switch molting sites between years. Change in the hydrology and ecology of molting lakes, as during droughts, influenced molt site selection of Barrow's Goldeneyes (Hogan 2012). Fidelity to and movement between molting sites need to be better quantified as it has implications on population

structure, design of monitoring programs, and management concerns (Esler et al. 2006, Pearce and Talbot 2006).

FOOD RESOURCES

Areas used by molting waterfowl are believed to be selected, in part, for abundant, high-quality foods necessary to grow flight feathers (Salomonsen 1968). Changes in quality and quantity of prey in relation to timing or duration of remigial molt are not described for any species of sea duck. For many species, food resources used during remigial molt have not been quantified or even identified (Table 9.3). Few studies in marine habitats have included diet analyses of birds before, during, and after the remigial molt. In one such study (Petersen 1981), Steller's Eiders modified their diet during molt, going from a mixed diet of crustaceans and mussels to only mussels during their flightless period, thereby minimizing foraging costs and maximizing energy balance. The eiders reincorporated amphipods in their diet after the flightless period (Petersen 1981). The nutrient quality of forage in relation to food selection of sea ducks remains to be evaluated at a variety of molting locations to confirm this pattern.

The return of individuals to traditional molting sites depends on the annual variability of abundance and quality of forage. Few studies in North America have investigated annual variability of important prey items. In Iceland, however, Einarsson and Gardarsson (2004) found that numbers of Barrow's Goldeneyes and Red-breasted Mergansers at molting locations were correlated with variability in densities of invertebrates. The authors suggested that the choice of molting areas was influenced by food availability at a known site vs. the risk of migrating to a different site with unknown food availability. Most major sea duck molting areas are used consistently from year to year reflecting the general stability of these sites in terms of producing necessary conditions for molt.

FORAGING BEHAVIOR AND ECOLOGY

Foods eaten by sea ducks during the flightless period vary among species, but forage usually includes the dominant species available within the molting area. Eiders and scoters, when molting in marine and estuarine waters, forage on molluscs, particularly blue mussels and clams (Table 9.3). Similarly, sea ducks molting in freshwater commonly consume

TABLE 9.3
Primary forage items eaten by sea ducks during their flightless period by habitat type compared to winter foods.

Species	Primary forage items in winter	Primary forage items during the flightless period			Sources
		Marine habitats	Estuarine habitats	Freshwater habitats	
Steller's Eider	Bivalves, gastropods, crustaceans	Unknown	Bivalves, polychaetes, gastropods, crustaceans	N/A	1–4
Spectacled Eider	Molluscs (bivalves, gastropods), crustaceans	Likely molluscs, clams, mussels, crustaceans	N/A	N/A	5–7
King Eider	Molluscs, crustaceans, echinoderms, polychaetes	Bivalves, polychaetes, gastropods, cockles, crustaceans, amphipods	N/A	Unknown	8–10
Common Eider	Molluscs, *Mya* clams, *Mytilus*, polychaetes, sea urchins	Molluscs, *Mytilus*, crustaceans, amphipods	Unknown	N/A	11–15
Harlequin Duck	Crustaceans, molluscs, insects, echinoderms, fish, amphipods, isopods	Likely crustaceans, molluscs, insects, echinoderms, fish	N/A	N/A	13,16
Surf Scoter	Bivalves, polychaetes, eelgrass, epifauna, crustaceans, herring spawn	Molluscs, *Mytilus*, clams	N/A	N/A	17–19
White-winged Scoter	Molluscs, bivalves, gastropods, pelecypods, crustaceans	Molluscs, bivalves, clams, Chironomidae	Unknown	Amphipods	17,19–21
Black Scoter	Molluscs, bivalves, gastropods, crustaceans, *Mytilus*, *Macoma*, sea urchins	*Mytilus*, *Astarte*, Diptera, Chironomidae, crustaceans	Unknown	N/A	13,17,22,23
Long-tailed Duck	Diverse, clams, mussels, snails, mysids, isopods, crabs, fish, fish eggs, amphipods	Unknown	Mysids, amphipods	Crustaceans, chironomid larvae, midge larvae, fingernail clams	13,24–26
Bufflehead	Shrimps, crabs, amphipods, isopods, bivalves, snails, fish	N/A	N/A	Likely insects and molluscs	27,28
Common Goldeneye	Saltwater, molluscs, *Mytilus*, crustaceans, fish; freshwater, fish, molluscs, crustaceans; Trichoptera larvae	*Mytilus*	Fish, insects, crustaceans, molluscs	Fish, insects, crustaceans, Trichoptera larvae	17,22,29,30

(Continued)

TABLE 9.3 (Continued)
Primary forage items eaten by sea ducks during their flightless period by habitat type compared to winter foods.

Species	Primary forage items in winter	Primary forage items during the flightless period			Sources
		Marine habitats	Estuarine habitats	Freshwater habitats	
Hooded Merganser	Fish, crustaceans, insects	Unknown	Unknown	Unknown	32
Common Merganser	Fish	Crustaceans, insects, fish	Fish	Fish	17,33
Red-breasted Merganser	Crustaceans, fish	Crustaceans, fish	Fish	Fish (sticklebacks)	17,34,35

(1) Petersen (1980), (2) Petersen (1981), (3) Metzner (1993), (4) Fredrickson (2001), (5) Petersen et al. (1998), (6) Petersen et al. (2000), (7) Lovvorn et al. (2003), (8) Frimer (1995a), (9) Frimer (1997), (10) Suydam (2000), (11) Cantin et al. (1974), (12) Frimer (1995b), (13) Goudie and Ankney (1986), (14) Goudie et al. (2000), (15) Merkel et al. (2007), (16) Robertson and Goudie (1999), (17) Reed et al. (1996), (18) Savard et al. (1998), (19) Anderson et al. (2008), (20) Dau (1987), (21) Brown and Fredrickson (1997), (22) Nilsson (1972), (23) Bordage and Savard (1995), (24) Johnson (1984), (25) Taylor (1986), (26) Robertson and Savard (2002), (27) Erskine (1971), (28) Gauthier (1993), (29) Jepsen (1976), (30) Eadie et al. (1995), (31) Eadie et al. (2000), (32) Dugger et al. (1994), (33) Mallory and Metz (1999), (34) Titman (1999), (35) Craik et al. (2011).

various species of crustaceans and insects, of which chironomids are frequently eaten (Table 9.3). Jepsen (1976), Johnson (1984), and Taylor (1986) determined the species and abundance of potential forage items within various sea duck molting areas. Although there was some selection by size and species, in general, they found that flightless birds took the most abundant, accessible food.

As with other molting waterfowl (Bailey 1985, Austin 1987), most sea ducks spend more time resting during the remigial molt than at any other time during the nonbreeding period (Frimer 1994b, Adams et al. 2000, Langlois 2006, O'Connor 2008, Hogan et al. 2013a). Similarly, molting Red-breasted Mergansers (Craik 2009), Harlequin Ducks (Langlois 2006), and Surf Scoters (O'Connor 2008) fed less during their flightless period than at other life stages. However, Surf and White-winged Scoters molting in British Columbia and Alaska had similar foraging rates during molt as in winter (Dickson 2011, Dickson et al. 2012). The time devoted to comfort movements is greater during molt than at any other time of the year (Frimer 1994b, Langlois 2006). Captive Common Eiders, fed *ad libitum*, spent 55% of their time resting during remigial molt compared to only 19% of their time during nonmolt periods (Portugal et al. 2010). Time spent foraging by captive eiders also decreased during remigial molt dropping to 2% from 18% (Portugal et al. 2010). Also, under certain circumstances, some birds may supplement exogenous energy with depletion of fat stores

(Fox et al. 2014). Changes in behavior and source of nutrients may be related to the need to protect fragile growing feathers and to minimize heat loss or energy expenditure. New or growing primaries (Figure 9.1) are relatively unprotected and fragile, which may impose constraints on feeding as most sea ducks partially use their wings when swimming underwater. Piersma (1988) suggested that diving birds reduce foraging during remigial molt to protect newly emerging feathers. In Common Eiders (Guillemette et al. 2007), time spent in diving was shortest at the beginning of molt but increased from the middle to the end of the molt, supporting the idea that diving birds may reduce foraging at the onset of remigial molt to protect emerging feathers and as well reduce energetic demands related to diving with reduced wing loading.

Surf Scoters that molted remiges along the Labrador coast fed mostly in the early morning and evening in coastal waters and spent the late morning and afternoon resting in tight flocks nearly 2 km offshore (O'Connor 2008). King Eiders (Frimer 1994b, 1995b), Common Eiders (Frimer 1995b), and Harlequin Ducks (Langlois 2006) used a similar foraging schedule (day–evening) during the flightless period. During the flightless period, Steller's Eiders continued to feed on the low tides but changed their foraging behavior from dabbling and diving to almost exclusively dabbling and head dipping when molting (Petersen 1981). Such behavior suggests that food is abundant on these molting sites. Diving is the second most energy-demanding activity after flying (Butler 1991), and it may be advantageous to reduce diving frequency during remigial molt. However, molting sea ducks do not seem nutritionally stressed or limited, and energy conservation may not play a significant role in regulating foraging time. Conversely, because thermal and activity costs are greatly reduced in molting birds and the daily costs of molt are spread over a relatively long period, birds simply may not need to forage intensively. Foraging intensity likely increases as birds rebuild muscle mass and undergo lipogenesis for fall migration, but this physiological change has yet to be quantified.

Molting eiders usually fed in densely packed rafts of 50–70 birds and dove synchronously at Disko Bay, Greenland (Frimer 1994a). Similarly, molting flocks of scoters often dive in a synchronous fashion. Synchronous foraging may be advantageous because of foraging efficiency or predator risk (Schenkeveld and Ydenberg 1985), but little is known of the social and foraging behaviors of molting sea ducks.

PREDATION RISK DURING REMIGIAL MOLT

Molting sea ducks likely select molt sites in part based on low predation risk as flightless birds may be less able to evade predators. There are a few reported cases of flightless Common Eiders being preyed upon by killer whales (Booth and Ellis 2006, Smith 2006) or by seals (Moore 2001). However, in a limited number of studies, survival of sea ducks has been shown to be high during remigial molt, suggesting that appropriate site selection is effective in reducing predation risk. For example, survival of female Harlequin Ducks during the flightless period was as high or higher than during other periods of their life cycle (Iverson and Esler 2007); scoter survival was higher during molt than during breeding or wintering (Anderson et al. 2012); and survival of Barrow's Goldeneyes was high during remigial molt (Hogan et al. 2013c).

One way to reduce predation risk is to shorten the flightless period. A shorter duration can be accomplished by quickly growing feathers or being able to fly before flight feather growth is complete. Waterfowl can fly before their remiges have reached their final length, although the

proportional length at which flight is achieved varies by species (Table 9.2; Sjöberg 1988, Hohman et al. 1992, Brown and Saunders 1998, Dickson et al. 2012). Given the low rate of predation during molt, selective pressures to shorten the flightless period due to predation risk may not be strong. However, regaining flight capacities would enable birds to seek new feeding areas if needed.

ENVIRONMENTAL CONSTRAINTS

Ice

Birds at high northern latitudes must complete remigial feather growth before ice forms or risk being unable to depart on southbound migration. In areas with a short ice-free period, adult female sea ducks may either molt during brood rearing or leave the area before becoming flightless. If adult females become flightless later in the season, they may move prior to molt to lower latitudes (King Eiders, Phillips et al. 2006), to coastal waters (Spectacled Eiders *Somateria fischeri*, Petersen et al. 1999; Harlequin Duck, Brodeur et al. 2008a), or to large lakes (Barrow's Goldeneyes, Hogan et al. 2011; Common Merganser, J. M. Pearce, unpubl. data). Coastal waters usually freeze later than freshwater lakes and ponds; the movement of birds to coastal regions or more southern latitudes may result in a reduced probability of ducks being iced in before they complete their remigial molt and migrate to wintering areas.

Ross (1983) suggested that few subadult Black Scoters (*M. nigra*) migrate to James and Hudson bays to molt remiges and rectrices because habitats with extensive ice are unavailable early in the season when subadults typically undergo remigial molt; the ice is gone by the time adult males arrive to molt. However, this pattern remains to be confirmed because substantial ice-free areas are present in James Bay in late June and subadults and adult males of other scoter species molt almost at the same time (Dickson et al. 2012). Recent satellite telemetry studies indicate that nonbreeding adult males migrate to James Bay to molt (S. G. Gilliland, unpubl. data). Portenko (1981) suggested that only male King Eiders are able to molt flight feathers at the most northern latitudes because the ice-free period is short despite a high abundance of invertebrates. Hohman et al. (1992) hypothesized that in northern latitudes, time is too short for adult females to molt their flight feathers on the breeding grounds after their young fledge. Thus, birds move to coastal or more southern molt sites to undergo remigial molt. However, this pattern remains poorly documented and may not be that common. Indeed, in several northern molting locations, birds remain on their general molting area several weeks after having regrown their flight feathers suggesting that, at least in coastal settings, ice formation may not be a constraining factor. The importance of ice as a constraining factor in selection of molting locations may have been overstated.

Day Length

When molting in areas with longer day length, birds have a greater opportunity for diurnal foraging. It has been hypothesized that King Eiders move to more southern latitudes to undergo the flightless period in response to the quickly decreasing daylight at high latitudes (Frimer 1994b, 1995a). However, remigial molting areas of male and subadult sea ducks are not consistently at lower latitudes. Birds at high northern latitudes tend to undergo remigial molt early in the summer when day length is not likely a factor restricting foraging time and, as indicated in the following, molting sea ducks spend relatively little time foraging. Day length, however, may become a factor influencing the selection of molting locations for later-molting adult females (Frimer 1995a). Colder temperatures, ice formation, and reduced day length occur simultaneously and are not independent; therefore, it is difficult to determine how each influences decisions to migrate to more southern latitudes to undergo the flightless period. However, some sea ducks can feed at night (Hogan et al. 2013a), and this behavior may be more common than currently documented. As most sea ducks remain on their northern molting areas several weeks after having regrown their remiges (Brodeur et al. 2002, Robert et al. 2002), day length is not likely a constraining factor.

COMPETITION

Molt migration of males and subadults away from nesting areas eliminates potential competition for food resources with brood-rearing females and ducklings. Many subadult sea ducks do not migrate to breeding areas and thus do not compete for forage with other age classes. Males and subadults of all sea ducks are rarely found with broods; however, some males and subadults of *Bucephala* and *Mergus* molt their flight feathers within the general nesting region (Erskine 1971). In several species of sea ducks, some subadult females visit their natal area prior to their first remigial molt but do not molt there. Examples include Bufflehead and Goldeneyes (Eadie and Gauthier 1985), White-winged Scoters (S. Slattery, Ducks Unlimited, pers. comm.), Common Mergansers (J. Pearce, pers. comm.), and King Eiders (A. N. Powell, pers. comm.). Food of young usually differs from that of adults, which may partially explain why males leave the female at the onset of incubation and undertake a molt migration towards secluded areas rich in preferred foods. For example, adult scoters feed mostly on shellfish, whereas their young feed on freshwater invertebrates (Brown and Fredrickson 1986, Savard et al. 1998). In the Common Eider, coastal habitats used by broods tended by females differ from those used by unsuccessful females and males (Cantin et al. 1974, Diéval 2006). By moving away from breeding areas, males may optimize their feeding opportunities by moving to more productive or safer habitats, which may lead to reduced competition and enhance foraging opportunities for their offspring.

In North American sea ducks, nonbreeders and females that are failed breeders leave nesting areas several weeks after males but before successfully breeding females (Petersen et al. 1999, Savard et al. 2007). Failed breeding females may molt their flight feathers near the breeding grounds (*Bucephala, Mergus, Clangula*) or migrate to distant molting areas (*Somateria, Melanitta, Bucephala, Clangula*). Buffleheads and goldeneyes sometimes molt at breeding areas but use lakes that are larger than most brood-rearing ponds. Similarly, larger lakes are used more often by molting Long-tailed Ducks than by pairs (Taylor 1986). However, when on the same lake, broods tend to stay closer to shore than molting birds, which group and forage offshore (J.-P. L. Savard, pers. comm.). Although these adjustments result in reduced competition, competition may not have been the overriding factor in this pattern; the ability to move to better foraging locations and opportunities for molt was likely more important. Also, small breeding ponds cannot sustain large molting flocks, and as we have seen, groups afford greater protection during the flightless period (Hamilton 1971).

Competition for high-quality foods is reduced when different ages and sexes either use the same molting area at different times or use different molting areas. Many flightless birds are found in huge flocks, and Hohman et al. (1992) suggested that there may be little competition for forage on molting areas. Some scoter molting sites are also used during spring and fall staging and even during winter (Savard et al. 1999, Rail and Savard 2003, Fox et al. 2008, Dickson et al. 2012), which suggests these habitats are resource rich and that competition may not limit food availability. As most male sea ducks are larger than females, dimorphism may reduce to some degree food competition between the sexes. For example, male Barrow's Goldeneyes take larger blue mussels (Mytilus edulis) than females during winter (Koehl et al. 1984).

RESPONSE TO DISTURBANCE

Molting waterfowl have increased vulnerability when flightless and react strongly to disturbance (Mosbech and Boertmann 1999, O'Connor 2008). Barrow's Goldeneyes molting on lakes of the Old Crow Flats reacted to human disturbance by forming closely knit flocks and moving away from the shore (van de Wetering 1997). Long-tailed Ducks and Surf Scoters reacted to aircraft disturbances by moving offshore from their resting sites, and flock sizes increased as disturbance persisted; Surf Scoters were more affected by aircraft than Long-tailed Ducks (Gollop et al. 1974). In Greenland, molting flocks of King Eiders reacted to an airplane 4 km away by becoming alert and often dove underwater and dispersed when the airplane was within 1 km (Mosbech and Boertmann 1999). Dispersal behavior was most prevalent in areas subjected to greater human disturbance from subsistence hunting than in remote areas. Hikers and kayakers often forced molting Harlequin Ducks from their resting rocks (J.-P. L. Savard, pers. obs.).

Avoidance behaviors were more frequent when birds were resting in midafternoon than when they were foraging in the morning. Foraging King Eiders stopped feeding when disturbed and often did not return to the feeding area until 6–8 h later (Frimer 1994a). On the Labrador coast, flightless Surf Scoters interrupted their feeding and moved several kilometers offshore following disturbances (O'Connor 2008). Molting Long-tailed Ducks moved between lagoons in response to

aircraft, boat, and human disturbances (Johnson 1982), but underwater seismic surveys did not seem to affect them (Lacroix et al. 2003).

The escape behavior of flightless birds differs among species. Flightless sea ducks escape disturbance by diving and scattering underwater (Figure 9.2). Common Eiders, scoters, and Barrow's Goldeneyes regroup when disturbed, but Common Eiders form tight groups and are much easier to drive into traps than scoters that have a greater tendency to scatter (J.-P. L. Savard, pers. comm.). Molting Red-breasted Mergansers were nearly impossible to drive because birds respond strongly to disturbance and flocks dissolve quickly. Flightless Surf and White-winged Scoters, when released after banding, swam away underwater for periods of 40–60 s and resurfaced between dives with only their bill out of the water (J.-P. L. Savard, pers. comm.). This type of reaction to disturbance could be costly energetically if disturbances at molt sites occur frequently. Comeau (1923, 1954) reported instances of hundreds of flightless scoters drowning following the passage of a big steamship during foggy conditions (likely killed by the propeller) and of hunters forcing flightless molting scoters to dive until they drowned, supporting the idea that birds are more vulnerable during molt.

MOLT HABITATS

Salomonsen (1968) and others (Hohman et al. 1992; Frimer 1994a, b, 1995a) characterized molting areas as having the following characteristics: (1) abundant escape cover or few predators, (2) abundant accessible food resources, (3) little or no ice during the flightless period, and (4) sufficient daylight or foraging time to regrow flight feathers. In addition, by migrating to these molting areas, individuals avoid competition with breeding birds and ducklings. These characteristics are not limited to molting areas. Scoters molt sometimes at sites used during spring and fall migration (Savard et al. 1999) and even winter (Dickson et al. 2012). Habitats used by molting sea ducks are quite diverse and range from large freshwater wetlands and lakes to a variety of coastal areas (Table 9.4). Important molting areas for sea ducks in North America include primarily northern coastal waters, bays, and estuaries (Figure 9.4), large inland lakes, and coastal and

TABLE 9.4
Habitat characteristics of primary sites used by sea ducks in North America during remigial molt and winter.

Species	Primary winter sites; Water depth	Primary molting sites; Water depth	Microhabitat during molt	Sources[a]
Steller's Eider	Inshore and nearshore marine bays, coastal waters; <10 m	Estuaries, nearshore marine bays and coastal waters; <6 m	Sandy with pebbles, sandy–gravelly, eel grass	1–3
Spectacled Eider	Offshore marine waters; 40–90 m	Coastal offshore waters, large bays; <18 m	Sandy–gravelly, sandy	4–10
King Eider	Coastal waters; 15–40 m	Nearshore marine waters, bays, fjords; 0–30 m	Silt bottom, soft bottom, or hard bottom bays and coastal regions	11–15
Common Eider	Nearshore/inshore marine coastal waters, intertidal; ≤25 m	Inshore marine coastal waters, bays, estuaries, intertidal; <15 m	Rocky or sandy bays and coastal regions, reefs	12,16–20
Harlequin Duck	Inshore marine bays and coastal waters; <6 m	Inshore marine bays and coastal waters; <6 m	Rocky, sandy with pebbles/cobbles	21–24
Surf Scoter	Nearshore marine bays and coastal waters; <9 m	Large bays, coastal inshore, nearshore; 1–6 m	Mud, sandy, sandy with pebbles, rocky, near shoals	17,26–29
White-winged Scoter	Nearshore marine bays and coastal waters; 5–20 (<5) m	Large bays, coastal inshore, nearshore, large freshwater lakes; <10 m	Sandy, sandy with pebbles, rocky, near shoals	17,26,28,30
Black Scoter	Nearshore marine bays and coastal waters; <10 m	Large bays, coastal inshore, nearshore; <10 m	Mud, sandy, sandy with pebbles, near shoals, rocky	3,17,25,26,31
Long-tailed Duck	Nearshore marine bays and coastal waters; <22 m	Freshwater lakes, wetlands, estuaries, bays; 2–3 m	Silty, sandy	32–40
Common Goldeneye	Inshore marine and brackish waters, estuaries, freshwater lakes; <7 m	Shallow boreal lakes, bays, rivers, estuaries; *very shallow*	Eelgrass beds, sand and sand–gravel, detritus, vegetated	28,33,34, 41–43
Barrow's Goldeneye	Inshore marine and brackish waters, estuaries, freshwater lakes; <4 m	Large, productive shallow boreal and subboreal lakes, estuaries;	Lake bottom	32,33,44–47
Bufflehead	Inshore marine waters, estuaries, freshwater wetlands and lakes; <3 m	Large, productive shallow boreal and subboreal lakes, alkaline lakes	Lake bottom	42,48

(Continued)

TABLE 9.4 *(Continued)*
Habitat characteristics of primary sites used by sea ducks in North America during remigial molt and winter.

Species	Primary winter sites; Water depth	Primary molting sites; Water depth	Microhabitat during molt	Sources[a]
Hooded Merganser	Shallow freshwater and brackish bays, estuaries, tidal creeks, ponds; <1.5 m	Freshwater; <1.5 m	Lake bottom unknown, mud/sand, boulder screen	21,28,42,49
Common Merganser	Inshore marine and brackish waters, freshwater lakes and rivers; <4 m	Freshwater lakes, reservoirs, estuaries, bays, inshore marine	Rocky, sandy, near shoals, eelgrass beds	28,42
Red-breasted Merganser	Coastal waters, inshore marine and brackish waters; ~15 m	Coastal, bays, estuaries, freshwater lakes	Rocky, sandy, near shoals, eelgrass beds	28,34,42,50,51

[a] (1) Petersen (1980), (2) Fredrickson (2001), (3) M. R. Petersen (unpubl. data), (4) Feder et al. (1994a), (5) Feder et al. (1994b), (6) Petersen et al. (1998), (7) Petersen et al. (1999), (8) Petersen et al. (2000), (9) Lovvorn et al. (2003), (10) Petersen and Douglas (2004), (11) Frimer (1995a), (12) Frimer (1995b), (13) Frimer (1997), (14) Mosbech et al. (2006a), (15) Phillips et al. (2006), (16) Salomonsen (1968), (17) Gill et al. (1981), (18) Petersen and Flint (2002), (19) Mosbech et al. (2006b), (20) J.-P. L. Savard (unpubl. data), (21) Palmer (1976), (22) Robertson et al. (1999), (23) Robertson and Goudie (1999), (24) Brodeur et al. (2002), (25) Ross (1983), (26) Dau (1987), (27) Johnson and Richardson (1981), (28) Reed et al. (1996), (29) Savard et al. (1998), (30) Brown and Fredrickson (1997), (31) Bordage and Savard (1995), (32) King (1963), (33) King (1973), (34) Nilsson (1972), (35) Derksen et al. (1981), (36) Johnson (1984), (37) Taylor (1986), (38) Johnson and Herter (1989), (39) Noel et al. (1999), (40) Petersen et al. (2003), (41) Jepsen (1976), (42) Bellrose (1980), (43) Eadie et al. (1995), (44) van der Wetering and Cooke (2000), (45) Robert et al. (2002), (46) Savard and Robert (2013), (47) W. S. Boyd (unpubl. data), (48) Erskine (1971), (49) Dugger et al. (1994), (50) Craik et al. (2009), (51) Craik et al. (2011).

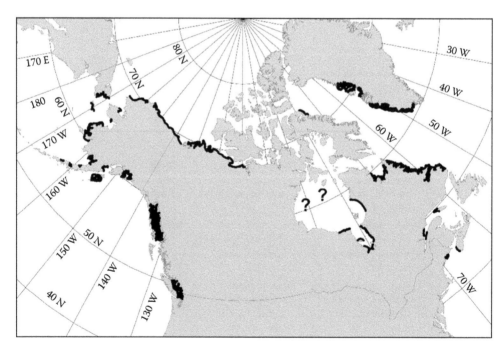

Figure 9.4. Major coastal molting areas of sea ducks in North America (Gauthier 1993; Dugger et al. 1994; Bordage and Savard 1995; Eadie et al. 1995, 2000; Brown and Fredrickson 1997; Savard et al. 1998; Mallory and Metz 1999; Robertson and Goudie 1999; Titman 1999; Goudie et al. 2000; Petersen et al. 2000; Suydam 2000; Fredrickson 2001; Robertson and Savard 2002; Powell and Suydam 2012). Question marks (?) indicate areas not surveyed.

inland wetlands (Figures 9.3 through 9.5). Few sea ducks molt below 45°N, most molting at more northern latitudes (Figure 9.4).

Steller's Eiders, Common Eiders, and eastern Harlequin Ducks undergo remigial molt only in shallow coastal waters (Figure 9.5a and b): Steller's Eiders in productive estuarine waters, Common Eiders over rocky subtidal substrates, and Harlequin Ducks along rocky foreshores (Petersen 1981, Robertson and Goudie 1999, Goudie et al. 2000). Long-tailed Ducks are perhaps the most diverse; they molt in marine estuaries, freshwater lakes both inland and in coastal deltas, freshwater wetlands, and coastal lagoons (Figure 9.5c). Scoters use mostly shallow coastal marine waters that are usually <10 m depth (Figure 9.5d), although some are known to molt inland (R. Dickson, unpubl. data). *Bucephala* and mergansers also molt in coastal marine waters and on freshwater lakes, but there is less information available on their molting distribution. Data for the small eastern population of Barrow's Goldeneyes

(<2,500 pairs) illustrate the diversity in the type and location of molting areas; birds molt in a variety of inland lakes and coastal waters (Figure 9.3, Robert and Savard 2006, Savard and Robert 2013). However, Barrow's Goldeneyes from western North America molt exclusively on shallow inland lakes and larger wetlands that seldom support larger bodied fish species and which are typically eutrophic to hypereutrophic in productivity (van de Wetering 1997; Hogan et al. 2011; Hogan 2012; W. S. Boyd, unpubl. data).

Most species of sea ducks molt in large flocks of hundreds to thousands of birds, especially when in coastal waters. Species molting on inland lakes such as goldeneyes, Buffleheads, and Common Mergansers form smaller flocks of tens to hundreds of birds. However, the size of molting flocks is variable among species, habitats, and activities—resting and roosting flocks are usually larger than foraging flocks.

Most sea ducks occur in monospecific flocks during the molting period although mixed flocks are

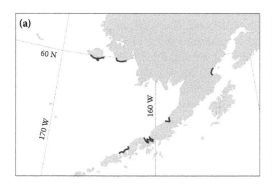

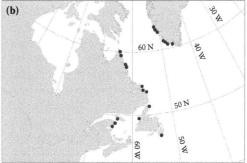

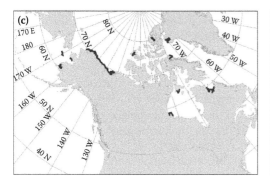

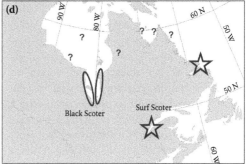

Figure 9.5. Major molting areas of (a) Steller's Eiders (Petersen 1981; Dau 1987; B. J. McCaffery, in litt. to USFWS), (b) eastern Harlequin Ducks (Brodeur et al. 2002, Chubbs et al. 2008, Robert et al. 2008, Thomas et al. 2008), (c) Long-tailed Ducks (King 1973, Palmer 1976, Johnson and Herter 1989, Robertson and Savard 2002, Petersen et al. 2003), and (d) eastern Black and Surf Scoters (Ross 1983, Bordage and Savard 1995, Savard et al. 1998). Question marks (?) indicate areas not yet surveyed.

sometimes observed; however, one species usually clearly dominates the flock. Common and King Eiders molting in Greenland forage at different depths and over different substrates (Frimer 1995b). In the St. Lawrence estuary (J.-P. L. Savard, unpubl. data) and on the northern Pacific Coast (Dickson 2011), Surf and White-winged Scoters are often found in the same area when molting over sandy substrates and can form either single species or multispecies flocks. When in a mixed flock, each species tends to be clumped within the flock. Within a species, molting habitats can vary considerably. Large inland shallow arctic lakes often support several species of molting waterfowl and thousands of birds. On those lakes, dabbling, diving, and sea ducks often cohabit during molt (King 1963, 1973). At Old Crow Flats in the Yukon as well as in NW Alberta, Barrow's Goldeneyes often molt with Canvasbacks and scaup (van de Wetering 1997, Hogan 2012).

SEASONAL DIFFERENCES OF HABITATS

Habitats used by birds during the flightless period often differ from those used during the remainder of the annual cycle (Table 9.4). Buffleheads molt on freshwater lakes yet are often found in marine waters in winter. Spectacled Eiders molt in relatively shallow (<18 m) nearshore waters (within 38 km from shore) yet winter at sea (>90 km from the nearest island) in waters more than twice that depth (>40 m, Petersen et al. 1999, Petersen and Douglas 2004). In contrast, some Harlequin Ducks often molt remiges and winter at the same locations and use the same habitats (Robertson et al. 1999).

For some species, the habitat used by males and subadults during the flightless period differs from that of adult females. For example, female Red-breasted Mergansers that undergo molt while rearing a brood use the same freshwater habitats as their ducklings, whereas adult males and subadults migrate to coastal marine waters to molt (Craik et al. 2009). Similarly, a few adult female King Eiders were flightless with their broods and molted in freshwater or brackish water lakes and ponds (Powell and Suydam 2012), whereas subadults, males, and failed breeding females molted in marine waters. Unfortunately, data are lacking on the impact of molting habitat on molt dynamics, bird condition, and potential survival.

Corresponding to differences in winter and molting habitats are differences in forage conditions. For example, the diet of Barrow's Goldeneyes wintering in marine waters differed from that of birds molting on freshwater lakes (Eadie et al. 2000). In Denmark, Common Goldeneyes had a more diverse diet of small fish, insects, crustacean, and molluscs when molting in brackish water (Jepsen 1976), but consumed mainly molluscs and crustaceans in marine waters during winter (Madsen 1954). Long-tailed Ducks molting in marine estuaries ate primarily mysids and some amphipods (Johnson 1984, 1985; although see Robertson and Savard 2002), whereas ducks molting in northern freshwater lakes consumed primarily crustaceans, insects, and fingernail clams (Taylor 1986). A better understanding of the quantity and quality of forage available and used during remigial molt and pre- and postremigial molt is needed to more fully understand the dynamic process of feather replacement in sea ducks.

DATA GAPS

The behavioral and physiological requirements and responses to molt, of both body and flight feathers of sea ducks, are still poorly understood, especially concerning the molt of body feathers. An understanding of these and other concepts have implications for conservation of many species. Indeed, habitat and nutritional requirements related to molt are still poorly documented, and differences in timing or speed of molt among different sexes and age classes may have fitness and survival consequences (Gehrold and Köhler 2013). Dynamics of organ and muscle size changes during the remigial molt has yet to be studied in most sea duck species. Current literature suggests that changes are quick and their timing may vary according to individual condition. For example, Fox et al. (2008) showed that, in Common Scoter, pectoral muscles were already undergoing increase in mass by the middle of remigial molt.

Site fidelity and movements among molting habitats need to be better quantified as it potentially affects population structure, monitoring schemes, and management concerns (Esler et al. 2006, Pearce and Talbot 2006). A better understanding of the diet of molting birds and of the quantity and quality of foraging conditions in different habitats before, during, and after the molt is needed to more fully understand the dynamic process of feather replacement in sea ducks and the features of molt habitats they select. Also included in understanding the quantity and quality of forage are the timing of prey availability; the interactions of birds in relation to their forage items, including the potential for prey depletion such as in mollusc-feeding species; and the impact of climate and other environmental conditions on prey populations.

Interactions among molting birds are poorly understood. Molting flocks are highly dynamic and can vary in size during and among days. Factors that determine the size of molting flocks are unknown, but dispersion and type of food resources as well as habitat configuration and size at local and landscape scales probably play a role. Molting in groups has not only predator avoidance benefits (Hamilton 1971), but also may enhance synchrony in timing of molt (Palmer 1972, Leafloor et al. 1996). It is not known whether some molting sites are used only or mainly by a single age or sex group and if so how they differ from other sites.

The increase in human disturbance at molting sites is an emerging concern for the conservation of sea ducks. Human populations continue to expand as does resource extraction into many areas where sea ducks molt their flight feathers (Hogan 2012). Some studies have shown that birds change their behavior in response to disturbance (O'Connor 2008), yet the physiological effects and the effects on their subsequent habitat use and site fidelity are poorly understood. An understanding of these and topics such as the energetic of molt will play important roles in the conservation and management of sea ducks. Understanding relationships among breeding, molting, and wintering locations is essential for a holistic management approach, especially in the context of harvested species. We are only slowly becoming able to unravel some of these relationships with satellite telemetry and genetic tools, and much remains to be done.

INTERNATIONAL CONCERNS

Several populations of sea ducks breed, molt, and winter in different countries: some Harlequin Ducks breed in Canada but molt in Greenland and winter in the United States; several species of ducks breed in Canada and then molt or winter in the United States or Mexico (Common Eiders, Surf Scoters, Black Scoters, Harlequin Ducks). In North America, some populations are shared by Russia and Alaska (Long-tailed Ducks, Steller's Eiders) and some by Greenland and Canada (Common Eider, King Eider, Harlequin Duck). Broad geographic distributions emphasize the need for international collaboration to insure a holistic and comprehensive approach to the management of sea ducks.

While cooperation between the United States and Canada is well established with the Migratory Bird Convention, efforts are needed with Russia and Greenland. The protection of Harlequin Ducks in Greenland has likely contributed to the slight recovery of the eastern North American population as a major portion of the North American population molt and winter there (Robertson et al. 2008). Research efforts on Common Eiders have spurred changes in hunting regulations in Greenland and Canada (Mosbech et al. 2006b, Gilliland et al. 2009, Savard et al. 2011). Protection of important molting sites is crucial for ensuring the health of sea duck populations. A need for a holistic approach to sea duck management is obvious but challenging.

ACKNOWLEDGMENTS

We thank R. Elliot, Science and Technology, and D. V. Derksen and J. Geiselman, Alaska Science Center, U.S. Geological Survey, for the opportunity to write this chapter. W. S. Boyd and J. M. Pearce shared their unpubl. data. We especially thank J. Terenzi and S. McCloskey for the preparation of the figures. We thank D. Esler, J. Thompson, A. J. Fox, and J. Hupp for reviewing this book. Their comments were quite pertinent and useful in improving the book. The use of trade names is for descriptive purposes only and does not imply endorsement by the US government.

LITERATURE CITED

Abraham, K. F., and G. H. Finney. 1986. Eiders of the eastern Canadian Arctic. Pp. 55–73 in A. Reed (editor), Eider ducks in Canada. Canadian Wildlife Service Report Series, No. 47. Canadian Wildlife Service, Ottawa, ON.

Adams, P. A., G. J. Robertson, and I. L. Jones. 2000. Time-activity budgets of Harlequin Ducks molting in the Gannet Islands, Labrador. Condor 102:703–708.

Alison, R. M. 1976. Oldsquaw brood behaviour. Bird-Banding 47:210–213.

Anderson, E. M., D. Esler, W. S. Boyd, J. R. Evenson, D. R. Nysewander, D. H. Ward, R. D. Dickson, B. D. Uher-Koch, C. S. VanStratt, and J. W. Hupp. 2012. Predation rates, timing, and predator composition for scoters (Melanitta spp.) in marine habitats. Canadian Journal of Zoology 90:42–50.

Anderson, E. M., J. R. Lovvorn, and M. T. Wilson. 2008. Reevaluating marine diets of Surf and White-winged Scoters: interspecific differences and the importance of soft-bodied prey. Condor 110:285–295.

Ankney, C. D. 1979. Does the wing molt cause nutritional stress in Lesser Snow Geese? Auk 96:68–72.

Ankney, C. D. 1984. Nutrient reserve dynamics of breeding and molting Brant. Auk 101:361–370.

Austin, J. E. 1987. Activities of postbreeding Lesser Scaup in southwestern Manitoba. Wilson Bulletin 99:448–456.

Bailey, R. O. 1981. The postbreeding ecology of the Redhead Duck (Aythya americana) on Long Island Bay, Lake Winnipegosis, Manitoba. Dissertation, McGill University, Montréal, QC.

Bailey, R. O. 1985. Protein reserve dynamics in post-breeding adult male Redheads. Condor 87:23–32.

Beard, E. B. 1964. Duck brood behavior at the Seney National Wildlife Refuge. Journal Wildlife Management 28:492–521.

Bellrose, F. C. (editor). 1980. Ducks, geese and swans of North America, 3rd ed. Wildlife Management Institute, Stackpole Books, Harrisburg, PA.

Benoit, R., M. Robert, C. Marcotte, G. Fitzgérald, and J.-P. L. Savard. 2001. Étude des déplacements du Garrot d'Islande dans l'est du Canada à l'aide de la télémétrie satellitaire. Technical Report Series, No. 360, Canadian Wildlife Service, Québec region, QC (in French).

Bluhm, C. K. 1988. Temporal patterns of pair formation and reproduction in annual cycles and associated endocrinology in waterfowl. Current Ornithology 5:123–185.

Booth, C. J., and P. Ellis. 2006. Common Eiders and Common Guillemots taken by killer whales. British Birds 99:533–535.

Bordage, D., and J.-P. L. Savard. 1995. Black Scoter (Melanitta nigra). in A. Poole and F. Gill (editors), The Birds of North America, No. 177. The Academy of Natural Sciences, Philadelphia, PA.

Breault, A. M., and J.-P. L. Savard. 1999. Philopatry of Harlequin Ducks moulting in southern British Columbia. Pp. 41–44 in R. I. Goudie, M. R. Petersen, and G. J. Robertson (editors), Behaviour and ecology of sea ducks. Canadian Wildlife Service, Occasional Paper, No. 100. Canadian Wildlife Service, Ottawa, ON.

Bridge, E. S. 2011. Mind the gaps: what's missing in our understanding of feather molt. Condor 113:1–4.

Brodeur, S., J.-P. L. Savard, M. Robert, A. Bourget, G. Fitzgerald, and R. D. Titman. 2008a. Abundance and movements of Harlequin Ducks breeding on rivers of the Gaspé Peninsula, Québec. Waterbirds 31:122–129.

Brodeur, S., J.-P. L. Savard, M. Robert, P. Laporte, P. Lamothe, R. D. Titman, S. Marchand, S. Gilliland, and G. Fitzgerald. 2002. Harlequin Duck Histrionicus histrionicus population structure in eastern Nearctic. Journal of Avian Biology 33:127–137.

Brodeur, S., J.-P. L. Savard, M. Robert, R. D. Titman, and G. Fitzgerald. 2008b. Failure time and fate of Harlequin Ducks implanted with satellite transmitters. Waterbirds 31:183–187.

Brown, P. W., and M. A. Brown. 1981. Nesting biology of the White-winged Scoter. Journal of Wildlife Management 45:38–45.

Brown, P. W., and L. H. Fredrickson. 1986. Food habits of breeding White-winged Scoters. Canadian Journal of Zoology 64:1652–1654.

Brown, P. W., and L. H. Frederickson. 1989. White-winged Scoter populations and nesting at Redberry Lake, Saskatchenan. Canadian Field-Naturalist 103:240–247.

Brown, P. W., and L. H. Fredrickson. 1997. White-winged Scoter (Melanitta fusca). in A. Poole and F. Gill (editors), The Birds of North America, No. 274. The Academy of Natural Sciences, Philadelphia, PA.

Brown, R. E., and D. K. Saunders. 1998. Regulated changes in body mass and muscle mass in Blue-winged Teal for an early return to flight. Canadian Journal of Zoology 76:26–32.

Bustnes, J. O., and K. E. Erikstad. 1991. Parental care in the Common Eider (Somateria mollissima): factors affecting abandonment and adoption of young. Canadian Journal of Zoology 69:1538–1545.

Butler, P. J. 1991. Exercise in birds. Journal of Experimental Biology 160:233–262.

Cantin, M., J. Bédard, and H. Milne. 1974. The food and feeding of Common Eiders in the St. Lawrence estuary in summer. Canadian Journal of Zoology 52:319–334.

Chubbs, T. E., P. G. Trimper, G. W. Humphries, P. W. Thomas, L. T. Elson, and D. K. Laing. 2008. Tracking seasonal movements of adult male Harlequin Ducks from central Labrador using satellite telemetry. Waterbirds 31:173–182.

Comeau, N. A. 1923. Notes on the diving of loons and ducks. Auk 50:525.

Comeau, N. A. 1954. Life and sport on the North Shore of the lower St. Lawrence and Gulf. Quebec Telegraph Printing Company, Quebec, QC.

Cooke, F., G. J. Robertson, R. I. Goudie, and W. S. Boyd. 1997. Molt and the basic plumage of male Harlequin Ducks. Condor 99:83–90.

Craik, S. R. 2009. Habitat use by breeding and molting Red-breasted Mergansers in the Gulf of St. Lawrence. Dissertation, McGill University, Montreal, QC.

Craik, S. R., J.-P. L. Savard, M. J. Richardson, and R. D. Titman. 2011. Foraging ecology of flightless male Red-breasted Mergansers in the Gulf of St. Lawrence, Canada. Waterbirds 34:280–288.

Craik, S. R., J.-P. L. Savard, and R. D. Titman. 2009. Wing and body molts of male Red-breasted Mergansers in the Gulf of St. Lawrence, Canada. Condor 111:71–80.

Dau, C. P. 1974. Nesting biology of the Spectacled Eider *Somateria fischeri* (Brandt) on the Yukon-Kuskokwim Delta, Alaska. M.Sc. thesis, University of Alaska, Fairbanks, AK.

Dau, C. P. 1987. Birds in nearshore waters of the Yukon-Kuskokwim Delta, Alaska. Murrelet 68:12–23.

Dau, C. P., P. L. Flint, and M. R. Petersen. 2000. Distribution of recoveries of Steller's Eiders banded on the lower Alaska Peninsula, Alaska. Journal of Field Ornithology 71:541–548.

Dean, W. R. J. 1978. Moult seasons of some Anatidae in the Western Transvaal. Ostrich 49:76–84.

Delacour, J. 1959. The waterfowl of the world (vol. 3). Country Life Limited, London, U.K.

Derksen, D. V., T. C. Rothe, and W. D. Eldridge. 1981. Use of wetland habitats by birds in the National Petroleum Reserve-Alaska. U.S. Fish and Wildlife Service, Resource Publication, No. 141. U.S. Fish and Wildlife Service, Washington, DC.

Dickson, D. L. 2012. Movements of King Eiders from breeding grounds on Banks Island, NWT, to moulting and wintering areas. Technical Report Series, No. 516. Canadian Wildlife Service, Edmonton, AB.

Dickson, R. 2011. Postbreeding ecology of White-winged Scoters (*Melanitta fusca*) and Surf Scoters (*M. perspicillata*) in western North America: wing moult phenology, body mass dynamics and foraging behaviour. Thesis, Simon Fraser University, Vancouver, BC.

Dickson, R. D., D. Esler, J. W. Hupp, E. M. Anderson, J. R. Evenson, and J. Barrett. 2012. Phenology and duration of remigial molt in Surf Scoters (*Melanitta perspicillata*) and White-winged Scoters (*Melanitta fusca*) on the Pacific coast of North America. Canadian Journal of Zoology 90:932–944.

Diéval, H. 2006. Répartition de l'Eider à Duvet pendant les périodes d'élevage des jeunes et de mue des adultes le long du fleuve Saint-Laurent. Thesis, University of Québec, Montréal, QC (in French).

Donald, P. F. 2007. Adult sex ratios in wild bird populations. Ibis 149:671–692.

DuBowy, P. J. 1985. Seasonal organ dynamics in postbreeding male Blue-winged Teal and Northern Shovelers. Comparative Biochemistry and Physiology 82A:899–906.

Dugger, B. D., K. M. Dugger, and L. H. Fredrickson. 1994. Hooded Merganser (*Lophodytes cucullatus*). in The Birds of North America, No. 98. The Academy of Natural Sciences, Philadelphia, PA.

Dwight Jr., J. 1914. The moults and plumage of the scoters—Genus Oidemia. Auk 31:293–308.

Eadie, J. M., and G. Gauthier. 1985. Prospecting for nest sites by cavity-nesting ducks of the genus *Bucephala*. Condor 87:528–534.

Eadie, J. M., M. L. Mallory, and H. G. Lumsden. 1995. Common Goldeneye (*Bucephala clangula*). in A. Poole and F. Gill (editors), The Birds of North America, No. 170. The Academy of Natural Sciences, Philadelphia, PA.

Eadie, J. M., J.-P. L. Savard, and M. L. Mallory. 2000. Barrow's Goldeneye (*Bucephala islandica*). in A. Poole and F. Gill (editors), The Birds of North America, No. 548. The Academy of Natural Sciences, Philadelphia, PA.

Einarsson, Á. and A. Gardarsson. 2004. Moulting diving ducks and their food supply. Aquatic Ecology 38:297–307.

Erskine, A. J. 1961. Nest-site tenacity and homing in the Bufflehead. Auk 78:389–396.

Erskine, A. J. 1971. Buffleheads. Canadian Wildlife Service, Monograph Series, No. 4. Environment Canada, Ottawa, ON.

Erskine, A. J. 1972. Populations, movements and seasonal distribution of mergansers. Canadian Wildlife Service Report Series, No. 17. Canadian Wildlife Service, Ottawa, ON.

Esler, D., S. A. Iverson, and D. J. Rizzolo. 2006. Genetic and demographic criteria for defining population units for conservation: the value of clear messages. Condor 108:480–483.

Feder, H. M., N. R. Foster, S. C. Jewett, T. J. Weingartner, and R. Baxter. 1994a. Mollusks in the northeastern Chukchi Sea. Arctic 47:145–163.

Feder, H. M., A. S. Naidu, S. C. Jewett, J. M. Hameedi, W. R. Johnson, and T. E. Whitledge. 1994b. The northeastern Chukchi Sea: benthos-environmental interactions. Marine Ecology Progress Series 111:171–190.

Ferrell, C. L. 1988. Contribution of visceral organs to animal energy expenditures. Journal of Animal Science 66:23–34.

Flint, P. L., J. B. Grand, J. A. Morse, and T. F. Fondell. 2000a. Late summer survival of adult female and juvenile Spectacled Eiders on the Yukon-Kuskokwim Delta, Alaska. Waterbirds 23:292–297.

Flint, P. L., D. L. Lacroix, J. A. Reed, and R. B. Lanctot. 2004. Movements of flightless Long-tailed Ducks during wing molt. Waterbirds 27:35–40.

Flint, P. L., M. R. Petersen, C. P. Dau, J. E. Hines, and J. D. Nichols. 2000b. Annual survival and site fidelity of Steller's Eiders molting along the Alaska Peninsula. Journal of Wildlife Management 64:261–268.

Fondell, T. F., P. L. Flint, J. A. Schmutz, J. L. Schamber, and C. A. Nicolai. 2013. Variation in body mass dynamics among moult sites in Brant Geese *Branta bernicla nigricans* supports adaptivity of mass loss during moult. Ibis 155:593–604.

Fox, A. D., P. L. Flint, W. L. Hohman, and J.-P. L. Savard. 2014. Waterfowl habitat use and selection during the remigial moult period in the northern hemisphere. Wildfowl 4 (Special Issue):131–168.

Fox, A. D., P. Hartmann, and I. K. Petersen. 2008. Changes in body mass and organ size during remigial moult in Common Scoter *Melanitta nigra*. Journal of Avian Biology 39:35–40.

Fox, A. D., and J. Kahlert. 2005. Changes in body mass and organ size during wing moult in non-breeding Greylag Geese, *Anser anser*. Journal of Avian Biology 36:538–548.

Fredrickson, L. H. 2001. Steller's Eider (*Polysticta stelleri*). in A. Poole and F. Gill (editors), The Birds of North America, No. 571. The Academy of Natural Sciences, Philadelphia, PA.

Frimer, O. 1994a. Autumn arrival and moult in King Eiders (*Somateria spectabilis*) at Disko, West Greenland. Arctic 47:137–141.

Frimer, O. 1994b. The behaviour of moulting King Eiders *Somateria spectabilis*. Wildfowl 45:176–187.

Frimer, O. 1995a. Adaptations by the King Eider *Somateria spectabilis* to its moulting habitat: review of a study at Disko, West Greenland. Ornithologisk Forenenigs Tidsskrift 89:135–142.

Frimer, O. 1995b. Comparative behaviour of sympatric moulting populations of Common Eider *Somateria mollissima* and King Eider *Somateria spectabilis* in central West Greenland. Wildfowl 46:129–139.

Frimer, O. 1997. Diet of moulting King Eiders *Somateria spectabilis* at Disko Island, West Greenland. Ornis Fennica 74:187–194.

Gaunt, A. S., R. S. Hikida, J. R. Jehl Jr., and L. Fenbert. 1990. Rapid atrophy and hypertrophy of avian flight muscle. Auk 107:649–659.

Gauthier, G. 1993. Bufflehead (*Bucephala albeola*). in A. Poole and F. Gill (editors), The Birds of North America, No. 67. The Academy of Natural Sciences, Philadelphia, PA.

Gehrold, A., and P. Köhler. 2013. Wing-moulting waterbirds maintain body condition under good environmental conditions: a case study of Gadwalls. Journal of Ornithology 154:783–793.

Geldenhuys, J. N. 1983. Morphological variation in wing-moulting South African Shelducks. Ostrich 54:19–25.

Gill Jr., R. E., M. R. Petersen, and P. D. Jorgensen. 1981. Birds of the northcentral Alaska Peninsula, 1976–1980. Arctic 34:286–306.

Gilliland, S. G., H. G. Gilchrist, R. F. Rockwell, G. J. Robertson, J.-P. L. Savard, F. Merkel, and A. Mosbech. 2009. Evaluating the sustainability of fixed-number harvest of northern Common Eiders in Greenland and Canada. Wildlife Biology 15:24–36.

Gin, H. B., and D. S. Melville. 1983. Moult in birds. BTO Guide 19. British Trust for Ornithology, Tring, U.K.

Gollop, M. A., J. R. Goldsberry, and R. A. Davis. 1974. Aircraft disturbance to moulting sea ducks, Hershel Island, Yukon Territory, August, 1972. Pp. 202–230 in W. W. H. Gunn and J. A. Livingston (editors), Disturbance to birds by gas compressor noise simulators, aircraft and human activity in the Mackenzie Valley and the North Slope, 1972. Arctic Gas Biological Report Series (vol. 14). Canadian Arctic Gas Study Ltd., Calgary, AB.

Goudie, R. I., and C. D. Ankney. 1986. Body size, activity budgets, and diets of sea ducks wintering in Newfoundland. Ecology 67:1475–1482.

Goudie, R. I., G. J. Robertson, and A. Reed. 2000. Common Eider (*Somateria mollissima*). in A. Poole and F. Gill (editors), The Birds of North America, No. 546. The Academy of Natural Sciences, Philadelphia, PA.

Gowans, B., G. J. Robertson, and F. Cooke. 1997. Behaviour and chronology of pair formation by Harlequin Ducks *Histrionicus histrionicus*. Wildfowl 48:135–146.

Guillemette, M., and J.-F. Ouellet. 2005. Temporary flightlessness in pre-laying Common Eiders *Somateria mollissima*: are females constrained by excessive wing-loading or by minimal flight muscle ratio? Ibis 147:293–300.

Guillemette, M., D. Pelletier, J.-M. Grandbois, and P. J. Butler. 2007. Flightlessness and the energetic cost of wing molt in a large sea duck. Ecology 88:2936–2945.

Hamilton, W. D. 1971. Geometry for selfish herd. Journal of Theoretical Biology 31:295–311.

Hawkins, G. L. 2011. Molts and plumages of ducks (Anatinae): an evaluation of Pyle (2005). Waterbirds 34:481–494.

Hochbaum, H. A. 1944. The Canvasback on a prairie marsh. American Wildlife Institute, Washington, DC.

Hochbaum, H. A. 1955. Travels and traditions of waterfowl. University of Minnesota Press, Minneapolis, MN.

Hogan, D. 2012. Postbreeding ecology of Barrow's Goldeneyes in northwestern Alberta. Thesis, Simon Fraser University, Vancouver, BC.

Hogan, D. E., D. Esler, and J. E. Thompson. 2013a. Body mass and foraging dynamics of Barrow's Goldeneyes: strategies for managing risk of simultaneous remigial molt. Auk 130:313–322.

Hogan, D. E., D. Esler, and J. E. Thompson. 2013b. Duration and phenology of remigial molt of Barrow's Goldeneye. Condor 115:762–768.

Hogan, D. E., J. E. Thompson, and D. Esler. 2013c. Survival of Barrow's Goldeneyes during remigial molt and fall staging. Journal of Wildlife Management 77:701–706.

Hogan, D. H., J. E. Thompson, D. Esler, and W. S. Boyd. 2011. Discovery of important postbreeding sites for Barrow's Goldeneye in the boreal transition zone of Alberta. Waterbirds 34:261–388.

Hohman, W. L. 1993. Body composition dynamics of Ruddy Ducks during wing moult. Canadian Journal of Zoology 71:2224–2228.

Hohman, W. L., C. D. Ankney, and D. H. Gordon. 1992. Ecology and management of postbreeding waterfowl. Pp. 128–189 in B. D. J. Batt, A. D. Afton, M. G. Anderson, C. D. Ankney, D. H. Johnson, J. A. Kadlec, and G. L. Krapu (editors), Ecology and management of breeding waterfowl. University of Minnesota Press, Minneapolis, MN.

Hohman, W. L., and R. D. Crawford. 1995. Molt in the annual cycle of Ring-necked Ducks. Condor 97:473–483.

Howell, M. D. 2002. Molt dynamics of male Long-tailed Ducks on the Beaufort Sea. Thesis, Auburn University, Auburn, AL.

Howell, M. D., J. B. Grand, and P. L. Flint. 2003a. Body molting of male Long-tailed Ducks in the nearshore waters of the North Slope, Alaska. Wilson Bulletin 115:170–175.

Howell, S. N. G. 2010. Molt in North American birds. Houghton Mifflin Hartcourt, Boston, MA.

Howell, S. N. G., C. Corbin, P. Pyle, and D. I. Rogers. 2003b. The first basic problem: a review of molt and plumage homologies. Condor 105:635–653.

Humphrey, P. S., and K. C. Parkes. 1959. An approach to the study of molts and plumages. Auk 76:1–31.

Iverson, S. A., and D. Esler. 2007. Survival of female Harlequin Ducks during wing molt. Journal of Wildlife Management 71:1220–1224.

Iverson, S. A., D. Esler, and W. S. Boyd. 2003. Plumage characteristics as an indicator of age class in the Surf Scoter. Waterbirds 26:56–61.

Iverson, S. A., B. D. Smith, and F. Cooke. 2004. Age and sex distributions of wintering Surf Scoters: implications for the use of age ratios as an index of recruitment. Condor 106:252–262.

Jehl Jr., J. R. 1990. Aspects of the molt migration. Pp. 102–113 in E. Gwinner (editor), Bird migration. Springer-Verlag, Berlin, Germany.

Jehl Jr., J. R. 1997. Cyclical changes in body composition in the annual cycle and migration of the Eared Grebe Podiceps nigricollis. Journal of Avian Biology 28:132–142.

Jepsen, P. U. 1973. Studies of the moult migration and wing-feather moult of the Goldeneye (Bucephala clangula) in Denmark. Danish Review of Game Biology 8:1–23.

Jepsen, P. U. 1976. Feeding ecology of Goldeneye (Bucephala clangula) during the wing-feather moult in Denmark. Danish Review of Game Biology 10:1–23.

Joensen, A. H. 1973. Moult migration and wing-feather moult of seaducks in Denmark. Danish Review of Game Biology 8:1–42.

Johnson, S. R. 1982. Continuing investigations of Oldsquaws (Clangula hyemalis L.) during the molt period in the Alaskan Beaufort Sea. Pp. 547–563 in Environmental assessment of Alaskan Continental Shelf, Final Report (vol. 23). NTSI PB 85–212595/As, Bureau of Land Management/National Oceanic and Atmospheric Administration, Outer Continental Shelf Environment Assessment, Juneau, AK.

Johnson, S. R. 1984. Prey selection by Oldsquaws in a Beaufort Sea lagoon, Alaska. Pp. 12–19 in D. N. Nettleship, J. A. Sanger, and P. F. Springer (editors), Marine birds: their feeding ecology and commercial fisheries relationships. Canadian Wildlife Service Special Publication for the Pacific Seabird Group, Dartmouth, NS.

Johnson, S. R. 1985. Adaptations of the Long-tailed Duck (Clangula hyemalis L.) during the period of molt in arctic Alaska. Proceedings International Ornithological Congress 18:530–540.

Johnson, S. R., and D. R. Herter. 1989. The birds of the Beaufort Sea. BP Exploration (Alaska) Inc., Anchorage, AK.

Johnson, S. R., and W. J. Richardson. 1981. Beaufort Sea barrier island-lagoon ecological process studies: final report, Simpson Lagoon. Pp. 109–383 in Environmental assessments. Alaskan Continental Shelf, Final Report of Principal Investigators (vol. 7). Bureau of Land Management and National Oceanic and Atmospheric Administration, Outer Continental Shelf Environmental Assessment Program, Boulder, CO.

Johnson, S. R., and W. J. Richardson. 1982. Waterbird migration near the Yukon and Alaskan Coast of the Beaufort Sea: II: molt migration of seaducks in summer. Arctic 35:291–301.

Kear, J. (editor). 2005. Ducks, geese and swans (vol. 2). Oxford University Press, Oxford, U.K.

Kellett, D. K. 1999. Causes and consequences of variation in nest success of King Eiders (Somateria spectabilis) at Karrak Lake, Northwest Territories. Thesis, University of Saskatchewan, Saskatoon, SK.

Kimball, R. T., and J. D. Ligon. 1999. Evolution of avian plumage dichromatism from a proximate perspective. American Naturalist 154:182–193.

King, J. G. 1963. Duck banding in arctic Alaska. Journal of Wildlife Management 27:356–362.

King, J. G. 1973. A cosmopolitan duck moulting resort; Takslesluk Lake Alaska. Wildfowl 24:103–109.

King, J. R. 1974. Seasonal allocation of time and energy resources in birds. Pp. 4–85 in R. A. Paynter Jr. (editor), Avian energetics. Nuttall Ornithological Club, Cambridge, MA.

King, J. R. 1981. Energetics of avian molt. Proceedings International Ornithological Congress 17:312–317.

Kjellen, N. 1994. Moult in relation to migration in birds—a review. Ornis Svecica 4:1–24.

Koehl, P. S., T. C. Rothe, and D. V. Derksen. 1984. Winter food habits of Barrow's Goldeneyes in southeast Alaska. Pp. 1–5 in D. N. Nettleship, G. A. Sanger, and P. F. Springer (editors), Marine birds: their feeding ecology and commercial fisheries relationships. Canadian Wildlife Service, Dartmouth, NS.

Lack, D. 1974. Evolution illustrated by waterfowl. Harper & Row, New York.

Lacroix, D. L., R. B. Lanctot, J. A. Reed, and T. L. McDonald. 2003. Effect of underwater seismic surveys on molting male Long-tailed Ducks in the Beaufort Sea, Alaska. Canadian Journal of Zoology 81:1862–1875.

Langlois, A. 2006. Écologie de la mue et de la migration automnale chez l'Arlequin Plongeur (Histrionicus histrionicus). Thesis, Laval University, Québec, QC (in French).

Leafloor, J. O., C. D. Ankney, and R. W. Risi. 1996. Social enhancement of wing moult in female Mallards. Canadian Journal of Zoology 74:1376–1378.

Little, B., and R. W. Furness. 1985. Long-distance moult migration by British Goosanders Mergus merganser. Ringing and Migration 6:77–82.

Lovvorn, J. R., S. E. Richman, J. M. Grebmeier, and L. W. Cooper. 2003. Diet and body condition of Spectacled Eiders wintering in pack ice of the Bering Sea. Polar Biology 26:259–267.

Madsen, F. J. 1954. On the food habits of the diving ducks in Denmark. Danish Review of Game Biology 2:157–266.

Mallory, M., and K. Metz. 1999. Common Merganser (Mergus merganser). in A. Poole and F. Gill (editors), The Birds of North America, No. 442. The Academy of Natural Sciences, Philadelphia, PA.

McGraw, K. J., J. Dale, and E. A. Mackillop. 2003. Social environment during molt and the expression of melanin-based plumage pigmentation in male House Sparrows (Passer domesticus). Behavioural Ecology and Sociobiology 53:116–122.

Merkel, F. R., S. E. Jamieson, K. Falk, and A. Mosbech. 2007. The diet of Common Eiders wintering in Nuuk, southwest Greenland. Polar Biology 30:227–234.

Metzner, K. A. 1993. Ecological strategies of wintering Steller's Eiders on Izembek Lagoon and Cold Bay, Alaska. Thesis, University of Missouri, Columbia, MO.

Moore, P. G. 2001. Concerning grey seals killing eider ducks in the Clyde Sea area. Journal of Marine Biological Association 81:1067–1068.

Mosbech, A. and D. Boertmann. 1999. Distribution, abundance and reaction to aerial surveys of post-breeding King Eiders (Somateria spectabilis) in western Greenland. Arctic 52:188–203.

Mosbech, A., R. S. Danø, F. Merkel, C. Stone, G. Gilchrist, and A. Flagstad. 2006a. Use of satellite telemetry to locate key habitats for King Eiders Somateria spectabilis in west Greenland. Pp. 769–776 in G. C. Boere, C. A. Galbraith and D. A. Stroud (editors), Waterbirds around the world. The Stationary Office, Edinburgh, U.K.

Mosbech, A., G. Gilchrist, F. Merkel, C. Sonne, A. Flagstand, and H. Nyegaard. 2006b. Year-round movements of northern Common Eiders Somateria mollissima borealis breeding in arctic Canada and west Greenland followed by satellite telemetry. Ardea 94:651–665.

Munro, J., and J. Bédard. 1977. Crèche formation in the Common Eider. Auk 94:759–771.

Murphy, M. E. 1996. Energetics and nutrition of moult. Pp. 158–198 in C. Carey (editor), Avian energetics and nutritional ecology. Plenum Press, New York, NY.

Murphy, M. E., and J. R. King. 1982. Amino acid composition of the plumage of the White-crowned Sparrow. Condor 84:435–438.

Nilsson, L. 1972. Habitat selection, food choice, and feeding habits of diving ducks in coastal waters of south Sweden during the non-breeding season. Ornis Scandinavica 3:55–78.

Noel, L. E., S. R. Johnson, and P. F. Wainwright. 1999. Aerial surveys of molting waterfowl in the barrier island-lagoon system between the Stockton Islands and Flaxman Island, Alaska, 1998. Final Report. BP Exploration (Alaska) Inc., Anchorage, AK.

O'Connor, M. 2008. Surf Scoter (Melanitta perspicillata) ecology on spring staging grounds and during the flightless period. Thesis, McGill University, Montreal, QC.

Oring, L. W. 1966. Breeding biology and molts of the Gadwall, Anas strepera Linnaeus. Dissertation, University of Oklahoma, Norman, OK.

Öst, M. 1999. Within-season and between-year variation in the structure of Common Eider broods. Condor 101:598–606.

Owen, M., and R. King. 1979. The duration of the flightless period in free-living Mallard. Bird Study 26:267–269.

Palmer, R. S. 1972. Patterns of molting. Pp. 65–102 in D. S. Farner, J. R. King, and K. C. Parkes (editors), Avian biology (vol. 2). Academic Press, New York, NY.

Palmer, R. S. 1976. Handbook of North American birds (vol. 3). Waterfowl (Part 2). Yale University Press, New Haven, CT.

Payne, R. B. 1972. Mechanisms and control of molt. Avian Biology 2:103–155.

Pearce, J. M., J. M. Eadie, J.-P. L. Savard, T. K. Christensen, J. Berdeen, E. Taylor, S. Boyd, A. Einarsson, and S. L. Talbot. 2014. Comparative population structure of cavity-nesting sea ducks. Auk 131:195–207.

Pearce, J. M. and S. L. Talbot. 2006. Demography, genetics, and the value of mixed messages. Condor 108:474–479.

Pearce, J. M., D. Zwiefelhofer, and N. Maryanski. 2009. Mechanisms of population heterogeneity among molting Common Mergansers on Kodiak Island, Alaska: implications for genetic assessments of migratory connectivity. Condor 111:283–293.

Pehrsson, O. 1987. Effects of body condition on molting in Mallards. Condor 89:329–339.

Petersen, M. R. 1980. Observations of wing-feather moult and summer feeding ecology of Steller's Eiders at Nelson Lagoon, Alaska. Wildfowl 31:99–106.

Petersen, M. R. 1981. Populations, feeding ecology and molt of Steller's Eiders. Condor 83:256–262.

Petersen, M. R., J. O. Bustnes, and G. H. Systad. 2006. Breeding and moulting locations and migration patterns of the Atlantic population of Steller's Eiders Polysticta stelleri as determined from satellite telemetry. Journal of Avian Biology 37:58–68.

Petersen, M. R., and D. C. Douglas. 2004. Winter ecology of Spectacled Eiders: environmental characteristics and population change. Condor 106:79–94.

Petersen, M. R., D. C. Douglas, H. M. Wilson, and S. E. McCloskey. 2012. Effects of sea ice on winter site fidelity of Pacific Common Eiders (Somateria mollissima v-nigrum). Auk 129:399–408.

Petersen, M. R., and P. L. Flint. 2002. Population structure of Pacific Common Eiders breeding in Alaska. Condor 104:780–787.

Petersen, M. R., J. B. Grand, and C. P. Dau. 2000. Spectacled Eider (Somateria fischeri). in A. Poole and F. Gill (editors), The Birds of North America, No. 547. The Academy of Natural Sciences, Philadelphia, PA.

Petersen, M. R., W. W. Larned, and D. C. Douglas. 1999. At-sea distribution of Spectacled Eiders: a 120-year-old mystery resolved. Auk 116:1009–1020.

Petersen, M. R., B. J. McCaffery, and P. L. Flint. 2003. Post-breeding distribution of Long-tailed Ducks Clangula hyemalis from the Yukon-Kuskokwim Delta, Alaska. Wildfowl 54:103–113.

Petersen, M. R., J. F. Piatt, and K. A. Trust. 1998. Foods of Spectacled Eiders Somateria fischeri in the Bering Sea, Alaska. Wildfowl 49:124–128.

Phillips, L. M., and A. N. Powell. 2006. Evidence for wing molt and breeding site fidelity in King Eiders. Waterbirds 29:148–153.

Phillips, L. M., A. N. Powell, and E. A. Rexstad. 2006. Large-scale movements and habitat characteristics of King Eiders throughout the nonbreeding period. Condor 108:887–900.

Pielou, E. C. 1991. After the ice age: the return of life to glaciated North America. University of Chicago Press, Chicago, IL.

Piersma, T. 1988. Breast muscle atrophy and constraints on foraging during the flightless period of wing moulting Great Crested Grebe. Ardea 76:96–106.

Piersma, T., and R. E. Gill Jr. 1998. Guts don't fly: small digestive organs in obese Bar-tailed Godwits. Auk 115:196–203.

Piersma, T., and Å. Lindström. 1997. Rapid reversible changes in organ size as a component of adaptive behaviour. Trends in Ecology and Evolution 12:134–138.

Ploeger, P. L. 1968. Geographical differentiation in arctic Anatidae as a result of isolation during the last glacial. Ardea 56:1–155.

Portenko, L. A. 1981. Birds of the Chukchi Peninsula and Wrangel Island (vol. 1). Amerind Publishing, New Delhi, India (Translated from Russian).

Portugal, S. J., R. Isaac, K. L. Quinton, and S. J. Reynolds. 2010. Do captive waterfowl alter their behaviour patterns during their flightless period of moult? Journal of Ornithology 151:443–448.

Powell, A. N., and R. S. Suydam. [online]. 2012. King Eider (Somateria spectabilis). in A. Poole (editor), The Birds of North American. Cornell Lab of Ornithology, Ithaca, NY. <http:/bna.birds.cornell.edu/bna/species/491> (10 February 2014).

Pöysä, H. 1992. Variation in parental care of Common Goldeneye (Bucephala clangula) females. Behaviour 123:247–260.

Pöysä, H., J. Virtanen, and M. Milonoff. 1997. Common Goldeneyes adjust maternal effort in relation to prior brood success and not current brood size. Behavioral Ecology and Sociobiology 40:101–106.

Pyle, P. 2005. Molts and plumages of ducks (Anatinae). Waterbirds 28:208–219.

Rail, J.-F., and J.-P. L. Savard. 2003. Identification des aires de mue et de repos au printemps des macreuses (Melanitta sp.) et de l'Eider à Duvet (Somateria mollissima) dans l'estuaire et le golfe du Saint-Laurent. Technical Report Series, No. 408. Canadian Wildlife Service, Quebec Region, QC (in French).

Reed, A., R. Benoit, R. Lalumière, and M. Julien. 1996. Duck use of the coastal habitats of northeastern James Bay. Canadian Wildlife Service, Occasional Paper, No. 90. Canadian Wildlife Service, Ottawa, ON.

Robert, M., R. Benoit, and J.-P. L. Savard. 2002. Relationship among breeding, molting, and wintering areas of male Barrow's Goldeneyes (Bucephala islandica) in eastern North America. Auk 119:676–684.

Robert, M., G. H. Mittelhauser, B. Jobin, G. Fitzgerald, and P. Lamothe. 2008. New insights on Harlequin Duck population structure in eastern North America as revealed by satellite telemetry. Waterbirds 31:159–172.

Robert, M., and J.-P. L. Savard. 2006. The St. Lawrence River Estuary and Gulf: a stronghold for Barrow's Goldeneyes wintering in eastern North America. Waterbirds 29:437–450.

Robertson, G. J., F. Cooke, R. I. Goudie, and W. S. Boyd. 1997. The timing of arrival and moult chronology of Harlequin Ducks Histrionicus histrionicus. Wildfowl 48:147–155.

Robertson, G. J., F. Cooke, R. I. Goudie, and W. S. Boyd. 1998. Moult speed predicts pairing success in male Harlequin Ducks. Animal Behaviour 55:1677–1684.

Robertson, G. J., F. Cooke, R. I. Goudie, and W. S. Boyd. 1999. Within-year fidelity of Harlequin Ducks to a moulting and wintering area. Pp. 45–51 in R. I. Goudie, M. R. Petersen, and G. J. Robertson (editors), Behaviour and ecology of sea ducks. Canadian Wildlife Service, Occasional Paper, No. 100. Canadian Wildlife Service, Ottawa, ON.

Robertson, G. J., and R. I. Goudie. 1999. Harlequin Duck (Histrionicus histrionicus). in A. Poole and F. Gill (editors), The Birds of North America, No. 547. The Academy of Natural Sciences, Philadelphia, PA.

Robertson, G. J., G. H. Mittelhauser, T. E. Chubbs, P. G. Trimper, R. I. Goudie, P. Thomas, S. Brodeur, M. Robert, S. G. Gilliland, and J.-P. L. Savard. 2008. Morphological variation among Harlequin Ducks in the Northwest Atlantic. Waterbirds 31:194–203.

Robertson, G. J., and J.-P. L. Savard. 2002. Long-tailed Duck (Clangula hyemalis). in A. Poole and F. Gill (editors), The Birds of North America, No. 651. The Academy of Natural Sciences, Philadelphia, PA.

Rodway, M. S. 2007. Timing of pairing in waterfowl II: testing the hypotheses with Harlequin Ducks. Waterbirds 30:506–520.

Rohwer, S., R. E. Ricklefs, V. G. Rohwer, and M. M. Copple. 2009. Allometry of the duration of flight feather molt in birds. PLoS Biology 7:e1000132.

Ross, R. K. 1983. An estimate of the Black Scoter, Melanitta nigra, population moulting in James and Hudson bays. Canadian Field-Naturalist 97:147–150.

Salomonsen, F. 1949. Some notes on the moult of the Long-tailed Duck (Clangula hyemalis). Avicultural Magazine 55:59–62.

Salomonsen, F. 1968. The moult migration. Wildfowl 19:5–24.

Sargeant, A. B., and D. G. Raveling. 1992. Mortality during the breeding season. Pp. 396–422 in B. D. J. Batt, A. D. Afton, M. G. Anderson, C. D. Ankney, D. H. Johnson, J. A. Kadlec, and G. L. Krapu (editors), Ecology and management of breeding waterfowl. University of Minnesota Press, Minneapolis, MN.

Savard, J.-P. L. 1987. Causes and functions of brood amalgamation in Barrow's Goldeneye and Bufflehead. Canadian Journal of Zoology 65:1548–1553.

Savard, J.-P. L. 1988. A summary of current knowledge on the distribution and abundance of moulting sea-ducks in the coastal waters of British Columbia. Technical Report Series, No. 45. Canadian Wildlife Service, Pacific and Yukon Region, BC.

Savard, J.-P. L., J. Bédard, and A. Nadeau. 1999. Spring and early summer distribution of scoters and eiders in the St. Lawrence River Estuary. Pp. 59–64 in R. I. Goudie, M. R. Petersen, and G. J. Robertson (editors), Behaviour and ecology of sea ducks. Occasional Paper, No. 100. Canadian Wildlife Service, Ottawa, ON.

Savard, J.-P. L., D. Bordage, and A. Reed. 1998. Surf Scoter (Melanitta perspicillata). in A. Poole and F. Gill (editors), The Birds of North America, No. 363. The Academy of Natural Sciences, Philadelphia, PA.

Savard, J.-P. L., L. Lesage, S. G. Gilliland, G. Gilchrist, and J.-F. Giroux. 2011. Molting, staging and wintering locations of Common Eiders breeding in the Gyrfalcon Archipelago, Ungava Bay. Arctic 64:197–206.

Savard, J.-P. L., A. Reed, and L. Lesage. 2007. Chronology of breeding and molt migration in Surf Scoters (Melanitta perspicillata). Waterbirds 30:223–229.

Savard, J.-P. L., and M. Robert. 2013. Relationships among breeding, molting and wintering areas of adult female Barrow's Goldeneyes (Bucephala islandica) in eastern North America. Waterbirds 36:34–42.

Schenkeveld, L. E., and R. C. Ydenberg. 1985. Synchronous diving by Surf Scoter flocks. Canadian Journal of Zoology 63:2516–2519.

Sjöberg, K. 1988. The flightless period of free-living male Teal Anas crecca in northern Sweden. Ibis 130:164–171.

Smith, C., F. Cooke, and R. I. Goudie. 1998. Ageing Harlequin Duck Histrionicus histrionicus drakes using plumage characteristics. Wildfowl 49:245–248.

Smith, W. E. 2006. Moulting Common Eiders devoured by killer whales. British Birds 99:264.

Soloviea, D. 1997. Timing, habitat use and breeding biology of Steller's Eider in the Lena Delta, Russia. Pp. 35–39 in Proceeding from Steller's Eider Workshop. Seaduck Specialists Group Bulletin, No. 7. Wetlands International, Ede, the Netherlands.

Stresemann, V. 1948. Eclipse plumage and nuptial plumage in the Old Squaw, or Long-tailed Duck (*Clangula hyemalis*). Avicultural Magazine 54:188–194.

Suydam, R., L. Quakenbush, M. Johnson, J. C. George, and J. Young. 1997. Migration of King and Common Eiders past Point Barrow, Alaska, in spring 1987, spring 1994, and fall 1994. in D. L. Dickson (editor), King and Common Eiders of the western Canadian Arctic. Occasional Paper, No. 94. Canadian Wildlife Service, Ottawa, ON.

Suydam, R. S. 2000. King Eider (*Somateria spectabilis*). in A. Poole, and F. Gill (editors), The Birds of North America, No. 491. The Academy of Natural Sciences, Philadelphia, PA.

Suydam, R. S., L. T. Quakenbush, D. L. Dickson, and T. Obritschkewitsch. 2000. Migration of King, *Somateria spectabilis*, and Common, *S. mollissima v-nigra*, eiders past Point Barrow, Alaska, during spring and summer/fall 1996. Canadian Field-Naturalist 114:444–452.

Taylor, E. J. 1986. Foods and foraging ecology of Oldsquaws (*Clangula hyemalis* L.) on the Arctic Coastal Plain of Alaska. Thesis, University of Alaska, Fairbanks, AK.

Thomas, P. W., G. H. Mittelhauser, T. E. Chubbs, P. G. Trimper, R. I. Goudie, G. J. Robertson, S. Brodeur, M. Robert, S. G. Gilliland, and J.-P. L. Savard. 2008. Movements of Harlequin Ducks in eastern North America. Waterbirds 31:188–193.

Thompson, C. W. 2004. Determining evolutionary homologies of molts and plumages: a commentary on Howell et al. (2003). Condor 106:199–206.

Thompson, D. Q., and R. A. Pearson. 1963. The eider pass at Point Barrow, Alaska. Journal of Wildlife Management 27:348–356.

Thompson, J. E., and R. D. Drobney. 1996. Nutritional implications of molt in male Canvasbacks: variation in nutrient reserves and digestive tract morphology. Condor 98:512–526.

Tickell, W. L. N. 2003. White plumage. Waterbirds 26:1–12.

Timson, R. S. 1976. Late summer migration at Barrow, Alaska. Pp. 364–399 in Environmental assessment of the Alaskan Continental Shelf, Principal Investigators' reports, April–June 1976 (vol. 1). National Oceanic and Atmospheric Administration, Boulder, CO.

Titman, R. D. 1999. Red-breasted Merganser (*Mergus serrator*). in A. Poole and F. Gill (editors), The Birds of North America, No. 443. The Academy of Natural Sciences, Philadelphia, PA.

Turcek, F. J. 1966. On plumage quality in birds. Ekologia Polska, Series A 14:617–633.

van de Wetering, D. 1997. Moult characteristics and habitat selection of postbreeding male Barrow's Goldeneye (*Bucephala islandica*) in northern Yukon. Technical Report Series, No. 296. Canadian Wildlife Service, Pacific and Yukon Region, BC.

van de Wetering, D. and F. Cooke. 2000. Body weight and feather growth of male Barrow's Goldeneye during wing molt. Condor 102:228–231.

Voitkevich, A. A. 1966. The feathers and plumage of birds. October House, New York, NY.

Walsberg, G. E., G. S. Campbell, and J. R. King. 1978. Animal coat color and radiative heat gain: a re-evaluation. Journal of Comparative Physiology B 126:211–222.

Welty, J. C. 1962. The life of birds. W. B. Saunders, Philadelphia, PA.

Willoughby, E. J. 2004. Molt and plumage terminology of Howell et al. (2003) still may not reflect homologies. Condor 106:191–196.

Young, D. A., and D. A. Boag. 1982. Changes in the physical condition of male Mallards (*Anas platyrhynchos*) during moult. Canadian Journal of Zoology 60:3220–3226.

CHAPTER TEN

Site Fidelity, Breeding Habitats, and the Reproductive Strategies of Sea Ducks*

Mark L. Mallory

Abstract. Reproductive strategies describe how species allocate resources to produce offspring and include time and energy invested in nest site choice, egg production, incubation, nest defense, and chick rearing. Sea ducks exhibit considerable variation in all of these parameters but typically have high annual survival, high natal and breeding site fidelity, and variable annual reproductive success compared to other groups of waterfowl. Collectively, these traits make sea ducks susceptible to anthropogenic threats including habitat change, and many populations are in decline.

Key Words: breeding propensity, brood rearing, clutch size, egg size, incubation, nest site selection, nest success, parental care, philopatry, reproductive allocation, site fidelity, survival.

The production of offspring is both a gamble and a balancing act for any bird. The gamble comes in that breeding individuals place themselves and their survival at higher risk than nonbreeding individuals, because they advertise their presence with courtship behavior, reduce their effective mobility by carrying eggs, and stay in one location for an extended period during nesting, all activities that increase the chance of predation. A balancing act results if the gamble undertaken can be reduced or increased depending on adjustments made by the breeding organism to expose itself to this risk. In the case of breeding birds, the management of risk can include choice of nesting location; adjustments to clutch size, egg size, or incubation constancy; defense of the nest; and time spent rearing young. This suite of traits describes the reproductive strategy of the organism.

Reproductive strategies are fascinating to study, because they are an integral component of the life-history strategy of an organism, which describes the anatomical, physiological, and behavioral adaptations that control how organisms invest in reproduction and self-maintenance to maximize lifetime reproductive success in response to environmental conditions (Williams 1966). The reproductive strategy component describes the manner in which a species allocates time and energy resources to produce viable offspring. Allocation can be divided into the energy to produce those offspring as well as parental care. Across species, strategies represent a spectrum,

* Mallory, M. L. 2015. Site fidelity, breeding habitats, and the reproductive strategies of sea ducks. Pp. 337–364 in J.-P. L. Savard, D. V. Derksen, D. Esler, and J. M. Eadie (editors). Ecology and conservation of North American sea ducks. Studies in Avian Biology (no. 46), CRC Press, Boca Raton, FL.

from organisms that produce many offspring with little or no parental care (often referred to as an r-strategy) to those that produce a few offspring requiring much parental care (K-strategy). Sea ducks are generally considered to be on the K-strategy side of this spectrum (Chapter 11, this volume).

Collectively, the waterfowl tribe Mergini (sea ducks) exhibit high variation in their reproductive strategies, both within and among species, making them an ideal group for study. Nesting locations are diverse, and some sea ducks produce among the largest of waterfowl clutches, while others are typically small (Bellrose 1980). Body size and lifespan also vary greatly, with the smallest species living shorter lives similar to dabbling and diving ducks, whereas larger species such as eiders often live and breed for nearly a decade and may live for >20 years (Goudie et al. 2000). Lifespan has important implications on lifetime reproductive success, with longer-lived females producing more offspring than shorter-lived ones (goldeneyes; Dow and Fredga 1984). This chapter focuses on the reproductive strategies of 15 sea duck species found in North America, which represent most of tribe Mergini with the exception of Smew (*Mergellus albellus*). Alisauskas and DeVink (Chapter 5, this volume) consider reproductive energetics of sea ducks such as income versus capital breeding strategies, whereas Eadie and Savard (Chapter 11, this volume) review pre- and posthatch brood amalgamation and within-species variation in parental care (Chapter 11, this volume). Here, my focus is on life-history variation in nesting microhabitats, reproductive effort, and reproductive success.

SEA DUCK REPRODUCTION COMPARED TO OTHER WATERFOWL

There are 34 species of ducks that commonly nest in North America (Bellrose 1980) and another 10 species of geese and swans. Of this waterfowl diversity, 15 species (34%; 44% of the ducks) are sea ducks. Much of the variation in reproductive strategies found among waterfowl in family Anatidae can also be found among sea ducks in tribe Mergini.

Sea ducks commonly breed either in forested areas or tundra (Table 10.1) and winter along marine coastlines or large, inland lakes and reservoirs (Chapter 13, this volume). The group is uncommon or absent as breeders from the grassland, prairie, and agricultural regions of North America, which are dominated by dabbling ducks (tribe Anatini) and diving ducks (tribe Aythyini). In contrast, goldeneyes and mergansers are ubiquitous in lakes, ponds, and rivers of forested regions of North America, particularly the boreal forest, whereas eiders and Long-tailed Ducks are commonly found along Arctic and northern coastlines. As a group, sea ducks breed across ~55° of latitude, from Chihuahua, Mexico (Common Merganser, *Mergus merganser*; Mallory and Metz 1999), to the northern tip of Ellesmere Island (King Eider, *Somateria spectabilis*, Suydam 2000; Long-tailed Duck, *Clangula hyemalis*, Robertson and Savard 2002). Sea ducks breed 10° farther north than any of the diving or dabbling ducks—farthest north are Northern Pintails (*Anas acuta*) at ~73°N on Banks Island. The northern breeding distribution of sea ducks, particularly into the tundra, is more similar to geese than most other ducks. Breeding sea ducks are distributed in the westernmost sites in coastal Alaska to easternmost sites in coastal Newfoundland across North America (Common Goldeneye, *Bucephala clangula*, Eadie et al. 1995; Red-breasted Merganser, *Mergus serrator*, Titman 1999; Common Merganser, Mallory and Metz 1999; Common Eider, *Somateria mollissima*, Goudie et al. 2000), comparable to the broad distribution of commonly known species like Mallard (*Anas platyrhynchos*; Drilling et al. 2002), Green-winged Teal (*A. crecca*; Johnson 1995), and Canada Goose (*Branta canadensis*, Mowbray et al. 2002).

Compared with other major waterfowl tribes, the Mergini have the highest proportion of cavity-nesting species (Figure 10.1a). At least half of the sea duck species exhibit nest parasitism (Table 10.1; Mallory and Weatherhead 1990, Birds of North America [BNA] accounts), a higher proportion than geese and dabbling ducks but lower than rates found among diving ducks and stifftails (Figure 10.1b; Chapter 11, this volume).

Despite broad distributions and often large populations, we know comparatively less about the ecology of sea ducks than geese, dabbling ducks, or diving ducks. It was only 1999, when scientists discovered where Spectacled Eiders (*Somateria fischeri*) spend the winter! Since the 1990s, much new research has been undertaken to understand aspects of the breeding biology that influence the reproductive strategies of sea ducks (Krementz et al. 1997; Flint et al. 1998, 2000; Kellett and

TABLE 10.1
Mean egg characteristics, nest characteristics, and nesting behavior of sea ducks.

Scientific name	Egg characteristics			Nest characteristics					
	Mass (g)	Shape index (Length/Breadth)	Shell thickness (mm)	Habitat	Location	Coloniality	Concealment	Substrate	Behavior
Polysticta stelleri	58	1.475	0.31	Tundra	Ground	Single	Exposed	T	NP
Somateria mollissima	108	1.502	0.35	Tundra	Ground	Colonial	Exposed	T/R	P
Somateria spectabilis	73	1.506	0.30	Tundra	Ground	Single	Exposed	T/R	NP[b]
Somateria fischeri	73	1.481	0.32	Tundra	Ground	Single	Exposed	T/G	NP
Histrionicus histrionicus	53	1.396	0.25	Forest/Tundra	Ground	Single	Concealed	S	NP[b]
Clangula hyemalis	43	1.398	0.26	Tundra	Ground	Single[a]	Exposed	G/T	NP[b]
Melanitta nigra	72	1.467	0.32	Tundra	Ground	Single	Concealed	G	NP[b]
Melanitta perspicillata	62	1.437	0.35	Tundra	Ground	Single	Concealed	S/G	NP
Melanitta fusca	77	1.432	0.38	Tundra	Ground	Single[a]	Concealed	S/G	NP[b]
Bucephala albeola	37	1.399	0.36	Forest	Cavity	Single	Concealed		P
Bucephala islandica	70	1.375	0.43	Forest	Cavity	Single	Concealed		P
Bucephala clangula	57	1.390	0.39	Forest	Cavity	Single	Concealed		P
Lophodytes cucullatus	60	1.210	0.64	Forest	Cavity	Single	Concealed		P
Mergus serrator	72	1.439	0.34	Forest	Ground	Single[a]	Concealed	S/G	P
Mergus merganser	70	1.433	0.35	Forest	Cavity	Single	Concealed		P

NOTES: Summary from Mallory and Weatherhead (1990) and BNA accounts. Note that Yom-Tov (2001) identifies several other species as parasitic in that they may lay their eggs parasitically, but for the purposes of this review, and based on a review of the literature and BNA accounts, I considered species as parasitic where this tactic was a common or regular feature of their reproductive strategy.

T, tundra; R, rock; G, grass; S, shrub; NP, non-parasitic; P, parasitic.

[a] Locally high densities found when nesting on islands or in association with colonial seabirds.

[b] Yom-Tov (2001) identifies this species as parasitic.

Alisauskas 2000; Pearce et al. 2005, 2008). Despite recent efforts, we still know little about reproduction among some species in North America such as Black Scoters (*Melanitta nigra*; Bordage and Savard 1995), including species that are common and widely distributed such as Common Mergansers (Mallory and Metz 1999).

BODY SIZE, CLUTCH SIZE, AND EGG SIZE

Among North American waterfowl, female body mass overlaps among the waterfowl tribes Anserini (geese), Anatini (dabbling ducks), Aythyini (diving ducks), and Mergini (sea ducks; Table 10.2 and Figure 10.2). Using the phylogeny of Rohwer

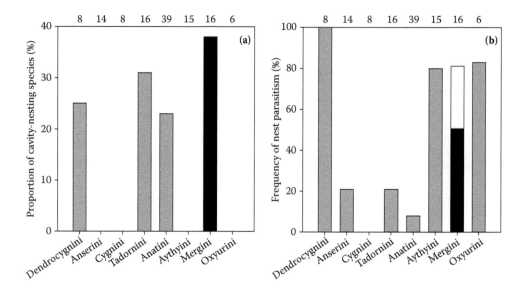

Figure 10.1. Propensity for cavity nesting and nest parasitism, shown as (a) proportion of cavity-nesting species within the main waterfowl tribes globally and (b) frequency of species within each tribe that exhibit nest parasitism (numbers across top are number of species in each tribe). Data from Mallory and Weatherhead (1990), and adjusted from BNA accounts. Black bar highlights data from the tribe Mergini. White bar above black bar in panel (b) reports proportion of sea duck species that are considered parasitic according to Yom-Tov (2001).

(1988), the Cygnini (swans) are all heavier than the sea ducks. Excluding the swans, female body mass differs significantly among the four other tribes (ANOVA on log-transformed female masses; $F_{3,33} = 11.2$, $P < 0.001$; phylogeny not controlled), with geese heavier than the ducks (Tukey–Kramer multiple comparison test, $P < 0.05$). However, there are no significant differences in mean female body mass among the different tribes of ducks (Figure 10.2; $P > 0.05$). Nonetheless, the sea ducks are noteworthy among North American ducks because the group includes the largest duck in the Northern Hemisphere, the Common Eider (*Somateria mollissima*), which lays the largest eggs of any species in this region (>100 g; Goudie et al. 2000). Tribe Mergini also includes the second smallest species of North American waterfowl, the Bufflehead (*Bucephala albeola*; ~300 g), only 6% heavier than the Green-winged Teal. Among cavity-nesting waterfowl in North America, the largest, nontropical species that nests in tree holes is a sea duck (Common Merganser, *M. merganser*; ~1200 g), which is at least 50% heavier than Wood Ducks (*Aix sponsa*), goldeneyes, or whistling ducks (*Dendrocygna* spp.; Rohwer 1988). Buffleheads are the smallest cavity-nesting duck in North America (Table 10.2).

Across waterfowl species, heavier species lay heavier eggs (Rohwer 1988), and the sea ducks conform to this general pattern (updated here with more recent data, Figure 10.3a; $F_{1,13} = 29.5$, $P < 0.001$, $r_{15} = 0.83$). However, sea duck species that are large-bodied females do not lay larger clutches (Figure 10.3b; $r_{s15} = -0.3$, $P = 0.27$), even if eiders are excluded ($r_{s12} = 0.37$, $P = 0.24$). Eiders are unusual as extreme capital breeders that must balance use of endogenous reserves to produce eggs against retaining sufficient endogenous reserves maintain high incubation constancy and completing incubation (Sénéchal et al. 2011). Overall, total clutch mass is greater for sea duck species with heavier body mass (excluding the eiders; Figure 10.3c; $r_{s12} = 0.82$, $P = 0.0011$), but clutch mass declines as a proportion of female body mass (Figure 10.3d; including eiders with $r_{s15} = -0.75$, $P = 0.0013$; excluding eiders with $r_{s12} = -0.53$, $P = 0.08$).

Egg morphology of sea ducks follows patterns found across all waterfowl, in that large-bodied species of waterfowl tend to have more oval eggs with thicker eggshells, even after controlling for phylogeny (Mallory and Weatherhead 1990). Eggs with thicker shells tend to be stronger, a necessity to prevent crushing when heavier females

TABLE 10.2

Body mass, lifespan, and nesting parameters of sea ducks.

Species	Body mass (g)		Breeding parameters							
Scientific name	Female	Male	Clutch size	Incubation constancy (%)	Nesting success (%)	Duckling survival (%)	Annual survival (%)	Mean lifespan (years)	Maximum longevity (years)[b]	Sources
Polysticta stelleri	852	887	5.4	—	15	—	90	—	23.0	1–3
Somateria mollissima	1760[a]	1870[a]	3.5	99.5	44	14	88.5	7.4	22.6	4–13
Somateria spectabilis	1717	1836	4.6	98	40	10	94	—	18.9	14–20
Somateria fischeri	1623	1494	4.7	90	48	49	78	>4	>18.0	21,22
Histrionicus histrionicus	570	642	5.8	—	74	44	70	—	18.2	23–25
Clangula hyemalis	618	800	7.1	89	33	10	74	3.1	15.6	26–28
Melanitta nigra	987	1117	7.5	—	13	22	77	—	18.1	29–31
Melanitta perspicillata	985	1059	7.6	—	70	40	—	—	9.5	32
Melanitta fusca	1450	1524	8.8	84	71	2	78	3.9	10.3	33–36
Bucephala albeola	337	465	7.8	89	79	45	63	3.3	18.7	37
Bucephala islandica	835	1160	6.5	—	59	56	60	3.4	15.3	38,39
Bucephala clangula	710	1120	7.1	82	64	48	67	3.7	20.4	40–44
Lophodytes cucullatus	554	680	8.9	85	73	—	72	—	14.5	45,46
Mergus serrator	998	1134	9.5	91	57	35	—	—	9.1	47–49
Mergus merganser	1185	1589	8.5	88	94	70	49	>2.3	13.4	50–52

NOTES: Where possible, body mass represents females measured after spring migration or early in the breeding season. Average annual survival based on estimates for females. Nesting success, duckling survival, and annual survival are averages from multiple studies, or multiple years of study, and are intended to provide context relative to other species. Because of variation in the methodology of gathering these data among studies, I provide only the average without measures of variation (available from the primary literature).

(1) Fredrickson (2001), (2) Flint et al. (2000), (3) Quackenbush et al. (2004), (4) Bellrose (1980), (5) Goudie et al. (2000), (6) Déscamps et al. (2009), (7) Kats (2007), (8) Wilson et al. (2007), (9) Hario et al. (2009), (10) Hoover et al. (2010), (11) Mendenhall and Milne (1985), (12) Bolduc and Guillemette (2003), (13) Flint et al. (1998), (14) Suydam (2000), (15) Kellett and Alisauskas (1997), (16) Oppel and Powell (2010), (17) Bentzen et al. (2008), (18) Mehl and Alisauskas (2007), (19) Kellett et al. (2003), (20) Bentzen et al. (2010), (21) Petersen et al. (2000), (22) Flint et al. (2006), (23) Robertson and Goudie (1999), (24) Robertson (2008), (25) Mittelhauser (2008), (26) Kellett et al. (2005), (27) Robertson and Savard (2002), (28) Schamber et al. (2009), (29) Bordage and Savard (1995), (30) Fox et al. (2003), (31) Schamber et al. (2010), (32) Savard et al. (1998), (33) Brown and Fredrickson (1997), (34) Krementz et al. (1997), (35) Traylor et al. (2004), (36) Traylor (2003), (37) Gauthier (1993), (38) Eadie et al. (2000), (39) Boyd et al. (2009), (40) Eadie et al. (1995), (41) Ludwichowski et al. (2002), (42) Paasivaara and Pöysä (2007), (43) Wayland and McNicol (1994), (44) Schmidt et al. (2006), (45) Dugger et al. (1994), (46) Pearce et al. (2008), (47) Bengtson (1971), (48) Bengtson (1972), (49) Craik and Titman (2009), (50) Mallory and Metz (1999), (51) Erskine (1971), (52) Pearce et al. (2005).

[a] borealis race.

[b] Longevity records from Lutmerding and Love (2014).

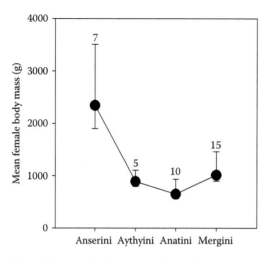

Figure 10.2. Mean female body mass (±SE) of four waterfowl tribes for which masses are known to overlap. (Data from Rohwer 1988.)

incubate eggs. However, greater strength is also conferred when eggs are rounder (Picman 1989). Within the sea ducks, the cavity-nesting species such as goldeneyes and mergansers have significantly rounder and thicker shells for their egg size than expected, possibly because nests in cavities may be more irregularly shaped, and thus eggs may get jostled more when females enter and exit the nest, thereby requiring greater eggshell strength to prevent egg damage. Interestingly, the Hooded Merganser (*Lophodytes cucullatus*) has the roundest egg and proportionally thickest eggshell of any living species of waterfowl, two characteristics that are typical of obligate brood parasites (Picman 1989). To date, no one has explained why this species produces eggs with such unusual characteristics, but production or emergence from such eggs does not appear to confer costs in terms of reduced clutch size, hatching success, or chick survival (Mallory and Weatherhead 1990).

In their summary of egg sizes and laying patterns, Alisauskas and Ankney (1992) noted that all sea ducks have an interval of >24 h between laying of consecutive eggs, which is the longest laying interval among the waterfowl except for the swans. Recent technological advances in nest monitoring suggest that egg-laying intervals may be shorter than previously thought (e.g., 21–25 h; Quackenbush et al. 2004), although this may reflect nest parasitism as host females appear to lay once every two days (Åhlund 2005). Alisauskas and Ankney (1992) also noted that many Mergini have variable laying intervals, which may shorten

as the clutch increases or the breeding season progresses (Morse et al. 1969, Erskine 1972, Gauthier 1993).

It is well established across bird species that there is an adaptive seasonal decline in clutch size, which may be related to condition-dependent reproductive effort of females or to lower reproductive value of later-hatched offspring due to reduced survival and recruitment (Klomp 1970, Drent and Daan 1980, Rowe et al. 1994). Sea ducks follow this avian trend. Female sea ducks lay smaller clutches if their nest is initiated relatively later in the season, because females have smaller endogenous reserves to commit to late clutches, or offspring from later clutches have lower reproductive value (Alisauskas and Ankney 1992). Recent research has suggested that nest parasitism may be more prevalent and undetected than previously recognized, even for host females because some females may only nest parasitically in a given year, whereas others may lay eggs parasitically before producing their own clutch (Åhlund and Andersson 2001). Consequently, the relationship between declining clutch size and increasing laying date requires more investigation.

Due to relatively slow egg-laying rates, long incubation periods, often high requirements for endogenous nutrient reserves, and relatively short breeding seasons, sea ducks seldom renest unless a clutch is lost early in laying (Afton and Paulus 1992, Alisauskas and Ankney 1992). Sea duck populations also exhibit a high proportion of nonbreeding individuals. Coulson (1984) showed that up to 65% of Common Eiders may skip breeding in a given year, a pattern also observed in Harlequin Ducks (Robertson and Goudie 1999), Barrow's Goldeneyes (Eadie et al. 2000), and Red-breasted Mergansers (Bengtson 1972). Åhlund and Andersson (2001) and Milonoff et al. (2004) also point out that apparent years of skipped breeding may be undetected parasitic laying. For example, a third of goldeneye females may lay only parasitically within a given season (Åhlund and Andersson 2001). A myriad of factors may influence a female's decision not to breed in a given year, including high predation pressure, poor climatic conditions, poor body condition due to harsh wintering conditions, or carryover effects from the previous year's breeding (Chapters 3, 5, 7, and 11, this volume).

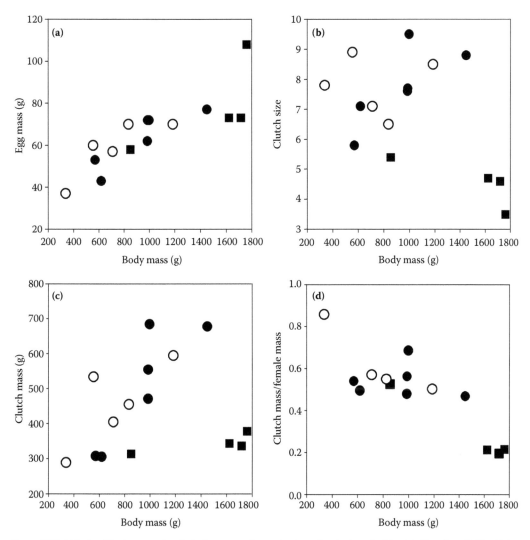

Figure 10.3. Relationships between female body mass and reproductive effort in the tribe Mergini (Data from Tables 10.1 and 10.2): (a) egg mass increases with female body mass; (b) clutch size shows no clear pattern in relation to body mass; (c) clutch mass generally increases in relation to body mass, when the eiders are excluded; and (d) clutch mass as a proportion of female mass decreases with increasing female mass. Eiders are depicted as filled squares, cavity-nesting sea ducks as unfilled circles, and the remaining ground-nesting sea ducks as filled circles.

ANNUAL SURVIVAL AND BREEDING SITE FIDELITY

Life-history theory predicts that in the balancing act of trying to maximize lifetime reproductive success, long-lived individuals should produce smaller clutches and have relatively lower annual reproductive success than short-lived individuals because lifetime reproductive success is enhanced principally through high adult survival, which allows many reproductive attempts (Williams 1966, Ricklefs 1977). Among sea ducks, we find a substantial range in reproductive metrics (Table 10.2), but which

generally conform to predictions of life-history theory. I compiled estimates of survival (sexes pooled) from the literature (Table 10.2) and from species accounts for dabbling ducks, diving ducks, and geese in the BNA series (hereafter, BNA accounts). Mean annual adult survival differs among tribes ($F_{3,29}$ = 6.5, P = 0.002). Sea ducks have an average annual survival of 73% ± 12% (SD; n = 13 species), similar to North American geese (77% ± 10%, n = 6) or diving ducks (66% ± 11%, n = 5; Tukey–Kramer tests, P > 0.05) but higher than dabbling ducks (55% ± 9%, n = 9; Tukey–Kramer test, P < 0.05). King and Common Eiders have the highest annual survival of any species of waterfowl in North America. The estimates of annual survival include mortality losses to hunting, which can be substantial for some populations of waterfowl (Chapter 12, this volume). Fewer sea ducks tend to be harvested annually (550,000 birds) than geese (3,200,000), dabbling ducks (11,500,000), or diving ducks (900,000; U.S. Fish and Wildlife Service 2014).

Breeding site fidelity is the tendency for an individual to return to the same breeding site annually and may be defined as return to the same nest site, island, or lake (Greenwood 1982, Clobert et al. 2001, Öst et al. 2011). Many species with high annual survival also exhibit high breeding site fidelity such as seabirds (>90%, Hamer et al. 2002). High fidelity is considered advantageous in relatively stable habitats as individuals may use prior experience with their breeding habitat or neighbors to improve their breeding success or chance of survival (Öst et al. 2011). In contrast, if environments suddenly change, such as the proliferation of a new predator or a rapid degradation of breeding sites, breeding site fidelity may prove an ecological trap as birds lack the behavioral plasticity to seek new breeding areas (Ekroos et al. 2012).

In addition to breeding site fidelity, some of sea ducks exhibit female-biased natal philopatry, including Common Eiders and the highly territorial species of *Bucephala* spp. In these species, more female ducklings return to their natal area than males, potentially resulting in competition between mothers and daughters for resources (Dow and Fredga 1983, Savard and Eadie 1989, Gauthier 1990, Pöysä et al. 1997a,b, McKinnon et al. 2006, Waldeck et al. 2008). Indeed, the advantages of inheriting maternal quality appear sufficiently high in goldeneyes that females in better condition produce female-biased broods that have higher survival than broods from smaller females (Jaatinen et al. 2012). Consequently, breeding site fidelity can have a substantial influence on phylogeography and population dynamics and demography of sea ducks (Chapters 2 and 3, this volume).

Among waterfowl, sea ducks have the highest mean female breeding site fidelity (data from table 11.1 in Anderson et al. 1992; $F_{2,45} = 13.0$, $P < 0.0001$). The average rate of breeding site fidelity among female sea ducks (66% $\pm$ 18%, n = 12 studies) was higher than either dabbling ducks (32% $\pm$ 15%, n = 26) or diving ducks (42% $\pm$ 26%, n = 10; Tukey–Kramer tests, $P < 0.05$).

NEST SITE SELECTION BY SEA DUCKS

The choice of where a bird builds a nest reflects a trade-off. On one hand, a bird needs to choose a site that minimizes the risk of predation for the eggs, chicks, and incubating parent but also keeps the eggs warm and sheltered from adverse weather. Simultaneously, the nest site must be accessible and be sufficiently close to a food source that the adult can replenish body reserves during incubation recesses without leaving the nest exposed for too long or allowing the eggs to cool (Welty and Baptista 1990). In the case of waterfowl, the nest site must also be situated close enough to water that chicks can reach brood-rearing sites while minimizing predation risk of overland travel (Eriksson 1979). As a widely distributed group of birds, it is unsurprising that sea ducks use a diverse array of nesting habitats, which tend to covary with other aspects of their reproductive strategy (Table 10.1).

NESTING MICROHABITAT

The nesting microhabitats of sea ducks vary within and among species, but some generalizations can be applied (Table 10.1). About half of the species nest in forested areas, and the other half in tundra or taiga habitats (Chapter 13, this volume). All tundra-nesting species make their nests on the ground, whereas five of the seven species found in forested habitats are cavity nesters, provided the area supports trees large enough for nesting cavities (Vaillancourt et al. 2009). Only one species, the Common Eider, is truly colonial (Goudie et al. 2000), although high nesting densities similar to colonies can be found in some island-nesting populations of Long-tailed Ducks (Robertson and Savard 2002), White-winged Scoters (Brown and Fredrickson 1997), and Red-breasted Mergansers (Titman 1999).

Eiders and Long-tailed Ducks have their nests in exposed locations on the tundra. Females employ a strategy of remaining motionless on the nest and relying on their cryptic coloration to minimize the risk of predation. Nests are typically placed among moss and grassy vegetation close to freshwater ponds or the marine coastline. In contrast, the scoters, Harlequin Duck, goldeneyes, mergansers, and non-Arctic nesting eiders select sites to conceal their nests from predators, including sites under shrubs (Schamber et al. 2010), among tall graminoid vegetation (Traylor et al. 2004, Craik and Titman 2009), in cavities among boulders

(Robertson and Goudie 1999), or in trees (Eadie and Gauthier 1985, Evans et al. 2002, Robert et al. 2010). Females of these species lack the mottled, cryptic coloration of the eiders.

Nesting microhabitat will vary for some species across their range, depending on the dominant land cover. Along southern Hudson Bay, Canada, Long-tailed Ducks generally nest near or under shrub cover (Alison 1975), whereas in Alaska they nest in somewhat more exposed locations along pond edges or in available shrub cover at inland sites (Schamber et al. 2009). In the High Arctic, Long-tailed Ducks nest in completely exposed moss pockets (M. L. Mallory, unpubl. data). Harlequin Ducks will nest under shrubs along forested streams in Western Canada but will use rock cavities in treeless regions of Eastern Canada and Iceland (Robertson and Goudie 1999). Even the colonial Common Eider exhibits variation in nest microhabitat across its range, from exposed nests on gravel in the High Arctic to forested islands in the Gulf of St. Lawrence (Goudie et al. 2000, Bolduc et al. 2005, Öst and Steele 2010). Thus, sea ducks exhibit some flexibility in nest site choice, depending on the type of habitat near brood-rearing ponds.

DISTANCE FROM NESTS TO NEAREST WATER

An important factor influencing where sea ducks can nest is the proximity of suitable nesting sites from brood-rearing ponds, lakes, or marine shorelines. Offspring survival may be affected by distance between nesting and brood-rearing habitat or the structure of navigable corridors between habitat patches, with reduced survival associated with longer distances traveled in unsuitable habitat (Eriksson 1979, Wayland and McNicol 1994, Pöysä and Paasivaara 2006).

Most ground-nesting sea ducks nest close to the water's edge. Distances are usually <30 m, and the vast majority of species nest within 200 m of the shore (BNA accounts). Mean distance (±SE) from the nest to water for nine species of ground-nesting sea ducks was 12 ± 3 m. White-winged Scoters typically nest the farthest from water among the ground-nesting sea ducks, with an average nest-to-water distance of 96 m and a maximum distance of 800 m (Brown and Fredrickson 1997). Nesting on islands reduces the risk from mammalian predators and allows females to nest farther from shore than sites on the mainland. In Arctic Russia, Spectacled Eiders nesting on the mainland had their nests

0.84 ± 0.09 m from the water, but on islands, they nested 75.4 ± 8.9 m away (Pearce et al. 1998).

In contrast to ground-nesting sea ducks, species nesting in tree cavities have fewer options for selecting nest locations and may experience nest site limitation (Pöysä and Pöysä 2002). For these species (Table 10.1), the typical distance between nests and the water depends on the distribution of suitable nesting cavities (excluding artificial nest boxes), which in turn is affected greatly by anthropogenic activities such as agriculture, forestry, and urban development (Bellrose 1980). Nonetheless, natural nests for Common and Hooded Mergansers as well as Buffleheads may be 500 m from water (Gauthier 1993, Dugger et al. 1994, Mallory and Metz 1999), whereas the goldeneyes may nest 2 km from the nearest open water (Eadie et al. 1995, 2000). Hence, cavity-nesting sea ducks can nest farther away from brood-rearing ponds than ground-nesting sea ducks.

Distances reported for nest to water refer to movements to the initial brood-rearing site, but not necessarily to the location where broods spend the majority of their time. For goldeneyes and Buffleheads, broods may move several kilometers from initial rearing sites (Gauthier 1987, Wayland and McNicol 1994, Eadie et al. 2000). In fact, Pöysä and Paasivaara (2006) and Paasivaara and Pöysä (2008) demonstrated that different habitat characteristics governed nest site versus brood site selection and thus that ultimately breeding site selection by goldeneyes includes a dynamic shift between nesting and brood-rearing habitats. Nest site selection and subsequent movements may also be influenced by the need for females and broods to join crèches once they leave the nest site (Chapter 11, this volume). For example, Common Merganser broods amalgamate and leave forested ponds and creeks for larger rivers, moving downstream and eventually rearing broods on larger lakes (Erskine 1972, Wood 1985). Common Eiders also form crèches and may move 80 km from their natal area prior to fledging (Goudie et al. 2000).

NEST BOXES

All cavity-nesting sea ducks will use artificial nest boxes (Table 10.1). Much work has gone into determining species-specific preferences in nest box dimensions so that wildlife managers or local conservation groups can attract these species or augment existing sites to increase local populations (Lumsden et al. 1980, 1986). After nest boxes are erected in a region, proportional occupancy by cavity-nesting sea ducks generally increases, but reproductive efficiency, measured as the number of eggs hatched per number of eggs laid, often decreases (Morse et al. 1969, Eriksson and Niittylä 1985, McNicol et al. 1997). The negative effects of nest box placement in high-visibility sites or high-density arrays and deleterious behavioral responses of cavity-nesting waterfowl have been recently investigated in comprehensive studies (Eadie and Fryxell 1992; Eadie et al. 1998; Chapter 11, this volume). Nest boxes are typically highly visible and have broad level floors. Sea ducks nesting in artificial boxes may have larger clutch sizes and higher rates of nest parasitism and abandonment than likely occurs in natural nests (Gauthier 1993; Eadie et al. 1995, 2000; Evans et al. 2002).

EFFECTS OF NESTING ON THE LOCAL ENVIRONMENT

In general, most nesting sea ducks have an imperceptible effect on their local environment. Goldeneyes, Bufflehead, and mergansers are all secondary cavity nesters that do not alter trees to make nesting sites. Harlequin Ducks, scoters, and most of the eiders typically nest at low densities across the landscape, making discreet nests except for nesting down. Colonial nesting eiders and possibly other species of island-nesting sea ducks have two distinct effects on their breeding habitats. First, the contribution of feces serves as fertilizer to the local environment. Freshwater ponds and low, terrestrial areas near these ponds are relatively eutrophic at eider colonies (Mallory et al. 2006), which may in turn make the habitat more attractive for other breeding birds such as Long-tailed Ducks. Ponds often develop algal conditions, and terrestrial areas produce lush moss or grassy vegetation (Michelutti et al. 2010). Second, eiders carry a variety of contaminants in their tissues and, through defecation and egg, chick, or adult mortality, concentrate these contaminants at their nesting colonies (Michelutti et al. 2010; Chapter 6, this volume), similar to the effects of other colonial marine birds. Concentrated contaminants can then move into the local terrestrial food web (Choy et al. 2010a,b).

NESTING SUCCESS

Many factors influence nesting success and duckling survival (Afton and Paulus 1992), including female body condition (Paasivaara and Pöysä 2007), disease (Korschgen et al. 1978, Descamps et al. 2009), contaminants (Franson et al. 1995), climatic conditions (Boyd 1996), nest location (Kellett et al. 2003), predation levels (Quackenbush et al. 2004, Hoover et al. 2010), and interactions among one or more of these factors. These first three factors (body condition, disease, contaminants) are covered in previous chapters (Chapters 4-6, this volume). Here, my focus is on the relationships between nesting success and nest location, climate, and predation.

EFFECTS OF CLIMATE ON SEA DUCK NESTING SUCCESS

Climatic conditions affect sea duck nesting success in different ways, including harsh conditions that lead to reduced nesting attempts or nest abandonment (Fredga and Dow 1983, Mallory et al. 2009), or indirectly by affecting food supplies or access to those resources, and consequently body condition (Robertson 1995a).

Local climate on the breeding grounds affects the timing of sea duck nesting (Bluhm 1992, Love et al. 2010; BNA accounts). In most cases, field studies have shown that later nesting corresponds to reduced reproductive effort with smaller clutch sizes (Dow and Fredga 1984, Alisauskas and Ankney 1992, Robertson 1995b, Grand and Flint 1997, Quackenbush et al. 2004, Traylor et al. 2004). The degree to which late winters lead to late nesting as a population-level effect and how this relates to reduced reproductive effort as individual-level effect is unclear—indirect, carryover, or wintering effects must be considered (Lehikoinen et al. 2006). In sea ducks, late winters can lead to reduced breeding propensity with fewer ducks attempting to breed (Bolduc et al. 2008, Love et al. 2010). For species nesting in the Arctic, a normal year offers a short window of time in which waterfowl can migrate north, select a nest site, lay and incubate eggs, and rear their chicks to fledging before migrating south in advance of freeze-up. In some years, low temperatures and extremely late ice breakup can result in reproductive failure across many species (Ganter and Boyd 2000). Late ice breakup can also cause massive mortality of migrating eiders that starve waiting for open water near their breeding areas (Fournier and Hines 1994). Starvation may in part reflect use of endogenous nutrient reserves that would have been used for egg production (Chapter 5, this volume) but instead are used for self-maintenance, waiting for suitable conditions to initiate the nest, or also could be related to reduced or late access to exogenous food supplies (Grand and Flint 1997, Love et al. 2010). At a broader level, harsh winters may reduce female overwintering body condition or delay breeding, which may have a negative influence on clutch size and even fledging success of ducklings (Lehikoinen et al. 2006).

In general, reports of nest failure due to direct effects of bad weather are uncommon among the Mergini and are certainly less common than reports among dabbling ducks, diving ducks, or geese (Sargeant and Raveling 1992, BNA Accounts). For example, Petersen et al. (2000) reported occasional losses of Spectacled Eider eggs or chicks during storm tides, and some Long-tailed Ducks will abandon nests during intense snowstorms in the High Arctic (M. L. Mallory, pers. obs.). Nonetheless, nesting sea ducks appear tolerant of cold and wet conditions, as would be expected for species largely breeding in northern regions, and are less susceptible to flooding than geese. Incubating eiders will remain on their nest even when covered by snow and ice or to protect their eggs from occasional periods of high heat (Fast et al. 2007). Naturally, nests in cavities are largely sheltered from weather extremes that might harm eggs or incubating females.

In contrast, weather can have deleterious effects on young sea ducks, particularly during their first week of life (Savard et al. 1991, 1998). Koskimies and Lahti (1964) found that most sea duck ducklings had better thermoregulatory capacity than dabbling or diving ducklings, at least until their energetic reserves were depleted, which may explain in part why larger ducklings have higher survival rates (Flint and Grand 1997, Flint et al. 2006). Losses of ducklings can be high during periods of cold, rainy conditions (Erskine 1971, Ganter and Boyd 2000, Schmidt et al. 2006). Bengtson (1972) found 80% of Red-breasted Merganser duckling mortality was due to snowstorms, and Boyd (1996) showed that fewer juvenile Long-tailed Ducks were shot by hunters in years of colder Arctic conditions, suggesting that reproductive effort or success was lower in years

with harsher weather. Bolduc et al. (2008) showed that low production of broods among Common Goldeneyes in northern Quebec was related to the cold, late spring, and wet conditions during early brood rearing. Other studies have found that weather has a relatively smaller, direct effect on duckling survival when compared with the effect of body condition of the attending female (Paasivaara and Pöysä 2007). Hildén (1964) suggested that higher gull predation occurred on scoter ducklings during poor weather. Similarly, Mendenhall and Milne (1985) suggested that poor weather conditions make eider ducklings more susceptible to predation because they have to spend more time feeding, less time brooding, and thus relatively less time vigilant for predators. In the Arctic, Barry (1968) reported an occasion where thousands of King Eiders died while flightless, ostensibly due to flash freezing.

PREDATORS OF NESTING SEA DUCKS

From an evolutionary perspective, predation is a critical factor driving natural selection (Darwin 1859) and consequently development of avian clutch size (Perrins 1977) and reproductive strategies (Lack 1954, Martin 1991). Sea duck reproductive strategies have evolved in partial response to predation pressure in diverse habitats like deciduous and coniferous forests, low taiga or treeless tundra, or rocky, unvegetated alluvial islands. In general, predation levels decrease with increasing latitude (McKinnon et al. 2010), but the options for hiding from predators also decrease as trees and shrubs also disappear at high latitudes.

Eggs and breeding females of Arctic-nesting birds experience avian predation pressure from gulls (Larus spp.), jaegers (Stercorarius spp.), Snowy Owls (Bubo scandiacus), and corvids (Alison 1975, Suydam 2000, Kellett et al. 2003, Quackenbush et al. 2004, Allard 2006, Bentzen et al. 2008, Hoover et al. 2010, Stein et al. 2010), as well as Sandhill Cranes (Grus canadensis, Hoover et al. 2010) and falcons (Falco spp., Mallory et al. 2009). Predation attempts by avian predators often result in partial clutch loss (Hoover et al. 2010, Öst and Steele 2010).

Greater reproductive failure is usually associated with mammalian predation on arctic-nesting birds. Polar bear (Ursus maritimus), grizzly bear (U. arctos horribilis), wolverine (Gulo gulo), red fox (Vulpes vulpes), and Arctic fox (V. lagopus) can easily consume entire clutches (Bengtson 1972, Suydam

2000, Flint et al. 2006, Mallory et al. 2009, Hoover et al. 2010, Schamber et al. 2010), and can cause dramatic reproductive failure of eider colonies (Goudie et al. 2000).

For sea ducks nesting in forested areas or regions south of the Arctic, the main predators differ (Pöysä et al. 1997a). Predation on sea duck eggs by other birds is relatively uncommon, except for southern eider colonies, because nests are typically more concealed or protected in cavities or shrubs. Nonetheless, various woodpeckers may destroy eggs, and clutches may be lost due to nest site competition with other bird species (Gauthier 1993; Dugger et al. 1994; Eadie et al. 1995, 2000). Once ducklings depart the nest and reach brood-rearing sites, however, avian predators include gulls, eagles, owls, and some larger hawks, as well as occasional predation by loons and herons (Gauthier 1993; Dugger et al. 1994; Eadie et al. 1995, 2000; Brown and Fredrickson 1997; Mawhinney et al. 1999; Titman 1999). Large predatory fish may also eat ducklings (Eadie et al. 1995, 2000; Paasivaara and Pöysä 2007). Similarly, mammalian predators differ from arctic latitudes and include black bear (*Ursus americanus*), red fox, striped skunk (*Mephitis mephitis*), raccoon (*Procyon lotor*), mink (*Mustela vison*), pine marten (*Martes americana*), and red squirrel (*Tamiasciurus hudsonicus*). In some areas, nesting success can be significantly influenced by marten predation (up to 58% of nests; Johnsson 1993). The Hooded Merganser may also experience depredation by black rat snakes (*Elaphe obsoleta*; Dugger et al. 1994).

INCUBATION

A female sea duck sitting on her nest is probably more at risk of predation than at any other time of the year. Some studies have suggested that the majority of predation on waterfowl nests occurs during incubation recesses (Swennen et al. 1993). If a predator traps a bird in her nest cavity, she has no escape option and will likely be killed. Even ground-nesting sea ducks can be cornered and killed when nesting in dense cover (Fast et al. 2007, Öst and Steele 2010). Thus, natural selection should favor behaviors that minimize the risk of predation during incubation, which includes taking few incubation recesses to avoid making noise or motion that attracts predators, maximizing nest attentiveness to reduce the incubation period,

and distraction displays to lure predators' attention away from the clutch (Afton and Paulus 1992, Mallory et al. 1998).

Afton and Paulus (1992) showed that diving ducks and other waterfowl nesting close to the water take more recesses per day than other ducks, presumably because they can slip into the water relatively undetected by predators. For sea ducks nesting on the ground or in cavities, taking fewer recesses reduces the risk of signaling the location of the nest to predators (Afton and Paulus 1992). Taking recesses during the day, when tree-climbing, nocturnal predators may be inactive, may serve to reduce predation risk on cavity-nesting ducks (Mallory and Weatherhead 1993). However, many other factors influence the timing and duration of incubation recesses such as egg-cooling rates (Flint and Grand 1999, Quackenbush et al. 2004) and need to be considered when evaluating the behavioral response of nesting sea ducks to predation risk.

Among waterfowl, heavier species have higher incubation constancy and greater nest attentiveness. Most sea ducks exhibit high incubation constancy (>85%) and take relatively few recesses per day (Afton and Paulus 1992). This behavior probably serves a similar function as in geese, where higher attentiveness is associated with lower losses of eggs to predators. Updating the formula in Afton and Paulus (1992) using data from Table 10.2, incubation constancy among sea ducks is described by the equation

$$\text{Constancy } (\%) = 82.393 + 0.006626 \text{ (Mass)},$$
$$F_{1,8} = 4.6, \quad P = 0.06, \quad r = 0.6,$$

as shown in Figure 10.4a. The main outlier to this pattern is the White-winged Scoter, which has relatively low nest attentiveness for its body mass. The incubation period measured as duration of time from start of incubation to date of hatching follows the opposite pattern, becoming shorter for heavier sea duck species (Figure 10.4b) as follows:

$$\text{Period (days)} = 32.486 - 0.0042 \text{ (Mass)},$$
$$F_{1,12} = 8.0, \quad P = 0.015.$$

Consequently, sea ducks with higher nest attentiveness have shorter incubation periods, as shown in Figure 10.4c:

$$\text{Period (days)} = 62.431 - 0.3835 \text{ (Constancy)},$$
$$F_{1,12} = 4.8, \quad P = 0.06.$$

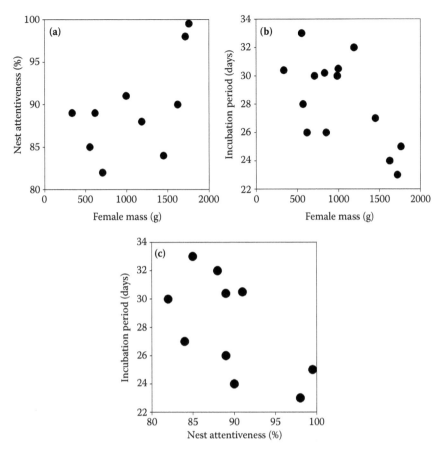

Figure 10.4. Relationships between incubation and female size in sea ducks: (a) nest attentiveness (or incubation constancy, %) of female sea ducks increases with female body mass; (b) incubation period decreases with increasing female body mass; and (c) incubation period decreases with increasing nest attentiveness (from references in BNA Accounts).

Natural selection has probably favored the evolution of larger body size in eiders, to allow them to store large amounts of endogenous nutrient reserves and thus maintain high nest attentiveness to reduce their incubation period and thus risk of predation (Thompson and Raveling 1987, Criscuolo et al. 2002, Bottitta et al. 2003).

NEST LOCATION

Typical nesting success of sea ducks varies with their nesting strategy. For example, nests in cavities are protected from extremes of precipitation and hot sun and from ground-based or avian predators (Martin and Li 1992). Cavity-nesting birds, including ducks, experience lower levels of nest predation (Fontaine et al. 2007). Correspondingly, cavity-nesting ducks tend to have relatively larger clutches for their body size than ground-nesting ducks (Figure 10.3c and d). Therefore, not

surprisingly, average annual nesting success among cavity-nesting sea ducks tends to be relatively high and similar across years (Figure 10.5a, Fredga and Dow 1983). However, nesting in cavities is associated with other costs that affect reproduction. For example, available cavities of suitable size may be limited in number (Pöysä and Pöysä 2002) or too distant from brood-rearing locations in some areas. Second, intense competition for available nest sites is associated with dump nesting or prehatch brood amalgamation (Chapter 11, this volume), negative density-dependent effects on production, and consequently reduced overall nesting efficiency (per capita number of eggs hatched or young fledged per egg laid; Gauthier 1987, Savard 1988, Eadie and Fryxell 1992, McNicol et al. 1997, Eadie et al. 1998, Pöysä and Pöysä 2002). Last, although predation levels may be lower than for ground nesters, cavity nests are still susceptible to predation, and breeding females from some species, such as the

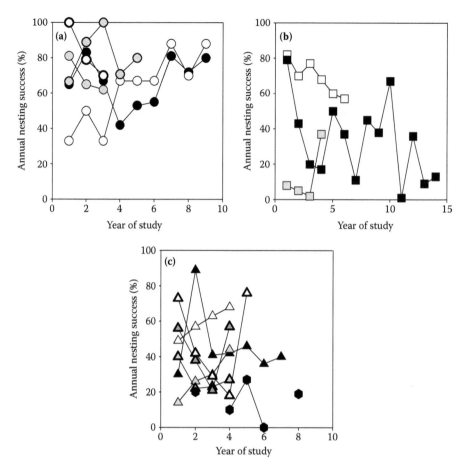

Figure 10.5. Variation in annual nesting success in sea ducks, as shown for (a) cavity-nesting species, which typically have higher annual success; (b) ground-nesting species excluding eiders, which show high variation among years; and (c) eiders, which have moderate but variable annual success (see also Table 10.2). Each symbol/fill combination in each graph represents data from a different study species. Data from Morse et al. (1969), McNicol et al. (1997), Grand and Flint (1997), Kellett et al. (2003), Traylor et al. (2004), Sénéchal et al. (2008), and Schamber et al. (2009, 2010).

Common Goldeneye, appear capable of evaluating relative safety of different nests, which is reflected as increased parasitic laying in safer nest sites the following year (Pöysä 2006).

In contrast, ground-nesting sea ducks are probably less limited in where they can nest but lack the protection of cavities, and thus their nests are at higher risk of predation or effects of bad weather. Ground-nesting sea ducks often select or have greater reproductive success at nest sites with greater cover, presumably to reduce the risk of detection by predators (Traylor et al. 2004, Öst et al. 2008b, Safine and Lindberg 2008, Hoover et al. 2010, Schamber et al. 2010, but see Bolduc et al. 2005). Despite selecting more concealed nest sites if available, overall success of sea ducks nesting on the ground tends to be lower and more variable (Figure 10.5b), as the annual levels of predation

or inclement weather can vary markedly. Based on a compilation of estimates (Figure 10.5), nesting success differed across the groups of sea ducks ($F_{2,83} = 20.3$, $P < 0.0001$). Mean annual nesting success for cavity-nesting sea ducks was 69.8% ± 3.1% (n = 29 site-years), significantly higher than that of eiders (38.6% ± 3.6%, n = 33) or other ground-nesting species (38.8% ± 5.5%, n = 24; Tukey–Kramer multiple comparison tests, both P < 0.001), but eider and ground-nesting sea duck success was similar.

PREDATOR AVOIDANCE BY NESTING ON ISLANDS

Many sea ducks nest on islands—why? The reason for this pattern probably lies in minimizing the risk of predation. There are relatively few options available to ground-nesting sea ducks to reduce the risk of predation by avian predators (Swennen 1983, Kay and Gilchrist 1998, Goudie et al. 2000) because predators can access the same sites and habitats as breeding ducks. However, mammalian predators have a far greater effect on reproductive success of sea ducks, particularly at the nesting stage. One tactic to reduce the risk of predation is for open, ground-nesting ducks to choose nest sites on small islands where they are available (Alison 1975, Clark and Shutler 1999, Kellett et al. 2003), because these sites tend to be inaccessible to ground-based, mammalian predators like foxes (Larson 1960).

Common Eiders nest almost exclusively on islands (Goudie et al. 2000). In the Arctic, eiders nest on islands that are not joined to the mainland by ice bridges, which can be traversed by foxes and other nest predators (Mallory and Gilchrist 2003, 2005). Traylor et al. (2004) found that almost all of the White-winged Scoters at Redberry Lake, Saskatchewan, nested on islands and in dense cover at those sites. Alison (1975) found higher numbers of nesting attempts of Long-tailed Ducks on islands than on the nearby mainland, as did Kellett et al. (2003) for King Eiders. However, loss of nests to predators was similar on island and mainland habitats in Alison's (1975) study, whereas for King Eiders, nest success was markedly higher on islands (Kellett et al. 2003). Hoover et al. (2010) found that Common Eider colonies located on islands in lakes had higher nesting success than eiders nesting on marine islands, presumably because the latter had little vegetative cover, whereas willows were present on freshwater islands. Schmutz et al. (1983) showed that eiders nesting on islands in Hudson Bay preferred sites where they had overhead shrub cover. Last, Fast et al. (2007) demonstrated that female eiders nesting on islands in more sheltered locations derived energetic benefits during incubation from a more moderated thermal regime, but these sites may be more risky or vulnerable if females are cornered during predation attempts. Like many long-lived birds, reproductive performance improves with age and experience in sea ducks (Goudie et al. 2000, Milonoff et al. 2002), and thus characteristics of the nesting female may override expected relationships with habitat quality (Bolduc et al. 2005, Öst and Steele 2010).

PREDATOR AVOIDANCE BY NESTING WITH OTHER SPECIES

Some birds are thought to preferentially nest in close proximity to other more aggressive or territorial bird species, such as Snow Geese (*Chen caerulescens*) nesting near Snowy Owls (Gauthier et al. 2004). By nesting within the territory of the defensive species, the nests of the geese are effectively defended against predators by the territorial owls. Some sea ducks nest in association with aggressive species, but the evidence of benefits to nesting is equivocal. Long-tailed Ducks often nest within the limits of the colonies of terns (*Sterna* spp.; Robertson and Savard 2002), because terns mob potential predators and thus may reduce the risk of predation, at least from the air. However, it is unclear whether nesting success is enhanced in these situations (Alison 1975). Kellett et al. (2003) found no evidence of improved reproductive success by King Eiders nesting near gulls (*Larus* spp.) or terns, and Bentzen et al. (2008) did not find that King Eiders selected these nesting associations. In contrast, Red-breasted Mergansers had higher nesting success by choosing sites near gulls and terns because predation levels were lower in the presence of these marine birds (Craik and Titman 2009). Schamel (1977) found a higher density of nesting Common Eiders within the territory of nesting Glaucous Gulls (*Larus hyperboreus*) than elsewhere on islands in Alaska. Whether these associations represent nest site choice by sea ducks to reduce predation risk or simply choice for similar habitats that reduce the risk of mammalian predators requires further study.

VARIATION IN PREDATION INTENSITY

Annual variation in predator numbers can lead to marked differences in the intensity of predation on nesting sea ducks and thus also influence annual breeding success. To some extent, numbers of predators may be linked to other available prey across a region (particularly for Arctic regions), often in a cyclical manner—a pattern referred to as the alternative prey hypothesis (Angelstam et al. 1984, Korpimäki et al. 1990). In its simplest form, this hypothesis posits that predators cue in and consume an abundant prey species, which drives its numbers down, and consequently predators switch to less abundant prey during times when the main prey numbers are low. Prey switching allows the population of that main prey to build again, and the cycle is generated.

Effects of the alternative prey hypothesis are supported in the Arctic, with predation levels on geese influenced by the well-established lemming cycle (Underhill et al. 1993, Bêty et al. 2002). Data are few, but sea duck nesting effort or nesting success can be higher in years with high lemming abundance, presumably because predation levels on ground-nesting ducks are reduced (Sittler et al. 2000, Bêty et al. 2002, Quackenbush et al. 2004).

Of course, there are also stochastic predation events that occur, such as polar bears accessing eider colonies and resulting in total colony failure (Goudie et al. 2000). More recently, global warming has resulted in earlier ice breakup, and changing seasonal events have been linked to more occurrences of polar bears foraging in bird colonies (Smith et al. 2010, Iverson et al. 2014), as well as local reports of more bears in eider colonies specifically (M. L. Mallory, unpubl. data).

ENVIRONMENTAL CUES INFLUENCING REPRODUCTIVE EFFORT

As relatively long-lived birds, often with delayed breeding maturity, it is probable that prebreeding females can gather information about the suitability of future nesting sites. Most of the sea ducks exhibit moderate to high levels of breeding site fidelity. Fidelity is particularly common among the cavity nesters (Eriksson and Niittylä 1985, Savard and Eadie 1989, Gauthier 1990, Zicus 1990), as well as in the colonial Common Eider (Swennen 1990). Among the cavity nesters, where competition for nest sites can be high, young, inexperienced birds often travel in groups to visit potential nesting cavities in a breeding area (Eadie and Gauthier 1985), and thus selection has favored strong abilities to find potential nest sites, including sites relatively exposed or concealed (Pöysä et al. 1999).

Younger, inexperienced birds may be able to gather information on the suitability of nest sites by using information from the remains of active or old nests to determine whether previous nests were successful or not (Dow and Fredga 1985, Pöysä et al. 2001), which may explain in part why some nest sites are used more than others, even if the same female does not return. However, prospecting visits by nonbreeding females to active nests where a resident female is incubating can have deleterious reproductive consequences during skirmishes at the nest site (Eadie et al. 1995; M. L. Mallory, pers. obs.).

Among colonial Common Eiders, some females, presumably nonbreeders or subadults, behave as aunts, traveling in groups through a colony and attending incubating hens (Goudie et al. 2000). These females may be gathering information on various nesting sites and behaviors for future breeding attempts. Even within the same breeding season, information from other nests and breeding site fidelity can improve reproductive success—Fast et al. (2010) showed that female eiders may use cues from remains of previous nests to select nest sites and initiate current nests earlier in the season.

Climatic conditions can have a significant effect on sea duck reproductive success (Boyd 1996, Ganter and Boyd 2000), but there is recent evidence that Common Eiders have some plasticity in responding to a variable environment. Love et al. (2010) showed that eiders use climatic cues in the early season to adjust the timing of their reproduction to enhance reproductive success. Similarly, Quackenbush et al. (2004) suggested that Steller's Eiders may use cues of lemming abundance to attempt nesting in years of high lemming numbers because there is abundant, alternate prey for foxes and other predators and hence reduced predation on eider nests. Moreover, Snowy Owls and Pomarine Jaegers also nest in these years and might provide local protection for nesting eiders.

BROOD-REARING BEHAVIOR

Brood-rearing strategies of sea ducks exhibit interspecific and even intraspecific variation (Öst et al. 2003a,b). Some species rear their young as a small family group, some travel great distances with their broods, and others amalgamate broods with other females, forming large crèches. Decisions by females on how to rear their brood may vary across years, influenced by the female's own physical condition. The adaptive significance of posthatch brood amalgamation is described by Eadie and Savard (Chapter 11, this volume).

While brood-rearing behaviors by sea ducks have received considerable research attention (Chapter 4, this volume), specific aspects of duckling survival are often difficult to determine for this tribe. In cavity nesters and Harlequin Ducks, broods may travel long distances and thus can be difficult to follow (Gauthier 1993; Dugger et al. 1994; Eadie et al. 1995, 2000; Robertson and Goudie 1999; Pöysä and Paasivaara 2006). Broods often merge into crèches in eiders, mergansers (with the possible exception of the Hooded Merganser), Long-tailed Ducks, and scoters, and following the success of individual ducklings or broods is challenging (Brown and Fredrickson 1997, Savard et al. 1998, Mallory and Metz 1999, Titman 1999, Goudie et al. 2000, Robertson and Savard 2002, Mehl and Alisauskas 2007). Despite these limitations, some consistent patterns have been documented. Duckling survival is generally higher in the cavity-nesting species and Harlequin Ducks and lower in the eiders, scoters, and Long-tailed Ducks (Table 10.2). All sea duck ducklings appear capable of feeding themselves soon after hatch; no substantiated reports exist of females feeding young. Sea ducks tend to have strong thermoregulatory capabilities at an early age (Koskimies and Lahti 1964), particularly when compared to more temperate nesting species, and Arctic-nesting species tend to be independent at an earlier age than temperate-nesting species (Kear 1970, Afton and Paulus 1992). Females in better body condition at the time of the hatching of their eggs are more likely to attend an amalgamated brood and remain with that brood for a longer duration, while females in poorer condition are more likely to attend larger groups of females and young (Bustnes and Erikstad 1991; Öst et al. 2003b; Chapter 11, this volume). Furthermore, evidence is mixed for whether larger broods afford enhanced survival for ducklings or not (Munro and Bédard 1977; Savard 1987; Kehoe 1989; Milonoff et al. 1995; Öst et al. 2003b, 2008a; Paasivaara and Pöysä 2007).

For goldeneyes and Buffleheads, females depart their nest cavities and move to brood-rearing ponds or shorelines, where they defend feeding territories for their young (Savard et al. 1991). Young are highly independent: broods actively feed with the female at varying distances away, and sometimes not present at all (Pöysä 1992). Strong, aggressive interactions may occur when brood-rearing females encounter each other, and young may be lost to other females or simply abandoned (Pöysä 1992, Eadie and Lyon 1998), and duckling survival is lower in situations of high breeding densities (Savard et al. 1991).

Common Eider females lead broods to marine waters shortly after hatch. Larger amalgamations (crèches) usually form near breeding colonies;

Munro and Bédard (1977) showed that crèches often form in response to predation. Nonbreeding females or failed breeders often join crèches and may occasionally brood the young (Goudie et al. 2000). Some females may stay with these crèches for the entire prefledging period, while others leave broods soon after crèche formation, in response to the female's physical condition (Bustnes and Erikstad 1991, Öst et al. 2003b). King Eiders and probably Long-tailed Ducks appear to follow a similar brood-rearing pattern (Suydam 2000, Robertson and Savard 2002), but female Steller's Eiders remain with broods until fledging (Fredrickson 2001).

Less is known about brood rearing in the scoters, consistent with our general lack of knowledge on these species. All three species form crèches, but it appears that females may abandon their brood relatively earlier than in eiders or goldeneyes. In White-winged Scoters, females usually abandon their brood by 3 weeks of age (Brown and Fredrickson 1997).

Rearing broods incurs costs to females. Milonoff et al. (2004) showed that female goldeneyes that raised a brood in a given year initiated nests later and laid smaller clutches in the following year. Hanssen et al. (2003) found that some female eiders in poorer initial body condition during breeding have suppressed immunity and were more likely to abandon their brood during rearing.

INTERANNUAL REPRODUCTIVE SUCCESS ACROSS THE TRIBE

Natural selection favors heritable traits that maximize the number of offspring produced (Darwin 1859). Individuals must appropriately allocate their limited resources through their lifespan to meet the competing demands of reproduction and survival (Horn and Rubenstein 1984, Stearns 1992). Life-history theory predicts that for waterfowl species facing environmental constraints that lead to lower annual reproductive success, natural selection would favor reduced annual reproductive effort such as smaller clutch size, delayed breeding maturity, and longer lifespan to maximize the number of reproductive attempts for the species (Lack 1968, Trivers 1972).

What are the environmental constraints that influence annual reproductive success of sea ducks and consequently the evolution of reproductive strategies? As I have outlined, annual variation in the direct or indirect effects of climatic conditions can markedly influence annual nesting success for most sea duck species, affecting breeding propensity, timing of nesting, reproductive effort in clutch or egg size, and juvenile survival. Annual variation in predation intensity, particularly as it may be related to availability of lemmings or other alternative prey for predators, can have dramatic effects on ground-nesting waterfowl nesting success, particularly for survival of nests and young.

Depending on the nesting habitat and environmental conditions, sea ducks face markedly different selective pressures that will influence many aspects of their ecology and behavior, which will affect adult or nest survival. Overall, the interspecific patterns follow predictions from life-history theory. Sea duck species with larger clutches and higher annual reproductive effort have lower annual female survival (Figure 10.6), as described by the following equation:

$$\text{Female survival (\%)} = 104.35 - 4.59 \text{ (Clutch size)};$$
$$F_{1,11} = 7.1, \quad P = 0.02, \quad r = -0.63.$$

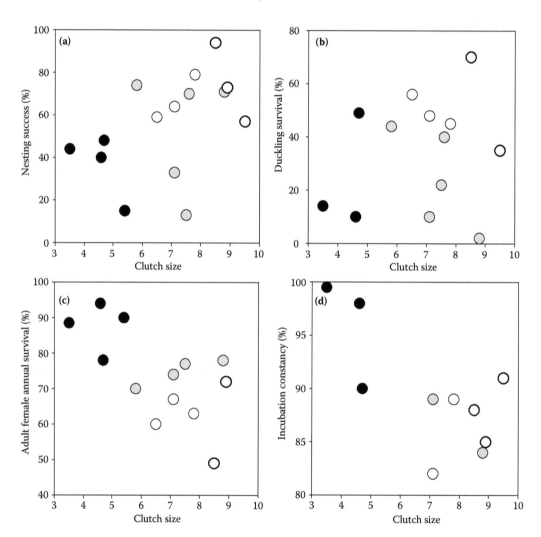

Figure 10.6. Relationships between mean clutch size for sea ducks and reproductive behavior or success. (Data from Table 10.2.) Eiders are depicted as black circles, goldeneyes and Buffleheads are white circles, mergansers are white circles with thick borders, and the remaining ground-nesting species are gray circles. (a) Annual nesting success typically increases with clutch size and is low for eiders, high for goldeneyes and mergansers, and variable for scoters, Harlequin Ducks, and Long-tailed Ducks; (b) duckling survival is typically low for eiders, high for goldeneyes and mergansers, and variable for scoters, Harlequin Ducks, and Long-tailed Ducks; (c) adult female annual survival is high for eiders, lower for mergansers and goldeneyes, and moderate for the other species, proportional to their mean clutch sizes; and (d) incubation constancy decreases with increasing clutch size, being high in eiders and lower in the other sea ducks.

The tradeoff not only extends across species (Table 10.2) but has been found among individuals in long-term studies of Common Eiders (Descamps et al. 2009) and Common Goldeneyes (Dow and Fredga 1984, Milonoff et al. 2002).

As a waterfowl tribe, the reproductive strategies of the sea ducks can be categorized by four different taxa (Figure 10.6). Colonial, ground-nesting Common Eiders have a low reproductive rate, apparently the outcome of frequent breeding failures, which is attributable often to die-offs of ducklings (Milne 1974, Swennen 1983, Mawhinney et al. 1999, Hario and Rintala 2006, Descamps et al. 2009), as well as a high proportion of nonbreeders in some years (Coulson 1984). In their reproductive strategy, which favors longevity in females, behavioral adaptations include crèching of young, which positively influences female survival (Öst et al. 2008a), as well as improved reproductive success with increasing female age and experience (Öst and Steele 2010). Collectively, reproductive success of Common Eiders oscillates substantially among years (Hario and Rintala 2006): during 71 site-years in northwestern Europe, eiders averaged 0.3 fledglings per female, with good years producing ≥0.6 but the more common poor years producing only 0.08. As a group, the eiders tend to have low annual nesting success (Figure 10.6a), low duckling survival (Figure 10.6b), and small clutches but exhibit high incubation constancy (Figure 10.6d) and high adult survival (Figure 10.6c). This reproductive strategy accords with a species that is susceptible to high rates of breeding failure in some years.

In contrast, the cavity-nesting Common Goldeneye has a higher reproductive rate, with clutches that are twice as large as eiders, and high duckling survival (Eadie et al. 1995) but lower adult female survival (Savard and Eadie 1989, Eadie et al. 1995, Ludwichowski et al. 2002). An average female goldeneye lives approximately half as long as an average female eider. However, nest site limitation by available cavities and strong, female-biased nest site fidelity may limit breeding propensity in some years and potentially result in alternative, often deleterious reproductive tactics such as nest parasitism (Eadie and Fryxell 1992, Pöysä et al. 1997a, Pöysä and Pöysä 2002). Furthermore, strong brood territoriality can result in reduced duckling survival (Pöysä 1992, Eadie and Lyon 1998, Ruusila and Pöysä 1998), as it does in Barrow's Goldeneye and Bufflehead (Savard et al. 1991).

From the data available to date, the mergansers generally appear to have similar reproductive strategies to goldeneyes and Buffleheads (Figure 10.6). Scoters, Long-tailed Ducks, and Harlequin Ducks, all ground-nesting species, appear to have strategies that are intermediate between eiders and cavity nesters. In particular, the scoters show strong preferences for nest concealment in their breeding strategies (Traylor et al. 2004, Schamber et al. 2010). More information is required on these species.

There are clearly some common patterns in reproductive strategies among the sea ducks that have been observed or should be expected with further research. For all sea duck species where it has been assessed, there is high breeding site fidelity (Anderson et al. 1992). This is not surprising for long-lived species or species nesting in stable environments such as cavity nesters in lake regions, because familiarity with a breeding location leads to better ability to assess predation risk, find food supplies, and move to brood-rearing habitats. Consequently, in many of the sea ducks, reproductive performance improves with age and breeding experience (Milonoff et al. 2002, Öst and Steele 2010). Natal philopatry is high in many of the species, notably the cavity nesters and colonial Common Eiders (Goudie et al. 2000, Ruusila et al. 2001), attributable to the benefits of experience with a breeding site under nest site limitation. Last, for many of the sea ducks, complex social interactions during brood rearing can influence duckling survival (Savard et al. 1991; Brown and Fredrickson 1997; Milonoff et al. 2002; Öst et al. 2008a; Chapter 11, this volume), which is an important component of their reproductive strategies.

INTERANNUAL REPRODUCTIVE SUCCESS: ANTHROPOGENIC EFFECTS

Anthropogenic activities can also lead to local adjustments in annual nesting success. Hunter harvest is relatively low on most North American sea ducks but can be a significant factor for some eider populations (Gilliland et al. 2009), reducing the number of birds that can return to breed. Logging and other forestry practices that remove nesting trees can lead to declines in nesting density of cavity-nesting sea ducks and probably contributed to population declines of some sea duck species in the nineteenth and twentieth centuries (Bellrose 1980; Gauthier 1993; Eadie et al. 1995, 2000; Mallory and Metz 1999). In the short term, provisioning of artificial nest boxes can increase cavity-nesting sea duck densities, but may not increase overall nesting success, and in fact may lead to reduced productivity (Eadie and Fryxell 1992, Pöysä and Pöysä 2002). Less direct factors can also influence survival, nesting propensity, and consequently annual nesting success. For example, improvements or restoration of the normal aquatic food chain conditions in lakes recovering for acidification are probably contributing to improved brood-rearing habitat for goldeneyes, mergansers, and other sea ducks in Eastern North America (McNicol et al. 1995). Introduction of zebra mussels (Dreissena spp.) in the Great Lakes of North America may be influencing both wintering locations and overwintering survival of species like Long-tailed Ducks (Schummer et al. 2008). At a regional level, annual reproductive success of sea ducks may respond to human activities, and sound management of sea duck populations must consider complex spatial and behavioral needs of sea ducks in habitat management (Chapter 14, this volume).

SUMMARY

Collectively, sea ducks exhibit a range of reproductive strategies that span the range found across most of the waterfowl and that follow the predictions of life-history theory. Although all of the sea duck species tend toward a K-strategy of reproduction, those species that have the highest annual survival tend to have delayed maturity, smaller clutches, lower duckling survival, and lower annual reproductive success (Table 10.2). Breeding site fidelity is common among the sea ducks, while renesting is uncommon, and delays in breeding and reductions in clutch size have a relatively strong effect on annual reproductive success. Stability of breeding populations is particularly sensitive to adult survival, and for most sea ducks, breeding success improves with age and experience (Coulson 1984, Hario and Rintala 2006, Hario et al. 2009), in part due to aspects of parental care (Chapter 11, this volume). The reproductive strategies of sea ducks, combined with their reliance on certain habitats or key locations at different stages of their annual cycle (Savard and Robert 2013; Chapter 13, this volume) and their susceptibility to particular anthropogenic threats (Chapter 14, this volume), suggest that many sea duck populations are vulnerable to decline, and recovery from low population numbers may be slow.

INFORMATION GAPS

Despite the tremendous increase in research on sea ducks from the 1990s onward, there are still substantial information gaps on reproductive strategies for some species. Compared to geese and dabbling ducks, we still know comparatively little about most sea ducks. Here, I highlight several key information gaps:

1. *Little-known species*—In contrast to the Common Eider and the goldeneyes, many aspects of the breeding biology of the mergansers and the scoters remain either unknown or understudied (Table 10.2). Field studies of breeding ecology have greatly improved our understanding of scoters (Traylor et al. 2004, Schamber et al. 2010), but additional studies are clearly needed. There is a clear dearth of recent work on any of the mergansers.

2. *Brood biology*—Factors influencing brood survival have been studied recently in some species (Öst et al. 2003a,b; Pöysä and Paasivaara 2006), but in general, the ecological correlates of brood survival in many of the sea ducks remain largely unknown. In particular, studies of the brood biology of Hooded Mergansers are lacking, outside of dietary research (Dugger et al. 1994).

3. *Long-term studies for long-lived birds*—Long-term studies have yielded great insights into factors influencing annual reproductive success and ultimately population demographics in some species (Dow and Fredga 1984, Descamps et al. 2009, Hario et al. 2009). Given the remarkable variation in reproductive strategies among sea ducks (Table 10.2), the extent to which these results can be applied reliably to other species is unclear. A significant gap in our information exists for species-specific data on reproductive success and survival across multiple years for some sea ducks. This information gap is particularly acute if we consider management needs for species at a time when we are still trying to identify or predict the effects of environmental stressors on sea duck populations (Descamps et al. 2009, 2010; Chapter 14, this volume).

4. *Regional variation*—Many sea duck species have broad breeding ranges, but our

measures of reproductive performance and survival often come from only a few locations or in some cases a single site. A significant knowledge gap remains in our limited geographic range of studies on species; the reliability of extrapolating to species-wide inferences from single or few locations is unclear. Recent studies on several sea duck species suggest that there are substantial differences in reproductive parameters associated with different geographic or environmental conditions (Alison 1975, Wayland and McNicol 1994, Brown and Fredrickson 1997, Schmidt et al. 2006, Safine and Lindberg 2008, Schamber et al. 2009).

LITERATURE CITED

Afton, A. D., and S. L. Paulus. 1992. Incubation and brood care. Pp. 62–108 in B. D. J. Batt, A. D. Afton, M. G. Anderson, C. D. Ankney, D. H. Johnston, J. A. Kadlec, and G. L. Krapu (editors), Ecology and management of breeding waterfowl. University of Minnesota Press, Minneapolis, MN.

Åhlund, M. 2005. Behavioural tactics at nest visits differ between parasites and hosts in a brood-parasitic duck. Animal Behaviour 70:433–440.

Åhlund, M., and M. Andersson. 2001. Female ducks can double their reproduction. Nature 414:600–601.

Alisauskas, R. T., and C. D. Ankney. 1992. Egg-laying and nutrient reserves. Pp. 30–61 in B. D. J. Batt, A. D. Afton, M. G. Anderson, C. D. Ankney, D. H. Johnston, J. A. Kadlec, and G. L. Krapu (editors), Ecology and management of breeding waterfowl. University of Minnesota Press, Minneapolis, MN.

Alison, R. M. 1975. Breeding biology and behavior of the Oldsquaw (Clangula hyemalis L.). Ornithological Monographs 18:1–52.

Allard, K. A. 2006. Foraging ecology of an avian predator, the Herring Gull and its colonial eider duck prey. Dissertation, University of New Brunswick, Fredericton, NB.

Anderson, M. G., J. M. Rhymer, and F. C. Rohwer. 1992. Philopatry, dispersal, and the genetic structure of waterfowl populations. Pp. 365–395 in B. D. J. Batt, A. D. Afton, M. G. Anderson, C. D. Ankney, D. H. Johnston, J. A. Kadlec, and G. L. Krapu (editors), Ecology and management of breeding waterfowl. University of Minnesota Press, Minneapolis, MN.

Angelstam, P., E. Lindström, and P. Widén. 1984. Role of predation in short-term populations fluctuations of some birds and mammals in Fennoscandia. Oecologia 62:199–208.

Barry, T. W. 1968. Observations on natural mortality and native use of eider ducks along the Beaufort Sea coast. Canadian Field-Naturalist 82:140–144.

Bellrose, F. C. 1980. Ducks, geese and swans of North America, 3rd ed. Wildlife Management Institute, Stackpole Books, Harrisburg, PA.

Bengtson, S.-A. 1971. Variations in clutch-size in ducks in relation to the food supply. Ibis 113:523–526.

Bengtson, S.-A. 1972. Reproduction and fluctuations in the size of duck populations at Lake Myvatn, Iceland. Oikos 23:35–58.

Bentzen, R. L., A. N. Powell, L. M. Phillips, and R. S. Suydam. 2010. Incubation behavior of King Eiders on the coastal plain of northern Alaska. Polar Biology 33:1075–1082.

Bentzen, R. L., A. N. Powell, and R. S. Suydam. 2008. Factors influencing nesting success of King Eiders on northern Alaska's coastal plain. Journal of Wildlife Management 72:1781–1789.

Bêty, J., G. Gauthier, E. Korpimäki, and J.-F. Giroux. 2002. Shared predators and indirect trophic interactions: lemming cycles and arctic-nesting geese. Journal of Animal Ecology 71:88–98.

Bluhm, C. K. 1992. Environmental and endocrine control on waterfowl reproduction. Pp. 323–364 in B. D. J. Batt, A. D. Afton, M. G. Anderson, C. D. Ankney, D. H. Johnston, J. A. Kadlec, and G. L. Krapu (editors), Ecology and management of breeding waterfowl. University of Minnesota Press, Minneapolis, MN.

Bolduc, F., and M. Guillemette. 2003. Incubation constancy and mass loss in the Common Eider Somateria mollissima. Ibis 145:329–332.

Bolduc, F., M. Guillemette, and R. D. Titman. 2005. Nesting success of Common Eiders Somateria mollissima as influenced by nest-site and female characteristics in the Gulf of St. Lawrence. Wildlife Biology 11:273–279.

Bolduc, F., S. Lapointe, and B. Gagnon. 2008. Common Goldeneye breeding in the eastern Canadian boreal forest: factors affecting productivity estimates. Waterbirds 31:42–51.

Bordage, D., and J.-P. L. Savard. 1995. Black Scoter (Melanitta nigra). in A. Poole and F. Gill (editors), The Birds of North America, No. 177. The Academy of Natural Sciences, Philadelphia, PA.

Bottitta, G. E., E. Nol, and H. G. Gilchrist. 2003. Effects of experimental manipulation of incubation length on behavior and body mass of Common Eiders in the Canadian Arctic. Waterbirds 26:100–107.

Boyd, H. 1996. Arctic temperatures and the Long-tailed Ducks shot in eastern North America. Wildlife Biology 2:113–117.

Boyd, W. S., B. D. Smith, S. A. Iverson, M. R. Evans, J. E. Thompson, and S. Schneider. 2009. Apparent survival, natal philopatry, and recruitment of

Barrow's Goldeneyes (*Bucephala islandica*) in the Cariboo-Chilcotin region of British Columbia, Canada. Canadian Journal of Zoology 87:337–345.

Brown, P. W., and L. H. Fredrickson. 1997. White-winged Scoter (*Melanitta fusca*). in A. Poole and F. Gill (editors), The Birds of North America, No. 274. The Academy of Natural Sciences, Philadelphia, PA.

Bustnes, J. O., and K. E. Erikstad. 1991. Parental care in the Common Eider *Somateria mollissima*: factors affecting abandonment and adoption of young. Canadian Journal of Zoology 69:1538–1545.

Choy, E. S., M. Gauthier, M. L. Mallory, J. P. Smol, D. Lean, and J. M. Blais. 2010a. An isotopic investigation of mercury accumulation in terrestrial food webs adjacent to an Arctic seabird colony. Science of the Total Environment 408:1858–1867.

Choy, E. S., L. E. Kimpe, M. L. Mallory, J. P. Smol, and J. M. Blais. 2010b. Biotransport of marine pollutants to a terrestrial food web: spatial patterns of persistent organic pollutants adjacent to a seabird colony in Arctic Canada. Environmental Pollution 158:3431–3438.

Clark, R. G., and D. Shutler. 1999. Avian habitat selection: pattern from process in nest-site use by ducks? Ecology 80:272–287.

Clobert, J., E. Danchin, A. A. Dhont, and J. D. Nichols. 2001. Dispersal. Oxford University Press, Oxford, U.K.

Coulson, J. C. 1984. The population dynamics of the Eider Duck *Somateria mollissima* and evidence of extensive nonbreeding by adult ducks. Ibis 126:525–543.

Craik, S. R., and R. D. Titman. 2009. Nesting ecology of Red-breasted Mergansers in a common tern colony in eastern New Brunswick. Waterbirds 32:282–292.

Criscuolo, F., G. W. Gabrielsen, and J.-P. Gendner. 2002. Body mass regulation during incubation in female Common Eiders *Somateria mollissima*. Journal of Avian Biology 33:83–88.

Darwin, C. D. 1859. The Origin of Species. Penguin, London, U.K.

Descamps, S., H. G. Gilchrist, J. Bety, E. I. Buttler, and M. R. Forbes. 2009. Costs of reproduction in a long-lived bird: large clutch size is associated with low survival in the presence of a highly virulent disease. Biology Letters 5:278–281.

Descamps, S., N. G. Yoccoz, J.-M. Gaillard, H. G. Gilchrist, K. E. Erikstad, S. A. Hanssen, B. Cazelles, M. R. Forbes, and J. Bêty. 2010. Detecting population heterogeneity in effects of North Atlantic Oscillations on seabird body condition: get into the rhythm. Oikos 119:1526–1536.

Dow, H., and S. Fredga. 1984. Factors affecting reproductive output of the Goldeneye Duck *Bucephala clangula*. Journal of Animal Ecology 53:679–692.

Dow, H., and S. Fredga. 1985. Selection of nest sites by a hole-nesting duck, the Goldeneye *Bucephala clangula*. Ibis 127:16–30.

Drent, R. H., and S. Daan. 1980. The prudent parent: energetic adjustment in avian breeding. Ardea 68:225–252.

Drilling, N., R. Titman, and F. McKenney. 2002. Mallard (*Anas platyrhynchos*). in A. Poole and F. Gill (editors), The Birds of North America, No. 658. The Academy of Natural Sciences, Philadelphia, PA.

Dugger, B. D., K. M. Dugger, and L. H. Fredrickson. 1994. Hooded Merganser (*Lophodytes cucullatus*). in A. Poole and F. Gill (editors), The Birds of North America, No. 98. The Academy of Natural Sciences, Philadelphia, PA.

Eadie, J. M., and J. Fryxell. 1992. Density-dependence, frequency-dependence, and alternative nesting behaviors in goldeneyes. American Naturalist 140:621–641.

Eadie, J. M., and G. Gauthier. 1985. Prospecting for nest sites by cavity-nesting ducks of the genus *Bucephala*. Condor 87:528–534.

Eadie, J. M., and B. Lyon. 1998. Cooperation, conflict and crèching behavior in goldeneye ducks. American Naturalist 151:397–408.

Eadie, J. M., M. L. Mallory, and H. G. Lumsden. 1995. Common Goldeneye (*Bucephala clangula*). in A. Poole and F. Gill (editors), The Birds of North America, No. 170. The Academy of Natural Sciences, Philadelphia, PA.

Eadie, J. M., J.-P. L. Savard, and M. L. Mallory. 2000. Barrow's Goldeneye (*Bucephala islandica*). in A. Poole and F. Gill (editors), The Birds of North America, No. 548. The Academy of Natural Sciences, Philadelphia, PA.

Eadie, J. M., P. Sherman, and B. Semel. 1998. Conspecific brood parasitism, population dynamics, and the conservation of cavity-nesting birds. Pp. 306–340 in T. Caro (editor), Behavioral Ecology and Conservation Biology. Oxford University Press, Oxford, U.K.

Ekroos, J., M. Öst, P. Karell, K. Jaatinen, and M. Kilpi. 2012. Philopatric predisposition to predation-induced ecological traps: habitat-dependent mortality of breeding eiders. Oecologia 170:979–986.

Eriksson, K., and J. Niittylä. 1985. Breeding performance of the Goosander *Mergus merganser* in the archipelago of the Gulf of Finland. Ornis Fennica 62:153–157.

Eriksson, M. O. G. 1979. Aspects of the breeding biology of the Goldeneye *Bucephala clangula*. Holarctic Ecology 2:186–194.

Erskine, A. J. 1971. Growth, and annual cycles in weights, plumages and reproductive organs in Goosanders in eastern Canada. Ibis 113:42–58.

Erskine, A. J. 1972. Populations, movements and seasonal distribution of mergansers. Canadian Wildlife Service Report Series, No. 17. Canadian Wildlife Service, Ottawa, ON.

Evans, M. R., D. B. Lank, W. S. Boyd, and F. Cooke. 2002. A comparison of the characteristics and fate of Barrow's Goldeneye and Bufflehead nests in nest boxes and natural cavities. Condor 104:610–619.

Fast, P. L., H. G. Gilchrist, and R. G. Clark. 2007. Experimental evaluation of nest shelter effects on weight loss in incubating Common Eiders *Somateria mollissima*. Journal of Avian Biology 38:205–213.

Fast, P. L., H. G. Gilchrist, and R. G. Clark. 2010. Nest-site materials affect nest-bowl use by Common Eiders (*Somateria mollissima*). Canadian Journal of Zoology 88:214–218.

Flint, P. L., and J. B. Grand. 1997. Survival of Spectacled Eider adult females and ducklings during brood rearing. Journal of Wildlife Management 61:217–221.

Flint, P. L., and J. B. Grand. 1999. Incubation behavior of Spectacled Eiders on the Yukon-Kuskokwim delta, Alaska. Condor 101:413–416.

Flint, P. L., C. L. Moran, and J. L. Schamber. 1998. Survival of Common Eider *Somateria mollissima* adult females and ducklings during brood rearing. Wildfowl 49:103–109.

Flint, P. L., J. A. Morse, J. B. Grand, and C. L. Moran. 2006. Correlated growth and survival of juvenile Spectacled Eiders: evidence of habitat limitation? Condor 108:901–911.

Flint, P. L., M. R. Petersen, C. P. Dau, J. E. Hines, and J. D. Nichols. 2000. Annual survival and site fidelity of Steller's Eiders molting along the Alaska Peninsula. Journal of Wildlife Management 64:261–268.

Fontaine, J. J., M. Martel, H. M. Markland, A. M. Niklison, K. L. Decker, and T. E. Martin. 2007. Testing ecological and behavioral correlates of nest predation. Oikos 116:1887–1894.

Fournier, M. A., and J. E. Hines. 1994. Effects of starvation on muscle and organ mass of King Eiders *Somateria spectabilis* and the ecological and management implications. Wildfowl 45:188–197.

Fox, A. D., A. Petersen, and M. Frederiksen. 2003. Annual survival and site-fidelity of breeding female Common Scoter *Melanitta nigra* at Myvatn, Iceland, 1925–1958. Ibis 145:E94–E96.

Franson, J. C., M. R. Petersen, C. U. Meteyer, and M. R. Smith. 1995. Lead poisoning of Spectacled Eiders (*Somateria fischeri*) and of a Common Eider (*Somateria mollissima*) in Alaska. Journal of Wildlife Diseases 31:268–271.

Fredga, S., and H. Dow. 1983. Annual variation in the reproductive performance of goldeneyes. Wildfowl 34:120–126.

Fredrickson, L. H. 2001. Steller's Eider (*Polysticta stelleri*). in A. Poole and F. Gill (editors), The Birds of North America, No. 571. The Academy of Natural History, Philadelphia, PA.

Ganter, B., and H. Boyd. 2000. A tropical volcano, high predation pressure, and the breeding biology of Arctic waterbirds: a circumpolar review of breeding failure in the summer of 1992. Arctic 53:289–305.

Gauthier, G. 1987. Brood territories in Buffleheads: determinants and correlates of territory size. Canadian Journal of Zoology 65:1402–1410.

Gauthier, G. 1990. Philopatry, nest site fidelity and reproductive performance in Buffleheads. Auk 106:568–573.

Gauthier, G. 1993. Bufflehead (*Bucephala albeola*). in A. Poole and F. Gill (editors), The Birds of North America, No. 67. The Academy of Natural Sciences, Philadelphia, PA.

Gauthier, G., J. Bêty, J.-F. Giroux, and L. Rochefort. 2004. Trophic interactions in a high Arctic Snow Goose colony. Integrative and Comparative Biology 44:119–129.

Gilliland, S. G., H. G. Gilchrist, R. F. Rockwell, G. J. Robertson, J.-P. L. Savard, F. Merkel, and A. Mosbech. 2009. Evaluating the sustainability of harvest among northern Common Eiders *Somateria mollissima borealis* in Greenland and Canada. Wildlife Biology 15:24–36.

Goudie, R. I., G. J. Robertson, and A. Reed. 2000. Common Eider (*Somateria mollissima*). in A. Poole and F. Gill (editors), The Birds of North America, No. 546. The Academy of Natural Sciences, Philadelphia, PA.

Grand, J. B., and P. L. Flint. 1997. Productivity of nesting Spectacled Eiders on the lower Kashunuk River, Alaska. Condor 99:926–932.

Greenwood, P. J. 1982. The natal and breeding dispersal of birds. Annual Review of Ecology and Systematics 13:1–21.

Hamer, K. C., E. A. Schreiber, and J. Burger. 2002. Breeding biology, life histories, and life history-environment interactions in seabirds. Pp. 217–261 in E. A. Schreiber and J. Burger (editors), Biology of Marine Birds. CRC Press, New York, NY.

Hanssen, S. A., I. Folstad, and K. E. Erikstad. 2003. Reduced immunocompetence and cost of reproduction in Common Eiders. Oecologia 136:457–464.

Hario, M., M. Mazerolle, and P. Saurola. 2009. Survival of female Common Eiders *Somateria m. mollissima* in a declining population of the northern Baltic Sea. Oecologia 159:747–756.

Hario, M., and J. Rintala. 2006. Fledgling production and population trends in Finnish Common Eiders (*Somateria mollissima mollissima*)—evidence for density dependence. Canadian Journal of Zoology 84:1038–1046.

Hildén, O. 1964. Ecology of duck populations in the island group of Valassaaret, Gulf of Bothnia. Annals Zoologica Fennica 1:153–279.

Hoover, A. K., D. L. Dickson, and K. W. Dufour. 2010. Survival and nesting success of the Pacific Eider (*Somateria mollissima v-nigrum*) near Bathurst Inlet, Nunavut. Canadian Journal of Zoology 88:511–519.

Horn, H. S., and D. I. Rubenstein. 1984. Behavioural adaptations and life history. Pp. 279–298 in J. R. Krebs and N. B. Davies (editors), Behavioural ecology: an evolutionary approach, 2nd ed. Blackwell Scientific Publications, Oxford, U.K.

Iverson, S. A., H. G. Gilchrist, P. A. Smith, A. J. Gaston, and M. R. Forbes. 2014. Longer ice-free seasons increase the risk of nest depredation by polar bears for colonial breeding birds in the Canadian Arctic. Proceedings of the Royal Society B 281:art20133128.

Jaatinen, K., M. Öst, P. Gienapp, and J. Merilä. 2012. Facultative sex allocation and sex-specific offspring survival in Barrow's Goldeneyes. Ethology 118:1–10.

Johnson, K. 1995. Green-winged Teal (*Anas crecca*). in A. Poole and F. Gill (editors), The Birds of North America, No. 193. The Academy of Natural Science, Philadelphia, PA.

Johnsson, K. 1993. The Black Woodpecker *Dryocopus martius* as a keystone species in forest. Dissertation, Swedish University of Agricultural Sciences, Uppsala, Sweden.

Kats, R. K. H. 2007. Common Eiders *Somateria mollissima* in the Netherlands: the rise and fall of breeding and wintering populations in relation to stocks of shellfish. Dissertation, Rijksuniversiteit Groningen, Groningen, Netherlands.

Kay, M. F., and H. G. Gilchrist. 1998. Distraction displays made by female Common Eiders (*Somateria mollissima borealis*) in response to human disturbance. Canadian Field-Naturalist 112:529–531.

Kear, J. 1970. The adaptive radiation of parental care in waterfowl. Pp. 357–392 in J. H. Crook (editor), Social behavior in birds and mammals. Academic Press, New York, NY.

Kehoe, F. P. 1989. The adaptive significance of creching behaviour in the White-winged Scoter (*Melanitta fusca deglandi*). Canadian Journal of Zoology 67:406–411.

Kellett, D. K., and R. T. Alisauskas. 1997. Breeding biology of King Eiders nesting on Karrak Lake, Northwest Territories. Arctic 50:47–54.

Kellett, D. K., and R. T. Alisauskas. 2000. Body-mass dynamics of King Eiders during incubation. Auk 117:812–817.

Kellett, D. K., R. T. Alisauskas, and K. R. Mehl. 2003. Nest-site selection, interspecific associations, and nest success of King Eiders. Condor 105:373–378.

Kellett, D. K., R. T. Alisauskas, K. R. Mehl, K. L. Drake, J. J. Traylor, and S. L. Lawson. 2005. Body mass of Long-tailed Ducks (*Clangula hyemalis*) during incubation. Auk 122:313–318.

Klomp, H. 1970. The determinants of clutch size in birds: a review. Ardea 58:1–125.

Korpimäki, E., K. Huhtaln, and S. Sukkavn. 1990. Does the year-to-year variation in the diet of Eagle and Ural Owls support the alternative prey hypothesis? Oikos 58:47–54.

Korschgen, C. E., H. C. Gibbs, and H. L. Mendall. 1978. Avian cholera in Eider Ducks in Maine. Journal of Wildlife Diseases 14:254–258.

Koskimies, J., and L. Lahti. 1964. Cold-hardiness of the newly hatched young in relation to ecology and distribution in ten species of European ducks. Auk 81:281–307.

Krementz, D. G., P. W. Brown, F. P. Kehoe, and C. S. Houston. 1997. Population dynamics of White-winged Scoters. Journal of Wildlife Management 61:222–227.

Lack, D. 1954. The natural regulation of animal numbers. Clarendon, Oxford, U.K.

Lack, D. 1968. Ecological adaptations for breeding in birds. Methuen, London, U.K.

Larson, S. 1960. On the influences of the arctic fox *Alopex lagopus* on the distribution of arctic birds. Oikos 11:276–305.

Lehikoinen, A., M. Kilpi, and M. Öst. 2006. Winter climate affects subsequent breeding success of Common Eiders. Global Change Biology 15:1355–1365.

Love, O. P., H. G. Gilchrist, S. Déscamps, C. A. D. Semeniuk, and J. Bêty. 2010. Pre-laying climatic cues can time reproduction to optimally match offspring hatching and ice conditions in an Arctic marine bird. Oecologia 164:277–286.

Ludwichowski, I., R. Barker, and S. Bräger. 2002. Nesting area fidelity and survival of female Common Goldeneyes *Bucephala clangula*: are they density-dependent? Ibis 144:452–460.

Lumsden, H. G., R. E. Page, and M. Gauthier. 1980. Choice of nest boxes by Common Goldeneyes in Ontario. Wilson Bulletin 92:497–505.

Lumsden, H. G., J. Robinson, and R. Hartford. 1986. Choice of nest boxes by cavity nesting ducks. Wilson Bulletin 98:167–168.

Lutmerding, J. A., and A. S. Love. 2014. Longevity records of North American birds (Ver. 2014.1). Patuxent Wildlife Research Center, Bird Banding Lab, Laurel, MD.

Mallory, M. L., A. J. Fontaine, P. A. Smith, M. O. Wiebe Robertson, and H. G. Gilchrist. 2006. Water chemistry of ponds on Southampton Island, Nunavut, Canada: effects of habitat and ornithogenic inputs. Archive fur Hydrobiologie 166:411–432.

Mallory, M. L., A. J. Gaston, and H. G. Gilchrist. 2009. Sources of breeding season mortality in Canadian Arctic seabirds. Arctic 62:333–341.

Mallory, M. L., and H. G. Gilchrist. 2003. Marine birds breeding in Penny Strait and Queens Channel, Nunavut, Canada. Polar Research 22:399–403.

Mallory, M. L., and H. G. Gilchrist. 2005. Marine birds of the Hell Gate Polynya, Nunavut, Canada. Polar Research 24:87–94.

Mallory, M. L., D. K. McNicol, R. A. Walton, and M. Wayland. 1998. Risk-taking by incubating Common Goldeneyes and Hooded Mergansers. Condor 100:694–701.

Mallory, M. L., and K. Metz. 1999. Common Merganser (*Mergus merganser*). in A. Poole and F. Gill (editors), The Birds of North America, No. 442. The Academy of Natural Sciences, Philadelphia, PA.

Mallory, M. L., and P. J. Weatherhead. 1990. Effects of nest parasitism and nest location on eggshell strength in waterfowl. Condor 92:1031–1039.

Mallory, M. L., and P. J. Weatherhead. 1993. Incubation rhythms and weight loss of Common Goldeneyes. Condor 95:1049–1059.

Martin, T. E. 1991. Interaction of nest predation and food limitation in reproductive strategies. Current Ornithology 9:163–197.

Martin, T. E., and P. Li. 1992. Life history traits of open- vs. cavity-nesting birds. Ecology 73:579–592.

Mawhinney, K., A. W. Diamond, P. Kehoe, and N. Benjamin. 1999. Status and productivity of Common Eiders in relation to the status of Great Black-backed Gulls and Herring Gulls in the southern Bay of Fundy and the northern Gulf of Maine. Waterbirds 22:253–262.

McKinnon, L., H. G. Gilchrist, and K. T. Scribner. 2006. Genetic evidence for kin-based female social structure in Common Eiders (*Somateria mollissima*). Behavioral Ecology 17:614–621.

McKinnon, L., P. A. Smith, E. Nol, J. L. Martin, F. I. Doyle, K. F. Abraham, H. G. Gilchrist, R. I. G. Morrison, and J. Bêty. 2010. Lower predation risk for migratory birds at high latitudes. Science 327:326–327.

McNicol, D. K., R. K. Ross, M. L. Mallory, and L. A. Brisebois. 1995. Trends in waterfowl populations: evidence of recovery from acidification. Pp. 203–217 in J. M. Gunn (editor), Environmental restoration and recovery of an industrial region. Springer-Verlag, New York, NY.

McNicol, D. K., R. A. Walton, and M. L. Mallory. 1997. Monitoring nest box use by cavity-nesting ducks on acid-stressed lakes in Ontario, Canada. Wildlife Biology 3:1–12.

Mehl, K. R., and R. T. Alisauskas. 2007. King Eider (*Somateria spectabilis*) brood ecology: correlates of duckling survival. Auk 124:606–618.

Mendenhall, V. M., and H. Milne. 1985. Factors affecting duckling survival of Eiders *Somateria mollissima* in northeast Scotland. Ibis 127:148–158.

Michelutti, N., J. Brash, J. Thienpont, J. M. Blais, L. Kimpe, M. L. Mallory, M. S. V. Douglas, and J. P. Smol. 2010. Trophic position influences the efficacy of seabirds as contaminant biovectors. Proceedings of the National Academy of Sciences USA 107:10543–10548.

Milne, H. 1974. Breeding numbers and reproductive rate of eiders at the Sands of Forvie National Nature Reserve, Scotland. Ibis 116:135–154.

Milonoff, M., H. Pöysä, and P. Runko. 2002. Reproductive performance of Common Goldeneye *Bucephala clangula* females in relation to age and lifespan. Ibis 144:585–592.

Milonoff, M., H. Pöysä, P. Runko, and V. Ruusila. 2004. Brood rearing costs affect future reproduction in the precocial Common Goldeneye *Bucephala clangula*. Journal of Avian Biology 35:344–351.

Milonoff, M., H. Pöysä, and J. Virtanen. 1995. Brood size-dependent offspring mortality in Common Goldeneye reconsidered: fact or artifact? American Naturalist 146:967–974.

Mittelhauser, G. H. 2008. Apparent survival and local movements of Harlequin Ducks wintering at Isle au Haut, Maine. Waterbirds 31:138–146.

Morse, T. E., J. L. Jakabosky, and V. P. McCrow. 1969. Some aspects of the breeding biology of the Hooded Merganser. Journal of Wildlife Management 33:596–640.

Mowbray, T. B., C. R. Ely, J. S. Sedinger, and R. E. Trost. 2002. Canada Goose (*Branta canadensis*). in A. Poole and F. Gill (editors), The Birds of North America, No. 682. The Academy of Natural Sciences, Philadelphia, PA.

Munro, J., and J. Bédard. 1977. Gull predation and crèching behaviour in the Common Eider. Journal of Animal Ecology 46:799–810.

Oppel, S., and A. N. Powell. 2010. Age-specific survival estimates for King Eiders derived from satellite telemetry. Condor 112:323–330.

Öst, M., A. Lehikoinen, K. Jaatinen, and M. Kilpi. 2011. Causes and consequences of fine-scale breeding dispersal in a female-philopatric species. Oecologia 166:327–336.

Öst, M., B. D. Smith, and M. Kilpi. 2008a. Social and maternal factors affecting duckling survival in Eiders *Somateria mollissima*. Journal of Animal Ecology 77:315–325.

Öst, M., and B. Steele. 2010. Age-specific nest-site preference and success in eiders. Oecologia 162:59–69.

Öst, M., M. Wickman, E. Matulionis, and B. Steele. 2008b. Habitat-specific clutch size and cost of incubation in eiders reconsidered. Oecologia 158:205–216.

Öst, M., R. Ydenberg, M. Kilpi, and K. Lindström. 2003a. Condition and coalition formation by brood-rearing Common Eider females. Behavioral Ecology 14:311–317.

Öst, M., R. Ydenberg, K. Lindström, and M. Kilpi. 2003b. Body condition and the grouping behavior of brood-caring female Common Eiders (*Somateria mollissima*). Behavioral Ecology and Sociobiology 54:451–457.

Paasivaara, A., and H. Pöysä. 2007. Survival of Common Goldeneye *Bucephala clangula* ducklings in relation to weather, timing of breeding, brood size, and female condition. Journal of Avian Biology 38:144–152.

Paasivaara, A., and H. Pöysä. 2008. Habitat-patch occupancy in the Common Goldeneye (*Bucephala clangula*) at different stages of the breeding cycle: implications to ecological processes in patchy environments. Canadian Journal of Zoology 86:744–755.

Pearce, J. M., P. Blums, and M. S. Lindberg. 2008. Site fidelity is an inconsistent determinant of population structure in the Hooded Merganser (*Lophodytes cucullatus*): evidence from genetic, mark-recapture, and comparative data. Auk 125:711–722.

Pearce, J. M., D. Esler, and A. G. Degtyarev. 1998. Nesting ecology of Spectacled Eiders *Somateria fischeri* on the Indigirka River Delta, Russia. Wildfowl 49:110–123.

Pearce, J. M., J. A. Reed, and P. L. Flint. 2005. Geographic variation in survival and migratory tendency among North American Common Mergansers. Journal of Field Ornithology 76:109–118.

Perrins, C. M. 1977. The role of predation in the evolution of clutch size. Pp. 181–191 in B. Stonehouse and C. M. Perrins (editors), Evolutionary ecology. Macmillan, London, U.K.

Petersen, M. R., J. B. Grand, and C. P. Dau. 2000. Spectacled Eider (*Somateria fischeri*). in A. Poole and F. Gill (editors), The Birds of North America, No. 547. The Academy of Natural Sciences, Philadelphia, PA.

Picman, J. 1989. Mechanisms of increased puncture resistance of eggs of Brown-headed Cowbirds. Auk 106:577–583.

Pöysä, H. 1992. Variation in parental care of Common Goldeneye (*Bucephala clangula*) females. Behaviour 123:247–260.

Pöysä, H. 2006. Public information and conspecific nest parasitism in goldeneyes: targeting safe nests by parasites. Behavioral Ecology 17:459–465.

Pöysä, H., M. Milonoff, V. Ruusila, and J. Virtanen. 1999. Nest-site selection in relation to habitat edge: experiments in the Common Goldeneye. Journal of Avian Biology 30:79–84.

Pöysä, H., M. Milonoff, and J. Virtanen. 1997a. Nest predation in hole-nesting birds in relation to habitat edge: an experiment. Ecography 20:329–335.

Pöysä, H., and A. Paasivaara. 2006. Movements and mortality of Common Goldeneye *Bucephala clangula* broods in a patchy environment. Oikos 115:33–42.

Pöysä, H., and S. Pöysä. 2002. Nest-site limitation and density dependence of reproductive output in the Common Goldeneye *Bucephala clangula*: implications for the management of cavity-nesting birds. Journal of Applied Ecology 39:502–510.

Pöysä, H., P. Runko, and V. Ruusila. 1997b. Natal philopatry and the local resource competition hypothesis: data from the Common Goldeneye. Journal of Avian Biology 28:63–67.

Pöysä, H., V. Ruusila, M. Milonoff, and J. Virtanen. 2001. Ability to assess nest predation risk in secondary hole-nesting birds: an experimental study. Oecologia 126:201–207.

Quackenbush, L., R. Suydam, T. Obritschkewitsch, and M. Deering. 2004. Breeding biology of Steller's Eiders (*Polysticta stelleri*) near Barrow, Alaska, 1991–1999. Arctic 57:166–182.

Ricklefs, R. E. 1977. On the evolution of reproductive strategies in birds: reproductive effort. American Naturalist 111:453–478.

Robert, M., M.-A. Vaillancourt, and P. Drapeau. 2010. Characteristics of nest cavities of Barrow's Goldeneyes in eastern Canada. Journal of Field Ornithology 81:287–293.

Robertson, G. J. 1995a. Annual variation in Common Eider egg size: effects of temperature, clutch size, laying date, and laying sequence. Canadian Journal of Zoology 73:1579–1587.

Robertson, G. J. 1995b. Factors affecting nest site selection and nesting success in the Common Eider *Somateria mollissima*. Ibis 137:109–115.

Robertson, G. J. 2008. Using winter juvenile/adult ratios as indices of recruitment in population models. Waterbirds 31:152–158.

Robertson, G. J., and R. I. Goudie. 1999. Harlequin Duck (*Histrionicus histrionicus*). in A. Poole and F. Gill (editors), The Birds of North America, No. 466. The Academy of Natural Sciences, Philadelphia, PA.

Robertson, G. J., and J.-P. L. Savard. 2002. Long-tailed Duck (*Clangula hyemalis*). in A. Poole and F. Gill (editors), The Birds of North America, No. 651. The Academy of Natural Sciences, Philadelphia, PA.

Rohwer, F. C. 1988. Inter- and intraspecific relationships between egg size and clutch size in waterfowl. Auk 105:161–176.

Rowe, L., D. Ludwig, and D. Schluter. 1994. Time, condition, and the seasonal decline of avian clutch size. American Naturalist 143:698–722.

Ruusila, V., and H. Pöysä. 1998. Shared and unshared parental investment in the precocial goldeneye (Aves: Anatidae). Animal Behaviour 55:307–312.

Ruusila, V., H. Pöysä, and P. Runko. 2001. Costs and benefits of female-biased natal philopatry in the Common Goldeneye. Behavioral Ecology 12:686–690.

Safine, D. E., and M. S. Lindberg. 2008. Nest habitat selection of White-winged Scoters on Yukon Flats, Alaska. Wilson Journal of Ornithology 120:582–593.

Sargeant, A. B., and D. G. Raveling. 1992. Mortality during the breeding season. Pp. 396–422 in B. D. J. Batt, A. D. Afton, M. G. Anderson, C. D. Ankney, D. H. Johnston, J. A. Kadlec, and G. L. Krapu (editors), Ecology and management of breeding waterfowl. University of Minnesota Press, Minneapolis, MN.

Savard, J.-P. L. 1987. Causes and functions of brood amalgamation in Barrow's Goldeneye and Bufflehead. Canadian Journal of Zoology 65:1548–1553.

Savard, J.-P. L. 1988. Winter, spring and summer territoriality in Barrow's Goldeneye: characteristics and benefits. Ornis Scandinavica 19:119–128.

Savard, J.-P. L., D. Bordage, and A. Reed. 1998. Surf Scoter (Melanitta perspicillata). in A. Poole and F. Gill (editors), The Birds of North America, No. 363. The Academy of Natural Sciences, Philadelphia, PA.

Savard, J.-P. L., and J. M. Eadie. 1989. Survival and breeding philopatry in Barrow's and Common Goldeneyes. Condor 91:198–203.

Savard, J.-P. L., and M. Robert. 2013. Relationships among breeding, molting and wintering areas of adult female Barrow's Goldeneyes (Bucephala islandica) in eastern North America. Waterbirds 36:34–42.

Savard, J.-P. L., G. E. J. Smith, and J. N. M. Smith. 1991. Duckling mortality in Barrow's Goldeneye and Bufflehead broods. Auk 108:568–577.

Schamber, J. L., F. J. Broerman, and P. L. Flint. 2010. Reproductive ecology and habitat use of Pacific Black Scoters (Melanitta nigra americana) nesting on the Yukon-Kuskokwim Delta, Alaska. Waterbirds 33:129–139.

Schamber, J. L., P. L. Flint, J. B. Grand, H. M. Wilson, and J. A. Morse. 2009. Population dynamics of Long-tailed Ducks breeding on the Yukon-Kuskokwim Delta, Alaska. Arctic 62:190–200.

Schamel, D. 1977. Breeding of the Common Eider on the Beaufort Sea coast of Alaska. Condor 79:478–485.

Schmidt, J. H., E. J. Taylor, and E. A. Rexstad. 2006. Survival of Common Goldeneye ducklings in Interior Alaska. Journal of Wildlife Management 70:792–798.

Schmutz, J. K., R. J. Robertson, and F. Cooke. 1983. Colonial nesting of the Hudson Bay Eider Duck. Canadian Journal of Zoology 61:2424–2433.

Schummer, M. L., S. A. Petrie, and R. C. Bailey. 2008. Dietary overlap of sympatric diving ducks during winter on northeastern Lake Ontario. Auk 125:425–433.

Sénéchal, É., J. Bêty, and H. G. Gilchrist. 2011. Interactions between lay date, clutch size, and postlaying energetic needs in a capital breeder. Behavioral Ecology 22:162–168.

Sénéchal, H., G. Gauthier, and J.-P. L. Savard. 2008. Nesting ecology of Common Goldeneyes and Hooded Mergansers in a boreal river system. Wilson Journal of Ornithology 120:732–742.

Sittler, B., G. Olivier, and T. B. Berg. 2000. Low abundance of King Eider nests during low lemming years in Northeast Greenland. Arctic 53:53–60.

Smith, P. A., K. H. Elliott, A. J. Gaston, and H. G. Gilchrist. 2010. Has early ice clearance increased predation on breeding birds by polar bears? Polar Biology 33:1149–1153.

Stearns, S. C. 1992. The evolution of life histories. Oxford University Press, Oxford, U.K.

Stein, J., N. G. Yoccoz, and R. A. Ims. 2010. Nest predation in declining populations of Common Eiders Somateria mollissima: an experimental evaluation of the role of Hooded Crows Corvus cornix. Wildlife Biology 16:123–134.

Suydam, R. S. 2000. King Eider (Somateria spectabilis). in A. Poole and F. Gill (editors), The Birds of North America, No. 491. The Academy of Natural Sciences, Philadelphia, PA.

Swennen, C. 1983. Reproductive output of Eiders Somateria mollissima in the southern border of its breeding range. Ardea 71:245–254.

Swennen, C. 1990. Dispersal and migratory movements of Eiders Somateria mollissima breeding in the Netherlands. Ornis Scandinavica 21:17–27.

Swennen, C., J. C. H. Ursem, and P. Duiven. 1993. Determinate laying and egg attendance in Common Eiders. Ornis Scandinavica 24:48–52.

Thompson, S. C., and D. G. Raveling. 1987. Incubation behavior of Emperor Geese compared to other geese: interactions of predation, body size and energetics. Auk 104:707–716.

Titman, R. D. 1999. Red-breasted Merganser (Mergus serrator). in A. Poole and F. Gill (editors), The Birds of North America, No. 443. The Academy of Natural Sciences, Philadelphia, PA.

Traylor, J. J. 2003. Nesting and duckling ecology of White-winged Scoters (Melanitta fusca deglandi) at Redberry Lake, Saskatchewan. Thesis, University of Saskatchewan, Saskatoon, SK.

Traylor, J. J., R. T. Alisauskas, and F. P. Kehoe. 2004. Nesting ecology of White-winged Scoters (Melanitta fusca deglandi) at Redberry Lake, Saskatchewan. Auk 121:950–962.

Trivers, R. L. 1972. Parental investment and sexual selection. Pp. 136–179 in B. Campbell (editor), Sexual selection and the descent of Man 1871–1971. Aldine, Chicago, IL.

Underhill, L. G., R. P. Prys-Jones, E. E. Syroechkovski Jr., N. M. Groen, V. Karpov, H. G. Lappo, M. W. J. van Roomen, A. Rybkin, H. Schekkerman, H. Spiekman, and R. W. Summers. 1993. Breeding of waders (Charadrii) and Brent Geese Branta bernicla bernicla at Pronchishcheva Lake, northeastern Taimyr, Russia, in a peak and decreasing lemming year. Ibis 135:277–292.

U.S. Fish and Wildlife Service. [online]. 2014. U.S. harvest, 2009 data. Flyways.us. <http://www.flyways.us/regulations-and-harvest/harvest-trends> (24 January 2014).

Vaillancourt, M. P., R. M. Drapeau, and S. Gauthier. 2009. Origin and availability of large cavities for Barrow's Goldeneye, a species at risk inhabiting the eastern Canadian boreal forest. Avian Conservation and Ecology 4:art6.

Waldeck, P., M. Andersson, M. Kilpi, and M. Öst. 2008. Spatial relatedness and brood parasitism in a female-philopatric bird population. Behavioral Ecology 19:67–73.

Wayland, M., and D. K. McNicol. 1994. Movements and survival of Common Goldeneye broods near Sudbury, Ontario, Canada. Canadian Journal of Zoology 75:1252–1259.

Welty, J. C., and L. Baptista. 1990. The life of birds, 4th ed. Harcourt Publishers, New York, NY.

Williams, G. C. 1966. Natural selection, the costs of reproduction, and a refinement of Lack's principle. American Naturalist 100:687–692.

Wilson, H. M., P. L. Flint, C. L. Moran, and A. N. Powell. 2007. Survival of breeding Pacific Common Eiders on the Yukon–Kuskokwim Delta, Alaska. Journal of Wildlife Management 71:403–410.

Wood, C. C. 1985. Food-searching behaviour of the Common Merganser (Mergus merganser) II: choice of foraging location. Canadian Journal of Zoology 63:1271–1279.

Yom-Tov, Y. 2001. An updated list and some comments on the occurrence of intraspecific nest parasitism in birds. Ibis 143:133–143.

Zicus, M. C. 1990. Nesting biology of Hooded Mergansers using nest boxes. Journal of Wildlife Management 54:637–643.

CHAPTER ELEVEN

Breeding Systems, Spacing Behavior, and Reproductive Behavior of Sea Ducks*

John M. Eadie and Jean-Pierre L. Savard

Abstract. Sea ducks in tribe Mergini exhibit a wide range of spacing, breeding, and brood-rearing behaviors and have provided important insights in both theoretical and applied behavioral ecology. The strength, timing, and duration of pairing vary among species. Long-term pair bonds of more than one year are common but the proportion of birds that re-pair annually is unknown. At least two pairing periods have been documented—one in fall, likely involving birds that are reuniting, and one in spring representing new pair formation. Courtship behaviors are diverse and spectacular but their ontogeny from juveniles to adult birds has not been described. Sea ducks have one of the largest gradients in pair spacing behavior among aquatic birds, ranging from a lack of pronounced spacing in the colonial Common Eider to extremely well-developed intra- and interspecific territorial behaviors in goldeneyes and Buffleheads. Brood spacing behavior ranges from the crèching behavior of Common Eiders to the highly developed territorial system of goldeneyes. The frequency of conspecific brood parasitism (CBP) is high in species that are cavity nesting or that nest in high densities, but uncommon in species nesting at low densities in dispersed ground nests. Strong female natal and breeding philopatry results in various levels of kinship between parasites and hosts suggesting that CBP may also constitute cooperation among generations of closely related females. We review the evidence supporting various models and hypotheses attempting to explain CBP. There is increasing evidence that nest-site quality or host quality may be important factors influencing how and where females lay their eggs. Interspecific brood parasitism is found in the same species for which CBP occurs, suggesting that there may be common ecological factors that influence both behaviors. Brood amalgamation (BA) after hatch is more frequent in cavity-nesting sea ducks and ground-nesting species that nest in high densities and rare or occasional in ground-nesting species that nest at low densities. Interestingly, BA occurs more frequently in those species in which CBP is also more common. Whether BA is the result of accidental mixing or has an adaptive basis is still debated, but studies suggest differences in benefits or costs of BA among populations and years. As for CBP, kinship may also play a role

* Eadie, J. M. and J.-P. L. Savard. 2015. Breeding systems, spacing behavior, and reproductive behavior of sea ducks. Pp. 365–415 in J.-P. L. Savard, D. V. Derksen, D. Esler, and J. M. Eadie (editors). Ecology and conservation of North American sea ducks. Studies in Avian Biology (no. 46), CRC Press, Boca Raton, FL.

in BA but evidence is equivocal. Unfortunately, we still know little about many of these species and our understanding of their social and reproductive ecology is based on detailed studies of less than half of the species in the tribe. The great behavioral diversity of ducks within the tribe Mergini offers a rich opportunity to explore the threads of physiology, ecology, and evolution that underlie a complex and intriguing spectrum of reproductive behaviors.

Key Words: brood amalgamation, conspecific brood parasitism, crèching, egg dumping, interspecific parasitism, kinship, pair formation, parental care, selfish herd, spacing behavior, territoriality.

Despite close phylogenetic affinities among sea ducks and a shared distribution as northern breeders (>40° latitude), the clade exhibits a remarkable degree of diversity in breeding systems, spacing behavior during breeding and nonbreeding seasons, and reproductive behavior. Sea ducks are unique in being one of the few groups of waterfowl in which some species are highly territorial in the classic sense of defending a fixed area and excluding all other conspecifics and even heterospecifics. Several species, including Barrow's Goldeneye (*Bucephala islandica*) and Buffleheads (*B. albeola*), have been the subject of detailed studies of territoriality (Savard 1982, 1984, 1988a; Gauthier 1987a,b, 1988, 1993). Conspecific brood parasitism (CBP, also known as nest parasitism or egg dumping) is common among goldeneyes, mergansers, and some of the eiders. Studies of these species in North America and Europe have provided new insight into the ecology of this unusual nesting behavior and the role of female philopatry and kinship in facilitating these behaviors within Mergini and Anseriformes in general (Andersson and Eriksson 1982; Eadie 1989, 1991; Pöysä 1999, 2003a,b, 2006; Andersson and Åhlund 2000, 2001; Åhlund and Andersson 2001; Pöysä and Pesonen 2007; Jaatinen et al. 2009, 2011a,b). Merging or mixing of broods after hatch (crèching or post-hatch brood amalgamation [BA]) also occurs in several species of sea ducks, and long-term studies of species such as Common Eiders (*Somateria mollissima*), in which this behavior is especially prominent, have illustrated a rich complexity underlying the patterns of parental care (Savard 1987; Bustnes and Erikstad 1991a,b; Eadie and Lyon 1998; Kilpi et al. 2001; Bustnes et al. 2002; Öst et al. 2003a,b, 2005, 2007a,b).

It is intriguing that these unusual reproductive behaviors are common in sea ducks relative to other waterfowl and other groups of birds, suggesting a common evolutionary and ecological origin. Nonetheless, there is considerable diversity among the 15 species in North America in their spacing systems and patterns of parental care, leading to ongoing discussion and debate about the proximate and ultimate basis for these behaviors. It may be that the underlying mechanisms differ among the species in the Mergini, such that there is no single unifying basis, but recent research has identified some common themes that resonate through the group. Clearly, there is much yet to be learned and the variation observed even among this one tribe of waterfowl offers considerable scope for comparative analysis. Our goal here is to provide a framework for further analyses of the reproductive behaviors of sea ducks.

In this chapter, we seek to synthesize and integrate our current understanding of the rich diversity of breeding systems, spacing behavior, and reproductive behavior of the Mergini. For a few species, we have relatively extensive data and these species have become model systems with which to explore territoriality, CBP and BA, or crèching. For many of the sea ducks, however, few data exist and information is anecdotal or descriptive; for these species, our review will serve instead to highlight gaps in our knowledge and to outline opportunities for further study.

Our chapter builds on the basic body of knowledge of reproductive strategies summarized by Mallory (Chapter 10, this volume) and can be viewed as a companion chapter to his ideas and themes. Our focus here is more restricted; specifically, we synthesize our current understanding in three key areas of sea duck reproductive behavior: (1) breeding systems, including the types and duration of pair bonds, sex ratios, evidence of extra-pair mating or forced copulations, and patterns of male and female breeding and non-breeding philopatry; (2) spacing behavior, focusing on breeding and pair territories (including interspecific territoriality), brood spacing behaviors, and winter spacing strategies and territories; and (3) brood parasitism and BA, whereby females provide parental care for the young of other females during the pre-hatching or post-hatching periods. As a group, studies of sea ducks have contributed greatly to our understanding of these systems in waterfowl; our hope is that this review will further highlight the important insights provided by research on this small group of birds on issues of broad interest and significance in avian behavioral ecology.

BREEDING SYSTEMS

A variety of breeding systems have evolved in birds in relation to variation in habitat and sex ratios (Oring 1982), but monogamy is by far the most widespread social system (90%; Lack 1968). Within the tribe Mergini, species vary in type and strength of pair bonds; timing of pairing and most species are highly philopatric.

Delayed Breeding

All sea ducks exhibit delayed maturity, and most females do not establish their first nest until they are 2 or even 3 years of age (Table 11.1). Several of the cavity-nesting species prospect for nest sites a year in advance of the next breeding season, as do failed breeders and even females who have successfully fledged a brood (Eadie and Gauthier 1985, Zicus and Hennes 1989). Prospecting a year in advance has not been recorded in ground-nesting ducks, but in some species, subadult females do return to their natal area where they could explore sites (Oppel and Powell 2010). However, breeding females search actively for nest sites upon arrival on the breeding grounds.

Mating Systems

Social monogamy is the norm among waterfowl and three types have been defined (Oring and Sayler 1992): perennial, when pairs stay together throughout the year as in geese and swans; annual, when new pairs are formed each year as in most dabbling ducks; and annual, with reuniting that is common in several species of sea ducks (Anderson

et al. 1992). Long-term pair bonds of more than 1 year are common among sea ducks and have been documented in Harlequin Ducks (*Histrionicus histrionicus*; Bengston 1972a, Smith et al. 2000), Long-tailed Ducks (*Clangula hyemalis*; Alison 1975), Common Eiders (Spurr and Milne 1976), Surf Scoters (*Melanitta perspicillata*; Savard et al. 1998a), Buffleheads (Gauthier 1987c), and Barrow's Goldeneyes (Savard 1985; Table 11.1). The proportion of birds that re-pair is unknown and has yet to be quantified.

Pair Bonds

In geese and swans, both parents raise young together and pair bonds are for life. Conversely, only the female raises the young in dabbling ducks and pair bonds are annual. In sea ducks, only the female raises the young but pair bonds range from annual to long term (Owen and Black 1990). In Harlequin Ducks, pair reunion occurs at wintering areas and not the remigial molting areas (Smith et al. 2000). In Barrow's Goldeneyes (Savard 1985), Harlequin Ducks (Robertson et al. 1998), and Common Eiders (Spurr and Milne 1976), pairing occurs soon after birds have returned to their wintering area in late fall. The reunion of a pair on the wintering areas was described for Barrow's Goldeneyes (Savard 1985): the male arrived on October 29 on its wintering area; the female was sighted on November 8, 4 km away from the male; and on November 9, she had joined her mate and spent all winter with him in the same territory they had used the previous winter. The same pair was seen together at the same wintering location for two more years. Male Harlequin Ducks, Barrow's Goldeneyes, and Buffleheads that lost their mate have been seen unpaired back at their previous breeding location indicating the difficulty of finding new mates and thus a potential benefit of reuniting with previous mates (Savard 1985, Gauthier 1987c, Smith et al. 2000). However, in sea ducks, most unpaired males do not migrate to breeding areas but rather migrate directly to molting areas located away from breeding areas (Robert et al. 2002). In scoters, unpaired males are rarely seen at freshwater breeding areas and rather migrate directly to their marine molting areas (Savard et al. 2007).

Sex ratios are male-biased in most adult sea ducks, heavily so in some species (Sargeant and Raveling 1992, Donald 2007; Table 11.1). Sex ratio

TABLE 11.1

Summary of key reproductive behaviors and characteristics of the sea ducks.

Common name	Latin name	Breeding system	Sex ratio	EPCs	Breeding philopatry	Winter/Molt philopatry	Pair formation	Pair bond	Density/Colonial	Pair spacing
Common Eider	*Somateria mollissima*	Seasonal monogamy	Equal; male-biased	Rare	High female	High	Winter	Seasonal and some long term	Colonial, high density	Colonial; female defense
King Eider	*Somateria spectabilis*	Seasonal monogamy	Equal	Multiple mating	—	—	Late winter/spring	Seasonal	Solitary, low density; can be colonial	Female defense; males aggressive
Spectacled Eider	*Somateria fischeri*	Seasonal monogamy	Equal	None reported	High female	High	Winter	Seasonal	Low density; dispersed, some clumps	Female defense
Steller's Eider	*Polysticta stelleri*	Seasonal monogamy	Variable	None reported	Female(?)	Moderate	Late winter	Seasonal(?)	Low density	Female defense; pairs drive others away
Harlequin Duck	*Histrionicus histrionicus*	Long-term monogamy	Male-biased	Rare	Both sexes	High	Winter	Long term	Low density—rivers	Female defense
Long-tailed Duck	*Clangula hyemalis*	Serial or long-term monogamy	Male-biased	None reported	Both sexes	Moderate	Winter	Long term	Low density; can nest in small clump	Territorial; site specific
Black Scoter	*Melanitta nigra*	Seasonal monogamy	Equal; male-biased	—	—	Moderate	Winter	Seasonal(?)	Low density; dispersed	Female defense
Surf Scoter	*Melanitta perspicillata*	Seasonal or long-term monogamy	Male-biased	Multiple mating	—	Moderate	Late winter/spring	Seasonal and long term	Low density	Female defense
White-winged Scoter	*Melanitta fusca*	Seasonal or long-term monogamy	Male-biased	Rare	High female	Moderate	Late winter/spring	Seasonal and long term	Can be high densities; islands	Female defense
Bufflehead	*Bucephala albeola*	Seasonal or long-term monogamy/rare polygyny	Male-biased	Multiple mating	High female	Moderate	Winter	Seasonal and long term	Med density—cavities	Territorial; site specific

Barrow's Goldeneye	*Bucephala islandica*	Long-term monogamy/rare polygyny	Slight male-biased	Multiple mating	Both sexes	High	Late winter/spring	Long term	Med density—cavities	Territorial; site specific
Common Goldeneye	*Bucephala clangula*	Seasonal or long-term monogamy	Male-biased	None reported	High female	—	Late winter/spring	Seasonal and long term	Low-medium density—cavities	Territorial; site specific
Hooded Merganser	*Lophodytes cucullatus*	Seasonal monogamy	Slight male-biased	—	High female	—	Winter	Seasonal	Med density—cavities	—
Common Merganser	*Mergus merganser*	Seasonal monogamy/rare polygyny	Variable	None reported	High female	High	Late winter/spring	Seasonal	Low density cavities	Not territorial; gregarious
Red-breasted Merganser	*Mergus serrator*	Seasonal monogamy	Variable	Rare	High female	—	Late winter/spring	Seasonal	Colonial Islands	Female defense

Common name	Nest	Prospects	Reuse old sites?	CBP	IBP	Post-hatch BA	Brood territories
Common Eider	Ground	Female; before laying	Yes	Variable; increases with nest density	Occasional; BRAN, CAGO, SNGO, ABDU	Frequent; aunts and large crèches (150 ducklings)	No—aunts
King Eider	Ground	Female; before laying	Low (7%)	Can be frequent (16.2% nests)	Not observed	Occasional; large (100) broods with 2–50 females	No
Spectacled Eider	Ground	Female; before laying	Yes (35%)	Infrequent (3/280; <2% nests)	Rare (3.7%); CAGO, BRAN, GRSC	Infrequent; temporary mixing; no females abandon young	No
Steller's Eider	Ground	Female; before laying	Yes (unknown if same female)	Infrequent (1/5 broods DNA)	—	Rare; nonbreeders may accompany broods	No

TABLE 11.1 (*Continued*)

Summary of key reproductive behaviors and characteristics of the sea ducks.

Common name	Nest	Prospects	Reuse old sites?	CBP	IBP	Post-hatch BA	Brood territories
Harlequin Duck	Ground	Female; before laying	Yes (9 of 20)	Yes (Iceland)	Rare; COME	Occasional when density high; nonbreeders accompany	No
Long-tailed Duck	Ground	Female; before laying	Yes (3 females)	Rare (1 nest)	Occasional; GRSC, RBME, LTDU	Frequent; large amalgamations (30–50 ducklings); multiple females	No
Black Scoter	Ground	Female; before laying	—	Rare	—	Occurs often on lakes and after females leave brood	No
Surf Scoter	Ground	Female; before laying	—	—	—	Frequent, especially at high density; 9–33 young	No
White-winged Scoter	Ground	Female; before laying	Yes—frequent	Frequent at high densities	Occasional; GADW, LESC, BWTE, MALL	Frequent; large amalgamations (150 ducklings); 1–7 females	No
Bufflehead	Cavity	Female; some year before	Yes—frequent	Frequent (5%–8% nests)	Occasional; BAGO, COGO	Frequent; 34% broods; 1 female	Yes
Barrow's Goldeneye	Cavity	Female; some year before	Yes—frequent (72%)	Very frequent (30%–64% nests)	Frequent; COGO, BUFF, HOME, WODU, RBME	Frequent; 22% broods; 1 female	Yes

Common Goldeneye	Cavity	Female; some year before	Yes—frequent (63%)	Very frequent (23%–54% nests)	Frequent; BAGO, BUFF, HOME, WODU, RBME	Frequent; 34% broods; 1 female	Yes
Hooded Merganser	Cavity	Female; some year before	Yes—frequent (29%)	Very frequent (34%–45% nests)	Frequent; COGO, COME, WODU	—	No
Common Merganser	Cavity	Female; some year before	Yes	Very frequent (20% nests)	Occasional; COGO, HOME	Frequent when young >7 days; 40+ young with 1 or more females	No
Red-breasted Merganser	Ground	Female; before laying	?	Very frequent (64% nests) on island; 4%–5% mainland	Frequent; BAGO, HADU, MALL, GADW, LESC	Frequent when nests grouped; 26–100 young; super-broody females	No

Common name	Clutch size	Max CS/ Median	Age first breed	Max lifespan recorded	Sources	Birds of North America account number
Common Eider	3–5 (1–14)	3.5	2–3+	22	Goudie et al. (2000)	546
King Eider	4–5 (1–16)	3.6	2–3+	15	Suydam (2000)	491
Spectacled Eider	4–5 (1–11)	2.4	2+	—	Petersen et al. (2000)	547
Steller's Eider	5–7 (3–10)	1.7	2–3+	21	Fredrickson (2001)	571

(Continued)

TABLE 11.1 (Continued)

Summary of key reproductive behaviors and characteristics of the sea ducks.

Common name	Clutch size	Max CS/ median	Age first breed	Max lifespan recorded	References	Birds of north america account number
Harlequin Duck	5–7 (3–9)	1.5	2–3+	10	Robertson and Goudie (1999)	466
Long-tailed Duck	6–8	—	2+	—	Robertson and Savard (2002)	651
Black Scoter	8–9 (5–11)	1.3	2+	8+	Bordage and Savard (1995)	177
Surf Scoter	7–8 (6–9)	1.2	2–3+	—	Savard et al. (1998a)	363
White-winged Scoter	8–10 (6–16)	1.8	2–3	18	Brown and Fredrickson (1997)	274
Bufflehead	7–9 (4–17)	2.1	2	14	Gauthier (1993)	067
Barrow's Goldeneye	7–9 (4–23)	2.9	2+	18	Eadie et al. (2000)	548
Common Goldeneye	7–9 (4–20)	2.5	2	12	Eadie et al. (1995)	170
Hooded Merganser	9–11 (5–44)	4.4	2–3+	11	Dugger et al. (1994)	098
Common Merganser	9–11 (6–17)	1.7	2+	14	Mallory and Metz (1999)	442
Red-breasted Merganser	9–10 (5–24)	2.5	2+	10	Titman (1999)	443

Most data are distilled from the Birds of North America species accounts. Question marks (?) indicate knowledge gaps.

BA, Brood amalgamation; BAGO, Barrow's Goldeneye; BRAN, Brant; BUFF, Bufflehead; BWTE, Blue-winged Teal; CAGO, Canada Goose; CBP, Conspecific brood parasitism; COGO, Common Goldeneye; COME, Common Merganser; CS, Clutch size; EPC, Extrapair copulations; GADW, Gadwall; GRSC, Greater Scaup; HADU, Harlequin Duck; HOME, Hooded Merganser; IBP, Interspecific brood parasitism; LESC, Lesser Scaup; LTDU, Long-tailed Duck; MALL, Mallard; RBME, Red-breasted Merganser; SNGO, Snow Goose; WODU, Wood Duck.

bias should favor pair reunion in males as competition for mates is likely fierce (Robertson et al. 1998). Also, the sexual dimorphism present in most sea ducks with males being larger and heavier than females likely reflects intense competition among males for females (Rodway 2007a). Advantages of pair reunion over the formation of new pair bonds include familiarity with the partner, familiarity of the male with the breeding location, reduction of time spent in courtship, and dominance over unpaired individuals (Spurr and Milne 1976, Rowley 1983, Hepp and Hair 1984, Black 1996). Greater reproductive output for birds reuniting than for those forming a new pair has been found in Barnacle Geese (*Branta leucopsis*, Owen et al. 1988) and Blue Ducks (*Hymenolaimus malacorhynchos*, Williams and McKinney 1996). Similar advantages may occur in sea ducks, although the question has not been well studied. Reunion of pairs has also been documented occasionally in dabbling ducks (Dwyer et al. 1973, Blohm 1978, Blohm and Mackenzie 1994, Mitchell 1997), but does not seem to be frequent. Unlike sea ducks, dabbling ducks are relatively short-lived and opportunities for pair reunion are likely not as frequent. In addition, the wintering habitats of dabbling ducks may not be as predictable as those of sea ducks (Anderson and Ohmart 1988, Fredrickson and Heitmeyer 1988), which would in turn limit any advantages to being philopatric to wintering sites.

In polygynous social systems, a male forms a prolonged pair bond with two or more females with overlapping nesting cycles. In theory, polygyny should be prevalent in waterfowl because males do not assist the female in brood rearing and young are highly precocial (Lack 1968). However, natal philopatry is female-biased in waterfowl, an unusual pattern that differs from most other bird groups where males are usually more philopatric than females (Rohwer and Anderson 1988). Female-biased philopatry in combination with pairing on wintering areas has likely countered the evolution of polygyny in waterfowl (Baldassarre and Bolen 2006). Indeed, females return to breed at their natal area with their mate. It is difficult for males to establish pair bonds with more than one female on wintering areas. Sea ducks exhibit female natal and breeding philopatry, and in some species, males are also philopatric to their breeding sites. Male and female polygyny has been documented in

territorial sea ducks and forms on the breeding areas after migration. Philopatry and attachment to territories and nest sites by females, loss of mate, or failure to establish a territory and familiarity between the birds involved lead occasionally to polygyny in Barrow's Goldeneyes (2% of 222 pairs of Barrow's Goldeneyes, Savard 1986) and Buffleheads (Gauthier 1986, 1993; Savard 1986).

Timing of Pairing

Most migratory bird species pair at their breeding areas. Waterfowl are unique in pairing at their wintering areas and this behavior has important implications. All sea ducks follow this general pattern (Table 11.1), although the timing of pairing is still unclear for some species (Rohwer and Anderson 1988, Rodway 2007a). Within a species, the timing of pairing varies with age (Spurr and Milne 1976), the timing of nuptial plumage acquisition, resource availability, physical condition, and winter severity (Baldassarre and Bolen 2006). In sea ducks, confusion arises because of the frequency of reuniting by pairs that occurs in late fall. In mergansers, pairing chronology varies between species, with Hooded Mergansers (*Lophodytes cucullatus*) pairing early, whereas pairs form later in Red-breasted Mergansers (*Mergus serrator*, Kahlert et al. 1998). In Harlequin Ducks, pairing begins early in fall after birds have regained their nuptial plumage (Gowans et al. 1997), and continues throughout winter (Robertson et al. 1998). Some Common Goldeneyes (*Bucephala clangula*) pair in December (Afton and Sayler 1982).

Several hypotheses have been proposed to explain timing of pairing in waterfowl (Rodway 2007a,b). The male costs and female benefit hypothesis poses that early pairing benefits females by protection from disturbance and harassment by other males, but Torres et al. (2002) and Rodway (2007a,b) did not find much support for this hypothesis. In Common Eiders, paired birds spent more time feeding and less time in aggressive interactions than unpaired birds (Ashcroft 1976). Rodway (2007a) argues that winter pairing in waterfowl evolved because of the benefits of prolonged periods for mate assessment and for improving mate coordination. He favors the mutual choice hypothesis, which attributes a greater role to the female in determining the timing of pairing (Rodway 2007a). However, these two hypotheses are not mutually exclusive, and more studies are needed to better quantify advantages of early pairing in sea ducks. Common Eider females rely solely on their body reserves for both laying and incubation (Milne 1976, Hario and Öst 2002) and most of these reserves originate from wintering or spring staging areas (Hario and Hollmén 2004). Thus, the role of the male may vary between wintering and breeding areas. While his role may improve feeding conditions for the female in winter, this does not necessarily translate to a similar function on breeding areas. Ashcroft (1976) showed that feeding rate of female Common Eiders was greater when their mate was close by, but Hario and Hollmén (2004) showed experimentally that females whose mate had been removed could feed undisturbed. The defense of an area around its mate on breeding areas is likely related to safeguarding paternity rather than providing undisturbed feeding conditions for their mate. In other species or populations where females obtain the nutrients for egg production on breeding areas, guarding by the male is likely a combination of both reducing disturbance and mate guarding, but detailed studies are lacking for most species. In Surf Scoters, males follow their mate and act aggressively toward other scoters; they do not defend a specific area but apparently a small moving area around the female. However, the aggressiveness is not constant and pairs can be seen occasionally resting together (Morrier et al. 2008).

Two pairing periods have been described in Common Eiders, one in fall likely involving reuniting birds and one in spring representing new pair formation (Gorman 1974, Spurr and Milne 1976). During these two periods, androgen production increased in males indicating that hormonal concentrations are related to courtship behaviors. Seasonal changes in the rate of displays were due in part to changes in the production of androgens by the testis (Gorman 1974), and triggered by day length (Lofts and Murton 1968). Resurgence of sexual behavior in fall has been reported in several species of birds and may be a common occurrence (Nice 1937, Jenkins et al. 1963). In sea ducks, where repairing is frequent, a dual pattern of pair formation in fall and spring is likely the norm. Recent observations of subadult White-winged Scoter females paired with adult males at breeding sites (S. Slattery, pers. obs.) suggest that some pairs may form earlier than

thought. Clearly, in sea ducks, timing of pair formation may be more variable within and between species than previously recognized and needs further quantification.

Forced Copulations

Forced copulations are common in waterfowl and in some species have been considered as a secondary mating tactic (McKinney et al. 1983, Gauthier 1988). However, the frequency of forced copulation attempts varies greatly among species. Forced copulations and extra-pair copulations (EPCs) have not been frequently documented in sea ducks (McKinney et al. 1983, McKinney and Evarts 1997). The degree of territoriality of a species and the frequency of forced copulations are inversely related (McKinney et al. 1983). Indeed, it has never been observed in territorial Buffleheads or Barrow's Goldeneyes (Savard 1985, Gauthier 1986). Common and Barrow's Goldeneye males take several minutes before mounting a prone female, a delay that makes forced copulation difficult (Afton and Sayler 1982, Savard 1985). In Common Eiders, females whose mate had been removed were not harassed by other males but fed undisturbed. Some females solicited copulations from males but most attempts were declined (Hario and Hollmén 2004). Rare instances of males mating with more than one female have been reported to occur in Common Eiders (Goudie et al. 2000), King Eiders (*Somateria spectabilis*; Suydam 2000), White-winged Scoters (*Melanitta fusca*; Brown and Fredrickson 1997), Buffleheads (Gauthier 1986), and Common Mergansers (*Mergus merganser*; Mallory and Metz 1999), and mixed paternity has been recorded in two broods of Red-breasted Mergansers (Titman 1999), but none of these instances were identified as forced matings (Table 11.1).

Courtship Behavior

McKinney (1992) listed four complex tasks associated with pair formation: demonstrating interest, attracting and maintaining attention, establishment of a bond, and countering interference from rivals. Pairs are formed through complex courtship behaviors specific to each species (Johnsgard 1965, McKinney 1992). Complex displays undoubtedly help isolate species from each other but are not foolproof as suggested by the occasional presence of hybrids (Martin and Di Labio 1994, Eadie et al. 1995, Livezey 1995). Sea duck courtship behaviors are spectacular and have been well described (Myres 1959a,b; Johnsgard 1965; Palmer 1976; Cramp and Simmons 1977). In most sea ducks, courtship is most prevalent in late winter and spring when small courting parties of multiple males courting a single female are common. Courtship parties often occur at the periphery of large flocks of already paired birds raising the question as to when most pairs formed. Courtship behavior is minimal during repairing (Savard 1988a, Gowans et al. 1997) but quite intense during the formation of new pair bonds. The ontogeny of courtship behaviors from juveniles to adult birds and the respective role of the male and the female in mate selection are two future research areas.

Philopatry and Breeding Propensity

In most land birds, natal philopatry is stronger in males than females, but in waterfowl, it is the reverse (Anderson et al. 1992). Female-biased natal and breeding philopatry is pronounced in every species of Mergini for which data are available (Table 11.1). In some populations of Common Eiders, as many as 98% of females return to their natal area and 94%–100% return to their previous breeding site such as a section of breeding island (Goudie et al. 2000). Fidelity to a specific nest bowl is much less common and seems to be associated with successful hatching. Bustnes and Erikstad (1993) found that only 25% of the successful females reused their nest bowl the following year and none of the unsuccessful ones did. In all five species of cavity-nesting sea ducks, subadult females return to their breeding area before they are sexually mature to prospect for nest sites (Table 11.1; Eadie and Gauthier 1985). Subadult females of a few ground-nesting sea ducks also are known to return to their natal areas (Robertson and Goudie 1999, Oppel and Powell 2010). Recent studies on White-winged Scoters documented the presence of subadult females (possibly not yet sexually mature) paired with adult males on the breeding areas (S. Slattery, pers. comm.). This report is intriguing and deserves to be quantified through further studies. If a common event among sea ducks, such pairings would make it difficult to interpret the result of spring pair surveys. Breeding propensity or the proportion of adult females that nest has

been identified as an important parameter in the dynamics of sea duck populations, but it has yet to be adequately quantified in most species. Estimates of breeding propensity averaged about 75% in a Common Eider population (Coulson 1984). Nonbreeding is thought to occur in Steller's Eiders (Quakenbush and Suydam 1999) and appears to be related to weather and predator density. However, it was not clear whether some birds did breed elsewhere in some years. Bond et al. (2008) found a high breeding propensity (92%) in a population of Harlequin Ducks. In general, breeding propensity is difficult to adequately measure.

Philopatry to molting and wintering areas is not as well documented as breeding philopatry but is likely strong as well (Breault and Savard 1999; Chapters 8 and 9, this volume). Recent satellite telemetry studies have confirmed site fidelity to molting and wintering areas in several species of sea ducks (Savard and Robert 2013, Sea Duck Joint Venture 2014). However, some birds change molting and wintering locations between years and causes are still unclear but are likely related in part to breeding status, reproductive success, and bird condition.

SPACING BEHAVIOR

Spacing behavior is common in waterfowl and has been the subject of much speculation about its function (McKinney 1965, Anderson and Titman 1992). Spacing strategies are related to intraspecific competition for resources such as food, nesting sites, and mates and vary considerably among species. Spacing helps disperse nesting efforts, reducing nest vulnerability to predators (McKinney 1965), and may contribute to regulation of local populations (Amat 1983, Savard 1988a,b, Pöysä and Pöysä 2002).

Spring Spacing Behavior and Pair Territories

In spring, most adult sea ducks are paired and flocks of paired birds are the norm at staging areas. Unpaired courting birds often form small peripheral and ephemeral groups. Such groups are composed of a female and a small number of adult males. Paired males are often seen defending a small area around their mate, protecting her from unpaired males. Unpaired males and subadults often form distinct flocks. No territorial defense has been described on any spring staging area.

At breeding areas, spacing strategies vary between species. For example, in North American dabbling ducks, spacing ranges from simple mate defense in Northern Pintails (*Anas acuta*) to strong territorial behavior in Northern Shovelers (*A. clypeata*, McKinney 1965, Titman and Seymour 1981). A similar gradient is found in sea ducks, from a lack of pronounced spacing in the colonial Common Eider to extremely well-developed intra- and interspecific territorial behaviors in goldeneyes and Buffleheads (Table 11.1; Savard 1982, 1984, 1988a).

Many definitions of territory have been proposed (Anderson and Titman 1992). In birds, the territory has been variously defined as any defended area (Noble 1939), to a fixed, actively defended area and exclusive with respect to rivals (Brown and Orians 1970). Among the Mergini, goldeneyes are considered to be the most aggressive as they are highly territorial, both intra- and interspecifically. Pairs establish breeding territories from which they exclude conspecifics and congeners. Territories are fiercely defended by the male, even when the female is absent, and provide the female with an undisturbed feeding site. Males exclude all other goldeneyes but their mate from the territory (Savard 1982). Territorial males display less vigorously against neighboring pairs than strange conspecifics (Savard 1982, Einarsson 1990), which is a pattern common in territorial birds (Weeden and Falls 1959, Wunderle 1978). Einarsson (1990) studied the settlement pattern of Barrow's Goldeneye pairs in Iceland and discovered that superior feeding areas were settled faster than inferior ones; defensibility was an important component of territory quality; territory size decreased with the proportion of territory boundary exposed to intrusion; and birds in small territories spent more time feeding than those in large territories. Gauthier (1987b) showed that in Buffleheads, food did not influence territory size, but density and conspecific intruder pressure did.

McKinney (1965) indicated that pair territoriality in waterfowl was primarily associated with stable habitats and, as suggested by Nudds and Ankney (1982), was a function of a bird's ability to defend the resource. Similarly, Einarsson (1990) found defensibility to be an important component in the settlement pattern of Barrow's Goldeneye pairs. All three species of *Bucephala* are cavity nesters, but this is unlikely the unique cause of territorial behavior as other cavity-nesting species such

as Wood Duck (*Aix sponsa*), Common Merganser, and Hooded Merganser are not territorial. Savard (1982, 1988a) and Einarsson (1988, 1990) suggested that the main role of pair territories in Barrow's Goldeneye and other *Bucephala* species is to provide food resources. Wood Ducks, Common Mergansers, and Hooded Mergansers exploit different habitats and food than goldeneyes and Buffleheads, and these resources may not have been conducive to the development of territoriality (Dugger et al. 1994, Hepp and Bellrose 1995, Mallory and Metz 1999). The origin and development of territoriality in waterfowl has yet to be fully understood (Titman and Seymour 1981, Savard 1988a).

Gauthier (1987a, 1988) presented an alternative viewpoint; he argued that territoriality in Buffleheads served to provide the female with undisturbed feeding time and protect her from harassment by other males. Protection of the nest site by competitors could be an additional benefit of the site-specific nature of the territory and explain why a female defense system evolved into defense of a fixed site (Gauthier 1993). Gauthier (1988) contended that the female was the most important resource to the male. He developed a simple graphical model to illustrate how factors that increase variance of reproductive success in stable versus variable habitats could influence the relative advantage of territorial behavior versus extra-pair mating (or forced copulation) as a secondary mating tactic. Gauthier's model predicted that EPC should be rare and mate attendance high in stable habitats, as found for most species of sea ducks. It is possible that pair territories in the genus *Bucephala* serve multiple functions, protecting both critical and defendable resources such as food and possibly nest sites, but also protecting the female and providing undisturbed foraging time. It is intriguing that mate guarding has been suggested to play such an important role in species such as Buffleheads, given that so few instances of EPCs or forced copulations have been reported. Perhaps male defense of the female is so effective that such opportunities are rare, or perhaps extra-pair mating may be more common than observed behaviorally. Mating systems of sea ducks present an interesting opportunity for future detailed parentage analyses; several studies have used molecular genetic methods to examine multiple maternity in sea ducks, but there has yet to be a comparable effort to consider multiple paternity. In most studies of multiple maternity, techniques limited consideration to only the maternal genotype such as protein fingerprinting (Andersson and Åhlund 2000, 2001), or the study was restricted to samples of only possible mothers and males were not sampled (Jaatinen et al. 2009, 2011a). Interestingly, the assumption of little or no multiple paternity could confound estimates of multiple maternities and potentially affect conclusions of kinship among host–parasite pairs. The role of territoriality and mate guarding in *Bucephala* offers potential for additional research.

Interspecific territoriality is quite rare in birds but is highly developed among goldeneyes and Buffleheads (Donaghey 1975, Savard 1984, Gauthier 1987b, Savard and Smith 1987). For example, Barrow's Goldeneyes are as aggressive toward conspecifics as against Common Goldeneyes and Buffleheads, and the territories of these species are mutually exclusive (Savard 1982, 1988a). Adjacent interspecific territories sometimes occur on highly productive ponds and territorial disputes are frequent and spectacular (Savard 1984). Interactions between congeneric Buffleheads and Common and Barrow's Goldeneyes are similar to intraspecific interactions suggesting that the primary role of the territory is for the defense of food resources. If the territory were solely related to providing an undisturbed area for the female for feeding and to escape sexual harassment then Buffleheads would not be excluded from the territory. Territorial Barrow's Goldeneyes are often aggressive toward other diving ducks and even dabbling ducks with the level of aggression related to the degree of diet overlap (Savard 1984). However, even if food and feeding opportunities are the basis of interspecific behavior, it does not ensure that territory size is directly adjusted to food resources but rather, the proximal factor seems to be pair density (Savard 1982, Einarsson 1990). Indeed, when a pair is removed from an area, the adjacent pair expands its territory (Savard 1982; Gauthier 1986, 1987a,b). In Buffleheads, when males of a pair are removed, they are replaced by previously unmated males, or the territory is taken over by another pair (Gauthier 1986).

In all three species of scoters and in Common Eiders, paired males defend only a moving area around their mate and are not strictly territorial (Bordage and Savard 1995, Brown and Fredrickson 1997, Savard et al. 1998a, Goudie et al. 2000, Fredrickson 2001). All three mergansers and

Harlequin Ducks also have a similar mate defense system and do not defend territories at breeding areas (Dugger et al. 1994, Mallory and Metz 1999, Titman 1999). Long-tailed Ducks, however, defend pair territories from which they exclude conspecifics (Alison 1975). Contrary to the behavior of goldeneyes and Buffleheads, Long-tailed Ducks are apparently not aggressive to other waterfowl species. However, their territorial behavior needs better documentation.

Brood Spacing Behaviors and Brood Territories

Sea ducks rear only a single brood per year and renesting is rare. In most waterfowl, females with broods exhibit some form of spacing behavior, but the establishment of territories is relatively uncommon, occurring usually in species in which both males and females attend the young (Common Shelduck, *Tadorna tadorna*, Patterson 1982) and in species that maintain territories all year (African Black Duck, *Anas sparsa*, Ball et al. 1978; steamer ducks, *Tachyeres* spp., Weller 1976). In sea ducks, females attend the brood alone and breeding and wintering areas are separated so that territories cannot be maintained all year.

Brood spacing behavior in sea ducks is diversified, ranging from the crèching behavior of Common Eiders (Munro and Bédard 1977a; Bustnes and Erikstad 1991a,b; Öst 1999) to the highly developed territorial system of goldeneyes (Savard 1982, 1988a) and Buffleheads (Table 11.1, Donaghey 1975, Gauthier 1987a). Scoters do not have a well-developed brood spacing strategy as females are only slightly aggressive toward other females and unfamiliar young (Savard et al. 1998b). BA occurs in nearly all species of sea ducks, even those that are territorial (Savard 1987, Savard et al. 1998b). Unlike in Common Eiders, it is rare to observe more than one female associated with a brood in Surf Scoters (Savard et al. 1998b) or White-winged Scoters (Brown 1977, Kehoe 1989, Traylor et al. 2008), even though female scoters without broods may be present (Savard et al. 1998b, Savard et al. 2007). The behavior of unsuccessful females has been documented in a few species and is likely similar in most species. In Surf Scoters, unsuccessful females remain for a while at the breeding area before eventually departing for molting areas (Savard et al. 2007). Similar patterns have been documented for unsuccessful Common Eider females (Diéval et al. 2011).

Brood territories established by *Bucephala* are unique among sea ducks. Females defend distinct brood territories and aggressive territorial encounters are frequent between females on adjacent territories. Brood territories do not usually coincide with the pair territory and are often on different ponds (Gauthier 1987a, Savard 1988a, Paasivaara and Pöysä 2008). Gauthier (1987a, 1988) in a detailed study of brood territoriality in Buffleheads suggested that stable habitat and defensibility were necessary components but not always sufficient to explain brood territoriality because several other species of waterfowl found in similar habitats were not territorial. In Buffleheads, territory size is inversely related to both food and brood density but the later association is stronger suggesting that brood density is the proximal factor influencing territory size (Gauthier 1987a). Highly productive ponds support more broods, but the importance of brood density in determining territory size is understandable. If a single brood is found on a water body, no territory boundaries are present. However, if there is more than one brood, female motivation or aggression levels related to body condition, brood size, or lake productivity and chance events will determine the fate of brood encounters and sometimes lead to establishment of adjacent brood territories.

In Barrow's Goldeneyes and Buffleheads, securing a food supply for ducklings is an important function of brood territories (Savard 1984, Gauthier 1987a, Einarsson 1988). Territorial behavior has an indirect spacing effect with late-hatching broods actively avoiding established broods, often moving to adjacent water bodies. Encounters among broods can yield different outcomes including complete or partial BA, establishment of a defended territory boundary, brood expulsion from the breeding pond, and sometimes death of ducklings (Savard 1987, Savard and Smith 1987).

Female *Bucephala* with young exclude other waterfowl species from their brood territory. In areas of sympatry, female Barrow's Goldeneyes dominate female Bufflehead and exclude their broods with the same intensity as conspecific broods (Savard 1987). In larger-bodied Barrow's Goldeneye, territorial behavior may negatively affect the survival of other waterfowl broods (Einarsson 1985, Savard 1988a,b).

Winter Spacing Strategies and Winter Territories

Flocking behavior varies greatly among sea ducks. Wintering Common Eiders form flocks of several thousand birds (Bordage et al. 1998), whereas Buffleheads occur in small, dispersed flocks (Erskine 1971). In general, winter spacing is related to food distribution and abundance. Eiders feeding on dense mussel beds (*Mytilus* spp.) form some of the largest flocks (Guillemette et al. 1993, Bordage et al. 1998), although scoters and Long-tailed Ducks also occur in large flocks. Species feeding on fish schools, such as the Red-breasted Merganser, are found in medium-size flocks of a few hundred birds (Titman 1999). Climate conditions affect flocking behavior, with species wintering in frozen areas and associated with polynyas, open water surrounded by sea ice, can exhibit some of the greatest number of birds per unit area (Spectacled Eider, *Somateria fischeri*, Petersen et al. 1995; Common Eider, Goudie et al. 2000).

The distribution and flock size of wintering Common Eiders in the Gulf of St. Lawrence vary with ice conditions with larger flocks found in the coldest part of the winter (Guillemette et al. 1993). In British Columbia, Surf Scoters form larger and denser flocks when feeding on blue mussels along rocky shorelines than when feeding on clams in mud and sandy habitats (Iverson 2002). Herring spawns (*Clupea* spp.) create some of the most spectacular concentrations of sea ducks, attracting thousands of birds each year (Munro and Clemens 1931, Vermeer et al. 1997, Sullivan et al. 2002, Rodway 2003).

Little is known about the behavior of birds within a flock but aggressive interactions are common and their functions unknown. One would expect a dominance hierarchy within flocks, with larger males dominating smaller females and adult birds dominating yearlings or subadults. Such dominance may explain some of the segregation of ages and sexes observed in the winter distributions of some species (Hepp and Hair 1984, Rodway et al. 2003, Iverson et al. 2004).

First-year male Surf Scoters increasingly associate with other first-year males as flock sizes increase, and the proportion of females is higher in large flocks (Iverson et al. 2004). In most species of sea ducks, winter spacing behavior appears mostly dictated by distribution of food resources (Iverson 2002). Among Harlequin Ducks wintering in the Strait of Georgia, British Columbia, group sizes were smaller in winter than spring and nearly 10 times larger during herring spawning events (Rodway 2003). Mate defense is a common occurrence even in flocks. Spacing then takes mostly the form of a moving area around the female. The transition in flock composition from unpaired birds in early winter to mostly paired birds in late winter and spring has not been studied.

Most species of sea ducks form roosting flocks at night (Linsell 1969, Jones 1979), and movements to roosting sites can be impressive. Wintering Long-tailed Ducks feed in offshore waters of Nantucket Sound but return each evening by the thousands to these protective waters to roost. Common Goldeneyes and Common Mergansers also form roosting aggregations (Breckenridge 1953, Reed 1971). Similar nocturnal concentrations occur in scoters, eiders, and Harlequin Ducks (Robertson and Goudie 1999, Rodway 2003).

Spacing in flocks and between pairs in winter does not involve defense of specific areas and is relatively passive with limited aggression and is linked to the immediate area around a given individual or its mate. Defense of winter territories has so far been only documented in Barrow's Goldeneye (Savard 1988a). The importance of habitat stability and defensibility for territoriality is illustrated by the behavior of wintering Barrow's Goldeneyes in eastern and Western Canada. In Western Canada, Barrow's Goldeneyes winter in ice-free waters along stable rocky shores. There, several Barrow's Goldeneye pairs can maintain winter territories where they exclude other goldeneyes and even other competing waterfowl (Savard 1988a). On the east coast, Barrow's Goldeneyes winter in ice-dominated waters and contend with shifting ice associated with tidal regime. Accordingly, habitat conditions are highly unstable resulting in the absence of territorial behavior in winter. Winter territorial behavior has not been reported in Common Goldeneyes or Buffleheads wintering on the west coast, possibly because they use more estuarine habitats that harbor fewer sessile organisms and are less predictable.

BROOD PARASITISM

Brood parasitism occurs when females lay their eggs in the nests of other individuals and the recipient or host female provides all care thereafter for the eggs and resulting offspring. The best-known examples include obligate interspecific brood parasites such as the European Cuckoo (*Cuculus canorus*) and the North American Brown-headed Cowbird (*Molothrus ater*; Payne 1977, Rothstein 1990, Davies 2000). These species, along with other parasitic cuckoos, cowbirds, and finches, rely entirely on other species to raise their young. However, facultative parasitism that is only used by a subset of females or on a limited range of occasions can also occur, and it is the most frequent mode of parasitism in waterfowl (Yom-Tov 1980, 2001; Eadie et al. 1988; Rohwer and Freeman 1989; Sayler 1992; Reichart et al. 2010). Facultative parasitism occurs both within species as CBP and among species as interspecific brood parasitism (IBP), although CBP appears to be much more common of the two. Facultative parasitism is difficult to detect and is likely to be underreported. Of the 236 species of birds in which CBP has been recorded (Yom-Tov 2001), over 30% (74 species) are waterfowl and CBP occurs in 45% of all waterfowl species (166 species). Among sea ducks, the proportion is even higher because 9 of 15 species (60%) have been reported to exhibit at least occasional or frequent CBP (Table 11.1). No information is known for Black Scoters (*Melanitta americana*) or Surf Scoters and the proportion could be even higher. Accordingly, CBP is disproportionally common among the sea ducks, making this clade of special interest in understanding this intriguing reproductive behavior (Chapter 10, this volume).

Conspecific Brood Parasitism

Occurrence and Frequency

CBP within the Mergini is most common in goldeneyes, Buffleheads, and mergansers (Table 11.1). Parasitism also occurs frequently in Common Eiders, and possibly in King Eiders, but appears to be infrequent in Spectacled (Petersen et al. 2000) and Steller's Eiders (*Polysticta stelleri*, Fredrikson 2001). Likewise, CBP is frequent in White-winged Scoters in some populations, but has not been reported in Black and Surf Scoters although Yom-Tov (2001, citing Bengtson 1972b) included Common Scoters (*Melanitta nigra*) in his list of conspecific brood parasites. CBP occurs in Harlequin Ducks in Iceland but has not been reported in North America, and there is one record of CBP in Long-tailed Ducks in Manitoba, Canada (Robertson and Savard 2002). Clearly, there is interesting variation among the sea ducks in the extent to which CBP occurs, although it is noteworthy that at least some evidence of CBP has been reported (even if rare) in all 13 species for which there are data (Table 11.1). This pattern suggests that CBP is a behavior that potentially can be expressed by any member of the tribe depending on ecological and environmental conditions.

The frequency of CBP is usually measured as the percentage of active nests that are identified as being parasitized but varies considerably among populations and years within species of sea ducks (Table 11.2, Figures 11.1 and 11.2). Assessing the true frequency of CBP within any population can be difficult, given the similarity of eggs within a species and the challenges of correctly identifying parasitic eggs. Most sea ducks lay eggs at longer intervals than many other ducks, often one egg every other day (Bellrose 1976, Palmer 1976, BNA Accounts, Table 11.1). Thus, criteria used to identify parasitism included two or more eggs laid per day, new eggs laid in the nest after the onset of incubation, long breaks in egg laying, large clutch sizes exceeding an expected number of eggs, and differences in egg size, shape, or color (Eriksson and Andersson 1982). Eadie (1989), Pöysä et al. (2001, 2010), and Eadie et al. (2010) examined differences in egg morphology as a method to detect parasitic eggs in goldeneyes. Differences were measured as the maximum Euclidean distances of the most dissimilar eggs in a nest based on length, width, and egg mass. Both studies found that differences in egg morphology could be used to identify parasitic eggs in nests of goldeneyes. In contrast, Ådahl et al. (2004) reported that this technique was not successful for Common Goldeneyes in Germany, although repeatability for egg measurements within females in that study was low, unlike in Pöysä et al. (2001, 2010) and Eadie et al. (2010). Egg morphology may be a useful technique to assess parasitism in sea ducks but would need to be validated for any species or population.

Not unexpectedly, the use of different criteria can give rise to different estimates of CBP, although few studies have examined this systematically. Eriksson and Andersson (1982) found that estimated frequencies of CBP in Common Goldeneyes in Sweden varied from 23% to 41% of nests depending on three criteria used: >1 egg laid per 24 h, skips in egg laying >3 days, and eggs laid after incubation. Eadie et al. (2010) further

evaluated these criteria and three other criteria to assess the frequency of CBP within populations of Barrow's Goldeneyes and Common Goldeneyes in British Columbia, Canada. The additional criteria were egg morphology, direct observations of two females laying in the same nest box, and clutch sizes exceeding 12 eggs (11 being the maximum number of postovulatory follicles in laying birds). Their results confirmed Eriksson and Andersson's (1982) findings that different criteria can yield different estimates of CBP. Using combinations of criteria and recursive partitioning analyses, Eadie et al. (2010) found that three criteria were effective at identifying 75% of parasitic nests: two eggs laid per day, eggs laid after incubation, and clutch sizes >12 eggs. While some parasitized nests would be missed using only these criteria, such an approach would allow at least a minimum estimate of the frequency of CBP in a population.

With the advent of molecular genetic techniques, the ability to estimate the frequency of CBP in natural populations has increased considerably. Several recent studies have applied molecular techniques to detect multiple maternity in nests of sea ducks (Table 11.2). Andersson and Åhlund (2000, 2001) developed the novel technique of protein fingerprinting, whereby a small sample of albumin is extracted from an egg and, using isoelectric focusing electrophoresis in immobilized pH gradients, protein banding patterns are detected that are unique to individual females. An advantage of this technique is that eggs can be sampled when freshly laid, whereas DNA-based methods require blood from newly hatched ducklings or from embryos that have been incubated long enough that sufficient DNA can be acquired. Eggs can disappear or be abandoned before incubation and these nests would be missed in a DNA-based sample. Protein fingerprinting alleviates this concern, although assignment of maternity can be more difficult if banding patterns are similar among related females in a nesting colony. Using this technique, Andersson and Åhlund (2001) found that CBP occurred in 68% (13 of 19) of the Common Goldeneye nests examined, with a range of 1–5 parasitic females identified per nest. Of the 234 eggs analyzed, 36% were assigned to females other than the incubating female. The proportion of the clutch that comprised parasitic eggs ranged from 8% to 67%.

Waldeck et al. (2004, 2011) used protein fingerprinting to examine CBP in several populations of

TABLE 11.2

Estimates of the frequency of CBP (percentage of nests and percentage of eggs) within species of sea ducks.

Species	Location	Years	Percentage of nests parasitized (n)	Percentage of parasitic eggs (n)	Number of parasitic eggs (or maximum clutch size observed)	Method	Sources
Common Eider	Spitsbergen, Norway	1989–1990	13% (232)–21% (198)	7.5% (year: 1989) 13.0% (year: 1990)	1 (10%), 2 (1%–4%), 3 (1%–3%), 5 (2%)	Observational; timing, egg morphology	Bjørn and Erikstad (1994)
Common Eider	La Perouse Bay, MB, Canada	1991–1993	20%–42%			Observational; timing, egg morphology	Robertson (1998)
Common Eider	Finland Baltic Sea, Finland	2001–2002	19% (70)–21% (94) 4%–55% among 9 sites	6% (754 eggs) 1%–14% among 9 sites	1 (67%), 2 (27%), 3 (6%); max CS: 8	Protein fingerprinting	Waldeck et al. (2004)
Common Eider	La Perouse Bay, MB, Canada	2002	31% (86)	8% (416 eggs)	$CS_{non-parasitized}$: 4.58 ± 0.68 $CS_{parasitized}$: 6.64 ± 1.01	Protein fingerprinting	Waldeck and Andersson (2006)
Common Eider	Prince Heinrich Island, Svalbard	2007–2008	19% (63)	6% (670 eggs)		Protein fingerprinting	Waldeck et al. (2011)
Common Eider	Christiansø, Denmark	1998–2001	38% (63)	17% (243 ducklings)	2–4	DNA microsatellite markers (7)	Tiedemann et al. (2011)
Common Eider	Akureyri, Iceland	2000–2001	30% (77)	42% (263 ducklings)	2–4	DNA microsatellite markers (7)	Tiedemann et al. (2011)
Common Eider	Gulf of Finland	2001–2003	51% (37)	34% (152 eggs)	1–2 (58% mixed clutches)	DNA microsatellite markers (7)	Hario et al. (2012)
King Eider	Karrak Lake, NWT, Canada	1995	16% (37)		Max CS: 10	Observational	Kellet and Alisauskas (1997)
Spectacled Eider	YK Delta, Alaska, USA	1995	1% (280)		Max CS: 11	Observational; timing, egg morphology	Petersen et al. (2000)
Bufflehead	British Columbia, Canada	1982–1986; 1957–1962	5% (75); 7.5% (264)		Max CS: 20	Observational; clutch size	Gauthier (1987b, 1993) and Erskine (1990)

(Continued)

TABLE 11.2 (*Continued*)

Estimates of the Frequency of CBP (percentage of nests and percentage of eggs) within species of sea ducks.

Species	Location	Years	Percentage of nests parasitized (n)	Percentage of parasitic eggs (n)	Number of parasitic eggs (or maximum clutch size observed)	Method	Sources
Barrow's Goldeneye	Iceland		21.5% (228)		Max CS: 23	Observational; clutch size	Erskine (1990)
Barrow's Goldeneye	British Columbia, Canada	1981–1984	7%–20% among years (409)			Observational; clutch size	Savard (1988b)
Barrow's Goldeneye	British Columbia, Canada	1984–1994	31%–68% among years; 0%–65.2% among lakes	24.3% (2758 eggs)	3.35 ± 0.22 SE; max CS: 28	Observational; timing, egg morphology	Eadie (1989) and Eadie et al. (2000)
Barrow's Goldeneye	British Columbia, Canada	2006		29% (288 ducklings)		DNA microsatellite markers (19)	Jaatinen et al. (2009)
Barrow's Goldeneye	British Columbia, Canada	2006–2007		27% min (183 ducklings); 9% min (122 ducklings)		DNA microsatellite markers (19)	Jaatinen et al. (2011a)
Common Goldeneye	Svartedalen, Sweden	1971–1980	41% (37) during egg-laying; 23% (35) during incubation		$CS_{non-parasitized}$: 7.6 ± 0.6 (12) $CS_{non-parasitized}$: 8.4 ± 0.5 (28) $CS_{parasitized}$: 10.6 ± 1.1 (7) $CS_{parasitized}$: 9.6 ± 0.9 (7)	Observational; timing, egg morphology	Eriksson and Andersson (1982)
Common Goldeneye	Göteborg, Sweden		68% (19)	36% (234)		Protein fingerprinting	Andersson and Åhlund (2001)

Species	Location	Years	Parasitism		Clutch size	Method	Reference
Common Goldeneye	British Columbia, Canada	1984–1994	32%–70% among years; 0%–78% among lakes	21.5% (1594 eggs)	3.6 ± 0.4 SE Max CS: 24	Observational; timing, egg morphology	Eadie (1989) and Eadie et al. (1995)
Common Goldeneye	Göteborg, Sweden	1989–1991	64% (22)	44.9% (187 eggs)	1–11	Protein fingerprinting	Åhlund (2005)
Common Goldeneye	Maaninka, Finland	1999–2005	46%–53% among years (551 successful nests)			Observational; timing, egg morphology	Paasivaara et al. (2010)
Hooded Merganser	Minnesota, USA	1981–1985	45% (91)		Max CS: 26	Observational; clutch size	Zicus (1990)
Hooded Merganser	Missouri, USA		34% (429)				Dugger et al. (1994)
Common Merganser	Finland		20%		Max CS: 18	Observational; clutch size	Mallory and Metz (1999)
Red-breasted Merganser	New Brunswick, Canada	1984	64% (125)		$CS_{non-parasitized}$: 9.9 ± 0.03 SD; $CS_{parasitized}$: 11.6 ± 1; Max CS: 24	Observational; clutch size	Young and Titman (1988)
Red-breasted Merganser	Europe		4.5%; 21%		Max CS: 56	Observational; clutch size	Hildén (1964) and Bengston (1972b)
Red-breasted Merganser	Iceland		45% (356)			Observational; clutch size	Bengston (1972b)

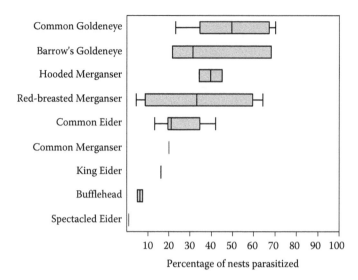

Figure 11.1. Variation in the estimated frequency of CBP among populations/years for several species of sea ducks, measured as the percentage of nests parasitized. Box and whisker plots indicate the median (central vertical line), 25%–75% quartiles (shaded box), and 10%–90% quantiles (whiskers). Data and references in Table 11.2.

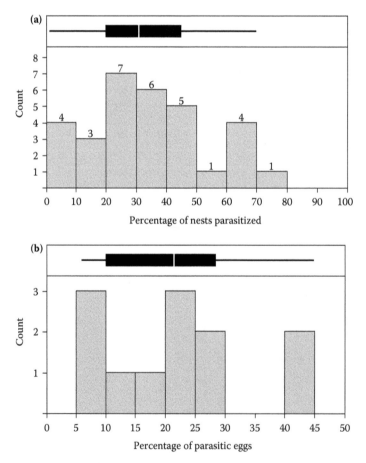

Figure 11.2. Variation in the estimated frequency of CBP within populations of sea ducks, measured as (a) the percentage of nests parasitized and (b) percentage of parasitic eggs. Box and whisker plots indicate the median (central vertical line), 25%–75% quartiles (shaded box), and 10%–90% quantiles (whiskers). Data and references in Table 11.2.

Common Eiders. In a population near the central Baltic Sea, parasitic eggs occurred in 20%–22% of nests and 6% of eggs were laid by females other than the host. In Hudson Bay, Canada, 31% of nests were the result of more than one female and 8% of the eggs were laid by parasitic females (Waldeck and Andersson 2006). In the high arctic near Svalbard, 19% of nests contained eggs of more than one female and 6% of eggs were parasitic (Waldeck et al. 2011; Table 11.2). Thus, frequencies of CBP detected by protein fingerprinting in several different populations of Common Eiders appear to be relatively consistent (20%–30% of nests and 6%–8% of eggs).

Other studies have used microsatellite DNA methods to genotype females and offspring and assign maternity in sea ducks (Jaari et al. 2009). Jaatinen et al. (2009) used 22 polymorphic microsatellite DNA loci for the Barrow's Goldeneye to evaluate maternity of young hatched in nest boxes in British Columbia. Of 288 ducklings sampled, 29.5% (85) were apparently parasitic. Jaatinen et al. (2009) were able to assign maternity to other females in the population for half (43) of these ducklings. In a follow-up study, Jaatinen et al. (2011a) analyzed 455 ducklings, of which they were able to assign 13% (61) as parasitic, while another 20% (89) were ambiguous (low likelihood or not assigned to a female). Tiedemann et al. (2011) used seven polymorphic microsatellite loci to genotype Common Eider ducklings and found that 30%–38% of the nests in two locations contained ducklings that were not assigned to the incubating host female and were parasitic, and the proportion of parasitic ducklings across all parasitized and unparasitized nests was 16%–17%.

We summarized available estimates of the frequency of CBP within sea ducks that were determined by both observational and molecular techniques (Table 11.2), and present the percentage of parasitized nests and parasitic eggs within species—pooled over species and populations (Figure 11.2). Although the database is slowly growing, it is clear that most studies have focused on only four to five species and data are limited or lacking for the remainder of the tribe. Based on the data available, CBP is most frequent in the goldeneyes and mergansers, Common Eiders, and to a lesser extent King Eiders (Figure 11.1). The percentage of parasitized nests within the Mergini ranges from 1% to 80% with a median estimate of 31% (n = 31 observations; Figure 11.2a). Estimates

of the percentage of parasitic eggs range 6%–45% with a median of 22% (n = 12 observations; Figure 11.2b). Future studies using molecular and observational methods for several of the other species would be extremely informative.

Ecological Correlates

The frequency of CBP, both within and among species of sea ducks, correlates most strongly with two ecological factors: cavity-nesting and high nesting densities, either in colonies as in Common Eiders or in clusters on islands or peninsulas where nests are in close proximity such as Red-breasted Mergansers, King Eiders, and even Harlequin Ducks in Iceland. Indeed, the only sea ducks for which CBP is rare are those species nesting at low densities in dispersed ground nests (Figure 11.3). Among King Eiders, CBP was only recorded as being frequent when the nests were clustered on islands and peninsulas at Karruk Lake Northwest Territories, Canada. It is intriguing that almost all species of sea ducks will reuse previous nest sites. Reuse is particularly common among cavity-nesting species but also occurs for ground nesters that reuse previous nest bowls or even take over the nest bowls of other females (Table 11.1).

All of these observations suggest that the occurrence of CBP in the Mergini is tied to the availability of suitable (or preferred) nest sites and/or the density of nesting females. However, there are many reasons why a relationship between CBP and nest density or nest location might arise:

1. *Nest-site availability*—suitable sites might be limited such that females without a site may have little recourse but to lay eggs parasitically or forgo breeding altogether (Eadie et al. 1988, Sayler 1992). Nest-site limitation would be more likely in cavity-nesting ducks, especially birds needing large, relatively scarce cavities such as Barrow's Goldeneyes, Common Goldeneyes, or Common Mergansers. However, nest sites might also be limited for ground-nesting ducks if only some sites are suitable such as among arctic breeding species when few snow-free sites are available prior to snowmelt.

2. *Host availability*—the occurrence of CBP may be dictated instead by the availability of active host nests to parasitize (Rohwer and

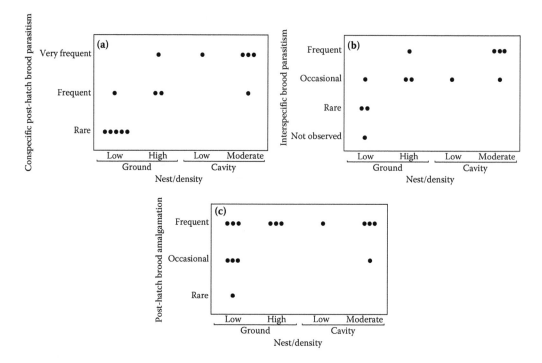

Figure 11.3. Frequency of (a) CBP, (b) IBP, and (c) post-hatch BA as a function of nest type and nesting density in the sea ducks. Each point represents a single species.

Freeman 1989). High densities of nests provide more opportunities and reduced search efforts for parasitic females.

3. *Nest competition*—parasitism might result simply as an inadvertent consequence of two females contesting the same site; both females might lay eggs in the site with the intent to incubate the clutch, but only one female succeeds (possibly the first to begin incubation), such that the second female is evicted and becomes an accidental parasite (Erskine 1990, Semel and Sherman 2001). Higher densities of females attempting to procure the same nest sites could lead to higher rates of inadvertent parasitism.

4. *Nest-site quality*—the quality (rather than the quantity) of sites or the hosts might vary; females might lay some or all of their clutch parasitically in these nests if the increment in the quality of the site or the experience or ability of the host female leads to greater reproductive success in sites that are safer from predation (Pöysä 1999, 2003a, 2006; Pöysä and Pesonen 2007).

5. *Nest loss*—if females lose their nest to predation during egg laying or early

incubation, they may be able to lay some additional eggs or may have eggs in development but lack sufficient reserves to establish a new nest or clutch (see Hanssen and Erikstad 2013). However, if nest or host densities are high, parasitism may provide those females who have lost their nest early in the breeding season with an opportunity to lay at least a few eggs.

6. *Information*—a final possibility is that high densities of nest sites or fixed locations of cavities or other nest sites afford females with different levels of information. Information could influence CBP in a number of ways. For example, some females may simply have more information about the location of possible sites, perhaps due to greater age or experience. Females lacking such information would be more constrained in their choices, which could lead them to turn to parasitism. This idea is a variant of the nest limitation hypothesis, although it is not strictly the number of nests that is limited, but rather the information about the location of suitable sites. Alternatively, higher densities of

nests or fixed cavity locations could provide females with more opportunities to
prospect and learn about the quality of nest
sites and thereby allow parasites to more
effectively target their egg laying as a variant of the nest site/host quality hypothesis.

Other explanations for CBP exist and the previous list is not intended to be comprehensive.
However, the list clearly illustrates that even a
simple and apparently strong pattern between the
frequency of CBP and ecological correlates such as
nest density or cavity locations may have multiple
interpretations. Recent work on several species of
sea ducks is helping to illuminate which of these
hypotheses might apply.

Conceptual Framework

A number of reviews have summarized the hypotheses on the adaptive basis for CBP (Andersson 1984,
Eadie et al. 1988, Petrie and Møller 1991, Sorenson
1991, Sayler 1992, Lyon 1993). Typically, these
hypotheses have classified CBP into three types:

1. *Best of a bad job* (BOBJ)—females lay
   parasitically when environmental or
   phenotypic factors limit the ability to
   breed otherwise (constraint) or when
   environmental conditions are unfavor-
   able such that the prospects for successful
   reproduction by nesting are low (restraint).
   A variant of this hypothesis focuses on nest
   loss whereby females lay parasitically after
   their nest is destroyed during laying.

2. *Professional or lifelong specialist parasites*—females
   never raise their own young and only lay in
   the nests of other females. Presumably, these
   females have higher lifetime fitness if rare in
   the population, because they are freed from
   the costs of parental care and so can invest
   in enhanced production of eggs. When the
   relative frequency of such individuals is
   high in the population, they do less well
   because they need nesting host females to
   parasitize. Negative frequency-dependent
   selection is proposed to act such that the
   frequencies of nesting and parasite females
   in the population are stabilized in a mixed
   evolutionary stable strategy. This category
   has yet to be found in the tribe Mergini or
   in any other species of conspecific brood
   parasite (Lyon and Eadie 2008).

3. *Fecundity enhancement*—nesting females also
   lay some additional eggs parasitically
   and increase fitness beyond that possible
   through nesting alone, presumably by
   bypassing some of the constraints or costs
   of raising the additional eggs/young on
   their own.

Lyon and Eadie (2008) suggested a revision to
the traditional set of hypotheses based on a conceptual framework derived from Sorenson's
(1991) reproductive decision model modified by
Lyon and Eadie (2008). According to this model,
females either nest or not, depending on ecological and phenotypic conditions. Females of either
type (nesters or non-nesters) might also pursue
parasitism, but the context differs. For non-nesting females, parasitic egg laying allows them to
better match their investment to the probability
of success, rather than attempting to nest (above
optimum) or not breeding at all (below optimum). Parasitism by nesting females, in contrast,
allows females to adjust their reproductive effort
upward when conditions are good, beyond that
which might be expended in a single nest.

The conceptual framework unifies all four possible nesting options (not breeding, parasitize
only, nest only, parasitize and nest) as part of a
single continuum that varies from low to high
investment and low to high expected fitness
benefits. The framework also captures the variation found both within and among species of
sea ducks (Tables 11.1 and 11.2, Figures 11.1 and
11.2). Indeed, not all options may be realized in
all species. Further, brood parasitism can be combined with nesting in various ways over a female's
lifetime to provide a flexible life history—females
adjust their reproductive investment and options
to prevailing ecological and social conditions and
since these conditions likely vary among females,
years, or locations, so too should their reproductive allocation. Of particular importance in the
Sorenson (1991) model is the distinction between
parasites with and without their own nests
because these two contexts likely involve different
constraints, and different hypotheses may apply
(summarized by table 1 of Lyon and Eadie 2008).

Relative Role of Relatives

One additional factor must be considered before
the adaptive basis of CBP can fully be explored

for sea ducks—the potential influence of kinship among females in a population and particularly between parasites and hosts. Thirty years ago, Andersson and Eriksson (1982) raised the possibility that hosts and parasites might be related, fundamentally changing the nature of CBP from a parasitic interaction to a potential form of cooperative breeding. It is no accident that Andersson and Eriksson (1982) originally proposed this idea based on their work with Common Goldeneyes, although they suggested that the same principle could apply to many species of waterfowl. The key is that females are the philopatric sex in waterfowl, both in terms of natal philopatry (returning to natal areas where they hatched) and breeding site fidelity (returning to an area where they previously nested; reviewed by Anderson et al. 1992). Female philopatry in sea ducks is even more prominent than in other waterfowl—high levels of female philopatry have been observed in all species (Table 11.1)—and in some species, breeding site fidelity of males is strong as well (Savard 1985, Gauthier 1987c). In some species, both sexes may also be strongly faithful to winter areas (Limpert 1980) and retain long-term pair bonds (Table 11.1). The consequence is that populations of sea ducks may exhibit strong local genetic structure (Chapter 2, this volume), and females may be returning to nest on breeding sites with close relatives present in the population.

Andersson and Eriksson (1982) proposed that elevated levels of relatedness among females could result in apparently parasitic females laying eggs in the nests of kin. Andersson (1984) explored this further and showed that indeed relatedness between host and parasites could facilitate the evolution of CBP. Sixteen years later, a flurry of theoretical papers spurred new interest in this idea (Zink 2000, Andersson 2001, Lopez-Sepulcre and Kokko 2002), with the emerging consensus that kinship could facilitate CBP, provided that the costs to host were low and some degree of kin recognition existed (Lyon and Eadie 2000). These theoretical advances, in turn, inspired several empirical studies using both protein fingerprinting and microsatellite DNA analyses to directly test the hypothesis that hosts and parasites were related (review in Eadie and Lyon 2011). Several of these field studies were conducted on sea ducks, perhaps not surprisingly given the high levels of female philopatry. In particular, three species— Common Goldeneyes, Barrow's Goldeneyes, and

Common Eiders—have been the focus of much research, not only on the role of relatives but also the causes, consequences, and potential adaptive basis of CBP in the Mergini. The work on Common Goldeneye has been particularly extensive and provides one of the best-studied examples of CBP in a precocial bird.

Case Studies of Goldeneyes

Studies of CBP in the Common Goldeneye date to Grenquist (1963). However, Andersson and Eriksson's (1982) pivotal paper provided one of the first attempts to explain why CBP has evolved mainly among ducks; specifically, (1) suitable nest sites are often rare, (2) nests are relatively easy to locate, (3) female philopatry makes it likely that host and parasite are genetically related, and (4) ducks do not defend their nests during the laying period. Andersson (1984) expanded these ideas, including the presentation of his original model of kin selection and CBP. This work clearly set the stage for the interest in CBP that has occurred in the intervening 30 years.

Several studies of CBP in Common Goldeneyes followed, notably an extensive series by Pöysä (1999, 2003a,b, 2004, 2006) and Pöysä et al. (2010) in Finland; further work in Sweden by Andersson and Åhlund (2000, 2001, 2012), Åhlund and Andersson (2001), and Åhlund (2005); and work in British Columbia done in conjunction with studies of Barrow's Goldeneyes (Eadie 1989, 1991; Eadie and Fryxell 1992; Eadie et al. 1998).

Pöysä (1999, 2003a,b, 2004, 2006) and Pöysä et al. (2010) focused on the potential role of predation in facilitating the evolution of CBP in goldeneyes and the degree to which parasites can gain information about the potential safety of nest sites. In a series of experimental and observational studies, Pöysä showed that (1) predation risk varies considerably between nest sites in Finland and that nests were more likely to be parasitized if they were located in sites that were not depredated during the nesting attempt the year before (Pöysä 1999); (2) parasites prefer newly erected nest boxes on lakes where real nest predation risk is low, but they do not discriminate among nests with evidence of experimental predation (broken eggs) versus control nests (Pöysä 2003a); (3) females prospect for nest sites at the end of the nesting season by exploring areas with active nest sites more frequently than sites without nests, and sites visited more frequently by females were more likely to be parasitized the following year (Pöysä 2006); and (4) the occurrence of CBP is better explained by nest-site characteristics than host traits, implying that parasitic females target a given nest based on factors related to the nest site rather than the host (Paasivaara et al. 2010). These observations led Pöysä and his colleagues to argue that "... more attention should be paid to factors associated with nest site attractiveness and quality when studying laying decisions of parasites and the occurrence of CBP in general" (Paasivaara et al. 2010:662).

Pöysä and Pesonen (2007) developed a model to reexamine the potential role of risk spreading by laying eggs in multiple nests to reduce the risk of any single egg being depredated. Earlier analyses indicated that the conditions under which such within-generation bet hedging would evolve were restrictive and an unlikely explanation for CBP (Bulmer 1984). However, when Pöysä and Pesonen (2007) revisited Bulmer's (1984) model by including variation among nest sites in nest safety, and the fact that parasite females might preferentially select safe sites, the fitness advantages of CBP were more evident: "the use of risk assessing, instead of random risk spreading, makes parasitic laying evolutionarily advantageous" (Pöysä and Pesonen 2007:94). Pöysä's work on Common Goldeneyes in Finland suggests that nest predation and nest-site quality may be a more important determinant of CBP than previously thought. The influence of nest predation in other populations needs to be further explored.

Andersson and Åhlund (2000, 2001) followed up earlier work on the Swedish Common Goldeneye population, using a newly developed protein fingerprinting technique along with detailed nest video recordings. Most interestingly, Åhlund and Andersson (2001) followed the egg-laying patterns of individual Common Goldeneye females and found that females employed one of the three different tactics: nonparasitic nesting, pure parasitism with egg laying only in the nests of other females, and a combination of parasitism followed by nesting. All three tactics were equally common, although the reproductive payoffs varied considerably. Females that laid parasitically and then produced their own clutch produced 1.5 times more offspring than nonparasites and twice more than pure parasites—by combining parasitism with normal nesting, some females were able to double their reproductive output. Eadie (1989) found similar patterns in Common and Barrow's Goldeneyes in British Columbia. Åhlund and Andersson (2001) and Eadie (1989) found that females switched tactics between years, indicating that the tactics are flexible. Moreover, Åhlund and Andersson (2001) suggested that nesting parasites may be in their prime reproductive years, in agreement with the predictions of the model of Sorenson (1991).

In earlier work, Erskine (1990:53) presented a different view of how CBP might arise in the genus *Bucephala* (goldeneyes and Buffleheads). He proposed the following hypothesis:

> Most *Bucephala* ducks, when using natural cavities in the wild, initiate laying to incubate the eggs that are in the nest when they complete their clutches. They continue laying even if another bird lays there during the same period, unless the disturbance or conflict involved exceeds their tolerance level. Either bird may finish laying and start incubating before the other is ready to do so. The evolutionary pressures acting on both birds are the same, and they differ little from those on a female with no competition for its nest-site.

Thus, Erskine (1990) argued that there might be little that is "intentional about CBP (joint-laying)," but rather "joint laying need not be deliberate, but may arise from competition for nest-sites".

Eadie (1989) and Pöysä (2003b) argued against this hypothesis based on observations that parasitic egg laying occurred in experimental decoy egg nests (new or empty nests with experimental eggs added, but without a host female), and these eggs were not incubated, implying that the parasitism was intentional and not a result of a female being excluded by another host female. Åhlund's (2005) detailed study of the behavioral tactics of host and parasite Common Goldeneyes argues further against the accidental parasitism hypotheses. He found striking difference between hosts and parasites in timing of egg laying, duration of time and behavior on the nest, deposition of down, covering eggs on departure, and time spent on the nest as the egg-laying sequence progressed. Åhlund concluded that while accidental joint laying does occur, competition for nests was not a major cause of brood parasitism in his study population.

Nonetheless, it may be premature to discard the accidental parasitism hypothesis for all cases of CBP. Certainly, there are observations—even in Åhlund's (2005) study—where it does appear that two females both lay eggs in a nest with the intention of incubating the nest. The fact that females will lay in nests with decoy eggs and then proceed to incubate the nest suggests that at least some species are not deterred from using nests with eggs already present (Eadie 1989, Pöysä 2003b; see Odell and Eadie 2010 for Wood Ducks). Semel and Sherman (2001) invoked Erskine's (1990) hypothesis to account for a large proportion of CBP observed in their study of Wood Ducks, although they also suggested that young females laid eggs as intentional parasites without any intent to incubate. Given extensive evidence that female Common Goldeneyes select for safe nest sites in Finland (Pöysä 1999, 2003a,b, 2004, 2006), it remains possible that laying eggs in a high-quality nest site—even if there is a risk of eviction prior to clutch completion—could yield greater reproductive success than laying more eggs in a lower-quality site. Hence, parasitism could still be an unintended consequence of competition for high-quality nest sites. Even if a female is evicted, she benefits through her eggs that are now tended by the unintentional host, and the evicted female could yet lay eggs in another nest or establish one of her own. It is not clear that such a mechanism would apply to other non-cavity-nesting sea ducks or other waterfowl, but this remains an intriguing possibility given the frequent use of old nest bowls and perhaps even selection for sites with down or safe sites in Common Eiders (Ruxton 1999, Fast et al. 2010, but see Robertson 1998, Lusignan et al. 2010).

Research on Barrow's Goldeneyes has been less extensive than on Common Goldeneyes, but similar patterns have emerged. Based on detailed observational and experimental studies of a population of Barrow's Goldeneyes in British Columbia, Canada, Eadie (1989, 1991) found that parasitism was not simply an inadvertent consequence of two females choosing the same nest site by chance—the frequency of parasitism was significantly higher than that expected if females were distributed randomly among nest sites. Moreover, parasites did not incubate nests into which they had laid eggs, even if given the chance when the host female was experimentally removed. The frequency of parasitism increased significantly when population densities were high and nest sites were less available. Experimental manipulation of nest-site densities on four lakes confirmed the strong effect of nest availability on parasitic behavior (Eadie 1991). However, low levels of parasitism occurred even when nest sites were abundant, and some females laid parasitically in addition to establishing a nest of their own. Of the 38 females followed throughout egg laying, 21% acted only as parasites in a given year, 55% attended their own nest, while 24% laid some eggs parasitically in addition to establishing nests of their own. In all of these last cases, females laid parasitically in a neighboring nest located within 50–500 m on the same lake or pond. Females also switched behaviors between years suggesting that professional parasites did not exist in this population. Females that pursued both parasitism and nesting in the same year as dual nesters laid more eggs than females that only nested, while the parasite-only females laid the fewest eggs, consistent with Åhlund and Andersson's (2001) results for Common Goldeneye. Parasite-only females tended to be younger, while nesting females and dual nesters tended to be older. Consistent with the model of Sorenson (1991), age and nest availability influence parasitism by inexperienced birds, whereas parasitism by older birds may be a mechanism of fecundity enhancement (Lyon and Eadie 2008).

Jaatinen et al. (2009, 2011a) used molecular genetic techniques to study CBP in a second population of Barrow's Goldeneye in British Columbia. In these studies, Jaatinen et al. were particularly interested in the role of relatedness among females and spatial proximity of nests as determinants of host–parasite interactions. Jaatinen et al. (2009) found that the number of parasitic eggs laid in host nests increased with decreasing distance between nests and increased with the degree of relatedness between the host and the parasite. In a follow-up study, Jaatinen et al. (2011a) found even more remarkable patterns with regard to kinship. Using molecular genetic techniques, they were able to assign parasitic offspring to parasitic females. Parasites included females both with and without their own nests, as observed by Eadie (1989, 1991) and Åhlund and Andersson (2001) and consistent with Sorenson's model (Lyon and Eadie 2008). However, Jaatinen et al. (2011a) further observed that the nesting status of the parasite (with or without their own nest) played an important role in how parasites chose host nests. As in their first study, the probability of parasitism increased with the relatedness to the host and with the spatial proximity to host nests. However, females that both nested and parasitized laid more parasitic eggs as their degree of relatedness to the host increased, while no such relationship existed for non-nesting parasites. Non-nesting parasites laid fewer eggs in total. The study by Jaatinen et al. (2011a) is intriguing because it suggests that nesting and non-nesting parasites in the same population use different host selection criteria and the influence of kinship varies for the two types of parasites. More work remains to be done, but it is clear that kinship, spatial proximity, and differences in the tactics of hosts and parasites influence CBP in both Barrow's and Common Goldeneyes, offering intriguing possibilities for further research.

Case Studies of Common Eiders

CBP has been found in almost every population of Common Eider (Table 11.2). Studies of CBP in the Common Eider are interesting because it is one of the few species of ground-nesting sea ducks in which this behavior has been well studied and the mechanisms that facilitate this behavior may be different from those that influence CBP in cavity-nesting sea ducks. The intensity of CBP appears to be lower in Common Eiders than for other sea ducks in that only 1–3 parasitic eggs are usually laid (Bjørn and Erikstad 1994, Robertson 1998, Waldeck et al. 2004) and large dump nests of >20 eggs are rare, although occasionally observed (Goudie et al. 2000). Parasitic eggs tend to be laid early in the host's laying sequence (Bjørn and Erikstad 1994); Robertson (1998) found that >75% of parasitic eggs were laid before the host laid her third egg. The occurrence of CBP (% of nests parasitized) is positively correlated with breeding density in several different populations (Bjørn and Erikstad 1994, Robertson 1998, Waldeck et al. 2004). Robertson (1998) found that 20%–40% of nests were parasitized in a colony near Hudson Bay, with the highest frequency in the year when nest density was highest. Similarly, Waldeck et al. (2004), using protein fingerprinting techniques, found that the frequency of parasitized nests on nine islands in the central Baltic Sea ranged from 4% to 55% of nests, and there was a significant positive correlation of CBP with nest density.

The correlation of parasitic egg-laying and nest density suggests that either availability of nest sites or hosts may play some role in this species. Several authors have suggested that CBP is not a result of limited nest sites, as there are often unused sites available such as empty nest bowls or sites that have been used in previous years (Robertson 1998, Waldeck et al. 2004, Waldeck and Andersson 2006). This observation has led to the suggestion that perhaps CBP arises in Common Eiders as a mechanism to procure safe nest sites, perhaps similar to the mechanism proposed by Pöysä (2003a,b, 2004, 2006) for Common Goldeneyes (Ruxton 1999, Waldeck and Anderson 2006, Fast et al. 2010).

This idea is supported by the observation that Common Eiders will reuse old nest sites—nest bowls are well established, reuse of old sites is the norm, and in some populations, nest bowls may be reused for many years, perhaps representing hundreds of years of occupation (Fast et al. 2010). Predation on Common Eider nests can be severe, particularly from arctic foxes (*Alopex lagopus*), red foxes (*Vulpes vulpes*), polar bears (*Ursus maritimus*), gulls (*Larus* spp.), Common Ravens (*Corvus corax*), American Crows (*C. brachyrhynchos*), and jaegers (*Stercorarius* spp.). Nesting success or the percentage of nests that successfully hatch can vary widely among locations, ranging from as little as 9% to more than 90% (Goudie et al. 2000). Similarly, hatching success, or the percentage of eggs that hatch, ranges from 25% to over 80%. Predation appears to fall more heavily on the first one or two eggs laid, possibly because the females are not in attendance and the eggs are more vulnerable (Robertson 1998). For example, Waldeck and Andersson (2006) found that partial predation occurred in 22% of 86 nests and the first 1 or 2 eggs disappeared before the female began to incubate in many of those nests. In addition, the authors found a large number of nests where the first egg disappeared and the nest was subsequently abandoned. The results indicated there is substantial predation early in the laying sequence (see also Robertson 1998, Andersson and Waldeck 2006). Given this pattern of high nest predation and frequent reuse of some nest sites, it is perhaps not unreasonable that selection for safe nest sites could be a factor facilitating CBP in Common Eiders.

Fast et al. (2010) tested this idea experimentally by placing vegetation or feather down in nest bowls prior to nesting. The authors found no difference in the likelihood of successful nest establishment between control and experimental nests; however the onset of incubation occurred 2–3 days earlier in nest bowls that contained feather down versus those with little nesting material. Fast et al. (2007) speculated that nest bowls containing feathers or vegetation may be selected first if the materials resulted in increased nest survival, possibly by enhancing egg concealment during the early laying period when predation is high and females are not present at the nest.

Robertson (1998) presented a different hypothesis. He reported that in 31% of parasitized nests, the putative parasitic egg was laid before the host laid her own first egg. The host female thus began laying in a nest in which another egg was already present, and then subsequently completed her clutch and incubated that nest. Robertson (1998) proposed that CBP was a result of nest takeover and egg adoption. In those nests where females took over nests, predation on the first eggs was lower than in comparable nests produced by only a single female. Robertson suggested that the presence of an egg in a nest would indicate that the site was a safe nest location since the egg had not been depredated, and therefore the nest may be attractive to another female. The benefit of obtaining a safe nest site could outweigh the potential cost of caring for the additional eggs.

Ruxton (1999) extended this hypothesis by suggesting that females can detect variation in the predation risk associated with different nests, and use this information to target nests with low predation risk as sites for laying parasitically. Ruxton (1999) further suggested that such parasitism would pay if eggs laid parasitically in the nest of highly attentive females have a higher chance of survival than if they were laid in the females own nest. Waldeck and Andersson (2006) used protein fingerprinting to explore this further for the population of Common Eiders studied earlier by Robertson (1998). They found that in 41% of mixed clutches, another female laid before the host started laying, corroborating that takeover of nests accounts for many of the mixed clutches in that population. Hario et al. (2012) used DNA microsatellites and found that 58% of parasitic eggs in a Common Eider population in Finland were the first or second laid egg. Waldeck and Andersson (2006) also found differences in predation rates during the early egg laying period:

there was no predation on nests that were taken over, whereas 5 of 59 (9%) single female clutches were depredated. When combined with a larger 3-year sample of nests from Robertson (1998), the proportion of nests that escape predation before incubation starts was 80% for mixed female nests but only 57% for single female nests, a highly significant difference. Waldeck and Andersson (2006) concluded that taken-over nests have higher early survival than other nests, consistent with the hypothesis that at least some CBP in Common Eiders is driven by selection for safe nest sites.

However, the pattern does not appear to be ubiquitous. Lusignan et al. (2010) tested the risk assessment hypothesis and whether parasites preferentially lay their eggs in safe nest sites in a population of Common Eiders nesting in Labrador, Canada. They used protein fingerprinting to quantify the frequency and distribution of CBP among three habitats that varied in exposure to avian predators. Nest-site safety did not explain patterns of CBP among habitats and parasitized and nonparasitized nests did not differ in their probability of nest survival. Nests in dense woody vegetation had the highest rates of survival but the lowest frequency of CBP. In contrast, levels of CBP were higher in nest shelters. Shelters were a relatively safe nesting site, but Lusignan et al. (2010) suggested that shelters were used more by parasites because "… they (1) are highly visible and easily found and observed for host activity, (2) provide protection against avian nest predators and increase the probability that a parasite will find a host nest during laying, and (3) protect evidence of previous nesting attempts from being scattered by wind and precipitation, making shelters attractive to both normal nesters and parasites" (Lusignan et al. 2010:770). Accordingly, Lusignan et al. (2010) argued that nest visibility and effects on detectability were a more important determinant of CBP than nest safety for their study population.

A full understanding of CBP in Common Eiders cannot be obtained without considering the role of kinship. As for other sea ducks, females are highly philopatric, resulting in strong genetic structure in local populations (Tiedemann et al. 1998, McKinnon et al. 2006, Waldeck et al. 2008, Sonsthagen et al. 2010). In several different populations, females nesting in close proximity are more closely related than birds nesting apart (McKinnon et al. 2006, Waldeck et al. 2008, Sonsthagen et al. 2010), although this is not always the case (Andersson and Waldeck 2007, Tiedemann et al. 2011, Hario et al. 2012). However, in all cases where it has been examined—with protein fingerprinting or DNA microsatellite analyses—there is a significant nonrandom pattern of greater relatedness between hosts and parasites, and this remains true even after accounting for local microgeographic genetic structure. Andersson and Waldeck (2007) found that females laying eggs in the same nest were, on average, closer kin than nesting neighbor females in a Hudson Bay population. Likewise, in a Baltic Sea population, Waldeck et al. (2008) found that average pairwise relatedness between hosts and parasites was significantly greater than that expected by chance, even after accounting for decreasing relatedness among females with increasing distances among nest sites. Last, in a Danish population of Common Eiders, Tiedemann et al. (2011) also found that relatedness between hosts and parasites was higher than expected by chance. They further discovered that parasitized females were older than nonparasitized females, the percentage of nests parasitized and the number of parasitic eggs increases with the age of nesting females, and the number of host eggs in the nest of parasitized females decreases significantly with age. The authors suggested that CBP "… may allow those older females unable to produce as many eggs as they can incubate to gain indirect fitness without impairing their direct fitness: genetically related females specialize in their energy allocation, with young females producing more eggs than they can incubate and entrusting these to their older relatives" (Tiedermann et al. 2011:3237). In other words, rather than a form of parasitism, CBP may constitute cooperation among generations of closely related females ("help from Grandma"). Remarkably, Tiedemen et al. (2011) found three females incubating nests with none of their own biological offspring, and all of these females were >15 years old. High levels of kinship could explain why some females take over nests in which other females have already laid eggs; if those eggs were laid by a relative, the "host" female who takes over the nest may instead be providing parental care to increase the fitness of a relative and thereby enhancing her own indirect fitness.

Several consistent patterns emerge from case studies of goldeneyes and Common Eiders. First,

CBP does not appear to be simply an artifact of competition for nest sites, given that it occurs so frequently in many different species and populations under different circumstances, and often when nest sites are readily available. The accidental parasitism hypothesis is not well supported as a general explanation for this widespread behavior, although there may well be instances of CBP that arise in this manner. Second, there is increasing evidence that nest-site quality or host quality may be important factors influencing how and where females lay their eggs. Whether this is a primary selective force driving CBP or just one of the many factors that influence the frequency and fitness tradeoffs of this behavior remains to be determined. Third, a growing body of research with a variety of molecular techniques has demonstrated that hosts and parasites are often related, implicating a role for kinship in CBP as proposed by Andersson (1984). The mechanisms of kin bias and kin recognition are poorly understood. Moreover, parasitism occurs frequently between non-kin so other factors must be involved. Nonetheless, a growing suite of molecular tools, accompanied by newly emerging telemetry and remote sensing technology, could help to resolve some of these uncertainties in the next decade. With the wealth of information already available, Common Goldeneyes, Barrow's Goldeneyes and Common Eiders provide ideal model systems with which to pursue these inquiries.

Costs to Host and Population Dynamics

Surprisingly, for a behavior that has long been referred to as "parasitism," remarkably few studies have looked at the costs of parasitism to the host in sea ducks, or many other species of waterfowl. This gap in information is important for two reasons. First, it is important to know if there are costs to the hosts to determine whether this is truly a parasitic interaction. Parasitism may not be costly. Waterfowl have precocial young and the cost of raising additional young may not be large for many species. Indeed, some studies have even suggested that there could be a benefit to the host, obtained through increased survival of their own young via dilution or selfish-herd effects in enlarged broods (Anderson 1984, Eadie and Lumsden 1985). Second, costs of parasitism figure importantly in theoretical models of kinship and CBP. If parasitism is costly to the host, recent models predict that parasites should avoid relatives; conversely, if there is little cost to the host, parasites should instead prefer to lay in nests of a relative, and hosts should be willing to accept the eggs of kin (Zink 2000, Andersson 2001, Lopez-Sepulcre and Kokko 2002).

Andersson and Eriksson (1982) used egg additions of 1, 4, or 7 eggs to nests of Common Goldeneyes to examine responses of host females to experimental parasitism. Host responses differed among treatments. Nests were deserted (even though egg-laying continued) when seven eggs were added, but not when 1–4 eggs were added. Further, when four eggs were added early in the hosts' laying sequence, the host reduced her own clutch size by the number of eggs added, whereas no reduction occurred if the experimental eggs were added late. Andersson and Eriksson (1982) developed an optimality model that predicted that hosts should reduce their own clutch size by one-half the number of parasitic eggs added. The timing of parasitism during early versus late egg-laying may be an important determinant of how hosts respond. Host clutch size is reduced for each egg added during early parasitism, whereas no reductions occur during late parasitism.

Eadie (1989) repeated Andersson and Ericksson's (1982) experiments for Common and Barrow's Goldeneyes in British Columbia. In contrast to Andersson and Ericksson's results, hosts did not reduce their own clutch size when 2, 4, or 6 parasitic eggs were added experimentally. The only response by hosts was when large numbers of parasitic eggs were added (8–12 eggs). In those cases, hosts were more likely to abandon the nest altogether. In contrast, Jaatinen et al. (2009) using molecular genetic assignment techniques, found that Barrow's Goldeneye hosts did reduce their own clutch size when parasitized, but surprisingly, the effect was greater when parasites were more closely related to the host. Further work on the costs of parasitism is clearly warranted. It is important to note that studies should be based on experimental, rather than observational methods. Assessments of the cost of parasitism will not be representative if parasitized hosts are a nonrandom subset of the population: low-quality females unable to defend their nest or, conversely, high-quality older females willing to accept additional eggs.

One area of interest for future studies of CBP is to explore the link between parasitism and population dynamics. A number of theoretical papers have suggested that CBP could influence population dynamics (Eadie and Fryxell 1992, Nee and May 1993, Eadie et al. 1998, de Valpine and Eadie 2008), resulting in population cycles or even local population crashes. Eadie and Fryxell (1992) presented evidence that parasitism could lead to fluctuating population dynamics in Barrow's Goldeneyes. Similar links have been suggested for Wood Ducks—high levels of CBP may lead to local population crashes. Semel et al. (1988) and Semel and Sherman (1995, 2001) argued that these crashes might be the result of intense nest box management programs and the provision of high densities of nest boxes, leading to abnormal frequencies of CBP and nest failure. In contrast, addition of nest boxes does not seem to exacerbate CBP in Barrow's Goldeneyes, possibly because goldeneyes are territorial and high population densities and extreme levels of CBP are limited (Savard 1984, 1988a,b; Eadie et al. 1998). Nonetheless, the linkages between CBP and population dynamics remain to be explored further, and additional empirical field studies are needed. Field experiments may be important if well-intended nest box management programs exacerbate CBP and lead to less stable or more variable populations.

Future Research Needs

We have focused on CBP because the study system offers much potential for future research. We suggest six areas in which further investigations will be informative:

1. Develop a better understanding of the variation in female reproductive nesting behaviors. Are there different types of parasitism as described by the model of Sorensen (1991)? Goldeneyes exhibit alternative laying strategies with parasite-only, nesting-only, and dual parasitism and nesting. What factors influence which of these strategies a female might adopt?

2. What is the role of nest-site quality or host quality on CBP and nest-site selection by conspecific parasites? Could CBP arise because of competition for high-quality sites, either as an adaptive mechanism favoring parasitic egg-laying (Pöysä 1999, 2003a,b, 2004, 2006) or as an inadvertent consequence of competition among females for these nest sites (Erskine 1990, Semel and Sherman 2001)?

3. What are the fitness cost and benefits of CBP to parasites and hosts, and particularly for females that are following different life-history trajectories? Investigating this question will require following females throughout their lifetime rather than single-season snapshots.

4. What is the role of kinship? Hosts and parasites appear be related in several species, but not all. It is unclear whether kinship is a key driver in the evolution of this behavior or just one of the several factors that could shape the frequency of CBP within and among populations.

5. What is the influence of CBP on population dynamics? Theoretical models suggest that CBP could influence population variation or stability, but empirical evidence is lacking. CBP could have both frequency-dependent and density-dependent effects and therefore it is rich for game theory analysis. It is important to learn how parasitic behavior impacts population dynamics, and hence how management programs might influence both behavior and population viability. How does CBP interact with high densities of artificial nest sites or creation of nesting islands where nesting densities may be high?

6. What factors determine whether a female lays parasitically one year and not the next? Are there fixed, different trajectories for young and experienced breeders? Do the behaviors have a heritable basis or does CBP vary among years or locations simply based on environmental and habitat conditions?

Interspecific Brood Parasitism

Occurrence and Frequency

Brood parasitism also occurs between species within the tribe Mergini, although it has been much less studied than CBP. IBP has been recorded in at least 11 species of sea ducks; the only species in which it has not been observed or reported are King Eiders, Steller's Eiders, Black Scoters, and Surf Scoters (Table 11.1). IBP is most frequent in cavity-nesting sea ducks such as Barrow's Goldeneye, Common Goldeneye and Hooded Merganser, as well as in dense colonial ground nests of Red-breasted Mergansers. Much of the parasitic egg-laying occurs among the nests of the other cavity-nesting ducks. Barrow's Goldeneyes, Common Goldeneyes and Buffleheads often lay eggs in each other's nests (Table 11.1). Interspecific parasitism is also common for the Common Merganser, Hooded Merganser and Common Goldeneye where the species co-occur. In eastern North America, mixed clutches of Wood Ducks, Hooded Mergansers and Common Goldeneyes are frequent (Table 11.1).

Among ground-nesting sea ducks, IBP occurs occasionally in Common Eiders, Long-tailed Ducks, and White-winged Scoters. The identity of eggs of other species in these mixed nests includes three species of geese (Black Brant *Branta bernicla*, Canada Goose B. *canadensis*, and Snow Goose *Chen caerulescens*) with Common Eiders, Greater Scaup (*Aythya marila*) with Long-tailed Ducks, Lesser Scaup (*A. affinis*) with White-winged Scoters, and three species of dabbling ducks (Gadwall, *Anas strepera*; Mallard, *A. platyrhynchos*; and Blue-winged Teal, *A. discors*) with White-winged Scoters (Table 11.1).

Ecological Correlates

IBP appears to be found in the same species for which CBP occurs suggesting that there may be common ecological factors that facilitate both behaviors. For eight species in which CBP is occasional or frequent, IBP also occurs occasionally or frequently (Table 11.1). Conversely, for four species where IBP is rare or not recorded, evidence of CBP is also limited. Indeed, there is a statistically detectable association between CBP and IBP ($G_w = 4.61$, $P < 0.05$). The association may be driven by common ecological factors—specifically, cavity nesting, use of nest boxes, and relatively high densities of nesting females in colonies or on islands (Figure 11.3).

At least two hypotheses may explain these patterns. First, IBP may be a carryover from factors that promote CBP. Hence, IBP may have little adaptive value and may be an inadvertent consequence of different species choosing the same nest sites or attempting to parasitize other nests regardless of host identity. Second, IBP may be an adaptive behavior in its own right. If host species are sufficiently similar such that nesting, incubation, and brood-rearing patterns are suitable and the parasitic young are able to hatch and survive, then parasitism of a closely related species could increase the availability of potential hosts. IBP may be more common in sea ducks and other waterfowl simply because, with precocial young, the need for a host species with similar patterns of parental care and diet may be less critical. This explanation might be the case for closely related species in the genus *Bucephala*.

However, a number of factors may cause IBP to be a less successful reproductive tactic than CBP. IBP may be disadvantageous when diet and habitat use of the species differ, such between Hooded Mergansers and Wood Ducks. In those cases, IBP may be more directly related to competition and overlap in use of cavity nest sites and nest boxes. A second potential cost of IBP is that the young of a different species may not respond appropriately or as quickly to alarm calls or brooding behaviors of the foster female and may be more susceptible to predation or mortality. Indeed, Eadie and Lumsden (1985) speculated that differential mortality could benefit the host female if predation fell disproportionately on the parasite young of the other species (the cannon fodder hypothesis). Last, an additional underexplored cost of IBP relates to species sexual imprinting. Young raised in a brood of another species may become sexually imprinted on the foster species, impairing mating and reproductive success of the parasitic young as adults. IBP may lead to observations of hybridization between species such as between Common and Barrow's Goldeneyes (Martin and Di Labio 1994). To date, systematic study of the causes and consequences of IBP in sea ducks and other waterfowl has been limited and much remains to be learned.

BROOD AMALGAMATION

Occurrence and Frequency

BA occurs when broods combine posthatch after leaving the nest. A variety of terms have been used for this behavior, including crèching, gang brooding, brood mixing, adoption, and kidnapping (Eadie et al. 1988). Some of the terms such as adoption or kidnapping imply function or causality of the behavior, whereas other terms have been used to refer to specific types of BA such as that in Common Eiders where multiple females or "aunts" attend the combined brood. Bédard and Munro (1977), Munro and Bédard (1977a,b), and Gorman and Milne (1972) used the term crèche (French meaning nursery) to refer to "a group containing any number of adult female(s) and duckling(s), two or more of which are parentally unrelated" (Bédard and Munro 1977:223). Because the function of the behavior is often not known, Eadie et al. (1988) suggested the neutral

term *posthatch BA*—shortened here simply to BA. There are many possible parallels between CBP and BA as both involve exchange of young among females and, in most cases, there is a donor and a recipient, although relationships can be less well defined in cases of BA. BA has potential costs and benefits for both donors and recipients, but different viewpoints persist because it is not entirely clear that BA consistently has an adaptive basis.

BA occurs in nearly all species of sea ducks (Table 11.1) and is particularly common in Common Eiders (Beauchamp 2000, Goudie et al. 2000), White-winged Scoters (Brown and Fedrickson 1997, Traylor et al. 2008), Surf Scoters (Savard et al. 1998b), and all three species of *Bucephala* (Gauthier 1993; Eadie et al. 1995, 2000), although possibly for different reasons. The only species for which BA has not been recorded is the Hooded Merganser, although it is also rare or infrequent in Steller's Eiders and appears to be only a transitory behavior in Spectacled Eiders (Table 11.1). The frequency of BA varies considerably among and within species (Table 11.3). The percentage of amalgamated broods that contain foreign young can be as low as 3% in Surf Scoters to as high as 60% in Common Eiders (Table 11.3). Likewise, the percentage of foreign young in amalgamated broods ranges 0%–65%, although this has not been well estimated in many species without following marked young (and because of high brood loss before mixing; Table 11.3).

The structure of amalgamated broods also varies widely among the Mergini (Tables 11.1 and 11.3). In some species, amalgamated broods can grow extremely large with over 100–150 ducklings in a single brood as in Common and King Eiders (Suydam 2000) and White-winged Scoters (Brown and Brown 1981, Brown and Fredrickson 1997). Combined broods of 30–50 ducklings are found frequently in Long-tailed Ducks, Surf Scoters, and Common and Red-breasted Mergansers (Tables 11.1 and 11.3). In most of these species, multiple females (1–7) may be in attendance although this is most extreme in Common Eiders where up to 17 females might attend one brood (Bédard and Munro 1977). In other species, such as Buffleheads and goldeneyes, combined broods are smaller and often the result of two broods intermixing, typically with only one female in attendance (Gauthier 1993; Eadie et al. 1995, 2000). Only a single female attended mixed broods of Surf Scoters in Québec (Savard et al. 1998b).

TABLE 11.3

Estimates of the frequency of post-hatch BA within species of sea ducks.

Species	Location	Years	Percentage of broods amalgamated (n broods)	Percentage of young in amalgamated (n broods)	Number attending females	Range of brood sizes in mixed broods (Min–Max)	Method	Sources
Common Eider	Ythan estuary Scotland	1965–1969?			1–85	2–470	Observational; marked females and ducklings	Gorman and Milne (1972)
Common Eider	St. Lawrence River, Québec, Canada	1971			1–17	1–60; 80–120 sometimes observed	Observational; marked females; some ducklings	Bédard and Munro (1977)
Common Eider	St. Lawrence River, Québec, Canada	1973–1974	27 h after hatch: 50% broods (75); 30–40 h: 35% (68); 41–65 h: 33% (12); 1 week: 17% (49)		1–5 (2 most frequent)	1–60 (7–9 most frequent)	Observational; marked females; some ducklings	Munro and Bédard (1977a,b)
Common Eider	Grindøya, Norway	1987–1989	42% of females (126) abandoned broods; 58% tended broods	47% (n = 76 tagged ducklings) with foster mother		1–10	Observational; marked females and ducklings	Bustnes and Erikstad (1991a)
Common Eider	sw Finland, Baltic Sea	1996–1999	Créched broods: 1996: 23.8% (21) 1997: 44.7% (38) 1998: 60.0% (50) 1999: 43.5% (46)				Observational; marked females	Klipi et al. (2001)
Common Eider	sw Finland, Baltic Sea	1995–1998	25% (n = 124) of lone female tended broods (1996); 38% (63) (1997)		1 (29%–44% broods); 2+ (56%–71%)	8–12	Observational	Öst (1999)
King Eider	Canadian Arctic				2–50	Several to >100	Observational	Suydam (2000)

(Continued)

TABLE 11.3 (*Continued*)
Estimates of the frequency of post-hatch BA within species of sea ducks.

Species	Location	Years	Percentage of broods amalgamated (n broods)	Percentage of young in amalgamated (n broods)	Number attending females	Range of brood sizes in mixed broods (Min–Max)	Method	Sources
Spectacled Eider	Canadian Arctic				2–50	Several to >100	Observational	Suydam (2000)
Surf Scoter	Québec, Canada	1994–1996	3% (33) (1995) 18% (11) (1994) 34% (29) (1995) 25% (12) (1996)	0%–53% (1994) 18%–65% (1995) 0%–64% (1995)	1	9–33 (1994) 9–40 (1995) 9–17 (1995)	Observational; some marked females	Savard et al. (1998b)
White-winged Scoter	Alberta, Saskatchewan, Canada	1976	61% (13/21) marked hens tended mixed broods (1984); 59% (16/27; 1985)		1–7	Up to 55 (Alberta); Up to 150 (Sask)	Observational; some marked females	Brown and Brown (1981)
White-winged Scoter	Saskatchewan, Canada	1984–1985	37.2% (n = 35) of females were "donors"; 5.7% acceptors (2000); 18.9% (n = 58) "donors"; 3.5% acceptors; 3.5% both (2001)		1	Largest 64	Observational; marked females and ducklings	Kehoe (1989)
White-winged Scoter	Saskatchewan, Canada	2000–2001	38% ± 4% (456) 32% (84) (1980) 27% (84) (1981) 42% (85) (1982) 41% (95) (1983) 49% (108) (1984)		1 (99%; n = 293) 2 (n = 1) 3 (n = 2)	Largest 39 (2000); 20 (2001)	Observational; marked females and ducklings	Traylor et al. (2008)

Species	Location	Years	Percentage	No.	Values	Method	Reference
Barrow's Goldeneye	British Columbia, Canada	1980–1984		1	4–28 9 (1980) 6 (1981) 11–28 (1982) 4–20 (1983) 4–23 (1985)	Observational; some marked females	Savard (1987)
Barrow's Goldeneye	British Columbia, Canada	1984–1986	23% (87) 14% (14) (1984) 17% (30) (1985) 30% (43) (1986)	1		Observational; some marked females	J. Eadie, unpubl. data
Common Goldeneye	British Columbia, Canada	1984–1986	29% (59) 13% (16) (1984) 33% (16) (1985) 33% (27) (1986)	1		Observational; some marked females	J. Eadie, unpubl. data
Bufflehead	British Columbia, Canada	1980–1984	34% ± 6% (370) 22% (67) (1980) 25% (73) (1981) 45% (78) (1982) 52% (73) (1983) 28% (79) (1984)	1			Savard (1987)
Long-tailed Duck	Churchill Manitoba, Canada	1970		1–4	32 (single female); 52 (4 females)		Robertson and Savard (2002)

Ecological Correlates

Several ecological variables appear to be important factors influencing the occurrence and frequency of BA in many species of Mergini. Perhaps foremost among these variables is the number of birds nesting in close proximity or at high densities. BA is common in colonial species such as the Common Eider (Beauchamp 2000, Goudie et al. 2000) and the White-winged Scoter (Brown and Fredrickson 1997, Traylor et al. 2008), especially when they nest in high densities. Under these conditions, brood encounters are more frequent and there is a greater potential opportunity of amalgamation. In contrast, BA is infrequent or only occasional in King and Spectacled Eiders, which do not nest colonially (Flint et al. 1998, Mehl and Alisauskas 2007, Phillips and Powell 2009). In Surf Scoters, BA is rare in the center of their breeding range where they nest at low density on small lakes but relatively frequent when they nest on large lakes at high density (Savard et al. 1998b). BA is also observed in river specialist species such as Common Mergansers and Harlequin Ducks that live in a relatively narrow confined habitat. BA also occurs more frequently in species or populations where newly hatched broods are subjected to high predation rates—such as gulls (*Larus* spp.) depredating young of Common Eiders or White-winged Scoters. BA is also common in species where females defend brood territories (*Bucephala*), and aggressive interactions among females and young may lead to brood displacement or expulsions. Overall, the general pattern is that BA is more frequent in cavity-nesting sea ducks and ground-nesting species that nest in high densities and rare or occasional in ground-nesting species that nest at low densities (Figure 11.3c), as found for CBP and IBP.

Interestingly, BA also tends to occur more frequently in those species in which CBP is also more common (both BA and CBP are common in 7 of 13 species and both uncommon in 3 species [Table 11.1]; $G_w = 8.85$, $P = 0.09$). This pattern suggests that similar ecological factors may influence both behaviors or life-history traits that predispose some species to CBP also promote BA after hatch.

Evolutionary Explanations: Accident or Adaption?

Our understanding of the proximate and ultimate factors that influence the ecology and evolution of BA is much less developed than that for CBP. Most research conducted to date on BA has focused on a few species and especially the Common Eider, perhaps because of its well-known habit of crèching (Gorman and Milne 1972; Bédard and Munro 1977; Munro and Bédard 1977a,b; Bustnes and Erikstad 1991a,b). Papers by Öst (1999), Kilpi et al. (2001), Öst et al. (2002, 2003a,b, 2005, 2007a,b, 2008), Jaatinen et al. (2011b, 2012), and Jaatinen and Öst (2011, 2013) represent the most comprehensive analysis of this behavior to date. A handful of other studies have been conducted on White-winged Scoters (Brown and Brown 1981; Kehoe 1989; Brown and Fredrickson 1997; Traylor et al. 2004, 2008), Surf Scoters (Savard et al. 1998b), and Barrow's Goldeneyes and Buffleheads (Savard 1987, Eadie and Lyon 1998). Nonetheless, despite its widespread occurrence among the Mergini, much of our current information comes from a few species.

Despite the lack of extensive data (or perhaps because of it), two contrasting explanations for BA have emerged, and these ideas remain highly contested. One school of thought holds that BA is simply a result of accidental mixing of broods with little adaptive basis (Savard 1987, Savard et al. 1998b). A second school of thought holds that BA is an adaptive behavior, although the adaptive mechanisms may vary and multiple hypotheses exist (Eadie et al. 1988; Bustnes and Erikstad 1991a,b; Beauchamp 1997).

Accidental Mixing

In their extensive field studies of Common Eiders in the St. Lawrence River estuary in Québec, Munro and Bédard (1977a) concluded … "In our opinion, the only evolved trait of this behavior is the spontaneous tendency for females to regroup their ducklings in the face of predation with concurrent lowering of normal aggressiveness among them. The contingent stabilization of groups formed following encounters (be these due to predation, chance, or alarm) seems to be a fortuitous consequence of overcrowding at the precise period when family links tend to crystallize" (Munro and Bédard 1997a:768–769). Munro and Bédard (1977a) suggested that newly hatched young were incompletely imprinted on their parent and therefore crèches formed simply because young followed the most broody female. Thus, Munro and Bédard (1977a,b) and Bédard and Munro (1977) argued that brood mixing was largely the result of disruptions caused by gull attacks shortly after hatch when mother–young bonds were not well established.

Savard (1987) extended these ideas to Buffleheads and Barrow's Goldeneyes, suggesting that brood mixing resulted inadvertently from territorial disputes among females with newly hatched broods. When territorial females with broods encounter other broods, the possible outcomes are establishment of territorial boundaries, expulsion of one of the broods from the lake, death of young when Barrow's Goldeneye females kill unfamiliar young (Savard 1987, Eadie and Lyon 1998), or complete or partial BA. Savard (1987) suggested that a number of proximate factors could influence whether BA occurred. These factors include the following:

1. *Female aggressiveness*—dominant or more aggressive females may chase away other broods, possibly to the point where the subordinate female is excluded from the area leaving her young behind. Observations of Barrow's Goldeneyes indicated that the resident female always dominated newly hatched broods.

2. *Age of young*—if the young of two broods are more similar in age, they are more likely to mix easily and the females' ability to distinguish their own young from others is diminished. Lack of discrimination may be especially true for newly hatched broods.

3. *Strength of the female–duckling and duckling–duckling bonds*—at least in cavity-nesting ducks, and likely many waterfowl, imprinting on the female and siblings is largely auditory until after nest exodus; visual imprinting is weak (Gottlieb 1971) and the female–duckling bond is not strong. Hence, newly hatched young may be more likely to mix with other ducklings or follow another female. Aggressive or broody females might even prove to be more attractive to ducklings while larger broods of ducklings could provide a greater stimulus.

4. *Chance, location, and brood density*—the topography of the area and the density of broods determine the likelihood that two or more broods encounter each other. Encounter rates will be greater when densities are high, when broods converge on the same small lakes or inlets after hatch, if broods hatch when territories are being established, or where newly hatched broods must traverse areas with established resident broods after leaving the nest.

Savard argued that there was little evidence of intentionality for the behavior or a heritable basis in goldeneyes and Buffleheads. He also postulated that if BA was adaptive, there should be higher survival of ducklings in larger, merged broods yet no such relationship was evident in his study.

Savard et al. (1998b) offered a similar explanation for BA in Surf Scoters in Québec. In that study, BA was frequent, especially when brood densities were high. Females were aggressive toward unfamiliar young and no evidence suggested voluntary abandonment of the young. Savard et al. (1998b) concluded that BA in that population occurred accidentally, favored by crowding in local areas and weak female–young bonds. In reviewing BA in other Mergini, Savard et al. (1998b) distilled the accidental mixing hypothesis (AMH) to four key elements:

1. Close proximity of one or more broods disrupted by predation, aggressive encounters, or territorial disputes among females.

2. Weak mother–young bonds at hatching.

3. Strong attraction among young of similar age.

4. Variable levels of aggressiveness or dominance among females, resulting in exclusion of other females.

Savard et al. (1998b) felt that these four factors were sufficient to explain most, if not all, occasions of BA in the Mergini, and more complex adaptive explanations were unnecessary.

Brood Amalgamation as Adaptive Behavior

In contrast to the AMH, a number of researchers have argued that BA provides fitness benefits to the donor, recipient, or both individuals (reviews in Eadie et al. 1988, Bustnes and Erikstad 1991a, Beauchamp 1997). The salvage strategy hypothesis (SSH) focuses on the potential benefits to the donor stating that females in poor condition, young, inexperienced, or having suffered high brood mortality and reduced brood size might abandon their brood. Females may abandon their young to the care of other female(s), or the abandoned young might secondarily find their way into another brood. The alternatives prove to be hard to disentangle given that the young rapidly join mixed broods without detailed behavioral data on the sequence of events. The central premise of all of these explanations is that females who abandon their young or leave them to the care of another female are emancipated from further parental investment; while her young may fare poorly, the female is able to salvage some reproductive success while improving the chances of her own survival and future reproduction as a trade-off between current and future reproduction. Advantages of merging broods are thought to include benefits of increased vigilance achieved through multiple female care such as in eiders, reduced duckling mortality in enlarged broods through dilution or selfish-herd effects, increased foraging efficiency of females who leave their young in a mixed brood to feed in areas better suited to adult foraging needs, and possibly indirect fitness benefits of caring for offspring of related females (Gorman and Milne 1972; Eadie et al. 1988; Kehoe 1989; Bustnes and Erikstad 1991a,b; Kilpi et al. 2001).

Several studies have found support for elements of the SSH. Gorman and Milne (1972) found that the crèche system of Common Eiders on the Ythan Estuary in Scotland "was "manned" by a constant turnover of breeding females who arrived with their young, stayed for a few days, and then abandoned their young and left the crèche system" (Gorman and Milne 1972:23). In this population, blue mussels and other high-quality foods needed by adults were spatially segregated from areas used for brood rearing. Gorman and Milne (1972:26) argued that "crèching evolved in eiders where the parent females have to leave their young in order to recover body weight lost during incubation."

In a Norwegian population of Common Eiders, Bustnes and Erikstad (1991a) found that females that abandoned their young laid smaller clutches and had a lower body weight at hatching than brood- and crèche-tending females. Poor body condition of females that abandoned their young was consistent with energetic stress and the SSH. Bustnes and Erikstad (1991a:1543) concluded that when "body condition is poor, abandonment might benefit lifetime reproduction by increasing the likelihood of the female's own survival."

Likewise, Kilpi et al. (2001) found that Common Eider females in poor body condition abandoned their broods more frequently than females in better condition. Females that attended a brood alone (lone tenders) were in the best condition at hatching, followed by females that remained with the merged brood for the brood-rearing period (permanent crèche tenders), females that joined and then left the merged brood (transient crèche tenders), and finally failed breeders. Kilpi et al. (2001) suggested that abandoning females may be able to replenish body reserves more efficiently and thereby ensure future breeding attempts. Females that abandoned their young used sites with abundant mussels, whereas females tending young broods were limited to feed on lower-value food items in the shallow littoral zone where ducklings are reared (Öst and Kilpi 1999).

One other factor that could influence whether a female stays with or abandons her young is the actual or expected mortality rate for the current brood. Bustnes and Erikstad (1991a) found that females who abandoned their young were in poor body condition and the survival rate of abandoned young was lower than those cared for by their own mother. In commenting on this study, Pöysä (1995) suggested that females might assess the survival prospects of their young versus investing in parental care. The brood success hypothesis posits that females should reduce the amount or duration of care "when offspring are surviving less than average ... because high offspring mortality indicates low benefits from future investment in the same brood" (Pöysä 1995:1576). Based on parental investment theory (Trivers 1972, Carlisle 1982), the hypothesis assumes a trade-off such that reduced investment in current reproduction enhances the prospect for future reproductive success. Pöysä (1995) further argued that females that adopt or tend broods often have large broods of their own (and even larger broods after amalgamation) and so should invest heavily in the care of the brood, based on the prospects of a high fitness return for continued investment. Bustnes and Erikstad (1995:1578) did not dispute this idea, but questioned whether a female has "the capacity to forecast the future benefits of her brood before the young have formed a close, unbreakable bond with her." Bustnes and Erikstad (1995) felt that because brood abandonment and adoption in their Common Eider population happened so quickly over a period of only a few days with some females giving up their young within a few minutes of reaching the water, females did not have sufficient time to assess the survival prospects of young. Bustnes and Erikstad (1995:1577) argued instead that females abandoned their young "because of depleted energy reserves and a great need to feed, presumably to increase their own chance of survival," which is consistent with the SSH. We note, as did both sets of authors, that the debate has been more about proximate factors influencing continued parental investment—energy reserves versus brood mortality rates—than about the adaptive significance of abandonment or adoption.

Pöysä et al. (1997) studied a population of Common Goldeneyes to further test the brood success hypothesis and to consider an alternative hypothesis. The brood size hypothesis predicts that females should adjust parental investment based on brood size. Pöysä et al. (1997) used observational data on brood survival rates, brood sizes, and duration of maternal care and reported that previous mortality predicts future mortality within broods and that the rate of offspring mortality was associated with subsequent maternal effort; broods that experienced a higher mortality rate were deserted earlier, consistent with the brood success hypothesis. Pöysä et al. (1997) did not find evidence in support of the brood size hypothesis. Eadie and Lyon (1998) experimentally tested the brood size hypothesis by manipulating brood sizes of Barrow's Goldeneyes at nest departure and found that brood desertion depended strongly on brood size at hatch with higher desertion rates among smaller broods. Pöysä and Milonoff (1999) argued that brood survival could have still contributed to this pattern. Nonetheless, both studies suggest that goldeneye females adjust maternal investment based on the relative value of continued care for the current brood and requirements of parental care could play some role in BA.

Eadie and Lyon (1998) also found that deserted ducklings joined other broods, but their success in doing so was strongly dependent on the availability of potential host broods and on the age of the recipient broods. Ducklings were accepted into young broods (<10 days old) but were always rejected from older broods. Discrimination is consistent with Savard's (1987) suggestion that age of ducklings is an important factor mitigating the outcome of brood encounters. Bustnes and Erikstad (1995) also noted that age differences among young can strongly influence adoption or rejection. They reported that "during 5 years of study of more than 150 broods and crèches with marked females, we have only observed one case of obvious age differences between ducklings accompanying the same female. When a bond is established, aggression toward foreign ducklings that intermingle with the brood is common … It is also very rare to see eider ducklings younger than 9–10 weeks without brood caring females, indicating that abandonment without adoption is very uncommon" (Bustnes and Erikstad 1995:1578). It remains to be determined whether differences in the age of ducklings impose some unmeasured cost to brood-tending females or whether it is simply a case of not being able to distinguish between young of similar age, especially in the first few days after hatch when female–duckling bonds are formed.

Why do females accept young of other females? The potential benefits of enlarged broods (dilution, selfish-herd effects) or shared parental care (multifemale groups such as eiders) are often cited, but evidence is mixed. Eadie and Lyon (1998) found no benefits or costs of brood adoption to the brood-tending females that were recipients of young in Barrow's Goldeneye, although the strong aggression against older foreign ducklings implied some cost of accepting additional young, possibly because of increased competition for food in the brood territory. Savard (1987) found no relation between brood size and survival in broods of Barrow's Goldeneyes or Buffleheads. Bustnes and Erikstad (1991a) noted that ducklings in crèches cared for by two females had higher survival than broods tended by single females, but not significantly so. Kehoe (1989) found that duckling survival was higher in larger broods of White-winged Scoters in 1 of 2 years. Traylor et al. (2008) also found that duckling survival increased with initial brood size in the same population of scoters studied several years later. Munro and Bédard (1977b) reported that survival of Common Eider ducklings was higher in larger broods when faced by single gull attacks, but mortality rate increased with brood size when multiple gulls attacked. Larger crèches suffered more attacks than small ones but offered much higher duckling survival in the face of attack by lone gulls. Even though Munro and Bédard (1977b:809) argued that brood mixing was accidental, they stated that "crèching remains advantageous even when the overall results of the two types of predation [single or multiple gull attacks] are combined" as multiple gull attacks are much more successful.

Öst et al. (2002) found that BA might directly benefit the females attending a crèche; in their study, the proportion of time each female spent feeding increased, and the proportion of time spent vigilant decreased as the number of females attending the brood increased. Moreover, the collective vigilance of multifemale groups was at least 20% higher than the vigilance of single females brooding alone (lone tenders), suggesting benefits for the young. Öst et al. (2008) confirmed this in a later study—variation in survival of ducklings from multifemale broods was influenced nonlinearly by the number of tending females, with the probability of survivorship highest for groups attended by two to three females and declining with larger groups.

Recent work on Common Eiders in the Baltic Sea has attempted to integrate many of these factors and provide a conceptual theoretical framework, at least for species with multifemale brood groups. Öst et al. (2003a) constructed a model in which females, based on their body condition and the structure of the joint brood, assess the fitness consequences of joining a multifemale brood-tending group—termed a coalition—versus tending for young alone. Their model rests on four main assumptions: (1) body condition affects the intensity of care that females are able to provide; (2) two females are better able to provide care than one; (3) females cannot care equally well for all the offspring in a brood; and (4) the fitness gain from entering a coalition for any individual female depends on how young are distributed spatially in the brood array—close to their mother or not, central or peripheral in the flock. Assumptions about proximity are based on the premise that ducklings closer to their mother receive better care because they have better access to food, better defense, or both. The model of Öst et al. (2003a) showed that the range of acceptable brood arrays in a coalition (the likelihood of joining) decreased with increasing condition of the female, so females tending alone should be in better condition than multifemale tenders. The prediction was supported by their data (Kilpi et al. 2001).

The coalition-formation model of Öst et al. (2003a) is related to models of reproductive skew (Vehrencamp 1983, Johnstone 2000, Nonacs and Hager 2011). In skew models, social groups are comprised of dominant and subordinate individuals, and the models address the question of how much of reproduction or other group resources the dominant individual should concede to ensure that subordinates stay in the group. The degree of control, prospects for solitary breeding, value of help provided by the subordinates, and levels of relatedness (indirect versus direct fitness) all factor into the predicted outcomes. Öst et al. (2007a) extended their coalition model to formally include the notion of reproductive skew. Using a bidding game approach, they developed a partner effort model to determine the evolutionary stable efforts of partners tending a coalition. As with their earlier model, the predicted outcome depends on body condition of the partners and the efforts extended by each partner; the benefits of shared care are unequally divided (skew). The model makes three predictions: (1) effort (vigilance) in a coalition will be lower than when alone, controlling for parental condition; (2) effort in a coalition should be positively correlated with body condition; and (3) effort in a coalition should depend on partner condition. Effort increases when partner condition decreases and decreases when partner condition increases. The first two predictions were supported by their data, while the third received partial support.

Öst et al. (2007a:84) concluded that the natural history of BA in Common Eiders fit well with the structure of their model.

> For a few days after hatch, eider hens are found in groups of up to more than 10 females, during which they form enduring coalitions with just one or two hens (Öst et al. 2003a), take up lone tending, or abandon their ducklings. Intense socializing and fighting are common during this time but occur only rarely later. Aggression may provide a means to influence the reproductive skew within the group, since a female's aggression frequency predicts her centrality in brood rearing coalitions (Öst et al. 2007b) and predation of ducklings by gulls is edge biased ... We hypothesize that these behaviors are related to the search for suitable coalition partners and that hens bargain with potential partners about the effort each will allocate and the skew of the joint brood.

In a series of subsequent studies, Öst and colleagues found further support for many of the assumption of their model: (1) brood structure is nonrandom with a mother's own ducklings closer to her and the level of female aggression is related to the presence of unrelated ducklings (Öst and Bäck 2003); (2) aggressive females and their ducklings occupy more central positions in the coalition, female age is positively correlated with aggression, and central females are more vigilant, all translating into potentially higher survival of a central female's own young (Öst et al. 2007b); and (3) older females form smaller coalitions, maximum body condition in a coalition increases with the age of the oldest female, and younger females in good condition are potentially more likely to form coalitions with older experienced females (Jaatinen and Öst 2011). Last, further extensions of the bidding game model by Jaatinen et al. (2011b) predicted that under conditions of heightened nest predation risk to females, cooperative brood care should become more prevalent and coalition group sizes should become larger, both of which were supported by their data.

Reconciling the Viewpoints

All of these case studies suggest that, rather than a mere result of accidental mixing, BA in some species such as Common Eiders might instead comprise a nuanced and subtle series of negations, bids, and choice of partners, functioning to increase a female's prospects for successful reproduction, balancing her own body condition, that of her potential partners, and her own current and future survival prospects and those of her young. The degree of complexity and sophistication implied in the most recent models is a polar opposite of the simplicity of the accidental mixing hypotheses. How do we reconcile these viewpoints? Perhaps the appeal of the AMH is its simplicity. It requires no finely honed adaptive behavior—merely a series of particular circumstances (high densities and chance brood encounters) and constraints (incomplete formation of parent offspring bonds after hatch) to result in at least occasional BA. Conversely, the apparent rigor of ever sophisticated adaptive models may cause us to tip the balance in their favor not because they are necessarily correct (or even fully testable) but because of the elegance and novelty of the theory. There is as much sociology as there is biology in the long-standing debates about the two schools of thought on the ecology and functions of BA in waterfowl, and perhaps for this reason alone, it remains an intriguing field for further inquiry.

It is likely that no single hypothesis suffices for all species or even for all populations of the same species. Indeed, Traylor et al. (2008:1085), in their rigorous multistate modeling analysis of the ecological correlates of duckling adoption in the White-winged Scoter, concluded, "adoption in this population is consistent with both the salvage strategy hypothesis and the accidental mixing hypothesis." The authors found support for the SSH in 1 year when females that were of small size and in poor nutritional condition produced offspring that were adopted by foster females. Traylor et al. (2008) thought that adoption was the result of abandonment decisions by female scoters. Consistent with Eadie and Lyon (1998), abandoned ducklings seek parental care and adopted individuals realized higher survival. In a second year, Traylor et al. (2008) found support for the AMH when they observed high levels of brood mixing and adoption after a period of severe cold and windy weather; high waves caused broods to congregate in protected inlets soon after hatch where accidental mixing occurred. Ducklings in poor condition were often stragglers and so became separated from their mother and joined foster broods.

Debates over the AMH and SSH may also be confounded by levels of explanation. Many of the mechanisms posited by the AMH represent proximate behavioral factors that could easily facilitate or mediate BA. Constraints on the development of family bonds, scattering in response to predation, attraction by ducklings for other ducklings, and variation in the level of aggression by brooding females all may play an important role. However, the SSH and other adaptive hypothesis instead address the ultimate evolutionary forces that shape these proximate mechanisms. Many species of waterfowl do not amalgamate broods (Eadie et al 1988, Beauchamp 1997), yet they too have similar mechanisms of family bonding, duckling attraction, and intense predation especially in the first few weeks. Why is it that species in the Mergini show such frequent evidence of BA and in numerous populations on different continents? If BA is indeed purely an accidental behavior with deleterious consequences, sea duck species might express behaviors to ensure the same degree of brood separation and isolation as found for many other species of ducks. Instead, perhaps it is a feature of northern climates, the greater need for ducklings to remain with a brood, and the strong energetic demands of egg production and incuba-tion in females that cause BA (and CBP and IBP) to be of greater adaptive value among sea ducks. Proximate mechanisms and ultimate explanations need not be in conflict unless we seek to have only one explanation for the behavior, but this reductionism does not fit modern thinking in the behavioral sciences (Bolhuis and Giraldeau 2005). For example, the weak mother–offspring bonds presented as a factor leading to accidental mixing could also be interpreted as a mechanism facilitat-ing more rapid BA because of the adaptive value to the female or ducklings—reduced strength of par-ent–offspring bonds could be a consequence of the value of amalgamating broods, rather than a cause. In any case, it is clear that additional oppor-tunities remain to further explore and clarify the proximate and ultimate factors that underlie BA in the Mergini and other waterfowl. Indeed, the two coauthors of this chapter have happily shared divergent viewpoints on the topic since graduate school and look forward to many more years of animated debate.

Relative Role of Relatives Revisited

The strong patterns of female-biased philopatry among Mergini and other waterfowl and the high degree of nest-site fidelity raise the possibility of kin selection among females that merge broods shortly after hatch or accept young of other females, as found in several cases of CBP described earlier. Tests of this hypothesis are fewer than for CBP but the potential remains. McKinnon et al. (2006) found significantly higher levels of relatedness relative to background levels among Common Eider females within groups depart-ing the colony with ducklings. Departing groups included pairs of females with levels of relatedness equivalent to full siblings. In contrast, Öst et al. (2005) used six to eight microsatellite markers to measure relatedness of co-tending eider females in enduring coalitions but found no pattern of non-random relatedness among coalition partners. In a later study of the same population with a larger number of microsatellite loci (19), Jaatinen et al. (2012) found a more complex pattern. Median relatedness values between coalition-forming females showed marked annual fluctuations; in 1 of 6 years (2010), relatedness was significantly higher than expected under random assortment of females, while a marginally significant trend was

detected in another year. Median intragroup relat-edness was elevated in small eider brood-rearing coalitions but decreased with increases in group size of females. Jaatinen et al. (2012) thought that in some years elevated relatedness observed in small brood-rearing coalitions may arise as simply a by-product of demography and partner choice. Older females prefer smaller coalitions but also have more relatives in the population by virtue of their longevity and accumulation of surviving relatives. However, in two years, levels of relat-edness exceeded these expectations, suggesting a more active mechanism of kin discrimination. Overall, the results suggest that patterns of relat-edness and benefits of associating with kin vary among years for as yet unknown reasons. The role of kinship may be more complex than previously thought and the value of rearing with relatives may indeed be relative!

CONCLUSIONS

In this chapter, we have reviewed and highlighted the considerable diversity of breeding systems, spacing behavior, and reproductive strategies of the sea ducks. A remarkable range of mating, social, and reproductive behaviors exists within tribe Mergini. It is astonishing that such varia-tion should exist within one monophyletic group and suggests that strong ecological pressures have driven these differences. Consequently, this one lineage of Order Anseriformes offers many opportunities to explore the threads of physiol-ogy, ecology, and evolution that have facilitated the radiation and maintenance of a complex and intriguing spectrum of reproductive behaviors.

Unfortunately, we still know little about many of these species and our understanding of their social and reproductive ecology is based on detailed studies of less than half of the species in the tribe. We have much yet to discover and it will be essential to expand our knowledge, not only to improve scientific understanding but also to develop sound management and conservation plans. The arctic and boreal environments where this unique group of birds live may be impacted more in the next several decades than many other biomes on earth. Understanding the reproduc-tive diversity of tribe Mergini, and the factors that promote successful reproduction of these birds breeding in some of the most extreme environ-ments in North America and Eurasia, will be essential to managing and mitigating the potential impacts of anthropogenic influences on sea ducks (Chapters 13 and 14, this volume).

ACKNOWLEDGMENTS

We thank D. Derksen, V. Skean, C. Amundson, and J. Bustnes for their many helpful comments on the earlier draft of this chapter. Financial support to assist in preparation of this chapter was provided by the Dennis G. Raveling Waterfowl Endowment at UC Davis (JME).

LITERATURE CITED

Ådahl, E., J. Lindstrom, G. D. Ruxton, K. E. Arnold, and T. Begg. 2004. Can intraspecific brood parasitism be detected using egg morphology only? Journal of Avian Biology 35:360–364.

Afton, A. D., and R. D. Sayler. 1982. Social courtship and pair bonding of Common Goldeneyes *Bucephala clangula*, wintering in Minnesota. Canadian Field-Naturalist 96:295–300.

Åhlund, M. 2005. Behavioural tactics at nest visits differ between parasites and hosts in a brood-parasitic duck. Animal Behaviour 70:433–440.

Åhlund, M., and M. Andersson. 2001. Brood parasitism—female ducks can double their reproduction. Nature 414:600–601.

Alison, R. M. 1975. Breeding biology and behavior of the Oldsquaw (*Clangula hyemalis* L.). Ornithological Monographs 18:1–52.

Amat, J. A. 1983. Pursuit flights of Mallard and Gadwall under different environmental conditions. Wildfowl 34:14–19.

Anderson, M. G., and R. D. Ohmart. 1988. Structure of the winter duck community on the lower Colorado River: patterns and processes. Pp. 191–236 in M. W. Weller (editor), Waterfowl in winter. University of Minnesota Press, Minneapolis, MN.

Anderson, M. G., J. M. Rhymer, and F. C. Rohwer. 1992. Philopatry, dispersal, and the genetic structure of waterfowl populations. Pp. 365–395 in B. D. J. Batt, A. D. Afton, M. G. Anderson, C. D. Ankney, D. H. Johnson, J. A. Kadlec, and G. L. Krapu (editors), Ecology and management of breeding waterfowl. University of Minnesota Press, Minneapolis, MN.

Anderson, M. G., and R. D. Titman. 1992. Spacing patterns. Pp. 251–289 in B. D. J. Batt, A. D. Afton, M. G. Anderson, C. D. Ankney, D. H. Johnson, J. A. Kadlec and G. L. Krapu (editors), Ecology and management of breeding waterfowl. University of Minnesota Press, Minneapolis, MN.

Andersson, M. 1984. Brood parasitism within species. Pp. 195–228 in C. J. Barnard (editor), Producers and scroungers: strategies of exploitation and parasitism. Croom Helm, London, U.K.

Andersson, M. 2001. Relatedness and the evolution of conspecific brood parasitism. American Naturalist 158:599–614.

Andersson, M., and M. Åhlund. 2000. Host-parasite relatedness shown by protein fingerprinting in a brood parasitic bird. Proceedings of the National Academy of Sciences USA 97:13188–13193.

Andersson, M., and M. Åhlund. 2001. Protein fingerprinting: a new technique reveals extensive conspecific brood parasitism. Ecology 82:1433–1442.

Andersson, M., and M. Åhlund. 2012. Don't put all your eggs in one nest: spread them and cut time at risk. American Naturalist 180:354–363.

Andersson, M., and M. O. G. Eriksson. 1982. Nest parasitism in Goldeneye *Bucephala clangula*: some evolutionary aspects. American Naturalist 120:1–16.

Andersson, M., and P. Waldeck. 2006. Reproductive tactics under severe egg predation: an eider's dilemma. Oecologia 148:350–355.

Andersson, M., and P. Waldeck. 2007. Host-parasite kinship in a female-philopatric bird population: evidence from relatedness trend analysis. Molecular Ecology 16:2797–2806.

Ashcroft, R. E. 1976. A function of the pair bond in the Common Eider. Wildfowl 27:101–105.

Baldassare, G. A. and E. G. Bolen. 2006. Waterfowl ecology and management. Krieger, Malabar, FL.

Ball, J. G., P. G. H. Frost, W. R. Siegfried, and F. McKinney. 1978. Territories and local movements of African Black Ducks. Wildfowl 29:61–69.

Beauchamp, G. 1997. Determinants of intraspecific brood amalgamation in waterfowl. Auk 114:11–21.

Beauchamp, G. 2000. Parental behaviour and brood integrity in amalgamated broods of the Common Eider. Wildfowl 51:169–179.

Bédard, J., and J. Munro. 1977. Brood and crèche stability in the Common Eider of the St. Lawrence estuary. Behaviour 60:221–236.

Bellrose, F. C. 1976. Ducks, geese and swans of North America. Stackpole, Harrisburg, PA.

Bengtson, S.-A. 1972a. Breeding ecology of the Harlequin Duck (*Histrionicus histrionicus*) in Iceland. Ornis Scandinavica 3:1–19.

Bengtson, S.-A. 1972b. Reproduction and fluctuations in the size of duck populations at Lake Mývatn, Iceland. Oikos 23:35–58.

Bjørn, T. H., and K. E. Erikstad. 1994. Patterns of intraspecific nest parasitism in the high arctic Common Eider (*Somateria mollissima borealis*). Canadian Journal of Zoology 72:1027–1034.

Black, J. M. 1996. Partnership in birds: the study of monogamy. Oxford University Press, Oxford, U.K.

Blohm, R. J. 1978. Migrational homing of male Gadwalls to breeding grounds. Auk 95:763–766.

Blohm, R. J., and K. A. MacKenzie. 1994. Additional evidence of migrational homing by a pair of Mallards. Journal of Field Ornithology 65:476–478.

Bolhuis, J. J., and L.-A. Giraldeau. 2005. The behavior of animals: mechanisms, function, and evolution. Wiley-Blackwell, Oxford, U.K.

Bond, J. C., D. Esler, and T. D. Williams. 2008. Breeding propensity of female Harlequin Ducks. Journal of Wildlife Management 72:1388–1393.

Bordage, D., N. Plante, A. Bourget, and S. Paradis. 1998. Use of ratio estimators to estimate the size of Common Eider populations in winter. Journal of Wildlife Management 62:185–192.

Bordage, D., and J.-P. L. Savard. 1995. Black Scoter (Melanitta nigra). in A. Poole and F. Gill (editors), The Birds of North America, No. 177. The Academy of Natural Sciences, Philadelphia, PA.

Breault, A. M. and J.-P. L. Savard. 1999. Philopatry of Harlequin Ducks moulting in southern British Columbia. Pp. 41–44 in R. I. Goudie, M. R. Petersen, and G. J. Robertson (editors), Behaviour and ecology of sea ducks. Canadian Wildlife Service Occasional Paper, No. 100, Canadian Wildlife Service, Ottawa, ON.

Breckenridge, W. J. 1953. Night rafting of American Golden-eyes on the Mississippi River. Auk 70:201–204.

Brown, J. L., and G. H. Orians. 1970. Spacing patterns in mobile animals. Annual Review of Ecology and Systematics 1:239–263.

Brown, P. W. 1977. Breeding biology of the White-winged Scoter (Melanitta fusca deglandi). Thesis, University of Missouri, Columbia, MO.

Brown, P. W., and M. A. Brown. 1981. Nesting biology of the White-winged Scoter Melanitta fusca deglandi. Journal of Wildlife Management 45:38–45.

Brown, P. W., and L. H. Fredrickson. 1997. White-winged Scoter (Melanitta fusca). in A. Poole and F. Gill (editors), The Birds of North America, No. 274. The Academy of Natural Sciences, Philadelphia, PA.

Bulmer, M. G. 1984. Risk avoidance and nesting strategies. Journal of Theoretical Biology 106:529–535.

Bustnes, J. Ø., and K. E. Erikstad. 1991a. Parental care in the Common Eider (Somateria mollissima): factors affecting the abandonment and adoption of young. Canadian Journal of Zoology 69:1538–1545.

Bustnes, J. Ø., and K. E. Erikstad. 1991b. The role of failed nesters and brood abandoning females in the crèching system of the Common Eider Somateria mollissima. Ornis Scandinavica 22:335–339.

Bustnes, J. Ø., and K. E. Erikstad. 1993. Site-fidelity in breeding Common Eider Somateria mollissima females. Ornis Fennica 70:11–16.

Bustnes, J. Ø., and K. E. Erikstad. 1995. Brood abandonment in Common Eiders: a reply to H. Pöysä. Canadian Journal of Zoology 73:1577–1578.

Bustnes, J. Ø., K. E. Erikstad, and T. H. Bjorn. 2002. Body condition and brood abandonment in Common Eiders breeding in the high Arctic. Waterbirds 25:63–66.

Carlisle, T. R. 1982. Brood success in variable environments: implications for parental care allocation. Animal Behaviour 30:824–836.

Coulson, J. C. 1984. The population dynamics of the Eider Duck Somateria mollissima and evidence of extensive non breeding adult ducks. Ibis 126:525–543.

Cramp, S., and K. L. Simmons (editors). 1977. Handbook of the birds of Europe, the Middle East and North Africa: the birds of the western Palearctic (vol. 1). Oxford University Press, Oxford, U.K.

Davies, N. B. 2000. Cuckoos, cowbirds and other cheats. T. & A. D. Poyser, London, U.K.

de Valpine, P., and J. M. Eadie. 2008. Conspecific brood parasitism and population dynamics. American Naturalist 172:547–562.

Diéval, H., J.-F. Giroux, and J.-P. L. Savard. 2011. Distribution of Common Eiders Somateria mollissima during the brood-rearing and moulting periods in the St. Lawrence Estuary, Canada. Wildlife Biology 17:124–134.

Donaghey, R. H. 1975. Spacing behaviour of breeding Buffleheads (Bucephala albeola) on ponds in the southern boreal forest. Thesis, University of Alberta, Edmonton, AB.

Donald, P. F. 2007. Adult sex ratios in wild bird populations. Ibis 149:671–692.

Dugger, B. D., K. M. Dugger, and L. H. Fredrickson 1994. Hooded Merganser (Lophodytes cucullatus). in A. Poole and G. Gill (editors), The Birds of North America, No. 98. The Academy of Natural Sciences, Philadelphia, PA.

Dwyer, T. S., S. R. Derrickson and D. S. Gilmer. 1973. Migrational homing by a pair of Mallards. Auk 90:687.

Eadie, J. M. 1989. Alternative female reproductive tactics in a precocial bird: the ecology and evolution of brood parasitism in goldeneyes. Dissertation, University of British Columbia, Vancouver, BC.

Eadie, J. M. 1991. Constraint and opportunity in the evolution of brood parasitism in waterfowl. International Ornithological Congress 20:1031–1040.

Eadie, J. M., and J. M. Fryxell. 1992. Density dependence, frequency-dependence, and alternative nesting strategies in goldeneyes. American Naturalist 140:621–641.

Eadie, J. M., and G. Gauthier. 1985. Prospecting for nest sites by cavity-nesting ducks of the genus *Bucephala*. Condor 87:528–534.

Eadie, J. M., F. P. Kehoe, and T. D. Nudds. 1988. Pre-hatch and post-hatch brood amalgamation in North-American Anatidae—a review of hypotheses. Canadian Journal of Zoology 66:1709–1721.

Eadie, J. M., and H. G. Lumsden. 1985. Is nest parasitism always deleterious to goldeneyes. American Naturalist 126:859–866.

Eadie, J. M., and B. E. Lyon. 1998. Cooperation, conflict, and crèching behavior in goldeneye ducks. American Naturalist 15:397–408.

Eadie, J. M., and B. E. Lyon. 2011. The relative role of relatives in conspecific brood parasitism. Molecular Ecology 20:5114–5118.

Eadie, J. M., M. L. Mallory, and H. G. Lumsden. 1995. Common Goldeneye (*Bucephala clangula*). in A. Poole and F. Gill (editors), The Birds of North America, No. 170. The Academy of Natural Sciences, Philadelphia, PA.

Eadie, J. M., J.-P. L. Savard, and M. L. Mallory. 2000. Barrow's Goldeneye (*Bucephala islandica*). in A. Poole and F. Gill (editors), The Birds of North America, No. 548. The Academy of Natural Sciences, Philadelphia, PA.

Eadie, J., B. Semel, and P. W. Sherman. 1998. Conspecific brood parasitism, population dynamics, and the conservation of cavity-nesting birds. Pp. 306–340 in T. Caro (editor), Behavioral ecology and conservation biology. Oxford University Press, Oxford, U.K.

Eadie, J. M., J. N. M. Smith, D. Zadworny, U. Kuhnlein, and K. Cheng. 2010. Probing parentage in parasitic birds: an evaluation of methods to detect conspecific brood parasitism using Goldeneyes *Bucephala islandica* and Bl. *clangula* as a test case. Journal of Avian Biology 41:163–176.

Einarsson, A. 1985. Use of space in relation to food in Icelandic Barrow's Goldeneye (*Bucephala islandica*). Dissertation, University of Aberdeen, Aberdeen, Scotland.

Einarsson, A. 1988. Distribution and movements of Barrow's Goldeneye *Bucephala islandica* young in relation to food. Ibis 130:153–163.

Einarsson, A. 1990. Settlement into breeding habitats by Barrow's Goldeneyes *Bucephala islandica*: evidence for temporary oversaturation of preferred habitat. Ornis Scandinavica 21:7–16.

Eriksson, M. O. G., and M. Andersson. 1982. Nest parasitism and hatching success in a population of Goldeneyes *Bucephala clangula*. Bird Study 29:49–54.

Erskine, A. J. 1971. Buffleheads. Canadian Wildlife Service Monograph Series, No. 4. Environment Canada, Ottawa, ON.

Erskine, A. J. 1990. Joint laying in *Bucephala* ducks parasitism or nest-site competition. Ornis Scandinavica 21:52–56.

Fast, P. L. F., H. G. Gilchrist, and R. G. Clark. 2010. Nest-site materials affect nest-bowl use by Common Eiders (*Somateria mollissima*). Canadian Journal of Zoology 88:214–218.

Flint, P. L., C. L. Moran, and J. L. Schamber. 1998. Survival of Common Eider *Somateria mollissima* adult females and ducklings during brood rearing. Wildfowl 49:103–109.

Fredrickson, L. H. 2001. Steller's Eider (*Polysticta stelleri*). in A. Poole and F. Gill (editors), The Birds of North America, No. 571. The Academy of Natural Sciences, Philadelphia, PA.

Fredrickson, L. H., and M. E. Heitmeyer. 1988. Waterfowl use of forested wetlands of the southern United States: an overview. Pp. 307–324 in M. W. Weller. Waterfowl in winter. University of Minnesota Press, Minneapolis, MN.

Gauthier, G. 1986. Experimentally-induced polygyny in Buffleheads: evidence for a mixed reproductive strategy? Animal Behaviour 34:300–302.

Gauthier, G. 1987a. Brood territories in Buffleheads: determinants and correlates of territory size. Canadian Journal of Zoology 65:1402–1410.

Gauthier, G. 1987b. The adaptive significance of territorial behaviour in breeding Bufflehead: a test of three hypotheses. Animal Behaviour 35:348–360.

Gauthier, G. 1987c. Further evidence of long-term pair bonds in ducks of the genus *Bucephala*. Auk 104:521–522.

Gauthier, G. 1988. Territorial behaviour, forced copulations, and mixed reproductive strategy in ducks. Wildfowl 39:102–114.

Gauthier, G. 1993. Bufflehead (*Bucephala albeola*). in A. Poole and F. Gill (editors), The Birds of North America, No. 67. The Academy of Natural Sciences, Philadelphia, PA.

Gorman, M. L. 1974. The endocrine basis of pair formation behaviour in the male Eider *Somateria mollissima*. Ibis 116:451–465.

Gorman, M. L., and H. Milne. 1972. Crèche behavior in the Common Eider *Somateria mollissima mollissima*. Ornis Scandinavica 3:21–26.

Gottlieb, G. 1971. Development of species identification in birds: an inquiry into the prenatal determinants of perception. University of Chicago Press, Chicago, IL.

Goudie, R. I., G. J. Robertson, and A. Reed. 2000. Common Eider (*Somateria mollissima*). in A. Poole and F. Gill (editors), The Birds of North America, No. 546. The Academy of Natural Sciences, Philadelphia, PA.

Gowans, B., G. J. Robertson and F. Cooke. 1997. Behaviour and chronology of pair formation by Harlequin Ducks *Histrionicus histrionicus*. Wildfowl 48:135–146.

Grenquist, P. 1963. Hatching losses of Common Goldeneye in the Finnish Archipelago. Proceedings of the International Ornithological Congress 13:685–689.

Guillemette, M., J. Himmelman, C. Barette, and A. Reed. 1993. Habitat selection by Common Eiders in winter and its interaction with flock size. Canadian Journal of Zoology 71:1259–1266.

Hanssen, S. A., and K. E. Erikstad. 2013. The long-term consequences of egg predation. Behavioral Ecology 24:564–569.

Hario, M., and T. E. Hollmén. 2004. The role of mate guarding in pre-laying Common Eiders *Somateria m. mollissima* in the northern Baltic Sea. Ornis Fennica 81:119–127.

Hario, M., M.-L. Koljonen, and J. Rintala. 2012. Kin structure and choice of brood care in a Common Eider (*Somateria m. mollissima*) population. Journal of Ornithology 153:963–973.

Hario, M., and M. Öst. 2002. Does heavy investment in foraging implicate low food acquisition for female Common Eiders *Somateria mollissima*? Ornis Fennica 79:111–120.

Hepp, G. R., and F. C. Bellrose. 1995. Wood Duck (*Aix sponsa*). in A. Poole and F. Gill (editors), The Birds of North America, No. 169. The Academy of Natural Sciences, Philadelphia, PA.

Hepp, G. R., and J. D. Hair. 1984. Dominance in wintering waterfowl (Anatini): effects on distribution of sexes. Condor 86:251–257.

Hildén, O. 1964. Ecology of duck populations in the island group of Valassaaret, Gulf of Bothnia. Annales Zoologica Fennica 1:153–279.

Iverson, S. A. 2002. Recruitment and the spatial organization of Surf Scoter (*Melanitta perspicillata*) populations during winter in the Strait of Georgia, British Columbia. Thesis, Simon Fraser University, Burnaby, BC.

Iverson, S. A., B. D. Smith, and F. Cooke. 2004. Age and sex distributions of wintering Surf Scoters: implications for the use of age ratios as an index of recruitment. Condor 106:252–262.

Jaari, S., K. Jaatinen, and J. Merila. 2009. Isolation and characterization of 22 polymorphic microsatellite loci for the Barrow's Goldeneye (*Bucephala islandica*). Molecular Ecology Resources 9:806–808.

Jaatinen, K., S. Jaari, R. B. O'Hara, M. Öst, and J. Merila. 2009. Relatedness and spatial proximity as determinants of host–parasite interactions in the brood parasitic Barrow's Goldeneye (*Bucephala islandica*). Molecular Ecology 18:2713–2721.

Jaatinen, K., K. Noreikiene, J. Merila, and M. Öst. 2012. Kin association during brood care in a facultatively social bird: active discrimination or by-product of partner choice and demography? Molecular Ecology 21:3341–3351.

Jaatinen, K., and M. Öst. 2011. Experience attracts: the role of age in the formation of cooperative brood-rearing coalitions in eiders. Animal Behaviour 81:1289–1294.

Jaatinen, K., and M. Öst. 2013. Brood size matching: a novel perspective on predator dilution. American Naturalist 181:171–181.

Jaatinen, K., M. Öst, P. Gienapp, and J. Merila. 2011a. Differential responses to related hosts by nesting and non-nesting parasites in a brood-parasitic duck. Molecular Ecology 20:5328–5336.

Jaatinen, K., M. Öst, and A. Lehikoinen. 2011b. Adult predation risk drives shifts in parental care strategies: a long-term study. Journal of Animal Ecology 80:49–56.

Jenkins, D., A. Watson, and G. R. Miller. 1963. Population studies on Red Grouse, *Lagopus lagopus scoticus* (Lath.) in north-east Scotland. Journal of Animal Ecology 32:317–376.

Johnsgard, P. A. 1965. Handbook of waterfowl behavior. Cornell University Press, Ithaca, NY.

Johnstone, R. A. 2000. Models of reproductive skew: a review and synthesis. Ethology 106:5–26.

Jones, P. H. 1979. Roosting behaviour of Long-tailed Ducks in relation to possible oil pollution. Wildfowl 30:155–158.

Kahlert, J., M. Coupe, and F. Cooke. 1998. Winter segregation and timing of pair formation in Red-breasted Merganser *Mergus serrator*. Wildfowl 49:161–172.

Kehoe, F. P. 1989. The adaptive significance of crèching behaviour in the White-winged Scoter (*Melanitta fusca deglandi*). Canadian Journal of Zoology 67:406–411.

Kellett, D. K., and R. T. Alisauskas. 1997. Breeding biology of King Eiders nesting on Karrak Lake, Northwest Territories. Arctic 50:47–54.

Kilpi, M., M. Öst, K. Lindström, and H. Rita. 2001. Female characteristics and parental care mode in the crèching system of eiders *Somateria mollissima*. Animal Behaviour 62:527–534.

Lack, D. 1968. Ecological adaptations for breeding in birds. Methuen and Company, London, U.K.

Limpert, R. J. 1980. Homing success of adult Buffleheads to a Maryland wintering site. Journal of Wildlife Management 44:905–908.

Linsell, S. E. 1969. Pre-dusk and nocturnal behaviour of goldeneye, with notes on population composition. Wildfowl 20:75–77.

Livesey, B. C. 1995. Phylogeny and evolutionary ecology of modern seaducks (Anatidae: Mergini). Condor 97:233–255.

Lofts, B., and R. K. Murton. 1968. Photoperiodic and physiological adaptations regulating avian breeding cycles and their ecological significance. Journal of Zoology 155:327–394.

Lopez-Sepulcre, A., and H. Kokko. 2002. The role of kin recognition in the evolution of conspecific brood parasitism. Animal Behaviour 64:215–222.

Lusignan, A. P., K. R. Mehl, I. L. Jones, and M. L. Gloutney. 2010. Conspecific brood parasitism in Common Eiders (Somateria mollissima): do brood parasites target safe nest sites? Auk 127:765–772.

Lyon, B. E. 1993. Conspecific brood parasitism as a flexible female reproductive tactic in American Coots. Animal Behaviour 46:911–928.

Lyon, B. E., and J. M. Eadie. 2000. Family matters: kin selection and the evolution of conspecific brood parasitism. Proceedings of the National Academy of Sciences USA 97:12942–12944.

Lyon, B. E., and J. M. Eadie. 2008. Conspecific brood parasitism in birds: a life-history perspective. Annual Review of Ecology, Evolution and Systematics 39:343–363.

Mallory, M., and K. Metz. 1999. Common Merganser (Mergus merganser). in A. Poole and F. Gill (editors), The Birds of North America, No. 442. The Academy of Natural Sciences, Philadelphia, PA.

Martin, P. R., and M. Di Labio. 1994. Natural hybrids between the Common Goldeneye, Bucephala clangula, and Barrow's Goldeneye, B. islandica. Canadian Field-Naturalist 108:195–198.

McKinney, F. 1965. Spacing and chasing in breeding ducks. Wildfowl trust 16th Annual Report 92–106.

McKinney, F. 1992. Courtship, pair formation, and signal systems. Pp. 214–250 in B. D. J. Batt, A. D. Afton, M. G. Anderson, C. D. Ankney, D. H. Johnson, J. A. Kadlec, and G. L. Krapu (editors), Ecology and management of breeding waterfowl. University of Minnesota Press, Minneapolis, MN.

McKinney, F., and S. Evarts. 1997. Sexual coercion in waterfowl and other birds. Ornithological Monographs 49:163–195.

McKinney, F., S. R. Derrickson, and P. Mineau. 1983. Forced copulation in waterfowl. Behaviour 86:250–294.

McKinnon, L., H. G. Gilchrist, and K. T. Scribner. 2006. Genetic evidence for kin-based female social structure in Common Eiders (Somateria mollissima). Behavioral Ecology 17:614–621.

Mehl, K. R., and R. T. Alisauskas. 2007. King Eider (Somateria spectabilis) brood ecology: correlates of duckling survival. Auk 124:606–618.

Milne, H. 1976. Body weights and carcass composition of the Common Eider. Wildfowl 27:115–122.

Mitchell, C. 1997. Re-mating in migratory Widgeon Anas penelope. Ardea 85:275–277.

Morrier, A., L. Lesage, A. Reed, and J.-P. L. Savard. 2008. Étude sur l'écologie de la Macreuse à Front Blanc au lac Malbaie, Réserve des Laurentides, 1994–1995. Technical Report Series, No. 301, Canadian Wildlife Service, Quebec Region, QC (in French).

Munro, J., and J. Bédard. 1977a. Crèche formation in the Common Eider. Auk 94:759–771.

Munro, J., and J. Bédard. 1977b. Gull predation and crèching behavior in the Common Eider. Journal of Animal Ecology 46:799–810.

Munro, J. A., and W. A. Clemens. 1931. Waterfowl in relation to the spawning of herring in British Columbia. Bulletin of the Biological Board of Canada 17:1–46.

Myres, M. T. 1959a. Display behavior of Bufflehead, scoters and goldeneyes at copulation. Wilson Bulletin 71:159–168.

Myres, M. T. 1959b. The behavior of the sea-ducks and its value in the systematics of the tribes Mergini and Somateriini, of the family Anatidae. Dissertation, University of British Columbia, Vancouver, BC.

Nee, S., and R. M. May. 1993. Population-level consequences of conspecific brood parasitism in birds and insects. Journal of Theoretical Biology 161:95–109.

Nice, M. M. 1937. Studies in the life history of the Song Sparrow (vol. II). The behavior of the Song Sparrow and other passerines. Transactions of the Linnaean Society of New York, Dover, New York, NY.

Noble, G. K. 1939. The role of dominance in the social life of birds. Auk 56:263–273.

Nonacs, P., and R. Hager. 2011. The past, present and future of reproductive skew theory and experiments. Biological Reviews 86:271–298.

Nudds, T. D., and D. Ankney. 1982. Ecological correlates of territory and home range size in North American dabbling ducks. Wildfowl 33:58–62.

Odell, N. S., and J. M. Eadie. 2010. Do Wood Ducks use the quantity of eggs in a nest as a cue to the nest's value? Behavioral Ecology 21:794–801.

Oppel, S., and A. N. Powell. 2010. Age-specific survival estimates of King Eiders derived from satellite telemetry. Condor 112:323–330.

Oring, L. W. 1982. Avian mating systems. Pp. 1–92 in D. S. Farner, J. R. King, and K. C. Parkes (editors), Avian biology (vol. VI). Academic Press, New York, NY.

Oring, L. W., and R. D. Sayler. 1992. The mating system of waterfowl. Pp. 190–213 in B. D. J. Batt, A. D. Afton, M. G. Anderson, C. D. Ankney, D. H. Johnson, J. A. Kadlec, and G. L. Krapu (editors), Ecology and management of breeding waterfowl. University of Minnesota Press, Minneapolis, MN.

Öst, M. 1999. Within-season and between-year variation in the structure of Common Eider broods. Condor 101:598–606.

Öst, M., and A. Bäck. 2003. Spatial structure and parental aggression in eider broods. Animal Behaviour 66:1069–1075.

Öst, M., C. W. Clark, M. Kilpi, and R. Ydenberg. 2007a. Parental effort and reproductive skew in coalitions of brood rearing female Common Eiders. American Naturalist 169:73–86.

Öst, M., K. Jaatinen, and B. Steele. 2007b. Aggressive females seize central positions and show increased vigilance in brood-rearing coalitions of eiders. Animal Behaviour 73:239–247.

Öst, M., and M. Kilpi. 1999. Parental care influences the feeding behaviour of female Eiders *Somateria mollissima*. Annales Zoologici Fennici 36:195–204.

Öst, M., L. Mantila, and M. Kilpi. 2002. Shared care provides time-budgeting advantages for female eiders. Animal Behaviour 64:223–231.

Öst, M., B. D. Smith, and M. Kilpi. 2008. Social and maternal factors affecting duckling survival in Eiders *Somateria mollissima*. Journal of Animal Ecology 77:315–325.

Öst, M., E. Vitikainen, P. Waldeck, L. Sundstrom, K. Lindström, T. Hollmén, J. C. Franson, and M. Kilpi. 2005. Eider females form non-kin brood-rearing coalitions. Molecular Ecology 14:3903–3908.

Öst, M., R. Ydenberg, M. Kilpi, and K. Lindström. 2003a. Condition and coalition formation by brood-rearing Common Eider females. Behavioral Ecology 14:311–317.

Öst, M., R. Ydenberg, K. Lindström, and M. Kilpi. 2003b. Body condition and the grouping behavior of brood-caring female Common Eiders (*Somateria mollissima*). Behavioral Ecology and Sociobiology 54:451–457.

Owen, M., and J. M. Black. 1990. Waterfowl ecology. Chapman & Hall, New York, NY.

Owen, M., J. M. Black, and H. Liber. 1988. Pair bond duration and timing of its formation in Barnacle Geese *Branta leucopsis*. Pp. 23–28 in M. W. Weller (editor), Waterfowl in winter. University of Minnesota Press, Minneapolis, MN.

Paasivaara, A., and H. Pöysä. 2008. Habitat-patch occupancy in the Common Goldeneye (*Bucephala clangula*) at different stages of the breeding cycle: implications to ecological processes in patchy environments. Canadian Journal of Zoology 86:744–755.

Paasivaara, A., J. Rutila, H. Pöysä, and P. Runko. 2010. Do parasitic Common Goldeneye *Bucephala clangula* females choose nests on the basis of host traits or nest site traits? Journal of Avian Biology 41:662–671.

Palmer, R. S. 1976. Handbook of North American birds (vol. 3). Yale University Press, New Haven, CT.

Patterson, I. J. 1982. The Shelduck. Cambridge University Press, Cambridge, U.K.

Payne, R. B. 1977. Ecology of brood parasitism in birds. Annual Review of Ecology and Systematics 8:1–28.

Petersen, M. R., D. C. Douglas, and D. M. Mulcahy. 1995. Use of implanted satellite transmitters to locate Spectacled Eiders at-sea. Condor 97:276–278.

Petersen, M. R., J. B. Grand, and C. P. Dau. 2000. Spectacled Eider (*Somateria fischeri*). in A. Poole and F. Gill (editors), The Birds of North America, No. 547. The Academy of Natural Sciences, Philadelphia, PA.

Petrie, M., and A. P. Møller. 1991. Laying eggs in others' nests: intraspecific brood parasitism in birds. Trends in Ecology and Evolution 6:315–320.

Phillips, L. M., and A. N. Powell. 2009. Brood rearing ecology of King Eiders on the north slope of Alaska. Wilson Journal of Ornithology 121:430–434.

Pöysä, H. 1995. Factors affecting abandonment and adoption of young in Common Eiders and other waterfowl: a comment. Canadian Journal of Zoology 73:1575–1577.

Pöysä, H. 1999. Conspecific nest parasitism is associated with inequality in nest predation risk in the Common Goldeneye (*Bucephala clangula*). Behavioral Ecology 10:533–540.

Pöysä, H. 2003a. Parasitic Common Goldeneye (*Bucephala clangula*) females lay preferentially in safe neighbourhoods. Behavioral Ecology and Sociobiology 54:30–35.

Pöysä, H. 2003b. Low host recognition tendency revealed by experimentally induced parasitic egg laying in the Common Goldeneye (*Bucephala clangula*). Canadian Journal of Zoology 81:1561–1565.

Pöysä, H. 2004. Relatedness and the evolution of conspecific brood parasitism: parameterizing a model with data for a precocial species. Animal Behaviour 67:673–679.

Pöysä, H. 2006. Public information and conspecific nest parasitism in goldeneyes: targeting safe nests by parasites. Behavioral Ecology 17:459–465.

Pöysä, H., K. Lindblom, J. Rutila, and J. Sorjonen. 2010. Response of parasitically laying goldeneyes to experimental nest predation. Animal Behaviour 80:881–886.

Pöysä, H., and M. Milonoff. 1999. Processes underlying parental care decisions and crèching behaviour: clarification of hypotheses. Annales Zoologici Fennici 36:125–128.

Pöysä, H., and M. Pesonen. 2007. Nest predation and the evolution of conspecific brood parasitism: from risk spreading to risk assessment. American Naturalist 169:94–104.

Pöysä, H., and S. Pöysä. 2002. Nest-site limitation and density dependence of reproductive output in the Common Goldeneye (*Bucephala clangula*): implications for the management of cavity-nesting birds. Journal of Applied Ecology 39:502–510.

Pöysä, H., P. Runko, V. Ruusila, and M. Milonoff. 2001. Identification of parasitized nests by using egg morphology in the Common Goldeneye: an alternative to blood sampling. Journal of Avian Biology 32:79–82.

Pöysä, H., J. Virtanen, and M. Milonoff. 1997. Common Goldeneyes adjust maternal effort in relation to prior brood success and not current brood size. Behavioral Ecology and Sociobiology 40:101–106.

Quakenbush, L., and R. Suydam. 1999. Periodic non breeding of Steller's Eiders near Barrow, Alaska, with speculations on possible causes. Pp. 34–40 in R. I. Goudie, M. R. Petersen, and G. J. Robertson (editors), Behaviour and ecology of sea ducks. Occasional Paper, No. 100, Canadian Wildlife Service, Ottawa, ON.

Reed, A. 1971. Pre-dusk rafting flights of wintering goldeneyes and other diving ducks in the Province of Quebec. Wildfowl 22:61–62.

Reichart, L. M., S. Anderholm, V. Muñoz-Fuentes, and M. S. Webster. 2010. Molecular identification of brood-parasitic females reveals an opportunistic reproductive tactic in Ruddy Ducks. Molecular Ecology 19:401–413.

Robert, M., R. Benoit, and J.-P. L. Savard. 2002. Relationship among breeding, molting, and wintering areas of male Barrow's Goldeneyes (*Bucephala islandica*) in eastern North America. Auk 119:676–684.

Robertson, G. J. 1998. Egg adoption can explain joint egg-laying in Common Eiders. Behavioral Ecology and Sociobiology 43:289–296.

Robertson, G. J., F. Cooke, R. I. Goudie, and W. S. Boyd. 1998. The timing of pair formation in Harlequin Ducks. Condor 100:551–555.

Robertson, G. J., and R. I. Goudie. 1999. Harlequin Duck (*Histrionicus histrionicus*). in A. Poole and F. Gill (editors), The Birds of North America, No. 466. The Academy of Natural Sciences, Philadelphia, PA.

Robertson, G. J., and J.-P. L. Savard. 2002. Long-tailed Duck (*Clangula hyemalis*). in A. Poole and F. Gill (editors), The Birds of North America, No. 651. The Academy of Natural Sciences, Philadelphia, PA.

Rodway, M. S. 2003. Timing of pairing in Harlequin Ducks: interaction of spacing behaviour, time budgets, and the influx of herring spawn. Dissertation, Simon Fraser University, Burnaby, BC.

Rodway, M. S. 2007a. Timing of pairing in waterfowl I: reviewing the data and extending the theory. Waterbirds 30:488–505.

Rodway, M. S. 2007b. Timing of pairing in waterfowl II: testing the hypothesis with Harlequin Ducks. Waterbirds 30:506–520.

Rodway, M. S., H. R. Regehr, and F. Cooke. 2003. Sex and age differences in distribution, abundance, and habitat preferences of wintering Harlequin Ducks: implications for conservation and estimating recruitment. Canadian Journal of Zoology 81:492–503.

Rohwer, F. C., and M. G. Anderson. 1988. Female biased philopatry, monogamy, and the timing of pair formation in migratory waterfowl. Current Ornithology 5:187–221.

Rohwer, F. C., and S. Freeman. 1989. The distribution of conspecific nest parasitism in birds. Canadian Journal of Zoology 67:239–253.

Rothstein, S. I. 1990. A model system for coevolution—avian brood parasitism. Annual Review of Ecology and Systematics 21:481–508.

Rowley, I. 1983. Remating in birds. Pp. 331–360 in P. Bateson (editor), Mate choice. Cambridge University Press, Cambridge, U.K.

Ruxton, G. D. 1999. Are attentive mothers preferentially parasitised? Behavioral Ecology and Sociobiology 46:71–72.

Sargeant, A. B., and D. G. Raveling. 1992. Mortality during the breeding season. Pp. 396–392 in B. D. J. Batt, A. D. Afton, M. G. Anderson, C. D. Ankney, D. H. Johnson, J. A. Kadlec, and G. L. Krapu. Ecology and Management of Breeding Waterfowl. University of Minnesota Press, Minneapolis, MN.

Savard, J.-P. L. 1982. Intra- and inter-specific competition between Barrow's Goldeneye (*Bucephala islandica*) and Bufflehead (*Bucephala albeola*). Canadian Journal of Zoology 60:3439–3446.

Savard, J.-P. L. 1984. Territorial behaviour of Common Goldeneye, Barrow's Goldeneye and Bufflehead in areas of sympatry. Ornis Scandinavica 15:211–216.

Savard, J.-P. L. 1985. Evidence of long-term pair bonds in Barrow's Goldeneye (*Bucephala islandica*). Auk 102:389–391.

Savard, J.-P. L. 1986. Polygyny in Barrow's Goldeneye. Condor 88:250–252.

Savard, J.-P. L. 1987. Causes and functions of brood amalgamation in Barrow's Goldeneye and Bufflehead. Canadian Journal of Zoology 65:1548–1553.

Savard, J.-P. L. 1988a. Winter, spring and summer territoriality in Barrow's Goldeneye: characteristics and benefits. Ornis Scandinavica 19:119–128.

Savard, J.-P. L. 1988b. Use of nest boxes by Barrow's Goldeneyes: nesting success and effect on the breeding population. Wildlife Society Bulletin 16:125–132.

Savard, J.-P. L., D. Bordage, and A. Reed. 1998a. Surf Scoter (*Melanitta perspicillata*). in A. Poole and F. Gill (editors), The Birds of North America, No. 363. The Academy of Natural Sciences, Philadelphia, PA.

Savard, J.-P. L., A. Reed, and L. Lesage. 1998b. Brood amalgamation in Surf Scoters Melanitta perspicillata and other Mergini. Wildfowl 49:129–138.

Savard, J.-P. L., A. Reed, and L. Lesage. 2007. Chronology of breeding and molt migration in Surf Scoters (Melanitta perspicillata). Waterbirds 30:223–229.

Savard, J.-P. L., and M. Robert. 2013. Relationships among breeding, molting and wintering areas of adult female Barrow's Goldeneyes (Bucephala islandica) in eastern North America. Waterbirds 36:34–42.

Savard, J.-P. L., and J. N. M. Smith. 1987. Interspecific aggression by Barrow's Goldeneye: a descriptive and functional analysis. Behaviour 102:168–184.

Sayler, R. D. 1992. Ecology and evolution of brood parasitism in waterfowl. Pp. 290–322 in B. J. D. Batt, A. D. Afton, M. G. Anderson, C. D. Ankney, D. H. Johnson, J. A. Kadlec, and G. L. Krapu (editors), Ecology and Management of Breeding Waterfowl. University of Minnesota Press, Minneapolis, MN.

Sea Duck Joint Venture. [online]. 2014. Atlantic and Great Lakes sea duck migration study: progress report February 2014. <http://seaduckjv.org/atlantic_migration_study.html> (3 April 2014).

Semel, B., and P. W. Sherman. 1995. Alternative placement strategies for Wood Duck nest boxes. Wildlife Society Bulletin 23:463–471.

Semel, B., and P. W. Sherman. 2001. Intraspecific parasitism and nest-site competition in Wood Ducks. Animal Behaviour 61:787–803.

Semel, B., P. W. Sherman, and S. M. Byers. 1988. Effects of brood parasitism and nest-box placement on Wood Duck breeding ecology. Condor 90:920–930.

Smith, C. M., F. Cooke, G. J. Robertson, R. I. Goudie, and S. Boyd. 2000. Long-term pair bonds in Harlequin Ducks. Condor 102:201–205.

Sonsthagen, S. A., S. L. Talbot, R. B. Lanctot, and K. G. McCracken. 2010. Do Common Eiders nest in kin groups? Microgeographic genetic structure in a philopatric sea duck. Molecular Ecology 19:647–657.

Sorenson, M. D. 1991. The functional significance of parasitic egg laying and typical nesting in Redhead ducks: an analysis of individual behaviour. Animal Behaviour 42:771–796.

Spurr, E., and H. Milne. 1976. Adaptive significance of autumn pair formation in the Common Eider (Somateria mollissima L.). Ornis Scandinavica 7:85–89.

Sullivan, T. M., R. W. Butler, and S. W. Boyd. 2002. Seasonal distribution of waterbirds in relation to spawning Pacific herring, Clupea pallasi, in the Strait of Georgia, British Columbia. Canadian Field-Naturalist 116:366–370.

Suydam, R. S. 2000. King Eider (Somateria spectabilis). in A. Poole and F. Gill (editors), The Birds of North America, No. 491. The Academy of Natural Sciences, Philadelphia, PA.

Tiedemann, R., and H. Noer. 1998. Geographic partitioning of mitochondrial DNA patterns in European Eider Somateria mollissima. Hereditas 128:159–166.

Tiedemann, R., K. B. Paulus, K. Havenstein, S. Thorstensen, A. Petersen, P. Lyngs, and M. C. Milinkovitch. 2011. Alien eggs in duck nests: brood parasitism or a help from Grandma? Molecular Ecology 20:3237–3250.

Titman, R. D. 1999. Red-breasted Merganser (Mergus serrator). in A. Poole and F. Gill (editors), The Birds of North America, No. 443. The Academy of Natural Sciences, Philadelphia, PA.

Titman, R. D., and N. R. Seymour. 1981. A comparison of pursuit flights by six North American ducks of the genus Anas. Wildfowl 32:11–18.

Torres, R., F. Cooke, G. J. Robertson and S. W. Boyd. 2002. Pairing decisions in the Harlequin Duck: costs and benefits. Waterbirds 25:340–347.

Traylor, J. J., R. T. Alisauskas, and F. P. Kehoe. 2004. Multistate modeling of brood amalgamation in White-winged Scoters Melanitta fusca deglandi. Animal Biodiversity and Conservation 27:369–370.

Traylor, J. J., R. T. Alisauskas, and F. P. Kehoe. 2008. Ecological correlates of duckling adoption among White-winged Scoters Melanitta fusca: strategy, epiphenomenon, or combination? Behavioral Ecology and Sociobiology 62:1085–1097.

Trivers, R. L. 1972. Parental investment and sexual selection. Pp. 136–179 in B. Campbell (editor), Sexual Selection and the Descent of Man. Aldine Publishing, Chicago, IL.

Vehrencamp, S. L. 1983. Optimal degree of skew in cooperative societies. American Zoologist 23:327–335.

Vermeer, K., M. Bentley, K. H. Morgan, and G. E. J. Smith. 1997. Association of feeding flocks of Brant and sea ducks with herring spawn at Skidegate Inlet. Pp. 102–107 in K. Vermeer and K. H. Morgan (editors), The ecology, status, and conservation of marine and shoreline birds of the Queen Charlotte Islands. Occasional Paper, No. 93, Canadian Wildlife Service, Ottawa, ON.

Waldeck, P., and M. Andersson. 2006. Brood parasitism and nest takeover in Common Eiders. Ethology 112:616–624.

Waldeck, P., M. Andersson, M. Kilpi, and M. Öst. 2008. Spatial relatedness and brood parasitism in a female-philopatric bird population. Behavioral Ecology 19:67–73.

Waldeck, P., J. I. Hagen, S. A. Hanssen, and M. Andersson. 2011. Brood parasitism, female condition and clutch reduction in the Common Eider *Somateria mollissima*. Journal of Avian Biology 42:231–238.

Waldeck, P., M. Kilpi, M. Öst, and M. Andersson. 2004. Brood parasitism in a population of Common Eider (*Somateria mollissima*). Behaviour 141:725–739.

Weeden, J. S., and J. B. Falls. 1959. Differential responses of male Ovenbirds to recorded songs of neighbouring and more distant individuals. Auk 76:342–351.

Weller, M. W. 1976. Ecology and behaviour of steamer-ducks. Wildfowl 27:45–53.

Williams, M., and F. McKinney. 1996. Long-term monogamy in a river specialist—the Blue Duck. Pp. 73–90 in J. M. Black (editor), Partnership in birds: the study of monogamy. Oxford University Press, Oxford, U.K.

Wunderlee Jr., J. M. 1978. Differential response of territorial Yellow-throats to the songs of neighbors and non-neighbors. Auk 95:389–395.

Yom-Tov, Y. 1980. Intraspecific nest parasitism in birds. Biological Reviews 55:93–108.

Yom-Tov, Y. 2001. An updated list and some comments on the occurrence of intraspecific nest parasitism in birds. Ibis 143:133–143.

Young, A. D., and R. D. Titman. 1988. Intraspecific nest parasitism in Red-breasted Mergansers. Canadian Journal of Zoology 66:2454-2458.

Zicus, M. C. 1990. Nesting biology of Hooded Mergansers using nest boxes. Journal of Wildlife Management 54:637–643.

Zicus, M. C., and S. K. Hennes. 1989. Nest prospecting by Common Goldeneyes. Condor 91:807–812.

Zink, A. G. 2000. The evolution of intraspecific brood parasitism in birds and insects. American Naturalist 155:395–405.

CHAPTER TWELVE

Harvest of Sea Ducks in North America*

A CONTEMPORARY SUMMARY

*Thomas C. Rothe, Paul I. Padding, Liliana C. Naves,
and Gregory J. Robertson*

Abstract. Sea ducks present unique challenges to waterfowl harvest management because the species have relatively low intrinsic population growth rates and varied population structure and harvest occurs under a diversity of rangewide hunting traditions. Sea duck harvest occurs throughout North America, ranging from inland harvest of widely distributed species, such as goldeneyes and mergansers, to specialized harvest of eiders and scoters in coastal and northern regions. Harvest of widely distributed species is well represented in continental waterfowl harvest monitoring programs. More localized harvests, such as those in coastal and remote areas, have proven challenging to monitor, and some special surveys have been implemented. Sea duck harvest regulations have evolved over the decades according to changes in population levels, management philosophies, and improvements in harvest information. Hunting of goldeneyes and Buffleheads has usually been regulated within general bag limits for ducks. Regulations for large mergansers have been liberal, but limits for Hooded Mergansers have remained conservative. Harvest regulations for eiders, scoters, Long-tailed Ducks, and Harlequin Ducks have recently become more restrictive, subject to special seasons

and bag limits in primary coastal hunting areas. With a few exceptions, harvest of widely distributed species and most species along the Pacific Coast is considered sustainable. Common Eider harvest in the Atlantic Flyway is a management concern given fluctuations in eider populations, high harvest pressure, and the presence of two subspecies. Sea ducks are important subsistence resources in the North; eiders are harvested by coastal communities and scoters by inland communities. Harvest estimates are now available for most northern jurisdictions, and management is undertaken in cooperation with First Nations and Inuit organizations in Canada and subsistence management bodies in Alaska. Additional information on the delineation and demography of sea duck populations is essential, along with improved harvest estimation techniques, to inform collaborative harvest management and to ensure sustainable harvest.

Key Words: age ratio, bag limit, eiders, egg collecting, ethnotaxonomy, harvest management, harvest surveys, HIP surveys, hunting seasons, regulations, scoters, sex ratio, subsistence harvest, wing survey, wounding losses.

* Rothe, T. C., P. I. Padding, L. C. Naves, and G. J. Robertson. 2015. Harvest of sea ducks in North America: A contemporary summary. Pp. 417–467 in J.-P. L. Savard, D. V. Derksen, D. Esler, and J. M. Eadie (editors). Ecology and conservation of North American sea ducks. Studies in Avian Biology (no. 46), CRC Press, Boca Raton, FL.

arvest management of sea duck populations has focused on concerns that sea duck recruitment is lower than in other taxa of ducks and that the effects of harvest may be largely additive and density independent (Goudie et al. 1994). Since harvest can be regulated, bag limits and seasons should be conservative as a primary means of sustaining populations (Chapter 3, this volume). In contrast, the tone of popular sporting literature often emphasizes opportunities to harvest exotic species of sea ducks during late seasons, under relatively liberal bag limits, and that sea ducks are an underutilized resource (Gillilan 1988). The divergent viewpoints highlight the need for research into the role of harvest in population dynamics so managers can develop appropriate harvest strategies.

Ideally, the influence of harvest on sea duck populations ought to be assessed in the context of comprehensive population models driven by reliable parameter datasets. Recent attention to declines in sea ducks has stimulated the development of models for some populations (Krementz et al. 1997, Gilliland et al. 2009, Schamber et al. 2009, Iverson and Esler 2010, Bentzen and Powell 2012, Wilson et al. 2012). However, most sea duck populations are poorly defined and have not been investigated with harvest models, in part because data on demographic parameters are often scarce or inadequate.

For most sea duck species, harvest assessment is constrained by the lack of critical information on (1) population structure and delineation of cohesive and manageable population units; (2) appropriate geographic management scales; (3) seasonal ranges, migration patterns, and fidelity to sites; and (4) reliable estimates of population size, productivity, and sources of nonhunting mortality (Sea Duck Joint Venture Management Board 2008, Sea Duck Joint Venture 2013). In addition, major challenges exist for documenting hunter participation and activity, the size and composition of harvest, and the rangewide distribution of harvest. In light of these uncertainties, it is challenging to design and implement appropriate harvest management actions, particularly for populations with declining or unknown trends in bird numbers. Identifying the information required to guide harvest management has been identified as a priority for the Sea Duck Joint Venture (2013), which has initiated efforts to evaluate harvest potential of several populations.

There is no question that in situations where mortality is a limiting factor for a population changes in harvest regulations, especially those affecting harvest of adult females, can provide survival benefits to improve population trajectories (Gilliland et al. 2009, Merkel 2010). However, if the limiting factors are primarily related to productivity, conventional regulatory prescriptions for managing harvest may have to be applied over many years and may have little effect on population dynamics.

Our objectives in this chapter are to provide (1) a synopsis of harvest management authorities and sources of harvest information, (2) an overview of sea duck harvesting traditions and regulation history along the Atlantic and Pacific Coasts and in the northern regions where hunting of sea ducks is most prevalent, and (3) a summary of the magnitude and composition of sea duck harvest in North America.

HARVEST MANAGEMENT AUTHORITIES

Regulation of waterfowl hunting has largely been a responsibility of the federal governments in the United States and Canada, particularly after ratification of the 1916 Migratory Bird Convention (Treaty) between the United States and Great Britain (for Canada) for the Protection of Migratory Birds. The treaty established a framework to protect shared bird populations by actions including regulation of hunting. Both the United States and Canada codified the treaty in federal laws: in the United States as the Migratory Bird Treaty Act of 1918 and in Canada as the Migratory Birds Convention Act of 1917. The concept of managing waterfowl on the basis of migratory corridors or flyways was developed in the 1940s and formally established in 1952. Since then, waterfowl hunting framework regulations in the United States have been developed annually through consultation between the U.S. Fish and Wildlife Service (USFWS) and state wildlife agencies associated with four Flyway Councils. Hunting rules are implemented through state regulations that are at least as restrictive as federal frameworks. In Canada, the Canadian Wildlife Service (CWS) consults with the provincial and territorial wildlife agencies to develop final waterfowl hunting regulations, which are then established in federal law. The federal regulatory processes in the United States and Canada remain separate, but representatives of the Canadian provinces, territories, and federal agencies collaborate with the Flyway Councils to exchange information and develop cooperative management programs.

SURVEYS TO ASSESS FALL–WINTER HARVEST

The main sources of information on hunter activity in fall and winter and sea duck harvest are national surveys that the USFWS and the CWS have conducted for decades. Sample frames, stratification, sampling procedures, data collection, and analytical methods have been described in detail elsewhere (Martin and Carney 1977, Cooch et al. 1978, Geissler 1990, Padding et al. 2006a, Johnson et al. 2012). Here, we focus on aspects of those surveys that specifically address sea ducks. Some US states have conducted other sea duck harvest surveys, and their results allow comparisons with the USFWS survey that we will examine.

Traditional surveys have not covered a large portion of North American harvest of sea ducks, namely, subsistence harvest in the north that occurs mostly from spring through early fall. Subsistence harvests have not been well documented historically; most were technically illegal but largely not enforced until relatively recent times. In this chapter, we make general distinctions between traditional *subsistence harvest*, defined by nutritional and cultural aspects, and conventional *fall and winter harvest* that includes recreation, food harvesting, and other values under more regulated conditions.

It is important to recognize, from a regulatory standpoint and for interpretation of harvest information, that coastal species such as eiders (*Polysticta* and *Somateria* spp.), scoters (*Melanitta* spp.), Harlequin Ducks (*Histrionicus histrionicus*), Long-tailed Ducks (*Clangula hyemalis*), and Barrow's Goldeneye (*Bucephala islandica*) are treated as a subset of the 15 taxonomic sea duck species (Mergini) in North America because they are subject to special hunting regulations. Widely distributed species (hereafter *ubiquitous sea ducks*) such as Common Goldeneye (*Bucephala clangula*), Bufflehead (*Bucephala albeola*), and mergansers (*Lophodytes* and *Mergus* spp.) are harvested across the continent and are usually regulated under general duck hunting seasons and bag limits.

U.S. Fish and Wildlife Service Harvest Survey: 1952–2001

From 1952 through 2001, the sample frame for the annual national survey consisted of every person who purchased a Migratory Bird Hunting and Conservation Stamp (or federal duck stamp). All waterfowl hunters 16 years of age or older are required to buy a federal duck stamp, which is valid for one year. A sample of duck stamp purchasers were asked to document their waterfowl hunting activity and harvest throughout the hunting season and report it on a survey form (Martin and Carney 1977).

Initially, hunters were asked to report duck and goose harvest by species, but it soon became apparent that some hunters were unable to identify species. To address this problem, the Waterfowl Parts Collection Survey (hereafter, wing survey) was developed to allow identification of the species, sex, and age of harvested ducks based on wing plumage (Carney 1984). A modified harvest survey form asked about hunting and harvest of general waterfowl categories (ducks, sea ducks, and geese) rather than species, and the wing survey generated data to partition harvest estimates by species, sex, and age (Martin and Carney 1977).

The harvest survey design allowed estimation of sea duck harvest separately from other species of ducks in the Atlantic Flyway, where special sea duck hunting seasons for eiders, scoters, and Long-tailed Ducks were held in all coastal states except Florida. In all other states, sea duck harvest reports and wings were combined with other duck data for harvest estimation. The Atlantic Flyway sea duck harvest analysis assumed that (1) hunters were able to identify eiders, scoters, and Long-tailed Ducks, and that (2) they knew that only those species were considered sea ducks for regulatory purposes.

The mechanism of hunter sampling from the harvest survey relied on cooperation by postal clerks and other vendors of federal duck stamps, and response rates declined in the 1980s when cooperation deteriorated (Tautin et al. 1989). In the 1990s, concerns about increasing bias from nonresponses in the waterfowl harvest survey (Barker et al. 1992) and a long-standing need to establish a national sample frame of all migratory bird hunters (Tautin et al. 1989) led the USFWS and state wildlife agencies to develop the Harvest Information Program (HIP; Elden et al. 2002). Under this program, the state agencies use their hunting license systems to provide the USFWS with an annual list of all licensed migratory bird hunters, forming the sample frame for annual harvest surveys. Before discontinuing the federal duck stamp–based waterfowl harvest survey, the USFWS conducted the duck stamp survey and the HIP waterfowl harvest survey concurrently for a 3-year period from 1999 to 2001 (Johnson et al. 2012).

U.S. Fish and Wildlife Service Harvest Survey: 1999 to Present

The HIP survey integrates 49 state-specific sample frames, treated as strata (Padding et al. 2006b). In the United States, issuing annual hunting licenses is a state purview; each individual state determines who must obtain a license to hunt in the state and, conversely, which hunters are exempt from the hunting license requirement. Eligibility varies from state to state, but groups of hunters most commonly exempted from hunting license requirements are young hunters (typically ≤16 years of age), landowners hunting on their own property, senior hunters (typically ≥60–65 years of age), and disabled veterans (Sheriff et al. 2002). The HIP sample frame does not include most license-exempt hunters, whereas only young hunters were previously excluded from the duck stamp survey sample frame. However, this difference apparently had little effect on survey results because the annual duck stamp–based and HIP estimates of active waterfowl hunters for the 1999–2001 overlap period were nearly the same (U.S. Fish and Wildlife Service, unpubl. data).

Under HIP, state hunting license vendors are required to ask migratory bird hunters a series of questions about what species they hunted the previous year and how many birds were harvested. The answers to those questions enable the USFWS to sample primarily duck and goose hunters for the waterfowl harvest survey, woodcock hunters for the woodcock harvest survey, and so on, and to sample hunters at different rates based on their previous reported harvest. In coastal Atlantic Flyway states, license vendors also specifically ask migratory bird hunters if they hunted sea ducks (eiders, scoters, or Long-tailed Ducks) during the previous hunting season. Persons that did are considered likely sea duck hunters and constitute a separate sampling stratum. California and Alaska identify likely sea duck hunters by similar means, except that only scoters and Harlequin Ducks are considered sea ducks in California, whereas eiders, scoters, Harlequin Ducks, Long-tailed Ducks, and mergansers are included in Alaska's special sea duck bag limits. Hunters must purchase a separate permit to hunt scoters in Oregon. In Washington

State, a separate permit is required for hunting Harlequin Ducks, scoters, Long-tailed Ducks, and goldeneyes. As in the duck stamp–based survey, respondents are asked to report their sea duck hunting activity and harvest separately from other duck hunting and harvest.

Like the previous system, the HIP survey system consists of a questionnaire survey that asks hunters to report their harvest of ducks, sea duck, geese, and brant. The wing survey is unchanged from the previous system and provides estimates of species, sex, and age composition for the bag. However, the HIP questionnaire asks hunters to report their harvest for each hunting trip, including the county and state in which they hunted.

Some hunters mistakenly report diving ducks or other ducks as sea ducks, so the USFWS uses the county information to determine whether the reported sea duck harvest could have occurred in special sea duck zones in the Atlantic Flyway or coastal areas of the Pacific Flyway, because nearly all of the sea duck harvest occurs in those two flyways. All sea duck harvest reported in other counties is added to other duck harvest for estimation purposes. Likewise, sea duck wings received are separated according to the counties where the birds were shot, so that wings from birds shot in coastal counties are used to estimate the species composition of the reported sea duck harvest and wings from sea ducks shot in other counties are used to estimate sea ducks as a proportion of other ducks harvested. For example, the sea duck harvest in New York is the sum of the reported sea duck harvest in the special sea duck hunting zone near Long Island, combined with sea ducks harvested in the rest of the state in areas such as the Finger Lakes and Lake Ontario. In addition to correcting some reported sea ducks harvested that were actually other diving ducks, sorting harvest and wings by county enables a better assessment of the impact of special sea duck hunting regulations.

Alaska is treated separately because sea duck hunting regulations define the group more broadly to include Common Mergansers (*Mergus merganser*), Red-breasted Mergansers (*M. serrator*), and Harlequin Ducks, in addition to eiders, scoters, and Long-tailed Ducks. Alaska has no special sea duck zones, and harvest is not necessarily restricted to the state's coastal areas because scoters, Long-tailed Ducks, and Harlequin Ducks can be found inland. Thus, sea duck harvest and species composition is estimated statewide for Alaska. In all other states, reported sea duck harvest and sea duck wings are combined with other ducks for analyses that estimate species-specific harvest.

Strengths and Weaknesses of the U.S. Fish and Wildlife Service Harvest Surveys

Sample Frames and Sampling

The primary strength of both survey systems is their sample frames and sampling designs. Both versions of the harvest survey are based on sample frames of nearly all waterfowl hunters and provide representative samples, including sea duck hunters. The main goal of implementing HIP was to increase the accuracy of state and flyway harvest estimates, so HIP does not adequately address questions about regional and local harvest of sea ducks and other species related to management areas, habitat units, or particular seasonal aggregations of waterfowl. More detailed harvest surveys related to specific sea duck population units would require intensive and expensive efforts.

State hunting license vendors do not always ask migratory bird hunters the questions required under the HIP (Moore et al. 2002), including questions designed to identify likely sea duck hunters and create a separate sampling stratum for them. Thus, each year some sea duck hunters are assigned to the wrong stratum in the HIP sample frame. However, all strata are sampled every year, albeit at different rates, and any misclassifications result in reduced precision but do not otherwise affect sea duck hunter activity and harvest estimates. Furthermore, all state HIP sample frames include nonresident (out-of-state and alien) license holders. Assuming that response rates for resident and nonresident hunters are similar, sea duck harvest by nonresident hunters and individuals who hunt with professional guides is captured in the HIP questionnaire survey, even if those hunters are misclassified with regard to the sea duck stratum.

Harvest Estimates

The duck stamp–based survey's sampling mechanism deteriorated over time but apparently that did not affect the accuracy of duck harvest estimates. Both survey systems were used from 1999 to 2001, and the annual duck harvest estimates derived from the two separate systems were similar (Padding and Royle 2012). Padding and Royle (2012) found that the duck stamp–based survey and HIP survey overestimated goose harvest by factors of 1.50 (SE = 0.02) and 1.63 (SE = 0.04), respectively, and both surveys overestimated duck harvest by a factor of 1.37 (SE = 0.02). The authors did not investigate the accuracy of sea duck harvest estimates, but it seems likely that those values were also overestimated to a similar degree.

Wing Surveys

Wing survey participants consist of two groups: (1) hunters who participated in the wing survey the previous year, and (2) a sample of hunters who participated in the questionnaire survey the previous year and reported harvesting at least one duck or goose. Hunters in the first group are removed from the sample after three years of participation. In a typical year, 6,000–8,000 hunters participate in the wing survey (usually ≤1% of all waterfowl hunters), roughly apportioned by state of residence based on the proportion of the national waterfowl harvest that usually is taken in that state (U.S. Fish and Wildlife Service, unpubl. data). Most of the primary states for sea duck harvest account for a small proportion of the total national harvest of waterfowl so hunter sample sizes are usually small for those states. Because many individual hunters probably exhibit the same hunting patterns from year to year and the sample always includes some hunters who have participated for 1 or 2 previous years, the small sample of sea duck wings received from some states may not be representative of sea duck harvest in those states. Thus, annual estimates of species- and date-specific sea duck harvest at the state level are often imprecise and autocorrelated with previous years' estimates. We report survey results at those levels as 10-year averages.

Nonresident hunters who do not reside in the state where the harvest occurred do not pose problems in estimating sea duck harvest but do potentially affect estimates of species composition. Nonresident hunters, especially those who hunt sea ducks with professional guides, probably do not hunt sea ducks year after year. Therefore, even if they are selected for the wing survey as a result of reporting sea duck harvest on the previous year's questionnaire, they are less likely to hunt sea ducks again while they are participants in the wing survey program. This issue would not affect the accuracy of species-specific harvest estimates if the species composition of sea ducks that resident and nonresident hunters harvest is the same, which is likely the case for the most part. However, harvest of species that nonresident hunters select for taxidermy would be underestimated if wings from those targeted species were not always submitted.

Canadian Wildlife Service Harvest Survey: 1967 to Present

The sample frame for Canada's national migratory bird harvest survey is based on a federal permit that is sold primarily at post offices and is required of all migratory bird hunters. The sample frame is stratified according to (1) which province, or zone within province, the hunter purchased the permit; (2) hunter experience, based on whether the hunter purchased a permit the previous year or the previous two years; and (3) whether or not the permit purchaser is a resident of Canada (Cooch et al. 1978). The survey began in 1967 and several refinements were made from 1967 to 1974 (Johnson et al. 2012).

Like the US system, the Canadian harvest survey system consists of a questionnaire survey that asks hunters to report their harvest of ducks, geese, and other species or species groups and a wing survey called the Species Composition Survey that provides species, sex, and age information. The questionnaire survey does not ask hunters to report sea duck hunting and harvest separately, but it does ask them to report the date and location of each of their hunts (Cooch et al. 1978, Johnson et al. 2012). Questionnaire survey results are used to estimate duck hunter activity and the total harvest of ducks, and the wing survey is used to estimate the species, sex, and age composition of the duck harvest, including species of sea ducks. The wing sample typically overrepresents the early part of the hunting season, so the date of harvest provided by both the questionnaire and wing surveys is used to estimate species composition by time period, and species-specific harvest estimates are summed across time periods (Johnson et al. 2012). This calculation ensures that harvest of species typically hunted later in the hunting season, such as sea ducks, is fully represented in estimates of total duck harvest.

Strengths and Weaknesses of the CWS Harvest Survey

The main strength of the survey system is the complete sample frame of all migratory bird hunters that the national permit provides, enabling the selection of stratified representative samples to maximize efficiency. However, the questionnaire survey response rate is typically only about 40% (Padding et al. 2006b), and nonresponse in harvest surveys is thought to result in inflated estimates (Barker 1991).

Hunters do not participate in the wing survey for more than two consecutive years, so impacts associated with multiyear participation do not affect the Canadian species composition and temporal and geographic distribution estimates as much as they do the US results. The annual sample is about 10,000 duck wings, which, like the US wing survey sample, is usually ≤1% of the total duck harvest. The proportion of the sea duck harvest that is sampled is small enough that resulting harvest estimates are imprecise; therefore, we report estimated harvest in Canada as 10-year averages.

The CWS wing survey excludes hunters who are not Canadian residents because administrative difficulties make it unreasonable to ask hunters to ship bird parts across international borders. This exclusion requires the assumption that Canadian and nonresident hunters both harvest the various species in the same proportions. However, this assumption is likely reasonable with regard to sea duck harvest. The proportion of nonresident hunters (mainly from the United States) has grown significantly since the mid-1990s in the prairie provinces of Alberta, Saskatchewan, and Manitoba (Alisauskas 2011), but nonresidents represent only <1% of the people who purchase migratory bird hunting permits in coastal provinces where most of the sea duck hunting occurs (M. Gendron, pers. comm.).

Several provinces in Atlantic Canada, including Newfoundland and Labrador, New Brunswick, and Nova Scotia, have special late sea duck hunting seasons in January and February. Harvest during those months may be underrepresented in the questionnaire survey because some hunters probably return their survey responses before that season occurs.

Wendt and Sileff (1986) documented issues with harvest estimates for sea ducks in Newfoundland, where a special survey was sent to hunters to report the kill of sea ducks and murres during the entire hunting season (September–March) in the late 1970s. Estimates from the special survey indicated that the national survey underestimated sea duck harvest by four to seven times. The authors attributed this difference to the timing of large harvests that occur late in the season and the wing survey, which concluded by November. More efforts have been made in recent years to ensure coverage of the national harvest survey and the wing survey for the entire hunting season. Similar special surveys were conducted in the late 1990s and showed the same bias, but it was much less pronounced (special surveys were ~1.6 times higher, S. Gilliland, Environment Canada, unpubl. data). Other issues indicated by Wendt and Sileff (1986) include the low number of sea duck hunters, their low response rates in some areas, a highly clumped distribution, and a strongly skewed distribution of harvest by individual hunters. All of these factors lead to imprecise and usually underestimated harvest of sea ducks.

State Surveys

Several states conduct independent efforts to estimate sea duck harvest, including Maryland and Washington. The Maryland Department of Natural Resources (MDNR) annually surveys a random sample of 7% of the state's hunting license purchasers to estimate hunting activity and harvest of all game species in the state (W. F. Harvey, pers. comm.). The method is a mail questionnaire survey that uses the HIP sample frame for Maryland, so it is generally comparable to the HIP survey. Annual estimates from the two surveys do not always agree (especially 2010 and 2011; Table 12.1), but the average estimated annual harvest over the entire period from the HIP surveys (16,700 birds) is similar to the average from the MDNR surveys (14,200 birds, Maryland Department of Natural Resources, unpubl. data).

To gain better information about sea duck harvest and hunting activity for the state, the Washington Department of Fish and Wildlife (WDFW) began requiring all sea duck hunters hunting in western Washington to obtain a special permit and harvest card in 2004 (Washington Department of Fish and Wildlife, unpubl. report). Permittees are required to immediately record harvest in the field and provide a report of their sea duck hunting after the season. Harvest estimates are expanded to account for the total number of permittees, with a correction for nonresponse.

TABLE 12.1

Comparisons of annual state-specific sea duck harvest (scoters, Long-tailed Ducks, and Harlequin Ducks) as estimated from federal (USFWS) and state (MDNR) waterfowl harvest surveys and mandatory hunter reports (WDFW).

	Maryland		Western Washington	
Year	Federal survey	State survey	Federal survey	State survey
1999	12,000	13,600		
2000	9,900	9,800		
2001	16,900	10,500		
2002	13,100	10,100		
2003	18,700	15,900		
2004	20,400	11,600	2,433	2,275
2005	20,400	23,000	2,383	1,928
2006	27,500	13,700	2,452	3,007
2007	17,900	16,800	3,325	2,594
2008	16,600	17,300	5,055	2,447
2009	16,100	13,800	8,963	3,903
2010	9,300	19,200	1,404	2,182
2011	18,200	9,500	4,552	1,577

The mandatory reporting system yields a harvest estimate of coastal sea duck species (scoters, Long-tailed Ducks, Harlequin Ducks, and goldeneyes) that can be compared with federal survey estimates for the state. Washington's annual sea duck permittee list has not been utilized in the federal survey; consequently, the federal harvest estimates are based on reported harvest and wings received for all ducks, rather than using a separate stratum of prospective sea duck hunters. The Washington survey relies on species identification by hunters, rather than the parts collection survey, and misclassification has been documented but not accurately quantified (D. Kraege, pers. comm.).

The federal harvest estimates for sea ducks in western Washington were greater than annual estimates from WDFW's mandatory reports in 6 of 8 years from 2004 through 2011 (Table 12.1), and the average of the federal estimates (3,798) was about 50% higher than the average of the WDFW counts (2,489). The difference in estimates suggests that the federal survey overestimated Washington sea duck harvest, that some permittees did not submit the mandatory report, or both.

The 2009 federal estimate for Washington illustrates a consequence of relying solely on the wing survey to differentiate between sea duck harvest and harvest of other ducks. That year, a single avid scoter hunter selected for the wing survey submitted a large number of scoter wings, resulting in an unusually high estimate of scoter harvest. Beginning in 2013, the HIP survey corrected this problem by using Washington's sea duck permit sales information to sample sea duck hunters and estimate sea duck harvest separately, as in the Atlantic Flyway and the other coastal Pacific Flyway states.

Wounding Loss

Wounding loss occurs when birds are injured or killed but not retrieved by hunters. Losses are probably greater in sea ducks than other waterfowl because (1) sea ducks are usually shot over open water at longer ranges than decoying dabbling ducks, (2) most sea ducks are large birds with tough feathers and skin that is difficult to penetrate, and (3) they are strong swimmers and divers that have a good chance of escaping retrieval (Bellrose 1953, Hochbaum and Walters

1984). To reduce the loss of wounded sea ducks, Atlantic Flyway hunters in the United States are allowed to shoot wounded ducks from motorboats under power in designated offshore sea duck hunting zones; this method of take is prohibited elsewhere in the United States and in Canada.

In the late 1960s, the Maryland DNR conducted studies of sea duck hunters in Chesapeake Bay that yielded estimates of wounding rates (L. J. Hindman, pers. comm.). The studies consisted of (1) observations recorded by MDNR personnel during hunts in which they participated and (2) surreptitious observations of other hunting parties by trained MDNR personnel, modeled after the hunter performance studies of the 1960s and 1970s (Kimball et al. 1971). Hunters observed during these studies were unable to retrieve about 28% of downed scoters, eiders, and Long-tailed Ducks (Maryland Department of Natural Resources, unpubl. data).

The annual HIP survey asks hunters to report the number of sea ducks they wounded and lost. During the 1999–2003 hunting seasons, reports of wounding loss [birds wounded/ (birds wounded + birds killed and retrieved)] averaged 0.18 for sea ducks and 0.12 for other ducks (Padding et al. 2006b, Moore et al. 2007). However, these data are difficult to interpret because (1) some wounding cannot be readily detected, (2) reactions of birds to near misses may be perceived as wounding, and (3) some hunters may be reluctant to report wounding loss. Martin and Carney (1977) concluded that although harvest surveys probably provide reliable indices of wounding loss, they likely underestimate loss rates. In harvest management and population models, wounding loss is typically considered to be 0.20 of the total duck kill (Anderson and Burnham 1976, Johnson et al. 1997). Assuming that hunter-reported sea duck wounding loss is 1.5 times greater than for other ducks, 0.30 or higher is a comparable estimate for sea ducks, depending on hunting methods. Hunter reports of wounding of ducks and geese have declined in the 1990s and 2000s (Schulz et al. 2006; U.S. Fish and Wildlife Service, unpubl. data), so perhaps actual wounding rates are lower now.

Another difficulty in assessing wounding loss is that the ultimate fate of a wounded bird is unknown; the bird could (1) die of its wounds or because of increased vulnerability to predation,

(2) recover enough to survive but not enough to reproduce as part of the breeding population, or (3) recover fully (Van Dyke 1981). Several studies have shown that substantial numbers of live ducks and geese had one or more shotgun pellets embedded in their tissue, indicating that they had survived hunting wounds (Elder 1955, Peterson and Ellarson 1975, Perry and Geissler 1980, Kirby et al. 1981, Madsen and Riget 2007). Hicklin and Barrow (2004) captured and examined 1,005 incubating female Common Eiders (*Somateria mollissima*) in eastern Canada and found that 29% of them carried at least one embedded pellet. Merkel et al. (2006) showed that embedded shots affected the body condition of juvenile Common Eiders but not that of subadults and adults. Exposure to hunting may be substantial for some species, but actual mortality due to wounding could be lower than currently accepted wounding loss estimates.

HARVEST OF UBIQUITOUS SPECIES (BUFFLEHEADS, COMMON GOLDENEYES, MERGANSERS)

Hunting Regulations: Past and Present

Migration routes and wintering areas of Buffleheads, Common Goldeneyes, and the mergansers are broadly distributed in North America, whereas eiders, scoters, Harlequin Ducks, and Long-tailed Ducks are primarily hunted in coastal areas and the Great Lakes. Consequently, harvest management of ubiquitous species has usually been implemented at national rather than regional scales. Ubiquitous species can be harvested under conventional fall and winter harvest regulations as well as by subsistence hunters.

For the most part, hunting Buffleheads and goldeneyes has only been allowed during general duck hunting seasons, and the species have been included in the aggregate daily bag limit for ducks, pooling any combination of species. However, shooting Buffleheads was prohibited in the United States from 1932 through 1937 in response to a presumed population decline. Bruette (1934) noted that Bufflehead numbers had decreased, although he characterized the species as quite abundant. In 1938, US hunters were allowed a maximum of three Buffleheads per day as part of a 10-duck daily bag limit, and in 1944, the restriction on Buffleheads was removed.

Canada offered some additional hunting opportunity from Québec west to Alberta in the 1960s and early 1970s by allowing take of two bonus goldeneyes in addition to the full regular daily bag limit of ducks. Later, as concerns mounted over the status of western Barrow's Goldeneyes, the daily bag limit for goldeneyes was restricted to two birds in British Columbia in 1990 and western Washington in 2010. Concerns were raised about the status of eastern Barrow's Goldeneyes where a small breeding population was centered in Québec and wintered in the northern Atlantic states (Robert et al. 2000), which also led to harvest restrictions in the east.

Common and Red-breasted Mergansers

Anglers and fisheries managers have not held mergansers in high regard over the years. Mergansers have been viewed as major fish predators since the early 1900s (Beach 1936, White 1957, Erskine 1972, Anderson et al. 1985), and most fish and wildlife agencies supported reduction of mergansers. In general, Common Mergansers were considered more egregious fish predators than Red-breasted Mergansers; Hooded Mergansers (*Lophodytes cucullatus*) were not a concern (see Munro and Clemens 1932, 1937, 1939; Salyer and Lagler 1940). Thus, it is not surprising that hunting regulations were established that encouraged hunting of mergansers, first in Canada and later in the United States.

The bag limit for all merganser species was removed entirely in Ontario in the mid-1930s, a few years later in Prince Edward Island, and in Newfoundland in 1965. Beginning in 1950, there was no bag limit for Common and Red-breasted Mergansers in Québec. When the no limit policy was discontinued in 1977, extra mergansers were still allowed in addition to the general duck bag limits in Newfoundland and Québec, and mergansers were also included in the additional seasons for eiders, scoters, mergansers, and Long-tailed Ducks in the coastal waters of New Brunswick and Nova Scotia. Additional mergansers in the daily bag were allowed in Québec until 1997 and were still allowed in Newfoundland and Labrador as of 2013.

Beginning in 1944, hunters in the United States were allowed to take up to 25 Common and Red-breasted Mergansers daily in addition to the general duck bag limit. The special bag limit for mergansers, over and above limits for other ducks, was based on the rationale of protecting economic fisheries of salmon, trout, and herring. In 1954, the additional 25-bird daily bag limit for Common and Red-breasted Mergansers was eliminated nationwide in the United States. In 1957, the separate additional bag limit for mergansers was reestablished in all four flyways with a daily bag limit of five (10 in possession), of which only one bird could be a Hooded Merganser. All merganser species have been included in the Pacific Flyway's general duck limits since 1980, but states in the other three flyways still allow the additional merganser limits under the federal frameworks. Some states have elected to forego this option: in 2012, Connecticut, Maryland, and New York in the Atlantic Flyway, Missouri in the Mississippi Flyway, and Montana and New Mexico in the Central Flyway opted to include mergansers in their general bag limits for ducks.

Hooded Merganser

Conservation concerns about Hooded Mergansers in the United States were recognized in the early 1900s based on population declines in California (Grinnell et al. 1918), the northeastern and Mid-Atlantic states (Phillips 1926), as well as other parts of their continental range. As regulation of waterfowl hunting evolved, the sparseness of Hooded Mergansers compared to the Mergus species led to protective measures nationwide in the United States when the bag limit for Hooded Mergansers was reduced to one bird per day and one in possession from 1953 through 1962, and one bird per day and two in possession through 1978. The restriction on Hooded Mergansers in Pacific Flyway states was lifted in 1979, and harvested birds have been included in the general duck limits since 1980. Low levels of harvest in the Pacific Flyway since the 1960s, typically ≤5,000 of about three million ducks harvested annually (S. M. Olson and R. E. Trost, U.S. Fish and Wildlife Service, unpubl. report; R. E. Trost and M. S. Drut, U.S. Fish and Wildlife Service, unpubl. report), suggest there is low interest in merganser hunting there and hunters probably take most Hooded Mergansers incidentally while hunting other ducks. In the other three flyways, however, the restriction remains in place, although the daily bag and possession limits were

raised to two and four birds, respectively, in 2006 in the Mississippi and Central Flyways and 2007 in the Atlantic Flyway.

Canada did not impose similar limitations on harvest of Hooded Mergansers, and the species was included under regulations for other species of mergansers or as part of the general duck bag limit (Québec, Manitoba, and the Central and Pacific Flyway provinces). No bag limit was set for Hooded Mergansers in Ontario until 1971, and in Newfoundland and Prince Edward Island until 1974, at which time the species was included as part of the general duck bag limits in those provinces.

Harvest

As expected from their broad distribution during hunting seasons, Buffleheads, Common Goldeneyes, and Hooded Mergansers are harvested throughout the continent (Figures 12.1 and 12.2). The Great Lakes region is a particularly important harvest area for these species, due to both bird abundance and large numbers of waterfowl hunters. The most commonly harvested sea ducks in North America (excluding subsistence harvest) are Buffleheads with annual continental harvest averaging about 211,000 birds, Common Goldeneyes averaging 103,000 birds, and Hooded Mergansers averaging 102,000 birds/year during the decade from 2002 to 2011.

The three species are all considered late migrants, but Buffleheads undertake fall migration earlier than the rest (Bellrose 1980). The difference in timing is reflected in the seasonal harvest in the United States, where about half of the Bufflehead harvest occurs before December compared with the other species (Figure 12.3). These duck species are not highly sought after by hunters, but goldeneye and Bufflehead decoys are fairly common in some regions, and wing survey receipts indicate that at least a few hunters specifically target mergansers (U.S. Fish and Wildlife Service, Branch of Harvest Surveys, unpubl. data). Overall, annual harvest of these species appears to track annual estimates of total duck harvest closely in both Canada and the United States (Figures 12.4 and 12.5), suggesting that most of the birds are taken opportunistically during hunts primarily targeting diving (Aythya spp.) or dabbling ducks (Anas spp.).

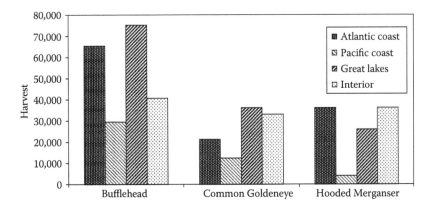

Figure 12.1. Regional distribution of the estimated average annual harvest of Buffleheads, Common Goldeneyes, and Hooded Mergansers in the United States and Canada from 2002 through 2011.

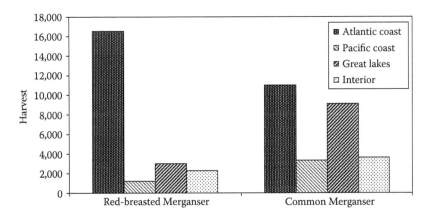

Figure 12.2. Regional distribution of the estimated average annual harvest of Red-breasted and Common Mergansers in the United States and Canada from 2002 through 2011.

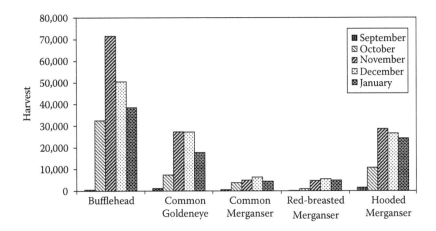

Figure 12.3. Seasonal distribution of the estimated average annual harvest of Buffleheads, Common Goldeneyes, and mergansers in the United States from 2002 through 2011.

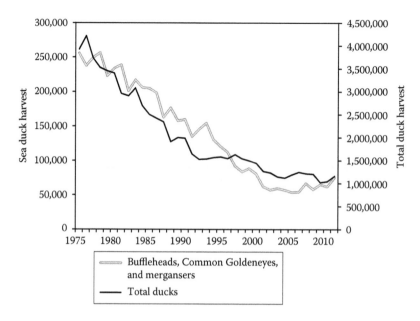

Figure 12.4. Estimated annual combined harvest of Buffleheads, Common Goldeneyes, and mergansers compared to total duck harvest in Canada from 1975 through 2011.

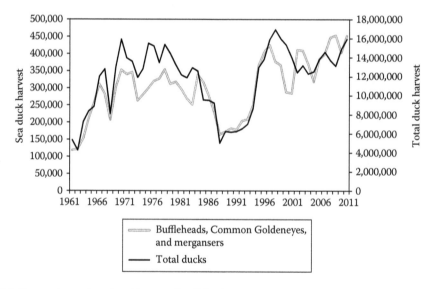

Figure 12.5. Estimated annual combined harvest of Buffleheads, Common Goldeneyes, and mergansers compared to total duck harvest in the United States from 1961 through 2011.

Sex and Age Composition

Fall and winter plumages of most adult male sea ducks are striking compared to immature birds and adult females, and many US hunters likely target adult males as a result. Thus, sex ratios of harvested birds are influenced by hunter selectivity and do not necessarily reflect the standing sex ratio of wild populations. Nonetheless, adult males

outnumber females in populations of goldeneyes, Buffleheads, and most mergansers with the exception of the Common Merganser (Table 12.2). The female-biased sex ratio of Common Merganser may reflect the continental population's sex composition but could also be the result of different migration patterns among the different age and sex classes (Anderson and Timken 1972), which may make adult males less vulnerable to hunting.

TABLE 12.2
Adult sex ratios (males–females) and age ratios (immature birds of both sexes–adult females) of goldeneyes, Buffleheads, and mergansers harvested in Canada and the United States from 2002 through 2011.

Species	Adult sex ratio		Age ratio		Wings examined	
	Canada	United States	Canada	United States	Canada	United States
Common Goldeneye	2.34	2.10	4.00	2.75	2,698	7,236
Barrow's Goldeneye	1.03	2.64	1.41	2.36	106	588
Bufflehead	2.79	2.95	4.85	3.65	1,801	15,197
Common Merganser	0.82	0.97	3.28	2.06	759	2,019
Red-breasted Merganser	2.71	1.41	6.05	2.17	579	1,795
Hooded Merganser	2.44	2.34	3.40	3.22	1,343	6,103

Harvest age ratios of most species have a higher ratio of immature birds in Canada than the United States (Table 12.2), suggesting that immature birds of these species become more wary after their initial exposure to hunting in Canada. Age ratios of birds harvested in the United States imply that recent annual production rates are relatively high for Common Goldeneyes, Buffleheads, and Hooded Mergansers but are less robust for Barrow's Goldeneyes and Common and Red-breasted Mergansers, suggesting lower harvest potential for the latter group of species.

ATLANTIC COAST FALL–WINTER TRADITIONS

The Atlantic coast of North America has a rich sea duck hunting tradition, from subsistence hunting by aboriginal peoples, to market hunting in the 1800s and early 1900s, to sport hunting. Commercial waterfowl hunting was still an accepted practice on the Atlantic coast until the late 1930s (Smith 1985), and sea ducks were hunted despite their less favorable reputation among gourmands. Market prices for Buffleheads and goldeneyes ($0.30–$0.50 per pair), scoters ($0.50), and Long-tailed Ducks ($0.70–$0.90) were much lower than the $5–$7 per pair that Canvasback (*Aythya valisineria*) commanded as a highly regarded species but were similar to prices for Northern Pintails (*Anas acuta*) and American Wigeon (*A. americana*, Walsh 1971). Even as market hunting faded into history, traditional fall and winter sea duck concentration areas remained the favored haunts of hunters.

Bays, sounds, inlets, river mouths, and other Atlantic coastal waters are the primary wintering grounds for several sea duck populations, and over the years, decades, and even centuries, those areas have provided ample opportunity for the hardy hunters who pursued these birds. The 1909 painting of Winslow Homer, *Right and Left*, depicts two goldeneyes falling to the gun and is an iconic image of Atlantic coast sea duck hunting on a rough day. Similar scenes still occur in hundreds of locales in the Gulf of St. Lawrence and the Bay of Fundy, along the coasts of Maine and Cape Cod, and in Long Island Sound, Chesapeake Bay, and Pamlico Sound. In 2013, more than 70 guide services from Maine to North Carolina offered specialized sea duck hunts, indicating that interest in hunting sea ducks along the Atlantic coast remains high to this day.

Further north, sea duck hunting is the main waterfowl harvest; when hunters in Newfoundland *are going at the ducks*, the main quarry is Common Eiders and not American Black Ducks (*Anas rubripes*) or Mallards (*A. platyrhynchos*). Attitudes of many hunters in Newfoundland and Labrador toward sea duck hunting are more similar to those of northerners than recreational hunters. The hunt is largely considered a subsistence harvest and has long traditions and cultural value. Hunting is often a community-based activity, and hunters and nonhunters alike participate in cleaning and processing sea ducks for consumption. Unfortunately, the perception of a traditional right to hunt sea ducks and other birds has been taken to extreme levels by a few individuals, and large-scale illegal harvest and selling of sea ducks still occurs despite considerable effort of enforcement officials (Chardine et al. 2008).

Harvest of sea ducks also continues with a relatively small but passionate group of sea duck hunters in the islands of St. Pierre and Miquelon, a small overseas collectivity of France located off the south coast of Newfoundland. The Migratory Bird Treaty/Convention between Canada and the United States does not apply in this area; hunting regulations have been set by local authorities, and they have traditionally allowed relatively liberal access to sea ducks and seasons extending into April. These late seasons have been challenged recently to align with European Union regulations on hunting birds.

Hunting Regulations: Past and Present

Eiders, Scoters, and Long-tailed Ducks

Market hunting had nearly extirpated American Common Eiders (*S. m. dresseri*) by the end of the 1800s (Goudie et al. 2000), and the Migratory Bird Treaty Act of 1916 stipulated that the United States and Canada would both prohibit eider hunting entirely. However, the Migratory Bird Convention Act (MBCA) did not apply in Newfoundland and Labrador because the region was a British colony that did not join confederation with Canada until 1949. Eider hunting, largely for subsistence purposes, continued in coastal waters. Likewise, subsistence hunters in the Canadian North continued taking eiders, as the MBCA was generally not enforced in the north. The ban on eider hunting in the rest of Canada and the United States remained until 1932, when eiders were included as part of the daily bag limit during the regular duck season.

Special seasons to provide additional sea duck hunting opportunity were first implemented in 1938, in Maine, New Hampshire, Connecticut, Massachusetts, and Rhode Island. Hunters there were allowed to take 10 scoters/day in open coastal waters from September 15 until the beginning of the general duck hunting season, at which point scoters were included along with other ducks as part of the duck bag limit. After expanding to include Long Island in 1940, the additional scoter season was replaced with a separate 92-day sea duck season in 1949, with a daily bag limit of seven eiders and scoters in any combination; Long-tailed Ducks were added the following year. In 1963, the season was increased to 107 days, the maximum season allowed under the 1916 Migratory Bird Treaty Act, and by 1971, all Atlantic coast states except Florida had special sea duck seasons. For more information, Caithamer et al. (2000) provide a detailed description of the evolution of sea duck hunting regulations in Atlantic coast states.

In Canada, an additional 30-day season for eiders and scoters was allowed in the coastal waters of New Brunswick and Nova Scotia beginning in 1948 and the following year in Newfoundland when it became a province. In 1950, Long-tailed Ducks were added to the list of species that could be taken during the additional sea duck seasons, and beginning in 1952, the additional season was also provided in the coastal waters of Québec. The additional seasons were retained in New Brunswick and Nova Scotia but were replaced with separate seasons for eider, scoter, and Long-tailed Duck hunting that were independent of general duck hunting seasons in Newfoundland (1958) and Québec (1963). The seasons provided hunting opportunity late in the winter, when most other ducks have long since migrated south but many sea ducks remain along Canada's east coast. Daily bag limits during both additional and separate seasons were the same as they were during general duck hunting seasons except in Labrador, where hunters could take 25 eiders, scoters, and Long-tailed Ducks daily in any combination. In 1967, separate bag limits of 10 birds in any combination were established for the additional seasons in New Brunswick and Nova Scotia, and Québec hunters were allowed two eiders, scoters, or Long-tailed Ducks in addition to the general daily duck bag limit. The daily bag limit for the separate season on the island of Newfoundland was increased to 12 birds in 1969.

The special sea duck seasons remained unchanged in most of the Atlantic coastal states and provinces during the 1970s and most of the 1980s. When regular duck hunting season lengths and daily bag limits were significantly reduced in the late 1980s, hunters in prime sea duck areas such as Chesapeake Bay increasingly turned to the special sea duck season for more hunting opportunity. Concerned about the possible impacts of this increased hunting pressure on scoters and Long-tailed Ducks, the state of Maryland reduced the daily bag limit to five sea ducks in 1989, and in 1993, the USFWS reduced the daily bag limit to four scoters in all Atlantic Flyway states. Similarly, concerns about the status of Common Eiders and scoters led CWS to reduce New Brunswick's additional-season daily bag limit in 1990 to six birds, of which no more than four could be eiders or scoters. That year, eider harvest was also limited to six birds daily in Newfoundland and seven birds daily in Labrador. The 10-bird limit was maintained in Nova Scotia, but in 1994, a restriction on scoter harvest was implemented with limits of no more than 4 birds/day, and a few years later, Nova Scotia's sea duck bag limit was reduced from 10 to 5 birds.

By 1997, eider harvest in Newfoundland had declined by 55% below the levels taken in the mid-1980s, and examination of females for embedded shot indicated that eiders migrating through or wintering around Newfoundland were exposed to high levels of hunting pressure (Hicklin and Barrow 2004). Confronted with declining harvest, evidence of heavy hunting pressure, and widespread expressions of public concern for eider populations, the CWS implemented harvest restrictions in 1998. The bag limit for eiders, scoters, and Long-tailed Ducks combined was reduced from 12 to 6 birds throughout the province, only three eiders could be taken daily in February, and the season was closed for the first 10 days of March. Subsequently, a survey that covered the wintering range of northern Common Eiders estimated that over 200,000 northern eiders wintered in Canada. The harvest assessment developed using this new information suggested the population in Canada could sustain additional harvest and the February restriction was lifted (Gilliland et al. 2009).

In the United States, as sea duck hunting continued to gain popularity, additional states restricted bag limits more than federal frameworks required. Maine maintained the aggregate seven-bird daily bag limit but restricted eider harvest to five birds daily in 1999, and further reduced the limit to four birds in 2009. In Massachusetts in 1999 and New Hampshire in 2000, the two states reduced the eider and Long-tailed Duck limit to four birds. Further, hunters in Massachusetts were allowed to take only one female eider per day. In 2004, Connecticut reduced the aggregate sea duck bag limit to five birds, of which no more than four birds could be Long-tailed Ducks, and Rhode Island also reduced the aggregate bag limit to five birds in 2008.

Barrow's Goldeneye

Uncertainty about the status of Barrow's Goldeneyes in British Columbia and interest in understanding the impacts of harvest on the western population of this species led to investigations of the ecology of the species in the 1980s (Savard 1987, Savard and Eadie 1989). The research led to increased scrutiny of the small eastern population and concern for its well-being (Savard and Dupuis 1999). The first harvest restrictions for eastern Barrow's Goldeneye, including closures and reduced bag limits, were established in Québec in 1995. The eastern birds were listed as a population of special concern in Canada (Committee on the Status of Endangered Wildlife in Canada 2000). Subsequently, the bag limit for Barrow's Goldeneye was reduced to one bird daily and two in possession in 2007 in Newfoundland and Labrador, New Brunswick, Nova Scotia, Ontario, and Prince Edward Island and in Québec the following year. A complete ban on harvest of Barrow's Goldeneyes was considered, but given the difficulty in identifying goldeneye species under typical hunting conditions, a one-bird daily limit was put in place. In 2012, the possession limit was reduced to one bird in those provinces. The US federal frameworks for the Atlantic Flyway have no specific restrictions on goldeneyes, but the state of Maine elected to close the season on this species in 2008.

Harlequin Duck

The species has probably never been numerous in eastern North America (Vickery 1988) but likely has declined since the early 1900s (Goudie 1989). As it became apparent that the eastern population numbered <1,000 birds (Goudie 1989), Canada closed the hunting season on Harlequin Ducks in Newfoundland and Labrador and Prince Edward Island in 1987, Nova Scotia in 1988, New Brunswick in 1989, and Ontario and Québec in 1990. That year, the eastern population of Harlequin Ducks was also listed as endangered in Canada. The United States closed the season in the Atlantic Flyway in 1989. The population has rebounded through the 2000s but remains relatively small at ≤2,000 birds (Mittelhauser 2008, Thomas 2008). Hunting of Harlequin Ducks is still prohibited in eastern North America.

Hunting Activity

The number of active waterfowl hunters in eastern Canada (the Atlantic Provinces and Quebec) declined steadily from the mid-1970s to the 2000s and has apparently stabilized at about 38,000 hunters

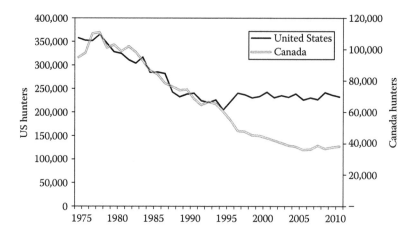

Figure 12.6. Number of active waterfowl hunters in Atlantic coastal states and provinces of the United States and Canada from 1975 through 2011.

per year in recent years (Figure 12.6). In contrast, although waterfowl hunters in Atlantic coast states have also declined since the 1970s, they stabilized about a decade earlier, in the mid-1990s (Figure 12.6). Estimates of sea duck hunter numbers are not available for the same time period, but harvest estimates suggest that they tracked waterfowl hunter numbers in Canada. In the 1980s, when regular duck hunting seasons were shortened, bag limits were reduced, and additional restrictions were imposed on harvest of American Black Ducks, the attention of waterfowl hunters turned to sea ducks in some areas. In Maine, for example, many guide services for sea duck hunting were established at this time (B. Allen, pers. comm.). The number of sea duck hunting guide services in Maine has declined since the 1990s (B. Allen, pers. comm.), indicating that interest in sea duck hunting there has waned recently. Estimates of active sea duck hunters (people who hunted eiders, scoters, and Long-tailed Ducks) are only available for Atlantic coast states for 1999 onward, and the estimates for that period are fairly stable, averaging about 11,500 hunters annually (range 9,800–13,700) in 1999–2011.

Fall and Winter Harvest

Fall and winter harvest of Common Eiders is about evenly split between Canada and the United States, but scoters and Long-tailed Ducks are taken primarily in the United States (Figure 12.7). Harvest of Common Eiders is about the same in both countries, but Canada mostly takes northern

Common Eiders, whereas US harvest is focused almost exclusively on American Common Eiders (Reed and Erskine 1986). Most of the eider, scoter, and Long-tailed Duck fall and winter harvest in both countries occurs along the Atlantic coast, although the Great Lakes region (>20% of the average annual harvest of Long-tailed Ducks) and the Pacific coast (>10% of the average annual harvest of Surf Scoters [*Melanitta perspicillata*]) are also significant harvest areas (Figure 12.8). A few King Eiders (*Somateria spectabilis*) are taken in either country in the sport harvest; from 2000 through 2011, the estimated mean annual harvest was 124 birds in Canada (all in eastern Canada) and 135 in the United States (85 on the Atlantic coast and 50 in Alaska). Estimated harvest may be biased low because Gilliland and Robertson (2009) reported that about 10% of the thousands of eiders harvested in northern Newfoundland were King Eiders and mainly juveniles. However, even taking into account a potential bias, harvest of King Eiders likely amounts to no more than a few hundred birds annually in each country.

Harvest of scoters and Long-tailed Ducks in eastern Canada dropped precipitously from the 1970s to the 2000s (Figure 12.9). Despite a similar decrease in hunter numbers (Figure 12.6), the decline in Common Eider harvest was much less pronounced. In Atlantic Canada, eiders are the targeted species in coastal sea duck hunts, with other species being taken somewhat opportunistically. The steepest declines in Long-tailed Duck harvest were in Ontario and Québec, suggesting shifts in the species distribution or changing

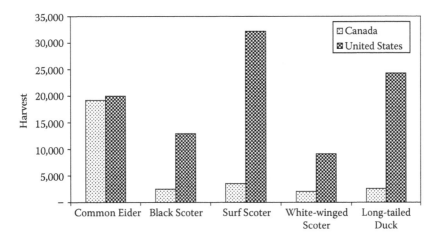

Figure 12.7. Estimated average annual harvest of Common Eiders, Black Scoters, White-winged Scoters, Surf Scoters, and Long-tailed Ducks in the United States and Canada from 2002 through 2011.

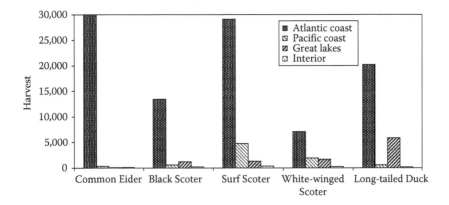

Figure 12.8. Regional distribution of the estimated average annual harvest of Common Eiders, Black Scoters, White-winged Scoters, Surf Scoters, and Long-tailed Ducks in the United States and Canada from 2002 through 2011.

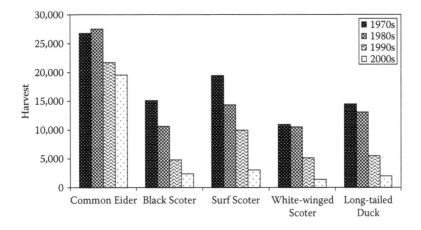

Figure 12.9. Estimated average annual harvest of several sea duck species in eastern Canada by decade from the 1970s through the 2000s.

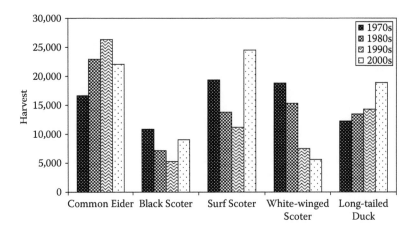

Figure 12.10. Estimated average annual harvest of several sea duck species in Atlantic coast states of the United States by decade from the 1970s through the 2000s.

hunting traditions in these provinces. Temporal patterns were different in the United States during the same period and varied more among species (Figure 12.10). Long-tailed Duck harvest increased steadily from the 1970s to the 2000s, whereas the declining harvest of White-winged Scoters (*Melanitta fusca*) was similar to the decline that Canada experienced. White-winged Scoter harvest on the Atlantic coast has decreased from an average of >25,000 birds/year in the 1970–1980s to about 7,000 birds/year in the 2000s, suggesting that this species is less available to hunters there than in the past. Population size and distribution data are insufficient to determine whether changes in harvest are due to a population decrease, a shift in geographic distribution, or some other causes.

In the United States, most harvest of Long-tailed Ducks and Surf Scoters occurs from November through January along the Atlantic coast, whereas harvest of Common Eiders, White-winged Scoters, and Black Scoters (*Melanitta americana*) peaks in November and declines thereafter (Figure 12.11). The temporal patterns may be due to differences in wintering areas and differential exposure to hunting pressure. Common Eiders and White-winged Scoters winter primarily from Maine to Long Island (Silverman et al. 2013), and the cold temperatures of December and January at those latitudes may deter some hunters. On the other hand, wintering Surf Scoters and Long-tailed Ducks are more broadly distributed with concentrations in Chesapeake Bay (Silverman et al. 2013),

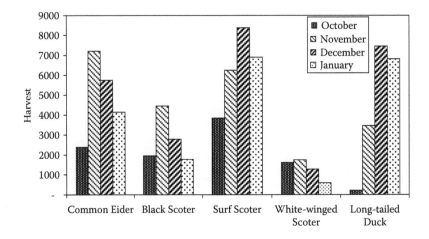

Figure 12.11. Seasonal distribution of the estimated mean annual harvest of Common Eiders, Black Scoters, White-winged Scoters, Surf Scoters, and Long-tailed Ducks in Atlantic coast states from 2002 through 2011.

which is heavily hunted throughout the comparatively mild winter. Black Scoters winter from Maine to northern Florida, with significant concentrations along the coasts of North and South Carolina and Georgia (Silverman et al. 2013), where wing survey receipts indicate there are few sea duck hunters.

The largest Common Eider harvest in Canada occurs in Newfoundland, where harvest reflects the seasonal abundance of birds in different parts of the province (Gilliland and Robertson 2009). In general, the harvest on the northern coasts peaks in November and December and is focused mainly on northern Common Eiders. In southern areas of the province, the harvest is distributed throughout the winter months of November through February, with a peak in December, and is split evenly between the northern and American Common Eiders (Gilliland and Robertson 2009). Patterns of harvest are driven by seasonal dynamics of sea ice, which push coastal species to southern waters as the winter progresses.

Sex and Age Composition of the Harvest

The sex and age composition of a population can provide insights about aspects of population status and ecology, such as effective size of the breeding population (adult sex ratio) and productivity (age ratio). In practice, sex and age information on sea ducks is obtained from samples of harvested birds via wing surveys rather than samples from entire populations. Thus, harvest sex and age ratios should be considered indices and not estimates of population composition.

Adult male sea ducks are readily identifiable from their plumage during fall and winter, making them easy for hunters to target (Metz and Ankney 1991). Consequently, sex ratios of harvested adult birds only provide indices of population sex ratios and not direct measures. Interestingly, adult sex ratios (males to females) of harvested Common Eiders and scoters are lower in Canada than in the United States (Table 12.3). It is thought that Canadian eider hunters do not preferentially target adult males and that sex ratios in the harvest reflect the population at large (Gilliland and Robertson 2009), whereas in the United States, hunters target adult males. However, it is also possible that adult male Common Eiders and scoters are less available to hunters in Canada, perhaps due to different migration patterns of adult males compared to adult females and birds that are young of the year. Assuming that most of the harvest of eiders, scoters, and Long-tailed Ducks in the United States consists of wintering birds, that hunters in Canada do not target males but US hunters target them, and that adult males and females winter together, it appears that male scoters and Long-tailed Ducks significantly outnumber females, whereas the Common Eider harvest sex ratio may be due primarily to hunters selecting adult males (Table 12.3).

Adult male sea ducks are obvious to most hunters, but plumage differences between adult females and immature birds of both sexes are subtle. To reduce the effects of hunter selectivity on harvest age ratio estimates, we excluded adult males and present the ratios as immatures to adult females (Table 12.3). However, immature birds are typically more vulnerable to hunting than adults. Thus, although they are correlated with population age structure, these age ratios are indices rather than direct measures (Martin and Carney 1977). With the exception of the White-winged Scoter

TABLE 12.3

Adult sex ratios (males–females) and age ratios (immature birds of both sexes–adult females) of eiders, scoters, and Long-tailed Ducks harvested in Canada and the United States from 2002 through 2011.

Species	Adult sex ratio		Age ratio		Wings examined	
	Canada	United States	Canada	United States	Canada	United States
Common Eider	0.94	2.24	1.65	0.63	1,890	2,485
Black Scoter	2.22	2.93	4.12	2.66	310	1,022
Surf Scoter	0.89	3.55	5.39	2.47	447	2,021
White-winged Scoter	1.24	3.04	3.77	5.36	237	783
Long-tailed Duck	2.85	2.65	6.49	2.22	284	1,528

(but note small sample sizes), harvest age ratios of eiders, scoters, and Long-tailed Ducks are greater in Canada than in the United States (Table 12.2), indicating that the initially naïve immature birds gain experience as they are exposed to hunting during migration to their wintering grounds.

The sex- and age-specific harvest rates that are needed to estimate vulnerability differences among sex and age cohorts are scarce for sea ducks. Joensen (1974) estimated that immature Common Eiders were two to four times more vulnerable to hunters in Denmark than adults. Merkel (2004a) found that eider age ratios were fairly even in Greenland wintering birds, but that harvest was heavily skewed (75%–95%) toward immature birds. In Newfoundland, Gilliland and Robertson (2009) estimated that 60% of King Eiders harvested were immature birds and that immature Common Eiders may have been five times more vulnerable to hunters than adults. However, reliable estimates of age-specific vulnerability are absent for most populations of sea ducks in North America.

Immature female midcontinent Mallards are about 1.75 times more likely to be shot than adult females (Runge et al. 2002). If we assume that vulnerability of immature sea ducks relative to that of adult females is of similar magnitude in the United States, recent productivity of Black Scoters, Surf Scoters, and Long-tailed Ducks is about 1.2–1.6 fledged young of both sexes per adult female. In contrast, the same assumptions yield a low estimate of the annual production rate for Atlantic Common Eiders: 0.36 total fledged young or about 0.18 fledged young females per adult female. The implications of such low productivity with regard to harvest management depend on whether production is density dependent or density independent (Chapter 3, this volume), but the answer to that question is presently unknown.

PACIFIC COAST FALL–WINTER TRADITIONS

Hunting traditions and harvest of sea ducks on the Pacific Coast are greatest in Alaska where breeding and wintering sea ducks are most abundant, less common in British Columbia and Washington, and primarily opportunistic take during hunts of diving ducks in Oregon, California, and Mexico. Undoubtedly, there are local traditions of waterfowl hunting on all bays and estuaries along the Pacific Coast, but popular and technical literature are scarce (Hagerbaumer 1998, Kramer 2003).

Some notable historical hunting areas in California with populations of migrant and wintering sea ducks include San Francisco Bay, Tomales Bay, Bodega Bay, and Humboldt Bay. The Oregon Coast has fewer areas with sea ducks, including Coos Bay, Alsea Bay, Yaquina Bay, Siletz Bay, Netarts Bay, Tillamook Bay, and the mouth of the Columbia River. The principal sea duck areas in Washington include numerous sites in Puget Sound and outer coastal bays such as Willapa Bay and Grays Harbor.

Hunting Regulations: Past and Present

Pacific Flyway

In 1948, representatives of western state wildlife agencies, British Columbia, and the federal agencies initiated a collaborative waterfowl program (today, the Pacific Flyway Study Committee) to expand data collection and create a forum for discussing issues about management and harvest of shared populations in the west (Bartonek 1984). With expansion of the flyway concept nationwide, the Pacific Flyway Council was established in 1951 to provide policy direction and coordination of programs, as well as formulating state recommendations on US waterfowl hunting regulations. Representatives from the Yukon Territory, Northwest Territories, and British Columbia have worked with the Pacific Flyway Council at the technical level to exchange information and develop complementary management programs.

Unlike the Atlantic Coast where sea duck hunting has been more traditional, historic waterfowl hunting regulations in the Pacific Coast states and British Columbia seldom included special provisions for sea ducks. The one exception was special liberal bag limits for mergansers, based on the rationale of protecting economically valuable fisheries for salmon, trout, and herring. Since 1990, conservation concerns in British Columbia and Washington stimulated bag limit restrictions for Barrow's Goldeneyes, Harlequin Ducks, scoters, and Long-tailed Ducks.

Alaska

Sea duck hunting has been traditional in Alaska, originally as part of year-round subsistence economies of aboriginal groups that depended on diverse resources and later for pioneering immigrants. After Alaska became a territory of the

United States in 1867, early territorial laws were prompted partly in response to a fabricated story in the early 1890s that large numbers of wild bird eggs were being commercially exported from Alaska. Though "The Great Duck Egg Fake" was thoroughly refuted in an 1895 issue of *Forest and Stream* (Sherwood 1977), federal territorial laws adopted in 1900–1902 prohibited the collection and possession of "eggs of any crane, wild duck, brant, or goose" and established the first game bird hunting regulations (Cameron 1929); "Indians, Eskimos, miners, and travelers" in need of food were exempted.

In 1925, Congress passed the Alaska Game Act that established the federally appointed Alaska Game Commission, which adopted waterfowl regulations consistent with the Migratory Treaty Act and prevailing national management regime; sea ducks were included in general duck bag limits. During and after World War II, Alaska experienced unprecedented population growth, including many hunters, adding harvest management challenges. Alaska administratively became part of the Pacific Flyway in 1952 and gained full parity with other states at statehood in 1959.

From the 1940s through the 1960s, many adjustments were made to waterfowl regulations, but the rationale and justifications are largely lost to history. General duck season lengths varied in Alaska, from 40 days in 1948 to 94 days by 1960. In 1950–1953, the first special sea duck seasons (only scoters and eiders) allowed 6–51 days in addition to the general duck season, varying among several regions. From 1954 through 1960, seasons for sea ducks were extended to 105 days, varying from 11 to 30 days longer than the general duck season, and included all *regulatory* sea ducks, including scoters, eiders, Harlequin Ducks, Long-tailed Ducks, and mergansers. From 1950 through 1960, Alaska hunters were allowed to take 10 sea ducks in addition to general limits of 5–7 ducks and an additional limit of mergansers. The merganser limit was 25 birds of all three species in 1950–1952; 25 mergansers, but only one Hooded Merganser in 1953–1956; and five birds daily, 10 birds in possession, but only one Hooded Merganser only in southern Alaska in 1957–1959. Mergansers have been incorporated into the special Alaska sea duck limit since 1960.

Since the first waterfowl regulations were established, deference has been given to limited hunting opportunities in Alaska, similar to accommodations in northern Canada. Seasons cannot be opened before September 1, and most ducks and geese migrate south by late October, leaving only 40–60 days of hunting before freeze-up in most parts of the state. After October, hunting for Mallards and sea ducks settling into coastal wintering areas provides harvest opportunities through the latest seasons. Since 1961, Alaska has had the longest allowable seasons for all ducks (105 days, reinterpreted to 107 days in 1974), running through January 22. A special sea duck limit of 15 daily, 30 in possession, was established in 1961 in further recognition of the limited hunting opportunity in Alaska. In 1999, out of general concern for apparent liberal limits, the federal framework regulations reduced the special sea duck bag limits from 15 birds daily and 30 in possession, to 10 birds daily with 20 in possession.

Hunting Activity

Over the long term, the number of waterfowl hunters has declined in the Pacific Flyway (Figure 12.12), similar to trends across the United States and Canada. The number of active hunters in Pacific states has been relatively stable since 1999, averaging about 140,000 in the past 7 years. In British Columbia and Yukon Territory, permit sales have declined below 7,000 and active waterfowl hunters have averaged <4,000 during the past 10 years.

Implementation of HIP in the United States allowed a first systematic effort to identify and enumerate sea duck hunters for harvest survey sampling, providing the first estimates in 1999 from coastal states that had special sea duck seasons or limits, or required sea duck hunters to obtain special permits. In the Pacific Flyway, sea duck hunter activity and harvest have been estimated in Alaska since 1999. In Washington, HIP has not sampled hunters from a stratum of sea duck hunters, but a requirement to obtain a permit to hunt sea ducks in western Washington has provided hunter and harvest information since 2004. Stratified HIP sampling for sea duck hunters was expanded to Oregon in 2006 and California in 2008. Numbers of hunters hunting sea ducks are not estimated in the Canadian harvest survey.

There are likely <3,000 active sea duck hunters annually in the Pacific Flyway states (Table 12.4), with few in California and Oregon, and perhaps

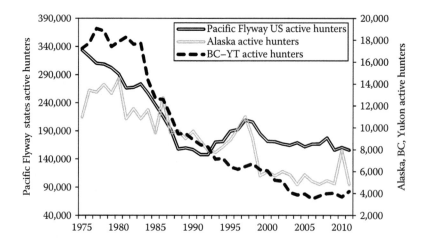

Figure 12.12. Trends in the number of active waterfowl hunters in the lower Pacific Flyway states, Alaska, and British Columbia and Yukon, 1975–2011.

TABLE 12.4

Estimated number of active sea duck hunters in Pacific coast states.

| Year | Active sea duck hunters | | | Permits[a] |
	Alaska	California	Oregon	Washington
1999	600			
2000	900			
2001	500			
2002	800			
2003	1,200			
2004	900			906
2005	900			1,359
2006	800		<50	1,861
2007	1,100		<50	1,941
2008	1,100	900	100	2,340
2009	1,100	400	200	2,129
2010	1,300	1,500	200	2,848
2011	600	100	100	1,757
2012	1,200	100	<50	1,895
AVG	929	600	150	1,893

SOURCE: Data from Washington State are holders of western Washington sea duck permits (Washington Department of Fish and Wildlife, unpubl. data).

[a] About 45% of Washington sea duck permittees were active hunters (D. Kraege, pers. comm.).

800–900 hunters in Washington (assuming about 45% of permittees hunt each year; Washington Department of Fish and Wildlife, unpubl. data). In Alaska, an average of about 1,000 hunters reported hunting sea ducks annually since the 1999 inception of HIP. Over the past 10 years, the number of active sea duck hunters has increased by 30%–40%, while the number of total active waterfowl hunters has remained stable around 5,500. Of Alaskan hunters that reported taking sea ducks, over 40% were residents of the Gulf Coast Zone that includes Anchorage, 28% were from southeast Alaska, and 15% resided in Kodiak. Fewer than 200 nonresident hunters reported taking sea ducks (Division of Wildlife Conservation, Alaska Department of Fish and Game, unpubl. HIP enrollment data).

In 2002–2011, Pacific Flyway hunters averaged about 8 days afield annually hunting waterfowl. Hunters in coastal states who indicated they hunted sea ducks spent fewer days hunting than general waterfowlers (California, 4.3 days vs. 10.1 days; Oregon, 1.9 days vs. 8.3 days; Washington, 2–3 days vs. 7.7 days; Alaska, 4.5 days vs. 5.1 days). In these coastal states, sea duck hunters averaged 6.9 sea ducks per season in Alaska, about 4.0 in Washington, 3.6 in California, and 3.4 in Oregon.

Fall and Winter Harvest

Annual HIP estimates of total duck harvest in the Pacific Flyway (including Alaska) averaged 3.04 million ducks in 2002–2011, including 94,000 sea ducks (Table 12.5). Sea duck taxa contributed an average of about 4.3% of total duck harvest in the 11 contiguous Pacific Flyway states. As expected, sea ducks make up 17% of total duck harvest in Alaska where sea ducks are numerous throughout fall and winter, but most dabbling ducks have departed by late October. The flyway harvest of all sea duck

TABLE 12.5

Average annual harvest of sea duck species in the Pacific flyway, 2002–2011.

State	COEI	KIEI	BLSC	SUSC	WWSC	LTDU	HARD	BUFF	COGO	BAGO	COME	RBME	HOME	Total
Alaska	324	60	531	1,418	818	309	1,844	1,416	1,331	2,078	860	981	7	11,977
Arizona	—	—	—	4	—	6	—	932	580	8	233	4	82	1,849
California	—	—	—	1,018	15	25	—	10,136	4,649	141	393	31	965	17,373
Colorado	—	—	—	9	—	—	—	71	1,095	21	98	—	48	1,342
Idaho	—	—	—	8	—	55	14	2,425	7,289	861	340	20	512	11,524
Montana	—	—	—	9	—	—	—	373	1,882	335	350	7	168	3,124
Nevada	—	—	—	7	6	—	—	471	307	8	89	13	72	973
New Mexico	—	—	—	—	—	—	—	13	103	—	70	—	6	192
Oregon	—	—	—	257	30	7	—	10,688	2,203	560	1,395	82	1,803	17,025
Utah	—	—	—	24	10	8	—	2,570	6,683	199	890	312	216	10,912
Washington	—	—	36	2,073	1,075	260	46	6,822	3,962	695	650	109	1,245	16,973
Wyoming	—	—	—	—	—	—	—	211	465	74	14	—	7	771
Coastal states	—	—	36	3,348	1,120	292	46	27,646	10,814	1,396	2,438	222	4,013	51,371
Inland states	—	—	—	61	16	69	14	7,066	18,404	1,506	2,084	356	1,111	30,687
PF Lower 48	—	—	36	3,409	1,136	361	60	34,712	29,218	2,902	4,522	578	5,124	82,058
Pac. Flyway	324	60	567	4,827	1,954	670	1,904	36,128	30,549	4,980	5,382	1,559	5,131	94,035
BC and YT	—	—	—	4	2	—	—	457	214	182	21	—	55	935

species was distributed primarily in California, Washington, and Oregon, each with 18% and Alaska with 13%. Idaho and Utah each harvested 12% of the flyway's sea ducks, largely on the occurrence of Buffleheads, goldeneyes, and mergansers.

Buffleheads and goldeneyes occur widely across North America and constituted 76% of sea ducks harvested in the Pacific Flyway. The harvest of these species creates a seeming paradox where sea ducks made up greater proportions of total ducks in Wyoming (9.1%), Colorado (5.2%), and Idaho (5%) than only 1.2% of California's large harvest of mostly dabbling ducks. In Alaska, 40% of sea ducks harvested were Buffleheads and goldeneyes.

Coastal-oriented species (eiders, scoters, Long-tailed Ducks, and Harlequin Ducks) subject to special regulations in Alaska and western Washington made up only 11% of the Pacific Flyway sea duck harvest. Across individual states, coastal species comprised 44% of sea duck harvest in Alaska, 21% in Washington, and <10% in California and Oregon. Alaska had the highest harvest of each coastal species, except that more Surf and White-winged Scoters were taken in Washington and California (Figure 12.13). Scoters have the greatest harvest among the coastal species, making up over 90% of this group in Washington, Oregon, and California. In Alaska, where sea duck availability and diversity are greater, scoters made up 23% of the sea duck harvest and 52% of coastal species taken. Harlequin Ducks are abundant in Alaska during winter and are managed separately from harlequins in the Pacific Northwest. Harlequins are relatively easy to hunt and made

up 15% of the Alaska sea duck harvest and 41% of the Alaska coastal species harvest (Table 12.5).

The HIP survey was designed to accurately measure harvest at the statewide level but provides little information on harvest locations and seasonality. Alaska Department of Fish and Game (ADF&G) conducted mail questionnaire surveys in 1971–1996 that collected data specifically on sea duck harvest and hunting locations. Though the sampling was not robust, 11 years of data indicated that fall and winter sea duck harvest occurred primarily on Kodiak Archipelago (27%), Cook Inlet (26%), and Southeast Alaska (21%) where the majority of wintering sea ducks occur and most of the state's population resides. In addition, sea duck hunting is prevalent along the western Alaska Peninsula, Aleutian Islands, Pribilof Islands, and Saint Lawrence Island, but information on hunting activity and harvest from state and HIP surveys is not sufficient for harvest estimates.

Regional Harvest Issues and Management Responses

Overall, given the healthy status of most sea duck species along the Pacific Coast in fall and winter and relatively low levels of harvest, few concerns have been raised about the effects of hunting on sea ducks at the level of flyway populations.

Barrow's Goldeneye

During the 1980s, research and surveys in British Columbia provided new information on the

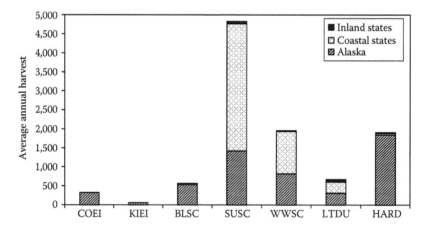

Figure 12.13. Distribution of average annual harvest of coast-oriented sea duck species in the Pacific Flyway across inland states, coastal states, and Alaska, 2002–2011.

numbers, distribution, and biology of Barrow's Goldeneyes in British Columbia and raised interest in the effects of contemporary harvest in the Pacific Flyway. Until recently, it was thought that up to 60% of the global population bred and wintered in British Columbia and that the Pacific population numbered up to 150,000 birds (Eadie et al. 2000). Limited banding data indicated that harvests in British Columbia and perhaps in Washington were derived from birds breeding in the Pacific Northwest (McKelvey and Smith 1990) and the regional harvest of western Barrow's was considered sustainable (Savard 1987).

As geographic coverage of winter waterfowl surveys expanded in Alaska, it has become apparent that more than 165,000 Barrow's Goldeneyes winter from the Gulf of Alaska westward through the Alaska Peninsula and that these birds constitute 85%–95% of all goldeneyes wintering in Alaska (Forsell and Gould 1981; Hodges et al. 2008; B. A. Agler et al., U.S. Fish and Wildlife Service, unpubl. report; A. McKnight et al., U.S. Fish and Wildlife Service, unpubl. report). Collectively, winter surveys indicate that there may be 250,000–300,000 Barrow's Goldeneyes in the western population from Alaska to California.

Recent telemetry work indicates that the Barrow's Goldeneyes breeding in interior British Columbia molt in northwestern Alberta and as far north as Great Bear Lake, NWT, then winter along the coasts of British Columbia and Washington (Boyd et al. 2013). These wintering birds probably define the southern extent of the species range in the west. Concurrent work in Alaska indicates that Barrow's Goldeneyes wintering in Prince William Sound breed in Interior Alaska, with males molting northeast only as far as Old Crow Flats in the Yukon. Barrow's Goldeneyes that were radio-marked near Juneau, Alaska, in spring of 2012 showed a breeding distribution in between the central Alaska birds and the British Columbia birds. Taken together, these studies indicate several regional affiliations of western Barrow's Goldeneyes, with only rare mixing of British Columbia and Washington birds with the abundant Alaska-wintering birds. Recent analyses of genetics and band recoveries indicate shallow structuring among western Barrow's Goldeneyes and low interchange between Alaska and British Columbia birds (Pearce et al. 2014).

As a precaution for regional Barrow's Goldeneyes, bag limits in British Columbia were restricted in 1990 to no more than two goldeneyes daily and four birds in possession within a general duck limit of eight ducks per day. In 2010, the aggregate bag limit for goldeneyes in western Washington was reduced to two birds per day as part of a broad sea duck harvest strategy that recognized the small number of Barrow's wintering in the state (Washington Department of Fish and Wildlife, unpubl. report).

The harvest of western Barrow's Goldeneyes is difficult to assess because, like many sea duck species, the number of wing samples submitted to the national harvest surveys is low and harvest estimates are variable. Since 1999, the HIP survey has estimated an average of 700 Barrow's Goldeneyes harvested in the entire state of Washington, and estimates based on state permit reports indicate a harvest of about 110 in western Washington. Several hundreds of Barrow's Goldeneyes are harvested annually in Idaho, Montana, Oregon, and Utah.

In western Canada, western Barrow's Goldeneyes are harvested in Yukon Territory, British Columbia, and Alberta; harvest of the species has declined substantially in each of these jurisdictions. Prior to 1990, combined harvest averaged 2,800 birds, with 85% occurring in British Columbia. The most recent 10-year average harvest of Barrow's Goldeneyes in the western provinces is <400 birds. Much of the decline in harvest likely reflects a steep drop in active waterfowl hunters in western Canada since the early 1980s (Figure 12.12). In British Columbia, however, implementation of a 2-bird bag limit in 1990 also may have contributed to harvest reduction. The most recent 10-year average harvest of western Barrow's Goldeneyes in British Columbia is about 155 birds, <50% of the species total of western provinces.

Harlequin Duck

Significant declines in Harlequin Ducks in eastern North America stimulated several investigations and surveys of the species on the Pacific Coast from British Columbia to Oregon during the 1990s (Robertson and Goudie 1999). Losses of many Harlequin Ducks during the T/V Exxon Valdez oil spill in the spring of 1989 led to several studies of breeding ecology, food habits, and seasonal distribution of Harlequin Ducks along the Gulf of Alaska coast.

Banding and survey programs on western Harlequin Ducks produced substantial evidence of a Rocky Mountain–Northwest Coast (RMNWC) component that breeds from British Columbia, Washington, and Oregon eastward to Colorado, Wyoming, and Alberta and winters along the British Columbia and Washington coasts (Pacific Flyway Study Committee, unpubl. report). Incomplete surveys of these birds suggest about 15,000–20,000 Harlequin Ducks wintering south of Alaska to Oregon (Robertson and Goudie 1999). Banding and telemetry data provide little evidence of exchange between these birds and the nearly 250,000 Harlequin Ducks wintering in Alaska (Forsell and Gould 1981; Byrd et al. 1992; Hodges et al. 2008; B. A. Agler et al., unpubl. report; C. P. Dau and E. J. Mallek, U.S. Fish and Wildlife Service, unpubl. report; A. McKnight et al., unpubl. report).

Concerns about low abundance, low productivity, and potential vulnerability to hunting of RMNWC Harlequin Ducks in winter stimulated regulatory restrictions, first in British Columbia with daily limits of harlequins reduced from eight to six concurrent with goldeneye reductions. In 1998, Washington adopted Harlequin Duck limits of one bird per day and one in possession within the seven-duck general limit. British Columbia followed suit in 2000 with reduced harlequin limits of two birds per day and four in possession within a daily limit of eight ducks. Washington further reduced the Harlequin Duck limit to only one bird per season in 2004 (Washington Department of Fish and Wildlife, unpubl. report). In 2012, possession limits in British Columbia were increased to three times daily limits (up to six Harlequin Ducks).

The former mail questionnaire survey and HIP surveys have registered only trace estimates of Harlequin Duck harvest in a few Pacific Flyway states, including rare records in Idaho, Montana, and Colorado. Since 1999, HIP estimated a harvest of only 70 birds in Washington State. Western Washington permit hunt data for 2004–2012 show an average of 134 Harlequin Ducks were taken per year under the 1-bird limit. Harlequin Ducks wintering in Puget Sound have been relatively stable at 3,000–4,000 birds since 1995 (Washington Department of Fish and Wildlife, unpubl. report).

Northwest Scoters and Long-tailed Ducks

Over the past 20 years, the WDFW, USFWS, CWS, and the province of British Columbia increasingly have collaborated to assess the status of wintering sea ducks from Georgia Strait south through Puget Sound, broadening conservation programs beyond initial focus on Harlequin Ducks and Barrow's Goldeneyes. The efforts were supported by cooperative international programs initiated to address increasing regional environmental and economic issues, including human encroachment on coastal areas and environmental hazards (Mahaffy et al. 1994). As part of the Puget Sound Ambient Monitoring Program, later renamed Puget Sound Ecosystem Monitoring Program, the WDFW has conducted aerial and boat surveys of wintering sea ducks since 1992 (D. R. Nysewander et al., Washington Department of Fish and Wildlife, unpubl. report). The resulting survey program was integrated with prescriptive harvest guidelines and other management actions in 2010 and updated in *Sea Duck Management Strategies* (Washington Department of Fish and Wildlife, unpubl. report) focused on sustaining the number of wintering scoters in the sound and avoiding redirection of harvest to Barrow's Goldeneyes and Long-tailed Ducks.

In 1998, Washington State established special sea duck restrictions, limiting daily bag limits to no more than four birds each of scoters, Long-tailed Ducks, and Harlequin Ducks within the general daily limit of seven ducks in western Washington. In 2004, in order to develop better harvest information, Washington Fish and Wildlife Commission required sea duck hunters in western Washington to obtain a license permit to identify participants and to report their sea duck harvest. The 1998 bag limit restrictions for scoters prevailed until the 2010–2011 season, when wintering scoters had dropped below the management threshold of 75,000; daily species limits were reduced to two scoters and two Long-tailed Ducks. In addition, goldeneyes were added as a restricted species with two birds per day.

The number of scoters wintering in Puget Sound has declined by over 50% since 1995 to about 50,000 birds (Washington Department of Fish and Wildlife, unpubl. report). Since 2004, state survey data indicate that average scoter harvest declined in western Washington from 2,300 under the 4-bird limit to about 1,300 in 2010–2012 under a 2-bird limit. Since 2010, the harvest of goldeneyes also declined, perhaps by half, to <400 birds (~60% Common Goldeneyes).

Alaska Sea Duck Conservation Concerns

Alaska has abundant sea ducks year-round, during migration, breeding, and wintering periods, and in general, harvest of sea ducks has been low in relation to the number of birds found throughout the state. A few situations have raised conservation concerns resulting in restrictions of sea duck hunting. The T/V Exxon Valdez oil spill killed large numbers of scoters, goldeneyes, Harlequin Ducks, and Long-tailed Ducks during spring migration through Prince William Sound and later near the Kodiak Archipelago, in Lower Cook Inlet, and westward along the Alaska Peninsula (Piatt et al. 1990, Piatt and Ford 1996).

The most prominent concern was the resulting reduction of resident breeding Harlequin Ducks in Prince William Sound and potential impacts of additional mortality from harvest. The prevailing sea duck season had opened September 1 with bag limits of 15 birds/day and 30 in possession. From 1991 through 1998, the Alaska Board of Game delayed the opening of Harlequin Duck season by 1 month to October 1 and imposed a limit on Harlequin Ducks of two birds per day and six in possession in Prince William Sound. The restrictions were deemed necessary to promote restoration of Harlequin Ducks breeding in the sound, but pre- and postspill harvest surveys lacked scope and accuracy to detect effects of these regulations on harvest. State and federal harvest survey data suggest that prespill harvest of Harlequin Ducks from fall and winter aggregations may have been <200 birds in the entire Gulf Coast Zone, based on an average annual harvest of 500 sea ducks, including about 15% Harlequin Ducks (Alaska Department of Fish and Game, unpubl. data; U.S. Fish and Wildlife Service, harvest survey).

In 1990, a petition was filed with USFWS to list Spectacled Eiders (Somateria fischeri) and Steller's Eiders (Polysticta stelleri) under the Endangered Species Act. Hunting seasons for both species were closed statewide in 1991. In May 1993, Spectacled Eiders were listed as threatened rangewide (U.S. Fish and Wildlife Service 1993). In August 1993, the USFWS determined that rangewide listing was not warranted for Steller's Eiders because substantial numbers breed in Russia and winter along the Alaska Peninsula. In 1997, Alaska-breeding Steller's Eiders were listed as threatened, based on a significant reduction in range in western Alaska (U.S. Fish and Wildlife Service 1997).

Evaluation of the role of harvest in historical declines or during recovery is difficult for these species. Harvest has not been implicated as a primary cause of critical population declines (U.S. Fish and Wildlife Service, unpubl. reports), but minimizing harvest is an important part of recovery programs for the listed populations. Historical harvest data from USFWS harvest surveys are sparse; there have been no records of Spectacled Eider wings in the wing survey, which have been conducted in Alaska since 1965, and therefore no means to detect or estimate harvest among surveyed hunters. Steller's Eider wings have been recorded in only 10 years since 1965, and harvest estimates average <100 birds/year. Since the 1991 season closure, an annual average of 0.1 wings of Steller's Eiders have been submitted to the wing survey, providing a projected total statewide harvest of 3.7 birds. These data illustrate the difficulty in estimating harvest of birds that occur in remote areas where access is limited for nonlocal hunters. Harvest of Spectacled and Steller's Eiders in Alaska is documented primarily from subsistence hunters that have access to these birds but are rarely sampled in the federal and state mail questionnaire surveys linked to hunting license sales.

Federal reductions in Alaska sea duck bag limits were implemented in 1999 and arose from general caution about harvest of sea duck species that have relatively low recruitment rates and the appearance of liberal limits. In an effort to reduce impacts for Alaska residents, the Alaska Board of Game set regulations for nonresident hunters to be a season limit of 20 sea ducks, including no more than four birds of each species. In 2001, the state capped Harlequin Duck and Long-tailed Duck limits for resident hunters at 6 birds/day and 12 in possession.

Similar to situations in other coastal areas where community development, seafood harvesting, and public recreation have expanded, hunting has stirred conflicts among user groups and with local residents in a few areas of the southern Alaska Coast. Kachemak Bay, within a 4 h drive from populous Anchorage and the Matanuska-Susitna Valley, has been a traditional sea duck hunting area for many years. Over the past 20 years, controversy has continued about potentially increasing numbers of duck hunters and hunting guides, the nature of sea duck hunting practices, and the importance of local winter aggregations of sea ducks to manageable population units. In 2010, the Alaska Board of Game reduced daily bag limits in Kachemak Bay to one eider, two Harlequin Ducks, and two Long-tailed Ducks. The ADF&G has conducted multiyear surveys to assess changes in wintering sea duck abundance in the bay, but conventional harvest surveys cannot provide insights on the effects of local regulations. The restriction of one eider per day is not likely to affect harvest because King and Common Eiders are rare in Kachemak Bay but 2-bird limits are likely to reduce the local harvest of Harlequin and Long-tailed Ducks.

Changing Coastal Environments and Hunting

Concerns are growing about human encroachment, commercial exploitation of coastal resources, habitat degradation, and hunting pressure in some staging and wintering areas. Expanding residential and commercial development, more intensive recreational activities, and increasingly diverse uses of coastal lands and waters not only affect habitat quality for sea ducks but also have resulted in steady losses of accessible hunting areas and local harvest traditions. As a result of broad biological uncertainties about sea duck populations, provincial and state management agencies and public interest groups have taken steps to maintain levels of winter aggregations in many areas through zoning of hunting and other activities and through conservative harvest regulations for sensitive areas such as Kachemak Bay, the Fraser River Delta, and Puget Sound.

NORTHERN SUBSISTENCE TRADITIONS

Subsistence Economies in the North

Historically and to this day, the harvest of birds is an important activity for subsistence cultures in Alaska and Canada (Figure 12.14). Birds constitute a small proportion of the total subsistence harvest in Alaska (Wolfe and Walker 1987), which is estimated to be 3% of 1.2 million edible pounds/year (J. A. Fall, unpubl. report), but the timing of harvest is important. In spring, when food supplies are depleted, the arrival of migratory birds brings relief until other subsistence resources such as fish and caribou become available. Fall and winter harvest is also important in some areas. Birds are a special treat and bring diversity to the subsistence diet, which tends to be monotonous. Besides nutritional aspects, bird harvesting also has cultural and social importance for these communities. Subsistence harvest is widely shared in the communities; numbers of users of these resources are much larger than numbers of harvesters.

Originally, Arctic and sub-Arctic peoples lived in small, nomadic groups moving across landscapes following the seasonality of biological resources. In the early 1900s, demographic and socioeconomic developments such as construction of schools and trading posts and population reductions caused by disease led to aggregation of people in villages or rural communities. The seasonal rounds of hunting, fishing, and gathering that still are a main characteristic of life in these communities date back thousands of years, although the congregation in villages likely affected original patterns of wildlife uses. In recent decades, most subsistence bird hunting is done with shotguns and aluminum skiffs, although some older harvest methods are still used in small scale. Hunting gear is owned and operated by family groups.

Some subsistence bird hunts are specialized, but bird hunting is commonly a supplemental and opportunistic activity done in conjunction with pursuits such as whaling, seal and walrus hunting, berry picking, travelling, or wood gathering. In most areas, bird hunting decreases or stops during summer because of other subsistence activities such as fishing and because of traditions of letting birds alone to breed. In contrast to most fall and winter hunting that includes recreation and food gathering, subsistence hunting is based on needs and optimizes harvest efficiency. Most bird parts are considered edible including head, feet, gizzard, heart, liver, and brain. For instance, the fatty bill knobs of male King Eiders are especially appreciated by some people (Wolfe and Paige 1995). Patterns of subsistence uses of birds, including sea ducks, are shaped by species distribution and seasons of occurrence and interactions with use patterns of other subsistence resources. Sea ducks are widely accessible as subsistence resources to northern peoples. In some regions, sea ducks represent the bulk of harvest: eiders on the coast and scoters inland.

History of Regulations and Public Involvement

In the seventeenth and eighteenth centuries, British, French, Spanish, and Russian explorers brought changes to indigenous cultures and

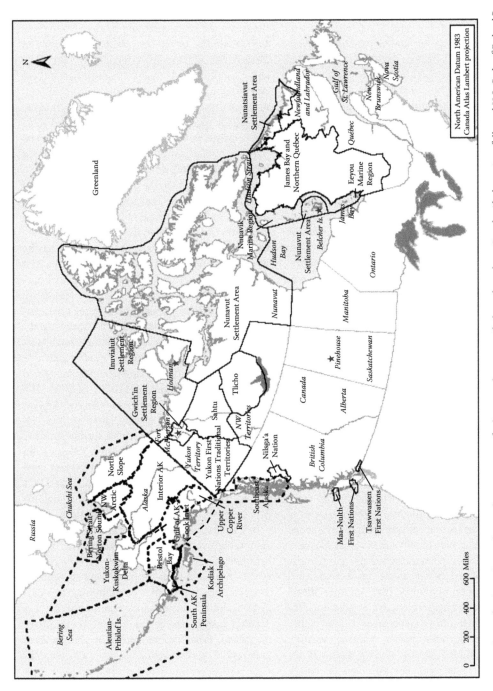

Figure 12.14. Regions relevant for the understanding and management of subsistence harvest in Alaska and northern Canada. Alaska regions followed U.S. Code of Federal Regulations (Title 50 Part 92.5). Canada regions were based on modern land claims agreements. (http://www.landclaimscoalition.ca/, accessed 22 January 2014; Map prepared by the ADF&G Division of Subsistence, Anchorage.)

introduced large-scale commercial harvest of fish and wildlife. In Alaska, indigenous peoples, trappers, whalers, miners, and immigrants harvested waterfowl and other wildlife for food and commerce. There was little regulation of hunting and fishing, if any, until 1900 when federal laws prohibited egging and destructive harvesting. The earliest US federal laws barely recognized the extensive waterfowl harvest by Alaska Native peoples, granting only that Indians, Eskimos, and travelers could take what they needed for food.

The 1916 Convention between the United States and Canada guided regulation of hunting, but provisions for subsistence hunting in the north were narrow and did not match customary and traditional practices. The Convention set an annual closed hunting season on migratory game birds between March 10 and September 1 for their protection during the breeding season. However, the closed season banned waterfowl hunting when most traditional harvest occurred: spring migration when food storages were depleted, summer molt migrations and aggregations of flightless birds, and early fall migrations. Perceptions of species that were taken for food and deemed appropriate for harvest were narrowly defined in Article II stating that Eskimos and Indians could take scoters and several species of nongame seabirds and their eggs at any time for food. Article IV provided specific protections for eiders by a 5-year closed season, establishing refuges and other rules to restrict harvest. The treaty recognized the importance of scoters, eiders, and some seabirds in the seasonal rounds of subsistence activities across northern North America, but it outlawed harvest of cranes, swans, geese, 19 common species of ducks (including seven species of sea ducks), and dozens of other migratory species that have been traditionally taken across Alaska.

Federal laws implementing the treaty in Canada and the United States made most traditional spring–summer migratory bird hunting illegal and set increasingly restrictive regulations for the fall–winter hunts (September 1–March 10). Despite the spring–summer hunting closure, harvest continued as an activity essential for subsistence. Decades of law enforcement issues, community hardships, and the inability to assess and manage these harvests prompted considerations to create legal, managed seasons involving subsistence users in the management process. In Canada, a Supreme Court decision (R. v. Sparrow, 1990 Can LII 104 SCC) affirmed that subsistence harvest of migratory birds by aboriginal peoples is an assured right under section 35(1) of the Constitution Act of 1982. In 1995, the United States and Canada agreed on a protocol to amend the treaty, which was ratified by the US Senate in 1997. A similar treaty between United States and Mexico was subsequently amended to be consistent with the Canada treaty. The purposes of these amendments were to sustain migratory bird populations through guiding conservation principles and legally recognize traditional spring and summer subsistence harvest, including constitutional rights in Canada and authorization of regulated spring–summer subsistence hunts in Alaska.

Implementation of the amended treaty in the United States directs that subsistence hunting in Alaska is to be incorporated in national management processes, provided that (1) subsistence harvest remains at traditional levels relative to the size of bird populations, (2) subsistence harvest data are integrated with flyway and national harvest management programs, and (3) regulatory processes for all migratory bird hunting are inclusive to users and responsive to conservation needs. Incorporation of Alaska's indigenous inhabitants into the management process was established through the formation of the Alaska Migratory Bird Co-Management Council (AMBCC) in 2000. The AMBCC includes representatives from the USFWS, ADF&G, and regional Alaska Native entities (FR 65(60): 16405–16409). An Alaska spring–summer subsistence season (April 2–August 31) has been authorized annually since 2003. The fall and winter migratory bird hunt (starting September 1) is managed under 50 Code of Federal Regulations Part 20, covering all states including Alaska.

In Canada, wildlife harvest is a component of comprehensive settlement agreements, which have been finalized or are under negotiation with First Nations and Inuit peoples. These agreements recognize the importance of traditional and cultural wildlife harvest and commonly include detailed harvest studies as part of their implementation. First Nations and Inuit peoples have preferential access rights to natural resources and wildlife within their jurisdictions, and their needs must be considered first before allocations for other hunters are set in conventional regulations. Wildlife management in First Nations and Inuit jurisdictions is generally undertaken within a comanagement board structure (Berkes 2009).

Subsistence Harvest of Sea Ducks in Northern Canada

Data Sources for Subsistence Harvest in Canada

Aboriginal peoples (First Nations, Inuit, and Métis) are not required to obtain a migratory bird hunting permit when harvesting waterfowl in traditional territories; therefore, their harvest is not captured in the national harvest survey. Reports of bands recovered in subsistence harvest are sporadic and depend on individual and community attitudes toward marking birds. Attempts to quantify subsistence use of birds and eggs in northern Canada have employed harvest logbooks or calendars, where hunters are asked to report their harvest of all fish and wildlife species over a period of time. Some studies have used interviews to validate or complement the data. However, these studies have been primarily designed to assess harvest of large game, such as caribou and marine mammals involving few species and of which a small number of animals are taken annually by individual hunters. On the other hand, a diversity of bird species may be hunted in relatively large numbers, which makes species identification and recall issues more prominent in bird harvest assessments.

Subsistence Harvest Patterns in Canada

Of the sea ducks present in coastal regions of Canada's north, the Common Eider is the species harvested in highest numbers. Large body size and relatively high abundance at certain times of the year make eiders an important subsistence resource. King Eiders make significant contributions to harvest only in the western portions of the Arctic. Long-tailed Ducks are taken in small numbers.

Generally, the first opportunity to harvest eiders is during spring migration. Spring hunting traditions occur in areas where eiders reliably congregate in open water patches such as near the community of Holman, on Victoria Island (Fabijan et al. 1997, Byers and Dickson 2001). Egging is important in communities close to breeding colonies, such as those on Belcher Islands, Hudson Strait, and coastal Labrador (Reed 1986). Adult birds are taken during the breeding season as well. Fall harvest is important in some regions, especially where migration routes take birds close to shore. Communities in the Belcher Islands have the opportunity to hunt Common Eiders, King Eiders, and Long-tailed Ducks during winter in polynyas and shore leads (Gilchrist and Robertson 2000). At inland sites, scoters are an important supplement to spring and summer diets. In the western Northwest Territories, most scoter harvest occurs in spring (May, Gwich'in Renewable Resource Board 2009). On the north shore of the Gulf of St. Lawrence, scoters are harvested by Innu people in spring when birds stage in marine waters before continuing migration toward inland breeding grounds. Scoters then move inland into Québec and Labrador, where Innu hunt scoters and other waterfowl at traditional early open water sites known as *Ashkui* (Sable et al. 2006). The following paragraphs refer to the harvest data available for several regions, from east to west along the coast, and then for the interior of northern Canada.

In a comprehensive recall harvest survey conducted in 2007 in Nunatsiavut (the Inuit region of Labrador), almost all households in Inuit communities were surveyed (Natcher et al. 2011). The Common Eider (2,608 birds/year), Surf Scoter (745 birds/year), Black Scoter (615 birds/year), and White-winged Scoter (86 birds/year) were among the bird species harvested in the largest numbers and were taken mainly in spring (Natcher et al. 2011). Egg harvest estimates in 2007 included 4,019 eggs from the nests of Common Eiders (Natcher et al. 2012).

In Nunavik (the Inuit region of northern Québec), estimates are available for 1973–1980 (James Bay and Northern Québec Native Harvesting Research Committee 1988). The average estimated duck harvest (species combined) was 12,970 birds/year (range 8,258–14,851), and was comprised of eiders (79%, 10,246 birds/year), scoters (10%, 1,297 birds/year), and mergansers (6%, 778 birds/year). The average estimated harvest of duck eggs was 35,421 eggs/year (range 11,189–111,322). Egg harvest was not identified to species, but eggs of Common Eiders likely represented the vast majority of the take: the subspecies *borealis* is available to communities in the Hudson Strait, while Common Eider eggs taken by communities in eastern Hudson Bay are mostly of the *sedentaria* subspecies.

The Nunavut Wildlife Harvest Survey provides estimates for all Inuit harvest in the territory of Nunavut in 1996–2001 (Priest and Usher 2004). The annual average eider harvest was 6,000 birds/year (range 5,004–6,387) including mostly Common Eiders and a small proportion of King Eiders. Other sea duck species were taken in low numbers: Common Mergansers (117 birds/year), Long-tailed Ducks (100 birds/year), Surf Scoters (11 birds/year), White-winged Scoters (9 birds/year), Red-breasted Mergansers (9 birds/year), and Black Scoters (5 birds/year). The estimated egg harvest included 7,909 eider eggs/year (range 4,446–11,669, Priest and Usher 2004). Some estimates from this survey seem low when compared to other studies. For example, 1982 harvest estimates included 6,000 eiders alone in the Belcher Islands, 554 eiders in the High Arctic, and 8,067 eiders in the Low Arctic (Donaldson 1984, Reed and Erskine 1986).

In the Inuvialuit Settlement Region (western Canadian Arctic), eider harvest averaged 3,446 birds/year in 1988–1994 (range 1,804–5,013, Fabijan et al. 1997). The community of Holman accounted for most of this harvest, composed of King Eiders taken in June (96%) and Pacific Common Eiders (v-nigra). In 1996–1998, a follow-up study was conducted in Holman to provide a detailed assessment of the harvest (Byers and Dickson 2001). Total harvest mortality (including wounding losses) ranged from 2,517–2,801 King Eiders and 19–29 Common Eiders annually, which corresponds to 3.7%–6.9% and 0.3%–0.9% of the populations migrating past Holman in those years. Wounding losses were low in years with extensive sea ice (3.2%–9.1%) but increased in one year with extensive open water (13.0%–20.0%). Sex ratios in the harvest reflected those in the population (Byers and Dickson 2001).

At inland sites south of the tree line, First Nations peoples take species such as scoters, goldeneyes, and mergansers. Harvest data for these areas are sparse, but in 1995–2001, a comprehensive study was conducted in the Gwich'in Settlement Area (far northwestern Northwest Territories and south of the Inuvialuit Settlement Region, Gwich'in Renewable Resource Board 2009). White-winged and Surf Scoters, collectively referred to as "black ducks" by Gwich'in people, were taken in the largest numbers (717 birds/year, range 452–1,002) and the majority of the harvest occurred at Fort McPherson. Other sea ducks harvested were goldeneyes (14 birds/year, range 0–61) and Long-tailed Ducks (7 birds/year, range 0–34). Harvest of sea ducks occurred mostly in May (Gwich'in Renewable Resource Board 2009).

A 1983–1984 study in the Cree community of Pinehouse (central Saskatchewan) likely represents typical annual harvest levels of communities in boreal forests of the prairie provinces. Harvest included 332 unspecified scoters, 179 Common Goldeneyes, 142 Surf Scoters, 139 Red-breasted Mergansers, 119 Common Mergansers, 36 Buffleheads, 16 unspecified mergansers, 13 White-winged Scoters, and 3 Black Scoters (Tobias and Kay 1994).

Subsistence harvest studies in the Hudson Bay and James Bay Lowlands of Ontario and Québec partitioned duck harvest into only three species: Mallards, American Black Ducks, and Northern Pintails. Other species, including sea ducks, were considered of minor importance and likely had low levels of harvest. Bird harvest by the Cree people in this area was largely composed of geese (Berkes et al. 1994).

Regional Management Topics

Subsistence harvest of Common Eiders in northern Canada is substantial, and harvest of the different subspecies needs to be considered in studies of population ecology. The borealis subspecies of Common Eiders are also subject to large fall and winter harvest in southern Canada and to recreational, subsistence, and commercial harvest in Greenland (Gilliland et al. 2009), where a significant portion of the Canadian breeding population winters (Mosbech et al. 2006). The Greenlandic harvest of borealis Common Eiders became a major management concern because population trends and modeling suggested the harvest was not sustainable (Merkel 2004b, Gilliland et al. 2009). Harvesting in Greenland is different than in northern North America, as the major harvest is done by Greenlandic harvesters that are allowed to sell their harvest in community markets (Merkel and Christensen 2008). Therefore, harvest regulations not only impact harvesting opportunities but also cash income in these isolated communities. Harvest limitations were implemented in 2002, substantially reducing the harvest from 52,000 to 84,000 eiders harvested annually between 1993 and 2001 to between 18,000 and 27,000 eider taken annually in 2002–2010 (Merkel and Christensen 2008, Greenland Home Rule 2013). Reduced levels of harvest were projected to be sustainable and even lead to population growth (Gilliland et al.

2009), and the Greenlandic breeding population of Common Eiders has shown signs of recovery (Merkel 2010, Burnham et al. 2012). Changes in regulations should lead to reduced harvest pressure on Canadian breeding eiders wintering in Greenland, but detecting a positive response from this reduced harvest pressure was not possible as large-scale outbreak of avian cholera in the early 2000s overwhelmed the dynamics of northern Common Eider populations in eastern Canada (Descamps et al. 2012). Egg collecting has also led to the apparent decline of eider colonies close to some communities (Cooch 1986), and so efforts have been made to manage the harvest through education and promotion of egging practices that minimize impact to colonies.

The harvest and relationships that the Inuit of the Belcher Islands have with the *sedentaria* population that spends the entire year in that region are a special case. The *sedentaria* subspecies occurs only in Hudson Bay and harvests by other communities are minimal. Therefore, the management of this subspecies can be done at a relatively local level. Harvest of eiders can be significant at the Belcher Islands (Reed and Erskine 1986), but is likely sustainable unless winter kill or other factors cause additional mortality (Robertson and Gilchrist 1998). Recent aerial surveys indicate that large numbers of eiders winter in ice leads well beyond the reach of the community of Sanikiluaq creating a natural refuge for this population (Gilchrist et al. 2006).

Important harvest in the western Arctic of Pacific Common Eiders (*v-nigra*) and King Eiders has generated management concerns for these populations (Suydam et al. 2000). Harvest of King Eiders that migrate past Holman (Victoria Island) may take 4%–7% of the regional subpopulation but is thought to be sustainable (Byers and Dickson 2001). This population of eiders is not hunted elsewhere in Canada (Fabijan et al. 1997) but is hunted in Alaska's North Slope (Braund 1993a,b) and Yukon–Kuskokwim Delta (C. Wentworth, unpubl. report) and also in eastern Russia, where harvest levels are unknown but suspected to be relatively high (E. Syroechkovski, Jr. and K. B. Klokov, IPEE RAN, unpubl. report).

Given the overall concerns for scoter populations in North America, subsistence harvest needs consideration. However, the overall take of scoters in the northern subsistence harvest in Canada is likely small, on the order of 10,000 annually, and probably not a major factor driving population decreases.

Subsistence Harvest of Sea Ducks in Alaska

Data Sources for Subsistence Harvest in Alaska

Until the mid-1980s, research on subsistence harvest was qualitative and focused on methods of harvest and uses, seasonal rounds, and patterns of sharing. Quantification of subsistence harvest developed as management systems for biological resources emerged and became more complex, and economic development initiatives progressively overlapped subsistence regions. Until the 1990s, surveys referred to categories of birds (ducks, geese, seabirds) and some subcategories are still used (eiders, scoters). Wolfe et al. (1990) compiled the first statewide subsistence bird harvest estimates referring to bird categories, and Paige and Wolfe (1997, 1998) provided estimates at the species level. In the mid-1990s, the estimated harvest was 63,301 sea ducks/year and represented 32% of the total duck harvest and 17% of the total bird harvest in the state (Table 12.6).

Annual harvest monitoring for waterfowl in 1980–2002 started in the context of the Yukon–Kuskokwim Delta Goose Management Plan (Copp and Roy 1986; Pamplin 1986; Zavaleta 1999; C. Wentworth, unpubl. report). The AMBCC harvest monitoring program started in 2004 to meet the intentions of the amended Migratory Bird Treaty and relies on collaboration among USFWS, ADF&G, and Alaska Native partners (Reynolds 2007; Naves et al. 2008; Naves 2010, 2012). The AMBCC survey covers 193 rural communities in 10 regions (population of 89,481; U.S. Census Bureau 2011). Regions have been surveyed depending on annual management priorities and funding availability. Harvest reports are completed by face-to-face interviews conducted by local surveyors. Survey seasons are spring (April 2–June 30), summer (July 1–August 31), and fall (September 1–October 31; not done in the North Slope, because birds out-migrate in late summer). Winter surveys (November 1–March 9) are done in southern coastal Alaska (Gulf of Alaska–Cook Inlet, Kodiak Archipelago, Aleutian–Pribilof Islands, and South Alaska Peninsula), a wintering area for many species, including sea ducks.

Harvest estimates presented in this chapter were based on AMBCC data (2004–2012; 368 community-years) complemented by data generated by the ADF&G Division of Subsistence (1993–2011; 50 community-years; Table 12.7).

TABLE 12.6
Alaska subsistence harvest of sea ducks and fall and winter harvest in the Pacific flyway.

	Alaska subsistence harvest			Fall–winter harvest[a] (2002–2011 average)	
	Mid-1980s–Early 1990s	1996[b]	2011[c]	Alaska	Pacific flyway, Lower 48
Steller's Eider	313[d]	438	230	0	0
Spectacled Eider	896[d]	1,127	222	0	0
King Eider	11,138[d]	16,469	16,203	60	0
Common Eider	4,204[d]	6,919	4,460	324	0
Harlequin Duck	—	2,217	2,080	1,844	60
Surf Scoter	—	967	2,765	1,418	3,409
White-winged Scoter	—	3,506	7,538	818	1,136
Black Scoter	—	8,451	11,617	531	36
Scoter (unidentified)	—	4,689	0	0	0
Long-tailed Duck	—	10,341	4,020	309	361
Bufflehead	—	3,916	3,782	1,416	34,712
Goldeneye	—	6,973	7,252	3,400	32,000
Merganser	—	1,977	1,556	1,900	5,000
Total sea ducks	—	63,301	61,725	11,977	94,035
Total ducks	210,448[e]	197,577	—	70,138	2,970,620
Total birds	307,242[e]	371,223	342,778[f]	79,931	3,370,600

NOTES: Total birds in "[a]," "[b]," and "[e]" did not include resident grouse and ptarmigan. —, data not available.

[a] S. M. Olson and R. E. Trost, U.S. Fish and Wildlife Service, unpubl. report; U.S. Fish and Wildlife Service Harvest Information Program (HIP) report series. Available online http://www.fws.gov/migratorybirds/NewReportsPublications/HIP/hip.htm. Accessed January 13, 2014.
[b] Paige and Wolfe (1998).
[c] Present study.
[d] Wolfe and Paige (1995).
[e] Wolfe et al. (1990).
[f] 2004–2010 average based on AMBCC data, including resident grouse and ptarmigan (L. C. Naves, Alaska Department of Fish and Game Division of Subsistence, unpubl. data).

The analyses are intended to portray current sea duck harvest levels and therefore did not include older surveys, except for 19 community-years (CSIS, 1993–1997) poorly represented in more recent data. The regions most represented in the dataset were Yukon–Kuskokwim Delta, Bering Strait–Norton Sound, and North Slope, and the regions least represented were Northwest Arctic, Kodiak Archipelago, and Aleutian–Pribilof Islands. Studies including harvest surveys may have different objectives and methods: sampling methods, species categories, definition of seasons, availability of seasonal estimates, and geographic scale for reporting. Studies combining data from different sources, such as estimates presented here, are constrained by compatibility issues and usually include only a subset of the data potentially available.

We present annual average harvest estimates at region and statewide levels, and used them to estimate harvest for the reference year 2011. Data were not available to assess and account for wounding losses in harvest estimates. Average estimates were generated for communities with >1 year of data. Within subregions, average community estimates were expanded to nonsurveyed communities by the following equation:

Subregion harvest

= Sum of average community harvest

$$\times \left(\frac{\text{Total households in subregion}}{\text{Households in surveyed communities}} \right)$$

where the number of occupied households per community was based on results of the U.S.

TABLE 12.7
Dataset used to generate sea duck subsistence harvest estimates for Alaska.

	Gulf of AK–Cook Inlet	Kodiak Archipelago	Aleutian–Pribilof Is.	Bristol Bay	Y–K Delta	Bering Strait–Norton Sound	NW Arctic	North Slope	Interior AK	Upper Copper River	Total
1993	—	3	—	—	—	—	—	—	—	—	3
1994	—	—	2	—	—	—	1	—	—	—	3
1996	—	—	4	—	—	—	2	—	—	—	6
1997	—	2	—	—	—	—	5	—	—	—	7
2004	4	—	—	13	16	11	—	—	18	6	68
2005	1	—	3	15	24	9	—	7	9	—	68
2006	2	4	—	1	24	—	4	—	19	—	54
2007	—	—	1	12	20	11	—	4	8	5	61
2008	—	—	4	8	14	—	—	4	2	—	32
2009	—	—	—	—	18	2	—	3	—	1	24
2010	2	5	—	1	15	7	—	—	20	2	52
2011	—	—	—	6	13	2	1	1	14	—	37
2012	—	—	—	—	—	2	1	—	—	—	3
Community-years	9	14	14	56	144	44	14	19	90	14	418
Communities in region	5	12	12	27	47	16	11	8	43	8	189
Data year per community	1.8	1.2	1.2	2.1	3.1	2.8	1.3	2.4	2.1	1.8	2.2

Census Bureau (2011). Similarly, within regions, subregion estimates were expanded to nonsurveyed subregions by

Regional harvest

= Sum of subregion harvest

$$\times \left(\frac{\text{Total households in region}}{\text{Households in surveyed subregions}} \right)$$

Harvest estimates were presented at the species level except for mergansers and goldeneyes, which included all species occurring in different areas of the state. In five community-years, estimates of 2–17 unknown eiders and scoters were omitted in calculations of region estimates causing negligible underestimation of harvest.

Considerations on Species Identification

Bird species identification in most harvest surveys is subject to hunters' abilities to correctly identify and report species harvested. A limited proportion of subsistence and sport hunters may develop the advanced bird identification skills necessary to identify some species. In the US fall–winter harvest surveys, species composition of harvest estimates is derived from biological sampling and verification by wings and tails provided by hunters. The subsistence harvest survey has not included species verification through parts collection or bag checks because of cultural sensitivities and of logistic difficulties of operating in these remote areas. This fact further increases the need to test and fine-tune harvest survey materials to accurately represent local systems of species identification and naming, especially regarding species of management concern. Mismatches between ethnotaxonomy and scientific taxonomy and confusion related to English bird names must be carefully considered to minimize species identification issues. Some potential sea duck species identification issues require further assessment and must be considered when interpreting subsistence harvest data.

It is unknown whether Native cultures commonly distinguish between Common and Barrow's Goldeneyes, or Common and Red-breasted Mergansers. Potential issues arise for species identification when English names include the word "common", which may be misunderstood as the locally most common species (Naves and Zeller 2013). Harvest of species with "common" in their English name may be overestimated, while harvest of other species in the area may be underestimated. This issue creates problems for separation of different species of mergansers and goldeneyes in subsistence harvest surveys. Estimates of harvest for Common Eiders may also be affected by this issue.

Female eiders may be difficult to tell apart, and Alaska Native words for undefined female eiders suggest that these birds may be treated as a category at least in some circumstances. Female scoters also may be difficult to tell apart by species. Locally, scoters are commonly referred to as "black ducks". Therefore, a tendency to report all species of scoters combined as Black Scoters could lead to overestimation of Black Scoter harvest and underestimation of other scoter species locally available. A small number of goldeneye and Bufflehead eggs were reported as harvested. These species are obligate tree cavity nesters, and it is unknown whether subsistence harvesters indeed encounter harvest opportunities or if

harvest reports are due to species identification issues. Last, the Long-tailed Duck is sometimes locally called "pintail", which may lead to confusion with the Northern Pintail, but the extent of this potential issue is unknown.

Subsistence Harvest Patterns in Alaska

The 2011 estimated subsistence harvest of sea ducks in Alaska was 61,725 birds/year; this value corresponds to about 18% of the statewide subsistence harvest of all migratory and resident birds (Table 12.6). Regionally, sea ducks represented 71%–78% of the total bird harvest in the North Slope, Kodiak Archipelago, and Gulf of Alaska–Cook Inlet; 24% of the harvest in Interior Alaska; and 6%–15% in the remaining regions of the state (2004–2010 average; L. C. Naves, unpubl. data). Compared to the 1995 estimates (Paige and Wolfe 1998), the main differences were lower harvest of Common Eiders (−2,459 birds/year) and Long-tailed Ducks (−6,321 birds/year) and higher harvest of scoters (+4,307 birds/year, species combined). The seasonal distribution of the estimated sea duck harvest was 54% spring, 17% summer, and 29% fall–winter (Figure 12.15). A large proportion of summer harvests (68%) were King and Common Eiders in the North Slope, likely harvested during the postbreeding migration. Excluding this region, the seasonal breakdown of harvest was 57% spring, 7% summer, and 37% fall–winter.

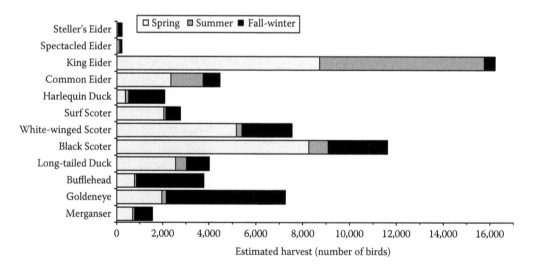

Figure 12.15. Seasonality of subsistence harvest of sea duck species in Alaska. Harvest during the postbreeding migration of King and Common Eiders in the North Slope accounts for a large proportion of the summer harvest of these species.

King Eiders (16,023 birds/year) and Black Scoters (11,617 birds/year) were harvested in the largest numbers statewide, followed by White-winged Scoters (7,538 birds/year) and goldeneyes (7,252 birds/year, Table 12.6, Figures 12.16 and 12.17). Eiders were harvested mostly in the North Slope and Bering Strait–Norton Sound, but King Eiders were also harvested in relatively large numbers in the Yukon–Kuskokwim Delta. Scoters and Long-tailed Ducks were harvested mostly in the Yukon–Kuskokwim Delta, Northwest Arctic, and Interior Alaska. Harlequin Ducks, Buffleheads, and goldeneyes were mostly harvested in southern coastal regions although relatively large harvest of Buffleheads and goldeneyes also occurred in the Yukon–Kuskokwim Delta and Interior Alaska. Merganser harvest was distributed statewide.

The estimated harvest of sea duck eggs (5,794 eggs/year) was largely comprised of eggs of Common Eiders (60%), Long-tailed Ducks (18%), and King Eiders (16%). The regions accounting for most of the egg harvest were Bering Strait–Norton Sound (84%; mostly Common Eiders, Long-tailed Ducks, and King Eiders) and North Slope (7%; mostly Common and King Eiders).

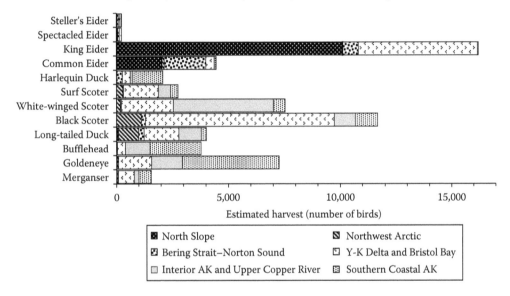

Figure 12.16. Distribution of subsistence harvest of sea duck species by region of Alaska.

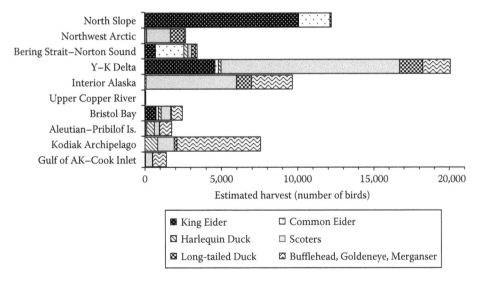

Figure 12.17. Species composition of subsistence sea duck harvest by region of Alaska.

Steller's Eider

The estimated Alaska subsistence harvest of Steller's Eider was 230 birds/year. Most harvest was in the Northwest Arctic, Bering Strait–Norton Sound, and Yukon–Kuskokwim (53–72 birds/year each), while the North Slope, Bristol Bay, and Aleutian–Pribilof Islands accounted for 13–20 birds/year each. More than half (65%) of bird harvest reports occurred in fall–winter and 24% occurred in spring. The estimated harvest of Steller's Eider eggs was 50 eggs/year divided between Bering Strait–Norton Sound (76%) and Yukon–Kuskokwim Delta (24%).

Spectacled Eider

The estimated harvest of Spectacled Eiders was 222 birds/year divided among four regions. Summer harvest (26% of the annual total) occurred in the North Slope, Bering Strait–Norton Sound, and Yukon–Kuskokwim Delta. Harvest in the Bristol Bay region occurred largely in fall–winter (57%), while harvest in the Yukon–Kuskokwim Delta (68%) and Bering Strait–Norton Sound (52%) was largely in spring. The estimated harvest of Spectacled Eider eggs totaled 25 eggs/year and occurred only in the Bering Strait–Norton Sound.

King Eider

The estimated harvest of King Eiders in Alaska was 16,203 birds/year, mostly in the North Slope (62%) and Yukon–Kuskokwim Delta (28%). In all regions south of the North Slope, most harvest occurred in spring (82%), likely during the prebreeding migration. In the North Slope, there was a tendency for a small harvest of King Eiders in spring (37%), with the primary harvest (63%) in summer during the postbreeding migration through the North Slope coast to molting areas in Russia (Woodby and Divoky 1982, Dickson et al. 1997).

The estimated harvest of King Eider eggs was 925 eggs/year. Harvest reports of King Eider eggs in known breeding areas in the North Slope and Northwest Arctic represented 27% of the annual estimated harvest. Harvest reports in the Bering Strait–Norton Sound represented 69% of annual estimates. Out of 44 community-years representing the Bering Strait–Norton Sound, harvest reports of King Eider eggs occurred in 10 community-years. Although species identification may

be an issue, harvest of King Eider eggs seems to regularly occur in this region, indicating that the breeding range of the species extends farther south than shown in most sources (Bellrose 1980).

Common Eider

The estimated harvest of Common Eiders in rural Alaska was 4,460 birds/year. Most of the harvest occurred in the North Slope (46% of the species annual harvest) and in the Bering Strait–Norton Sound (42%). In the North Slope, harvest occurred in spring (59%) and in summer (41%). In the Bering Strait–Norton Sound, harvest occurred in all seasons, but mostly in spring (46% spring, 22% summer, 32% fall–winter). The estimated harvest of Common Eider eggs was 3,496 eggs/year, of which 91% were in the Bering Strait–Norton Sound and 5% in the North Slope.

Harlequin Duck

The estimated harvest of Harlequin Ducks was 2,080 birds/year. More than half (64%) of the harvest occurred in the Kodiak Archipelago and Aleutian–Pribilof Islands; 75% occurred in fall–winter. Breeding Harlequin Ducks are sparsely distributed on high gradient streams and rivers throughout southern, interior, and western Alaska. The majority (92%) of the small summer harvest occurred in the Bering Strait–Norton Sound, Interior Alaska, and Kodiak Archipelago. Harvest in the Bristol Bay and Yukon–Kuskokwim Delta typically occurred in spring and fall. Harvest of Harlequin Duck eggs was not documented in the harvest surveys, probably because their nests are difficult to find.

Surf, White-winged, and Black Scoters

The Surf Scoter estimated harvest totaled 2,765 birds/year. About half of the harvest (51%) occurred in the Yukon–Kuskokwim Delta; 74% occurred in spring. The estimated harvest of Surf Scoter eggs was 15 eggs/year and occurred only on the Yukon–Kuskokwim Delta.

The estimated harvest of White-winged Scoters was 7,541 birds/year, and most harvest occurred in Interior Alaska (59%) and Yukon–Kuskokwim Delta (30%). In Interior Alaska, White-winged Scoter harvest occurred mostly in spring (68%) and fall (28%). In the Yukon–Kuskokwim Delta, most harvest occurred in spring (83%). The estimated harvest of White-winged Scoter eggs was 47 eggs/year and occurred in interior and western Alaska.

The Black Scoter estimated harvest was 11,617 birds/year. More than half of the harvest (69%) occurred in the Yukon–Kuskokwim Delta; 71% occurred in spring. The estimated harvest of Black Scoter eggs was 78 eggs/year and occurred in interior and western Alaska and North Slope.

The scoter species composition in subsistence harvest requires consideration because of potential for species misidentification. The contribution of Black Scoters to the statewide harvest may be overestimated because of potential confusion with the name "black scoter". In Alaska, Surf and White-winged Scoters are generally more abundant than Black Scoters and breed more commonly in interior boreal forest. However, most of Alaska scoter harvest occurs in high-harvest, western coastal areas, which are the primary breeding range of Black Scoters. Scoter migration ecology is not well known, and several factors may influence regional scoter harvest composition across seasons, including variation in scoter availability and factors affecting amount and distribution of harvest effort.

Long-tailed Duck

The estimated harvest of this species was 4,020 birds/year. Most harvest occurred in Yukon–Kuskokwim Delta (38%), Interior Alaska (25%), and Northwest Arctic (22%). The estimated harvest of Long-tailed Duck eggs was 1,027 eggs/year, of which 93% were in the Bering Strait–Norton Sound, 5% in the Yukon–Kuskokwim Delta, and 2% in the North Slope.

Bufflehead

The estimated harvest of Buffleheads was 3,782 birds/year. Most harvest occurred in the Kodiak Archipelago (55%) and Interior Alaska (29%). Harvest in the Kodiak Archipelago occurred mostly in fall–winter (94%), while harvest in Interior Alaska was divided between spring (40%) and fall (58%). The estimated harvest of Bufflehead eggs was 62 eggs/year and was in interior and western Alaska, mostly in the Yukon–Kuskokwim Delta (48 eggs/year). Buffleheads are obligate tree cavity nesters, and it is unknown whether subsistence harvesters indeed encounter egg harvest opportunities or if harvest reports are due to species misidentification.

Goldeneyes

The estimated harvest of goldeneyes was 7,252 birds/year. Harvest in the Aleutian–Pribilof Islands, western Alaska, and Interior Alaska (48%) likely refers to Common Goldeneye and was divided between spring (48%) and fall–winter (46%). Harvest in the Gulf of Alaska-Cook Inlet, Kodiak Archipelago, and Upper Copper River (52%) likely refers to both Common and Barrow's Goldeneyes and occurred mostly in fall–winter (92%). The estimated harvest of goldeneye eggs was 17 eggs/year. Goldeneyes are obligate tree cavity nesters, and egg harvest reports may be due to species misidentification.

Mergansers

The estimated harvest of mergansers was 1,556 birds/year. Most of the harvest (95%) occurred in regions of the state where both Common and Red-breasted Mergansers occur (southern coast, Yukon–Kuskokwim Delta, Bristol Bay, and interior) and was divided between spring (46%) and fall–winter (49%). The estimated harvest of merganser eggs was 52 eggs/year, of which 77% occurred in the Kodiak Archipelago and Aleutian–Pribilof Islands.

Regional Distribution of Harvest

The Yukon–Kuskokwim Delta ranked first in the number of sea ducks harvested (20,170 birds/year) and had a diverse species composition. The species harvested most often were Black Scoter (40%), King Eider (23%), and White-winged Scoter (11%). The North Slope ranked second in the number of sea ducks harvested but had the least diverse species composition (two species represented 99% of the harvest). King and Common Eiders represented most of the sea duck harvest in the Bering Strait–Norton Sound (72% of 3,579 birds/year) and North Slope (99% of 12,300 birds/year, Figure 12.17). Interior Alaska ranked third in the number of sea ducks harvested (9,663 birds/year), 46% of which were White-winged Scoters. Sea duck harvest in the Northwest Arctic region was largely composed of Black Scoters (40%) and Long-tailed Ducks (33%) and differed from its neighboring regions to the north and south (Figure 12.17).

The Bering Strait–Norton Sound (4,888 eggs/year) and the North Slope (409 eggs/year) had the highest harvest of sea duck eggs. North Slope harvest was largely composed of eggs of King Eiders (45%) and Common Eiders (45%). Bering Strait–Norton Sound harvest was composed of eggs of Common Eiders (65%), Long-tailed Ducks (20%), and King Eiders (13%). In these regions, Common Eiders were represented in more egg harvest (45% in North Slope, 65% in Bering Strait–Norton Sound) than bird harvest (17% in North Slope, 52% in Bering Strait–Norton Sound). The difference may arise from (1) preference for harvest of Common Eider eggs, (2) indirect factors leading to selective harvest such as overlap between breeding areas and harvest effort, and (3) difficulty in identifying the species of incubating females, nests, and eggs, leading to a tendency to report unknown eider eggs as "common eiders" on the survey form.

Regional Management Topics

Steller's and Spectacled Eiders are closed to harvest in Alaska, but some birds and eggs are taken in subsistence harvest. Russian breeding birds likely comprise most of Alaska harvest occurring during fall–winter (65% of Steller's Eider and 32% of Spectacled Eider annual harvest estimates) and may be part of spring harvest (24% of Steller's Eider and 42% of Spectacled Eider annual harvest estimates, Table 12.8). Information on the status of Russian populations is sparse, but there are concerns about potential rangewide declines of Steller's Eiders and the security of wintering Spectacled Eiders. Recent statewide harvest estimates suggest a reduction in harvest of Steller's and Spectacled Eiders compared to the mid-1980s through the early 1990s and 1996 (Table 12.7). The current lower harvest estimates may be related to (1) outreach and communication work focusing on conservation and harvest closure, (2) law enforcement efforts, (3) reduced reporting rate because of law enforcement action, or (4) reduced availability of these species at least in some areas in the last 25 years. It is difficult to estimate harvest of species taken in low numbers because only few data points are available for generalization over a large geographic area. The AMBCC recognizes the need for continued outreach and communication work among all stakeholders to address management and conservation concerns.

Final Considerations for Subsistence Harvest in Canada and Alaska

In general, Common Eiders are a primary resource for coastal communities in eastern North America, and King Eiders are important for west coastal communities. Common Eiders represent a special challenge to management because of strong population structure with four subspecies in North America and many hybrid zones (Sonsthagen et al. 2011). Moreover, different population segments are subject to various combinations of subsistence and fall–winter harvest (Gilliland et al. 2009). Eiders have been sensitive to overexploitation but, on the other hand, also have shown a surprising ability to quickly recover when harvest is managed at appropriate levels (Merkel 2010).

Wildlife harvests, including sea ducks, are poorly known or estimated for large regions in interior Canada. Scoters are important subsistence resources in coastal and interior regions and in the Arctic and sub-Arctic (21,981 scoters/year in all regions of Alaska). All harvests of scoters from Alaska to Mexico need to be considered in management plans for these species. Harvests in eastern Russia are poorly documented but are important for understanding population dynamics of shared populations of eiders, Long-tailed Ducks, and other species (E. Syroechkovski Jr. and K. B. Klokov, IPEE RAN, unpubl. report). Well-considered regional management plans and effective collaboration across jurisdictions will help to ensure that harvests of key species, such as eiders and scoters, are sustainable into the future.

New information provided and older information assembled in this chapter allow better understanding of sea duck harvest patterns and quantification of harvest demand. While progress is made in delineating and estimating population sizes, information on harvest demand is important to set minimum management and conservation objectives.

Subsistence harvest is challenging to monitor or manage because harvest occurs in remote areas and in a particular cultural context. Additionally, data obtained may include species identification issues and issues resulting from difficulty in implementing standard data collection methods.

TABLE 12.8

Alaska average annual subsistence harvest of sea ducks by region.

	North Slope	NW Arctic	Bering Strait–Norton Sound	Y–K Delta	Interior AK	Upper Copper River	Bristol Bay	Aleutian–Pribilof Is.	Kodiak Archipelago	Gulf of AK–Cook Inlet	Total
Steller's Eider, birds	20	58	72	54	b	b	13	13	0	0	230
Spring	12	0	8	22	b	b	12	0	0	0	54
Summer	8	0	13	5	b	b	0	0	0	0	26
Fall–winter	c	58	51	27	b	b	1	13	0	0	150
Steller's Eider, eggs	0	0	38	12	b	b	0	0	0	0	50
Spectacled Eider, birds	55	0	73	31	b	b	63	b	b	b	222
Spring	8	0	38	21	b	b	27	b	b	b	94
Summer	47	0	9	1	b	b	0	b	b	b	57
Fall–winter	c	0	26	9	b	b	36	b	b	b	71
Spectacled Eider, eggs	0	0	25	0	b	b	0	b	b	b	25
King Eider, birds	10,087	32	707	4,598	b	b	740	38	1	0	16,203
Spring	3,695	22	448	3,814	b	b	702	24	1	0	8,706
Summer	6,392	9	104	529	b	b	0	0	0	0	7,034
Fall–winter	c	1	155	255	b	b	38	14	0	0	463
King Eider, eggs	185	61	643	15	b	b	3	18	0	0	925
Common Eider, birds	2,045	99	1,862	224	b	b	148	82	0	0	4,460
Spring	1,198	60	859	148	b	b	72	0	0	0	2,337
Summer	847	21	402	15	b	b	47	75	0	0	1,407
Fall–winter	c	18	601	61	b	b	29	7	0	0	716
Common Eider, eggs	185	15	3,188	41	b	b	0	67	0	0	3,496
Harlequin Duck, birds	b	7	261	170	92	0	184	504	832	30	2,080
Spring	b	4	77	72	18	0	118	24	54	5	372
Summer	b	0	98	2	22	0	0	5	14	4	145
Fall–winter	c	3	86	96	52	0	66	475	764	21	1,563
Harlequin Duck, eggs	b	0	0	0	0	0	0	0	0	0	0
Surf Scoter, birds	0	277	27	1,417	543	0	163	5	214	119	2,765

(Continued)

TABLE 12.8 (*Continued*)
Alaska average annual subsistence harvest of sea ducks by region.

	North Slope	NW Arctic	Bering Strait–Norton Sound	Y–K Delta	Interior AK	Upper Copper River	Bristol Bay	Aleutian–Pribilof Is.	Kodiak Archipelago	Gulf of AK–Cook Inlet	Total
Spring	0	251	22	1,277	323	0	70	4	62	29	2,038
Summer	0	9	1	33	41	0	12	0	0	0	96
Fall–winter	c	17	4	107	179	0	81	1	152	90	631
Surf Scoter, eggs	0	0	0	15	0	0	0	0	0	0	15
White-winged Scoter, birds	1	165	43	2,257	4,432	24	80	171	300	65	7,538
Spring	0	89	29	1,870	3,033	8	47	4	50	26	5,156
Summer	1	9	2	61	161	8	0	0	0	4	246
Fall–winter	c	67	12	326	1,238	8	33	167	250	35	2,136
White-winged Scoter, eggs	13	0	6	10	18	0	0	0	0	0	47
Black Scoter, birds	5	1,101	184	7,989	920	21	399	144	570	284	11,617
Spring	5	926	150	6,182	520	0	287	4	46	137	8,257
Summer	0	64	0	681	62	13	8	5	0	0	833
Fall–winter	c	111	34	1,126	338	8	104	135	524	147	2,527
Black Scoter, eggs	9	0	16	18	35	0	0	0	0	0	78
Long-tailed Duck, birds	86	904	249	1,533	991	0	29	28	185	15	4,020
Spring	35	784	160	800	722	0	14	0	28	10	2,553
Summer	51	62	27	250	83	0	0	0	0	0	473
Fall–winter	c	58	62	483	186	0	15	28	157	5	994
Long-tailed Duck, eggs	15	0	959	53	0	0	0	0	0	0	1,027
Bufflehead, birds	b	0	35	295	1,106	8	59	147	2,091	41	3,782
Spring	b	0	35	134	442	5	20	0	121	10	767
Summer	b	0	0	66	22	0	0	3	0	4	95
Fall–winter	c	0	0	95	642	3	39	144	1,970	27	2,920
Bufflehead, eggs[a]	b	0	7	48	0	2	0	0	5	0	62
Goldeneyes, birds	b	41	38	1,166	1,374	11	313	509	3,212	588	7,252
Spring	b	32	10	723	675	7	226	0	113	170	1,956

(Continued)

TABLE 12.8 (*Continued*)
Alaska average annual subsistence harvest of sea ducks by region.

	North Slope	NW Arctic	Bering Strait–Norton Sound	Y–K Delta	Interior AK	Upper Copper River	Bristol Bay	Aleutian–Pribilof Is.	Kodiak Archipelago	Gulf of AK–Cook Inlet	Total
Summer	b	0	0	94	76	1	12	1	11	0	195
Fall–winter	c	9	28	349	623	3	75	508	3,088	418	5,101
Goldeneyes, eggs[a]	b	0	6	8	0	0	0	0	4	0	17
Merganser, birds	0	44	30	371	203	0	346	135	159	268	1,556
Spring	0	0	8	206	144	0	193	15	19	105	690
Summer	0	0	6	21	41	0	9	8	0	0	85
Fall–winter	c	44	16	144	18	0	144	112	140	163	781
Merganser, eggs	1	0	0	5	0	0	6	16	24	0	52

[a] Obligate tree cavity nesting, reported egg harvest may need further assessment.
[b] Species unlikely to occur in this region has not been included in harvest surveys and has not been reported as harvested.
[c] AMBCC surveys not conducted in North Slope in fall because birds migrate out of this region starting in late summer.

Better understanding of sea duck ethnotaxonomy likely will help to minimize issues with species identification. However, ethnotaxonomy may include biological units defined by plumage, age, or other elements and may confound integration with the species-based scientific taxonomy. Implementation of a species verification system remains a main challenge in subsistence bird harvest monitoring.

Despite difficulties, subsistence harvest surveys are important for conservation of sea duck populations and also to ensure sustainable hunting opportunities, through the values they bring to resource management: (1) Documentation of subsistence harvest is important in resource allocation among user groups; subsistence communities depend on these resources and by law aboriginal peoples in Canada and qualified rural residents in Alaska have priority of access. (2) Subsistence harvest accounts for the bulk of the take for some sea duck populations, and continued harvest monitoring generates key information for management, especially for species with declining populations. (3) Harvest surveys are a main channel of communication between subsistence users and resource management agencies. Surveys create opportunities for stakeholders to work together and engage subsistence users in management and conservation of the resources they depend on for food and also to maintain a lifestyle that supports their social and cultural well-being.

Subsistence harvest differs from recreational harvest in fundamental ways, including the socioeconomic context and seasonal timing primarily as a spring hunt. It is difficult to fit subsistence harvest in the current framework of the Migratory Bird Convention and associated hunting regulations in Canada and the United States. Measures to monitor and, where needed, regulate subsistence harvest are still being developed and implemented in comanagement systems including subsistence users. Management of subsistence harvest in North America is evolving toward a workable framework recognizing the importance of harvest for northern peoples and honoring the intent of the Migratory Bird Treaty: to maintain viable populations for future generations through management that accounts for the seasonal travels of birds across multiple jurisdictions. Compared to a generation ago, managers now have better tools to estimate subsistence harvest, and partnerships are evolving for effective communication and comanagement of birds, including sea ducks. Care will be needed in the future; sea duck populations can be overexploited and climate change may bring extensive modifications to habitats and to their patterns of use by both people and sea ducks.

Changes to environments and human activities in the north will present both challenges and opportunities to sea ducks and other wildlife.

CONTEMPORARY SEA DUCK HARVEST ISSUES AND INFORMATION GAPS

The annual surveys that monitor fall and winter sea duck harvest in Canada and the United States are adequate to track the relative magnitude, distribution, and species composition of harvest over time at a continental scale. However, current harvest surveys and resultant datasets are not reliable and detailed enough to assess regional harvest or evaluate impacts on defined populations. Information needed to fill this gap includes (1) improved harvest sampling methods, (2) improved understanding of the structure of sea duck populations and how harvest is linked to breeding populations, and (3) larger sample sizes in wing surveys in both countries or alternate methods to assess species, sex, and age composition.

Comprehensive demographic data that provide context for annual harvest estimates and guide harvest management are unavailable for nearly all sea duck species. Harvest assessment goals cannot be achieved without fundamental definition of cohesive sea duck population units, including interpretation of seasonal structuring and the validity of management at regional scales; once populations are defined, the effects of harvest can be assessed with associated rates of recruitment and survival. For example, sex- and age-specific harvest rate estimates are needed in combination with harvest sex and age ratios to provide reliable estimates of population sex ratios and productivity. Development of population models that are parameterized with improved demographic information and recent harvest data will ensure that resources are focused on more effective population surveys, research, and appropriate hunting regulations.

Subsistence harvest estimates are not comprehensive, particularly for northern Canada, primarily due to (1) difficulties in obtaining complete sample frames of subsistence hunters, (2) logistical challenges with conducting surveys in remote areas, and (3) sociocultural barriers to participation in surveys. Government wildlife agencies and organizations representing subsistence hunters recognize these challenges and are collaborating to develop and improve methods to estimate subsistence harvest. These joint efforts are more fully describing the extent and value of sea duck harvest in subsistence economies and a more complete understanding of waterfowl harvest in North America.

The role of harvest in sea duck population dynamics is uncertain (Chapter 3, this volume), but current harvest levels appear to be sustainable for most populations. Managers have reduced or curtailed hunting of small populations that were at greatest risk, such as the eastern populations of Harlequin Duck and Barrow's Goldeneye, as well as Steller's and Spectacled Eiders in Alaska. High levels of harvest of Common Eiders in Greenland were substantially reduced through regulation and international cooperation, and the population appears to be recovering. The most heavily harvested sea duck species are abundant and seem to have stable (Bufflehead and Common Goldeneye) or increasing population trends (Hooded Merganser) since the mid-1900s, and productivity indices based on age ratios among harvested birds are relatively high.

Harvest of scoters, Long-tailed Ducks, and Common and Red-breasted Mergansers are much lower. Based on harvest age ratios, these species are apparently productive enough to sustain current harvest levels. Limited band recovery data also suggest that these species are harvested at low rates. For example, adult male Surf Scoters banded in eastern Canada in 2004–2008 were harvested at about a 2%–3% rate (Gilliland et al. 2011). The other scoter species and Long-tailed Ducks have similar temporal and spatial distribution along the Atlantic coast during the hunting season and likely experience similarly low harvest rates. Atlantic Common Eiders may be an exception. Despite high adult survival rates and low harvest rates (Krementz et al. 1996), eiders can experience widespread reproductive failure, sometimes for prolonged periods. Productivity is low in this population, especially along the Maine coast where nearly all eider ducklings are killed by predators, primarily gulls (B. Allen, pers. comm.).

The Sea Duck Joint Venture's Continental Technical Team is currently employing the *potential biological removal* analytical method to determine harvest potential for scoters, Atlantic Common Eiders, and Long-tailed Ducks (Wade 1998, Runge et al. 2004). The method allows estimation of allowable take or levels of sustainable harvest despite uncertainty about the magnitude and variability of the demographic parameters used in the analyses. Results of this work will provide an initial indication of whether current harvest levels are indeed sustainable. More importantly, these analyses will show researchers and managers what demographic information is needed to most effectively inform and ensure sound harvest management of sea ducks in North America.

ACKNOWLEDGMENTS

The authors recognize the valuable contributions of many wildlife administrators, managers, and researchers who have raised interest in the role of harvest in sea duck population management and produced many stimulating studies and analyses over the past 20 years. We especially thank K. Wilkins and B. Raftovich (USFWS) and M. Gendron (CWS) for providing the hunter activity and harvest data for the United States and Canada that were used throughout this chapter. W. Harvey and L. Hindman (Maryland DNR), D. Kraege (Washington DFW), and B. Allen (Maine DIFW) gave us state survey data on sea duck hunting and insights into the traditions and current status of sea duck hunting in the Atlantic and Pacific Flyways. We are grateful to M. Robertson (CWS) for valuable insights on sources of subsistence harvest data in Canada and to J. Fall (Alaska DF&G, Division of Subsistence) for his continuous guidance and expertise on all matters related to subsistence harvest. The authors commend the AMBCC and its subsistence harvest survey, supported by a collaboration of national wildlife refuges, village councils, local surveyors, and Alaska Native organizations and funded by USFWS; this evolving survey is providing crucial harvest data for migratory bird management, especially for sea ducks. The authors appreciate critical reviews by D. Kraege (Washington DFW), K. Dickson (CWS), S. Gilliland (CWS), and J.-P. L. Savard, who provided important insights and comments to greatly improve the manuscript. Last, but not least, we are indebted to all of the hunters, households, and communities who have voluntarily provided information on their harvest and local knowledge upon which we depend to fulfill science-based stewardship of North American sea ducks.

LITERATURE CITED

Alaska Department of Fish and Game Division of Subsistence. 2014. Community Subsistence Information System (CSIS). <http://www.adfg.alaska.gov/sb/CSIS/> (13 January 2014).

Alisauskas, R. T. 2011. Does species composition of arctic geese recovered in prairie Canada vary by hunter residency? Wildlife Society Bulletin 35:85–92.

Anderson, B. W., and R. L. Timken. 1972. Sex and age ratios and weights of Common Mergansers. Journal of Wildlife Management 36:1127–1133.

Anderson, D. R., and K. P. Burnham. 1976. Population ecology of the Mallard VI: the effect of exploitation on survival. Resource Publication 128. U.S. Fish and Wildlife Service, Washington, DC.

Anderson, J. M., K. Schiefer, and D. B. Arrier. 1985. A study of merganser predation and its impact on Atlantic salmon stocks in the Restigouche River system, 1984. Atlantic Salmon Federation, St. Andrews, NB.

Barker, R. J. 1991. Nonresponse bias in New Zealand waterfowl harvest surveys. Journal of Wildlife Management 55:126–131.

Barker, R. J., P. H. Geissler, and B. A. Hoover. 1992. Sources of nonresponse to the federal waterfowl hunter questionnaire survey. Journal of Wildlife Management 56:337–343.

Bartonek, J. C. 1984. Pacific flyway. Pp. 395–403 in A. S. Hawkins, R. C. Hanson, H. K. Nelson, and H. M. Reeves (editors), Flyways: pioneering waterfowl management in North America. U.S. Department of the Interior, Fish and Wildlife Service, Washington, DC.

Beach, U. S. 1936. The destruction of trout by fish ducks. Transactions of the American Fisheries Society 66:338–341.

Bellrose, F. C. 1953. A preliminary evaluation of cripple losses in waterfowl. Transactions of the North American Wildlife Conference 18:337–360.

Bellrose, F. C. 1980. Ducks, geese and swans of North America, 3rd ed. Stackpole Books, Harrisburg, PA.

Bentzen, R. L., and A. N. Powell. 2012. Population dynamics of King Eiders breeding in northern Alaska. Journal of Wildlife Management 76:1011–1020.

Berkes, F. 2009. Evolution of co-management: role of knowledge generation, bridging organizations and social learning. Journal of Environmental Management 90:1692–1702.

Berkes, F., P. J. George, R. J. Preston, A. Hughes, J. Turner, and B. D. Cummins. 1994. Wildlife harvesting and sustainable regional native economy in the Hudson and James Bay Lowland, Ontario. Arctic 47:350–360.

Boyd, S., D. Esler, T. Bowman, J. E. Thompson, and J. Schamber. 2013. Migration and habitat use of Pacific Barrow's Goldeneye homepage. <http://www.sfu.ca/biology/wildberg/CWESeaducksfolder/BAGOwebpage/BAGOMigrationHome.html> (11 January 2014).

Braund, S. R. 1993a. North Slope subsistence study Barrow, 1987, 1988 and 1989. Technical Report No. 149, Minerals Management Service, U.S. Department of the Interior, Anchorage, AK.

Braund, S. R. 1993b. North Slope subsistence study Wainwright 1987 and 1988. Technical Report No. 147, Minerals Management Service, U.S. Department of the Interior, Anchorage, AK.

Bruette, W. 1934. American duck, goose, and brant shooting. Charles Scribner's Sons, New York, NY.

Burnham, K. K., J. A. Johnson, B. Konkel, and J. L. Burnham. 2012. Nesting Common Eider (Somateria mollissima) population quintuples in Northwest Greenland. Arctic 65:458–464.

Byers, T., and D. L. Dickson. 2001. Spring migration and subsistence hunting of King and Common Eiders at Holman, Northwest Territories, 1996–98. Arctic 54:122–134.

Byrd, G. V., J. C. Williams, and A. Durand. 1992. The status of Harlequin Ducks in the Aleutian Islands, Alaska. Pp. 14–32 in Proceedings of the Harlequin Duck Symposium. Northwest Section of the Wildlife Society, Moscow, ID.

Caithamer, D. F., M. Otto, P. I. Padding, J. R. Sauer, and G. H. Haas. 2000. Sea ducks in the Atlantic Flyway: population status and a review of special hunting seasons. U.S. Fish and Wildlife Service, Laurel, MD.

Cameron, J. 1929. Bureau of Biological Survey; its history, activities and organization. The Johns Hopkins Press, Baltimore, MD.

Carney, S. M. 1984. Estimating the harvest. Pp. 256–259 in A. S. Hawkins, R. C. Hanson, H. K. Nelson, and H. M. Reeves (editors), Flyways: pioneering waterfowl management in North America. U.S. Fish and Wildlife Service, Washington, DC.

Chardine, J. W., G. J. Robertson, and H. G. Gilchrist. 2008. Seabird harvest in Canada. Pp. 20–29 in F. Merkel and T. Barry (editors), Seabird harvest in the Arctic. Circumpolar Seabird Group (CBird). CAFF Technical Report No. 16, CAFF International Secretariat, Akureyri, Iceland.

Committee on the Status of Endangered Wildlife in Canada. 2000. COSEWIC assessment and status report on the Barrow's Goldeneye Bucephala islandica, Eastern population, in Canada. Committee on the Status of Endangered Wildlife in Canada, Ottawa, ON.

Cooch, F. G. 1986. The numbers of nesting northern eiders on the West Foxe Is., NWT, in 1956 and 1976. Pp. 114–118 in A. Reed (editor), Eider ducks in Canada. Canadian Wildlife Service Report Series No. 47. Canadian Wildlife Service, Ottawa, ON.

Cooch, F. G., S. Wendt, G. E. J. Smith, and G. Butler. 1978. The Canada Migratory Game Bird Hunting Permit and associated surveys. Pp. 8–39 in H. Boyd and G. H. Finney (editors), Migratory game bird hunters and hunting in Canada. Canadian Wildlife Service Report Series No. 43, Ottawa, ON.

Copp, J. D., and G. M. Roy. 1986. Results of the 1985 survey of waterfowl hunting on the Yukon-Kuskokwim Delta, Alaska. Report by the Oregon State University, Corvallis to the U.S. Fish and Wildlife Service, Division of Migratory Bird Management, Anchorage, AK.

Descamps, S., S. Jenouvrier, H. G. Gilchrist, and M. R. Forbes. 2012. Avian cholera, a threat to the viability of an Arctic seabird colony? PLoS One 7:e29659.

Dickson, D. L., R. C. Cotter, J. E. Hines, and M. F. Kay. 1997. Distribution and abundance of King Eiders in the western Canadian Arctic. Pp. 29–39 in D. L. Dickson (editor), King and Common Eiders of the western Canadian arctic. Canadian Wildlife Service Occasional Paper No. 94. Canadian Wildlife Service, Ottawa, ON.

Donaldson, J. 1984. Wildlife harvest statistics for the Baffin region, Northwest Territories. Technical Report No. 2, Baffin Region Inuit Association, Yellowknife, NWT.

Eadie, J. M., J.-P. L. Savard, and M. L. Mallory. 2000. Barrow's Goldeneye (Bucephala islandica). In A. Poole and F. Gill (editors), The birds of North America, no. 548. The Academy of Natural Sciences, Philadelphia, PA.

Elden, R. C., W. V. Bevill, P. I. Padding, J. E. Frampton, and D. L. Shroufe. 2002. A history of the development of the Harvest Information Program. Pp. 7–16 in J. M. VerSteeg and R. C. Elden (compilers), Harvest Information Program: evaluation and recommendations. International Association of Fish and Wildlife Agencies, Migratory Shore and Upland Game Bird Working Group, Ad Hoc Committee on HIP, Washington, DC.

Elder, W. H. 1955. Fluoroscopic measurement of hunting pressure in Europe and North America. Transactions of the North American Wildlife Conference 20:298–322.

Erskine, A. J. 1972. Populations, movements and seasonal distribution of mergansers in northern Cape Breton Island. Canadian Wildlife Service Report Series 17. Canadian Wildlife Service, Report 17, Ottawa, ON.

Fabijan, M., R. Brook, D. Kuptana, and J. E. Hines. 1997. The subsistence harvest of King and Pacific Eiders in the Inuvialuit Settlement Region, 1988–1984. Pp. 67–73 in D. L. Dickson (editor), King and Pacific Eiders of the Western Canadian Arctic. Canadian Wildlife Service Occasional Paper No. 94. Canadian Wildlife Service, Ottawa, ON.

Forsell, D. J., and P. J. Gould. 1981. Distribution and abundance of marine birds and mammals wintering in the Kodiak area of Alaska. FWS/OBS-81/13. U.S. Fish and Wildlife Service, Washington, DC.

Geissler, P. H. 1990. Estimation of confidence intervals for federal waterfowl harvest surveys. Journal of Wildlife Management 54:201–205.

Gilchrist, H. G., J. Heath, L. Arrangutainaq, G. Robertson, K. Allard, S. Gilliland, and M. L. Mallory. 2006. Combining science and local knowledge to study Common Eider ducks wintering in Hudson Bay. Pp. 284–303 in R. Riewe and J. Oakes (editors), Climate change: linking traditional and scientific knowledge. Aboriginal Issues Press, University of Manitoba, Winnipeg, MB.

Gilchrist, H. G., and G. J. Robertson. 2000. Observations of marine birds and mammals wintering at polynyas and ice edges in the Belcher Islands, Nunavut, Canada. Arctic 53:61–68.

Gillilan, G. H. 1988. Gunning for sea ducks. Tidewater Publishers, Centerville, MD.

Gilliland, S. G., H. G. Gilchrist, R. F. Rockwell, G. J. Robertson, J.-P. L. Savard, F. Merkel, and A. Mosbech. 2009. Evaluating the sustainability of harvest among northern Common Eiders Somateria mollissima borealis in Greenland and Canada. Wildlife Biology 15:24–36.

Gilliland, S. G., E. Reed, C. Lepage, L. Lesage, J.-P. L. Savard, and K. McAloney. 2011. Harvest and survival of eastern North American Surf Scoters. P. 74 in Abstracts of the Fourth International Sea Duck Conference, September 12–16, 2011, Seward, AK.

Gilliland, S. G., and G. J. Robertson. 2009. Composition of eiders harvested in Newfoundland. Northeastern Naturalist 16:501–518.

Goudie, R. I. 1989. Historical status of Harlequin Ducks wintering in eastern North America—a reappraisal. Wilson Bulletin 101:112–114.

Goudie, R. I., S. Brault, B. Conant, A. V. Kondratyev, M. R. Petersen, and K. Vermeer. 1994. The status of sea ducks in the North Pacific Rim: towards their conservation and management. Transactions of the North American Wildlife Natural Resources Conference 59:27–49.

Goudie, R. I., G. J. Robertson, and A. Reed. 2000. Common Eider (Somateria mollissima). In A. Poole (editor), The Birds of North America online. Cornell Lab of Ornithology, Ithaca, New York, NY.

Greenland Home Rule. 2013. Harvest survey homepage. Department of Fisheries, Hunting and Agriculture, Nuuk, Greenland. Piniarneq 2013. <http://vintage.nanoq.gl/Emner/Erhverv/Erhvervsomraader/Fangst_og_Jagt/PINIARNEQ.aspx> (11 January 2014) (in Greenlandic).

Grinnell, J., H. C. Bryant, and T. I. Storer. 1918. The game birds of California. University of California, Berkeley, CA.

Gwich'in Renewable Resource Board. 2009. Gwich'in harvest study final report. Gwich'in Renewable Resource Board, Inuvik, NWT.

Hagerbaumer, D. 1998. Waterfowling these past fifty years: especially brant. Sand Lake Press, Amity, OR.

Hicklin, P. W., and W. R. Barrow. 2004. The incidence of embedded shot in waterfowl in Atlantic Canada and Hudson Strait. Waterbirds 27:41–45.

Hochbaum, G. S., and C. J. Walters. 1984. Components of hunting mortality in ducks. Canadian Wildlife Service Occasional Paper No. 52, Environment Canada, Ottawa, ON.

Hodges, J. I., D. J. Groves, and B. P. Conant. 2008. Distribution and abundance of waterbirds near shore in southeast Alaska, 1997–2002. Northwestern Naturalist 89:85–96.

Iverson, S. A., and D. Esler. 2010. Harlequin Duck population injury and recovery dynamics following the 1989 Exxon Valdez oil spill. Ecological Applications 20:1993–2006.

James Bay and Northern Québec Native Harvesting Research Committee. 1988. Final report: research to establish present levels of harvesting for the Inuit of northern Québec, 1976–1980. James Bay Northern Québec Native Harvesting Research Committee, Québec City, QC.

Joensen, A. H. 1974. Populations and shooting utilization of migratory ducks in Denmark, with particular reference to the Eider Duck (Somateria mollissima). Pp. 252–261 in Proceeding of the XIII International Congress of Game Biologists, Stockholm, Sweden.

Johnson, F. A., C. T. Moore, W. L. Kendall, J. A. Dubovsky, D. F. Caithamer, J. R. Kelley Jr., and B. K. Williams. 1997. Uncertainty and the management of Mallard harvests. Journal of Wildlife Management 61:203–217.

Johnson, M. A., P. I. Padding, M. H. Gendron, E. T. Reed, and D. A. Graber. 2012. Assessment of harvest from conservation actions for reducing midcontinent light geese and recommendations for future monitoring. Pp. 46–94 in J. O. Leafloor, T. J. Moser, and B. D. J. Batt (editors), Evaluation of special management measures for midcontinent Lesser Snow Geese and Ross's Geese. Arctic Goose Joint Venture Special Publication, U.S. Fish and Wildlife Service, Washington, DC.

Kimball, C. F., F. A. Bishop, C. D. Crider, J. H. Dunks, R. M. Hopper, and D. D. Kennedy. 1971. Analysis of the 12-state point regulation test, 1970, based on the Hunter Performance Survey. Administrative Report 206, U.S. Fish and Wildlife Service, Migratory Bird Population Station, Laurel, MD.

Kirby, R. E., J. H. Riechman, and T. W. Schoenfelder. 1981. Recuperation from crippling in ducks. Wildlife Society Bulletin 9:150–153.

Kramer, G. 2003. A Ducks Unlimited guide to hunting diving and sea ducks. Ducks Unlimited, Memphis, TN.

Krementz, D. G., P. W. Brown, F. P. Kehoe, and C. S. Houston. 1997. Population dynamics of White-winged Scoters. Journal of Wildlife Management 61:222–227.

Krementz, D. G., J. E. Hines, and D. F. Caithamer. 1996. Survival and recovery rates of American Eiders in eastern North America. Journal of Wildlife Management 60:855–862.

Land Claims Agreements Coalition. 2014. Map of Modern Treaty Territories. Ottawa, ON. http://www.landclaimscoalition.ca/ (13 January 2014).

Madsen, J., and F. Riget. 2007. Do embedded shotgun pellets have a chronic effect on body condition of Pink-footed Geese? Journal of Wildlife Management 71:1427–1430.

Mahaffy, M. S., D. R. Nysewander, K. Vermeer, T. R. Wahl, and P. E. Whitehead. 1994. Status, trends and potential threats related to birds in the Strait of Georgia, Puget Sound, and the Strait of Juan de Fuca. P. 256 In R. C. Wilson, R. J. Beamish, F. Aitkens, and J. Bell (editors), Review of the marine environment and biota of Strait of Georgia, Puget Sound and Juan de Fuca Strait: Proceedings of the British Columbia/ Washington Symposium on the Marine Environment, January 13–14, 1994. Canadian Technical Report of Fisheries and Aquatic Science, No. 1948, Victoria, BC.

Martin, E. M., and S. M. Carney. 1977. Population ecology of the Mallard IV: a review of duck hunting regulations, activity, and success, with special reference to the Mallard. Resource Publication 130. U.S. Fish and Wildlife Service, Washington, DC.

McKelvey, R., and G. E. J. Smith. 1990. The distribution of waterfowl banded or returned in British Columbia, 1951–1985. Technical Report Series 79. Canadian Wildlife Service, Pacific and Yukon Region, Delta, BC.

Merkel, F. R. 2004a. Impact of hunting and gillnet fishery on wintering eiders in Nuuk, Southwest Greenland. Waterbirds 27:469–479.

Merkel, F. R. 2004b. Evidence of population decline in Common Eiders breeding in western Greenland. Arctic 57:27–36.

Merkel, F. 2010. Evidence of recent population recovery in Common Eiders breeding in Western Greenland. Journal of Wildlife Management 74:1869–1874.

Merkel, F., and T. Christensen. 2008. Seabird harvest in Greenland. Pp. 41–49 in F. Merkel and T. Barry (editors), Seabird harvest in the Arctic. Circumpolar Seabird Group (CBird), CAFF Technical Report No. 16. CAFF International Secretariat, Akureyri, Iceland.

Merkel, F. R., K. Falk, and S. E. Jamieson. 2006. Effect of embedded lead shot on body condition of Common Eiders. Journal of Wildlife Management 70:1644–1649.

Metz, K. J., and C. D. Ankney. 1991. Are brightly coloured male ducks selectively shot by hunters? Canadian Journal of Zoology 69:279–282.

Mittelhauser, G. H. 2008. Harlequin Ducks in the eastern United States. Waterbirds 31:58–66.

Moore, M. T., K. D. Richkus, P. I. Padding, E. M. Martin, S. S. Williams, and H. L. Spriggs. 2007. Migratory bird hunting activity and harvest during the 2001 and 2002 hunting seasons: final report. U.S. Fish and Wildlife Service, Washington, DC.

Moore, M. T., S. L. Sheriff, and D. T. Cobb. 2002. Do the Harvest Information Program screening questions provide the information necessary to stratify the survey as envisioned? Pp. 17–30 in J. M. VerSteeg and R. C. Elden (compilers), Harvest Information Program: evaluation and recommendations. International Association of Fish and Wildlife Agencies, Migratory Shore and Upland Game Bird Working Group, Ad Hoc Committee on HIP, Washington, DC.

Mosbech, A., H. G. Gilchrist, F. R. Merkel, C. Sonne, and A Flagstad. 2006. Year-round movements of Northern Common Eider *Somateria mollissima borealis* breeding in Arctic Canada and West Greenland followed by satellite telemetry. Ardea 94:651–665.

Munro, J. A., and W. A. Clemens. 1932. Food of the American Merganser (*Mergus merganser americanus*) in British Columbia: a preliminary paper. Canadian Field-Naturalist 46:166–168.

Munro, J. A., and W. A. Clemens. 1937. The American Merganser in British Columbia and its relation to the fish population. Bulletin 55. Biological Board of Canada, Ottawa, ON.

Munro, J. A., and W. A. Clemens. 1939. The food and feeding habits of the Red-breasted Merganser in British Columbia. Journal of Wildlife Management 3:46–53.

Natcher, D., L. Felt, K. Chaulk, A. Procter, and Nunatsiavut Government. 2011. Monitoring the domestic harvest of migratory birds in Nunatsiavut Labrador. Arctic 64:362–366.

Natcher, D., L. Felt, K. Chaulk, and A. Procter. 2012. The harvest and management of migratory bird eggs by Inuit in Nunatsiavut, Labrador. Environmental Management 50:1047–1056.

Naves, L. C. 2010. Alaska migratory bird subsistence harvest estimates, 2004–2007, Alaska Migratory Bird Co-Management Council. Technical Paper No. 349. Alaska Department of Fish and Game, Division of Subsistence, Anchorage, AK.

Naves, L. C. 2012. Alaska migratory bird subsistence harvest estimates, 2010, Alaska Migratory Bird Co-Management Council. Technical Paper No. 376. Alaska Department of Fish and Game, Division of Subsistence, Anchorage, AK.

Naves, L. C., D. Koster, M. G. See, B. Easley, and L. Olson. 2008. Alaska Migratory Bird Co-Management Council migratory bird subsistence harvest survey: assessment of the survey methods and implementation. Special Publication SP2008–005. Alaska Department of Fish and Game, Division of Subsistence, Anchorage, AK.

Naves, L. C., and T. K. Zeller. 2013. Saint Lawrence Island subsistence harvest of birds and eggs, 2011–2012, addressing Yellow-billed Loon conservation concerns, Alaska Migratory Bird Co-Management Council. Technical Paper No. 384. Alaska Department of Fish and Game, Division of Subsistence, Anchorage, AK.

Padding, P. I., J.-F. Gobeil, and C. Wentworth. 2006b. Estimating waterfowl harvest in North America. Pp. 849–852 in G. C. Boere, C. A. Galbraith, and D. A. Stroud (editors), Waterbirds around the world. The Stationery Office, Edinburgh, U.K.

Padding, P. I., M. T. Moore, K. D. Richkus, E. M. Martin, S. S. Williams, and H. L. Spriggs. 2006a. Migratory bird hunting activity and harvest during the 1999 and 2000 hunting seasons: final report. U.S. Fish and Wildlife Service, Washington, DC.

Padding, P. I., and J. A. Royle. 2012. Assessment of bias in US waterfowl harvest estimates. Wildlife Research 39:336–342.

Paige, A. W., and R. J. Wolfe. 1997. The subsistence harvest of migratory birds in Alaska: compendium and 1995 update. Technical Paper No. 228. Alaska Department of Fish and Game, Division of Subsistence, Juneau, AK.

Paige A. W., and R. Wolfe. 1998. The subsistence harvest of migratory birds in Alaska, 1996 update. Draft report for U.S. Fish and Wildlife Service. Alaska Department of Fish and Game, Division of Subsistence, Juneau, AK.

Pamplin Jr., W. L. 1986. Cooperative efforts to halt population declines of geese nesting on Alaska's Yukon-Kuskokwim Delta. Transactions of the North American Wildlife and Natural Resources Conference 51:487–506.

Pearce, J. M., J. M. Eadie, J.-P. L. Savard, T. K. Christensen, J. Berdeen, E. J. Taylor, S. Boyd, A. Einarsson, and S. L. Talbot. 2014. Comparative population structure of cavity-nesting sea ducks. Auk 131:195–207.

Perry, M. C., and P. H. Geissler. 1980. Incidence of embedded shot in Canvasbacks. Journal of Wildlife Management 44:888–894.

Peterson, S. R., and R. S. Ellarson. 1975. Incidence of body shot in Lake Michigan Oldsquaws. Journal of Wildlife Management 39:217–219.

Phillips, J. C. 1926. A natural history of the ducks (vols. 1–4). Houghton Mifflin, Boston, MA. Reprinted (1986) as Volumes 1–2, Dover Publications, Mineola, New York, NY.

Piatt, J. F., and R. G. Ford. 1996. How many seabirds were killed by the Exxon Valdez oil spill? Pp. 712–719 in S. D. Rice, R. B. Spies, D. A. Wolfe, and B. A. Wright (editors), Proceedings: 1993 Exxon Valdez Oil Spill Symposium. American Fisheries Society Symposium 18, Bethesda, MD.

Piatt, J. F., C. J. Lensink, W. Butler, M. Kendziorek, and D. R. Nysewander. 1990. Immediate impact of the Exxon Valdez oil spill on marine birds. Auk 107:387–397.

Priest, H., and P. J. Usher. 2004. The Nunavut wildlife harvest study. Nunavut Wildlife Management Board, Iqaluit, NU.

Reed, A. 1986. Eiderdown harvesting and other uses of Common Eiders in spring in summer. Pp. 138–146 in A. Reed (editor), Eider ducks in Canada. Canadian Wildlife Service Report Series No. 47. Canadian Wildlife Service, Ottawa, ON.

Reed, A., and A. J. Erskine. 1986. Populations of the Common Eider in eastern North America: their size and status. Pp. 156–162 in A. Reed (editor), Eider ducks in Canada. Canadian Wildlife Service Report Series 47. Canadian Wildlife Service, Ottawa, ON.

Reynolds, J. H. 2007. Investigating the impact of sampling effort on annual migratory bird subsistence harvest survey estimates. Solutions Statistical Consulting report for U.S. Fish and Wildlife Service Division of Migratory Bird Management Order No. 701812M816. U.S. Fish and Wildlife Service, Anchorage, AK.

Robert, M., R. Benoit, and J.-P. L Savard. 2000. COSEWIC status report on the Barrow's Goldeneye Bucephala islandica in Canada. Committee on the Status of Endangered Wildlife in Canada, Ottawa, ON.

Robertson, G. J., and H. G. Gilchrist. 1998. Evidence for population declines among Common Eiders breeding in the Belcher Islands, Northwest Territories. Arctic 51:378–385.

Robertson, G. J., and R. I. Goudie. 1999. Harlequin Duck (Histrionicus histrionicus). In A. Poole and F. Gill (editors), The birds of North America, no. 466. The Academy of Natural Sciences, Philadelphia, PA.

Runge, M. C., F. A. Johnson, J. A. Dubovsky, W. L. Kendall, J. Lawrence, and J. Gammonley. 2002. A revised protocol for the adaptive harvest management of mid-continent Mallards. U.S. Fish and Wildlife Service, Washington, DC.

Runge, M. C., W. L. Kendall, and J. D. Nichols. 2004. Exploitation. Pp. 303–328 in W. J. Sutherland, I. Newton, and R. E. Green (editors), Bird ecology and conservation: a handbook of techniques. Oxford University Press, Oxford, U.K.

Sable, T., G. Howell, D. Wilson, and P. Penashue. 2006. Ashkui Project; linking western science and indigenous knowledge. Pp. 109–127 in P. Sillitoe (editor), Local science vs. global knowledge: approaches to indigenous knowledge in international development. Berghahn Books, New York, NY.

Salyer II, J. C., and K. F. Lagler. 1940. The food and habits of the American Merganser during winter in Michigan, considered in relation to fish management. Journal of Wildlife Management 4:186–219.

Savard, J.-P. L. 1987. Status report on Barrow's Goldeneye. Technical Report Series No. 23. Canadian Wildlife Service, Pacific and Yukon Region, Delta, BC.

Savard, J. P. L., and P. Dupuis. 1999. A case for concern? The eastern population of Barrow's Goldeneyes (Bucephala islandica). Pp. 65–76 in R. I. Goudie, M. R. Petersen, and G. J. Robertson (editors), Behavior and ecology of sea ducks. Occasional Paper 100, Canadian Wildlife Service Ottawa, ON.

Savard, J. P. L., and J. M. Eadie. 1989. Survival and breeding philopatry in Barrow's and Common Goldeneyes. Condor 91:198–203.

Schamber, J. L., P. L. Flint, J. B. Grand, H. M. Wilson, and J. A. Morse. 2009. Population dynamics of Long-tailed Ducks breeding on the Yukon-Kuskokwim Delta, Alaska. Arctic 62:190–200.

Schulz, J. H., P. I. Padding, and J. J. Millspaugh. 2006. Will Mourning Dove crippling rates increase with nontoxic-shot regulations? Wildlife Society Bulletin 34:861–865.

Sea Duck Joint Venture. 2013. Sea Duck Joint Venture Implementation Plan for April 2013 through March 2016. Report of the Sea Duck Joint Venture. U.S. Fish and Wildlife Service, Anchorage, AK.

Sea Duck Joint Venture Management Board. 2008. Sea Duck Joint Venture Strategic Plan 2008–2012. U.S. Fish and Wildlife Service, Anchorage, AK.

Sheriff, S. L., J. H. Schulz, B. D. Bales, M. T. Moore, P. I. Padding, and D. A. Shipes. 2002. The current reliability of Harvest Information Program surveys. Pp. 51–68 in J. M. VerSteeg and R. C. Elden (compilers), Harvest Information Program: evaluation and recommendations. International Association of Fish and Wildlife Agencies, Migratory Shore and Upland Game Bird Working Group, Ad Hoc Committee on HIP, Washington, DC.

Sherwood, M. 1977. The great duck egg fake. Alaska Journal 7:88–94.

Silverman, E. D., D. T. Saalfeld, J. B. Leirness, and M. D. Koneff. 2013. Wintering sea duck distribution along the Atlantic coast of the United States. Journal of Fish and Wildlife Management 4:178–198.

Smith, W. N. 1985. Marsh tales: market hunting, duck trapping, and gunning. Tidewater Publishers, Centreville, MD.

Sonsthagen, S. A., S. L. Talbot, K. T. Scribner, and K. G. McCracken. 2011. Multilocus phylogeography and population structure of Common Eiders breeding in North America and Scandinavia. Journal of Biogeography 38:1368–1380.

Suydam, R. S., D. L. Dickson, J. B. Fadely, and L. T. Quakenbush. 2000. Population declines of King and Common Eiders of the Beaufort Sea. Condor 102:219–222.

Tautin, J., S. M. Carney, and J. B. Bortner. 1989. A national migratory gamebird harvest survey: a continuing need. Transactions of the North American Wildlife and Natural Resources Conference 54:545–551.

Thomas, P. W. 2008. Harlequin Ducks in Newfoundland. Waterbirds 31:44–49.

Tobias, T. N., and J. J. Kay. 1994. The bush harvest in Pinehouse, Saskatchewan, Canada. Arctic 47:207–221.

U.S. Census Bureau. 2011. Profiles of general demographic characteristics, Alaska: 2010. U.S. Department of Commerce, Washington, DC.

U.S. Fish and Wildlife Service. 1993. Endangered and threatened wildlife and plants; threatened status for the Spectacled Eider: final rule. Federal Register 58:24474–27480.

U.S. Fish and Wildlife Service. 1997. Endangered and threatened wildlife and plants; threatened status for the Alaska breeding population of the Steller's Eider: final rule. Federal Register 62:31748–31757.

U.S. Fish and Wildlife Service. 2014. Harvest Information Program (HIP) migratory bird harvest report series. http://www.fws.gov/migratorybirds/NewReportsPublications/HIP/hip.htm (13 January 2014).

Van Dyke, F. 1981. Mortality in crippled Mallards. Journal of Wildlife Management 45:444–453.

Vickery, P. D. 1988. Distribution and population status of Harlequin Ducks (Histrionicus histrionicus) wintering in eastern North America. Wilson Bulletin 100:119–126.

Wade, P. R. 1998. Calculating limits to the allowable human-caused mortality of cetaceans and pinnipeds. Marine Mammal Science 14:1–37.

Walsh, H. M. 1971. The outlaw gunner. Tidewater Publishers, Centreville, MD.

Wendt, J. S., and E. Sileff. 1986. The kill of eiders and other sea ducks in eastern Canada. Pp. 147–154 in A. Reed (editor), Eider ducks in Canada. Canadian Wildlife Service Report Series 47. Canadian Wildlife Service, Ottawa, ON.

White, H. C. 1957. Food and natural history of mergansers on salmon waters in the Maritime Provinces of Canada. Bulletin of the Fisheries Research Board of Canada 116, Ottawa, ON.

Wilson, H. M., P. L. Flint, A. N. Powell, J. B. Grand, and T. L. Moran. 2012. Population ecology of breeding Pacific Common Eiders on the Yukon-Kuskokwim Delta, Alaska. Wildlife Monographs 182:1–28.

Wolfe, R. J., A. W. Paige, and C. L. Scott. 1990. The subsistence harvest of migratory birds in Alaska. Technical Paper No. 197. Alaska Department of Fish and Game, Division of Subsistence, Juneau, AK.

Wolfe, R. J., and A. W. Paige. 1995. The subsistence harvest of Black Brant, Emperor Geese, and Eider Ducks in Alaska. Technical Paper No. 234. Alaska Department of Fish and Game, Division of Subsistence, Juneau, AK.

Wolfe, R. J., and R. J. Walker. 1987. Subsistence economies in Alaska: productivity, geography, and development impacts. Arctic Anthropology 24:56–81.

Woodby, D. A., and G. J. Divoky. 1982. Spring migration of eiders and other waterbirds at Point Barrow, Alaska. Arctic 35:403–410.

Zavaleta, E. 1999. The emergence of waterfowl conservation among Yup'ik hunters in the Yukon-Kuskokwim Delta, Alaska. Human Ecology 27:231–266.

CHAPTER THIRTEEN

Habitats of North American Sea Ducks*

Dirk V. Derksen, Margaret R. Petersen, and Jean-Pierre L. Savard

Abstract. Breeding, molting, fall and spring staging, and wintering habitats of the sea duck tribe Mergini are described based on geographic locations and distribution in North America, geomorphology, vegetation and soil types, and freshwater and marine characteristics. The dynamics of habitats are discussed in light of natural and anthropogenic events that shape areas important to sea ducks. Strategies for sea duck habitat management are outlined and recommendations for international collaboration to preserve key terrestrial and aquatic habitats are advanced.

Key Words: anthropogenic effects, brood rearing, climatic conditions, competition with fish, diet, food resources, freshwater, marine, molt migration, nest sites, spring staging, stopover sites, winter habitat.

We follow the definition of habitat advanced by Odum (1971), which is the place or space where an organism lives. Weller (1999) emphasized that habitats for waterbirds require presence of sufficient resources such as food, water, cover, and space for different stages of their annual cycle. Habitats exploited by North American sea ducks are diverse and widespread across the continent and adjacent marine waters, and until recently, most were only superficially known. A 15-year-long effort funded research on sea duck habitats through the Sea Duck Joint Venture and the Endangered or Threatened Species programs of the United States and Canada. Nevertheless, important gaps still remain in our understanding of key elements required by some species during various life stages. Many significant habitats, especially staging and wintering sites, have been and continue to be destroyed or altered by anthropogenic activities. The goal of this chapter is to develop a comprehensive summary of marine, freshwater, and terrestrial habitats and their characteristics by considering sea duck species with similar needs as groups within tribe Mergini. Additionally, we examine threats and changes to sea duck habitats from human-caused and natural events. Last, we evaluate conservation and management programs underway or available for maintenance and enhancement of habitats critical for sea ducks.

HABITATS OF EIDERS

Eiders (*Polysticta* and *Somateria*) are the most marine of the North American sea ducks. The birds nest at high latitudes and are often found in polynyas, areas of the pack ice that remain open in winter. In many remote, inaccessible regions, the molting,

* Derksen, D. V., M. R. Petersen, and J.-P. L. Savard. 2015. Habitats of North American sea ducks. Pp. 469–527 in J.-P. L. Savard, D. V. Derksen, D. Esler, and J. M. Eadie (editors). Ecology and conservation of North American sea ducks. Studies in Avian Biology (no. 46), CRC Press, Boca Raton, FL.

staging, and wintering areas of eiders were first located by the use of satellite telemetry (Petersen et al. 1995, 1999; Petersen and Flint 2002; Phillips et al. 2006) and then described with environmental data obtained from satellite imagery, remote sensors, and other sources. Remotely generated characteristics of the environment such as ice extent, temperature, chlorophyll, salinity, sea surface temperature (SST), and water depth can provide insight into the habitat of a smaller region such as a bay or island group or even at broader geographical scales of the Bering Sea, Chukchi Sea, Beaufort Sea, or the Atlantic coast of North America (Petersen and Douglas 2004; Mosbech et al. 2006a; Phillips et al. 2006; Oppel et al. 2009a,b; Zipkin et al. 2010). Analyses from remote sensing complement habitat descriptions from land-based studies that provide a much more detailed analysis of areas used by birds (Goudie et al. 2000, Petersen et al. 2000, Suydam 2000, Fredrickson 2001). The following habitat descriptions consider both regional and local attributes.

Breeding Habitats

Eiders nest in arctic and subarctic wetlands, where the birds are often found on the shores and islands of ponds and lakes. Vegetation in these tundra habitats consists of short grasses, sedges, dwarf alders (Alnus crispa), and willows (Salix spp.). Eiders tend to be spatially segregated during the breeding season, although there is much overlap in distribution and variability among regions. On the Arctic Coastal Plain, King Eiders are generally located further inland than Spectacled Eiders (Suydam 2000; W. W. Larned et al., unpubl. report). Among the four eider species, the North American distribution of Steller's Eider populations is the most restricted. Subtle differences in nesting distribution and brood-rearing habitat separate Steller's, Spectacled, and King Eiders (N. A. Rojek, unpubl. report; D. E. Safine, unpubl. reports). These eiders use features of arctic and subarctic habitats that are much different than those used by Common Eiders.

In North America and elsewhere, the presence or absence of avian and mammalian predators and alternative prey are key features of breeding habitats of eiders (Goudie et al. 2000, Petersen et al. 2000, Suydam 2000, Fredrickson 2001). Quakenbush and Suydam (1999) and Quakenbush et al. (2004) suggest that Steller's Eiders nest primarily in years

with abundant lemmings; Chaulk et al. (2007) and others concluded that Common Eiders may delay nest initiation or switch nesting islands when mammalian predators such as red fox (Vulpes vulpes) or arctic fox (V. lagopus) are present. Eiders that nest in association with pugnacious territorial birds, such as geese, gulls, terns, jaegers, and Snowy Owls, may gain protection from avian as well as mammalian predators (Schamel 1977, van Dijk 1986, Blomqvist and Elander 1988, Robertson 1995, Kellett and Alisauskas 1997, Pearce et al. 1998, Quakenbush et al. 2004). However, Bentzen et al. (2009) found little evidence that King Eiders selected nest sites in gull territories, and Kellett et al. (2003) found no increase in nest survival when King Eiders nested near gulls or terns. Little evidence indicates that nest success of Common Eiders increases (and may decrease) when birds nest in mixed colonies or adjacent to gull colonies. Gulls often take Common Eider eggs from nests or depredate ducklings when females leave the nesting colony with newly hatched young (Bourget 1973, McAloney 1973, Schmutz et al. 1983, Gilliland 1990).

Steller's Eider

The Pacific population of Steller's Eiders nesting in North America is now almost exclusively found in the tundra wetlands of the Arctic Coastal Plain near Barrow, Alaska (Kertell 1991, Quakenbush et al. 2004), and with only occasional nests found in western Alaska (Flint and Herzog 1999). The following descriptions of habitats and habitat use on the Arctic Coastal Plain are from the 21-year period, 1991–2011 (Quakenbush et al. 2004; L. T. Quakenbush et al., unpubl. report; N. A. Rojek, unpubl. report; D. E. Safine, unpubl. reports). Habitat use in Russia is summarized by Fredrickson (2001) and differs little from the breeding grounds at Barrow. In early spring, Steller's Eiders visit wetlands such as flooded low-centered polygons, moats around the perimeter of lake basins, and water-filled ice-wedge troughs, as well as adjacent upland tundra. Features of these wetlands include tundra flooded from snow melt and shallow and medium depth ponds with emergent stands of pendent grass (Arctophila fulva) and sedges (Carex aquatilis). Nests and broods of Steller's Eiders, however, were found almost exclusively near permanent, shallow ponds dominated by A. fulva and C. aquatilis, although some birds used deeper ponds and flooded tundra and were near streams.

Spectacled Eider

In North America, breeding Spectacled Eiders are restricted to broad, coastal wetlands in tundra habitats of northern and western Alaska. Breeding habitats include delta islands covered primarily by grasses and arctic and subarctic wetlands with low vegetation, which is underlain with continuous or discontinuous permafrost (Petersen et al. 2000). In northern Alaska, Anderson et al. (1999) found nests of Spectacled Eiders on shores and islands of ponds, on river delta islands, and on dry spots in wetlands with periodic incursion of marine waters, nonpatterned wet meadows, and low- and high-relief moist meadows. Common features of breeding habitats of Spectacled Eiders on the North Slope included large, wind-oriented thaw lakes, pingos, and polygon ridges (Anderson et al. 1999, Petersen et al. 2000, Bart and Earnst 2005). In western Alaska, the breeding habitat of Spectacled Eiders in the vegetated intertidal zone is a combination of coastal salt marshes that grade into higher freshwater wetlands and thaw lakes; the area is inundated by seawater with periodic storm-driven surge tides (King and Dau 1981). Nest sites of Spectacled Eiders in both western and northern Alaska are located primarily near ponds and lakes and on shorelines, peninsulas, and islands (Dau 1974, Anderson et al. 1999, Petersen et al. 2000, Bart and Earnst 2005).

In northern Alaska, female Spectacled Eiders raise broods in deep, freshwater ponds that are often fringed with *A. fulva* and on shallow ponds with *Carex* spp. (D. E. Safine, unpubl. report), and in western Alaska on freshwater and brackish tundra ponds (Flint et al. 2000a, 2006). In western Alaska, habitat characteristics of areas used by Spectacled Eider broods include ponds in dwarf shrub–graminoid meadow, high-graminoid meadow, high-sedge meadow, mixed high-graminoid meadow and dwarf shrub upland, and low-sedge meadow (Flint et al. 2006). At Hock Slough on the Yukon–Kuskokwim Delta, young broods made extensive movements up to 8 km from their nests to other wetlands, but little movement was detected at a study site on Kigigak Island (Flint et al. 2006). On the North Slope near Barrow, Spectacled Eider broods moved <2.5 km during the brood-rearing period (D. E. Safine, unpubl. report).

King Eider

King Eiders nest primarily in northern arctic ecosystems in freshwater habitats and raise their young in both fresh and marine waters. Nests of King Eiders are frequently found in polygonal wetlands, grass and sedge meadows, and closed graminoid meadows (Barry 1986). King Eiders arrive on their breeding grounds before nesting sites are available and feed and roost in meltwater ponds before laying eggs (Barry 1986, Holcroft-Weerstra and Dickson 1997, Suydam 2000, Oppel et al. 2011). Characteristics of nest sites are similar to those of Steller's and Spectacled Eiders, although King Eiders will nest in more upland tundra habitats in the vicinity of lakes and ponds (Abraham and Finney 1986, Dickson et al. 1997, Suydam 2000). Detailed analyses of nest site characteristics of King Eiders suggested that on the North Slope of Alaska, female King Eiders selected sites near water, on islands, and in areas with high willow cover (Bentzen et al. 2009). At Karrak Lake, Nunavut, Canada, King Eider nest success improved when nests were placed on isolated islands (Kellett et al. 2003). In general, King Eider ducklings are raised in large, deeper, freshwater ponds near nesting areas and often move to marine habitats before fledging (Barry 1986, Suydam 2000).

Common Eider

The common features of nesting areas for all subspecies of Common Eiders include suitable nest sites on islands or other coastal regions with protection from predators and access to marine waters. Colonies of Common Eiders have been found on low barrier islands, islands in bays and estuaries, and coastal marine islands, as well as mainland areas (Prach et al. 1986, Goudie et al. 2000). In mainland areas, Common Eiders are associated with brackish ponds within a few km of the Bering Sea coast (Flint et al. 1998); in the Canadian Arctic, Common Eiders were found on the mainland within 12 km of the coast and up to 46 km inland (Cornish and Dickson 1997) and on islands in freshwater lakes (Chapdelaine et al. 1986, Nakashima 1986, Prach et al. 1986, Cornish and Dickson 1997, Falardeau et al. 2003).

Nest site selection by Common Eiders on islands includes sand, gravel, and boulders; sites next to logs and other debris deposited during extreme storms; sites in or next to small clumps of beach rye (Elymus mollis) or other plants; open areas with grasses, sedges, mosses, and other low vegetation; sites resembling a burrow (but not subterranean) that are completely surrounded and covered by vegetation; sites under bushes; and forested clumps (Schamel 1977, Schmutz et al. 1983, Barry 1986, van Dijk 1986, Reed et al. 1996, Cornish and Dickson 1997, Goudie et al. 2000, The Joint Working Group on the Management of the Common Eider 2004, Noel et al. 2005b). In the Aleutian Archipelago and Alaska Peninsula, Common Eiders nest at a large variety of sites on fox-free islands; these nest locations include steep, Elymus-covered slopes, old midden sites; debris of derelict buildings; sandy, gravel, and rocky beaches; cliff faces and stacks; and islands and shores of lakes (Murie 1959). Few Common Eider nests were found on islands with foxes, and the nests that were present were primarily on ledges of cliff faces, stacks, rocky islets, and islands on lakes (Murie 1959). In the high arctic when on islands between Ellesmere and Devon islands, Nunavut, Canada, Common Eiders nested on old beach ridges up to 200 m from shore, cliff ledges, bases of cliffs or near large rocks, tops of cliffs or boulders, man-made structures, talus slopes, caves or pockets in rock massives, and boulder fields (Prach et al. 1986). Common Eiders readily use man-made structures designed to provide protection of nests from predators (The Joint Working Group on the Management of the Common Eider 2004).

Islands with colonies of Common Eiders generally have no resident mammalian predators; however, polar bears (Ursus maritimus) and foxes will destroy most or all nests if they obtain access to these islands (Murie 1959, Petersen 1982, Quinlan and Lehnhausen 1982, Nakashima 1986, Noel et al. 2005b). Common Eiders will delay or refrain from nesting if ice bridges occur between the mainland and nesting islands; ice may be an important habitat feature of islands, particularly if it provides predators with access to nests (Ahlén and Andersson 1970, Schamel 1977, Chaulk et al. 2007). Chaulk and Mahoney (2012) found that in years with increased ice surrounding islands, nest initiation date of Common Eiders was delayed.

After hatching, female Common Eiders move their broods from nesting sites on islands to marine waters or from terrestrial habitats to the sea via freshwater lakes, streams, and rivers. Flint

et al. (1998) found that most females moved their broods to saltwater by 15 days of age. Unlike other eiders, Common Eiders rear their broods exclusively in coastal marine habitats but may be found in freshwater or brackish water ponds early in the brood-rearing period. Broods remain in shallow marine waters when the ducklings are young, then move to deeper waters that are a few meters in depth as the young mature (McAloney 1973, Öst and Kilpi 1999). In the waters along the coast in northeastern James Bay, foraging habitats used by broods of Common Eiders included open water areas, over boulder-strewn tidal flats, and eelgrass beds (Reed et al. 1996). Blinn et al. (2008) found that foraging areas with a gradual slope supported more Common Eider ducklings than areas with a steep slope, and rockweed (Ascophyllum nodosum) was important in predicting duckling density in the Bay of Fundy, New Brunswick, Canada.

Invertebrates eaten by Common Eider ducklings at the St. Lawrence River estuary include Littorina spp., amphipods, and insects, with increasing amounts of Littorina as the ducklings aged (Cantin et al. 1974). Somateria mollissima dresseri ducklings in Nova Scotia ate similar types of molluscs with Lacuna vincta (65%), Mytilus edulis (19%), and Littorina spp. (9%) as the main prey species with comparable variation among habitats (McAloney 1973). The prey size of Common Eider ducklings increased as ducklings grew and foraged in deeper water (McAloney 1973). In the St. Lawrence River estuary and elsewhere, feeding activity of Common Eider broods varied with the tide and feeding occurred when food was most available (McAloney 1973, Cantin et al. 1974, Minot 1980). In New Brunswick, Minot (1980) also reported that broods of Common Eiders foraged early and late in the day.

Remigial Molting Habitats

Waterfowl replace flight feathers (remiges) with simultaneous molt and become flightless in remote, inaccessible locations in marine habitats (Chapter 9, this volume). Studies at a few sites in North America and Greenland have included direct observations of birds during remigial molt (Petersen 1980, 1981; Frimer 1993, 1994a,b, 1995a,b, 1997). Habitat characteristics based on data gathered remotely have provided insight into key areas used during remigial molt (Phillips and Powell 2006, Phillips et al. 2006). With the exception of a few female King Eiders (Knoche 2004), eiders undergo remigial molt exclusively in shallow, marine habitats. In general, female eiders migrate to the same areas used by males of the same species (Petersen 1980, 1981; Petersen et al. 1999; Flint et al. 2000b; Phillips et al. 2006; Chapter 9, this volume).

Steller's Eider

In North America, habitats used during the flightless period of the Pacific population of Steller's Eiders consist of shallow waters a few meters or less in depth (at low tide) in estuaries and lagoons and low barrier islands and shoals. Detailed studies of Steller's Eiders during the flightless period have only been conducted along the north side of the Alaska Peninsula at Izembek and Nelson Lagoons. Both areas are shallow estuarine systems but differ substantially in habitat structure. Nelson Lagoon contains sand flats and gravel areas in which mussels, clams, amphipods, and polychaete worms are taken by Steller's Eiders, although mussels are consumed more when birds are flightless (Petersen 1981). A major feature of Izembek Lagoon is the vast expanse of eelgrass (*Zostera marina*). The community of invertebrates associated with eelgrass beds consists of many species of marine worms, amphipods, snails, clams, and other small marine life (over 30 taxa); *Macoma balthica* and polychaetes dominate the sand flats. Crustacea (Amphipods), bivalves (*M. balthica* and *Turtonia minuta*), gastropods, and polychaetes are common prey items of Steller's Eiders in Izembek Lagoon (Metzner 1993). Failed and successful breeding females, males, and subadults molt remiges in the same lagoons, although not necessarily at exactly the same locations (Flint et al. 2000b, but see Petersen 1981), and differ in the timing of remigial molt (Petersen 1980, 1981; Flint et al. 2000b). Petersen (1981) detected no differences in habitat types used by different sexes or ages during the remigial molt. The benthic characteristics of areas with flightless Steller's Eiders elsewhere have not been described.

Spectacled Eider

In general, areas of Spectacled Eiders where birds are flightless are in waters <20 m in depth in coastal regions of western and northern Alaska, the Chukotka Peninsula in western Russia, and offshore between the deltas of the Indigirka and Kolyma rivers in the Siberian Sea (Petersen et al. 1999). Spectacled Eiders were found at an average of 7.1 km (both sexes) offshore at Mechigmenski Bay, Russia; 16.9 km (males) offshore between the Indigirka and Kolyma River deltas, Russia; 34.5 km (males) and 37.9 km (females) offshore of Ledyard Bay, Alaska; and 26.7 km (females) from the land in Norton Sound, Alaska (Petersen et al. 1999). Both sexes of Spectacled Eiders undergo remigial molt in the same locations but temporal segregation occurs because males and subadults are flightless before adult females (Petersen et al. 1999). Unlike other eiders, there is some evidence of spatial segregation of breeding female Spectacled Eiders; all marked breeding females from the Yukon–Kuskokwim Delta undergo remigial molt in eastern Norton Sound (no marked males moved there), and a large proportion of nesting females from the North Slope used Ledyard Bay during the flightless period (Petersen et al. 1999).

King Eider

Phillips et al. (2006) found that models including distance to shore, salinity, chlorophyll, interactions between chlorophyll and salinity, and water depth best explained variation in habitat characteristics of King Eider molting areas from random sites along the Alaska coast south to Bristol Bay and the eastern Siberian coasts to the Kamchatka Peninsula. Schamber et al. (2010b) found that King Eiders in Bristol Bay, Alaska, used sites farther from shore and in deeper waters than Black Scoters. Male King Eiders showed some degree of site fidelity to the region where they molted during the previous year (Phillips and Powell 2006, Knoche et al. 2007), which suggests little change in the types of habitats used by individuals among years. At Disko Island in west Greenland, molting King Eiders feed on benthic invertebrates (primarily molluscs, Frimer 1995a, 1997) in waters 15–25 m deep and over soft substrates (Frimer 1995a,b; 1997). Based on aerial survey data over a broad region along the west Greenland coast, Mosbech and Boertmann (1999) reported the highest mean densities of flightless King Eiders in coastal waters with unconsolidated, fine-grained sediments. Molting areas of King Eiders in North America and west Greenland are used by both sexes (Frimer 1994a,b, 1995a,b, 1997; Phillips et al. 2006). A few female King Eiders molt remiges at the breeding grounds while attending broods (Knoche 2004), thus the molting habitat of this segment of the population is identical to sites used during brood rearing.

Common Eider

Common Eiders undergo molt in shallow waters close to shore. Characteristics of molting habitats of Common Eiders in North America are described only from locations in the eastern Canadian Arctic and to the east along the Atlantic coastal region. Although birds may roost on exposed rocks in the intertidal zone, in James Bay, male and female Common Eiders occur primarily in open water shoal areas during the flightless period (Reed et al. 1996). In the eastern Canadian Arctic and the St. Lawrence River estuary, both sexes of flightless Common Eiders may occur in the same general area although frequently in separate flocks (Abraham and Finney 1986, Savard et al. 1999). Although present at the same time, brood flocks remain separate from flightless adult and subadult Common Eiders (Diéval et al. 2011). In west Greenland, Frimer (1995b) found molting *Somateria mollissima borealis* foraging in waters <15 m in depth and with a hard bottom. In the St. Lawrence Estuary and Gulf, molting male Common Eiders were associated with mussel beds, molting females without young were associated with mussels and gammarids, and females with young at sites with periwinkles (*Littorina littorea*) (Diéval et al. 2011).

Female Common Eiders (*S. m. borealis*) molted remiges within 50 km of their nesting area at Ungava Bay, Canada, (Savard et al. 2011), and <100 km (10 of 18 females) from breeding sites in eastern Canada and west Greenland (Mosbech et al. 2006b). In the St. Lawrence Estuary and Gulf, some females molted flight feathers close to their breeding islands, whereas others molted remiges hundreds of kilometers away from nesting areas (J.-P. L. Savard, unpubl. data). Barry (1986) reported that female Common Eiders (*Somateria mollissima v-nigrum*) remained near the breeding area to molt remiges. Based on movement data from transmitters, M. R. Petersen (unpubl. data) found female eiders (*S. m. v-nigrum*) that were likely flightless

used sites within 25 km of their nest locations at five study areas in Alaska. Data from male Common Eiders marked with satellite transmitters is sparse, but one male *S. m. v-nigrum* may have undergone remigial molt >1,000 km from his nesting region (M. R. Petersen, unpubl. data). Mosbech et al. (2006b) found that all of their sample of male *S. m. borealis* marked with satellite transmitters molted remiges >100 km from their nesting area. Based on aerial surveys, Abraham and Finney (1986) found flightless *Somateria mollissima sedentaria* males in the general vicinity of nesting female Common Eiders.

Fall and Spring Staging Habitats

Staging or stopover locations of eiders are in marine habitats, although the places used by a particular species or population may vary among years and between seasons. Ice is a major factor influencing the location, timing of movement, and residence time of eiders during the stopover period, especially in spring (McLaren and McLaren 1982; Barry 1986; W. W. Larned, unpubl. report). During molt migration (Chapter 8, this volume), long-distance migrants often stage at locations along their route between breeding and molting grounds. Male eiders of all species stage in marine habitats during the period after they leave the breeding grounds until they begin remigial molt (Petersen et al. 1999, 2006; Phillips et al. 2007; D. Rosenberg, unpubl. data).

Steller's Eider

Spring staging areas used by the Pacific population of Steller's Eiders are known from aerial surveys in coastal regions on the north side of the Alaska Peninsula and the southeast portion of the Bering Sea (W. W. Larned, unpubl. report). When free of ice, W. W. Larned (unpubl. report) found large numbers of Steller's Eiders in the shallow lagoons and bays along the north side of the Alaska Peninsula; Steller's Eiders moved north as the ice disappeared and migrated to and staged in the shallow flats and shoals of Kuskokwim Bay. In Alaska, stopover locations are shallow with large areas exposed at low tide. Staging areas of Steller's Eiders in Alaska and Russia during spring migration are also visited by Steller's Eiders during the molt and fall migrations (Petersen 1980, 1981; Metzner 1993; U.S. Fish and Wildlife Service

2002; D. Rosenberg, unpubl. data). Steller's Eiders staged during the molt and fall migrations in shallow, protected waters in Vankarem Lagoon, Kolyuchin Bay, and elsewhere along the northeast coast of the Chukotka Peninsula (A. A. Kochnev et al., unpubl. report).

Spectacled Eider

The stopover locations and habitats used by Spectacled Eiders in spring are unknown. However, Dau and Kistchinski (1977) reviewed previous reports of Spectacled Eiders in spring and, based on those data combined with their own observations, concluded that the availability of open water and ice conditions in the northern Bering Sea may be important factors influencing the time of arrival of paired birds on their nesting grounds. The authors also concluded that changing ice conditions also may dictate the distribution and abundance of staging Spectacled Eiders. This hypothesis is consistent with other species of eiders in regions dominated by sea ice.

During molt migration from the three breeding areas in western Alaska, northern Alaska, and arctic Russia, Spectacled Eiders used marine coastal waters and occasionally large, freshwater lakes (Petersen et al. 1999; D. Troy, unpubl. report). In the Beaufort Sea of Alaska, a few male and most female Spectacled Eiders staged in marine coastal waters during molt migration; these birds were associated with open water at mouths of rivers and other ice-free areas (Petersen et al. 1999; D. Troy, unpubl. report). In the Chukchi Sea, Bering Sea, and Russia, major staging areas used by Spectacled Eiders during molt and fall migrations are also key molting areas (Petersen et al. 1999). As with Steller's Eiders, A. A. Kochnev et al. (unpubl. report) observed Spectacled Eiders staging during molt migration and in fall in shallow, protected waters in Vankarem Lagoon, Kolyuchin Bay, and elsewhere along the northeast coast of the Chukotka Peninsula.

King Eider

Spring staging areas of King Eiders are partly restricted by presence of ice in their most northern habitats (McLaren and McLaren 1982, Barry 1986, Dickson and Gilchrist 2002). Important spring staging areas in the eastern Beaufort Sea have annually reoccurring open water and shallow (<50 m), clear waters that enhance King Eider foraging efficiency (Alexander et al. 1997). These habitat conditions are similar to sites used by King Eiders during spring staging in eastern Lancaster Sound and western Baffin Bay, Canada (McLaren and McLaren 1982). Oppel et al. (2009a) found that King Eiders staged in the eastern Chukchi Sea during spring in waters averaging 23 ± 7 (SD) m in depth, and in autumn found adults in waters 20 ± 10 (SD) m and juveniles in waters 10 ± 7 (SD) m deep. During spring staging in the Beaufort Sea, King Eiders primarily used flaw lead habitats and selected narrower sections usually within 2 km of the ice edge; birds were rarely found in land-fast ice leads, and up to a third of the King Eiders were in the pack ice. King Eiders foraged in waters with an average depth of 30 m often close to the pack ice edges (Dickson and Smith 2013). In winters with extremely cold conditions, some polynyas may not have open water available in spring for staging King Eiders, and these birds stage elsewhere (McLaren and McLaren 1982). Many King and some Common Eiders died during unusually severe spring storms when the open waters in which birds staged froze and eiders starved (Barry 1968, Fournier and Hines 1994). Mortality events are irregular and not well documented but can be catastrophic; Barry (1968) reported as many as 100,000 eiders dying along the Canadian Beaufort Sea coast and vicinity.

Common Eider

Many breeding female Common Eiders have a short molt migration (1–3 days, <100 km) and are not found at distant staging areas (Savard et al. 2011; M. R. Petersen unpubl. data, but see Mosbech et al. 2006b). As with King Eiders, during spring migration, timing of arrival, duration of stay, and locations of stopover regions used by Common Eiders in the most northern habitats are influenced by the extent and movement of ice (Barry 1986, Alexander et al. 1997, Mosbech et al. 2006b, Petersen 2009). Alexander et al. (1997) found Common Eiders at spring staging areas closer to shore and in shallower waters (<20 m deep) than King Eiders (<50 m deep) and similarly in clear, nonturbid waters. In spring, staging Common Eiders in the southeast Beaufort Sea preferred flaw lead habitats and avoided pack ice; birds were primarily present in shallower (22 m) waters than King Eiders (Dickson and Smith

2013). Characteristics of stopover areas used by Common Eiders in fall (excluding areas also used by molting and wintering birds) have not been described.

Wintering Habitats

Wintering habitats vary within and among eider species and range from ice-free waters, coastal waters often influenced by ice, and northern marine waters subject to extreme weather, cold temperatures, and limited day lengths. Food resources affect the abundance and distribution of wintering eiders. Bustnes et al. (2013) showed that Common and King Eiders wintering in northern Norway aggregated into a few large flocks at the beginning of winter, but as winter progressed and prey density declined, birds separated into smaller flocks, changed their distribution, and subsequently remained at fixed locations.

Steller's Eider

Wintering habitats of the Pacific population of Steller's Eider are described from studies on the Alaska Peninsula (Metzner 1993, Laubhan and Metzner 1999, Fredrickson 2001) and near Dutch Harbor on Unalaska Island, west of the tip of the Alaska Peninsula (Reed and Flint 2007). In Nelson Lagoon, Steller's Eiders remained throughout winter unless forced out by ice and were most commonly found in inshore and nearshore waters (Gill et al. 1981). Molting, staging, and wintering Steller's Eiders used the same habitats each season: generally sand flats and nearshore areas with gravel bottoms. Steller's Eiders remained throughout the winter in the relatively shallow Izembek Lagoon and departed only when ice forced birds to move to deeper, open waters in the adjacent Cold Bay (Metzner 1993, Laubhan and Metzner 1999). Sediments of Izembek Lagoon consist of oozing mud, especially in eelgrass beds, to loose, coarse, volcanic sand. Izembek Lagoon is a vast expanse of eelgrass beds that harbor many species of amphipods, snails, and other small invertebrates; *Macoma* spp. and polychaetes dominate the sand flats. By contrast, Cold Bay is an unvegetated, deep-water embayment in which birds forage on infauna, epibenthos, and mobile fauna. The depth of Cold Bay reaches 6 m within 500 m from shore and the bottom is likely mud with rock projections.

Spectacled Eider

Spectacled Eiders winter in the Bering Sea south and southwest of St. Lawrence Island, Alaska (Petersen et al. 1995, 1999). In early winter, Spectacled Eiders were located 52.0 km (median, 32.1–71.9 km interquartile range) south of St. Lawrence Island, then in midwinter, 108.2 km (median, 90.8–125.5 km interquartile range) southwest of the island. The winter habitat of Spectacled Eiders south of St. Lawrence Island is summarized by Petersen and Douglas (2004). The region includes waters 40–90 m in depth within a large, broad, undulating basin with gentle slopes and distinct openings at both ends. Ice is a dominant feature in winter beginning in December and extending into April; although the amount and characteristics of ice vary considerably among years, on average, the core area was ≥90% ice-covered for >50% of the time and otherwise, ice was largely absent (<5% ice) or dense (≥85% ice). Spectacled Eiders instrumented with implanted satellite transmitters were found in waters 60.9 ± 0.5 (SD) m in depth (range 49–68 m) in depth.

Key features of the benthic communities in the Bering Sea used by Spectacled Eiders in winter include bivalves, gastropods, and crustaceans, although community structure in this area has changed over the decades from previously being dominated by *Macoma calcarea* to currently *Nuculana radiata* (Sirenko and Koltun 1992). During late winter (March 2001), Spectacled Eiders collected by Lovvorn et al. (2003) in the Bering Sea south of St. Lawrence Island had eaten clams, primarily *N. radiata*. Spectacled Eiders collected in nearshore waters at St. Lawrence Island had eaten gastropods, bivalves, and crustaceans; the predominant species group was *Macoma* clams (Petersen et al. 1998).

Ice and weather conditions in the Bering Sea wintering area of the Spectacled Eider south of St. Lawrence Island can have an effect on the population index for the number of adult Spectacled Eiders returning to the Yukon–Kuskokwim Delta the subsequent spring (Petersen and Douglas 2004). Based on their model, Lovvorn et al. (2009) showed that the availability of ice for roosting is an important component in energy budgets of Spectacled Eiders and lack of ice could reduce the amount of viable wintering habitat available to this eider species.

King Eider

The wintering areas of King Eiders breeding in North America are often remote sites off of coastal regions (Suydam 2000, Phillips et al. 2006, Oppel et al. 2009b). Some birds winter near more populated regions but are often too far offshore for direct observations of habitat use. In the north Pacific and eastern Atlantic, sea ice is a factor influencing King Eider distribution in winter (Gilchrist and Robertson 2000, Mosbech et al. 2006a, Phillips et al. 2006, Oppel et al. 2009b). In the Pacific region, when compared to random sites within this potential wintering area, King Eiders were located closer to shore and in shallower areas with lower salinity and percent ice cover (Phillips et al. 2006). Large concentrations of King Eiders were found in winter in open areas in the ice over offshore, shallow (<50 m in depth) banks in the southwest Greenland Open Water Area (Mosbech and Johnson 1999).

Common Eider

Land-based studies of winter habitats used by Common Eiders from North America have been conducted primarily in southwest Greenland, coastal eastern North America, and Hudson Bay near towns and villages. Comparatively little information is available for wintering Common Eiders in western North America and the Pacific Ocean. Wintering habitats and foraging areas of Common Eiders in North America and Greenland are characterized as areas close to shore and in rocky, shallow waters (Goudie et al. 2000). Based on aerial surveys in Greenland, S. m. borealis were found dispersed in rocky coast waters during winter (Mosbech and Johnson 1999); however, birds wintering at Disko Island, western Greenland, ate benthic invertebrates more common to softer substrates (Merkel et al. 2007). In southeastern Newfoundland, Common Eiders were found feeding primarily next to bedrock slabs with boulders, cobble, and sand (Goudie and Ankney 1988). Hamilton (2000) studied wintering Common Eiders in New Brunswick, Canada, in lower, intertidal habitats. The substrate of intertidal habitats was primarily rocks and small boulders over soft sediment; blue mussels (M. edulis) were the dominant prey species, and barnacles (Semibalanus balanoides) and limpets (Collisella testudinalis) were common (Hamilton 2000).

In winter, the Gulf of St. Lawrence, Canada, is typified by low temperatures, little daylight, and variable sea ice. Guillemette et al. (1993) found birds in the Gulf of St. Lawrence primarily using waters 0–6 m and rarely in waters 24–42 m in depth. Soft-bottomed substrates were available, but Common Eiders wintering in the Gulf of St. Lawrence preferred reef habitats with kelp patches and urchin barrens. Common Eiders also foraged in a mixed habitat of bedrock, boulders, cobles, and soft bottom; however, soft-bottom habitat associated with deeper waters (>24 m) was avoided (Guillemette et al. 1993).

Bird responses to effects of winter ice were variable depending on location. In the Gulf of St. Lawrence, Guillemette et al. (1993) found that Common Eiders remained at their study area as long as some open water was available. In the Bering Sea, Petersen et al. (2012) noted that S. m. v-nigrum remained within the same polynya despite extensive ice formation. In the Belcher Islands of Nunavut, Canada, Gilchrist and Robertson (2000) examined S. m. sedentaria use of polynyas and land-fast ice edges and incorporated observations and traditional knowledge from local Inuit. Common Eiders used shallow, ice-free waters of polynyas and adjacent land-fast ice and did not leave the area as preferred ice-free waters froze. Gilchrist and Robertson (2000) describe a die-off of Common Eiders (S. m. sedentaria) when a wintering area froze near the Belcher Islands, Hudson Bay. Robertson and Gilchrist (1998) suggested that following this severe winter, numbers of nesting S. m. sedentaria were substantially reduced.

Heath et al. (2006, 2007) and Heath and Gilchrist (2010) reported that Common Eiders (S. m. sedentaria) in Hudson Bay foraged in relatively shallow waters (11.3 m deep) and were restricted by the speed of tidal currents; birds foraged less when the tide was the strongest. Using depth recorders, Guillemette et al. (2004) determined that Common Eiders dove deeper in winter (<6 m) than in summer (<3 m). Foraging behavior of Common Eiders varied between fjord and outer coastal habitat in Southwest Greenland (Merkel and Mosbech 2008). In the coastal area, Common Eiders foraged at twilight and during the day (rarely at night) in waters near (0.5–1 km) to shore, and in the fjord, Common Eiders foraged close (<50 m) to shore and almost exclusively at twilight and night. Merkel and Mosbech (2008) hypothesized that nighttime feeding of Common

Eiders in the fjords was an avoidance tactic to prevent predation by White-tailed Eagles (*Haliaeetus albicilla*) that forage during the day. Zipkin et al. (2010) analyzed the distribution data of Common Eiders from the eastern Atlantic coast (United States and Canada) between 1991 and 2002 in relationship to latitude, the North Atlantic Oscillation, water depth and slope, SST, and if birds were in a bay or not. The authors discovered that Common Eiders were most prevalent at more northern latitudes; higher densities occurred nearshore during wet, cold winters; and this species was often found in areas with steeper, more rugged bottoms.

Foraging by Common Eiders in winter can have a substantial impact on benthic communities. In New Brunswick, Canada, Hamilton (2000) used predator exclusion cages in the intertidal zone and found that predation of blue mussels by Common Eiders directly reduced abundance of mussels, which ultimately resulted in a change of community structure. In Greenland, Blicher et al. (2011) estimated that Common and King Eiders can take 58%–81% of the total production of macrobenthos to meet their energetic demands during winter. Changes in numbers of Common Eiders on two study areas in the Baltic Sea were correlated with annual variation in biomass of the benthic community (particularly blue mussels) in waters 0–6 m in depth (Larsen and Guillemette 2000). Common Eiders exploited food resources in relationship to the availability of foods (Larsen and Guillemette 2000). Based on their models, Larsen and Guillemette (2000) predicted that in a winter with low abundance of food and no blue mussels, Common Eiders consumed 64% biomass of the forage and in a winter of higher food abundance consumed 38% of the available benthos and 58% of blue mussel biomass. When feeding on mussels, Guillemette et al. (1996) estimated that Common Eiders removed 48%–69% of the biomass of their foraging area in winter and removed 3%–6% of the biomass when feeding on urchins.

HABITATS OF HARLEQUIN DUCKS

There are two primary groups of Harlequin Ducks (*Histrionicus histrionicus*) breeding in North America: a Pacific population that occurs in northwest North America, and an Atlantic population in northeast North America (Robertson and Goudie 1999). Some birds from the Atlantic population molt and winter along the coast line of west Greenland and potentially mix with the Greenland population (Robertson and Goudie 1999, Brodeur et al. 2002, Robert et al. 2008b), and some birds that winter in Alaska likely breed in Russia (D. Rosenberg, pers. comm.). Within the Atlantic population, different segments of the breeding population are demographically distinct (Brodeur et al. 2002, Chubbs et al. 2008, Robert et al. 2008b); the occurrence of distinct breeding segments within the Pacific population is under debate (Lanctot et al. 1999, Esler et al. 2006, Pearce and Talbot 2006).

Breeding Habitats

Breeding Harlequin Ducks exploit rapidly flowing rivers and streams in forested, alpine, arctic, and subarctic habitats, as well as forested streams (Cassirer et al. 1993; Robertson and Goudie 1999; Brodeur et al. 2008; E. F. Cassirer and C. R. Groves, unpubl. report). Harlequin Ducks nest exclusively along rivers flowing into coastal waters or inland watersheds (Robertson and Goudie 1999). Freshwater habitats are often used by Harlequin Ducks from their arrival in spring until the broods fledge and leave the river systems in fall. Rivers and streams are especially important to Harlequin Ducks in spring; during that period, females acquire the resources needed for egg production (Hunt and Ydenberg 2000, Bond et al. 2007). Many female Harlequin Ducks nest and raise broods on the same rivers and streams they occupied in spring and may exhibit year-to-year site fidelity (Bengtson 1972; Kuchel 1977; Wright et al. 2000; E. F. Cassirer and C. R. Groves, unpubl. report). Harlequin Duck broods are also found at the inlets or outlets of lakes or ponds associated with rivers (Gardarsson 2008, Goudie and Gilliland 2008) and on lakes (Hunt and Ydenberg 2000). At coastal areas in spring and summer, Harlequin Ducks will move to the deltas of streams along the coast (Dzinbal and Jarvis 1984). Harlequin Ducks nesting along short streams or rivers that flow directly into coastal waters will move their young into and out of these waters during the brood-rearing period (Dzinbal 1982, Crowley 1999, Gardarsson 2008).

Physical characteristics of habitats may be similar, but the occurrence and abundance of breeding pairs can vary both in relationship to predators and changes in the availability of invertebrates in rivers and streams. The number of breeding Harlequin Ducks can be substantially influenced by predation of adult females by aerial predators such as cliff-nesting raptors (Heath and Montevecchi 2008). Similarly, the quality of the breeding habitat for Harlequin Ducks may be reduced by antipredator behavior of insects in the presence of fish (LeBourdais et al. 2009). Streams and rivers are subject to periodic flooding or high water, which may change the species composition or abundance of invertebrates, thus affecting the distribution of broods (Kuchel 1977) and abundance or success of adult breeding birds (Wright et al. 2000, but see Goudie and Jones 2005).

Rivers, streams, and inlets or outlets of lakes in all freshwater areas used by Harlequin Ducks feature currents of variable rates from turbulent, cascading reaches with considerable elevation gradients to slow-moving flows at the confluence of streams and lakes. In Labrador, streams used by Harlequin Ducks were narrow, had high pH, and were warmer than unused streams; streams that were used also had large boulders and less frequently sand or gravel, steeper shorelines, and more vegetation cover on islands and shores than unused streams (Rodway 1998a). Benthic substrates in which birds forage are primarily shallow with gravel, stones, boulders, and bedrock that support high densities of bottom-dwelling invertebrates; however, densities and dominant invertebrate species may vary by year, location, and season (Bengtson and Ulfstrand 1971; Bengtson 1972; Wright et al. 2000; Gardarsson 2006, 2008). Foraging areas on streams and rivers tended to be associated with rapids and small waterfalls (Bengtson 1972); however, birds also foraged in slower moving water (Rodway 1998b, Robert and Cloutier 2001). Soon after their arrival to the breeding grounds in spring (April), Hunt and Ydenberg (2000) found birds congregating at small side channels to eat eggs of longnose suckers (*Catostomus catostomus*).

Due to low nesting densities and secretive behavior, few Harlequin Duck nests have been found in North America (Hunt 1998, Rodway et al. 1998, Crowley 1999, Brodeur et al. 2008, Goudie and Gilliland 2008, Savard et al. 2008, but see Bond et al. 2007). Nest sites in North America include crevices and cavities on cliffs, on the ground on islands, and at the base of cliffs and against trees. In British Columbia, 42 breeding records included 29 nests on inland streams and

rivers, seven on lakes, and six on coastal islands or mainland (Savard 1991). Bengtson (1972) described locations of 98 nests in Iceland: 67% were on islands and 33% were on riverbanks. Nesting cover included areas beneath shrubs, moorland open vegetation and grasses, and areas next to rocks and in rock cavities.

Migration, Molt, and Wintering Habitats

Some Harlequin Ducks remain at the same coastal location throughout the flightless period, winter, and spring and exhibit fidelity to the same location within and among years (Breault and Savard 1999, Cooke et al. 2000, Iverson et al. 2004, Iverson and Esler 2006, Thomas and Robertson 2008). Other Harlequin Ducks can be found at separate although similar coastal staging, molting, and wintering areas (Robertson et al. 1999, Brodeur et al. 2002, Gilliland et al. 2002, Chubbs et al. 2008, Robert et al. 2008b, Savard et al. 2008). Nonbreeders often remain along the coast throughout the year, and breeding males and females return to the coast after leaving rivers and streams (Brodeur et al. 2002, 2008).

Harlequin Ducks are found along exposed, rocky seacoasts once they leave the breeding grounds (Robertson and Goudie 1999, Gardarsson 2008). Rodway et al. (2003b) examined habitat preferences of wintering Harlequin Ducks and provided more detailed descriptions of the wintering area in the Strait of Georgia, Canada. The authors found that Harlequin Duck densities were highest along shorelines with intertidal habitat >100 m wide consisting of cobble–gravel or bedrock–boulder substrates, although birds preferred shorelines with small islets and reefs. The habitat features described by Rodway et al. (2003b) are generally similar to what Esler et al. (2000a) found in Prince William Sound, Alaska. Esler et al. (2000a) noted that Harlequin Duck densities increased where there were offshore reefs within 500 m of shore and bird density was more variable among islands depending on the type of benthic substrate, yet positively associated with streams. In eastern North America, Goudie and Ankney (1988) found that wintering Harlequin Ducks used sunkers or submerged rocks, and were close to shore in shallower waters. In Maine, Mittelhauser (2000) found higher densities of wintering Harlequin Ducks along shorelines exposed to greater wave action and with granite ledges as intertidal substrate. Similarly, in Iceland, Gardarsson (2008) concluded that Harlequin Ducks are only found on exposed rocky coasts often with shallow rock shelves and indented exposed shorelines.

When along the coast, the depth of the water in which Harlequin Ducks feed can vary by several meters within a day with tidal activity; Harlequin Ducks typically forage in shallow intertidal and to some extent subtidal waters, thus the availability of their forage and their foraging behavior can vary throughout the tide cycle (Goudie 1999, Fisher and Griffin 2000, Heath et al. 2008), although foraging is not always influenced by tide changes (Goudie and Ankney 1986, Mittelhauser 2000).

Harlequin Ducks of both sexes will move to quieter, shoreline waters at the mouths of streams in response to availability of salmon roe in summer (July to August; Dzinbal and Jarvis 1984); birds also aggregate at locations with herring roe on beaches and shallow, subtidal habitats during spring/late winter (Vermeer et al. 1997, Sullivan et al. 2002, Rodway et al. 2003a, Bond and Esler 2006). Reed and Flint (2007) also reported that wintering Harlequin Ducks increased their feeding activity at locations in Alaska with seafood processing or municipal sewage effluent, likely in response to high concentrations of invertebrates.

HABITATS OF SCOTERS

Breeding Habitats

The three species of scoters occurring in North America have distinct breeding distributions, but the species overlap in the northern open taiga (Bordage and Savard 1995, Brown and Fredrickson 1997, Savard et al. 1998). Surf Scoters are endemic to North America and primarily breed in the boreal forest and to some extent in the open taiga, which is the main breeding habitat of Black Scoters (Perry et al. 2006). Common Scoters (Melanitta nigra nigra) have been elevated to species status based on behavioral traits (M. nigra, Sangster 2009, Chesser et al. 2010) but likely select similar habitats as Black Scoters (Melanitta americana). White-winged Scoters occur in Europe, Asia, and North America where birds breed on the prairie and boreal forest, but rarely in taiga habitats (Brown and Fredrickson 1997). However, coastal ponds of the lowlands adjacent to eastern James Bay support the largest known breeding concentration of White-winged Scoters in Quebec (Benoit et al. 1993, 1994, 1996).

Black and Surf Scoters primarily use shallow (<5 m depth), small (<30 ha and even <10 ha) rocky lakes and avoid rivers and large deep lakes (Savard and Lamothe 1991, Décarie et al. 1995). At a landscape scale (1 km²), occurrence of broods of both species is correlated with the presence and number of small lakes 10–100 ha (Décarie et al. 1995). Lakes used by Black and Surf Scoters tend to be high in saturated oxygen, clear, low in conductivity, and lack emergent vegetation (Bordage and Savard 1995, Savard et al. 1998). Décarie et al. (1995) determined that broods of Black Scoters used small bogs and fen more frequently and lakes 10–100 ha less frequently than Surf Scoter broods. Common Scoters in northern Scotland (Fox et al. 1989) and Black Scoters in northern Quebec (DesGranges and Houde 1989) are associated with productive, nonacidic lakes with islands. Haszard (2004) similarly found that in the Mackenzie Delta region, Surf and White-winged Scoters used wetlands with more abundant food. On Lake Malbaie, Quebec, Surf Scoter broods preferred shallow (<2 m), protected waters and total brood home ranges averaged 95 ha (95% confidence area; range: 28–173; n = 7) or 12 ha (50% confidence area; range: 3–22 ha; Lesage et al. 2008).

At the Yukon–Kuskokwim Delta of Alaska, Black Scoters nest in shrub edge habitats, preferring well-concealed nesting sites and avoiding dry upland tundra (Schamber et al. 2010a). Shrub cover is also used by breeding Common Scoters in Iceland (Bengtson 1970). However, birds also nest in tall grass habitats that provide good lateral cover (Bordage and Savard 1995). At Lake Malbaie, Quebec, a 664 ha basin of shallow water (<7 m) supported >60 pairs of Surf Scoters and the highest known nesting density of the species. Nests were found on two islands ranging from 6 to 27 m from shore and all were well concealed in the vegetation or under fallen trees (n = 17, Morrier et al. 2008). Black, Common, and Surf Scoters do not usually nest close to shore and nests have been found up to 50 m from shore (Bordage and Savard 1995, Savard et al. 1998). Aerial surveys in northern Ontario found scoter pairs associated with areas that averaged 18% of wetland land cover, whereas areas without scoters averaged only 6% wetland cover; scoters also avoided large lakes as well as fen areas (Brook et al. 2012).

White-winged Scoters are more diverse in their selection of breeding lakes and often nest at high densities on brackish alkaline shallow lakes with shrubby islands in the prairies and aspen parkland as well as the boreal forest (Brown 1977, Brown and Fredrickson 1997). As with Surf Scoters, young White-winged Scoter ducklings forage in shallower waters than older ducklings, but most used waters <2 m in depth (Brown and Fredrickson 1997, Lesage et al. 2008). The species seems to nest farther from shore than Black and Surf Scoters with nests in Saskatchewan averaging 96 m from the water edge and being often located in thorny shrubs and on islands (Brown and Brown 1981, Brown and Fredrickson 1989, Traylor et al. 2004). At the Yukon Flats of Alaska, nesting White-winged Scoters preferred dwarf tree scrub and avoided graminoid habitat. Birds also selected nest sites with good overhead and lateral cover and closer to edge (<10 m) and water (Safine and Lindberg 2008). In northern Quebec, on the eastern coast of James Bay, scoters selected alkaline (pH ranging from 7.2 to 9.2), shallow (<4 m) coastal ponds with abundant submerged vegetation (Benoit et al. 1993, 1996). Lakes used by broods had high densities of the amphipod *Hyalella azteca* (Benoit et al. 1994, 1996), which dominates the diet of ducklings in Alberta (Brown and Fredrickson 1997). In Finland, White-winged Scoters nest far from shore, well into the upland, and use well-concealed shrub sites (Hilden 1964). White-winged Scoters also nest on islands of coastal archipelagos with Common Eiders (Hilden 1964). Coastal nesting habitat is apparently not used by the species in North America. However, White-winged Scoters nest in lowlands adjacent to James Bay where they use small coastal ponds (Reed et al. 1996).

Molting Habitats

Major molting sites are all located in coastal waters in areas rich in shellfish resources, especially clams and mussels (Bordage and Savard 1995, Brown and Fredrickson 1997, Savard et al. 1998). The three species of scoters sometimes molt together, but birds tend to form species-specific groups most of the time. In the St. Lawrence Estuary, Surf and White-winged Scoters often molt together over large, sandy subtidal substrates usually <10 m deep (Rail and Savard 2003). However, even when molting together, scoters tend to form species-specific groups. In eastern North America, most Black Scoters molt within James Bay, and some do so along the southwestern shore of Hudson Bay. Within this large area, distinct molting sites have been identified but have yet to be characterized (Ross 1983, Bordage and Savard 1995). Surf and White-winged Scoters also molt in the region, but in smaller numbers. Reed et al. (1996) indicated that in James Bay, all scoters fed mostly on bivalves but that White-winged Scoters avoided mussels, which were the preferred prey of the other two species. In James Bay, all three species molted over coarse substrates in relatively open, shallow (1–6 m) waters with strong tidal currents and often over underwater shallow shelves around islands (Reed et al. 1996).

On the Pacific coast, scoters molt in nearshore habitats that are similar to sites in eastern North America (Savard 1988b). Molting sites in Frederick Sound, southeast Alaska, were characterized by shallow gravel beaches and submerged glacial moraines at the entrance to inlets (Butler 1998). Optimal shellfish habitat is likely correlated with dynamic water flow. As in the St. Lawrence River, Surf and White-winged Scoters were often found molting together (Butler 1998), but Surf Scoters are the most numerous species (Hodges et al. 2008). In southern British Columbia, both scoter species molt at the Fraser River Delta and in Boundary Bay, which is an area characterized by intertidal mudflats and eelgrass beds with a few rocky outcrops (Dickson et al. 2012). In the western Palearctic, Common and Velvet Scoters molt over sandy substrates rich in bivalves and usually at depths <10 m (Fox 2003).

Some female scoters that successfully raised broods may molt on freshwater lakes, along with some males (Benoit et al. 1994; J. Thompson, pers. comm.). King (1963) captured several hundreds of White-winged Scoters in molting flocks at Ohtig Lake, Alaska, documenting that some birds molt inland. However, the proportion of scoters molting on inland lakes is probably relatively small given the large numbers of molting birds recorded in the marine environment (Savard 1988b, Benoit et al. 1994, Brown and Fredrickson 1997).

Fall and Spring Staging Habitats

In scoters, spring staging is more spectacular than fall staging, especially along the Pacific coast where birds often concentrate by the thousands at herring spawning sites to consume roe and take advantage of ephemeral, abundant polychaetes (Sullivan et al. 2002, Lacroix et al. 2005, Lewis et al. 2007). Lok et al. (2008) showed that Surf and White-winged Scoters, which are quite sedentary during winter, moved considerable distances during the herring spawning season to take advantage of ephemeral but highly nutrient-rich resources. Surf Scoters tended to move greater distances than White-winged Scoters. On the Pacific coast, the chronology of Surf Scoter spring migration and spawning herring is coincident, which indicates an association between Surf Scoters and spawning herring (Lok et al. 2012). However, Surf Scoters do not always respond to herring spawn in Puget Sound (De la Cruz et al. 2009). During spring migration on the Atlantic coast, scoters do not seem to be closely tied with herring spawn, possibly because of the timing of spawn, which occurs mostly in late summer and fall (Graham et al. 1984). However, Surf Scoters may exploit herring spawn in the fall in the St. Lawrence Estuary (J.-P. Savard, unpubl. data). Herring spawn may be more important on the west coast as Surf Scoters migrate through wintering areas (De la Cruz et al. 2009), whereas this is not the case on the east coast where there are no wintering scoters in important spring staging areas.

Several areas used during molt are also occupied as spring and fall staging habitats, indicating their high productivity (Rail and Savard 2003, Schamber et al. 2010b). Molting Black Scoters remain in James Bay until late October through early November (Sea Duck Joint Venture, unpubl. data), and it is likely that young birds may also stage at the site before migrating to the Atlantic coast. The St. Lawrence Estuary is an important fall staging area that attracts mixed flocks of adult and young Surf Scoters in early October. Most birds utilize large, sandy shores and to some extent, the same habitats used during molt.

Wintering Habitats

Surf and Black Scoters winter almost exclusively in marine waters, whereas some White-winged Scoters also winter in freshwater sites such as Lake Ontario. In the marine and freshwater environment, scoters winter in small to large flocks in a variety of habitats supporting bivalves such as clams and mussels (Bordage and Savard 1995, Brown and Fredrickson 1997, Savard et al. 1998).

On the Pacific coast, species are segregated by habitat or feeding preferences and mixed flocks are rare. Surf Scoters can be found either over clam or mussel beds and birds exploit mussel beds more than the other species, thus Surf Scoters are most prevalent over rocky substrates (Savard 1989). Surf Scoters are by far the most numerous scoters wintering in the inlets and fjords of British Columbia and southeastern Alaska (Vermeer 1981, Hodges et al. 2008). Surf Scoters form denser flocks over mussel habitat than over clam beds likely reflecting the different spacing of their prey. Indeed, the distribution of mussels is often highly clumped and localized. Also, dense flocks of Surf Scoters may facilitate exploitation of tightly packed and strongly attached mussel beds as scoters benefit from the openings in the mussel clumps created by feeding congeners (Lacroix 2001). Lacroix (2001) found that Surf Scoters ate mostly mussels that were just over a year old and that birds could completely consume the younger mussels of a bed. Foraging areas were quickly recolonized by mussels but it took 2 years after depletion for mussels to grow to a size favored by scoters. Thus, sites were not exploited each year. Also, if scoters did not use a site for a few years, mussels could grow to a size too large for Surf Scoters (Lacroix 2001). On the west coast, because of the absence of winter ice cover, young mussels can colonize tidally exposed rocks and grow to an optimal size for scoters. Conditions differ along the north Atlantic coast where ice prevents overwinter survival of these mussels. Thus, ice conditions as well as different tidal regimes contribute to different mussel bed dynamics between the two coasts. It is not clear how regional variation in ecological conditions may affect sea ducks.

Surf Scoters, like White-winged Scoters, use sandy shores where they feed on clams. On the Atlantic coast, scoters were associated with the flattest subtidal areas, in contrast to Common Eiders that occurred over the steepest areas. Black Scoters tended to occupy areas of shallower depth than those used by White-winged Scoters (Silverman et al. 2013). In Narragansett Bay, Rhode Island, the highest densities of wintering Surf Scoters were found over sand substrates with homogenous assemblages of infaunal prey (Loring et al. 2013). In northern New England, Surf Scoters were also associated with sandy substrates (Stott and Olson 1973). Lewis et al. (2008) found that in Baines Sound in British Columbia, the foraging efforts of both species did not vary significantly with clam density and that the habitat was not overexploited, suggesting that clam beds may be more difficult to overexploit than mussel beds. Kirk et al. (2007) found that Surf Scoter foraging efforts did not vary between habitats but did increase throughout the winter as prey depletion occurred. The authors also found that scoters did not increase foraging efforts in depleted habitats, but instead redistributed to less depleted habitats. Surf Scoters were more sedentary and more faithful to foraging sites in clam than in mussel habitats (Kirk et al. 2008). In Baynes Sound, British Columbia, Surf and White-winged Scoters occurred at higher densities in areas with sand sediments, more intertidal area, and higher clam densities (Žydelis et al. 2006).

On the Pacific coast, Black Scoters are mostly found along gravel and cobble shores and do not frequent the deep inlets used by Surf Scoters (Savard 1988b, Bordage and Savard 1995). Vermeer and Bourne (1982) indicated that all three species of scoters were found over sand–mud and cobble substrates, White-winged Scoters over gravel beds, and Surf Scoters over rocky substrates. On the Atlantic coast, Black Scoters avoided rocky shores, preferring sandy substrates (Stott and Olson 1973). Similar use of sandy substrates by Black Scoters also was documented in Puget Sound on the west coast (Hirsch 1980). In southern New England, Black Scoters were associated with nearshore subtidal habitats and preferred benthic habitats with coarse sand substrate to substrate with finer sand (Loring 2012, Loring et al. 2013). In western Europe, Common Scoters were associated with shallow (<10 m) sandy habitats that had high densities of bivalves (Degraer et al. 1999, Fox 2003). In Newfoundland, Black Scoters used the same habitats as Harlequin Ducks, including waters with cobbles and bedrock ledges (Goudie 1984). A similar association between Black Scoter and Harlequin Duck distribution was found in the Georgia Strait, British Columba (Savard 1989). There, the distribution of Black Scoters was associated with White-winged Scoters, but not with Surf Scoters (Savard 1989).

HABITATS OF LONG-TAILED DUCKS

The Long-tailed Duck (*Clangula hyemalis*) is a circumpolar species of northern environments (Dement'ev and Gladkov 1952, Kistchinski 1973, Robertson and Savard 2002) and is listed as vulnerable on the IUCN Red List (BirdLife International 2012). The distribution of Long-tailed Ducks appears to show a broad east–west divide in North America (Robertson and Savard 2002), without connectivity among breeding, molting, and wintering areas (Petersen et al. 2003; Braune et al. 2005; M. R. Petersen, unpubl. data). Relatively little research has focused on breeding Long-tailed Ducks in North America (Alison 1975, Schamber et al. 2009), but some information is available from short notes, unpubl. reports, theses, and ecological surveys. Most studies in North America have addressed wintering ecology and distribution, with the majority of work on habitat relationships on the east coast of North America.

Breeding Habitats

Long-tailed Ducks are widespread throughout the wetlands of arctic and subarctic habitats. Salter et al. (1980) reported they were the most numerous duck species on the Arctic Coastal Plain between Cache Creek and Firth River, Canada. Long-tailed Ducks are considered generalists and will use a diversity of wetlands and uplands within any location. Thus, descriptions of nest sites and brood-rearing locations are limited by the wetland and upland types available in study areas. Descriptions of nesting locations from a few study areas do not necessarily represent a complete picture of nesting habitats of Long-tailed Ducks throughout the species' breeding range. Birds may nest solitarily or in colonies on barrier islands, offshore islands, or the mainland. Bailey (1925) found that Long-tailed Duck nests along the shore of Lopp Lagoon, Alaska, including little islets, and on small islets and in pits in the sand with and without grasses; and on one little island, he found at least 100 nesting pits.

At Churchill, Manitoba, Alison (1975) described his study area as tundra, dry upland, marshland, and scrubland habitats that included ponds and lakes. Long-tailed Duck nests were located at the shorelines, islands, and peninsulas of freshwater ponds and lakes, on dry upland, and in scrubland; nests were solitary or clumped. During early brood rearing, females with ducklings remained in the same tundra pond or lake near where they hatched then moved to nearby and larger ponds with older young (Alison 1976). Derksen et al. (1981) also found that on the North Slope of Alaska, female Long-tailed Ducks with broods moved to larger water bodies later in the season. Evans (1970) reported a positive association between nesting Long-tailed Ducks and Arctic Terns (*Sterna paradisaea*). Alison (1975) also noted that Long-tailed Ducks and Arctic Terns nested together but concluded that the association was a result of similar habitat preferences.

In Beaufort Sea coastal wetlands, Long-tailed Ducks used open water as it became available in spring (Howard 1974), particularly moats of large lakes and deep-*Arctophila* wetlands (Bergman et al. 1977, Derksen et al. 1981). Adult Long-tailed Ducks in spring and females with broods were found in shallow-*Carex*, shallow-*Arctophila*, and deep-*Arctophila* wetlands as well as deep-open basins, basin-complex wetlands, and beaded streams (Bergman et al. 1977, Derksen et al. 1981). Thus, during the breeding season on the Arctic Coastal Plain, Long-tailed Ducks exploit a diversity of wetland types, with the exception of flooded tundra and coastal estuarine waters.

An important aspect of breeding habitats concerns the potential impacts of severe weather, snow, and ice on nesting Long-tailed Ducks. Boyd (1996) examined the ratios of first-winter to adult female Long-tailed Ducks shot in eastern North America in relationship to summer and autumn temperatures on the breeding grounds and found that hunter kill of first-winter birds was correlated with seasonal temperature in the Arctic tundra. Alison (1973, 1975) found that nest initiation of Long-tailed Ducks was delayed in a cold May, due to snow on potential nest sites and ice on ponds, and was interrupted by snow and cold weather during nest initiation. Similarly, in northern Sweden, Pehrsson (1986) found that breeding phenology of Long-tailed Ducks was delayed due to snow and adverse weather before or during egg laying. The numbers of ducklings fledging is also mediated by predation. In northern Sweden, the productivity of Long-tailed Ducks was positively correlated with small rodent abundance, likely a result of increased nest success due to the abundance of alternate prey for potential nest predators (Pehrsson 1986).

The presence or absence of preferred or alternate prey of ducklings in ponds is a key factor of breeding habitats of Long-tailed Ducks. On the Arctic Coastal Plain, Alaska, Long-tailed Duck broods were located in ponds and lakes with high densities of invertebrates (Howard 1974, Bergman et al.

1977, Derksen et al. 1981). In northern Sweden, Long-tailed Duck broods were closely associated with the presence and abundance of slow-swimming crustacea (Pehrsson 1974). Long-tailed Ducks likely move their broods to ponds with greater food resources as invertebrates become depleted (Pehrsson 1974, Alison 1976).

Molting Habitats

Throughout their range, in North America and elsewhere, Long-tailed Ducks molt their flight feathers while in barrier island lagoon systems, in coastal marine habitats, in large freshwater lakes, and in wetlands in arctic and subarctic environments (King 1973, Bergman et al. 1977, Derksen et al. 1981, McLaren and McLaren 1982, Taylor 1986, Dickson and Gilchrist 2002, Robertson and Savard 2002, Petersen et al. 2003, Noel et al. 2005a). Adult female Long-tailed Ducks marked as breeding birds in western Alaska remained on the breeding grounds, moved to larger lakes in western Alaska, or migrated to molting locations in other regions of North America (Petersen et al. 2003). Flightless adult females marked in a lagoon in northern Alaska either returned to the same lagoon to molt the following year or molted elsewhere in Alaska or coastal Russia (M. R. Petersen, unpubl. data). Thus, molting habitats of Long-tailed Ducks may vary substantially throughout the northern portion of North America, as well as the Northern Hemisphere, and individuals may molt at different locations in different years. Detailed studies in North America of Long-tailed Duck use of molting habitats have only been conducted in a barrier island lagoon system of the Beaufort Sea (Johnson and Richardson 1981; Johnson 1984a,b, 1985) and a freshwater system on the Alaska North Slope (Bergman et al. 1977, Derksen et al. 1981, Taylor 1986). Similar studies at other locations in North America are needed to more clearly define the broad array of molting habitats used by Long-tailed Ducks.

During surveys along the western Beaufort Sea coast, Fischer and Larned (2004) found the highest densities of Long-tailed Ducks in shallow marine waters <10 m in depth. Long-tailed Ducks are widespread in the western Beaufort Sea (Johnson and Richardson 1982, Johnson and Herter 1989, Fischer and Larned 2004), and a detailed examination of coastal habitats used by molting ducks has been conducted around Simpson Lagoon, Jones

Islands (Johnson and Richardson 1981, 1982). At this site, Long-tailed Ducks fed in shallow waters (<2.5 m) on epibenthic invertebrates; foods were primarily mysids (58.7%–68.5%) and amphipods (14.2%–15.5%), which are the primary species in the epibenthos in the lagoon system (Griffiths and Dillinger 1981; Johnson 1984a,b). Flightless Long-tailed Ducks in Beaufort Lagoon (western Beaufort Sea) ate primarily amphipods (30.3%), gastropods (16.5%), and mysids (12.8%) (Brackney and Platte 1986), which suggests invertebrate fauna may differ between these two lagoon systems. Johnson (1984b) concluded that, based on their survey data of Long-tailed Ducks and an analysis of weather data, the birds preferentially roosted and fed near lee shores of barrier islands. Flint et al. (2004) also suggested that movement patterns of flightless Long-tailed Ducks monitored via radio telemetry were likely influenced by geography of barrier islands that provided protection from wind and variation in weather patterns that determined prevailing winds and wave dynamics.

Taylor (1986) studied habitat use of Long-tailed Ducks during the breeding, postbreeding, and molting periods in a freshwater system in northern Alaska. During the molt, Taylor (1986) and Bergman et al. (1977) found most birds in comparatively deep-open lakes. The authors defined these wetlands as class V, deep-open lakes with waters up to 1.1 m in depth and have abrupt shores, sublittoral shelves, and a deep central zone. Derksen et al. (1981) also found molting birds in deep-Arctophila ponds (class IV wetlands; large ponds or lakes exceeding 0.4 m in depth, characterized by no emergent plants in the central zone, Arctophila stands near the shore; abrupt shores; and flat or gently sloping bottoms; Bergman et al. 1977). Long-tailed Ducks collected by Taylor (1986) in lakes without least cisco (Coregonus sardinella) ate primarily Daphniidae (57%), with about equal proportions (13%–14%) of Chironomidae larvae and small crustaceans (Anostraca and Copepoda). Without ciscos in lakes, Long-tailed Ducks consumed zooplankton in greater amounts and used smaller quantities of the more abundant Chironomidae larvae and Oligochaeta (Taylor 1986). When fish were present, Long-tailed Ducks primarily consumed larvae, larval cases, and adults of Chironomidae (80%) and Pelecypoda (17%) in proportions similar to their availability (Taylor 1986). Thousands of Long-tailed Ducks, mostly females (61%), molted on the large (114 km^2),

shallow (<5 m), productive Takslesluk Lake of the Yukon Delta (King 1973).

Fall and Spring Staging Habitats

Most studies on habitat use of staging Long-tailed Ducks have been conducted during spring migration when birds are in polynyas and leads in the ice near their most northern breeding areas. Little research has been conducted on stopover sites used by Long-tailed Ducks after spring migration begins or during fall migration at northern sites, and data are limited for the staging habitats of southern populations of breeding birds, including western Alaska. Little information is available on fall stopover habitats used by Long-tailed Ducks from the time they leave their molting area until they arrive to wintering or staging sites at lower latitudes. It is often difficult to separate characteristics of fall stopover from wintering habitats since many Long-tailed Ducks stage at the same locations where they occur in winter (Portenko 1972, Robertson and Savard 2002, Petersen et al. 2003, Mallory et al. 2006).

In spring, in more southern regions along the west coast of British Columbia, Long-tailed Ducks can be found feeding on herring spawn in bays and beaches (Vermeer et al. 1997, Sullivan et al. 2002). Similarly, during spring migration in Norway, Gjøsæter and Sætre (1974) reported Long-tailed Ducks feeding on eggs of capelin (Mallotus villosus). Characteristics of stopover sites are likely highly variable as Long-tailed Ducks migrated along the coast, crossed sea ice with leads and broken ice, and flew over land (Richardson and Johnson 1981, Mallory et al. 2006).

Aerial surveys in the western Canadian Arctic have been conducted in spring before birds disperse to breeding areas (Alexander et al. 1997, Dickson and Gilchrist 2002). During those surveys, Long-tailed Ducks were found in leads along the land-fast ice edges of the mainland. The highest densities of Long-tailed Ducks were located between the Ballie Islands off the coast of the Bathurst Peninsula and Cape Dalhousie (Alexander et al. 1997). Alexander et al. (1997) concluded that the reoccurrence of open water, water depth, and water turbidity influenced waterfowl distribution in the western Canadian Arctic. Long-tailed Ducks were consistently found in the Cape Bathurst polynya and in deeper waters as well as the shallow waters used by eiders. Long-tailed Ducks occur in the more turbid waters as well as clear-water areas. Alexander et al. (1997) concluded that the variable habitats used by Long-tailed Ducks during spring were a result of their ability to dive in deeper waters (to 73 m) and their flexible food selection for different invertebrates associated with ice, the water column, or on the sea floor.

Aerial surveys of waterfowl in the eastern Canadian Arctic during spring staging focused primarily along the northwest coast of Baffin Bay (McLaren and McLaren 1982). The distribution of Long-tailed Ducks varied among years, yet most observations of this duck were usually within 1 km of land or shore-fast ice. Variability in spring distribution patterns of Long-tailed Ducks may have been influenced by climate conditions during migration and on staging areas.

In many areas, Long-tailed Ducks initiate fall migration and leave molting or brood-rearing areas by September through November (McLaren and McLaren 1982; Petersen et al. 2003; Mallory et al. 2006; M. R. Petersen, unpubl. data), often when insufficient light is available for surveys and winter weather has begun. For Long-tailed Ducks migrating from most northern locations, information is limited for stopover areas after October and during late fall until the birds arrive in more southerly regions.

Wintering Habitats

Long-tailed Ducks show ecological plasticity in both foods and habitats selected during winter (Žydelis and Ruškytė 2005), and are often described as a generalist species (Stott and Olson 1973, Bustnes and Systad 2001, Robertson and Savard 2002). As with breeding locations, wintering areas of Long-tailed Ducks are restricted by the variability in habitats within a study area (Stott and Olson 1973). Despite a broad selection of forage items and habitats, birds may be not in better body condition if eating one set of forage items in one habitat versus another (Žydelis and Ruškytė 2005). In the Baltic Sea, the diet of Long-tailed Ducks varied by age and sex, but Stempniewicz (1995) concluded that these differences were partly related to winter distribution, feeding grounds, and feeding efficiency. Stempniewicz (1995) also found that foods changed over the winter and early spring. Thus, habitats described below should be considered in light of the large variability of habitats used by Long-tailed Ducks,

the time during winter when data were collected, the depth at which birds foraged, and age and sex of individuals in the sample.

In a study of foraging sea ducks wintering in Newfoundland, Goudie and Ankney (1988) determined that Long-tailed Ducks used bedrock ledge and cobble beach habitats. In eastern North America, Stott and Olson (1973) reported that Long-tailed Ducks had a generalized diet and were found in all habitats. Habitats at their study area included sandy beach, sand with rock, half sand and half rock, rock with sand, and rocky headland along the coast in harbors. In the Nantucket Shoals, White et al. (2009) recorded Long-tailed Ducks up to 70 km offshore, generally in waters <20 m in depth, and their diet was primarily amphipods (Gammarus annulatus). In contrast, Mackay (1892) observed Long-tailed Ducks eating a wider variety of foods including fish, mussels, and crabs. In Kachemak Bay, Alaska, Sanger and Jones (1984) found Long-tailed Ducks mostly over sand and mud and in waters <20 m deep, where they captured a wide variety of infauna and epibenthic prey. In Greenland, Merkel et al. (2002) conducted coastal surveys that were 15–20 km from shore in waters <50 m and located most Long-tailed Ducks in nearshore coastal waters and rarely in fjords. Thousands of Long-tailed Ducks concentrate to feed occasionally at the upwelling offshore of the junction of the Saguenay and St. Lawrence rivers (Savard 1990, 2009). In winter, Long-tailed Ducks use subtidal and offshore waters of the St. Lawrence River and are rarely found in intertidal waters (Savard 2009).

Zipkin et al. (2010) analyzed long-term data from the Atlantic Flyway Sea Duck Surveys from eastern North America to evaluate the potential effects of both local and broad environmental factors that may influence sea duck distribution. At a broad scale, they determined that the SST had the strongest influence on expected counts of Long-tailed Ducks in nearshore waters with numbers decreasing with decreasing SST. At a more local scale, Long-tailed Ducks may prefer marine areas with steeper, more rugged bottoms (Zipkin et al. 2010). The authors concluded that the expected survey estimates and distribution of Long-tailed Ducks and other sea ducks were likely influenced by both broad environmental and localized habitat factors, and Long-tailed Ducks were more abundant at northern latitudes and had sharply delineated latitudinal distribution breaks.

Long-tailed Ducks also winter in polynyas in Hudson Bay (Gilchrist and Robertson 2000, Robertson and Savard 2002); dive in the broken ice in Kachemak Bay, Alaska (Sanger and Jones 1984); occur in the Sireniki polynya, Chukotka Peninsula; and are found in other polynyas and leads in the ice, as well as more southern ice-free regions (Konyukhov et al. 1998; Robertson and Savard 2002; M. R. Petersen, unpubl. data).

Over 100,000 Long-tailed Ducks winter on the Great Lakes (Robertson and Savard 2002) where the birds are found offshore in open waters and close to shore in bays and harbors (Schorger 1951; Ellarson 1956; Schummer et al. 2008a,b, 2012). Long-tailed Ducks were previously reported to dive to 37–58 m (Scott 1938, Schorger 1951), and Ellarson (1956) found that this duck was commonly captured in gill nets at 47 m in Lake Michigan. Foods of Long-tailed Ducks were measured as a proportion of total volume and were dominated by crustaceans, primarily the amphipod *Monoporeia affinis* (formerly *Pontoporeia affinis*; Ellarson 1956, Peterson and Ellarson 1977). The amphipod is primarily a benthic organism found in high concentrations in the bottom fauna of large, deep lakes in shallow waters 0–5 m in depth. Densities decline by about half at 15–20 m and increase again in 25–30 m (Rawson 1953), with concentration zones between 35 and 50 m in Lake Michigan (Eggleton 1937). Additional foods of Long-tailed Ducks include molluscs (pelecypods and gastropods), fish and fish eggs, insects, and a variety of other small animals (Hull 1914, Ellarson 1956, Peterson and Ellarson 1977, Ross et al. 2005). Diet of Long-tailed Ducks varies among sample locations and lakes (Eggleton 1937, Rawson 1953).

Nonnative species of zebra (*Dreissena polymorpha*) and quagga mussels (*D. bugensis*) invaded the Great Lakes during the 1980s and resulted in changes in food habitats of Long-tailed Ducks (Hamilton and Ankney 1994, Ross et al. 2005, Schummer et al. 2008a). Long-tailed Ducks consumed mussels in Lake Ontario, but amphipods and chironomids continued to be the primary forage items while mussels dominated the benthos (Schummer et al. 2008a,b). As during the period before the mussel invasion in the Great Lakes, amphipods and chironomid numbers continued to vary among sampling locations and seasons (Schummer et al. 2008b). One consequence of expansion of *Dreissena* mussels is that they provide a microclimate favorable to amphipods.

Thus, densities of amphipods have increased in areas of poor habitat and locations where amphipods occurred in the past (Wisenden and Bailey 1995, Schummer et al. 2008b).

In general, body mass declined, and foraging activity of Long-tailed Ducks increased as winter progressed in the Great Lakes (Peterson and Ellarson 1979, Schummer et al. 2012). During winter on Lake Ontario, models developed by Schummer et al. (2012) that best described foraging effort included two variables: date and temperature on the day before collection. Changes in ice cover and effects on forage availability were correlated with time and temperature but were not a factor influencing fat levels in Long-tailed Ducks at Lake Ontario. Winter fat levels in skin, abdominal, and visceral fat depots were related to date and temperature as primary explanatory factors (with sex as a covariate), but substantial variation in the data remained unexplained (Schummer et al. 2012). Other habitat factors changing through time and influenced by temperature such as forage species abundance, distribution, and quality were not examined but may influence foraging effort and body condition.

During unusual years, ice can be a factor influencing survival and condition of Long-tailed Ducks within the Great Lakes. Long-tailed Ducks in Lake Michigan and Lake Erie have been found dead and dying on the ice or in holes in the ice during extremely cold weather (Gromme 1936, Trautman et al. 1939). During midwinter, a few Long-tailed Ducks have been found in the open water of a harbor in Lake Michigan feeding on minnows (Hull 1914). Mass mortality of Long-tailed Ducks on the Great Lakes due to excessive ice occurs on an irregular basis (Ellarson 1956). Schummer et al. (2012) concluded that Long-tailed Ducks were tolerant of increased ice cover primarily because of their ability to exploit a variety of species of potential foods and to move to deeper, ice-free water when ice forms at shallow foraging areas.

HABITATS OF GOLDENEYES AND BUFFLEHEADS

Breeding Habitats

Goldeneyes and Buffleheads are obligate cavity nesters and require suitable cavities to lay and incubate their eggs. All species start breeding at 2–3 years of age but 1-year-old females may return to their natal area to prospect for nesting cavities (Eadie and Gauthier 1985, Boyd et al. 2009). In North America, Buffleheads readily use cavities excavated by Northern Flickers (*Colaptes auratus*), whereas goldeneyes are limited to cavities of Pileated Woodpeckers (*Dryocopus pileatus*, Bonar 2000, Evans et al. 2002, Martin et al. 2004). However, the ducks also nest in broken snags and cavities created by natural weathering (Skinner 1937, Prince 1968, Vaillancourt et al. 2009). In British Columbia, Barrow's Goldeneyes selected cavity trees with a diameter at breast height (dbh) >35 cm, whereas trees used by Buffleheads had a dbh >25 cm (Evans 2003). Cavities used by Barrow's Goldeneyes were 10–20 cm in diameter, whereas sites occupied by Buffleheads measured 5–12 cm, suggesting little overlap between the species. Avoidance of large entrance cavities by Buffleheads may reduce confrontations with the more dominant goldeneyes (Savard 1984, Evans 2003). Most nests are located in trees, but Buffleheads and goldeneyes have also been recorded nesting in rock crevices, under tree stumps, and in a marmot burrow (Harris et al. 1954; Eadie et al. 1995, 2000). Breeding cavities are usually immediately adjacent to suitable water bodies, although some sites have been 1–2 km away.

All three *Bucephala* species differ in their breeding habitat preferences but can often coexist in the most productive habitats. The birds defend breeding pair and brood territories, but the sites rarely coincide (Savard 1984, Gauthier 1993, Eadie et al. 2000, Paasivaara and Pöysä 2008). The nest site is sometimes adjacent to the pair territory, but most are not and often are on different lakes. Goldeneye and Bufflehead broods can travel a few kilometers before settling onto a territory (Evans 2003, Paasivaara and Pöysä 2008). The distance covered depends on the location of the nest site in relation to suitable brood-rearing lakes and also on the abundance and location of already settled broods (Erskine 1972a, Pöysä and Paasivaara 2006). Both pair and brood territories are defended and territory size is a function of food abundance and density of congeners (Gauthier 1987a,b; Savard 1984; Einarsson 1990; Thompson and Ankney 2002). Strong intra- and interspecific competition likely affects local settlement patterns, so that some birds are displaced to suboptimal habitats. A general association of goldeneyes with clusters

of ponds likely reflects their spacing needs during brood rearing and also provides alternate habitats for raising young (Robert et al. 2008a).

Common Goldeneyes have the widest breeding distribution and frequent oligotrophic boreal lakes and ponds as well as slow-flowing river systems (Prince 1968, Rajala and Ormio 1970, Eadie et al. 1995). Small lakes <30 ha in size with little or no emergent or submerged vegetation are preferred by Common Goldeneyes (DesGranges and Darveau 1985, McNicol et al. 1987, Nummi and Pöysä 1993, Wayland and McNicol 1994). Food abundance is an important factor in selecting brood-rearing lakes, and female Common Goldeneyes often move their ducklings from the hatching lake to a basin with more invertebrates (Eriksson 1978, Pöysä and Virtanen 1994). Common Goldeneyes can breed successfully near lakes with fish but prefer lakes where fish are absent (Eriksson 1983, Mallory et al. 1994, Eadie et al. 1995, Nummi et al. 2012).

Barrow's Goldeneyes appear to be more restricted to highly productive breeding lakes or ponds than are Common Goldeneyes (Savard 1984, Eadie et al. 2000). In western North America, Barrow's Goldeneyes are most often found breeding on alkaline ponds without fish (Munro 1939, Savard et al. 1994), whereas in eastern North America, the species is largely confined to small (<10 ha), high elevation lakes without fish (Robert et al. 2000, 2008a). By contrast, in Iceland, Barrow's Goldeneyes are restricted to highly productive rivers and lakes where birds settle in relation to food abundance (Einarsson 1988). In the Columbia Valley, British Columbia, Barrow's Goldeneyes are found almost exclusively on the alkaline ponds of the adjacent plateau, whereas Common Goldeneyes are most numerous on the less productive valley ponds (Savard 1984). Similar segregation is found in the Ste-Marguerite Valley, Quebec, where Common Goldeneyes use the valley bottom, and Barrow's Goldeneyes exploit headwater ponds at higher elevations (>500 m; Sénéchal 2003, 2008; Robert et al. 2008a). In British Columbia, Barrow's Goldeneye pairs used lakes with greater abundance of invertebrates and adjacent to potential nesting cavities. Females with broods selected lakes with more invertebrates and without emergent vegetation. Evans (2003) showed that Barrow's Goldeneye duckling survival, mass, and return rates as 1-year-olds were positively associated with food abundance at their natal lake.

Buffleheads are a small-bodied species and sympatric with Barrow's and Common Goldeneyes in western Canada. Buffleheads prefer ponds with little or no emergent vegetation (Erskine 1972a, Gauthier 1993, Evans 2003) and are especially numerous on productive alkaline ponds without fish but also use smaller ponds than goldeneyes (J.-P. Savard, unpubl. data). Buffleheads are rarely found on ponds under 1 ha (Erskine 1972a). In central Alberta, the species uses small and shallow ponds or lakes with no outlets or just seasonal outflow (Erskine 1972a, Donaghey 1975). Buffleheads reach higher densities and have higher duckling survival rates on lakes with high invertebrate density (Gauthier 1987a). Like goldeneyes, Buffleheads occasionally move overland to reach more productive ponds (Gauthier 1987a). In sites where Buffleheads cohabit with Barrow's Goldeneyes, lake selection by breeding pairs or females with broods was not related to food abundance, which may be partially due to their avoidance of Barrow's Goldeneyes (Evans 2003).

Molting Habitats

Little is known about molting habitats of Common Goldeneyes in North America. In the St. Lawrence River, shallow estuarine areas are used as molting sites and birds are known to molt at the mouth of some of the fjords and inlets of the Labrador coast and Ungava Bay (Todd 1963; Canadian Wildlife Service, unpubl. data). Some birds molt at beds of eelgrass (*Z. marina*) in intertidal areas of James Bay (Benoit et al. 1994, Reed et al. 1996). In Alaska, some males molt on large, shallow lakes (King 1973). In Denmark and Estonia, Common Goldeneyes molt in shallow (<2 m), sheltered, fresh, or brackish waters with no emergent vegetation (Jepsen 1973, Kumari 1979). However, more birds molt in brackish waters of fjords than on freshwater lakes; freshwater lakes are mostly used by females (Jepsen 1973).

Molting habitats of Barrow's Goldeneyes are better documented. The species uses a variety of habitats, all of which are characterized by shallow, productive waters. In Alaska, several thousand birds molt at Ohtig Lake, a large, shallow basin in the Yukon River watershed that is also used by molting Buffleheads, scaup (*Aythya* spp.), and American Wigeon (*Anas americana*) (King 1963). More than 5,000 males molt at the Old Crow Flats, northern Yukon, Canada, where density

of molting goldeneyes was correlated with lake productivity, suggesting that food abundance was an important factor in lake selection (Van de Wetering 1997). Recently, several molting lakes were located in central and northern Alberta, and most were shallow, productive lakes (Hogan et al. 2011). Most lakes supported <100 birds but a few had over 3,000 molting males. To date, no coastal sites in western North America have been identified as molting locations for either Barrow's or Common Goldeneyes. On the East Coast, the small population of Barrow's Goldeneyes molts in a variety of habitats from freshwater lakes to shallow estuaries and coastal brackish waters (Benoit et al. 2001, Robert et al. 2002). In North America, molting habitats of Barrow's Goldeneyes are located hundreds and even thousands of km from breeding areas (Robert et al. 2002, Hogan et al. 2011). However, in Iceland, birds molt in the same area where they breed (Gardarsson 1978), including Lake Mývatn, a large, spring-fed volcanic basin. Females often molt on lakes close to their breeding areas but also at some areas where males undergo remigial molt (Hogan et al. 2011, Savard and Robert 2013).

Buffleheads molt mostly on shallow lakes and are sometimes found on lakes used by goldeneyes (King 1963, 1973). Females often regroup on large, shallow lakes within their breeding area (Erskine 1972a). There is no published information indicating that Buffleheads use coastal waters during molt.

Fall and Spring Staging Habitats

Some goldeneyes fly along a direct pathway to their breeding areas from their wintering locations and do not stage for an extended length of time inland during spring migration (Robert et al. 2002, Savard and Robert 2013). Other individuals, especially birds wintering further south, use spring and fall staging areas. In the spring, close to their breeding areas, goldeneyes sometimes use lake and river outlets when breeding lakes are still frozen. During spring migration, Common Goldeneyes occupy intertidal areas of James Bay and especially sites with eelgrass (Reed et al. 1996). In the St. Lawrence River estuary, Barrow's Goldeneyes stage for a few weeks along the south shore before flying to breeding areas just a few hundred kilometers north (Bourget et al. 2007, Savard and Robert 2013). Buffleheads stage in the

Klamath Basin in northern California and southern Oregon for a few weeks each spring and autumn where they prefer seasonally flooded wetlands to permanently flooded ones (Gammonley and Heitmeyer 1990).

Satellite telemetry studies have indicated that Barrow's Goldeneyes remain at their molting sites until late fall and then fly directly to winter habitats (Robert et al. 2002, Savard and Robert 2013). Hogan et al. (2013) showed that survival of Barrow's Goldeneyes was lower during fall staging than during molting on lakes of the boreal transition zone of northwestern Alberta, mostly because of hunting mortality. Common Goldeneyes and Buffleheads likely have similar behavior. It is thus difficult to distinguish staging and wintering habitats. In the St. Lawrence River, Barrow's and Common Goldeneyes use different locations in winter than during spring and fall, but more or less similar habitats. The south shore of the estuary is not accessible in winter and forces goldeneyes to winter along the north shore. The south shore may be used in spring before their northward migration, but some goldeneyes remain on the north shore (Bourget et al. 2007). If goldeneyes remain at molting areas until late fall, they could be considered as staging birds, and they are indeed molting their body feathers during that time to revert from eclipse to the nuptial plumage.

Wintering Habitats

All three species frequent relatively shallow and protected waters and are especially common over intertidal areas, but are absent from open water areas (Vermeer 1982, Conant et al. 1988). However, they are rarely found in similar habitats. Barrow's Goldeneyes are closely associated with rocky foreshores (Savard 1989, Esler et al. 2000b) and are especially numerous in the coastal inlets of British Columbia and Alaska where the birds feed on mussels and associated invertebrates (Mitchell 1952, Koehl et al. 1982, Vermeer 1982, Hodges et al. 2008). In Alaska, Barrow's Goldeneyes capture most of their prey within 2 m of the water surface (Koehl et al. 1982). Common Goldeneyes avoid coastal inlets and prefer shallow estuarine conditions in North America and Europe (Pehrsson 1976, Vermeer 1982, Vermeer and Morgan 1992). In the St. Lawrence River, Quebec, Barrow's Goldeneyes winter in saltwater, whereas Common Goldeneyes winter in both

fresh and marine waters (Robert and Savard 2006, Drolet 2007). In saltwater habitats, both species prefer shallow depths and intertidal areas (Ouellet et al. 2010). In freshwater, goldeneyes are often associated with riverine systems (Sayler and Afton 1981, Eggeman 1986).

In the St. Lawrence River estuary, Barrow's and Common Goldeneyes are restricted to ice-free intertidal areas but segregate within these locations; Barrow's Goldeneyes are associated with rocky foreshores with boulders, whereas Common Goldeneyes are found in more sandy and estuarine conditions (Robert and Savard 2006, Bourget et al. 2007, Ouellet et al. 2010). In Iceland, Barrow's Goldeneyes winter in volcanic spring-fed freshwaters that are similar to their breeding habitats (Gardarsson 1978). Ice conditions influence the wintering strategies of Common Goldeneyes and Buffleheads as lipid reserves decline by >50% as the percent of ice cover on Lake Ontario increased from 0% to >40% (Schummer et al. 2012). Foraging effort increased as temperatures decreased, but where Common Goldeneyes could adjust foraging effort to compensate for lipid use, Buffleheads could not compensate because they foraged at full capacity for most of the winter (Schummer et al. 2012).

Buffleheads avoid rocky shorelines and are associated with estuarine environments. Several thousand birds winter in Chesapeake Bay, Maryland (Limpert 1980), and are numerous in most estuaries of Puget Sound, Washington (Hirsch 1980), British Columbia (Vermeer 1982, Butler et al. 1989), and Alaska (Hodges et al. 2008). Along the Atlantic coast, birds are mostly associated with soft, sandy, or mud substrates (Stott and Olson 1973).

As observed in other sea ducks (Iverson et al. 2003), winter habitats used by goldeneyes differ by age, sex, and pairing status (Nilsson 1970, Sayler and Afton 1981, Eggeman 1986). Pairs of Common Goldeneyes used shallower waters than unpaired birds (Eggeman 1986). In British Columbia, some pairs of Barrow's Goldeneyes defended winter territories in shallow water, whereas unpaired birds were found in deeper waters (Savard 1988a).

Goldeneyes regroup at dusk at their wintering areas and spend the night offshore in large groups and redistribute themselves each morning (Breckenridge 1953, Reed 1971, Sayler and Afton 1981). Buffleheads also roost in groups and depart after sunset to their roosting area, possibly to minimize predation (Finlay 2007). Size of roosting flocks and locations vary with temperature and local ice conditions; larger flocks are more common during cold nights and in ice-free areas. Nocturnal gathering may be an adaptation to conserve heat at night because it is most prevalent in winter.

HABITATS OF MERGANSERS

A primary feature of habitats used by mergansers (also known as sawbills or fish ducks) is the presence of fish and crustaceans. Mergansers occur in fresh and marine waters, including wetland ponds, lakes, rivers, estuaries, and nearshore coastal waters. Larger-bodied species of mergansers are known to take fry, parr, and smolts of trout and salmon. Thus, management actions to reduce or eliminate mergansers are often conducted in rivers where salmonid fish are present. Many of the studies on breeding ecology and food habits of mergansers have been conducted on rivers popular for sport and commercial fishing because of the real or perceived reduction in the numbers of salmon returning as adults.

Breeding Habitats

In North America, the breeding range of mergansers extends from the high arctic above tree line (>79°N) for Red-breasted Mergansers to the northern Gulf coast for Hooded Mergansers. Nest sites of North American mergansers vary substantially among species. Hooded and Common Mergansers are obligate cavity-nesting species in freshwater habitats, whereas Red-breasted Mergansers are an exclusively ground-nesting species in both freshwater and marine habitats.

Common Merganser

Common Mergansers nest primarily in cavities of trees or artificial nest boxes as a similar habitat but can occasionally be found on the ground in partially enclosed areas such as small caves, cracks in rocks, and under shrubs (Munro and Clemens 1937, White 1957, Bengtson 1970, Mallory and Metz 1999). Because of their need for cavities, the primary nesting habitat of Common Mergansers is in the western mountains to the edge of the boreal forest (Bellrose 1976, Palmer 1976). In northwest California and the central

Coast Range of Oregon, nesting habitats along rivers include coastal redwoods (*Sequoia sempervirens*) and riparian areas dominated by Douglas fir (*Pseudotsuga menziesii*), red alder (*Alnus rubra*), oaks (*Quercus* spp.), Pacific madrone (*Arbutus menziesii*), bigleaf maple (*Acer macrophyllum*), and western red cedar (*Thuja plicata*, Foreman 1976, Loegering and Anthony 1999).

Lemelin et al. (2010) used data from aerial surveys in the boreal forest of Quebec and examined numbers of indicated breeding pairs (IBPs) or the number of males in breeding plumage observed either singly or in groups of up to four individuals including females and unsexed birds. The authors concluded that Common Mergansers used nearshore waters of lakes and occupied both isolated and connected ponds as well as rivers and avoided offshore waters of lakes. Within those habitats, Common Mergansers selected shorelines of open wetlands and small islands in open water and avoided shorelines in flooded swamps and forests. Common Mergansers preferentially used streams in open wetlands and shrub, flooded swamps, and water bodies >8 ha. Rempel et al. (1997) described the habitats associated most frequently with pairs of Common Mergansers in spring on the breeding grounds and broods in Ontario. Based on analyses of observed and expected use of habitats, Rempel et al. (1997) showed that pairs of Common Mergansers differentially used large river riparian wetlands and open water in wetland complexes, ponds, and fens.

Brood requirements have been investigated at several sites. Broods of Common Mergansers were present in wetlands with moving water associated with beaver ponds and large lakes with marshy shores (Rempel et al. 1997). Broods of Common Mergansers were more abundant in larger than smaller streams (Loegering and Anthony 1999) and gradually moved from small streams near their nests to larger streams, lakes, and bays downriver (Erskine 1972b, Marquiss and Duncan 1994b, Loegering and Anthony 1999, Mallory and Metz 1999). As in Quebec and Ontario, Common Mergansers during summer in northern Wisconsin preferred deeper lakes with high water clarity (Newbrey et al. 2005). Wood (1985, 1986) examined pair and brood densities on salmon-producing streams on Vancouver Island, Canada, and found that Common Mergansers concentrated in areas where the density of prey was high. In summer, Newbrey et al. (2005) examined the presence or absence of Common Mergansers in relation to characteristics of small lakes in Michigan (<200 ha), including maximum depth, surface area, pH, water clarity, macrophytes, trees, boats, houses, and boat access (public or private) and type (gasoline or nongasoline powered). The presence of Common Mergansers was positively associated with clarity of water and maximum lake depth and surface area.

Red-breasted Merganser

Breeding Red-breasted Mergansers are widely distributed across North America at high latitudes (79°30′N; Hofmann et al. 1997) and northern regions of the United States (Titman 1999). Red-breasted Mergansers nest primarily in the tundra and boreal forests and islands in marine waters and large, freshwater lakes (Trauger and Bromley 1976, Titman 1999). Breeding habitats are highly variable and descriptions differ among locations. Unlike cavity-nesting species such as Common and Hooded Mergansers, Red-breasted Mergansers nest exclusively on the ground, often under vegetation, in dense cover, and in holes and crevices (Bengtson 1970, Trauger and Bromley 1976, Titman 1999, Craik and Titman 2009) and sometimes in association with gulls or terns (Young and Titman 1986, Titman 1999, Craik and Titman 2009). Based on IBP surveys in the boreal forest of Quebec, Lemelin et al. (2010) concluded that breeding pairs of Red-breasted Mergansers avoided offshore waters of open lakes and isolated ponds but selected nearshore waters of lakes and connected ponds. Red-breasted Mergansers frequently used shorelines of open wetlands and were found less than expected on forested shorelines and were most often in ponds >8 ha (Lemelin et al. 2010). In spring, the densities of Red-breasted Mergansers decreased with increasing distance upstream when the width of the river decreased, at higher elevations, and an increased slope of river bed (Marquiss and Duncan 1993). Råd (1980), Sjöberg (1989), and others have suggested that the movement patterns of Red-breasted Mergansers are tied to the activity patterns and movements of small, abundant fishes.

In river habitats during spring and summer, Red-breasted Mergansers tend to occupy wide stream sections and prefer slower, smoother river sections (White 1957, Gregory et al. 1997, Titman 1999). In some river systems, female Red-breasted

Mergansers move their ducklings downstream to the mouths of rivers or estuaries (Marquiss and Duncan 1993, Titman 1999). Red-breasted Mergansers in freshwater areas in western Norway were found breeding in lakes >15 ha and in the lower portion of rivers with calm and slow-moving water (Råd 1980). Duckling production of Red-breasted Mergansers in the river North Esk, Scotland, was lowest in upstream regions and highest in the downstream regions (Marquiss and Duncan 1993).

Craik and Titman (2008) evaluated brood movements and habitat use of Red-breasted Mergansers in the marine environment. The authors found that broods with young ducklings were led by females from nesting islands into intertidal estuaries or to the mouths of tidal streams. Red-breasted Merganser ducklings from 1 to 45 days of age were found primarily at the mouths of tidal streams, barrier island estuarine intertidal zones, and mainland estuarine intertidal waters and rarely in estuarine subtidal habitats (Titman 1999, Craik and Titman 2009). In general, Red-breasted Merganser broods used shallow (<0.5 m), nearshore waters (<50 m) with submerged eelgrass.

In a large lake complex in Iceland, recently hatched Red-breasted Merganser ducklings were likely restricted to areas where larvae of aquatic insects were abundant (Bengtson 1971a, b). As the ducklings became larger and older (>9 days), most Red-breasted Merganser females moved their broods to more open areas in the lake where there were higher densities of sticklebacks and little emergent aquatic vegetation (Bengtson 1971a, b). Broods of all ages roosted on shores with short grass and scattered shrubs.

Hooded Merganser

Hooded Mergansers nest in tree cavities and nest boxes throughout their breeding range and are almost exclusively restricted to wetlands that contain or are bordered by live and dead trees (Dugger et al. 1994, Sénéchal et al. 2008). Morse et al. (1969) found that if provided alternatives, Hooded Mergansers used nest boxes closest to water. In a study of nest box occupancy in Minnesota, Hooded Mergansers nested in coniferous forests, deciduous forest, and grasslands and were most common in regions having the highest lake and stream densities and where forest cover was most extensive (Zicus and Hennes

1988). Sénéchal et al. (2008) found that in the boreal forest of Quebec, nest box use by Hooded Mergansers tended to be a greater distance from trees ≥10 cm in diameter and with greater tree canopy than nest boxes that were not used.

Based on locations of breeding pairs of Hooded Mergansers observed during IBP surveys in Quebec, Lemelin et al. (2010) discovered that Hooded Mergansers avoided forested wetlands. Pairs of Hooded Mergansers selected ponds ≤8 ha; preferentially used connected and isolated ponds; were found on shorelines of open wetlands and flooded swamps, yet rarely at small islands or water bodies in forests; and frequented stream shorelines in open wetlands, shrub swamps, and flooded swamps (Lemelin et al. 2010). In contrast, Dugger et al. (1994) reported that Hooded Mergansers in Missouri seemed to select nest boxes placed along slow moving, heavily forested rivers and lakes, and nests were found in dense buttonbush/cypress swamps. During summer months of May–August, Newbrey et al. (2005) found that the presence of Hooded Mergansers on small lakes in Michigan (<200 ha) was positively influenced by the number of islands on a lake. At a finer spatial scale, McNicol et al. (1997) examined the presence of nesting Hooded Mergansers in Ontario in relationship to fish density, pH, lake area, number of wetlands within 500 m, and connectivity in smaller ponds and wetlands (<8 ha, range = 0.6–86 ha). Nesting attempts and reproductive success of Hooded Mergansers were not related to any of the variables measured, and the authors concluded these habitat characteristics were not important for nest site selection but might be important during brood rearing.

Broods of Hooded Mergansers can be found in all suitable habitats within a few kilometers of nests (Dugger et al. 1994). Suitable habitats vary from location to location within and among biomes. Characteristics of brood habitats, as summarized by Dugger et al. (1994), include "emergent marshes, small lakes, ponds, beaver wetlands, forested creeks and rivers, and swamps." In concordance with an examination of nest sites near ponds in the boreal forest in Ontario, McNicol and Wayland (1992) found that Hooded Merganser broods showed no preference for the presence of fish. The occurrence of Hooded Mergansers in lakes was not related to invertebrate species occurrences, the abundance or groupings of invertebrates, acidity, or other features (McNicol and Wayland 1992). In Ontario, Rempel et al. (1997) compared habitats in which pairs of Hooded Mergansers were seen in spring to habitats used by broods during summer. The habitat strata *Riparian: Beaver Pond Marsh* was used almost exclusively by pairs and broods, but pairs were also located in semienclosed fens with 5%–25% open water (Rempel et al. 1997). Hooded Merganser broods avoided deep, open water wetlands and small stream riparian wetlands.

Habitats used by broods in Wisconsin included a variety of types of rivers, streams, and marshes and was dominated by mixed upland and lowland hardwoods and black spruce, dead timber stands in the water, sedge meadows, and areas of lowland brush and mixed hardwoods adjacent to lakes, rivers, and streams (Kitchen and Hunt 1969). Kitchen and Hunt (1969) found that broods were located primarily in rivers rather than lakes and were more common in the faster currents (0.3 m/s) in wider rivers (12–18 m). Pine River was the primary location where broods were found and contained the highest number of fish and crayfish per km and had the greatest density of potential nest cavities per hectare (Kitchen and Hunt 1969). Pine River was relatively shallow (~0.6 m) with rocky bottom instead of a mud or sand bottom (Kitchen and Hunt 1969); broods were observed foraging in rocky areas of other rivers as well.

In Maine, Longcore et al. (2006) examined habitat use by Hooded Mergansers and other waterfowl in relation to water chemistry, invertebrate species and density, presence or absence of fishes in ponds, and other wetland characteristics including water area and vegetation features. The density of pairs of Hooded Mergansers tended to be higher on beaver ponds than glacial basins, but the difference was not significant. The authors concluded that water chemistry and macrophyte diversity affected macroinvertebrate abundance, which in turn influenced water bird use among wetlands. Thus, the lack of fish and resulting changes in invertebrate populations in some wetlands may have been beneficial to Hooded Mergansers.

Molting Habitats

Molting habitats of mergansers are known from only a few locations within their broad distribution. Little is known about molting habitats of Hooded Mergansers, primarily because large, easily located flocks of flightless birds are rarely found and detailed descriptions of molting locations are few. Molting habitats of Common and Red-breasted Mergansers in North America are better described (Reed et al. 1996, Craik et al. 2011).

Common Merganser

Like most other sea ducks, Common Mergansers undertake a molt migration. Based on observations of male Common Mergansers marked with satellite transmitters, marked birds molted over 1,000 km northeast to southwest of the area from likely breeding sites or nonbreeding sites used for spring staging (Stiller 2011). Male Common Mergansers used a variety of habitats during migration including streams, lakes, coastal areas, and estuaries. Molting sites were diverse but included coastal areas of James Bay and of northeastern Hudson Bay, small northern inland lakes, and a northern riverine system (Stiller 2011).

On the River Dee, Scotland, male Common Mergansers are often reported to move down river from nesting areas to river mouths and then on to molting areas (Marquiss and Duncan 1994a). Male Common Mergansers in North America molt in large freshwater lakes and marine bays and estuaries (Bellrose 1976, Pearce et al. 2009). On the River Dee, Scotland, female Common Mergansers molted on the river, and a few flightless males were found at the mouth of the river, but Marquiss and Duncan (1994b) concluded that most locally breeding male and female Common Mergansers molted elsewhere. Flightless female Common Mergansers are also found in flocks on large lakes in Alaska (J. M. Pearce, USGS Alaska Science Center, pers. comm.). Molting female Common Mergansers are also found in shallow, protected estuaries (Hatton and Marquiss 2004). Flocks of flightless male mergansers (*Mergus* spp.) have been located in coastal habitats of James Bay, Canada (Reed et al. 1996). Males of Red-breasted and Common Mergansers are generally indistinguishable in eclipse plumage (Mallory and Metz 1999, Titman 1999), and it is especially difficult to differentiate these species during aerial surveys. Reed et al. (1996) reported molting birds as mergansers but separated the two species whenever possible. During the molt, Reed et al. (1996) found that mergansers tended to congregate in open water habitats.

Red-breasted Merganser

Craik et al. (2011) evaluated habitat characteristics of a marine molting area used by Red-breasted Mergansers in the shallow waters around Anticosti Island in the Gulf of St. Lawrence. Characteristics of the molting habitat included clear, shallow waters (≤12 m in depth), near shores (<850 m), and rock–sand substrate with rockweed and kelp (Craik et al. 2011). Craik et al. (2011) found that molting male Red-breasted Merganser flocks tended to forage closer to shore (<850 m) and in shallow waters (<4 m in depth). At Anticosti Island, most feeding of flightless Red-breasted Mergansers occurred at low tide in the morning or evening (Craik et al. 2011). Foraging areas at Anticosti Island included vegetation, but flightless Red-breasted Mergansers primarily ate small fish (Craik et al. 2011).

Hooded Merganser

Little is known about the habitats used by Hooded Mergansers during remigial molt. Bellrose (1976) stated that "after leaving the female at the start of incubation, the male vanishes into limbo." Dugger et al. (1994) do not provide information on the ecology of molting Hooded Mergansers during the flightless period. During the molting period at James Bay, Reed et al. (1996) reported that almost all of the few Hooded Mergansers observed were in freshwater habitats. In northern Quebec, molting Hooded Mergansers were found in small groups on slow, meandering sloughs and rivers throughout the boreal forest and taiga (Savard 1977).

Fall and Spring Staging Habitats

Staging areas of mergansers through mainland and along coastal North America include all waters with suitable habitat with sufficient open water for takeoff. Stopover locations that birds use for more than a few days require appropriate foods in sufficient abundance for mergansers to gain reserves to continue migration. Detailed descriptions of the water chemistry, forage availability, and diet at staging areas are limited for any species of merganser.

Common Merganser

It is difficult to differentiate staging from molting, wintering, or breeding for Common Mergansers in regions where birds occur throughout the year. Males and some females will remain as far north as open water in rivers, lakes, and estuaries allow (Erskine 1972b, Marquiss and Duncan 1994b, Mallory and Metz 1999). In North America, migration tendency and distance of migration varies among Common Mergansers throughout their range (Pearce et al. 2005). Staging Common Mergansers have been reported in estuaries and nearshore waters (Vermeer and Morgan 1992) but have primarily been found at larger, freshwater lakes and reservoirs (Meissner and Niklewska 1993, Leslie et al. 1994, Mallory and Metz 1999, Spence and Bobowski 2003). At James Bay, Reed et al. (1996) found that staging mergansers foraged primarily in sparsely vegetated eelgrass beds, open water, and boulder-strewn tidal flats, although they were also found in tidal flats associated with salt marshes. Munro and Clemens (1937) indicate that mergansers concentrated at spawning sites of kokanee on lakes in British Columbia during August to November where birds stayed until the fish had completed the spawning season.

Red-breasted Merganser

Large flocks of Red-breasted Mergansers are found in freshwater habitats in the midwest, but most major stopover locations are in marine and estuarine habitats (Palmer 1976, Vermeer and Morgan 1992, Meissner and Niklewska 1993, Titman 1999). Red-breasted Mergansers foraged in relatively shallow waters (<9 m) in Lake Erie during migration (Bur et al. 2008). Red-breasted Mergansers were not found on the Colorado River during migration, but after the river was dammed, Red-breasted Mergansers used the newly created Lake Powell in Arizona and Utah during migration (Spence and Bobowski 2003). Additional detailed descriptions and use of habitats during migration are needed to fully understand the habitat relationships of Red-breasted Mergansers.

Hooded Merganser

Descriptions of habitats of spring and fall stopover sites used by Hooded Mergansers during migration are scarce and general in nature. Dugger et al. (1994) report that during migration, Hooded Mergansers are found in many habitats including open waters of rivers and lakes, brackish coastal bays, tidal creeks, seasonally flooded forest, dead timber, and scrub/shrub wetlands. Single individuals or small flocks of Hooded Mergansers with <13 individuals have been reported during migration at Lake Powell (Spence and Bobowski 2003).

Wintering Habitats

Mergansers can be found wintering in freshwater and marine habitats with sufficient food resources and open water. Wintering areas are often different for subadults, adult females, and adult males, with adult males wintering in areas north of those used by adult females. A difference in wintering location implies differences in habitats, although comparisons among wintering habitats between sexes and ages needs further examination.

Common Merganser

In winter, Common Mergansers almost exclusively occur in freshwater habitats on large lakes, reservoirs, lagoons, and rivers, and less commonly in brackish waters (Bellrose 1976, Palmer 1976, Heitmeyer and Vohs 1984, Leslie et al. 1994, McCaw et al. 1996, Mallory and Metz 1999). Common and Red-breasted Mergansers winter in fjords in northern marine waters (Vermeer and Morgan 1992); Common Mergansers were found in the highest density in estuaries without bays, whereas the highest densities of Red-breasted Mergansers were in estuaries with bays. In the Gulf of Gdańsk in the Baltic Sea, Red-breasted Mergansers preferred open seacoasts, whereas Common Mergansers were found more often in harbors and the mouths of rivers (Meissner and Niklewska 1993). Ice and floods can affect the distribution of Common Mergansers in marine and freshwater habitats (Erskine 1972b, Palmer 1976, Żydelis 2001, Faragó and Hangya 2012).

Numbers of Common Mergansers increased dramatically at Caballo Reservoir in the Rio Grande River, New Mexico, from a low of 300 birds in the winter of 1960–1961 to 29,000 birds in the winter of 1989–1990. During 1980–1994, numbers varied among years with more birds found in years with a smaller pool size (McCaw et al. 1996). At Caballo Reservoir and elsewhere, fluctuations in water levels among winters can have an effect on the abundance of fish and numbers of wintering Common Mergansers (McCaw et al. 1996). Both Common and Red-breasted Mergansers will winter in the lower reaches of rivers, which remain ice-free in winter (Marquiss and Duncan 1994b, Cosgrove et al. 2004). Several thousand Common Mergansers winter in the St. Lawrence River near Montreal in open waters below a hydroelectric dam apparently feeding on fish that pass through the dam (Reed and Bourget 1977).

Red-breasted Merganser

Although found in freshwater habitats, Red-breasted Mergansers primarily winter in marine environments (Titman 1999). When wintering together, habitat preferences of Red-breasted Mergansers differ slightly from Common Mergansers. A few Red-breasted Mergansers now winter in Lake Powell during some years, where birds were not seen before the dam was built on the Colorado River (Spence and Bobowski 2003). In northeast Scotland, Richner (1988) analyzed Red-breasted Merganser movements along a gradient between freshwater and seawater within an estuary. The greatest number of Red-breasted Mergansers was 2–4 km above the estuary mouth but the salinity of these areas was not determined. Month and time of day explained the total variation of numbers of Red-breasted Mergansers within the estuary. Richner (1988) also found an increase in numbers of Red-breasted Mergansers with the rising tide and early in the morning; daily tidal cycles, time of day, month, or difference in tidal ranges did not influence the abundance of Red-breasted Mergansers in the different sections of the river from freshwater to marine habitats. Richner (1988) suggests that the section of the estuary that was preferred had a greater diversity and abundance of foods as a result of nutrients supplied by sewage water influx from a nearby village.

Holm and Burger (2002) examined the numbers of Red-breasted Mergansers in relationship to areas with strong tidal currents in nearshore waters of Vancouver Island, Canada. Red-breasted Mergansers were generally found in slack water <10 m in depth, with relatively few birds in eddies and none in turbulent water (Holm and Burger 2002). In the North Sea, Red-breasted Mergansers were often associated with floating seaweeds and debris and likely fed on the macrofauna and small fishes beneath floating vegetation and objects (Vandendriessche et al. 2007). Of all the sea ducks, Red-breasted Mergansers have the widest range of wintering areas in terms of latitude and temperature. In eastern North America, the species commonly winters from the St. Lawrence River estuary and Gulf to Cuban coastal waters (J.-P. L. Savard, unpubl. data).

Hooded Merganser

Hooded Mergansers winter in a wide variety of freshwater and marine habitats. In coastal areas, Hooded Mergansers are found in shallow waters of freshwater and brackish bays, estuaries, and tidal creeks and ponds (Dugger et al. 1994). In forested wetlands, Hooded Mergansers variously used permanently flooded sloughs, backwaters, rivers, dead tree, beaver pond, scrub/shrub, and sites with overcup oak (*Quercus lyrata*), likely in relationship to crayfish abundance and availability (Fredrickson and Heitmeyer 1988). During aerial surveys of wetlands during winter in Oklahoma, Heitmeyer and Vohs (1984) found Hooded Mergansers on reservoirs and in riverine habitats. Holm and Burger (2002) found wintering Hooded Mergansers in nearshore marine waters and almost exclusively in slack waters ≤5 m in depth. Birds were not seen on turbulent water and only rarely in eddies and the main flow.

DYNAMICS OF HABITATS

Habitats are not static, which requires sea ducks to either adapt to new conditions in areas of traditional use or move to alternative places. Changes in critical habitats that are exploited by sea ducks result from natural events and human activities. Habitat degradation and loss are considered limiting factors for some species of sea ducks in North America (Sea Duck Joint Venture Management Board 2008). Here, we will explore some of the physical and biological processes that affect breeding, postbreeding, and wintering habitats of sea ducks in North America, and we draw also on supporting evidence from studies conducted outside the continent.

Natural Events

Natural events are usually considered to be environmental conditions that are not associated with human activities. Examples may include drought and other climatic phenomena, geological processes such as earthquakes and volcanic eruptions, and wildfires. Some factors that result in alteration of ecosystems and sea duck habitats can be either natural, man caused, or a combination. Fire can be the result of lightning strikes, but many forest and prairie fires are also started by humans, and both can result in demonstrable changes to ecosystems. In this section, we consider several natural conditions and events that shape habitats and influence the response and population dynamics of sea ducks.

Water Variability

Water levels and quality are especially critical in ensuring that freshwater and marine habitats support sea ducks. Many species of sea ducks use coastal and riparian habitats during the reproductive period, and nest losses can be high during seasonal flooding associated with storm events. In western Alaska on the Yukon–Kuskokwim Delta, tidal flooding is often associated with coastal storms, but the frequency and extent of nest failure of sea ducks as a result of these events is not well understood (Mickelson 1975, Petersen et al. 2000).

Drought is a well-known limiting factor for nesting dabbling ducks, especially in prairie and parkland wetlands of North America where birds nest in wet years (Weller and Spatcher 1965). Many ducks may overfly these areas to more northern areas in years when basins are dry (Hansen and McKnight 1964, Smith 1970, Derksen and Eldridge 1980). Common Goldeneye, Bufflehead, White-winged Scoter, and other sea duck species associated with these biomes may be negatively affected by increasing aridity associated with climate change in prairie and parkland ecosystems. Larson (1995) modeled dynamics of northern prairie wetlands to examine potential effects of temperature and precipitation changes on basins holding water and compare sensitivities of wetlands across a broad geographic area of the prairie pothole region (PPR). The models explained more than 60% of the variation in number of wet basins throughout the PPR study area, with parkland wetlands much more vulnerable to increased temperatures than wetlands in either Canadian or United States grasslands. Larson (1995) concluded that anticipated climate changes may result in both fewer and lower-quality wetlands for waterfowl production in the PPR. Johnson et al. (2005) developed a wetland simulation model to demonstrate that the most productive habitat for breeding waterfowl would shift to the eastern and northern fringes of the PPR under a predicted dryer climate.

By contrast to the prairie and parkland biomes, the boreal forest and associated streams, lakes, and wetlands of North America provide significant breeding and molting habitats for several species of sea ducks. The boreal forest in Canada alone contains 25% of the world's wetlands, and combined with the boreal forest of Alaska, this North American biome includes about 1.5 million lakes and many millions of wetlands (U.S. Fish and Wildlife Service 2013). Water bodies in boreal ecosystems are increasingly subject to evaporation and low precipitation associated with recent climate warming, which results in closed basins if water levels drop below the threshold for stream outflow. Loss of connectivity leads to increased salinity, which can cause gradual changes to more salt-tolerant taxa of plants and animals (Schindler and Lee 2010). Furthermore, boreal bog and fen plant communities will likely evolve in response to regional or global climate warming and/or alterations in water-table elevation (Weltzin et al. 2003), with uncertain consequences to sea ducks, especially during the nesting period.

Forest Fires

An especially important habitat for breeding sea ducks in North America is the boreal forest, which is dominated by coniferous trees, peatlands, and a diversity of wetlands. The boreal forest covers a broad, contiguous expanse of land from Newfoundland and Labrador westward across the central and northern Canadian provinces and through the Yukon Territory to interior and western Alaska. Among the Mergini, only the four species of tundra-nesting eiders do not utilize the boreal forest during the breeding period. Fire is a critical process in the boreal forest as these periodic events organize plant communities, successional patterns, and ecosystem structure and function across landscapes (Rowe and Scotter 1973, Weber and Stocks 1998). In Canada, fire occurrence has increased steadily from ~6,000 fires per year in the 1930–1960 period to about 10,000 fires per year in the 1980s, with an ongoing average of ~10,000 fires per year from 1990 through 1998 (Weber and Stocks 1998). A similar trend has occurred in Alaskan boreal forests where half of the largest fire years during a 56-year record occurred since 1990 (Chapin et al. 2008).

Kasischke and Turetsky (2006) calculated that the area burned across the entire North American boreal forest biome tripled from the 1960s to the 1990s because of an increased frequency of fire. Kasischke et al. (2006) suggest that increased fire activity over the past four decades is a result of an increase in average ambient temperatures in the boreal forest region of North America. General circulation models (GCMs) predict future increases in fire levels with warming conditions in these forests (Stocks et al. 1998, 2000). Populations of cavity-nesting sea ducks, such as goldeneyes, Bufflehead, Common Mergansers, and Hooded Mergansers, could potentially be negatively affected at a local level in boreal areas where nesting trees have been reduced by fire (Eadie et al. 1995). No field studies have been conducted yet to quantify such impacts. Some populations of cavity-nesting sea ducks might be reestablished before recovery of burned areas through nest box programs (Eadie et al. 1995), but such projects are labor intensive, subject to logistical limitations in remote areas, and artificial boxes require periodic maintenance to be fully effective. Eadie et al. (2000) noted that such efforts could be detrimental by exposing breeding females to a greater risk of predation if nest boxes are easier to locate or access than natural cavities. Haszard (2004) found that Surf and White-Winged Scoters avoided wetlands in recently burned areas, but the transitory response lasted only 2 years until ground vegetation had regrown, suggesting that the absence of nesting cover is important in the selection of a wetland. Clearly, we need a much better understanding of how increasing fire activity associated with escalating ambient temperatures affects population status of cavity-nesting species and other sea ducks, especially across the boreal forest biome.

Role of Fish in Sea Duck Habitat Selection

The three merganser species in North America have evolved to exploit freshwater and marine fishes throughout their life cycles; their primary breeding, staging, and wintering habitats include ponds and lakes, fluvial systems, and oceans that support an abundance and diversity of fish species. Other sea ducks, such as Long-tailed Ducks, have varied diets that may include a small amount of fishes or fish roe in some seasons. However, relatively little is known about habitat relationships between fish versus sea ducks that are nonpiscivorous or consume fish and fish eggs as a minor portion of their diet. Eriksson (1979) was the first study to examine the question of whether fish reduce the availability of invertebrate foods preferred by Common Goldeneye. Eriksson's experimental research on Swedish lakes demonstrated that some aquatic insect groups proved sensitive to predation by perch (*Perca fluviatilis*) and roach (*Leuciscus rutilus*) and that availability of these invertebrate foods was negatively affected for Common Goldeneyes. Eadie and Keast (1982) tested Eriksson's (1979) hypothesis that Common Goldeneye and yellow perch (*Perca flavescens*) compete for invertebrate prey in lakes of Ontario, Canada. The authors also found a high degree of diet overlap between perch and goldeneye and an inverse relationship between fish and duck abundances within the study lakes and potentially across large geographic areas in North America where the species overlap. Taylor (1986) evaluated diets of Long-tailed Ducks and least cisco (*C. sardinella*) in tundra lakes on the North Slope of Alaska. He compared stomach contents of cisco and Long-tailed Ducks to lake samples of zooplankton and found that size selective predation by cisco prevented large zooplankton species such as *Daphnia middendorffiana*, *D. pulex*, *Branchinecta paludosa*, and *Polyartemia hazeni* from inhabiting or becoming an important component of the community in basins with planktivorous fish. In Finland, Pöysä et al. (1994) documented the disappearance of fish in lakes subject to acidification from industrial pollution. Their research showed that acid-induced disappearance of perch (*P. fluviatilis*) in small forest lakes may benefit breeding pairs and broods of Common Goldeneyes who were not subject to competition for key aquatic invertebrates. DesGranges and Gagnon (1994) evaluated the potential for a reduction in invertebrate food abundance resulting from lake acidification and implications on trophic relationships in Quebec, Canada. Their model indicated that lake productivity, lake acidity, and competition between ducklings of Common Goldeneyes and American Black Duck (*Anas rubripes*) with brook trout (*Salvelinus fontinalis*) affected prey availability and the sharing of food resources between ducklings and fish. Robert et al. (2008a) explored habitat preferences of Barrow's Goldeneyes breeding in eastern North America and found a clear negative relationship between occurrence of goldeneyes and brook trout. The negative relationship probably reflects a positive linkage between these ducks and highly productive aquatic ecosystems where there is a lack of competition with fishes for aquatic insects (Robert et al. 2008a). Similarly, LeBourdais et al. (2009) determined that the presence of fish, such as introduced rainbow trout (*Oncorhynchus mykiss*), lowers the quality of streams and rivers for breeding by Harlequin Ducks in British Columbia, Canada. By contrast, Epners et al. (2010) found that Common Goldeneye densities in the boreal transition zone of western Canada were greater in lakes with large-bodied fish such as yellow perch, than lakes without fish or small-bodied species, which the authors conclude may be due to abundance of foods that would preclude competition between waterfowl and fishes. In summary, nonpiscivorous sea ducks select breeding habitats devoid of fish competitors in oligotrophic systems in eastern North America or, alternatively, highly productive systems that provide adequate aquatic invertebrates to support both fish and ducks.

Anthropogenic Activities

Human activities have played an important role in altering terrestrial, freshwater, and marine habitats critical to sea ducks, not only in North America but across the circumpolar arctic, subarctic, and temperate zones used by these species during breeding and postbreeding periods. A few man-caused changes to landscapes may have resulted in opportunities for sea ducks to expand their range or otherwise increase survival.

Shellfish Industry

Oyster, clam, and mussel harvest and aquaculture in marine habitats have expanded on the Atlantic and Pacific coasts of North America and northern Europe. Aquaculture activities may affect molting, staging, and wintering areas by destroying natural habitats and by directly excluding molluscivorous sea ducks from areas of traditional use (Savard et al. 1998), or alternatively, these fisheries may provide novel food resources for sea ducks. Shellfish operations result in modifications to nearshore habitats and thus may have an impact on sea ducks that depend on these areas for foods.

Coastal British Columbia has experienced modifications to nearshore habitats, especially in the Baynes Sound area, as a result of the commercial shellfish aquaculture industry. To understand the potential impact of the shellfish industry on wintering Surf and White-Winged Scoters, Žydelis et al. (2006) quantified habitat use by sea ducks in relation to natural environmental attributes and habitat modifications associated with shellfish aquaculture. Despite extensive clam and oyster farming in their study area, densities of scoters were related primarily to natural environmental attributes, particularly intertidal area, clam density, and sediment type. While current intensities and practices of shellfish aquaculture in Baynes Sound, British Columbia, have not had a strong negative impact on habitat use by these two scoter species, Žydelis et al. (2006) cautioned that further industrialization in British Columbia could eventually lead to detrimental effects if the level of habitat change approaches that associated with wild shellfish harvest (Camphuysen et al. 2002, Oosterhuis and van Dijk 2002, Atkinson et al. 2003). Further investigations into possible effects of the shellfish aquaculture industry on sea duck habitats in British Columbia by Kirk et al. (2007) revealed especially interesting and unexpected results. These authors compared density and morphology of wild bay mussels (*Mytilus trossulus*) growing naturally on shellfish farming rafts, buoys, and associated lines to that of mussels from adjacent intertidal areas. Mussel density on shellfish farming structures greatly exceeded that of areas free of structures. Importantly, farm mussels were larger, had lower shell mass, and had weaker byssal attachments, attributes that make these mussels more profitable for foraging sea ducks. Furthermore, Kirk et al. (2007) showed that Surf Scoters and Barrow's Goldeneyes exploited mussels on aquaculture structures, which in turn suggest management options that could enhance marine foraging habitats for sea ducks that depend on bivalves as a primary source of food.

In eastern North America, large-scale harvest of blue mussels, as well as sea urchins and marine algae, has resulted in degradation of traditional winter habitats in Maine, New Brunswick, and Nova Scotia (Goudie et al. 2000). Aquaculture of molluscs along the eastern seaboard has led to exclusion of sea ducks from prime intertidal habitat and, in some cases, depredation of mollusc farms by these birds because sites overlap with traditionally used littoral ranges.

Sewage and Seafood-Processing Effluent

Sewage lagoons can attract waterfowl, including some sea duck species, because they are phosphorus- and nitrogen-enriched habitats with high densities of invertebrates (Baldassarre and Bolen 2006). In the Aleutian Archipelago of Alaska, Steller's Eiders and Harlequin Ducks were attracted to the outfall area associated with a seafood processing plant and municipal sewage system at Dutch Harbor, likely because of improved feeding opportunities associated with invertebrate abundance enhanced by eutrophication (Reed and Flint 2007). Increased abundance of invertebrates provided enhanced foraging opportunities for these two sea ducks at Dutch Harbor, but Reed and Flint (2007) documented an important negative consequence to birds concentrated at outfall sites—birds were subject to a sustained, unnaturally high level of predation by a large population of Bald Eagles (*Haliaeetus leucocephalus*). Other limiting factors for sea ducks attracted to outfall areas included exposure to contaminants and pathogens (Skei et al. 2000). Research conducted by Hollmén et al. (2011) at the Dutch Harbor, Alaska, outfall site showed that prevalence of *Escherichia coli* in fecal samples from Steller's Eiders was 16% compared to 2% presence in birds sampled at a control location far from the industrialized site. *E. coli* levels in Harlequin Ducks from the outfall area at Dutch Harbor were much higher (67%) than levels found in Steller's Eiders. Most importantly, Hollmén et al. (2011) found evidence of avian pathogenic *E. coli* in both species and detected strains that carried virulence genes associated with mammals in Harlequin Ducks. What is not yet understood is the extent to which sea ducks may be affected by nearshore microbial exposure associated with outfalls, but Hollmén et al. (2011) concluded that at their study site, ~5% of the local wintering population of Steller's Eiders was subject to impacts from exposure to *E. coli*.

Contaminants

A vast array of contaminants has been isolated from breeding, staging, and wintering habitats used by sea ducks (Chapter 6, this volume). We now have documented levels for many heavy metals, trace elements, pesticides, persistent aromatic hydrocarbons, fertilizers, and petroleum hydrocarbons in several species of sea ducks and have an improved understanding of how these contaminants affect duck physiology and survival. Less attention has been devoted to how various contaminants alter freshwater wetland, oceanic, and terrestrial habitats of sea ducks. Nevertheless, there are studies that focus on these questions and provide useful insights into some of the mechanisms and significance of changes that have occurred. For example, the Canadian Shield lakes in southern Quebec experienced a lowering of pH levels as a result of acid precipitation caused by airborne pollutants, especially high-sulfur fuels. Acidification occurs because these waters pass through areas where the substrate is formed of minerals with a low carbonate content, which reduces buffering (Bobée et al. 1982, DesGranges and Darveau 1985). Acidification of waters affects the availability of inorganic nutrients and carbon for macrophytes, and increases in H^+ ions may also directly affect composition of aquatic plants in lakes with some species eliminated, while others tolerant of acid conditions respond favorably (Farmer 1990). Highly acidic lakes (pH < 4.5) can significantly reduce or eliminate macroinvertebrates from the water column and even benthic organisms making these basins unattractive to aquatic birds (DesGranges and Darveau 1985). Experimental acidification of a small lake demonstrated significant changes in the lake's food web including alteration of the phytoplankton complex, cessation of fish reproduction, disappearance of benthic crustaceans, and appearance of filamentous algae (Schindler et al. 1985). Interestingly, Common Goldeneyes in the Canadian Shield lakes area benefited from limited lake acidification because they avoided basins with yellow perch (*P. flavescens*), which are absent from highly acidic lakes. Absence of fish likely provides reduced competition for goldeneyes for the available invertebrate foods (Eriksson 1984, DesGranges and Darveau 1985). By contrast, a study of the relative abundance of Common Goldeneyes in Sweden and Finland showed no correlation with lake pH (Elmberg et al. 1994). Potential impacts of lake acidification and changes in zooplankton and fish on other sea duck species is unknown (Beamish 1976, Havens et al. 1993), but clearly worthy of further investigation and modeling (DesGranges and Gagnon 1994).

Clear-Cutting and Other Silviculture Methods

Boreal forests are a primary source for an ever-increasing demand of wood fiber, and large areas of commercially valuable timberlands are scheduled for cutting. Clear-cutting is the most common harvest method employed in the boreal forests of North America, especially in Canada. While the practice is widespread and much debated, impacts on forest dynamics and environmental change are poorly understood (Dussart and Payette 2002). Potential ecological effects of clear-cutting and other forest management practices on habitats important to sea ducks include change in tree and other vegetation diversity from preharvest communities. Fragmentation of mature forests into smaller areas result in an increase in the proportion of forest-edge habitat (Andrén 1994, Schmiegelow and Mönkkönen 2002); reduced water quality due to elevated flows, silt loads, and chemical concentrations (Schindler 1998b); diminished snag density and suitable natural cavities (Vaillancourt et al. 2009); and reduced soil fertility (McRae et al. 2001). Many of these and other potential impacts to nesting sea ducks remain unstudied, but some recent investigations have addressed responses of several species of cavity-nesting sea ducks to forest habitat change associated with clear-cutting.

Pierre et al. (2001) examined the effects of habitat destruction and fragmentation on nest predation of cavity-nesting sea ducks and other waterfowl in the boreal mixed-wood forest in north central Alberta, Canada. The experimental research demonstrated that predation of nests in artificial cavities was significantly lower in cutblocks than in uncut forest 150 m from lakeshores. At harvested and unharvested lakes, predation did not differ significantly in uncut forest at 50 m versus 150 m from the lakeshore (Pierre et al. 2001). More importantly, it is the reduced stem density and loss of natural cavities as potential nesting sites in harvested areas of the boreal forest in Quebec, Canada, that are likely most limiting for Barrow's Goldeneye and other cavity-nesting sea ducks (Vaillancourt et al. 2009). The authors found highly detectable cavities in clear-cuts but concluded that an absence of recruitment in dead trees will eventually lead to a deficit in cavities and potential cavity trees as decaying trees collapse. Landscape-scale conversion of the eastern Canadian boreal forest from a mixed to a deciduous cover has resulted in dramatic changes to avian communities. New growth of deciduous trees, if allowed to mature, may eventually provide cavities large enough for nesting sea ducks. Drapeau et al. (2009) argue for ecosystem-oriented management approaches in eastern boreal forests where live and dead tree retention should be paramount in forestry practices to ensure maintenance of key ecological processes and conservation of snag-dependent avian species. More comprehensive knowledge on the overall effects of forest harvesting on breeding habitats is needed, and especially for cavity-nesting sea ducks (Lemelin et al. 2007).

Wind Turbine Facilities

Worldwide, wind power installations have increased in number and size in recent years as demand for renewable and cleaner sources of energy has escalated in response to concerns about climate change associated with burning carbon-based fuels (Chapter 14, this volume). What do we understand about potential changes to offshore marine habitats important to sea ducks as a result of construction of wind farms? Fox et al. (2006) evaluated European wind farms in terms of habitat loss, modification, or gain. The authors categorized the effects of at-sea construction as: (1) destruction of feeding habitats under turbine foundations or anti-scour structures (boulders), and (2) creation of novel habitats on turbine foundations or anti-scour structures. Initial evidence from Denmark wind farms suggests that the extent of physical loss to turbine foundations and to anti-scour protection provisions is at a relatively small scale of not more than 2% of the total area of a wind farm (Fox et al. 2006). Direct habitat loss resulting from at-sea placement of turbines is thought to be small, but Drewitt and Langston (2006) note that effects could be more widespread if development disrupts geomorphological processes in offshore habitats, leading to environmental changes such as increased erosion. At present, no evidence indicates that marine wind farms in Denmark have created new habitats or food resources for sea ducks (Fox et al. 2006). The authors suggested that potential changes in habitat could be indirectly measured by continuing to monitor bird densities at marine wind farms.

An indirect measure of habitat loss associated with offshore wind farms is avoidance behavior of sea ducks to the presence of turbine fields. Masden et al. (2009) evaluated avian avoidance responses during pre- and post-construction periods at an offshore wind farm in Denmark. The authors used surveillance radar to measure flight trajectories of about 200,000 migrating Common Eiders. The curvature of eider trajectories was greatest post-construction and within 500 m of the wind farm, with a median curvature significantly greater than preconstruction, suggesting that the eiders adjusted their flight paths in the presence of the wind farm (Masden et al. 2009). At another Danish wind farm, Larsen and Guillemette (2007) conducted a series of controlled experiments to determine whether avoidance responses of Common Eiders was due to the standing towers or the revolving rotor blades of the turbines. The results showed that the avoidance behavior of eiders was caused by the presence of the structures rather than movement or noise of rotors. Available evidence from these projects and other European studies indicates that avoidance behavior of sea ducks can result in reduction in habitat availability within and around at-sea wind farms (Fox et al. 2006). This issue could become an important limiting factor for sea duck populations as wind farms proliferate in marine habitats around the world, although Stewart et al. (2007) argue that the evidence base for assessment of wind farm impacts on birds from current studies is poor and more long-term impact assessments are required.

Hydroelectric Development

By the mid-1990s, over 52% of the large river systems in northern North America were moderately or strongly affected by fragmentation of their river channels due to dams and water regulation resulting from reservoir operation, interbasin diversion, and irrigation (Dynesius and Nilsson 1994). With the exception of the Yellowstone River in Montana, all rivers greater than 1,000 km in length in the contiguous 48 states of the United States have been severely altered for hydropower or navigation (Benke 1990). New hydroelectric projects have been proposed and several completed in boreal and tundra biomes, especially in northern Quebec. Collectively, these developments have raised concerns about potential

impacts to breeding habitats of sea ducks including Harlequin Ducks (Robertson and Goudie 1999), Black Scoters (Bordage and Savard 1995), and Surf Scoters (Savard et al. 1998). Broad-scale trends of habitat loss and loss of ecosystem function associated with hydropower dams are not well documented (Benke 1990), although recent research offers some insights into impacts that are relevant to sea ducks, especially those that breed in boreal habitats.

Flooding of forested habitat from hydroelectric megaprojects removes nesting areas exploited by sea ducks. An example is the James Bay Project in Quebec, Canada, which dammed nine free-flowing rivers on the east side of James Bay and created reservoirs that cover an area of more than 13,300 km². A massive impounded lake system replaced important boreal forest and wetland breeding habitats of Common Goldeneye, Bufflehead, Harlequin Duck, Black Scoter, Surf Scoter, Red-breasted Merganser, and Common Merganser, but no estimates are available for the significance of lost habitats to the status of affected populations.

In addition to direct loss of forest and wetland nesting and brood-rearing habitats due to flooding, hydroelectric dams affect watersheds in other important ways. Consider the potential consequences of impoundments to small tributaries upstream of a dammed river. Fast-flowing streams in riparian, subalpine, and coastal habitats are favored breeding sites for Harlequin Duck, but altered flow dynamics associated with damming may render these fluvial systems unsuitable (Robertson and Goudie 1999), because of changes to the biological and physical characteristics of channels (Bednarek 2001). An example of such a transformation that would impact Harlequin Ducks on breeding areas, where their diet is predominantly aquatic insects (Robertson and Goudie 1999), include modifications of aquatic insect emergence timing and subsequent changes in prey availability (Jonsson et al. 2012a,b). Englund and Malmqvist (1996) compared macroinvertebrate abundance and diversity in regulated and unregulated Swedish rivers. Where flow patterns were altered by dams, total abundance and species richness of invertebrates were lower than on streams that had natural flow regimes, which would have important implications for sea ducks and other predators dependent on these food resources. Another consideration of hydroelectric projects concerns potential effects of river regulation on vegetation of river margins. Nilsson et al. (1991) documented lower plant cover and species richness in a regulated river compared to a free-flowing stream in Sweden. Jansson et al. (2000) compared riparian vegetation at eight boreal rivers in Sweden; four that were without dams and four strongly regulated for hydroelectric purposes. Plant species richness and cover were reduced in run-of-river impoundments and storage reservoirs due to water-level fluctuations (Jansson et al. 2000). In addition to alteration of riparian and aquatic habitats resulting from variation in flow regimes below dams (Collier et al. 1996), other downstream effects include loss or alteration of coastal and delta habitats as a result of reduction of sediments that are trapped in reservoirs (U.S. Department of the Interior 1995). Results of these studies should be considered when evaluating potential impacts to sea ducks, especially those species dependent on riparian and delta habitats during nesting and brood-rearing periods.

Creation of large impoundments as a result of damming rivers in some geographic areas of North America has resulted in new habitats of value to sea ducks. Reservoirs are known to attract breeding (Duda et al. 2008), migrating (Thompson 1973), and wintering populations of sea ducks (Hobaugh and Teer 1981, Heitmeyer and Vohs 1984). River sections that are downstream of dam sites can provide habitat for wintering populations of goldeneyes and mergansers because heavy flows and heated water discharged from electrical generating plants maintain open water (Sayler and Afton 1981, Jones and Drobney 1986). As Baldassarre and Bolen (2006) note, reservoirs are of growing interest to managers because as natural wetlands continue to decline, these artificial habitats take on greater importance as waterfowl habitat. However, positive aspects of reservoirs as habitats for goldeneyes, mergansers, and Bufflehead during the nonbreeding season should not be considered mitigation for important losses to breeding habitats in more northern areas of North America.

As demand for alternatives to fossil-fuel-fired generating facilities escalates, it is expected that additional large-scale hydroelectric projects will be developed in sea duck habitats in arctic and subarctic areas of North America. Understanding the significance of these developments to the habitats and dynamics of sea duck populations should be a critical component of the preconstruction review process, which has not previously been addressed.

Alien or Invasive Species

Alien species are defined in several ways in the literature. For our discussion about dynamics of sea duck habitats, we define an alien or invasive species as one that occurs outside its normal distribution, and because the species cannot reach this new location by its own means, human activity of some kind is involved in moving or introducing the species concerned across a biogeographical barrier (Shine et al. 2000). Worldwide, species invasions are considered important threats to conservation and have caused numerous extinctions of native biota. Invasive species can alter the evolutionary pathway of native species by competitive exclusion, niche displacement, hybridization, introgression, predation, and ultimately extinction (Mooney and Cleland 2001). Ecosystem modification is considered the most significant impact of invasive species because such changes are likely to affect most of the originally resident species, and physical structure of the habitat can be changed as a result (Simberloff 2010). Importantly, nonnative species are not the only factor controlling the ecosystems that they invade; factors such as disturbance, hydrology, and weather conditions may interact strongly with the invading species (Strayer et al. 2006). Is there evidence that ecosystems and habitats important to breeding, staging, or wintering sea ducks have been compromised by the invasion of nonnative plants or animals? If so, what are the consequences to the dynamics of populations of sea ducks?

The introduction of the nonnative zebra mussel (D. polymorpha) into North America from Europe likely occurred as a result of discharge of larvae from ship ballast water. Zebra mussels were first discovered in 1988 in Lake St. Clair but may have been present as early as 1986 (Griffiths et al. 1991). Thereafter, Dreissena spread rapidly throughout the Great Lakes region and into large navigable rivers including the Detroit, Mississippi, Tennessee, Cumberland, Ohio, Arkansas, Illinois, St. Lawrence, Niagara, and Hudson. Zebra mussels have spread to other water bodies in at least 28 states and 2 provinces (Benson 2011). Strayer (1991) predicted that this exotic bivalve would spread over much of North America with only the extreme northern and southern parts of the continent unlikely to be colonized. Invasion of *Dreissena* into North American aquatic ecosystems has already resulted in significant abiotic and biotic changes (MacIsaac 1996), including declines in algae abundance and biovolume (Nichols and Hopkins 1993, Idrisi et al. 2001), reduction in chlorophyll-*a* (Fahnenstiel et al. 1993, Higgins et al. 2011), declines of phytoplankton biomass (Caraco et al. 1997), reduced levels of dissolved organic carbon, (Johengen et al. 1995), degradation of oxygen resources (Effler et al. 1996), declines of zooplankton (MacIsaac et al. 1995), and reductions or extirpation of native clams in the family Unionidae (Mackie 1991, Ricciardi et al. 1995, Schloesser et al. 1996) and Sphaeriidae (Lauer and McComish 2001). Zebra mussels also affect the dynamics of polychlorinated biphenyls (PCBs) through biomagnification (Bruner et al. 1994). The issue is a concern because PCBs are hazardous chemicals and changes in transfer pathways could lead to increases in the chemical body burdens of aquatic biota (Morrison et al. 1998).

Densities of zebra mussels up to 800,000/m² have been reported in the Great Lakes region (reviewed by Effler et al. 1996). Several species of sea ducks have responded to this newly available and abundant food source. Custer and Custer (1996) evaluated the diet of bay ducks and sea ducks on Lake Erie and Lake St. Clair and found that zebra mussels were the dominant food consumed by Common Goldeneyes (79.2% as an aggregate percentage) but were less important for Buffleheads (23%). However, the authors caution that it is unclear whether this invasive mussel provides adequate winter forage for waterfowl because *Dreissena* is a low-energy food by comparison to other high-energy forage, including rapidly declining native bivalves. Ross et al. (2005) suggested that dietary shifts from nonfilter-feeding gastropods to filter-feeding zebra mussels may be linked to elevated contaminant burdens in Lesser Scaup (*Aythya affinis*) and Greater Scaup (*A. marila*) in the lower Great Lakes. Indeed, Mazak et al. (1997) clearly demonstrated this pattern in western Lake Erie where concentrations of PCBs were significantly higher in scaup and Bufflehead that consumed *Dreissena* than in individuals that ate mainly macrophytes, although it is unknown whether any of the sampled waterfowl that exploited *Dreissena* experienced adverse health effects, such as impairment of reproductive performance (Fox

1993). Custer and Custer (2000) examined four species of waterfowl in Lake Erie, Lake St. Clair, and Lake Michigan for organochlorine contaminant and trace element analyses. Concentrations of PCBs and a metabolite of DDT (DDE) in scaup were below known effect levels, although selenium in Lesser Scaup, Bufflehead, and Common Goldeneye were in the range of elevated or potentially harmful levels.

We still know little about the potential impacts of zebra mussel invasions in aquatic ecosystems and habitats important to sea ducks in North America. Mackie (1991) hypothesized that *D. polymorpha* invasions in Lake Erie will result in an increase in the size of the euphotic zone, a decline in primary production, and reductions in pelagic autotrophs, heterotrophs, herbivorous zooplankton, and planktivores, ultimately resulting in alteration of the entire pelagic–benthic energy balance. Despite the considerable ecosystem level changes that have been documented following invasion of zebra mussels into North American waters, no evidence indicates that these alterations have resulted in overall population declines of sea ducks. However, because population size of the invasive species and environmental factors typically vary over time, the full effects of the invader might not be seen for a considerable time after the initial invasion (Strayer et al. 2006).

Long-term, ecosystem-based research on the impact of invasive species is needed, including their effects on the population dynamics of sea ducks (Strayer et al. 2006). We need to be able to distinguish invaders with minor and major effects to be able to effectively prioritize management recommendations (Parker et al. 1999).

Role of Climate Events in Evolution of Habitats

Arctic ecosystems are undergoing dramatic changes as a result of increasing ambient and ocean temperatures, increasing length of growing seasons, decreasing thickness and extent of ice in the Arctic Ocean, decreasing annual extent of terrestrial snow cover, reduction in glacial area and volume, accelerated coastal erosion, and increasing soil temperatures and degradation of permafrost (Arctic Council 2005, Intergovernmental Panel on Climate Change 2007, Post et al. 2009, Derksen et al. 2012, McLennan et al. 2012). Abiotic events linked to climate warming are well documented, but the ecological consequences of climate change in the Arctic are comparatively underreported (Post et al. 2009), including effects on habitats and populations of sea ducks.

Sea duck species that occur largely in the North American Arctic include Steller's Eider, Spectacled Eider, King Eider, Common Eider, and Long-tailed Duck. Other members of the tribe Mergini, including Common Goldeneye, Bufflehead, Harlequin Duck, Black Scoter, White-Winged Scoter, Surf Scoter, and Red-breasted Merganser, also use habitats north of the Arctic Circle, especially during the breeding season. Collectively, these 12 species depend on a diversity of high-latitude terrestrial, freshwater, and marine habitats for migratory staging, breeding, molting, and wintering periods. Limited information is available concerning direct impacts of climate events on sea ducks, but recent studies have begun to address this pressing need identified by managers in state, provincial, and federal governments in the United States and Canada. Next, we consider specific regional and continental scale examples where we now know, or can reasonably predict, potential responses of sea ducks to changing Arctic habitats.

Spectacled Eiders are uniquely adapted to extreme Arctic conditions and environments including sea ice and the tundra biome. The species depends on access to polynyas and open leads in fast ice of the Bering Sea during winter (Petersen et al. 1999). Aggregations of Spectacled Eiders are correlated with high densities of benthic food resources at their main winter sites south of St. Lawrence Island, Alaska (Cooper et al. 2013). Sea ice that surrounds polynyas and leads functions as substrate for resting and roosting Spectacled Eiders during intervals when they are not diving for bivalves and other marine invertebrates, and it also attenuates wave height that may contribute to increased energy expenditure and have negative consequences on the thermodynamics of eiders diving for food (Petersen and Douglas 2004). The authors evaluated population dynamics of breeding Spectacled Eiders on the Yukon–Kuskokwim Delta, Alaska, in relation to sea ice dynamics in the Bering Sea and found that population indices of eiders were negatively correlated with extreme ice concentration during time scales spanning 1957–2002 and 1988–2002. More recent trends of ice dynamics in the Bering Sea show considerable variation with ice extent near a record minimum during 2000–2005 but near

a record maximum between the years 2007 and 2010 (Douglas 2010). Despite considerable variability, GCMs predict reduction of ice cover in the Bering Sea; retreat is expected to be largest in the summer, resulting in a delayed freeze-up and thinner sea ice during winter relative to the present climate (Walsh 2008). Based on evaluation of 18 different GCM models, Douglas (2010) demonstrated that median March ice extent in the Bering Sea is projected to be about 25% less than the 1979–1988 average by the middle of the twenty-first century and 60% less by the end of the century, and the ice-free season is expected to increase from its contemporary average of 5.5 months to a median of about 8.5 months by the end of the century. In concert with reduced sea ice cover, good evidence indicates that northern Bering Sea ecosystems are changing from arctic to subarctic conditions in regions where Spectacled Eiders winter. Bering Sea ecosystems are shifting away from high water column and sediment carbon production and tight pelagic–benthic coupling of organic production (Grebmeier et al. 2006). Additionally, biological communities have shown recent significant changes including juxtaposition of dominance of two bivalves exploited by wintering Spectacled Eiders (Lovvorn et al. 2003) and reduction of benthic prey populations (Grebmeier and Dunton 2000). Responses of biological systems within northern Bering Sea polynyas to climate warming may range from an increase in microbial carbon remineralization to higher energy allocation to growth versus reproduction beneath polynyas (Grebmeier and Barry 2007). The ice-dominated, shallow ecosystem of the northern Bering Sea that favored benthic communities and bottom-feeding Spectacled Eiders is being replaced by an ecosystem dominated by pelagic fish (Grebmeier et al. 2006). Consequently, we would expect that wintering Spectacled Eider populations would be negatively affected by these documented and predicted climate-related changes to ecosystems and bivalve prey populations in the Bering Sea pack ice.

Are numbers of sea ducks affected by changing climate events that alter marine habitats? Flint (2013) addressed this complex and challenging question by examining changes in North American breeding populations of sea ducks from 1957 to 2011 in relation to regime shifts in the North Pacific Ocean that occurred in 1977, 1989, and 1998. Flint

(2013) modeled survey data and climate data for six groups of sea ducks, including eiders, scoters, Long-tailed Ducks, mergansers, goldeneyes, and Buffleheads. He found that four of six groups experienced population-level effects of the 1977 shift, five of six responded to the 1989 shift, and one group demonstrated effects of the 1998 shift.

At regional and site-specific scales, evidence has demonstrated how contemporary climate events affect habitats and sea duck populations during the breeding period. Drever et al. (2012) conducted a retrospective, comparative analysis of population-level responses of waterfowl, including Surf and White-Winged Scoters, to changes in snow cover duration in the western boreal forest of North America during a 34-year period from 1973 to 2007. These authors used a time series of snow extent and snow cover duration derived from satellite data and found a significant trend to shorter spring snow cover duration. Scoter population growth rates were positively linked to spring snow cover duration; larger populations resulted after springs with long snow cover duration than following springs with short snow cover duration. The total Surf and White-Winged Scoter populations decreased strongly over the period most likely because they are late-nesting species with reduced flexibility in their timing of breeding (Drever et al. 2012).

Several long-term studies of Common Eiders provide additional understanding of impacts of climate change on breeding habitats and consequently on the dynamics of these high-latitude nesting populations. D'Alba et al. (2010) examined a 30-year population data set on over 2000 Common Eiders at a colony in southwest Iceland to model response of this species to climate fluctuations. Ambient temperatures at this eider colony generally increased between 1977 and 2006; nesting females laid their clutches earlier following mild winters, number of nests increased over time, and year-to-year population fluctuations were correlated positively with temperatures during the breeding season two years previously when the new recruits to the population hatched (D'Alba et al. 2010). Jónsson et al. (2009) evaluated two nesting colonies of Common Eiders in northwest Iceland over 29-year (1978–2007) and 55-year (1953–2007) periods to understand if climatic conditions had effects on breeding numbers, arrival dates, and clutch size. The results were variable between the two colonies. At one location, numbers of nests increased following warm, wet winters and first nests were produced later following windy and wet winters, while windy conditions tended to be followed by earlier female arrivals in spring. By contrast, warm, wet springs were positively correlated with larger clutch sizes, and clutch sizes decreased following especially wet and warm autumns at the longer-studied colony. Likewise, Iles et al. (2013) found that cold and wet and warm and dry conditions in early spring were correlated with decreased nest success in Common Eiders, whereas warm and wet conditions in late spring increased eider nest success in La Pérouse Bay, Manitoba, Canada. Clearly, spring snow cover is an important aspect of nesting habitat for many arctic-nesting species of waterfowl, including sea ducks, and relatively recent climate change has resulted in variable responses, but longer-term data are necessary to understand consequences to the population status of sea ducks.

In addition to mainland breeding sites in arctic Alaska and Canada, Pacific Common Eiders and Long-tailed Ducks nest on barrier islands, and birds molt in shallow waters protected by these islands in the Chukchi and Beaufort seas (Goudie et al. 2000, Robertson and Savard 2002). These habitats are expected to be particularly vulnerable to changes linked to amplified climate warming. Storm surges result in extensive coastline erosion (Jones et al. 2009), reconfiguration, and migration of barrier islands (Reimnitz and Maurer 1979, Mars and Houseknecht 2007). With rising levels in the Beaufort Sea (Manson and Solomon 2007), landward retreat and loss of surface area of barrier islands is predicted (Shaw et al. 1998), which would directly impact the value of these mostly predator-free habitats for nesting sea ducks.

We encourage additional research that employs retrospective analyses focused on examining correlations between time series of climate metrics and sea duck populations, as well as application of climate models to predict habitat conditions and population response of these species, many of which depend on arctic and subarctic habitats where climate changes are most dramatic. Many challenges are present in detection of ecological cause-and-effect relationships in climate change studies. Fundamental considerations include presentation of a specific, detailed, mechanistic hypothesis and the conduct of critical testing by means of observational experiments (Krebs and Berteaux

2006). Further, Seavy et al. (2008) address questions associated with model uncertainties, climatic conditions that fall outside the historical range of variability, shifting vegetation communities, and other aspects to predicting how bird populations will respond to changing climates. The authors recommend that climate modeling efforts consider regional summaries of climate projections to aid in understanding the magnitude of effects projected for more specific study sites.

STRATEGIES FOR HABITAT MANAGEMENT AND PROTECTION

Our discussion of natural and anthropogenic changes in sea duck habitats in North America suggests that major international efforts are needed to reduce impacts to and stabilize breeding, molting, migratory staging, and wintering areas critical for these species. Arctic tundra and boreal forest are the two biomes in North America that represent the primary breeding habitats for most sea ducks including eiders, Long-tailed Duck, scoters, Bufflehead, goldeneyes, and two of three merganser species. Tundra and boreal forest are vast geographic areas, not yet densely populated by humans, but subject to increasing human-related activities and other stressors. Considering just these two biomes, the boreal forest is perhaps of most immediate concern because significant changes are occurring as a result of climate events coupled with exploitation of natural resources, especially water, timber, and hydrocarbons (Schindler 1998a). Schindler and Lee (2010) proposed a catchment-scale conservation approach where entire watersheds of thousands of square kilometers would need to be protected from resource development to maintain ecosystem health and biodiversity and provide adequate habitats to sustain viable populations of aquatic and terrestrial species, including sea ducks. Most of the Canadian boreal forest, which accounts for one-quarter of the intact, original forest remaining in the world, is without permanent protection from resource development. In Alaska, significant portions of the boreal forest biome are free from industrial development through six National Parks, and an additional seven National Wildlife Refuges encompass large tracts of boreal forest where commercial timber harvest is precluded. Nevertheless, large areas of the boreal forest managed by the State of Alaska and private corporations are under development and subject to future timber harvest, mining, and hydropower

projects. To maintain adequate sea duck breeding habitats in the boreal forest of North America, we recommend establishment of additional large, interconnected conservation units such as those currently under consideration in the provinces of Ontario and Quebec through the Canadian Boreal Initiative.

Marine habitats and especially estuaries are increasingly subject to a great diversity of threats (reviewed by Kennish 2002). In 2009, the Sea Duck Joint Venture (2012) implemented a multiyear satellite telemetry study of sea ducks in the Atlantic Flyway to enhance our understanding of the locations of the most important at-sea and nearshore wintering, staging, and molting areas for Long-tailed Duck, Black Scoter, White-Winged Scoter, and Surf Scoter. Results from these and other telemetry studies in the Arctic, Pacific, and Atlantic oceans may be employed, in concert with other ecological data, to identify a network of marine conservation units designed to provide protection for sea ducks during the postbreeding period, as has been advanced for conservation of ocean fisheries and maintenance of marine biodiversity (Botsford et al. 2003, Gerber et al. 2003). Kennish (2002) emphasized the need for a global estuarine and marine management framework to address the documented degradation of these habitats, which would be especially crucial for conservation of sea duck species, such as eiders, that depend on coastal and marine areas in both North America and abroad.

ACKNOWLEDGMENTS

This manuscript benefited from insightful reviews by J. E. Thompson, D. Esler, and J. M. Pearce. V. Skean checked the references for accuracy and provided expert guidance on other editorial considerations. We thank the U.S. Geological Survey, Alaska Science Center, and Environment Canada for support.

LITERATURE CITED

Abraham, K. F., and G. H. Finney. 1986. Eiders of the eastern Canadian Arctic. Pp. 55–73 in A. Reed (editor), Eider ducks in Canada. Canadian Wildlife Service, Report Series, No. 47. Canadian Wildlife Service, Ottawa, ON.

Ahlén, I., and A. Andersson. 1970. Breeding ecology of an eider population on Spitsbergen. Ornis Scandinavica 1:83–106.

Alexander, S. A., D. L. Dickson, and S. E. Westover. 1997. Spring migration of eiders and other waterbirds in offshore areas of the western Arctic. Pp. 6–20 in D. L. Dickson (editor), King and Common Eiders of the Western Canadian Arctic. Canadian Wildlife Service, Occasional Paper, No. 94. Canadian Wildlife Service, Ottawa, ON.

Alison, R. M. 1973. Delayed nesting in Oldsquaws. Bird-Banding 44:61–62.

Alison, R. M. 1975. Breeding biology and behavior of the Oldsquaw (Clangula hyemalis L.). Ornithological Monographs 18:1–52.

Alison, R. M. 1976. Oldsquaw brood behavior. Bird-Banding 47:210–213.

Anderson, B. A., C. B. Johnson, B. A. Cooper, L. N. Smith, and A. A. Stickney. 1999. Habitat associations of nesting Spectacled Eiders on the Arctic Coastal Plain of Alaska. Pp. 27–33 in R. I. Goudie, M. R. Petersen, and G. J. Robertson (editors), Behaviour and ecology of sea ducks. Canadian Wildlife Service, Occasional Paper, No. 100. Canadian Wildlife Service, Ottawa, ON.

Andrén, H. 1994. Effects of habitat fragmentation on birds and mammals in landscapes with different proportions of suitable habitat: a review. Oikos 71:355–366.

Arctic Council. 2005. Arctic Climate Impact Assessment-Scientific Report, 2005. Cambridge University Press, New York, NY.

Atkinson, P. W., N. A. Clark, M. C. Bell, P. J. Dare, J. A. Clark, and P. L. Ireland. 2003. Changes in commercially fished shellfish stocks and shorebird populations in the Wash, England. Biological Conservation 114:127–141.

Bailey, A. M. 1925. A report on the birds of northwestern Alaska and regions adjacent to Bering Strait. Part IV. Condor 27:164–171.

Baldassarre, G. A., and E. G. Bolen. 2006. Waterfowl ecology and management, 2nd ed. Krieger, Malabar, FL.

Barry, T. W. 1968. Observations on natural mortality and native use of eider ducks along the Beaufort Sea coast. Canadian Field-Naturalist 82:140–144.

Barry, T. W. 1986. Eiders of the western Canadian Arctic. Pp. 74–80 in A. Reed (editor), Eider ducks in Canada. Canadian Wildlife Service, Report Series, No. 47. Canadian Wildlife Service, Ottawa, ON.

Bart, J., and S. L. Earnst. 2005. Breeding ecology of Spectacled Eiders Somateria fischeri in northern Alaska. Wildfowl 55:85–100.

Beamish, R. J. 1976. Acidification of lakes in Canada by acid precipitation and the resulting effects on fishes. Water, Air, and Soil Pollution 6:501–514.

Bednarek, A. T. 2001. Undamming rivers: a review of the ecological impacts of dam removal. Environmental Management 27:803–814.

Bellrose, F. C. 1976. Ducks, Geese and Swans of North America, 2nd ed. The Wildlife Management Institute, Stackpole Books, Harrisburg, PA.

Bengtson, S.-A. 1970. Location of nest-sites of ducks in Lake Mývatn area, north-east Iceland. Oikos 21:218–229.

Bengtson, S.-A. 1971a. Food and feeding of diving ducks breeding at Lake Mývatn, Iceland. Ornis Fennica 48:77–92.

Bengtson, S.-A. 1971b. Habitat selection of duck broods in Lake Mývatn Area, north-east Iceland. Ornis Scandinavica 2:17–26.

Bengtson, S.-A. 1972. Breeding ecology of the Harlequin Duck Histrionicus histrionicus (L.) in Iceland. Ornis Scandinavica 3:1–19.

Bengtson, S.-A., and S. Ulfstrand. 1971. Food resources and breeding frequency of the Harlequin Duck Histrionicus histrionicus in Iceland. Oikos 22:235–239.

Benke, A. C. 1990. A perspective on America's vanishing streams. Journal of American Benthological Society 9:77–88.

Benoit, R., R. Lalumière, and A. Reed. 1993. Étude de la sauvagine sur la côte nord-est de la Baie James-1992. Report for the Société d'Énergie de la Baie James, Direction Ingénierie et Environnement, Service Écologie. Groupe Environnement Shooner, QC (in French).

Benoit, R., R. Lalumière, and A. Reed. 1994. Étude de la sauvagine sur la côte nord-est de la Baie James-1993. Report for the Société d'Énergie de la Baie James, Direction Ingénierie et Environnement, Service Écologie. Groupe Environnement Shooner, QC (in French).

Benoit, R., R. Lalumière, and A. Reed. 1996. Étude de la sauvagine sur la côte nord-est de la Baie James-1995. Report for the Société d'Énergie de la Baie James, Direction Ingénierie et Environnement, Service Écologie. Groupe Environnement Shooner, QC (in French).

Benoit, R., M. Robert, C. Marcotte, G. Fitzgérald, and J.-P. L. Savard. 2001. Étude des déplacements du Garrot d'Islande dans l'est du Canada à l'aide de la télémétrie satellitaire. Technical Report Series, No. 360, Canadian Wildlife Service, QC (in French).

Benson, A. J. 2011 [online]. Zebra mussel sightings distribution. U.S. Geological Survey website: <http://nas.er.usgs.gov/taxgroup/mollusks/zebramussel/zebramusseldistribution.aspx> (31 July 2013).

Bentzen, R. L., A. N. Powell, and R. S. Suydam. 2009. Strategies for nest-site selection by King Eiders. Journal of Wildlife Management 73:932–938.

Bergman, R. D., R. L. Howard, K. F. Abraham, and M. W. Weller. 1977. Water birds and their wetland resources in relation to oil development at Storkersen Point, Alaska. U.S. Fish and Wildlife Service, Resource Publication, No. 129. U.S. Fish and Wildlife Service, Washington, DC.

BirdLife International. [online]. 2012. *Clangula hyemalis.* in IUCN 2013, IUCN Red List of Threatened Species. Version 2013.1. <www.iucnredlist.org> (31 July 2013).

Blicher, M. E., L. M. Ramussen, M. K. Sejr, F. R. Merkel, and S. Rysgaard. 2011. Abundance and energy requirements of eiders (*Somateria* spp.) suggest high predation pressure on macrobenthic fauna in a key wintering habitat in SW Greenland. Polar Biology 34:1105–1116.

Blinn, B. M., A. W. Diamond, and D. J. Hamilton. 2008. Factors affecting selection of brood-rearing habitat by Common Eiders (*Somateria mollissima*) in the Bay of Fundy, New Brunswick, Canada. Waterbirds 31:520–529.

Blomqvist, S., and M. Elander. 1988. King Eider (*Somateria spectabilis*) nesting in association with Long-tailed Skua (*Stercorarius longicaudus*). Arctic 41:138–142.

Bobée, B., Y. Grimard, M. Lachance, and A. Tessier. 1982. Nature et étendue de l'acidification des lacs du Québec. Rapport scientifique, No. 140. Ministére de l'Environnement de Québec. Service de la qualité des eaux, QC (in French).

Bonar, R. L. 2000. Availability of Pileated Woodpecker cavities and use by other species. Journal of Wildlife Management 64:52–59.

Bond, J. C., and D. Esler. 2006. Nutrient acquisition by female Harlequin Ducks prior to spring migration and reproduction: evidence for body mass optimization. Canadian Journal of Zoology 84:1223–1229.

Bond, J. C., D. Esler, and K. A. Hobson. 2007. Isotopic evidence for sources of nutrients allocated to clutch formation by Harlequin Ducks. Condor 109:698–704.

Bordage, D., and J.-P. Savard. 1995. Black Scoter (*Melanitta nigra*). in A. Poole and F. Gill (editors), The Birds of North America, No. 177. The Academy of Natural Sciences, Philadelphia, PA.

Botsford, L. W., F. Michell, and A. Hastings. 2003. Principles for the design of marine reserves. Ecological Applications 13:S25–S31.

Bourget, A. A. 1973. Relation of eiders and gulls nesting in mixed colonies in Penobscot Bay, Maine. Auk 90:809–820.

Bourget, D., J.-P. L. Savard, and M. Guillemette. 2007. Distribution, diet, and dive behaviour of Barrow's and Common Goldeneyes during spring and autumn in the St. Lawrence estuary. Waterbirds 30:230–240.

Boyd, H. 1996. Arctic temperatures and the Long-tailed Ducks shot in eastern North America. Wildlife Biology 2:113–117.

Boyd, W. S., B. D. Smith, S. A. Iverson, M. R. Evans, J. E. Thompson, and S. Schneider. 2009. Apparent survival, natal philopatry, and recruitment of Barrow's Goldeneyes (*Bucephala islandica*) in the Cariboo-Chilcotin region of British Columbia, Canada. Canadian Journal of Zoology 87:337–345.

Brackney, A. W., and R. M. Platte. 1986. Habitat use and behavior of molting Oldsquaw on the coast of the Arctic National Wildlife Refuge, 1985. in G. W. Garner and P. E. Reynolds (editors), Arctic National Wildlife Refuge Progress Report, No. FY86–17. Arctic National Wildlife Refuge Coastal Plain Resource Assessment. U.S. Fish and Wildlife Service, Anchorage, AK.

Braune, B. M., K. A. Hobson, and B. J. Malone. 2005. Regional differences in collagen stable isotope and tissue trace element profiles in populations of Long-tailed Duck breeding in the Canadian Arctic. Science of the Total Environment 346:156–168.

Breault, A. M., and J.-P. L. Savard. 1999. Philopatry of Harlequin Ducks moulting in southern British Columbia. Pp. 41–44 in R. I. Goudie, M. R. Petersen, and G. J. Robertson (editors), Behaviour and ecology of sea ducks. Canadian Wildlife Service, Occasional Paper, No. 100. Canadian Wildlife Service, Ottawa, ON.

Breckenridge, W. J. 1953. Night rafting of American Goldeneyes on the Mississippi River. Auk 70:201–204.

Brodeur, S., J.-P. L. Savard, M. Robert, A. Bourget, G. Fitzgerald, and R. D. Titman. 2008. Abundance and movements of Harlequin Ducks breeding on rivers of the Gaspé Peninsula, Québec. Waterbirds 31:122–129.

Brodeur, S., J.-P. L. Savard, M. Robert, P. Laporte, P. Lamothe, R. D. Titman, S. Marchand, S. Gilliland, and G. Fitzgerald. 2002. Harlequin Duck *Histrionicus histrionicus* population structure in eastern Nearctic. Journal of Avian Biology 33:127–137.

Brook, R. W., K. F. Abraham, K. R. Middel, and R. K. Ross. 2012. Abundance and habitat selection of breeding scoters (*Melanitta* spp.) in Ontario's Hudson Bay Lowlands. Canadian Field-Naturalist 126:20–27.

Brown, P. W. 1977. Reproductive ecology of the White-winged Scoter (*Melanitta fusca deglandi*). Thesis, University of Missouri, Columbia, MO.

Brown, P. W., and M. A. Brown. 1981. Nesting biology of the White-winged Scoter. Journal of Wildlife Management 45:38–45.

Brown, P. W., and L. H. Fredrickson. 1989. White-winged Scoter populations and nesting at Redberry Lake, Saskatchewan. Canadian Field-Naturalist 103:240–247.

Brown, P. W., and L. H. Fredrickson. 1997. White-winged Scoter (*Melanitta fusca*). in A. Poole and F. Gill (editors), The Birds of North America, No. 274. The Academy of Natural Sciences, Philadelphia, PA.

Bruner, K. A., S. W. Fisher, and P. F. Landrum. 1994. The role of zebra mussel, *Dreissena polymorpha*, in contaminant cycling: I. The effect of body size and lipid content on the bioconcentration of PCBs and PAHs. Journal of Great Lakes Research 20:725–734.

Bur, M. T., M. A. Stapanian, G. Bernhardt, and M. W. Turner. 2008. Fall diets of Red-breasted Merganser (*Mergus serrator*) and walleye (*Sander vitreus*) in Sandusky Bay and adjacent waters of western Lake Erie. American Midland Naturalist 159:147–161.

Bustnes, J. O., and G. H. Systad. 2001. Comparative feeding ecology of Steller's Eider and Long-tailed Ducks in winter. Waterbirds 24:407–412.

Bustnes, J. O., G. H. Systad, and R. C. Ydenberg. 2013. Changing distribution of flocking sea ducks as non-regenerating food resources are depleted. Marine Ecology Progress Series 484:249–257.

Butler, R. W. 1998. Moulting sites of sea ducks and other marine birds in Frederick Sound, southeast Alaska. Canadian Field-Naturalist 112:346–347.

Butler, R. W., N. K. Dawe, and D. E. C. Trethewey. 1989. The birds of estuaries and beaches in the Strait of Georgia. Pp. 142–147 in K. Vermeer and R. W. Butler (editors), The ecology and status of marine and shoreline birds in the Strait of Georgia, British Columbia. Special Publication, Canadian Wildlife Service, Ottawa, ON.

Camphuysen, C. J., C. M. Berrevoets, H. J. W. M. Cremers, A. Dekinga, R. Dekker, B. J. Ens, T. M. van der Have et al. 2002. Mass mortality of Common Eiders (*Somateria mollissima*) in the Dutch Wadden Sea, winter 1999/2000: starvation in a commercially exploited wetland of international importance. Biological Conservation 106:303–317.

Cantin, M., J. Bédard, and H. Milne. 1974. The food and feeding of Common Eiders in the St. Lawrence estuary in summer. Canadian Journal of Zoology 52:319–334.

Caraco, N. F., J. J. Cole, P. A. Raymond, D. L. Strayer, M. L. Pace, S. E. G. Findlay, and D. T. Fischer. 1997. Zebra mussel invasion in a large, turbid river: phytoplankton response to increased grazing. Ecology 78:588–602.

Cassirer, E. F., G. Schirato, F. Sharpe, C. R. Groves, and R. N. Anderson. 1993. Cavity nesting by Harlequin Ducks in the Pacific Northwest. Wilson Bulletin 105:691–694.

Chapdelaine, G., A. Bourget, W. B. Kempt, D. J. Nakashima, and D. J. Murray. 1986. Population d'Eider à Duvet près des côtes du Québec septentrional. Pp. 39–50 in A. Reed (editor), Eider ducks in Canada. Canadian Wildlife Service, Report Series, No. 47. Canadian Wildlife Service, Ottawa, ON (in French).

Chapin III, F. S., S. F. Trainor, O. Huntington, A. L. Lovecraft, E. Zavaleta, D. C. Natcher, A. D. McGuire et al. 2008. Increasing wildfire in Alaska's boreal forest: pathways to potential solutions of a wicked problem. BioScience 58:531–540.

Chaulk, K. G., and M. L. Mahoney. 2012. Does spring ice cover influence nest initiation date and clutch size in Common Eiders? Polar Biology 35:645–653.

Chaulk, K. G., G. J. Robertson, and W. A. Montevecchi. 2007. Landscape features and sea ice influence nesting Common Eider abundance and dispersion. Canadian Journal of Zoology 85:301–309.

Chesser, R. T., R. C. Banks, F. K. Barker, C. Cicero, J. L. Dunn, A. W. Kratter, I. J. Lovette et al. 2010. Fifty-first supplement to the American Ornithologists' Union Check-List of North American Birds. Auk 127:726–744.

Chubbs, T. E., P. G. Trimper, G. W. Humphries, P. W. Thomas, L. T. Elson, and D. K. Laing. 2008. Tracking seasonal movements of adult male Harlequin Ducks from central Labrador using satellite telemetry. Waterbirds 31:173–182.

Collier, M., R. H. Webb, and J. C. Schmidt. 1996. Dams and rivers: a primer on the downstream effects of dams. U.S. Geological Survey Circular 1126, Tucson, AZ.

Conant, B., J. G. King, J. L. Trapp, and J. I. Hodges. 1988. Estimating populations of ducks wintering in Southeast Alaska. Pp. 541–551 in M. W. Weller (editor), Waterfowl in winter. University of Minnesota Press, Minneapolis, MN.

Cooke, F., G. J. Robertson, C. M. Smith, R. I. Goudie, and W. S. Boyd. 2000. Survival, emigration, and winter population structure of Harlequin Ducks. Condor 102:137–144.

Cooper, L. W., M. G. Sexson, J. M. Grebmeier, R. Gradinger, C. W. Mordy, and J. R. Lovvorn. 2013. Linkages between sea-ice coverage, pelagic-benthic coupling, and the distribution of Spectacled Eiders: observations in March 2008, 2009, and 2010, Northern Bering Sea. Deep-Sea Research Part II: Topical Studies in Oceanography 91:31–43.

Cornish, B. J., and D. L. Dickson. 1997. Common Eiders nesting in the western Canadian Arctic. Pp. 40–50 in D. L. Dickson (editor), King and Common Eiders of the western Canadian Arctic. Canadian Wildlife Service, Occasional Paper, No. 94. Canadian Wildlife Service, Ottawa, ON.

Cosgrove, P. J., J. R. A. Butler, and R. L. Laughton. 2004. Canoe and walking surveys of wintering Goosanders, Red-breasted Mergansers, Great Cormorants and Common Goldeneyes on the River Spey, 1994–2003. Scottish Birds 24:1–10.

Craik, S. R., J.-P. L. Savard, M. J. Richardson, and R. D. Titman. 2011. Foraging ecology of flightless male Red-breasted Mergansers in the Gulf of St. Lawrence, Canada. Waterbirds 34:280–288.

Craik, S. R., and R. D. Titman. 2008. Movements and habitat use by Red-breasted Merganser broods in eastern New Brunswick. Wilson Journal of Ornithology 120:743–754.

Craik, S. R., and R. D. Titman. 2009. Nesting ecology of Red-breasted Mergansers in a Common Tern colony in eastern New Brunswick. Waterbirds 32:282–292.

Crowley, D. W. 1999. Productivity of Harlequin Ducks breeding in Prince William Sound, Alaska. Pp. 14–20 in R. I. Goudie, M. R. Petersen, and G. J. Robertson (editors), Behaviour and ecology of sea ducks. Canadian Wildlife Service, Occasional Paper, No. 100. Canadian Wildlife Service, Ottawa, ON.

Custer, C. M., and T. W. Custer. 1996. Food habits of diving ducks in the Great Lakes after the zebra mussel invasion. Journal of Field Ornithology 67:86–99.

Custer, C. M., and T. W. Custer. 2000. Organochlorine and trace element contamination in wintering and migrating diving ducks in the southern Great Lakes, USA, since the zebra mussel invasion. Environmental Toxicology and Chemistry 19:2821–2829.

D'Alba, L., P. Monaghan, and R. G. Nager. 2010. Advances in laying date and increasing population size suggest positive responses to climate change in Common Eiders Somateria mollissima in Iceland. Ibis 152:19–28.

Dau, C. P. 1974. Nesting biology of the Spectacled Eider Somateria fischeri (Brandt) on the Yukon-Kuskokwim Delta, Alaska. Thesis, University of Alaska, Fairbanks, AK.

Dau, C. P., and S. A. Kistchinski. 1977. Seasonal movements and distribution of the Spectacled Eider. Wildfowl 28:65–75.

Décarie, R., F. Morneau, D. Lambert, S. Carrière, and J.-P. L. Savard. 1995. Habitat use by brood-rearing waterfowl in subarctic Québec. Arctic 48:383–390.

Degraer, S., M. Vincx, P. Meire, and H. Offringa. 1999. The macrozoobenthos of an important wintering area of the Common Scoter (Melanitta nigra). Journal of the Marine Biological Association of the United Kingdom 79:243–251.

De la Cruz, S. E. W., J. Y. Takekawa, M. T. Wilson, D. R. Nysewander, J. R. Evenson, D. Esler, W. S. Boyd, and D. H. Ward. 2009. Spring migration routes and chronology of Surf Scoters (Melanitta perspicillata): a synthesis of Pacific coast studies. Canadian Journal of Zoology 87:1069–1086.

Dement'ev, G. P., and N. A. Gladkov (editors). 1952. Birds of the Soviet Union (vol. 4). Publishing House Sovetskaya Nauka, Moscow, USSR (English translation by the Israel Program for Scientific Translation, Jerusalem, Israel, in 1967).

Derksen, C., S. L. Smith, M. Sharp, L. Brown, S. Howell, L. Copland, D. R. Mueller et al. 2012. Variability and change in the Canadian cryosphere. Climatic Change 115:59–88.

Derksen, D. V., and W. D. Eldridge. 1980. Drought-displacement of pintails to the arctic coastal plain, Alaska. Journal of Wildlife Management 44:224–229.

Derksen, D. V., T. C. Rothe, and W. D. Eldridge. 1981. Use of wetland habitats by birds in the National Petroleum Reserve-Alaska. U.S. Fish and Wildlife Service, Resource Publication, No. 141. U.S. Fish and Wildlife Service, Washington, DC.

DesGranges, J.-L., and M. Darveau. 1985. Effect of lake acidity and morphometry on the distribution of aquatic birds in southern Quebec. Holarctic Ecology 8:181–190.

DesGranges, J.-L., and C. Gagnon. 1994. Duckling response to changes in the trophic web of acidified lakes. Hydrobiologia 279/280:207–221.

DesGranges, J.-L., and B. Houde. 1989. Effects of acidity and other environmental parameters on the distribution of lacustrine birds in Quebec. Pp. 7–41 in J.-L. DesGranges (editor), Studies on the effects of acidification on aquatic wildlife in Canada: lacustrine birds and their habitats in Quebec. Canadian Wildlife Service, Occasional Paper, No. 67. Canadian Wildlife Service, Ottawa, ON.

Dickson, D. L., R. C. Cotter, J. E. Hines, and M. F. Kay. 1997. Distribution and abundance of King Eiders in the western Canadian Arctic. Pp. 29–39 in D. L. Dickson (editor), King and Common Eiders of the Western Canadian Arctic. Canadian Wildlife Service, Occasional Paper, No. 94. Canadian Wildlife Service, Ottawa, ON.

Dickson, D. L., and H. G. Gilchrist. 2002. Status of marine birds of the southeastern Beaufort Sea. Arctic 55:46–58.

Dickson, D. L., and P. A. Smith. 2013. Habitat used by Common and King Eiders in spring in the southeast Beaufort Sea and overlap with resource exploration. Journal of Wildlife Management 77:777–790.

Dickson, R. D., D. Esler, J. W. Hupp, E. M. Anderson, J. R. Evenson, and J. Barrett. 2012. Phenology and duration of remigial moult in Surf Scoters (Melanitta perspicillata) and White-winged Scoters (Melanitta fusca) on the Pacific coast of North America. Canadian Journal of Zoology 90:932–944.

Diéval, H., J.-F. Giroux, and J.-P. Savard. 2011. Distribution of Common Eiders *Somateria mollissima* during the brood-rearing and moulting periods in the St. Lawrence Estuary, Canada. Wildlife Biology 17:124–134.

Donaghey, R. H. 1975. Spacing behaviour of breeding Buffleheads (*Bucephala albeola*) on ponds in the southern boreal forest. Thesis, University of Alberta, Edmonton, AB.

Douglas, D. C. 2010. Arctic sea ice decline: projected changes in timing and extent of sea ice in the Bering and Chukchi Seas. U.S. Geological Survey Open-File Report 2010–1176. U.S. Geological Survey, Reston, VA.

Drapeau, P., A. Nappi, L. Imbeau, and M. Saint-Germain. 2009. Standing deadwood for keystone bird species in the eastern boreal forest: managing for snag dynamics. Forestry Chronicle 85:227–234.

Drever, M. C., R. G. Clark, C. Derksen, S. M. Slattery, P. Toose, and T. D. Nudds. 2012. Population vulnerability to climate change linked to timing of breeding in boreal ducks. Global Change Biology 18:480–492.

Drewitt, A. L., and R. H. W. Langston. 2006. Assessing the impacts of wind farms on birds. Ibis 148:29–42.

Drolet, C. 2007. Structure des groupes et comportement d'alimentation des Garrots à œil d'or hivernant sur le fleuve Saint-Laurent. Thesis, McGill University, Montréal, QC (in French).

Duda, J. J., J. E. Freilich, and E. G. Schreiner. 2008. Baseline studies in the Elwha River ecosystem prior to dam removal: introduction to the special issue. Northwest Science 82:1–12.

Dugger, B. D., K. M. Dugger, and L. H. Fredrickson. 1994. Hooded Merganser (*Lophodytes cucullatus*). in A. Poole and F. Gill (editors), The Birds of North America, No. 99. The Academy of Natural Sciences, Philadelphia, PA.

Dussart, E., and S. Payette. 2002. Ecological impact of clear-cutting on black spruce-moss forests in southern Québec. Ecoscience 9:533–543.

Dynesius, M., and C. Nilsson. 1994. Fragmentation and flow regulation of river systems in the northern third of the world. Science 266:753–762.

Dzinbal, K. A. 1982. Ecology of Harlequin Ducks in Prince William Sound, Alaska, during summer. Thesis, Oregon State University, Corvallis, OR.

Dzinbal, K. A., and R. L. Jarvis. 1984. Coastal feeding ecology of Harlequin Ducks in Prince William Sound, Alaska, during summer. Pp. 6–10 in D. N. Nettleship, G. A. Sanger, and P. F. Springer (editors), Marine birds: their feeding ecology and commercial fisheries relationships. Canadian Wildlife Service, Ottawa, ON.

Eadie, J. M., and G. Gauthier. 1985. Prospecting for nest sites by cavity-nesting ducks of the genus *Bucephala*. Condor 87:528–534.

Eadie, J. M., and A. Keast. 1982. Do goldeneye and perch compete for food? Oecologia 55:225–230.

Eadie, J. M., M. L. Mallory, and H. G. Lumsden. 1995. Common Goldeneye (*Bucephala clangula*). in A. Poole and F. Gill (editors), The Birds of North America, No. 170. The Academy of Natural Sciences, Philadelphia, PA.

Eadie, J. M., J.-P. L. Savard, and M. L. Mallory. 2000. Barrow's Goldeneye (*Bucephala islandica*). in A. Poole and F. Gill (editors), The Birds of North America, No. 548. The Academy of Natural Sciences, Philadelphia, PA.

Effler, S. W., C. M. Brooks, K. Whitehead, B. Wagner, S. M. Doerr, M. Perkins, C. A. Siegfried, L. Walrath, and R. P. Canale. 1996. Impact of zebra mussel invasion on river water quality. Water Environment Research 68:205–214.

Eggeman, D. R. 1986. Influence of environmental conditions on distribution and behavior of Common Goldeneye wintering in Maine. Thesis, University of Maine, Orono, ME.

Eggleton, F. E. 1937. Productivity of the profundal benthic zone in Lake Michigan. Michigan Academy of Science, Arts and Letters 22:593–611.

Einarsson, A. 1988. Distribution and movements of Barrow's Goldeneye (*Bucephala islandica*) young in relation to food. Ibis 130:153–163.

Einarsson, A. 1990. Settlement into breeding habitats by Barrow's Goldeneyes *Bucephala islandica*: evidence for temporary oversaturation of preferred habitat. Ornis Scandinavica 21:7–16.

Ellarson, R. S. 1956. A study of the Old-squaw duck on Lake Michigan. Thesis, University of Wisconsin, Madison, WI.

Elmberg, J., K. Sjöberg, P. Nummi, and H. Pöysä. 1994. Patterns of lake acidity and waterfowl communities. Hydrobiologia 279/280:201–206.

Englund, G., and B. Malmqvist. 1996. Effects of flow regulation, habitat area and isolation on the macroinvertebrate fauna of rapids in north Swedish rivers. Regulated Rivers: Research and Management 12:433–445.

Epners, C. A., S. E. Bayley, J. E. Thompson, and W. M. Tonn. 2010. Influence of fish assemblage and shallow lake productivity on waterfowl communities in the boreal transition zone of western Canada. Freshwater Biology 55:2265–2280.

Eriksson, M. O. G. 1978. Lake selection by goldeneye ducklings in relation to the abundance of food. Wildfowl 29:81–85.

Eriksson, M. O. G. 1979. Competition between freshwater fish and Goldeneyes *Bucephala clangula* (L.) for common prey. Oecologia 41:99–107.

Eriksson, M. O. G. 1983. The role of fish in the selection of lakes by nonpiscivorous ducks: Mallard, Teal, and Goldeneye. Wildfowl 34:27–32.

Eriksson, M. O. G. 1984. Acidification of lakes: effects on waterbirds in Sweden. Ambio 13:260–262.

Erskine, A. J. 1972a. Buffleheads. Canadian Wildlife Service, Monograph Series, No. 4. Canadian Wildlife Service, Ottawa, ON.

Erskine, A. J. 1972b. Populations, movements and seasonal distribution of mergansers. Canadian Wildlife Service, Report Series, No. 17. Canadian Wildlife Service, Ottawa, ON.

Esler, D., T. D. Bowman, T. A. Dean, C. E. O'Clair, S. C. Jewett, and L. L. McDonald. 2000a. Correlates of Harlequin Duck densities during winter in Prince William Sound, Alaska. Condor 102:920–926.

Esler, D., T. D. Bowman, C. E. O'Clair, T. A. Dean, and L. L. McDonald. 2000b. Densities of Barrow's Goldeneyes during winter in Prince William Sound, Alaska in relation to habitat, food, and history of oil contamination. Waterbirds 23:425–431.

Esler, D., S. A. Iverson, and D. J. Rizzolo. 2006. Genetic and demographic criteria for defining population units for conservation: the value of clear messages. Condor 108:480–483.

Evans, M. R. 2003. Breeding habitat selection by Barrow's Goldeneye and Bufflehead in the Cariboo-Chilcotin region of British Columbia: nest-sites, brood-rearing habitat, and competition. Thesis, Simon Fraser University, Burnaby, BC.

Evans, M. R., D. B. Lank, W. S. Boyd, and F. Cooke. 2002. A comparison of the characteristics and fate of Barrow's Goldeneye and Bufflehead nests in nest boxes and natural cavities. Condor 104:610–619.

Evans, R. M. 1970. Oldsquaw nesting in association with Arctic Terns at Churchill, Manitoba. Wilson Bulletin 82:383–390.

Fahnenstiel, G. L., T. B. Bridgeman, G. A. Lang, M. J. McCormik, and T. F. Nalepa. 1993. Phytoplankton productivity in Saginaw Bay, Lake Huron: effects of zebra mussel (Dreissena polymorpha) colonization. Journal of Great Lakes Research 21:465–475.

Falardeau, G., J.-F. Rail, S. Gilliland, and J.-P. Savard. 2003. Breeding survey of Common Eiders along the west coast of Ungava Bay, in summer 2000, and a supplement on other nesting aquatic birds. Technical Report Series, No. 405, Canadian Wildlife Service, Québec City, QC.

Faragó, S., and K. Hangya. 2012. Effects of water level on waterbird abundance and diversity along the middle section of the Danube River. Hydrobiologia 697:15–21.

Farmer, A. M. 1990. The effects of lake acidification on aquatic macrophytes—a review. Environmental Pollution 65:219–240.

Finlay, J. K. 2007. Offshore flight of Buffleheads, Bucephala albeola, after twilight in winter: an anti-predation tactic? Canadian Field-Naturalist 121:375–378.

Fischer, J. B., and C. R. Griffin. 2000. Feeding behavior and food habits of wintering Harlequin Ducks at Shemya Island, Alaska. Wilson Bulletin 112:318–325.

Fischer, J. B., and W. W. Larned. 2004. Summer distribution of marine birds in the western Beaufort Sea. Arctic 57:143–159.

Flint, P. L. 2013. Changes in size and trends of North American sea duck populations associated with North Pacific oceanic regime shifts. Marine Biology 160:59–65.

Flint, P. L., J. B. Grand, J. A. Morse, and T. F. Fondell. 2000a. Late summer survival of adult female and juvenile Spectacled Eiders on the Yukon-Kuskokwim Delta, Alaska. Waterbirds 23:292–297.

Flint, P. L., and M. P. Herzog. 1999. Breeding of Steller's Eiders, Polysticta stelleri, on the Yukon-Kuskokwim Delta, Alaska. Canadian Field-Naturalist 113:306–308.

Flint, P. L., D. L. Lacroix, J. A. Reed, and R. B. Lanctot. 2004. Movements of flightless Long-tailed Ducks during wing molt. Waterbirds 27:35–40.

Flint, P. L., C. M. Moran, and J. L. Schamber. 1998. Survival of Common Eider Somateria mollissima adult females and ducklings during brood rearing. Wildfowl 49:103–109.

Flint, P. L., J. A. Morse, J. B. Grand, and C. L. Moran. 2006. Correlated growth and survival of juvenile Spectacled Eiders: evidence of habitat limitation? Condor 108:901–911.

Flint, P. L., M. R. Petersen, C. P. Dau, J. E. Hines, and J. D. Nichols. 2000b. Annual survival and site fidelity of Steller's Eiders molting along the Alaska Peninsula. Journal of Wildlife Management 64:261–268.

Foreman, L. D. 1976. Observations of Common Merganser broods in northwestern California. California Fish and Game 62:207–212.

Fournier, M. A., and J. E. Hines. 1994. Effects of starvation on muscle and organ mass of King Eiders Somateria spectabilis and the ecological and management implications. Wildfowl 45:188–197.

Fox, A. D. 2003. Diet and habitat use of scoters Melanitta in the western Palearctic—a brief overview. Wildfowl 54:163–182.

Fox, A. D., M. Desholm, J. Kahlert, T. K. Christensen, and I. K. Petersen. 2006. Information needs to support environmental impact assessment of the effects of European marine offshore wind farms on birds. Ibis 148:129–144.

Fox, A. D., N. Jarrett, H. Gitay, and D. Paynter. 1989. Late summer habitat selection by breeding waterfowl in northern Scotland. Waterfowl 40:106–114.

Fox, G. A. 1993. What have biomarkers told us about the effects of contaminants on the health of fish-eating birds in the Great Lakes? The theory and a literature review. Journal of Great Lakes Research 19:722–736.

Fredrickson, L. H. 2001. Steller's Eider (*Polysticta stelleri*). in A. Poole and F. Gill (editors), The Birds of North America, No. 571. The Academy of Natural Sciences, Philadelphia, PA.

Fredrickson, L. H., and M. E. Heitmeyer. 1988. Waterfowl use of forested wetlands of the southern United States: an overview. Pp. 307–323 in M. W. Weller (editor), Waterfowl in winter. University of Minnesota Press, Minneapolis, MN.

Frimer, O. 1993. Occurrence and distribution of King Eiders *Somateria spectabilis* and Common Eiders *S. mollissima* at Disko, west Greenland. Polar Research 12:111–116.

Frimer, O. 1994a. Autumn arrival and moult in King Eiders (*Somateria spectabilis*) at Disko, West Greenland. Arctic 47:137–141.

Frimer, O. 1994b. The behaviour of moulting King Eiders *Somateria spectabilis*. Wildfowl 45:176–187.

Frimer, O. 1995a. Adaptations by the King Eider *Somateria spectabilis* to its moulting habitat: review of a study at Disko, West Greenland. Dansk Ornithologisk Forenings Tidsskrift 89:135–142.

Frimer, O. 1995b. Comparative behaviour of sympatric moulting populations of Common Eider *Somateria mollissima* and King Eider *S. spectabilis* in central West Greenland. Wildfowl 46:129–139.

Frimer, O. 1997. Diet of moulting King Eiders *Somateria spectabilis* at Disko Island, West Greenland. Ornis Fennica 74:187–194.

Gammonley, J. H., and M. E. Heitmeyer. 1990. Behavior, body condition, and foods of Buffleheads and Lesser Scaups during spring migration through the Klamath Basin, California. Wilson Bulletin 102:672–683.

Gardarsson, A. 1978. Distribution and numbers of the Barrow's Goldeneye (*Bucephala islandica*) in Iceland. Natturufraedingurinn 48:62–191.

Gardarsson, A. 2006. Temporal processes and duck populations: examples from Mývatn. Hydrobiologia 567:98–100.

Gardarsson, A. 2008. Harlequin Ducks in Iceland. Waterbirds 31:8–14.

Gauthier, G. 1987a. Brood territories in Buffleheads: determinants and correlates of territory size. Canadian Journal of Zoology 65:1402–1410.

Gauthier, G. 1987b. The adaptive significance of territorial behaviour in breeding Buffleheads: a test of three hypotheses. Animal Behaviour 35:348–360.

Gauthier, G. 1993. Bufflehead (*Bucephala albeola*). in A. Poole and F. Gill (editors), The Birds of North America, No. 67. The Academy of Natural Sciences, Philadelphia, PA.

Gerber, L. R., L. W. Botsford, A. Hastings, H. P. Possingham, S. D. Gaines, S. R. Palumbi, and S. Andelman. 2003. Population models for marine reserve design: a retrospective and prospective synthesis. Ecological Applications 13:S47–S64.

Gilchrist, H. G., and G. J. Robertson. 2000. Observations of marine birds and mammals wintering at polynyas and ice edges in the Belcher Islands, Nunavut, Canada. Arctic 53:61–68.

Gill Jr., R. E., M. R. Petersen, and P. D. Jorgensen. 1981. Birds of the northcentral Alaska Peninsula, 1976–1980. Arctic 34:286–306.

Gilliland, S. G. 1990. Predator prey relationships between Great Black-backed Gull and Common Eider populations on the Wolves Archipelago, New Brunswick: a study of foraging ecology. Thesis, University of Western Ontario, London, ON.

Gilliland, S. G., G. J. Robertson, M. Robert, J.-P. L. Savard, D. Amirault, P. Laporte, and P. Lamothe. 2002. Abundance and distribution of Harlequin Ducks molting in eastern Canada. Waterbirds 25:333–339.

Gjøsæter, J., and R. Sætre. 1974. Predation of eggs of capelin (*Mallotus villosus*) by diving ducks. Astarte 7:83–89.

Goudie, R. I. 1984. Comparative ecology of Common Eiders, Black Scoters, Oldsquaws, and Harlequin Ducks wintering in southern Newfoundland. Thesis, University of Western Ontario, London, ON.

Goudie, R. I. 1999. Behaviour of Harlequin Ducks and three species of scoters wintering in the Queen Charlotte Islands, British Columbia. Pp. 6–13 in R. I. Goudie, M. R. Petersen, and G. J. Robertson (editors), Behaviour and ecology of sea ducks. Canadian Wildlife Service, Occasional Paper, No. 100. Canadian Wildlife Service, Ottawa, ON.

Goudie, R. I., and C. D. Ankney. 1986. Body size, activity budgets, and diets of sea ducks wintering in Newfoundland. Ecology 67:1475–1482.

Goudie, R. I., and C. D. Ankney. 1988. Patterns of habitat use by sea ducks wintering in southeastern Newfoundland. Ornis Scandinavica 19:249–256.

Goudie, R. I., and S. G. Gilliland. 2008. Aspects of distribution and ecology of Harlequin Ducks on the Torrent River, Newfoundland. Waterbirds 31:92–103.

Goudie, R. I., and I. L. Jones. 2005. Feeding behavior of Harlequin Ducks (*Histrionicus histrionicus*) breeding in Newfoundland and Labrador: a test of the food limitation hypothesis. Bird Behavior 17:9–18.

Goudie, R. I., G. J. Robertson, and A. Reed. 2000. Common Eider (*Somateria mollissima*). in A. Poole and F. Gill (editors), The Birds of North America, No. 546. The Academy of Natural Sciences, Philadelphia, PA.

Graham, J. J., B. J. Joule, C. L. Crosby, and D. W. Townsend. 1984. Characteristics of the Atlantic herring (Clupea harengus L.) spawning population along the Maine coast, inferred from larval studies. Journal of Northwest Atlantic Fishery Science 5:131–142.

Grebmeier, J. M., and J. P. Barry. 2007. Benthic processes in polynyas. Elsevier Oceanography Series 74:363–390.

Grebmeier, J. M., and K. H. Dunton. 2000. Benthic processes in the northern Bering/Chukchi Seas: status and global change. in H. P. Huntington (editor), Impacts of changes in sea ice and other environmental parameters, the Arctic. U.S. Marine Mammal Commission, Washington, DC.

Grebmeier, J. M., J. E. Overland, S. E. Moore, E. V. Farley, E. C. Carmack, L. W. Cooper, K. E. Frey, J. H. Helle, F. A. McLaughlin, and S. L. McNutt. 2006. A major ecosystem shift in the northern Bering Sea. Science 311:1461–1464.

Gregory, R. D., S. P. Carter, and S. R. Baillie. 1997. Abundance, distribution and habitat use of breeding Goosanders Mergus merganser and Red-breasted Mergansers Mergus serrator on British Rivers. Bird Study 44:1–12.

Griffiths, R. W., D. W. Schloesser, J. H. Leach, and W. P. Kovalak. 1991. Distribution and dispersal of the zebra mussel (Dreissena polymorpha) in the Great Lakes Region. Canadian Journal of Fisheries and Aquatic Sciences 48:1381–1388.

Griffiths, W. B., and R. E. Dillinger. 1981. Beaufort Sea barrier island-lagoon ecological process studies: final report, Simpson Lagoon. Part 5. Invertebrates. U.S. Department of Commerce, NOAA, OCSEAP Final Report, No. 8, Boulder, CO.

Gromme, O. J. 1936. Effect of extreme cold on ducks in Milwaukee Bay. Auk 53:324–325.

Guillemette, M., J. H. Himmelman, C. Barette, and A. Reed. 1993. Habitat selection by Common Eiders in winter and its interaction with flock size. Canadian Journal of Zoology 71:1259–1266.

Guillemette, M., A. Reed, and J. H. Himmelman. 1996. Availability and consumption of food by Common Eiders wintering in the Gulf of St. Lawrence: evidence of prey depletion. Canadian Journal of Zoology 74:32–38.

Guillemette, M., A. J. Woakes, V. Henaux, J.-M. Grandbois, and P. J. Butler. 2004. The effect of depth on the diving behaviour of Common Eiders. Canadian Journal of Zoology 82:1818–1826.

Hamilton, D. J. 2000. Direct and indirect effects of predation by Common Eiders and abiotic disturbance in an intertidal community. Ecological Monographs 70:21–43.

Hamilton, D. J., and C. D. Ankney. 1994. Consumption of zebra mussels Dreissena polymorpha by diving ducks in Lakes Erie and St. Clair. Wildfowl 45:159–166.

Hansen, H. A., and D. E. McKnight. 1964. Emigration of drought-displaced ducks to the arctic. Transactions of the North American Wildlife and Natural Resources Conference 29:119–129.

Harris, S. W., C. L. Buechele, and C. F. Yocom. 1954. The status of Barrow's Golden-eye in eastern Washington. Murrelet 35:33–38.

Haszard, S. 2004. Habitat use by White-winged Scoters (Melanitta fusca) and Surf Scoters (Melanitta perspicillata) in the Mackenzie Delta Region, Northwest Territories. Thesis, University of Saskatchewan, Sakatoon, SK.

Hatton, P. L., and M. Marquiss. 2004. The origins of moulting Goosanders on the Eden estuary. Ringing and Migration 22:70–74.

Havens, K. E., N. D. Yan, and W. Keller. 1993. Lake acidification: effects on crustacean zooplankton populations. Environmental Science and Technology 27:1621–1624.

Heath, J. P., and H. G. Gilchrist. 2010. When foraging becomes unprofitable: energetics of diving in tidal currents by Common Eiders wintering in the Arctic. Marine Ecology Progress Series 403:279–290.

Heath, J. P., H. G. Gilchrist, and R. C. Ydenberg. 2006. Regulation of stroke patterns and swim speed across a range of current velocities: diving by Common Eiders wintering in polynyas in the Canadian arctic. Journal of Experimental Biology 209:3974–3983.

Heath, J. P., H. G. Gilchrist, and R. C. Ydenberg. 2007. Can dive cycle models predict patterns of foraging behavior?: diving by Common Eiders in an Arctic polynya. Animal Behaviour 73:877–884.

Heath, J. P., and W. A. Montevecchi. 2008. Differential use of similar habitat by Harlequin Ducks: trade-offs and implications for identifying critical habitat. Canadian Journal of Zoology 86:419–426.

Heath, J. P., W. A. Montevecchi, and G. J. Robertson. 2008. Allocating foraging effort across multiple time scales: behavioral responses to environmental conditions by Harlequin Ducks wintering at cape St. Mary's Newfoundland. Waterbirds 31:71–80.

Heitmeyer, M. E., and P. A. Vohs Jr. 1984. Distribution and habitat use of waterfowl wintering in Oklahoma. Journal of Wildlife Management 48:51–62.

Higgins, S. N., M. J. Vander Zanden, L. N. Joppa, and Y. Vadeboncoeur. 2011. The effect of dreissenid invasions on chlorophyll and the chlorophyll: total phosphorus ratio in north-temperate lakes. Canadian Journal of Fisheries and Aquatic Sciences 68:319–329.

Hildén, O. 1964. Ecology of duck populations in the island group of Valassaaret, Gulf of Bothnia. Annales Zoology Fennici 1:153–279.

Hirsch, V. K. 1980. Winter ecology of sea ducks in the inland marine waters of Washington. Thesis, University of Washington, Seattle, WA.

Hobaugh, W. C., and J. G. Teer. 1981. Waterfowl use characteristics of flood-prevention lakes in north-central Texas. Journal of Wildlife Management 45:16–26.

Hodges, J. I., D. J. Groves, and B. P. Conant. 2008. Distribution and abundance of waterbirds near shore in southeast Alaska, 1997–2002. Northwestern Naturalist 89:85–96.

Hofmann, T., J. W. Chardine, and H. Blokpoel. 1997. First breeding record of Red-breasted Merganser, *Mergus serrator*, on Axel Heiberg Island, Northwest Territories. Canadian Field-Naturalist 11:308–309.

Hogan, D., J. E. Thompson, and D. Esler. 2013. Survival of Barrow's Goldeneyes during remigial molt and fall staging. Journal of Wildlife Management 77:701–706.

Hogan, D., J. E. Thompson, D. Esler, and W. S. Boyd. 2011. Discovery of important post-breeding sites for Barrow's Goldeneye in the boreal transition zone of Alberta. Waterbirds 34:261–288.

Holcroft-Weerstra, A. C., and D. L. Dickson. 1997. Activity budgets of King Eiders on the nesting grounds in spring. Pp. 58–66 in D. L. Dickson (editor), King and Common Eiders of the western Canadian Arctic. Canadian Wildlife Service, Occasional Paper, No. 94. Canadian Wildlife Service, Ottawa, ON.

Hollmén, T. E., C. DebRoy, P. L. Flint, D. E. Safine, J. L. Schamber, A. E. Riddle, and K. A. Trust. 2011. Molecular typing of *Escherichia coli* strains associated with threatened sea ducks and near-shore marine habitats of south-west Alaska. Environmental Microbiology Reports 3:262–269.

Holm, K. J., and A. E. Burger. 2002. Foraging behavior and resource partitioning by diving birds during winter in areas of strong tidal currents. Waterbirds 25:312–325.

Howard, R. L. 1974. Aquatic invertebrate-waterbird relationships on Alaska's Arctic Coastal Plain. Thesis, Iowa State University, Ames, IA.

Hull, E. D. 1914. Habits of the Old-squaw (*Harelda hyemalis*) in Jackson Park, Chicago. Wilson Bulletin 26:116–123.

Hunt, B., and R. Ydenberg. 2000. Harlequins *Histrionicus histrionicus* in a Rocky Mountain watershed I: background and general breeding ecology. Wildfowl 51:155–168.

Hunt, W. A. 1998. The ecology of Harlequin Ducks (*Histrionicus histrionicus*) breeding in Jasper National Park, Canada. Thesis, Simon Fraser University, Burnaby, BC.

Idrisi, N., E. L. Mills, L. G. Rudstam, and D. J. Stewart. 2001. Impact of zebra mussels (*Dreissena polymorpha*) on the pelagic lower trophic levels of Oneida Lake, New York. Canadian Journal of Fisheries and Aquatic Sciences 58:1430–1441.

Iles, D. T., R. F. Rockwell, P. Matulonis, G. J. Robertson, K. F. Abraham, J. C. Davies, and D. N. Koons. 2013. Predators, alternative prey and climate influence annual breeding success of a long-lived sea duck. Journal of Animal Ecology 82:683–693.

Intergovernmental Panel on Climate Change. 2007. Climate Change 2007—The Physical Science Basis. in S. Solomon, D. Qin, M. Manning, Z. Chen, M. Marquis, K. B. Averyt, M. Tignor, and H. L. Miller (editors), Fourth Assessment Report of the Intergovernmental Panel on Climate Change. Cambridge University Press, New York, NY.

Iverson, S. A., and D. Esler. 2006. Site fidelity and the demographic implications of winter movements by a migratory bird, the Harlequin Duck *Histrionicus histrionicus*. Journal of Avian Biology 37:219–228.

Iverson, S. A., D. Esler, and W. S. Boyd. 2003. Plumage characteristics as an indicator of age class in the Surf Scoter. Waterbirds 26:56–61.

Iverson, S. A., D. Esler, and D. J. Rizzolo. 2004. Winter philopatry of Harlequin Ducks in Prince William Sound, Alaska. Condor 106:711–715.

Jansson, R., C. Nilsson, M. Dynesius, and E. Andersson. 2000. Effects of river regulation on river-margin vegetation: a comparison of eight boreal rivers. Ecological Applications 10:203–224.

Jepsen, P. U. 1973. Studies of the moult migration and wing-feather moult of the Goldeneye (*Bucephala clangula*) in Denmark. Danish Review of Game Biology 8:1–24.

Johengen, T. H., T. F. Nalepa, G. L. Fahnenstiel, and G. Goudy. 1995. Nutrient changes in Saginaw Bay, Lake Huron, after the establishment of the zebra mussel (*Dreissena polymorpha*). Journal of Great Lakes Research 21:449–464.

Johnson, S. R. 1984a. Prey selection by Oldsquaws in a Beaufort Sea lagoon, Alaska. Pp. 12–19 in D. N. Nettleship, G. A. Sanger, and P. F. Springer (editors), Marine birds: their feeding ecology and commercial fisheries relationships. Canadian Wildlife Service, Ottawa, ON.

Johnson, S. R. 1984b. Continuing investigations of Oldsquaws (*Clangula hyemalis* L.) during the molt period in the Alaskan Beaufort Sea. Pp. 547–635 in U.S. Department of Commerce, NOAA, OCSEAP Final Report 23, Juneau, AK.

Johnson, S. R. 1985. Adaptations of the Long-tailed Duck (*Clangula hyemalis* L.) during the period of molt in arctic Alaska. Proceedings of the International Ornithological Congress 18:530–540.

Johnson, S. R., and D. R. Herter. 1989. The birds of the Beaufort Sea. BP Exploration (Alaska), Anchorage, AK.

Johnson, S. R., and W. J. Richardson. 1981. Beaufort Sea barrier island-lagoon ecological process studies: final report, Simpson Lagoon. Pp. 109–383 in Environmental assessment of the Alaskan Continental Shelf, final reports of principal investigators (vol. 7). Biological Studies. NOAA, Office of Marine Pollution Assessment, Boulder, CO.

Johnson, S. R., and W. J. Richardson. 1982. Waterbird migration near the Yukon and Alaskan coast of the Beaufort Sea: II. Moult migration of sea ducks in summer. Arctic 35:291–301.

Johnson, W. C., B. V. Millett, T. Gilmanov, R. A. Voldseth, G. R. Guntenspergen, and D. E. Naugle. 2005. Vulnerability of northern prairie wetlands to climate change. BioScience 55:863–872.

Joint Working Group on the Management of the Common Eider. 2004. Quebec management plan for the Common Eider *Somateria mollissima dresseri*. A special publication of the Joint Working Group on the Management of the Common Eider, QC.

Jones, B. M., C. D. Arp, M. T. Jorgenson, K. M. Hinkel, J. A. Schmutz, and P. L. Flint. 2009. Increase in the rate and uniformity of coastline erosion in arctic Alaska. Geophysical Research Letters 36:L03503.

Jones, J., and R. D. Drobney. 1986. Winter feeding ecology of Scaup and Common Goldeneye in Michigan. Journal of Wildlife Management 50:446–452.

Jónsson, J. E., A. Gardarsson, J. A. Gill, A. Petersen, and T. G. Gunnarsson. 2009. Seasonal weather effects on the Common Eider, a subarctic capital breeder, in Iceland over 55 years. Climate Research 38:237–248.

Jonsson, M., P. Deleu, and B. Malmqvist. 2012a. Persisting effects of river regulation on emergent aquatic insects and terrestrial invertebrates in upland forests. River Research and Applications 29:537–547.

Jonsson, M., D. Strasevicius, and B. Malmqvist. 2012b. Influences of river regulation and environmental variables on upland bird assemblages in northern Sweden. Ecological Research 27:945–954.

Kasischke, E. S., T. S. Rupp, and D. L. Verbyla. 2006. Fire trends in the Alaskan boreal forest region. Pp. 285–301 in F. S. Chapin III, M. A. Oswood, K. Van Cleve, L. A. Viereck, and D. L. Verbyla (editors), Alaska's changing boreal forest. Oxford University Press, New York, NY.

Kasischke, E. S., and M. R. Turetsky. 2006. Recent changes in the fire regime across the North American boreal region-spatial and temporal patterns of burning across Canada and Alaska. Geophysical Research Letters 33:L09703.

Kellett, D. K., and R. T. Alisauskas. 1997. Breeding biology of King Eiders nesting on Karrak Lake, Northwest Territories. Arctic 50:47–54.

Kellett, D. K., R. T. Alisauskas, and K. R. Mehl. 2003. Nest-site selection, interspecific associations, and nest success of King Eiders. Condor 105:373–378.

Kennish, M. J. 2002. Environmental threats and environmental future of estuaries. Environmental Conservation 29:78–107.

Kertell, K. 1991. Disappearance of the Steller's Eider from the Yukon-Kuskokwim Delta, Alaska. Arctic 44:177–187.

King, J. G. 1963. Duck banding in arctic Alaska. Journal of Wildlife Management 27:356–362.

King, J. G. 1973. A cosmopolitan duck moulting resort: Takslesluk Lake Alaska. Wildfowl 24:103–109.

King, J. G., and C. P. Dau. 1981. Waterfowl and their habitat in the eastern Bering Sea. Pp. 739–753 in D. W. Hood and J. A. Calder (editors), The Eastern Bering Sea Shelf: oceanography and resources (vol. 2). University of Washington Press, Seattle, WA.

Kirk, M., D. Esler, S. A. Iverson, and W. S. Boyd. 2008. Movements of wintering Surf Scoters: predator responses to different prey landscapes. Oecologia 155:859–867.

Kirk, M. K., D. Esler, and W. S. Boyd. 2007. Foraging effort of Surf Scoters (*Melanitta perspicillata*) wintering in a spatially and temporally variable prey landscape. Canadian Journal of Zoology 85:1207–1215.

Kistchinski, A. A. 1973. Waterfowl in north-east Asia. Wildfowl 24:88–102.

Kitchen, D. W., and G. S. Hunt. 1969. Brood habitat of the Hooded Merganser. Journal of Wildlife Management 33:605–609.

Knoche, M. J. 2004. King Eider wing molt: inferences from stable isotope analyses. Thesis, University of Alaska, Fairbanks, AK.

Knoche, M. J., A. N. Powell, L. T. Quakenbush, M. J. Wooller, and L. M. Phillips. 2007. Further evidence for site fidelity to wing molt locations by King Eiders: integrating stable isotope analyses and satellite telemetry. Waterbirds 30:52–57.

Koehl, P., T. C. Rothe, and D. V. Derksen. 1982. Winter food habits of Barrow's Goldeneyes in southeast Alaska. Pp. 1–5 in D. N. Nettleship, G. A. Sanger, and P. F. Springer (editors), Marine birds: their feeding ecology and commercial fisheries relationships. Canadian Wildlife Service, Darmouth, NS.

Konyukhov, N. B., L. S. Bogoslovskaya, B. M. Zvonov, and T. I. Van Pelt. 1998. Seabirds of the Chukotka Peninsula, Russia. Arctic 51:315–329.

Krebs, C. J., and D. Berteaux. 2006. Problems and pitfalls in relating climate variability to population dynamics. Climate Research 32:143–149.

Kuchel, C. R. 1977. Some aspects of the behavior and ecology of Harlequin Ducks breeding in Glacier National Park, Montana. Thesis, University of Montana, Missoula, MT.

Kumari, E. 1979. Moult and moult migration of waterfowl in Estonia. Wildfowl 30:90–98.

Lacroix, D. L. 2001. Foraging impacts and patterns of wintering Surf Scoters feeding on bay mussels in coastal Strait of Georgia, British Columbia. Thesis, Simon Fraser University, Vancouver, BC.

Lacroix, D. L., S. B. Boyd, D. Esler, M. Kirk, T. Lewis, and S. Lipovsky. 2005. Surf Scoters Melanitta perspicillata aggregate in association with ephemerally abundant polychaetes. Marine Ornithology 33:61–63.

Lanctot, R., B. Goatcher, K. Scribner, S. Talbot, B. Pierson, D. Esler, and D. Zwiefelhofer. 1999. Harlequin Duck recovery from the Exxon Valdez oil spill: a population genetics perspective. Auk 116:781–791.

Larsen, J. K., and M. Guillemette. 2000. Influence of annual variation in food supply on abundance of wintering Common Eiders Somateria mollissima. Marine Ecology Progress Series 201:301–309.

Larsen, J. K., and M. Guillemette. 2007. Effects of wind turbines on flight behavior of wintering Common Eiders: implications for habitat use and collision risk. Journal of Applied Ecology 44:516–522.

Larson, D. L. 1995. Effects of climate on numbers of Northern Prairie wetlands. Climate Change 30:169–180.

Laubhan, M. K., and K. A. Metzner. 1999. Distribution and diurnal behavior of Steller's Eiders wintering on the Alaska Peninsula. Condor 101:694–698.

Lauer, T. E., and T. S. McComish. 2001. Impact of zebra mussels (Dreissena polymorpha) on fingernail clams (Sphaeriidae) in extreme southern Lake Michigan. Journal of Great Lakes Research 27:230–238.

LeBourdais, S. V., R. C. Ydenberg, and D. Esler. 2009. Fish and Harlequin Ducks compete on breeding streams. Canadian Journal of Zoology 87:31–40.

Lemelin, L.-V., M. Darveau, L. Imbeau, and D. Bordage. 2010. Wetland use and selection by breeding waterbirds in the boreal forest of Quebec, Canada. Wetlands 30:321–332.

Lemelin, L.-V., L. Imbeau, M. Darveau, and D. Bordage. 2007. Local, short-term effects of forest harvesting on breeding waterfowl and Common Loon in forest-dominated landscapes of Quebec. Avian Conservation and Ecology 2:art10.

Lesage, L., A. Reed and J.-P. L. Savard. 2008. Duckling survival and use of space by Surf Scoter (Melanitta perspicillata) broods. Écoscience 15:81–88.

Leslie Jr., D. M., W. J. Stancill, and R. F. Raskevitz. 1994. Use of an old multipurpose reservoir by migrating and wintering non-dabbling ducks. Proceedings of the Oklahoma Academy of Science 74:21–24.

Lewis, T. L., D. Esler, and W. S. Boyd. 2007. Foraging behaviors of Surf Scoters and White-winged Scoters during spawning of Pacific herring. Condor 109:216–222.

Lewis, T. L., D. Esler, and W. S. Boyd. 2008. Foraging behaviour of Surf Scoters (Melanitta perspicillata) and White-winged Scoters (M. fusca) in relation to clam density: inferring food availability and habitat quality. Auk 125:149–157.

Limpert, R. J. 1980. Homing success of adult Buffleheads to a Maryland wintering site. Journal of Wildlife Management 44:905–908.

Loegering, J. P., and R. G. Anthony. 1999. Distribution, abundance, and habitat association of riparian-obligate and -associated birds in the Oregon Coast Range. Northwest Science 73:168–185.

Lok, E. K., D. Esler, J. Y. Takekawa, S. W. De La Cruz, W. S. Boyd, D. R. Nysewander, J. R. Evenson, and D. H. Ward. 2012. Spatiotemporal associations between Pacific herring spawn and Surf Scoter spring migration: evaluating a 'silver wave' hypothesis. Marine Ecology Progress Series 457:139–150.

Lok, E. K., M. Kirk, D. Esler, and W. S. Boyd. 2008. Movements of pre-migratory Surf and White-winged Scoters in response to Pacific herring spawn. Waterbirds 31:385–393.

Longcore, J. R., D. G. McAuley, G. W. Pendelton, C. R. Bennatti, T. M. Mingo, and K. L. Stromborg. 2006. Macroinvertebrate abundance, water chemistry, and wetland characteristics affect use of wetlands by avian species in Maine. Hydrobiologia 567:143–167.

Loring, P. H. 2012. Phenology and habitat use of scoters along the southern New England continental shelf. Thesis, University of Rhode Island, Kingston, New York, NY.

Loring, P. H., P. W. C. Patton, S. R. McWilliams, R. A. McKinney, and C. A. Oviatt. 2013. Densities of wintering scoters in relation to benthic prey assemblages in a North Atlantic estuary. Waterbirds 36:144–155.

Lovvorn, J. R., J. M. Grebmeier, L. W. Cooper, J. K. Bump, and S. E. Richman. 2009. Modeling marine protected areas for threatened eiders in a climatically changing Bering Sea. Ecological Applications 19:1596–1613.

Lovvorn, J. R., S. E. Richman, J. M. Grebmeier, and L. W. Cooper. 2003. Diet and body condition of Spectacled Eiders wintering in pack ice of the Bering Sea. Polar Biology 26:259–267.

MacIsaac, H. J. 1996. Potential abiotic and biotic impacts of zebra mussels on the inland waters of North America. American Zoology 36:287–299.

MacIsaac, H. J., C. J. Lonnee, and J. H. Leach. 1995. Suppression of microzooplankton by zebra mussels: importance of mussel size. Freshwater Biology 34:379–387.

Mackay, G. H. 1892. Habits of the Oldsquaw (*Clangula hyemalis*) in New England. Auk 9:330–337.

Mackie, G. L. 1991. Biology of the exotic zebra mussel, *Dreissena polymorpha*, in relation to native bivalves and its potential impact in Lake St. Clair. Hydrobiologia 219:251–268.

Mallory, M., and K. Metz. 1999. Common Merganser (*Mergus merganser*). in A. Poole and F. Gill (editors), The Birds of North America, No. 442. The Academy of Natural Sciences, Philadelphia, PA.

Mallory, M. L., J. Akearok, N. R. North, D. V. Weseloh, and S. Lair. 2006. Movements of Long-tailed Ducks wintering on Lake Ontario to breeding areas in Nunavut, Canada. Wilson Journal of Ornithology 118:494–501.

Mallory, M. L., D. K. McNicol, and P. J. Weatherhead. 1994. Habitat quality and reproductive effort of Common Goldeneyes nesting near Sudbury, Canada. Journal of Wildlife Management 58:552–560.

Manson, G. K., and S. M. Solomon. 2007. Past and future forcing of Beaufort Sea coastal change. Atmosphere-Ocean 45:107–122.

Marquiss, M., and K. Duncan. 1993. Variation in the abundance of Red-breasted Mergansers *Mergus serrator* on a Scottish river in relation to season, year, river hydrography, salmon density and spring culling. Ibis 135:33–41.

Marquiss, M., and K. Duncan. 1994a. Diurnal activity patterns of Goosanders *Mergus merganser* on a Scottish river system. Wildfowl 45:209–221.

Marquiss, M., and K. Duncan. 1994b. Seasonal switching between habitats and changes in abundance of Goosanders *Mergus merganser* within a Scottish river system. Wildfowl 45:198–208.

Mars, J. C., and D. W. Houseknecht. 2007. Quantitative remote sensing study indicates doubling of coastal erosion rate in past 50 yr along a segment of the arctic coast of Alaska. Geology 35:583–586.

Martin, K., K. E. H. Aitken, and K. L. Wiebe. 2004. Nest sites and nest webs for cavity-nesting communities in interior British Columbia, Canada: nest characteristics and niche partitioning. Condor 106:5–19.

Masden, E. A., D. T. Haydon, A. D. Fox, R. W. Furness, R. Bullman, and M. Desholm. 2009. Barriers to movement: impacts of wind farms on migrating birds. ICES Journal of Marine Science 66:746–753.

Mazak, E. J., H. J. MacIsaac, M. R. Servos, and R. Hesslein. 1997. Influence of feeding habits on organochlorine contaminant accumulation in waterfowl on the Great Lakes. Ecological Applications 7:1133–1143.

McAloney, R. K. 1973. Brood ecology of the Common Eider (*Somateria mollissima dresseri*) in the Loscombe area of Nova Scotia. Thesis, Acadia University, Wolfville, NS.

McCaw III, J. H., P. J. Zwank, and R. L. Steiner. 1996. Abundance, distribution, and behavior of Common Mergansers wintering on a reservoir in southern New Mexico. Journal of Field Ornithology 67:669–679.

McLaren, P. L., and M. A. McLaren. 1982. Waterfowl populations in eastern Lancaster Sound and western Baffin Bay. Arctic 35:149–157.

McLennan, D. S., T. Bell, D. Berteaux, W. Chen, L. Copland, R. Fraser, D. Gallant et al. 2012. Recent climate-related terrestrial biodiversity research in Canada's Arctic national parks: review, summary, and management implications. Biodiversity 13:157–173.

McNicol, D. K., B. E. Bendell, and R. K. Ross. 1987. Studies of the effects of acidification on aquatic wildlife in Canada: waterfowl and trophic relationships in small lakes in northern Ontario. Canadian Wildlife Service, Occasional Paper, No. 62. Canadian Wildlife Service, Ottawa, ON.

McNicol, D. K., R. A. Walton, and M. L. Mallory. 1997. Monitoring nest box use by cavity-nesting ducks on acid-stressed lakes in Ontario, Canada. Wildlife Biology 3:1–12.

McNicol, D. K., and M. Wayland. 1992. Distribution of waterfowl broods in Sudbury area lakes in relation to fish, macroinvertebrates, and water chemistry. Canadian Journal of Fisheries and Aquatic Sciences 49:122–133.

McRae, D. J., L. C. Duchesne, B. Freedman, T. J. Lynham, and S. Woodley. 2001. Comparisons between wildfire and forest harvesting and their implications in forest management. Environmental Reviews 9:223–260.

Meissner, W., and I. Niklewska. 1993. Zimowanie szlachara (*Mergus serrator*), nurogsia (*Mergus merganser*) i bielaczka (*Mergus albellus*) na Zatoce Gdaskiej w sezonach 1984–1985–1986/1987. Notation Ornitologiczen 34:111–123 (in Polish with English summary).

Merkel, F. R., S. E. Jamieson, K. Falk, and A. Mosbech. 2007. The diet of Common Eiders wintering in Nuuk, southwest Greenland. Polar Biology 30:227–234.

Merkel, F. R., and A. Mosbech. 2008. Diurnal and nocturnal feeding strategies in Common Eiders. Waterbirds 31:580–586.

Merkel, F. R., A. Mosbech, D. Boartman, and L. Gronhahl. 2002. Winter seabird distribution and abundance off south-western Greenland, 1999. Polar Research 21:17–36.

Metzner, K. A. 1993. Ecological strategies of wintering Steller's Eiders on Izembek Lagoon and Cold Bay, Alaska. Thesis, University of Missouri, Columbia, MO.

Mickelson, P. J. 1975. Breeding biology of Cackling Geese and associated species on the Yukon-Kuskokwim Delta, Alaska. Wildlife Monographs 45.

Minot, E. O. 1980. Tidal, diurnal and habitat influences on Common Eider rearing activities. Ornis Scandinavica 11:165–172.

Mitchell, G. J. 1952. A study of the distribution of some members of the Nyrocinae wintering on the coastal waters of southern British Columbia. Thesis, University of British Columbia, Vancouver, BC.

Mittelhauser, G. H. 2000. The winter ecology of Harlequin Ducks in coastal Maine. Thesis, University of Maine, Orono, ME.

Mooney, H. A., and E. E. Cleland. 2001. The evolutionary impact of invasive species. Proceedings of the National Academy of Sciences 98:5446–5451.

Morrier, A., L. Lesage, A. Reed, and J.-P. L. Savard. 2008. Étude sur l'écologie de la Macreuse à Front Blanc au Lac Malbaie, Réserve des Laurentides, 1994–1995. Série de rapports techniques, No. 301. Région du Québec, Service canadien de la faune, Section Conservation des populations, Environnement Canada, Québec, QC (in French).

Morrison, H. A., F. A. P. C. Gobas, R. Lazar, D. M. Whittle, and G. D. Haffner. 1998. Projected changes to the trophodynamics of PCBs in the western Lake Erie ecosystem attributed to the presence of zebra mussels (Dreissena polymorpha). Environmental Science and Technology 32:3862–3867.

Morse, T. E., J. L. Jakabosky, and B. P. McCrow. 1969. Some aspects of the breeding biology of the Hooded Merganser. Journal of Wildlife Management 33:596–604.

Mosbech, A., and D. Boertmann. 1999. Distribution, abundance and reaction to aerial surveys of post-breeding King Eiders (Somateria spectabilis) in western Greenland. Arctic 52:188–203.

Mosbech, A., R. S. Danø, F. Merkel, C. Sonne, G. Gilchrist, and A. Flagstad. 2006a. Use of satellite telemetry to locate key habitats for King Eiders Somateria spectabilis in West Greenland. Pp. 769–776 in G. C. Boere, C. A. Galbraith, and D. A. Stroud (editors), Waterbirds around the world. The Stationary Office, Edinburg, U.K.

Mosbech, A., G. Gilchrist, F. Merkel, C. Sonne, A. Flagstad, and H. Nyegaard. 2006b. Year-round movements of northern Common Eiders Somateria mollissima borealis breeding in Arctic Canada and West Greenland followed by satellite telemetry. Ardea 94:651–665.

Mosbech, A., and S. R. Johnson. 1999. Late winter distribution and abundance of sea-associated birds in south-western Greenland, the Davis Strait and southern Baffin Bay. Polar Research 18:1–17.

Munro, J. A. 1939. Studies of water-fowl in British Columbia, Barrow's Golden-eye, American Golden-eye. Transactions Royal Canadian Institute 24:259–318.

Munro, J. A., and W. A. Clemens. 1937. The American Merganser in British Columbia and its relation to the fish population. Bulletin of the Biological Board of Canada 55:1–50.

Murie, O. J. 1959. Fauna of the Aleutian Islands and Alaska Peninsula. North American Fauna, No. 61. U.S. Fish and Wildlife Service, Washington, DC.

Nakashima, D. J. 1986. Inuit knowledge of the ecology of the Common Eider in northern Quebec. Pp. 102–113 in A. Reed (editor), Eider Ducks in Canada. Canadian Wildlife Service, Report Series, No. 47. Canadian Wildlife Service, Ottawa, ON.

Newbrey, J. L., M. A. Bozek, and N. D. Niemuth. 2005. Effects of lake characteristics and human disturbance on the presence of piscivorous birds in northern Wisconsin, USA. Waterbirds 28:478–486.

Nichols, K. H., and G. J. Hopkins. 1993. Recent changes in Lake Erie (north shore) phytoplankton: cumulative impacts of phosphorus loading reductions and the zebra mussel introduction. Journal of Great Lakes Research 19:637–647.

Nilsson, C., A. Ekblad, M. Gardfjell, and B. Carlberg. 1991. Long-term effects of river regulation on river-margin vegetation. Journal of Applied Ecology 28:963–987.

Nilsson, L. 1970. Local and seasonal variation in sex-ratios of diving ducks in south Sweden during the non-breeding season. Ornis Scandinavica 1:115–128.

Noel, L. E., S. R. Johnson, and G. M. O'Doherty. 2005a. Long-tailed Duck, Clangula hyemalis, eider, Somateria spp., and scoter, Melanitta spp., distributions in central Alaska Beaufort Sea Lagoons, 1999–2002. Canadian Field-Naturalist 119:181–185.

Noel, L. E., S. R. Johnson, G. M. O'Doherty, and M. K. Butcher. 2005b. Common Eider (Somateria mollissima v-nigrum) nest cover and depredation on Central Alaskan Beaufort Sea barrier islands. Arctic 58:129–136.

Nummi, P., and H. Pöysä. 1993. Habitat associations of ducks during different phases of the breeding season. Ecography 16:319–328.

Nummi, P., V.-M. Vaananen, M. Rask, K. Nyberg, and K. Taskinen. 2012. Competitive effects of fish in structurally simple habitats: perch, invertebrates, and goldeneye in small boreal lakes. Aquatic Sciences 74:343–350.

Odum, E. P. 1971. Fundamentals of ecology. Saunders, Philadelphia, PA.

Oosterhuis, R., and K. van Dijk. 2002. Effect of food shortage on the reproductive output of Common Eiders *Somateria mollissima* breeding at Griend (Wadden Sea). Atlantic Seabirds 4:29–38.

Oppel, S, D. L. Dickson, and A. N. Powell. 2009a. International importance of the eastern Chukchi Sea as a staging area for migrating King Eiders. Polar Biology 32:775–783.

Oppel, S., A. N. Powell, and M. G. Butler. 2011. King Eider foraging effort during the pre-breeding period in Alaska. Condor 113:52–60.

Oppel, S., A. N. Powell, and D. L. Dickson. 2009b. Using an algorithmic model to reveal individually variable movement decisions in a wintering sea duck. Journal of Animal Ecology 78:524–531.

Ost, M., and M. Kilpi. 1999. Parental care influences feeding behavior of female Eiders *Somateria mollissima*. Annales Zoologici Fennici 36:195–204.

Ouellet, J.-F., M. Guillemette, and M. Robert. 2010. Spatial distribution and habitat selection of Barrow's and Common Goldeneyes wintering in the St. Lawrence marine system. Canadian Journal of Zoology 88:306–314.

Paasivaara, A., and H. Pöysä. 2008. Habitat-patch occupancy in the Common Goldeneye (*Bucephala clangula*) at different stages of the breeding cycle: implications to ecological processes in patchy environments. Canadian Journal of Zoology 86:744–755.

Palmer, R. S. (editor). 1976. Handbook of North American birds (vol. 3), Waterfowl (Part 2). Yale University Press, New Haven, CT.

Parker, I. M., D. Simberloff, W. M. Lonsdale, K. Goodell, M. Wonham, P. M. Kareiva, M. H. Williamson, B. Von Holle, P. B. Moyle, J. E. Byers, and L. Goldwasser. 1999. Impact: toward a framework for understanding the ecological effects of invaders. Biological Invasions 1:3–19.

Pearce, J. M., D. Esler, and A. G. Degtyarev. 1998. Nesting ecology of Spectacled Eiders *Somateria fischeri* on the Indigirka River Delta, Russia. Wildfowl 49:110–123.

Pearce, J. M., J. A. Reed, and P. L. Flint. 2005. Geographic variation in survival and migratory tendency among North American Common Mergansers. Journal of Field Ornithology 76:109–118.

Pearce, J. M., and S. L. Talbot. 2006. Demography, genetics, and the value of mixed messages. Condor 108:474–479.

Pearce, J. M., D. Zwiefelhofer, and N. Maryanski. 2009. Mechanisms of population heterogeneity among molting Common Mergansers on Kodiak Island, Alaska: implications for genetic assessments of migratory connectivity. Condor 111:283–293.

Pehrsson, O. 1974. Nutrition of small ducklings regulating breeding area and reproductive output in the Long-tailed Duck, *Clangula hyemalis*. XI International Congress of Game Biologists, National Swedish Environment Protection Board, Stockholm, Sweden, 3–7 September 1973.

Pehrsson, O. 1976. Food and feeding grounds of the Goldeneye *Bucephala clangula* (L.) on the Swedish west coast. Ornis Scandinavica 7:91–112.

Pehrsson, O. 1986. Duckling production of the Oldsquaw in relation to spring weather and small-rodent fluctuations. Canadian Journal of Zoology 64:1835–1841.

Perry, M. C., D. M. Kidwell, A. M. Wells, E. J. R. Lohnes, P. C. Osenton, and S. H. Altmann. 2006. Characterization of breeding habitats for Black and Surf Scoters in the eastern boreal forest and subarctic regions of Canada. Pp. 80–89 in A. Hanson, J. Kerekes, and J. Paquet (editors), Limnology and aquatic birds: abstracts and selected papers from the fourth conference of the Societas Internationalis Limnologiae. Aquatic Birds Working Group. Canadian Wildlife Service Technical Report Series, No. 474. Canadian Wildlife Service, Sackville, NB.

Petersen, M. R. 1980. Observations of wing-feather moult and summer feeding ecology of Steller's Eiders at Nelson Lagoon, Alaska. Wildfowl 31:99–106.

Petersen, M. R. 1981. Populations, feeding ecology and molt of Steller's Eiders. Condor 83:256–262.

Petersen, M. R. 1982. Predation on sea birds by red foxes at Shaiak Island, Alaska. Canadian Field-Naturalist 96:41–45.

Petersen, M. R. 2009. Multiple spring migration strategies in a population of Pacific Common Eiders. Condor 111:59–70.

Petersen, M. R., J. O. Bustnes, and G. H. Systad. 2006. Breeding and moulting locations and migration patterns of the Atlantic population of Steller's Eiders *Polysticta stelleri* as determined from satellite telemetry. Journal of Avian Biology 37:58–68.

Petersen, M. R., and D. C. Douglas. 2004. Winter ecology of Spectacled Eiders: environmental characteristics and population change. Condor 106:79–94.

Petersen, M. R., D. C. Douglas, and D. M. Mulcahy. 1995. Use of implanted satellite transmitters to locate Spectacled Eiders at-sea. Condor 97:276–278.

Petersen, M. R., D. C. Douglas, H. M. Wilson, and S. E. McCloskey. 2012. Effects of sea ice on winter site fidelity of Pacific Common Eiders (*Somateria mollissima v-nigrum*). Auk 129:399–480.

Petersen, M. R., and P. L. Flint. 2002. Population structure of Pacific Common Eiders breeding in Alaska. Condor 104:780–787.

Petersen, M. R., J. B. Grand, and C. P. Dau. 2000. Spectacled Eider (*Somateria fischeri*). in A. Poole, and F. Gill (editors), The Birds of North America, No. 547. The Academy of Natural Sciences, Philadelphia, PA.

Petersen, M. R., W. W. Larned, and D. C. Douglas. 1999. At-sea distributions of Spectacled Eiders (*Somateria fischeri*): a 120 year-old mystery resolved. Auk 116:1009–1020.

Petersen, M. R., B. J. McCaffery, and P. L. Flint. 2003. Post-breeding distribution of Long-tailed Ducks *Clangula hyemalis* from the Yukon-Kuskokwim Delta, Alaska. Wildfowl 54:103–113.

Petersen, M. R., J. F. Piatt, and K. A. Trust. 1998. Foods of Spectacled Eiders *Somateria fischeri* in the Bering Sea, Alaska. Wildfowl 49:124–128.

Peterson, S. R., and R. S. Ellarson. 1977. Food habits of Oldsquaws wintering on Lake Michigan. Wilson Bulletin 89:81–91.

Peterson, S. R., and R. S. Ellarson. 1979. Changes in Oldsquaw carcass weight. Wilson Bulletin 91:288–300.

Phillips, L. M., and A. N. Powell. 2006. Evidence for wing molt and breeding site fidelity in King Eiders. Waterbirds 29:148–153.

Phillips, L. M., A. N. Powell, and E. A. Rexstad. 2006. Large-scale movements and habitat characteristics of King Eiders throughout the nonbreeding period. Condor 108:887–900.

Phillips, L. M., A. N. Powell, E. J. Taylor, and E. A. Rexstad. 2007. Use of the Beaufort Sea by King Eiders breeding on the North Slope of Alaska. Journal of Wildlife Management 71:1892–1898.

Pierre, J. P., H. Bears, and C. A. Paszkowski. 2001. Effects of forest harvesting on nest predation in cavity-nesting waterfowl. Auk 118:224–230.

Portenko, L. A. 1972. Birds of the Chukchi Peninsula and Wrangel Island (vol. 1). Nauka Publishers, Leningrad, Russia (English translation published by Amerind Publishing Co., New Delhi, India, 1981).

Post, E., M. C. Forchhammer, S. Bret-Harte, T. V. Callaghan, T. R. Christensen, B. Elberling, A. D. Fox et al. 2009. Ecological dynamics across the Arctic associated with recent climate change. Science 325:1355–1358.

Pöysä, H., and A. Paasivaara. 2006. Movements and mortality of Common Goldeneye *Bucephala clangula* broods in a patchy environment. Oikos 115:33–42.

Pöysä, H., M. Rask, and P. Nummi. 1994. Acidification and ecological interactions at higher trophic levels in small forest lakes: the perch and the Common Goldeneye. Annales Zoologici Fennici 31:397–404.

Pöysä, H., and J. Virtanen. 1994. Habitat selection and survival of Common Goldeneye (*Bucephala clangula*) broods—preliminary results. Hydrobiologia 279/280:289–296.

Prach, R. W., A. R. Smith, and A. Dzubin. 1986. Nesting of the Common Eider near the Hell Gate—Cardigan Strait polynya, 1980–81. Pp. 127–135 in A. Reed (editor), Eider ducks in Canada. Canadian Wildlife Service, Report Series, No. 47. Canadian Wildlife Service, Ottawa, ON.

Prince, H. H. 1968. Nest sites used by Wood Ducks and Common Goldeneyes in New Brunswick. Journal of Wildlife Management 32:489–500.

Quakenbush, L., and R. Suydam. 1999. Periodic non-breeding of Steller's Eiders near Barrow, Alaska, with speculations with possible causes. Pp. 34–40 in R. I. Goudie, M. R. Petersen, and G. J. Robertson (editors), Behavior and ecology of sea ducks. Canadian Wildlife Service, Occasional Paper, No. 100. Canadian Wildlife Service, Ottawa, ON.

Quakenbush, L., R. Suydam, T. Obritschkewitsch, and M. Deering. 2004. Breeding biology of Steller's Eiders (*Polysticta stelleri*) near Barrow, Alaska, 1991–99. Arctic 57:166–182.

Quinlan, S. E., and W. A. Lehnhausen. 1982. Arctic fox, *Alopex lagopus*, predation on nesting Common Eiders, *Somateria mollissima*, at Icy Cape, Alaska. Canadian Field-Naturalist 96:462–466.

Råd, O. 1980. Breeding distribution and habitat selection of Red-breasted Mergansers in freshwater in western Norway. Wildfowl 31:53–56.

Rail, J.-F., and J.-P. Savard. 2003. Identification des aires de mue et de repos au printemps des macreuses (*Melanitta* sp.) et de l'Eider à Duvet (*Somateria mollissima*) dans l'estuaire et le golfe du Saint-laurent. Technical Report Series, No. 408. Canadian Wildlife Service, Quebec, QC (in French).

Rajala, P., and T. Ormio. 1970. On the nesting of the Goldeneye *Bucephala clangula* (L.), in the Meltaus Game Research Area in Northern Finland, 1959–1966. Riistatieteellisia Julkaisuja 31:3–9.

Rawson, D. S. 1953. The bottom fauna of Great Slave Lake. Journal of Fisheries Research Board Canada 10:486–520.

Reed, A. 1971. Pre-dusk rafting flights of wintering goldeneyes and other diving ducks in the Province of Quebec. Wildfowl 22:61–62.

Reed, A., R. Benoit, R. Lalumière, and M. Julien. 1996. Duck use of the coastal habitats of northeastern James Bay. Canadian Wildlife Service, Occasional Paper, No. 90. Canadian Wildlife Service, Ottawa, ON.

Reed, A., and A. Bourget. 1977. Distribution and abundance of waterfowl wintering in southern Quebec. Canadian Field-Naturalist 91:1–7.

Reed, J. A., and P. L. Flint. 2007. Movements and foraging effort of Steller's Eiders and Harlequin Ducks wintering near Dutch Harbor, Alaska. Journal of Field Ornithology 78:124–132.

Reimnitz, E., and D. K. Maurer. 1979. Effects of storm surges on the Beaufort Sea coast, northern Alaska. Arctic 32:329–344.

Rempel, R. S., K. F. Abraham, R. R. Gadawski, S. Garbor, and R. K. Ross. 1997. A simple wetland habitat classification for boreal forest waterfowl. Journal of Wildlife Management 61:746–757.

Ricciardi, A., J. B. Rasmussen, and F. G. Whoriskey. 1995. Predicting the intensity and impact of *Dreissena* infestation on native unionid bivalves from *Dreissena* field density. Canadian Journal of Fisheries and Aquatic Sciences 52:1449–1461.

Richardson, W. J., and S. R. Johnson. 1981. Waterbird migration near the Yukon and Alaskan coast of the Beaufort Sea: I. timing, routes and numbers in spring. Arctic 34:108–121.

Richner, H. 1988. Temporal and spatial patterns in the abundance of wintering Red-breasted Mergansers *Mergus serrator* in an estuary. Ibis 130:73–78.

Robert, M., R. Benoit, and J.-P. L. Savard. 2000. COSEWIC status report on the Barrow's Goldeneye (*Bucephala islandica*) eastern population in Canada. Committee on the status of endangered wildlife in Canada. Canadian Wildlife Service, Ottawa, ON.

Robert, M., R. Benoit, and J.-P. L. Savard. 2002. Relationship among breeding, molting and wintering areas of male Barrow's Goldeneyes (*Bucephala islandica*) in eastern North America. Auk 119:676–684.

Robert, M., and L. Cloutier. 2001. Summer food habits of Harlequin Ducks in eastern North America. Wilson Bulletin 113:78–84.

Robert, M., B. Drolet, and J.-P. L. Savard. 2008a. Habitat features associated with Barrow's Goldeneyes breeding in eastern Canada. Wilson Journal of Ornithology 120:320–330.

Robert, M., G. H. Mittelhauser, B. Jobin, G. Fitzgerald, and P. Lamothe. 2008b. New insights on Harlequin Duck population structure in eastern North America as revealed by satellite telemetry. Waterbirds 31:159–172.

Robert, M., and J.-P. L. Savard. 2006. The St. Lawrence River Estuary and Gulf: a stronghold for Barrow's Goldeneyes wintering in eastern North America. Waterbirds 29:437–450.

Robertson, G. J. 1995. Factors affecting nest site selection and nesting success in the Common Eider *Somateria mollissima*. Ibis 137:109–115.

Robertson, G. J., F. Cooke, R. I. Goudie, and W. S. Boyd. 1999. Within-year fidelity of Harlequin Ducks to a moulting and wintering area. Pp. 45–51 in R. I. Goudie, M. R. Petersen, and G. J. Robertson (editors), Behaviour and ecology of sea ducks. Canadian Wildlife Service, Occasional Paper, No. 100. Canadian Wildlife Service, Ottawa, ON.

Robertson, G. J., and H. G. Gilchrist. 1998. Evidence of population declines among Common Eiders breeding in the Belcher Islands, Northwest Territories. Arctic 51:378–385.

Robertson, G. J., and R. I. Goudie. 1999. Harlequin Duck (*Histrionicus histrionicus*). in A. Poole and F. Gill (editors), The Birds of North America, No. 466. The Academy of Natural Sciences, Philadelphia, PA.

Robertson, G. J. and J.-P. L. Savard. 2002. Long-tailed Duck (*Clangula hyemalis*). in A. Poole and F. Gill (editors), The Birds of North America, No. 651. The Academy of Natural Sciences, Philadelphia, PA.

Rodway, M. S. 1998a. Habitat use by Harlequin Ducks breeding in Hebron Fjord, Labrador. Canadian Journal of Zoology 76:897–901.

Rodway, M. S. 1998b. Activity patterns, diet, and feeding efficiency of Harlequin Ducks breeding in northern Labrador. Canadian Journal of Zoology 76:902–909.

Rodway, M. S., J. W. Gosse Jr., I. Fong, and W. A. Montevecchi. 1998. Discovery of a Harlequin Duck nest in eastern North America. Wilson Bulletin 110:282–285.

Rodway, M. S., H. M. Regehr, J. Ashley, P. V. Clarkson, R. I. Goudie, D. E. Hay, C. M. Smith, and K. G. Wright. 2003a. Aggregative response of Harlequin Ducks in herring spawning in the Strait of Georgia, British Columbia. Canadian Journal of Zoology 81:504–514.

Rodway, M. S., H. M. Regehr, and F. Cooke. 2003b. Sex and age differences in distribution, abundance, and habitat preferences of wintering Harlequin Ducks: implications for conservation and estimating recruitment rates. Canadian Journal of Zoology 81:492–503.

Ross, R. K. 1983. An estimate of the Black Scoter, *Melanitta nigra*, population moulting in James and Hudson bays. Canadian Field-Naturalist 97:147–150.

Ross, R. K., S. A. Petrie, S. S. Badzinski, and A. Mullie. 2005. Autumn diet of Greater Scaup, Lesser Scaup, and Long-tailed Ducks on eastern Lake Ontario prior to zebra mussel invasion. Wildlife Society Bulletin 33:81–91.

Rowe, J. S. and G. W. Scotter. 1973. Fire in the boreal forest. Quaternary Research 3:444–464.

Safine, D. E., and M. S. Lindberg. 2008. Nest habitat selection of White-winged Scoters on Yukon Flats, Alaska. Wilson Journal of Ornithology 120:582–593.

Salter, R. E., M. A. Gollop, S. R. Johnson, W. R. Koski, and C. E. Tull. 1980. Distribution and abundance of birds on the Arctic Coastal Plain of northern Yukon and adjacent Northwest Territories, 1971–1976. Canadian Field Naturalist 94:219–238.

Sanger, G. A., and R. D. Jones Jr. 1984. Winter feeding ecology and tropic relationships of Oldsquaws and White-winged Scoters on Kachemak Bay, Alaska.

Pp. 20–28 in D. N. Nettleship, G. A. Sanger, and P. F. Springer (editors), Marine birds: their feeding ecology and commercial fisheries relationships. Proceedings of the Pacific Seabird Group Symposium, Seattle, Washington. Special Publication, Canadian Wildlife Service, Ottawa, ON.

Sangster, G. 2009. Acoustic differences between the scoters Melanitta nigra nigra and M. n. americana. Wilson Journal of Ornithology 121:696–702.

Savard, J.-P. L. 1977. Étude de la faune avienne dans les basins de la Grande rivière de la baleine et de la Petite rivière (été 1976). Technical report for the Environment Direction of Hydro-Québec, Direction Environment. Éco-recherches, Inc., GB-BIOP-ECO-77-3 (in French).

Savard, J.-P. L. 1984. Territorial behaviour of Common Goldeneye, Barrow's Goldeneye and Bufflehead in areas of sympatry. Ornis Scandinavica 15:211–216.

Savard, J.-P. L. 1988a. Winter, spring and summer territoriality in Barrow's Goldeneye: characteristics and benefits. Ornis Scandinavica 19:119–128.

Savard, J.-P. L. 1988b. A summary of current knowledge on the distribution and abundance of moulting seaducks in the coastal waters of British Columbia. Technical Report Series, No. 45. Canadian Wildlife Service, Pacific and Yukon Region, Vancouver, BC.

Savard, J.-P. L. 1989. Birds of rocky coastlines and pelagic waters in the Strait of Georgia. Pp. 132–141 in K. Vermeer and R. W. Butler (editors), The ecology and status of marine and shoreline birds in the Strait of Georgia, British Columbia. Special Publication, Canadian Wildlife Service, Ottawa, ON.

Savard, J.-P. L. 1990. Population de sauvagine hivernant dans l'estuaire du saint-Laurent: Écologie, distribution et abondance. Technical Report Series, No. 89. Canadian Wildlife Service, Quebec region, QC (in French).

Savard, J.-P. L. 1991. Status report on the distribution and ecology of Harlequin Ducks in British Columbia. Technical Report Series, No. 110. Canadian Wildlife Service, Pacific and Yukon Region, Vancouver, BC.

Savard, J.-P. L. 2009. Diversité, abondance et répartition des oiseaux aquatiques hivernant dans les eaux côtières et pélagiques du Parc marin Saguenay–Saint-Laurent. Journal of Water Science 22:353–371 (in French).

Savard, J.-P. L., J. Bédard, and A. Nadeau. 1999. Spring and early summer distribution of scoters and eiders in the St. Lawrence River estuary. Pp. 60–65 in R. I. Goudie, M. R. Petersen, and G. J. Robertson (editors), Behaviour and ecology of sea ducks, Canadian Wildlife Service Occasional Paper, No. 100. Canadian Wildlife Service, Ottawa, ON.

Savard, J.-P. L., D. Bordage, and A. Reed. 1998. Surf Scoter (Melanitta perspicillata). in A. Poole and F. Gill (editors), The Birds of North America, No. 363. The Academy of Natural Sciences, Philadelphia, PA.

Savard, J.-P. L., W. S. Boyd, and G. E. J. Smith. 1994. Waterfowl-wetland relationships in the aspen parkland of British Columbia: comparison of analytical methods. Hydrobiologia 279/280:309–325.

Savard, J.-P. L., and P. Lamothe. 1991. Distribution, abundance and aspects of breeding ecology of Black Scoters and Surf Scoters in Northern Quebec. Canadian Field-Naturalist 105:488–496.

Savard, J.-P. L., L. Lesage, S. G. Gilliland, H. G. Gilchrist, and J.-R. Giroux. 2011. Molting, staging, and wintering locations of Common Eiders breeding in the Gyrfalcon Archipelago, Ungava Bay. Arctic 64:197–206.

Savard, J.-P. L., and M. Robert. 2013. Relationships among breeding, molting and wintering areas of adult female Barrow's Goldeneyes (Bucephala islandica) in eastern North America. Waterbirds 36:34–42.

Savard, J.-P. L., M. Robert, and S. Brodeur. 2008. Harlequin Duck in Quebec. Waterbirds 31:19–31.

Sayler, R. D., and A. D. Afton. 1981. Ecological aspects of Common Goldeneyes Bucephala clangula wintering on the upper Mississippi River. Ornis Scandinavica 12:99–108.

Schamber, J. L., F. J. Boerman, and P. Flint. 2010a. Reproductive ecology and habitat use of Pacific Black Scoters (Melanitta nigra americana) nesting on the Yukon-Kuskokwim Delta, Alaska. Waterbirds 33:129–171.

Schamber, J. L., P. L. Flint, J. B. Grand, H. M. Wilson, and J. A. Morse. 2009. Population dynamics of Long-tailed Ducks breeding on the Yukon-Kuskokwim Delta, Alaska. Arctic 62:190–200.

Schamber, J. L., P. L. Flint, and A. N. Powell. 2010b. Patterns of use and distribution of King Eiders and Black Scoters during the annual cycle in northeastern Bristol Bay, Alaska. Marine Biology 157:2169–2176.

Schamel, D. 1977. Breeding of the Common Eider (Somateria mollissima) on the Beaufort Sea coast of Alaska. Condor 79:478–485.

Schindler, D. W. 1998a. A dim future for boreal waters and landscapes. BioScience 48:157–164.

Schindler, D. W. 1998b. Sustaining aquatic ecosystems in boreal regions. Conservation Ecology 2:art18.

Schindler, D. W., and P. G. Lee. 2010. Comprehensive conservation planning to protect biodiversity and ecosystem services in Canadian boreal regions under a warming climate and increasing exploitation. Biological Conservation 143:1571–1586.

Schindler, D. W., K. H. Mills, D. F. Malley, D. L. Findlay, J. A. Shearer, I. J. Daview, M. A. Turner, G. A. Linsey, and D. R. Cruikshank. 1985. Long-term ecosystem stress: the effects of years of experimental acidification on a small lake. Science 228:1395–1401.

Schloesser, D. W., T. F. Nalepa, and G. L. Mackie. 1996. Zebra mussel infestation of unionid bivalves (Unionidae) in North America. American Zoologist 36:300–310.

Schmiegelow, F. K. A., and M. Mönkkönen. 2002. Habitat loss and fragmentation in dynamic landscapes: avian perspectives from the boreal forest. Ecological Applications 12:375–389.

Schmutz, J. K., R. J. Robertson, and F. Cooke. 1983. Colonial nesting of the Hudson Bay Eider Duck. Canadian Journal of Zoology 61:2424–2433.

Schorger, A. W. 1951. Deep diving of the Old-squaw. Wilson Bulletin 63:112.

Schummer, M. L., S. A. Petrie, and R. C. Bailey. 2008a. Dietary overlap of sympatric diving ducks during winter on northeastern Lake Ontario. Auk 125:425–433.

Schummer, M. L., S. A. Petrie, and R. C. Bailey. 2008b. Interaction between macroinvertebrate abundance and habitat use by diving ducks during winter in northeastern Lake Ontario. Journal of Great Lakes Research 34:54–71.

Schummer, M. L., S. A. Petrie, R. C. Bailey, and S. S. Badzinski. 2012. Factors affecting lipid reserves and foraging activity of Buffleheads, Common Goldeneyes, and Long-tailed Ducks during winter at Lake Ontario. Condor 114:62–74.

Scott, W. E. 1938. Old-squaws taken in gill-nets. Auk 55:668.

Sea Duck Joint Venture. 2012 [online]. Atlantic and Great Lakes sea duck migration study. Progress Report, Washington, DC. <http://seaduckjv.org> (7 August 2013).

Sea Duck Joint Venture Management Board. 2008. Sea Duck Joint Venture Strategic Plan 2008–2012. U.S. Fish and Wildlife Service, Anchorage, AK.

Seavy, N. E., K. E. Dybala, and M. A. Snyder. 2008. Climate models and ornithology. Auk 125:1–10.

Sénéchal, H. G. 2003. Étude comparative de l'écologie de nidification et d'élevage des couvées de Garrot à œil d'or (Bucephala clangula) et de Harle Couronné (Lophodytes cucullatus) dans un habitat de rivière. Thesis, Université Laval, Québec, QC (in French).

Sénéchal, H., G. Gauthier, and J.-P. L. Savard. 2008. Nesting ecology of Common Goldeneyes and Hooded Mergansers in a boreal river system. Wilson Journal of Ornithology 120:732–742.

Shaw, J., R. B. Taylor, S. Solomon, H. A. Christian, and D. L. Forbes. 1998. Potential impacts of global sea-level rise on Canadian coasts. Canadian Geographer 42:365–379.

Shine, C., N. Williams, and L. Gündling. 2000. A guide to designing legal and institutional frameworks on alien invasive species. Environmental and Law Policy Paper, No. 40. International Union for Conservation of Nature and Nature Resources. Gland, Switzerland; Cambridge, U.K.; and Bonn, Germany.

Silverman, E. D., D. T. Saalfeld, J. B. Leirness, and M. D. Koneff. 2013. Wintering sea duck distribution along the Atlantic Coast of the United States. Journal of Fish and Wildlife Management 4:178–198.

Simberloff, D. 2010. Invasive species. Pp. 131–152 in N. S. Sodhi and P. R. Ehrlich (editors), Conservation biology for all. Oxford University Press, Oxford, U.K.

Sirenko, B. L., and V. M. Koltun. 1992. Characteristics of benthic biocenoses of the Chukchi and Bering Seas. Pp. 251–261 in P. A. Nagel (editor), Results of the third joint US-USSR Bering and Chukchi Seas expedition (BERPAC), summer 1988. U.S. Fish and Wildlife Service, Washington, DC.

Sjöberg, K. 1989. Time-related predator/prey interactions between birds and fish in a northern Swedish river. Oecologia 80:1–10.

Skei, J., P. Larsson, R. Rosenberg, P. Jonsson, M. Olsson, and D. Broman. 2000. Eutrophication and contaminants in aquatic ecosystems. Ambio 29:184–194.

Skinner, M. P. 1937. Barrow's Golden-eye in the Yellowstone National Park. Wilson Bulletin 59:3–11.

Smith, R. I. 1970. Response of Pintail breeding populations to drought. Journal of Wildlife Management 34:943–946.

Spence, J. R., and B. R. Bobowski. 2003. 1994–1997 Water bird surveys of Lake Powell, a large oligotrophic reservoir on the Colorado River, Utah and Arizona. Western Birds 34:133–148.

Stempniewicz, L. 1995. Feeding ecology of the Long-tailed Duck Clangula hyemalis wintering in the Gulf of Gdańsk (southern Baltic Sea). Ornis Svecica 5:133–142.

Stewart, F. B., A. S. Pullin, and C. F. Coles. 2007. Poor evidence-base for assessment of windfarm impacts on birds. Environmental Conservation 34:1–11.

Stiller, J. C. 2011. Effects of Common Merganser on hatchery-reared brown trout and spring movements of adult males in southeastern New York, USA. Thesis, State University of New York, Syracuse, NY.

Stocks, B. J., M. A. Fosberg, T. J. Lynham, L. Mearns, B. M. Wotton, Q. Yang, J.-Z. Jin, K. Lawrence, G. R. Hartley, J. A. Mason, and D. W. McKenney. 1998. Climate change and forest fire potential in Russian and Canadian boreal forests. Climatic Change 38:1–13.

Stocks, B. J., M. A. Fosberg, B. M. Wotton, T. J. Lynham, and K. C. Ryan. 2000. Climate change and forest fire activity in North American boreal forests. Pp. 368–376 in E. S. Kasischke and B. J. Stocks (editors), Fire, climate change, and carbon cycling in the North American boreal forest. Springer-Verlag, New York, NY.

Stott, R. S., and D. P. Olson. 1973. Food-habitat relationship of sea ducks on the New Hampshire coastline. Ecology 54:996–1007.

Strayer, D. L. 1991. Projected distribution of the zebra mussel, Dreissena polymorpha, in North America. Canadian Journal of Fisheries and Aquatic Sciences 48:1389–1395.

Strayer, D. L., V. T. Eviner, J. M. Jeschke, and M. L. Pace. 2006. Understanding the long-term effects of species invasions. Trends in Ecology and Evolution 21:645–651.

Sullivan, T. M., R. W. Butler, and W. S. Boyd. 2002. Seasonal distribution of waterbirds in relation to spawning Pacific herring, Clupea pallasi, in the Strait of Georgia, British Columbia. Canadian Field-Naturalist 116:366–370.

Suydam, R. S. 2000. King Eider (Somateria spectabilis). in A. Poole and F. Gill (editors), The Birds of North America, No. 491. The Academy of Natural Sciences, Philadelphia, PA.

Taylor, E. J. 1986. Foods and foraging ecology of Oldsquaws (Clangula hyemalis L.) on the Arctic Coastal Plain of Alaska. Thesis, University of Alaska, Fairbanks, AK.

Thomas, P. W., and G. J. Robertson. 2008. Apparent survival of male Harlequin Ducks molting at the Gannet Islands, Labrador. Waterbirds 31:147–151.

Thompson, D. 1973. Feeding ecology of diving ducks on Keokuk Pool, Mississippi River. Journal of Wildlife Management 37:367–381.

Thompson, J. E., and C. D. Ankney. 2002. Role of food in territoriality and egg production of Buffleheads and Barrow's Goldeneyes. Auk 119:1075–1090.

Titman, R. D. 1999. Red-breasted Merganser (Mergus serrator). in A. Poole and F. Gill (editors), The Birds of North America, No. 443. The Academy of Natural Sciences, Philadelphia, PA.

Todd, W. E. C. 1963. Birds of the Labrador Peninsula and adjacent areas. A distributional list. University of Toronto Press, Toronto, ON.

Trauger, D. L., and R. G. Bromley. 1976. Additional bird observations on the West Mirage Islands, Great Slave Lake, Northwest Territories. Canadian Field-Naturalist 90:114–122.

Trautman, M. B., W. E. Bills, and E. L. Wickliff. 1939. Winter losses from starvation and exposure of waterfowl and upland game birds in Ohio and other northern states. Wilson Bulletin 51:86–104.

Traylor, J. J., R. T. Alisauskas, and F. P. Kehoe. 2004. Nesting ecology of White-winged Scoters Melanitta fusca deglandi at Redberry Lake, Saskatchewan. Auk 121:950–962.

U.S. Department of the Interior. 1995. Final environmental impact statement: Elwha River ecosystem restoration. Olympic National Park, Washington, DC.

U.S. Fish and Wildlife Service. 2002. Steller's Eider recovery plan. Fairbanks, AK.

U.S. Fish and Wildlife Service. [online]. 2013. International migratory bird day. <http://www.fws.gov/birds/imbd/> (8 August 2013).

Vaillancourt, M.-A., P. Drapeau, M. Robert, and S. Gauthier. 2009. Origin and availability of large cavities for Barrow's Goldeneye (Bucephala islandica), a species at risk inhabiting the eastern Canadian boreal forest. Avian Conservation and Ecology 4:art6.

Vandendriessche, S., E. W. M. Stienen, M. Vincx, and S. Degraer. 2007. Seabirds foraging at floating seaweeds in the northeast Atlantic. Ardea 95:289–298.

Van de Wetering, D. 1997. Moult characteristics and habitat selection of post-breeding male Barrow's Goldeneye (Bucephala islandica) in northern Yukon. Canadian Wildlife Service Technical Report Series, No. 296. Pacific and Yukon Region, Vancouver, BC.

van Dijk, B. 1986. The breeding biology of eiders at Ile aux Pommes, Quebec. Pp. 119–135 in A. Reed (editor), Eider ducks in Canada. Canadian Wildlife Service, Report Series, No. 47. Canadian Wildlife Service, Ottawa, ON.

Vermeer, K. 1981. Food and populations of Surf Scoters in British Columbia. Wildfowl 33:107–116.

Vermeer, K. 1982. Food and distribution of three Bucephala species in British Columbia waters. Wildfowl 33:22–30.

Vermeer, K., M. Bentley, K. H. Morgan, and G. E. J. Smith. 1997. Association of feeding flocks of brant and sea ducks with herring spawn at Skidegate Inlet. Pp. 102–107 in K. Vermeer and K. H. Morgan (editors), The ecology, status, and conservation of marine and shoreline birds of the Queen Charlotte Islands. Canadian Wildlife Service, Occasional Paper, No. 93. Canadian Wildlife Service, Ottawa, ON.

Vermeer, K., and N. Bourne. 1982. The White-winged Scoter diet in British Columbia waters: resource partitioning with other scoters. Pp. 30–38 in D. N. Nettleship, G. A. Sanger, and P. F. Springer (editors), Marine birds: their feeding ecology and commercial fisheries relationships. Canadian Wildlife Service, Darmouth, NS.

Vermeer, K., and K. H. Morgan. 1992. Marine bird populations and habitat use in a fjord on the west coast of Vancouver Island. Pp. 86–96 *in* K. Vermeer, R. W. Butler, and K. H. Morgan (editors), The ecology, status, and conservation of marine and shoreline birds on the west coast of Vancouver Island. Canadian Wildlife Service, Occasional Paper, No. 75. Canadian Wildlife Service, Ottawa, ON.

Walsh, J. E. 2008. Climate of the arctic marine environment. Ecological Applications 18:S3–S22.

Wayland, M., and D. K. McNicol. 1994. Movements and survival of Common Goldeneye broods near Sudbury, Ontario, Canada. Canadian Journal of Zoology 72:1252–1259.

Weber, M. G., and B. J. Stocks. 1998. Forest fires and sustainability in the boreal forests of Canada. Ambio 27:545–550.

Weller, M. W. 1999. Wetland birds: habitat resources and conservation implications. Cambridge University Press, Cambridge, U.K.

Weller, M. W., and C. S. Spatcher. 1965. Role of habitat in the distribution and abundance of marsh birds. Iowa State University Agricultural and Home Economics Experiment Station Special Report, No. 43. Iowa State University, Ames, IA.

Weltzin, J. F., S. D. Bridgham, J. Pastor, J. Chen, and C. Harth. 2003. Potential effects of warming and drying on peatland plant community composition. Global Change Biology 9:141–151.

White, H. C. 1957. Food and natural history of mergansers on salmon waters in the Maritime Provinces of Canada. Fisheries Research Board of Canada, Bulletin No. 116.

White, T. P., R. R. Veit, and M. C. Perry. 2009. Feeding ecology of Long-tailed Ducks *Clangula hyemalis* wintering on the Nantucket Shoals. Waterbirds 32:293–299.

Wisenden, P. A., and R. C. Bailey. 1995. Development of macroinvertebrate community structure associated with zebra mussel (*Dreissena polymorpha*) colonization of artificial substrates. Canadian Journal of Zoology 73:1438–1443.

Wood, C. C. 1985. Aggregative response of Common Mergansers (*Mergus merganser*): predicting flock size and abundance on Vancouver Island salmon streams. Canadian Journal of Fisheries and Aquatic Sciences 42:1259–1271.

Wood, C. C. 1986. Dispersion of Common Merganser (*Mergus merganser*) breeding pairs in relation to the availability of juvenile Pacific salmon in Vancouver Island streams. Canadian Journal of Zoology 64:756–765.

Wright, K. K., H. Bruner, J. L. Li, R. Jarvis, and S. Dowlan. 2000. The distribution, phenology, and prey of Harlequin Ducks, *Histrionicus histrionicus*, in a Cascade mountain stream, Oregon. Canadian Field-Naturalist 114:187–195.

Young, A. D., and R. D. Titman. 1986. Costs and benefits to Red-breasted Mergansers nesting in tern and gull colonies. Canadian Journal of Zoology 64:2339–2343.

Zicus, M. C., and S. K. Hennes. 1988. Cavity nesting waterfowl in Minnesota. Wildfowl 39:115–123.

Zipkin, E. F., B. Gardner, A. T. Gilbert, A. F. O'Connell Jr., J. A. Royle, and E. D. Silverman. 2010. Distribution patterns of wintering sea ducks in relation to the North Atlantic Oscillation and local environmental characteristics. Oecologia 163:893–902.

Žydelis, R. 2001. Some remarks on effect of climatic parameters on wintering waterbirds in the Eastern Baltic. Acta Zoologica Lituanica 11:303–308.

Žydelis, R., D. Esler, W. W. Boyd, D. Lacroix, and M. Kirk. 2006. Habitat use by wintering Surf and White-winged Scoters: effects of environmental attributes and shellfish aquaculture. Journal of Wildlife Management 70:1754–1762.

Žydelis, R., and D. Ruškytė. 2005. Winter foraging of Long-tailed Ducks (*Clangula hyemalis*) exploiting different benthic communities in the Baltic Sea. Wilson Bulletin 117:133–141.

CHAPTER FOURTEEN

Conservation of North American Sea Ducks*

W. Sean Boyd, Timothy D. Bowman,
Jean-Pierre L. Savard, and Rian D. Dickson

Abstract. Several species of North American sea ducks have experienced population declines in the last century and, in most cases, the causes remain unknown. Of primary concern for conservation is the fact that research on the tribe Mergini has lagged far behind other waterfowl groups, leading to a poor understanding of the key factors that may limit sea duck populations. We lack basic information on the demography and ecology of most species, including patterns of distribution and abundance, habitat associations, demographic rates, and population structure. Moreover, sea ducks face a wide range of potential threats, from large-scale environmental drivers such as climate change that may have long-lasting effects on many species to specific resource development projects that may affect some species at local geographic areas over short time periods. Further efforts are required to assess the individual and cumulative impacts of these threats to sea ducks before we can formulate effective conservation strategies.

Key Words: climate change, disturbance, habitat requirements, harvest, invasive species, mismatch in timing, parasites, pathogens, population delineation, population trends, resource development, satellite telemetry, trophic interactions.

At the turn of the twenty-first century, 10 of the 15 North American species of sea ducks (Tribe Mergini) were thought to be declining (Caithamer et al. 2000, Sea Duck Joint Venture 2003), and several species were considered species of conservation concern and listed as *at risk* or *threatened*. The eastern population of Harlequin Ducks (*Histrionicus histrionicus*) was listed as endangered in Canada in 1986 (later downgraded to species of special concern; Thomas and Robert 2001) and as a threatened species in Maine since 1997 (Maine Department of Inland Fisheries and Wildlife 2013). In 2000, the eastern population of Barrow's Goldeneyes (*Bucephala islandica*) was designated as a species of special concern in Canada (Robert et al. 2000) and is currently considered threatened in Maine (Maine Department of Inland Fisheries and Wildlife 2013). Spectacled Eiders (*Somateria fischeri*) were listed as a federally threatened species in the United States in 1993 (U.S. Fish and Wildlife Service 1996), and the Alaska-breeding population of Steller's Eider (*Polysticta stelleri*) was added to the same list in 1997 (U.S. Fish and Wildlife

* Boyd, W. S., T. D. Bowman, J.-P. L. Savard, and R. D. Dickson. 2015. Conservation of North American sea ducks. Pp. 529–559 in J.-P. L. Savard, D. V. Derksen, D. Esler, and J. M. Eadie (editors). Ecology and conservation of North American sea ducks. Studies in Avian Biology (no. 46), CRC Press, Boca Raton, FL.

Service 2002). Apparent declines in abundance and subsequent listing decisions have generated considerable concern for the welfare of sea ducks in North America.

In 1986, the North American Waterfowl Management Plan (NAWMP) was signed by the United States and Canada to address concerns related to declining waterfowl populations and management of shared waterfowl resources. A combination of favorable climatic conditions, habitat enhancement and conservation programs, and harvest restrictions resulted in population increases of some waterfowl species in subsequent years (Williams et al. 1999). However, most of the species exhibiting positive population trajectories were dabbling ducks, while many sea duck populations remained below historic levels (North American Waterfowl Management Plan 1998). The purpose of our chapter is to highlight some of the most pressing issues facing North American sea ducks. Of primary concern for conservation is the fact that research on the tribe Mergini has lagged far behind other waterfowl groups, leading to a poor understanding of the key factors limiting population numbers of sea ducks. Moreover, sea ducks face a wide range of potential threats, from large-scale environmental drivers such as climate change that may have long-lasting effects on many species to specific resource development projects that may affect some species at local geographic areas over short time periods. We begin this chapter with a discussion of knowledge gaps followed by an examination of specific conservation challenges. Throughout the chapter, we emphasize the necessity of having reliable ecological information to guide conservation efforts.

KNOWLEDGE GAPS

Our ability to conserve sea ducks depends on having relevant biological and ecological information at both the species and population levels. Harvest management and habitat protection efforts have been identified as important conservation priorities (Sea Duck Joint Venture 2012). To address these topics, managers need a better understanding of how sea duck populations are structured, as well as their respective abundance and distribution patterns, trends, demographic vital rates, and harvest levels (Chapters 2, 3, and 12, this volume).

Demography, Population Models, and Limiting Factors

Sea ducks are generally long-lived, with delayed sexual maturity and variable rates of nonbreeding (Goudie et al. 1994). Annual reproductive success is low on average, but can be highly variable, and population growth is dependent on intermittent years of relatively high duckling production (Goudie et al. 1994, Schamber et al. 2009). Small changes in adult survival, relative to other population parameters, can affect population trajectories (Gilliland et al. 2009; Coulson 2010; Bentzen and Powell 2012; Wilson et al. 2012; Chapter 3, this volume). For example, declines in adult survival were correlated with reduced population viability of northern Common Eiders (Somateria mollissima) in Greenland and Canada (Gilliland et al. 2009), but following hunting restrictions implemented in 2001, clear evidence of population recovery was documented (Merkel 2010). In western Alaska, toxic effects of ingestion of lead shot were a key contributing factor in increased mortality rates of Spectacled Eiders, and restrictions on lead shot are a potential target for management (Flint et al. 1997, 2000; Grand et al. 1998). At the same time, demographic data collected for nesting Pacific Common Eiders were used to develop a population model, which could be contrasted with declining populations of Spectacled Eiders at the same breeding grounds (Wilson et al. 2012). Population studies have noted that eiders are most sensitive to declines in adult survival, but practical ways to influence this parameter were limited (Bentzen and Powell 2012, Wilson et al. 2012). Nest survival, duckling survival, and other reproductive parameters may be more likely to influence population trajectories because components of reproduction are inherently more variable and are more feasible demographic parameters to target with management actions.

Population models can formalize assessments of population-level processes and provide a means to determine the relative contributions of different parameters to population growth. Models can also identify gaps in existing knowledge and facilitate development of hypotheses to prioritize research and assess potential outcomes of different management prescriptions. Useful models require robust estimates of age-specific reproduction and survival rates, which in turn require reasonably good estimates of breeding propensity, clutch size, nest success, and juvenile dispersal. It is necessary to know the scale at which the population of concern is demographically independent from other populations. Further, parameter estimates are most relevant if derived from a representative set of sites throughout the geographic range of the population.

Population models are especially useful for identifying when and where constraints on growth are occurring, and this information can be used to target research and conservation planning (Gilliland et al. 2009, Schamber et al. 2009, Wilson et al. 2012). Where models lack data on critically sensitive parameters, efforts should be made to improve parameter estimates. Managing populations based on model-generated priorities alone does not address the logistical challenges of changing vital rates (Mills et al. 1999). Some parameters may be difficult, if not impossible, for managers to address such as the potential for mismatches in timing due to climate change or changes in sea ice. A diversified approach is likely to be most efficient (Wilson et al. 2012). For example, policy and legislation can be used to simultaneously limit the effects of harvest, habitat degradation, and disturbance, and more localized efforts such as focused predator control may have concurrent, positive effects on multiple aspects of reproduction and survival.

Estimation of vital rates is likely to become an increasing priority for some sea duck species, particularly for discrete population units. A population study of White-winged Scoters (*Melanitta deglandi*) in the Northwest Territories of Canada determined that apparent breeding propensity of paired females, as measured by blood assays, was highly variable from year to year. Many of the paired but nonbreeding females were subadults, implying that surveys counting pairs may overestimate local breeding populations (S. Slattery, unpubl. data). Studies of Harlequin Ducks on the east and west coasts have indicated that demographic information for this species can be obtained relatively easily, and one study on the Pacific Coast provided estimates of subadult dispersal rates (Regehr 2003). One obvious need is for a population study of the intensively hunted population of Common Eiders on the east coast of North America; large numbers of adults and young could be captured and banded, permitting the quantification of important demographic parameters (Joint Working Group on the

Management of the Common Eider 2004). Cavity-nesting sea ducks include Buffleheads (*Bucephala albeola*), Common Goldeneyes (*B. clangula*), Barrow's Goldeneyes, Hooded Mergansers (*Lophodytes cucullatus*), and Common Mergansers (*Mergus merganser*). Many of these species will readily use artificial nest boxes, as well as natural cavities, enabling capture and monitoring of females during the breeding period to estimate reproductive rates (Dugger et al. 1999, Schmidt et al. 2006).

Unfortunately, for most sea duck populations, it is difficult to band sufficient numbers and obtain an adequate number of recoveries to permit detailed analyses of survival, so conservation efforts may be forced to proceed without complete demographic data and models (Heppell et al. 2000). A multifaceted approach may be needed, where several measures of population growth are compared and contrasted and serve to validate one another, rather than relying on a single estimate alone (Eberhardt 2002, Sandercock and Beissinger 2002, Wilson et al. 2012). Integration of multiple measures of population change including trends in abundance, survival estimates from mark–recapture studies, and projected population modeling would improve our understanding of population dynamics and the efficacy of management actions (Wilson et al. 2012).

Population Delineation

The identification of demographically independent populations is a necessary precursor for almost all other information needs in wildlife management. Population dynamics can be influenced by intrinsic factors, such as recruitment and survival, and by movements of individuals among different segments of the population. The interpretation of data on trends and harvest rates must take into account the scale at which the population or subpopulation of interest is demographically independent, which requires an understanding of patterns of connectivity among breeding, molting, and wintering areas, and levels of individual philopatry and site fidelity to those areas. The key factors limiting population growth cannot be assessed without these basic types of information.

A variety of approaches can be used to delineate sea duck populations, including satellite telemetry, banding studies, stable isotopes, and population genetics. Each technique provides resolution at varying spatial and temporal scales, and each has a unique set of assumptions and potential biases. Satellite telemetry has been used in recent decades and results have been generally successful but variable among species, location, and time of year. Telemetry underscored the importance of wintering and molting areas in Greenland for Northern Common Eiders (*borealis* subsp., Mosbech et al. 2006b) and eastern Harlequin Ducks that breed in Canada (Brodeur et al. 2002, Savard et al. 2011). A recent project marked birds of all age and sex classes in Pacific Barrow's Goldeneyes, and successfully identified a key molting lake for males in Alberta (Hogan et al. 2011), and provided insights into demographically independent subpopulations ranging from British Columbia to Alaska (W. S. Boyd, unpubl. data). Pacific Surf Scoters (*Melanitta perspicillata*) marked on wintering areas from Mexico to Alaska generated movement data to describe migration routes, timing, and affiliations between wintering and breeding areas (De La Cruz et al. 2009). Telemetry studies of Black Scoters (*M. americana*) confirmed the presence of two independent populations in North America—one that breeds and winters in the eastern part of the continent and one with a western distribution (Alaska Department of Fish and Game, Canadian Wildlife Service, U.S. Fish and Wildlife Service, and U.S. Geological Survey, unpubl. data). Satellite telemetry confirmed earlier stable isotope work by Mehl et al. (2005) and indicated that King Eiders (*Somateria spectabilis*) breeding in Alaska and the western Canadian Arctic migrate west to winter along the coasts of Alaska and Russia, whereas birds breeding in the eastern Canadian Arctic migrate east to winter in western Greenland and Maritime Canada (Mosbech et al. 2006a, Powell 2009, Dickson 2012).

Advances in satellite transmitter technology enable the tracking of individual birds for more than one annual cycle, allowing researchers to address questions about annual variability in movements, seasonal habitat use patterns, and site fidelity. Understanding the level of site fidelity should be a priority for all sea duck species but is particularly important for managing hunted species. Unusually high rates of harvest in combination with high site fidelity or low natal dispersal rates can lead to local extirpation.

Satellite telemetry has helped solve many mysteries about the migration ecology of sea ducks but it is an expensive technology, and as a result,

sample sizes tend to be small. Uncertainties remain about the potential effects of implanted transmitters on aspects of migratory behavior such as timing, duration, direction of travel, and components of reproduction and productivity. Implanted transmitters have been shown to cause malformed eggs in captive diving ducks (G. Olsen, unpubl. data). Black Scoters that received implants at spring staging areas delayed migration toward breeding or molting areas in the year they were implanted (S.G. Gilliland, unpubl. data). Fast et al. (2011), in their study on Common Eiders, noted the need to assess the impacts of implant transmitters on the flight behavior of sea ducks. Transmitters negatively affected nesting propensity and apparent survival of Black-tailed Godwits (*Limosa limosa*), but migration patterns were unaffected in this large-bodied shorebird (Hooijmeijer et al. 2014).

Leg bands, stable isotopes, and genetic markers have also been used to describe population-level connections among geographically separate regions (Mehl et al. 2005; Pearce et al. 2009b; Chapter 2, this volume). Pearce et al. (2008, 2009a) argued for the use of multiple data sources to describe population structure, and Pearce et al. (2014) used a combination of banding and genetic analysis to delineate populations of Buffleheads, Common Goldeneyes, and Barrow's Goldeneyes.

Population Size and Trend

Monitoring is fundamental to understanding population dynamics and evaluating management actions, yet most species of sea ducks are poorly surveyed. The Waterfowl Breeding Population and Habitat Survey (WBPHS) is flown each spring and is used as a basis for setting population goals for many North American waterfowl. However, the survey does not cover the core breeding ranges of most sea duck species and is not timed to capture peak counts of breeding sea ducks, which generally nest later than dabbling ducks. Some groups of sea ducks are not differentiated to species during the survey and may be pooled as scoters, goldeneyes, or mergansers. Monitoring sea duck populations therefore will require new surveys or modifications to WBPHS as an existing survey program to ensure adequate geographic coverage, timing, and precision. The Sea Duck Joint Venture partnership has helped test survey techniques and designs that have led to one operational survey for breeding Black Scoters in Alaska and tests of feasibility for several others that could become operational with adequate funding. Nest counts in colonies of Common Eiders are providing one of the most useful measures of trends for breeding sea duck populations in eastern North America (Joint Working Group on the Management of the Common Eider 2004). Additional details on population trends and monitoring programs for sea ducks in North America are important (Chapter 1, this volume).

Harvest

Accurate measurements of sport and subsistence harvest of sea ducks are difficult to obtain (Chapter 12, this volume). National harvest surveys do not provide reliable estimates, although the implementation of the U.S. Harvest Information Program in 2001 has improved our ability to estimate harvest metrics (total birds, sex and age ratios of the bag) and model population effects of regulatory changes. In Canada, the National Harvest Survey has provided wildlife managers with data on hunter activity and harvest of migratory game species since 1967. Despite these long-standing surveys, estimates of sea duck harvest remain relatively poor because the number of sea duck hunters participating in the surveys is low, survey response rates are poor, important hunting areas for sea ducks are clustered, and the sampling period tends to miss late season periods when the majority of sea ducks are harvested. In addition, data from parts collection programs are often inadequate to allow reliable assessment of numbers, species, sex, and age composition (Chapter 12, this volume).

Recent satellite telemetry studies have confirmed that some populations of sea ducks winter along the coasts of Greenland (Brodeur et al. 2002, Mosbech et al. 2006b) and northern Russia (Oppel et al. 2009a). While harvest data from Greenland are generally good, the commercial harvest and sale of sea ducks, particularly Common Eiders, have resulted in significant population declines in the past (Hansen 2002), highlighting a need to closely monitor harvest (Gilliland et al. 2009). Recent restrictions in harvest quotas have resulted in local increases in breeding Common Eiders, indicating that sustainable harvest is possible in Greenland (Merkel 2010, Burnham et al. 2012).

Sea duck harvest data from Russia are sparse or nonexistent, although harvest of some North American species may be significant during winter.

Sea ducks are harvested for subsistence in several countries (Cooch 1965, Bartonek 1986, Nakashima 1986, Reed 1986, Merkel 2004), but this use has not been thoroughly assessed, and monitoring and management of subsistence harvest is particularly challenging (Mitchell 1986; Natcher et al. 2011, 2012). Given the relative sensitivity of sea duck populations to changes in adult survival, better data are needed on timing and levels of sport and subsistence harvest (Gilliland et al. 2009; Chapter 12, this volume).

Habitat

Sea ducks occupy a broad range of habitat types throughout the annual cycle, mostly in northern and coastal regions where ecological information is limited. Many species travel to inland freshwater areas to breed and molt, to coastal sites to molt and winter, and to inland or coastal staging sites during migration. Common Eiders breed and molt in marine waters, and large numbers of scoters, Long-tailed Ducks (*Clangula hyemalis*), and Harlequin Ducks also molt in marine environments. From fall through spring, most sea ducks congregate at coastal sites that offer reliable food resources and stable environments. Sea ducks may be resident at coastal sites for 8–9 months of the year, underscoring the importance of these habitats.

Information on habitat associations of sea ducks is limited (Chapter 13, this volume). Investigations of the relationships between habitat and sea duck productivity and survival are a fundamental need, particularly for remote environments. Principal information needs include (1) seasonal distribution and abundance in relation to habitat types; (2) ecological characterization of breeding, brood rearing, molting, staging, and wintering habitats; and (3) seasonal habitat requirements in relation to food and other factors.

In addition to basic ecological information, understanding the potential effects of natural and anthropogenic alterations to sea duck habitats should be a management priority (Chapter 13, this volume). At northern breeding and molting areas, climate change may already be having dramatic effects on tundra and boreal forest environments.

In some areas, resource extraction for oil and gas, mining, and timber harvest are impinging on sea duck habitats. Coastal staging, molting, and wintering areas are subject to other challenges, such as urbanization and industrial development, shipping and commerce of petroleum and other hazardous cargo, harvest or aquaculture of shellfish and finfish, nearshore wind energy development, and increases in recreational activity. Sea ducks concentrate at specific geographic areas throughout the year and especially during winter, which leads to a pressing need for (1) identification of key, vulnerable sites and habitat types, (2) research on development-related impacts to assess population-level effects, (3) strategies to protect important habitats and mitigate impacts, and (4) assessment of habitat conservation efforts over large geographic scales.

Parasites, Pathogens, and Contaminants

Biotic and anthropogenic agents can influence the viability of sea duck populations, either through direct mortality or indirectly by lowering reproductive potential. Relatively little information exists on population-level impacts of parasites, diseases, and contaminants. Nevertheless, considerable documentation exists regarding the presence of a diversity of internal and external parasites, diseases, and chemical toxins among different sea ducks in the tribe Mergini (Chapter 4, this volume).

Parasites known to occur in sea ducks include protozoans, trematodes, cestodes, nematodes, acanthocephalans, hematozoans, mallophagans, and siphonapterans, and several sea duck die-offs have been attributed to these agents. For example, renal coccidiosis, which is caused by protozoan parasites that infect the kidneys and associated tissues, accounted for 25%–40% of mortality of 1–2-week-old ducklings of Common Eiders (Mendenhall and Milne 1985). A holarctic species of spiny-headed worm (Acanthocephala) has been identified as the cause of epizootics and subsequent mortality in Common Eiders (Clark et al. 1958). Die-offs of Red-breasted Mergansers (*Mergus serrator*) have been attributed to a larval nematode (Locke et al. 1964, Friend and Franson 1999).

The impact of diseases can be significant as the loss of mature adults tends to have the greatest effect on population dynamics (Goudie et al. 1994, Gilliland et al. 2009, Schamber et al. 2009,

Bentzen and Powell 2012, Wilson et al. 2012). Avian cholera, one of the most destructive diseases, is important in the dynamics of colonially nesting Common Eiders (Descamps et al. 2012, Tjornlov et al. 2013) and should be closely monitored. Cholera outbreaks mostly affect colonial breeding birds (Gershman et al. 1964, Reed and Cousineau 1967), but also occur in populations of physiologically stressed birds and appear to be moving northward (Korschgen et al. 1978, Harms 2011). Unfortunately, little is known of the factors leading to cholera outbreaks and whether climate warming will increase its occurrence and frequency (Harms 2011). Cholera has caused large die-offs among Common Eiders in the St. Lawrence River estuary population, including an estimated mortality of 7,000–10,000 birds in 2002 (Joint Working Group on the Management of the Common Eider 2004). The disease seriously threatens the continued persistence of the largest breeding colony of northern Common Eiders in the Canadian Arctic (Descamps et al. 2009, 2011, 2012; Buttler et al. 2011). An outbreak of avian cholera in Chesapeake Bay may have killed as many as 200,000 birds, mostly Long-tailed Ducks (80%) and scoters (10%, Montgomery et al. 1979). Adenoviruses and reoviruses have been linked to mortality events in both Long-tailed Ducks and Common Eiders (Hollmén et al. 2002, 2003). In Alaska, four species of eiders, Long-tailed Ducks, Harlequin Ducks, and Buffleheads, tested positive for avian influenza, although die-offs have not been attributed to low pathogenic forms of the influenza A viruses (Ip et al. 2008; Hill et al. 2010; Ramey et al. 2010, 2011).

The impact of contaminants on sea ducks is poorly understood (Henny et al. 1995). Screening for contaminants in sea ducks has revealed a diversity of persistent organic pollutants (e.g., chlorinated pesticides, polychlorinated biphenyls, polycyclic aromatic hydrocarbons), heavy metals (e.g., arsenic, cadmium, copper, chromium, lead, zinc, mercury), and trace elements (e.g., selenium). Lead toxicosis due to the consumption of spent shot has been identified as a limiting factor in the recovery of Spectacled Eiders in Alaska (Grand et al. 1998). In fact, lead poisoning was the most common form of toxicity in a review of sea duck necropsies conducted at the U.S. Geological Survey–National Wildlife Health Center (Skerratt et al. 2005). One of the most important contaminants affecting sea ducks is petroleum products, especially when spilled into marine environments. Oil spills have caused mortality of thousands of sea ducks (Piatt et al. 1990, Fowler and Flint 1997), but the precise cause of death is often unrecorded and could be either oil toxicosis or hypothermia due to oiled plumage. Overall, the prevalence of oil toxicosis in sea ducks is difficult to assess and likely underestimated (Skerratt et al. 2005).

Further research is needed on the susceptibility of sea ducks to parasites, diseases, and contaminants. Areas of emphasis should include avian cholera, viral diseases, and other agents that might infect seasonal aggregations of sea ducks (Chapter 4, this volume). New studies are also needed to document mobility and fate of contaminants, specific histological and physiological effects, and critical exposure rates, as well direct or indirect impacts on survival and reproduction (Chapter 6, this volume).

CLIMATE CHANGE

Climate change is one of the most challenging conservation issues, as it will manifest in a variety of ways, both negative and positive (Anderson and Sorenson 2001). Impacts will differ among species, locations, and seasons, and some responses may be delayed in long-lived species such as sea ducks.

Large-Scale Regime Shifts

Most sea ducks are associated with cold environments and are likely to be negatively affected by warming waters and temperatures. In North America, the greatest forecasted changes in temperature are in northern areas where most sea ducks breed and molt (Serreze et al. 2000). Climatic conditions are affected by large-scale oceanic–atmospheric systems such as the Pacific Decadal Oscillation, El Niño Southern Oscillation, and the North Atlantic Oscillation (NAO; Ottersen et al. 2001, Mantua and Hare 2002, Stenseth et al. 2003). These regional systems involve moisture budgets and ocean temperatures that directly affect marine and terrestrial systems (Ottersen et al. 2001). Climate change will affect sea ducks through its impacts on the patterns and intensities of such systems (Forchhammer and Post 2004).

The NAO reflects winter climate conditions from the eastern US seaboard to Siberia (Hurrell

1995, Hurrell and Deser 2009). In eastern North America, a positive NAO index indicates mild winter conditions on the U.S. Atlantic coast and cold harsh conditions in the Gulf of St. Lawrence and eastern Greenland, whereas a negative index reflects the opposite. The NAO also influences northeastern Europe with warm conditions in the Baltic Sea during positive indices and cold icy conditions during negative indices. Negative NAO indices were associated with greater ice coverage in winter in the Baltic Sea and ice cover negatively affected Velvet Scoters (*Melanitta fusca*, Hartman et al. 2013). Similar climatic fluctuations can also have significant ecosystem effects (Ottersen et al. 2001, Stenseth et al. 2003, Briers et al. 2004). In Europe, NAO is positively correlated with body condition of Common Eiders, which in turn increases reproductive success (Lehikoinen et al. 2006). The NAO cycles have the opposite effect on Common Eiders wintering in Greenland (Descamps et al. 2010) and Iceland (Jonsson et al. 2013). Thus, if the frequency of positive NAO increases with climate change, it may result in lower reproductive success for eiders breeding in northern Canada. However, for sea ducks wintering along the Atlantic coast in the United States, a positive NAO implies warmer conditions, and species reactions may differ with some species doing better during positive NAO and others during negative NAO (Zipkin et al. 2010).

The North Pacific has experienced several broadscale oceanic regime shifts in the half century since the 1950s that appear to be correlated with trends in the breeding populations of several species of North American sea ducks (Flint 2013). The regime shifts are associated with changes in the size and intensity of the Aleutian low-pressure system, as well as temperature and precipitation patterns in western North America. Eiders, scoters, and Long-tailed Ducks followed similar population trajectories with overall decreases from 1950 to 1989, but stable or slightly increasing numbers in recent years. Conversely, Bufflehead, goldeneyes, and mergansers have increased continentally since 1950. It remains unclear whether regime shifts affected survival rates, reproductive output, or both, but the effects appear to differ among species. Flint (2013) speculated that oceanic regime shifts might affect food resources used by sea ducks, with consequences for survival and productivity. Since marine and terrestrial climatic patterns are linked, the effects of oceanic regime shifts may be mediated through conditions at sea or at breeding areas. In Prince William Sound, Alaska, changes in marine bird abundance were associated with a shift in climatic regime. Following a regime shift in the North Pacific in 1976–1977, population numbers of White-winged Scoters, Surf Scoters, and mergansers increased, whereas Harlequin Ducks and goldeneyes decreased in numbers (Agler et al. 1999).

Climate change will affect ecosystems in ways that are difficult to predict. The magnitude of weather extremes will likely increase the frequency of harmful algal blooms (Glibert et al. 2005), which can kill sea ducks (Jessup et al. 2009). Some sea duck species, such as Bufflehead and Barrow's Goldeneyes, breed and molt in alkaline wetlands that rely on snowmelt for water and may be affected if climate warming leads to lower precipitation levels. In some cases, effects on trophic relationships in aquatic ecosystems could result in lower productivity and decreased duckling survival (Winder and Schindler 2004). Takekawa et al. (2011) showed that the breeding distribution of Surf Scoters in northwestern Canada followed the snowline at the time of their spring arrival. Thus, breeding distribution of Surf Scoters may shift northward in response to climate warming.

Climate warming is resulting in sea level rise and an increase in the frequency of storm events (U.S. Fish and Wildlife Service, unpubl. report). Changes in storm events will affect low-lying coastal areas such as the Alaskan barrier islands used by nesting Common Eiders and important nearshore foraging areas used by sea ducks. In some areas, rising sea levels may remove land bridges, restricting access for mammalian predators and leading to increased breeding success (Chaulk 2006, Chaulk et al. 2006). Warming of oceanic waters along the eastern seaboard will likely trigger a shift in shellfish distribution and thus affect the distribution patterns of several species of sea ducks. Common Eiders are especially vulnerable as the species is tightly associated with mussel distribution (Larsen and Guillemette 2000), a shellfish adapted to cold waters (Newell 1989). In the Baltic Sea, warmer temperatures are linked to lower proportions of tissue in mussels, decreasing their nutritional value for sea ducks (Waldeck and Larsson 2013).

Extreme weather conditions associated with climate change may affect survival rates of sea ducks. For example, mass mortality of King Eiders has been related to freeze-up of polynyas during spring migration (Barry 1968, Fournier and Hines 1994). The nonmigratory population of Common Eiders in Hudson Bay is sensitive to ice conditions, and high mortality has been observed during unusually cold winters (Robertson and Gilchrist 1998, Heath and Gilchrist 2010). The distribution of sea ducks wintering in the Baltic Sea is controlled by the extent of ice cover that concentrates birds in a few ice-free areas and increases competition for food, leading to overexploitation of food resources in some years (Vaitkus 2001). On the other hand, Oppel et al. (2008, 2009b) found that King Eiders may benefit from the presence of ice because it dampens wave actions, can be used as resting sites, and may allow some areas to recover from intensive exploitation of prey.

Changes in ice cover could have significant impacts on the distribution and ecology of sea ducks, which may be negative or positive, depending on the season and species. The distribution of sea ducks wintering in the St. Lawrence estuary and gulf is greatly influenced by ice conditions (Bourget et al. 2007; Canadian Wildlife Service, unpubl. data). Grebmeier et al. (2006) suggested that stocks of benthic invertebrates in the Bering Sea are affected by warming trends that could negatively impact wintering King Eiders (Phillips et al. 2006). The authors also speculated that as sea ice cover declines, King Eiders may winter further north. The breeding dynamics of Common Eiders in the north are linked to ice conditions and earlier springs will likely improve reproductive success (Chaulk et al. 2006, 2007; Chaulk and Mahoney 2012; Hanssen et al. 2013). D'Alba et al. (2010) indicated that earlier breeding of Common Eiders in Iceland corresponded to an increase in population size, suggesting a positive response to climate change. However, Iverson et al. (2014) found that longer ice-free seasons increase the risk of nest predation by polar bears in colonial nesting Common Eiders.

Mismatch in Timing

A mismatch in timing between peak food production vs. arrival of adults at northern breeding areas or peak hatch of young could negatively affect productivity and survival of sea ducks (McKinnon et al. 2012). Charmantier et al. (2008) suggested that birds breeding at higher latitudes are already adapted to respond to interannual variability in phenology. Sea ducks have some flexibility in their migration timing and could adjust to earlier seasons but responses may vary among species (Drever et al. 2012). Common Eiders breeding in the Arctic track spring phenology and adjust their breeding timing to spring temperatures (Love et al. 2010), but late-nesting scoters may have less phenological flexibility, and a timing mismatch with important prey could have negative effects (Takekawa et al. 2011, Drever et al. 2012). Sea ducks are generally long-lived and genetic responses to climate change may be slow to develop (Parmesan 2006; Gienapp et al. 2007, 2008).

Timing mismatches could also occur throughout the food chain with indirect effects on sea ducks (Durant et al. 2007, Van der Putten et al. 2010). Increasing sea temperatures have been shown to negatively affect the recruitment of bivalves by creating a mismatch between timing of bivalve reproduction (responding to temperature) vs. blooms of phytoplankton (responding to light conditions, Philippart et al. 2003). Under this scenario, scoters, eiders, and other species of sea ducks that feed heavily on bivalves could face declining food resources.

Predator–Prey Interactions

Climate change can affect predator–prey interactions through shifts in the distribution of predators or the abundance and availability of alternative prey species. With a warming climate, new predators may enter an ecosystem, such as northward expansion of red foxes (*Vulpes vulpes*, Hickling et al. 2006). Early breakup of sea ice in the eastern Canadian Arctic makes it difficult for polar bears (*Ursus maritimus*) to hunt seals, prompting bears to seek alternative food sources (Smith et al. 2010). Since 2003, polar bear presence has increased dramatically at two Common Eider nesting colonies in northern Hudson Bay (Iverson et al. 2014). Polar bears target larger colonies, reducing nest success through direct consumption of eggs and by disturbance that facilitates additional egg predation by gulls. The combined effects of bear and gull predation can result in total reproductive failure for affected colonies of Common Eiders, and current rates of nest predation could lead to precipitous population declines.

Climate variation may alter predator–prey dynamics in subtle ways. Iles et al. (2013) found that cold and wet conditions in early spring led to decreased nest survival for Common Eiders, but warm and wet conditions in late spring were beneficial. Increased precipitation prevented mammalian predators from accessing nesting islands, whereas higher temperatures promoted vegetative growth that concealed nests from avian predators. Predators can also affect habitat use frequency and patterns (Heath et al. 2006, Cresswell 2008). In boreal and arctic ecosystems, small mammal cycles influence the reproductive success of several species of sea ducks as predators tend to focus on birds and their eggs following declines in small mammal numbers (Pehrsson 1986, Berg et al. 2000, Sittler et al. 2000, Quakenbush et al. 2004, Iles et al. 2013). Climate change may influence the magnitude and frequency of these cycles (Kausrud et al. 2008, Gilg et al. 2009), and thus patterns of prey abundance, with possible consequences for sea ducks.

HABITAT

Sea ducks use a variety of habitats, from Arctic islands to urban harbors and from pack ice leads in the Bering Sea to coastal lagoons in Baja California (Chapter 13, this volume). Many species, particularly King Eiders, scoters, and Long-tailed Ducks, are widely dispersed across the tundra and boreal forest where they nest at low densities. Conversely, during the nonbreeding season, many species congregate in large, dense flocks at molting locations, migratory staging sites, and especially wintering areas. Sites that offer reliable food resources and stable environments are often used consistently from year to year, but distributions can vary annually and seasonally as sea ducks take advantage of ephemeral but rich foods such as herring spawn and blooms of marine organisms (Munro and Clemens 1931, Lacroix et al. 2005, Lok et al. 2008). An understanding of spatial distribution patterns, habitat associations, and annual variability is needed to assess how these factors influence productivity, survival, and vulnerability to environmental and anthropogenic changes.

Tracking individual ducks provides fine-scale temporal and spatial details on specific habitat features that correspond to individual home ranges, nest site locations, foraging areas, and molting locations. Telemetry with very-high-frequency (VHF) radio transmitters can be used to follow daily movements of individuals and has revealed that some species will move from inshore coastal feeding areas to nocturnal offshore roosting locations (Lewis et al. 2005). Satellite telemetry is rapidly expanding our understanding of continental-scale movement patterns, generating data over the entire annual cycle, and linking movements among breeding, molting, and wintering areas. Satellite telemetry also provides movement and settling data that can be used to identify specific sites and habitat types used by large numbers of birds (Petersen et al. 1999, Hogan et al. 2011). Managers can use spatial analyses to link sea duck distribution data with biotic and abiotic features to understand specific habitat requirements for different species. In conjunction with formal surveys, local ecological knowledge from rural or native communities can be a valuable source of information, particularly in remote areas of northern Canada and Alaska where scientific information may be limited (Gilchrist et al. 2005, Mallory et al. 2006).

Habitats or sites used consistently by high numbers of sea ducks are obvious candidates for protection, as are sites used by multiple species. For example, the entire global population of Spectacled Eiders winters in a relatively small region of the Bering Sea (Petersen et al. 1999). Large proportions of the eastern populations of Long-tailed Ducks, White-winged Scoters, Common Eiders, and Black Scoters regularly winter or stage in the Cape Cod and Nantucket area off the Massachusetts coast. If negative impacts are imminent, then the conservation value of an area should be evaluated and appropriate actions taken to mitigate risk.

Basic research is needed on the ecological requirements of species using special or unique habitats. Examples include cavity-nesting ducks (Bufflehead, Common and Barrow's Goldeneyes, mergansers); island nesters (Common Eider, White-winged Scoter, Red-breasted Mergansers); river specialists (Harlequin Duck, mergansers); species that use nearshore sites for wintering or molting (Harlequin Duck, Surf Scoter, White-winged Scoter and Black Scoter, Bufflehead, goldeneyes, mergansers); species that require deep, offshore waters for wintering or molting (Spectacled and King Eiders, Long-tailed Duck); and shellfish specialists (Common Eider, Surf Scoter, White-winged Scoter, Black Scoter).

Sea ducks are protected in Canada under the Migratory Birds Convention Act and in the United States under the Migratory Bird Treaty Act. However, these acts do not confer protection to habitat, applying only to birds, their nests, and eggs. Protection of critical habitat is, however, legally required for species if they are listed as threatened or endangered under the Canadian Species at Risk Act or the U.S. Endangered Species Act. Steller's and Spectacled Eiders are listed as threatened in the United States and critical habitat has been designated for both species (U.S. Fish and Wildlife Service 2001a,b). The eastern populations of Harlequin Ducks and Barrow's Goldeneyes are listed as species of special concern in Canada, but critical habitat requirements do not apply to this listing status.

The U.S. National Wildlife Refuge System protects more than 60 million hectares of wildlife habitat in Refuges, Wetland Management Districts, and Waterfowl Production Areas. In Canada, national wildlife areas (>1 million hectares) and migratory bird sanctuaries (>10 million hectares) have been established to protect wildlife of national concern, particularly migratory birds and species at risk. National wildlife areas exist on federal lands and offer the strongest level of habitat protection available to Environment Canada (Canadian Nature Federation 2002). Migratory bird sanctuaries, which may be on federal, provincial, territorial or private lands, do not provide such stringent habitat protection, but do prohibit hunting of migratory birds. Hunting is allowed in certain national wildlife areas (Canada) and National Wildlife Refuges (United States), but harvest is more easily controlled than in non-protected areas.

An important milestone in continental waterfowl conservation was the development of the NAWMP (1986). Since 1986, NAWMP partners have conserved or restored over 6 million hectares of waterfowl habitat on both publicly and privately owned lands. Some non-waterfowl-specific programs have contributed to significant conservation of wetlands and other waterfowl habitat, such as the Clean Water Act and the Farm Bill (Conservation Reserve Program and the Wetlands Reserve Program) in the United States and the Environmental Farm Planning, National Farm Stewardship, and Greencover Canada programs developed by Agriculture Canada (North American Waterfowl Management Plan 2012). However, most of these targeted habitat conservation efforts have focused on the Prairie Pothole region of Canada and the United States, the so-called "duck factory", and not at breeding areas of sea ducks in the boreal forest, taiga, tundra, or coastal areas of Alaska and northern Canada. Many of the northern landscapes important to sea ducks are still relatively undeveloped, but forestry, mining, and oil and gas development are quickly expanding throughout the region. Effective habitat conservation should focus on permanently protecting naturally functioning ecosystems but should also consider the consequences of large-scale changes such as climate change or oceanic regime shifts (North American Waterfowl Management Plan 2012). More than 40% of Alaska is managed for some level of conservation (Smith et al. 2008), whereas in Canada, <6% of the boreal region is permanently protected from industrial development (Canadian Boreal Initiative 2005). However, Canada's Boreal Leadership Council has committed to protecting at least half of the boreal region in a network of large protected areas (Canadian Boreal Initiative 2012).

Overall, the conservation of marine ecosystems has lagged behind terrestrial efforts. With signing of the international Convention on Biological Diversity, the global community recognized the importance of marine conservation efforts (United Nations 1992). In 2010, ratifying countries agreed to protect at least 10% of their marine areas by 2020 (United Nations Environmental Programme 2010). In Canada, the responsibility for protection lies with Fisheries and Oceans Canada, Parks Canada, and Environment Canada with the latter agency being responsible for protecting habitat for wildlife, including migratory birds. In 1997, with the creation of the Oceans Act, Canada committed to establish a national network of marine protected areas (MPAs), but this network has not been fully developed and only about one percent of Canada's marine area is protected (Vaughan 2012). In 1994, Canada's Wildlife Act was amended to establish marine wildlife areas (MWAs). To date, no protected areas have been established, although the Scott Islands MWA in British Columbia is nearing completion (Thompson 2008, Environment Canada 2013). The United States has not ratified the U.N. Convention on Biological Diversity but a national system of MPAs has been created. An impressive 40% of US waters are included in MPAs, but only ~8% of the total area are sites

focused on conservation of natural or cultural resources (Wenzel et al. 2012). In Canada, most marine protection is based on ecological reasons, and over half of the protected area is classified by the International Union for Conservation of Nature (IUCN) as protection of biodiversity (category Ia), nature reserve or ecosystem protection (category Ib), or natural areas that allow recreation (category II, Government of Canada 2010). The selection of MPAs is still in progress in both countries and offers a potential opportunity to conserve important molting, staging, and wintering areas for sea ducks.

DISTURBANCE

Disturbance by commercial or recreational activities can take many forms, from indirect impacts and displacement caused by resource exploitation to direct disturbance of birds in their daily activities. Several species of sea ducks congregate in large flocks throughout the year or are associated with specialized habitats, rendering them susceptible to disturbance. Colonial nesters and flightless molting birds are especially vulnerable. Common Eider breeding colonies are particularly sensitive to disturbance during nest initiation and early incubation (Bolduc and Guillemette 2003, Bourgeon et al. 2006, Wilson et al. 2012), and eider broods are easily disturbed in the first weeks following hatching. Boat traffic can reduce survival of eider ducklings by increasing their vulnerability to gull predation (Mendenhall and Milne 1985). Red-breasted Mergansers nesting on islands may desert nests when disturbed (Craik and Titman 2009). Harlequin Ducks, Common Mergansers, and, occasionally, Red-breasted Mergansers are disturbed by rafting, canoeing, kayaking, and sport fishing, which may force broods into suboptimal habitats, disrupt brood cohesion, and increase duckling mortality (Wallen 1987, Diamond and Finnegan 1993, Mallory and Metz 1999, Titman 1999).

Disturbance to sea ducks, especially molting and wintering birds, may increase in the future as human populations continue to expand. Boat traffic is a concern at staging (Korschgen et al. 1985, Laursen et al. 2005), molting (Thiel et al. 1992), and wintering areas (Kaiser et al. 2006, Laursen and Frikke 2008, Merkel et al. 2009). In the Baltic Sea, molting sea ducks avoided areas with heavy boat traffic (Thiel et al. 1992); along the Labrador coast, molting Surf Scoters stopped feeding and swam offshore when disturbed (O'Connor 2008); and molting King Eiders reacted strongly to approaching aircraft by diving and dispersing (Mosbech and Boertmann 1999). A recent threat is the advent of high-speed ferries in sea duck concentration areas, which can exclude birds from foraging areas (Larsen and Laubek 2005). Kaiser et al. (2006) indicated that Common Scoters (*Melanitta nigra*) avoided areas with high boat traffic even if these areas had high prey biomass. Increased boat traffic in northern coastal waters in response to a warming climate and ice-free shipping routes will likely increase disturbance in important breeding, molting and staging areas. In southern British Columbia, male Harlequin Ducks have recently deserted a few traditional molting sites, possibly because of increasing harassment or predation risk from a growing population of Bald Eagles (*Haliaeetus leucocephalus*), combined with increasing levels of disturbance from boaters, kayakers, and paddle-boarders in local waters (W. S. Boyd, unpubl. data). Measures to minimize disturbance can include creation of boat exclusion areas or the use of buffer zones to protect foraging and resting waterfowl (Madsen 1998, Rodgers and Schwikert 2002). Birds were displaced from preferred foraging areas when motorboats were used to hunt wintering Common Eiders in the Danish Wadden Sea, but a subsequent ban on use of motorboats resulted in increased numbers of eiders (Laursen and Frikke 2008).

Vulnerability to disturbance is often quantified by measuring behavioral changes, such as avoidance of human activity. However, it is unclear whether avoidance or other behavioral changes result in demographic consequences (Gill et al. 2001). It is difficult to determine the relationship between disturbance and mortality or recruitment, but effects can be predicted by using models that include such parameters as rate of disturbance, time to return after disturbance, energy costs of increased flight, and reductions in prey intake (Stillman et al. 2007).

PREDATORS

Thriving gull and eagle populations can reduce the annual production of some sea duck species (Mawhinney et al. 1999, Bolduc and Guillemette 2003) and affect the breeding distribution of others (Heath et al. 2006). Harlequin Ducks are less numerous on streams with cliff-breeding raptors, suggesting that predation danger influences their distribution (Heath and Montevecchi 2008). Eagles may also cause behavioral changes in sea ducks; Common Eiders wintering in Greenland fjords feed nocturnally, apparently in reaction to predation risk from White-tailed Eagles (*Haliaeetus albicilla*, Merkel 2004, Jamieson et al. 2006). Expanding populations of White-tailed Eagles in Europe are affecting the behavior of breeding Common Eiders (Kilpi and Öst 2002, Jaatinen et al. 2011), and increasing populations of sea eagles (*Haliaeetus* spp.) may be causing declines of marine birds at many locations throughout the northern hemisphere (Hipfner et al. 2012).

Gull predation on Common Eider ducklings can be important locally and, in some cases, can greatly reduce reproductive success (Gilliland 1990, Mawhinney 1999). Gull predation on eggs can be significant, especially when combined with human disturbance at colonies and particularly at sites with little vegetative cover (Bourget 1973, Gotmark and Ahlund 1988, Ahlund and Gotmark 1989, Bolduc and Guillemette 2003). Increased nest cover can reduce avian predation and has been associated with increased nest success in areas with few or no mammalian predators (Choate 1967, Milne and Reed 1974). On islands with poor cover, nesting shelters can reduce avian predation (Woolaver 1997, Joint Working Group on the Management of the Common Eider 2004), improve female body condition (Fast et al. 2007, D'Alba et al. 2009), and in some cases increase breeding densities (Clark 1968). On islands where mammalian predators occur, Common Eiders may avoid concealed nest sites (Divoky and Suydam 1995, Noel et al. 2005). It is therefore important to assess the predator community before undertaking nest shelter programs (Fast et al. 2007).

Predator control can enhance productivity (Côté and Sutherland 1997). Control of nest predators can be an efficient way of increasing local populations, as successful females are more likely to breed in the same area in subsequent years (Hanssen and Erikstad 2013). Hanssen et al. (2013) found a three- to fourfold increase in density in Common Eider colonies if predators were controlled. Gull control may be successful in increasing reproductive success of some sea ducks (Mawhinney 1999), but it is often controversial (Blodget 1988, Coulson 1991). The relationship between nesting eiders and gulls is complex and may be positive in some cases. For example, the presence of breeding gulls can indicate islands that are free of terrestrial predators, and in some cases, nesting eiders may benefit from the aggressive nature of gulls, which can deter other avian predators (van Dijk 1986).

Occasionally, foxes, polar bears, or other mammalian predators can reach islands and destroy entire colonies of sea duck nests (Quinlan and Lehnhausen 1982, Smith et al. 2010). On small islands, the presence of foxes is often dictated by movement of ice floes and the formation of ice bridges. On larger islands, predation may last for a few years as foxes may overwinter by subsisting on alternate prey (Joint Working Group on the Management of the Common Eider 2004). The incidence of predators on breeding islands is also increasing in the north, possibly because of warmer conditions in the Arctic (Smith et al. 2010). If mammalian predators only occasionally colonize islands, then control of these invaders can be an efficient management technique (Joint Working Group on the Management of the Common Eider 2004).

Introduced predators can have drastic impacts on populations. Escaped American mink (*Mustela vison*) from fur farms have caused considerable problems in Iceland and Scandinavia (Nordstrom and Korpimaki 2004). Common Eiders were almost extirpated from several Aleutian Islands where introduced arctic foxes (*Vulpes lagopus*) depredated adults and eggs, drastically reducing eider populations (U.S. Fish and Wildlife Service, unpubl. report). Predator control or removal has proven successful in the recovery of some populations. White-winged Scoters and Red-breasted Mergansers responded positively to removal of introduced American mink in Finland but the larger-bodied Common Eider did not (Nordstrom et al. 2002). In Alaska, a few colonies of Common Eiders have been reestablished following arctic fox removal, and sustained fox control has been established in key nesting areas to aid recovery of Steller's Eiders (U.S. Fish and Wildlife Service, unpubl. reports).

INTRODUCED SPECIES

Sea ducks can be affected by nonpredatory, introduced species through competition and changes in prey availability. Nonnative fish and shellfish have been introduced into freshwater and marine environments throughout North America. The food quality of these invasive species may differ from native species and lead to complex trophic interactions within ecosystems (Richman and Lovvorn 2003, 2004).

Common Goldeneyes compete for food with fish (Eriksson 1979, Eadie and Keast 1982, Rask et al. 2001), and numbers of goldeneye broods on small boreal lakes declined after introductions of Eurasian perch (*Perca fluviatilis*, Nummi et al. 2012). Barrow's Goldeneyes in eastern North America generally breed on high-elevation lakes without fish, and stocking lakes with trout can lead to reductions in prey availability (Robert et al. 2008). Logging operations opened road access to remote lakes, and may have led to widespread stocking, but there is now a moratorium on fish stocking throughout much of the goldeneye breeding range in Quebec (Committee on the Status of Endangered Wildlife in Canada 2011). Density of Harlequin Ducks on breeding streams is negatively related to the presence of fish, likely due to indirect competition for invertebrate prey (LeBourdais et al. 2009). Many fish species, including rainbow trout (*Oncorhynchus mykiss*), have been introduced to watersheds throughout the breeding range of Pacific Harlequin Ducks, and it is hypothesized that fish stocking may have contributed to long-term population declines in duck numbers (LeBourdais et al. 2009).

Several species of introduced bivalves have become important food sources for sea ducks. Zebra mussels (*Dreissena polymorpha*), introduced to the Great Lakes basin in the 1980s, became a highly profitable food (Hamilton et al. 1994, Custer and Custer 1996). Diving duck numbers increased dramatically after zebra mussels became established (Petrie and Knapton 1999), suggesting large-scale shifts in distribution and possibly increased survival (Ross et al. 2005). Zebra mussels and their larvae are easily transported inadvertently via boats, trailers, fishing equipment, and ballast water and are now well established throughout the Great Lakes, the St. Lawrence Seaway, the Mississippi River, and other rivers in northeast United States. Similarly, varnish clams (*Nuttallia obscurata*), accidentally introduced near Vancouver, British Columbia, around 1990, have radiated rapidly throughout the region (Dudas et al. 2007). By 2001, this bivalve species had already become a major prey species of scoters in the Strait of Georgia, British Columbia (Lewis et al. 2007), and may be contributing to changes in winter distributions (D. Esler, pers. comm.). In San Francisco Bay, the recently introduced Asian clam (*Potamocorbula amurensis*) has become extremely abundant and is the primary food consumed by Surf Scoters in some areas of the bay (Lovvorn et al. 2013).

If invasive bivalves become superabundant, high densities may cause shifts in community structure, and opportunistic predators such as sea ducks may be affected. Contamination issues, however, may outweigh the benefits of an increased food supply. As filter feeders, bivalves accumulate contaminants that are subsequently passed on to sea ducks, with possible adverse effects on reproductive output or survival (Schummer et al. 2009; Chapter 6, this volume). Moreover, some invasive bivalves may concentrate contaminants at higher rates than other invertebrates (Linville et al. 2002, Petrie et al. 2007). Thus, Bufflehead and Common Goldeneye in the Great Lakes had elevated levels of PCBs and selenium, likely because of foraging on contaminated zebra mussels (Custer and Custer 2000).

RESOURCE EXPLOITATION

Resource development can negatively affect sea ducks through mechanisms such as habitat destruction or degradation, disturbance, and reduced prey availability. Most development activities have short-term, local effects but their cumulative impact can be important. We consider six types of resource development that can potentially affect sea ducks: oil and gas, hydroelectric, offshore wind energy, forestry, fisheries and aquaculture, and harvest of eider down.

Oil and Gas

Oil and gas development may take different forms, from onshore wells to offshore drilling to mining of oil sands and oil shale deposits. Adverse effects can result from activities during exploration and development, but greater risks are posed by oil spills once a site has been developed. The severity of damage to wildlife depends on many factors, including the size and type of oil spill, timing and location of the event, the sensitivity of exposed species to oil, prevailing winds and other climatic conditions, and physical characteristics of the environment. Sea ducks often concentrate in nearshore and offshore marine areas, and oil spills in sensitive habitats pose an especially grave risk.

Oil drilling and oil spills pose potential threats to sea ducks in both direct and indirect ways. Direct effects include collision with exploration and drilling structures or vessels and mortality via toxicity or through disruption of feather waterproofing leading to hypothermia. Indirect effects include disturbance and displacement of birds that lead to habitat loss, contamination at discharge sites or from oil wells, sublethal effects of ingestion or contact with oil, and alteration of predator communities associated with anthropogenic food sources or nesting structures (Minerals Management Service 2007). Ingesting oil may result in debilitating effects that include gastrointestinal irritation, pneumonia, dehydration, red blood cell damage, impaired osmoregulation, suppressed immune system, hormonal imbalance, inhibited reproduction, retarded growth, and abnormal parental behavior (Hartung and Hunt 1966, Szaro et al. 1981, Albers 1983, Jenssen et al. 1985, Leighton 1991, Eppley 1992, Burger and Fry 1993, Fowler et al. 1993, Jenssen 1994, Briggs et al. 1997, Walton et al. 1997). Adverse effects may lead to reduced fitness and survival, particularly in harsh arctic environments. Changes in adult survival have a relatively high impact on population growth rates of sea ducks, and mortality events caused by oil spills could have a major effect on populations of sea ducks.

Major oil spills in marine environments pose an obvious risk to sea ducks because the birds live in marine environments and often forage in intertidal areas. One of the best-studied events was the *Exxon Valdez* oil spill that occurred in Prince William Sound, Alaska, in 1989. Harlequin Ducks were particularly vulnerable to this spill because much of the oil was deposited in shallow, intertidal areas where the birds tend to forage. In the aftermath of the spill, Harlequin Duck numbers decreased by an estimated 25%, and survival remained depressed for 6–9 years after the spill (Iverson and Esler 2010). Evidence of exposure to residual oil from the *Exxon Valdez* was found in Harlequin Ducks and Barrow's Goldeneyes up to 20 years after the spill (Trust et al. 2000; Esler et al. 2002, 2010, 2011). Iverson and Esler (2010) predicted a timeline to recovery for Harlequin Ducks of ~24 years, which is much longer than previously thought.

Oil development in terrestrial areas can be expansive, as evidenced by the massive development of oil sand deposits in western Canada, mostly in Alberta. As conventional oil production becomes increasingly constrained, and technology for alternative sources has improved, more effort has gone into production of fuels from low-quality hydrocarbon stores such as bitumen. The extraction of Alberta bitumen has been described as the biggest oil development scheme in the history of North America (Gawthrop 1999). As of 2012, more than 500 km^2 of boreal forest habitat had been lost as potential breeding habitat for sea ducks, most notably Surf Scoters and White-winged Scoters. In addition, the extraction and processing of these reserves require huge amounts of water, much of which ends up contaminated and stored in tailings ponds.

The impacts of chronic oiling on sea ducks may be even more significant than occasional large oil spills (Weise and Ryan 2003). Chronic oiling may result from illegal discharges from tank washings, and dirty ballast water, and bilge pumping from vessels. Weise and Ryan (2003) reported on chronic marine oiling off the coast of Newfoundland and found that Common Eiders were highly vulnerable. Chronic oiling is likely to become more of a concern in Arctic areas as sea ice melts due to climate change and northern shipping corridors through the Northwest Passage lead to an increase in traffic of marine vessels.

Mitigation measures to reduce the potential mortality of sea ducks include avoidance of drilling in areas of high importance to sea ducks, particularly offshore areas; siting of drilling pads, pipelines, pump stations, roads, and other infrastructure in nonsensitive areas; using bird-friendly lighting on structures to lessen the probability of collision mortalities (Poot et al. 2008); increasing enforcement of environmental laws regarding oil pollution; and ensuring that oil spill response vessels and equipment are on-site with capability for rapid deployment. Response efforts assume that the technology to address a major oil spill exists, which arguably has not been demonstrated for offshore events such as the recent British Petroleum oil spill in Gulf of Mexico. Responses may be particularly challenging in northern areas if a spill occurs under ice pack where oil biodegrades slowly, greatly complicating containment, cleanup, and long-term recovery. The selection of appropriate clean-up techniques and effectiveness of operations can influence the amount of damage and the ability of the ecosystem to recover.

Oil exploration and development in the Arctic will likely increase as the climate warms. To predict and mitigate the negative effects of activities associated with this industry, basic information is needed, such as the location of key sea duck habitats and seasonal use patterns and the indirect and direct effects of oil or associated contaminants. Further, cumulative impacts need to be considered in addition to the effects of individual projects.

Hydroelectric

Large-scale hydroelectric power projects provide over 60% of the total electricity in Canada (Environment Canada 2010) and 20% in Alaska (Fay et al. 2012). Construction of dams and reservoirs for power generation can flood huge areas and dramatically alter downstream characteristics of waterways and fragment associated habitats with roads and transmission lines. Many large hydroelectric projects are located in remote and northern areas, and land flooded to form reservoirs may reduce breeding habitat for some sea duck species. Major hydraulic changes have occurred over the past several decades at rivers along the north shore of Québec and in the watersheds of Hudson Bay and James Bay. Downstream effects from large installations and smaller run-of-the-river projects are of particular concern for Harlequin Ducks and mergansers, which breed and raise young on fast-flowing streams (Robertson and Goudie 1999).

Hydroelectric developments can also affect estuarine and marine environments. In north temperate regions, high water flows during spring are trapped in reservoirs and stored until winter, when energy demands increase and water is released to generate power. An altered hydrological cycle can lower spring primary productivity leading to negative effects on invertebrates and fish and may even alter ocean currents and climate patterns (Rosenberg et al. 1997). In fact, Inuit people of the Belcher Islands have claimed that hydroelectric projects alter sea ice conditions in eastern Hudson Bay, possibly affecting wintering Common Eiders (Heath and the community of Sanikiluaq 2011). Projects near the northeast coast of James Bay have been identified as possible causes of recent changes in eelgrass beds (*Zostera marina*) and other subtidal habitats used by molting scoters, Common Goldeneye, and mergansers (Reed et al. 1996).

Offshore Wind Energy

Offshore wind energy projects are increasing globally and a better understanding of impacts on sea ducks is urgently needed. Preferred sea duck foraging habitats are relatively shallow coastal waters that provide abundant benthic foods (Zipkin et al. 2010), but these same habitats are often optimal for offshore wind energy development. The primary concerns include behavioral avoidance, habitat alteration or loss, mortality losses to collisions with above-water structures, and cumulative, long-term impacts from multiple wind farm projects.

Studies from Europe have demonstrated that some sea ducks avoid wind power facilities (Tulp et al. 1999, Guillemette and Larsen 2002, Desholm and Kahlert 2005, Larsen and Guillemette 2007). In certain instances, avoidance appears to be temporary as birds increased their use of the area in subsequent years, but it is unclear whether behavioral responses were due to habituation or changing food resources (Petersen and Fox 2007). Altered behavior of sea ducks in response to offshore structures has also been documented (Hicklin and Bunker-Popma 2001, Bunker-Popma 2006). Tulp et al. (1999) showed that scoters and eiders sometimes move at night with an increased risk of collision with above water structures, but evidence from Europe suggests that such collisions are not a major concern (Desholm and Kahlert 2005). In North America, no offshore wind farms are currently operational, but several projects have been proposed or are under development along the Atlantic and Pacific coasts and in the Great Lakes. Unfortunately, baseline data for sea duck use of the proposed development sites are limited in most cases. To evaluate the cumulative population-level or landscape-level impacts from multiple offshore wind farms, new information is needed on the location and types of habitats used by sea ducks throughout their winter range; seasonal patterns of use of key habitats; direction, elevation, and other aspects of flight behavior; and whether habitat availability is a limiting factor. Appropriate techniques to address these questions will vary by site, although intensive surveys, digital imagery, satellite telemetry, and radar studies will be useful tools to document spatial and temporal abundance of sea ducks and identify high-priority habitats. Data from these studies can then be incorporated into risk assessments and planning processes to ensure that offshore wind energy development minimizes impacts to sea ducks.

Forestry

Four species of sea ducks (Buffleheads, Common and Barrow's Goldeneyes, and Hooded Mergansers) are obligate cavity nesters, while two other species sometimes use cavities (Common and Red-breasted Mergansers). Species relying on tree cavities for nesting are especially sensitive to forestry operations as logging reduces the availability of suitable cavities and increases intra- and interspecific competition for remaining sites (Vaillancourt et al. 2009). Logging can also result in increased predation rates as cavities become scarce and predators use remnant habitats as corridors (Gilliam and Fraser 2001). Timber harvesting can alter riparian habitat, disrupt stream flow, and increase siltation rates, thereby affecting breeding habitat conditions for Harlequin Ducks and mergansers (Dugger et al. 1994, Mallory and Metz 1999, Robertson and Goudie 1999). Forestry may also reduce habitat availability for species such as White-winged Scoters, which nest at low densities in the boreal forest (Brown and Fredrickson 1997).

Provision of artificial nesting structures may mitigate some of the impacts of timber harvesting, as cavity-nesting sea ducks will use artificial nest boxes for successful breeding (Lumsden et al. 1980, 1986; Gauthier 1988; Savard 1988; Savard and Robert 2007). Nest boxes have been used for centuries to manage goldeneye populations in Europe and Iceland (Siren 1951, Dennis and Dow 1984). However, Pöysä and Pöysä (2002) showed that nest boxes led to an increase in the number of Common Goldeneye pairs but not in brood numbers or reproductive success. For Barrow's Goldeneye, reproductive success was slightly lower in nest boxes than in natural cavities because of greater desertion and predation rates in the more conspicuous nest boxes (Evans et al. 2002). Furthermore, adding nest boxes in British Columbia increased numbers of Barrow's Goldeneye, which had a negative impact on Bufflehead broods (Savard 1987). Trophic interactions among sea ducks indicate a need to consider the potential impact of specific management prescriptions on the entire waterfowl community.

Despite the potential problems noted earlier, nest boxes can be used to establish new populations of cavity-nesting waterfowl (Doty and Kruse 1972, Dennis and Dow 1984, Corrigan et al. 2011). In addition, the use of artificial nesting structures may be warranted in intensively exploited areas to maintain breeding populations of cavity-nesting sea ducks if the retention and recruitment of trees with natural cavities are inadequate (Vaillancourt et al. 2009).

Commercial Fisheries and Aquaculture

Commercial fisheries can cause direct mortality to sea ducks and can have indirect effects such as disturbance and competition for food resources. Sea ducks are susceptible to accidental entanglement in gillnets but quantitative data on bycatch are limited (Žydelis et al. 2009a, Bellebaum et al. 2013). In Greenland, gillnet fisheries for lumpfish (*Cyclopterus lumpus*) have drowned thousands of Common and King Eiders per year (Merkel 2004). Hundreds of Red-breasted Mergansers and Barrow's Goldeneyes have drowned in fishing nets on Lake Myvatn, Iceland (Bengtson 1971), and thousands of Long-tailed Ducks have drowned in gillnet fisheries on the Great Lakes (Robertson and Savard 2002). In the Baltic and North Seas, tens of thousands of Long-tailed Ducks and thousands of eiders, scoters, and goldeneyes drown each year in gillnets (Žydelis et al. 2009a, Bellebaum et al. 2013). Molting sea ducks may be particularly vulnerable as they form large, concentrated flocks and use relatively small areas while flightless. Bycatch rates are affected by factors such as the spatial and temporal overlap between fishing activity and foraging sea ducks, fishing gear characteristics, and water clarity (Tasker et al. 2000, Žydelis et al. 2009a). Other relationships between sea ducks and fisheries can be complex and need to be closely monitored (Nehls and Ruth 1994). For example, some fisheries may cause changes in the availability of key prey species used by sea ducks. In the Netherlands, increased harvesting of shellfish may have led to high rates of starvation for wintering eiders (Camphuysen et al. 2002, Swart and Andel 2007).

Aquaculture or the controlled production of finfish and shellfish is one of the fastest expanding sectors of food production in the world (Food and Agriculture Organization 2010). From 1980 to 2010, world fish aquaculture increased almost 12 times, and by 2010, aquaculture provided 47% of the global food fish supply (Food and Agriculture Organization 2012). At a global scale, aquaculture is dominated by freshwater finfish production, while marine aquaculture accounts for 32% of total production (Food and Agriculture Organization 2010). The potential for interactions between sea ducks and aquaculture operations is high. Most species of sea ducks feed to some extent on shellfish or finfish, and large numbers of sea ducks use productive, sheltered coastal habitats that are also preferred for aquaculture production (Žydelis et al. 2009b).

Several studies have documented neutral or beneficial effects of shellfish aquaculture (Caldow et al. 2003; Connolly and Colwell 2005; Žydelis et al. 2006, 2009b; Kirk et al. 2007a). Aquaculture facilities can provide profitable foraging sites for sea ducks, as prey tend to be concentrated and contained, easy to locate, readily available in shallow water, and energetically or nutritionally rich (e.g., cultivated mussels have higher tissue/shell ratio, Kirk et al. 2007b, Žydelis et al. 2009a). In many cases, sea ducks learn to exploit food sources provided directly by aquaculture operations (Ross and Furness 2000, Thompson 2003, Žydelis et al. 2009b). Some operations may indirectly benefit sea ducks by contributing to population growth of farmed species in surrounding areas as seeded clams or their larvae disperse away from shellfish lease areas (Žydelis et al. 2006). Increasing aquaculture production could reduce commercial exploitation of wild fish and shellfish and lead to higher prey availability for sea ducks. At moderate levels, some aquaculture operations may be compatible with sea duck conservation (Žydelis et al. 2006). Sea ducks can benefit without causing conflict if they consume insignificant amounts of product, forage on noncultured prey, or forage on prey populations that have been enhanced by aquaculture operations (Žydelis et al. 2009b). At oyster farms in British Columbia, Surf Scoters and Barrow's Goldeneyes were attracted to wild mussels growing on rafts and lines (Kirk et al. 2008, Žydelis et al. 2009b). The sea ducks did not feed on the cultured oysters, and shellfish farmers benefited from their gleaning of unwanted mussels off structures (Žydelis et al. 2009b).

Negative interactions with aquaculture operations are also possible (Dankers and Zuidema 1995, Bendell 2011), but depend on bird species, type and intensity of aquaculture, and habitat type (Žydelis et al. 2009b). Conflicts between sea ducks and aquaculture have increased as aquaculture production has expanded. Eiders, scoters, goldeneyes, and Long-tailed Ducks commonly feed on mussel farms (Ross and Furness 2000, Dionne et al. 2006, Kirk et al. 2007a), and Common Mergansers have been reported as predators at fish stocking operations (Wood 1985a,b; Wood and Hand 1985). In Prince Edward Island, Canada, predation by sea ducks reduced harvestable yields of mussels by 10–20% (Thompson and Gillis 2001, Thompson 2003). Feltham (1995) reported that mergansers consumed ~38% of their body mass per day (~400–522 g), suggesting a potential conflict with fisheries. In southeastern New York, brown trout (*Salmo trutta*) were released as yearlings, and Common Mergansers consumed 51% of 14,000 fish during a 3-month period (Stiller 2011).

Aquaculture operations can exclude sea ducks from important foraging habitat and disturbance may disrupt activity patterns (Žydelis et al. 2009b). Deterrence efforts to reduce damage by sea ducks can lead to further displacement, disturbance, and direct mortality (Rueggeberg and Booth 1989, Burnett et al. 1994, Ross and Furness 2000, Dionne 2004, Dionne et al. 2006, Žydelis et al. 2009b). Deterrence efforts have had limited success as sea ducks can quickly become habituated to methods such as noise cannons and scarecrows (Ross et al. 2001; S. E. Richman, pers. comm.). Persecution and other interactions can also lead to negative public perception of sea ducks, which may impair conservation efforts.

Harvest of Eiderdown

Eiderdown is used locally for clothing by the Inuit peoples of Hudson Bay, Nunavut, Nunavik, and Greenland (Nakashima 1986, Reed 1986, Oakes 1999, Bédard et al. 2008). In North America, commercial harvest of eiderdown currently occurs only in the St. Lawrence estuary and in northern Quebec and represents a symbiosis between industry and conservation (Reed 1986, Bédard et al. 2008). Commercial harvest of eiderdown is particularly well developed in Iceland, where Common Eiders are closely managed by farmers

who provide nesting shelters and exclude or control predators.

Impacts of eiderdown harvest on nest survival are difficult to assess but appear to be relatively small (Reed 1986, Melhum et al. 1991, Kristjansson and Jonsson 2011). However, the practice may have contributed to the historic decline of the breeding population of Common Eiders in Greenland when both adults and down were collected and disturbance at breeding colonies was high (Boertmann et al. 1996). Techniques for harvest of eiderdown and impacts need to be better quantified in northern Quebec, but harvest in the St. Lawrence is highly regulated (Bédard et al. 2008). Down is harvested each year from colonies and predators and disturbance are intensively managed. To minimize the impact on nest success, down is collected late in the incubation period when nest desertion is less likely (Bolduc and Guillemette 2003), and nests are counted during harvest of down so that the population is closely monitored. The industry is dependent on economic conditions and the price of eiderdown fluctuates greatly from year to year (Bédard et al. 2008). However, because eiderdown is still used by the Inuit, the commercial use provides a good opportunity to enhance conservation of northern Common Eiders. When properly managed, down can be collected with minimal impact.

CONCLUSIONS

Sea ducks are facing increasing threats at every stage of the annual cycle. Addressing these challenges will require an international effort and far more resources than are currently available through existing conservation programs. With increasing pressures on sea duck habitats, quantitative assessments are needed to prioritize and direct limited resources toward protecting the most important and most vulnerable habitats. A better understanding of the key factors that limit sea duck populations is urgently needed but requires completion of relevant demographic studies and population delineations. Last, the effects of environmental and human perturbations, as well as the effectiveness of management actions, cannot be evaluated without robust information on trends and population numbers of sea ducks.

As we learn to assess the multitude of potential threats facing sea ducks, the next step is to consider their relative importance in the holistic context of annual cycles and life spans. More studies have identified cross-seasonal effects as being important (Gardarsson 2006; DeVink et al. 2008; Chapter 5, this volume). Conditions in a particular season and location can have lag effects that impact reproductive success, survival, and other demographic parameters at future times and places. We must also evaluate risks created by cumulative and combined impacts of stressors. Synergistic interactions among such varied factors as contaminant exposure, disturbance levels, changing climatic conditions, and food availability could lead to larger negative impacts on sea ducks that would not be predicted if each factor were viewed independently.

ACKNOWLEDGMENTS

We thank the editors for giving us the opportunity to contribute this chapter on the Conservation of Sea Ducks. We also thank D. Derksen, A. Reed, V. Skean, and H. Wilson for providing helpful comments on a previous version of the manuscript. Preparation of this work was supported by the Environment Canada, Science and Technology Branch, U.S. Fish and Wildlife Service, and U.S. Geological Survey.

LITERATURE CITED

Agler, B. A., S. J. Kendall, D. B. Irons, and S. P. Klosiewski. 1999. Declines in marine bird populations in Prince William Sound, Alaska coincident with a climatic regime shift. Waterbirds 22:98–103.

Ahlund, M., and F. Gotmark. 1989. Gull predation on Eider ducklings Somateria mollissima: effects of human disturbance. Biological Conservation 48:115–127.

Albers, P. H. 1983. Effects of oil on avian reproduction: a review and discussion. Pp. 78–96 in D. Rosie and S. N. Barnes (editors), The effects of oil on birds: a multi-discipline symposium. Tri-State Bird Rescue and Research, Newark, DE.

Anderson, M. G., and L. G. Sorenson. 2001. Global climate change and waterfowl: adaptation in the face of uncertainty. Transactions of the North American Wildlife and Natural Resources Conference 66:300–319.

Barry, T. W. 1968. Observations on natural mortality and native use of eider ducks along the Beaufort Sea coast. Canadian Field-Naturalist 82:140–144.

Bartonek, J. C. 1986. Waterfowl management and subsistence harvests in Alaska and Canada: an overview. Transactions of the North American Wildlife and Natural Resource Conference 51:459–463.

Bédard, J., A. Nadeau, J.-F. Giroux, and J.-P. L. Savard. 2008. Eiderdown: characteristics and harvesting procedures. Société Duvetnor Ltée and Canadian Wildlife Service, Environment Canada, Québec Region, QC.

Bellebaum, J., B. N. Schirmeister, N. Sonntag, and S. Garthe. 2013. Decreasing but still high: bycatch of seabirds in gillnet fisheries along the German Baltic coast. Aquatic Conservation: Marine and Freshwater Ecosystems 23:210–221.

Bendell, L. I. 2011. Sea ducks and aquaculture: the cadmium connection. Ecotoxicology 20:474–478.

Bengtson, S.-A. 1971. Food and feeding of diving ducks breeding at Lake Myvatn, Iceland. Ornis Fennica 48:77–92.

Bentzen, R. L., and A. N. Powell. 2012. Population dynamics of King Eiders breeding in northern Alaska. Journal of Wildlife Management 76:1011–1020.

Berg, T. B., O. Gilg, and B. Sittler. 2000. Low abundance of King Eider nests during low lemming years in northeast Greenland. Arctic 53:53–60.

Blodget, B. G. 1988. The half century battle for gull control. Massachusetts Wildlife 38:4–10.

Boertmann, D., A. Mosbech, K. Falk, and K. Kampp. 1996. Seabird colonies in western Greenland. Technical Report, No. 170. National Environmental Research Institute, Roskilde, Denmark.

Bolduc, F., and M. Guillemette. 2003. Human disturbance and nesting success of Common Eiders: interaction between visitors and gulls. Biological Conservation 110:77–83.

Bourgeon, S., F. Griscuolo, F. Bertile, T. Raclot, G. W. Gabrielsen, and S. Massemin. 2006. Effects of clutch sizes and incubation stage on nest desertion in the female Common Eider Somateria mollissima nesting in the high Arctic. Polar Biology 29:358–363.

Bourget, A. A. 1973. Relation of eiders and gulls nesting in mixed colonies in Penobscot Bay, Maine. Auk 90:809–820.

Bourget, D., J.-P. L. Savard, and M. Guillemette. 2007. Distribution, diet, and dive behaviour of Barrow's and Common Goldeneyes during spring and autumn in the St. Lawrence estuary. Waterbirds 30:230–240.

Briers, R. A., J. H. R. Gee, and R. Geoghegan. 2004. Effects of the North Atlantic Oscillation on growth and phenology of stream insects. Ecography 27:811–817.

Briggs, K. T., M. E. Gershwin, and D. W. Anderson. 1997. Consequences of petrochemical ingestion and stress on the immune system of seabirds. ICES Journal of Marine Science 54:718–725.

Brodeur, S., J.-P. L. Savard, M. Robert, P. Laporte, P. Lamothe, R. D. Titman, S. Marchand, S. Gilliland, and G. Fitzgerald. 2002. Harlequin Duck *Histrionicus histrionicus* population structure in eastern Nearctic. Journal of Avian Biology 33:127–137.

Brown, P. W., and L. H. Fredrickson. 1997. White-winged Scoter (*Melanitta fusca*). in A. Poole and F. Gill (editors), The Birds of North America, No. 274. The Academy of Natural Sciences, Philadelphia, PA.

Bunker-Popma, K. 2006. Scoter, *Melanitta* spp., migrations interrupted by Confederation Bridge: an update. Canadian Field-Naturalist 120:232–233.

Burger, A. E., and D. M. Fry. 1993. Effects of oil pollution on seabirds in the northeast Pacific. Pp. 254–263 in K. Vermeer, K. T. Briggs, K. H. Morgan, and D. Siegel-Causey (editors), The status, ecology and conservation of marine birds in the North Pacific. Canadian Wildlife Service Special Publication, Ottawa, ON.

Burnett, J. A., W. Lidster, P. Ryan, and C. Baldwin. 1994. Saving cultured mussels and waterfowl in Newfoundland. Canadian Wildlife Service, Environmental Conservation Branch, Environment Canada, Atlantic Region, Sackville, NB.

Burnham, K. K., J. A. Johnson, B. Konkel, and J. L. Burnham. 2012. Nesting Common Eider (*Somateria mollissima*) population quintuples in Northwest Greenland. Arctic 65:456–464.

Buttler, E. I., H. G. Gilchrist, S. Descamps, M. R. Forbes, and C. Soos. 2011. Handling stress of female Common Eiders during avian cholera outbreaks. Journal of Wildlife Management 75:283–288.

Caithamer, D. F., M. Otto, P. I. Padding, J. R. Sauer, and G. H. Haas. 2000. Sea ducks in the Atlantic flyway: population status and a review of special hunting seasons. Office of Migratory Bird Management, U.S. Fish and Wildlife Service, Laurel, MD.

Caldow, R. W. G., H. A. Beadman, S. McGrorty, M. J. Kaiser, J. D. Goss-Custard, K. Mould, and A. Wilson. 2003. Effects of intertidal mussel cultivation on bird assemblages. Marine Ecology Progress Series 259:173–183.

Camphuysen, C. J., C. M. Berrevoets, H. J. W. M. Cremers, A. Dekinga, R. Dekker, B. J. Ens, T. M. van der Have et al. 2002. Mass mortality of Common Eiders (*Somateria mollissima*) in the Dutch Wadden Sea, winter 1999/2000: starvation in a commercially exploited wetland of international importance. Biological Conservation 106:303–317.

Canadian Boreal Initiative. 2005. The boreal in the balance: securing the future of Canada's boreal region. Canadian Boreal Initiative, Ottawa, ON.

Canadian Boreal Initiative. 2012. Canadian boreal forest conservation framework. Canadian Boreal Initiative, Ottawa, ON.

Canadian Nature Federation. 2002. Conserving wildlife on a shoestring budget—opportunities and challenges for Canada's National Wildlife Areas, Migratory Bird Sanctuaries and Marine Wildlife Areas. Canadian Nature Federation, Ottawa, ON.

Charmantier, A., R. H. McCleery, L. R. Cole, C. Perrins, L. E. B. Kruuk, and B. C. Sheldon. 2008. Adaptive phenotypic plasticity in response to climate change in a wild bird population. Science 320:800–803.

Chaulk, K. G. 2006. Spatial and temporal ecology of a colonial waterbird: the distribution and abundance of nesting Common Eiders (*Somateria mollissima*) in Labrador. Dissertation, Memorial University of Newfoundland, St. John's, NL.

Chaulk, K. G., and M. L. Mahoney. 2012. Does spring ice cover influence nest initiation date and clutch size in Common Eiders? Polar Biology 35:645–653.

Chaulk, K. G., G. J. Robertson, and W. A. Montevecchi. 2006. Extinction, colonization, and distribution patterns of Common Eider populations nesting in a naturally fragmented landscape. Canadian Journal of Zoology 84:1402–1408.

Chaulk, K. G., G. J. Robertson, and W. A. Montevecchi. 2007. Landscape features and sea ice influence nesting Common Eider abundance and dispersion. Canadian Journal of Zoology 85:301–309.

Choate, J. S. 1967. Factors influencing nesting success of eiders in Penobscot Bay, Maine. Journal of Wildlife Management 31:769–777.

Clark, G. M., D. O'Meara, and J. W. Van Weelden. 1958. An epizootic among eider ducks involving an acanthocephalid worm. Journal of Wildlife Management 22:204–205.

Clark, S. H. 1968. The breeding ecology and experimental management of the American Eider in Penobscot Bay, Maine. Thesis, University of Maine, Orono, ME.

Committee on the Status of Endangered Wildlife in Canada. 2011. Committee on the Status of Endangered Wildlife in Canada status appraisal summary on the Barrow's Goldeneye *Bucephala islandica* eastern population in Canada. Committee on the Status of Endangered Wildlife in Canada. Ottawa, ON.

Connolly, L. M., and M. A. Colwell. 2005. Comparative use of longline oysterbeds and adjacent tidal flats by waterbirds. Bird Conservation International 15:237–255.

Cooch, F. G. 1965. The breeding biology and management of the Northern Eider (Somateria mollissima borealis) in the Cape Dorset Area, Northwest Territories. Canadian Wildlife Service, Wildlife Management Bulletin, Series 2, Number 10, Ottawa, ON.

Corrigan, R. M., G. J. Scrimgeour, and C. Paszkowski. 2011. Nest boxes facilitate local-scale conservation of Common Goldeneye (Bucephala clangula) and Bufflehead (Bucephala albeola) in Alberta, Canada. Avian Conservation and Ecology 6:art1.

Côté, I. M., and W. J. Sutherland. 1997. The effectiveness of removing predators to protect bird populations. Conservation Biology 11:395–405.

Coulson, J. C. 1991. The population dynamics of culling Herring Gulls and Lesser Black-backed Gulls. Pp. 479–497 in C. M. Perrins, J.-D. Lebreton, and G. J. M. Hirons (editors), Bird population studies: relevance to conservation and management. Oxford University Press, New York, NY.

Coulson, J. C. 2010. A long term study of the population dynamics of Common Eiders Somateria mollissima: why do several parameters fluctuate markedly? Bird Study 57:1–18.

Craik, S. R., and R. D. Titman. 2009. Nesting ecology of Red-breasted Mergansers in a common tern colony in Eastern New Brunswick. Waterbirds 32:282–292.

Cresswell, W. 2008. Non-lethal effects in birds. Ibis 150:3–17.

Custer, C. M., and T. W. Custer. 1996. Food habits of diving ducks in the Great Lakes after the zebra mussel invasion. Journal of Field Ornithology 67:86–99.

Custer, C. M., and T. W. Custer. 2000. Organochlorine and trace element contamination in wintering and migrating diving ducks in the southern Great Lakes, USA, since the zebra mussel invasion. Environmental Toxicology and Chemistry 19:2821–2829.

D'Alba, L., P. Monaghan, and R. G. Nager. 2009. Thermal benefits of nest shelter for incubating female eiders. Journal of Thermal Biology 34:93–99.

D'Alba, L., P. Monaghan, and R. G. Nager. 2010. Advances in laying date and increasing population size suggest positive response to climate change in Common Eiders Somateria mollissima in Iceland. Ibis 152:19–28.

Dankers, N., and D. R. Zuidema. 1995. The role of the mussel (Mytilus edulis L.) and mussel culture in the Dutch Wadden Sea. Estuaries 18:71–80.

De La Cruz, S. E. W., J. Y. Takekawa, M. T. Wilson, D. R. Nysewander, J. R. Evenson, D. Esler, W. S. Boyd, and D. H. Ward. 2009. Spring migration routes and chronology of Surf Scoters (Melanitta perspicillata): a synthesis of Pacific Coast studies. Canadian Journal of Zoology 87:1069–1086.

Dennis, R. H., and H. Dow. 1984. The establishment of a population of Goldeneyes Bucephala clangula breeding in Scotland. Bird Study 31:217–222.

Descamps, S., M. R. Forbes, H. G. Gilchrist, H. G. Love, and J. Bêty. 2011. Avian cholera, post-hatching survival and selection on hatch characteristics in a long-lived bird. Journal of Avian Biology 42:39–48.

Descamps, S., H. G. Gilchrist, J. Bêty, E. I. Buttler, and M. R. Forbes. 2009. Cost of reproduction in a long-lived bird: large clutch size reduces survival in the presence of a highly virulent disease. Biology Letters 5:278–281.

Descamps, S., S. Jenouvrier, H. G. Gilchrist, and M. R. Forbes. 2012. Avian cholera, a threat to the viability of an Arctic seabird colony? PLoS One 7:e29659.

Descamps, S., N. G. Yoccoz, J.-M. Gaillard, H. G. Gilchrist, K. E. Erikstad, S. A. Hanssen, B. Cazelles, M. R. Forbes, and J. Bêty. 2010. Detecting population heterogeneity in effects of North Atlantic Oscillations on seabird body condition: get into the rhythm. Oikos 119:1526–1536.

Desholm, M., and J. Kahlert. 2005. Avian collision risk at an offshore wind farm. Biology Letters 1:296–298.

DeVink, J.-M., R. G. Clark, S. M. Slattery, and A. M. Scheuhammer. 2008. Cross-seasonal association between winter trophic status and breeding ground selenium levels in boreal White-winged Scoters. Avian Conservation and Ecology 3:art3.

Diamond, S., and P. Finnegan. 1993. Harlequin Duck ecology on Montana's Rocky Mountain Front. U.S. Forest Service, Rocky Mountain District, Choteau, MT.

Dickson, D. L. 2012. Seasonal movement of King Eiders breeding in Western Arctic Canada and Northern Alaska. Canadian Wildlife Service Technical Report Series, No. 520. Canadian Wildlife Service, Edmonton, AB.

Dionne, M. 2004. Relationships between diving ducks and mussel aquaculture in Prince Edward Island, Canada. Thesis, University of New Brunswick, Fredericton, NB.

Dionne, M., J.-S. Lauzon-Guay, D. J. Hamilton, and M. A. Barbeau. 2006. Protective socking material for cultivated mussels: a potential non-disruptive deterrent to reduce losses to diving ducks. Aquaculture International 14:595–615.

Divoky, G. J., and R. Suydam. 1995. An artificial nest site for arctic nesting Common Eiders. Journal of Field Ornithology 66:270–276.

Doty, H. A., and A. D. Kruse. 1972. Techniques for establishing local breeding populations of Wood Ducks. Journal of Wildlife Management 36:428–435.

Drever, M. C., R. G. Clark, C. Derksen, S. M. Slattery, P. Toose, and T. D. Nudds. 2012. Population vulnerability to climate change linked to timing of breeding in boreal ducks. Global Change Biology 18:480–492.

Dudas, S. E., J. F. Dower, and B. R. Anholt. 2007. Invasion dynamics of the varnish clam (*Nuttallia obscurata*): a matrix demographic modeling approach. Ecology 88:2084–2093.

Dugger, B. D., K. M. Dugger, and L. H. Fredrickson. 1994. Hooded Merganser (*Lophodytes cucullatus*). in A. Poole and F. Gill (editors), The Birds of North America, No. 98. The Academy of Natural Sciences, Philadelphia, PA.

Dugger, K. M., B. D. Dugger, and L. H. Fredrikson. 1999. Annual survival rates of female Hooded Mergansers and wood ducks in southeast Missouri. Wilson Bulletin 111:1–6.

Durant, J. M., D. O. Hjermann, G. Ottersen, and N. C. Stenseth. 2007. Climate and the match or mismatch between predator requirements and resource availability. Climate Research 33:271–283.

Eadie, J. M., and A. Keast. 1982. Do goldeneye and perch compete for food? Oecologia 55:225–230.

Eberhardt, L. L. 2002. A paradigm for population analysis of long-lived vertebrates. Ecology 83:2841–2854.

Environment Canada. [online]. 2010. Hydro power. <http://www.ec.gc.ca/energie-energy/default.asp?lang = En&n = A0D4263A-1> (24 September 2013).

Environment Canada. [online]. 2013. The Scott Islands: a proposed Marine National Wildlife Area. <http://www.ec.gc.ca/ap-pa/default.asp?lang=En&n=90605DDB-1> (24 September 2013).

Eppley, Z. A. 1992. Assessing indirect effects of oil in the presence of natural variation: the problem of reproductive failure in South Polar Skuas during the *Bahia Paraiso* oil spill. Marine Pollution Bulletin 25:9–12.

Eriksson, M. O. G. 1979. Competition between freshwater fish and Goldeneyes *Bucephala clangula* (L.) for common prey. Oecologia 41:99–107.

Esler, D., B. E. Ballachey, K. A. Trust, S. A. Iverson, J. A. Reed, A. K. Miles, J. D. Henderson et al. 2011. Cytochrome P4501A biomarker indication of the timeline of chronic exposure of Barrow's Goldeneyes to residual *Exxon Valdez* oil. Marine Pollution Bulletin 62:609–614.

Esler, D., T. D. Bowman, K. A. Trust, B. E. Ballachey, T. A. Dean, S. C. Jewett, and C. E. O'Clair. 2002. Harlequin Duck population recovery following the *Exxon Valdez* oil spill: progress, process and constraints. Marine Ecology Progress Series 241:271–286.

Esler, D., K. A. Trust, B. E. Ballachey, S. A. Iverson, T. L. Lewis, D. J. Rizzolo, D. M. Mulcahy et al. 2010. Cytochrome P4501A biomarker indication of oil exposure in Harlequin Ducks up to 20 years after the *Exxon Valdez* oil spill. Environmental Toxicology and Chemistry 29:1138–1145.

Evans, M. R., D. B. Lank, W. S. Boyd, and F. Cooke. 2002. A comparison of the characteristics and fate of Barrow's Goldeneye and Bufflehead nests in nest boxes and natural cavities. Condor 104:610–619.

Fast, L. F., H. G. Gilchrist, and R. G. Clark. 2007. Experimental evaluation of nest shelter effects on weight loss in incubating Common Eiders *Somateria mollissima*. Journal of Avian Biology 38:205–213.

Fast, P. L., M. Fast, A. Mosbech, C. Sonne, H. G. Gilchrist, and S. Deschamps. 2011. Effects of implanted satellite transmitters on behavior and survival of female Common Eiders. Journal of Wildlife Management 75:1553–1557.

Fay, G., A. V. Meléndez, and C. West. 2012. Alaska energy statistics: 1960–2011 preliminary report. Institute of Social and Economic Research, University of Alaska Anchorage. Anchorage, AK.

Feltham, M. J. 1995. Consumption of Atlantic salmon smolts and parr by Goosanders: estimates from doubly-labelled water measurements of captive birds released on two Scottish rivers. Journal of Fish Biology 46:273–281.

Flint, P. L. 2013. Changes in size and trends of North American sea duck populations associated with North Pacific oceanic regime shifts. Marine Biology 160:59–65.

Flint, P. L., J. B. Grand, J. A. Morse, and T. F. Fondell. 2000. Late summer survival of adult female and juvenile Spectacled Eiders on the Yukon–Kuskokwim Delta, Alaska. Waterbirds 23:292–297.

Flint, P. L., M. R. Petersen, and J. B. Grand. 1997. Exposure of Spectacled Eiders and other diving ducks to lead in western Alaska. Canadian Journal of Zoology 75:439–443.

Food and Agriculture Organization of the United Nations Fisheries and Aquaculture Department. 2010. The state of the world fisheries and aquaculture 2010. Food and Agriculture Organization of the United Nations, Rome, Italy.

Food and Agriculture Organization of the United Nations Fisheries and Aquaculture Department. 2012. The state of the world fisheries and aquaculture 2012. Food and Agriculture Organization of the United Nations, Rome, Italy.

Forchhammer, M. C., and E. Post. 2004. Using large-scale climate indices in climate change ecology studies. Population Ecology 46:1–12.

Fournier, M. A., and J. E. Hines. 1994. Effects of starvation on muscle and organ mass of King Eiders *Somateria spectabilis* and the ecological and management implications. Wildfowl 45:188–197.

Fowler, A. C., and P. L. Flint. 1997. Persistence rates and detection probabilities of oiled King Eider carcasses on St. Paul Island, Alaska. Marine Pollution Bulletin 34:522–526.

Fowler, S. W., J. W. Readman, B. Oregioni, J.-P. Villeneuve, and K. McKay. 1993. Petroleum hydrocarbons and trace metals in nearshore Gulf sediments and biota before and after the 1991 war. Marine Pollution Bulletin 27:171–182.

Friend, M., and J. C. Franson. 1999. Field manual of wildlife diseases. Field procedures and disease of birds. Biological Resources Division, Information and Technology Report 1999–001. U.S. Department of the Interior and U.S. Geological Survey, Washington, DC.

Gardarsson, A. 2006. Temporal processes and duck populations: examples from Myvatn. Hydrobiologia 567:89–100.

Gauthier, G. 1988. Factors affecting nest-box use by Buffleheads and other cavity-nesting birds. Wildlife Society Bulletin 16:160–169.

Gawthrop, D. 1999. Vanishing halo—saving the boreal forest. Greystone Books, Vancouver, BC.

Gershman, M., J. F. Witter, H. E. Spencer Jr., and A. Kalvaitis. 1964. Case report: epizootic of fowl cholera in the Common Eider duck. Journal of Wildlife Management 28:587–589.

Gienapp, P., R. Leimu, and J. Merila. 2007. Responses to climate change in avian migration time—microevolution versus phenotypic plasticity. Climate Research 35:25–35.

Gienapp, P., C. Teplitsky, J. S. Alho, J. A. Mills, and J. Merila. 2008. Climate change and evolution: disentangling environmental and genetic responses. Molecular Ecology 17:167–178.

Gilchrist, G., M. Mallory, and F. Merkel. 2005. Can local ecological knowledge contribute to wildlife management? Case studies of migratory birds. Ecology and Society 10:art20.

Gilg, O., B. Sittler, and I. Hanski. 2009. Climate change and cyclic predator–prey population dynamics in the high Arctic. Global Change Biology 15:2634–2652.

Gill, J. A., K. Norris, and W. J. Sutherland. 2001. Why behavioural responses may not reflect the population consequences of human disturbance. Biological Conservation 97:265–268.

Gilliam, J. F., and D. F. Fraser. 2001. Movement in corridors: enhancement by predation threat, disturbance and habitat structure. Ecology 82:258–273.

Gilliland, S. G. 1990. Predator prey relationships between Great Black-backed Gull and Common Eider populations on the Wolves Archipelago, New Brunswick: a study of foraging ecology. Thesis, University of Western Ontario, London, ON.

Gilliland, S. G., H. G. Gilchrist, R. F. Rockwell, G. J. Robertson, J.-P. L. Savard, F. Merkel, and A. Mosbech. 2009. Evaluating the sustainability of fixed-number harvest of northern Common Eiders in Greenland and Canada. Wildlife Biology 15:24–36.

Glibert, P., D. Anderson, P. Gentien, E. Granéli, and K. Sellner. 2005. The global, complex phenomena of harmful algal blooms. Oceanography 18:136–147.

Gotmark, F., and M. Ahlund. 1988. Nest predation and nest site selection among Eiders *Somateria mollissima*: the influence of gulls. Ibis 130:111–123.

Goudie, R. I., S. Brault, B. Conant, A. V. Kondratyev, M. R. Petersen, and K. Vermeer. 1994. The status of sea ducks in the north Pacific Rim: toward their conservation and management. Transactions of the North American Wildlife and Natural Resources Conference 59:27–49.

Government of Canada. 2010. Spotlight on marine protected areas in Canada. Fisheries and Oceans Canada, Ottawa, ON.

Grand, J. B., P. L. Flint, M. R. Petersen, and C. L. Moran. 1998. Effect of lead poisoning on Spectacled Eider survival rates. Journal of Wildlife Management 62:1103–1109.

Grebmeier, J. M., J. E. Overland, S. E. Moore, E. V. Farley, E. C. Carmarck, L. W. Cooper, K. E. Frey, J. H. Helle, F. A. McLaughlin, and S. L. McNutt. 2006. A major shift in the northern Bering Sea. Science 311:1461–1464.

Guillemette, M., and J. K. Larsen. 2002. Post development experiments to detect anthropogenic disturbances: the case of sea ducks and wind parks. Ecological Applications 12:868–877.

Hamilton, D. J., C. D. Ankney, and R. C. Bailey. 1994. Predation of zebra mussels by diving ducks: an exclosure study. Ecology 75:521–531.

Hansen, K. 2002. A farewell to Greenland's wildlife. Gads Forlag et Narayana Press, Gylling, Copenhagen, Denmark.

Hanssen, S. A., and K. E. Erikstad. 2013. The long term consequences of egg predation. Behavioral Ecology 24:564–569.

Hanssen, S. W., B. Moe, B.-J. Bardsen, F. Hanssen, and G. W. Gabrielsen. 2013. A natural antipredation experiment: predator control and reduced sea ice increases colony size in a long-lived duck. Ecology and Evolution 3:3554–3564.

Harms, N. J. 2011. Avian cholera in the eastern Canadian Arctic: investigating disease origins and reservoirs. Arctic 64:501–505.

Hartman, G., A. Kolzch, K. Larsson, M. Nordberg, and J. Hoglund. 2013. Trends and population dynamics of a Velvet Scoter (Melanitta fusca) population: influence of density dependence and winter climate. Journal of Ornithology 154:837–847.

Hartung, R., and G. S. Hunt. 1966. Toxicity of some oils to waterfowl. Journal of Wildlife Management 30:564–570.

Heath, J., and the community of Sanikiluaq. 2011. People of a feather. Sanikiluaq Running Pictures, Vancouver, BC.

Heath, J. P., and H. G. Gilchrist. 2010. When foraging becomes unprofitable: energetics of diving in tidal currents by Common Eiders wintering in the Arctic. Marine Ecology Progress Series 403:279–290.

Heath, J. P., and W. A. Montevecchi. 2008. Differential use of similar habitat by Harlequin Ducks: trade-offs and implications for identifying critical habitat. Canadian Journal of Zoology 86:419–426.

Heath, J. P., G. J. Robertson, and W. A. Montevecchi. 2006. Population structure of breeding Harlequin Ducks and the influence of predation risk. Canadian Journal of Zoology 84:855–864.

Henny, C. J., D. D. Rudis, T. J. Roffe, and E. Robinson-Wilson. 1995. Contaminants and sea ducks in Alaska and the circumpolar region. Environmental Health Perspectives 103:41–49.

Heppell, S. S., H. Caswell, and L. B. Crowder. 2000. Life histories and elasticity patterns: perturbation analysis for species with minimal demographic data. Ecology 81:654–665.

Hicklin, P., and K. Bunker-Popma. 2001. The spring and fall migrations of scoters, Melanitta spp., at Confederation Bridge in the Northumberland Strait between New Brunswick and Prince Edward Island. Canadian Field-Naturalist 115:436–445.

Hickling, R., D. B. Roy, J. K. Hill, R. Fox, and C. D. Thomas. 2006. The distributions of a wide range of taxonomic groups are expanding polewards. Global Change Biology 12:450–455.

Hill, N. J., J. Y. Takekawa, C. J. Cardona, J. T. Ackerman, A. K. Schultz, K. A. Spragens, and W. M. Boyce. 2010. Waterfowl ecology and avian influenza in California: do host traits inform us about viral occurrence? Avian Diseases 54:426–432.

Hipfner, J. M., L. K. Blight, R. W. Lowe, S. I. Wilhelm, G. J. Robertson, R. T. Barrett, T. Anker-Nilssen, and T. P. Good. 2012. Unintended consequences: how the recovery of sea eagle Haliaeetus spp. populations in the northern hemisphere is affecting seabirds. Marine Ornithology 40:39–52.

Hogan, D., J. E. Thompson, D. Esler, and W. S. Boyd. 2011. Discovery of important post breeding sites for Barrow's Goldeneye in the boreal transition zone of Alberta. Waterbirds 34:261–388.

Hollmén, T. E., J. C. Franson, P. L. Flint, J. B. Grand, R. B. Lanctot, D. E. Docherty, and H. M. Wilson. 2003. An adenovirus linked to mortality and disease in Long-tailed Ducks (Clangula hyemalis) in Alaska. Avian Diseases 47:1434–1440.

Hollmén, T., J. C. Franson, M. Kilpi, D. E. Docherty, W. R. Hansen, and M. Hario. 2002. Isolation and characterization of a reovirus from Common Eiders (Somateria mollissima) from Finland. Avian Diseases 46:478–484.

Hooijmeijer, J. C. E. W., R. E. Gill Jr., D. M. Mulcahy, T. L. Tibbitts, R. Kentie, F. J. Gerritsen, L. W. Bruinzeel, D. C. Tijssen, C. M. Harwood, and T. Piersma. 2014. Abdominally implanted satellite transmitters affect reproduction and survival rather than migration of large shorebirds. Journal of Ornithology 155:447–457.

Hurrell, J. W. 1995. Decadal trends in the North Atlantic Oscillation: regional temperatures and precipitation. Science 269:676–679.

Hurrell, J. W., and C. Deser. 2009. North Atlantic climate variability: the role of the North Atlantic Oscillation. Journal of Marine Systems 78:28–41.

Iles, D. T., R. F. Rockwell, P. Matulonis, G. J. Robertson, K. F. Abraham, J. C. Davies, and D. Koons. 2013. Predators, alternative prey and climate influence annual breeding success of a long-lived sea duck. Journal of Animal Ecology 82:683–693.

Ip, H. S., P. L. Flint, J. C. Franson, R. J. Dusek, D. V. Derksen, R. E. Gill, C. R. Ely et al. 2008. Prevalence of influenza A viruses in wild migratory birds in Alaska: patterns of variation in detection at a crossroads of intercontinental flyways. Virology Journal 5:71.

Iverson, S. A., and D. Esler. 2010. Harlequin Duck population injury and recovery dynamics following the 1989 Exxon Valdez oil spill. Ecological Applications 20:1993–2006.

Iverson, S. A., H. G. Gilchrist, P. A. Smith, A. J. Gaston, and M. R. Forbes. 2014. Longer ice-free seasons increase the risk of nest depredation by polar bears for colonial breeding birds in the Canadian Arctic. Proceedings of the Royal Society B 281:art20133128.

Jaatinen, K., M. Öst, and A. Lehikoinen. 2011. Adult predation risk drive shifts in parental care strategies: a long-term study. Journal of Animal Ecology 80:49–56.

Jamieson, S. E., H. G. Gilchrist, F. R. Merkel, A. W. Diamond, and K. Falk. 2006. Endogenous reserve dynamics of northern Common Eiders wintering in Greenland. Polar Biology 29:585–594.

Jenssen, B. M. 1994. Review article: effects of oil pollution, chemically treated oil, and cleaning on the thermal balance of birds. Environmental Research 20:425–444.

Jenssen, B. M., M. Ekker, and C. Bech. 1985. Thermoregulation in a naturally oil-contaminated Thick-billed Murre *Uria aalge*. Bulletin of Environmental Contamination and Toxicology 35:9–14.

Jessup, D. A., M. A. Miller, J. P. Ryan, H. M. Nevins, H. A. Kerkering, A. Mekebri, D. B. Crane, T. A. Johnson, and R. M. Kudela. 2009. Mass stranding of marine birds caused by a surfactant-producing red tide. PLoS One 4:e4550.

Joint Working Group on the Management of the Common Eider. 2004. Quebec management plan for the Common Eider *Somateria mollissima dresseri*. A special publication of the Joint Working Group on the Management of the Common Eider, Quebec City, QC.

Jonsson, J. E., A. Gardarsson, J. A. Gill, U. K. Pétusdottir, A. Petersen, and T. G. Gunnarsson. 2013. Relationships between long-term demography and weather in a sub-arctic population of Common Eider. PLoS One 8:e67093.

Kaiser, M. J., M. Galanidi, D. A. Showler, A. J. Elliott, R. W. G. Caldow, E. I. S. Rees, R. A. Stillman, and W. J. Sutherland. 2006. Distribution and behaviour of Common Scoter *Melanitta nigra* relative to prey resources and environmental parameters. Ibis 148:110–128.

Kausrud, K. L., A. Mysterud, H. Steen, J. O. Vik, E. Østbye, B. Cazelles, E. Framstad, A. M. Eikeset, I. Mysterud, T. Solhøy, and N. C. Stenseth. 2008. Linking climate change to lemming cycles. Nature 456:93–97.

Kilpi, M., and M. Öst. 2002. The effect of White-tailed Sea Eagle predation on breeding eider females off Tvär-minne, Western Gulf of Finland. Suomen Riista 48:27–33.

Kirk, M. K., D. Esler, and W. S. Boyd. 2007a. Foraging effort of Surf Scoters (*Melanitta perspicillata*) wintering in a spatially and temporally variable prey landscape. Canadian Journal of Zoology 85:1207–1215.

Kirk, M. K., D. Esler, and W. S. Boyd. 2007b. Morphology and density of mussels on natural and aquaculture structure habitats: implications for sea duck predators. Marine Ecology Progress Series 346:179–187.

Kirk, M., D. Esler, S. A. Iverson, and W. S. Boyd. 2008. Movements of wintering Surf Scoters: predator responses to different prey landscapes. Oecologia 155:859–867.

Korschgen, C. E., L. S. George, and W. L. Green. 1985. Disturbance of diving ducks by boaters on a migrational staging area. Wildlife Society Bulletin 13:290–296.

Korschgen, C. E., H. C. Gibbs, and H. L. Mendall. 1978. Avian cholera in eider ducks in Maine. Journal of Wildlife Diseases 14:254–258.

Kristjansson, T. O., and J. E. Jonsson. 2011. Effects of down collection on incubation temperature, nesting behaviour and hatching success of Common Eiders (*Somateria mollissima*) in west Iceland. Polar Biology 34:985–994.

Lacroix, D. L., W. S. Boyd, D. Esler, M. Kirk, T. Lewis, and S. Lipovsky. 2005. Surf Scoters *Melanitta perspicillata* aggregate in association with ephemerally abundant polychaetes. Marine Ornithology 33:61–63.

Larsen, J. K., and M. Guillemette. 2000. Influence of annual variation in food supply on abundance of wintering Common Eiders *Somateria mollissima*. Marine Ecology Progress Series 201:301–309.

Larsen, J. K., and M. Guillemette 2007. Effects of wind turbines on flight behaviour of wintering Common Eiders: implications for habitat use and collision risk. Journal of Applied Ecology 44:516–522.

Larsen, J. K., and B. Laubek. 2005. Disturbance effects of high-speed ferries on wintering sea ducks. Wildfowl 55:99–116.

Laursen, K., and J. Frikke. 2008. Hunting from motorboats displaces Wadden Sea Eiders *Somateria mollissima* from their favoured feeding distribution. Wildlife Biology 14:423–433.

Laursen, K., J. Kahlert, and J. Frikke. 2005. Factors affecting escape distances of staging waterbirds. Wildlife Biology 11:13–19.

LeBourdais, S. V., R. C. Ydenberg, and D. Esler. 2009. Fish and Harlequin Ducks compete on breeding streams. Canadian Journal of Zoology 87:31–40.

Lehikoinen, A., M. Kilpi, and M. Öst. 2006. Winter climate affects subsequent breeding success of Common Eiders. Global Change Biology 12:1355–1365.

Leighton, F. A. 1991. The toxicity of petroleum oils to birds: an overview. Pp. 43–57 in J. White, L. Frink, T. M. Williams, and R. W. Davis (editors), The effects of oil on wildlife. Sheridan Press, Hanover, PA.

Lewis, T. L., D. Esler, and W. S. Boyd. 2007. Effects of predation by sea ducks on clam abundance in soft-bottom intertidal habitats. Marine Ecology Progress Series 329:131–144.

Lewis, T. L., D. Esler, W. S. Boyd, and R. Žydelis. 2005. Nocturnal foraging behavior of wintering Surf Scoters and White-winged Scoters. Condor 107:637–647.

Linville, R. G., S. N. Luoma, L. Cutter, and G. A. Cutter. 2002. Increased selenium threat as a result of invasion of the exotic bivalve *Potamocorbula amurensis* into the San Francisco Bay-Delta. Aquatic Toxicology 57:51–64.

Locke, L. N., J. B. DeWitt, C. M. Menzie, and J. A. Kerwin. 1964. A merganser die-off associated with larval *Eustrongylides*. Avian Diseases 8:420–427.

Lok, E. K., M. Kirk, D. Esler, and W. S. Boyd. 2008. Movements of pre-migratory surf and White-winged Scoters in response to Pacific herring spawn. Waterbirds 31:385–393.

Love, O. P., H. G. Gilchrist, S. Descamps, C. A. D. Semeniuk, and J. Bêty. 2010. Pre-laying climatic cues can time reproduction to optimally match offspring hatching and ice conditions in an arctic marine bird. Oecologia 164:277–286.

Lovvorn, J. R., S. E. De La Cruz, J. Y. Takekawa, L. E. Shaskey, and S. E. Richman. 2013. Niche overlap, threshold food densities, and limits to prey depletion for a diving duck assemblage in an estuarine bay. Marine Ecology Progress Series 476:251–268.

Lumsden, H. G., R. E. Page, and M. Gauthier. 1980. Choice of nest boxes by Common Goldeneyes in Ontario. Wilson Bulletin 92:497–505.

Lumsden, H. G., J. Robinson, and R. Hartford. 1986. Choice of nest boxes by cavity-nesting ducks. Wilson Bulletin 98:167–168.

Madsen, J. 1998. Experimental refuges for migratory waterfowl in Danish wetlands: I. Baseline assessment of the disturbance effects of recreational activities. Journal of Applied Ecology 35:386–397.

Maine Department of Inland Fisheries and Wildlife. [online]. 2013. State list of endangered and threatened species. <http://www.maine.gov/ifw/wildlife/endangered/listed_species_me.htm> (17 January 2014).

Mallory, M., and K. Metz. 1999. Common Merganser (*Mergus merganser*). in A. Poole and F. Gill (editors), The Birds of North America, No. 442. The Academy of Natural Sciences, Philadelphia, PA.

Mallory, M. L., A. J. Fontaine, J. A. Akearok, and V. H. Johnston. 2006. Synergy of local ecological knowledge, community involvement and scientific study to develop marine wildlife areas in eastern Arctic Canada. Polar Record 42:205–216.

Mantua, N. J., and S. R. Hare. 2002. The Pacific decadal oscillation. Journal of Oceanography 58:35–44.

Mawhinney, K. 1999. Factors affecting adult female crèche attendance and duckling survival of Common Eiders in the southern Bay of Fundy and northern Gulf of Maine. Dissertation, University of New Brunswick, Fredericton, NB.

Mawhinney, K., A. W. Diamond, P. Kehoe, and N. Benjamin. 1999. Status and productivity of Common Eiders in relation to the status of Great Black-backed Gulls and Herring Gulls in the southern Bay of Fundy and the northern Gulf of Maine. Waterbirds 22:253–263.

McKinnon, L., M. Picotin, E. Bolduc, C. Juillet, and J. Bêty. 2012. Timing of breeding, peak food availability, and effects of mismatch on chick growth in birds nesting in the high arctic. Canadian Journal of Zoology 90:961–971.

Mehl, K., R. T. Alisauskas, K. A. Hobson, and F. R. Merkel. 2005. Linking breeding and wintering areas of King Eiders: making use of polar isotope gradients. Journal of Wildlife Management 69:1297–1304.

Mehlum, F., L. Nielsen, and I. Gjertz. 1991. Effect of down harvesting on nesting success in a colony of the Common Eider *Somateria mollissima* in Svalbard. Norsk Polarinstitut Skrifter 195:47–50.

Mendenhall, V. M., and H. Milne. 1985. Factors affecting duckling survival of Eiders *Somateria mollissima* in northeast Scotland. Ibis 127:148–158.

Merkel, F. R. 2004. Impact of hunting and gillnet fishery on wintering eiders in Nuuk, southwest Greenland. Waterbirds 27:469–479.

Merkel, F. R. 2010. Evidence of recent population recovery in Common Eiders breeding in western Greenland. Journal of Wildlife Management 74:1869–1874.

Merkel, F. R., A. Mosbech, and F. Riget. 2009. Common Eider *Somateria mollissima* feeding activity and the influence of human disturbances. Ardea 97:99–107.

Mills, L. S., D. F. Doak, and M. J. Wisdom. 1999. Reliability of conservation actions based on elasticity analysis of matrix models. Conservation Biology 13:815–829.

Milne, H., and A. Reed. 1974. Annual production of fledged young from the eider colonies of the St. Lawrence Estuary. Canadian Field-Naturalist 88:163–169.

Minerals Management Service. 2007. Biological opinion for Chukchi Sea planning area oil and gas lease sale 193 and associated seismic surveys and exploratory drilling. Consultation with Minerals Management Service—Alaska OCS Region, Anchorage, AK.

Mitchell, D. C. 1986. Native subsistence hunting of migratory waterfowl in Alaska: a case study demonstrating why politics and wildlife management don't mix. Transactions of the North American Wildlife and Natural Resource Conference 51:527–534.

Montgomery, R. D., G. Stein Jr., V. D. Stotts, and F. H. Settle. 1979. The 1978 epornitic of avian cholera on the Chesapeake Bay. Avian Diseases 23:966–978.

Mosbech, A., and D. Boertmann. 1999. Distribution, abundance and reaction to aerial surveys of post-breeding King Eiders (*Somateria spectabilis*) in western Greenland. Arctic 52:188–203.

Mosbech, A., R. S. Danø, F. Merkel, C. Sonne, G. Gilchrist, and A. Flagstad. 2006a. Use of satellite telemetry to locate key habitats for King Eiders *Somateria spectabilis* in west Greenland. Pp. 769–776 in G. C. Boere, C. A. Galbraith, and D. A. Stroud (editors), Waterbirds around the world. The Stationary Office, Edinburgh, U.K.

Mosbech, A., G. Gilchrist, F. Merkel, C. Sonne, A. Flagstand, and H. Nyegaard. 2006b. Year-round movements of northern Common Eiders *Somateria mollissima borealis* breeding in Arctic Canada and west Greenland followed by satellite telemetry. Ardea 94:651–665.

Munro, J. A., and W. A. Clemens. 1931. Waterfowl in relation to the spawning of herring in British Columbia. Bulletin of the Biological Board of Canada 17:1–46.

Nakashima, D. J. 1986. Inuit knowledge of the ecology of the Common Eider in northern Quebec. Pp. 102–113 in A. Reed (editor), Eider ducks in Canada. Canadian Wildlife Service, Report Series. Ottawa, ON.

Natcher, D. C., L. Felt, K. Chaulk, and A. Procter. 2011. Monitoring the domestic harvest of migratory birds in Nunatsiavut, Labrador. Arctic 64:362–366.

Natcher, D. C., L. Felt, K. Chaulk, and A. Procter. 2012. The harvest and management of migratory bird eggs by Inuit in Nunatsiavut, Labrador. Environmental Management 50:1047–1056.

Nehls, G., and M. Ruth. 1994. Eiders, mussels and fisheries—continuous conflicts or relaxed relations? Ophelia 6:263–278.

Newell, R. I. E. 1989. Species profiles: life histories and environmental requirements of coastal fishes and invertebrates (North and Mid-Atlantic)—blue mussel. U.S. Fish and Wildlife Service. Biological Report 82 and U.S. Army Corps of Engineers, TR E1-82-4. U.S. Fish and Wildlife Service, Slidell, LA.

Noel, L. E., S. R. Johnson, G. M. O'Doherty, and M. K. Butcher. 2005. Common Eider (*Somateria mollissima v-nigrum*) nest cover and depredation on Central Alaskan Beaufort Sea Barrier Islands. Arctic 58:129–136.

Nordström, M., J. Högmander, J. Nummelin, J. Laine, N. Laanetu, and E. Korpimäki. 2002. Variable responses of waterfowl breeding populations to long-term removal of introduced American mink. Ecography 25:385–394.

Nordström, M., and E. Korpimäki. 2004. Effects of island isolation and feral mink removal on bird communities on small islands in the Baltic Sea. Journal of Animal Ecology 73:424–433.

North American Waterfowl Management Plan. [online]. 1986. North American waterfowl management plan: a strategy for cooperation. U.S. Department of the Interior and Environment Canada, Washington, DC. <http://nawmp.ca/> (24 September 2013).

North American Waterfowl Management Plan. [online]. 1998. Expanding the vision: 1998 update North American waterfowl management plan. North American Waterfowl Management Plan Committee. <http://www.nawmp.ca/eng/pub_e.html> (24 September 2013).

North American Waterfowl Management Plan. [online]. 2012. North American waterfowl management plan 2012: people conserving waterfowl and wetlands. North American Waterfowl Management Plan Committee. <http://www.nawmp.ca> (24 September 2013).

Nummi, P., V. M. Väänänen, M. Rask, K. Nyberg, and K. Taskinen. 2012. Competitive effects of fish in structurally simple habitats: perch, invertebrates, and goldeneye in small boreal lakes. Aquatic Sciences 74:343–350.

Oakes, J. 1999. Vêtements d'eider. Aboriginal Issues Press, Winnipeg, MB.

O'Connor, M. 2008. Surf Scoter (*Melanitta perspicillata*) ecology on spring staging grounds and during the flightless period. Thesis, McGill University, Montreal, QC.

Oppel, S., D. L. Dickson, and A. N. Powell. 2009a. International importance of the eastern Chukchi Sea as a staging area for migrating King Eiders. Polar Biology 32:775–783.

Oppel, S., A. N. Powell, and D. L. Dickson. 2008. Timing and distance of King Eider migration and winter movements. Condor 110:296–305.

Oppel, S., A. N. Powell, and D. L. Dickson. 2009b. Using an algorithmic model to reveal individually variable movement decisions in a wintering sea duck. Journal of Animal Ecology 78:524–531.

Ottersen, G., B. Planque, A. Belgrano, E. Post, P. C. Reid, and N. C. Stenseth. 2001. Ecological effects of the North Atlantic Oscillation. Oecologia 125:1–14.

Parmesan, C. 2006. Ecological and evolutionary response to recent climate change. Annual Review of Ecology, Evolution, and Systematics 37:637–669.

Pearce, J. M., P. Blums, and M. S. Lindberg. 2008. Site fidelity is an inconsistent determinant of population structure in the Hooded Merganser (*Lophodytes cucullatus*): evidence from genetic, mark-recapture, and comparative data. Auk 125:711–722.

Pearce, J. M., J. M. Eadie, J.-P. L. Savard, T. K. Christensen, J. Berdeen, E. J. Taylor, W. S. Boyd, Á. Einarsson, and S. L. Talbot. 2014. Comparative population structure of cavity-nesting sea ducks. Auk 131:195–207.

Pearce, J. M., K. G. McCracken, T. K. Christensen, and Y. N. Zhuravlev. 2009a. Migratory patterns and population structure among breeding and wintering Red-breasted Mergansers (*Mergus serrator*) and Common Mergansers (*M. merganser*). Auk 126:784–798.

Pearce, J. M., D. Zwiefelhofer, and N. Maryanski. 2009b. Mechanisms of population heterogeneity among molting Common Mergansers on Kodiak Island, Alaska: implications for genetic assessments of migratory connectivity. Condor 111:283–293.

Pehrsson, O. 1986. Duckling production of the Oldsquaw in relation to spring weather and small rodent fluctuations. Canadian Journal of Zoology 64:1835–1841.

Petersen, I. K., and A. D. Fox. 2007. Changes in bird habitat utilization around the Horns Rev 1 offshore wind farm, with particular emphasis on Common Scoter. National Environmental Institute, University of Aarhus, Aarhus, Denmark.

Petersen, M. R., W. W. Larned, and D. C. Douglas. 1999. At-sea distribution of Spectacled Eiders: a 120-year-old mystery resolved. Auk 116:1009–1020.

Petrie, S. A., S. S. Badzinski, and K. G. Drouillard. 2007. Contaminants in Lesser and Greater Scaup staging on the lower Great Lakes. Archives of Environmental Contamination and Toxicology 52:580–589.

Petrie, S. A., and R. W. Knapton. 1999. Rapid increase and subsequent decline of zebra and quagga mussels in Long Point Bay, Lake Erie: possible influence of waterfowl. Journal of Great Lakes Research 25:772–782.

Philippart, C. J. M., H. M. van Aken, J. J. Beukema, O. G. Bos, G. C. Cadée, and R. Dekker. 2003. Climate-related changes in recruitment of the bivalve *Macoma balthica*. Limnology and Oceanography 48:2171–2185.

Phillips, L. M., A. N. Powell, and E. A. Rexstad. 2006. Large-scale movements and habitat characteristics of King Eiders throughout the nonbreeding period. Condor 108:887–900.

Piatt, J. F., C. J. Lensink, W. Butler, M. Kendziorek, and D. R. Nysewander. 1990. Immediate impact of the 'Exxon Valdez' oil spill on marine birds. Auk 107:387–397.

Poot, J., B. J. Ens, H. de Vries, M. A. H. Donners, M. R. Wernand, and J. M. Marquenie. 2008. Green light for nocturnally migrating birds. Ecology and Society 13:art47.

Powell, A. N. [online]. 2009. Satellite telemetry of King Eiders from northern Alaska 2002–2009. OBIS-SEAMAP. <http://seamap.env.duke.edu/data-sets/detail/487> (27 January 2010).

Pöysä, H., and S. Pöysä. 2002. Nest-site limitation and density dependence of reproductive output in the Common Goldeneye (*Bucephala clangula*): implications for the management of cavity-nesting birds. Journal of Applied Ecology 39:502–510.

Quakenbush, L., R. Suydam, T. Obritschkewitsch, and M. Deering. 2004. Breeding biology of Steller's Eiders (*Polysticta stelleri*) near Barrow, Alaska, 1991–99. Arctic 57:166–182.

Quinlan, S. E., and W. A. Lehnhausen. 1982. Arctic fox, *Alopex lagopus*, predation on nesting Common Eiders, *Somateria mollissima*, at Icy Cape, Alaska. Canadian Field-Naturalist 96:462–466.

Ramey, A. M., J. M. Pearce, C. R. Ely, L. M. S. Guy, D. B. Irons, D. V. Derksen, and H. S. Ip. 2010. Transmission and reassortment of avian influenza viruses at the Asian–North American interface. Virology 406:352–359.

Ramey, A. M., J. M. Pearce, A. B. Reeves, J. C. Franson, M. R. Petersen, and H. S. Ip. 2011. Evidence for limited exchange of avian influenza viruses between seaducks and dabbling ducks at Alaska Peninsula coastal lagoons. Archives of Virology 156:1813–1821.

Rask, M., H. Pöysä, P. Nummi, and C. Karppinen. 2001. Recovery of the perch (*Perca fluviatilis*) in an acidified lake and subsequent responses in macroinvertebrates and the Goldeneye (*Bucephala clangula*). Water, Air, and Soil Pollution 130:1367–1372.

Reed, A. 1986. Eiderdown harvesting and other uses of Common Eiders in spring and summer. Pp. 138–146 in A. Reed (editor), Eiders ducks in Canada. Canadian Wildlife Service, Report Series, No. 47, Ottawa, ON.

Reed, A., R. Benoit, R. Lalumière, and M. Julien. 1996. Duck use of the coastal habitats of northeastern James Bay. Canadian Wildlife Service, Occasional Paper, No. 90. Canadian Wildlife Service, Ottawa, ON.

Reed, A., and J.-G. Cousineau. 1967. Epidemics involving the Common Eider (*Somateria mollissima*) at Ile Blanch, Quebec. Naturaliste Canadien 94:327–334.

Regehr, H. M. 2003. Movement patterns and population structure of Harlequin Ducks wintering in the Strait of Georgia, British Columbia. Dissertation, Simon Fraser University, Vancouver, BC.

Richman, S. E., and J. R. Lovvorn. 2003. Effects of clam species dominance on nutrient and energy acquisition by Spectacled Eiders in the Bering Sea. Marine Ecology Progress Series 261:283–297.

Richman, S. E., and J. R. Lovvorn. 2004. Relative foraging value to Lesser Scaup ducks of native and exotic clams from San Francisco Bay. Ecological Applications 14:1217–1231.

Robert, M., R. Benoit, and J.-P. L. Savard. 2000. COSEWIC status report on the Barrow's Goldeneye (*Bucephala islandica*) eastern population in Canada. Committee on the Status of Endangered Wildlife in Canada, Canadian Wildlife Service, Ottawa, ON.

Robert, R., B. Drolet, and J.-P. L. Savard. 2008. Habitat features associated with Barrow's Goldeneye breeding in eastern Canada. Wilson Journal of Ornithology 120:320–330.

Robertson, G. J., and H. G. Gilchrist. 1998. Evidence of population declines among Common Eiders breeding in the Belcher Islands, Northwest Territories. Arctic 51:378–385.

Robertson, G. J., and R. I. Goudie. 1999. Harlequin Duck (*Histrionicus histrionicus*). in A. Poole and F. Gill (editors), The Birds of North America, No. 466. The Academy of Natural Sciences, Philadelphia, PA.

Robertson, G. J., and J.-P. L. Savard. 2002. Long-tailed Duck (*Clangula hyemalis*). in A. Poole and F. Gill (editors), The Birds of North America, No. 651. The Academy of Natural Sciences, Philadelphia, PA.

Rodgers, J. A., and S. T. Schwikert. 2002. Buffer-zone distances to protect foraging and loafing waterbirds from disturbance by personal watercraft and outboard-powered boats. Conservation Biology 16:216–224.

Rosenberg, D. M., F. Berkes, R. A. Bodaly, R. E. Hecky, C. A. Kelly, and J. W. Rudd. 1997. Large-scale impacts of hydroelectric development. Environmental Reviews 5:27–54.

Ross, B. P., and R. W. Furness. 2000. Minimizing the impact of eider ducks on mussel farming. Ornithology Group, University of Glasgow, Glasgow, Scotland.

Ross, B. P., J. Lien, and R. W. Furness. 2001. Use of underwater playback to reduce the impact of eiders on mussel farms. ICES Journal of Marine Science 58:517–524.

Ross, R. K., S. A. Petrie, S. S. Badzinski, and A. Mullie. 2005. Autumn diet of Greater Scaup, Lesser Scaup, and Long-tailed Ducks on eastern Lake Ontario prior to zebra mussel invasion. Wildlife Society Bulletin 33:81–91.

Rueggeberg, H., and J. Booth. 1989. Marine birds and aquaculture in British Columbia: assessment and management of interactions, phase 3 report: preventing predation by scoters on a west coast mussel farm. Technical Report Series, No. 74. Canadian Wildlife Service, Pacific and Yukon Region, BC.

Sandercock, B. K., and S. R. Beissinger. 2002. Estimating rates of population change for a neotropical parrot with ratio, mark-recapture, and matrix methods. Journal of Applied Statistics 29:589–607.

Savard, J.-P. L. 1987. Causes and functions of brood amalgamation in Barrow's Goldeneye and Bufflehead. Canadian Journal of Zoology 65:1548–1553.

Savard, J.-P. L. 1988. Use of nest boxes by Barrow's Goldeneyes: nesting success and effect on the breeding population. Wildlife Society Bulletin 16:125–132.

Savard, J.-P. L., L. Lesage, S. C. Gilliland, H. G. Gilchrist, and J.-F. Giroux. 2011. Molting, staging and wintering locations of Common Eiders breeding in the Gyrfalcon Archipelago, Ungava Bay. Arctic 64:197–206.

Savard, J.-P. L., and M. Robert. 2007. Use of nest boxes by goldeneyes in eastern North America. Wilson Journal of Ornithology 119:28–34.

Schamber, J. L., P. L. Flint, J. B. Grand, H. M. Wilson, and J. A. Morse. 2009. Population dynamics of Long-tailed Ducks breeding on the Yukon–Kuskokwim Delta, Alaska. Arctic 62:190–200.

Schmidt, J., E. J. Taylor, and E. A. Rexstad. 2006. Survival of Common Goldeneye ducklings in interior Alaska. Journal of Wildlife Management 70:792–798.

Schummer, M. L., S. S. Badzinski, S. A. Petrie, Y.-W. Chen, and N. Belzile. 2009. Selenium accumulation in sea ducks wintering at Lake Ontario. Archives of Environmental Contaminants and Toxicology 58:854–862.

Sea Duck Joint Venture. [online]. 2003. Species status reports. U.S. Fish and Wildlife Service, Anchorage, Alaska, and Environment Canada, Sackville, NB. <http://www.seaduckjv.org/meetseaduck/species_status_summary.pdf> (24 September 2013).

Sea Duck Joint Venture. [online]. 2012. Sea Duck Joint Venture Implementation Plan for April 2012 through March 2015. Report of the Sea Duck Joint Venture. U.S. Fish and Wildlife Service, Anchorage, Alaska, and Environment Canada, Sackville, NB. <http://seaduckjv.org/sdjv_implementation_plan.pdf> (24 September 2013).

Serreze, M. C., J. E. Walsh, and F. S. Chapin. 2000. Observational evidence of recent change in the northern high-latitude environment. Climatic Change 46:159–207.

Siren, M. 1951. Increasing the goldeneye population with nest boxes. Suomen Riista 6:83–101.

Sittler, B., O. Gill, and T. B. Berg. 2000. Low abundance of King Eider nests during low lemming years in northeast Greenland. Arctic 53:53–60.

Skerratt, L. F., J. C. Franson, C. U. Meteyer, and T. E. Hollmén. 2005. Causes of mortality in sea ducks (Mergini) necropsied at the USGS-National Wildlife Health Center. Waterbirds 28:193–207.

Smith, C., S. Feirer, R. Hagenstein, A. Couvillion, and S. Leonard. 2008. Assigning conservation management status to Alaska's Lands. Gap Analysis Program Bulletin 15:1–8.

Smith, P. A., K. H. Elliot, A. J. Gaston, and H. G. Gilchrist. 2010. Has early ice clearance increased predation on breeding birds by polar bears? Polar Biology 33:1149–1153.

Stenseth, N. C., G. Ottersen, J. W. Hurrell, A. Mysterud, M. Lima, K.-S. Chan, N. G. Yoccoz, and B. Adlansvik. 2003. Studying climate effects on ecology through the use of climate indices: the North Atlantic Oscillation, El Nino Southern Oscillation and beyond. Proceedings of the Royal Society B 270:2087–2096.

Stiller, J. C. 2011. Effects of Common Merganser on hatchery-reared brown trout and spring movements of adult males in southeastern New York, USA. Thesis, University of New York, Syracuse, NY.

Stillman, R. A., A. D. West, R. W. Caldow, and S. D. Durell. 2007. Predicting the effect of disturbance on coastal birds. Ibis 149:73–81.

Swart, J. A. A., and J. van Andel. 2007. Rethinking the interface between ecology and society: the case of the cockle controversy in the Dutch Wadden Sea. Journal of Applied Ecology 45:82–90.

Szaro, R. C., G. Hensler, and G. H. Heinz. 1981. Effects of chronic ingestion of No. 2 fuel oil on Mallard ducklings. Journal of Toxicology and Environmental Health 7:789–799.

Takekawa, J. Y., S. W. De La Cruz, M. T. Wilson, E. C. Palm, J. Yee, D. R. Nysewander, J. R. Evenson, J. M. Eadie, D. Esler, W. S. Boyd, and D. H. Ward. 2011. Breeding distribution and ecology of Pacific Coast Surf Scoters. Pp. 41–64 in J. V. Wells (editor), Boreal birds of North America. Studies in Avian Biology (vol. 41), University of California Press, Berkeley, CA.

Tasker, M. L., C. J. Camphuysen, J. Cooper, S. Garthe, W. A. Montevecchi, and S. J. Blaber. 2000. The impacts of fishing on marine birds. ICES Journal of Marine Science 57:531–547.

Thiel, M., G. Nehls, S. Brager, and J. Meisner. 1992. The impact of boating on the distribution of seals and molting ducks in the Wadden Sea of Schleswig-Holstein. Netherlands Institute for Sea Research, Publication Series, No. 20:221–233.

Thomas, P. W., and M. Robert. 2001. Updated COSEWIC status report of the Eastern Canada Harlequin Duck (Histrionicus histrionicus). Committee on the Status of Endangered Wildlife in Canada, Canadian Wildlife Service, Ottawa, ON.

Thompson, R. 2003. Sea ducks and mussel aquaculture interactions in Prince Edward Island October 2001–January 2003. Technical Report, No. 230. Prince Edward Island Department of Fisheries, Aquaculture and Environment, Fisheries and Aquaculture Division, Charlottetown, PE.

Thompson, R. 2008. Federal protected areas and wildlife. Chapter 4 Pp. 1–20 in Status report of the Commissioner of the Environment and Sustainable Development to the House of Commons. Office of the Auditor General of Canada, Ottawa, ON.

Thompson, R., and B. Gillis. 2001. Sea ducks and mussel aquaculture interactions in Prince Edward Island October 2000–January 2001. Technical Report, No. 227. Prince Edward Island Department of Fisheries, Aquaculture and Environment, Fisheries and Aquaculture Division, Charlottetown, PE.

Titman, R. D. 1999. Red-breasted Merganser (Mergus serrator). in A. Poole and F. Gill (editors), The Birds of North America, No. 443. The Academy of Natural Sciences, Philadelphia, PA.

Tjornlov, R. S., J. Humaidan, and M. Fredriksen. 2013. Impacts of avian cholera on survival of Common Eiders Somateria mollissima in a Danish colony. Bird Study 60:321–326.

Trust, K. A., D. Esler, B. R. Woodin, and J. J. Stegeman. 2000. Cytochrome P4501A induction in sea ducks inhabiting nearshore areas of Prince William Sound, Alaska. Marine Pollution Bulletin 40:397–403.

Tulp, I., H. Schekkerman, J. K. Larsen, J. van der Winden, R. J. W. van de Haterd, P. van Horssen, S. Dirksen, and A. L. Spaans. 1999. Nocturnal flight activity of sea ducks near the windfarm Tunø Knob in the Kattegat. Bureau Waardenburg, proj. nr. 98.100, report nr. 99.64. Novem, Utrecht, the Netherlands (in Dutch).

United Nations. [online]. 1992. Convention on biological diversity. <www.cbd.int/convention>.

United Nations Environment Programme. 2010. Decision adopted by the Conference of the Parties to the Convention on Biological Diversity: The Strategic Plan for Biodiversity 2011–2020 and the Aichi Biodiversity Targets. Conference of the Parties to the Convention on Biological Diversity at Its 10th Meeting, Nagoya, Japan.

U.S. Fish and Wildlife Service. 1996. Spectacled Eider recovery plan. U.S. Fish and Wildlife Service, Anchorage, AK.

U.S. Fish and Wildlife Service. 2001a. Endangered and threatened wildlife and plants: final determination of critical habitat for the Alaska-breeding population of the Steller's Eider. 50 CFR Part 17, RIN 1018-AF95. Department of the Interior. Federal Register 66:8850–8884.

U.S. Fish and Wildlife Service. 2001b. Endangered and threatened wildlife and plants: final determination of critical habitat for the Spectacled Eider. 50 CFR Part 17, RIN 1018-AF92. Department of the Interior. Federal Register 66:9146–9185.

U.S. Fish and Wildlife Service. 2002. Steller's Eider recovery plan. U.S. Fish and Wildlife Service, Fairbanks, AK.

Vaillancourt, M.-A., P. Drapeau, M. Robert, and S. Gauthier. 2009. Origin and availability of large cavities for Barrow's Goldeneye (Bucephala islandica), a species at risk inhabiting the eastern Canadian boreal forest. Avian Conservation and Ecology 4:art6.

Vaitkus, G. 2001. Ecological adaptations of seabirds to the gradient of winter climatic conditions in the Baltic Sea Region. Acta Zoologica Lituanica 11:280–287.

Van der Putten, W. H., M. Macel, and M. E. Visser. 2010. Predicting species distribution and abundance responses to climate change: why is it essential to include biotic interactions across trophic levels. Philosophical Transactions of the Royal Society B 365:2025–2034.

Van Dijk, B. 1986. The breeding biology of eiders at Ile aux Pommes, Quebec. Pp. 119–126 in A. Reed (editor), Eider Ducks in Canada. Canadian Wildlife Service, Report Series, No. 47. Canadian Wildlife Service, Ottawa, ON.

Vaughan, S. 2012. The commissioner's perspective main points. Chapters 1 to 4 Pp. 1–28 in the Fall Report of the Commissioner of the Environment and Sustainable Development. Office of the Auditor General of Canada, Ottawa, ON.

Waldeck, P., and K. Larsson. 2013. Effects of winter water temperature on mass loss in Baltic blue mussels: implications for foraging sea ducks. Journal of Experimental Marine Biology and Ecology 444:24–30.

Wallen, R. L. 1987. Habitat utilization by Harlequin Ducks in Grand Teton National Park. Thesis, Montana State University, Bozeman, MT.

Walton, P., C. M. R. Turner, G. Austin, M. D. Burns, and P. Monaghan. 1997. Sub-lethal effects of an oil pollution incident on breeding Kittiwakes (Rissa tridactyla). Marine Ecology Progress Series 155:261–268.

Weise, F. K., and P. C. Ryan. 2003. The extent of chronic marine oil pollution in southeastern Newfoundland waters assessed through beached bird surveys 1984–1999. Marine Pollution Bulletin 46:1090–1101.

Wenzel, L., M. D'Iorio, and K. Yeager. 2012. Analysis of United States MPAs. Office of Ocean and Coastal Resource Management, NOAA Ocean Service, Silver Spring, MD.

Williams, B. K., M. D. Koneff, and D. A. Smith. 1999. Evaluation of waterfowl conservation under the North American Waterfowl Management Plan. Journal of Wildlife Management 63:417–440.

Wilson, H. M., P. L. Flint, A. N. Powell, J. B. Grand, and C. L. Moran. 2012. Population ecology of breeding Pacific Common Eiders on the Yukon–Kuskokwim Delta, Alaska. Wildlife Monographs 182:1–28.

Winder, M., and D. E. Schindler. 2004. Climate change uncouples trophic interactions in an aquatic ecosystem. Ecology 85:2100–2106.

Wood, C. C. 1985a. Food-searching behaviour of the Common Merganser (Mergus merganser) II: choice of foraging location. Canadian Journal of Zoology 63:1271–1279.

Wood, C. C. 1985b. Aggregative response of Common Mergansers (Mergus merganser): predicting flock size and abundance on Vancouver Island salmon streams. Canadian Journal of Fisheries and Aquatic Sciences 42:1259–1271.

Wood, C. C., and C. M. Hand. 1985. Food-searching behaviour of the Common Merganser (Mergus merganser) I: functional responses to prey and predator density. Canadian Journal of Zoology 63:1260–1270.

Woolaver, L. G. 1997. Habitat and artificial shelter use by American Eider (Somateria mollissima dresseri Sharpe) nesting on the eastern shore of Nova Scotia. Thesis, Acadia University, Wolfville, NS.

Zipkin, E. F., B. Gardner, A. T. Gilbert, A. F. O'Connell Jr., J. A. Royle, and E. D. Silverman. 2010. Distribution patterns of wintering sea ducks in relation to the North Atlantic Oscillation and local environmental characteristics. Oecologia 163:893–902.

Žydelis, R., J. Bellebaum, H. Österblom, M. Vetemaa, B. Schirmeister, A. Stipniece, M. Dagys, M. van Eerden, and S. Garthe. 2009a. Bycatch in gillnet fisheries—an overlooked threat to waterbird populations. Biological Conservation 142:1269–1281.

Žydelis, R., D. Esler, W. S. Boyd, D. Lacroix, and M. Kirk. 2006. Habitat use by wintering Surf and White-winged Scoters: effects of environmental attributes and shellfish aquaculture. Journal of Wildlife Management 70:1754–1762.

Žydelis, R., D. Esler, M. Kirk, and W. S. Boyd. 2009b. Effects of off-bottom shellfish aquaculture on winter habitat use by molluscivorous sea ducks. Aquatic Conservation: Marine and Freshwater Ecosystems 19:34–42.

CHAPTER FIFTEEN

Conclusions, Synthesis, and Future Directions*

UNDERSTANDING SOURCES
OF POPULATION CHANGE

*Daniel Esler, Paul L. Flint, Dirk V. Derksen, Jean-Pierre L. Savard,
and John M. Eadie*

Abstract. The chapters in this volume of *Studies in Avian Biology* reflect the burgeoning interest in sea ducks, both as study species with compelling and unique ecological attributes and as taxa of conservation concern. In this review, we provide perspective on the current state of sea duck knowledge by highlighting key findings in the preceding chapters that are of particular value for understanding or influencing population change. We also introduce a conceptual model that characterizes links among topics covered by individual chapters and places them in the context of demographic responses. Last, we offer recommendations for areas of future research that we suggest will have importance for understanding and managing population dynamics of sea ducks.

Key Words: abundance, conservation, density-dependence, life-history traits, population model, vital rates.

S ea ducks differ from other waterfowl, and other birds, in their unique combinations of life-history traits, habitats used, and conservation concerns (Introduction, this volume). Many of these attributes have been highlighted in the preceding chapters of our volume and provide the first major compilation of ecological data for tribe Mergini. Among waterfowl, sea ducks are a diverse group (Chapter 2, this volume). Eiders and other species have life-history characteristics similar to geese in terms of high survival rates and low levels of fecundity, whereas mergansers are more similar to dabbling ducks with high fecundity and lower survival (Chapter 3, this volume). Sea ducks have goose-like life histories, but mating systems are similar to other ducks, with females providing all direct parental care. Further, some species of sea ducks are among the best examples of capital breeders where most nutrients and energy for egg laying and incubation come from stored reserves (Korschgen 1977). Other aspects of reproductive behavior, especially pre- and posthatch brood amalgamation (Chapter 11, this volume), and the high proportion of species that are cavity nesters, are unusual and the sea duck tribe has been an important study system for understanding

* Esler, D., P. L. Flint, D. V. Derksen, J.-P. L. Savard, and J. M. Eadie. 2015. Conclusions, synthesis, and future directions:
Understanding sources of population change. Pp. 561–568 in J.-P. L. Savard, D. V. Derksen, D. Esler, and J. M. Eadie (editors).
Ecology and conservation of North American sea ducks. Studies in Avian Biology (no. 46), CRC Press, Boca Raton, FL.

these phenomena. As highlighted throughout this volume of *Studies in Avian Biology*, the ecology of sea ducks is a fascinating and important field of inquiry that has grown substantially in recent years. For the purposes of this concluding chapter, we emphasize assimilation of the information presented in our volume as it pertains to understanding variation in demographic attributes and subsequently population dynamics, which provides a contextual structure for considering the various aspects of sea duck ecology throughout the annual cycle and is directly relevant for conservation of sea duck populations.

CONCEPTUAL MODEL

To fully consider population dynamics of sea ducks, and the underlying factors leading to demographic variation, we needed to link together the various features of sea duck ecology described in the preceding chapters. To conceptualize these relationships, we developed a model that includes ecological or anthropogenic drivers (e.g., habitat conditions, harvest) that influence individual performance and vital rates at different stages of the annual cycle (e.g., adult survival and recruitment),

which subsequently drive population dynamics (Figure 15.1). In developing the model, we considered habitat conditions in a broad sense, pooling a suite of factors that might influence individuals or populations, including food availability and quality, temperature, precipitation, winds, ice conditions, as well as the functional and numerical responses of predators.

Our conceptual model illustrates a number of key relationships. First, the model demonstrates that ecological and anthropogenic drivers can influence vital rates through multiple pathways, both direct and indirect. Inclusion of indirect pathways fits with a growing appreciation for the importance of cross-seasonal effects on vital rates (Chapter 5, this volume), through indirect relationships where drivers of demographic change are imposed at different stages of the annual cycle than when the demographic effects are realized. Our model also captures the idea that multiple demographic attributes may be influenced by the same factor, leading to temporal correlation among vital rates. The model also accommodates the potential for life-history trade-offs between current reproductive effort and subsequent survival. The conceptual end point of this model is

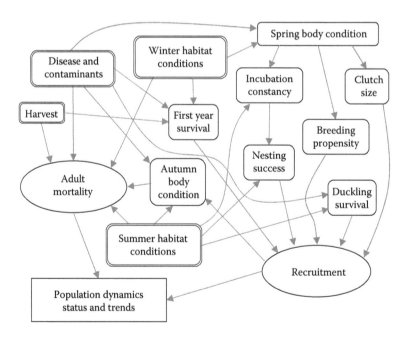

Figure 15.1. Conceptual model describing sea duck demography through the full annual cycle. The model outcome is *population dynamics*, defined as abundance and trends in abundance, which is shown in the lower left rectangle. Drivers of population dynamics, *adult survival* and *recruitment*, are shown as ovals. Measures of demographic or energetic performance are indicated with solid lined boxes. Double-lined boxes represent anthropogenic or ecological drivers influencing performance. Arrows between boxes represent cause and effect influences.

population size and, equally importantly, change in population size, which are often the metrics used for indicating conservation status and gauging efficacy of management activities.

In some cases, the relationships described by the model have been quantified such as linkages among ecological or anthropogenic factors, individual performance, and population vital rates. In other cases, effect sizes are unknown. For example, we know that disease can affect individual survival and body condition, but disease effects on vital rates at a population level and hence effects on population dynamics are largely unknown. Some demographic parameters in the model are easily and frequently measured such as clutch size, whereas others including breeding propensity are not, complicating our understanding of the relative importance of different pathways for affecting vital rates and population dynamics.

The conceptual model allows consideration of hypotheses for broad-scale or long-term population change. For example, consider the hypothesis that broad-scale variation in sea duck population status and trends results from variation in winter habitat conditions. Based on the model, winter habitat variation would be expected to result in alteration of survival for both first-year and adult birds, as well as changes in spring body condition and resulting effects on breeding propensity, clutch size, and incubation constancy. This example illustrates the complexity of ascribing sea duck population dynamics to specific driving mechanisms and factors. Many changes in vital rates are not due to a single driving factor. For example, adult mortality could be affected by habitat conditions in winter, summer, or both seasons. Nor are effects of any ecological or anthropogenic driver necessarily unique—the potential negative effects of degraded winter habitat conditions would be similar to negative effects of disease and contaminants. As such, considering effects of specific drivers individually could lead to spurious conclusions.

ANTHROPOGENIC AND ECOLOGICAL DRIVERS

Historical and contemporary harvest of sea ducks varies among species (Chapter 12, this volume). While commercial harvest of most waterfowl species ended >50 years ago, there was an extensive commercial harvest of sea ducks as recently as the late 1990s (Merkel 2010). Some species of sea ducks have essentially no harvest; others cannot be harvested due to their remote distributions, while some populations have relatively high harvest rates. For example, some merganser populations have band recovery rates that are comparable to Mallards (*Anas platyrhynchos*, Pearce et al. 2005). In general, harvest rates for sea ducks are difficult to estimate and monitor, and management of sea duck harvest has received less attention than dabbling ducks because banding effort is limited and recoveries are few. However, one of the clearest examples of harvest effects on population dynamics for any species of waterfowl comes from management of commercial exploitation of Common Eiders (*Somateria mollissima*) in Greenland, where reductions in harvest resulted in an immediate population increase (Merkel 2010).

Contaminants have been relatively well studied in sea ducks, compared to other waterfowl (Chapter 6, this volume). As top-level predators in nearshore marine systems, sea ducks tend to feed on long-lived marine invertebrates such as bivalves, which can concentrate contaminants in their tissues. Sea ducks are generally more susceptible to bioaccumulation than dabbling ducks that feed on ephemeral invertebrates and seeds or herbivorous geese that feed on vegetation. However, sea ducks may tolerate higher concentrations of some contaminants, particularly metals and trace elements, than freshwater species of waterfowl. Many studies have identified presence of contaminants in sea duck tissues, although the most useful research considers links between contaminant exposure and metrics of breeding performance or survival of individuals or population demography (Chapter 6, this volume).

Sea ducks use terrestrial and marine habitats and are, therefore, exposed to a diversity of disease-causing viruses, bacteria, and parasites during the annual cycle, with potential effects on vital rates and population dynamics (Figure 15.1). Adenoviruses have been associated with mortality events of Common Eiders and Long-tailed Ducks (*Clangula hyemalis*) in marine habitats of Europe and North America during the prebreeding and molting periods (Chapter 4, this volume). Avian cholera, which is more typically associated with freshwater habitats, has caused die-offs of nesting and wintering sea ducks in a variety of locations. The mechanisms of exposure of sea ducks to pathogens at diverse habitat types are not fully understood, but patterns of habitat use by sea ducks may facilitate redistribution of bacteria, viruses, and parasites across the marine–freshwater interface. Parasites and pathogens of sea ducks have been well described, but less work has focused on the impacts of these agents at the population level, despite a growing appreciation for the potential role of disease for influencing population dynamics (Chapter 4, this volume).

Anthropogenic changes to important sea duck habitats throughout the annual cycle, particularly in nearshore marine systems that serve as important molting, migration, and wintering habitats, are likely to increase (Chapters 13 and 14, this volume). In general, we have an improved understanding of the effects of local- to regional-scale changes to habitats on sea ducks, such as shellfish aquaculture, offshore wind power development, oil and gas development and oil spills. These and other forms of anthropogenic habitat change will continue and require additional evaluation of effects on populations. The annual cycle model provides context for understanding how effects of habitat change may be manifested (Figure 15.1). Additional work on trophic interactions is needed to improve understanding of the importance of variation in nearshore marine systems, particularly of benthic prey, and potential for subsequent effects on sea duck population dynamics (Chapter 7, this volume). In addition to bottom-up effects of food resources, sea ducks can have strong top-down effects on nearshore marine systems in some circumstances. Numerous studies have documented significant depletion of invertebrate prey by sea ducks (Chapter 7, this volume). The work has tended to be at local scales and the broad applicability and generality of these top-down effects is an area that warrants additional research. Estimates of prey depletion and associated effects on populations would be a key step in assessing density-dependent relationships. Flint (2013) noted that sea duck species that preferred nearshore habitats appeared to be less influenced by oceanic regime shifts than their more pelagic counterparts. Thus, nearshore ecosystems that are likely to be influenced by anthropogenic effects may function differently than offshore areas, and sea ducks may play a more substantial part in ecosystem function than previously assumed. Quantifying the role of sea ducks within nearshore communities should be an area of high priority research, both for understanding sea duck

ecology and for identifying limiting factors that might be influenced by anthropogenic effects.

VITAL RATES

Compared with most groups of waterfowl, studies of demographic rates are not common for sea ducks, with the exception of measures of variation in productivity (Chapters 10 and 11, this volume). Few studies have examined survival or dispersal (Chapter 3, this volume), and fewer yet have considered the complete annual cycle for any population (Hario and Rintala 2006, Hario et al. 2009, Coulson 2010). Accordingly, use of population models for understanding dynamics of sea duck abundance has been limited (Chapter 3, this volume). We need additional studies that can demonstrate the range of annual variation in these vital rates. Importantly, these studies need to measure the full range of parameters to allow assessment of correlations among these rates.

RESEARCH PRIORITIES AND CHALLENGES FOR THE FUTURE

Sea ducks will continue to be taxa of research and conservation interest for a number of reasons. First, North American populations of Spectacled Eiders (*Somateria fischeri*) and Steller's Eiders (*Polysticta stelleri*) remain listed under the U.S. Endangered Species Act, and eastern populations of Harlequin Ducks (*Histrionicus histrionicus*) and Barrow's Goldeneyes (*Bucephala islandica*) are species of concern in Canada. Continental numbers of some sea duck species remain well below historical levels (Chapter 1, this volume). On an encouraging note, downward trends in scoters, eiders, Long-tailed Ducks, and other species of sea ducks have abated over the last decade. Nevertheless, the trends remain somewhat disconcerting as we do not understand why these populations declined or why these trends have reversed (Flint 2013). Some groups such as Buffleheads (*Bucephala albeola*), goldeneyes, and mergansers have shown persistent numerical increases at a continental scale. Continued documentation of these patterns, and refinement in the methods and resolution of detecting population trends will be important. Developing mechanistic understanding of underlying causes of population change is critical for advancing our field, but documenting change is only the first step toward gaining knowledge that can be used to understand and influence population dynamics.

In addition to general conservation interest in sea duck populations, numerous specific issues affect conservation of the birds of tribe Mergini, at spatial scales ranging from local to global (Chapter 14, this volume). Climate change is certain to have large effects on animal abundance and distributions. Given the distribution of sea ducks at high latitudes, and the greater effects of climate change observed and predicted for Arctic environments (Post et al. 2009), sea ducks will almost certainly be affected (D'Alba et al. 2010). The size, direction, and mechanisms of these effects will be of interest in terms of conservation of sea ducks and as a valuable metric of ecosystem change in the range of sea ducks. The growing body of literature on sea ducks reflected in our volume is already contributing to understanding these issues, and we foresee continuing need for work in this arena.

As a field, we have been successful at conducting research that addresses local- or regional-scale perturbations to habitats or populations of sea ducks. In most cases, we still cannot identify specific causes of broad-scale, long-term population change. However, correlations have been detected between population change and a variety of climatic and oceanographic signals (Flint 2013). These linkages are a critical advancement, as they offer potential explanatory value to understand continental-scale changes in abundance that have been observed in most sea duck species (Chapter 1, this volume). Still missing from the correlations is an understanding of the functional relationships that lead from climate variation to changes in vital rates and subsequently to variation in population dynamics. We think that understanding broad-scale underlying mechanisms leading to population change should be a key element of continued sea duck research, particularly given the relevance for conservation. To date, studies of Common Eiders at Coquet Island, England (Coulson 1999, 2010, 2013), and the Söderskär Archipelago, Finland (Hario and Rintala 2006, Hario et al. 2009), are perhaps the best examples of long-term research that have consistently collected such comprehensive data. To fully understand sea duck population dynamics, we recommend long-term research that measures the full range of demographic parameters. Such studies can then compare relationships between annual variation in life-history parameters and environmental conditions to begin to assess the relationships outlined in the overall model. Further, Flint (2013) demonstrated population changes associated with environmental shifts in oceanic regimes but noted that the actual vital rates that may have changed are unknown. Long-term studies would have the potential to contrast life-history parameters in a before-and-after design when future regime shifts happen. Natural experiments may provide the only tractable method toward a better understanding of population regulation for sea ducks. The key to understanding sea duck population dynamics will be to collect long-term data to assess patterns of variation in vital rates in relation to associated variation in external drivers. Under this scenario, each year provides a single sample for evaluating relative importance of anthropogenic and ecological factors as causes of demographic change and variation in vital rates and bird numbers as the population response, which emphasizes a need for long-term and replicated studies.

For migratory birds generally, including sea ducks, there is a growing realization that an understanding of population processes and effective conservation requires consideration of the full annual cycle. Like most Northern Hemisphere birds, research on sea ducks tends to be skewed toward studies of breeding biology, although studies during the breeding season only slightly outnumber those from other stages of the annual cycle. A balance in research focus is an encouraging finding, because it provides the basis for full life-cycle evaluations, which in turn can identify stages of the annual cycle that may serve as limiting factors to population stability or growth. Factors during one season can have carryover effects that influence performance during another season (Figure 15.1; Chapter 5, this volume). Carryover or cross-seasonal effects may be one mechanism by which varying ocean conditions could influence individual variation in breeding performance and, in turn, population dynamics. Continued consideration of the full annual cycle and cross-seasonal effects is a research area with great value for understanding population dynamics of sea ducks.

A central theme throughout this chapter is that quantifying demographic variation is critical for understanding sea duck population change and directing conservation efforts. Many species of sea ducks breed at low densities across large geographic areas. Selection of a study site at random, which would be most appropriate for generating unbiased estimates at the population scale, would almost certainly result in a small sample size for assessment of life-history parameters. Field sites selected at random would likely have difficult and expensive field logistics. Therefore, researchers tend to select areas with relatively high densities of birds or ease of field logistics, despite potential for results that are not representative at a population scale. Long-term work on Common Eiders by Coulson (2013) is one potential example, as Coquet Island is a site at the southern edge of the species distribution (Kear 2005). Yet the relatively easy field logistics is likely the only feasible way that such comprehensive, valuable, long-term data could be collected. Thus, available data represent a trade-off between random or representative samples and our ability to collect useful field data. We encourage both spatial and temporal replication of such studies to validate the application of results at population or landscape scales.

Studies of foraging, behavior, migration, and other topics in this volume have considerable value in their own right. Nevertheless, field studies that have linked these aspects of sea duck ecology to demographic performance have had the greatest influence on our understanding of how and why sea duck populations change. Research that can estimate the effects denoted by the arrows in our conceptual model will be most useful in understanding cause and effect relationships (Figure 15.1). Some aspects of sea duck demography remain a mystery. For example, variation in breeding propensity is almost surely a demographic response to environmental change in habitat conditions, but few studies have quantified this critical parameter for long-lived sea ducks (Coulson 2013). As another example, juvenile survival between fledging and age of first reproduction is essentially unknown, but may be a demographic parameter that would be highly responsive to changing conditions. Integrating and understanding the importance of various demographic attributes will require detailed population modeling. Recent

deployment of satellite transmitters in concert with phylogenetic, phylogeographic, and population genetics research has revolutionized our understanding of sea duck migration, providing insights into population delineation, relationships between breeding, molting, and wintering areas, and location of spring and fall staging areas (Chapters 2, 8, and 9, this volume). A greater resolution of sea duck movements will aid in defining populations, which is an essential step in developing and applying population models.

A significant, and rarely addressed, question in terms of understanding sea duck population dynamics is the role of density dependence (Hario and Rintala 2006). A lack of understanding is not surprising given that the role of density is poorly understood even for species that are well studied (Jamieson and Brooks 2004, Viljugrein et al. 2005, Murray et al. 2010). In the broad sense, density dependence represents the relationship between habitat conditions and population size that results in changes to vital rates and subsequent changes in population size. The most common way to assess density dependence is through examination of autocorrelation of population counts. However, the use of counts bypasses the underlying demographic processes that are driven by changes in vital rates and has numerous statistical issues (Lebreton 2009). Assessment of the relationship between population size and vital rates or correlations among vital rates and habitat characteristics is more direct but can be misleading if carrying capacity is not static. Carrying capacity for sea ducks may vary annually in a largely stochastic or irregular fashion (e.g., as influenced by ice conditions), cyclically in response to climate-driven cycles (i.e., El Niño), or as a trend associated with longer-term directional environmental change (e.g., global climate change). Also, the portion of the life cycle where density dependence occurs may vary between years and between populations adding another level of complexity. Under these conditions, detection of density dependence is difficult. Further, migratory species may be less likely to demonstrate density-dependent effects as they have more options than sedentary species. When habitat conditions are heterogeneous, birds may disperse from areas with poor conditions and thereby avoid direct density-dependent effects (Zipkin et al. 2010).

Flint (2013) demonstrated a correlation between oceanic regime shifts and population responses, implying a link between habitat conditions and population dynamics, which is a prerequisite for density-dependent regulation. Given that many sea duck populations have declined, one could assume that populations are well below carrying capacity and as such density dependence should not apply. Alternatively, the large-scale population declines may have been caused by broad-scale habitat changes and as such, the observed declines would represent a reduction in carrying capacity. Understanding the role of density dependence is essential for interpreting population change and defining management options. For example, if density dependence is occurring, then some fraction of harvest may be compensatory. Conversely, in the absence of density dependence, harvest is functionally additive. Further, management actions targeted at increasing demographic rates will not likely result in population change if density-dependent limitation is occurring during other portions of the annual cycle (Pehrsson and Nyström 1988). If density dependence is occurring, then management actions likely need to target habitat improvement during the limiting period. That is, under density-dependent regulation, population size can only increase through increased carrying capacity. Thus, understanding the role of density dependence for sea ducks is needed to decide if management should focus on habitat improvement or direct alteration of demographic parameters, such as decreased harvest or increased productivity through predator manipulation.

In sum, key research needs include large-scale, complete annual cycle studies with demographic and modeling components. The geographic, logistical, and technical scope of that type of research will require collaboration among sea duck researchers throughout North America, as well as direction of funding from a diversity of sources. Collaboration among research scientists working with sea ducks in North America, Asia, and Europe also will lead to benefits for populations that are shared among continents or flyways and for developing new insights from the rapid advancements in studies of sea duck biology that are occurring across the Northern Hemisphere.

LITERATURE CITED

Coulson, J. C. 1999. Variation in clutch size of the Common Eider: a study based on 41 breeding seasons on Coquet Island, Northumberland, England. Waterbirds 22:225–238.

Coulson, J. C. 2010. A long-term study of the population dynamics of Common Eiders Somateria mollissima: why do several parameters fluctuate markedly? Bird Study 57:1–18.

Coulson, J. C. 2013. Age-related and annual variation in mortality rates of adult female Common Eiders (Somateria mollissima). Waterbirds 36:234–239.

D'Alba, L., P. Monaghan, and R. G. Nager. 2010. Advances in laying date and increasing population size suggest positive responses to climate change in Common Eiders Somateria mollissima in Iceland. Ibis 152:19–28.

Flint, P. L. 2013. Changes in size and trends of North American sea duck populations associated with North Pacific oceanic regime shifts. Marine Biology 160:59–65.

Hario, M., M. J. Mazerolle, and P. Saurola. 2009. Survival of female Common Eiders Somateria m. mollissima in a declining population of the northern Baltic Sea. Oecologia 159:747–756.

Hario, M., and J. Rintala. 2006. Fledgling production and population trends in Finnish Common Eiders (Somateria mollissima mollissima)—evidence for density dependence. Canadian Journal of Zoology 84:1038–1046.

Jamieson, L. E., and S. P. Brooks. 2004. Density dependence in North American ducks. Animal Biodiversity and Conservation 27:113–128.

Kear, J. 2005. Ducks, geese and swans (vol. 2). Oxford University Press, Oxford, U.K.

Korschgen, C. E. 1977. Breeding stress of female Common Eiders in Maine. Journal of Wildlife Management 41:360–373.

Lebreton, J.-D. 2009. Assessing density-dependence: where are we left? Pp. 19–42 in D. L. Thomson, E. G. Cooch, and M. J. Conroy (editors), Modeling demographic processes in marked populations. Environmental and Ecological Statistics Series (vol. 3), Springer, New York, NY.

Merkel, F. R. 2010. Evidence of recent population recovery in Common Eiders breeding in western Greenland. Journal of Wildlife Management 74:1869–1874.

Murray, D. L., M. G. Anderson, and T. D. Steury. 2010. Temporal shift in density dependence among North American breeding duck populations. Ecology 91:571–581.

Pearce, J. P., J. A. Reed, and P. L. Flint. 2005. Geographic variation in survival and migratory tendency among North American Common Mergansers. Journal of Field Ornithology 76:109–118.

Pehrsson, O., and K. G. K. Nyström. 1988. Growth and movements of Oldsquaw ducklings in relation to food. Journal of Wildlife Management 52:185–191.

Post, E., M. C. Forchhammer, M. S. Bret-Harte, T. V. Callaghan, T. R. Christensen, B. Elberling, A. D. Fox et al. 2009. Ecological dynamics across the arctic associated with recent climate change. Science 325:1355–1358.

Viljugrein, H., N. C. Stenseth, G. W. Smith, and G. H. Steinbakk. 2005. Density dependence in North American ducks. Ecology 86:245–254.

Zipkin, E. F., B. Gardner, A. T. Gilbert, A. F. O'Connell Jr., J. A. Royle, and E. D. Silverman. 2010. Distribution patterns of wintering sea ducks in relation to the North Atlantic Oscillation and local environmental characteristics. Oecologia 163:893–902.

APPENDIX A: NORTH AMERICAN SEA DUCKS

Figure A.1. **(See color insert.)** (a) Barrow's Goldeneyes, (b) Common Goldeneyes, and (c) Buffleheads are cavity nesters that exhibit strong intra- and inter-specific territorial behavior, and differ in their distribution in North America. (d) Harlequin Ducks are river-breeding specialists that molt and forage along rocky, marine shores in late summer and winter. Photographs by (a) Gary Kramer, (b) Dirk Derksen, (c) Ian Routley, and (d) Jeff Coats.

Figure A.2. **(See color insert.)** All four eider species winter and molt in marine waters. (a) King, (b) Spectacled, and (c) Steller's Eiders breed on freshwater lakes; (d) the Common Eider is the only true colonial sea duck. Photographs by (a) Ken Wright, (b) Ted Swem, (c) Gary Kramer, and (d) Jeff Coats.

(a)   (b)

(c)   (d)

Figure A.3. **(See color insert.)** Most scoters breed on freshwater lakes and molt and winter in salt waters where they form large flocks feeding mostly on clams and mussels. (a,b) The Surf Scoter is endemic to North America, (c) the Black Scoter is now recognized as a full species distinct from the Common Scoter of Europe, and (d) the White-winged Scoter occurs in both Europe and North America. Photographs by (a) Brian Uher-Koch, (b) Milo Burcham, (c) Ryan Askren, and (d) Gary Kramer.

(a)   (b)

(c)   (d)

Figure A.4. **(See color insert.)** Mergansers differ by their habitat preferences: (a) the Red-breasted Merganser is associated with salt water, whereas (b) Hooded and (c) Common Mergansers exploit fresh waters for breeding. All mergansers forage primarily on fishes. (d) The Long-tailed Duck breeds in northern ponds and coastal regions and is pelagic throughout the winter. It is the only sea duck with distinct winter and breeding plumages. Photographs by (a,b) Gary Kramer, (c) Ian Routley, and (d) Ryan Askren.

INDEX

A

acanthocephalans, 111–112
adenoviruses, 100–101
age ratios, 428, 429 (table)
age-structured population model, 66, 67 (fig.)
Alaska Arctic Coastal Plain Breeding Waterfowl
        survey, 6
Alaska Migratory Bird Co-Management Council (AMBCC),
        446
Alaska, subsistence harvest
    Bufflehead, 455
    considerations, 456, 459–460
    data sources, 449–451, 450 (table), 451 (table)
    Duck, Harlequin, 454
    Duck, Long-tailed, 455
    Eider, Common, 454
    Eider, King, 454
    Eider, Spectacled, 454
    Eider, Steller's, 454
    Goldeneyes, 455
    Mergansers, 455
    patterns, 450 (table), 452–453, 452 (fig.), 453 (fig.)
    regional distribution, 455–456
    regional management topics, 456, 457–459 (table)
    Scoters, Black, 454–455
    Scoters, Surf, 454–455
    Scoters, White-winged, 454–455
    species identification, 451–452
algal toxins, 113–114
alien/invasive species, 503–504
allometry, 255–256, 256 (fig.)
annual flight paths, 287–288 (fig.) (table)
annual survival and breeding site fidelity, 343–344
anthropogenic activities
    alien/invasive species, 503–504
    clear-cutting and other silviculture methods, 501
    contaminants, 500
    ecological drivers, 563–564
    hydroelectric development, 502–503
    sewage and seafood-processing effluent, 500
    shellfish industry, 499
    wind turbine facilities, 501–502
anticholinesterase pesticides, 221
apparent realized fitness gradient, 86, 87 (fig.)
Arctic Monitoring and Assessment Programme, 204
Asian clam (*Potamocorbula amurensis*), 542
aspergillosis, 109
Atlantic coast fall-winter traditions, 429–430
    Ducks, Harlequin, 431
    Ducks, Long-tailed, 430–431
    Eiders, 430–431
    fall and winter harvest, 432–435, 432 (fig.), 433 (fig.),
        434 (fig.)
    Goldeneye, Barrow's, 431
    hunting activity, 431–432
    Scoters, 430–431
    sex and age composition, 435–436, 435 (table)
Audubon Christmas Bird Counts, 6
avian cholera, 106–108
avian influenza virus, 102–104
avian malaria, 111
Avipoxvirus, 104

B

bacteria
    avian cholera, 106–108
    chlamydiosis, 108
    colibacillosis, 108
    description, 105
    Erysipelothrix rhusiopathiae, 108–109
    Mycobacterium avium, 109
basal metabolic rate (BMR), 251
biochemical markers, 248
biological toxins
    algal toxins, 113–114
    avian botulism, 112–113
biomagnification, 170
bioremediation, 205
blood metabolites, 152
body composition analysis, 148–149

body mass, disadvantages, 145
body reserves, 250–251
body size, 339–343
botulism, 112–113
breeding, capital, 144
breeding habitats
    Buffleheads, 487–489
    Duck, Long-tailed, 483–484
    Ducks, Harlequin, 478–479
    Eiders, Common, 471–472
    Eiders, King, 471
    Eiders, Spectacled, 471
    Eiders, Steller's, 470–471
    Goldeneyes, 487–489
    Mergansers, Common, 491–492
    Mergansers, Hooded, 492–493
    Mergansers, Red-breasted, 492
    Scoters, 480–481
breeding, income, 144
breeding propensity, 368–372 (table), 375
breeding systems
    courtship behavior, 374–375
    delayed breeding, 367, 369–372 (table)
    forced copulations, 369–372 (table), 374
    mating systems, 367, 369–372 (table)
    pair bonds, 367, 369–372 (table), 373
    philopatry and breeding propensity, 369–372 (table),
        375
    timing of pairing, 369–372 (table), 373–374
British Columbia Coastal Waterbird survey, 6
brood amalgamation
    accidental mixing, 400–401
    adaptive behavior, 401–405
    ecological correlates, 386 (fig.), 400
    occurrence and frequency, 369–372 (table), 396,
        397–399 (table)
    relative role of relatives, 406
    viewpoints, 405–406
brood biology, 356
brood parasitism
    conspecific
        conceptual framework, 369–372 (table), 381–383
            (table), 386 (fig.), 387
        ecological correlates, 369–372 (table), 384 (fig.),
            385–387
        Eiders, Common, 390–393
        Goldeneyes, 388–391
        host and population dynamics, 393–394
        occurrence and frequency, 369–372 (table),
            379–380, 381–383 (table), 384 (fig.), 385
        relative role of relatives, 369–372 (table),
            387–388
        research needs, 394–395
    interspecific
        ecological correlates, 369–372 (table), 386 (fig.),
            395–396
        occurrence and frequency, 369–372 (table), 395
brood-rearing behavior, 352–353, 470–471
brood spacing behaviors and brood territories, 377–378
Brown trout (Salmo trutta), 546
Bucephala
    B. albeola (see Buffleheads (Bucephala albeola))
    B. angustipes, 44
    B. fossilis, 44
    B. ossivalis, 44

Buffleheads (Bucephala albeola), 31, 564
    abundance, 4
    breeding habitats, 487–489
    cavity-nesting, 531
    dive depth and duration, 242
    duck plague, 99
    eggs, 134
    fall and spring staging habitats, 489–490
    flightless period, 310
    lead poisoning, 174
    molting habitats, 489
    mtDNA haplotype networks, 44–45 (fig.)
    reproduction, 340
    timing of spring migration, 269
    wintering habitats, 490–491

C

cadmium, 201–203
calcium, 147
Canada, subsistence harvest
    data sources, 447
    patterns, 447–448
    regional management topics, 448–449
Canadian Wildlife Service (CWS) harvest survey
    1967 to present, 422
    strengths and weaknesses, 422–423
capital breeding, 144
carbohydrates, 147
cavity-nesting sea ducks, 531
CBP. See conspecific brood parasitism (CBP)
chlamydiosis, 108
Chlamydophila psittaci, 108
chlordane, 210
chromium, 203
climate change, 565
    large-scale regime shifts, 535–536
    mismatch in timing, 536–537
    predator–prey interactions, 537
Clostridium botulinum, 112–113
clutch production
    clutch mass, 134, 136, 137 (fig.)
    clutch size, 133 (fig.)
    North American sea ducks, 130–131 (table)
coccidia, 110
colibacillosis, 108
commercial fisheries and aquaculture, 545–546
compound-specific stable isotope analysis (CSIA), 248
conspecific brood parasitism (CBP), 366
    conceptual framework, 387
    ecological correlates, 369–372 (table), 385–387, 386
        (fig.)
    Eiders, Common, 390–393
    Goldeneyes, 388–391
    host and population dynamics, 393–394
    occurrence and frequency, 369–372 (table), 379–383,
        384 (fig.), 385
    relative role of relatives, 387–388
    research needs, 394–395
contaminants, 534–535
    heavy metals, 158
    organic, 158–159
    trace elements, 158
copper, 204
Coquet Island, 565–566

cost of flight, 252–253, 253 (fig.)
cost of foraging, 254, 255 (fig.)
courtship behavior, 374–375
cross-seasonal nutritional effects
    climatic variation, 127
    oceanic conditions, 127
Cuckoo, European (*Cuculus canorus*), 379

D

dabbling ducks, 373
daily energy expenditure (DEE), 251
daily energy requirements (DER), 255–256, 256 (fig.)
delayed breeding, 367, 369–372 (table)
delta-aminolevulinic acid dehydratase (ALAD) assay, 173
density-dependent effects, 87–89, 566
deterministic sea duck matrix population model, 66–68
dichlorodiphenyldichloroethane (DDD), 210
diet composition, 246–247, 247 (fig.)
disturbance, 539–540
diving behavior. *See* foraging ecology
double-observer methods, 3
duck factory, 539
duckling–duckling bonds, 401
duckling survival
    brood size, 403
    bursal infections and immunosuppression, 102
    definition, 68
    fertility variation, 81
    first-year survival, 78
    growth rate, 77
    lower-level elasticity estimates, 74
    residual body reserves, 128
duck plague, 99–100
Ducks, Blue (*Hymenolaimus malacorhynchos*), 373
Ducks, Harlequin (*Histrionicus histrionicus*), 31
    Atlantic coast fall-winter traditions, 431
    breeding habitats, 478–479
    breeding population, 18–20
    dive depth and duration, 242
    EROD activity, 207–208
    origins, 47–49
    Pacific coast fall-winter traditions, 441–442
    population, 529
    spring migration routes, 272
    wintering habitats, 479–480
Ducks, Labrador (*Camptorhynchus labradorius*), 31, 51
Ducks, Long-tailed (*Clangula hyemalis*), 31
    adenoviruses, 563
    Atlantic coast fall-winter traditions, 430–431
    breeding habitats, 483–484, 533
    breeding population, 16–18
    dive depth and duration, 242
    fall and spring staging habitats, 485–486
    molting habitats, 484–485
    natural history and contaminants, 171–172
    origins, 49–50
    Pacific coast fall-winter traditions, 442
    spring migration routes, 272, 533
    wintering habitats, 486–487
Ducks, Muscovy (*Cairina moschata*), 46
Ducks, Tufted (*Aythya fuligula*), 103
Ducks, Wood (*Aix sponsa*)
    avian influenza virus, 103
    cavity-nesting species, 376

renal pathology, 202
reproduction, 340
dynamics of habitats
    alien/invasive species, 503–504
    clear-cutting and silviculture methods, 501
    contaminants, 500
    forest fires, 497–498
    hydroelectric development, 502–503
    natural events, 496–497
    role of climate events, 504–507
    role of fish, 498–499
    sewage and seafood-processing effluent, 500
    shellfish industry, 499
    water variability, 497
    wind turbine facilities, 501–502

E

Eastern birds, 49
Eastern Waterfowl Breeding Survey, 4–5
eclipse plumage, 308
egg drop syndrome virus, 101
egg formation, 128
egg production costs, 128–129
    composition of egg, 134, 135 (table)
    egg-laying rates, 134, 136 (fig.)
    egg mass and mean female body mass, 132, 133 (fig.)
    eiders and non-eiders, 129
    North American sea ducks, 130–131 (table)
    nutrient production, 137
    nutrient requirements, 134
    RFG, 129
egg size, 339–343
Eiderdown, 546
Eiders
    Atlantic coast fall-winter traditions, 430–431
    Common, 41 (fig.)
    King, 40–41
    origins, 39–40
    Polysticta and *Somateria*, 33
    Spectacled, 40
    Steller's, 41–42
Eiders, Common (*Somateria mollissima*), 272
    breeding habitats, 471–472
    breeding population, 9–14
    cadmium concentration, 203
    conspecific brood parasitism, 390–393
    copper concentration, 204
    dive depth and duration, 242
    eggs, 340
    fall and spring staging habitats, 475–476
    harvest reduction, 563
    molting habitats, 474
    Polycyclic aromatic hydrocarbons (PAHs), 206
    population model, 84, 85 (fig.)
    population viability, 530
    selenium concentration, 197–198
    wintering habitats, 477–478
Eiders, King (*Somateria spectabilis*), 32
    breeding habitats, 471
    breeding population, 14
    delayed reproduction cost, 269
    dive depth and duration, 242
    duck plague, 100
    fall and spring staging habitats, 475

molting habitats, 473–474
satellite telemetry, 532
wintering habitats, 477
Eiders, Spectacled (*Somateria fischeri*), 7–8, 273
breeding habitats, 471
breeding population, 7–8
dive depth and duration, 242
fall and spring staging habitats, 475
federally threatened species, 529
incubation constancy, 145
molting habitats, 473
North American populations, 564
wintering habitats, 476
Eiders, Steller's (*Polysticta stelleri*)
breeding habitats, 470–471
breeding population, 8–9
fall and spring staging habitats, 474–475
federally threatened species, 529
molting habitats, 473
North American populations, 564
wintering habitats, 476
elasticity, population dynamics
lower-level, 72
vs. sensitivity, 70–71, 73–74
endogenous nutrient, 152–155, 154 (table), 155 (fig.)
energetic requirements, breeding, 128
energy and nutrient supply assessment
techniques
blood metabolites, 152
body composition analysis, 148–149
fatty acid analysis, 151–152
stable-isotope analysis, 149–151, 150 (table)
energy balance approach, 248, 249 (fig.)
energy cost elements
cost of flight, 252–253, 253 (fig.)
cost of foraging, 254, 255 (fig.)
daily energy requirements and allometry,
255–256, 256 (fig.)
individual-based modeling, 257–258
maintenance energy costs, 245 (table), 251–252,
252 (fig.), 253 (table)
thermoregulation, 254–255
threshold prey densities, 256–257
energy gain elements
body reserves, 250–251
energy intake, 249–250
environment
changes, 294
day length, 321
ice, 320–321
migration, 291–292
Erysipelothrix rhusiopathiae, 108–109
Escherichia coli, 108, 500
ethnotaxonomy, 459
7-ethoxyresorufin-O-deethylase (EROD), 206
Eurasian perch (*Perca fluviatilis*), 541
evolutionary significant units (ESUs), 51
extra-pair copulations (EPCs), 374
Exxon Valdez oil spill, 207–208

F

fall and spring staging habitats
Buffleheads, 489–490
Ducks, Long-tailed, 485–486
Eider, Common, 475–476
Eider, King, 475
Eider, Spectacled, 475
Eider, Steller's, 474–475
Goldeneyes, 489–490
Merganser, Common, 494–495
Merganser, Hooded, 495
Merganser, Red-breasted, 495
Scoters, 481–482
fall migration
flight behavior, 286
interannual constancy, 286
routes, 285–286
stopover behavior, 286
timing, 285
variation, sexes and age groups, 286–287
fatty acid analysis, 151–152
fecundity enhancement, 387
female aggressiveness, 401
female-biased philopatry, 373
female–duckling bond strength, 401
fish spawning events, 241
flight feathers, 305–308, 311, 314, 317, 320–322, 326, 327
food habits, 98
food resources, 317
forage items, 317, 318–319 (table)
foraging ecology
behavior, 317–320
biochemical markers, 248
diet composition, 246–247, 247 (fig.)
diet studying methods, 247–248
dive depth and duration, 242–243, 243 (fig.)
foraging effort, 243–245, 244 (fig.), 245
(table)
nocturnal foraging, 245–246
foraging energetics, 248–249, 249 (fig.)
foraging habitat requirements, 258
forced copulations, 369–372 (table), 374
forestry, 544–545
fungi, 109

G

Geese, Barnacle (*Branta leucopsis*), 373
Geese, Egyptian (*Aloplochen aegyptiaca*), 106
Geese, Pink-Footed (*Anser brachyrhynchus*), 291
Geese, Spur-winged (*Plectropterus gambensis*), 106
generic sea duck, 68
glutathione metabolism, 196
glycogen, 147
Godwits, Black-tailed (*Limosa limosa*), 532
Goldeneyes, Barrow's (*Bucephala islandica*), 21, 31
Atlantic coast fall-winter traditions, 431
breeding population, 20–21
duck plague, 99
Eastern populations, 564
Pacific coast fall-winter traditions, 440–441
phylogenetic reconstruction, 34
special concern species, 529
spring migration routes, 272
Goldeneyes, Common (*Bucephala clangula*),
44–45 (fig.)
accidental parasitism hypothesis, 390
breeding habitats, 487–489
breeding population, 20–21

egg-laying sequence, 389
fall and spring staging habitats, 489–490
fecundity enhancement, 390
female philopatry, 388
frequency of parasitism, 390
incubation constancy, 145
molecular genetic techniques, 390
molting habitats, 489
nest-site characteristics, 389
non-nesting parasites, 391
predation, 388
protein fingerprinting technique, 389
spring stopover behavior, 274
timing of pairing, 373
wintering habitats, 490–491
Greater Scaup (*Aythya marila*), 44

H

habitat
    biotic and abiotic features, 538
    cavity-nesting ducks, 538
    conservation planning, 259
    duck factory, 539
    herring spawn, 537
    management and protection, 507
    marine wildlife areas, 539
    northern and coastal regions, 533–534
    protection, 530
    telemetry, 537
    U.S. National Wildlife Refuge System, 538
Harriers, Marsh (*Circus aeruginosus*), 291
harvest management
    Atlantic coast fall-winter traditions, 429–436
    authorities, 418
    Canadian Wildlife Service Harvest survey,
        422–423
    issues and information gaps, 460–461
    northern subsistence traditions, 444–460, 445 (fig.),
        450 (table), 451 (table), 452 (fig.), 453 (fig.),
        457–459 (table)
    Pacific coast fall-winter traditions, 436–444, 439
        (table), 440 (fig.)
    state surveys, 423–424, 423 (table)
    ubiquitous species, 425–429
    U.S. Fish and Wildlife Service Harvest survey,
        419–422
    wounding loss, 424–425
Hawk, Sharp-shinned (*Accipiter striatus*), 35
helminths, 111
hemolytic anemia, 206
hemoparasites, 110–111
heterozygosity, 38
homozygosity, 38
host availability, 385–386
hydroelectric power projects, 543–544
hypothetical fitness gradient, 86 (fig.)

I

income breeding, 144
incubation, 348–349
    clutch mass, 138, 142 (fig.)
    constancy, 144

endogenous nutrient use during, 152–155,
        154 (table), 155 (fig.)
    incubation constancy, 139, 143 (fig.)
    interspecific patterns of, 139 (fig.), 140–141 (table)
indicated breeding pairs (IBPs), 491
individual variation, 290–291
infectious bursal disease virus (IBDV), 102
inter- and intra-annual stopover site fidelity,
        288–290 (fig.)
International Union for Conservation of Nature
        (IUCN), 539
interspecific brood parasitism
    ecological correlates, 395–396
    occurrence and frequency, 395
ionizing radiation, 222

L

lead poisoning
    delta-aminolevulinic acid dehydratase, 173
    shotgun pellets, 172, 174
    suspected sea duck death, 176
Leslie style matrix model, 68–69, 68 (table)
lipid reserves, 250–251
little-known species, 356
long-lived birds, 356, 356.
lower-level elasticities, 72, 72 (fig.)
lowest observed adverse effect level (LOAEL), 220

M

Mallard (*Anas platyrhynchos*), 46
    band recovery rates, 562
    duck plague, 99
    moisture content, 171
    nickel concentration, 203
    reproduction, 338
    zinc poisoning, 204
management units (MUs), 51
marine wildlife areas (MWAs), 539
Maryland Department of Natural Resources
        (MDNR), 423
mating systems, 367, 369–372 (table)
Merganser, Auckland Islands (*Mergus australis*), 31
Mergansers
    breeding population, 21–22
    Common, 46–47
    fossils, 45
    Hooded, 47
    origins, 45–46
    Red-breasted, 47
Mergansers, Common (*Mergus merganser*), 22
    breeding habitats, 491–492
    cavity-nesting, 531
    diet composition, 246
    duck plague, 99
    fall and spring staging habitats, 494–495
    flightless adult female, 309
    hunting regulations, 425–426
    molting habitats, 494
    phylogenetic analysis, 33
    population model, 84–85, 86 (fig.)
    spring migration variation, 277
    wintering habitats, 495–496
    zinc poisoning, 204

Mergansers, Hooded (*Lophodytes cucullatus*), 22
  breeding habitats, 492–493
  cavity-nesting, 531
  dive depth and duration, 242
  duck plague, 100
  environmental changes, 294
  fall and spring staging habitats, 495
  hunting regulations, 426
  molting habitats, 494
  pairing chronology, 373
  roundest eggs, 342
  wintering habitats, 496
Mergansers, Red-breasted (*Mergus serrator*), 22
  breeding habitats, 492
  diet composition, 246
  duck plague, 100
  fall and spring staging habitats, 495
  hunting regulations, 425–426
  molting habitats, 494
  molt timing, 308
  spring migration routes, 272
  wintering habitats, 496
Mergansers, Scaly-sided (*Mergus squamatus*), 45
Mergini, 30, 35, 48, 338, 340
  brood amalgamation, 400–401,
      405–406
  CBP, 379, 385, 388
  forest fires, 497
  goldeneyes, 376
  molecular data, 33–34
  morphology and behavior, 31–33
  mortalities, 109
  uniqueness, 293
methylmercury, 199
migration
  environment, 291–292
  sea ducks, 268–269
  speed and duration, 292–293
  strategy, 267–268
  waterfowl, 268
Migratory Birds Convention Act, 538
Migratory Bird Treaty Act, 538
molt habitats
  breeding birds and ducklings, 322
  coastal molting areas, 324–325
  Ducks, Harlequin, 325
  flocks, 325
  freshwater wetlands and lakes, 322
  large inland shallow arctic lakes, 326
  Scoters, 325
  seasonal differences, 326
  winter, 322–324
molting flocks, 327
molt migration, 315–317
  flight behavior, 283
  interannual constancy, 284
  routes, 278–283 (fig.) (table)
  stopover behavior, 283
  timing, 277–278
  variation, sexes and age, 284–285
monotypic sea ducks
  Duck, Harlequin, 47–49
  Duck, Long-tailed, 49–50
Mussel, Zebra (*Dreissena polymorpha*), 294
Mycobacterium avium, 109

N

natal philopatry, 355
nesting
  boxes, 345–346
  climate effects, 346–347
  competition, 386
  islands, 350–351
  local environment, 346
  location, 349–350
  microhabitats, 344–345
  nearest water, 345
  other species, 351
  predators, 347–348
  site selection, 344
  success, 346
nest loss, 386
nest-site availability, 385
nest-site quality, 386
nickel, 203–204
nocturnal foraging, 245–246
North American Waterfowl Management Plan (NAWMP),
    259, 530
North Atlantic Oscillation, 127
Northern Gadwall (*Anas strepera*), 308
Northern Pintails (*Anas acuta*), 103, 291, 338, 376, 429
northern subsistence traditions
  history of regulations and public involvement,
      444–446
  subsistence economies, 444, 445 (fig.)
  subsistence harvest, Alaska, 449–456, 450 (table),
      451 (table), 452 (fig.), 453 (fig.), 457–459 (table)
  subsistence harvest, Canada, 447–449
Northwest Scoters, 442
Nunavut Wildlife Harvest Survey, 446
nuptial plumage, 308
nutrient/energy sources
  calcium, 147
  carbohydrates, 147
  high body mass, disadvantages, 145
  incubation constancy, 144
  limitations, 157
  protein, 146–147
  somatic fat, 146
  timing and location, 155–157
nutrition, 126

O

oceanic regime shifts, 567
offshore wind energy, 544
oil and gas resources, 542–543
organophosphorus and carbamate pesticides, 221
ornithosis, 108
orthoreovirus, 101–102
oviduct, 129

P

Pacific coast fall-winter traditions
  Alaska, 436–437
  coastal environments and hunting, 444
  Duck, Harlequin, 441–442
  Ducks, Long-tailed, 442
  fall and winter harvest, 438–440, 439 (table)

Goldeneye, Barrow's, 440–441
    hunting activity, 437–438, 438 (fig.), 438 (table)
    Northwest Scoters, 442
    Pacific Flyway, 436
    sea duck conservation concerns, 443–444
Pacific Flyway, 436
Pacific herring (*Clupea pallasi*), 274, 291
pair bonds, 367–373
paleontology, 31, 35
parasites, 534–535
    acanthocephalans, 111–112
    definition, 109
    helminths, 111
    protozoan, 110–111
parental care, 338
Pasteurella multocida, 106–107
pathogens, 534–535
Penguin, Emperor (*Aptenodytes forsteri*), 151
peripheral populations, 38
persistent organic pollutants (POPs)
    characteristics, 209
    PCBs, 210
    PCDDs and PCDFs, 210–211
    sources, 209
    tissues of sea ducks, 212–218
petroleum
    adverse effects, 205–206
    EROD activity, 206–207
    external oiling, 206
    oil spills, 205
philopatry, 369–372 (table), 375
phylogenetics
    molecular, 33–34 (fig.)
    relationships, 30–31 (fig.)
phylogeography
    effective population size, 38
    general genetic patterns, 34–37 (fig.)
    genetic structure, 37–38
    population structure, 38
Pleistocene, Early, 45
plumages, 306–307
Pochard (*Aythya ferina*), 44, 103
polychlorinated dibenzofurans (PCDFs), 210–211
polychlorinated dibenzo-p-dioxins (PCDDs), 210–211
polycyclic aromatic hydrocarbons (PAHs), 206
polygynous social systems, 373
POPs. *See* persistent organic pollutants (POPs)
population delineation, 531–532
population dynamics, 562
    balance equation, 66
    definition, 63
    density-dependent mechanism, 87–89, 91
    deterministic sea duck matrix population model, 66–68
    Eiders, Common, 84
    elasticity, 69–71
    female-only model, 65
    lambda, 65
    lower-level elasticities, 72
    management scenarios, 82–83, 91
    Mergansers, Common, 84–85
    model parameters, 68–69
    negative correlations, 79–80
    nonlinear responses, 83
    positive correlations, 77–79
    prospective population projection, 66

response to management actions, 90–91
    retrospective population modeling, 80–82
    sea duck management, 89–90
    sensitivity, 72–73
    stochastic matrix population model, 74–77
    two-sex model, 65–66
    uses, 65
population relationship
    Buffleheads, 44–45
    Eiders, 39–42
    Goldeneyes, 44–45
    Mergansers, 45–47
    Scoters, 42–44
population status and trends
    Alaska Arctic Coastal Plain Breeding Waterfowl
        survey, 6
    American Common Eider, 10–12
    Atlantic Black Scoter, 16
    Audubon Christmas Bird Counts, 6
    British Columbia Coastal Waterbird survey, 6
    Buffleheads, 21
    Duck, Harlequin, 18–20
    Duck, Long-tailed, 16–18
    Eastern Waterfowl survey, 4–5
    Eiders, Common, 9
    Eiders, King, 14
    Eiders, Spectacled, 7–8
    Eiders, Steller's, 8–9
    Goldeneyes, 20–21
    Hudson Bay Common Eider, 12–13
    Mergansers, 21–22
    Northern Common Eider, 13–14
    Pacific Black Scoter, 6, 16
    Pacific Common Eiders, 9–10
    Scoter, Black, 16
    Scoter, Surf, 16
    Scoter trends, 7
    Scoter, White-winged, 16
    trend estimate, 7
    Waterfowl Breeding Population and Habitat survey,
        3–4
    WBPHS trend analysis, 6–7
    Yukon–Kuskokwim Delta Nest survey, 5–6
population structure, 38
poxviruses, 104–105
pre- and posthatch brood amalgamation, 561
prebreeding phase, nutritional conditioning, 126
predation intensity variation, 351–352
predator–prey interactions, 537
predators, 540–541
prenesting phase, nutritional conditioning, 126
pre-rapid follicular growth (pre-RFG) stage, 128–129
prospective population projection, 66
protein, 146–147, 202
protoporphyrin assay, 173
psittacosis, 108

R

radiation, 221–223
Rainbow trout (*Oncorhynchus mykiss*), 541
rapid follicular growth (RFG), 128–129
Redheads (*Aythya americana*), 46
    neurological signs, 103
    somatic fat, 146

remigial molting
    Buffleheads, 489
    competition, 321–322
    data gaps, 326–327
    definition, 305
    Ducks, Long-tailed, 484–485
    Eider, Common, 474
    Eider, King, 473–474
    Eider, Spectacled, 473
    Eider, Steller's, 473
    energetics, 311–314
    energy management strategy, 314–315
    environmental constraints, 320–321
    flight feathers, 307–308
    food resources, 317
    foraging behavior and ecology, 317–320
    Goldeneyes, 489
    habitats (see molt habitats)
    international concerns, 327
    interspecific variation, 310
    intraspecific variation, 308–310
    length of period, 310–311
    length of stay, 311
    Merganser, Common, 494
    Merganser, Hooded, 494
    Merganser, Red-breasted, 494
    migration, 315
    molting areas vs. breeding and wintering locations,
        315–317
    movements and fidelity, 317
    nutritional demands, 305
    plumages, 306–307
    predation risk, 320
    response to disturbance, 322
    scoters, 481
reoviruses, 101
reproductive strategies, 337
resource exploitation
    commercial fisheries and aquaculture,
        545–546
    eiderdown, 546
    forestry, 544–545
    hydroelectric, 543–544
    offshore wind energy, 544
    oil and gas, 542–543
resting metabolic rate (RMR), 251
retrospective population modeling, 80–82

S

Sarcocystis sp., 110
satellite telemetry, 40, 46, 48–50, 531–533, 538, 544
Scaup, Lesser (Aythya affinis)
    contaminant burdens, 504
    dive costs and intake rates, 256
    selenium, 158
    WNV, 105
Scoters
    Atlantic coast fall-winter traditions, 430–431
    Black, 43–44
    breeding habitats, 480–481
    molting habitats, 481
    origins, 42–43
    White-winged and Surf, 43 (fig.)
    wintering habitats, 482–483

Scoters, Black (Melanitta americana), 32
    breeding population, 16
    breeding survey, 6
    fall migration, 286
Scoters, Common (Melanitta nigra)
    CBP, 379
    flight behavior, 274
    foraging areas, 540
Scoters, Surf (Melanitta perspicillata), 32
    breeding population, 16
    foraging effort, 244
    nocturnal foraging, 245
    population delineation, 532
    spring migration, 269
Scoters, Velvet (Melanitta fusca), 535
Scoters, White-winged (Melanitta fusca), 32
    breeding population, 16
    incubation constancy, 145
    migration, 273
    nocturnal foraging, 245
sea duck demography, 562
sea duck ecology, 562
sea duck reproduction
    environmental cues, 352
    interannual reproductive success
        anthropogenic effects, 355–356
        tribe, 353–355
    vs. other waterfowl
        breeding, 338–339
        Ducks, Long-tailed , 338
        Eiders, Spectacled, 338
        Mergini, 338, 340
        Scoters, Black, 339
seasonal movements
    fall migration, 285–287
    molt migration, 277–285
    spring migration, 269–277
selenium, 196–199
selenomethionine, 196
sensitivity, population dynamics
    changes in life history parameters, 72–73
    vs. elasticity, 70–71, 73–74
sewage and seafood-processing effluent, 500
sex ratios, 367, 368–372 (table), 428, 429 (table)
social monogamy, 367
somatic fat, 146
spacing behavior
    brood spacing behaviors and brood territories,
        377–378
    spring spacing behavior and pair territories, 375–377
    winter spacing strategies and winter
        territories, 378–379
spring migration
    flight behavior, 274–276 (table)
    interannual constancy, 276
    nutrient reserve acquisition and, 272–274
    routes and stopover sites, 272, 273 (fig.)
    stopover behavior, 274
    timing, 269–272 (table)
    variation, sexes and ages, 276–277
spring spacing behavior and pair territories
    Duck, Wood, 376
    interspecific territoriality, 377
    mating systems, 376
    Northern Pintails, 376

territoriality, 376
unpaired courting birds, 375
stable-isotope analysis
diet and egg components, 150
endogenous nutrients, 151
endogenous *vs.* exogenous sources, 150
limitations, 149
marine *vs.* freshwater sources, 151
technique, 149
stochastic matrix population model
average lambda, 76–77, 76 (fig.)
input values, 76 (table)
life history strategies, 75
nesting success, 75
process variance, 75
subsistence harvest
Alaska (*see* Alaska, subsistence harvest)
Canada, 447–449

T

taxonomy, 30–34 (fig.)
thorny-headed worms, 111
threshold prey densities, 256–257
timing of pairing, 369–372 (table), 373–374
trace elements, metals and, 177–195
cadmium, 201–203
chromium, 203
contaminants, 158
copper, 204
lead, 172–176
mercury, 199–201
nickel, 203–204
selenium, 196–199
zinc, 204
tributyltin (TBT), 204, 211, 219
trivalent chromium, 203
true yolk, 129
two-sex model, 65–66

U

U.S. Fish and Wildlife Service harvest survey (USFWS)
harvest estimates, 421
1952–2001, 419–420
1999 to present, 420–421
sample frames and sampling, 421
wing surveys, 421–422
U.S. Geological Survey–National Wildlife Health
Center, 534
U.S. National Wildlife Refuge System, 538

V

variance decomposition approach, 81, 82 (table)
Varnish clams (*Nuttallia obscurata*), 541
viruses
adenoviruses, 100–101
avian influenza, 102–104
duck plague, 99–100
IBDV, 102
orthoreovirus, 101–102
pathogenicity, 99
poxviruses, 104–105
Wellfleet Bay, 104
WNV, 105
vulnerability index model, 171–172

W

Washington Department of Fish and Wildlife (WDFW), 423
Waterfowl Breeding Population and Habitat Survey
(WBPHS), 3–4, 6–7, 532
Wellfleet Bay virus, 104
West Nile virus (WNV), 105
wing surveys, 421–422
wintering habitats
Buffleheads, 490–491
Ducks, Harlequin, 479–480
Ducks, Long-tailed, 486–487
Eider, Common, 477–478
Eider, King, 477
Eider, Spectacled, 476
Eider, Steller's, 476
Goldeneyes, 490–491
Mergansers, Common, 495–496
Mergansers, Hooded, 496
Mergansers, Red-breasted, 496
Scoters, 482–483
winter spacing strategies and winter territories, 378–379
WNV. *See* West Nile virus (WNV)
wounding loss, 424–425

Y

Yukon–Kuskokwim Delta Aerial Goose–Duck–Waterbird
Survey, 4–5
Yukon–Kuskokwim Delta Nest Survey, 5–6

Z

Zebra mussels (*Dreissena polymorpha*), 356, 541
zinc, 204

STUDIES IN AVIAN BIOLOGY

Series Editor: Brett K. Sandercock

1. Status and Distribution of Alaska Birds. Kessel, B., and D. D. Gibson. 1978

2. Shorebirds in Marine Environments. Pitelka, F. A., editor. 1979

3. Bird Community Dynamics in a Ponderosa Pine Forest. Szaro, R. C., and R. P. Balda. 1979

4. The Avifauna of the South Farallon Islands, California. DeSante, D. F., and D. G. Ainley. 1980.

5. Annual Variation of Daily Energy Expenditure by the Black-billed Magpie: A Study of Thermal and Behavioral Energetics. Mugaas, J. N., and J. R. King. 1981.

6. Estimating Numbers of Terrestrial Birds. Ralph, C. J., and J. M. Scott, editors. 1981.

7. Population Ecology of the Dipper (Cinclus mexicanus) in the Front Range of Colorado. Price, F. E., and C. E. Bock. 1983.

8. Tropical Seabird Biology. Schreiber, R. W., editor. 1984.

9. Forest Bird Communities of the Hawaiian Islands: Their Dynamics, Ecology, and Conservation. Scott, J. M., S. Mountainspring, F. L. Ramsey, and C. B. Kepler. 1986.

10. Ecology and Behavior of Gulls. Hand, J. L., W. E. Southern, and K. Vermeer, editors. 1987.

11. Bird Communities at Sea off California:1975 to 1983. Briggs, K. T., W. B. Tyler, D. B. Lewis, and D. R. Carlson. 1987.

12. Biology of the Eared Grebe and Wilson's Phalarope in the Nonbreeding Season: A Study of Adaptations to Saline Lakes. Jehl, J. R., Jr. 1988.

13. Avian Foraging: Theory, Methodology, and Applications. Morrison, M. L., C. J. Ralph, J. Verner, and J. R. Jehl, Jr., editors. 1990.

14. Auks at Sea. Sealy, S. G., editor. 1990.

15. A Century of Avifaunal Change in Western North America. Jehl, J. R., Jr., and N. K. Johnson, editors. 1994.

16. The Northern Goshawk: Ecology and Management. Block, W. M., M. L. Morrison, and M. H. Reiser, editors. 1994.

17. Demography of the Northern Spotted Owl. Forsman, E. D., S. DeStefano, M. G. Raphael, and R. J. Gutiérrez, editors. 1996.

18. Research and Management of the Brown-headed Cowbird in Western Landscapes. Morrison, M. L., L. S. Hall, S. K. Robinson, S. I. Rothstein, D. C. Hahn, and T. D. Rich, editors. 1999.

19. Ecology and Conservation of Grassland Birds of the Western Hemisphere. Vickery, P. D., and J. R. Herkert, editors. 1999.

20. Stopover Ecology of Nearctic–Neotropical Landbird Migrants: Habitat Relations and Conservation Implications. Moore, F. R., editor. 2000.

21. Avian Research at the Savannah River Site: A Model for Integrating Basic Research and Long-Term Management. Dunning, J. B., Jr., and J. C. Kilgo, editors. 2000.

22. Evolution, Ecology, Conservation, and Management of Hawaiian Birds: A Vanishing Avifauna. Scott, J. M., S. Conant, and C. van Riper, II, editors. 2001.

23. Geographic Variation in Size and Shape of Savannah Sparrows (Passerculus sandwichensis). Rising, J. D. 2001.

24. The Mountain White-crowned Sparrow: Migration and Reproduction at High Altitude. Morton, M. L. 2002.

25. Effects of Habitat Fragmentation on Birds in Western Landscapes: Contrasts with Paradigms from the Eastern United States. George, T. L., and D. S. Dobkin, editors. 2002.

26. Ecology and Conservation of the Willow Flycatcher. Sogge, M. K., B. E. Kus, S. J. Sferra, and M. J. Whitfield, editors. 2003.

27. Ecology and Conservation of Birds of the Salton Sink: An Endangered Ecosystem. Shuford, W. D., and K. C. Molina, editors. 2004.

28. Noncooperative Breeding in the California Scrub-Jay. Carmen, W. J. 2004.

29. Monitoring Bird Populations Using Mist Nets. Ralph, C. J., and E. H. Dunn, editors. 2004.

30. Fire and Avian Ecology in North America. Saab, V. A., and H. D. W. Powell, editors. 2005.

31. *The Northern Goshawk: A Technical Assessment of its Status, Ecology, and Management.* Morrison, M. L., editor. 2006.

32. *Terrestrial Vertebrates of Tidal Marshes: Evolution, Ecology, and Conservation.* Greenberg, R., J. E. Maldonado, S. Droege, and M. V. McDonald, editors. 2006.

33. *At-Sea Distribution and Abundance of Seabirds off Southern California: A 20-Year Comparison.* Mason, J. W., G. J. McChesney, W. R. McIver, H. R. Carter, J. Y. Takekawa, R. T. Golightly, J. T. Ackerman, D. L. Orthmeyer, W. M. Perry, J. L. Yee, M. O. Pierson, and M. D. McCrary. 2007.

34. *Beyond Mayfield: Measurements of Nest-Survival Data.* Jones, S. L., and G. R. Geupel, editors. 2007.

35. *Foraging Dynamics of Seabirds in the Eastern Tropical Pacific Ocean.* Spear, L. B., D. G. Ainley, and W. A. Walker. 2007.

36. *Status of the Red Knot (Calidris canutus rufa) in the Western Hemisphere.* Niles, L. J., H. P. Sitters, A. D. Dey, P. W. Atkinson, A. J. Baker, K. A. Bennett, R. Carmona, K. E. Clark, N. A. Clark, C. Espoz, P. M. González, B. A. Harrington, D. E. Hernández, K. S. Kalasz, R. G. Lathrop, R. N. Matus, C. D. T. Minton, R. I. G. Morrison, M. K. Peck, W. Pitts, R. A. Robinson, and I. L. Serrano. 2008.

37. *Birds of the US–Mexico Borderland: Distribution, Ecology, and Conservation.* Ruth, J. M., T. Brush, and D. J. Krueper, editors. 2008.

38. *Greater Sage-Grouse: Ecology and Conservation of a Landscape Species and Its Habitats.* Knick, S. T., and J. W. Connelly, editors. 2011.

39. *Ecology, Conservation, and Management of Grouse.* Sandercock, B. K., K. Martin, and G. Segelbacher, editors. 2011.

40. *Population Demography of Northern Spotted Owls.* Forsman, E. D., et al. 2011.

41. *Boreal Birds of North America: A Hemispheric View of Their Conservation Links and Significance.* Wells, J. V., editor. 2011.

42. *Emerging Avian Disease.* Paul, E., editor. 2012.

43. *Video Surveillance of Nesting Birds.* Ribic, C. A., F. R. Thompson, III, and P. J. Pietz, editors. 2012.

44. *Arctic Shorebirds in North America: A Decade of Monitoring.* Bart, J. R., and V. H. Johnston, editors. 2012.

45. *Urban Bird Ecology and Conservation.* Lepczyk, C. A., and P. S. Warren, editors. 2012.

46. *Ecology and Conservation of North American Sea Ducks.* Savard, J.-P. L., D. V. Derksen, D. Esler and J. M. Eadie, editors. 2015.

Milton Keynes UK
Ingram Content Group UK Ltd.
UKHW051538141024
449569UK00028B/1519